Environmental Science

EARTH AS A LIVING PLANET

Fourth Edition

• ## Daniel B. Botkin

Research Professor
Department of Ecology, Evolution and Marine Biology
University of California, Santa Barbara

President,
The Center for the Study of the Environment
Santa Barbara, California

• ## Edward A. Keller

Professor of Environmental Studies and
  Geological Sciences
University of California, Santa Barbara

with assistance from:
Dorothy B. Rosenthal
Mel S. Manalis
Harold Ward
Marc J. McGinnes
Valery A. Rivera

JOHN WILEY & SONS, INC.

Senior Acquisitions Editor   Keri Witman
Senior Development Editor   Ellen Ford
Freelance Developmental Editor   Beverly Peavler
Marketing Manager   Clay Stone
Senior Production Editor   Patricia McFadden
Production Services   Ingrao Associates
Senior Designer   Kevin Murphy
Text Designer   Laura C. Ierardi
Senior Photo Editor   Sara Wight
Photo Researcher   Alexandra Truitt
Illustration Editor   Sandra Rigby
Cover Photo   "Leadqueen Meadow, Purcell Range" © Larry Carver
Cover Designer   Lynn Rogan

This book was set in 10/12 Galliard by LCI Design
and printed and bound by Von Hoffmann.
The cover was printed by Von Hoffmann.

This book is printed on acid-free paper.

The paper in this book was manufactured by a mill whose forest management programs
include sustained yield harvesting of its timberlands. Sustained yield harvesting principles
ensure that the numbers of trees cut each year does not exceed the amount of new growth.

ISBN 0-471-38914-5

Printed in the United States of America

10 9 8 7 6 5 4 3

Dedications

- **FOR JANE O'BRIEN,**
  who, during her life, gave joy and spirit to my work.
  Dan Botkin

  and

- **FOR VALERY RIVERA**
  who contributed so much to this book and
  is a fountain of inspiration in our work and lives.
  Ed Keller

About the Authors

Daniel B. Botkin is President of The Center for the Study of Environment and Research Professor of Ecology, Evolution and Marine Biology, University of California, Santa Barbara. From 1978 to 1993, he was Professor of Biology and Environmental Studies at the University of California, Santa Barbara, serving as Chairman of the Environmental Studies Program from 1978 to 1985. For more than three decades, Professor Botkin has been active in the application of ecological science to environmental management. He is the winner of the Mitchell International Prize for Sustainable Development and the Fernow Prize for International Forestry, and he has been elected to the California Environmental Hall of Fame.

Trained in physics and biology, Professor Botkin is a leader in the application of advanced technology to the study of the environment. The originator of widely used forest gap-models, his research has involved endangered species, characteristics of natural wilderness areas, the study of the biosphere, and attempts to deal with global environmental problems. During his career, Professor Botkin has advised the World Bank about tropical forests, biological diversity, and sustainability; the Rockefeller Foundation about global environmental issues; the government of Taiwan about approaches to solving environmental problems; and the state of California on the environmental effects of water diversion on Mono Lake. He served as the primary advisor to the National Geographic Society for their centennial edition map on "The Endangered Earth." He recently directed a study for the states of Oregon and California concerning salmon and their forested habitats.

He has published many articles and books about environmental issues. His latest books are *Passage of Discovery: The American Rivers Guide to the Missouri River of Lewis and Clark* (Penguin-Putnam), *The Blue Planet* (Wiley), *Our Natural History: The Lessons of Lewis and Clark* (Putnam), *Discordant Harmonies: A New Ecology for the 21st Century* (Oxford University Press), and *Forest Dynamics: An Ecological Model* (Oxford University Press).

Professor Botkin was on the faculty of the Yale School of Forestry and Environmental Studies (1968–1974) and was a member of the staff of the Ecosystems Center at the Marine Biological Laboratory, Woods Hole, MA (1975–1977). He received a B.A. from the University of Rochester, an M.A. from the University of Wisconsin, and a Ph.D. from Rutgers University.

Edward A. Keller was chair of the Environmental Studies and Hydrologic Sciences Programs from 1993 to 1997 and is Professor of Geological Sciences at the University of California, Santa Barbara, where he teaches geomorphology, environmental geology, environmental science, river processes, and engineering geology. Prior to joining the faculty at Santa Barbara, he taught geomorphology, environmental studies, and earth science at the University of North Carolina, Charlotte. He was the 1982–1983 Hartley Visiting Professor at the University of Southampton and a Visiting Fellow in 2000 at Emmanuel College of Cambridge University, England.

Professor Keller has focused his research efforts into three areas: studies of Quaternary stratigraphy and tectonics as they relate to earthquakes, active folding, and mountain building processes; hydrologic process and wildfire in the chaparral environment of southern California; and physical habitat requirements for the endangered southern California steelhead trout. He is the recipient of various Water Resources Research Center grants to study fluvial processes and U.S. Geological Survey and Southern California Earthquake Center grants to study earthquake hazards.

Professor Keller has published numerous papers, and is the author of the textbooks *Environmental Geology, Introduction to Environmental Geology* and (with Nicholas Pinter) *Active Tectonics* (Prentice-Hall). He holds bachelors degrees in both geology and mathematics from California State University, Fresno; an M.S. in geology from the University of California; and a Ph.D. in geology from Purdue University.

Preface

A study of the environment in the 21st century is an exciting endeavor as we continue to move from confrontation to cooperative problem solving and place the study of the environment on a sound scientific basis. The enthusiasts of the 1960s have matured into today's environmental professionals: executives in alternative energy corporations, applied scientists who work on international projects to spread appropriate technologies to our inner cities and to developing nations, economists who calculate cost and benefits of pollution controls, environmental lawyers who mediate problems and help write laws to promote sustainable use of our resources, and other people in many related fields.

In recent years our understanding of many aspects of the environment has greatly increased. This has ranged from our understanding of the transport and fate of pollutants and toxins in the air and water, to appreciation of what is really necessary to save endangered species, to understanding how the Earth as a system operates.

Study of our environment has undergone tremendous change during the past four decades. In the 1960s and 1970s a grass roots movement to protect the environment began with the recognition of potential and real adverse effects of modern civilization on our environment. Views became polarized. Some environmentalists argued that everything about the environment was good and should be protected without change—that is, all development of natural resources was bad. Some environmentalists believed that the world would be destroyed if we did not change our approach to the environment. Opponents argued that these extreme "environmentalists" were opposed to progress and possibly to everything good stemming from civilization and technology. Nevertheless, during this period progress was made in dealing with environmental issues and problems. Enthusiasm for the environment remained high.

Environmentalism in the 1980s was characterized by a conscious shift from enthusiastic rhetoric to development of alternative ways to solve environmental problems associated with local, regional, and global issues such as human population, hazardous waste, acid precipitation, global warming, and stratospheric ozone depletion. People and institutions began putting more energy and resources into solving environmental problems and learning more about how the earth works as a system.

During the 1990s there was a return to considerable confrontation and emotionalism surrounding the environment. Some activists place the Earth's life support system, the biosphere, at the top of the moral pyramid, to be protected at all cost from adverse human interference. At the other extreme some people see environmentalists and environmentalism as a threat to private property rights and their "way of life." Finding solutions to environmental problems in the 21st century will require that a spectrum of potential solutions be carefully evaluated in light of our values and scientific knowledge, while at the same time ensuring social justice.

● GOALS OF BOOK

The purpose of *Environmental Science* is to provide an up-to-date introduction to the most important and useful concepts in the study of the environment. Information is presented from an analytical and interdisciplinary perspective from which we must view environmental issues in order to deal successfully with them. The goal is to teach the student *how* to think through environmental issues.

Critical Thinking

We must do more than simply identify and discuss environmental problems and solutions. To be effective, we must know what science is and is not. Then, we need to develop critical thinking skills. Critical thinking is so important that we have made it the focus of its own chapter, Chapter 2. With this in mind, we have also developed *Environmental Science* to present the material in a factual and unbiased format. Our goal is to help you think through the issues, not tell you what to think. To this purpose, at the end of each chapter, critical thinking exercises, called "Critical Thinking Issues," are provided. Critical thinking is further emphasized throughout the text in analytical discussions of topics, evaluation of perspectives, and integration of important Themes, which are described in detail later.

Interdisciplinary Approach

The approach of *Environmental Science* is interdisciplinary in nature. Environmental science integrates many disciplines and includes some of the most important topics of modern civilization as well as some of the oldest philosophical concerns of human beings—that of the nature of our relationship with our environment. Applied and basic aspects of environmental sciences require a solid foundation in the natural sciences, in addition to fields such as anthropology, economics, history, sociology, and philosophy of the environment. Not only do we need the best ideas and information to deal successfully with our environmental problems, but we also must be aware of the cultural and historical contexts in which we make decisions about the environment and understand ways in which choices are made and implemented. Thus, the field of environmental science integrates the natural sciences with environmental ethics, environmental economics, environmental law, environmental impact, and environmental planning. As a result, *Environmental Science* provides an introduction to the entire spectrum of relationships between people and the environment.

THEMES

Environmental Science is based on the philosophy that several threads of inquiry are of particular importance to environmental science. These key themes are woven throughout the book.

These five key themes or threads of inquiry are discussed in more detail in Chapter 1. They are also revisited at the end of each chapter where we discuss some of the pertinent material relative to these themes. They are further emphasized in the Closer Look boxes, each of which is highlighted by an icon suggesting the major underlying Theme of the discussion. In many cases, more than one Theme is relevant, again showing the inter-relationships at work.

Human Population

Underlying nearly all environmental problems is the rapidly increasing human population. Ultimately, we cannot expect to solve these other problems unless we can limit the total number of people on Earth to an amount the environment can sustain. We believe that education is important to solving the population problem. As people become more educated, and as the rate of literacy increases, population growth tends to decrease.

Sustainability

Sustainability is a term that has gained much popularity recently. Speaking generally, it means that a resource is used in such a way that it continues to be available. However, the term is used vaguely and it is something we are struggling to clarify. Some would define it as ensuring that future generations have equal opportunities to the resources that our planet offers. Others would argue that sustainability refers to types of developments that are economically viable, do not harm the environment, and are socially just. We all agree that we must learn how to sustain our environmental resources so that they continue to provide benefits for people and other living things on our planet.

A Global Perspective

Until recently we generally believed that human activity caused only local, or at most regional, environmental change. We now know that effects of human activity on Earth are of such an extent that we are involved in a series of unplanned planetary experiments. The main goal of the emerging science known as *Earth System Science* is to obtain basic understanding of how our planet works as a system. This understanding can then be applied to help solve global environmental problems. The emergence of Earth System Science has opened up a new area of inquiry for faculty and students. Understanding the relationships between biological and physical sciences requires interdisciplinary cooperation and education.

The Urban World

An ever-growing number of people are living in urban areas. Unfortunately our urban centers have long been neglected and the quality of the urban environment has suffered. It is here we experience air pollution, waste disposal problems, social unrest, and other stresses of the environment. In the past we have centered our studies of the environment more on wilderness than the urban environment. In the future we must place greater focus on towns and cities as livable environments.

Science and Values

Finding solutions to environmental problems involves more than simply gathering facts and understanding the scientific issues of a particular problem. It also has much to do with our systems of values and issues of social justice. To solve our environmental problems, we must understand what our values are, and which potential solutions are socially just. Then, we can apply scientific knowledge about specific problems and find acceptable solutions.

ORGANIZATION

We believe a real strength of *Environmental Science* is the systematic and in-depth coverage of the multitude of subjects that comprise the field of Environmental Science.

An important objective is to integrate physical and biological processes within a social framework. We recognize that environmental education is a life-long process and any one course may not be able to cover in depth all of the subjects presented in *Environmental Science*. The goal is to provide an instrument of learning useful to educators and students today so that future generations of students will be more informed and able to make judgments concerning the environment based upon sound scientific knowledge. We believe that the understanding of, and critical thinking about, environmental problems is much more important than the mere presentation of facts and information.

To support this goal, our text is divided into eight parts. *Part I* provides a broad overview of the key themes in *Environmental Science*, the scientific method, and thinking critically about the environment. *Part II* presents the study of the Earth as a system, emphasizing how systems work and the basic biochemical cycles of our planet. *Part III* focuses on life and the environment and includes subjects such as human population, ecosystems, biological diversity, biological productivity and energy flow, and restoration and recovery of ecosystem response to disturbance. *Part IV* presents living resources from a sustainability viewpoint, and topics covered include world food supply, agriculture and environment, plentiful and endangered species, forest ecology, conserving and managing life in the oceans, and environmental health and toxicology. *Part V* introduces and discusses a wide variety of topics related to energy including basics necessary for understanding energy, fossil fuels and environment, alternative energy, and nuclear energy. *Part VI* presents the water environment on Earth in terms of water supply use and management, and water pollution treatment. *Part VII* concerns the air environment, from global issues such as climate, global warming, and stratospheric ozone depletion to regional issues such as acid rain, to local issues including urban air pollution and indoor air pollution. *Part VIII* is concerned with relationships between environment and society. Topics include environmental economics, the urban environment, integrated waste management, minerals and the environment, environmental impact and planning, and how we might achieve sustainability.

● SPECIAL FEATURES

In writing *Environmental Science* we have designed a text that incorporates a number of special features that we believe will help teachers to teach and students to learn. These include:

- A **Case Study** introduces each chapter. The purpose is to interest the reader in the subject being discussed and to raise important questions on the subject matter. For example, in Chapter 15, which deals with Environmental Health and Toxicology, the Case Study introduces the problem of lead toxicity and asks the question, "Is lead in the urban environment contributing to antisocial (criminal) behavior?"

- **Learning Objectives** are introduced at the beginning of each chapter to help students focus on what is important in the chapter and what they should achieve after reading and studying the chapter.

- **A Closer Look** is the name of special learning modules presented in most of the chapters. The purpose here is to present more detailed information concerning a particular concept or issue. For example, A Closer Look 4.1 (Matter and Energy) discusses some basic physics and chemistry. Many of these special features contain figures and other data to enrich the reader's understanding, and relate back to the book themes.

- Near the end of each chapter, a **Critical Thinking Issue** is presented as a method of encouraging critical thinking about the environment and to help students understand how these issues may be studied and evaluated. For example, Chapter 20 presents the environmental issue of how wet is a wetland? The issue in Chapter 17 examines the important environmental question of whether or not we should raise the gasoline tax.

- Following the Summary, a special section, **Reexamining Themes and Issues**, reinforces the five major themes of the textbook. Here we also make value judgments concerning important environmental issues and questions.

- **Study Questions** for each chapter are provided. The purpose is to provide a study aid for students and utilize critical thinking skills.

- **Further Readings** are provided with each chapter so that students may expand their knowledge and reading through major sources of information (both print and electronic) on the environment.

- **References** cited in the text are provided at the end of the book as notes for each chapter. These are numbered according to their citation in the text. We believe it's very important that introductory textbooks carefully cite sources of information used in the writing. These are provided to recognize those scholars whose work we depend on, and so that the reader may draw upon these references as needed for additional reading and research.

● CHANGES IN THE FOURTH EDITION

Environmental Science is a rapidly developing set of fields. Populations grow; species become threatened or released from near-extinction; our actions change. To remain comtemporary, a textbook in Environmental Sciences requires frequent updating. For example, human populations continue to grow and the distribution changes, as does the rate of urbanization. Data have been brought up to date throughout. As information and ideas change, references also require updating, and this has been done throughout.

A major revision of the fourth edition is the re-organization of Parts III and IV on ecological processes, restoration ecology, and sustaining living resources. Biogeography and Biological Productivity have been split into separate chapters (8 and 9). There is a new Chapter 10 on the important subject of restoration ecology.

Other changes and special features in the fourth edition include:

- A new companion Web site (www.wiley.com/college/botkin) provides activities for students and resources for instructors, including on-line quizzing, Virtual Field Trips, Environmental Debates, additional cases and critical thinking articles, and Web links to important environmental data and research
- A totally new design and re-organized layout to provide a clearer understanding of the topics and issues
- Increased emphasis on critical thinking
- Increased emphasis on the "Themes" of Human Population, Sustainability, Global Perspective, Urban World, and Values and Science. All are integrated within every chapter, highlighted in Closer Look boxes, and revisited at the end of each chapter
- Many new photos and illustrations
- New and updated Case Studies, Closer Look boxes and Critical Thinking Issues
- Short chapters allow for syllabus flexibility and manageable reading assignments

Augmentation of Web Site References

Valid information is becoming increasingly available over the Web, and easy access to these data is of great value. Government data that used to take weeks of library search are available almost instantly over the Web. For this reason, we have greatly augmented the number of Web site references and have gathered them all on the book's companion Web site.

Integration of Themes

The fourth edition continues to integrate five major themes, adding more references in the text to the connection between a specific discussion and one of these themes. Closer Look boxes also emphasize specific themes through icons in the design.

New Case Studies

Each chapter begins with a case study that helps the student learn about the chapter's topic through a specific example. A major improvement in the fourth edition is the replacement of some older case studies with new ones that discuss current issues and are more closely integrated into the chapter. New case studies are listed following the Table of Contents.

New Critical Thinking Issues

Each chapter ends with a discussion of an environmental issue, with critical thinking questions for the students. This is one of the ways that the text is designed to help the student learn to think for oneself about the analysis of environmental issues. In the fourth edition, some older environmental issues have been replaced with new ones and these have been more closely integrated into the text. These are listed following the Table of Contents.

● SUPPLEMENTARY MATERIALS

Environmental Science, Fourth Edition, features a full line of teaching and learning resources developed to help professors create a more dynamic and innovative learning environment. For students, we offer tools to build their ability to think clearly and critically. For the convenience of both the professors and students, we provide teaching and learning tools in print, software, and on the Web.

For Students

Student Review Guide (0-471-21885-5)

Prepared by Joseph Luczkovich and David Knowles, of East Carolina University, the student review guide offers students a great way to review materials from the text and test their knowledge. Each chapter in the text has a corresponding chapter in the review guide. The following tools have been included to help students master the material: an overview feature called The Big Picture, Frequently Asked Questions, Ecology in Your Backyard, references to related resource materials, and an Ecotest designed for chapter review.

Take Note! (0-471-41602-9)

This helpful student ancillary is composed of the same images as the Transparency Acetates, enlarged, and with added space next to the images for note-taking.

Student Web Site (www.wiley.com/college/botkin)

A completely new, redesigned, content-rich Web site has been created to provide enrichment activities and resources for students. These features include review of Learning Objectives, on-line quizzing, Virtual Field Trips, interactive Environmental Debates, additional case studies and critical thinking readings, and Web links to important data and research in the field of environmental studies.

For Instructors

Instructor's Resource CD-ROM (0-471-41600-2)

The Instructor's Resource CD (IRCD) is a multi-platform CD-ROM. The Instructor's Resource CD includes the *Instructor's Resource Guide,* the *Test Bank,* the

Computerized Test Bank, PowerPoint Presentations, and jpeg files for all the *Illustrations* in the text. The IRCD is designed to address the needs of all teaching levels. These assets are also available to professors at the Botkin/Keller text Web site (www.wiley.com/college/botkin).

Instructor's Resource Guide

Prepared by Ann S. Causey of Auburn University. Provides useful tools to highlight for instructors key concepts from each chapter of the text. Each chapter includes the following topics: Lecture Lead-Ins, Outline/Lecture Notes, Critical Thinking Activities, and Suggested Readings. In addition, the author includes a section titled "What You Can Do" to help you with additional resource ideas for providing engaging lectures.

Test Bank

Prepared by Nicholas Pinter, Southern Illinois University. Available on both the Instructor's Resource CD and the Wiley Botkin/Keller website (www.wiley.com /college/botkin). The test bank includes approximately 2000 questions, in multiple-choice, short answer, and essay formats. The Test Bank is provided in a word.doc format for your convenience to use and edit for your individual needs. For this edition, the author has created many new questions, and has labeled the boxed applications according to the five themes and issues set forth in the text. In addition, the author has created questions for the theme boxes and emphasized the themes in many of the questions throughout the test bank.

Computerized Test Bank

The Computerized Test Bank (CTB) is a multi-platform CD-ROM (*Diploma/Exam*) which is also available on the Wiley Botkin/Keller website (www.wiley.com /college/botkin). The CTB includes all the files from the Test Bank, but, puts them into a dynamic computerized format. The easy-to-use test-generated program fully supports graphics, print tests, student answer sheets, and answer keys quickly and easily. The software's advanced features allow you to create an exam to your exact specifications, with an easy-to-use interface.

PowerPoint Slides

Available on the Wiley Botkin/Keller website (www.wiley.com/college/botkin) and Instructor's Resource CD, this classroom presentation tool incorporates all of the line art from the text which can be sequenced and customized by instructors to fit any lecture. Designed according to the organization of the material in the textbooks, this series of electronic transparencies can be used to illustrate concepts visually and graphically.

Transparency Acetates (0-471-23667-5)

Full-color traditional acetate transparencies of illustrations from the text are available for professors who are unable to use the slides available on the *Instructor's Resource CD-ROM*. Included are 100 images from the text, with large, boldface labels to aid instructors with classroom presentations.

Instructor's Web Site

The resources provided on the Instructor's Resource CD will also be available on the instructor section of the Wiley Botkin/Keller Web site (www.wiley.com/college /botkin).

Daniel B. Botkin

Edward A. Keller

Acknowledgments

Completion of this book was only possible due to the cooperation and work of many people. To all those who so freely offered their advice and encouragement in this endeavor, we offer our most sincere appreciation. We are indebted to our colleagues who made contributions: Dorothy A. Rosenthal for writing Chapter 2 and the environmental issues at the end of each chapter; Mel S. Manalis for assistance in developing the energy chapters, global warming, and stratospheric ozone depletion; Marc J. McGinnis for assistance in helping develop discussions concerning environmental law; Harold Ward for writing parts of Chapter 30, particularly relating to environmental legislation and review; Amy Selting for assistance in research; and Beverly Peavler for her excellent editing of the manuscript. Finally, we would like to thank Harold Morowitz for helpful suggestions on how to introduce the basic energy concepts.

We greatly appreciate our editors, Keri Witman and Ellen Ford, at John Wiley & Sons, for their support, encouragement, assistance, and professional work, as well as Beverly Peavler for her excellent editing of the manuscript. We also extend thanks to our production editors, Trish McFadden and Suzanne Ingrao, who did a great job and made important contributions in many areas; to Kevin Murphy, Laura Ierardi, and Lynn Rogan for a beautiful interior design and striking cover; Sara Wight for photo research; and Sandra Rigby for the illustration program. Thanks go out also for editorial assistance from Kelly Coaxum and Justin Bow. The extensive supplements package was enhanced through the efforts of Joan Petrokovsky, Cindy Taylor, and Art Pomponio.

Of particular importance to the development of the book were the individuals who read the book chapter by chapter and provided valuable comments and constructive criticism. This was a particularly difficult job given the wide variety of topics covered in the text and we believe that the book could not have been successfully completed without their assistance. These reviewers are offered our special gratitude:

Diana Anderson, Northern Arizona University
James Bartolome, University of California – Berkeley
Susan Beatty, University of Colorado – Boulder
Brian Beeder, Morehead State University

William Berry, University of California – Berkeley
Alan Bjorkman, North Park University
Charles Bomar, University of Wisconsin – Stout
Kelly Cain, University of Wisconsin – River Falls
Rosanna Cappellato, Emory University
W.B. Clapham Jr., Cleveland State University
William Davin, Berry College
Jerry Delsol, Modesto Junior College
Jean Dupon, Menlo College
Richard Feldman, Marist College
Deborah Freile, Berry College
Nancy Goodyear, Bainbridge College
Paul Grogger, University of Colorado
Herbert Grossman, Pennsylvania State University
Gian Gupta, University of Maryland
Lonnie Guralnick, Western Oregon University
Raymond Hames, University of Nebraska
John Harley, Eastern Kentucky University
Alan Holyoak, Manchester College
Donald Humphreys, Temple University
James Jensen, SUNY- Buffalo
David Johnson, Michigan State University
John Kinworthy, Concordia University
Thomas Klee, Hillsborough Community College
Mark Knauss, Shorter College
Ned Knight, Linfield College
Steven Kolmes, University of Portland
Matthew Laposata, Kennesaw State University
Timothy Lyon, Ball State University
Timothy McCay, Colgate University
Michele Morek, Brescia University
James Morris, University of South Carolina
Kathleen Nolan, St. Francis College
Nancy Ostiguy, Pennsylvania State University
John Pratte, Kennesaw State University
Carlton Rockett, Bowling Green State University
Robert Sanford, University of Southern Maine
Jeffrey Schneider, SUNY – Oswego
Jill Scheiderman, SUNY- Dutchess Community College
Peter Schwartzman, Knox College
Daniel Sivek, University of Wisconsin
Meg Stewart, Vassar College
Richard Stringer, Harrisburg Area Community College
Janice Swab, Meredith College

Jeffrey Tepper, Valdosta State University
Richard Vance, UCLA
Sarah Warren, North Carolina State University
William Winner, Oregon State
Bruce Wyman, McNeese State University
Richard Zingmark, University of South Carolina

Reviewers of Previous Editions

Marc Abrams, Pennsylvania State University
Michele Barker-Bridges, Pembroke State University (NC)
Susan Beatty, University of Colorado
David Beckett, University of Southern Mississippi
Mark Belk, Brigham Young University
Kristen Bender, California State University, Long Beach
Gary Booth, Brigham Young University
Grace Brush, Johns Hopkins University
Kelley Cain, University of Wisconsin, River Falls
John Campbell, Northwest Community College (WY)
Ann Causey, Prescott College (AZ)
Simon Chung, Northeastern Illinois State
Thomas B. Cobb, Bowling Green State University
Jim Dunn, University of Northern Iowa
Robert Feller, University of South Carolina
Andrew Friedland, Dartmouth College
Douglas Green, Arizona State University
James H. Grosklags, Northern Illinois University
Bruce Hayden, University of Virginia
David Hilbert, San Diego State University
Peter Kolb, University of Idaho
Henry Levin, Kansas City Community College
Hugo Lociago, University of California, Santa Barbara
Tom Lowe, Ball State University
Timothy Lyon, Ball State University
Mel Manalis, University of California, Santa Barbara
Earnie Montgomery, Tulsa Junior College, Metro Campus
Walter Oechel, San Diego State University
C. W. O'Rear, East Carolina University
Stephen Overmann, Southeast Missouri State University
Martin Pasqualetti, Arizona State University
David Pimental, Cornell University
Brian Reeder, Morehead State University
C. Lee Rockett, Bowling Green State University
Joseph Simon, University of South Florida

Lloyd Stark, Pennsylvania State University
Laura Tamber, Nassau Community College (NY)
Bruce Wyman, McNeese State University
Ann Zimmerman, University of Toronto
Marvin Baker, University of Oklahoma
Mary Benbow, University of Manitoba
Grady Blount, Texas A&M University-Corpus Christi
John Bounds, Sam Houston State University
Vincent Breslin, SUNY-Stony Brook
Bonnie Brown, Virginia Commonwealth University
Annina Carter, Adirondack Community College
Peter Colverson, Mohawk Valley Community College
Harry Corwin, University of Pittsburgh
Craig Davis, Ohio State University
Paul Grogger, University of Colorado
David Johnson, Michigan State University
S. B. Joshi, York University
Stephen Malcolm, Western Michigan University
James Melville, Mercy College
Chris Migliaccio, Miami-Dade Community College-Wolfson
Clayton Penniman, Central Connecticut State University
Jeffery Schneider, SUNY-Oswego

Completion of this project required an incredible amount of word processing and other administrative assistance. We are very grateful to Susan Leska for transcribing and typing many chapters of this text. We are indebted to Valery A. Rivera who reviewed chapters of this book with the careful eye and mind that only a dedicated teacher could bring to the subject. Valery made many important suggestions for improvement of both style and content in the fourth edition.

The University of California, Santa Barbara, and George Mason University provided the stimulating atmosphere necessary for writing. Parts of the text were also written while E. A. Keller was on leave at Emannual College, Cambridge University (UK) and La Laguna University (Spain). Fellowships to Daniel B. Botkin from the East–West Center Honolulu and Rockefeller Bellagio Study and Conference Center, Bellagio, Italy, and from the Woodrow Wilson International Center for Scholars, Washington, D.C., provided time and stimulating environments in which ideas in the text underwent growth and enrichment.

Daniel B. Botkin

Edward A. Keller

Brief Contents

Contents

Cases, Closer Look and Critical Thinking Issue Boxes

A Special Note to Students

We have written this book with the belief that students really want to be challenged to think about the important environmental problems that face the world today. We are attempting to bring the environment to the forefront of education. This is in response to tremendous public concern about the environment. Public opinion polls confirm that the environment is an important social and political issue. We also believe that scientific literacy and critical thinking are very important. As a result, we have chosen to discuss important environmental issues, some of which are multi-faceted, difficult, and challenging. We all need to recognize that solutions to the many environmental problems facing us today are not easy to find. There are differing opinions on which problems we should address and what solutions should be attempted.

This is something we cannot teach you. We can present information and arguments, but ultimately you must think for yourself and develop solutions based on your value system and understanding. If this appears as a vague concept—it is! You are responsible for future generations just as are your instructors. Making informed decisions requires a lot of study as well as a commitment to be responsible in helping to insure that our planet maintains a healthy productive environment. We believe that if our book, *Environmental Science,* helps you in that endeavor, then it has been successful.

If our book has helped you to think more critically about environmental problems, we would like to hear from you. Please share your ideas or questions with us at DBotkin@Wiley.com.

Daniel B. Botkin
Edward A. Keller

CHAPTER 1

Basic Issues in Environmental Sciences

The rich diversity of life on and around coral reefs is illustrated in this photo from the Red Sea, Egypt.

● LEARNING OBJECTIVES

Certain issues are basic to the study of environmental science. After reading this chapter, you should understand:

- Why rapid human population growth is *the* fundamental environmental issue.
- Why we must learn to sustain our environmental resources so that they will be available in the future.
- How human beings affect the environment of the entire planet and why we must take a global perspective on environmental problems.
- Why urban environmental issues and the effects of urban areas on environments elsewhere need to be given primary focus.
- Why developing solutions to environmental problems requires making value judgments based on knowledge of scientific facts.

Shrimp, Mangroves, and Pickup Trucks: Local and Global Connections Reveal Major Environmental Concerns

Maitri Visetak, who owns a small plot of land along the coast of southern Thailand, wanted to improve life for his family, and he succeeded. A growing demand for shrimp as a luxury food and overfishing of wild shrimp had fueled growth of the world market for farm shrimp from a $1.5 billion industry 30 years ago to an $8 billion business today. In the early 1990s, Mr. Visetak began farming shrimp in two small ponds (0.2 hectare or 0.5 acre; Figure 1.1a). Within two years, he had accumulated enough capital to purchase two pickup trucks—a clear indication of financial success in Thailand. By then, though, his ponds were contaminated with shrimp waste, antibiotics, fertilizers, and pesticides. Shrimp could no longer live in the ponds. And there was an even more widespread effect: Pollutants escaping from the ponds threatened survival of the area's mangrove trees (Figure 1.1b). Like thousands of other shrimp farmers in southeast Asia, India, Africa, and Latin America, Mr. Visetak considered abandoning these ponds and moving on to others.

Maitri Visetak is trying to feed his family in the best way he knows how, but along with thousands of other shrimp farmers in the world, he is unwittingly contributing to destruction of coastal mangroves, one of the world's valuable ecosystems. Half of the world's mangrove forests have been destroyed and with them a major source of food for local human populations and breeding grounds for much of the tropical world's sea life. The United Nations Environment Program has estimated that one-fourth of the destruction of mangroves can be traced to shrimp farming. Environmentalists have become alarmed, and in many areas, local people have staged protests against shrimp farming. With the world's population expected to increase from 6.1 billion to 9 billion by the middle of the twenty-first century, concern over the world's mangrove forests is growing.[1-6]

(a)

(b)

Figure 1.1 • Sustainability. (*a*) Shrimp farms such as this one threaten the survival of mangrove forests. (*b*) Mangroves on the banks of Indian River, Isle of Dominica, West Indies. Mangrove trees grow in coastal wetlands. Their specialized roots can survive immersion in ocean water at high tide and exposure to the drying sun at low tide. Swamps formed by mangroves provide habitat for many kinds of ocean life and are important for commercial fisheries in many parts of the world.

• *The story of Maitri Visetak illustrates several of the major themes of environmental science. First, human population increase is a major contributor to environmental problems. Second, industrial development and urbanization have serious environmental consequences. Third, unsustain-*

able use of resources must be replaced with sustainable practices. Fourth, local changes can have global effects. Finally, environmental issues involve values and attitudes as well as scientific understanding. The story of Maitri Visetak also illustrates important questions that we all must face: Which individual actions contribute to environmental degradation? What actions can people, both as individuals and as groups, take to limit environmental damage?

1.1 Key Themes

Today we stand at the threshold of a major change in our approach to environmental issues. Two paths lie before us. One is an approach developed some 40 years ago—an approach that has produced many advances but also many failures. This path has emphasized confrontation and emotionalism. It has been characterized by a lack of understanding of basic facts about the environment and how natural ecological systems function, as well as a willingness to base solutions on political ideologies and on ancient myths about nature.

The second path offers the potential for long-lasting, successful solutions to environmental problems. This approach seeks to move from confrontation to cooperative problem solving, from explaining the environment in terms of ancient myths to providing a sound scientific basis from which to view environmental issues. The purpose of this book is to take the student down the second pathway. A little historical explanation will help clarify what we seek to accomplish.

Before 1960, few people had ever heard the word *ecology*, and the term *environment* meant little as a political or social issue. Then came the publication of Rachel Carson's landmark book, *Silent Spring* (Houghton Mifflin, Boston, 1960, 1962). At about the same time, several major environmental events, such as oil spills along the coasts of southern California and Massachusetts, occurred, along with highly publicized threats of extinction to many species, including whales, elephants, and songbirds. The environment became a popular issue.

As with any new social or political issue, at first relatively few people recognized its importance. These people felt a need to stress the problems—to emphasize the negative—in order to bring public attention to environmental concerns. The early days of modern environmentalism were dominated by confrontations between those labeled environmentalists and those labeled antienvironmentalists.

These two approaches can be contrasted in very simplistic terms. Environmentalists believed that the world was in peril. To them, economic and social development meant the destruction of the environment and ultimately the end of civilization, the extinction of many species, and perhaps the extinction of human beings. Their solution was a new worldview that depended only secondarily on facts, understanding, and science.

The anti-environmentalists, on the other hand, believed that social and economic health and progress were necessary if people and civilization were to prosper. From their perspective, environmentalists represented a dangerous and extreme view, with a focus on the environment to the detriment of people—a focus that they thought would destroy the very basis of civilization and lead to the ruin of our modern way of life.

Adding to the limitations of the early approach to environmental issues was a lack of scientific knowledge and practical know-how. Environmental sciences were in their infancies. Some people even saw science as part of the problem.

Today, the situation has changed considerably. Public opinion polls repeatedly show that people around the world rank the environment among the most important social and political issues. There is no longer a need to prove that environmental problems are serious. In addition, most people recognize that we must seek sustainability not only of the environment but also of our economic activities so that humanity and the environment can persist together into the future. Among many, the dichotomy of the last generation is giving way to a new unity: the idea that a sustainable environment and a sustainable economy may be compatible.

Although our scientific understanding of the environment lags greatly behind our need to know, significant progress has been made in many areas of environmental science. Advances have also been made in the creation of legal frameworks for the management of the environment, thus providing a new basis for solving environmental issues. The time is now ripe to seek truly lasting, rational solutions to environmental problems.

The environment is complex and multifaceted. This book covers a wide range of topics, from environmental ethics to the chemistry of the ozone layer in the atmosphere. This wide range of topics is connected by five unifying themes. The purpose of this chapter is to acquaint you with these themes.

Human Population Problem

The rapidly increasing human population underlies all environmental problems. Ultimately, we cannot solve our environmental problems unless we can learn to limit the number of people on the earth to a number that the environment can sustain.

Sustainability

In the past, we have used living environmental resources such as forests, fish, and wildlife faster than they could replenish. We have extracted minerals, oil, and groundwater without concern for their limits or the need for recycling. As a result, many of these resources are no longer abundant. We must learn how to sustain our environmental resources so that they continue to provide benefits to us and to the larger environment of which we are a part.

A Global Perspective

Life on our planet is characterized by complex relationships among living things—the atmosphere, the water, and the land. Human activity has begun to change the environment on a global scale in ways harmful to people and other living things. As a result, the next generation of environmental scientists, as well as all citizens, must have a global perspective.

An Urban World

Most of us live in cities and towns, and it is in these urban settings that many solutions to environmental problems must be found. Cities and towns are increasing in size and spreading over the landscape, sometimes eliminating valuable farmland and natural areas. In the past, environmental activism usually centered on wilderness. In the future, however, it will be important to focus on towns and cities and to develop harmony between industrial and commercial activities and the quality of urban environments.

Science and Values

Although environmental issues have often been portrayed as simply a question of facts, central conflicts about the environment have to do with values and science. Knowledge, which includes knowing the scientific data as well as understanding the issues, tells us what solutions are possible. What we ultimately decide to do—which of the possible solutions we choose—depends on our values. A central theme of this book is how we value the envi-

ronment. Given the size of our population and our great technological power to change the environment, we can no longer avoid the connection between values and science. (See this chapter's Critical Thinking.)

These five issues—the human population problem, sustainability, a global perspective, an urban world, and values and science—are central to our understanding of environmental science and recur frequently throughout the text. They are the threads that tie together our ideas about the environment and provide the focus for our study of environmental science. The remainder of this chapter examines these issues in more detail.

1.2 Human Population Growth

The John Eli Miller Family

John Eli Miller was a farmer like many other farmers in the United States, except for one thing—when he died on his farm in Middlefield, Ohio, in the mid-twentieth century, he was the head of the largest family in the United States (see Figure 1.2). He was survived by 5 children, 61 grandchildren, 338 greatgrandchildren, and 6 greatgreatgrandchildren. Within his lifetime, John Miller witnessed a family population explosion. What was perhaps even more remarkable was that the explosion started with a family of just 7 children—not all that unusual for the nineteenth century United States.

During most of John Miller's life, his family was not unusually large. It is just that he lived long enough to find out what simple multiplication can do, and he lived in a time when the death rate among infants, children, and young adults was very low compared with typical death rates during the history of most human populations.[7] Of 7 children born to John Miller, 5 survived him; of 63 grandchildren, 61 survived him; and of 341 greatgrandchildren (born to 55 married grandchildren—an average of slightly more than 6 children per parent), 338 survived him.

John Miller's family emphasizes a major factor in our modern population explosion. Modern technology, modern medicine, and the supply of food, clothing, and shelter have decreased death rates and accelerated the net rate of growth. As a result, human population has increased greatly, threatening the environment.

The Population Bomb

In the last part of the twentieth century, the most dramatic increase in the history of the human population occurred. In merely the past 35 years, the human population of the world more than doubled, increasing from 2.5 billion to over 6.1 billion. Figure 1.3 illustrates the rapid explosion of the human population, sometimes referred to as the population bomb.[8] The figure breaks down the increases by region.

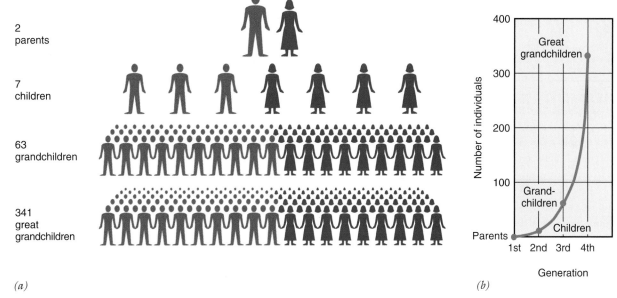

(a)

(b)

Figure 1.2 • The population bomb starts with little sparks. (*a*) A simplified family tree of four generations of the John Eli Miller family. (*b*) The population explosion of the John Eli Miller family shown in graphic form.

The human population issue is the underlying issue of the environment because most current environmental damage results from the very large number of people on the earth. Human population is discussed in detail in Chapter 5; an outline of the problem is given here

because it is so fundamental to an analysis of environmental issues.

As you will see in Chapter 5, for most of human history, the total population was small, with an extremely low long-term rate of increase compared with that of today's population.[9,10] In addition, the growth of the human population has not been a steady march. Although it is customary to think of the population as increasing continuously without declines or fluctuations, this has not been the case. For example, great declines occurred during the time of the Black Death and the recent famines in Africa. (See A Closer Look 1.1.)

African Famines

The world already has examples of what happens when a human population exceeds its environmental resources. Among these examples are major episodes of famine in Africa, which continue today. In the mid-1970s, following a drought in the Sahel region, 500,000 Africans starved to death, and several million more were permanently affected by malnutrition.[11] Starvation in African nations gained worldwide attention some 10 years later, in the 1980s.[12,13] In one year during that period, as many as 22 African nations suffered catastrophic food shortages, and 150 million Africans faced starvation.

Famine in Africa has had multiple, interrelated causes. One, as suggested, is drought. Although drought is not new to Africa, the size of the population affected by drought is new. Food shortages affect many more people. Because of food shortages, the control and destruction of food has sometimes been used as a weapon in political

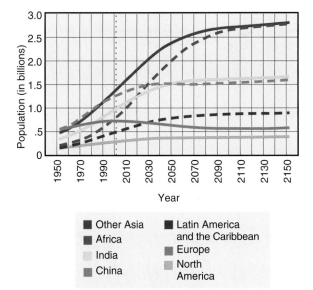

Figure 1.3 • Population change since 1950 projected to the year 2150 for major areas of the world, medium fertility scenario. The population of Africa will nearly quadruple. The only major area whose population is projected to drop over time is Europe—from 728 million to 595 million, a decline of 18% over 155 years. [*Source:* Population Division, Department of Economic and Social Affairs, United Nations Secretariat, *World Population Projections to 2150* (New York: United Nations), 1998].

The Black Death

The epidemic disease bubonic plague, commonly known as the Black Death, spread throughout Europe during the fourteenth century. The most severe episodes occurred between 1347 and 1351, but there were many recurrences throughout the century (Figure 1.4).[14] The disease is caused by the bacteria *Yersinia pestis* and spread by fleas that live on rodents. It was first recorded in Western history as a major human problem when it occurred in the seventh century in the Roman Empire and in North Africa. The plague probably first appeared in India in the seventh century and spread rapidly north and west. Not until the fourteenth century did another major epidemic occur. The plague again spread rapidly, reaching Italy in 1348 and Spain, France, Scandinavia, and central Europe within two years. In England, one-fourth to one-third of the population died within a single decade, although mortality varied widely by region. Entire towns were abandoned, and the production of food for the remaining population was jeopardized.

The Black Death had many environmental and economic consequences. For example, the great reduction in the labor force led to an increase in wages and is believed to have been a contributing factor to a subsequent increase in the standard of living. Much agricultural land was abandoned because no one was available to work it.

As this example illustrates, human populations have not always increased continuously but have suffered setbacks and declines. The bubonic plague is one of the best-recorded and best-known setbacks in human history. It is likely, though, that other such episodes, resulting from changes in climate, declines in the food supply, or environmental catastrophes, have happened many times in human history.

(a)

(b)

Figure 1.4 ● (*a*) The change in the population of Europe during the time of the Black Death. (*b*) During the fourteenth century, the Black Death killed many people in Europe. The reduction in human population had economic and environmental effects. This fourteenth century illustration of two plague victims is from a miniature from the Toggenburg Bible. Note the swellings over the bodies, characteristic of the plague.

disruptions (Figure 1.5). In addition, deserts in Africa appear to be spreading, in part because of changing climate but also because of human activities. Poor farming practices have increased erosion, and deforestation may be helping to make the environment drier.

Famines in Africa illustrate the fact that population crises feed on themselves. People affect the environment; the environment affects people. The environment affects agriculture, and agriculture also affects the environment. Human population growth in Africa has severely stretched the capacity of the land to provide food and has threatened its future productivity. This situation leads to a key question about values: *Which is more important, the survival of people alive today or the conservation of the environment, on which future food production and human life depend?*[15]

Answering this question demands value judgments but information and knowledge are necessary to make such value-based decisions. For example, we must determine whether we can continue to increase agricultural produc-

Values and science. • Social conditions affect the environment, and the environment affects social conditions. Political disruption in Somalia (illustrated by a Somalian boy with a gun, left photo) interrupted farming and food distribution, leading to starvation.

Overpopulation, climate change, and poor farming methods also lead to starvation, which in turn promotes social disruption. Famine has been common in parts of Africa since the 1980s, as illustrated by gifts of food from aid agencies in southern Sudan (right photo).

tion without destroying the very environment on which agriculture and, indeed, the persistence of life on Earth are based. This is a technical, scientific investigation. The result provides a basis for a judgment that in turn depends on our values.

The human population is doubling every few decades, but human effects on the environment are growing at an even faster rate. Human beings cannot escape the laws of population growth. The question is: What will we do about the increase in our own species and its impact on our planet and our future?

1.3 Sustainability and Carrying Capacity

One of the central environmental questions is this: What is the maximum number of people the earth can sustain? The answer depends on the rate at which we use our environmental resources.

There is little doubt that we are using our renewable environmental resources faster than they can be replenished. In general, we are using forests and fish faster than they can regrow and eliminating habitats of endangered species and wildlife faster than they can be replenished. We are extracting minerals, oil, and groundwater without sufficient concern for their limits or the need to recycle them. As a result, there is a present shortage of some resources and an expectation of more shortages in the future. Clearly, we must learn how to *sustain* our environmental resources so that they continue to provide benefits for people and other living things on our planet.

Defining Sustainability

Sustainability must be achieved, but we are unclear at present how to achieve it, in part, because the word is used to mean different things, often leading to confusion and to people talking cross-purposes. **Sustainability** refers to resources and their environment. In this book, sustainability has two scientific definitions: (1) *sustainable resource harvest*, such as a sustainable supply of timber, means that the same quantity of that resource can be harvested each year (or other harvest interval) for an unlimited or specified amount of time without decreasing the ability of the resources to produce the same harvest level. (2) A *sustainable ecosystem* is an ecosystem from which we are harvesting a resource that is still able to maintain its essential functions and properties.

Other kinds of sustainability refer to human societies. (3) A *sustainable economy* is an economy that maintains its level of activity over time in spite of its uses of environmental resources. (4) *Sustainable development* typically means that a society is able to continue to develop its economy and social institutions and maintain its environment for an indefinite time.

Carrying Capacity of the Earth

Carrying capacity is a concept related to sustainability. It is usually defined as *the maximum number of individuals of a species that can be sustained by an environment without decreasing the capacity of the environment to sustain that same amount in the future.* When we ask, "What is

Carrying Capacity of the Chinook Salmon

One approach to determining the carrying capacity and the sustainable harvest of a resource is to examine historic records and see if previous catch levels were maintained. One problem of determining carrying capacity using this method is illustrated by the history of commercial fishing of chinook salmon in the Columbia River of the Pacific Northwest. Figure 1.5 graphs the annual catch of salmon from 1866 (just after the Civil War, when commercial fishing began in earnest on the river) to 1966.[16] The catch increased rapidly from 1860 to 1880, then declined and fluctuated between 16 and 36 million pounds until 1920, when it declined again in a highly varying pattern.

From this history, what, if any, levels of chinook salmon are sustainable? The high catches between 1890 and 1920 are about the same, so a person in charge of managing the salmon fishery in 1920 might reasonably have concluded that the salmon had a sustainable catch of between 16 and 36 million pounds. However, a person who had only the information after 1950, when the catch represents the tail of a declining curve, would conclude that the approximate level of catch in the early 1960s was sustainable. This graph illustrates the difficulty in determining a truly sustainable harvest simply from historical records.

Chinook salmon landings by the
Columbia River commercial fishery
1866-1966

Figure 1.5 ● The commercial catch of chinook salmon on the Columbia River between 1866 and 1966. (*Source*: J. M. Van Hyning, "Factors Affecting the Abundance of Fall Chinook Salmon in the Columbia River," *Research Reports of the Fish Commission of Oregon* 4[1], Figure 14, p. 38.)

the maximum number of people that the earth can sustain?" we are asking about the Earth's carrying capacity, and we are also asking about sustainability.

For the human population, the carrying capacity depends in part on our values. Do we want those who follow us to live short lives in crowded surroundings without a chance to enjoy Earth's scenery and diversity of life? Or do we hope that our descendants will have a life of high quality and good health? Once we choose a goal for the quality of life, we can use scientific information to understand what the carrying capacity might be and how we might achieve it. (See A Closer Look 1.2.)

1.4 A Global Perspective

Because human actions have begun to change the environment at a global level, the next human generation must embrace a global perspective on environmental issues. The recognition that, worldwide, civilization can change the environment at a global level is relatively recent. As discussed in detail in later chapters, scientists now believe that emissions of modern chemicals are changing the ozone layer high in the atmosphere. Scientists also believe that the burning of fossil fuels increases the concentration of greenhouse gases in the atmosphere, which may change Earth's climate. These atmospheric changes suggest that the actions of many groups of people at many locations affect the environment on a global level.[17,18] Another new idea explored in later chapters is that nonhuman life affects the environment of our planet at a global level and has changed it over the course of several billion years. These two new ideas have profoundly affected our approach to environmental issues.

Awareness of the global interactions between life and the environment has led to the development of the **Gaia hypothesis**, originated by British chemist James Lovelock and American biologist Lynn Margulis.[17] The Gaia hypothesis (discussed in Chapter 3) proposes that the environment at a global level has been profoundly changed by life over the history of life on Earth and that these changes have tended to improve the chances for the continuation of life. Because life affects the environment at a global level, the environment of our planet is different from that of a lifeless one.

1.5 An Urban World

In part as a result of the rapid growth of the human population and in part as a result of changes in technology, we are becoming an urban species, and our effects on the environment are more and more the effects of urban life (Figure 1.6).[19] With economic development comes urbanization; people move from farms to cities and then perhaps to suburbs. Cities and towns increase in size. Because cities are commonly located near rivers and along coastlines, urban sprawl often overtakes the good agricultural land of river floodplains, as well as the coastal wetlands, which are an important habitat for many rare and endangered species. As urban areas expand, wetlands are filled in, forests cut, and soils covered over and removed from productive use.

In developed countries, about 75% of the population live in urban areas and 25% live in rural areas; but in developing countries, only 38% of the people are city dwellers.[20] By 2025, almost two-thirds of the population—5 billion people—will live in cities. Only a few urban areas had populations over 4 million in 1950. In 1999, Tokyo, Japan, was the world's largest city. In 2015, Tokyo will still be the world's largest city, with a population of 28.9 million. *The number of megacities (urban areas with at least 8 million inhabitants) increased from two (New York and London) in 1950 to 23 in 1995, with 17 of them in the developing world. By 2015, there will be 36 megacities; 23 of them will be located in Asia. Los Angeles and New York City are among the 10 largest megacities.[21,22]*

In the past, environmental organizations often focused their concern and their activities on non-urban areas—wilderness, endangered species, and natural resources, including forests, fisheries, and wildlife. Although these areas will remain important, in the future, we must place more emphasis on urban environments and on the effects of urban environments on the rest of the planet.

1.6 Science and Values

As previously mentioned, one theme of this book is that *decisions concerning solutions to environmental problems require both values and science.* We must choose what we want the environment to be. To make this choice, we must first know what is possible. That requires knowing and understanding the implications of the scientific data. Once we know our options, we can select from among them. What we choose is determined by our values. An example of a value judgment regarding the world's human environmental problem is the choice between the desire of an individual to have many children and the need to find a way to limit the human population worldwide.

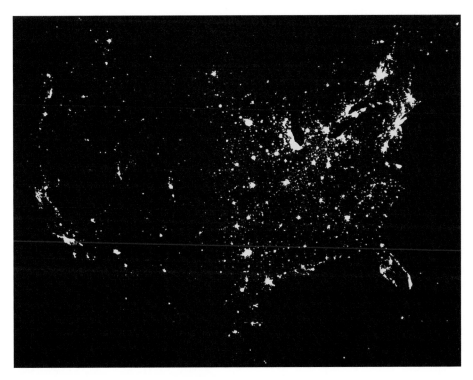

Figure 1.6 • An urban world and a global perspective. When the United States is viewed at night from space, the urban areas show up as bright lights. The number of urban areas reflects the urbanization of our nation.

Once we have chosen a goal based on knowledge and values, we have to find a way to attain that goal. This step also requires knowledge. The more technologically advanced and powerful our civilization, the more knowledge is required. For example, current fishing methods make it possible for us to harvest very large numbers of chinook salmon from the Columbia River, and demand for salmon encourages us to harvest as many as possible. To determine whether chinook salmon are sustainable, we must know how many there are now and how many there have been in the past. We must also understand the processes of birth and growth for this fish, its food requirements, its habitat, its life cycle, and so forth—all the factors that ultimately determine the abundance of salmon in the Columbia River. Consider, in contrast, the situation almost two centuries ago. When Lewis and Clark first made an expedition to the Columbia, they found many small villages of Native Americans who depended in large part on the fish in the river for food. The human population was small, and the methods of fishing were simple. The maximum number of fish the people could catch probably posed no threat to the salmon. These people could fish without scientific understanding of numbers and processes. (This example does not suggest that pre-scientific societies lacked an appreciation for the idea of sustainability. On the contrary, many so-called primitive societies held strong beliefs about the limits of harvests.)

Placing a Value on the Environment

How do we place a value on any aspect of our environment? How do we choose between two different concerns? Environmental values can be based on four categories of justification: utilitarian, ecological, aesthetic, and moral. A **utilitarian justification** sees some aspect of the environment as valuable because it provides individuals with economic benefit or is directly necessary to their survival. For example, fishermen obtain their livelihood from the ocean and need a supply of fish so that they can continue to earn a living. An **ecological justification** is based on the value of some factor that is essential to larger life-support functions, even though it may not benefit an individual directly. For example, as discussed in the opening case study, mangrove trees provide habitat for marine fish. Although we do not eat mangroves, we may eat the fish that depend on them. As another example, burning coal and oil adds greenhouse gases to the atmosphere, which may lead to a change in climate that could affect the entire earth. These ecological reasons form a basis for the conservation of nature that is essentially enlightened self-interest.

Aesthetic justification has to do with our appreciation of the beauty of nature. For example, many people find wilderness scenery beautiful and would rather live in a world with wilderness than without it. The aesthetic justification is gaining a legal basis. The state of Alaska acknowledges that sea otters have an important role related to recreation: People observe and photograph the otters and enjoy viewing them in a wilderness setting. Many examples illustrate the importance of the aesthetic values of the environment. When people grieve following the death of a loved one, they typically seek out places with grass, trees, and flowers, and thus we decorate our graveyards. Although popular discussions of environmental issues might make aesthetic justifications seem superficial, in fact, beauty in their surroundings is of profound importance to people. Frederick Law Olmsted, the great American landscape planner, argued that plantings of vegetation provide medical, psychological, and social benefits and are essential to city life.[18]

Moral justification has to do with the belief that various aspects of the environment have a right to exist and that it is our moral obligation to allow them to continue or help them to persist. Moral arguments have been extended to many nonhuman organisms, to entire ecosystems, and even to inanimate objects.[19] For example, the historian Roderick Nash has written an article entitled "Do Rocks Have Rights?" which discusses such moral justification.[20] And the United Nations General Assembly World Charter for Nature, signed in 1982, states that species have a moral right to exist.

The analysis of such issues is the province of a new discipline known as *environmental ethics*. Another concern of environmental ethics is our obligation to future generations: Do we have a moral obligation to leave the environment in good condition for our descendants, or are we at liberty to use environmental resources to the point of depletion within our own lifetimes?

Summary

- Five threads, or themes, run through this text: the urgency of the population issue, the importance of urban environments, the need for sustainability of resources, the importance of a global perspective, and the role of science and values in the decisions we face.

- The human population grew at a rate unprecedented in history in the twentieth century. Population growth is *the* environmental problem.

- When the impact of technology is combined with the impact of population, the impact on the environment is multiplied.

- In an increasingly urban world, we must focus much of our attention on the environments of cities and on the effects of cities on the rest of the environment.

- Determining the earth's carrying capacity for people and levels of sustainable harvests of resources is difficult but crucial if we are to plan effectively to meet our needs in the future.

How Can We Preserve the World's Coral Reefs?

Coral reefs are among the largest, oldest, most diverse, and most beautiful communities of plants and animals. Today, many coral reefs have been seriously damaged or are at risk. Scientists estimate that approximately 10% have already been destroyed, while another 30% are threatened. The major threats to reefs are direct or indirect results of human activities. *Almost 60% of the world's reefs are threatened by activities including coastal development, destructive fishing practices, overexploitation of resources, and marine pollution.*[23]

The pieces of coral that most people know from souvenir and jewelry shops are limestone skeletons secreted by colonies of living animals related to sea anemones and jellyfish. Like their relatives, these small individual coral animals, or polyps, use tentacles equipped with stinging cells to capture food. In addition, polyps obtain nourishment from photosynthetic algae that live in their cells. When polyps die, their skeletons remain while the next generation of individuals secrete new material. Thus, reefs grow slowly by accretion. Coral reefs that exist today are 5,000 to 10,000 years old. By taking the brunt of the force of waves, coral reefs protect coastlines from erosion, a function that has been estimated to have a value of $50,000 a year per square foot. In addition, *reefs can provide humans with living resources (fish) and services (tourism, coastal protection) worth $375 billion per year.*[24]

Coral reefs provide homes for a vast variety of plants and animals. Approximately 25% of all marine organisms, about 1 million species, are associated with coral reefs. Reef organisms are the source of many useful chemicals and medicines, and scientists are currently searching for others. The plant and animal species found around coral reefs are linked in intricate ways, so that removing only one or two key elements may cause a catastrophic collapse. For example, overfishing in the waters off the Cook Islands in the South Pacific in the 1980s removed most of the reef's parrot fish and sea urchins, both of which feed on algae. Soon, algae overgrew the reef, and the entire community of reef life collapsed.

Coral reefs have long been the main source of protein for tropical people, who today number approximately 1 billion. With modern transportation and preservation methods, fish and other food organisms from coral reefs are now eaten by many other people as well. In fact, reef fish constitute about 15% of the entire worldwide catch. Unfortunately, many of the world's reefs are being overfished, so that some species are now rare and endangered.

Recently, some consumers, especially in Asian countries, have placed a high premium on eating fish that are alive when they reach restaurants. The demand for fish for aquariums also fuels the demand for live fish. To obtain live fish, many fishermen use dynamite to stun fish or cyanide to poison them temporarily. Both methods can kill or damage other organisms, and dynamite can destroy the reef material itself. When fish are lodged in crevices in the reef, fishermen may use crowbars to pry apart the coral so they can reach the fish. Fishing is not the only threat to coral reefs. The limestone material that forms the bulk of a reef is sometimes mined to use as construction material. Millions of tourists from around the world who flock to reef areas to fish, swim, dive, and enjoy their beauty pose an additional threat. But perhaps the greatest threat to coral reefs comes from increasing population in the tropics. Population densities greater than that of the New Jersey coast (about 500 people per square kilometer) are found in parts of tropical Asia and the Caribbean. Half a billion people live within one kilometer of a coral reef.[25] In many areas, raw or inadequately treated sewage is released near shore. Runoff from land development and deforestation adds to the burden of sediment and pollution. Degradation of the coral reef in Kaneohe Bay in Hawaii due to sewage and other runoff was dramatically reversed in the 1980s when the sewage was diverted to the open ocean. But with increasing urbanization and population growth around the bay in the 1990s, recovery slowed, and perhaps even reversed.

Critical Thinking Questions

1. How does the current state of the world's coral reefs illustrate each of the five key themes of this textbook?
2. What are the utilitarian, ecological, aesthetic, and moral justifications for preserving coral reefs?
3. If Maitri Visetak were making his living from fishing rather than farming shrimp, how might he view the preservation of coral reefs? What arrangements could be made to meet his needs but at the same time preserve coral reefs in his area?
4. What things can you do in your everyday life to contribute to the preservation of coral reefs?

- Awareness of how people at a local level affect the environment globally gives credence to the Gaia hypothesis. Future generations will need a global perspective on environmental issues.
- Placing a value on various aspects of the environment

requires knowledge and understanding of the science involved but also depends on our judgments concerning the uses and aesthetics of the environment and on our moral commitments to other living things and to future generations.

This chapter lays out the five themes that run through this book. They are:

- Human population growth is the fundamental environmental issue.
- Sustainability is the key to future availability of resources and environmental qualities.

- Our modern civilization is having global environmental effects.
- Our world is becoming increasingly urban, and urban areas have their own kinds of environmental effects, both inside the city and on the surrounding countryside.
- Any solution to environmental issues requires consideration of both knowledge and values.

These lead to key questions, including:

Human Population: What is more important: the quality of life of a person alive today or the quality of life of future generations?

Sustainability: What is more important: abundant resources today—as much as we want and can obtain—or the persistence of these resources for future generations?

Global Perspective: What is more important: the quality of your local environment or the quality of the global environment—the environment of the entire planet?

Urban World: What is more important: human creativity and innovation, including arts, humanities, and science, or the persistence of certain endangered species? Must this always be a tradeoff, or are there ways to have both?

Values and Science: Does Nature know best, so that we never have to ask what environmental goal we should seek, or do we need knowledge about our environment, so that we can make the best judgments, given available information?

● KEY TERMS

aesthetic justification **10**	ecological justification **10**	moral justification **10**	utilitarian justification **10**
carrying capacity **7**	Gaia hypothesis **8**	sustainability **7**	

● STUDY QUESTIONS

1. Refer to Figure 1.5. How would you respond to the statement "the catch of salmon in the Columbia River is sustainable at 1960s levels"?

2. In what ways do the effects on the environment of a resident of a large city differ from the effects of someone living on a farm? In what ways are the effects similar?

3. Programs were established to supply food from Western nations to feed starving people in Africa. Some people argue that such food programs, which may have short-term benefits, actually increase the threat of starvation in the future. What are the pros and cons of international food relief programs?

4. What are the values involved in deciding whether to create an international food relief program? What are five kinds of information required to determine the long-term effects of such programs?

5. Which of the following are global environmental problems? Why?

   a. The growth of the human population.

   b. The furbish lousewort, a small flowering plant found in the state of Maine. It is so rare that it has been seen by few people and is considered endangered.

   c. The blue whale, listed as an endangered species under the U.S. Marine Mammal Protection Act.

d. A car that has air-conditioning.

e. Seriously polluted harbors and coastlines in major ocean ports.

6. How could you determine **(a)** the carrying capacity and **(b)** the sustainable yield of salmon in the Columbia River? (Refer to Figure 1.5.)

7. Is it possible that all the land on Earth will sometime in the future become one big city? If not, why not? To what extent does the answer depend on:

a. Global environmental considerations?

b. Scientific information?

c. Values?

● FURTHER READING

Botkin, D. B. 2000. *No Man's Garden: Thoreau and a New Vision for Civilization and Nature.* Discusses many of the central themes of this textbook, with special emphasis on values and science and an urban world. Henry David Thoreau's life and works illustrate approaches that can be beneficial in our dealing with modern environmental issues.

Botkin, D. B. 1990. *Discordant Harmonies: A New Ecology for the 21st Century.* New York: Oxford University Press. An analysis of the myths that lie behind attempts to solve environmental issues.

Kent, M. M. 1990. *World Population: Fundamentals of Growth.* Washington, D.C.: Population Reference Bureau. Facts and data about the growing human population.

Kessler, E., ed. 1992. "Population, Natural Resources and Development," *AMBIO* 21(1). A special issue of the journal *AMBIO*, addressing many problems concerning human population growth and its economic and environmental implications.

Leopold, A. 1949. *A Sand County Almanac.* New York: Oxford University Press. Perhaps, along with Rachel Carson's *Silent Spring*, one of the most influential books of the post–World War II and pre–Vietnam War era about the value of the environment. Leopold defines and explains the land ethic and writes poetically about the aesthetics of nature.

Lutz, W. 1994. *The Future of World Population.* Washington, D.C.: Population Reference Bureau. Summary of current information on population trends and future scenarios of fertility, mortality, and migration.

Nash, R. F. 1988. *The Rights of Nature: A History of Environmental Ethics.* Madison: University of Wisconsin Press. An introduction to environmental ethics.

2 Thinking Critically about the Environment

Forester recording measurements as part of collecting data which is part of the scientific method.

● LEARNING OBJECTIVES

Science is not a collection of facts to be memorized. It is a process of refining our understanding of nature through continual questioning and active investigation. After reading this chapter, you should understand:

- That thinking about environmental issues involves thinking scientifically.
- That scientific knowledge is acquired through observations of the natural world that can be tested through additional observations and experiments.
- The difference between deductive and inductive reasoning and how both are used in scientific thinking.
- That every measurement involves some degree of approximation—that is, uncertainty—and that a measurement without a statement about its degree of uncertainty is meaningless.
- That scientific discovery involves a number of processes, including the scientific method, and that science and scientists are too diverse to be described by just one method.
- That technology is not science but science and technology interact, stimulating growth in each other.
- That decision making about environmental issues involves society, politics, culture, economics, and values as well as scientific information.

The Case of the Mysterious Crop Circles

(a)

For 13 years, circular patterns appeared "mysteriously" in grain fields in southern England (Figure 2.1). Proposed explanations included aliens, electromagnetic forces, whirlwinds, and pranksters. The mystery generated a journal and a research organization headed by a scientist, as well as a number of books, magazines, and clubs devoted solely to crop circles. Scientists from Great Britain and Japan brought in scientific equipment to study the strange patterns. Then, in September 1991, two men confessed to having created the circles by entering the fields along paths made by tractors (to disguise their footprints) and dragging planks through the fields. When they made their confession, they demonstrated their technique to reporters and some crop circle experts.

Figure 2.1 • (*a*) A series of crop circles and other shapes at the Vale Pwesey in southern England in July 1990. (*b*) Crop circles seen from the air make distinctive patterns.

(b)

• *How is it that so many people, including some scientists, took the crop circles seriously? The answer is that they misunderstood the scientific method and engaged in fallacious reasoning. The failure of some people to think critically about crop circles resulted in no harm, but the same type of thinking applied to other, more serious environmental issues can have serious consequences.*

2.1 Thinking about Environmental Science

Thinking about the environment is as old as our first human ancestors. Their survival depended on knowledge of the environment. Today, the environment plays a crucial role in the development of each of us; normal human development does not occur in the absence of environmental stimuli. In modern technological society, the environment for most of us is what we encounter in our everyday lives.

Thinking *scientifically* about the environment is only as old as science itself. Science had its beginnings in the ancient civilizations of Babylonia and Egypt. There, observations of the environment were carried out primarily for practical reasons, such as planting crops, or for religious reasons, such as using the positions of the planets and stars to predict human events. These ancient practices differed from modern science in that they did not distinguish between science and technology or between science and religion.

These distinctions first appeared in classical Greek science. Because of their interest in ideas, the Greeks developed a more theoretical approach to science. Knowledge for its own sake, rather than for practical purposes, became the primary goal. At the same time, the Greeks' philosophical approach began to move science away from religion and toward philosophy.

Modern science is usually considered to have its roots in the end of the sixteenth and the beginning of the seventeenth centuries, with the development of the scientific method by Gilbert (magnets), Galileo (physics of motion), and Harvey (circulation of blood). Unlike earlier classical scientists who asked "why"? in the sense of "for what purpose?" they made important discoveries by asking "how"? in the sense of "how does it work?" Galileo also pioneered the use of numerical observations and mathematical models. The scientific method, which quickly proved to be very successful in advancing knowledge, was first described explicitly by Francis Bacon in 1620. Although not a practical scientist himself, Bacon recognized the importance of the scientific method, and his writings did much to promote scientific research.[3]

Our cultural heritage, then, gives us two ways of thinking about the environment: the kind of thinking we do in everyday life and the kind of thinking scientists do (see Table 2.1). There are many similarities between the two, which helps us to realize that thinking scientifically is something all of us can do. At the same time, there are crucial differences. Ignoring these differences can lead to invalid conclusions, in general, and to serious errors in decisions about the environment, in particular.[4]

The complexity of environmental sciences suggests several questions: How can we understand ecological phenomena? How can we take a scientific approach to the great, ancient questions about nature around us, the source of great wonder at the complexity of life and the marvelous adaptations of living things? How can we seek answers to practical questions about the effects of people on nature and the actions people should take in solving environmental problems?

To begin to tackle the problem of a scientific understanding of life and its environment, it is useful to review the fundamentals of the scientific method and consider in what ways ecology can fit that mold and in what ways it may need to cast new molds. The next sections of this chapter deal with the scientific method and its applications to environmental issues.

Science as Process

Science is a process, a way of knowing. It results in conclusions, generalizations, and sometimes scientific theories and even scientific laws. These comprise a body of beliefs. Often people confuse the *process* of science with a fixed set of beliefs—the *results*. But science does not lead so much to a fixed set of beliefs as to a set of beliefs that, at the present time, account for all known observations about a kind of phenomenon.

Table 2.1   Knowledge in Everyday Life Compared with Knowledge in Science

Parameter	*Everyday Life*	*Science*
Goal	To lead a satisfying life (implicit)	To know, predict, and explain (explicit)
Requirements	Context-specific knowledge; no complex series of inferences; can tolerate ambiguities and lack of precision	General knowledge; complex, logical sequences of inferences; must be precise and unambiguous
Resolution of questions	Through discussion, compromise, consensus	Through observation, experimentation, logic
Understanding	Acquired spontaneously through interacting with world and people; criteria not well defined	Pursued deliberately; criteria clearly specified
Validity	Assumed, no strong need to check; based on observations, common sense, tradition, authorities, experts, social mores, faith	Must be checked; based on replications, converging evidence, formal proofs, statistics, logic
Organization of knowledge	Network of concepts acquired through experience; local, not integrated	Organized, coherent, hierarchical, logical; global, integrated
Acquisition of knowledge	Perception, patterns, qualitative; subjective	Plus formal rules, procedures, symbols, statistics, mental models; objective
Quality control	Informal correction of errors	Strict requirements for eliminating errors and making sources of error explicit

Source: Based on F. Reif and J. H. Larkin, "Cognition in Scientific and Everyday Domains: Comparison and Learning Implications," *Journal of Research in Science Teaching*, 28(9), pp. 733–760. Copyright © 1991 by National Association for Research in Science Teaching. Reprinted by permission of John Wiley & Sons.

Science is a process of discovery—a continuing process whose essence is change in ideas. Often, the fact that scientific ideas change seems frustrating. Why cannot scientists agree on what is the best diet for people? Why is a chemical considered dangerous in the environment for a while and then determined not to be? Why do scientists in one decade believe that fire in nature is an undesirable disturbance and in a later decade decide that it is important and natural? Is there going to be global warming or not? Can't scientists just find out the truth for each of these questions, once and for all, and agree on it?

Rather than looking to science for answers to such questions, it is more accurate to think of science as a continuing adventure with increasingly better approximations about how the world works. Sometimes changes in ideas are small, and the major context remains the same. Sometimes a science undergoes a fundamental revolution in ideas.

 ## 2.2 Understanding What Science Is (and What It Isn't)

Science is one way of looking at the world. It begins with **observations** about the natural world. From these observations, scientists formulate hypotheses that can be tested. Modern science does not deal with things that cannot be tested by observation, such as the ultimate purpose of life or the existence of a supernatural being. Science also does not deal with questions that involve values, such as standards of beauty or issues of good and evil. The criterion by which we decide whether a statement is in the realm of science is whether it is possible, at least in principle, to disprove the statement.

Disprovability

It is generally agreed today that the essence of the scientific method is **disprovability**. *A statement can be said to be scientific if someone can state a method by which it could be disproven.* Thus, if you can think of a test that could disprove a statement, then that statement can be said to be scientific. If you cannot think of such a test, then the statement is said to be nonscientific. The distinction between a scientific statement and a nonscientific statement is not a value judgment—that is, the distinction is not meant to imply that science is the only "good" kind of knowledge. The distinction is simply a philosophical one involving kinds of knowledge and kinds of logic.

Although there are many other ways of looking at the world, such as the religious, aesthetic, and moral views, they are not science. The assertions made in those worldviews are not disprovable in the scientific sense; they are based ultimately on faith, beliefs, and cultural and personal values. To say that these other ways of looking at the world are not science is not to denigrate them. Each

way of viewing the world gives us a different way of perceiving and of making sense of our world, and each is valuable to us.

Assumptions of Science

Science makes certain assumptions about the natural world. It is important to be aware of these assumptions in order to understand what science is.

1. Events in the natural world follow patterns that can be understood through careful observation and analysis.
2. The basic patterns, or rules, that describe the behavior of events in the natural world are the same throughout the universe.
3. Science is based on a type of reasoning known as induction; it begins with specific observations about the natural world and extends to generalizations.
4. Generalizations can be subjected to tests that may disprove them. If such a test cannot be devised, then a generalization cannot be treated as a scientific statement.
5. Although new evidence can disprove existing scientific theories, science can never provide absolute proof of the truth of its theories.

The Nature of Scientific Proof

One source of serious misunderstanding about science is the use of the word *proof*, which most students encounter in mathematics, particularly in geometry.

Proof in mathematics and logic involves reasoning from initial definitions and assumptions. If a conclusion follows logically from these **premises**, the conclusion is said to be proved. This process is known as **deductive reasoning**.

An example of deductive reasoning is the following syllogism, or series of logically connected statements.

> *Premise: A straight line is the shortest distance between two points.*
> *Premise: The line from A to B is the shortest distance between points A and B.*
> *Conclusion: Therefore, the line from A to B is a straight line.*

Note that the conclusion in this syllogism follows directly from the premises.

Deductive proof does not require that the premises be true, only that the reasoning be foolproof. Logically valid but untrue statements can result from false premises, as in the following example:

> *Premise: Humans are the only toolmaking organisms.*
> *Premise: The woodpecker finch uses tools.*
> *Conclusion: Therefore, the woodpecker finch is a human being.*

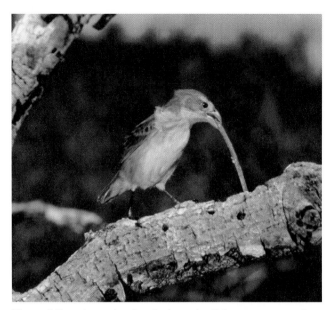

Figure 2.2 • A woodpecker finch on the Galapagos uses a twig to remove insects from a hole in a tree, demonstrating tool use by nonhuman animals.

In this case, the concluding statement *must* be true if both the preceding statements are true. However, we know that the conclusion is not only false but ridiculous (Figure 2.2). If the second statement is true (which it is), then the first cannot be true, and the conclusion must be false. The conclusion that a bird, the woodpecker finch, which uses cactus spines to remove insects from tree limbs, is human follows logically but defies common sense.

Because science is based on observations, its conclusions are only as true as the premises from which they are deduced. The rules of deductive logic govern only the process of moving from premises to conclusion. Science, in contrast, requires not only logical reasoning but also correct premises. To be scientific, the three statements should be expressed conditionally (with reservation):

> *If humans are the only toolmaking organisms*
>   *and*
> *the woodpecker finch is a toolmaker,*
>   *then*
> *the woodpecker finch is a human being.*

When we formulate generalizations based on a number of observations, we are engaging in **inductive reasoning**. To illustrate, let us define birds with particular characteristics as swans. If we always observe that such birds are white, we may make the *inductive statement* "All swans are white." What we really mean is, "All of the swans we have seen are white." We never know when our very next observation will turn up a bird that is like a swan in all ways except that it is black. Occasionally, mutations of black sheep occur, so it is not impossible that some swan, somewhere, may be black (Figure 2.3).

(a)

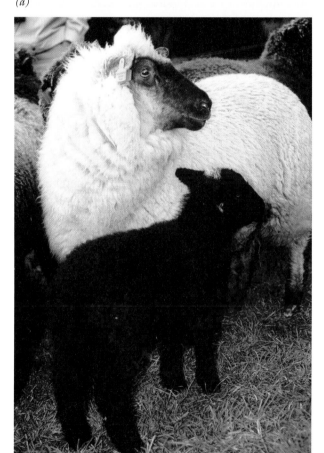

(b)

Figure 2.3 • (a) A white swan and offspring, and (b) a black lamb with its mother.

Proof in inductive reasoning is, therefore, very different from proof in deductive reasoning. In induction, we can only state that it is *usually* true that all swans are white; we cannot say so with absolute certainty. When we say something is proved in induction, what we really mean is that it has a very high degree of probability.

Probability is a way of expressing our certainty (or uncertainty)—our estimation of how good our observations are, how confident we are of our predictions.

When we have a fairly high degree of confidence in our conclusions in science, we often forget to state the degree of certainty or uncertainty. Instead of saying, "There is a 99.9% probability that...," we say, "It has been proved that...." Unfortunately, many people interpret this as a deductive statement, meaning the conclusion is absolutely true, which has led to much misunderstanding about science.

Although science begins with observations and, therefore, inductive reasoning, deductive reasoning is useful in helping scientists analyze whether conclusions based on inductions are logically valid. Scientific reasoning combines induction and deduction—different but complementary ways of thinking.[5]

What we have just described is the classic scientific method. Scientific advances, however, often begin with instances of insight—leaps of imagination, which are then subjected to the stepwise inductive process. Some scientists have made major advances by being in the right place at the right time and knowing the right things at the right time. For example, penicillin was discovered "by accident" in 1928 when Sir Alexander Fleming was studying the pus-producing bacteria *Staphylococcus aureus*. A culture of this bacteria was accidentally contaminated by the green fungus *Penicillium notatum*. Fleming noticed that the bacteria did not grow in those areas of a culture where the fungus grew. He isolated the mold, grew it in a fluid medium, and found that it produced a substance that killed many of the bacteria that caused diseases. Eventually, this discovery led other scientists to develop an injectable agent to treat diseases. *Penicillium notatum* is a common mold found on stale bread. No doubt many others had seen the mold, perhaps even noticing that other strange growths on bread did not overlap with *Penicillium notatum*. It took the knowledge that Fleming had and his observational ability for this piece of "luck" to occur.

2.3 Measurements and Uncertainty

People in general put more faith in the accuracy of measurements than do scientists. Scientists realize that *every measurement is only an approximation*. Measurements are limited; they depend on the instruments used and the people who use the instruments. Measurement uncertainties are inevitable; they can be reduced, but they can never be completely eliminated. Because all measurements have some degree of uncertainty, a measurement is meaningless unless it is accompanied by an estimate of its uncertainty.

Imagine a simplified scenario in which an engineer is given a rubber O-ring used to seal fuel gases in a space shuttle. She is asked to determine the flexibility of the O-rings under different temperature conditions to help answer the question, "At what temperature do the O-rings become brittle and subject to failure?" "At what temperature(s) is it unsafe to fire the shuttle?"

After doing some tests, the engineer says that the rubber becomes brittle at -1°C (30°F). Is it safe to fire the shuttle at 0°C (32°F)? At this point, you do not have enough information to answer the question. You assume that the temperature data may have some degree of uncertainty, but you have no idea how large it is. Is the uncertainty ±5°C, ±2°C, or ±0.5°C? In order to make a reasonably safe and economically sound decision about whether to fire the shuttle, you must know the uncertainty of the measurement.

Dealing with Uncertainties

There are two sources of uncertainty. One is the real variability of nature. The other is the fact that every measurement has some error. *Measurement* uncertainties and other errors that occur in experiments are called *experimental errors*. Errors that occur consistently, such as those resulting from incorrectly calibrated instruments, are *systematic errors*, in contrast to *random errors*, which occur with no apparent consistency.

Scientists traditionally include a discussion of experimental errors when they report results. Often, error analysis leads to greater understanding and sometimes even to important discoveries. For example, the existence of the eighth planet in our solar system, Neptune, was discovered when scientists investigated apparent inconsistencies in the orbit of the seventh planet, Uranus.

Measurement uncertainties can be reduced by improving the instruments used and by requiring a standard procedure in making measurements. They can also be reduced by using carefully designed experiments and appropriate statistical procedures. However, uncertainties are inherent in every measurement and can never be completely eliminated.

As difficult as it is for us to live with uncertainty, that is the nature of nature, as well as the nature of measurement and of science. We need to use our understanding of the uncertainties in measurements to read reports of scientific studies critically, whether they appear in science journals or in magazines and newspapers. (See A Closer Look 2.1: Measurement of Carbon Stored in Vegetation.)

Accuracy and Precision

Certain measurements have been made very carefully by many people over a long period of time, and accepted values have been determined. *Accuracy* is the extent to which a measurement agrees with the accepted value. *Precision* is the degree of exactness with which a quantity is measured. A very precise measurement may not be

Measurement of Carbon Stored in Vegetation

A number of people have suggested that a partial solution to global warming might be a massive worldwide program of tree planting. Trees take carbon dioxide (an important greenhouse gas) out of the air in the process of photosynthesis. Because they live a long time, trees can store carbon for decades, even centuries. But how much carbon can be stored? Many books and reports published during the past 20 years contained numbers representing the total stored carbon in the vegetation of the Earth, but all were presented without any estimate of error (Table 2.2). Without an estimate of that uncertainty, the figures are meaningless, yet important environmental decisions have been based on them.

Recent studies have reduced the error by replacing guesses and extrapolations with scientific sampling techniques, sim-ilar to the procedures used to predict the outcome of elections. Even these improved data would be meaningless, however, without an estimate of error. The new figures show that the earlier estimates were three to four times too large, grossly overestimating the storage of carbon in vegetation and, therefore, the contribution tree planting could make in offsetting global warming.

Table 2.2   Estimates of Above-Ground Biomass in the North American Boreal Forest

Source	Biomass[a] (kg/m²)	Carbon[b] (kg/m²)	Total Biomass[c] (10⁹ metric tons)	Total Carbon[c] (10⁹ metric tons)
This study[d]	4.2 ± 1.0	1.9 ± 0.4	22 ± 5	9.7 ± 2
Previous estimates[e]				
1	17.5	7.9	90	40
2	15.4	6.9	79	35
3	14.8	6.7	76	34
4	12.4	5.6	64	29
5	5.9	2.7	30	13.8

Source: D. B. Botkin and L. Simpson, 1990, "The First Statistically Valid Estimate of Biomass for a Large Region," *Biogeochemistry*, 9 (1990) 161–274. Reprinted by permission of Kluwer Academic, Dordrecht, The Netherlands.

[a]Values in this column are for total above-ground biomass. Data from previous studies giving total biomass have been adjusted using the assumption that 23% of the total biomass is in below-ground roots. Most references use this percentage; Leith and Whittaker use 17%. We have chosen to use the larger value to give a more conservative comparison.

[b]Carbon is assumed to be 45% of total biomass following R. H. Whittaker, *Communities and Ecosystems* (New York: Macmillan, 1974).

[c]Assuming our estimate of the geographic extent of the North American boreal forest: 5,126,427 km² (324,166 mi²).

[d]Based on a statistically valid survey; above-ground woodplants only.

[e]Lacking estimates of error: (1) G. J. Ajtay, P. Ketner, and P. Duvigneaud, "Terrestrial Primary Production and Phytomass," in *The Global Carbon Cycle*, ed. B. Bolin, E. T. Degens, S. Kempe, and P. Ketner (New York: Wiley, 1979), pp. 129–182. (2) R. H. Whittaker and G. E. Likens, "Carbon in the Biota," in *Carbon and the Biosphere*, eds. G. M. Woodwell and E. V. Pecan (Springfield, Va.: Nat. Tech. Infor. Center, 1973), pp. 281–300. (3) J. S. Olson, H. A. Pfuderer, and Y. H. Chan, *Changes in the Global Carbon Cycle and the Biosphere*, ORNL/EIS-109 (Oak Ridge, Tenn.: Oak Ridge National Laboratory, 1978). (4) J. S. Olson, I. A. Watts, and L. I. Allison, *Carbon in Live Vegetation of Major World Ecosystems*, ORNL-5862 (Oak Ridge, Tenn.: Oak Ridge National Laboratory, 1983). (5) G. M. Bonnor, *Inventory of Forest Biomass in Canada* (Petawawa, Ontario: Can. For. Serv., Petawawa Nat. For. Institute, 1985).

accurate. If the accepted value for the temperature of boiling water at sea level is 100.000°C (212.000°F) and you measure it as 99.875EC (211.775EF), your meas-urement is just as precise as the accepted value—that is, it is measured to the nearest 1/1000 degree—but it is still slightly inaccurate.

Although it is important to make measurements as pre-cisely as possible, it is equally important not to report measurements with more precision than they warrant. Doing so conveys a misleading sense of precision. A meas-urement based on a ruler divided into tenths of a cen-timeter should be precise to the nearest hundredth of a centimeter, but no more. A small bone that appears to end exactly at the 2.1-cm mark should be reported as 2.10 cm long, not as 2.100 cm. The latter would imply that the first zero was read directly, when in fact it was estimated.

Observations, the basis of science, may be made through any of the five senses or by instruments that measure beyond what we can sense. The accuracy of observations is checked by comparison of observations made by many scientists. When an observation is agreed on by all, or almost all, it is often called a **fact**.

An important distinction needs to be made between observations and **inferences**, or ideas based on observations. We might observe that a substance is a white, crystalline material with a sweet taste. We might infer that the substance is sugar. Before this inference can be accepted as fact, it must be subjected to further tests. Confusing observations with inferences and accepting untested inferences is the kind of sloppy thinking often described by the phrase "thinking makes it so."

When scientists wish to test an inference, they convert it into a statement that can be disproved. This type of statement is known as a **hypothesis**. A hypothesis continues to be accepted until it is disproved. A hypothesis that has not been disproved has not been proved to be true in the deductive sense. It has only been found to be probably true until, and unless, evidence to the contrary is found.

The most typical hypotheses take the form of if-then statements. For example, "If I apply more fertilizer, then the tomato plants will produce larger tomatoes." The statement relates two conditions: amount of fertilizer applied and size of tomatoes. Each condition is called a variable—that is, something that can vary. The size of tomatoes is called the **dependent variable** because it is assumed to depend on the amount of fertilizer, the **independent variable**. The independent variable is also sometimes called a **manipulated variable**, because the scientist deliberately changes, or manipulates, it. The dependent variable is then referred to as a **responding variable**—one that responds to changes in the manipulated variable.

In growing tomato plants, many variables may exist. Some, like the position of the North Star, can be assumed to be irrelevant. Others, like the duration of daylight, are potentially relevant. In testing a hypothesis, a scientist tries to keep all relevant variables constant, except for the independent and dependent variables. This practice is known as *controlling variables*. In a **controlled experiment**, the experimental is compared to a standard, or control—an exact duplicate of the experiment except for the condition of the one variable being tested (the independent variable). Any difference in outcome (dependent variable) between the experiment and the control can be attributed to the effect of the independent variable.

An important aspect of science, but one often overlooked in descriptions of the scientific method, is the need to define or describe variables in very exact terms that can be understood by all scientists. The least ambiguous way to define or describe a variable is in terms of what one would have to do to duplicate the variable's measurement.

Such definitions are called **operational definitions**. Cooking recipes are examples of operational definitions. Before carrying out an experiment, both the independent and dependent variables must be defined operationally. Operational definitions allow other scientists to repeat experiments exactly and to check on the results reported.

While performing an experiment, a scientist must record the values of the input (independent variable) and the output (dependent variable). These values are referred to as *data* (singular *datum*). They may be numerical, **quantitative data** or nonnumerical, **qualitative data**. In our example, qualitative data would be the size of tomatoes reported as small, medium, or large; quantitative data would be the mass in grams or diameter in centimeters.

Knowledge in an area of science grows as more hypotheses are supported. Scientists use accumulated knowledge to develop explanations, or models, that are consistent with currently accepted hypotheses. A *model* is "a deliberately simplified construct of nature."[6] It may be an actual working model, a pictorial model, a mental model, a computer model, a laboratory model, or a mathematical model. As new knowledge accumulates, models may have to be revised or replaced, with the goal of finding models more consistent with nature. Because hypotheses in science are continually being tested and evaluated by other scientists, science has a built-in self-correcting feedback system.[6]

Models may be *conceptual, mathematical,* or *numerical.* A conceptual model is a set of concepts expressed qualitatively. The phrase "The world is like a watch" is an example of a simple conceptual model. Often, in science, models are explicitly mathematical or numerical. A mathematical model is a set of equations describing how things work. A computer simulation is a numerical model. With a computer simulation, scientists can develop models of very complex systems. Because environmental systems are complex, computer simulations are becoming more and more useful in environmental science. Computer simulation of the atmosphere has become important in scientific analysis of the possibility of global warming. Computer simulation is becoming important for biological systems as well, such as simulations of forest growth (Figure 2.4). Models that offer broad, fundamental explanations of many observations are called **theories**.

The ideas discussed in this section are usually referred to as the scientific method and can be presented as a series of steps:

1. Make observations and develop a question about the observations.
2. Develop a tentative answer to the question—a hypothesis.
3. Design a controlled experiment to test the hypothesis (implies identifying and defining independent and dependent variables).
4. Collect data in an organized form, such as a table.
5. Interpret the data through graphic or other means.

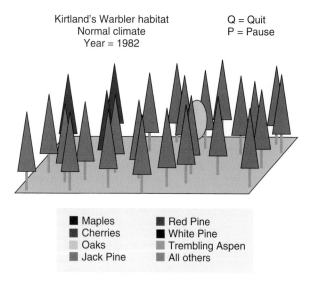

Kirtland's Warbler habitat
Normal climate
Year = 1982

Q = Quit
P = Pause

■ Maples   ■ Red Pine
■ Cherries ■ White Pine
□ Oaks   ■ Trembling Aspen
■ Jack Pine ■ All others

Figure 2.4 • A computer simulation of forest growth. Shown here is a screen display of individual trees, whose growth is forecast year by year, depending on environmental conditions. In this computer run, only three types of trees are present. This kind of model is becoming increasingly important in environmental sciences. (*Source:* From *JABOWA-II* by D. B. Botkin. Copyright © 1993 by D. B. Botkin. Used by permission of Oxford University Press, Inc.)

6. Draw a conclusion from the data.
7. Compare the conclusion with the hypothesis and determine whether the results support or disprove the hypothesis.
8. If the hypothesis is accepted, conduct further experiments to test it. If the hypothesis is rejected, make additional observations and construct a new hypothesis (Figure 2.5).

 2.5 Misunderstandings about Science

The scientific method does not take into account the differences among the various disciplines of science. The logic of research is not the same in physics, for example, as in biology. In fact, within biology itself, research on evolution differs in significant ways from research on ecology. It is much more realistic to speak of the methods of science than of the scientific method.

Theory in Science and Language

A common misunderstanding about science arises from confusion between the use of the word *theory* in science and its use in everyday language. A scientific theory is a grand scheme that relates and explains many observations and is supported by a great deal of evidence. In contrast, in everyday usage, a theory can be a guess, a hypothesis, a prediction, a notion, a belief. We often hear the phrase "It's *just* a theory." That may make sense in everyday lan-

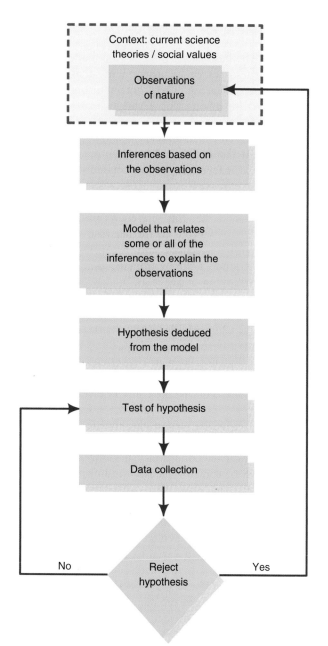

Figure 2.5 • Scientific investigation as a feedback process. (*Source:* Modified from C. M. Pease and J. J. Bull, *Bioscience* 42 [April 1992]:293–298.)

guage but not in the language of science. In fact, theories have tremendous prestige and are considered the greatest achievements of science.[7]

Further misunderstanding arises when scientists use the word *theory* in several different senses.[5] For example, we may encounter references to a currently accepted, widely supported theory, such as the *theory of evolution by natural selection;* a discarded theory, such as the *theory of inheritance of acquired characteristics;* a new theory, such as the *theory of evolution of multicellular organisms by symbiosis;* and a model dealing with a specific or narrow area of science, such as the *theory of enzyme action.*[7]

One of the most important misunderstandings about the scientific method is the relationship between research and theory. Although theory is usually presented as growing out of research, in fact, theories also guide research. The very observations a scientist makes occur in the context of existing theories. At times, discrepancies between observations and accepted theories become so great that a scientific revolution occurs; old theories are discarded and replaced with new or significantly revised theories.[8]

Science and Technology

Another misunderstanding about science occurs when science is confused with technology. Science is a search for understanding of the natural world, whereas technology is the control of the natural world for the benefit of humans. Science often leads to technological developments, just as new technologies lead to scientific discoveries. The telescope began as a technological device, but when Galileo used it to study the heavens, it became a source of new scientific knowledge. That knowledge stimulated the technology of telescope making so that better telescopes were produced, which in turn led to further advances in astronomy.

Science is limited by the technology available. Before the invention of the electron microscope, scientists were limited to magnifications of 1000 times and to studying objects about the size of 10ths of a micrometer. (A micrometer is 1/1,000,000 of a meter, or 1/1000 of a millimeter.) The electron microscope enabled scientists to view objects far smaller by magnifying more than 100,000 times. The electron microscope, a basis for new science, was also the product of science. Without prior scientific knowledge about electron beams and how to focus them, the electron microscope could not have been developed.

Most of us do not come in direct contact with science in our daily lives; instead, we come in contact with the products of science—technological devices such as cars, toasters, and microwave ovens. Thus, people tend to confuse the products of science with science itself. As you study science, it will help if you keep the distinction between science and technology in mind.

Science and Objectivity

One myth about science is the myth of objectivity, or value-free science—the idea that scientists are capable of complete objectivity independent of their personal values and the culture in which they live and that science deals only with objective facts. Objectivity is certainly a goal of scientists, but it is unrealistic to think that they can be totally free of influence from their social environments and personal values. A more realistic view is to admit that scientists do have biases and to try to identify these explicitly rather than ignore them. In some ways, this situation is similar to that of measurement error; it is inescapable, and we can best deal with it by recognizing it and estimating its effects.

We have only to look at recent controversies about environmental issues, such as whether or not to adopt more stringent automobile emission standards, to find examples of how personal and social values affect science. Genetic engineering, nuclear power, and the preservation of threatened or endangered species such as the spotted owl involve conflicts among science, technology, and society.

The idea that science is not entirely value free should not be taken to mean that fuzzy thinking is acceptable in science. It is still important to think critically and logically about science and related social issues. Without the high standards of evidence held up as the norm for science, we run the danger of accepting unfounded ideas about the world. When we confuse what we would like to believe with what we have the evidence to believe, we have a weak basis for making critical environmental decisions—decisions that could have far-reaching and serious consequences.

Science, Pseudoscience, and Frontier Science

Certain ideas are sometimes presented as scientifically valid but are in fact not scientific, because they are inherently untestable, lack empirical support, or are based on faulty reasoning or poor scientific methodology. Such ideas are referred to as **pseudoscientific** (*pseudo* means "false").

Pseudoscientific ideas arise from various sources, as shown in Figure 2.6. With more research, however, some of these frontier models may move into the realm of accepted science and new ideas will take their place at the advancing frontier.[9] Research may not support other

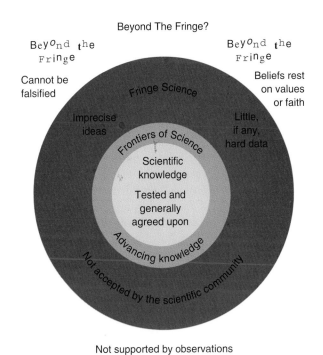

Figure 2.6 • Beyond the fringe?

hypotheses at the frontier, and these will be discarded by scientists.

Some people continue to believe in discarded scientific ideas, or pseudoscience. For example, although scientists long ago discarded the notion that the movements of the heavenly bodies could affect the personalities and fates of humans, a substantial number of Americans believe in or are not sure about astrology (in one study, 25% believed and 22% were not sure).[10] Many of the things that astrologers learned about the movements of the stars and planets were so accurate that they became the basis for astronomy, the scientific study of the heavens. Parts of astrology moved into accepted science, parts into pseudoscience.

Accepted science may merge into frontier science, which in turn may merge into more far-out ideas, or fringe science. Really wild ideas may be considered beyond the fringe.

Although scientists have no trouble distinguishing between accepted science and pseudoscience, they do have trouble identifying ideas at the frontier that will become accepted and those that will be relegated to pseudoscience. That trouble arises because science is a process of continual investigation. This ambiguity at the frontiers of science leads many people to accept some frontier science before it has been completely verified and to confuse pseudoscience with frontier science. (See the discussion of the Gaia hypothesis in Chapter 3.)

2.6 Environmental Questions and the Scientific Method

Environmental sciences deal with especially complex systems and include a relatively new set of sciences. Therefore, the process of study has not always proceeded neatly following the scientific method discussed earlier in this chapter. Often, observations are not used to develop formal hypotheses. Controlled laboratory experiments have been the exception rather than the rule. Much environmental research has been limited to field observations of processes and events that have been difficult to subject to controlled experiments.

Environmental research presents several obstacles to following the classic scientific method. The long time frame of many ecological processes relative to human lifetimes, professional lifetimes, and lengths of research grants poses problems for establishing statements that can, in practice, be subject to disproof. What do we do if a theoretical disproof through direct observation would take a century or more? Other obstacles include difficulties in setting up adequate experimental controls in field studies; in developing laboratory experiments that contain sufficient complexity; and in developing theory and models for complex systems. Throughout the text, we present differences between the "standard" scientific method and the actual approach that has been used in ecology. Here, we introduce an illustrative example.

An Example: The California Condor

As an example of an environmental question that has been difficult to investigate using the traditional scientific method, consider the problem of the California condor. In the 1970s, the total population of this species, America's largest bird in terms of wingspan, had declined to 22. The bird had never been abundant since European settlement of North America, but the numbers had declined over the years. Several suggestions were made in the early 1980s about what to do to save this species from extinction.

One suggestion was to remove all the condors from the wild and attempt to breed them in captivity in zoos. Then, once the numbers had increased, some of the birds could be released back into the wild. Another suggestion was to improve the habitat of the condors, especially by providing more food and returning more of their native range to grasslands through controlled burning. Grasslands had been more prevalent in the condors' habitat prior to the 1980s. Over much of the condors' natural range, fire had been suppressed during the twentieth century, and areas that had been comparatively open grasslands had reverted to dense shrublands called chaparral. It is difficult for the large condors to land and take off in dense shrubland and to see the carcasses of dead animals, on which they feed.

In theory, scientists could have taken half of the 22 remaining condors out of the wild and tried captive breeding and left the other half in the wild and modified their habitat. But 22 individuals was considered too few to take any chances with, and an experimental approach attempting two different kinds of treatment of the condors seemed too risky. The result was that all the condors were captured and a program of captive breeding begun (Figure 2.7). In the 1990s, enough condors had been bred that it was believed safe to try to reintroduce some of them into the wild. Some reintroductions have been attempted, but so far no condor reintroduced into the wild has become self-sufficient in finding food.

Perhaps the solution for the condor will be a combination of the two approaches discussed earlier: captive breeding and habitat restoration. But if so, these methods will occur in sequence rather than simultaneously. Perhaps some of the habitat modifications could be done as experiments in a way that would include controls (habitats not modified) and treatments (habitats subject to controlled burns and reintroduction of deer and other sources of food for the condors).

We will find that environmental sciences—young sciences dealing with complex phenomena—are filled with unanswered questions and that many of these questions do not seem open to the traditional scientific method. They require new approaches that retain the essential key of the scientific method: *the ability to disprove a statement*. The large number of unanswered questions and the difficulty in answering them may seem discouraging. But they

(a)

(b)

Figure 2.7 • (*a*) Condors in a captive breeding program. (*b*) Released condors in their natural wild habitat.

can also be viewed as an adventure in which doubt and uncertainty are a wonderful challenge. The famous Nobel-prize-winning physicist Richard Feynman expressed this idea when he said, "*You cannot understand science and its relation to anything else unless you understand and appreciate [science as] the great adventure of our time. You do not live in your time unless you understand that this is a tremendous adventure and a wild and exciting thing.*"[11] In this spirit, we can approach environmental science and seek a set of scientific methods that will allow us to answer its fascinating questions.

Some Alternatives to Direct Experimentation

So how have ecologists tried to answer these difficult questions? Several approaches have been taken, including use of historic records, modern catastrophes, and disturbances.

Historic Evidence

Ecologists have made use of both human and ecological historic records. A classic example is a study of the history of fire in the Boundary Waters Canoe Area (BWCA) of Minnesota, 1 million acres of boreal forests, streams, and lakes well known for recreational canoeing.

Murray (Bud) Heinselman had lived near the BWCA for much of his life and was instrumental in having it declared a wilderness area. A forest ecological scientist, Heinselman set out to determine the past patterns in fires in this wilderness. Those patterns are important to the maintenance of the wilderness. If the wilderness has been characterized by fires of a specific frequency, then it can be argued that this frequency is required to maintain the area in its most "natural" state.

Heinselman used three kinds of historic records: written records, tree-ring records, and buried records (fossil and pre-fossil organic deposits). Trees of the boreal forests, like most trees that are conifers or angiosperms (flowering plants), produce annual growth rings. If a fire burns through the bark of a tree, it leaves a scar, just as a serious burn leaves a scar on human skin. The tree grows over the scar, depositing a new growth ring for each year. (Figure 2.8 shows fire scars and tree rings on a cross section of a tree.) By examining cross sections of trees, it is possible to determine the date of each fire and the number of years between fires. From written and tree-ring records, Heinselman

Figure 2.8 • Cross section of a tree showing fire scars and tree rings.

found that since the 17th century, the BWCA forests had burned an average of once per century. However, the frequency of fires varied over time. Furthermore, buried charcoal dated using Carbon-14 revealed that fires could be traced back more than 30,000 years.

The three kinds of historic records developed important evidence about fire in the history of the BWCA. At the time Heinselman did his study, the standard hypothesis was that fires were bad for forests and should be suppressed. The historic evidence provided a disproof of this hypothesis. It showed fires to be a natural and integral part of the forest, and demonstrated that the forest persisted with fire for a very long time. Thus, the use of historic information meets the primary requirement of the scientific method—the ability to disprove a statement. *Historic evidence is a major source of data that can be used to test scientific hypotheses in ecology.*

Modern Catastrophes and Disturbances as Experiments

Sometimes, a large-scale catastrophe provides a kind of modern ecological experiment. The volcanic explosion of Mount St. Helens in 1980 supplied such an experiment, destroying vegetation and wildlife over a wide area (Figure 2.9). The recovery of plants, animals, and ecosystems following this explosion gave scientists insights into the dynamics of ecological systems. It provided some surprises. The main surprise was the rapidity with which vegetation recovered and wildlife returned to parts of the mountain. In other ways, the recovery followed expected patterns in ecological succession (see Chapter 9). Another recent example of a catastrophe that has become the subject of ecology research is the 1988 wildfire in Yellowstone National Park.

It is important to point out that the greater the quantity and the better the quality of ecological data *prior to* such a catastrophe, the more that can be learned from the response of ecological systems to the event. This calls for careful monitoring of the environment.

2.7 Science and Decision Making

Like the scientific method, the process of making decisions is sometimes presented as a series of steps:

1. Formulate a clear statement of the issue to be decided.
2. Gather the scientific information related to the issue.
3. List all alternative courses of action.
4. Predict the positive and negative consequences of each course of action and the probability that each consequence will occur.
5. Weigh the alternatives and choose the best solution.

Such a procedure is a good guide to rational decision making, but it assumes a simplicity not often found in real-world issues. It is difficult to anticipate all the consequences of a course of action, and unintended consequences are at the root of many environmental problems. Often the scientific information is incomplete and even controversial. For example, the insecticide DDT causes eggshells of birds that feed on insects to be so thin as to result in the death of unhatched birds. When DDT first came into use, this consequence was not predicted. Only when populations of species such as the brown pelican became seriously endangered did people become aware of it.

(a)

(b)

Figure 2.9 • (*a*) Mount St. Helens and its surroundings soon after the 1980 eruption, showing destruction of the forests. (*b*) Current state of some of the forests, showing recovery.

Evaluating Media Coverage

The following questions will help you evaluate a report on science[12]:

1. Where is the report? Is it in a scientific journal, a serious newspaper, a popular science magazine, or a tabloid?

2. Is the report based on observations of actual occurrences? Were those observations made by more than one person? More than a few?

3. Are the sources of the report identified specifically? Are the sources specific, named scientists, scientific journals, or scientific organizations?

4. Does the report provide evidence that the claims are supported by other members of the scientific community?

5. Does the evidence for the claims seem sufficient? Is there contradictory evidence that might offset the evidence given for them?

6. Do the claims follow logically from the evidence? Do the claims violate reason (for example, a 99-year-old woman gives birth to baby)? Is there a simpler explanation for the observations?

7. Is there a clear basis for suspecting bias on the part of the source or the writer of the report?

8. Can you describe the line of reasoning that leads from the evidence to the claims?

In the face of incomplete information, scientific controversies, conflicting interests, and emotionalism, how can environmental decisions be made? We need to begin with the scientific evidence from all relevant sources, with estimates of the uncertainties in each. Where scientists disagree about the interpretation of data, it may be possible to develop a consensus or a series of predictions based on the different interpretations. The impacts of the scenarios need to be identified and the risks associated with each analyzed in comparison with the benefits. Avoiding emotionalism and resisting slogans and propaganda are essential to developing sound approaches to environmental issues. Ultimately, however, environmental decisions are policy decisions negotiated through the political process. The makers of policies are rarely professional scientists. Rather, they are political leaders and ordinary citizens. The scientific education of those in government and business as well as of all citizens is therefore crucial.

 2.8 **Learning about Science**

Science is an open-ended process of finding out about the natural world. In contrast, science lectures and texts are usually summaries of the answers arrived at through this process, and science homework and tests are exercises in finding the right answer.

Therefore, students often perceive science as a body of facts to be memorized and lectures and texts as authoritative sources of absolute truths about the world. In contrast, scientists view scientific knowledge as *currently* accepted truth, always subject to change as new observations and interpretations are made. Students tend not to question the material in textbooks and lectures, whereas the essence of science is questioning and critical examination of accepted truths.

Viewing science as a collection of facts may lead students to see science as difficult to understand and remember. In contrast, scientists emphasize the collection of facts into coherent pictures of the world that have the power to explain many phenomena. These pictures (models, theories) are so powerful that scientists find them easy to remember and expect that students will also.

Learning science also frequently involves solving problems. Here, too, the attitudes of students and scientists differ. Where students may look to formulas, type problems, and algorithms to help solve problems, scientists look to general principles, critical thinking, and creativity.[13]

We encourage you to learn environmental science in an active mode, to be critical of what you hear, what you see, and what you read, including this textbook. We encourage you to try to understand what science is—its assumptions, limitations, and methods—and to apply this understanding to your study of environmental science. Most of all, we hope that you will see not isolated facts to be learned by rote but the connections among facts that reflect the interdependence and unity in the environment.

 2.9 **Science and Media Coverage**

Most media reports on scientific issues deal with new discoveries, fringe science, and pseudoscience. (See A Closer Look 2.2: Evaluating Media Coverage.) It is important to analyze the claims in such reports to decide in which category they should be placed. Critical reading and listening require thinking carefully about whether a statement is based on observations and data, an objective interpretation of data, an interpretation based on the experience of an expert in the area, or a subjective opinion.

How Do We Decide What to Believe about Environmental Issues?

When you read about an environmental issue in a newspaper or magazine, how do you decide whether to accept the claims made in the article? Are they based on scientific evidence, and are they logical?

Scientific evidence is based on observations, but media accounts often rely mainly on inferences (interpretations) rather than evidence. Distinguishing inferences from evidence is an important first step in evaluating articles critically.

Secondly, it is important to consider the source of a statement. Is the source a reputable scientific organization or publication? Does the source have a vested interest that might bias the claims? When sources are not named, it is impossible to judge the reliability of claims.

If a claim is based on scientific evidence presented logically from a reliable, unbiased source, it is appropriate to accept the claim tentatively, pending further information. Practice your critical evaluation skills by reading the article in the box and answering the critical thinking questions.

Critical Thinking Questions

1. What is the major claim made in the article?
2. What evidence does the author present to support the claim?
3. Is the evidence based on observations, and is the source of the evidence reputable and unbiased?
4. Is the argument for the claim, whether based on evidence or not, logical?
5. Would you accept or reject the claim?
6. Even if the claim were well supported by evidence based on good authority, why would your acceptance be only tentative?

CLUE FOUND IN DEFORMED FROG MYSTERY

BY MICHAEL CONLON
Reuters News Agency (as printed in the *Toronto Star*)
November 6, 1996

A chemical used for mosquito control could be linked to deformities showing up in frogs across parts of North America, though the source of the phenomenon remains a mystery. "We're still at the point where we've got a lot of leads that we're trying to follow but no smoking gun," says Michael Lannoo of Ball State University in Muncie, Ind. "There are an enormous number of chemicals that are being applied to the environment and we don't understand what the breakdown products of these chemicals are," says Lannoo, who heads the U.S. section of the world-wide Declining Amphibian Population Task Force.

He says one suspect chemical was Methoprene, which produces a breakdown product resembling retonoic acid, a substance important in development. "Retonoic acid can produce in the laboratory all or a majority of the limb deformities that we're seeing in nature," he says. "That's not to say that's what's going on. But it is the best guess as to what's happening." Methoprene is used for mosquito control, among other things, Lannoo says.

Both the decline in amphibian populations and the deformities are of concern because frogs and related creatures are considered "sentinel" species that can provide early warnings of human risk. The skin of amphibians is permeable and puts them at particular risk to agents in the water.

Lannoo says limb deformities in frogs had been reported as far back as 1750, but the rate of deformities showing up today was unprecedented in some species. Some were showing abnormalities that affected more than half of the population of a species living in certain areas, he adds. He says he doubted that a parasite believed to have been the cause of some deformities in frogs in California was to blame for similar problems in Minnesota and nearby states. Deformed frogs have been reported in Minnesota, Wisconsin, Iowa, South Dakota, Missouri, California, Texas, Vermont, and Quebec. The deformities reported have included misshapen legs, extra limbs, and missing or misplaced eyes.

The need for critical examination applies to statements made by scientists as well as by other sources. On the one hand, scientists' training in analyzing data may make them more qualified than the general public to decide certain complex issues; on the other, as discussed before, scientists' own values, interests, and cultural backgrounds may influence their interpretations of data and bias their pronouncements. The situation is complicated by the fact that some scientists become "media scientists," whose opinions are sought on a wide variety of topics, some of which they have not studied as scientists. The opinions of experts do need to be heard, but appealing to scientists as authorities outside their area of expertise is contrary to the antiauthoritarian nature of science.

Summary

- Science has a tradition of critical thinking about the natural world. Its goal is understanding how nature works. Decisions on environmental issues must begin with an examination of the relevant scientific evidence. However, environmental decisions also require careful analysis of economic, social, and political consequences. Solutions will reflect religious, aesthetic, and ethical values as well.

- Science begins with careful observations of the natural world, from which scientists formulate hypotheses. Whenever possible, scientists test hypotheses with controlled experiments.

- The process of formulating hypotheses based on observations or prior knowledge and testing the hypotheses through controlled experiments is referred to as the scientific method. Although the scientific method is often taught as a prescribed series of steps, it is better to think of it as a general guide to scientific thinking, with many variations.

- Scientific knowledge is acquired through inductive reasoning, in which general conclusions are based on specific observations. Conclusions arrived at through induction can never be proved with certainty. Because of the inductive nature of science, it is possible to disprove hypotheses, but it is not possible to prove them with 100% certainty.

- Science also makes use of deductive reasoning, in which conclusions follow necessarily from particular premises, provided that the rules of logic are followed. If the premises are true, then the conclusion is certainly true.

- Measurements are approximations that may be more or less exact, depending on the measuring instruments and the people who use the instruments. A measurement is meaningful when it is accompanied by an estimate of the degree of uncertainty, or error.

- Accuracy in measurement is the extent to which the measurement agrees with an accepted value. Precision is the degree of exactness with which a measurement is made. A precise measurement may not be accurate. The estimate of uncertainty provides information on the precision of a measurement.

- A general statement that relates and explains a great many hypotheses is called a theory. Theories are the greatest achievements of science. Like other scientific knowledge, however, they may be modified or discarded as new information is acquired. Existing theories provide a framework for the observations and hypotheses of science. New observations and hypotheses inconsistent with existing theories stimulate the development of new theories.

- Critical thinking can help us distinguish science from pseudoscience. It can also help us recognize possible bias on the part of scientists and in the media. Critical thinking involves questioning and synthesizing what is learned in order to achieve knowledge, rather than merely acquire information.

REEXAMINING THEMES AND ISSUES

This chapter summarizes the scientific method, which is essential to the analysis and solution of environmental problems. The scientific method is critical to understanding the **human population** problem and to developing sound approaches to **sustainability**. The **global perspective** on environment arises out of new findings in environmental science. The importance of our increasingly **urbanized world** is best understood with the assistance of scientific investigation.

Solutions to environmental problems require both **values** and **knowledge**. Understanding the scientific method is especially important if we are going to understand the connection between values and knowledge. Ultimately, environmental decisions are policy decisions, negotiated through the political process. Policymakers often lack sufficient understanding of the scientific method, leading to false conclusions.

Uncertainty is part of the nature of measurement and science. We must learn to accept uncertainty as part of our attempt to conserve and use our natural resources.

KEY TERMS

1. Which of the following are scientific statements and which are not? What is the basis for your decision in each case?

   a. The amount of carbon dioxide in the atmosphere is increasing.

   b. Picasso was the greatest painter of the twentieth century.

   c. Helping terminally ill people to die is wrong.

   d. The earth is flat.

   e. The good will reap their reward in the afterlife.

2. What is the logical conclusion of each of the following syllogisms? Which conclusions correspond to observed reality?

   a. All men are mortal.
   Socrates is a man.
   Therefore,

   b. All sheep are black.
   Mary's lamb is white.
   Therefore,

   c. All Amazons are women.
   No men are Amazons.
   Therefore,

   d. All elephants are animals.
   All animals are living beings.
   Therefore,

3. Which of the following statements are supported by deductive reasoning and which by inductive reasoning?

   a. The sun will rise tomorrow.

   b. The square of the hypotenuse of a right triangle is equal to the sum of the squares of the other two sides.

   c. Only male deer have antlers.

   d. If $A = B$ and $B = C$, then $A = C$.

   e. The net force acting on a body equals its mass times its acceleration.

4. The accepted value for the number of inches in a centimeter is 0.3937. Two students mark off a centimeter on a piece of paper and then measure the distance using a ruler (in inches). Student A finds the distance equal to 0.3827 in., and student B finds it equal to 0.39 in. Which measurement is more accurate? Which is more precise? If student B measured the distance as 0.3900 in., what would be your answer?

5. a. A teacher gives five students each a metal bar and asks them to measure the length. The measurements obtained are 5.03, 4.99, 5.02, 4.96, and 5.00 cm. How can you explain the variability in the measurements? Are these systematic or random errors?

   b. The next day, the teacher gives the students the same bars but tells them that the bars have contracted because they have been in the refrigerator. In fact, the temperature difference would be too small to have any measurable effect on the length of the bars. The students' measurements, in the same order as in part (a), are 5.01, 4.95, 5.00, 4.90, and 4.95 cm. Why are the students' measurements different from those of the day before? What does this illustrate about science?

6. Identify the independent and dependent variables in each of the following:

   a. Change in the rate of breathing in response to exercise.

   b. The effect of study time on grades.

   c. The likelihood that people exposed to smoke from other people's cigarettes will contract lung cancer.

7. a. Identify a technological advance that resulted from a scientific discovery.

   b. Identify a scientific discovery that resulted from a technological advance.

   c. What technological device have you used today? What scientific discoveries were necessary before the device could be developed?

8. What is fallacious about each of the following statements?

   a. A fortune cookie contains the statement, "A happy event will occur in your life." Four months later, you find a hundred dollar bill. You conclude that the fortune was correct.

   b. A person claims that aliens visited Earth in prehistoric times and influenced the cultural development of humans. As evidence, the person points to ideas among many groups of people about beings who came from the sky and performed amazing feats.

   c. A person observes that light-colored forms of an animal almost always occur on light surfaces, whereas dark forms of the same species occur on dark surfaces. The person concludes that the light surface causes the light color of the animals.

   d. A person knows three people who have had fewer colds since they began taking vitamin C on a regular basis. The person concludes that vitamin C prevents colds.

9. Find a newspaper article on a controversial topic. Identify some loaded words in the article—that is, words that convey an emotional reaction or a value judgment.

10. Identify some social, economic, aesthetic, and ethical issues involved in a current environmental controversy.

● FURTHER READING

American Association for the Advancement of Science (AAAS). 1989. *Science for All Americans*. Washington, D.C.: AAAS. This report focuses on the knowledge, skills, and attitudes a student needs to be scientifically literate.

Furnham, A. 1991 (January). "Hooked on Horoscopes," *New Scientist*, pp. 33–36.

Gould, S. J. 1996. *The Mismeasure of Man*. New York: Norton. The author takes on the controversial concept of biological determinism.

Grinnell, F. 1992. *The Scientific Attitude*. New York: Guilford Press. Examples from biomedical research are used to illustrate the processes of science (observing, hypothesizing, experimenting) and how scientists interact with each other and with society.

Kuhn, Thomas S., 1996. *The Structure of Scientific Revolutions*. Chicago: University of Chicago Press. A modern classic in the discussion of the scientific method, especially regarding major transitions in new sciences like environmental sciences.

McCain, G., and E. M. Segal. 1982. *The Game of Science*. Monterey, Calif.: Brooks/Cole Publishing. The authors present a lively look into the subculture of science.

Sagan, C. 1995. *The Demon-Haunted World*. New York: Random House. The author argues that irrational thinking and superstition threaten democratic institutions and discusses the importance of scientific thinking to our global civilization.

CHAPTER 3 Systems and Change

Amboseli National Park, Kenya, Africa. The mountain in the background is Mount Kilimanjaro.

● LEARNING OBJECTIVES

Changes in systems may occur naturally or may be induced by humans. Many complex and far-reaching interactions can result. After reading this chapter, you should understand:

● Why solutions to many environmental problems involve the study of systems and rates of change.
● How positive and negative feedback operate in a system.
● What are the implications of exponential growth and doubling time.
● That natural disturbances and changes in systems such as forests, rivers, and coral reefs are important to their continued existence.
● What an ecosystem is and why sustained life on Earth is a characteristic of ecosystems.
● What the Gaia hypothesis is and how life on Earth has affected Earth itself.
● What the principle of uniformitarianism is and how it can be used to anticipate future changes.
● Why the principle of environmental unity is important in studying environmental problems.
● How human activities amplify the effects of natural disasters.

Amboseli National Park

Environmental change is often caused by a complex web of interactions, and the most obvious answer to the question of what caused a particular change may not be the right answer. Amboseli National Park is a case in point. In the short span of a few decades, this park, located in southern Kenya at the foot of Mount Kilimanjaro, underwent a significant environmental change. An understanding of physical, biological, and human-use factors—and how these factors are linked—was needed to explain what happened.

Figure 3.1 shows the boundary of the park and the major geologic units. The park is centered on an ancient lake bed, remnants of which include the seasonally flooded Lake Amboseli and some swampland. Mount Kilimanjaro is a well-known volcano, composed of alternating layers of volcanic rock and ash deposits. Rainfall that reaches the slopes of Mount Kilimanjaro infiltrates the volcanic material (becomes groundwater) and moves slowly down the slopes to saturate the ancient lake bed, eventually emerging at springs in the swampy, seasonally flooded land. The groundwater becomes very saline (salty) as it percolates through the lake bed, since salt stored in the lake bed sediments dissolves easily when the sediments are wet.

Before the mid-1950s, the dominant vegetation in the park was fever tree woodlands, which provided habitat for mammalian species such as kudu, baboons, vervet monkeys, leopards, and impalas. Starting in the 1950s, and accelerating in the 1960s, these woodlands disappeared and were replaced by short grass and brush, which provided habitat for typical plains animals such as zebras and wildebeest. During these decades, the annual amounts of rainfall increased.

Loss of the woodland habitat was initially blamed on overgrazing of cattle by the Masai people (Figure 3.2) and damage to the trees from elephants (Figure 3.3). Environmental scientists eventually rejected these

Figure 3.1 • Generalized geologic and landform map of Amboseli National Park, southern Kenya, Africa, and Mount Kilimanjaro. (*Source:* After T. Dunn and L. B. Leopold, *Water in Environmental Planning* [San Francisco: W. H. Freeman, 1978.]

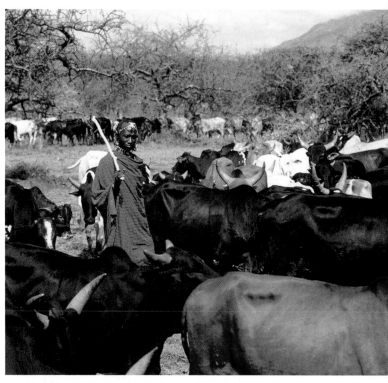

Figure 3.2 • Masai people grazing cattle in Amboseli National Park, Kenya. Grazing activities were prematurely blamed for loss of fever tree woodlands.

Figure 3.3 • Elephant feeding on a yellow-bark acacia tree. Elephant damage to trees is considered a factor in loss of woodland habitat in Amboseli National Park. However, elephants probably play a relatively minor role compared with oscillations in climate and groundwater conditions.

the woodlands that suffered the least decline had the highest density of people and cattle. These observations suggested that overgrazing by cattle was not responsible for loss of the trees. Elephant damage was thought to be a major factor because elephants had stripped bark from more than 83% of the trees in some areas and had pushed over some younger, smaller trees. However, researchers concluded that elephants played only a secondary role in changing the habitat. As the density of fever trees and other woodland plants decreased, the incidence of damage caused by elephants increased. In other words, elephant damage interacted with some other, primary factor in changing the habitat.

Research on rainfall, groundwater history, and soils suggested that the Amboseli National Park area is very sensitive to changing amounts of rainfall. During dry periods, the salty groundwater sinks lower in the earth, and soil near the surface has a relatively low salt content. The fever trees grow well in the nonsalty soil. During wet periods, the groundwater rises close to the surface, bringing with it salt, which invades the root zones of trees and kills them. The groundwater level rose as much as 3.5 m (11.4 ft) in response to unusually wet years in the 1960s. Analysis of the soils confirmed that the tree stands that suffered the most damage were those associated with highly saline soils.[1,2] As the trees died, they were replaced by salt-tolerant grasses and low brush.[1,2]

Evaluation of the historic record from early European explorers, accounts from Masai herders, and fluctuating lake levels in other East African lakes suggested that before 1890, there had been another period of above-normal rainfall and loss of woodland environment. Thus, the scientists concluded that cycles of greater and lesser rainfall change hydrology and soil conditions, which in turn change the plant and the animal life of the area.[1] Cycles of wet and dry periods can be expected to continue, and associated with these will be changes in the soils, distribution of plants, and abundance and types of animals present.[1]

hypotheses as the main causes of the environmental change. Their careful work showed that changes in rainfall and soils were the primary culprits, rather than people or elephants.[1,2]

Let us look at the evidence more closely to see why the scientists rejected the overgrazing and elephant-damage hypotheses. Investigators were surprised to note that most dead trees were in an area that had been free of cattle since 1961, which was before the major decline in the woodland environment. Furthermore, some of

• *The Amboseli story illustrates how environmental science attempts to work out sequences of events that follow a particular change. At Amboseli, rainfall cycles change hydrology and soil conditions, which in turn change the vegetation and animals of the area. In this chapter, we examine environmental systems and changes that occur as a result of natural and human-induced processes.*

3.1 Systems and Feedback

In environmental science at every level, we must deal with a variety of systems that range from simple to complex. We must therefore understand systems and how different parts of systems interact with one another. We begin our discussion of systems by defining what a system is and how it can operate.

System Defined

A **system** is a set of components or parts that function together to act as a whole. A single organism (such as your body) is a system, as is a sewage treatment plant, a city (Figure 3.4), a river (Figure 3.5), and your dorm room. On a much different scale, the entire earth is a system.

Systems may be open or closed. An **open system** is not generally contained within boundaries, and some energy or material (solid, liquid, or gas) moves into or out of the system. The ocean is an open system with regard to water, because water moves into the ocean from the atmosphere and out of the ocean to the atmosphere. Conversely, in a **closed system**, no such movements take place. Earth is a closed system (for all practical purposes) with regard to material.

Systems respond to inputs and have outputs. Your body, for example, is a complex system. If you are hiking and see a grizzly bear, the sight of the bear is an input. Your body reacts to that input: The adrenaline level in your blood goes up, your heart rate increases, and the hair on your head and arms may rise. Your response—perhaps moving slowly away from the bear—is an output. We call these responses feedback.

Feedback

Feedback occurs when the output of the system also serves as an input and leads to further changes in the system. A good example of feedback is human temperature regulation. If you go out in the sun and get hot, the increase in temperature affects your sensory perceptions (input). If you stay in the sun, your body responds physiologically: Your skin pores open, and you are cooled by evaporating water (you sweat). The cooling is output, and it is also input to your sensory perceptions. You may also respond behaviorally: Because you feel hot (input), you walk into the shade and your temperature returns to normal.

In our example, an increase in temperature is followed by a response that leads to a decrease in temperature. This is an example of **negative feedback**, in which an increase in output leads to a later decrease. Negative feedback is self-regulating, or stabilizing; it usually keeps a system in a relatively constant condition.

Positive feedback occurs when an increase in output leads to a further increase in the output. A fire starting in a forest provides an example of positive feedback. The

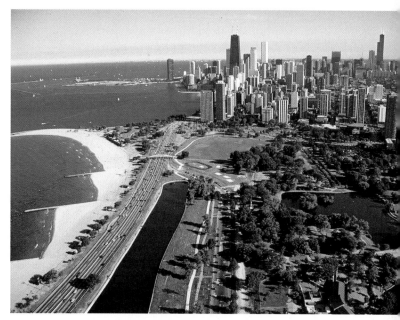

Figure 3.4 • Lake Michigan, Lincoln Park, and the city of Chicago make up an example of a large, complex urban system that includes air, water, and land resources. Urban systems are particularly important in environmental science because more and more people are living in urban areas.

wood may be slightly damp at the beginning and so may not burn well. Once a fire starts, wood near the flame dries out and begins to burn, which in turn dries out a greater quantity of wood and leads to a larger fire. The larger the fire, the more wood is dried, and the more rapidly the fire increases. Positive feedback, sometimes called "a vicious cycle," is destabilizing.

Figure 3.5 • The Owens River, on the east side of the Sierra Nevada in California, is a system that includes water, sediment, vegetation, and animals such as fish and insects that function together as a whole.

Environmental damage can occur when human use of the environment leads to positive feedback. Off-road vehicle use (Figure 3.6), for example, may cause positive feedback with respect to soil erosion. These vehicles' churning tires, designed to grip the earth, erode the soil and uproot plants, increasing the rate of erosion. As more soil is exposed to erosion by running water, ruts and gullies are carved. Drivers then avoid the ruts and gullies, driving on adjacent sections that are not as eroded, thus widening paths and further increasing erosion. The gullies themselves cause an increase in erosion because they concentrate runoff and have steep side slopes. Once formed, gullies tend to grow in length, width, and depth, causing additional erosion (Figure 3.7). Eventually, an area of intensive off-road vehicle use may become a wasteland of eroded paths and gullies. Positive feedback has made the situation get worse and worse.

Some situations can be examined in terms of both positive and negative feedback loops. Changes in human population in large cities present an example, as illustrated in Figure 3.8. Positive feedback, which increases the population in cities, may occur when people perceive greater opportunities in cities and hope for a higher standard of living. Negative feedback may result from air and water pollution, disease, crime, and discomfort, if these factors encourage some people to migrate from the cities to rural areas.

Practicing your critical thinking skills, you may ask, "Is negative feedback generally desirable, and is positive feedback generally undesirable?" Reflecting on this question, we can see that although negative feedback is self-regulating, it may in some instances not be desirable. The period of time over which the positive or negative feedback occurs is the important factor.

For example, suppose we are interested in restoring the ecology of Yellowstone National Park through the reintroduction of wolves. We will expect positive feedback in the population for a time as the number of wolves grows. In this case, positive feedback for a period of time is desirable because it produces a change we want. We might also envision a system in a state that is stable but undesirable. An example is a polluted stream in an urban environment. Urban runoff and its associated pollutants, such as oil and other chemicals from streets, entering the

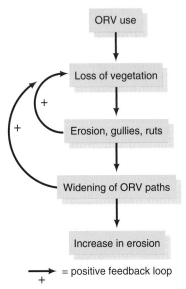

Figure 3.6 • How use of off-road vehicles (ORVs) produces positive feedback that increases erosion.

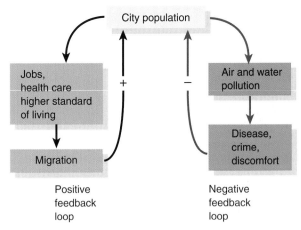

Figure 3.7 • Off-road vehicle damage of rare plants living on coastal dunes near San Luis Obispo, California. Note the tire tracks extending into the dune field.

Figure 3.8 • Potential positive and negative feedback loops for changes of human population in large cities. (*Source:* Modified from M. Maruyama. "The Second Cybernetics: Deviation-Amplifying Mutual Causal Processes," *American Scientist* 51 (1963):164–670. Reprinted by permission of *American Scientist*, magazine of Sigma Xi, The Scientific Research Society.)

stream's system may, through negative feedback mechanisms, reach a stable state between the water and the pollutants in the stream. However, this system would be considered by most people to be undesirable. A channel restoration project might be implemented to control pollutants by collecting and treating them before they enter the stream. As a result, the stream might reach a new state that is ecologically more desirable.

We can see that whether we view positive or negative feedback as desirable depends on the system and potential changes. Nevertheless, some of the major environmental problems we face today result from positive feedback mechanisms that are out of control. These include resource use and growth of human population, among others.

Interestingly, throughout most of human history, strong negative feedback cycles resulted in a very low growth in human population. Disease and limited capacity to produce food kept growth low. However, in the past hundred years, modern medicine, sanitation, and agricultural practices turned negative feedback into positive feedback, and rapid increase in human population occurred.

3.2 Exponential Growth

A particularly important example of positive feedback occurs with **exponential growth**. Simply stated, growth is exponential when it occurs at a constant *rate* per time period (rather than a constant *amount*). For instance, suppose you have $1,000 in the bank, and it grows at 10% per year. The first year, $100 in interest is added to your account. The second year, you earn more, because you earn 10% on the new total amount, $1,100. The greater the amount, the greater the interest earned, so the money (or the population, or some other quantity) increases by larger and larger amounts. When we plot data in which exponential growth is occurring, the curve we obtain is said to be "J" shaped. It looks like a skateboard ramp, starting out nearly flat and then rising steeply. (The actual shape depends on the scale of the units of the curve.) Figure 3.9 shows two typical exponential growth curves.

Calculating exponential growth involves two related factors: the rate of growth measured as a percentage and the doubling time in years. The **doubling time** is the time necessary for the quantity being measured to double. A useful rule is that the doubling time is approximately equal to 70 divided by the annual percentage growth rate. Working It Out 3.1 describes exponential growth calculations and explains why 70 divided by the annual growth rate is the doubling time. If you understand algebra and natural logarithms, you should find the example interesting. However, it is the general principles that are most important.

One significant principle of exponential growth is that it is incompatible with the concept of sustainability, which is

(a)   *(b)*

Figure 3.9 • *(a)* Idealized curve illustrating exponential growth. Growth rate is constant at 7%, and time necessary to double the quantity is constant at 10 years. Notice that growth is slow at first and much faster after several doubling times. For example, the quantity changes from 2 to 4 (absolute increase of 2) for the doubling from 10 to 20 years. It increases from 32 to 64 (absolute increase of 32) during the doubling from 50 to 60 years. *(b)* Human population increase for the last 2000 years. (*Source:* Data from U.S. Department of State.)

Exponential Growth

If the quantity of something (say, the number of humans on Earth) increases or decreases at a fixed fraction per unit of time, whose symbol is k (for example, $k = +0.02$ per year), then the quantity is changing exponentially. With positive k, we have exponential growth. With negative k, we have exponential decay.

Growth rate R is defined as the percent change per unit of time—that is, $k = R/100$. Thus, if $R = 2\%$ per year, then $k = +0.02$ per year.

The equation to describe exponential growth is

$$N = N_0\, e^{kt}$$

where N is the future value of whatever is being evaluated; N_0 is present value; e, the base of natural logarithms, is a constant 2.71828; k is as defined above; and t is the number of years over which the growth is to be calculated. This equation can be solved using a simple hand calculator, and a number of interesting environmental questions can be answered. For example, assume that we want to know what the world population is going to be in the year 2020, given that the population in 2000 is 6.2 billion and the population is growing at a constant rate of 1.36% per year ($k =$ 0.0136). We can estimate N, the world population for the year 2020, by applying the preceding equation:

$$
\begin{aligned}
N &= (6.2 \times 10^9) \times e^{(0.0136 \times 20)} \\
&= 6.2 \times 10^9 \times e^{0.272} \\
&= 6.2 \times 10^9 \times 2.71828^{0.272} \\
&= 8.14 \times 10^9, \text{ or } 8.14 \text{ billion persons}
\end{aligned}
$$

The doubling time for a quantity undergoing exponential growth (i.e., increase by 100%) can be calculated by the following equation:

$$2\,N_0 = N_0\, e^{k T_d}$$

where T_d is the doubling time.

Take the natural logarithm of both sides

$$\ln 2 = k T_d \text{ and } T_d = \ln 2 / k$$

Then remembering that $k = R/100$

$$
\begin{aligned}
T_d &= 0.693/(R/100) \\
&= 100\,(0.693)/R \\
&= 69.3/R, \text{ or about } 70/R
\end{aligned}
$$

This result is our general rule—that the doubling time is approximately 70 divided by the growth rate. For example, if $R = 10\%$ per year, then $T_d = 7$ years.

---

a long-term process (decades to hundreds of years or more). In fact, the term *sustainable growth* is an oxymoron—a self-contradiction. Even at modest growth rates, the number of whatever is growing will eventually reach extraordinary levels that are impossible to maintain.[3,4]

Exponential growth has interesting (and sometimes alarming) consequences, as illustrated in a fictional story by Albert Bartlett.[3] Imagine a hypothetical strain of bacteria in which each bacterium divides into two every 60 seconds (the doubling time is one minute). Assume that one bacterium is put in a bottle at 11:00 A.M. The bottle (its world) is full at 12:00 noon. When was the bottle half full? The answer is 11:59 A.M. If you were a bacterium in the bottle, at what time would you realize that you were running out of space? There is no single answer to this question. Consider, though, that at 11:58 the bottle was 75% empty, and at 11:57 it was 88% empty. Now assume that at 11:58 some farsighted bacteria realized that the population was running out of space and started looking around for new bottles. Let's suppose that they were able to find and move into three more bottles. How much time did they buy? Two additional minutes. They will run out of space at 12:02 P.M. If they had found 16 additional bottles, how much more time would they have?

The story of bacteria in bottles, while obviously hypothetical, illustrates the power of exponential growth. We look at exponential growth and doubling time again in Chapter 5, when we consider the growth of the human population. Here, we simply note that many systems in nature display exponential growth for some period of time, so it is important that we be able to recognize it. In particular, it is important to recognize exponential growth in a positive feedback cycle, as accompanying changes may be very difficult to control or stop.

3.3 Environmental Unity

Our discussion of positive and negative feedback set the stage for another fundamental concept in environmental science: **environmental unity**. Simply stated, environmental unity means that it is impossible to change only one thing; everything affects everything else. Of course, this is something of an overstatement. The concept is not absolutely true; the extinction of a species of snails in North America, for instance, is hardly likely to change the flow characteristics of the Amazon River. Many aspects of the natural environment are closely linked, however.

Changes in one part of a system often have secondary and tertiary effects within the system and effects on adjacent systems. Earth and its ecosystems are complex entities in which any action may have several or many effects. The case study of Amboseli National Park is a good example of the principle of environmental unity. We can find many other examples in both human-made and natural settings.

An Urban Example

Consider the changes occurring in midwestern U.S. cities such as Chicago and Indianapolis during a major shift in land use from forests or agricultural land to an urban development. Clearing the land for urban use increases runoff and the amount of sediment eroded from the land (soil erosion). Increased runoff from streets and sediment eroded from bare ground during construction affect the form and shape of the river channel. The river carries more sediment, and some of it is deposited on the bottom of the channel, reducing channel depth and increasing the flood hazard.

Eventually, as more land is paved, the amount of sediment eroded from the land decreases, runoff increases further, and the streams readjust to a lower sediment load (the amount of sediment carried by the stream) and more runoff. The readjustment is a form of negative feedback inherent in streams and rivers.

Urbanization is also likely to pollute the streams or otherwise change water quality. The increased fine sediment makes the water muddy, and chemicals from street and yard runoff pollute the stream. These changes affect the biological systems in the stream and adjacent banks. Thus, land-use conversion can set off a series of changes in the environment, and each change is likely to trigger additional changes.

A Forest Example

The interaction among forests, streams, and fish in the Pacific Northwest provides another example of environmental unity. In the redwood forests of northern California and southern Oregon, large pieces of woody debris, such as tree trunks and roots, are necessary to form and maintain nearly all the pool environments' small streams (Figure 3.10). Large pieces of redwood fall naturally into streams and partially block their flow, producing pools of deeper water. These pools provide much of the rearing habitat for young salmon, which spend part of their lives in the streams before migrating to the ocean.

It was formerly common practice to remove woody debris from streams because it was thought to block the migration of adult salmon attempting to return to spawning beds in the streams. We now know that this practice degrades the fish habitat. Stream restoration projects now

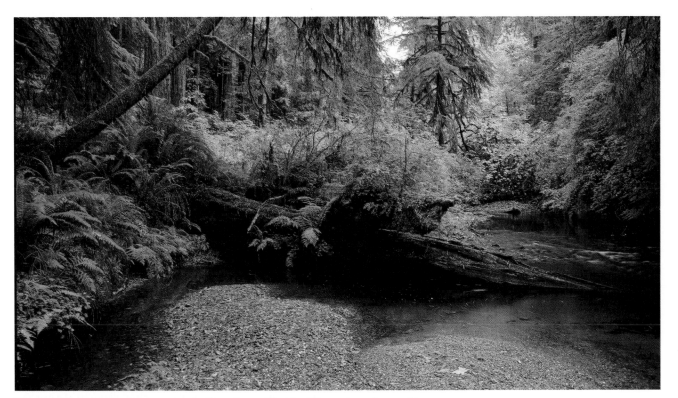

Figure 3.10 • Stream processes are significantly modified by fallen redwood trees. The large tree trunk in the central portion of the photograph produced a small pool, which is a good fish habitat. The site is Prairie Creek, Redwood State Park, California.

often place large woody debris into channels to improve the fish habitat. The role of large woody debris in stream processes and salmon habitat illustrates the value of studying relations between physical and biological systems to help provide for sustainable fish populations. Such studies are at the heart of environmental science.

3.4 Uniformitarianism

Earth and its life-forms have changed many times, but the processes necessary to sustain life and the environment for life have occurred throughout much of Earth's history. The principle that physical and biological processes presently forming and modifying Earth can help explain the geologic and evolutionary history of Earth is known as **uniformitarianism**. The principle may be more simply stated as "the present is the key to the past." For example, if a deposit of gravel and sand found at the top of a mountain is similar to stream gravels found today in an adjacent valley, we may infer by uniformitarianism that a stream once flowed in a valley where the mountaintop is now.

Uniformitarianism was first suggested in 1785 by the Scottish scientist James Hutton, known as the father of geology. Charles Darwin was impressed by the concept of uniformitarianism, and the concept pervades his ideas on biological evolution. Today, uniformitarianism is considered one of the fundamental principles of the biological and Earth sciences.

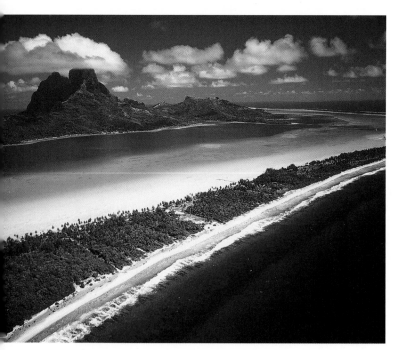

Figure 3.11 • This beach on the island of Bora Bora, French Polynesia, is an example of a geologically short-lived landform, vulnerable to rapid change from storms and other natural processes.

Uniformitarianism does not demand or even suggest that the magnitude and frequency of natural processes remain constant. Obviously, some processes do not extend back through all of geologic time. For example, the early Earth atmosphere did not contain free oxygen. However, for the past several billion years, the continents, oceans, and atmosphere have been similar to those of today. We assume that physical and biological processes that form and modify the Earth's surface have not changed significantly over this period. To be useful from an environmental standpoint, the principle of uniformitarianism has to be more than a key to the past. We must turn it around and say that a study of past and present processes is the key to the future. That is, we can assume that in the future the same physical and biological processes will operate, although the rates will vary as the environment is influenced by natural change and human activity. Geologically short-lived landforms such as beaches (Figure 3.11) and lakes will continue to appear and disappear in response to storms, fires, volcanic eruptions, and earthquakes. Extinctions of animals and plants will continue in spite of, as well as because of, human activity. We want to improve our ability to predict what the future may bring, and uniformitarianism can assist in this task.

3.5 Changes and Equilibrium In Systems

Uniformitarianism, then, suggests that changes in natural systems may be predictable. We turn next to an examination of how systems may change. This will involve looking at the relation of system inputs and outputs.

Where the input into a system is equal to the output (Figure 3.12a), there is no net change in the size of the reservoir (the amount of whatever is being measured), and the system is said to be in a **steady state**. The steady state is a dynamic equilibrium, because material or energy is entering and leaving the system in equal amounts. The opposing processes occur at equal rates. An approximate steady state may occur on a global scale, such as in the balance between incoming solar radiation and outgoing radiation from Earth, or on the smaller scale of a university, where new freshmen begin their studies and seniors graduate at about the same rate.

When the input into the system is less than the output (Figure 3.12b), the size of the reservoir declines. For example, if a resource, such as groundwater, is consumed faster than it can be naturally or humanly replaced, it may be used up. Conversely, in a system where input exceeds output (Figure 3.12c), the reservoir will increase. Examples are the buildup of heavy metals in lakes and the pollution of groundwater.

By using rates of change or input–output analysis of systems, we can derive an average residence time for objects or material moving through a system. The **average residence time** is the time it takes for a given part of

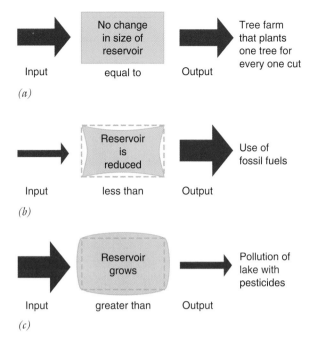

(a) Input — equal to — Output

No change in size of reservoir

Tree farm that plants one tree for every one cut

(b) Input — less than — Output

Reservoir is reduced

Use of fossil fuels

(c) Input — greater than — Output

Reservoir grows

Pollution of lake with pesticides

Figure 3.12 • Major ways in which a reservoir, or stock, of some material can change. (*Source:* Modified from P. R. Ehrlich, A. H. Ehrlich, and J. P. Holvren, *Ecoscience: Population, Resources, Environment*, 3rd ed. [San Francisco: W. H. Freeman, 1977].)

the total reservoir of a particular material to be cycled through the system. To compute the average residence time when the size of the reservoir and the rate of transfer are constant, we divide the total size of the reservoir by the average rate of transfer through that reservoir. For example, suppose the university mentioned above has 10,000 students. Each year 2,500 freshmen start, and 2,500 seniors graduate. The average residence time for students is 10,000 divided by 2,500, or 4 years.

Average residence time has important implications for environmental systems. A system such as a small lake with an inlet and an outlet and a high transfer rate of water has a short residence time for water. On the one hand, that makes the lake especially vulnerable to change if, for example, a pollutant is introduced. On the other hand, the pollutant soon leaves the lake. Large systems with a slow rate of transfer of water, such as oceans, have a long residence time and are much less vulnerable to quick change. However, once polluted, large systems with slow transfer rates are difficult to clean up.

Let us look more closely at system inputs and outputs. Inputs to systems may be thought of as causes and outputs or responses as effects. For example, we may add a nitrogen fertilizer to an orchard of orange trees. Adding the fertilizer is an input (or cause), and the output (or effect) is the number of oranges the tree produces.

If the relationship between a cause (input) and effect (output) is strictly proportional for all values, then we call this relationship *linear*. Input and output coupled with

feedback in a system may result in relationships between cause and effect that are *nonlinear*, and there may be *delays* in the response.[4] Some relationships in systems are linear over a particular range of input and then become nonlinear. For example, if you apply 0.25 kg of fertilizer per orange tree and the yield increases by 5% and you then apply 0.50 kg per tree and the yield increases by 10% and then 0.75 kg per tree increases the yield by 15%, the relationship is linear over these values of input of fertilizer. But what if you apply 50 kg per tree in the hope of increasing the yield by 1000%? You would probably damage or kill the tree and the yield would be zero!

Over the entire range from 0.25 to 50 kg, then, the relationship between cause and effect changes. You might also note delays in response. When you add fertilizer, for example, it takes time for it to enter the soil and be used by the tree. Many responses to environmental inputs (including human population change; pollution of the land, water and air; and use of resources) are nonlinear and may involve delays that must be recognized if we are to understand and solve environmental problems.

With an understanding of input and output, we have a framework for interpreting some of the changes that may affect systems. An idea that has been used and defended in the study of our natural environment is that natural systems that have not been affected by human activity tend toward some sort of steady state, or dynamic equilibrium. Sometimes this is called the *balance of nature*. Certainly, negative feedback operates in many natural systems and may tend to hold a system at equilibrium. Nevertheless, we need to ask how often the equilibrium model really applies.

If we examine natural systems in detail and perform our evaluation over a variety of time frames, it is evident that a steady state, or dynamic equilibrium, is seldom obtained or maintained for very long. Rather, systems are characterized not only by human-induced disturbances but also by natural disturbances (sometimes called natural disasters, such as flood and wildfires). Thus, changes over time can be expected. In fact, studies of such diverse systems as forests, rivers, and coral reefs suggest that disturbances due to natural events such as storms, floods, and fires are necessary for the maintenance of those systems.

The environmental lesson is that systems change naturally. If we are going to manage systems for the betterment of the environment, we need to gain a better understanding of the following:[5,6]

- Types of disturbances and changes that are likely to occur.
- Time periods over which changes occur.
- The importance of each change to the long-term productivity of the system.

These concepts are at the heart of understanding the principles of environmental unity and sustainability.

3.6 Earth and Life

We turn next from a general discussion of systems to a more direct focus on Earth as a living planet. Earth formed approximately 4.6 billion years ago when a cloud of interstellar gas known as a solar nebula collapsed, creating protostars and planetary systems. Life on Earth began approximately 1.6 billion years later (3 billion years ago) and since that time has profoundly affected the planet. Since the emergence of life, many kinds of organisms have evolved, flourished, and died, leaving only their fossils to record their place in history.

Several million years ago, there were evolutionary beginnings for the eventual dominance of humans on Earth. Eventually, the fossil record tells us, we too will disappear. The brief moment of humanity in Earth history may not be particularly significant. However, to us living now and to the human generations still to come, how we affect our environment is important.

Human activities increase and decrease the magnitude and frequency of some natural Earth processes. For example, rivers may rise and flood the surrounding countryside regardless of human activities, but the magnitude and frequency of flooding may be greatly increased or decreased by human activity. In order to predict the long-range effects of such processes as flooding, we must be able to determine how our future activities will change the rates of physical processes.

From a biological and geological point of view, we know that the ultimate fate of every species is extinction. However, humans have accelerated this fate for many species. We see in Figure 3.13 that as the human population has increased, a parallel increase in the extinction of species has occurred. These extinctions are closely related to land use change—to agricultural and urban uses that change the ecological conditions of an area. Some species are domesticated or cultivated, and their numbers grow; others are removed as pests.

Human activities now also affect Earth on a global scale, and these effects are increasing with technological advances. Our civilization has the potential to greatly alter our climate—the chemistry of our atmosphere, soil, and water—and even the chances that life will persist on Earth. We explore the interaction of life with its environment in the next section.

3.7 Earth as a Living System

Earth as a planet has been profoundly altered by the life that inhabits it. Earth's air, oceans, soils, and sedimentary rocks are very different from what they would be on a lifeless planet. In many ways, life helps control the make-up of the air, oceans, and sediments.

Life interacts with its environment on many levels. A single bacterium in the soil interacts with the air, water,

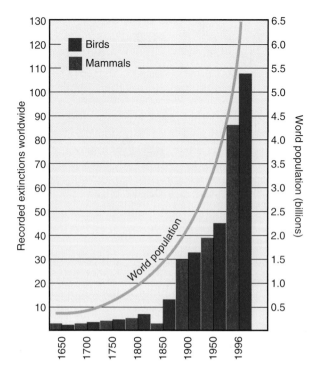

Figure 3.13 • An increase in human population paralleled by an increase in the extinction of birds and mammals. (*Source:* Modified from U. Ziswiler, *Extinction and Vanishing Species* [New York: Springer-Verlag, 1967], with 1993 data from B. Groombridge, ed., *IUCN Red List of Threatened Animals* [IUCN, World Conservation Union, 1996].)

and particles of soil around it within the space of a fraction of a cubic centimeter. A forest extending hundreds of square kilometers interacts with large volumes of air, water, and soil. All of the oceans, all of the lower atmosphere, and all of the near-surface part of the solid Earth are affected by life.

A general term, **biota**, is used to refer to all living things (animals and plants, including microorganisms) within a given area—from an aquarium to a continent to Earth as a whole. The region of Earth where life exists is known as the **biosphere**. It extends from the depths of the oceans to the summits of mountains. The biosphere includes all life as well as the lower atmosphere and the oceans, rivers, lakes, soils, and solid sediments that actively interchange materials with life. All living things require energy and materials. In the biosphere, energy is received from the sun and the interior of Earth and is used and given off as materials are recycled.

To understand what is required to sustain life, consider the following question: How small a part of the biosphere could be isolated from the rest and still sustain life? Suppose you put parts of the biosphere into a glass container and sealed it. What minimum set of contents would sustain life? If you placed a single green plant in the container along with air, water, and some soil, the plant could make sugars from water and from carbon dioxide in the air. It could also make many organic compounds, including proteins and woody tissue, from sugars and from inorganic compounds in the soil. But no green plant can decompose its own products and recycle the materials. Eventually, your green plant would die.

We know of no single organism, population, or species that both produces all its own food and completely recycles all its own metabolic products. For life to persist, there must be several species within an environment that includes fluid media—air and water—to transport materials and energy. Such an environment is an ecosystem, our next important topic of discussion.

3.8 Ecosystems

An **ecosystem** is a community of organisms and its local nonliving environment in which matter (chemical elements) cycles and energy flows. It is a fundamental principle that *sustained life on Earth is a characteristic of ecosystems*, not of individual organisms or populations or single species.

The Nature of Ecosystems

The term *ecosystem* is applied to areas of all sizes, from the smallest puddle of water to a large forest or the entire global biosphere. Ecosystems differ greatly in composition— that is, in the number and kinds of species, the kinds and relative proportions of nonbiological constituents, and the degree of variation in time and space. Sometimes the borders of an ecosystem are well defined, as in the transition from the ocean to a rocky coast or from a pond to the surrounding woods. Sometimes the borders are vague, as in the subtle gradation of forest to prairie in Minnesota and the Dakotas or from grasslands to savannas or forests in East Africa. What is common to all ecosystems is not physical structure—size, shape, variations of borders—but the existence of the processes we have mentioned—the flow of energy and the cycling of chemical elements.

Ecosystems can be natural or artificial. A pond constructed as part of a waste treatment plant is an artificial ecosystem. Ecosystems can be natural or managed, and the management can vary over a wide range of actions. Agriculture can be thought of as partial management of certain kinds of ecosystems.

Natural ecosystems carry out many public service functions for us. Wastewater from houses and industries is often converted to drinkable water by passage through natural ecosystems such as soils. Pollutants, such as those in the smoke from industrial plants or in the exhaust from automobiles, are often trapped on leaves or converted to harmless compounds by forests.

The Gaia Hypothesis

Our discussion of Earth as a system, life in its environment, the biosphere, and ecosystems leads us to the question of how much life on Earth has affected our planet. In recent years, the **Gaia hypothesis**—named for Gaia, the Greek goddess Mother Earth—has become a hotly debated subject.[7] The hypothesis states that life manipulates the environment for the maintenance of life. For example, some scientists believe that algae floating near the surface of the ocean influence rainfall at sea and the carbon dioxide content of the atmosphere, thereby significantly affecting the global climate. It follows, then, that the planet Earth is capable of physiological self-regulation.

According to James Lovelock, a British scientist who has been developing the Gaia hypothesis since the early 1970s, the idea of a living Earth is probably as old as humanity.[7] James Hutton, whose theory of uniformitarianism was discussed earlier, stated in 1785 that he believed Earth to be a superorganism and compared the cycling of nutrients from soils and rocks in streams and rivers to the circulation of blood in an animal.[7] In this metaphor, the rivers are the arteries and veins, the forests are the lungs, and the oceans are the heart of Earth.

The Gaia hypothesis is really a series of hypotheses. The first is that life, since its inception, has greatly affected the planetary environment. Few scientists would disagree. The second hypothesis asserts that life has altered Earth's environment in ways that have allowed life to persist. Certainly, there is some evidence that life has had such an effect on Earth's climate. A popularized extension of the Gaia hypothesis is that life *deliberately* (consciously) controls the global environment. Few scientists accept this idea.

The extended Gaia hypothesis may have merit in the future, however. Humans have become conscious of our effects on the planet, some of which influence future changes in the global environment. Thus, the concept that humans can consciously make a difference in the future of our planet is not as extreme a view as would once have been thought. The future status of the human environment may depend, in part, on actions we take now and in coming years. This aspect of the Gaia hypothesis exemplifies the key theme of thinking globally, which was introduced in Chapter 1.

The decisions we make in managing our global environment depend on our values as well as our understanding of how Earth works (another key theme from Chapter 1). With this in mind, we explore in greater depth how human processes are linked to environmental change.

3.9 Human Amplification of Environmental Change

As we have seen, the environment is changed in various ways by natural events. Natural change may rearrange the landscape almost overnight by earthquake, flood, volcanic eruption, or hurricane. Such changes from natural earth processes are becoming more damaging to human systems, and human activities may be in part responsible.

Events that were formerly hazardous are now becoming catastrophes. A **catastrophe** is an event in which damages to people, property, and society are sufficiently great that recovery is long and difficult. Two of the most damaging catastrophes resulting from natural hazards in 1998 were the flooding of the Yangtze River in China and Hurricane Mitch, which struck Central America. The floods on the Yangtze River caused nearly 4,000 deaths, while Hurricane Mitch killed approximately 11,000 people (Figure 3.14).

It has been suggested that the damage in China and Central America was particularly severe because of land-use changes that preceded these events. People had removed one-half of the forest of Honduras. (In addition, an 11,000-km² fire had damaged the region prior to the hurricane.) Hillsides stripped of vegetation washed away during the hurricane. Along with them went farms, homes, roads, and bridges. Much the same story emerged from the flood in Central China. The Yangtze River basin had lost about 85% of its forest to timber harvesting and the conversion of land to agricultural uses in recent years. Because much of the vegetation was gone, the runoff from storms was much greater, contributing to the flooding of the Yangtze River and greatly increased damage from the flood.[8]

The lesson (and a new principle) learned from Hurricane Mitch and the flooding of the Yangtze is that *human activities are increasing potential impacts of natural disasters to the extent that these disasters are producing catastrophes*. We are learning that if we wish to minimize damages from natural hazards in the future, we will need to consider land rehabilitation (restoration ecology) with the goal of achieving sustainable development based on maintenance of healthy ecosystems.[8] However, this will be a difficult task, because pressures of human population growth in many parts of the world will continue well into the 21st century. This fact emphasizes the need to address human population growth if we are to solve pressing environmental problems and reach the goal of sustainability for our environment.

Not all changes happen as rapidly as those associated with Hurricane Mitch and the Yangtze floods. Most changes brought on by human activity involve rather slow processes with cumulative effects. For example, our present global warming began with the Industrial Revolution, when people started burning massive amounts of fossil fuels. Levels of carbon dioxide in the atmosphere added by human activities have nearly doubled since the Industrial Revolution and are causing human-enhanced global warming. Certainly, we are aware that climate may change naturally and that such change may be rapid (see Chapter 22). However, present global warming is significant and results in part from human activity.

The decline of fisheries is another process that generally takes place over a relatively long time. In such cases, recognizing that irreversible damage may have been done often involves crossing a threshold that is difficult to identify in time to avoid problems. More commonly, we recognize that a threshold has been exceeded after we begin to identify consequences of a particular activity. For example, we have painfully come to the conclusion that the crash of the salmon population in the Pacific Northwest is the combined result of fishing, building of dams, and land-use practices such as timber harvesting. We did not recognize the crash until it occurred and we reflected on past activities. This example further supports the idea that environmental science problems are often complex, with linkages among various parts of ocean, river, land, and forest systems.

Change can also be chaotic. This occurs when some

Figure 3.14 ◦ Hurricane Mitch in 1998 killed about 11,000 people in Central America. Shown here is a disabled man being helped from mud deposited by floods that damaged Tegucigalpa, Honduras.

Is the Gaia Hypothesis Science?

According to the Gaia hypothesis, Earth and all living things form a single system, with interdependent parts, communication among these parts, and the ability to self-regulate. Are the Gaia hypothesis and its component hypotheses science, fringe science, or pseudoscience? Is the Gaia hypothesis anything more than an attractive metaphor? Does it have religious overtones? Answering these questions is more difficult than answering similar questions about, say, crop circles, as described in Chapter 2. Analyzing the Gaia hypothesis forces us to deal with some of our most fundamental ideas about science and life.

Critical Thinking Questions

1. What are the main hypotheses included in the Gaia hypothesis?
2. What kind of evidence would support each hypothesis?
3. Which of the hypotheses can be tested?
4. Is each hypothesis a science, fringe science, or pseudoscience?
5. Some scientists have criticized James E. Lovelock, who formulated the Gaia hypothesis, for using the term *Gaia*. Lovelock responds that it is better than referring to a "biological cybernetic system with homeostatic tendencies." What does this phrase mean?
6. What are the strengths and weaknesses of the Gaia hypothesis?

small change is amplified, resulting in complex and perhaps periodic activity or behavior. Chaos theory is a mathematical description of change that has been used to describe the behavior of a variety of systems, including fluctuations of populations and changes in circulation and patterns of air current in the atmosphere. An often-cited hypothetical example is a scenario in which the beating of the wings of a hummingbird in Brazil causes, through a series of amplified activities, a hurricane in Miami. Although this example seems unlikely to occur in nature, many surprises confront us when we consider natural earth systems. For example, changes in the temperature of the Pacific Ocean cause far-ranging changes in storms, floods, and other natural hazards on a global basis (see Chapter 22).

As stated, one of our goals in understanding the role of human processes in environmental change is to help manage our global environment. To accomplish this, we need to be able to predict changes before they occur. But as the examples above demonstrate, prediction presents great challenges. Although some changes are anticipated, others come as a surprise. As we learn to apply the principles of environmental unity and uniformitarianism more skillfully, we will be better able to anticipate changes that would otherwise have been surprises.

Summary

- A system is a set of components or parts that function together as a whole. Environmental studies deal with complex systems at every level, and solutions to environmental problems often involve understanding systems and rates of change.
- Systems respond to inputs and have outputs. Feedback is a special kind of system response. Positive feedback is destabilizing, whereas negative feedback tends to stabilize or encourage more constant conditions in a system.
- Relationships between the input (cause) and output (effect) of systems may be nonlinear and may involve delays.
- The principle of environmental unity, simply stated, holds that everything affects everything else. It emphasizes linkages among parts of systems.
- The principle of uniformitarianism can help predict future environmental conditions on the basis of the past and the present.
- A particularly important aspect of positive feedback is exponential growth, in which the increase per time period is a constant fraction or percentage of the current amount. Exponential growth involves two factors: the rate of growth and the doubling time.
- Changes in systems can be studied through input-output analysis. The average residence time is the average time it takes for the total reservoir of a particular material to be cycled through the system.
- Life on Earth began about 3 billion years ago and since that time has profoundly changed our planet. Sustained life on Earth is a characteristic not of individual organisms or populations but of ecosystems—local communities of interacting populations and their nonbiological environments.
- The general term *biota* refers to all living things, and the region of Earth where life exists is known as the *biosphere*.
- The Gaia hypothesis states that life on Earth, through a complex system of positive and negative feedback, regulates the planetary environment to help sustain life.
- Human activities are amplifying the effects of natural processes, especially those producing natural disasters.

Human Population:

The human population of Earth is experiencing a variety of positive-feedback mechanisms leading to an increasing population. Of particular concern are local or regional increases in population density (number of people per unit area), which strain resources and lead to human suffering and economic damage. For example, land-use changes linked to increases in human population have changed some disasters into catastrophes.

Sustainability:

Negative feedback is stabilizing. If we are to have a sustainable human population and use our resources sustainably, then we need to set in place a series of negative feedbacks within our agricultural, urban, and industrial systems.

Global Perspective:

This chapter introduced Earth as a system. One of the most fruitful areas for environmental research remains the investigation of relationships between physical and biological processes on a global scale. More of these relationships must be discovered if we are to solve environmental problems related to such issues as potential global warming, ozone depletion, and disposal of toxic waste.

Urban World:

The concepts of environmental unity and uniformitarianism are particularly appropriate in urban environments, where land-use changes result in a variety of changes that affect physical and biochemical processes.

Science and Values:

Our discussion of the Gaia hypothesis reminds us that we still know very little about how our planet works and how physical, biological, and chemical systems are linked. What we do know is that we need more scientific understanding. This understanding will be driven in part by the value that we place on our environment and on the well-being of other living things.

● KEY TERMS

average residence time **40**	closed system **35**	exponential growth **37**	positive feedback **35**
biosphere **42**	doubling time **37**	feedback **35**	steady state **40**
biota **42**	ecosystem **43**	negative feedback **35**	system **35**
catastrophe **44**	environmental unity **38**	open system **35**	uniformitarianism **40**

● STUDY QUESTIONS

1. How does the Amboseli National Park case history exemplify the principle of environmental unity?

2. What is the difference between positive and negative feedback in systems? Provide an example of each.

3. What is the main point concerning exponential growth? Is exponential growth good or bad?

4. Why is the idea of equilibrium in systems somewhat misleading in regard to environmental questions? Is the establishment of a balance of nature ever possible?

5. Why is the concept of the ecosystem so important in the study of environmental science? Should we be worried about disturbing ecosystems? Under what circumstances should we worry or not worry?

6. Is the Gaia hypothesis a true statement of how nature works, or is it simply a metaphor? Explain.

7. How might you use the principle of uniformitarianism to help evaluate environmental problems? Is it possible to use this principle to help evaluate the potential consequences of too many people on Earth?

● FURTHER READING

Abelson, P. H. 1990. "Global Change." *Science* 249:1085. This editorial outlines what the United States is doing to address the challenges of global change and what it can do in the future.

Bunyard, P., ed. 1996. *Gaia in Action: Science of the Living Earth.* Edinburgh: Floris Books. Investigations into implications of the Gaia hypothesis.

Ehrlich, P. R., A. H. Ehrlich, and J. P. Holdren. 1970. *Ecoscience.* San Francisco: W. H. Freeman. Although this is an older book, Chapter 2 provides a good overview of the physical world and how systems and changes may affect our environment.

Kirchner, J. W. 1989. "The Gaia Hypothesis: Can It Be Tested?" *Reviews of Geophysics* 27:223–235. The Gaia theories are evaluated both as a colorful metaphor and as a scientific hypothesis.

Lovelock, J. 1995. *The Ages of Gaia: A Biography of Our Living Earth.* New York: W. W. Norton. This small book explains the Gaia hypothesis, presenting the case that life very much affects our planet and in fact may regulate it for the benefit of life.

CHAPTER 4

The Biogeochemical Cycles

The city of Seattle, Washington is located between Puget Sound (right) and Lake Washington (left). View is to the south.

● LEARNING OBJECTIVES

Life is composed of many chemical elements, which exist in the right amounts, the right concentrations, and the right ratios to one another. If these conditions are not met, then life is limited. The study of chemical availability and biogeochemical cycles is important to the solution of many environmental problems. After reading this chapter, you should understand:

- What are the major biogeochemical cycles.
- What are the major factors and processes that control biogeochemical cycles.
- Why some chemical elements cycle quickly and some slowly.
- How each major component of Earth's global system (the atmosphere, waters, solid surfaces, and life) are involved and linked with biogeochemical cycles.
- How the biogeochemical cycles most important to life, especially the carbon cycle, generally operate.
- How humans affect biogeochemical cycles.

Lake Washington

Humans may unknowingly alter the natural cycling of chemicals to the detriment of the environment. Science can provide potential solutions, but our value systems dictate whether we are willing to pursue these solutions. The following story concerning Lake Washington shows that adverse effects of chemical cycling can be addressed and reduced.

The city of Seattle, Washington, lies between two major bodies of water—saltwater Puget Sound to the west and freshwater Lake Washington to the east (Figure 4.1a). Beginning in the 1930s, the freshwater lake was used for disposal of sewage. By 1959, 11 sewage treatment plants had been constructed along the lake. The plants removed disease-causing organisms and much of the organic matter that had formerly entered the lake. Each day, 76,000 m^3 per day (about 20 million gallons) of treated wastewater flowed from the plants into the lake. Smaller streams that fed the lake brought in small amounts of untreated sewage. The lake's response to the introduction of treated wastewater was a major bloom of undesirable algae and photosynthetic bacteria, which reduced water clarity and the general aesthetics of the lake.[1]

Seattle's mayor appointed an advisory committee to determine what might be done. Scientific research showed that phosphorus in the treated wastewater had stimulated the growth of algae and bacteria. Although the treatment process had eliminated disease-causing organisms, it had not removed phosphorus, which came from common laundry detergent and other urban sources. The phosphorus, when it entered the lake, acted as a potent fertilizer, posing a chemical problem for the lake's ecosystem. Puget Sound is a large body of ocean water with a rapid rate of exchange with the Pacific Ocean. Because of this rapid exchange, the phosphorus would be quickly diluted to a very low concentration.

To solve the problem, the committee advised the city to discharge wastewater from the sewage treatment plants into Puget Sound rather than into the lake.

The change was made by 1968, and the lake improved rapidly. The diversion of the phosphorus-rich wastewater greatly reduced the amount of phosphorus cycling through the lake's ecosystem. In terms of input-output analysis, introduced in Chapter 3, the input of phosphorus was greatly reduced, and so the concentration of the nutrient in the lake water was also reduced. By the time a year had passed, the unpleasant algae had decreased, and surface waters had become much clearer

(a)

(b)

Figure 4.1 • (a) Thematic image of the Seattle, Washington, region (about 24,000 km^2). Bodies of water (e.g., Puget Sound and Lake Washington) are dark blue/black; urban areas (e.g., Seattle) are pink/light-blue/gray; forested lands, are dark green; clear-cut (logged) lands are pink/yellow; and snow and ice on Mt. Rainier and other areas are baby blue. (b) Lake Washington in the summer of 1998 at a popular recreational beach on the eastern shoreline.

than they had been five years before. Oxygen concentrations in deep water increased immediately to levels greater than those observed in the 1930s, favoring an increase in fish.[1]

Urbanization around Lake Washington (Figure 4.1b) continues to be a concern, because urban runoff is a potential source of water pollution that may degrade the quality of streams or lakes it enters. Nevertheless, the water quality of Lake Washington in the early 21st century remains good for a large urban lake.

The story of Lake Washington illustrates the importance of understanding how chemicals cycle in ecosystems. The story confirms several key themes from Chapter 1. As the population in Seattle increased, so did its sewage. The scientific study of the lake and the discovery of the role of phosphorus reflected the value the people of Seattle placed on Lake Washington. Recognizing the need for a high-quality environment in an urban area, they developed a plan to sustain the water quality of their urban lake. They "acted locally" to solve a water pollution problem and provided a positive example of pollution abatement.

4.1 How Chemicals Cycle

Earth is a particularly good planet for life from a chemical point of view. The atmosphere contains plenty of water and oxygen, which we and other animals need to breathe. In many places, soils are fertile, containing the chemical elements necessary for plants to grow; and Earth's bedrock contains valuable metals and fuels. Of course, some parts of the Earth's surface are not perfect for life—deserts with little water, chemical deserts (such as the middle regions of the oceans) where nutrients necessary for life are not abundant, and certain soils in which some of the chemical elements required for life are lacking or others toxic to life are present.[2] Scientific questions reflecting our values for a quality environment are: What kinds of chemical processes benefit or harm the environment, ourselves, and other life forms? How can we manage chemicals in the environment to improve and sustain ecosystems from the local to the global level? To answer these questions, we need to know how chemical elements cycle, and this is our starting point.

Biogeochemical Cycles

The term *chemical* refers here to an element such as carbon (C) or phosphorus (P), or to a compound, such as water (H_2O). (See A Closer Look 4.1 for a discussion of matter and energy.) A **biogeochemical cycle** is the complete path a chemical takes through the four major components, or reservoirs, of Earth's system: atmosphere, hydrosphere (oceans, rivers, lakes, groundwaters, and glaciers), lithosphere (rocks and soils), and biosphere (plants and animals). A biogeochemical cycle is *chemical* because it is chemicals that are cycled, *bio-* because the cycle involves life, and *geo-* because a cycle may include atmosphere, water, rocks, and soils.

Consider as an example an atom of carbon (C) in carbon dioxide (CO_2) emitted from burning coal (which is made up of fossilized plants hundreds of millions of years old). The carbon atom is released into the atmosphere and then taken up by a plant and incorporated into a seed. The seed is eaten by a mouse. The mouse is eaten by a coyote, and the carbon atom is expelled following digestion as scat onto the soil. Decomposition of the scat allows our carbon atom to enter the atmosphere again. It may also enter another organism, such as an insect, which utilizes the scat as a resource.

Chemical Reactions

It is important to acknowledge in our discussion of how chemical cycles work that the emphasis is on chemistry. Many chemical reactions occur within and between the living and nonliving portions of ecosystems. A **chemical reaction** is a process in which new chemicals are formed from elements and compounds that undergo a chemical change. For example, a simple reaction between rainwater (H_2O) and carbon dioxide (CO_2) in the atmosphere produces weak carbonic acid (H_2CO_3):

$$H_2O + CO_2 \longrightarrow H_2CO_3.$$

This weak acid reacts with earth materials, such as rock and soil, to release chemicals into the environment. The released chemicals include calcium, sodium, magnesium, and sulfur, with smaller amounts of heavy metals such as lead, mercury, and arsenic. The chemicals appear in various forms, such as compounds and ions in solution.

Many other chemical reactions determine whether chemicals are available to life. For example, photosynthesis is a series of chemical reactions by which living green plants, with sunlight as an energy source, convert carbon

Matter and Energy

The universe as we know it consists of two entities: matter and energy. Matter is the material that makes up our physical and biological environments (you are composed of matter). Energy is the ability to do work. The first law of thermodynamics—also known as the law of conservation of energy or, the first energy law—states that energy cannot be created or destroyed but can change from one form to another. This law stipulates that the total amount of energy in the universe does not change.

Our sun produces energy through nuclear reactions at high temperatures and pressures that change mass (a measure of the amount of matter) into energy. At first glance, this may seem to violate the law of conservation of energy. However, this is not the case, because energy and matter are interchangeable. Albert Einstein first described the equivalence of energy and mass in his famous equation $E = mc^2$, where E is energy, m is mass, and c is the velocity of light in a vacuum, such as outer space (approximately 300,000 km/s or 186,000 mi/s). Because the velocity of light squared is a very large number, even a small amount of change in mass produces very large amounts of energy.[3]

Energy, then, may be thought of as a mathematical abstract quantity that is always conserved. This means that it is impossible to get something for nothing when dealing with energy; it is impossible to extract more energy from any system than the amount of energy that originally entered the system. In fact, the second law of thermodynamics states that you cannot break even. When energy is changed from one form to another, it always moves from a more useful form to a less useful one. Thus, as energy moves through a real system and is changed from one form to another, energy is conserved, but it becomes less useful. We return to the two energy laws when we discuss energy in ecosystems (Chapter 9) and in modern society (Chapter 16).

We turn next to a brief introduction to the basic chemistry of matter, which will help you in understanding biogeochemical cycles. An atom is the smallest part of a chemical element that can take part in a chemical reaction with another atom. An element is a chemical substance composed of identical atoms that cannot be separated by ordinary chemical processes into different substances. Each element is given a symbol. For

example, the symbol for the element carbon is C, and that for phosphorus is P.

A model of an atom (Figure 4.2) shows three subatomic particles: neutrons, protons, and electrons. The atom is visualized as a central nucleus composed of neutrons with no electrical charge and protons with positive charge. A cloud of electrons, each with a negative charge, revolves about the nucleus. The number of protons in the nucleus is unique for each element and is the atomic number for that element. For example, hydrogen, H, has one proton in its nucleus, and its atomic number is 1. Uranium has 92 protons in its nucleus, and its atomic number is 92. A list of known elements with their atomic numbers, called the Periodic Table, is shown in Figure 4.5.

Electrons in our model of the atom are arranged in shells (representing energy levels), and the electrons closest to the nucleus are bound tighter to the atom than those in the outer shells. Electrons have negligible mass compared with neutrons or protons; therefore, nearly the entire mass of an atom is in the nucleus.

The sum of the number of neutrons and protons in the nucleus of an atom is known as the atomic weight. Atoms of

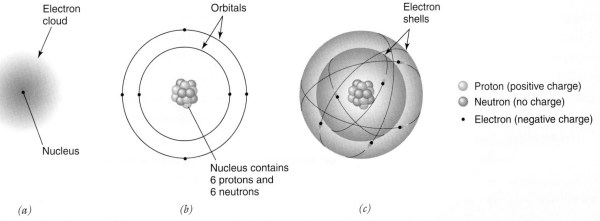

(a)　　　　(b)　　　　(c)

Figure 4.2 ● (a) Idealized diagram showing the basic structure of an atom as a nucleus surrounded by a cloud of electrons. (b) Conceptualized model of an atom of carbon with six protons and six neutrons in the nucleus and six orbiting electrons in two ener-

gy shells; and (c) a more realistic view of (b). Size of nucleus relative to size of electron shells is greatly exaggerated. After Press, F., and Siever, R. 1994. *Understanding Earth*. [New York: W. H. Freeman]) the same element always have the same atomic number (the same number of protons in the nucleus), but they can have different numbers of neutrons and therefore different atomic weights. Two atoms of the same element with different numbers of neutrons in their nuclei and different atomic weights are known as isotopes of that element. For example, two isotopes of oxygen are O-16 and O-18, where 16 and 18 are the atomic weights. Both isotopes have an atomic number of 8, but O-18 has two more neutrons than O-16. The study is proving very useful in studying how the earth works. For example, the study of oxygen isotopes has resulted in a better understanding of how the global climate has changed. This topic is beyond the scope of our present discussion but can be found in many basic textbooks on oceanography.

An atom is chemically balanced in terms of electric charge when the number of protons in the nucleus is equal to the number of electrons. However, an atom may lose or gain electrons, changing the balance in the electrical charge. An atom that has lost or gained electrons is called an ion. An atom that has lost one or more electrons has a net positive charge and is called a cation. For example, the potassium ion K^+ has lost one electron, and the calcium ion Ca^{2+} has lost two electrons. An atom that has gained electrons has a net negative charge and is called an anion. For example, O^{2-} is an anion of oxygen that has gained two electrons.

A compound is a chemical substance composed of two or more atoms of the same or different elements. The smallest unit of a compound is a molecule. For example, each molecule of water, H_2O,

contains two atoms of hydrogen and one atom of oxygen held together by chemical bonds. Minerals that form rocks are compounds, as are most chemical substances found in a solid, liquid, or gaseous state in the environment.

In the study of the environment, we are concerned with chemicals that move in and between the major components of the Earth system. An example is the element carbon, which moves through the atmosphere, hydrosphere, lithosphere, and biosphere in a large variety of compounds. These include carbon dioxide (CO_2) and methane (CH_4), which are gases in the atmosphere; sugar ($C_6H_{12}O_6$) in plants and animals; and complex hydrocarbons (compounds of hydrogen and carbon) in coal and oil deposits.

dioxide (CO_2) and water (H_2O) to sugar ($C_6H_{12}O_6$) and oxygen (O_2). The general chemical reaction for photosynthesis is

$$6CO_2 + 6H_2O \xrightarrow{\text{sunlight}} C_6H_{12}O_6 + 6O_2.$$

Photosynthesis produces oxygen as a byproduct, and that is why we have free oxygen in our atmosphere. We return to the topic of photosynthesis later in this chapter (Closer Look 4.3).

After considering the two chemical reactions and applying critical thinking, you may recognize that both reactions combine water and carbon dioxide, but the products are very different: carbonic acid in one combination and a sugar in the other. How can this be so? The answer lies in an important difference between the simple reaction in the atmosphere to produce carbonic acid and the production of sugar and oxygen in the series of reactions of photosynthesis. Green plants use the energy from the sun, which they absorbed through the chemical chlorophyll. Thus, active solar energy is converted to a stored chemical energy in sugar.

Perhaps the simplest way to think of a biogeochemical cycle is as a "box-and-arrow" diagram, which shows where a chemical is stored and the pathways along which it is transferred from one storage place to another. (See Figure 4.3 and A Closer Look 4.2, "A Biogeochemical Cycle."). We can consider biogeochemical cycles on any spatial scale of interest to us, from a single ecosystem to the whole earth. It is often useful to consider such a cycle from a global perspective. The problem of potential glob-

al warming, for example, calls for an understanding of the cycling of carbon transferred into and out of Earth's atmosphere. Sometimes, it is useful to consider a cycle at a more local level, as in the case study of Lake Washington. The key that unifies all these cycles is the involvement of the four principle components of the Earth system: lithosphere, atmosphere, hydrosphere, and biosphere. By their nature, chemicals in these four major components have different average times of storage. In general, the average residence time of chemicals is long in rocks, short in the atmosphere, and intermediate in the hydrosphere and biosphere.

 4.2 **Environmental Questions and Biogeochemical Cycles**

With a general idea of how chemicals cycle, we next consider some of the environmental questions that the science of biogeochemical cycles can help answer. These questions include the following:

Biological Questions
- What factors, including chemicals necessary for life, place limits on the abundance and growth of organisms and their ecosystems?
- What toxic chemicals might be present that adversely affect the abundance and growth of organisms and their ecosystems?
- How can people improve the production of a desired biological resource?

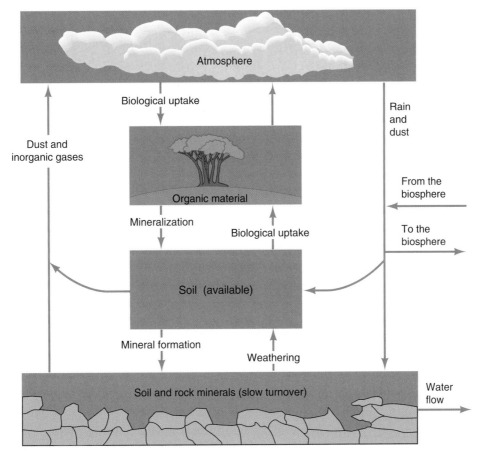

Figure 4.3 • A generalized cycling of a chemical in an ecosystem. Chemical elements cycle within an ecosystem or exchange between an ecosystem and the biosphere. Organisms exchange elements with the nonliving environment; some elements are taken up from and released to the atmosphere, and others are exchanged with water and soil or sediments. The parts of an ecosystem can be thought of as storage compartments for chemicals. The chemicals move among these compartments at different rates and remain within them for different average lengths of time. For example, the soil in a forest has an active part, which rapidly exchanges elements with living organisms, and an inactive part, which exchanges elements slowly (as shown in the lower part of the diagram). Generally, life benefits if needed chemicals are kept within the ecosystem and are not lost through geologic processes, such as erosion, that remove them from the ecosystem.

- What are the sources of chemicals required for life, and how might we make these more readily available?
- What problems occur when a chemical is too abundant, as was the case with phosphorus in Lake Washington?

Geological Questions
- What physical and chemical processes control the movement and storage of chemical elements in the environment?
- How are chemical elements transferred from the solid earth to the water, atmosphere, and life forms?
- How does the long-term storage of chemical elements (for thousands of years or longer) in rocks and soils affect ecosystems on local to global scales?

Atmospheric Questions
- What determines the concentrations of elements and compounds in the atmosphere?

- Where the atmosphere is polluted as the result of human activities, how might we alter a biogeochemical cycle to lower the pollution?

Hydrological Questions
- What determines whether a body of water will be biologically productive?
- When a body of water becomes polluted, how can we alter biogeochemical cycles to reduce the pollution and its effects?

4.3 **Biogeochemical Cycles and Life: Limiting Factors**

The first question on the preceding list concerns chemicals necessary for life and limits imposed by these chemicals. We consider this issue next.

A Biogeochemical Cycle

The simplest way to visualize a biogeochemical cycle is as a box-and-arrow diagram, where the boxes represent places where a chemical is stored (called storage compartments) and the arrows represent pathways of transfer (Figure 4.4*a*). The rate of transfer or flux is the amount per unit time of a chemical that enters or leaves a storage compartment. When a chemical enters a storage compartment from another compartment we say that the receiving compartment is a sink. For example, the forests of the world (which are a storage compartment for carbon) may function as a sink for carbon from the atmosphere, sequestering it in wood, leaves, and roots. The amount of carbon transferred from the atmosphere to the forest on a global basis is the flux, which can be measured in units, such as billions of metric tons of carbon per year.

A biogeochemical cycle is generally drawn for a single chemical element, but sometimes it is drawn for a compound—for example, water (H_2O). Figure 4.4*b* shows the basic elements of a biogeochemical cycle for water, which represents three parts of the hydrological cycle. Water is stored temporarily in a lake (compartment B). It enters the lake from the atmosphere (compartment A) as precipitation and from the land around the lake as runoff (compartment C). It leaves the lake through evaporation to the atmosphere or runoff to a surface stream or subsurface flows.

In each compartment, we can identify an average length of time that an atom is stored before it is transferred. This is called the residence time.

As an example, consider a salt lake with no transfer out except by evaporation. Assume that the lake contains 3,000,000 m^3 (106 million ft^3) of water and the evaporation is 3,000 m^3/day (106,000 ft^3/day). Surface runoff into the lake is also 3,000 m^3/day, so the volume of water in the lake remains constant. We can calculate the average residence time of the water in the lake as the volume of the lake divided by the evaporation rate, or 3,000,000 m^3 divided by 3,000 m^3/day, which is 1,000 days (or 2.7 years).

Another crucial aspect of a biogeochemical cycle is the set of factors or processes that controls the flow from one compartment to another. To understand a biogeochemical cycle, these factors and processes should be quantified and understood. For example, understanding how air temperature and wind velocity vary across a lake is important in understanding the rate of evaporation of water from the lake.

Figure 4.4 ● Basic parts of a biochemical cycle. (*a*) A and B are storage compartments. Chemicals flow from one compartment to another. (*b*) Some components of the hydrologic cycle.

All living things are made up of chemical elements, but of the 103 known elements, only 24 are required for life processes (Figure 4.5). These 24 are divided into the **macronutrients**, elements required in large amounts by all life, and **micronutrients**, elements required either in small amounts by all life or in moderate amounts by some forms of life and not at all by others.

The macronutrients, in turn, include the "big six," the elements that form the fundamental building blocks of life. These are carbon, hydrogen, nitrogen, oxygen, phosphorus, and sulfur. Each element plays a special role in organisms. Carbon is the basic building block of organic compounds. Along with oxygen and hydrogen, carbon forms carbohydrates. Nitrogen, along with these other three, makes proteins. Phosphorus is the "energy element"; it occurs in the compounds called ATP and ADP, important in the transfer and use of energy within cells.

In addition to the "big six," other macronutrients also play important roles. Calcium, for example, is the structure element, occurring in bones of vertebrates, shells of shellfish, and wood-forming cell walls of vegetation. Sodium and potassium are important to nerve signal transmission. Many of the metals required by living things are necessary for specific enzymes. (An enzyme is a complex organic compound that acts as a catalyst—it causes or speeds up chemical reactions such as digestion.)

For any form of life to persist, chemical elements must be available at the right times, in the right amounts, and in the right concentrations relative to each other. When this does not happen, a chemical can become a **limiting factor**, preventing the growth of an individual, population, or species, or even causing its local extinction. Limiting factors are discussed further in Chapters 5 and 11. Chemical elements may also be toxic to some life

Figure 4.5 • Periodic table of the elements.

forms and ecosystems. Mercury, for example, is toxic even in low concentrations. Copper and some other elements are required in low concentrations for life processes but are toxic when present in high concentrations.

4.4 General Concepts Central to Biogeochemical Cycles

Although there are as many biogeochemical cycles as there are chemicals, certain general concepts hold true for these cycles.

- Some chemical elements cycle quickly and are readily regenerated for biological activity. Oxygen and nitrogen are among these. Typically, these elements have a gas phase and are present in Earth's atmosphere, and/or they are easily dissolved in water and are carried by the hydrologic cycle.

- Other chemical elements are easily tied up in relatively immobile forms and are returned slowly, by geological processes, to where they can be reused by life. Typically, these elements lack a gas phase and are not

found in significant concentrations in the atmosphere. They also are relatively insoluble in water. Phosphorus is an example of this kind of chemical.

- Chemicals whose biogeochemical cycles include a gas phase and which are stored in the atmosphere tend to cycle rapidly. Those without an atmospheric phase are likely to end up as deep-ocean sediment and recycle slowly.

- Since life evolved, it has greatly altered biogeochemical cycles, and this alteration has changed our planet in many ways, such as in the development of the fertile soils on which agriculture depends.

- The continuation of processes that control biogeochemical cycles is essential to the long-term maintenance of life on Earth.

- Through modern technology, we have begun to transfer chemical elements between air, water, and soil at rates comparable to natural processes. These transfers can benefit society, as when they improve crop production, but they can also pose environmental dangers, as illustrated by the opening case study. To live wisely with our environment, we must recognize the positive and negative consequences of altering biogeochemical cycles. Then we must attempt to accentuate the positive and minimize the negative.

Discussion of biogeochemical cycles beyond the general concepts listed above requires an understanding of geologic and hydrologic cycles. Of particular importance are geologic processes linked to cycling of chemicals in the biosphere.

4.5   The Geologic Cycle

Throughout the 4.6 billion years of Earth's history, rocks and soils have been continuously created, maintained, changed, and destroyed by physical, chemical, and biological processes. Collectively, the processes responsible for formation and change of Earth materials are referred

to as the **geologic cycle** (Figure 4.6). The geologic cycle is best described as a group of cycles: tectonic, hydrologic, rock, and biogeochemical.

The Tectonic Cycle

The **tectonic cycle** involves creation and destruction of the solid outer layer of Earth, the lithosphere. The lithosphere is about 100 km (60 miles) thick and is broken into several large segments called plates, which are moving relative to one another (Figure 4.7). The slow movement of these large segments of Earth's outermost rock shell is referred to as **plate tectonics**. The plates "float" on denser material and move at rates of 2 to 15 cm/year (0.8 to 5.9 in/year), or about as fast as your fingernails grow. The tectonic cycle is driven by forces originating

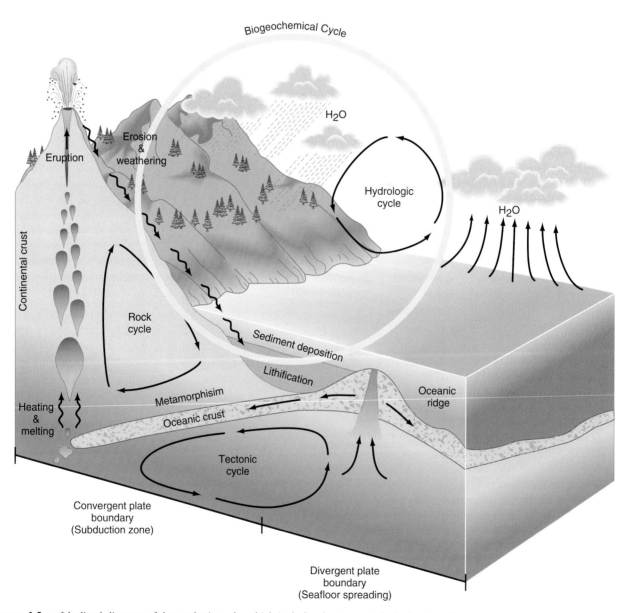

Figure 4.6 • Idealized diagram of the geologic cycle, which includes the tectonic, hydrologic, rock, and biogeochemical cycles.

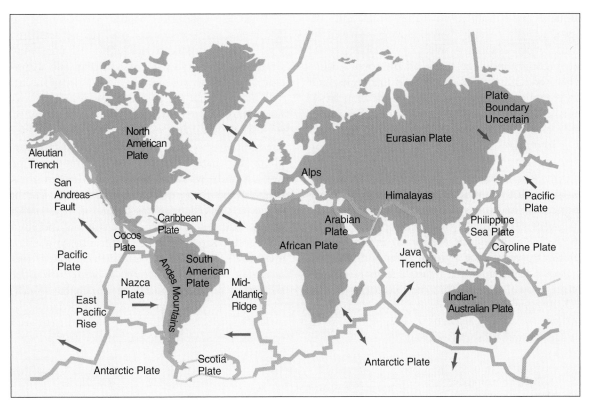

Figure 4.7 • Generalized map of Earth's lithospheric plates. Divergent plate boundaries are shown as heavy lines (for example, the mid-Atlantic ridge). Convergent boundaries are shown as barbed lines (for example, the Aleutian trench). Transform fault boundaries are shown as yellow thinner lines (for example, the San Andreas Fault). Arrows indicate directions of relative plate motions. (*Source:* Modified from B. C. Birchfiel, R. J. Foster, E. A. Keller, W. N. Melhorn, D. G. Brookins, L. W. Mintz, and H. V. Thurman, *Physical Geology: The Structures and Processes of the Earth* [Columbus, Ohio: Charles E. Merrill, 1982].)

deep within the earth. Closer to the surface, rocks are deformed by spreading plates, which produce ocean basins, and collisions of plates, which produce mountain ranges.

Plate tectonics has important environmental effects. Moving plates change the location and size of continents, altering atmospheric and ocean circulation and thereby altering climate. Plate movement has also created ecological islands by breaking up continental areas. When this happens, closely related life forms are isolated from one another for millions of years, leading to the evolution of new species. Finally, boundaries between plates are geologically active areas, and most volcanic activity and earthquakes occur there. Earthquakes occur when the brittle upper lithosphere fractures along faults. Movement of several meters between plates can occur within a few seconds or minutes, in contrast to the slow, deeper plate movement described above.

Three types of plate boundaries occur: divergent, convergent, and transform faults.

- A *divergent plate* boundary occurs at a spreading ocean ridge, where plates are moving away from one another and new lithosphere is produced. This process, known as *seafloor spreading*, produces ocean basins.

- A *convergent plate boundary* occurs when plates collide. When a plate composed of relatively heavy ocean basin rocks dives, or subducts, beneath the leading edge of a plate composed of lighter continental rocks, a subduction zone is present. Such a convergence may produce linear coastal mountain ranges, such as the Andes in South America. When two plates, each composed of lighter continental rocks, collide, a continental mountain range such as the Himalayas in Asia may form.[4,5]

- A *transform fault boundary* occurs where one plate slides past another. An example is the San Andreas Fault in California, which is the boundary between the North American and Pacific plates. The Pacific plate is moving north relative to the North American plate at about 5 cm/year (2 in/year). As a result, Los Angeles is moving slowly toward San Francisco about 500 km (300 mi) north. If this motion continues, in about 10 million years, San Francisco will be a suburb of Los Angeles.

The Hydrologic Cycle

The **hydrologic cycle** (Figure 4.8) is the transfer of water from the oceans to the atmosphere to the land and back to the oceans. The processes involved include evaporation of water from the oceans; precipitation on land; evaporation from land; and runoff from streams, rivers, and sub-surface groundwater. The hydrologic cycle is driven by solar energy, which evaporates water from oceans, fresh-water bodies, soils, and vegetation. Of the total 1.3 billion km³ water on Earth, about 97% is in oceans, and about 2% is in glaciers and ice caps. The rest is in fresh-water on land and in the atmosphere. Although it represents only a small fraction of the water on earth, the water on land is important in moving chemicals, sculpturing landscape, weathering rocks, transporting sediments, and providing our water resources. The water in the atmosphere—only 0.001% of the total on Earth—cycles quickly to produce rain and runoff for our water resources.

Annual rates of transfer from the storage compartments in the hydrologic cycle are shown in Figure 4.8. These rates of transfer define a global water budget. If we sum the arrows going up in the figure, and then sum the arrows going down, we find that the two sums are the same, 577,000 km³/year. Similarly, precipitation on land (119,000 km³/year) is balanced by evaporation from land plus surface and subsurface runoff.

Especially important from an environmental perspective is that rates of transfer on land are small relative to what's happening in the ocean. For example, most of the water that evaporates from the ocean falls again as precipitation into the ocean. On land, most of the water that falls as precipitation comes from evaporation of water from land. This means that regional land-use changes, such as the building of large dams and reservoirs, can change the amount of water evaporated into the atmosphere and change the location and amount of precipitation on land—water we depend on to raise our crops and supply water for our urban environments. Furthermore, as we pave over large areas of land in cities, storm water runs off quicker and in greater volume, thereby increasing flood hazards. Bringing water into semi-arid cities by pumping groundwater or transporting water from distant mountains through aqueducts may increase evaporation, thereby increasing humidity and precipitation in a region.

We can see from Figure 4.8 that approximately 60% of water that falls by precipitation on land each year evaporates to the atmosphere. A smaller component (about

Figure 4.8 • The hydrologic cycle, showing the transfer of water (thousands of km³/yr) from the oceans to the atmosphere to the continents and back to the oceans again. (*Source:* From P. H. Gleick, *Water in Crisis.* [New York: Oxford University Press, 1993].)

Storage Compartments of Water		
Compartment	Vol. (thousands of km³)	Percentage of Total Water
Ocean	1,230,000	97.2
Glaciers and ice caps	28,600	2.2
Shallow groundwater	4,000	0.3
Lakes	123	0.009
Atmosphere	12.7	0.001
Rivers	1.2	0.0001

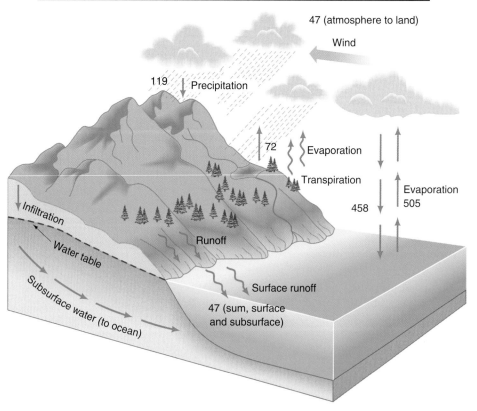

40%) returns to the ocean surface and subsurface runoff. This small annual transfer of water supplies resources for rivers and urban and agricultural lands. Unfortunately, distribution of water on land is far from uniform. As a result, water shortages occur in some areas. As human population increases, water shortages will become more frequent in arid and semi-arid regions, where water is naturally nonabundant.

At the regional and local level, the fundamental hydrological unit of the landscape is the **drainage basin** (also called a watershed or catchment). A drainage basin is the area that contributes surface runoff to a particular stream or river. The term *drainage basin* is usually used in evaluating the hydrology of an area, such as the stream flow or runoff from hill slopes. Drainage basins vary greatly in size, from less than a hectare (2.5 acres) to millions of square kilometers. A drainage basin is usually named for its main stream or river, such as the Mississippi River drainage basin.

The Rock Cycle

The **rock cycle** consists of numerous processes that produce rocks and soils. The rock cycle depends on the tectonic cycle for energy and on the hydrologic cycle for water. As shown in Figure 4.9, rock is classified as igneous, sedimentary, or metamorphic. These types of rock are involved in a worldwide recycling process. Internal heat from the tectonic cycle produces igneous rocks from molten material near the surface, such as lava from volcanoes. These new rocks weather when exposed at the surface. The freezing of water in cracks of rocks creates a physical weathering process. The water expands when it freezes, breaking the rocks apart. Physical weathering makes smaller particles of rocks from bigger ones, producing sediment such as gravel, sand, and silt. Chemical weathering occurs when the weak acids in water dissolve chemicals from rocks. The sediments and dissolved chemicals are then transported by water, wind, or ice (glaciers).

Weathered materials accumulate in depositional basins, such as the oceans. Sediments in depositional basins are compacted by overlying sediments and converted to sedimentary rocks by compaction and the cementing together of particles. After sedimentary rocks are buried to sufficient depths (usually tens to hundreds of kilometers), they may be altered by heat, pressure, or chemically active fluids and transformed to metamorphic rocks. Deeply buried rocks may be transported to the surface by an uplift associated with plate tectonics and then subjected to weathering.

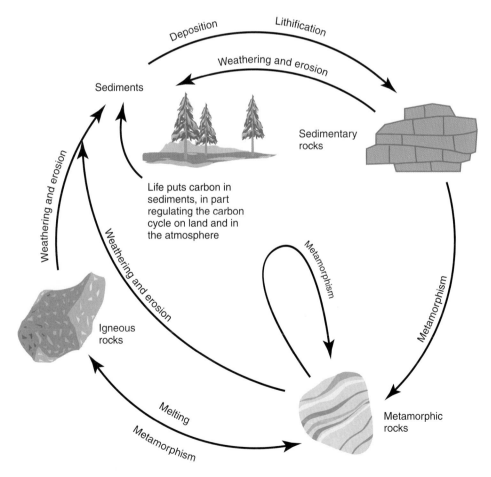

Figure 4.9 • The rock cycle and major paths of material transfer as modified by life.

You can see on Figure 4.8 that life processes play an important role in the rock cycle through the addition of organic carbon rocks. Through the addition of organic carbon, rocks such as limestone, which is mostly a calcium carbonate (the material of seashells and bones), as well as fossil fuel resources such as coal, are produced.

Plate tectonic processes of uplift and subsidence of rocks, along with erosion, produce the earth's varied topography. The spectacular Grand Canyon of the Colorado River in Arizona (Figure 4.10), sculptured from mostly sedimentary rocks, is one example. Another is the beautiful Tower Karst in China (Figure 4.11); these resistant blocks of limestone have survived chemical weathering and erosion that removed the surrounding rocks.

Our discussion of geologic cycles has emphasized tectonic, hydrologic, and rock-forming processes. We can now begin to integrate biogeochemical processes into the picture.

Figure 4.10 • In response to slow tectonic uplift of the region, the Colorado River has eroded through the sedimentary rocks of the Colorado plateau to produce the spectacular scenery of the Grand Canyon. The river in recent years has been greatly modified by dams and reservoirs above and below the Grand Canyon. Sediment once carried to the Gulf of California is now deposited in reservoirs. As a result of the Glen Canyon Dam, the water flowing through the Grand Canyon is clearer and colder, resulting in changes in the abundance of sandbars and species of fish. The presence of this upstream dam has changed the hydrology and environment of the Colorado River in the Grand Canyon.

 4.6 Biogeochemical Cycling in Ecosystems

When we ask questions about what chemicals might limit the abundance of a specific individual organism, population, or species, we look for the answer first at the ecosystem level. As explained in Chapter 3, an ecosystem is a community of different species and their nonliving envi-

Figure 4.11 • This landscape in the People's Republic of China features *Tower Karst*, steep-sided hills or pinnacles composed of limestone. The rock has been slowly dissolving through chemical weathering. The pinnacles and hills are remnants of the weathering and erosion processes.

ronment in which energy flows and chemicals cycle. The boundaries that we choose for this investigation may be somewhat arbitrary, selected for convenience of measurement and analysis. On land, we often evaluate biogeochemical cycles for a fundamental element of the landscape, usually a drainage basin. Freshwater bodies—lakes, ponds, and bogs—are also convenient landforms for analysis of ecosystems and biogeochemical cycling.

Chemical cycling in an ecosystem begins with inputs from outside. On land, chemical inputs to an ecosystem come from the atmosphere via rain, wind-transported dust (called dry fallout), and volcanic ash from eruptions and from adjoining land via stream flow from flooding and groundwater from springs. Ocean and freshwater ecosystems have the same atmospheric and land inputs (including large, nearshore submarine springs) and, in addition, have inputs from ocean currents and submarine vents (hot springs) at divergent plate boundaries.

Chemicals cycle internally within an ecosystem through air, water, rocks, soil, and food chains by way of physical transport and chemical reactions. With the death of individual organisms, decomposition through chemical reactions returns chemical elements to other parts of the ecosystem. In addition, living organisms release some chemical elements directly into an ecosystem. Defecation by animals and dropping of fruit by plants are examples.

An ecosystem can lose chemical elements to other ecosystems. For example, rivers transport chemicals from the land to the sea. An ecosystem that has little loss of chemical elements can function in its current condition for longer periods than a "leaky" ecosystem that loses chemical elements rapidly. All ecosystems, however, lose chemicals to some extent. Therefore, all ecosystems require some external inputs of chemicals.

Ecosystem Cycles of a Metal and a Nonmetal

Within an ecosystem, different chemical elements may have very different pathways, as illustrated in Figure 4.12 for calcium and in Figure 4.13 for sulfur. The calcium

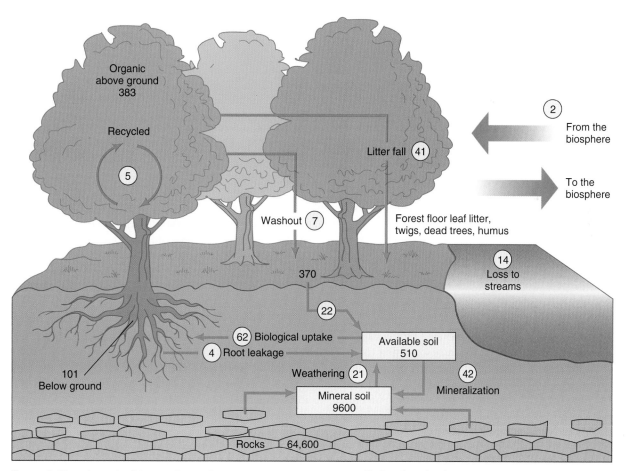

Figure 4.12 • Annual calcium cycle in a forest ecosystem. The circled numbers are flux rates in kilograms per hectare per year. The other numbers are the amounts stored in kilograms per hectare. Unlike sulfur, calcium does not have a gaseous phase, although it does occur in compounds as part of dust particles. Calcium is highly soluble in water in its inorganic form and is readily lost from land ecosystems in water transport. The information in this diagram was obtained from Hubbard Brook Ecosystem. (*Source:* G. E. Likens, F. H. Bormann, R. S. Pierce, J. S. Eaton, and N. M. Johnson, *The Biogeochemistry of a Forested Ecosystem*, 2nd ed. [New York: Springer-Verlag, 1995].)

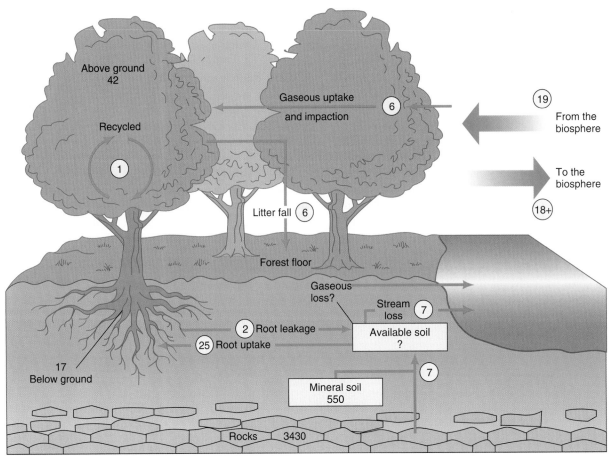

Figure 4.13 • Annual sulfur cycle in a forest ecosystem. The circled numbers are the flux rates in kilograms per hectare per year. The other numbers are the amounts stored in kilograms per hectare. Sulfur has a gaseous phase as H_2S and SO_2. The information in this diagram was obtained from Hubbard Brook Ecosystem. (*Source:* G. E. Likens, F. H. Bormann, R. S. Pierce, J. S. Eaton, and N. M. Johnson, *The Biogeochemistry of a Forested Ecosystem*, 2nd ed. [New York: Springer-Verlag, 1995].)

cycle is typical of a metallic element, and the sulfur cycle is typical of a nonmetallic element.

An important difference between these cycles is that calcium, like most metals, does not form a gas on the earth's surface. Therefore, calcium is not present as a gas in the atmosphere. In contrast, sulfur forms several gases, including sulfur dioxide (a major air pollutant and component of acid rain; see Chapter 23) and hydrogen sulfide (swamp or rotten egg gas, usually produced biologically).

Because sulfur has gas forms, it can be returned to an ecosystem more rapidly than can calcium. Annual input of sulfur from the atmosphere to a forest ecosystem has been measured to be 10 times that of calcium. For this reason, calcium and other elements without a gas phase are more likely to become limiting factors.

Chemical Cycling and the Balance of Nature

For life to be sustained indefinitely within an ecosystem, energy must be continuously added, and the store of essential chemicals must not decline. We have already discussed the common belief that, without human interference, life would be sustained indefinitely in a steady state, or "balance of nature." We have also discussed the belief that life tends to function to preserve an environment beneficial to itself. Both beliefs presume that chemical elements necessary for life exist in a dynamic steady state within an ecosystem.[6] These beliefs can be rephrased as a scientific hypothesis: "Without human disturbance, the net storage of chemical elements within an ecosystem will remain constant over time." However, as noted, inevitably, some fraction of the chemical elements stored in an ecosystem is lost and must be replaced. Are ecosystems, then, ever in a dynamic steady state in regard to chemical elements? Studies indicate that they are not, because rates of input and output of chemicals do not balance, and concentrations of some chemicals decrease over time.

4.7 Some Major Global Chemical Cycles

Earlier in this chapter, we asked what chemical elements limit the abundance of life. We pointed out that the chemical elements required by life are divided into two

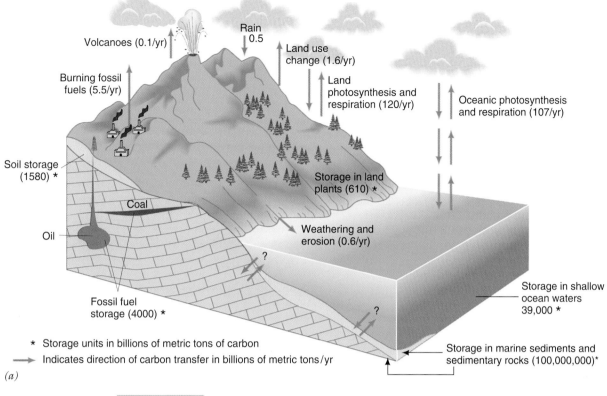

Storage in atmosphere (750 + 3/yr due to burning fossil fuels) *

Volcanoes (0.1/yr)

Rain 0.5

Land use change (1.6/yr)

Land photosynthesis and respiration (120/yr)

Oceanic photosynthesis and respiration (107/yr)

Burning fossil fuels (5.5/yr)

Soil storage (1580) *

Coal

Oil

Storage in land plants (610) *

Weathering and erosion (0.6/yr)

?

Storage in shallow ocean waters 39,000 *

Fossil fuel storage (4000) *

?

* Storage units in billions of metric tons of carbon

→ Indicates direction of carbon transfer in billions of metric tons/yr

Storage in marine sediments and sedimentary rocks (100,000,000)*

(a)

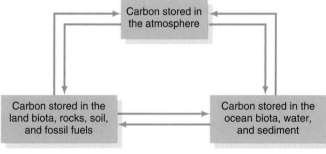

Carbon stored in the atmosphere

Carbon stored in the land biota, rocks, soil, and fossil fuels

Carbon stored in the ocean biota, water, and sediment

(b)

Figure 4.14 • (a) Generalized global carbon cycle. (b) Parts of the carbon cycle are simplified to illustrate the cyclic nature of the movement of carbon. [*Source:* Modified from G. Lambert, *La Recherche* 18 (1987):, 782–783, with some data from R. Houghton, *Bulletin of the Ecological Society of America* 74, no. 4 (1993): 355–356.]

major groups: macronutrients, which are required by all forms of life in large amounts, and micronutrients, which are either required by all forms of life in small amounts or required by only certain life forms. In this section, we consider the global cycles of three macronutrients—carbon, nitrogen, and phosphorus. We focus on these in part because they are three of the "big six"—the elements that are the basic building blocks of life. Each also is related to important environmental problems that have attracted attention in the past and will continue to do so in the future.

The Carbon Cycle

Carbon is the element that anchors all organic substances, from coal and oil to DNA (deoxyribonucleic acid), the compound that carries genetic information. Although of central importance to life, carbon is not one of the most abundant elements in the earth's crust. It contributes only 0.032% of the weight of the crust, ranking far behind oxygen (45.2%), silicon (29.5%), aluminum (8.0%), iron (5.8%), calcium (5.1%), and magnesium (2.8%.)[7,8]

The major pathways and storage reservoirs of the **carbon cycle** are shown in Figure 4.14. Notice that carbon has a gaseous phase as part of its cycle. It occurs in the earth's atmosphere as carbon dioxide (CO_2) and methane (CH_4), both greenhouse gases (see Chapter 22). Carbon enters the atmosphere through the respiration of living things, through fires that burn organic compounds, and by diffusion from the ocean. It is removed from the atmosphere by photosynthesis of green plants, algae, and photosynthetic bacteria. (See A Closer Look 4.3, "Photosynthesis and Respiration.") Over the past 3 billion years of Earth's history, the rate at which biological processes have removed carbon dioxide from the atmosphere has exceeded the rate of addition. Therefore, Earth's atmosphere has much less carbon dioxide than would occur on a lifeless Earth.

Photosynthesis and Respiration

Carbon dioxide enters biological cycles through photosynthesis, the process whereby the cells of living organisms (such as plants) convert energy from sunlight in a series of chemical reactions to chemical energy. In the process, carbon dioxide and water are combined to form organic compounds such as simple sugars and starches, with oxygen as a byproduct (Figure 4.15). Carbon leaves living biota through respiration, the process by which organic compounds are broken down to release gaseous carbon dioxide. For example, animals (including people) take in air, which has a relatively high concentration of oxygen. Oxygen is absorbed by blood in the lungs. Through respiration, carbon dioxide is released into the atmosphere. Figure 4.16 illustrates the role of photosynthesis and respiration, along with other processes, in the carbon cycle of a lake.

Energy from sun

CO$_2$

O$_2$

H$_2$O

At the cell level: chlorophyll (green plant absorbs sunlight)

Figure 4.15 ● Idealized diagram illustrating photosynthesis for a green plant (tree) and generalized reaction.

General Photosynthesis: chemical reaction

$$6CO_2 + 6H_2O \xrightarrow{\text{sunlight}} C_6H_{12}O_6 + 6O_2$$

$$\text{carbon + water} \xrightarrow{\text{sunlight}} \text{sugar (glucose) + oxygen}$$

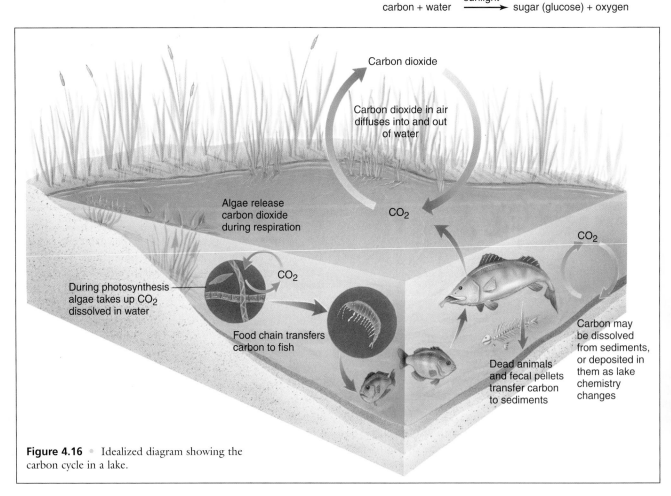

Carbon dioxide

Carbon dioxide in air diffuses into and out of water

CO$_2$

CO$_2$

Algae release carbon dioxide during respiration

During photosynthesis algae takes up CO$_2$ dissolved in water

CO$_2$

Food chain transfers carbon to fish

Dead animals and fecal pellets transfer carbon to sediments

Carbon may be dissolved from sediments, or deposited in them as lake chemistry changes

Figure 4.16 ● Idealized diagram showing the carbon cycle in a lake.

Carbon occurs in the ocean in several inorganic forms, including dissolved carbon dioxide as well as carbonate (CO_3^{2-}) and bicarbonate (HCO_3^-). It also occurs in organic compounds of marine organisms and their products, such as seashells ($CaCO_3$). Carbon enters the ocean from the atmosphere by simple diffusion of carbon dioxide. The carbon dioxide then dissolves and is converted to carbonate and bicarbonate. Marine algae and photosynthetic bacteria obtain the carbon dioxide they use from the water in one of these three forms. Carbon is transferred from the land to the ocean in rivers and streams as dissolved carbon, including organic compounds, and as organic particulates (fine particles of organic matter). Winds also transport small organic particulates from the land to the ocean. Transfer via rivers and streams makes up a relatively small fraction of the total global carbon flux to the oceans. However, on the local and regional scale, input of carbon from rivers to nearshore areas such as deltas and salt marshes, which are often highly biologically productive, is important.

Carbon enters the biota through photosynthesis and is returned to the atmosphere or waters by respiration or by wildfire. When an organism dies, most of its organic material decomposes to inorganic compounds, including carbon dioxide. Some carbon may be buried where there is not sufficient oxygen to make this conversion possible or where the temperatures are too cold for decomposition. In these locations, organic matter is stored. Over years, decades, and centuries, storage of carbon occurs in wetlands, including parts of floodplains, lake basins, bogs, swamps, deep-sea sediments, and near-polar regions. Over longer periods (thousands to several million years), some carbon may be buried with sediments that become sedimentary rocks. This carbon is transformed into fossil fuels such as natural gas, oil, and coal. Nearly all of the carbon stored in the lithosphere exists as sedimentary rocks. Most of this is in the form of carbonates such as limestone, much of which has a direct biological origin.

The cycling of carbon dioxide between land organisms and the atmosphere is a large flux. Approximately 15% of the total carbon in the atmosphere is taken up by photosynthesis and released by respiration annually. Thus, as noted, life has a large effect on the chemistry of the atmosphere.

The Missing Carbon Link

Because carbon forms two of the most important greenhouse gases—carbon dioxide and methane—much research has been devoted to understanding the carbon cycle. However, at a global level, some key issues remain unanswered. For example, monitoring of atmospheric carbon dioxide levels over the past 35 years suggests that of the approximately 7.1 units released per year by human activities into the atmosphere, approximately 3.2 units remain there. It is estimated that about 2 units diffuse into the ocean. This leaves 1.9 units unaccounted for.[9,10]

Several hundred or more million tons of carbon are burned each year from fossil fuel and end up somewhere unknown to science. Inorganic processes do not account for the fate of this "missing carbon sink." Either marine or land photosynthesis, or both, must provide the additional flux. At this time, however, scientists do not agree which processes dominate, nor in what regions of the earth this missing flux occurs.

Why can't scientists agree on these basic issues? Unfortunately, there are many uncertainties. For example, when uncertainties in measuring carbon are considered, the missing carbon sink has a flux of 1.9 ± 1.2 billion metric tons/yr, or 1.9 ± 1.2 Pg/yr (petagrams/yr). The uncertainty is over 50% of the suspected flux. Including estimated uncertainties yields the following, in units of billions of metric tons of carbon (GtC) per year.[11]

$$\begin{matrix} \text{Atmospheric} \\ \text{increase} \end{matrix} = \begin{matrix} \text{Emissions} \\ \text{from} \\ \text{fossil fuels} \end{matrix} + \begin{matrix} \text{Net emissions} \\ \text{from changes} \\ \text{in land use} \end{matrix} - \begin{matrix} \text{Oceanic} \\ \text{uptake} \end{matrix} - \begin{matrix} \text{Missing} \\ \text{carbon} \\ \text{sink} \end{matrix}$$

$$3.2 \pm 0.2 = 5.5 \pm 0.5 + 1.6 \pm 0.7 - 2.0 \pm 0.8 - 1.9 \pm 1.2$$

Since the beginning of the Industrial Revolution in the mid-19th century, the unidentified sink has apparently steadily increased in size (Figure 4.17). It is believed that a possible sink to account for part of the missing carbon (about 0.6 billion metric tons/yr) is the boreal forests of the world, which cover 16 to 20 million km^2 above 40° north latitude. However, fires and melting of frozen ground in the boreal forests also release carbon, and it's not certain whether the forests comprise a net sink or a net source of carbon to the atmosphere.[12] The missing carbon problem illustrates the complexity of biogeochemical cycles, especially ones in which the biota play an important role.

The carbon cycle will continue to be an important area of research because of its significance to global climate investigations, especially to global warming.[13,14] To better understand this issue, we turn next to the interactions between the carbon and tectonic cycles through the carbon-silicate cycle.

The Carbon-Silicate Cycle

Carbon cycles rapidly among the atmosphere, oceans, and life. However, over geologically long periods of time, the cycling of carbon becomes intimately involved with the cycling of silicon. The combined **carbon-silicate cycle** is therefore of geological importance to the long-term stability of the biosphere over periods that exceed one-half billion years.[15]

The carbon-silicate cycle begins when carbon dioxide in the atmosphere dissolves in the water to form weak carbonic acid (H_2CO_3) that falls as rain (Figure 4.18). As the mildly acidic water migrates through the ground, it weathers rocks and facilitates the erosion of the earth's abundant silicate-rich rocks. Among other products, weathering and

Figure 4.17 • Global flux of carbon, 1850–1990. Modified after Woods Hole, "The Missing Carbon Sink," 2000, http://www.whrc.org/science/carbon/carbon.htm, accessed July 5, 2000.

erosion release calcium ions (CA^{2+}) and bicarbonate (HCO_3^-) ions. These ions enter the ground and surface waters and eventually are transported to the ocean. Calcium and bicarbonate ions make up a major portion of the chemical load that rivers deliver to the oceans.

Tiny floating marine organisms in the ocean use the calcium and bicarbonate to construct their shells. When these organisms die, the shells sink to the bottom of the ocean, where they accumulate as carbonate-rich sediments. Eventually, carried by moving tectonic plates, they

Figure 4.18 • Idealized diagram showing the carbon-silicate cycle. (*Source:* Modified from J. E. Kasting, O. B. Toon, and J. B. Pollack, "How Climate Evolved on the Terrestrial Planets," *Scientific American* 258 (1988): 2.)

enter a subduction zone, where they are subjected to increased heat, pressure, and partial melting. The resulting magma releases carbon dioxide, which rises in volcanoes and is released into the atmosphere. This process provides a lithosphere-to-atmosphere flux of carbon.

The long-term carbonate-silicate cycle (Figure 4.18) and the short-term carbon cycle (Figure 4.14) interact to affect levels of CO_2 and O_2 in the atmosphere. For example, the burial of organic material in an oxygen-poor environment amounts to a net increase of photosynthesis (which produces O_2) over respiration (which produces CO_2). Thus, if burial of organic carbon in oxygen-poor environments increases, the concentration of atmospheric oxygen will increase. Conversely, if more organic carbon escapes burial and is oxidized to produce CO_2, then the CO_2 concentration in the atmosphere will increase.[16] Long-term changes in CO_2 and O_2 have been profoundly important in global change processes over geologic time, far beyond what we normally consider in environmental science.

The Nitrogen Cycle

Nitrogen is essential to life because it is necessary for the manufacture of proteins and DNA. Free nitrogen (N_2 uncombined with any other element) makes up approximately 80% of Earth's atmosphere. However, many organisms cannot use this nitrogen directly. Some, such as animals, require nitrogen in an organic compound.

Others, including plants, algae, and bacteria, can take up nitrogen either as the nitrate ion (NO_3^-) or the ammonium ion (NH_4^+). Because nitrogen is a relatively unreactive element, few processes convert molecular nitrogen to one of these compounds. Lightning oxidizes nitrogen, producing nitric oxide. In nature, essentially all other conversions of molecular nitrogen to biologically useful forms are conducted by bacteria.

The **nitrogen cycle** is one of the most important and most complex of the global cycles (Figure 4.19). The process of converting inorganic, molecular nitrogen in the atmosphere to ammonia or nitrate is called **nitrogen fixation**. Once in these forms, nitrogen can be used on land by plants and in the oceans by algae. Through chemical reactions, bacteria, plants, and algae then convert these inorganic nitrogen compounds into organic ones, and the nitrogen becomes available to ecological food chains. When organisms die, other bacteria convert the organic compounds containing nitrogen back to ammonia, nitrate, or molecular nitrogen, which enters the atmosphere. The process of releasing fixed nitrogen back to molecular nitrogen is called **denitrification**.

Nearly all organisms depend on nitrogen-converting bacteria. Some organisms have evolved symbiotic relationships with these bacteria. For example, the roots of the pea family have nodules that provide a habitat for the bacteria. The bacteria obtain organic compounds for food from the plants, and the plants obtain usable nitrogen.

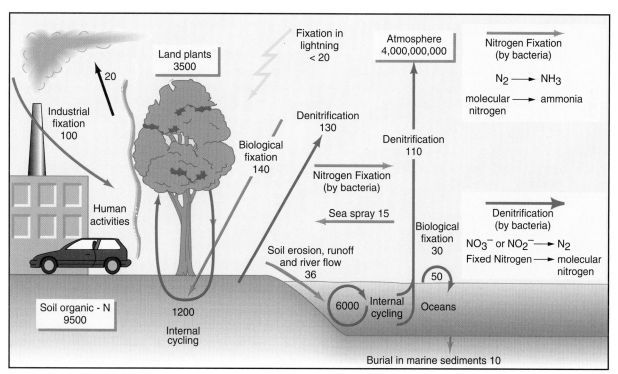

Figure 4.19 • The global nitrogen cycle. Numbers in boxes indicate amounts stored, and numbers with arrows indicate annual flux, in 10^{12} g N_2. Note that the industrial fixation of nitrogen is nearly equal to the global biological fixation. (*Source:* Data from R. Söderlund and T. Rosswall, in *The Handbook of Environmental Chemistry*, Vol. 1, Pt. B, ed. O. Hutzinger [New York: Springer-Verlag, 1982].)

Such plants can grow in otherwise nitrogen-poor environments. When these plants die, they contribute nitrogen-rich organic matter to the soil, thereby improving the soil's fertility. Alder trees, too, have nitrogen-fixing bacteria as symbionts in their roots. (Symbionts are organisms in symbiotic relationships.) These trees grow along streams, and their nitrogen-rich leaves fall into the streams and increase the supply of the element in a biologically usable form to freshwater organisms.

Nitrogen-fixing bacteria are also symbionts in the stomachs of some animals, particularly the cud-chewing animals. These animals, which include buffalo, cows, deer, moose, and giraffes, have a specialized four-chambered stomach. The bacteria provide as much as half of the total nitrogen needed by the animals, with the rest provided by protein in the green plants the animals eat.

In terms of availability for life, nitrogen falls somewhere between carbon and phosphorus. Like carbon, nitrogen has a gaseous phase and is a major component of Earth's atmosphere. Unlike carbon, however, it is not very reactive, and its conversion depends heavily on biological activity. Thus, the nitrogen cycle is not only essential to life but also is primarily driven by life.

In the early part of the twentieth century, scientists discovered that industrial processes could convert molecular nitrogen into compounds usable by plants. This greatly increased the availability of nitrogen in fertilizers. Today, industrial fixation of nitrogen is a major source of commercial nitrogen fertilizer. The amount of industrial fixed nitrogen is about 50% of the amount fixed in the biosphere. Nitrogen in agricultural runoff is a potential source of water pollution.

Nitrogen combines with oxygen in high-temperature atmospheres. As a result, many modern industrial combustion processes produce oxides of nitrogen. These processes include the burning of fossil fuels in gasoline and diesel engines. Thus, oxides of nitrogen, which are air pollutants, are indirect results of modern industrial activity and modern technology. Nitrogen oxides play a significant role in urban smog (see Chapter 23).

In summary, nitrogen compounds are a bane and a boon for society and for the environment. Nitrogen is required for all life, and its compounds are used in many technological processes and in modern agriculture. But nitrogen is also a source of air and water pollution.

The Phosphorus Cycle

Phosphorus, one of the "big six" elements required in large quantities by all forms of life, is often a limiting nutrient for plant and algal growth. However, if phosphorus is too abundant, it can cause environmental problems, as illustrated by the story of Lake Washington that opened this chapter.

Unlike carbon and nitrogen, phosphorus does not have a gaseous phase on the earth (Figure 4.20). Thus, the **phosphorus cycle** is significantly different from the carbon and nitrogen cycles. The rate of transfer of phosphorus in the earth's system is slow compared with that of carbon or nitrogen. Phosphorus exists in the atmosphere only in small particles of dust. In addition, phosphorus tends to form compounds that are relatively insoluble in water. Consequently, phosphorus is not readily weathered chemically. It does occur commonly in an oxidized state as phosphate, which combines with calcium, potassium, magnesium, or iron to form minerals.

Phosphorus enters the biota through uptake as phosphate by plants, algae, and some bacteria. In a relatively stable ecosystem, much of the phosphorus that is taken up by vegetation is returned to the soil when the plants die. Nevertheless, some phosphorus is inevitably lost to ecosystems. It is transported by rivers to the oceans, either in a water-soluble form or as suspended particles.

An important way in which phosphorus returns from the ocean to the land involves ocean-feeding birds, such as the Chilean pelican. These birds feed on small fish, especially anchovies, which in turn feed on tiny ocean plankton. Plankton thrive where nutrients such as phosphorus are present. Areas of rising oceanic currents known as upwellings are such places. Upwellings occur near continents where the prevailing winds blow offshore, pushing surface waters away from the land and allowing deeper waters to rise and replace those moved offshore. Upwellings carry nutrients, including phosphorus, from the depths of the oceans to the surface.

The fish-eating birds nest on offshore islands where they are protected from predators. Over time, their nesting sites become covered with their phosphorus-laden excrement, called guano. The birds nest by the thousands, and deposits of guano accumulate over centuries. In relatively dry climates, guano hardens into a rocklike mass that may be up to 40 m (130 ft) thick. The guano results from a combination of biological and nonbiological processes. Without the plankton, fish, and birds, the phosphorus would have remained in the ocean. Without the upwellings, the phosphorus would not have been available.

Guano deposits were once major sources of phosphorus for fertilizers. In the mid-1800s, as much as 9 million metric tons per year of guano deposits were shipped to London from islands near Peru. Today, most phosphorus fertilizers come from mining of phosphate-rich sedimentary rocks containing fossils of marine animals. The richest phosphate mine in the world is Bone Valley, 40 km east of Tampa, Florida. Between 10 and 15 million years ago, Bone Valley was the bottom of a shallow sea where marine invertebrates lived and died.[17] Through tectonic processes, Bone Valley was slowly uplifted, and in the 1880s and 1890s phosphate ore was discovered there. Today, Bone Valley provides more than one-third of the world's phosphate production and three-fourths of U.S. production.

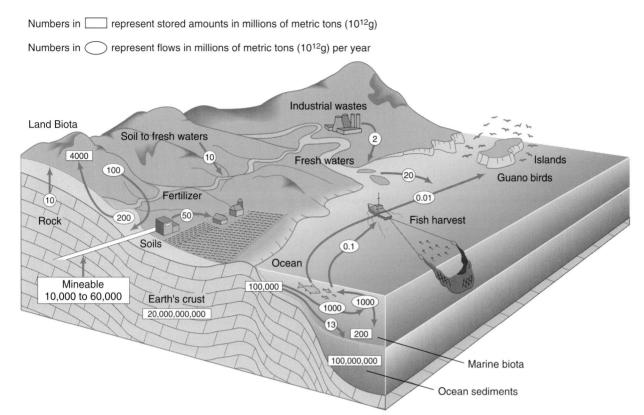

Numbers in ☐ represent stored amounts in millions of metric tons (10^{12}g)

Numbers in ◯ represent flows in millions of metric tons (10^{12}g) per year

Figure 4.20 • The global phosphorus cycle. Phosphorus is recycled to soil and land biota through geologic processes that uplift the land and erode rocks, by birds that produce guano, and by human beings. Although Earth's crust contains a very large amount of phosphorus, only a small fraction of it can be mined by conventional techniques. Therefore, phosphorus is an expensive resource to produce. Values of the amount of phosphorus stored or in flux are compiled from various sources. Estimates are approximate to the order of magnitude. (*Source:* Based primarily on C. C. Delwiche and G. E. Likens, "Biological Response to Fossil Fuel Combustion Products," in *Global Chemical Cycles and Their Alterations by Man*, ed. W. Stumm [Berlin: Abakon Verlagsgesellschaft, 1977], pp. 73–88; and U. Pierrou, "The Global Phosphorus Cycle," in *Nitrogen, Phosphorus and Sulfur—Global Cycles*, eds. B. H. Svensson and R. Soderlund [Stockholm: *Ecological Bulletin*, 1976, pp. 75–88.)

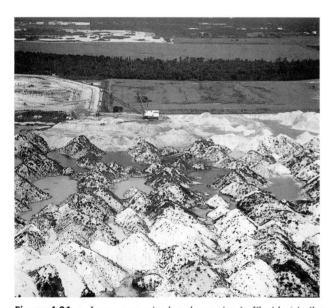

Figure 4.21 • Large open-pit phosphate mine in Florida (similar to Bone Valley), with piles of waste material. The land in the upper part of the photograph has been reclaimed and is being used for pasture. (*Source:* V. Smil, "Global Populations and the Nitrogen Cycle," *Scientific American*, July 1997, p. 77.)

Total U.S. reserves of phosphorus are estimated at 2.2 billion metric tons, a sufficient quantity to supply our needs for several decades. However, if the price of phosphorus increases as high-grade deposits are exhausted, phosphorus from lower-grade deposits can be mined at a profit. Florida is thought to have 8.1 billion metric tons of phosphorus that can be recovered with existing mining methods if the price is right.[17] Mining, of course, may have negative effects on the land and ecosystems. For example, in some phosphorous mines, huge pits and waste ponds have scarred the landscape, damaging biologic and hydrologic resources. Balancing the need for phosphorus with adverse environmental impacts of mining is a major environmental issue. Land disrupted by open-pit phosphate mining, shown on Figure 4.21, is reclaimed as mandated by law following phosphate extraction to pasture land.

Summary

• Biogeochemical cycles are the major way in which elements important to Earth processes and life are moved

How Are Human Activities Affecting the Nitrogen Cycle?

Scientists estimate that nitrogen deposition to Earth's surface will double in the next 25 years. What is causing this increase, and how will it affect the nitrogen cycle?

The natural rate of nitrogen fixation on land is estimated to be 140 teragrams (Tg) of nitrogen a year (1 teragram = 1 million metric tons). Human activities, such as use of fertilizers, draining of wetlands, clearing of land for agriculture, and burning of fossil fuels, are causing additional nitrogen to enter the environment. Currently, human activities are responsible for more than half of the fixed nitrogen that is deposited on land. Before the 20th century, fixed nitrogen was recycled by bacteria with no net accumulation. Since 1900, however, the use of commercial fertilizers has increased exponentially (see graph). Nitrates and ammonia from burning fossil fuels have increased 20% in the last decade. These inputs have overwhelmed the denitrifying part of the nitrogen cycle and the ability of plants to use fixed nitrogen.

Nitrate ions in the presence of soil or water may form nitric acid. With other acids in the soil, nitric acid can leach out chemicals important to plant growth, such as magnesium and potassium. When these chemicals are depleted, more toxic ones, such as aluminum, may be released, damaging tree roots. Acidification of soil by nitrate ions is also harmful to organisms. When toxic chemicals wash into streams, they can kill fish. Excess nitrates in rivers and along coasts can cause algae to overgrow, with effects similar to those described earlier for Lake Washington. High levels of nitrates in drinking water from streams or groundwater contaminated by fertilizers are a health hazard.

The nitrogen and carbon cycles are linked because nitrogen is a component of chlorophyll, the molecule plants use in photosynthesis. Because nitrogen is a limiting factor on land, it has been predicted that increasing levels of global nitrogen may increase plant growth.

Recent studies have suggested that a beneficial effect from increased nitrogen would be short-lived, however. As plants use additional nitrogen, some other factor will become limiting.

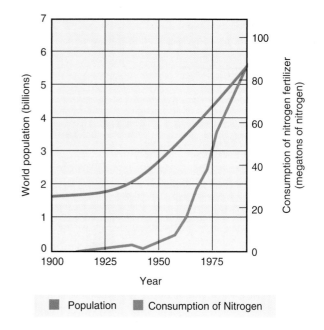

When that occurs, plant growth will slow, and so will the uptake of carbon dioxide. More research is needed to understand the interactions between carbon and nitrogen cycles and predict long-term effects of human activities.

Critical Thinking Questions

1. Compare the rate of human contributions to nitrogen fixation with the natural rate.
2. How does the change in fertilizer use relate to the change in world population? Why?
3. Develop a diagram to illustrate the links between the nitrogen and carbon cycles.
4. Make a list of ways in which human activities could be modified to reduce human contributions to the nitrogen cycle.

through the atmosphere, hydrosphere, lithosphere, and biosphere.

- Biogeochemical cycles can be described as a series of reservoirs, or storage compartments, and pathways, or fluxes, between reservoirs.
- In general, some chemical elements cycle quickly and are readily regenerated for biological activity. Elements

whose biogeochemical cycles include a gaseous phase in the atmosphere tend to cycle more rapidly.

- Our modern technology has begun to alter and transfer chemical elements in biogeochemical cycles at rates comparable to those of natural processes. Some of these activities are beneficial to society, but others pose dangers.
- To be better prepared to manage our environment, we

must recognize both positive and negative consequences of activities that transfer chemical elements and deal with them appropriately.

- Biogeochemical cycles tend to be complex, and Earth's biota has greatly altered the cycling of chemicals through the air, water, and soil. Continuation of these processes is essential to the long-term maintenance of life on Earth.

- Every living thing, plant or animal, requires a number of chemical elements. These chemicals must be available at the appropriate time and in the appropriate form and amount.

- Chemicals can be reused and recycled, but in any real ecosystem, some elements are lost over time and must be replenished if life in the ecosystem is to persist.

- Change and disturbance of natural ecosystems are the norm. A steady state, in which the net storage of chemicals in an ecosystem does not change with time, cannot be maintained.

- There are many uncertainties in measuring either the amount of a chemical in storage or the rate of transfer between reservoirs. For example, the global carbon cycle includes a large sink that science has not yet been able to locate.

● REEXAMINING THEMES AND ISSUES

Human Population:

Through modern technology, we are transferring some chemical elements through the air, water, soil, and biosphere at rates comparable to those of natural processes. As our population increases, so does our utilization of resources, and so do these rates of transfer. This is a potential problem because eventually the rate of transfer for a particular chemical may become so large that pollution of the environment results.

Sustainability:

If we are to sustain a high-quality environment, the major biogeochemical cycles must transfer and store the chemicals necessary to maintain healthy ecosystems. That is one reason why understanding biogeochemical cycles is so important. For example, the release of sulfur into the atmosphere is degrading air quality at local to global levels. As a result, the United States is striving to control these emissions.

Global Perspective:

The major biogeochemical cycles discussed in this chapter are presented from a global perspective. Through ongoing research, scientists are trying to better understand how major biogeochemical cycles work. For example, the carbon cycle and its relationship to the burning of fossil fuels and the storage of carbon in the biosphere and oceans is being intensely investigated. Results of these studies are assisting in the development of strategies to reduce carbon emissions. These strategies are implemented at the local level, at power plants and in cars and trucks that burn fossil fuels.

Urban World:

Our society has concentrated the use of resources in urban regions. As a result, the release of various chemicals into the biosphere, soil, water, and atmosphere is often greater in urban centers, resulting in biogeochemical cycles that cause pollution problems.

Science and Values:

Our understanding of biogeochemical cycles is far from complete. There are large uncertainties in the measurement of fluxes of chemical elements such as nitrogen, carbon, phosphorous, and others. We are studying biogeochemical cycles because we believe that understanding them will allow us to solve environmental problems. Which problems we address first will reflect the values of our society.

KEY TERMS

biogeochemical cycle **50**	drainage basin **59**	macronutrients **54**	plate tectonics **56**
carbon-silicate cycle **65**	geologic cycle **56**	micronutrients **54**	rock cycle **59**
carbon cycle **63**	hydrologic cycle **57**	phosphorous cycle **68**	tectonic cycle **56**
chemical reaction **50**	limiting factor **54**		

STUDY QUESTIONS

1. Why is an understanding of biogeochemical cycles important in environmental science? Explain your answer with two examples.

2. What are some of the general rules that govern biogeochemical cycles, especially the transfer of material?

3. Identify the major aspects of the carbon cycle and the environmental concerns associated with it.

4. What are differences in the geochemical cycles for phosphorus and nitrogen, and why are the differences important in environmental science?

5. How can the carbon-silicate cycle provide a negative-feedback mechanism to control the temperature of Earth's atmosphere?

FURTHER READING

Berner, R. A., and E. K. Berner. 1996. *Global Environment, Water, Air, and Geochemical Cycles.* Upper Saddle River, NJ: Prentice-Hall. A good discussion of environmental geochemical cycles, focusing on Earth's air and water systems.

Kasting, J. F., O. B. Toon, and J. B. Pollack. 1988. "How Climate Evolved on the Terrestrial Planets," *Scientific American* 258(2):90–97. This paper provides a good discussion of the carbonate-silicate cycle and why it is important in environmental science.

Lerman, A. 1990. "Weathering and Erosional Controls of Geologic Cycles," *Chemical Geology* 84:13–14. Natural transfer of elements from the continents to the oceans is largely accomplished by erosion of the land and transport of dissolved material in rivers.

Post, W. M., T. Peng, W. R. Emanual, A. W. King, V. H. Dale, and D. L. DeAngelis. 1990. "The Global Carbon Cycle," *American Scientist* 78:310–326. The authors describe the natural balance of carbon dioxide in the atmosphere and review why the global climate hangs in the balance.

Schlesinger, W. H. 1992. *Biogeochemistry: An Analysis of Global Change.* San Diego: Academic Press. This book provides a comprehensive and up-to-date overview of the chemical reactions on land, in the oceans, and in the atmosphere of Earth.

CHAPTER 5 The Human Population as an Environmental Problem

Crowded temporary housing at a gold rush site on Mt. Dimata, Davao Del Norte, Mindanao, Philippines, illustrates the impact of human beings on the environment and the relationship between valuable natural resources and a growing human population.

● LEARNING OBJECTIVES

The current human population represents something unprecedented in the history of the world. Never before has one species had such a great impact on the environment in such a short time and continued to increase at such a rapid rate. These qualities make human population *the* underlying environmental issue. After reading this chapter, you should understand:

- That ultimately, there can be no long-term solutions to environmental problems unless the human population stops increasing.
- That two major questions about the human population involve what controls its rate of growth and how many people the earth can sustain.
- That the rapid increase in the human population has occurred with little or no change in the maximum lifetime of an individual.
- That modern medical practices, as well as improvements in sanitation, control of disease-spreading organisms, and supplies of human necessities, have decreased death rates and accelerated the net rate of human population growth.
- That countries with a high standard of living have moved more quickly to a lower birth rate than have countries with a low standard of living.
- That although we cannot predict with absolute certainty what the future human carrying capacity of Earth will be, understanding of human population dynamics can help us make useful forecasts.

Death in Bangladesh

Bangladesh has 125.7 million people and an annual population growth rate of 1.8%.[1] This means that the annual increase in population is 2.26 million. Recently, more than 100,000 people died in Bangladesh when storms pushed ocean waters onshore, flooding the low-lying coastal lands that make up most of this nation (Figure 5.1).[2] This great human tragedy was made even more tragic by the fact that, given the average rate of population increase in Bangladesh, those 100,000 people were replaced in just two weeks. The growth curve of the population showed barely a ripple (Figure 5.2).[3]

Bangladesh is one of the poorest nations in the world, and this poverty affects human survival. According to World Bank standards for poverty, *over half the population (53%) is below the poverty line for poor,* *and 36% are below the poverty line for very poor. Nine out of ten children are malnourished, and 50% of children are stunted from lack of proper nourishment. Six hundred children die each day from malnutrition.* The average number of calories of food available per person is only 88% of that required for good health. Less than half the population (47%) living in urban areas has access to safe drinking water, and less than a fifth has access to adequate modern sanitation. Average life expectancy is about 59 years. With inadequate resources for each individual and a relatively rapid growth rate among nations, Bangladesh struggles to maintain even its existing poor standard of living.

Some have argued that the low-lying coastal areas that make up most of Bangladesh are fundamentally uninhabitable at high population densities for extended

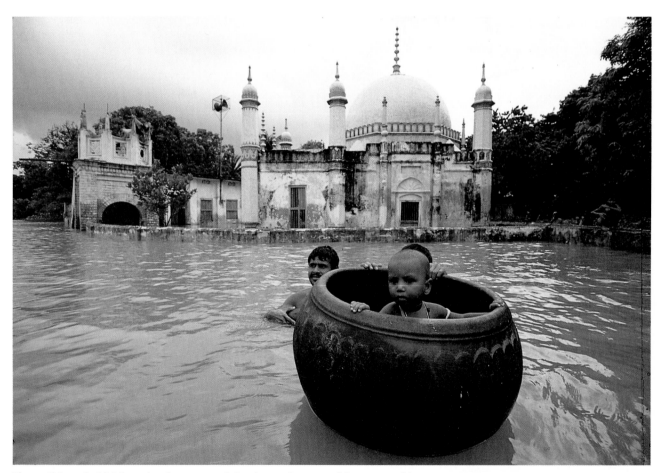

Figure 5.1 ● A child floats in a large urn and thereby survives one of the catastrophic floods in Bangladesh that have claimed hundreds of thousands of lives. Overcrowding in Bangladesh increases the number of deaths from these floods.

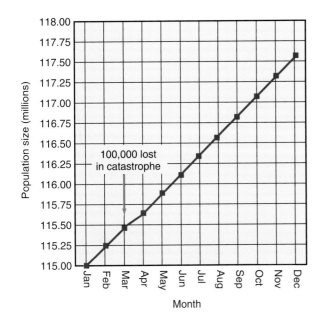

periods and are only inhabited now because of the huge numbers of people living on the Indian subcontinent.[4] For Bangladesh, it is difficult to talk about solving major environmental problems, conserving biological diversity, or optimizing production of fisheries and vegetation when people barely have sufficient resources to survive and the growth of the human population erases any advances.[5-8]

Figure 5.2 • Projected population growth in Bangladesh for the months shortly before and after a loss of 100,000 people in a single catastrophe. The population is so large that growth continues with little slowdown and rapidly recovers the number lost.

• *Progress is being made in Bangladesh in family planning and reducing the growth rate. However, as this case study illustrates, there can be no long-term solution to environmental issues in that nation, or elsewhere, as long as the population continues its momentous rise. As long as the human population continues to grow rapidly, the trend that leads to overwhelming the environment will continue. That is why human population growth is a major theme of this textbook.*

5.1 Basic Concepts

The rapid regrowth of Bangladesh's population following devastating storms vividly illustrates the human population's great capacity for growth and suggests the problems that this growth poses for the environment, perhaps even for the survival of our species on Earth. What can be done to reduce the rate of population growth in that country and others?

Knowledge of certain fundamental concepts about populations helps us to understand what approaches are open to us to solve the problem of population growth. In this chapter, we explain these concepts. We also describe how the human population has changed over time and how human beings function as ecological factors. The basic concepts of population growth and change discussed here serve as background for many discussions in later chapters.

Rates of Growth

By mid-2000, the world population numbered approximately 6 billion people and had an annual growth rate of approximately 1.4% (Figure 5.3a and b). At this rate, 84 million people are added to Earth's population in a single year, a number close to the 2002 population of Germany![1] Human population trends vary greatly among countries (Figure 5.3c). In developing countries, the average population growth rate in 1999 was 1.7%, whereas in the developed nations—those in Western Europe and North America, for example—the growth rate was less than 1% and in some cases much lower, even negative.[1] The growth rate of the population of the United States has declined (Figure 5.4).

Thus, there seems to be a correlation between poverty and population growth. The poorer a nation, the more

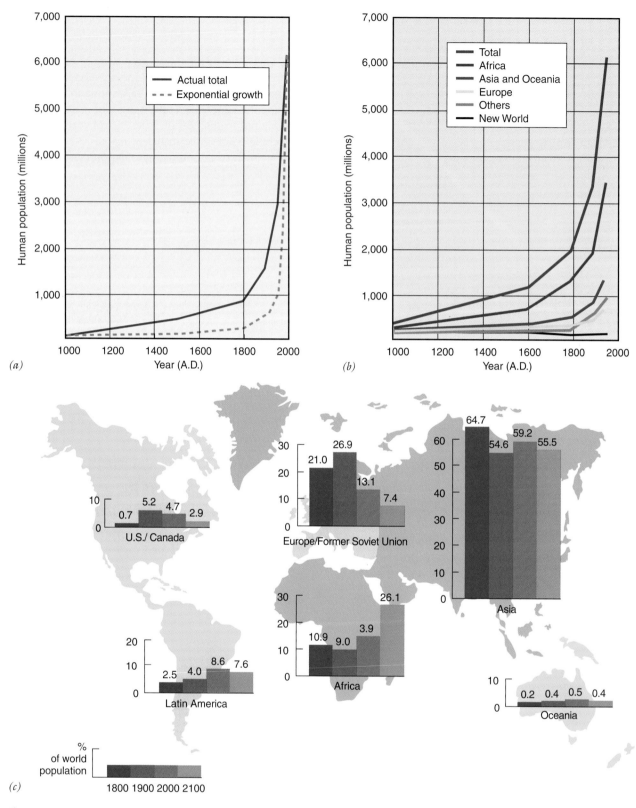

Figure 5.3 • World human population since 1000 A.D. (*a*) Actual increase compared with an exponential curve. Note that the rate of increase of the human population has exceeded an exponential rate. With an exponential curve, the net rate of increase is constant. In reality, the net rate of increase grew as a result of several factors—most importantly for our discussion, a decline in the death rate. (*b*) Human population from 1000 A.D. to 2000 A.D. by major geographic region. (*c*) Past and forecasted human population growth rate by geographic region and major nations. The main feature of this graph is that most growth in the 21st century will take place in developing nations. (*Source:* M. M. Kent and L. A. Crews, *World Populations: Fundamentals of Growth* [Washington, D.C.: Population Reference Bureau, 1990].)

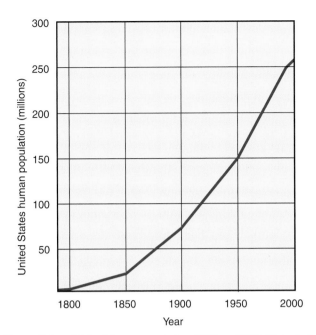

Figure 5.4 • United States population 1790 to 2000. Note that the rate of growth of the United States population appears to be slowing.

likely the population growth will be high; the wealthier a nation—and the higher the average per capita income—the lower the population growth rate. Since a high growth rate works against an increase in per capita income, poor nations are in danger of a positive feedback. The more people, the greater the growth rate; the greater the growth rate, the more people.

Population and Technology

The danger posed by the human population to the environment has two main factors: the number of people and the impact of each person on the environment. When there were few people on Earth and technology was limited, human impact was local. In that situation, the overuse of a local resource had few or no large or long-lasting effects. The fundamental problem now is that there are so many people and our technologies are so powerful that our effects are no longer local and unimportant.

Biologist and educator Paul Ehrlich has proposed that the simplest way to characterize the total impact of the human population on the environment is through a single relationship: The total environmental effect is the product of the impact per individual times the total number of individuals:[9]

$$T = P \times I$$

where P is the population size—the number of people—and I is the average environmental impact per person. Of course, in the real world, the situation is much more complicated. On the one hand, the impact per person who lives in the United States is on average much greater than the impact per person who lives in a low-technology society. On the other hand, even in a poor low-technology nation, like Bangladesh, the sheer number of people leads to large-scale environmental effects. In general, however, an increase in either the individual impact each of us has on the environment or the total number of people results in an increase in the total human effect on the environment. Almost 200 years ago, Thomas Malthus foresaw the human population problem (see A Closer Look 5.1, "The Prophecy of Malthus"). One major difference between our situation today and that foreseen by Malthus is that the expansion of technological power, although perhaps delaying a population crisis, has ultimately increased our effect on the environment.

Technology not only increases the use of resources but also enables us to affect the environment in many new ways, compared with hunters and gatherers or with people who farmed with simple wooden and stone tools. For example, before the invention of chlorofluorocarbons (CFCs), used as propellants in spray cans and coolants in refrigerators and air conditioners, we were not causing depletion of the ozone layer in the upper atmosphere. Similarly, before we started driving automobiles, there was much less demand for steel and little demand for oil—and much less air pollution.

An important consequence of the population-times-technology equation is that the addition of each new individual to the population of an industrialized nation leads, on average, to a greater effect on the environment than does the addition of each new individual in the population of a poor, undeveloped nation. Herein also lies a great irony involving two standard goals of international aid: improving the standard of living and slowing the overall human population growth. Improving the standard of living increases the total environmental impact, countering the environmental benefits of a decline in population growth.

The Human Population, the Quality of Life, and the Carrying Capacity of the Earth

How many people can live on Earth at the same time—that is, what is the carrying capacity of Earth with respect to people? This question has been asked throughout the twentieth century and continues to be controversial. There is no single answer, because the answer depends on what quality of life people desire and are willing to accept.

The usual approach to estimating the maximum number of people on Earth has involved two methods. One approach extrapolates from past growth to estimate what a likely future maximum might be. This approach, as discussed later in this chapter, assumes that the population will follow an S-shaped curve called the logistic growth curve and will gradually level off (Figure 5.5).

The Prophecy of Malthus

Almost 200 years ago, the English economist Thomas Malthus eloquently stated the human population problem. He based his argument on three simple premises:[10]

- Food is necessary for people to survive.
- "Passion between the sexes is necessary and will remain nearly in its present state" so that children will continue to be born.
- The power of population growth is "indefinitely greater than the power of the Earth to produce subsistence."

Malthus reasoned that it would be impossible to maintain a rapidly multiplying human population on a finite resource base. His projections of the ultimate fate of humankind were dire, as dismal a picture as that painted by the most extreme pessimists of today. The power of population growth is so great, he wrote, that "premature death must in some shape or other visit the human race. The vices of mankind are active and able ministers of depopulation, but should they fail, sickly seasons, epidemics, pestilence and plague, advance in terrific array, and sweep off their thousands and ten thousands." Should these fail, "gigantic famine stalks in the rear, and with one mighty blow, levels the population with the food of the world."

Critics of Malthus continue to point out that his predictions have yet to come true. Whenever things have looked bleak, technology has provided a way out, allowing us to live at greater densities. These critics have argued that our technologies will continue to save us from a Malthusian fate and that therefore we need not worry about the growth of the human population.

Who is correct? Ultimately, in a finite world, Malthus must be correct about the final outcome of unchecked growth. He may have been wrong about the timing; he did not anticipate the capability of technological changes to delay the inevitable. But although some people believe that Earth can support many more people than it does now, *in the long run there must be an upper limit.* The basic issue that confronts us is this: How can we achieve a constant world population, or at least halt the increase in the population, in a way that is most beneficial to most people? This is undoubtedly one of the most important questions that has ever faced humanity.

The second approach might be referred to as "the packing problem" approach. There has been a tendency to consider how many people might be packed onto the earth, not taking into sufficient account the need for lands and oceans to provide food, water, energy, construction materials, and scenic beauty and to maintain biological diversity. This has lead to very high estimates of the total number of people that might occupy Earth.

More recently, and at the other extreme, a philosophical movement has developed known as "Deep Ecology." This philosophy makes sustaining the biosphere the primary moral imperative for people. Its proponents argue

(a)

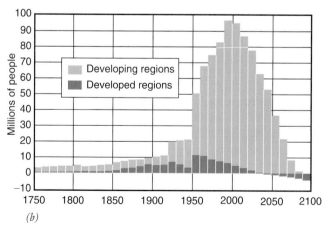
(b)

Figure 5.5 • Logistic growth curve. World population is shown as total numbers (*a*) and growth per decade (*b*), by development status, 1750–2100. For example, during the decade from 1980 to 1990, 82 million people were added each year, for a total addition of 820 million people.

that the whole earth is necessary to sustain life. Therefore, everything else must be sacrificed to the goal of sustaining the biosphere. People are considered active agents of destruction of the biosphere, and therefore the total number of people should be greatly reduced. The estimates of the number of people desirable vary greatly, from a few million to a few billion. At this time, we do not have enough scientific information to tell us what number of people and what rate of resource use will fall below the level of threatening the persistence of the biosphere. So at this time, no valid scientific number can be attached to the Deep Ecology goal.

In between the packing approach and Deep Ecology are a number of options. The options vary according to the quality of life for the average person. If all the people of the world were to live at the same level as those of the United States, with our high resource use, then the carrying capacity would be comparatively low. If all the people of the world were to live at the level of those in Bangladesh, with all of its risks as well as its poverty and its heavy drain on biological diversity and scenic beauty, the carrying capacity would be much higher.

Clearly, the acceptable carrying capacity is not simply a scientific issue; it is an issue combining *science and values*, one of the themes of this book. Science plays two roles. First, by leading to new knowledge that leads to new technology, it makes possible both a greater impact per individual on Earth's resources and a higher density of human beings. Second, scientific methods can be used to forecast a probable carrying capacity once a goal for the average quality of life, in terms of human values, is chosen. In this second use, science can tell us the implications of our value judgments, but it cannot provide those value judgments.

Human Demography

We began the chapter by asking what we can do about the rapid increase in our numbers. Understanding what causes the increases and learning how to calculate rates of increase are two key steps. People who study human population have defined a number of terms and concepts that help us to achieve this understanding. These include the following:

1. Age structure.
2. Demographic transition.
3. Total fertility.
4. Relationships between the human population and the environment.
5. Factors that increase the death rate, including why death rates increase with crowding and why death rates will increase if birth rates do not decrease.
6. Correlation of standard of living with birth rates, death rates, and growth rates.

The study of these and related concepts is called *human demography*. **Demography** means the study of populations. Demographers count a population (determine the actual size at some time) and attempt to project changes into the future.

The following sections introduce basic principles of population growth and change. These basic concepts form the study of **population dynamics**, which applies not only to human populations but also to nonhuman populations. The concepts discussed in this chapter therefore provide important background for later chapters.

5.2 Human Population Growth Rate

The history of the human population, shown in A Closer Look 5.2, "Growth of the Human Population," can be viewed as consisting of four major periods.

1. In the early period of hunters and gatherers, the world's total human population was probably less than a few million.
2. A second period, beginning with the rise of agriculture, allowed a much greater density of people and the first major increase in the human population.
3. The industrial revolution, with improvements in health care and the supply of food, led to a rapid increase in the human population.
4. Today, the rate of population growth has slowed in wealthy, industrialized nations but continues to increase rapidly in poorer, less-developed nations (Figures 5.3c and 5.5a).

It is also interesting to look at the total cumulative population over the history of humans, which is explored in A Closer Look 5.3, "How Many People Have Lived on Earth?"

Population Dynamics

Population dynamics deals with growth and change in populations. A **population** is a group of individuals of the same species living in the same area or interbreeding and sharing genetic information. A **species** is all individuals that are capable of interbreeding. A species is made up of populations. Basic equations of population change and examples of calculations are given in Working It Out 5.1, "Equations of Population Change."

How rapidly a population changes depends on the **growth rate**, which is the difference between the **birth rate** and the **death rate**. (See Table 5.1 for other useful terms.) The human population growth rate is usually expressed in the rate per 1,000, called the *crude rate*, rather than the more familiar percentage, which is the rate per 100. For example, in 1999, the *crude death rate* in the United States was 9, meaning that 9 of every 1,000

Growth of the Human Population

Stage 1. Hunters and Gatherers
From the first evolution of humans to the beginning of agriculture.[11]

Population density: About 1 person per 130–260 km² in the most habitable areas.

Total human population: As low as one-quarter million, less than the population of modern small cities like Hartford, Connecticut, and certainly less than a few million, which is fewer than now live in many of our largest cities.

Average rate of growth: The average annual rate of increase over the entire history of human population is less than 0.00011% per year.

Stage 2. Early, Preindustrial Agriculture
Beginning sometime between 9000 B.C. and 6000 B.C. and lasting approximately until the 16th century A.D.

Population density: With the domestication of plants and animals and the rise of settled villages, human population density increased greatly, to about 1 or 2 people/km² or greater, beginning a second period in human population history. (Even today, primitive people who practice agriculture have population densities greatly exceeding those of hunters and gatherers.)

Total human population: About 100 million by A.D. 1 and 500 million by A.D. 1600. (See Figure 5.3.)

Average rate of growth: Perhaps about 0.03%, which was large enough to increase the human population from 5 million in 10,000 B.C. to about 100 million in A.D. 1. The Roman Empire accounted for about 54 million. From A.D. 1 to A.D. 1000, the population increased to between 200 and 300 million.

Stage 3. The Machine Age
Some experts say that this time marked the transition from agricultural to literate societies, when better medical care and sanitation were factors in reducing the death rate.

Total human population: About 900 million in 1800, almost doubling in the next century and again (to 3 billion) by 1960.

Average rate of growth: By 1600, about 0.1% per year, with rate increases of about 0.1% every 50 years until 1950. This rapid increase occurred because of the discovery of causes of diseases, invention of vaccines, improvements in sanitation, other advances in medicine and health, and advances in agriculture that led to a great increase in the production of food, shelter, and clothing.

Stage 4. The Modern Era
Total human population: About 6 billion.

Average rate of growth: The growth rate of the human population reached 2% in the middle of the 20th century and has declined slightly, to 1.4%.[1]

● A CLOSER LOOK 5.3

How Many People Have Lived on Earth?

How many people have lived on Earth? Of course, before written history, there were no censuses. The first estimates of population in Western civilization were attempted in the Roman era. During the Middle Ages and the Renaissance, scholars occasionally estimated the number of people. The first modern census was taken in 1655 in the Canadian colonies by the French and British.[12] The first series of regular censuses taken by a country began in Sweden in 1750, and the United States has taken a census every decade since 1790. Most countries began much later. The first Russian census, for example, was taken in 1870. Even today, many countries do not take censuses or do not do so regularly. The population of China has only recently begun to be known with any accuracy. By studying modern primitive peoples and applying principles of ecology, however, we can gain a rough idea of the total number of people who may have lived on Earth. Summing all the values, including those since the beginning of written history, about 50 billion people are estimated to have lived on Earth.[13] Surprisingly, the 6 billion people alive today represent more than 10% of all of the people who have ever lived.

Equations of Population Change

Populations change in size through births, deaths, immigration (arrivals from elsewhere), and emigration (individuals leaving to go elsewhere). We can write a formula to represent population change:

$$P_2 = P_1 + (B - D) + (I - E)$$

where P_1 is the number of individuals in a population at time 1, P_2 is the number of individuals in that population at some later time 2, B is the number of births in the period from time 1 to time 2, D is the number of deaths from time 1 to time 2, I is the number entering as immigrants, and E is the number leaving as emigrants.

Ignoring for the moment immigration and emigration, we can state that how rapidly a population grows depends on the difference between the birth rate and the death rate. The birth rate (usually denoted by b) is the fraction or percentage of the population born in a unit of time. The death rate (d) is the fraction or percentage of the population that dies in a unit of time. The growth rate of a population is then

$$g = b - d$$

Note that in all these cases, the units are numbers per unit of time.

Recall from Chapter 3 that doubling time, the time for a population to reach twice its present size, can be estimated by the formula

$$T = 70/\text{annual growth rate}$$

where T is the doubling time and where the annual growth rate is expressed as a percentage. For example, a population growing at 2% per year would double in approximately 35 years.

Birth rates, death rates, and growth rates can be calculated from the number of events during some time period and the total population at some specific time during that period. Letting N equal the total number of individuals in the population, the birth rate (b) is the number of births per unit time (B) divided by the total population (N), or

$$b = B/N$$

The death rate is the number of deaths per unit time (D) divided by the total population (N), or

$$d = D/N$$

The growth rate (g) is the result of the number of births minus the number of deaths per unit time divided by the total number in the population (N), or

$$g = (B - D)/N \text{ or } g = G/N$$

In making these calculations, it is important to be consistent in using the population at the beginning, middle, or end of the period. Usually, the amount at the beginning or the middle is used.

Now we consider an example. There were 18,700,000 people in Australia in mid-1998, and there were 261,800 births from 1998 to 1999. The birth rate b calculated against the mid-1998 population was 261,800/18,700,000, or 1.4%. During the same period, there were 130,900 deaths; the death rate d was 130,900/18,700,000, or 0.7%. The growth rate g was (261,800 – 130,900)/18,700,000, or 0.7%, just under 1% per year.

---

people died. (The same information expressed as a percentage is a rate of 0.9%.) In 1999, the *crude birth rate* in the United States was 15.[1] The *crude growth rate* is the net change—the birth rate minus the death rate. Thus the *crude growth rate* in 1999 in the United States was 6. For every 1,000 people at the beginning of 1999, there were 1,006 at the end of the year.

The **maximum lifetime** is the genetically determined maximum possible age to which an individual of a species can live. **Life expectancy** is the average number of years an individual can expect to live *given the individual's present age*. (Often, the term is applied, without qualification, to mean the life expectancy of a newborn.)

A surprising aspect of the second and third periods in the history of human population is that *population growth*

occurred with little or no change in the maximum lifetime. What changed were birth rates, death rates, population growth rates, age structure, and average life expectancy.

Exponential Growth

As discussed in Chapter 3, a population experiencing exponential growth increases by a *constant percentage* per unit of time. The growth rate of the human population increased with a varying growth rate during the first part of the twentieth century, peaking during the years 1965 to 1970 at 2.1%, because of improvements in health, medicine, and food production. Thus, the human population has actually increased at a rate faster than the rate of exponential growth. This increase in the human

Table 5.1 • Common Terms

Crude birth rate: number of births per 1,000 individuals per year, "crude" because population age structure is not taken into account

Crude death rate: number of deaths per 1,000 individuals per year

Crude growth rate: net number added per 1,000 individuals per year; also equal to crude birth rate minus crude death rate

Fertility: pregnancy or the capacity to become pregnant or to have children

General fertility rate: number of live births expected in a year per 1,000 women aged 15 to 49 years, considered the childbearing years

Age-specific rate: number of births expected per year among a fertility-specific age group of women in a population

Total fertility rate (TFR): average number of children expected to be born to a woman through her child-bearing years

Cause-specific death rate: number of deaths from one cause per 100,000 total deaths

Incidence rate: number of people contracting a disease during a time period, usually measured per 100 people

Prevalence rate: number of people afflicted by a disease at a particular time

Case fatality rate: percentage of people who die once they contract a disease

Morbidity: general term meaning the occurrence of disease and illness in a population

Rate of natural increase (RNI): birth rate minus death rate, implying annual rate of population growth not including migration

Doubling time: number of years for a population to double assuming a constant rate of natural increase

Infant mortality rate: annual number of deaths of infants under age 1 per 1,000 live births

Life expectancy at birth: average number of years a newborn infant can expect to live under current mortality levels

GNP per capita: gross national product (GNP), which includes the value of all domestic and foreign output, per person

Source: C. Haub and D. Cornelius, *World Population Data Sheet* (Washington, D.C.: Population Reference Bureau, 1998).

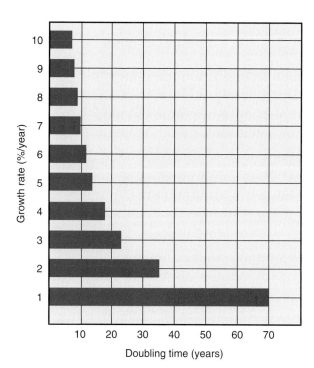

Figure 5.6 • Doubling time changes rapidly with the population growth rate. Because the world's population is increasing at a rate between 1% and 2%, we expect it to double within the next 35 to 70 years.

Doubling Time

Recall from Chapter 3 that *doubling time*, a concept used frequently in discussing human population growth, is the time required for a population to double in size (see Working It Out 5.1, "Equations of Population Change"). The standard way to calculate doubling time is to assume that the population is growing exponentially (has a constant growth rate). We can then estimate doubling time by dividing 70 by the annual growth rate stated as a percentage.

The doubling time based on exponential growth is very sensitive to growth rate; that is, it changes quickly as the growth rate changes (Figure 5.6). A few examples demonstrate this sensitivity. The U.S. population grew approximately 0.9% per year during the 1980s; the doubling time is 70 divided by 0.9, or 78 years. During the same period, the Ivory Coast's population grew at 3.8% per year; 70 divided by 3.8 is approximately 18. So, while the doubling time for the U.S. population was 78 years, the doubling time for the Ivory Coast's population was a little more than 18 years. Populations of Belgium, Denmark, Portugal, and Poland are growing at an annual rate of about 0.1%—or a doubling time of 700 years. The world's most populous country, China, has grown at an annual rate of 1.0%; its doubling time is 73 years. In the latter part of the twentieth century, the world population growth rate was 1.4%, which would lead to a doubling time of 49 years.[1]

growth rate has stopped, however, and the growth rate is declining globally. As mentioned, it is now approximately 1.4%.[1]

5.3 Projecting Future Population Growth

With human population growth such a central issue, it is important that we develop methods to forecast what will happen to our population in the future. One of the simplest approaches is to calculate the doubling time.

No population can sustain an exponential rate of growth indefinitely. Eventually the population will run out of food and space. A population of 100 increasing at 5% per year, for example, would grow to 1 billion in less than 325 years. If the human population had increased at this rate since the beginning of recorded history, it would now exceed all the known matter available in the universe. Clearly, then, *no population can grow exponentially forever.*

The Logistic Growth Curve

If a population cannot increase forever, what changes in the population can we expect over time? One of the first suggestions made about the human population is that it would follow a smooth S-shaped curve known as the **logistic growth curve**. The population would increase exponentially only temporarily. After that, the rate of growth would gradually decline (i.e., population would increase more slowly) until an upper population limit, called the **logistic carrying capacity**, was reached (Figure 5.5*a* and Figure 5.7). Once the logistic carrying capacity had been reached, the population would remain at that number.

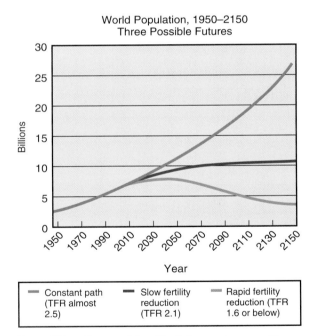

**World Population, 1950–2150
Three Possible Futures**

Legend:
- Constant path (TFR almost 2.5)
- Slow fertility reduction (TFR 2.1)
- Rapid fertility reduction (TFR 1.6 or below)

Figure 5.7 • Exponential and logistic growth curves. Three possible paths of future world population growth, as projected by the United Nations. The constant path assumes that the 1998 growth rate will continue unchanged, resulting in an exponential increase. The slow fertility reduction path assumes that the world's fertility will decline to reach replacement level by the year 2050 and that the world's population will stabilize at about 11 billion by the 22nd century. The rapid fertility reduction path assumes that the world's fertility will go into decline in the 21st century, peaking at 7.7 billion in 2050 and dropping to 3.6 billion by 2150. These are theoretical curves. (*Source:* United Nations Population Division, 1998.)

The logistic growth curve was first suggested in 1838 by a European scientist, P. F. Verhulst, as a theory for the growth of animal populations. It has been applied widely to the growth of many animal populations, including those important in wildlife management, endangered species, and fisheries (see Chapter 13), as well as the human population. Unfortunately, there is little evidence that human populations—or any animal populations, for that matter—actually follow this growth curve.

The logistic curve involves assumptions that are unrealistic for humans and for other mammals. These assumptions include a constant environment, a constant carrying capacity, and a homogeneous population (i.e., one in which all individuals are identical in their effects on each other). The logistic curve is especially unlikely if death rates continue to decrease from improvements in health, medicine, and food supplies. Once a human population has benefitted from these improvements, it must pass through what has become known as the demographic transition to achieve **zero population growth**, which leads to a stabilized population. The demographic transition is discussed later in this chapter.

Forecasting Human Population Growth Using the Logistic Curve

Although unrealistic, the logistic curve has been the method used for most forecasts of the long-term size of human populations in specific nations. This S-shaped curve first rises steeply upward and then changes slope, curving toward the horizontal carrying capacity (Figures 5.5*a* and 5.7). The point at which the curve changes is the *inflection point*. Until a population has reached the inflection point, we cannot project the final logistic size.

Unfortunately for those who want to make this calculation, the human population has not yet made the bend around the inflection point. Typically, forecasters have dealt with this problem by assuming that today's population is just reaching the inflection point. This standard practice leads invariably to a great underestimate of the maximum population. For example, one of the first projections of the upper limit of the U.S. population, made in the 1930s, assumed that the inflection point occurred then. That assumption resulted in an estimate that the final population of the United States would be approximately 200 million. That number has long since been exceeded; indeed, the U.S. population has passed 275 million.[14]

Future Population Trends Based on the Logistic Curve

The World Bank, an international organization that makes loans and provides technical assistance to developing countries, has made a series of projections based on current birth rates and death rates and assumptions about how these rates will change. These projections form the basis for the logistic curves presented in Figure 5.7. The projections are based on several critical assumptions: (1) mortality will fall everywhere and level off when female life expectancy reaches 82 years. (2) Fertility will reach replacement levels everywhere between 2005 and 2060. (3) There will be no major worldwide catastrophe. This approach projects an equilibrium world population of 10.1 to 12.5 billion.[15] Developed countries would only increase from 1.2 billion today to 1.9 billion, but developing countries would increase from 4.5 billion to 9.6 billion. Bangladesh would reach 257 million (in an area the size of Wisconsin); Nigeria, 453 million; and India, 1.86 billion. In these projections, the developing countries contribute 95% of the increase.[16]

The Demographic Transition

The **demographic transition** is a three-stage pattern of change in birth rates and death rates that has occurred during the process of industrial and economic development of Western nations. It leads to a decline in population growth.

A decline in the death rate is the first stage of the demographic transition (Figure 5.8).[11] In a nonindustrial country, birth rates and death rates are high, and the growth rate is low.[17] With industrialization, health and sanitation improve, and the death rate drops rapidly. The birth rate remains high, however, and the population enters stage II, a period with a high growth rate. Most European nations passed through this period in the eighteenth and nineteenth centuries.

As education and the standard of living increase and as family planning methods become more widely used, the population reaches stage III. The birth rate drops toward the death rate, and the growth rate also decreases, eventually to a low or zero growth rate. However, the decline in birth rate occurs only if families believe there is a direct connection between future economic well-being and funds spent on the education and care of their young. Such families have few children and put all their resources toward the education and well-being of those few.

Figure 5.8 • The demographic transition: (*a*) theoretical, including possible fourth and fifth stages that might take place in the future; (*b*) as has taken place for developed countries since 1775; and (*c*) as may be occurring in developing nations since 1775.

(*Source:* M. M. Kent and K. A. Crews, *World Population: Fundamentals of Growth* [Washington, D.C.: Population Reference Bureau, 1990]. Copyright 1990 by the Population Reference Bureau, Inc. Reprinted by permission.)

Historically, parents have preferred to have large families. Without other means of support, parents can depend on children for a kind of "social security" in their old age, and children help with many kinds of hunting, gathering, and low-technology farming. Unless a change in attitude occurs among parents—unless they see more benefits from a few well-educated children than many poorer children—nations face a problem in making the transition from stage II to stage III (Figure 5.8c).

Some developed countries are approaching stage III, but it is an open question whether other developing nations will make the transition before a serious population crash occurs. *The key point here is that the demographic transition will only take place if parents come to believe that a small family size is to their benefit.* Here we see again the connection between science and values. Scientific analysis can show the value of small families, but this knowledge must become part of cultural values to have an effect.

Potential Effects of Medical Advances on Human Population

Although the demographic transition is traditionally defined as consisting of three stages, advances in treating chronic health problems such as heart disease can lead a stage III country to a second decline in the death rate. This could bring about a second transitional phase of population growth (stage IV) in which the birth rate would remain the same as the death rate fell. A second stable phase of low or zero growth (stage V) would be achieved only when the birth rate declined to match the decline in the death rate. Thus, there is danger of a new spurt of growth, even in industrialized nations that have passed through the standard demographic transition.

Recent medical advances in the understanding of aging and the potential of new biotechnology techniques to increase both the average longevity and the maximum lifetime of human beings have major implications for the growth of the human population. If these medical advances take place, then the death rate will drop lower, and the growth rate will increase even more. Thus, a prospect that is positive from each individual's point of view—a longer, healthier, and more active life—could have negative effects on the environment. A decision therefore lies in the future for all of us. We have the following choices: stop medical research dealing with chronic diseases of old age and attempts to increase people's maximum lifetime; reduce the birth rate; or do neither and wait for Malthus's projections to come true—for famine, environmental catastrophes, and epidemic diseases to cause sporadic, large incidences of human death. The first choice seems inhumane, but the second is highly controversial. This is one of the most important issues concerning science and values that confront the people of the world.

Human Death Rates and the Rise of Industrial Societies

We return now to further consideration of the first stage in the demographic transition. We can get an idea of the first stage by comparing a modern industrialized country, such as Switzerland, which has a crude death rate of 9 per 1,000, with a developing nation, such as Niger, which has a crude death rate of 24.[1] Modern medicine has greatly reduced death rates from disease in Switzerland, with the first and greatest effect due to the reduction of death from acute or epidemic diseases.

An *acute* or *epidemic disease* appears in the population, affects a comparatively large percentage of it, and then declines or almost disappears for a while, only to reappear later. Epidemic diseases typically are rare but have occasional outbreaks during which a large proportion of the population is infected. Influenza, plague, measles, mumps, and cholera are examples of epidemic diseases. A *chronic disease*, in contrast, is always present in a population, typically occurring in a relatively small but relatively constant percentage of the population. Heart disease, cancer, and stroke are examples.

The great decrease in the percentage of deaths due to acute or epidemic diseases can be seen in a comparison of causes of deaths in Ecuador in 1987 and the United States in 1900, 1987, and 1998 (Figure 5.9).[18] In Ecuador, which is a developing nation, acute diseases and those listed as "all others" accounted for about 60% of mortality in 1987. In the United States in 1987, these accounted for only 20% of mortality. Chronic diseases accounted for about 70% of mortality in the modern United States. In contrast, these accounted for less than 20% of the deaths in the United States in 1900 and about 33% in Ecuador in 1987. Ecuador in 1987, then, resembled the United States of 1900 more than it resembled the United States of either 1987 or 1998.

Although outbreaks of epidemic diseases have declined greatly during the past century in industrialized nations, there is now concern that the incidence of these diseases might increase. Several factors contribute to the possibility that the incidence of epidemic diseases might rise. One is that, as the human population grows, people live in new habitats, where previously unknown diseases occur. Another is that strains of disease organisms have evolved to be resistant to antibiotics and other modern methods of control.

Contributing to the spread of diseases from one part of the world to another is the mobility of the human population; jet airliners daily carry vast numbers of people and goods around the world. Increasing density in populations of poor people also makes the spread of diseases more likely. AIDS is a factor in this increase, because individuals with AIDS lack resistance to diseases. Some 95% of the 33.6 million people with AIDS or HIV, the virus that causes AIDS, live in the developing world. Of these 33.6 million, 70% live in sub-Saharan Africa, which is home to 10% of the world's

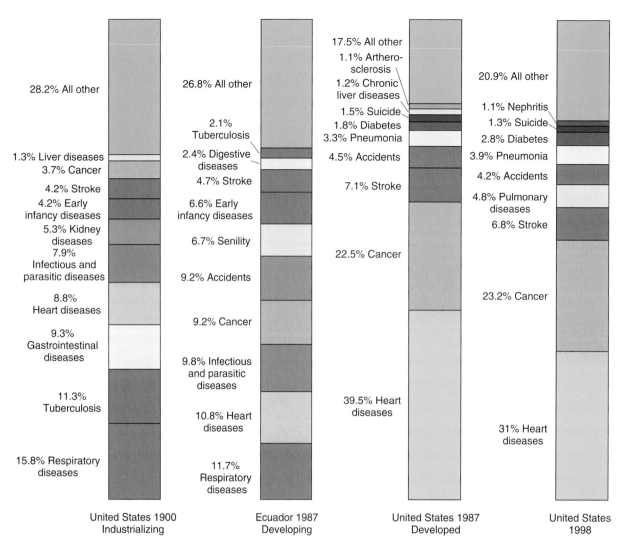

28.2% All other

1.3% Liver diseases
3.7% Cancer

4.2% Stroke
4.2% Early infancy diseases
5.3% Kidney diseases
7.9% Infectious and parasitic diseases

8.8% Heart diseases

9.3% Gastrointestinal diseases

11.3% Tuberculosis

15.8% Respiratory diseases

United States 1900
Industrializing

26.8% All other

2.1% Tuberculosis
2.4% Digestive diseases
4.7% Stroke

6.6% Early infancy diseases

6.7% Senility

9.2% Accidents

9.2% Cancer

9.8% Infectious and parasitic diseases

10.8% Heart diseases

11.7% Respiratory diseases

Ecuador 1987
Developing

17.5% All other
1.1% Arthero-sclerosis
1.2% Chronic liver diseases
1.5% Suicide
1.8% Diabetes
3.3% Pneumonia
4.5% Accidents

7.1% Stroke

22.5% Cancer

39.5% Heart diseases

United States 1987
Developed

20.9% All other

1.1% Nephritis
1.3% Suicide
2.8% Diabetes
3.9% Pneumonia
4.2% Accidents
4.8% Pulmonary diseases
6.8% Stroke

23.2% Cancer

31% Heart diseases

United States 1998

Figure 5.9 • Causes of mortality in industrializing, developing, and industrialized nations. (*Sources:* U. S. 1900, Ecuador 1987, and U.S. 1987 data from M. M. Kent and K. A. Crews, *World Population: Fundamentals of Growth* [Washington, D.C.: Population Reference Bureau, 1990]. Copyright 1990 by the Population Reference Bureau, Inc. Reprinted by permission. *National Vital Statistics Report* 48 [11], July 24, 2000.

population.[19] However, because fertility in African countries is high, populations will continue to grow.[20]

5.4 Limiting Factors

Eventually, human populations will be limited by some factor or combination of factors. We can group the limiting factors into those that affect a population during the year in which they become limiting (short-term factors), those whose effects are apparent after 1 year but before 10 years (intermediate-term factors), and those whose effects are not apparent for 10 years (long-term factors). Some factors fit into more than one category, having, say, both short-term and intermediate-term effects.

The most important short-term factor is something that disrupts the distribution of food within a country. This disruption may result from political events, includ-

ing wars, or a local loss of crops coupled with a shortage of energy needed to transport food. An abrupt change in weather, such as a drought, can cause such a disruption. Other short-term factors include major world catastrophes, such as thermonuclear war, the worldwide spread of a major toxic chemical, or the outbreak of a new disease or a new strain of a previously controlled disease.

Intermediate-term factors include certain climatic changes; energy shortages that affect food production and distribution; desertification (see Chapter 11); wide dispersal of certain pollutants, such as the spread of toxic metals into waters and fisheries; disruption in the supply of nonrenewable resources, such as rare metals used in making steel alloys for transportation machinery; and a decrease in the supply of firewood or other fuels for heating and cooking.

Long-term factors include soil erosion; a decline in groundwater supplies; disruption in the supply of nonrenewable resources; climatic changes, such as global warming, resulting from changes in atmospheric chemistry; and wide dispersal of certain pollutants, such as acid rain.

One way to view limiting factors is to consider the resources available per person, referred to as **per capita availability**. It has been a common belief that the quality of life will get better and better in the future as a result of improvements in our technology, but an analysis of the change in per capita resources suggests that we have already passed the peak in the amount of biological resources available per person. Some people argue, therefore, that we have exceeded the long-term carrying capacity of Earth for people. For example, the per capita production of wood peaked at 0.67 m³/person (0.88 yd³/person) in 1967, fish at 5.5 kg/person (12.1 lb./person) in 1970, beef at 11.81 kg/person (26.0 lb/person) in 1977, mutton at 1.92 kg/person (4.21 lb/person) in 1972, wool at 0.86 kg/person (1.9 lb/person) in 1960, and cereal crops at 342 kg/person (754.1 lb/person) in 1977.[21] Before these peaks, the per capita production of each resource had grown rapidly. Although part of the decline may be due to short-term economic forces, another part results from our nearing the limit of productive capacity with existing technology.

5.5 Age Structure

An important factor in population growth is the **population age structure**, which is the proportion of the population in each age class. The age structure of a population affects current and future birth rates, death rates, and growth rates; our impact on the environment; and current and future social and economic status.

Although age structure varies considerably by nation, we can discuss the general types (Figure 5.10). Kenya's pyramid-shaped age structure illustrates a rapidly growing population heavily weighted toward youth. In developing countries today, about 34% of the populations are under 15 years old. Such an age structure indicates that the population will grow very rapidly in the future. In contrast, the United States illustrates a population with slow growth, and Italy is an example of a nation with declining growth. Whereas elderly people make up a small percentage (3%) of Kenya's population, they make up a much larger percentage of the populations in the United States and Italy (13% and 17%, respectively).[1]

Age structure provides insight into a population's history, its current status, and its likely future. For example, the baby boom that occurred after World War II in the United States (a great increase in births from 1946 through 1964) forms a pulse in the population that can be seen as a bulge in the age structure, especially of those aged 40 to 50 in 2000. There is a secondary, smaller bulge as a result of a larger number of offspring from the first baby boom, which can be seen as a slight increase in the 5- to 15-year-olds. The second peak shows that baby boom pulse is moving through the age structure. At each life stage, the baby boom has put pressure on social and economic resources; for example, schools were crowded when the baby boomers were of primary and secondary school age.

Another enomic implication of age structure involves care of older people. In preindustrial societies, where average lifetimes are short, the young care for the old within the same family. In modern technological societies, family size is reduced, and care for older people is distributed through taxes, so that those who work pro-

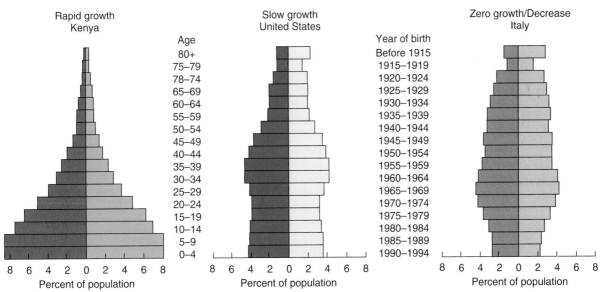

Figure 5.10 • Age structure of Kenya, the United States, and Italy, 1995. (*Sources:* U.S. Bureau of the Census, "U.S. Population Estimates by Age, Sex, and Race: 1990 to 1995," PPL-41, February 14, 1996; Council of Europe, *Recent Demographic Developments in Europe 1997*, Table 1–1; United Nations, *The Sex and Age Distribution of the World Populations—The 1996 Revision*, 500–1.)

vide funds to care for those who cannot. This provision is necessary in achieving zero population growth. Without it, parents would continue to rely on having many children to ensure their future well-being. However, a shift from a youthful age structure (like Kenya's) to an elderly age structure (like Italy's) means that a smaller percentage of the population work and therefore the taxes available for care for older people decrease.

This situation poses problems for a national government. The easiest way to increase tax income is to increase the percentage of young people and thereby promote rapid population growth. Thus, short-term economic pressures at a national level can lead to political policies supporting rapid population growth—not in the long-term interest of the nation.

5.6 Total Fertility Rate and the Lag in Response of Populations

Another important factor that can delay changes in population growth is the average number of children expected to be born to a woman during her lifetime; this is known as the **total fertility rate**, or TFR. The total fertility rate in the United States peaked at about 3.8 children per woman in the late 1950s and by 1999 was 2.0.[1,22] This means that a woman who was 15 years old in the late 1950s would have been expected to have about four children by the time she reached 44 years of age, whereas a 15-year-old woman in 1999 will be expected to have only about two children in her lifetime. A population that maintained a TFR of 3.8 over a long time would increase rapidly, whereas a population that maintained a TFR of 2.0 over a long time would decline.[1,22]

Replacement-level fertility. The TFR required for a population to remain constant in size is 2.1. From the day that a population achieves replacement-level fertility, however, that population will continue to grow for several generations, approximately 50 to 200 years. This phenomenon is called *population momentum* or *population-lag effect*. This time-lag effect is of great importance to human populations.

How does the effect occur? On the day that the women currently alive reach replacement-level fertility, the age structure is still weighted toward the young. Because previous fertility was greater than required for replacement levels, the population includes more women of childbearing age than would a completely constant population. These women continue to bear children throughout their childbearing years. Thus, the population continues to grow.

Eventually, if the population maintains replacement-level fertility, the population size will become constant, but the final size will be larger than the present size. Even though the U.S. TFR is below replacement level, for example, the total population is still growing.

A comparison of nations suggests that the TFR declines as income increases (Figure 5.11.) Bangladesh, which has

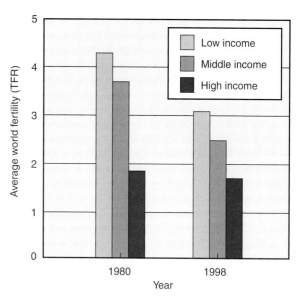

Figure 5.11 • Total fertility rate decreases as income increases. The graph shows average world values for 1980 and 1998. The total fertility rate is much higher among low-income families worldwide than among high-income families. In all cases, however, there has been some decrease in total fertility rate since 1980. (*Source:* World Development Indicators 2000 [http://www.worldbank.org/data/wdi2000/index.htm].)

an average income per person of $360 per year, has a high total fertility rate—approximately 3.3 children born to each woman. Brazil, with an average income of more than $4,790, has a TFR of 2.3.[1] Figure 5.12 shows that, worldwide, the TFR has steadily decreased since 1950.

Life Expectancy

Life expectancy, as stated earlier, is the estimated average number of years a person of a specific age can expect to live. Technically, it is an age-specific number; each age class within a population has its own life expectancy. For general comparison, however, the life expectancy at birth is used. Life expectancy differs by nation and by sex, age, and other factors. The life expectancy in a primitive hunter-gatherer society is short, for example; among the !Kung bushmen of Botswana, life expectancy at birth is 39 years.[23,24]

As mentioned earlier, the human population's great rate of increase has occurred despite the fact that life expectancy has not increased. In fact, the chances of a person 75 years old living to be age 90 were greater in ancient Rome than they are today in the United States. We can reconstruct the life expectancy of other periods in history by studying the dates on tombstones as has been done in Figure 5.13. The age at death can also be used to infer the survivorship curve of the population (Figure 5.13b). Such reconstructions suggest that death rates were much higher among young people in Rome than in 20th-century England. In ancient Rome, the life

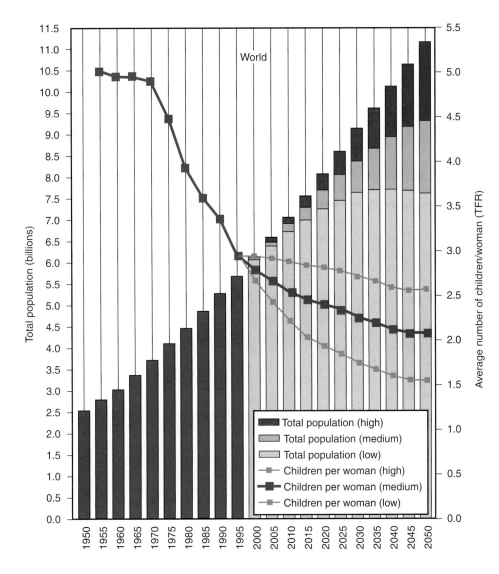

Figure 5.12 • Total fertility rate and the number of women of child-bearing age, 1950–2050, as observed until 1995 and projected from that year to 2050. The total fertility rate dropped between 1950 and 1990 as the world population increased. Projecting present trends into the future suggests that by 2050, a woman will have on average less than two children during her lifetime (statistically, 1.5 children on average per woman), a major decline from an average of greater than three at the present time and well below the replacement-level of 2.1. (*Source:* UN Population Division, *World Population Prospects, 1950–2050,* 1996 edition, Annexes I and II [1997]. Chart from G. K. Heilig, *IIASA-LUC,* 1996.)

(a)

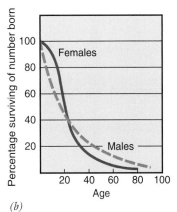

(b)

Figure 5.13 • (*a*) Life expectancy in ancient Rome and 20th-century England. This graph shows the average additional number of years one could expect to live having reached a given age. For example, a 10-year-old in England could expect to live about 55 more years; a 10-year-old in Rome could expect to live about 20 more years. Among the young, life expectancy was greater in 20th-century England than in ancient Rome. However, the graphs cross at about age 60. An 80-year-old Roman could expect to live longer than an 80-year-old Briton. The graph for Romans is reconstructed from ages given on tombstones. (*b*) Approximate survivorship curve for Rome for the first four centuries A.D. The percent surviving decreases rapidly in the early years, reflecting the high mortality rates for children in ancient Rome. Females had a slightly higher survivorship rate until age 20, after which males had a slightly higher rate. (*Source:* Modified from G. E. Hutchinson, *An Introduction to Population Ecology* [New Haven, Conn.: Yale University Press, 1978]. Copyright 1978 by Yale University Press. Used by permission.)

expectancy of a 1-year-old was about 22 years, whereas in 20th-century England it was about 50 years.

Notice that life expectancy in 20th-century England was greater than in ancient Rome for all ages until about age 55, when the life expectancy values appear higher for ancient Romans than for a 20th-century Briton. This would suggest that the hazards of modern life may be concentrated more on the aged. Pollution-induced diseases are one factor in this change.

 ## 5.7   How Can We Stop Population Growth?

We have surveyed several aspects of population dynamics. We now return to our earlier question: How can we stop population growth?

Age of First Childbearing

The simplest and one of the most effective means of slowing population growth is to delay the age of first childbearing by women.[6] As more women enter the workforce and as education levels and standards of living increase, this delay tends to occur naturally. Social pressures that lead to deferred marriage and childbearing can be very effective.

Countries with high growth rates have early marriages. In south Asia and in Africa south of the Sahara, about 50% of women marry between the ages of 15 and 19. In Bangladesh, women marry at age 16 on the average, whereas in Sri Lanka the average age for marriage is 25. The World Bank estimates that if Bangladesh adopted Sri Lanka's marriage pattern, families could average 2.2 fewer children.[16] Increases in the marriage age could account for 40% to 50% of the drop in fertility required to achieve zero population growth for many countries.

Age at marriage has increased in some countries, especially in Asia. For example, in Korea, the average marriage age went from 17 in 1925 to 24 in 1975. China passed laws fixing minimum marriage ages, first at 18 for women and 20 for men in 1950, then at 20 for women and 22 for men in 1980.[17] Between 1972 and 1985, China's birth rate dropped from 32 to 18 per thousand people, and the average fertility rate went from 5.7 to 2.1 children.

Birth Control in Developing Countries

Another simple means of decreasing birth rates is breast feeding, which can delay resumption of ovulation.[26] This is used consciously as a birth control method by women in a number of countries; in the mid-1970s the practice of breast feeding provided more protection against conception in developing countries than did family planning programs, according to the World Bank.[17]

Nevertheless, much emphasis is placed on the need for family planning.[27] Traditional methods range from abstinence to induction of sterility with natural agents.

Modern methods include the birth-control pill, which prevents ovulation through control of hormone levels; surgical techniques for permanent sterility; and mechanical devices. Contraceptive devices are used widely in many parts of the world, especially in eastern Asia, where data show that 78% of women use them. In Africa, the percentage is 18%, and in Central and South America it is 53% and 62%, respectively.[1]

Abortion is also widespread. Although now safe medically in most cases, abortion is one of the most controversial methods from a moral perspective. Ironically, it is also one of the most important birth control methods in terms of its effects on birth rates. Approximately 46 million abortions are performed each year.[28]

National Programs to Reduce Birth Rates

Reduction in birth rates requires a change in attitude, a knowledge of the means to control birth, and the ability to afford these means. As we have seen, a change in attitude can arise simply with an increase in the standard of living. In many countries, however, it has been necessary to provide formal family planning programs to explain the problems arising from rapid population growth and the benefits to individuals of reduced population growth. These programs also provide information about birth control methods and provide access to these methods.[22] The choice of population control methods is an issue that involves social, moral, and religious beliefs, which vary from country to country. It is difficult to generalize about what approach is best throughout the world.

The first country to adopt an official population policy was India in 1952. But few developing countries had official family planning programs before 1965. Since 1965, many such programs have been introduced. Over the last 30 years, the World Bank has lent $4.2 billion to over 80 countries to support reproductive health projects.[26] The United States contributed $372 million to international family planning funds through the U.S. Agency for International Development.[30] That amount was less than U.S. contributions of $547 million in 1995 and $385 million in 1997.[31]

Although most countries now have some kind of family planning program, the effectiveness varies greatly. A wide variety of approaches have been used, from simply providing more information, to promoting and providing means for birth control, to offering rewards and imposing penalties.

Penalties usually take the form of taxes. For example, Ghana, Malaysia, Pakistan, Singapore, and the Philippines have used a combination of methods, including limits on tax allowances for children and maternity benefits. Tanzania has taken another approach, restricting paid maternity leave for women to a frequency of once in three years. Singapore does not take family size into account in allocating government-built housing, so larger families

How Many People Can Earth Support?

In mid-1992, the world population reached 5.5 billion. By the end of 1999, it had reached 6 billion. Estimates of how many people the planet can support range from 2.5 billion to 40 billion. Why do the estimates vary so widely?

An estimate of 2.5 billion assumes that we maintain current levels of food production and that everyone eats as well as Americans do now—that is, 30% to 40% more calories than needed. The estimate of 40 billion assumes that all the remaining flat land of the world can be used to produce food, although in fact most of it is too cold or too dry to farm. What is a realistic carrying capacity? What factors need to be considered to answer this question?

Food Supply

World grain production has apparently leveled off since reaching its highest levels in the mid-1980s. The production of grain from 1984 to 1994 remained at approximately 1.7 billion tons, after rising from 631 million tons in 1950. The remarkable increase in productivity after 1950 resulted from the development of high-yielding varieties, use of chemical fertilizers, application of pesticides, and doubling of cropland acreage. If the present harvest were distributed evenly and everyone ate a vegetarian diet, it could support 6 billion people. As the world population has continued to grow, the per capita allotment of grain has been falling since 1984, when it stood at 346 kg per capita. By 1994, it had fallen to 311 kg per capita.

Land and Soil Resources

Almost all the usable agricultural land, approximately 1.5 billion ha (3.7 billion acres), is already being cultivated. An increase of 13% in agricultural lands is possible but would be costly. Per capita land area devoted to raising crops has dropped since 1950 to 1.7 ha (4.2 acres) and will continue to drop to approximately 1 ha (2.5 acres) by 2025 if present population projections hold. More soil is lost each year to erosion (about 26 billion metric tons) than is formed.

Water Resources

Water suitable for drinking and irrigation represents only a small proportion (less than 3%) of the water on Earth. Underground reservoirs are being depleted on the order of feet per year but are being replaced in inches or even fractions of inches per year. Per capita water consumption varies from 350–1,000 l (371–1,060 qt) a day in the developed countries to 2–5 l (2.1–5.3 qt) a day in rural areas, where people may obtain water directly from streams or primitive wells.

Net Primary Production

Human and domestic animals use about 4% of the net primary production of the world's land area and 2% of that of the oceans. (Primary production is discussed in Chapter 8.) Cultivated lands produce less than naturally vegetated areas. Net primary productivity on the planet has dropped about 13% since the 1950s.

Population Density

Population density varies greatly, from 3,076 people/km² on the tiny island of Malta to 66 people/km² in Africa as a whole. Bangladesh has 2,261 people/km², the Netherlands, 1,002/km², and Japan, 869/km².

Technology

Carrying capacity is not merely a matter of numbers of people. It also involves the impact they have on the world's resources—most critically on energy resources. Multiplying population by per capita energy consumption gives a relative measure of the impact people have on the environment. By that measure, each American has the impact of 35 people in India or 140 people in Bangladesh.

Critical Thinking Questions

1. Some people claim that agricultural technology can improve food production so that it continues to outpace, or at least keep pace with, population growth, thereby increasing carrying capacity above 6 billion. What evidence supports this point of view? What evidence refutes it?

2. The land area of the world is approximately 150 million km² (57.9 million mi²). What would the world population be if the average density were 400 people/km²? Is this a sound basis for determining carrying capacity? Explain.

3. In recent decades, some scientists have debated whether population or technology is the basic limitation on carrying capacity. What is your position, and how do you support it?

4. What factors, in addition to these six listed in this box, do you think should be considered in determining the planet's carrying capacity?

are more crowded. As an example of the use of rewards, Singapore increased the priority of school admission to children from smaller families.[15]

China has one of the oldest and most effective family planning programs. In 1978, China adopted an official policy to reduce the country's human population growth from 1.2% in that year to zero by the year 2000. Although in 2000 China's growth rate had slowed to 1.0% a year rather than to zero, the program has done much to curb the country's rapid population growth. The Chinese program places its emphasis on single-child families. The government uses education, a network of family planning that provides information and means for birth control, and a system of rewards and penalties. Women are given paid leave for abortions and for sterilization operations. Families with a single child receive benefits, including financial subsidies in some areas. In some parts of China, families that have a second child must return the bonuses received for the first. Other rewards and penalties vary from province to province.[15]

Some countries, including Bangladesh, India, and Sri Lanka, have paid people to be voluntarily sterilized. In Sri Lanka, this practice has applied only to families with two children, and a voluntary statement of consent must be signed.

Change in Values

People's values concerning how many children they wish to raise evidently change as a country becomes more prosperous (i.e., modern). For example, the TFR in Italy has fallen to 1.2 and in the rest of Western Europe and Japan it is about 1.5. These values are well below replacement-level fertility. In the United States, if immigration is excluded the TFR is just below the replacement-level and continues to fall. As the demographic transition occurs and the TFR drops, it often keeps falling below the replacement level. If this trend continues, the world population in two or three centuries could be less than 2 billion.[32]

Summary

- The human population issue is *the* environmental issue, because most current environmental damage results from the very high number of people on Earth and their great power to change the environment.

- Throughout most of our history, the human population and its average growth rate were small. The growth of the human population can be divided into four major phases. Although the population has increased in each phase, the current situation is unprecedented.

- Countries whose birth rates have decreased have experienced a demographic transition marked by a decrease in death rates followed by a decrease in birth rates. Many developing nations have experienced a great decrease in the death rate but still have a very high birth rate. It remains an open question whether some of these nations will be able to achieve a lower birth rate before reaching disastrously high population levels.

- The maximum population Earth can sustain and the population that will ultimately be attained by human beings are controversial questions. Standard estimates suggest that the human population will reach between 10 and 16 billion before stabilizing.

- As modernization, development, and the demographic transition occurs, TFR drops, often below the replacement level (2.1). World TFR is expected to be about 1.5 by 2050.

- Considerable lags in the responses of human populations to changes in birth rates and death rates occur because of the age structure of the population. A population that achieves replacement-level fertility will continue to grow for several generations.

● REEXAMINING THEMES AND ISSUES

Human Population: Our discussion in this chapter reemphasizes the point that there can be no long-term solution to our environmental problems unless the human population stops growing at its present rate. This makes the human population problem a top priority.

Sustainability: As long as the human population continues to grow, it is doubtful that our other environmental resources can be made sustainable.

Global Perspective: Although the growth rate of the human population varies from nation to nation, the overall environmental effects of the rapidly growing human population are global. For example, the increase in use of fossil fuels in Western nations since the beginning of the Industrial Revolution has affected the entire world. The growing demand for fossil fuels and their increasing use in developing nations are also having a global effect.

Urban World: One of the major patterns in the growth of the human population is the increasing urbanization of the world. Cities are not self-contained but are linked to the surrounding environment, depending on it for resources and affecting environments elsewhere. Urban development often leads to encroachment on highly valued natural areas, especially because cities are typically located where water, transportation, and material resources are readily available.

The human population problem epitomizes the connection between values and knowledge. Scientific and technological knowledge has helped us cure diseases, decrease death rates, and thereby increase the growth of the human population. Our present abilities to forecast human population growth provide much useful knowledge, but what we do with this knowledge is subject to intense debate around the world, because values are so important in relation to birth control and family size. We can only solve the human population problem by confronting the connection between the way we value human life and our scientific knowledge about human population and its environmental effects.

● KEY TERMS

birth rate **79**
death rate **79**
demographic transition **84**
demography **79**
growth rate **79**

life expectancy **81**
logistic carrying capacity **83**
logistic growth curve **83**
maximum lifetime **81**

per capita availability **87**
population **79**
population age structure **87**
population dynamics **79**

replacement-level fertility **88**
species **79**
total fertility rate **88**
zero population growth **83**

● STUDY QUESTIONS

1. What are the principal reasons that the human population has grown so rapidly in the twentieth century?

2. Given a world population in 1990 of 5.3 billion with a growth rate of 1.8%, what is the doubling time?

3. Why is it important to consider the age structure of a human population?

4. Three characteristics of a population are birth rate, growth rate, and death rate. How has each been affected by (a) modern medicine, (b) modern agriculture, and (c) modern industry?

5. What is meant by the statement "What is good for an individual is not always good for a population."?

6. Strictly from a biological point of view, why is it difficult for a human population to achieve a constant size?

7. What environmental factors are likely to increase the chances of an outbreak of an epidemic disease?

8. Why is it so difficult to predict the growth of Earth's human population?

9. Before the beginning of the scientific and industrial revolutions, what factors tended to decrease the size of the human population?

10. To which of the following can we attribute the great increase in human population since the beginning of the Industrial Revolution: changes in human (a) birth rates, (b) death rates, (c) longevity, or (d) death rates among the very old? Explain.

11. What is the demographic transition? When would one expect replacement-level fertility to be achieved—before, during, or after the demographic transition?

12. Present some arguments for or against the following statement: "With proper planning, we can achieve a constant carrying capacity for human beings on Earth."

13. Based on the history of human populations in various countries, how would you expect the following to change as per capita income increased: (a) birth rates, (b) death rates, (c) average family size, and (d) age structure of the population? Explain.

● FURTHER READING

Brown, L. R., and Kane, H. 1994. *Full House: Reassessing the Earth's Population Carrying Capacity.* New York: Norton. A discussion of factors affecting Earth's human carrying capacity and how they change.

Cohen, J. E. 1995. *How Many People Can the Earth Support?* New York: Norton. A detailed discussion of world population growth, Earth's human carrying capacity, and factors affecting both.

Kent, M. M., and Crews, K. A. 1990. *World Population: Fundamentals of Growth.* Washington, D.C.: Population Reference Bureau. A clear and concise introduction to population concepts and historical and present trends in the human population.

Kessler, E., ed. 1992. "Population, Natural Resources and Development," *AMBIO* 21(1). A special issue of the journal *AMBIO* addressing many problems concerning human population growth and its economic and environmental implications. It includes a description of several case studies.

McFalls, J. A. 1991. "Population: A Lively Introduction," *Population Bulletin* 46(2):1–43. A good introduction to human population dynamics.

World Bank. 1985. *Population Change and Economic Development.* Oxford: Oxford University Press. An excellent compilation of facts about human populations and methods for calculating population trends.

CHAPTER 6

Ecosystems and Ecosystem Management

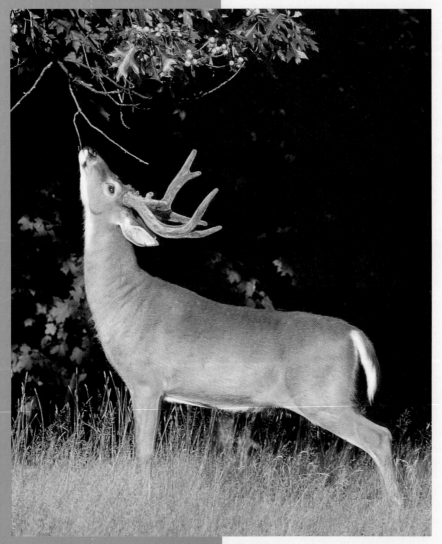

The linkages between white-tailed deer, Lyme disease, mice, and gypsy moths in eastern forests illustrate the complexities of ecosystems and ecological communities. The deer share the ticks that carry Lyme disease with mice and people.

● LEARNING OBJECTIVES

Life on Earth is sustained by ecosystems, which vary greatly but have certain attributes in common. After reading this chapter, you should understand:

- What basic characteristics of ecosystems allow them to sustain life.
- What are the basic concepts of the ecological communities and their processes within ecosystems.
- What are food chains, food webs, and trophic levels.
- What the concept of ecosystem management involves.
- Why the ecosystem is the basic system that supports life and allows it to persist.
- How conservation and management of the environment might be improved through ecosystem management.
- What a community level effect is.
- What a keystone species is.

The Acorn Connection

As young children, most people learn that acorns grow into oak trees. In fact, though, most acorns do not grow into trees but become food for mice, chipmunks, squirrels, and deer. In the woodlands of the northeastern United States, where oak trees abound, large crops of acorns (Figure 6.1*a*) are produced every three to four years. Acorns are rich in proteins and fats and an excellent source of nutrition. A steady supply of acorns would be an excellent food base for the woodland animals.

The production of acorns is affected by the amount of light and rain, the temperature patterns over the year, and the quality of the soil. Scientists reason that if oaks produced the same number of acorns each year, the populations of animals that feed on them would grow so large that very few acorns would survive.

In reality, the number of acorns produced varies from year to year, with "mast" years—years of high production—occurring occasionally. In the years between bumper crops of acorns, mice populations decline. With the next bumper crop, there are more acorns than can be eaten by the consumers of acorns, so many acorns survive to become oaks. Also, because of the abundance of food, the mice populations increase.

White-footed mice (Figure 6.1*b*) make up one such population. These mice also carry tick larvae (Figure 6.1*c*). As the ticks feed on the blood of the mice, they pick up the microorganisms responsible for Lyme disease. Mice populations are highest during the summer following a bumper crop of acorns, and so are tick larvae.

In later stages of their life cycle, ticks attach to other animals, including deer (Figure 6.1*d*). As deer brush against plants, ticks are deposited. The ticks can be picked up by people brushing against the plants as they walk past. If an infected tick bites a person, the person

(a) (b) (c)

(d) (e) (f)

Figure 6.1 • The tick that carries Lyme disease (*c*) feeds on both the white-footed mouse (*b*) and the white-tailed deer (*d*). Oak leaves (*f*) are an important food for the deer and for gypsy moths (*e*), while oak acorns (*a*) are important food for the mouse. But the mouse also eats the moths. The more mice, the fewer gypsy moths, but the more ticks.

may contract Lyme disease. As second-growth forest area has increased and deer populations have soared, Lyme disease has become the most common tick-borne disease in the United States.[1,2]

Why has forest area increased? Beginning in colonial times, the forests of the northeastern United States were cleared to make space for farming and settlements, to provide fuel, and to provide timber for commercial uses. As coal, oil, and gas replaced wood as a primary fuel, and as farming moved westward to the more fertile Great Plains, fields that had been cleared were abandoned. In many areas, the maximum clearing occurred around 1900. Since then, forests have grown back.

But let's return to the mice. In addition to feeding on acorns (and other grains), mice feed on insects, including larvae of the gypsy moth. Gypsy moth larvae (Figure 6.1e) feed on leaves of trees and are particularly fond of oak leaves. Studies suggest that in years when mice populations are low—the years between bumper crops of acorns—gypsy moth populations can increase dramatically. During these periodic outbreaks, gypsy moth larvae can virtually denude an area, stripping the leaves from the trees. Oaks that have lost most or all of their leaves may not produce bumper crops of acorns.

Once the leaves are off the trees, more light reaches the ground, and seedlings of many plants that could not do well in deep forest shade begin to grow. As a result, other species of trees may gain a foothold in the forest and change its species profile. Of course, the next generation of gypsy moth larvae find little to eat, and the population of gypsy moths begins to decline again.

Abundant acorns draw deer into the woods, where they browse on small plants and tree seedlings. Ticks drop off the deer and lay eggs in the leaf litter. When the eggs hatch, the larvae attach to mice, and the cycle of Lyme disease continues. Deer do not eat ferns, however, and in areas where deer populations are dense, many ferns but few wildflowers and tree seedlings are found.

Predators are also affected by the periodic nature of acorn crops. For example, birds that feed on gypsy moth larvae lose a food source when moth populations are low. When moth populations are high, however, bird nests are more exposed to predation because trees lose so many leaves.[2]

● *The acorn connection illustrates many of the basic characteristics of ecosystems and ecological communities. First, all of the living parts of the oak forest community depend on the nonliving parts of the ecosystem for their survival: water, soil, air, and the light that provides energy for photosynthesis. Second, the members of the ecological community affect the nonliving parts of the ecosystem. When gypsy moths denude an area, for example, more sunlight can reach the forest floor. Third, the living organisms in the ecosystem are connected in complex relationships that make it difficult to change one thing without changing many others. Fourth, the relationships among the members of the ecological community are dynamic and constantly changing. Many species are adapted to and benefit from a changing environment, as shown by the advantages provided oaks by varying acorn production. Fifth, the implication for human management of ecosystems is that any management practice involves tradeoffs. In this case, managing the forest to protect people against Lyme disease only results in more potential for gypsy moth damage.[1]*

The Ecosystem: Sustaining Life on Earth

We tend to associate life with individual organisms, for the obvious reason that it is individuals that are alive. But *sustaining* life on Earth requires more than individuals or even single populations or species. Life is sustained by the interactions of many organisms functioning together, in ecosystems, interacting through their physical and chemical environments. *Sustained life on Earth, then, is a characteristic of ecosystems, not of individual organisms or populations.*[3] To understand important environmental issues, such as conserving endangered species, sustaining renewable resources, and minimizing the effects of toxic substances, we must understand certain basic principles about occurring ecosystems.

Basic Characteristics of Ecosystems

Ecosystems have several fundamental characteristics.

- *Structure.* An ecosystem is made up of two major parts: nonliving and living. The nonliving part is the physical-chemical environment, including the local atmosphere, water, and mineral soil (on land) or other substrate (in water). The living part, called the **ecological community**, is the set of species interacting within the ecosystem.
- *Processes.* Two basic kinds of processes must occur in an ecosystem: a cycling of chemical elements and a flow of energy.
- *Change.* An ecosystem changes over time and can undergo development through a process called succession, which is discussed in Chapter 9.

The processes that occur in an ecosystem are necessary for the life of the ecological community, but no member of that community can carry out these processes alone. That is why we have said that sustained life on Earth is a characteristic of ecosystems rather than individuals or populations.

We can see this by looking at cycling in an ecosystem. As mentioned in Chapter 4, chemical cycling is complex. Each chemical element required for growth and reproduction must be made available to each organism at the right time, in the right amount, and in the right ratio relative to other elements. These chemical elements must also be recycled—converted to a reusable form. Wastes are converted into food, which is converted into wastes, which must be converted once again into food, with the cycling going on indefinitely, if the ecosystem is to remain viable.

For complete recycling of chemical elements to take place, several species must interact. In the presence of light, for example, green plants, algae, and photosynthetic bacteria produce sugar from carbon dioxide and water. From sugar and inorganic compounds, they make many other organic compounds, including proteins and woody tissue. But no green plant can decompose woody tissue back to its original inorganic compounds. Other forms of life—primarily bacteria and fungi—*can* decompose organic matter; but they cannot produce their own food. Instead, they obtain energy and chemical nutrition from the dead tissues on which they feed.

Theoretically, at its simplest, the ecological community in an ecosystem consists of at least one species that produces its own food from inorganic compounds in its environment and another species that decomposes the wastes of the first species, plus a fluid medium (air, water, or both). We turn next to a more detailed discussion of ecological communities—in particular, of food chains in ecological communities.

Ecological Communities and Food Chains

We have identified an ecological community as the set of interacting species that makes up the living part of an ecosystem. In practice, the term *ecological community* is defined by ecologists in two ways. One method is to define the community as a set of *interacting* species found in the same place and functioning together to make possible the persistence of life. That is essentially the definition we used earlier. A problem with this definition is that it often is difficult in practice to know the entire set of interacting species. Ecologists therefore may use a pragmatic or operational definition, in which the community consists of all the species found in an area, whether or not they are known to interact. Animals in different cages in a zoo could be called a community according to this definition.

One way in which individuals in a community interact is by feeding on one another. Energy, chemical elements, and some compounds are transferred from creature to creature along **food chains**, the linkage of who feeds on whom. In more complex cases, these linkages are called **food webs**.

Ecologists group the organisms in a food web into trophic levels. A **trophic level** consists of all those organisms in a food web that are the same number of feeding levels away from the original source of energy. The original source of energy in most ecosystems is the Sun. In other cases, it is the energy in certain inorganic compounds.

Green plants, algae, and certain bacteria produce sugars through the process of photosynthesis, using only the energy of the sun and carbon dioxide (CO_2) from the air, so they are grouped into the first trophic level. Organisms in the first trophic level, which make their own food and inorganic chemicals and a source of energy, are called **autotrophs**. *Herbivores*, organisms that feed on plants, algae, or photosynthetic bacteria, are members of the second trophic level. *Carnivores*, or meat-eaters, that feed directly on herbivores, make up the third trophic level. Carnivores that feed on third-level carnivores are in the fourth trophic level, and so on.

Food chains and food webs are often quite complicated and thus difficult to analyze. A detailed look at one of the simplest food chains is provided in A Closer Look 6.1, "Hot Springs Ecosystems in Yellowstone National Park." Next, we look briefly at several more complicated food chains.

Hot Spring Ecosystems in Yellowstone National Park

Perhaps the simplest natural ecosystem is a hot spring such as those found in geyser basins in Yellowstone National Park, Wyoming.[4] Few organisms can live in these hot springs because the environment is so severe. Water in parts of the springs is close to the boiling point. In addition, some springs are very acidic and others are very alkaline; either extreme makes a harsh environment. Some of the organisms that can live in hot springs are brightly colored and give these pools the striking appearance for which they are famous (Figure 6.2*a*).

Typically, the springs have a wide range of water temperatures, from almost boiling near the source to much cooler near the edges, especially in the winter, when there may be snow on the ground next to a spring. In a typical alkaline hot spring, the hottest waters, between 70° and 80°C (158°–176°), are colored bright yellow-green by photosynthetic blue-green bacteria, one of the few kinds of photosynthetic organisms that can survive in hot springs. In slightly cooler waters, 50° to 60°C (122–140°F), thick mats of bacteria and algae accumulate, some becoming 5 cm thick. These mats are formed by long strings of photosynthetic bacteria and algae. As the flowing springwater passes over the mats, the long strings of cells trap and hold single-cell algae.

First trophic level. Photosynthetic bacteria and algae make up the spring's first trophic level, which is composed of autotrophs—organisms that make their own food from inorganic chemicals and a source of energy. In the hot springs, as in most communities, the source of energy is sunlight (Figure 6.2*b*).

Second trophic level. Some flies, called *Ephydrid* flies, live in the cooler areas of the springs. One species, *Ephydra bruesi*, lays bright orange-pink egg masses on stones and twigs that project above the

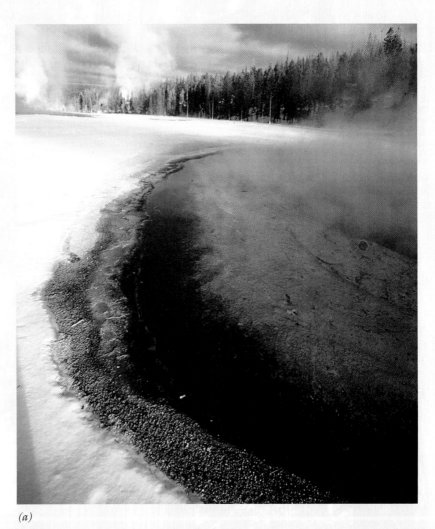

(a)

Figure 6.2 ● (*a*) One of the many hot springs in Yellowstone National Park. The bright greenish color comes from photosynthetic bacteria, one of the few kinds of organisms that can survive in the hot temperatures and chemical conditions of the springs. (*b*) Food web of a hot spring.

mat. Another species, *Aracoenia turbida*, lays white eggs in the mat. The fly larvae feed on the bacteria and algae. Since these flies eat only plants, they are herbivores, and they form the second trophic level.

Third trophic level. Another fly, called the *Dolichopopid* fly, is carnivorous and feeds on the eggs and larvae of the herbivorous flies. Dragonflies, wasps, spiders, tiger beetles, and one species of

bird, the killdeer, also feed on the herbivorous flies. The herbivorous *Ephydrid* flies also have a parasite, a red mite, which feeds on fly eggs and travels by attaching itself to the bodies of the adult flies. Another parasite, a small wasp, lays its eggs within the fly larvae. All these form the third trophic level.

Fourth trophic level. Wastes and dead organisms of all trophic levels are fed on

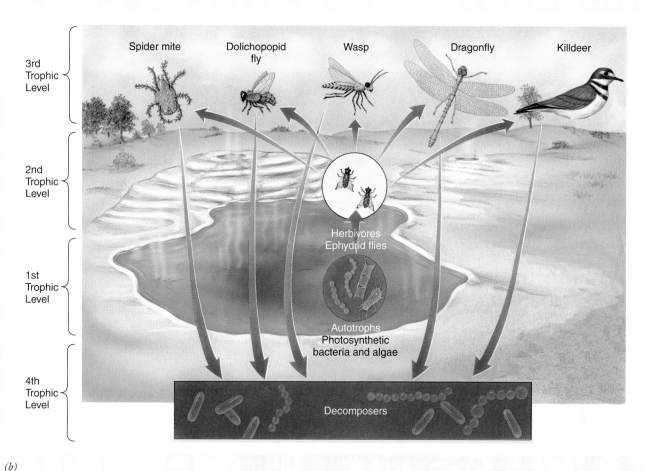

3rd Trophic Level: Spider mite, Dolichopopid fly, Wasp, Dragonfly, Killdeer

2nd Trophic Level

1st Trophic Level

Herbivores
Ephydrid flies

Autotrophs
Photosynthetic
bacteria and algae

4th Trophic Level: Decomposers

(b)

by decomposers, which in the hot springs are primarily bacteria. These form the fourth trophic level.

The entire hot springs community of organisms—photosynthetic bacteria and algae, herbivorous flies, carnivores, and decomposers—is maintained by two factors: (1) sunlight, which provides an input of usable energy for the organisms; and (2) a constant flow of hot water, which provides a continual new supply of chemical elements required for life and a habitat in which the bacteria and algae can persist.

Even though this is one of the simplest ecological communities in terms of the numbers of species, a fair number of species are found. About 20 species in all are important in this ecosystem. The ecological community they form has been sustained for long periods in these unusual habitats.

Another interesting aspect of the hot springs ecosystem is species dominance. Dominant species are those that are most abundant or otherwise most important in the community. (We discuss this in connection with diversity in Chapter 7.) As noted earlier, in the hot springs community, the species of photosynthetic bacteria or algae that is dominant changes with the temperature; one species dominates the hotter springs and hottest regions within a spring, and another species dominates cooler waters.

Because the algae are so brightly colored, this spatial patterning in dominance is readily apparent to visitors. It was striking to one of the earliest explorers of Yellowstone, a trapper named Osborne Russell, who visited the springs in the 1830s and 1840s. He wrote that one boiling spring about 100 m (328 ft) across had three distinct colors: "From the west side for one-third of the diameter it was white, in the middle it was pale red, and the remaining third on the east light sky blue."[5]

A Terrestrial Food Chain

An example of terrestrial food chains and trophic levels is shown in Figure 6.3. This north temperate woodland food web existed in North America before European settlement and includes human beings. The first trophic level, autotrophs, includes grasses, herbs, and trees. The second trophic level, herbivores, includes mice, an insect called the pine borer, and other animals (such as deer) not shown here. The third trophic level, carnivores, includes foxes and wolves, hawks and other predatory birds, spiders, and predatory insects. People are *omnivores* (eaters of both plants and animals) and feed on several trophic lev-

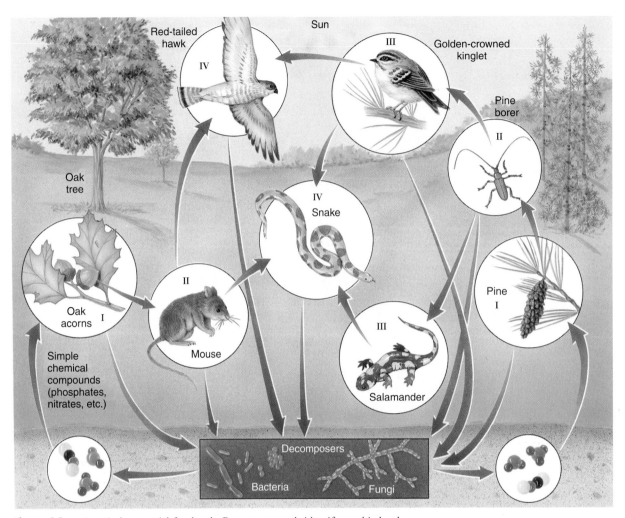

Figure 6.3 • A typical terrestrial food web. Roman numerals identify trophic levels.

els. In Figure 6.3, people would be included in the fourth trophic level, the highest level in which they would take part. **Decomposers**, such as bacteria and fungi, feed on wastes and dead organisms of all trophic levels. Decomposers are also shown here on the fourth level.

An Oceanic Food Chain

In oceans, food webs involve more species and tend to have more trophic levels than they do in the hot springs described in A Closer Look 6.1 or the terrestrial ecosystem just considered. In a typical oceanic ecosystem (Figure 6.4), microscopic single-cell planktonic algae and planktonic photosynthetic bacteria are in the first trophic level. Small invertebrates called *zooplankton* and some fish feed on the algae and photosynthetic bacteria, forming the second trophic level. Other fish and invertebrates feed on these herbivores and form the third trophic level. The great baleen whales filter seawater for food, feeding primarily on small herbivorous zooplankton (mostly crustaceans), and thus the baleen whales are also in the third

level. Some fish and marine mammals, such as killer whales, feed on the predatory fish and form higher trophic levels.

The Food Web of the Harp Seal

In the abstract, a diagram of a food web and its trophic levels seems simple and neat; but in reality, food webs are complex, because most creatures feed on several trophic levels. For example, consider the food web of the harp seal (Figure 6.5). The harp seal is shown at the fifth level.[6] It feeds on flatfish (fourth level), which feed on sand lances (third level), which feed on euphausiids (second level), which feed on phytoplankton (level 1). But the harp seal actually feeds at several trophic levels, from the second through the fourth, and it feeds on predators of some of its prey and thus is a competitor with some of its own food.[6] A species that feeds on several trophic levels typically is classified as belonging to the trophic level above the highest level from which it feeds. Thus, we place the harp seal on the fifth trophic level.

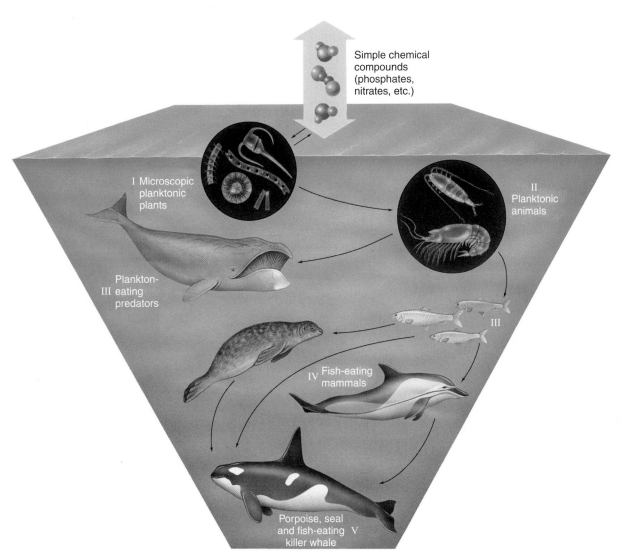

Figure 6.4 • An oceanic food web.

6.2 The Community Effect

Species can interact directly through food chains, as we have just seen. They also interact directly through symbiosis and competition, discussed in the next chapter. But a species can also affect other species *indirectly*, by affecting a third, a fourth, or many other species that, in turn, affect the second species. In addition, a species can affect the nonliving environment, which then affects a group of species in the community. Changes in that group affect another group. Such indirect and more complicated interactions are referred to as **community-level interactions**.

Interactions at the community level are illustrated by the sea otters of the Pacific Ocean. In fact, the community-level interactions of the sea otter are at the heart of some arguments in favor of conservation of this species. Sea otters feed on shellfish, including sea urchins and abalone (Figure 6.6*a*).

Sea otters originally occurred throughout a large area of the Pacific Ocean coasts, from northern Japan, northeastward along the Russian and Alaskan coasts, and southward along the coast of North America to Morro Hermoso in Baja California, to Mexico.[7] The otters were brought almost to extinction by commercial hunting for their fur during the 18th and 19th centuries; they have one of the finest furs in the world. By the end of the 19th century, there were too few otters left to sustain commercial exploitation, and there was concern that the species would become extinct.

A small population survived and has increased since then, so that today there are approximately 111,500 sea otters. According to the Marine Mammal Center, there are approximately 1,900 sea otters live along the coast of California, a few hundred in Washington and British Columbia, and 100,000 along the Aleutian Islands of Alaska. The sea otter population in Russian waters is

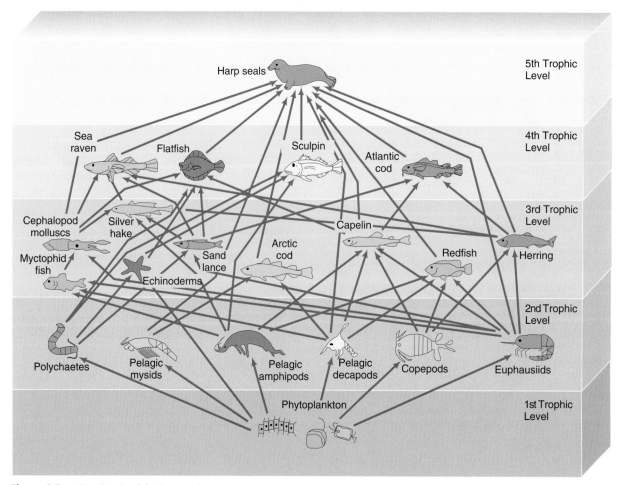

Figure 6.5 • Food web of the harp seal.

about 9,000.[8] Legal protection of the sea otter by the U.S. government began in 1911 and continues under the U.S. Marine Mammal Protection Act of 1972 and the Endangered Species Act of 1973.

The sea otter has been a focus of controversy and research. On the one hand, fishermen argue that the sea otter population has recovered—in fact recovered too much. In this view, there are too many sea otters today, and they interfere with commercial fishing because they take large amounts of abalone.[9] On the other hand, conservationists argue that sea otters have an important community-level role, necessary for the persistence of many oceanic species. They say that there are still too few sea otters for this role to be maintained at a sufficient level.

What is this important role? It is made up of many effects on the community that result from sea otters' feeding on sea urchins. Sea urchins are a preferred food of sea otters. Sea urchins, in turn, feed on kelp, large brown algae that form undersea forests and provide important habitat for many species. Sea urchins graze along the bottoms of the beds, feeding on the base of kelp, called *holdfasts,* which attach the kelp to the bottom. When holdfasts are eaten through, the kelp float free and die.

Where sea otters are abundant, as on Amchitka Island in the Aleutian Islands, kelp beds are abundant, and there are few sea urchins (Figure 6.6*b*). At nearby Shemya Island, which lacks sea otters, sea urchins are abundant, and there is little kelp (Figure 6.6*c*).[7] Experimental removal of sea urchins has led to an increase in kelp.[10]

Otters, then, affect the abundance of kelp, but the influence is indirect. Sea otters neither feed on kelp nor protect individual kelp plants from attack by sea urchins. Sea otters reduce the number of sea urchins. With fewer sea urchins, less kelp is destroyed. With more kelp, there is more habitat for many other species; so, sea otters indirectly increase the diversity of species.[9,11] Thus, sea otters have a community-level effect. This example shows that such effects can occur through food chains and can alter the distribution and abundance of individual species.

A species such as the sea otter that has a large effect on its community or ecosystem is called a **keystone species,** or a key species.[12] The balance of the entire system is keyed to the activities of this species. Its removal or a change in its role within the ecosystem changes the basic nature of the community.

(a)

(b)

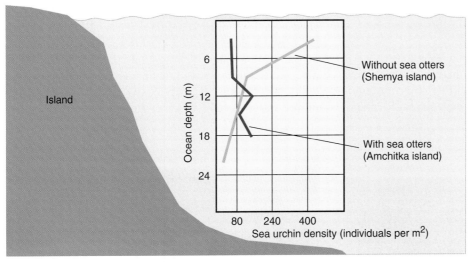

(c)

Figure 6.6 • The effect of sea otters on kelp.

The existence of community-level effects demonstrate the reality behind the concept of an ecological community; it shows us that certain processes can take place only because of the existence of a set of species interacting together. These effects also suggest that an ecological community is more than the sum of its parts—a perception called the *holistic view*. (Refer back to the discussion of environmental unity in Chapter 3.)

In spite of their importance, community-level interactions are often difficult to recognize. One difficulty is knowing when and how species interact. Community-level interactions are not always as clear as those involving the sea otter. Even there, considerable scientific research was required to understand the interactions. Adding to the complexity, the set of species that make up an ecological community is not fixed completely but varies within the same kind of ecosystem from time to time and place to place. This brings us to the question of how to identify ecosystems.

6.3 How Do You Know When You Have Found an Ecosystem?

An ecosystem is the minimal entity that has the properties required to sustain life. This implies that an ecosystem is real and important and therefore that we should be able to find one easily. However, ecosystems vary greatly in structural complexity and in the clarity of their boundaries. Ecosystems differ in size, from the smallest puddle of water to a large forest. Ecosystems and their communities differ in composition, from a few species in the small space of a hot spring to many species interacting over a large area of the ocean. Furthermore, ecosystems differ in the kinds and relative proportions of their nonbiological constituents and in their degree of variation in time and space.

Sometimes the borders of the ecosystem are well defined, such as the border between a lake and the surrounding countryside. But sometimes the transition from one ecosystem to another is gradual, as in the transition from desert to forest on the slopes of the San Francisco Mountains in Arizona (Figure 6.7) and in the subtle gradations from grasslands to savannas in East Africa and from boreal forest to tundra in the far north, where the trees thin out gradually (Figure 6.8).

A commonly used practical delineation of the boundary of an ecosystem on land is the watershed. A **watershed** is defined most simply as follows: Within a watershed, any drop of rain that reaches the ground flows out in the same stream. Topography (the lay of the land) determines the watershed. When a watershed is used to define the boundaries of an ecosystem, the ecosystem is unified in terms of chemical cycling. Some classic experi-

(a)

(b)

Figure 6.7 • (*a*) Sometimes the transition from one ecosystem to another is sharp and distinct, as in the transition from lake to forest at Lake Moraine in Banff National Park, Alberta, Canada. (*b*) Sometimes the transition is gradual and fuzzy, as in the transi- tions among vegetation types and their associated ecosystems on the slopes of Humphreys Peak, one of the San Francisco Peaks in Coconino National Forest near the Grand Canyon in Arizona.

of chemical elements. Ecological communities change over time, and it is the interactions among the species—a dynamic set of processes—that are the key to the community concept.

6.4 Ecosystem Management

Ecosystems can be natural or artificial or a combination of both. An artificial pond that is a part of a waste treatment plant is an example of an artificial ecosystem. Ecosystems can also be managed, and the management can vary over a large range of actions. Agriculture can be thought of as partial management of certain kinds of ecosystems (see Chapters 11 and 12), as can forests managed for timber production (see Chapter 13). Wildlife preserves are examples of partially managed ecosystems (see Chapter 13).

Sometimes, when we manage or domesticate individuals or populations, we separate them from their ecosystems. We also do this to ourselves (see Chapter 3, on urban ecosystems). When we do this, we must replace the ecosystem functions of energy flow and chemical cycling with our own actions. This is what happens in a zoo, where we must provide food and remove the wastes for individuals separated from their natural environments.

The ecosystem concept, then, lies at the heart of the management of natural resources. When we try to conserve species or manage natural resources so that they are sustainable, we must focus on their ecosystem and make sure that it continues to function. If it doesn't, we must replace or supplement ecosystem functions with our own actions. Ecosystem management, however, means more

Figure 6.8 • Often, the transition between boreal forest and tundra is subtle and occurs over a large area. Here we see thinly scattered, small trees interspersed among tundra vegetation, within the transition in Kluane Park, Yukon, Canada.

mental studies of ecosystems have been conducted on forested watersheds in U.S. Forest Service experimental areas, including the Hubbard Brook experimental forest in New Hampshire (Figure 6.9) and the Andrews experimental forest in Oregon.

What all ecosystems have in common is not a particular physical size or shape but the existence of the processes we have mentioned: the flow of energy and the cycling

Figure 6.9 • The V-shaped logged area in this picture is the famous Hubbard Brook ecosystem study. Here, a watershed defined the ecosystem, and the V shape is an entire watershed cut as part of the experiment.

How Are the Borders of an Ecosystem Defined?

The borders between ecosystems may be well defined or gradual. Those considered well defined include freshwater streams. Such ecosystems are often studied separately from surrounding ecosystems by researchers with different training and using different methods. Recent research on streams in southeast Alaska in which salmon spawn has raised questions about the practice of studying aquatic and terrestrial ecosystems separately.

Salmon are anadromous fish—fish that come from the ocean to spawn in freshwater streams. In southeast Alaska, enormous numbers of salmon spawn in over 5,000 streams. Although salmon are born in freshwater, they migrate to the ocean, where most of their growth occurs. After they return to their home streams, they spawn and die. In one sense, therefore, salmon are a means of transporting resources from the ocean to freshwater. Because of their large numbers, salmon have the potential of making significant contributions to organic and mineral content of streams.

Salmon have a high lipid content compared with many other fish and are thus a good energy source for the animals that prey on them. In addition, their decay adds nitrogen, phosphorus, carbon, and other inorganic elements to freshwater. In one lake in western Alaska, for example, its 24 million fish add 170 tons of phosphorus to the lake each year—an amount equal to or greater than recommended rates for applying fertilizers to trees. When the fish die, their carcasses decay and provide nourishment for algae, fungi, and bacteria. Invertebrates feed on these and on decaying bits of fish. Other fish feed on the invertebrates. Finally, bears and other carnivores eat salmon, both live and dead, during their upstream migration. In that way, nutrients derived from salmon pass into the soil and vegetation surrounding the streams.

Spawning fish have higher proportions of heavy isotopes of nitrogen and carbon (^{15}N and ^{13}C). These can be used to trace the relative contributions of anadromous fish to the nitrogen and carbon content of organisms in the food web. One such study showed that spawning salmon contributed 10.9% of the nitrogen found in invertebrate predators and 17.5% in the foliage of riparian plants. While it is not surprising to find aquatic invertebrates, which feed on salmon eggs and juveniles, with large amounts of nitrogen derived from salmon, researchers were surprised at the high levels in stream-side vegetation. When terrestrial mammals and birds feed on salmon, their feces and any uneaten salmon carcasses decay and add nutrients to the soil, where they can be taken up through the roots of plants. In southeast Alaska, over 40 species of mammals and birds feed on salmon. Salmon migrations attract large numbers of predators to streams and lakes. Salmon and other anadromous fish thus appear to link the ocean, freshwater, and land to an extent that is only beginning to be appreciated.

Critical Thinking Questions

1. Given the intricate connections between the aquatic and terrestrial ecosystems along salmon streams, how would you define the boundaries of the ecosystems?

2. When more numbers of adult salmon reach the spawning grounds than are needed to maintain the population, some are considered excess. How might the research described here affect that view?

3. Some biologists have called salmon a keystone species. Given what you know about keystone species, how would you argue for or against this designation?

4. In recent years, the numbers of anadromous fish along the Pacific Coast of North America have declined precipitously because of overfishing and habitat destruction. What effects would you predict this might have on the ecology of freshwater streams and their adjoining land areas?

5. What types of management decisions about fish, wildlife, and forests would follow from recognizing the connection between aquatic and terrestrial ecosystems?

than compensating for changes we make in ecosystems. It means managing and conserving life on Earth by considering chemical cycling, energy flow, community-level interactions, and the natural changes that take place within ecosystems.

Summary

- An ecosystem is the simplest entity that can sustain life. At its most basic, an ecosystem consists of several species and a fluid medium (air, water, or both). The ecosystem must sustain two processes, the cycling of chemical elements and the flow of energy.

- The living part of an ecosystem is the ecological community, a set of species connected by food webs and trophic levels. A food web or chain describes who feeds on whom. A trophic level consists of all the organisms that are the same number of feeding steps from the initial source of energy.

- Community-level effects result from indirect interactions between species, such as those that occur when sea otters influence the abundance of sea urchins.

- Ecosystems are real and important, but it is often difficult to define the limits of a system or to pinpoint all the interactions that take place.

- Ecosystem management is considered key to the successful conservation of life on Earth.

REEXAMINING THEMES AND ISSUES

Human Population: The human population depends on many ecosystems that are widely dispersed around the globe. Our modern technology may appear to make us independent of these natural systems. In fact, though, the more connections we establish through modern transportation and communication, the more kinds of ecosystems we depend on. Therefore, the ecosystem concept is one of the most important we will learn about in this book.

Sustainability: The ecosystem concept is at the heart of managing for sustainability. When we try to conserve species or manage living resources so that they are sustainable, we must focus on their ecosystem and make sure that it continues to function.

Global Perspective: Our planet has sustained life for approximately 3.5 billion years. To understand how Earth as a whole has sustained life for such a long time, we must understand the ecosystem concept, because the environment at a global level must meet the same basic requirements as any local ecosystem.

Urban World: Cities are embedded in larger ecosystems. But like any life-supporting system, a city must meet basic ecosystem needs. This is accomplished through connections between cities and surrounding environments. Together, these function as ecosystems or sets of ecosystems. To understand how we can create pleasant and sustainable cities, we must understand the ecosystem concept.

Science and Values: The introductory case study concerning acorns, mice, deer, and Lyme disease illustrates the interactions between values and scientific knowledge about ecosystems. Science can tell us how organisms interact like deer and mice. This knowledge confronts us with choices. Do we want to have many deer and mice and to live with Lyme disease? Do we want to invest in ecological and medical resources to find a better way to control that disease? Which choice we make depends on our values.

KEY TERMS

autotrophs **97**
community-level interactions **101**
decomposers **100**

ecological community **97**
food chains **97**
food webs **97**

keystone species **102**
trophic level **97**
watershed **104**

STUDY QUESTIONS

1. What is the difference between an ecosystem and an ecological community?

2. In what ways would an increase in the number of sea otters and a change in their geographic distribution benefit fishermen? In what ways would these changes be a problem for fishermen?

3. Based on the discussion in this chapter, would you expect a highly polluted ecosystem to have many species or few species?

4. Is our species a keystone species? Explain.

5. Which of the following are ecosystems? Which are ecological communities? Which are neither?
   a. Chicago
   b. A 1,000 ha farm in Illinois
   c. A sewage treatment plant
   d. The Illinois River
   e. Lake Michigan

Bilby, R. E., B. R. Fransen, and P. A. Bisson. 1996. "Incorporation of Nitrogen and Carbon from Spawning Coho Salmon into the Trophic System of Small Streams: Evidence from Stable Isotopes," *Canadian Journal of Fisheries and Aquatic Sciences* 53, pp. 164–173. A well-documented example of a community-level effect.

Bormann, F. H., and G. E. Likens. 1979. *Pattern and Process in a Forested Ecosystem.* New York: Springer-Verlag. A synthetic view of the northern hardwood ecosystem, including its structure, function, development, and relationship to disturbance.

Brown, D. E., F. Reichenbacher, and S. E. Franson, 1998. *A Classification of North American Biotic Communities/Maps.* Salt Lake City: University of Utah Press. A description of a hierarchical classification system for biotic communities occurring from the Arctic Circle through Central America.

Diamond, J., and T. J. Case, eds. 1986. *Community Ecology.* New York: Harper & Row. An examination of community ecology from several perspectives, including spatial and temporal scales, equilibrium theory, and community structuring mechanisms.

Duggins, D. O. 1980. "Kelp Beds and Sea Otters: An Experimental Approach," *Ecology* 61:447–453. A discussion of how experiments with sea urchin removal led to observations on kelp succession and diversity.

Pahl-Wostl, C. 1995. *The Dynamic Nature of Ecosystems: Chaos and Order Entwined.* New York: John Wiley & Sons. A book arguing that the tradeoff between the irregular, chaotic dynamics at the level of populations and the spatiotemporal organization at the level of the system as a whole shapes ecological systems.

Strong, D. R., D. Simberloff, L. G. Abele, and A. B. Thistle. eds. 1984. *Ecological Communities: Conceptual Issues and the Evidence.* Princeton, N.J.: Princeton University Press. A collection of papers that illuminates species interactions and the ecological effects of these interactions.

Willson, M. R., and K. C. Halupka. 1995 (June). "Anadromous Fish as Keystone Species in Vertebrate Communities." *Conservation Biology* 9(3):489–497. An article describing community-level effects of salmon.

7 Biological Diversity

Tropical rain forests have great biological diversity and are an increasing subject of research, sometimes involving unusual techniques.

● LEARNING OBJECTIVES

People have long wondered how the amazing diversity of life on Earth came to be. This diversity has developed through biological evolution and is affected by interactions among species and by the environment. After reading this chapter, you should understand:

● How the conservation of biological diversity involves an understanding of the intricate relationships among species and between species and their environments.
● Why people value biological diversity.
● What are the ecological functions of biological diversity.
● What major problems are associated with biological diversity.
● How mutation, natural selection, migration, and genetic drift lead to evolution of new species.
● Why so many species have been able to evolve and persist.
● How species interactions affect diversity.
● The concepts of the ecological niche and habitat.
● How people can affect biological diversity.

The Shrinking Mississippi Delta

Throughout the United States, coastal wetlands support a wide diversity of life, but these areas have decreased greatly in size in recent decades. The greatest coastal wetlands in the lower 48 states—the Mississippi Delta—have undergone such a decrease. This has led to a loss of many kinds of habitats, has reduced the abundance of many species, and poses threats to the diversity of life not only in the wetlands but also in the Gulf of Mexico.

Louisiana has 40% of all the coastal wetlands in the lower 48 states (Figure 7.1). Between 1956 and 1990, approximately 3,460 km² of coastal wetlands were lost—converted to open water. And even with conservation projects in place today, it is estimated that an additional 1,800 to 3,500 km² will be lost by the mid-21st century.[1]

Several factors contribute to the loss, ranging from a natural geological slumping of the land to the building of levees and navigation channels on the Mississippi River. The Mississippi River used to be the biggest land builder in the world, carrying sediment from the combined Missouri-Mississippi River system and depositing 400 million tons in the delta each year. Now, 13,000 km of canals in Louisiana's marshes transport the sediments through the wetlands and out into the Gulf of Mexico, rather than allowing the sediments to deposit in the delta. Shipping lanes alone have caused 30% of the wetland loss.[1]

Wetlands have their own special set of species, adapted to the wet climate; standing water and high water tables; and heavy, highly organic soils that develop in these habitats. The Mississippi Delta wetlands are also important as a breeding and feeding ground for many kinds of life in the Gulf of Mexico. So these wetlands not only have their own biological diversity but also affect the diversity of life over a much larger region. The continued major decline in these wetlands affects wetland plants, waterfowl, fin fish and shellfish, alligators, and a number of species of mammals.

A new program, called Coast 2050, is trying to improve the situation. This project would close some of the canals and a major navigation channel called the Mississippi River Gulf Outlet. Through construction of alternative waterways and closure of some existing ones, it would allow sediments carried by the Mississippi River to deposit once again in the delta. It would divert water flow to recreate some of the delta's former backwaters and marshes.

As part of the beginning of this project, a water flow of 240 m³/sec has been redirected to wetlands lying in an area adjacent to Breton Sound, where 400 ha of wetlands have been disappearing each year. In just three years since the project began, oysters increased by three orders of magnitude in this area (a thousand times more oysters),

(a)

(b)

Figure 7.1 • Mississippi Delta wetlands. (*a*) A wetland area. (*b*) Map of the wetlands area.

marsh plants increased in abundance by sevenfold, and there are many more muskrats, alligators, and waterfowl. The approach seems to work and to benefit the biological diversity of this once greatest of all coastal wetlands.

But the project is controversial. Expanding it would be very expensive. The prototype at Breton Sound has cost $26 million, and this is just a beginning. And fishermen who catch gulf shrimp, farmers on the lowlands, shipping interests, and towns and residents along the delta, which depend on the levees to prevent flooding, oppose the restoration. The more the wetlands are restored, the fewer levees and ship channels will be available and the greater the chance of floods.

In addition, some scientists believe that there are better methods to accomplish the same goals. For example, some scientists argue that simply removing many existing levees and channels would be enough and that spending money on new water diversions would be an unnecessary expense.

A major improvement in the wetlands thus awaits more funding and a resolution of conflicts over the different uses of the delta. Coastal 2050 shows that improvement is possible, however. It also shows that the condition of one set of habitats and ecosystems affects other, much larger ones. Finally, it shows that decisions affecting biological diversity involve one of the major themes of this book: science and values. Conservation of biological diversity in the Mississippi Delta wetlands is representative of the complexity involved in conserving biological diversity throughout most of the world.

● *The controversy over the Mississippi Delta's wetlands raises several basic questions: What is biological diversity? Why is it important? What do we need to conserve about biological diversity? How can we carry out such conservation? This chapter introduces basic concepts related to biological diversity and its ecological and environmental importance.*

7.1 Understanding Biological Diversity

Biological diversity attracts great public attention today. Frequently, articles in newspapers and discussions on television address issues associated with disappearing species around the world and the need to conserve these species.

Discussions of the value of biological diversity are complicated by the fact that people mean various things when they talk about the conservation of diversity. They may mean conservation of a variety of habitats, the number of genetic varieties, the number of species, the relative abundance of species, or a specific endangered species. All these concepts are interrelated, but each has a distinct meaning.

Another complication is the fact that, as the human population increases, human needs and desires come into increasing conflict with habitats and populations of endangered species. We saw an example of such conflict in the introductory case study. Clearly, then, decisions about what, where, when, and how to conserve biological diversity involve both values and science.

Before we can intelligently discuss the issues involved in conserving Earth's diversity of organisms, we must understand how this diversity came to be. This chapter first addresses the principles of biological evolution that lead to the evolution of new species. We then turn to biological diversity itself: its various meanings, how interactions between species increase or decrease diversity, and how the environment affects diversity.

7.2 Biological Evolution

Biological evolution is a key to understanding biological diversity. Outside biology, the term *evolution* is used broadly to mean the history and development of something. Within biology, however, the term has a more specialized meaning. For example, geologists talk about the evolution of Earth, which simply means Earth's history and geologic changes that have occurred over that history. Book reviewers talk about the evolution of the plot of a novel, meaning how the story unfolds as one reads the book. **Biological evolution** refers to the change in inherited characteristics of a population from generation to generation. It can result in new species—populations that can no longer reproduce with members of the original species. Along with self-reproduction, biological evolution is one of the features that distinguishes life from everything else in the universe.

Biological evolution is a one-way process. Once a species is extinct, it is gone forever. You can run a machine, like a mechanical grandfather clock, forward and backward. But when a new species evolves, it cannot evolve backward into its parents.

How does the evolution occur? Four processes lead to evolution: mutation, natural selection, migration, and genetic drift.

Mutation

Genes, contained in chromosomes within cells, are inherited—passed from one generation to the next. Genes are made up of a complex chemical compound called deoxyribonucleic acid (DNA). DNA in turn is made up of chemical building blocks that form a code, a kind of alphabet of information. The DNA alphabet consists of four letters (specific nitrogen-containing compounds, called bases), which are combined in pairs: (A) adenine, (C) cytosine, (G), guanine, and (T) thymine. How these letters are combined in long strands determines the "message" interpreted by a cell to produce specific compounds.

Sets of the four base pairs form a gene, which is a single piece of genetic information. The number of base pairs that make up a gene varies. To make matters more complex, some base pairs found in DNA are nonfunctional—they are not active and do not determine any chemicals that are produced by the cell. Another complication is that some genes affect the activity of others, turning those other genes on or off. And creatures such as ourselves have genes that limit the number of times a cell can divide, thereby determining a maximum number of cell divisions and therefore maximum longevity.

When a cell divides, the DNA is reproduced so that each new cell gets a copy. Sometimes there is an error or a failure in the reproduction of the DNA, resulting in a change in the DNA and therefore a change in inherited characteristics. At other times, an external environmental agent, such as radiation or a toxic organic chemical, comes in contact with DNA and alters it. Radiation, such as X rays and gamma rays, can break the DNA apart or change its chemical structure. Certain toxic chemicals can react with a DNA molecule to cause a chemical change. When DNA changes in any of these ways, the DNA is said to have undergone **mutation**.

In some cases, a cell or offspring with a mutation cannot survive. In some cases, the mutation simply adds variability to the inherited characteristics. But in some cases, individuals with mutations are so different from their parents that they cannot reproduce with the normal offspring. In this case, a new species has been created. Note that the change is not always for the better: *Mutation can result in a new species whether or not that species is better adapted than its parental species to the environment.*

Natural Selection

When there is variation within a species, some individuals may be better suited to the environment than others. Through the process of **natural selection**, organisms whose biological characteristics better fit them to the environment have relatively more offspring than those whose characteristics are less fit; therefore these organisms are better represented by descendants in future generations.

Natural selection is illustrated in A Closer Look 7.1, which describes how the mosquitoes that carry malaria developed resistance to DDT and how the species of microorganism that causes malaria developed a resistance to quinine. Such adaptations demonstrate the four primary characteristics of natural selection:

1. Inheritance of traits from one generation to the next and some variation in these traits—that is, genetic variability.
2. Environmental variability.
3. Differential reproduction that varies with the environment.
4. Influence of the environment on survival and reproduction.

When natural selection takes place over a very long time, a number of characteristics can change. The accumulation of these changes may be so great that the present generation can no longer reproduce with individuals with the original DNA structure, resulting in a new species. Sometimes, two populations of the same species become isolated completely from one another and cannot reproduce together for a long time. In such cases, the populations may change so much that they can no longer reproduce together even when they are brought back into contact. In this case, two new species have evolved from the original species. Thus, natural selection, combined with geographic separation of populations, is a mechanism by which the evolution of new species can take place. Evolution through geographic isolation is discussed in the next chapter.

Migration

A third process that can lead to evolution is **migration**. The migration of one population of a species into a habitat previously occupied by another population of that species can lead to changes in gene frequency. Even organisms that do not move themselves may have reproductive structures that migrate; seeds of flowering plants,

Natural Selection:
Mosquitoes and the Malaria Parasite

Malaria poses a great threat to 2.4 billion people—over one-third of the world's population—living in over 90 countries, most of them located in the tropics. In 1998 alone, this mosquito-borne infection killed 1.1 million people; an estimated 300 to 400 million more are infected each year.[2]

Once thought to be caused by filth or bad air (hence the name, *malaria*, from the Latin for "bad air)," malaria is actually caused by parasitic microbes (four species of *Plasmodium* protozoa). These microbes also affect and are carried by *Anopheles* mosquitoes.

One solution to the malaria problem, then, would be the eradication of *Anopheles* mosquitoes. By the end of World War II, scientists had discovered that the pesticide DDT was extremely effective against *Anopheles* mosquitoes. They had also found chloroquine was highly effective in killing *Plasmodium* parasites. (Chloroquine is an artificial derivative of quinine, a chemical from the bark of the quinine tree that was an early treatment for malaria.)

In 1957, the World Health Organization (WHO) began a $6 billion campaign to rid the world of malaria using a combination of DDT and chloroquine. At first, the strategy seemed successful. By the mid-1960s, malaria was nearly gone or had been eliminated from 80 percent of the target areas.

However, success was short-lived. The mosquitoes began to develop a resistance to DDT, and the protozoa became resistant to chloroquine. In many tropical areas, the incidence of malaria worsened. For example, as a result of the WHO program, the number of cases in Sri Lanka had dropped from 1 million to only 17 by 1963. However, by 1975, 600,000 cases had been reported, and the actual number is believed to be four times higher. Resistance among the mosquitoes to DDT became widespread, and resistance of the protozoa to chloroquine was found in 80% of the 92 countries where malaria was a major killer.[2,3]

The mosquitoes and the protozoa developed this resistance to the pesticide and the drugs through the process of natural selection. When the populations were exposed to the killing agent, the susceptible organisms died. The populations of mosquitoes and *Plasmodium* are very large, however, and within them is natural variation in DNA. Somewhere in this variation were characteristics that made some of the mosquitoes resistant to DDT and some of the *Plasmodium* resistant to chloroquine. *Only* these most resistant organisms survived. These individuals then passed on their resistant genes to their offspring through reproduction. Since the susceptible individuals died, they left few or no offspring; and any offspring they left were susceptible.

Thus, a change in the environment—the human introduction of DDT and chloroquine—led to a change in which genotype left more offspring. The resistant individuals became dominant in the populations. The development of resistance by both the mosquitoes and the protozoa illustrates the process of biological evolution by natural selection.

A practical lesson from this experience is that if we set out to eliminate a disease-causing species, we must attack it completely at the outset and destroy all the individuals before natural selection can lead to resistance. Sometimes this may be an impossible task, in part because of the natural genetic variation in the target species.

Since the drug chloroquine is generally ineffective now, new drugs have been developed to prevent malaria. However, these second- and third-line drugs will eventually become unsuccessful, too, as a result of the same process of biological evolution by natural selection. This process is speeded up by *Plasmodium*'s ability to rapidly mutate. In South Africa, for example, the protozoa became resistant to Mefloquine immediately after the drug became available as a treatment.

An alternative is to develop a vaccine for the *Plasmodium* protozoa. The advent of biotechnology has made possible the genetic mapping of these malaria-causing organisms. Scientists are currently mapping the genetic structure of *P. falciparum*, the most deadly of the protozoa, and expect to finish within several years. With this information, they expect to create a vaccine containing a variety of the species that is benign in human beings but that produces an immune reaction.[4-8]

In addition, scientists are mapping the genetic structure of *Anopheles gambiae*, the carrier mosquito. This project could show how to better control *Plasmodium* by providing insight concerning genes that could prevent development of the malaria parasite within the mosquito. In addition, it could identify genes associated with insecticide resistance and provide clues to developing a new pesticide.[4-8]

The development of resistance to DDT by mosquitoes and to chloroquine by *Plasmodium* is an example of the process of biological evolution in action today. With the aid of biotechnology, scientists are working to understand the specific chemical structure of the inheritance of characteristics.

for example, are blown by the wind and moved by animals. If seeds land in a new habitat, then the environment may be different enough to favor genotypes not as favored by natural selection in the parents' habitat. (The genotype is the genetic makeup of an individual or group.) Natural selection, combined with migration to new habitat, can thus lead to new dominant genotypes and eventually to new species.

Migration has been an important process over geologic time. For example, during intervals between recent ice ages and at the end of the last ice age, Alaska and Siberia were connected by a land bridge that permitted the migration of plants and animals. When the land bridge was closed off by rising sea water, populations that had migrated to the New World were cut off, and new species evolved. The same process happened much longer ago when marsupials reached Australia and were then cut off from other continents.

Genetic Drift

Another process that can lead to evolution is genetic drift. **Genetic drift** refers to changes in the frequency of a gene in a population as a result of chance rather than mutation, selection, or migration. Chance may determine which individuals become isolated in a small group from a larger population and thus which genetic characteristics are most common in that isolated population. The individuals may not be better adapted to the environment; in fact, they may be more poorly adapted or neutrally adapted.

Suppose, for example, that you begin a study of mosquitoes that carry the malaria parasite (A Closer Look 7.1). You want to study the resistance to DDT, and you go out and collect a sample of the mosquitoes. After you grow the population and obtain a number of generations of offspring, you begin your research. But you discover that by chance the ones you caught are not resistant to DDT. Therefore, none of the offspring are resistant, and the entire population is killed by your experimental use of DDT. This is an example of genetic drift, and it reveals the problem that genetic drift poses for rare and endangered species.

Genetic drift is generally considered to be an especially serious problem when populations become very small, as happens with endangered species. It can be a problem for a rare or endangered species for two reasons: (1) Characteristics that are less adapted to existing environmental conditions may dominate, making the chance of survival of the species less likely. (2) Genetic variability of the species is greatly reduced, meaning that its ability to adapt to future changes in the environment through natural selection has been reduced.

Genetic drift can occur in any small population and may also present problems when a small group is by chance isolated from the main population. For example, bighorn sheep live in the mountains of the desert southwest of the United States. These sheep feed high up in the mountains in the summer, where it is cooler and wetter and where more vegetation is available. Before human settlement of the region, the sheep could move freely and sometimes migrated from one mountain to another by descending into the valleys and crossing them in the win-

ter. With human settlement and the development of cattle ranches and other human activities, many populations of bighorn sheep could no longer migrate among the mountains by crossing the valleys. These sheep have become isolated in very small numbers—commonly, a dozen or so sheep. In such cases, genetic drift may pose a problem for the survival of these populations.

7.3 The Evolution of Life on Earth

In the case of the mosquitoes and the malaria parasite described in A Closer Look 7.1, evolution occurred relatively rapidly. Over most of the history of Earth, evolution seems to have proceeded on average much more slowly. We know about the history of the evolution of species in part through the study of fossils in rocks and in other geologic deposits. The story of evolution begins, as far as we know it today, with the oldest fossils of microorganisms, which appear to be ancestral forms similar to bacteria and what some microbiologists now call Archea.[9] The oldest known fossils of these organisms lived about 3.5 billion years ago. For 2 billion years, there was only such microbial life on Earth, but these organisms greatly changed the global environment, especially by altering the chemistry of the atmosphere.

Photosynthesis developed during that time, and early photosynthetic organisms removed carbon dioxide from Earth's atmosphere and added large amounts of oxygen (illustrating our ongoing assertion that life has always changed the environment on a global scale). With the high concentration of oxygen in the atmosphere, the ecological stage was set for the evolution of new forms of life. Free oxygen allowed the evolution of respiration, which paved the way for oxygen-breathing organisms, including, eventually, humans.

Genetic Engineering and Some New Issues about Biological Diversity

Our understanding of evolution today owes much to the science of molecular biology and the practice of genetic engineering, which are creating a revolution in how we think about and deal with species. At present, scientists have the complete DNA code for four species: the fruit fly (Drosphila), a nematode worm (a very small worm that lives in water, *C. elegans*); yeast; and a small weed plant, Thale cress (*Arabidopsis* thaliana). These species were selected either because they have been the subject of much genetic study or because they have relatively small chains of the base pairs that make up the genetic alphabet. The transcription of the DNA sequence for our own species is nearly completed.

The amount of information contained in DNA is huge. Thale cress's DNA consists of 125 million base pairs. And this is comparatively few. Important crop plants have large numbers of base pairs. Rice has 430 million and wheat more than 16 billion! The base pairs in Thale cress appear to make up approximately 25,000 genes, but many are duplicates; so there are probably about 15,000 unique genes that determine what Thale cress will be like. This is about the same number of genes in the nematode worm and the fruit fly, which have also been sequenced. In contrast, the number of human genes is estimated to be between 30,000 and 130,000.

Advances in genetic engineering reinforce the basis for confidence in the theory of evolution. Scientists can now manipulate DNA and can therefore manipulate inherited characteristics of crops, bacteria, and other organisms and give them new combinations of characteristics not found before, therefore demonstrating that characteristics are inherited and can be altered, as predicted by evolutionary theory. These new capabilities pose novel problems and hold new promise for biological diversity. On the one hand, we may be able to help the persistence of rare, endangered species by increasing genetic variability or by overcoming some of the less adaptive genetic characteristics resulting from genetic drift; on the other hand, we might inadvertently create super pests, predators, or competitors of endangered species. These new organisms can be to our benefit, but we must be careful not to release into the environment new strains that have great reproductive potential and could become unexpected pests. Genetic engineering poses new challenges for the environment, as will be discussed in later chapters. (Some environmental implications of genetic engineering are discussed in Chapters 11 and 13.)

Evolution as a Game

Biological evolution is so different from other processes that it is worthwhile to spend some extra time exploring this idea. There are no simple rules that species must follow to win or to stay in the game of life. Sometimes, when we try to manage species or to predict their future condition and probable persistence, we assume that evolution will follow simple rules. When we make these assumptions, we find that species play tricks on us; they adapt or fail to adapt over time in ways that we did not anticipate. Meanwhile, we repeatedly hope and plan as if life and its environment would follow simple rules. This is true even for the most recent work in genetic engineering.

Such unexpected outcomes result from our failure to fully understand how species have evolved in relation to their ecological situations. Complexity is a feature of evolution. Species have evolved many intricate and amazing adaptations that have allowed them to persist. It is essential to realize that these adaptations have evolved not in isolation but in the context of relationships to other organisms and to the environment. The environment sets up a situation or scene within which evolution, by natural selection, takes place. The ecologist G. E. Hutchinson referred to

this interaction in the title of one of his books, *The Ecological Theater and the Evolutionary Play.* Here, the ecological situation—the condition of the environment and other species—is the scenery against which natural selection occurs, and natural selection results in a story of evolution played out over the history of life on Earth.[10]

7.4 Basic Concepts of Biological Diversity

As suggested earlier, the term *biological diversity* is used loosely, even though it represents one of the most discussed of environmental problems. Sometimes, the term refers simply to the protection of a single endangered species, such as the African rhinoceros or the blue whale. At other times, the concern is with the conservation of the total number of species in a given area, regardless of what those species are. This is generally the case with the concern over the loss of tropical rain forests. Sometimes an endangered species is the focus for other goals, so that the concern may be less with the specific species as with its entire habitat or with a proposed action. For example, not long after the Endangered Species Act was passed, a debate erupted over the conservation of a previously little known fish, the snail darter. But the snail darter was a means used by some to try to stop what they believed was an unnecessary dam that would have many undesirable environmental effects. Those wanting the dam opposed the listing of the snail darter as endangered.

To develop workable policies for the conservation of biological diversity, we must be clear about the meaning of this concept. In terms of biological characteristics, then, **biological diversity** involves the following concepts:

1. *Genetic diversity:* the total number of genetic characteristics of a specific species, a subspecies, or a group of species. In terms of genetic engineering and our new understanding of DNA, this could mean: (a) the total base-pair sequences in DNA, (b) the total number of genes, active or not, or (c) the total number of active genes.
2. *Habitat diversity:* the diversity of habitats in a given unit area.
3. *Species diversity*, which, in turn, has three aspects:
   a. *Species richness*—the total number of species.
   b. *Species evenness*—the relative abundance of species.
   c. *Species dominance*—the most abundant species.

Let us look more closely at the differences among the three concepts that make up species diversity: species richness, species evenness, and species dominance. Imagine two ecological communities, each with 10 species and 100 individuals (Figure 7.2). In the first community, 82 individuals belong to a single species, and the remaining 9 species are represented by 2 individuals each. In the second commu-

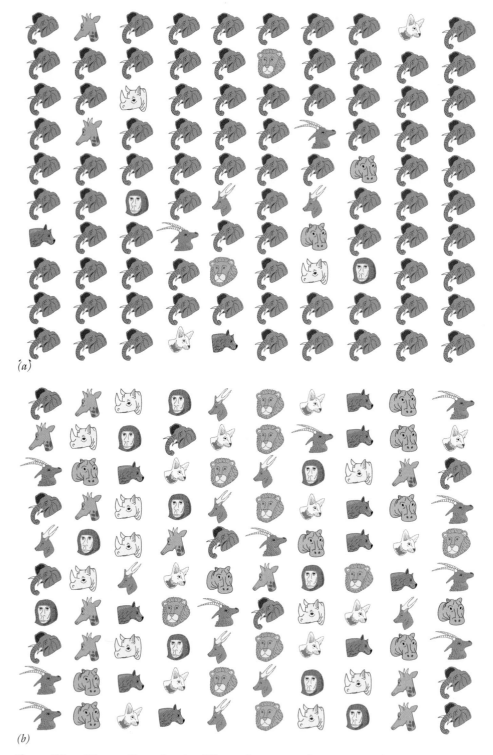

(a)

(b)

Figure 7.2 • Diagram illustrating the difference between species evenness and species richness. Figures (*a*) and (*b*) have the same richness but different evenness. Lay a ruler across each diagram and count the number of species the edge crosses. Do this several times, then determine which figure has greater evenness. See text for explanation of results.

nity, all the species are equally abundant; each therefore has 10 individuals. Which community is more diverse?

At first, one might think that the two communities have the same species diversity, because they have the same number of species. However, if you walked through both communities, the second would appear more diverse. As you walked through the first community, most of the time you would see individuals only of the dominant species (in the case of Figure 7.2, elephants); and there would be many species you probably would not see at all. In the second community, even a casual visitor would see many of the species in a short time. The first community would appear relatively low in diversity until it was subject to extremely careful study. You can test the differences in the probability of encountering a new species by laying a ruler down in any direction on Figures 7.2*a* and 7.2*b* and counting the number of species that it touches.

As this example suggests, merely counting the number of species is not enough to describe biological diversity. *Species diversity has to do with the relative chance of seeing species as much as it has to do with the actual number present.* For a fixed number of species, a community in which each species has the same number of individuals as all others would appear to be the most diverse, whereas a community in which most of the individuals belong to a single species would seem the least diverse. Ecologists divide these aspects of species diversity, referring to the total number of species in an area as species richness, the relative abundance of species as species evenness, and the most abundant species as dominant.

Whichever meaning of biological diversity is chosen, it can be said that people have greatly changed biological diversity, decreasing the number of species, increasing the rate of extinction of species, decreasing the genetic diversity of many species, decreasing the number of habitats, and greatly altering the geographic distribution of many species. In the balance of this chapter, we explore the factors that increase or decrease diversity. As we develop an understanding of how so many species came to be and why species become extinct, we will begin to understand what is involved in preserving this amazing diversity.

7.5 The Number of Species on Earth

No one knows the exact number of species on Earth, because new species are discovered all the time, especially in little-explored areas such as tropical rain forests. For example, since 1992, five new mammals have been discovered in Laos and two more have been "rediscovered." The rediscovered animals were known from a few specimens but had not been seen for a long time. The new species include: (1) the spindle-horned oryx (which is not only a new species but also represents a previously unknown genus); (2) the giant muntjak (muntjak, also known as "barking deer," are small deer; the giant muntjak is so called because it has large antlers; (3) the small black muntjak; (4) the striped hare (whose nearest relative lives in Sumatra); and (5) a new species of civet cat. The rediscovered species are: (1) a wild pig, previously known only from a skull found by a hunter in the 1890s; and (2) Roosevelt's muntjak, previously known from one specimen obtained in the 1920s.

That such a small country with a long history of human occupancy would have so many new mammal

species suggests how little we still know about the total biological diversity on Earth. Some 1.4 million species have been named, but biologists estimate that the total number is probably considerably higher. Estimates range from 3 million to as high as 100 million (see Table 7.1).

It is common to think in terms of two major kinds of life: animals and plants. Scientists, however, categorize life into major groups based on evolutionary relationships—a biological genealogy. In the recent past, scientists classified life into five kingdoms: animals, plants, fungi, protists, and bacteria. New evidence from the fossil record and studies in molecular biology suggest that it may be more appropriate to describe life as existing in three major domains, one called Eukaryotes or Eucarya, which includes animals, plants, fungi, and protists (mostly single-celled organisms); Bacteria; and Archaea.[11,12] Eucarya have a nucleus and other organelles; Bacteria and Archaea do not. Archaea used to be classified among Bacteria, but there are substantial molecular differences that suggest ancient divergence in heritage (Figure 7.3).

(a)

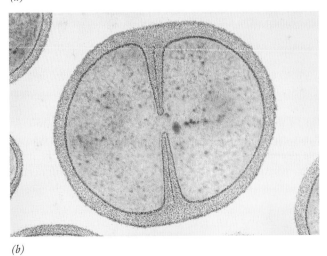

(b)

Figure 7.4 • Photomicrograph of (*a*) a eukaryote cell and (*b*) a bacterial (prokaryote) cell.

Table 7.1 Comparison of Species Estimates

Kingdom/Phylum Subphylum/Class	Mader, 1997	Campbell, 1996	Raven, 1999	Margulis, 1988	E. O. Wilson, 1988
Virus					1000
Monera				100000	4760 (+Archaea)
Archaebacteria			100		
Cyanobacteria			7500		
Fungi		100000	70000	100000	
Chytridiomycota			790		575
Zygomycota	600	600	1060	600	665
Ascomycota	30000	60000	32300	tens of thousands	28650
Basidiomycota	16000	25000	22300	25000	16000
Deuteromycota	25000			25000	
Protista		60000		thousands	
Ciliophora	8000		8000		
Myxomycota	560		700		500
Cryptophyta			200		
Dinophyta			2000–4000		
Oomycota	580		694		580
Euglenophyta	1000	800	900	800	800
Dinoflagellata	1000	1100		several thousand	1100
Sarcodina	40000				
Sporozoa	3600				
Chrysophyta	11000	850	1000		12500
Bacillariophyta		10000	100000	10000	
Phaeophyta	1500	1500	1500	1500	1500
Rhodophyta	4000	4000	4000–6000	4000	4000
Chlorophyta	7000	7000	17000	7000	7000
Plantae					
Bryophyta	12000	10000	9500	24000	16600
Hepatophyta	10000	6500	6000		
Anthocerophyta	100	100	100		
Psilophyta	several	10–13	several		9
Lycophyta	1000	1000	1000	1000	1275
Shenophyta		15	15	40	15
Filicinophyta				12000	10000
Peterophyta	12000	12000	11000		
Coniferophyta	550	550	550	550	
Cycadophyta	100	100	140	100	
Gneetophyta	70	70	70	70	
Anthophyta		235000	235000	230000	
Monocotyledones	65000				50000
Dicotyledones	170000				170000

Insects and plants make up most of the known species (Table 7.1). Many of the insects are tropical beetles in rain forests. Mammals, the group to which people belong, make up a comparatively small number of species, about 4,000. As more explorations are carried out, especially in tropical areas, the number of identified invertebrates and plants are most likely where the major increase in named species will occur.

7.6 Why Are There So Many Species?

Evolution is responsible for the development of species. But why are there so many? Species compete with one another for resources, and competition would seem to result in a reduction in the number of species. How, then, can so many different species survive? Part of the answer lies in the different ways in which organisms interact, and part of the answer lies with the idea of the ecological niche. The next section looks at the interactions between species in an attempt to explain this apparent paradox.

Interactions between Species

We can classify interactions between species according to three basic types: *competition*, in which the outcome is negative for both groups; *symbiosis*, which benefits both participants; and *predation-parasitism*, in which the outcome benefits one and is detrimental to the other. (Scientists use

Table 7.1 **Comparison of Species Estimates (continued)**

Kingdom/Phylum Subphylum/Class	Mader, 1997	Campbell, 1996	Raven, 1999	Margulis, 1988	E. O. Wilson, 1988
Animalia		>1,000,000			
Porifera	5000	9000		10000	5000
Cnidaria	9000	10000			9000
Platyhelminthes	13000	20000		15000	12200
Rotifera	2000	1800		2000	
Nematoda	12000	80000		80000	12000
Mollusca	110000	50000		110000	50000
Annelida	12000	15000		5400	12000
Arthropoda	>6 million	nearly 1,000,000		500000	989761
Cheliceriformes					
Uniramia					
Insecta					751000
Anoplura		2400			
Coleoptera		500000			
Dermaptera		1000			
Diptera		80000			
Hemiptera		55000			
Hymenoptera		90000			
Isoptera		2000			
Lepidoptera		140000			
Odonata		5000			
Orthoptera		30000			
Siphonaptera		1200			
Trichoptera		7000			
Crustacea		40000			
Brachiopoda		330		335	
Echinodermata	6550	7000		6000	6100
Chordata	45000			45000	
Vertebrata	44000				
Chondrichthyes	850	750			843
Osteichthyes	20000	30000			18150
Amphibia	3900	4000			4184
Reptilia	6000	7000			6300
Aves	9000	8600			9040
Mammalia	4500	4500			4000

Note: With ~15,000 new species described every year, a changing classification system, and no central data registry, it is not suprising that species estimates differ among publications and years.

Sources: Wilson, E. O. (ed.), 1988, *Biodiversity*, National Academy Press, Washington, D.C.; *Five Kingdoms: An Illustrated Guide to the Phyla of Life on Earth*, 1998, Lynn Margulis, Karlene V. Schwartz (contributor), Stephen Jay Gould, W.H. Freeman.

somewhat more detailed categories of interactions, as shown in Table 7.2.) Each type of interaction affects evolution, the persistence of species, and the overall diversity of life.

The discussion of malaria in A Closer Look 7.1; "Natural Selection: Mosquitoes and the Malaria Parasite", shows several of the ways in which species interact. The protozoan that causes malaria is a parasite of both mosquitoes and human beings; the mosquito is a parasite of human beings; for most of their lives, mosquitoes are herbivores, predators of primary producers.

The Competitive Exclusion Principle

A consequence of natural selection is that when two species compete, the more fit species will win and persist and the less fit species will lose and become extinct. This is known as the **competitive exclusion principle**, which states that *two species that have exactly the same requirements cannot coexist in exactly the same habitat.* Garrett Hardin expressed the idea most succinctly: "Complete competitors cannot coexist."[13]

An Illustration:
Gray and Red Squirrels in Great Britain

The American gray squirrel and the British red squirrel illustrate the competitive exclusion principle. The American gray squirrel was introduced into Great Britain because some people thought it was attractive and would be a pleasant addition to the landscape. Thus, this introduction was not accidental but intentional. In fact, about

Table 7.2   Types of Interactions between Species

Interaction	Species A	Species B	Result
Neutralism	0	0	Neither affects the other.
Competition	–	–	A and B compete for the same resource; each has a negative effect on the other. Example: hummingbirds compete for the same flower nectar.
Inhibition	0	–	A inhibits B; A is not affected. Example: bacteria in a cow's rumen release a chemical that inhibits other bacteria.
Parasitism	+	–	A, the parasite or predator, benefits. Example: heartworm in a reindeer.
Predation	+	–	A feeds on B, the host or prey, who suffers a direct negative effect. Example: wolves feeding on reindeer.
Symbiosis	+	+	A and B require each other to survive. Example: some bacteria and reindeer; hummingbirds and some flowers.
Commensalism	+	0	A requires B to survive; B is not affected significantly. Example: mistletoe on some trees.

Note: 0, no effect; +, positive effect on birth, growth, or survival; –, negative effect on birth, growth, or survival.

a dozen attempts at introduction were made, the first perhaps as early as 1830. By the 1920s, the American gray squirrel was well established in Great Britain, and its numbers expanded greatly in the 1940s and 1950s.

Today, the American gray squirrel is a problem; it competes with the native red squirrel and is winning the competition (Figure 7.4). The two species have almost exactly the same habitat requirements. At present, the red squirrel population is approximately 160,000. Although red squirrels used to be found in deciduous woodlands throughout the lowlands of central and southern Britain, they now are common only in Cambria, Northumberland, and Scotland,

with scattered populations in East Anglia, Wales, on the Isle of Wight, and on islands in Poole Harbor, Dorset.[14]

It is thought that one reason for the shift in the balance of these species is that the main source of food during winter for red squirrels is hazelnuts, whereas for gray squirrels it is acorns. Thus, red squirrels have a competitive advantage in areas with hazelnuts, and gray squirrels have the advantage in oak forests. When gray squirrels were introduced, oaks were the dominant mature tree in Great Britain; about 40% of the timber planted was oaks.

If present trends continue, the red squirrel may disappear from the British mainland in the next 20 years. The

(a)

(b)

Figure 7.4 • The American gray squirrel (*Sciurus carolinensis*) (*a*) was introduced into Great Britain, where it is displacing the native European red squirrel (*Sciurus vulgaris*) (*b*), illustrating the competitive exclusion principle.

introduction of the gray squirrel into Great Britain illustrates one of the major causes of modern threats to biological diversity: the introduction by people of species into new habitats. Introduced competitors frequently threaten native species. We look at this issue again in Chapter 8.

Why Hasn't the Competitive Exclusion Principle Limited Species Diversity?

According to the competitive exclusion principle, *complete competitors cannot coexist; one will always exclude the other.* Over a very long period of time, we might expect this to reduce the number of species over a large area. Taken to its logical extreme, we could imagine Earth with very few species, perhaps one green plant on the land, one herbivore to eat it, one carnivore, and one decomposer. If we added 4 species for the ocean and 4 for fresh waters, we would have only 12 species on our planet.

Being a little more realistic, we could allow for the necessity of adaptations to major differences in climate and other environmental aspects. Perhaps we could specify 100 major climatic or environmental categories: cold and dry, cold and wet, warm and dry, warm and wet, and so forth. Even so, we would still expect that within each environmental category, competitive exclusion would result in a few species. Allowing 4 species per major environmental category would result in only 400 species. Yet about 1.4 million have been named, and scientists speculate that many more millions may exist; the number is so great, we do not have even a good estimate of it. How can they all coexist?

Niches: How Species Coexist

Some classic studies of flour beetles (*Tribolium*), which live on wheat flour, help explain how species with similar requirements can coexist. Flour beetles make good experimental subjects because they require only small containers of wheat flour to live.

In a study, populations of two species of flour beetles are placed in a small container with flour. The number of individuals of each species is counted; the flour is passed through a sieve, separating the beetles, which are then identified by species and put back into the container with the flour. Whenever two species are placed in a container that has uniform environmental conditions, one species of beetle always wins. Competitive exclusion takes place. However, it is not always the same species that survives. How do we explain this?

The answer is that the two beetle species differ in rates of survival and reproduction under any set of environmental conditions. When it is warm and dry, one species has more offspring and wins. When it is cool and moist, the other species has more offspring and wins (Figure 7.5). When conditions are in between, sometimes one species wins and sometimes the other, seemingly randomly; but invariably one persists while the second becomes extinct.

In a mixed environment (an environment with a variety of conditions in different geographic areas), however, both species of beetle may survive. They do so by using separate parts of the habitat. The one that does well where it is wet and cool is found in those habitats, while the one that does well in warm and dry habitats is found there. In no location do the species coexist.

Here, then, is the key to the coexistence of many species. *Species that require the same resources can coexist by utilizing those resources under different environmental conditions.*[15] The species are said to have different ecological niches.

Professions and Places: The Ecological Niche and the Habitat

The ecological niche is an important concept. One way to understand it is to think of it as a species' profession. Where a species lives is its **habitat**, but what it does for a living (its profession) is its **ecological niche**.[16] The general profession of a flour beetle is eating flour. A squirrel's profession is eating seeds of trees. Suppose you have a neighbor who is a bus driver. Where your neighbor lives and works is your town; what your neighbor does is drive a bus. Your neighbor's niche is bus driving; your neighbor's habitat is your hometown. Similarly, if someone says, "Here comes a wolf," you think not only of a creature that inhabits the northern forests (its habitat) but also of a predator that feeds on large mammals (its niche).

Understanding the niche of a species is useful in assessing the impact of development or of changes in land use. Will the change remove an essential requirement for some species' niche? A new highway that makes car travel easier might eliminate your neighbor's bus route and therefore eliminate his niche. In the same way, cutting a forest may drive away prey and eliminate the niche of the wolf.

Measuring Niches

How can we measure the niche? When we think of the niche only in terms of a profession, it is difficult to quantify it—to determine exactly where it is and what its limits are for a given species. However, the niche can also be described as the set of all environmental conditions under which a species can persist.

The idea of a measured niche is known as the *Hutchinsonian niche*, after G. E. Hutchinson, who first suggested it.[17] It is illustrated by the distribution of two species of flatworm, a tiny worm that lives on the bottom of freshwater streams. A study was made of two species of these small worms in Great Britain, where it was found that some streams contained one species, some the other, and still others both.[18]

The stream waters are cold at their source in the mountains and become progressively warmer as they flow downstream. Each species of flatworm occurs within a specific range of water temperatures. In streams where species A occurs alone, it is found from 6° to 17°C

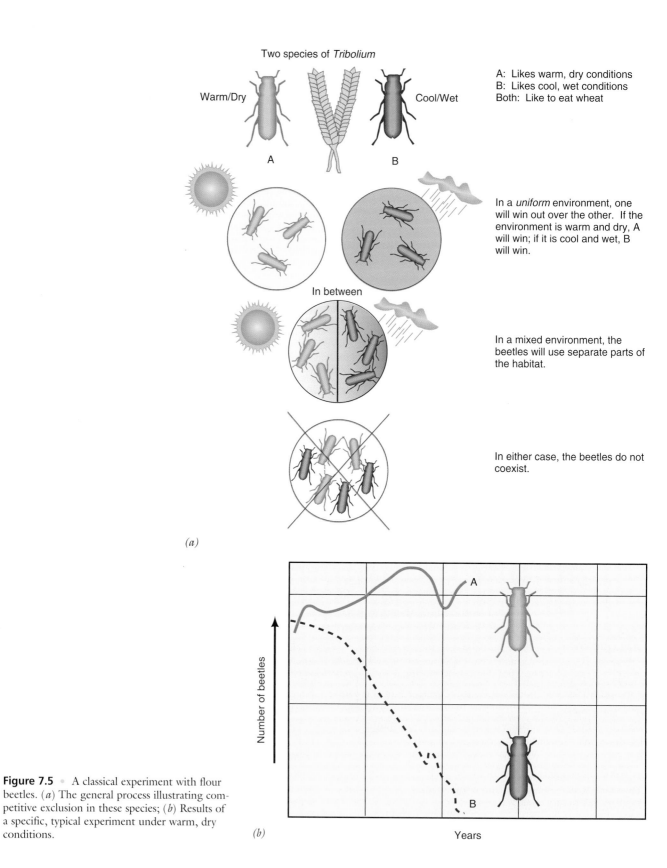

Two species of *Tribolium*

Warm/Dry   Cool/Wet

A   B

A: Likes warm, dry conditions
B: Likes cool, wet conditions
Both: Like to eat wheat

In a *uniform* environment, one will win out over the other. If the environment is warm and dry, A will win; if it is cool and wet, B will win.

In between

In a mixed environment, the beetles will use separate parts of the habitat.

In either case, the beetles do not coexist.

(*a*)

Number of beetles

A

B

Years

(*b*)

Figure 7.5 • A classical experiment with flour beetles. (*a*) The general process illustrating competitive exclusion in these species; (*b*) Results of a specific, typical experiment under warm, dry conditions.

(42.8°–62.6°F) (Figure 7.6*a*). Where species B occurs alone, it is found from 6° to 23°C (42.8°–73.4°F) (Figure 7.6*b*). When they occur in the same stream, their temperature ranges are much reduced. Species A occurs in the upstream sections, where the temperature ranges from 6° to 14°C (42.8°–57.2°F), and B occurs in downstream areas, where temperatures are warmer, from 14° to 23°C (57.2°–73.4°F) (Figure 7.6*c*).

(a)

(b)

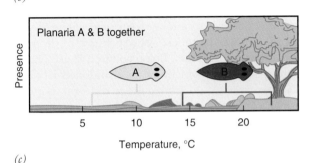

Temperature, °C

(c)

Figure 7.6 • The occurrence of freshwater flatworms in cold mountain streams in Great Britain. (*a*) The presence of species A in relation to temperature in streams where it occurs alone. (*b*) The presence of species B in relation to temperature in streams where it occurs alone. (*c*) The temperature range of both species in streams where they occur together. Inspect the three graphs; what is the effect of each species on the other?

The range of temperature over which species A occurs when it has no competition from B is called its *fundamental temperature niche*. The set of conditions under which it persists in the presence of B is called its *realized temperature niche*. The flatworms show that species divide up resources along environmental gradients in nature.

Of course, temperature is only one aspect of the environment. Flatworms also have requirements in terms of the acidity of the water and other factors. We could create graphs for each of these factors, showing the range under which A and B occurred. The collection of all those graphs would constitute the complete description of the niche of each species.

The Importance of the Niche in Conservation
From the discussion of the competitive exclusion principle and the ecological niche, we learn something important about the conservation of species. If we are to conserve a species in its native habitat, we must make sure

that all the requirements of its niche are present. Conservation of endangered species is more than a matter of placing many individuals of that species into an area; all the life requirements for that species must also be present.

Symbiosis

Our discussion up to this point might leave the impression that species interact mainly by interfering with one another. But this is not the case. **Symbiosis** is also very important. This term, derived from a Greek word meaning "living together," describes a relationship between two organisms that is beneficial to both. Each partner in symbiosis is called a **symbiont**. Through symbiosis, two species can benefit one another and enhance each other's chances of persisting.

Symbiosis is widespread and common; most animals and plants have symbiotic relationships with other species. Symbiosis therefore affects biological diversity. The widespread occurrence of symbiosis is illustrated by our own bodies. Microbiologists tell us that about 10% of a person's body weight is actually the weight of symbiotic microorganisms in the intestines. The resident bacteria help us in our digestion; we provide a habitat that supplies all their needs; both we and they benefit. We become aware of this intestinal community when it changes—for example, when we travel to a foreign country and ingest new strains of bacteria. Then we suffer a well-known traveler's malady, gastrointestinal upset.

Another important kind of symbiotic interaction occurs between certain mammals and bacteria. A reindeer on the northern tundra may appear to be alone but carries with it many companions. Like domestic cattle, the reindeer is a ruminant, with a four-chambered stomach (Figure 7.7) teeming with microbes (a billion per cubic centimeter). In this partially closed environment, the respiration of microorganisms uses up the oxygen ingested by the reindeer while eating. Other microorganisms digest cellulose, take nitrogen from the air in the stomach, and make proteins. The bacterial species that digest the parts of the vegetation that the reindeer cannot digest itself (in particular, the cellulose and lignins of cell walls in woody tissue) require a peculiar environment. They can survive only in an environment without oxygen. One of the few places on Earth's surface where such an environment exists is the inside of a ruminant's stomach.[19] The bacteria and the reindeer are symbionts, each providing what the other needs; and neither could survive without the other. They are therefore called **obligate symbionts**.

We can see that if we want to save a species from extinction, we must save not only its niche but also its symbionts. This suggests another important point that will become more and more evident throughout this text: The attempt to save a single species almost invariably

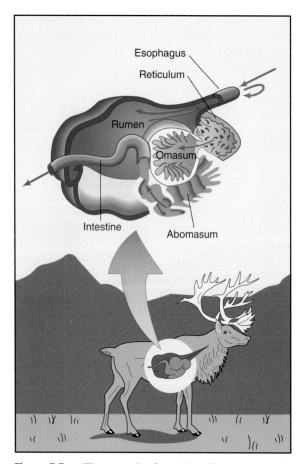

Figure 7.7 • The stomach of a reindeer illustrates complex symbiotic relationships. For example, in the rumen, bacteria digest woody tissue the reindeer could not otherwise digest. The result is food for the reindeer and food and a home for the bacteria, which could not survive in the local environment outside.

(Image labels: Esophagus, Reticulum, Rumen, Omasum, Intestine, Abomasum)

leads us to conserve a group of species, not just a single species nor a particular physical habitat.

Predation and Parasitism

Predation-parasitism is the third way in which species interact. Predation occurs when an organism (a predator) feeds on other, live organisms, usually of another species (prey). Parasitism is an association between organisms in which one (the parasite) lives on, in, or with the other (host) and depends on it for existence but makes no useful contribution to it and may harm it.

Predation can increase diversity of prey species by reducing the abundance of the dominant prey. In this way, predators can contribute to species evenness by keeping the dominant species from overwhelming others via competitive exclusion. For example, some studies have shown that a moderately grazed pasture has more species of plants than an ungrazed one. The same seems to be true about natural grasslands and savannas. Without graz-

ers and browsers, then, African grasslands and savannas might have fewer species of plants.

7.7 Environmental Factors That Influence Diversity

Species are not uniformly distributed over Earth's surface; diversity varies greatly from place to place. For instance, suppose you were to go outside and count all the species in a field or any open space near where you are reading this book (that would be a good way to begin to learn for yourself about biodiversity). The number of species would differ greatly depending on where you are. If you lived in northern Alaska or Canada, Scandinavia, or Siberia, you would probably find a significantly smaller number of species than if you lived in the tropical areas of Brazil, Indonesia, or central Africa. Variation in diversity is partially a question of latitude. In general, higher diversity occurs at lower latitudes. Many patterns of diversity also occur within local areas. If you counted species in the relatively sparse environment of an abandoned city lot, for example, you would find quite a different number than if you counted species in the rich environment of an old, long-undisturbed forest.

The large-scale geographic pattern in the distribution of species, called *biogeography*, is the topic of the next chapter. For now, we look more closely at some of the environmental factors that influence diversity. Table 7.3 summarizes several of these factors.

On a local scale on the land, the kinds of species and ecosystems that occur change with soil type and topographic characteristics: slope, aspect (the direction the slope faces), elevation, and relation to a drainage basin. These factors influence the number and types of plants. The plants, in turn, influence the number and types of animals. Some of the possible interrelationships are illustrated in Figure 7.8.[20]

Change in the relative abundance of a species over an area or a distance is referred to as an *ecological gradient*. Such a change can be seen with changes in elevation in mountainous areas. Figure 7.9 illustrates some of these changes for the Grand Canyon and the nearby San Francisco Mountains of Arizona. Although such patterns are most easily seen in vegetation, they occur for all organisms. For example, the pattern of distribution of representative African mammals on Mount Kilimanjaro is shown in Figure 7.10.

If environmental conditions were constant over time and space, the existing dominant species would become more abundant. They would tend to increase at the expense of rarer species, following the competitive exclusion principle. But a habitat that includes a variety of local environments is likely to offer a refuge to rarer species, thus leading to greater diversity. The more diverse habitat allows more niches, and more species persist.

Table 7.3	Some Major Factors That Increase and Decrease Biological Diversity

A. Factors that tend to increase diversity

1. A physically diverse habitat
2. Moderate amounts of disturbance (such as fire or storm in a forest or a sudden flow of water from a storm into a pond)
3. A small variation in environmental conditions (temperature, precipitation, nutrient supply, etc.)
4. High diversity at one trophic level increases the diversity at another trophic level. (Many kinds of trees provide habitats for many kinds of birds and insects.)
5. An environment highly modified by life (e.g., a rich organic soil)
6. Middle stages of succession
7. Evolution

B. Factors that tend to decrease diversity

1. Environmental stress
2. Extreme environments (conditions near the limit of what living things can withstand)
3. A severe limitation in the supply of an essential resource
4. Extreme amounts of disturbance
5. Recent introduction of exotic species (species from other areas)
6. Geographic isolation (being on a real or ecological island)

Some habitats harbor little diversity because they are so severe or stressful to life. A comparison of vegetation in two areas of Africa illustrates the point. In eastern and southern Africa, the well-drained, sandy soils support a diverse vegetation, including many species of *Acacia* and *Combretum* trees as well as many grasses. The well-drained soil is not stressful to plants. In contrast, on the very heavy clay soils of wet areas near rivers, such as occur along the Sengwa River in Zimbabwe, the woodlands are composed almost exclusively of a single species called *mopane*. In this case, the very heavy clay soils impose a stress on plants; these soils hold water and allow little

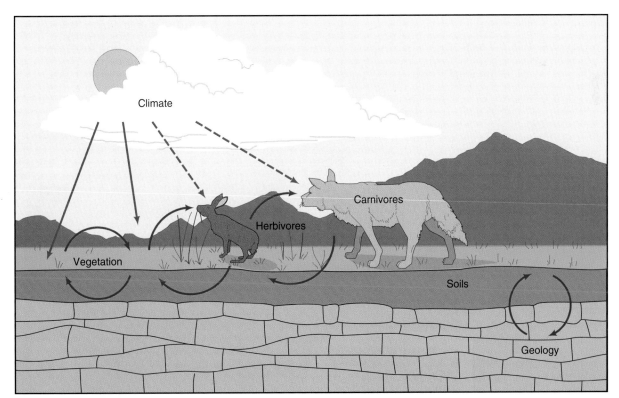

Figure 7.8 • Interrelationships among climate, geology, soil, vegetation, and animals. What lives where depends on many factors. Climate, geologic features (bedrock type, topographic features), and soils influence vegetation. Vegetation in turn influences soils and the kinds of animals that will be present. Animals affect the vegetation. Arrows represent a causal relationship; the direction is from cause to effect. A dashed arrow indicates a relatively weak influence, and a solid arrow a relatively strong influence.

North   San Francisco mountains   South

Figure 7.9 • Change in the relative abundance of a species over an area or a distance is referred to as an *ecological gradient*. Such a change can be seen with changes in elevation in mountainous areas. The altitudinal zones of vegetation in the Grand Canyon of Arizona and the nearby San Francisco Mountains are shown. (*Source:* From C. B. Hunt, *Natural Regions of the United States and Canada* [San Francisco: W. H. Freeman, 1974] Copyright 1974 by W. H. Freeman.)

oxygen to reach plant roots. Trees thus are restricted to species with very shallow roots.

Moderate environmental disturbance can also increase diversity. For example, fire is a common disturbance in many forests and grasslands. Occasional light fires produce a mosaic of recently burned and unburned areas; these patches favor different kinds of species and increase overall diversity.

Of course, people also affect diversity. In general, urbanization, industrialization, and agriculture decrease diversity, reducing the number of habitats and simplifying habitats. In addition, we intentionally favor specific species and manipulate populations for our own purposes, as when a person plants a lawn or when a farmer plants a single crop over a large area.

In the past, people rarely thought about cities as having any beneficial effects on biological diversity. Indeed, the development of cities tends to reduce biological diversity. This is especially true because cities have typically been located at good sites for travel, such as along rivers or near oceans, where biological diversity is often high. However, in recent years, there has been a growing

Figure 7.10 • Changes in the distribution of animals with elevation on a typical mountain in Kenya. (*Source:* From C. B. Cox, I. N. Healey, and P. D. Moore, *Biogeography* [New York: Halsted, 1973].)

Why Preserve Biodiversity?

Preserving a diversity of life on Earth has become an accepted goal for many people. But when that goal comes into conflict with other goals, such as economic development, the question becomes, "How much diversity and at what cost?" To find the answer, it is important to think carefully about the values of biological diversity and to separate those based on science from those based on other values, such as aesthetic, ethical, religious, or economic values.

In 1980, the International Union for Conservation of Nature and Natural Resources (IUCN) proposed a statement to form a basis for conserving biological diversity. That statement summarizes most of the arguments used as a rationale for conserving diversity. Read the statements reproduced below and answer the questions that follow.

An Ethical Basis for Preserving Biodiversity

1. The world is an interdependent whole made up of natural and human communities. The well-being and health of any one part depends on the well-being and health of the other parts.

2. Humanity is part of nature, and humans are subject to the same immutable ecological laws as are all other species on the planet.

3. All life depends on the uninterrupted functioning of natural systems that ensures the supply of energy and nutrients, so ecological responsibility among all people is necessary for the survival, security, equity, and dignity of the world's communities.

4. Human culture must be built on a profound respect for nature, a sense of being at one with nature, and a recognition that human affairs must proceed in harmony and in balance with nature.

5. The ecological limits within which we must work are not limits to human endeavor; instead, they give direction and guidance as to how human affairs can sustain environmental stability and diversity.

6. All species have an inherent right to exist. The ecological processes that support the integrity of the biosphere and its diverse species, landscapes, and habitats are to be maintained. Similarly, the full range of human cultural adaptations to local environments is to be enabled to prosper.

7. Sustainability is the basic principle of all social and economic development.

8. Personal and social values should be chosen to accentuate the richness of flora, fauna, and human experience. This moral foundation will enable the many utilitarian values of nature—for food, health, science, technology, industry, and recreation—to be equitably distributed and sustained for future generations.

9. The well-being of future generations is a social responsibility of the present generation. Therefore, the present generation should limit its consumption of nonrenewable resources to the level that is necessary to meet the basic needs of society and ensure that renewable resources are nurtured for their sustainable productivity.

10. All persons must be empowered to exercise responsibility for their own lives and for the life of Earth. They must therefore have full access to educational opportunities, political enfranchisement, and sustaining livelihoods.

11. Diversity in ethical and cultural outlooks toward nature and human life is to be encouraged by promoting relationships that respect and enhance the diversity of life, irrespective of the political, economic, or religious ideology in a society.

Critical Thinking Questions

1. Which of the statements are scientific—that is, which are based on repeatable observations, can be tested, and are supported by evidence?

2. Using the table that follows, identify which statement from the argument for preserving biodiversity is implied by each type of value. Write the number of the statement in the right-hand column. A statement may fit into more than one category.

Type of Value	Source of Value of Living Organisms	Statement Number
Ethical	The fact that they are alive	
Aesthetic	Their beauty and the rewards humans derive from their beauty	
Economic	The direct and indirect ways in which they benefit humans	
Ecological	Their contributions to the health of the ecosystem	
Intellectual	Their contributions to knowledge	
Emotive	The sense of awe and wonder they inspire in humans	
Religious	Having been created by a supernatural being or force	
Recreational	Sport, tourism, and other recreations	

3. How would you rank the types of values given?

4. Which type or types of values could be used to support preservation of smallpox viruses? All 1,200 (or more) species of beetles in the tropical rain forest? The northern spotted owl? Dolphins? Sharks?

realization that cities can contribute in important ways to the conservation of biological diversity. These topics are discussed in greater detail in Chapter 27.

The number of species per unit area varies over time as well as space. The changes occur over many time periods, from very short (a year), to moderate (years, decades), to even longer (centuries or more in a forest), and finally to very long geologic periods. As an example of the latter, diversity of animals in the early Paleozoic era (beginning about 500 million years ago) was much lower than in the late Mesozoic (beginning about 130 million years ago) and the Cenozoic (beginning about 75 million years ago) eras.[21]

Summary

- Biological evolution—the change in inherited characteristics of a population from generation to generation—is responsible for the development of the many species of life on Earth. Four processes that lead to evolution are mutation, natural selection, migration, and genetic drift.

- Biological diversity involves three concepts: genetic diversity (the total number of genetic characteristics), habitat diversity (the diversity of habitats in a given unit area), and species diversity. Species diversity, in turn, involves three ideas: species richness (the total number of species), species evenness (the relative abundance of species), and species dominance (the most abundant species).

- About 1.4 million species have been identified and named. Insects and plants make up most of the species. As more explorations are carried out, especially in tropical areas, the number of identified species, especially of invertebrates and plants, will increase.

- Species engage in three basic kinds of interactions: competition, in which the outcome is negative for both groups; symbiosis, which benefits both groups; and predation-parasitism, in which the outcome benefits one and is a disadvantage for the other. Each type of interaction affects evolution, the persistence of species, and the overall diversity of life. It is important to understand that organisms have evolved together so that predator, parasite, prey, competitor, and symbiont have adjusted to one another. Human interventions frequently upset these adjustments.

- The competitive exclusion principle states that two species that have exactly the same requirements cannot coexist in exactly the same habitat; one must win. The reason that more species do not die out from competition is that they have developed a particular niche and thus avoid competition.

- The number of species in a given habitat is affected by many factors, including latitude, elevation, topography, the severity of the environment, and the diversity of the habitat. Predation and moderate disturbances, such as fire, can actually increase the diversity of species. The number of species also varies over time. Of course, people affect diversity as well.

● REEXAMINING THEMES AND ISSUES

Human Population: The growth of human populations has led to decreases in biological diversity. As discussed in this chapter, people have had major effects on biological diversity. If the human population continues to increase, pressures will continue on endangered species, and maintaining existing biological diversity will be an ever-greater challenge.

Sustainability: Sustainability involves more than just having many individuals of a species. For a species to persist, its life requirements must be present, and its habitat must be in good condition. A diversity of habitats allows sustainability of more species.

Global Perspective: For several billion years, life has affected the environment on a global scale. These global effects have in turn affected biological diversity. Life added oxygen to the atmosphere and removed carbon dioxide, thereby making animal life possible.

Urban World: People have rarely thought about cities as having any beneficial effects on biological diversity. However, in recent years, there has been a growing realization that cities can contribute in important ways to the conservation of biological diversity. This topic will be discussed in Chapter 27.

Science and Values: Perhaps no environmental issue causes more debate, is more central to arguments over values, or has greater emotional importance to people than biological diversity. Concern with specific endangered species has been at the heart of many political controversies. The path to resolving these conflicts and debates involves a clear understanding of the values at issue as well as knowledge about species and their habitat requirements and the role of biological diversity in life's history on Earth.

● KEY TERMS

biological diversity **111**

biological evolution **112**

competitive exclusion principle **119**

ecological niche **121**

genetic drift **114**

habitat **121**

migration **112**

mutation **112**

natural selection **112**

obligate symbionts **123**

symbiont **123**

symbiosis **123**

● STUDY QUESTIONS

1. Why do introduced species often become pests?

2. On which of the following planets would you expect a greater diversity of species?

   a. A planet with intense tectonic activities.

   b. A tectonically dead planet.

   (Remember that tectonics refers to the geologic processes that involve the movement of tectonic plates and continents, processes that lead to mountain building and so forth.)

3. You are going to conduct a survey of national parks. What relationship would you expect to find between the number of species of trees and the size of the parks?

4. A city park manager has run out of money to buy new plants. How can the park labor force alone be used to increase the diversity of (a) trees and (b) birds in the parks?

5. A plague of locusts visits a farm field. Soon after, many kinds of birds arrive to feed on the locusts.

What changes occur in animal dominance and diversity? Begin with the time before the locusts arrive and end with the time after the birds have been present for several days.

6. What is the difference between dominance and diversity in (a) a zoo and (b) a natural wildlife preserve?

7. What is the difference between habitat and niche?

8. There are more than 600 species of trees in Costa Rica, most of which are in the tropical rain forests. What might account for the coexistence of so many species with similar resource needs?

9. Which of the following can lead to populations that are less adapted to the environment than were their ancestors?

   a. Natural selection.

   b. Migration.

   c. Mutation.

   d. Genetic drift.

● FURTHER READING

Darwin, C. A. 1859. *The Origin of Species by Means of Natural Selection, or the Preservation of Proved Races in the Struggle for Life.* London: Murray. Reprinted variously. A book that marked a revolution in the study and understanding of biotic existence.

Dawkins, R. 1996. *Climbing Mount Improbable*. New York: Viking. A discussion of some implications of modern discoveries in genetics and evolution.

Margulis, L., and D. Sagan. 1995. *What Is Life?* New York: Simon and Schuster. A beautifully illustrated and well-written introduction to the major forms of life on Earth and the effect of life's diversity on the global environment.

Pimentel, D. 1992. "Conserving Biological Diversity in Agricultural/Forestry Systems," *BioScience* 42:354–362. A journal article in which the author argues that most biological diversity exists in human-managed ecosystems and identifies means to conserve that diversity.

Strickberger, M. W. 1990. *Evolution*. Boston: Jones and Bartlett. A framework to explain current information and concepts in the study of evolution.

Wilson, E. O., ed. 1992. *The Diversity of Life*. New York: W. W. Norton. A book outlining the story of evolution of life on Earth, how species became diverse, and the scope of the current threat to that diversity.

CHAPTER 8 Biogeography

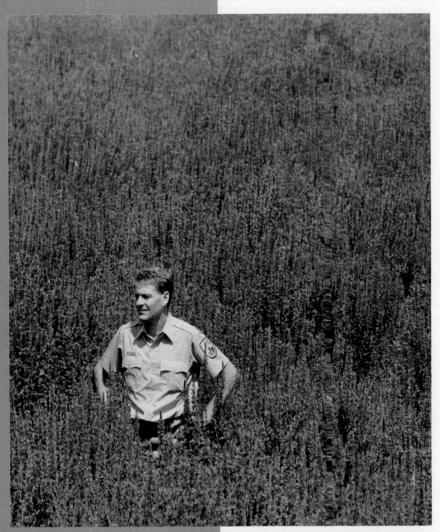

Invasion of a perennial wetland plant such as purple loosestrife has eliminated native plants that provide important wildlife habitat.

LEARNING OBJECTIVES

If we are to conserve biological diversity, we must understand the large-scale, global patterns of biogeography. After reading this chapter, you should understand:

- How large-scale global patterns and the environment affect biological diversity.
- How climate, bedrock, soils, and the geography of life are related to one another.
- What biotic provinces and biomes are and how they differ.
- How plate tectonics affects biogeography.
- What island biogeography is and what it implies for the general geography of life, especially the geography of biological diversity.
- What are the major patterns in the distribution of biomes on Earth and the major characteristics of each of the 17 biomes found on Earth.
- How people affect the geography of life.
- How the introduction of exotic species into new habitats typically affects the new habitats.
- What ecological islands are and how we can help to conserve their biological diversity.

Invasion of the Purple Loosestrife

Over Earth's history, species sometimes become isolated from one another by physical barriers. Introducing long-separated species from one part of the world into another, a common human practice, has many consequences. Without the natural enemies of their native habitats—that is, predators or parasites, including disease-causing organisms—introduced species can spread unchecked and become a threat to biodiversity in their new habitats.

Purple loosestrife, a perennial plant of wetlands, was brought to North America from Europe in the early 19th century. This introduction illustrates some of the negative consequences of introducing alien species (Figure 8.1).[1] It has been responsible for eliminating many of the native plants that provide food and cover for wildlife, including waterfowl, in wetlands of Canada and the United States.[2]

Purple loosestrife was introduced accidentally when ships from Europe released ballast water contaminated with loosestrife seeds in North American waters. It was also introduced deliberately as an herbal remedy for diarrhea, dysentery, bleeding ulcers, and sores. Without natural enemies and being highly tolerant of variations in moisture and temperature, purple loosestrife spread rapidly along the New England coast. Construction of canals and waterways in the 1880s helped it spread into more inland areas of New York and the St. Lawrence River Valley. By the mid-1990s, it occurred throughout southeastern Canada and the eastern and midwestern United States. It has started to appear in the western United States and southwestern Canada.

Purple loosestrife illustrates the great capacity of living things to grow and reproduce. A single plant can produce more than 2.5 million seeds each year.[3] The seeds, which can survive for long periods of time, are spread by water and in mud clinging to wildlife, livestock, and people. Disturbances such as fires or human disturbances of the environment, such as clearing of land, pro-vide exposed sites in which the seeds can germinate. As many as 20,000 seedlings can be found in one square meter (1.2 yd²). Few native species can compete with this productivity.

Once loosestrife matures, it shades other seedlings, and its relatively tall height among wetland herbs and shrubs (approximately 2 m or 6 ft) prevents other plants from growing. It has endangered such species as bulrushes in Massachusetts, dwarf spike rushes in New York, and bog turtles in the northeastern United States. In addition, livestock do not find it as palatable as native grasses and sedges, so it has reduced the carrying capacity of grazing areas and led to agricultural losses.

Roots of the purple loosestrife are exceptionally resistant and resilient when cut, burned, or sprayed with herbicides. The stems regrow readily from the roots. The only effective control is to remove every bit of the

(a)

(b)

(c)

Figure 8.1 ● (a) Purple loosestrife, a European plant introduced into North America, has caused many problems for the biological diversity of native plants. (b) This wetland is dominated by purple loosestrife, which has replaced native American vegetation. (c) An insect parasite of purple loosestrife, from its native habitat, may be useful in controlling this plant in North America.

plant, both above and below ground, which is expensive and time consuming. Currently, scientists from Canada and the United States are guiding research on the use of biological controls. The effort focuses on three species of European insects that are natural parasites of the loosestrife in its original habitat. The most promising of the insect species is being released with the goal of eliminating 75 to 80% of purple loosestrife in Canada and the United States (Figure 8.1*c*).

One lesson of the purple loosestrife case history is that we must be very careful about inadvertent introductions of exotic species. In many cases, such introductions threaten native species and cause other problems. In recent years, many problems have resulted from the introduction of exotic species by people, ranging from the introduction into South America of African killer bees, which have migrated north to the United States, to the introduction of cane toads into the wet tropics of Australia.

● *The story of the purple loosestrife illustrates biogeography and some of the environmental problems that result from our alterations of this geography. There is a geography of life, influenced by plate tectonics, climate, and ocean dynamics. Over the long time that life has existed on the Earth, groups of species have become isolated by geographic barriers. While the ability to invade new habitats is an essential and necessary quality of life—so that species can persist even though their environments may change—the introduction of species into new areas often creates problems for the species that are already there. This chapter explores the basic concepts of biogeography; introduces the major biomes—the kinds of ecosystems—found on the Earth and some of their characteristics; and discusses environmental problems that arise because of disruption of biogeographic patterns.*

8.1 The Geography of Life: Basic Terminology

The kinds of species as well as the number vary greatly from place to place on Earth, with great numbers of plant species in tropical rain forests and relatively few in the boreal forests of the far north. If we are interested in conserving biological diversity, it is important that we understand these large-scale, global patterns, which we refer to as patterns in **biogeography**.

Before beginning a discussion of biogeography, it is useful to define several terms related to the geographic distribution of species. As we have already seen, a species introduced into a new geographic area is called an **exotic species**. A species that is native to a particular area and not native elsewhere is called an **endemic species**. For example, Monterey pine is endemic to a portion of the California coast and exotic in New Zealand.

A species with a broad distribution, occurring all over the world wherever the environment is appropriate, is called a **cosmopolitan species**. The moose is found in both North America and Europe and is therefore a cosmopolitan species of the northern boreal forests. The house mouse is cosmopolitan because it occurs in many places where people provide a habitat for it.

Species that are found almost anywhere are **ubiquitous species**. Humans are ubiquitous, as are some bacteria. For example, *Escherichia coli*, the common bacterium of human intestines, can probably be found almost anywhere on the land and in fresh waters.

8.2 Wallace's Realms: Biotic Provinces

People have long been fascinated by the great geographic variability in species. Aristotle recorded some of the first principles of biogeography; he distinguished boreal, temperate, and tropical life zones. During the 19th century, global patterns became a subject of scientific study, partly because of explorations of the New World tropics,

(a)

(b)

(c)

Figure 8.2 • (*a*) The main biogeographic realms for animals are based on genetic factors. Within each realm, the vertebrates are in general more closely related to each other than to vertebrates filling similar niches in other realms. Herbivores from three realms are shown: (*b*) the eastern grey kangaroo from Australia; (*c*) the giraffe from Kenya, Africa; and (*d*) the agouti from South America.

(d)

South America, the South Pacific, Australia, and parts of Asia.

In 1876, the great British biologist and co-discoverer of the theory of biological evolution, A. R. Wallace, suggested that the world could be divided into six biogeographic regions on the basis of fundamental features of the animals found in those areas.[4] Wallace referred to these regions as **realms** and named them Nearctic (North America), Neotropical (Central and South America), Palaearctic (Europe, northern Asia, and northern Africa), Ethiopian (central and southern Africa), Oriental (the Indian subcontinent and Malaysia), and Australian. These have become known as *Wallace's realms* (Figure 8.2).

The idea of realms, as noted, involves a classification of animals. All living organisms are classified into groups called **taxa**, normally based on evolutionary relationships or similarity of characters. The hierarchy of these groups

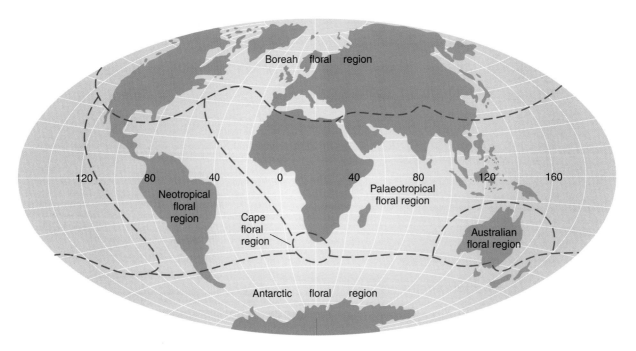

Figure 8.3 • The major vegetation realms are also based on genetic factors. Thus, for example, plants within a realm are more closely related to each other than to flowering plants of other realms.

(from largest and most inclusive to smallest and least inclusive) is domain or kingdom, phylum (in animals) or division (in plants), class, order, family, genus, and species. An order may contain several families, and a family may contain a few to many genera. Genera may contain one to many species.

In each biogeographic realm, certain families or orders of animals are dominant. Animals filling a particular ecological niche in one realm are of different genetic stock from those filling the same niche in the other realms. For example, bison and pronghorn antelope are among the large mammalian herbivores in North America. Rodents such as the capybara fill the same niches in South America, and kangaroos fill them in Australia. In central and southern Africa, many species, including giraffes and antelopes, fill these niches.

The basic concept of Wallace's realms is still considered valid and has been extended to all life-forms,[5] including plants (Figure 8.3)[6,7] and invertebrates. These regions are now referred to as biotic provinces.[8] A **biotic province** is a region inhabited by a characteristic set of taxa (species, families, orders), bound by barriers that prevent the spread of the distinctive kinds of life to other regions and the immigration of foreign species.[9] In a biotic province, organisms share a common genetic heritage but may live in a variety of environments as long as they are genetically isolated from other regions.

Wallace did not have the benefit of our modern understanding of geologic processes to explain how distinct biological groups could have evolved in major habitat regions. The modern explanation for the origin of Wallace's realms is that *continental drift*, caused by plate tectonics, has resulted in the periodic unification and separation of the continents (see the discussion in Chapter 4).[10,11] Unification of continents allowed genetic mixing, whereas separation imposed geographic isolation. Continental unification and land bridges spread genetic stock, which provided organisms with the potential to exploit new habitats. Continental separation led to genetic isolation and the evolution of new species.

8.3 Regional Patterns within Continents

The concept of continental drift leaves us with a picture of huge land masses moving ponderously over Earth's surface, periodically isolating and remixing groups of organisms and leading to an increase in the diversity of species. If each continent were a uniform plot of land in a uniform climate, however, then there would be fewer potential ecological niches (see Chapter 7), and therefore biological diversity within a continent would be low.

In fact, plate tectonics creates continents with complex topography, including mountain building and alterations in drainage patterns and the paths of rivers. These can create barriers to the migration of species, leading to geographic isolation *within* a continent. Thus, when the Rocky Mountains began to form 90 million years ago, they created a barrier to various kinds of nonmountain life forms. As a result of this—as well as climatic changes,

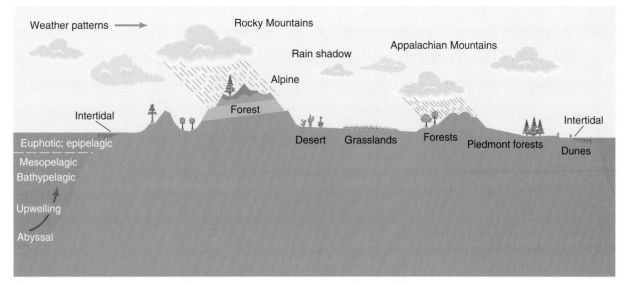

Figure 8.4 • Generalized cross section of North America showing weather, landforms, and the geography of life. The weather patterns move from west to east.

some of which in part result from the mountain building—California has vegetation quite distinct from that which occurs at similar elevations and latitudes east of the Rocky Mountains.

Other major factors that affect the geographic patterns of life within a continent include proximity to oceans or other large bodies of water, ocean currents near shore, position relative to mountain ranges, latitude, and longitude. As an example, Figure 8.4 shows the pattern of life from west to east across North America (see A Closer Look 8.1).

8.4 Evolution and Diversity in Biomes

Another way of looking at the biogeography of Earth is in terms of similarity of environments. Similar environments lead to the evolution of organisms similar in form and function (but not necessarily in genetic heritage or internal makeup) and to similar ecosystems. This is known as *the rule of climatic similarity* and leads to the concept of the biome.

A **biome** is a kind of ecosystem, such as a desert, tropical rain forest, or grassland. It is common and convenient to divide the biosphere into biomes. (Later in this chapter, we discuss the characteristics of the major biomes of the Earth.) Note that parts of a biome can be completely isolated geographically from other parts. Note also the distinction between a biome and a biotic province. A biological province is based on who is related to whom. Species within a biotic province, in general, are more closely related to each other than to species in other provinces. In two different biotic provinces, the same ecological niche will be filled with a species that performs a specific function and may look very similar to each

other, but have quite different genetic ancestry. In this way, a biotic province is an evolutionary unit. How a niche is filled in two different biotic provinces is a consequence of the concept of the biome.

A biome, as a kind of ecosystem, results from the pressures and opportunities presented by similar climates, topography, geological formations and structures, water, and soils in a region. Thus very cold areas with short growing seasons lead to the evolution of tundra plants. These plants occur near the pole and in high altitudes of mountains. The genetic ancestry can be quite different.

The strong relationship between climate and life suggests that if we know the climate of an area, we can predict what biome will be found there, what its approximate biomass (amount of living matter) and production will be, and what the shapes and forms of dominant organisms will be.[12,13] The general relationship between biome type and the two most important climatic factors—rainfall and temperature—is diagrammed in Figure 8.6.

We can look at deserts or grasslands that are widely separated (in different biotic provinces) to examine the idea that similar environmental conditions eventually lead to the evolution of species with similar adaptations and to similar ecosystems. For example, certain plants of the American and African deserts may look similar but are not closely related; they belong to different biological families (Figure 8.7). Geographically isolated for 180 million years, they have been subjected to similar climates, which imposed similar stresses and opened up similar ecological opportunities. In each continent, desert plants evolved to adapt to these stresses and potentials, and have come to look alike and prevail in like habitats, a process known as **convergent evolution**. The ancestral differences between these look-alike plants can be found in their flowers,

A Biogeographical Cross Section of North America

A generalized cross section of the United States shows the relationships among weather patterns, topography, and biota (Figure 8.4). Off the West Coast of the United States in the Pacific Basin occur the pelagic ecosystems: euphotic zones, which have sufficient light for photosynthesis and are occupied by small, mainly single-cell algae, and other zones, with too little light for photosynthesis, occupied by animals that depend for food on dead organisms that sink from above. Near the shore, particularly in areas of upwelling such as along the California coast, there are abundant algae, fish, birds, shellfish, and marine mammals. Where the tides and waves alternately cover and uncover the shore, a long, thin line of intertidal ecosystems is found, dominated by kelp and other large algae that are attached to the ocean bottom; by shellfish, such as mussels, barnacles, and abalone; by crabs and other invertebrates; and by shorebirds, such as sandpipers.

Weather systems move generally from west to east in the Northern Hemisphere. As air masses are forced over the coastal mountains and Rocky Mountains, they are cooled, and the moisture in them condenses to form clouds and rain. The West Coast is an area of moderate temperature, because water has a high capacity to store heat and the Pacific Ocean modulates the land's temperature. Annual amounts of rainfall increase with elevation on the western slopes of the mountains. In the south, moving east from the Southern California coast, rainfall remains low until the mountains force the air high enough to condense much of its moisture. In general, the colder, wetter heights of the mountains support coniferous forests.

In the north, along the coasts of Washington and Oregon, cool temperatures year round lead to heavy rains near the shore, producing an unusual temperate-climate rain forest. The most well-known example occurs in the Olympia National Forest on the northwestern edge of the state of Washington.

The eastern slopes of the coastal ranges form a so-called rain shadow. First, the air that passes over these eastern slopes gives up most of its moisture to the mountains; as a result, it is dry as it passes to the east. In addition, the air sinks to lower elevations, is warmed, and can hold more moisture. This dry air tends to take up moisture from the ground, producing the deserts of Utah, California, Arizona, and New Mexico. Whereas annual rainfall in the Olympic Peninsula of Washington reaches 375 cm/year (150 in/year), east of the Cascade Mountains it falls to 20 cm/year (8 in/year).

The same effect occurs in the Rockies. Less than 160 km (100 mi) west of Denver, in the Rocky Mountains, the annual rainfall is 100 cm (40 in). In the Great Plains, 160 km east of Denver, the annual rainfall is only 30–40 cm (12–16 in). Average annual rainfall increases steadily eastward: 50 cm (20 in) at Dodge City, Kansas; 70 cm (27.5 in) near Lincoln, Nebraska; and 90 cm (35.5 in) near Kansas City, Missouri.[14]

The biomes reflect these changes in rainfall. Just east of Denver are short-grass prairies, which become mixed-grass prairies (a mixture of short-grass and tall-grass prairies) and then tall-grass prairies as we continue eastward. Rainfall reaches levels sufficient to support forests farther east, near the South Dakota–Minnesota border in the north, where the annual rainfall reaches 50–64 cm (20–25 in). From there to the East Coast, the eastern deciduous forest (dominated by trees that lose their leaves during the winter) and the boreal forest of eastern North America predominate.

Sometimes these ecological borders, such as those between the short-grass and tall-grass prairies, are said to be subtle; but they were quite striking and visible to some of the early travelers in the West.

One such traveler, Josiah Gregg, wrote in his journal in 1831 that to the west of Council Grove, Kansas, at the border of the tall-grass and short-grass prairies, the "vegetation of every kind is more stinted—the gay flowers more scarce, and the scant timber of a very inferior quality," whereas to the east he found the prairies to have "a fine and productive appearance truly rich and beautiful."[14]

The patterns described for the United States occur worldwide. Changes with elevation from warm, dry-adapted woodlands to moist, cool-adapted woodlands are found in Spain, where beech and birch, characteristic of middle and northern Europe (Germany, Scandinavia), are found at high elevations and alpine tundra is found at the summits. Similar patterns are found in Venezuela, where a change in elevation from sea level to 5000 m (16,404 ft) at the summits of the Andes is equivalent to a latitudinal change from the Amazon basin to the southern tip of the South American continent. The seasonality of rainfall as well as the total amount often determines which ecosystems occur in an area.

Two other general concepts of biogeography, illustrated by both the latitudinal patterns from the arctic to the tropics and the altitudinal patterns from mountaintops to valleys, are (1) *the number of species declines as the environment becomes more stressful* and (2) *on land, the stature of vegetation decreases as the environment becomes more stressful* (Figure 8.5). These concepts apply to most stresses, including those that human beings impose by adding pollutants to the environment, decreasing the fertility of soils or otherwise impoverishing habitats, and increasing the rate of environmental disturbance. From these concepts, we can predict that highly polluted and disturbed landscapes and seascapes will have few species and that, on the land, the dominant species of plants will have small stature.

Figure 8.5 • Environmental stress and biogeography. Certain general patterns can be found as an environment becomes more stressful. This diagram shows the effect of increasing water stress. Where rainfall is plentiful, vegetation is abundant, and there are forests of tall trees of many species. As rainfall lessens, the size of the plants decreases to small trees, then shrubs and grasses, then scattered plants. The total biomass decreases; and in general, the number of species decreases. Similar changes accompany increases in other kinds of stress, including the stress of certain pollutants. (*Source:* Adapted from R. H. Whittaker, *Communities and Ecosystems,* 2nd ed. [New York: Macmillan, 1975].)

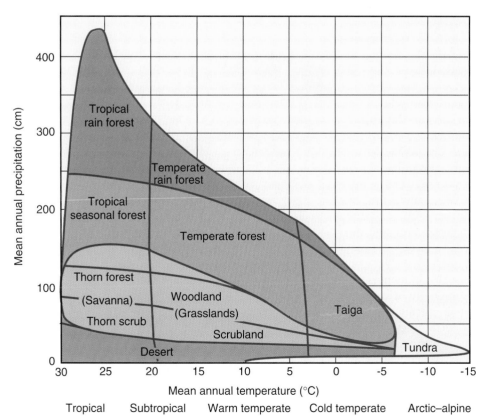

Figure 8.6 • A pattern of types of vegetation in relation to precipitation and temperature. Boundaries between types are approximate. For example, tropical rain forests occur in areas with an approximate annual rainfall of 250–450 cm/year and an approximate temperature range of 18°–28°C. Note that deserts occur over a wide range of annual temperatures, from 30°–5°C, as long as rainfall is less than about 50 cm/year. The warmer the climate, the more rainfall is required to move from desert to another biome. (*Source:* Adapted from R. H. Whittaker, *Communities and Ecosystems,* 2nd ed. [New York: Macmillan, 1975].)

(a)

(b)

(c)

Figure 8.7 • Convergent evolution. Given sufficient time and similar climates in different areas, species similar in shape and form will tend to occur. The Joshua tree (*a*) and saguaro cactus (*b*) of North America look similar to the giant euphorbia (*c*) of East Africa. All three are tall, have green succulent stems that replace the leaves as the major sites of photosynthesis, and have spiny pro- jections; but these plants are not closely related. The Joshua tree is a member of the agave family, the saguaro is a member of the cactus family, and the euphorbia is a member of the spurge family. Their similar shapes result from evolution under similar desert climates, a process known as convergent evolution.

fruits, and seeds, which are evolutionarily the most con- servative (least changing) organs and provide the best clues to the genetic history of a species.

People make use of convergent evolution when they move decorative and useful plants around the world. Cities that lie in similar climates in different parts of the world now share many of the same decorative plants. Bougainvillea, a spectacularly bright flowering shrub originally native to Southeast Asia, decorates cities as dis- tant as Los Angeles and the capital of Zimbabwe. In New York City and its outlying suburbs, one can find the Norway maple from Europe, the tree of heaven and the gingko tree from China, and such native species as sweet gum, sugar maple, and pin oak (Figure 8.8).

Another important concept process that influences the geography of life is divergent evolution. In **divergent evolution**, a population is separated, usually by geo- graphic barriers; the separated subpopulations evolve sep- arately but retain some common characteristics. It is now believed that the ostrich (native to Africa), the rhea

Figure 8.8 • A gingko tree.

(a)

(b)

Figure 8.9 • Divergent evolution. These three large flightless birds evolved from a common ancestor but are now found in widely separated regions: (*a*) the ostrich in Africa, (*b*) the rhea in South America, and (*c*) the emu in Australia.

(native to South America), and the emu (native to Australia) have a common ancestor but evolved separately (Figure 8.9). In open savannas and grasslands, a large bird that can run quickly but feed efficiently on small seeds and insects has certain advantages over other organisms seeking the same food. Thus, these species maintained the same characteristics in widely separated areas.

 8.5 Island Biogeography

Islands have a special fascination for naturalists. Darwin's visit to the Galapagos Islands gave him his most powerful insight into biological evolution.[15] There he found many species of finches that were related to a single species found elsewhere. On the Galapagos, each species was adapted to a different niche.[16] Darwin suggested that finches, isolated from other species on the continents, eventually separated into a number of groups, each adapted to a more specialized role. This process is called **adaptive radiation**. On the Hawaiian Islands, a finchlike ancestor evolved into several species, including fruit and seed eaters, insect eaters, and nectar eaters, each with a beak adapted for its specific food (Figure 8.10).[17]

(c)

We can make several generalizations about species diversity on islands. The primary generalizations are these:

1. The two sources of new species on an island are migration from the mainland and evolution of new species in place (as with Darwin's finches on the Galapagos).
2. Islands have fewer species than continents.

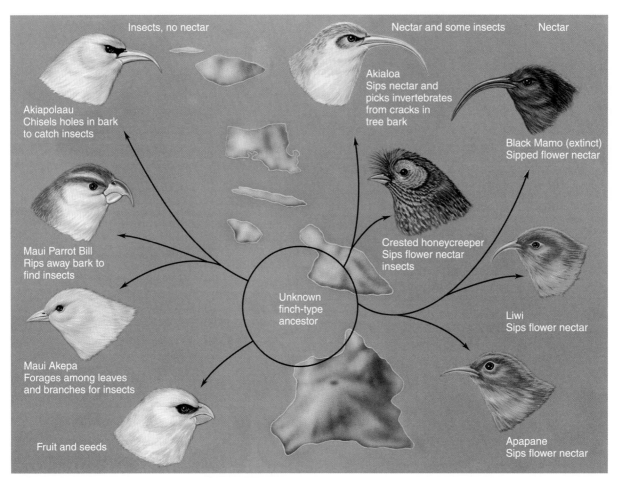

Insets, no nectar   Nectar and some insects   Nectar

Akiapolaau
Chisels holes in bark
to catch insects

Akialoa
Sips nectar and
picks invertebrates
from cracks in
tree bark

Black Mamo (extinct)
Sipped flower nectar

Maui Parrot Bill
Rips away bark to
find insects

Crested honeycreeper
Sips flower nectar
insects

Unknown
finch-type
ancestor

Liwi
Sips flower nectar

Maui Akepa
Forages among leaves
and branches for insects

Fruit and seeds

Apapane
Sips flower nectar

Figure 8.10 • Evolutionary divergence among honeycreepers in Hawaii. Sixteen species of birds, each with a beak specialized for its food, evolved from a single ancestor. Eight of the species are shown here. The species evolved to fit ecological niches that, on the North American continent, had previously been filled by other species not closely related to the ancestor. (*Source:* From C. B. Cox, I. N. Healey, and P. D. Moore, *Biogeography* [New York: Halsted, 1973].)

3. The smaller the island, the fewer the species, as can be seen in the number of reptiles and amphibians in various West Indian islands.

4. The farther the island from a mainland (continent), the fewer the species (Figure 8.11).[18]

Why are these generalizations so? With regard to size, small islands tend to have fewer habitat types. And some habitats on a small island may be too small to support a population large enough to have a good chance of surviving for a long time. A small population might be easily extinguished by a storm, flood, or other catastrophe or disturbance.

Concerning the distance of an island from the mainland, it makes sense that there will be fewer migrants if an island is far from the mainland, because the farther the island from the mainland, the harder it will be for an organism to travel the distance. In addition, the smaller the island, the smaller it is as a "target," so a small island far away from the mainland will be visited relatively rarely by individuals of any species.

Every species is subject to risk of extinction; extinctions can be caused by predation, disease (parasitism), competition, climatic change, or habitat alteration. *Generally, the smaller the population, the greater its risk of extinction. And the smaller the island, the smaller the population of a particular species that can be supported.*

A final generalization about island biogeography is that over a long time, an island tends to maintain a rather constant number of species, which is the result of the rate at which species are added minus the rate at which they become extinct. These numbers follow the curves shown in Figure 8.12. For any island, the number of species of a particular life-form can be predicted from the island's size and distance from the mainland.

The concepts of island biogeography apply to what are called "ecological islands" as well as to physical islands. An **ecological island** is a small habitat of a type that is separated from the major part of that habitat on a continent or in a region. For example, a pond within the woods of Michigan is an ecological island relative to the Great Lakes that border Michigan. A small stand of trees within a prairie

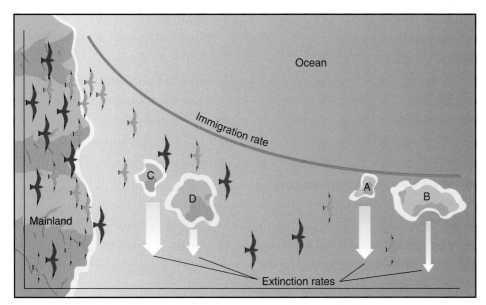

Figure 8.11 • Idealized relation of an island's size, distance from the mainland, and number of species. The nearer an island is to the mainland, the more likely it is to be found by an individual, and thus the higher the rate of immigration. The larger the island, the larger the population it can support, and the higher the chance of persistence of a species; small islands have a higher rate of extinction. The average number of species therefore depends on the rate of immigration and the rate of extinction. Thus, a small island near the mainland may have the same number of species as a large island far from the mainland. The thickness of the arrow represents the magnitude of the rate. (*Source:* Modified from R. H. MacArthur and E. O. Wilson, *The Theory of Island Biogeography* [Princeton, N.J.: Princeton University Press, 1967].)

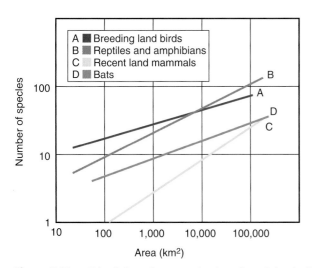

Figure 8.12 • Islands have fewer species than do mainlands. The larger the island, the greater the number of species. This general rule is shown by a graph of the number of species of birds, reptiles and amphibians, land mammals, and bats for islands in the Caribbean. (*Source:* Modified from Wilcox, B. (ed.), IUCN Red List of Threatened Animals, Gland, Switzerland, IUCN, 1988.)

forms a forest island. A city park is another example of an ecological island. When we consider whether an area such as a city park is large enough to support a population of a particular species, we can apply generalizations of island biogeography to forecast whether this is likely to be the case.

 ## 8.6 Biogeography and People

We have seen that biogeography affects biological diversity. Changes in biological diversity in turn affect people and the living resources on which we depend. These effects extend from individuals to civilizations. For example, the last ice ages had dramatic effects on plants and animals and thus on human beings.

Consider a specific example. Europe and Great Britain have fewer native species of trees than other temperate regions of the world. Only 30 tree species are native to Great Britain (that is, present prior to human settlement), although hundreds of species grow there today. Why are there so few native species in Europe and Great Britain?

The answer is the combined effects of climatic change and the geography of European mountain ranges. In Europe, major mountain ranges run east-west, whereas in North America and Asia, the major ranges run north-south. During the past 2 million years, Earth has experienced several episodes of continental glaciation, when glaciers several kilometers thick expanded from the arctic over the landscape. At the same time, glaciers formed in the mountains and expanded downward. Trees in Europe were caught between the ice from the north and the ice from the mountains and had few refuges; many species became extinct. In contrast, in North America and Asia, tree seeds could spread southward as the ice advanced, where they became established and produced new seeds.

Thus, the tree species "migrated" southward and survived each episode of glaciation.[19]

Since the rise of modern civilization, these ancient events have had many practical consequences. Starting in the 16th century, soon after the discovery of North America by Europeans, considerable effort was expended in bringing exotic species of trees and shrubs to Europe and Great Britain. These were used to decorate gardens, homesites, and parks and formed the basis of much of the commercial forestry in the region. For example, in the famous gardens of Alhambra in Granada, Spain, hedges that are grown and cut in elaborate shapes consist of Monterey cypress from North America. Douglas fir and Monterey pine are important commercial timber trees in Great Britain and Europe. As these examples show, knowledge of biogeography—being able to predict what will grow where based on climatic similarity—has been used for major economic benefits.

As mentioned in Chapter 7, people have altered biodiversity themselves. They have done so through three major actions: direct hunting, which causes extinction or serious decline in a species; direct disruption of habitat; and the introduction of exotic species into new habitats. The last is particularly relevant to this chapter.

The human introduction of exotics has had mixed results. On the one hand, the major foods of the world come from only a few species, and these species have been introduced widely by people. Without such introductions, people could not live in great abundances in most locations (see Chapter 12, on agriculture). Also, landscape beauty, including the landscapes in cities, has been improved through the use of exotic species of plants. Many pets that are important to people, including cats and dogs, are routinely introduced into new habitats by people. On the other hand, as this chapter's introductory Case Study and Critical Thinking Issue show, introductions of exotics into new habitats have often had disastrous ecological consequences.

Examination of the history of human introductions of exotics leads to some general rules: Unless there is a very good reason to introduce a species into a new habitat, don't do it. If you are going to do it, do it very carefully. Check first about the natural pests and parasites of the species you plan to introduce. Are some of them essential to keep the species from reaching undesirable abundances? Are some of them likely to cause problems themselves?

As the technology of travel increases, inadvertent introductions of exotics becomes more likely. We must therefore increase our awareness of the problems associated with these introductions, be on our guard to know when introductions happen, and prevent those that are clearly undesirable or whose desirability is unknown.

8.7 Earth's Biomes

Earth has 17 major biomes: tundra, taiga (boreal forests), temperate forests, temperate rain forests, temperate woodlands, temperate shrub lands, temperate grasslands, tropical rain forests, tropical seasonal forests and savannas, deserts, wetlands, fresh waters, intertidal areas, open oceans, benthos, upwellings, and hydrothermal vents. Figure 8.13 shows the major land biomes. Subtypes occur within these major types. Biomes are usually named for the dominant vegetation (for example, coniferous forests, grasslands); for the dominant shape and form, or

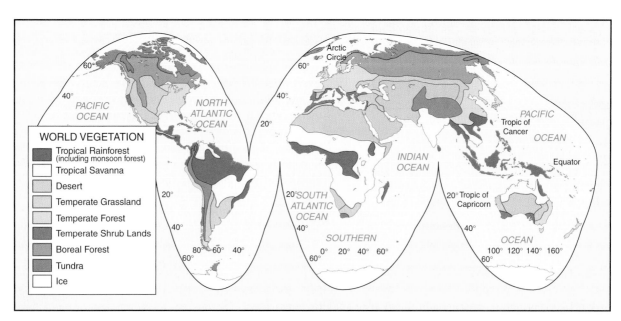

Figure 8.13 • Global distribution of the major land biomes. (*Source:* H. J. deBlij and P. O. Muller, *Physical Geography of the Global Environment* [New York: Wiley, 1996], Fig 27-1, p. 290.)

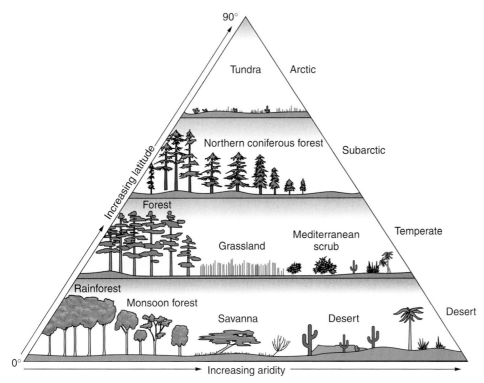

Figure 8.14 • Simplified diagram of the relationship between precipitation and latitude and Earth's major land biomes. Here, latitude serves as an index of average temperature, so latitude can be replaced by average temperature in this diagram. (Source: H. J. deBlij and P. O. Muller, *Physical Geography of the Global Environment* [New York: Wiley, 1996], Figure 27-4, p. 293.)

physiognomy, of the dominant organisms (forest, shrub land); or for the dominant climatic conditions (cold desert, warm desert).

Knowledge of basic characteristics of the world's major biomes and of life forms within each is important for planning, dealing with environmental issues, and determining beneficial introductions of new species. For example, we can use the correlation between geography of climate and geography of life to forecast possible effects of global warming on the geography of life.

As we have seen, similar environments lead to evolution of organisms similar in form and function but not necessarily in genetic heritage. (See the distinctions made earlier in this chapter between biotic provinces and *biomes*.) For example, deserts occur on all continents except Antarctica. Major desert plants of each continent often look similar but have different genetic heritages. The close relationship between environment and kinds of life-forms is illustrated in Figure 8.14.

The distribution of biomes (Figure 8.13) gives Earth a distinctive pattern as seen from space (Figure 8.15). Note the correspondence between this view from space and the average summer temperature as measured from space (Figure 8.16). For example, July average temperatures above 30°C (Figure 8.16) correspond with the occurrence of deserts (Figures 8.13 and 8.14) in the Americas

and in Africa. And the location of boreal forests corresponds with July average temperatures below 20°C.

Considering precipitation along with temperature leads to more refined correlations between climate and vegetation. Warm, wet climates favor high vegetation production. Biomes of these climatic zones tend to be highly productive as long as other factors, such as availability of chemical elements that plants require, are not limiting. Cold or dry climates do not permit high rates of vegetation growth. Therefore, different biomes are characterized by different average rates of vegetation production (Figure 8.17). Knowledge of climate can be used to predict what kinds of vegetation are likely to be found in a region. Vegetation is the form of life most visible from space, but other forms of life have similar geographic relationships.

Another major and intriguing geographic pattern is the change in biological diversity with latitude. In general, diversity decreases with latitude, so that the highest diversity typically occurs in the tropics. In North America, the diversity of plants decreases greatly with latitude. Mexico has 23,000 plant species. Brazil, Columbia, Peru, Madagascar, Mexico, India, China, Indonesia, and Australia contain about 60% of the world's named plant species.[20]

In the remainder of this chapter, we take a closer look at the 17 major biomes of Earth.

Figure 8.15 • NASA Landsat imagery of the vegetation of Earth from space. (From "The Endangered Earth," a map supplement to the 100th anniversary issue of *National Geographic* magazine, 1988. The same map can be obtained from F. G. Hall, Goddard Space Flight Center, Code 623, Greenbelt, MD.)

Tundra

Tundras are treeless plains that occur in the harsh climates of low rainfall and low average temperatures (Figure 8.18). The dominant vegetation includes grasses and their relatives (sedges), mosses, lichens, flowering dwarf shrubs, and mat-forming plants. As the environment becomes more harsh, vegetation grades from dwarf shrubs and grasslike plants to mosses and lichens and finally to bare rock surfaces with occasional lichens. The extreme tundra occurs in Antarctica, where the major

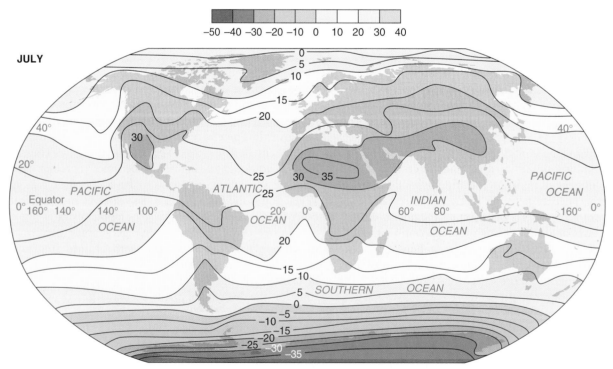

Figure 8.16 • Mean sea-level air temperatures (°C) for July. (*Source:* H. J. deBlij and P. O. Muller, *Physical Geography of the Global Environment* [New York: Wiley, 1996], Figure 8-11, p. 93.)

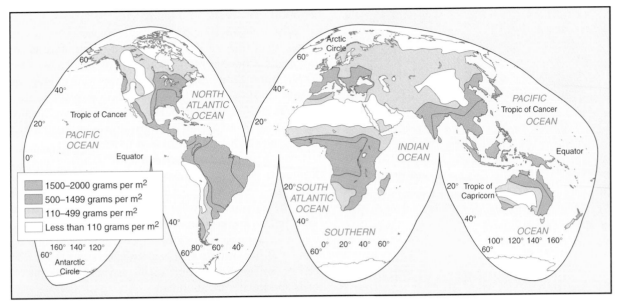

Figure 8.17 • Geography of vegetation production. The different biomes are characterized by different average rates of production of vegetation. Tropical forests, temperate rain forests, and temperate deciduous forests are highly productive. Here again we can see the relationship between climate and biomes. (*Source:* H. J. deBlij and P. O. Muller, *Physical Geography of the Global Environment* [New York: Wiley, 1996], Figure 26-3, p. 281.)

Legend:
- 1500–2000 grams per m^2
- 500–1499 grams per m^2
- 110–499 grams per m^2
- Less than 110 grams per m^2

Figure 8.18 • Tundra biome. Shown here is arctic tundra in Denali National Park, Alaska, with mountain avens in bloom.

land organism in some areas is a lichen that grows within rocks, just below the surface.

There are two kinds of tundra: arctic, which occurs at high latitudes, and alpine, which occurs at high elevations. The vegetation of both is similar, but the dominant kinds of animals are different. Arctic tundras typically have some large mammals, such as caribou of North America, as well as important small mammals, birds, and insects. In alpine tundras, dominant animals are small rodents and insects. This difference exists in part because alpine tundras occupy comparatively small, isolated areas, whereas arctic tundras cover the large territories required for populations of large mammals.

Parts of tundra have *permafrost*—permanently frozen ground—which is extremely fragile. When disturbed by such activities as road development, permafrost areas may be permanently changed or may take a very long time to recover.

Taiga, or Boreal Forests

The **taiga**, or boreal forest, biome includes the forests of the cold climates of high latitudes and high altitudes. Taiga forests are dominated by conifers, especially spruces, firs, and larches, and some pines. Aspens and birches are important flowering trees (Figure 8.19). Typically, boreal forests form dense stands of relatively small trees, typically usually less than 30 m in height. Although boreal forests cover very large areas, relatively few important species of trees occur—only about 20 major species in North American boreal forests, for example. Boreal forests are among the most economically

Figure 8.19 • The boreal forest and a moose, a characteristic mammal of this biome.

important biomes, as they are the source of much lumber and paper pulp.

Because the northern areas of North America and Eurasia have been connected by land bridges during past ice ages, the animals and vegetation of the boreal forest biome have been able to spread widely. Moose, for example, are found on both continents, as are small flowering plants, called *Saxifraga Flagellaris*. Thus, the boreal forests of North America and Eurasia share both genetic heritage (biotic provinces) and similar climates; the climate similarity has led to dominant life-forms similar in shape and form (biome characteristics). In this case, the biotic province and biome overlap. The dominant animals of boreal forests include a few large mammals (moose, deer, wolves, and bears), small rodents (squirrels and rabbits), small carnivores (foxes), many insects, and migratory birds, especially waterfowl and carnivorous land birds, such as owls and eagles.

Disturbances—particularly fires, storms, and outbreaks of insects—are common in the boreal forests. For example, the entire million acres of the Boundary Waters Canoe Area of Minnesota burns over (through numerous small fires) an average of once each century, and individual forest stands are rarely more than 90 years old.

Boreal forests contain some of Earth's largest remaining wilderness areas and are appreciated for conservation of wilderness and for various kinds of recreation. The fur-bearing animals of boreal and tundra biomes have long been of commercial value, although that value has declined greatly in western nations because of concerns with the animals' welfare.

Temperate Forests

Temperate forests occur in climates somewhat warmer than those of the boreal forest. These forests occur throughout North America, Eurasia, and Japan and have many genera in common. Dominant vegetation includes tall deciduous trees; common species are maples, beeches, oaks, hickories, and chestnuts, typically larger in stature than trees of the boreal forest. These forests are important economically for their hardwood trees, used for furniture. Temperate deciduous forests are among the biomes most changed by human beings because they occur in regions long dominated by civilization, including much of China, Japan, Western European, the United States, and the urbanized parts of Canada.

Large mammals are less common in the temperate deciduous forest than in the taiga; herbivores include deer. The low density of large mammals results in part from the deep shade of the forest interior; in such areas, there is less for ground-swelling mammals to eat. The dominant animals tend to be small mammals, including those that live in trees (such as squirrels) and those that feed on soil organisms and small plants (such as mice). Birds and insects are abundant.

There are few remaining uncut or unplowed stands of temperate forest, but some of those that remain are important nature preserves. As with the boreal forest biome, fire is commonly a natural, recurring feature, although the frequency of fire in many temperate forests appears to be lower than in the boreal forests. In this biome, hunting and gathering cultures appear to have contributed to the fire frequency in many regions.

Temperate Rain Forests

Temperate rain forests occur under moderate temperature regimes where the rainfall exceeds 250 cm/year. Such rain forests are rare but spectacular. These are the giant forests (Figure 8.20).

In the Northern Hemisphere, the dominant trees are evergreen conifers. Temperate rain forests of the Northern Hemisphere include the redwood forests of California and Oregon, where the tallest trees in the world exist, as well as forests of the state of Washington and adjacent Canada, dominated by such large trees as Douglas firs and western cedars. Trees are taller than 70 m and are long lived; Douglas firs live more than 400 years, redwoods several thousand.

Temperate rain forests also occur in the Southern Hemisphere. The best known of these are the forests of western New Zealand. The trees here are typically smaller than those in North America.

Temperate rain forests have comparatively low diversity of plants and animals, in part because the climatic conditions tend to favor specialized species and in part because the abundant growth of the dominant vegetation produces a very deep shade in which few other plants can grow to provide food for herbivores. The temperate rain forests of New Zealand contain a higher diversity of vegetation species than those of North America.

This biome is important economically; redwoods, Douglas firs, and western cedars are major North

Figure 8.20 • Temperate rain forest biome. Moss covers Sitka spruce in Olympic National Park, a famous example of the temperate rain forest of the Pacific Northwest of North America.

American timber crops. It is also important culturally. It is the home of Native American tribes who want to retain traditional practices in their use of forest resources. In addition, it is a focus of concern for biological conservation of the magnificent ancient forests, the many streams, and the associated species, such as salmon, the spotted owl, and the marbled murrelet, an oceanic bird that nests in these forests (Figure 8.21).

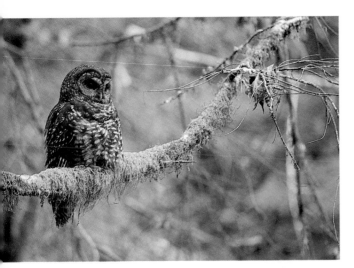

Figure 8.21 • The spotted owl, one of the animals of the temperate rain forest biome and a focus of recent debates about conservation of endangered species.

Temperate Woodlands

Where temperate woodlands occur, the temperature patterns are like those of deciduous forests but the climate is slightly drier. In North America, such areas occur from New England south to Georgia and the Caribbean islands. Temperate woodlands are dominated by small trees such as piñon pines and evergreen oaks. The stands tend to be open, with wide spaces between trees, allowing considerable light to reach the ground. These generally pleasant areas are often used for recreation. Fires are a common disturbance to which many species are adapted. Because the pines typical of these areas are often fast growing and produce wood good for timber, pulp, and paper, temperate woodlands are often economically valuable. Forest plantations occur in this biome. The pleasant combination of trees and grasses, along with the abundance of some large mammals such as mule deer in Ponderosa pine, have also made these a focus for biological conservation, once again bringing science and values to the fore.

Temperate Shrub Lands

Under still drier climates, temperate shrub lands called **chaparral** occur. These miniature woodlands are dominated by dense stands of shrubs rarely exceeding a few meters in height. Chaparrals occur in Mediterranean climates—climates with low rainfall, which is concentrated in the cool season. Chaparrals are found along the coast of California, in Chile, in South Africa, and in the Mediterranean region. While only about 5% of the Earth's land area is in this biome, it is among the most attractive to people because of its moderate, sunny climate. Therefore, it has been highly modified around the world by human action, and relatively few examples of native chaparral remain.

Typically, chaparral vegetation is distinctively aromatic; an example is sage. Some scientists believe the aromatic compounds are a means of competition among the plants. There are few large mammals; reptiles and small mammals are characteristic. The animals and plants of this biome have little economic value at present, but the biome is important for watersheds and erosion control.

The vegetation is adapted to fires; many species regenerate rapidly, and some promote fires by producing abundant fuel in the form of dead twigs and branches. As a result, stands are rarely more than 50 years old. When intense precipitation follows fire, erosion rates can increase to some of the highest known until renewed vegetation again protects the slopes. Managing fires in chaparral is especially important because housing is often near the fire-prone vegetation.

As suggested earlier, temperate shrublands tend to be among those most favored for human settlement, as illustrated by the location of ancient Greek and Roman civilizations in Mediterranean regions. Since fire is a persistent and natural feature of the vegetation, this creates risk to those who live within this biome. Vegetation from the world's temperate woodlands is widely circulated and used to decorate public and private gardens, streets, and other public areas in cities and towns in this biome throughout the world.

Temperate Grasslands

Temperate grasslands occur in regions too dry for forests and too moist for deserts. A major biome in terms of area covered, these temperate grasslands include the great North American prairies (Figure 8.22), which originally covered more land area than any other biome in the United States; the steppes of Eurasia; the plains of eastern and southern Africa; and the pampas of South America. Dominant species are grasses and other flowering plants, many of which are perennials with extensively developed roots.

Soils often have a deep organic layer, formed by the decaying deep roots of the grasses and the decaying stems and leaves of the prairie plants. The result is some of the best soils in the world for agriculture. By volume, most of the food of the world comes from this biome—all of the small grains, along with most of the large hoofed herbivores that provide meat for people, including cattle and American bison.

In North America, unplowed prairie is rare. Considerable effort has been made in recent years to restore prairies, a labor-intensive task. Prior to European settlement, the North America prairie extended from central

Figure 8.22 • Temperate grassland biome. The great American prairie was one of the world's largest areas of temperate grasslands before European settlement of North America.

Minnesota west to the Rocky Mountains and from Texas and Oklahoma northward into central Saskatchewan, Canada. Other areas of grasslands occurred in the high plains of northern California and in eastern Oregon and Washington.

Grasslands are home to the highest abundance and the greatest diversity of large mammals: the wild horses, asses, and antelopes of Eurasia; the once-huge herds of bison that roamed the prairies of the American West, along with pronghorn; the kangaroos of Australia; and the antelope and other large herbivores of Africa. Fossil evidence suggests that grasslands and grazing mammals evolved together, beginning about 60 million years ago. Grasslands plants are adapted to certain kinds of grazing, and animals are important in the transport of seeds. The animals, of course, require edible grasses and forbs (flowering plants that are not grasses, trees, or shrubs). Thus, there is a kind of symbiosis among the animals and plants of this biome which is essential for its continuation.

Temperatures are moderate, similar to those of the temperate deciduous forests, and rainfall is usually seasonal. Fire is a natural, recurring feature, and in most locations, both fire and grazing are necessary for the persistence of grasses and forbs. If fire and grazing are eliminated, the land tends to become desert shrubland in the drier regions and open woodlands in the wetter areas. In many grasslands, hunting and gathering cultures appear to have contributed to the fire frequency and in this way increased the area and persistence of this biome.

Tropical Rain Forests

Tropical rain forests occur where the average temperature is high and relatively constant throughout the year and where rainfall is high and relatively frequent throughout the year. In the wettest climates and in climates with rainfall well distributed throughout the year are found the tropical rain forests of South and Central America, northeastern Australia, Indonesia, the Philippines, Borneo, Hawaii, and parts of Malaysia. Tropical rain forests have long been home for hunting and gathering cultures, but relatively few civilizations have been able to persist in this biome.

Tropical rain forests are famous for their diversity of vegetation. Hundreds of species of trees may be found within a few square kilometers. Typically, some trees are very tall, and some, such as palm trees, remain relatively small. Some plants, like bromeliads and some ferns, grow on trees (Figure 8.23). Approximately two-thirds of the 300,000 known species of flowering plants occur in the tropics, mainly in tropical rain forests.[20] Many species of animals occur as well. Mammals tend to live in trees, but some are ground dwellers. Insects and other invertebrates are abundant and show a high diversity. Rain forests occur in some of the most remote regions of Earth and remain poorly known; many undiscovered species are believed to exist there.

Figure 8.23 • Tropical rain forest biome. The vegetation along the Segama River in Borneo illustrates lowland tropical rain forests.

Except for dead organic matter at the surface, soils tend to be very low in nutrients. Most chemical elements (nutrients) are held in the living vegetation, which has evolved to survive in this environment; otherwise, rainfall would rapidly remove many chemical elements necessary for life.

Tropical Seasonal Forests and Savannas

Tropical seasonal forests occur at low latitudes where the average temperature is high and relatively constant throughout the year and rainfall is high but very seasonal. Such forests are found in India and Southeast Asia, Africa, and South and Central America. In areas of even lower rainfall, tropical savannas—grasslands with scattered trees—are found. These include the savannas of Africa, which, along with the grasslands, have the greatest abundance of large mammals remaining anywhere in the world. The number of plant species is high as well.

Disturbances, including fires and the impact of herbivory on the vegetation, are common but may be necessary to maintain these areas as savannas; otherwise, they would revert to woodlands in wetter areas or to shrub lands in drier areas. Under still drier climates, savannas are replaced by shrub lands, characterized by small shrubs, a generally low abundance of vegetation, and a low density of vertebrate animals.

Deserts

Deserts occur in the driest regions where vegetation can survive, typically where the rainfall is less than 50 cm/year. Most deserts—such as the Sahara of North Africa and the deserts of the southwestern United States (Figure 8.24), Mexico, and Australia—occur at low lati-

Figure 8.24 • Desert biome. These desert lands are at White Sands National Monument, New Mexico. Deserts vary greatly in the amount of vegetation they contain; this desert is relatively sparse, and some have no vegetation.

tudes. However, cold deserts occur in the basin and range areas of Utah and Nevada and in parts of western Asia.

Most deserts have a considerable amount of specialized vegetation, as well as specialized vertebrate and invertebrate animals. Soils often have abundant nutrients but little or no organic matter and need only water to become very productive. Disturbances are common in the form of occasional fires, occasional cold weather, and sudden, infrequent, and intense rains that cause flooding.

Relatively few large mammals live in deserts. The dominant animals of warm deserts are nonmammalian vertebrates (snakes and reptiles). Mammals are usually small, like the kangaroo mice of North American deserts.

Wetlands

Wetlands include freshwater swamps, marshes, and bogs and saltwater marshes. All have standing water: The water table is at the surface, and the ground is saturated with water (Figure 8.25). Standing water creates a special soil environment with little oxygen, so decay takes place slowly, and only plants with specialized roots can survive. Bogs—wetlands with a stream input but no surface water outlet—are characterized by floating mats of vegetation. Swamps and marshes are wetlands with surface inlets and outlets.

Dominant plants are small, ranging from small trees—such as the mangroves of warm coastal wetlands and the black spruces and larches of the north—to shrubs, sedges, and mosses. Small changes in elevation make a great difference. On slight rises, roots can obtain oxygen, and small trees can grow; in lower areas are patches of open water with algae and mosses.

Although wetlands occupy a relatively small portion of Earth's land area, they are important in the biosphere. In the oxygenless soils, bacteria survive that cannot live in high-oxygen atmospheres. These bacteria carry out chemical reactions, such as the production of methane and hydrogen sulfide, that have important effects in the biosphere. In addition, over geologic time, wetland environments produced the vegetation that today is coal.

Saltwater marshes are important breeding areas for many oceanic animals and contain many invertebrates. Dominant animals include crabs and shellfish, such as clams. Saltwater marshes are therefore an important economic resource. Dominant animals in freshwater wetlands include many species of insects, birds, and amphibians; few mammals are exclusive inhabitants of this biome. The larger swamps of warm regions are famous for large reptiles and snakes, as well as for a relatively high diversity of mammals where topographic variation allows small upland areas.

Although, because of the high water table, people tend not to live within wetlands, this biome can produce many edible plants; plants useful for materials of human society, such as baskets and other containers; and animals, including fish, that provide food. They are often used for recreation and have been a favorite biome for many nat-

Figure 8.25 • Wetland biome. Wetlands include areas of open standing water and areas of herbs, grasses, shrubs, and trees that can withstand persistent or frequent flooding.

uralists and conservationists. Henry David Thoreau liked wetlands the best of all kinds of biomes.[21]

Freshwaters

Freshwater lakes, ponds, rivers, and streams make up a very small portion of Earth's surface but are critical for our water supply for homes, industry, recreation, and agriculture and play important ecological roles (Figure 8.26).

Figure 8.26 • Freshwater biome. The Great Lakes of North America are among the world's most famous examples of the freshwater biome. In this picture, at Miners Castle Picture Rocks National Lakeshore, the sharp transition between the waters of the lake and the adjacent forests is clearly illustrated.

Rivers and streams are also important in the biosphere as major transporters of materials from land to ocean.

Dominant freshwater plants are floating algae, referred to as *phytoplankton*. Along shores and in shallow areas are rooted flowering plants, such as water lilies. Animal life is often abundant. Open waters have many small invertebrate animals (collectively called *zooplankton*), both herbivores and carnivores. Many species of finfish and shellfish are found. Estuaries—areas at mouths of rivers where river water mixes with ocean waters—are rich in nutrients and usually support an abundance of fish. They are important as breeding sites for many commercially significant fish. Furthermore, many species that spend a greater portion of their life cycle in other biomes depend on freshwaters for reproduction or for food, as well as for drinking water. Freshwaters, then, are among the most important biomes for life's diversity.

Freshwaters are also among the areas most altered by human activities, especially by modern technological civilization. Throughout much of the history of civilization, the importance of freshwaters was recognized, but at the same time rivers and streams were heavily used as a way to get rid of wastes. Water power was one of the first sources of nonbiological energy, and rivers have long provided major transportation routes for people. Impoundment of streams and rivers (for example, through the building of dams and reservoirs), along with channelization to make transporation on them easier, has led to major changes in many freshwaters.

In recent years, the importance of freshwaters and their surrounding riparian and wetland areas has been recognized, and a major change has occurred in Western civilization's attitude toward rivers and streams. A human generation ago, water power was considered one of the "cleanest" and most environmentally friendly forms of energy production. Today, there is growing interest in the protection of free-flowing rivers and streams, whose complex channels, backwaters, meanders, floodplains, and seasonal ponds are important to fish, waterbirds, and other wildlife. Much conservation effort is now being expended to restore freshwaters.

Meanwhile, however, in developing nations, water power remains one of the least expensive sources of energy, and the development of major dams continues. We can expect freshwaters to be a major focus for environmental conflicts in the next decades.

Intertidal

The intertidal biome is made up of areas exposed alternately to air during low tide and ocean waters during high tide (Figure 8.27). Constant movement of waters transports nutrients into and out of these areas, which are usually rich in life and important to people as a direct source of food and a spawning and breeding ground for many important foods. As a result, they represent major economic resources. Large algae are found here, from

Figure 8.27 • Intertidal biome. A rocky intertidal biome on the Atlantic coast of North America is shown.

giant kelp of temperate and cold waters to algae of coral reefs in the tropics. Attached shellfish and birds are usually abundant and form a major part of economic resources. Near-shore areas are often important breeding grounds for many species of fish and shellfish, often of economic significance.

The near-shore part of the oceanic environment is most susceptible to pollution from land sources. It is often heavily polluted by human activities, because major cities and civilizations tend to develop at the mouths of major rivers and along productive intertidal coastlines. In addition, as a major recreational area, it is subject to considerable alteration by people. Some of the oldest environmental laws concern the rights to use resources of this biome, and today major legal conflicts continue about access to intertidal areas and harvesting of biological resources.

Disturbances are common in the intertidal biome. They include daily changes in sea level with the tides, seasonal changes in tidal minima and maxima, and ocean storms. Adaptation to these disturbances is essential to survival within the intertidal. Some of the most extreme variations in environmental conditions occur in this biome. Consider, for example, a barnacle or mussel that twice daily experiences a change from a cool or cold saline water environment to direct exposure to bright sunlight and the highly oxygenated atmosphere.

Open Ocean

Called the *pelagic* region, the open ocean biome includes open waters of much of the oceans. These vast areas tend to be low in nitrogen and phosphorus—chemical deserts with low productivity and low diversity of algae. Many species of large animals occur but at low density.

Benthos

The bottom portion of oceans is called the benthos (deeps). Primary input of food is dead organic matter that falls from above. The waters are too dark for photosynthesis, so no plants grow there.

Upwellings

Deep ocean waters are cold and dark, and life is scarce. However, these waters are nutrient rich because of numerous creatures who die in surface waters and sink. (See Chapter 9, for a discussion of energy flow in ocean ecosystems.) Upward flows, or upwellings, of deep ocean waters bring nutrients to the surface, allowing abundant growth of algae and animals that depend on algae. Upwellings occur off the west coast of North America, South America, West Africa, and near the Arctic and Antarctic ice sheets. In some areas, deeper waters are brought to the surface by winds that push coastal waters away from shore. These fertile upwelling zones are among the most important regions for the production of commercial fish.

Hydrothermal Vents

Hydrothermal vents, a recently discovered biome, occur in deep ocean, where plate tectonic processes create vents of hot water with a high concentration of sulfur compounds. These sulfur compounds provide an energy basis for chemosynthetic bacteria, which support giant clams, worms, and other unusual life-forms. (See Chapter 8.) Water pressure is high, and temperatures range from boiling point in waters of vents to the frigid (about 4°C) waters of deep ocean.

Summary

- If we are interested in conserving biological diversity, we must understand the large-scale global patterns of life known as biogeography.

- Geographic isolation leads to the evolution of new species. Wallace's realms, or biotic provinces, are major geographic divisions (generally continents) based on fundamental features of the animals found in them. Animals filling specific niches within one realm are of different stock from those filling the same niches in other realms.

- The rule of climatic similarity holds that similar environments lead to the evolution of biota and biological communities similar in external form and function but not in genetic heritage or internal makeup. Areas of climatic similarity with similar biota are known as biomes. A biome is a kind of ecosystem; examples are desert, grasslands, and rain forest.

- Convergent evolution occurs when two genetically dissimilar species that occur in geographically separate parts of a biome develop along similar lines and have similar external form and function. Divergent evolution occurs when several species evolve from a common ancestral species but develop separately because of geographic isolation.

- The study of island life has led to a theory of island biogeography, which includes several important concepts: Islands have fewer species than mainlands because of their smaller size and distance from the mainland. The smaller an island and the farther it is from the mainland, the fewer the species the island will contain.

- Ecological islands—habitats separated from the main part of a biome—show the same diversity characteristics as physical islands. The smaller the ecological island and the greater its distance from its "mainland," the fewer the species it can support.

Escape of an Exotic Species

About 20 years ago, a delicate seaweed named *Caulerpa taxifolia* (Figure 8.28) was brought from its native habitat in the Pacific Ocean to a zoo in Germany, where it was cultivated and used to embellish saltwater aquarium exhibits, a seemingly harmless action. The seaweed was such a success that samples were sent to other institutions, including the Oceanographic Museum in Monaco.

Within about five years of its introduction there, an unfortunate accident took place. The seaweed was inadvertently flushed into the Mediterranean when exhibit tanks were cleaned. This accident might seem to be innocuous, but considering it so would ignore the tremendous power of species to act as biological invaders.

Once freed in the Mediterranean, *Caulerpa* quickly changed its growth pattern and adapted to its new habitat along the southern coast of France. This may have occurred through a mutation, through hybridization with native seaweeds, or because its genetic code contained information that allowed for considerable plasticity. Whatever the exact genetic explanation, today *Caulerpa* grows about six times larger in the Mediterranean than it does in its native Pacific Ocean. It is also tolerant of colder temperatures, surviving in waters as low as 10°C, compared with a limit of 21°C in its home waters.

Within the past two or three years, *Caulerpa* has spread to the Adriatic and appears to threaten the entire Mediterranean with its ability to choke out competing algae. It grows on rocks, sand, and mud, unlike most seaweeds, which do best on only one kind of substrate. It grows so profusely that it blankets competing native seaweeds, excluding them; and it appears to be toxic to local animals that feed on seaweeds, such as sea urchins, which will not touch it.

Controlling this invader has proved very difficult. When mechanically removed, it soon recovers to form even denser stands. Scientists are seeking a biological control method and are currently studying an exotic snail for release.

Thus, a seemingly harmless algae that gracefully decorated indoor aquaria has become an invading monster, affecting algae and animal life in the Mediterranean Sea, with consequences for commerce, recreation, and aesthetics. The problem caused by

Figure 8.28 • *Caulerpa taxifolia.* This delicate seaweed brought from the Pacific Ocean has become a pest in the Mediterranean.

Caulerpa taxifolia is the result of the geography of life and the alteration of this geography by human action. The question that faces us today is: How can we control *Caulerpa taxifolia*?

Critical Thinking Questions

1. Suppose you were in charge of a program to control or eliminate this pest from the Mediterranean. Would you try to eliminate *Caulerpa* completely or reduce its abundance?

2. Develop a plan for the control of this algae, based on material in this and previous chapters. What kinds of biogeographical exploration might help you develop a plan to control this species?

3. If nothing is done, what would be a likely long-term outcome?

- Earth has 17 major biomes, each with its own characteristic dominant shapes and forms of life. Biomes vary in their importance to people, with some of great importance. Most biomes have been heavily altered by human actions. Understanding the major characteristics of these biomes is important to the conservation and sustainable use of their resources.

- People have long introduced exotic species into new

habitats, sometimes creating benefits, often causing new problems. From the study of biogeography, certain general rules can be set down concerning the introduction of exotic species by human actions. The primary rule is this: Unless there is a very clear and good reason to introduce an exotic species into a new habitat, don't do it; and take precautions to prevent such introductions from occurring inadvertently.

● REEXAMINING THEMES AND ISSUES

Human Population:

The spread of human beings around the Earth has been a major factor in the past several thousand years in altering the biogeography of the Earth. As long as people have traveled and migrated to new areas, they have brought with them animals and plants, and their fungal and bacterial companions. The introduction of these species into new areas has had major effects on native species, including the extinction of species. Human alteration of the landscape, increasingly an effect because of our advancing technology and growing population, also alters biogeography. Trade in products of species, such as rhinoceros horns and elephant ivory, has led to threatening species with extinction. Today, our species is one of the greatest causes of alterations of biogeography.

Sustainability:

How we alter the biogeography of species affects sustainability of human societies, of commercial production of living resources, and of the conservation of biological diversity. When we move species from one area to another, we can, and have, had major effects on the sustainability of native species. The guide to the future, in relation to biogeography, is caution. Do not introduce new species unless they seem essential, and do so only carefully, so that the introduction is tested to make sure that the species does not become a pest.

Global Perspective:

Biogeography is one of the fundamentally global qualities of life on the Earth. The abilities to invade and persist in new environments are essential properties of life. In this way, the ability to invade is a positive biological force. But the power of biological growth and invasiveness can also cause many problems. Because human beings are now spread around the Earth, we are a global factor with great influences on biogeography.

Urban World:

In our urbanized world we transport species, especially flowering plants, to decorate our cities and make them more livable. When done carefully, so that the introduced species do not become invasive pests, this can greatly improve urban life. Linnaeus understood this and led the early importation of plants from the New World to decorate urban gardens in the cities of Europe. Frederick Law Olmsted, the father of landscaping planning in America, thought of the vegetation of the world as a "palette" with which we could decorate our cities. Cities can also be a positive factor in the conservation of biological diversity. Some species, such as the peregrine falcon, have been introduced into cities where they survive, and this has helped in the recovery of such endangered species. Urban gardens, even in the small backyards of residents, can host plants that are rare and endangered and which in turn can provide food for threatened and endangered birds.

Science and Values:

The effects human beings have on biogeography are revealed by scientific research, as are the ways in which these effects are caused. But whether we choose to introduce a species into a new habitat or to allow the extinction of a native species where we live, depends on our values. We need science to tell us what can happen and warn of future potential effects on the distribution of species; with this knowledge, we can make decisions based on our values.

● KEY TERMS

adaptive radiation **139**
biogeography **132**

biome **135**
biotic province **134**

chaparral **147**
convergent evolution **135**

cosmopolitan species **132**
divergent evolution **138**

● STUDY QUESTIONS

1. What is a geologic barrier, and why is this concept important in the geography of living things?

2. What are the major factors that determine which species live in a particular location on a continent?

3. What are the consequences of geographic isolation?

4. In Jules Verne's classic novel *The Mysterious Island*, a group of Americans finds themselves on an isolated volcanic island inhabited by kangaroos and large rodents closely related to the agoutis of South America. Why is this situation unrealistic? What would make this co-occurrence possible?

5. What do we learn from biogeography that helps us understand why there are so many species on Earth? Make a list of the major factors that lead to high biodiversity.

6. What are three ways in which people have altered the distribution of living things?

7. From the perspective of biogeography, why do people attach so much importance to the conservation of tropical rain forests?

8. How can the concepts of island biogeography help us understand the conservation of parks and small nature preserves?

9. What ideas from the theory of island biogeography might explain why there are large mammals in arctic tundra but not in alpine tundra?

10. If you were to travel to Mars, which is dry and has great daily variations in temperature, and were to search for life, what kind of biome would you search for first?

11. Suppose you are going to build a spacecraft for long-term voyages, and you plan to use an ecological life support system—to use ecosystems to provide food; to recycle oxygen, carbon dioxide, and water; and to treat wastes. What biomes do you believe will be most important to take along? Would you create a "new" biome?

● FURTHER READING

Elton, C. S. 2000. *The Ecology of Invasions by Animals and Plants*. New York: Oxford University Press. (Reprinted with new introduction by Daniel Simberloff.) A classic work by one of the major ecologists of the 20th century.

Levin, S. A., ed. 2000. *Encyclopedia of Biodiversity*. San Diego: Academic Press. A new, broad-based review of biodiversity.

Noss, R. F., and A. Y. Cooperrider. 1994. *Saving Nature's Legacy: Protecting and Restoring Biodiversity*. Washington, D.C.: Island Press.

Reid, W. V., and K. R. Miller. 1989. *Keeping Options Alive: The Scientific Basis for Conserving Biodiversity*. Washington, D.C.: World Resources Institute.

Strahler, A. 1994. *Modern Physical Geography*, 4th ed. New York: John Wiley & Sons. A book long recognized as the leading introduction to geography, including an introduction to biogeography.

Wilson, E. O., ed. 1988. *Biodiversity*. Washington, D.C.: National Academy Press. A major overview by some leading scientists.

Wilson, E. O. 1999. *Conserving Earth's 1999 Biodiversity*. Washington, D.C.: Island Press. A CD-ROM that includes text of some of Wilson's writing. He is recognized as one of the leading authorities on biodiversity.

World Resources Institute. 1996. *National Biodiversity Planning: Guidelines Based on Early Experiences Around the World*. Washington, D.C.: World Resources Institute. An "illustrative biodiversity-planning process" based on the real-world experiences of 17 countries or regions—Australia, Canada, Chile, China, Costa Rica, Egypt, Germany, Indonesia, Kenya, Mexico, the Netherlands, Norway, the Philippines, Poland, the South Pacific, the United Kingdom, and Vietnam—already developing national strategies, plans, and programs.

9 Biological Productivity and Energy Flow

A moose feeding in wetland flowers participates in ecosystem energy flow, as do the flowering plants as they use sunlight in photosynthesis.

● LEARNING OBJECTIVES

To conserve and manage our biological resources wisely, we must understand the basic concepts of energy, energy flow in ecosystems, and biological production. After reading this chapter, you should understand:

- How energy flow determines the upper limit on the production of biological resources, including forests, fisheries, wildlife, and endangered species.
- Where energy comes from and how it is transferred from one living thing to another.
- How the first and second laws of thermodynamics affect energy and production.
- That energy flow is one way through the ecosystem.
- That a basic quality of life is its ability to create order from energy on a local scale.
- Why little of the energy available to an organism is fixed in new organic matter and how little of the energy available moves on to the next trophic level.

Harvesting Forests in Michigan and England

Life can be wondrously productive, but in many cases, the harvesting of wild living resources, from marine mammals to forests, has exceeded Earth's bounty. We seem to be in the process of destroying many resources by harvesting them faster than they can regrow, and we have been doing so for a long time.

In the United States, for example, between 1840 and 1920, 7.7 million ha (19 million acres) of white pine forests were logged in the state of Michigan alone. Those who started the logging believed that they would never run out of wood—that by the time the last acre had been cut the first acre logged would have regrown. But those first loggers underestimated the power of technology. Only 20 ha (49 acres) of uncut white pine[1] remain in Michigan, and only a small fraction of the original white pine forests have regenerated to anything like their original size and abundance.

To make matters worse, over much of the area, unused limbs, branches, twigs, and leaves were left on the ground and provided fuel for fires that were often started accidentally at sawmills. The complete clearing of the land removed seed trees, and the fires destroyed the organic matter in the soil. As a result of these practices, stump barrens—areas where large trees have never regrown—are common in northern Michigan (Figure 9.1).

In contrast, in medieval England, some small forested areas were harvested carefully and slowly. For example, in 1356, a survey of the estates of Bishop Ely stated that a "certain wood called Heylewode" was 32.4 ha (80 acres) in size. Every year the "underwood" (the shrubs and young, small trees) of Heylewode was harvested in 4.5 of the 32.4 ha (11 of the 80 acres), leaving mature trees everywhere and young trees on the remaining 27.9 ha (69 acres). This practice was continued "without causing waste or destruction"; in modern terminology, the practice was sustainable. Similar forests were managed in much the same way in other locations in England from the 14th century until World War I with little decline in the woodlands.[1] Today, we are seeing projects designed to redevelop these methods (Figure 9.2).

These two contrasting examples suggest that at certain levels of harvesting, forest production can be sustained, but once these levels are exceeded, forests will decline and may not recover. If they do recover, the recovery may take an exceedingly long time.

Figure 9.1 • Sometimes, forests that are clear-cut do not regenerate. That happened with some large areas of white pine forests in Michigan, called "stump barrens," including the area shown here, the Kingston Plains. These areas were logged and subject to repeated fires. Some believe that the logging practices of the time, along with climatic change, prevented regeneration.

Figure 9.2 • Some English woodlands have been managed since medieval times for sustainable forestry.

> • *Throughout the history of civilization, people have cut trees faster than trees have regrown. Beginning with the earliest civilizations, forests were cleared to make way for agriculture and to provide fuel and structural materials, the basis of early civilization.[2] The practice of clearing the land and eliminating forests has continued into our own time. Clearly, the amount of timber harvested cannot exceed forever the amount of new timber that grows between harvests. Eventually, the supply of timber will run out. With modern pressures for more and more use of wood and other biological resources, we need more than ever to understand the limits to the growth of timber—and to all biological production.*

9.1 How Much Can We Grow?

Determining how much organic matter can be produced in any time period is important to many environmental topics, especially those that concern biological resources. How many bushels of wheat can a farmer produce in a field in a year? What is the upper limit of food that can be produced for all the people on Earth? What is the limit on the number of whales in the ocean? What is the maximum production of forests that we can expect?

Many factors can limit the growth of living things, but the ultimate limit on production of organic matter is determined by energy flow. In order to estimate the actual and the maximum possible production of organic matter of any kind, it is necessary to understand basic concepts of energy, energy flow in ecosystems, and biological production.

9.2 Biological Production

The total amount of organic matter on Earth or in any particular ecosystem or area is called its **biomass**. This includes all living things plus all products of living things. Biomass is usually measured as the amount per unit surface area of Earth (for example, as grams per square meter [g/m^2] or metric tons per hectare [MT/ha]).

Biomass is increased through biological production (growth). Change in biomass over a given period of time is called **net production**. Biological production involves the capture of usable energy and the production of organic compounds in which that energy is stored. Three measures are used for biological production: biomass, energy stored, and carbon stored. We can think of these measures as the currencies of production. (General relationships for calculating production are given in Working It Out 9.1 and 9.2. Common units of measure of production are given in Table 9.1.)

Two Kinds of Biological Production

There are two kinds of biological production. Some organisms make their own organic matter from a source of energy and inorganic compounds. These organisms, introduced in the discussion of trophic levels in Chapter 6, are called **autotrophs** (meaning self-nourishing). The autotrophs include green plants (those containing chlorophyll), such as herbs, shrubs, and trees; algae, which are usually found in water but occasionally grow on land; and certain kinds of bacteria that grow in water.

The production carried out by autotrophs is called **primary production**. Most autotrophs make sugar from sunlight, carbon dioxide, and water in a process called

Table 9.1	Common Units of Measure for Biomass and Production	
Habitat	*Biomass*	*Production*
Units of Mass		
Land	g/cm^2	g/cm^2/d; g/cm^2/yr
	kg/ha	kg/ha/yr
	t/ha	t/ha/yr
Water	g/cm^3	g/cm^3/d
	kg/m^3	kg/m^3/yr
Units of Energy		
Land	kcal/cm^2	kcal/cm^2/d
	kcal/ha/yr	kcal/ha/yr
	kJ/cm^2	kJ/cm^2/d
	kJ/ha/yr	kJ/ha/yr
Water	kcal/m^3	kcal/m^3/yr
	kJ/m^3	kJ/m^3/yr

Note: Units of mass: 1.0 t (metric ton) = 1000 kg (kilogram) = 10^{6}g (gram). Units of area: 1.0 ha (hectare) = 10,000 square meters (m^2) = 10^8 square centimeters (cm^2). Units of time: yr (year, sometimes written as "a"); d (day). Units of energy: 1 kcal = 4.19 kJ.

Equations for Production, Biomass, and Energy Flow

We can write a general relation between biomass (B) and net production (NP):

$$B_2 = B_1 + NP$$

where B_2 is the biomass at the end of the time period, B_1 is the amount of biomass at the beginning of the time period, and NP is the change in biomass during the time period. Thus,

$$NP = B_2 - B_1$$

General production equations are given as

$$GP = NP + R$$
$$NP = GP - R$$

where GP is gross production, NP is net production, and R is respiration.

The three currencies of energy flow are biomass, energy content, and carbon content. The average of the energy in vegetation is approximately 21 kilojoules per gram (kJ/g). Energy content of organic matter varies. Ignoring bone and shells, woody tissue contains the least energy per gram, about 17 kJ/g; fat contains the most, about 38 kJ/g; and muscle contains approximately 21–25 kJ/g. Leaves and shoots of green plants have 21–23 kJ/g; roots have about 19 kJ/g.[3]

The kilojoule (1 kJ = 1000 J = 0.24 kcal) is the International System (SI) unit preferred in scientific notation for energy and work. It replaces the calorie or kilocalorie of earlier studies of energy flow. The kilocalorie is the amount of energy required to heat a kilogram of water 1 degree Celsius (from 15.5° to 16.5°C). (The calorie, which is one-thousandth of a kilocalorie, is the amount of energy required to heat a gram of water the same 1 degree Celsius.) Note that the calorie referred to in diet books is actually the kilocalorie, though for convenience it is referred to in popular literature as a calorie. To keep all this straight, just remember that almost nobody uses the "little" calorie, regardless of what they call it. To compare, an average apple contains about 419 kJ or 100 kcal. The calorie is typically used in studies of diets; the joule is used in physics and engineering.

photosynthesis, which releases free oxygen (see Working It Out 9.1 and Working It Out 9.2). Some autotrophic bacteria can derive energy from inorganic sulfur compounds; these bacteria are referred to as **chemoautotrophs**. Such bacteria have been discovered in deep-ocean vents, where they provide the basis for a strange ecological community. Chemoautotrophs are also found in muds of marshes, where there is no free oxygen.

Other kinds of life cannot make their own organic compounds from inorganic ones and must feed on other

Energy Equalities

For Those Who Make Their Own Food (autotrophs)

Photosynthesis, the process by which autotrophs make sugar from sunlight, carbon dioxide, and water, is defined as

$$6CO_2 + 6H_2O = C_6H_{12}O_6 + 6O_2$$

Chemosynthesis takes place in certain environments. In chemosynthesis, the energy in hydrogen sulfide (H_2S) is used by certain bacteria to make simple organic compounds. The reactions differ among species and depend on characteristics of the environment. Therefore, no single equation like that for photosynthesis is available for chemosynthesis.

Net production for autotrophs is given as

$$NPP = GPP - R_a$$

where NPP is net primary production, GPP is gross primary production, and R_a is the respiration of autotrophs.

For Those Who Do Not Make Their Own Food

Secondary production of a population is given as

$$NSP = B_2 - B_1$$

where NSP is net secondary production, B_2 is the biomass at time 2, and B_1 is the biomass at time 1. The change in biomass is the result of the addition of weight of living individuals, the addition of newborns and immigrants, and loss through death and emigration. The biological use of energy occurs through respiration, most simply expressed as

$$C_6H_{12}O_6 + 6O_2 = 6CO_2 + 6H_2O + Energy$$

living things. These are called **heterotrophs**. All animals, including human beings, are heterotrophs, as are fungi, many kinds of bacteria, and many other small forms of life. Production by heterotrophs is called **secondary production** because it depends on the production of autotrophic organisms. This dependency is the basis for the food web described in Chapter 6. (The associated energy flow is diagrammed in Figure 9.3.)

Once an organism has obtained new organic matter, it can use the energy in that organic matter to do things: to move, to make new kinds of compounds, to grow, to reproduce, and so forth. The use of energy in organic matter in most heterotrophic and autotrophic organisms is accomplished through **respiration**. In respiration, an organic compound is combined with oxygen to release energy and produce carbon dioxide and water (see Working It Out 9.2). The process is similar to the burning of organic compounds but takes place within cells at much lower temperatures through enzyme-mediated reactions. *Respiration is the use of biomass to release energy that can be used to do work.* Respiration returns to the environment the carbon dioxide that has been removed by photosynthesis.

Gross and Net Production

The production of biomass and its use as a source of energy by autotrophs includes three steps.

1. An organism produces organic matter within its body.
2. It uses some of this new organic matter as a fuel in respiration.
3. Some of the newly produced organic matter is stored for future use.

The first step, production of organic matter before use, is called **gross production**. The amount left after utilization is called **net production**. In general terms, net production is the amount added to and stored in an organism, population, species, trophic level, ecological community, or ecosystem. In these terms,

Net production = Gross production – Respiration.

Therefore, as mentioned earlier, net production is the change in biomass. This is a fundamental production relationship. The difference between gross and net production is like the difference between a person's gross and net income. Gross income is the amount you are paid. Net income is what you have left after money is deducted for taxes and other fixed costs. Respiration is like the necessary expenses that are required in order for you to do your work.

The gross production of a tree—or any other plant—is the total amount of sugar it produces by photosynthesis before any is used. Within living cells in a green plant, some

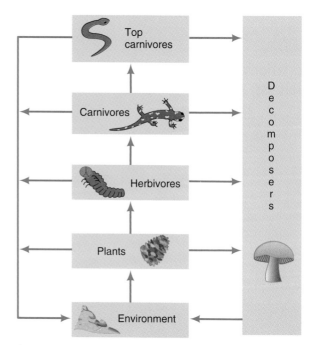

Figure 9.3 • Energy pathways through an ecosystem. Usable energy flows from the external environment (the sun) to the plants, then to the herbivores, carnivores, and top carnivores. Death at each level transfers energy to decomposers. Energy lost as heat is returned to the external environment.

of the sugar is oxidized in respiration. Energy is used to convert sugars to carbohydrates, carbohydrates to amino acids, amino acids to proteins and new leaf tissue, and so forth. Energy is also used to transport material within the plant. What is not used in such ways may itself be transported to roots, stems, flowers, and fruits or stored within the leaf cell. Some energy is lost as heat in the transfer. Other energy is used to make other organic compounds in other parts of the plant: cell walls, proteins, and so forth. Some is stored for later use. Net production of a green plant is energy contained in what is left at the end of the year. For woody plants like trees, it includes new wood laid down in the trunk, new buds that will develop into leaves and flowers the next year, and new roots.

9.3 Energy Flow

Energy is a difficult and abstract concept. When we buy electricity by the kilowatt-hour (1 kWh = 3.6 x 10³ kJ), what are we buying? We cannot see it or feel it, even if we have to pay for it.[4]

At first glance, energy flow seems simple enough: We take energy in and use it, just like a machine. But if we dig a little deeper into this subject, we discover its depth and philosophical importance. When we consider the role of energy in an ecosystem, we consider the heart of the matter of what it is that distinguishes life and life-containing systems from the rest of the universe.

Although most of the time, energy is invisible to us, with infrared film we can see the differences between warm and cold objects, and we can see some factors about energy flow that affect life. With infrared film, warm objects appear red and cool objects blue. Figure 9.4 shows birch trees in a New Hampshire forest. In Figure 9.4*a*, the trees are shown as we see them, using standard film; in Figure 9.4*b*, infrared film shows the leaves bright red, indicating that they have been warmed by the sun and are absorbing and reflecting energy, whereas the white birch bark remains cooler. The ability of tree leaves to absorb energy is essential; it is this source of energy that ulti-mately supports all life in a forest. Energy flows through life, and *energy flow* is a key concept.

All life requires energy. Energy is the ability to do work, to move matter. As anyone who has dieted knows, our weight is a delicate balance between the energy we take in through our food and the energy we use. What we do not use and do not pass on, we store. Our use of energy, and whether we gain or lose weight, follows the laws of physics. This is not only true for people, it is also true of all populations of living things, of all ecological communities and ecosystems, and of the entire bios-phere.

(a)

(b)

(c)

(d)

Figure 9.4 • Top: A birch forest in New Hampshire as we see it, using normal pho-tographic film (*a*) and the same forest photographed with infrared film (*b*). Red color means warmer temperatures; the leaves are warmer than the surroundings because they are heated by sunlight. Bottom: A nearby rocky outcrop as we see it, using normal photographic film (*c*) and the same rocky outcrop photographed with infrared film (*d*). Blue means that a surface is cool. The rocks appear deep blue, indicating that they are much cooler than the surrounding trees.

Ecosystem energy flow is the movement of energy through an ecosystem from the external environment through a series of organisms and back to the external environment. It is one of the fundamental processes common to all ecosystems. The flow of energy through ecosystems is one of the most profound of all topics discussed in this book; it tells us about the very nature of life.

At first glance energy flow might seem simple enough: An ecosystem takes energy in and uses it, just like a machine. But if we dig a little deeper into this subject, we discover its depth and philosophical importance. When we consider the role of energy in an ecosystem, we consider the heart of what distinguishes life and life-containing systems from the rest of the universe.

Energy enters an ecosystem by two pathways. One is the pathway already discussed: Energy is fixed by organisms that can produce their own food from energy (through photosynthesis) or by chemosynthesis from inorganic minerals. In the second pathway, heat energy is transferred by the air or water currents or by convection through soils and sediments and warms living things. For instance, when a warm air mass passes over a forest, heat energy is transferred from the air to the land and to the organisms.

What about production within an ecosystem? We saw earlier that net production equals gross production minus respiration. As stated, this relationship is true not only for an individual organism but also for an ecological community or an ecosystem (see Working It Out 9.3). In an ecosystem, the gross production is simply the gross production of all the autotrophs. The net ecosystem production is the amount of biomass added in some time period after all utilization, including the respiration of autotrophs and heterotrophs, has taken place.

9.4 The Ultimate Limit on the Abundance of Life

What ultimately limits the amount of organic matter that can be produced? How closely do ecosystems, species, populations, and individuals approach this maximum limit? The search for answers to these questions takes us to some of the most fundamental issues about life, issues that involve energy flow through ecosystems. As mentioned earlier, the relationship between energy and life is explained by fundamental laws of physics.

The Laws of Thermodynamics

Energy is subject to the *law of conservation of energy*, which states that in any physical or chemical change, energy is neither created nor destroyed but merely changed from one form to another. The law of conservation of energy is also called the *first law of thermodynamics* (discussed in A Closer Look 4.1, "Matter and Energy").

If the total amount of energy is always conserved—if it remains constant—then why can't we just recycle energy inside our bodies? Similarly, why can't energy be recycled in ecosystems and in the biosphere?

Let us imagine how that might work, say, with frogs and mosquitoes. Frogs eat insects, including mosquitoes. Mosquitoes suck blood from vertebrates, including frogs. Consider an imaginary closed ecosystem consisting of water, air, a rock for frogs to sit on, frogs, and mosquitoes. In this system, the frogs get their energy from eating the mosquitoes, and the mosquitoes get their energy from biting the frogs (Figure 9.5). Such a closed system would be a biological perpetual-motion machine: It could continue indefinitely without an input of any new material or energy. This sounds nice, but unfortunately, it is impossible.

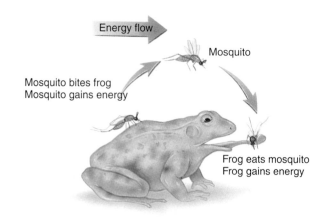

Figure 9.5 • An impossible ecosystem.

The Second Law of Thermodynamics

To better understand why we cannot recycle energy, imagine a closed system (a system that receives no input after the initial input) containing a pile of coal, a tank of water, air, a steam engine, and an engineer (Figure 9.6). Suppose that the engine runs a lathe that makes furniture. The engineer lights a fire to boil the water, creating steam to run the engine. As the engine runs, the heat from the fire gradually warms the entire system.

When all the coal is completely burned, the engineer will not be able to boil any more water, and the engine will stop. The average temperature of the system is now higher than the starting temperature. The energy that was in the coal is dispersed throughout the entire system, much of it as heat in the air.

Why can't the engineer recover all that energy, recompact it, put it under the boiler, and run the engine? The answer is found in the second law of thermodynamics. Physicists have discovered that no real use of energy can ever be 100% efficient. Whenever useful work is done, some energy is inevitably converted to heat. Collecting all the energy dispersed in this closed system would require more energy than could be recovered.

Our imaginary system begins in a highly organized state, with energy compacted in the coal. It ends in a less organized state, with the energy dispersed throughout the system as heat. The energy has been degraded, and the system is said to have undergone a decrease in order. The measure of the decrease in order (the disorganization of energy) is called **entropy**. The engineer did produce some furniture, converting a pile of lumber into nicely ordered tables and chairs. The system had a local increase of order (the furniture) at the cost of a general increase in disorder (the state of the entire system). All energy of all systems tends to flow toward states of increasing entropy.

Figure 9.6 • A system closed to the flow of energy.

Why? The general answer is found in the *second law of thermodynamics*, which addresses how energy changes in form.

Energy always changes from a more useful, more highly organized form to a less useful, disorganized form. That is, energy cannot be completely recycled to its original state of organized, high-quality usefulness. For this reason, the mosquito–frog system will eventually stop, when not enough useful energy is left. (There is also a more mundane reason: Only female mosquitoes require blood and then only in order to reproduce. Mosquitoes are otherwise herbivorous.)

From the discussion presented in A Closer Look 9.1, "The Second Law of Thermodynamics," we reach a new understanding of a basic quality of life.[3] *It is the ability to create order on a local scale that distinguishes life from its nonliving environment.* Having this ability requires obtaining energy in a usable form, and that is the essence of why we eat. This principle is true for every ecological level: individual, population, community, ecosystem, and biosphere. Energy must continually be added to an ecological system in a usable form. Energy is inevitably degraded into heat, and this heat must be released from the system. If it is not released, the temperature of the system will increase indefinitely. The *net flow of energy through an ecosystem, then, is a one-way flow.*

Based on what we have said about the energy flow through an ecosystem, we can see that an ecosystem must lie between a source of usable energy and a sink for degraded (heat) energy. The ecosystem is said to be an *intermediate* between the energy source and the energy sink. The energy source, ecosystem, and energy sink together form a **thermodynamic system**. The ecosystem can undergo an increase in order, called a *local increase*, as long as the entire system undergoes a decrease in order, called a *global decrease*. Creating local order involves the production of organic matter. Producing organic matter requires energy; organic matter stores energy.

Energy Efficiency and Transfer Efficiency

How efficiently do living things use energy? This is an important question for the management and conservation of all biological resources. We would like biological resources to be efficient in their use of energy—to produce much biomass from a given amount of energy.

The laws of physics and chemistry state that no system can be 100% efficient. As energy flows through a food web, it is degraded, and less and less is usable. Generally, the more energy an organism gets, the more it has for its own use. However, organisms differ in how efficiently they use the energy they obtain. A more efficient organism has an advantage over a less efficient one.

Efficiency can therefore be defined for both artificial and natural systems: machines, individual organisms, populations, trophic levels, ecosystems, and the biosphere.[5] Energy efficiency is defined as the ratio of output to input, and it is usually further defined as the amount of useful work obtained from some amount of available energy. Efficiency has different meanings to different users. A farmer thinks of an efficient corn crop as one that converts a great deal of solar energy to sugar and uses little of that sugar to produce stems, roots, and leaves. In other words, the most efficient crop is the one that has the most harvestable energy left at the end of the season. A truck driver views an efficient truck as one that uses as much energy as possible from its fuel and leaves as little energy as possible to exit in the exhaust. When we view organisms as food, we define efficiency as the farmer does, in terms of energy storage (net production from available energy). When we are energy users, we define efficiency as the truck driver does, in terms of how much useful work we accomplish with the available energy.

A common ecological measure of energy efficiency is called food chain efficiency, or **trophic-level efficiency**, which is the ratio of production of one trophic level to the production of the next lower trophic level. (See Chapter 6.) This efficiency is never very high. Green plants convert only 1 to 3% of the energy received from the sun during the year to new plant tissue. We would often like to increase trophic-level efficiency.

The efficiency with which herbivores convert the potentially available plant energy into herbivorous energy is usually less than 1%, as is the efficiency with which carnivores convert herbivores into carnivorous energy. It is frequently written in popular literature that the transfer is 10%—for example, that 10% of the energy in corn can be converted into energy in a cow. However, this is a managed ecological efficiency rather than the natural, trophic-level efficiency. In natural ecosystems, the organisms in one trophic level tend to take in much less energy than the potential maximum amount available to them, and they utilize more energy than they store for the next trophic level.

Consider an example. At Isle Royale National Park, an island in Lake Superior, wolves feed on moose in a natural wilderness. A pack of 18 wolves killed an average of one moose approximately every 2.5 days,[6] resulting in a trophic-level efficiency of wolves of about 0.01%. Wolves use most of the energy they take in from eating moose, especially for moving in the search for prey.[7] From the wolves' point of view, wolves are efficient, but from the point of view of someone who wants to feed on wolves, they appear inefficient.

The rule of thumb for ecological trophic energy efficiencies is that more than 90% (usually much more) of all energy transferred between trophic levels is lost as heat. Less than 10% (approximately 1% in natural ecosystems) is fixed as new tissue. In highly managed ecosystems, such as ranches, the efficiencies may be greater. Even in such highly managed systems, however, it takes an average of 3.2 kg (7 lb) of vegetable matter to produce 0.45 kg (1 lb) of edible meat. Cattle are among the least efficient producers, requiring around 7.2 kg (16 lb) of vegetable matter to produce 0.45 kg (1 lb) of meat. Chickens are much more efficient, using approximately 1.4 kg (3 lb) of vegetable matter to produce 0.45 kg (1 lb) of eggs or meat. Much attention has been paid to the idea that humans should eat at a lower trophic level in order to use resources more efficiently. (See The Critical Thinking Issues, "Should People Eat Lower on the Food Chain?")

Many other kinds of energetic efficiencies are widely used in ecological studies. Some of these are described in A Closer Look 9.2.[8]

 9.5 Some Examples of Energy Flow

We conclude the chapter with several examples of energy flow in ecosystems.

Ecological Efficiencies

Table 9.2 gives some values for growth efficiency, or gross production efficiency (*P/C*), which is the ratio of the material produced (*P* is the net production) by an organism or population to the material ingested or consumed (*C*). The amount consumed is normally much less than the maximum amount available. Estimates show, for example, that annually amounts of less than 1% to about 20% of the leaves available in forests and woodlands are consumed by leaf-eating insects.[9]

Table 9.2 also gives examples of net growth efficiency, or net production efficiency (*P/A*). This is the ratio of the material produced (*P*) to the material assimilated (*A*), which is less than the material consumed because some food taken in is discharged as waste and never used by an organism.

Table 9.2   Ecological Efficiencies for Animal Populations

	Ecological Efficiencies (%)	
Trophic Types	*Net Production Efficiency (P/A)*	*Gross Production Efficiency (P/C)*
Terrestrial Animals		
Microorganisms[a]	~40	
Invertebrates		
Herbivores	20–40	8–27
Carnivores	10–37	~34
Saprophages	17–40	5–8
Vertebrates		
Herbivores[a]	2–10	
Carnivores[a]	2–10	
Aquatic Animals		
Fishes[a,b]	—	1–7

Sources: T. Penczak, *Comp. Biochem. Physiology* 101 (1992): 791–798; D. E. Reichle, "The Role of Soil Invertebrates in Nutrient Cycling," in U. Lohm and T. Persson, eds., *Ecol. Bull.* (*Stockholm*) 25 (1977): 145–156; and M. Schaefer, "Secondary Production and Decomposition," in E. Rohrig and B. Ulrich, eds., *Temperate Deciduous Forests*, vol. 7 of *Ecosystems of the World* (Amsterdam: Elsevier, 1991).

[a] Data are based on characteristic values for trophic levels and populations.

[b] Populations in a tropical river.

Energy Flow in an Old Field Food Chain

In an old field in Michigan, meadow mice feed on grasses and herbs, and least weasels feed on mice (one of many food chains in this old field).[10] The first step in the flow of energy is the fixation of light energy by photosynthesis in the leaves of grasses, herbs, and shrubs. In this step, the energy in light is transferred to and stored in sugar or carbohydrates in the plants. Recall that this energy—the energy that is stored by autotrophs before any is used—is called gross primary production. Some of the energy is used immediately by the leaves to keep their own life processes going. As explained earlier, the amount stored by the autotrophs after using what is needed is net primary production.

As we have seen, only a small fraction of the energy available to each trophic level *is converted to net production of new tissue. A large fraction of the energy available to each trophic level is used in respiration.* In the old field in Michigan, about 15% of the vegetation's gross production is used in respiration; 68% of the energy taken up by the mice is used in respiration; and 93% of the energy taken up by the least weasel is used in respiration. Only a

part of the energy flow in the old field moves through the food chain of vegetation–mice–weasels. One reason is that mice eat only the seeds of the plant. Most of the energy remains in the vegetation until it is transferred from dead vegetation to animals, fungi, and bacteria by the decomposer food chain.

Energy Flow in a Stream or River

In most ecosystems, the original fixing of energy occurs within the ecosystem; however, some freshwater streams are an exception. In streams, photosynthesis of stream algae is often small compared with the amount of dead vegetation that falls in or flows in from the surrounding countryside (dead leaves and twigs fall into the stream or are washed into it during storms).[11] *Detritivores* (organisms that feed on dead organic material) are common in streams and feed mainly on this deposited vegetation. Some of the animals are shredders that tear up leaves; others feed on the smaller pieces.

Other grazing animals move along the surfaces of rocks and scrape off attached algae. Many stream predators are larvae of land-dwelling insects, such as dragon-

Should People Eat Lower on the Food Chain?

The energy content of a food chain is often represented by an *energy pyramid*, such as the one shown here for a hypothetical, idealized food chain. In an energy pyramid, each level of the food chain is represented by a rectangle whose area is more or less proportional to the energy content for that level. For the sake of simplicity, the food chain shown here assumes that each link in the chain has one and only one source of food.

Assume that if a 75-kg (165-lb) person ate frogs (and some people do!), he would need 10 a day, or 3,000 a year (approximately 300 kg or 660 lb). If each frog ate 10 grasshoppers a day, the 3,000 frogs would require 9,000,000 grasshoppers a year to supply their energy needs, or approximately 9,000 kg (19,800 lb) of grasshoppers. A horde of grasshoppers of that size would require 333,000 kg (732,600 lb) of wheat to sustain them for a year.

As the pyramid illustrates, energy content decreases at each higher level of the food chain. The result is that the amount of energy at the top of a pyramid is related to the number of layers the pyramid has. For example, if people fed on grasshoppers rather than frogs, each person could probably get by on 100 grasshoppers a day. The 9,000,000 grasshoppers could support 300 people for a year, rather than only one. If, instead of grasshoppers, people ate wheat, then 333,000 kg of wheat could support 666 people for a year.

This argument is often extended to suggest that people should become herbivores (vegetarians, in human parlance) and eat directly from the lowest level of all food chains, the autotrophs. Consider, however, that humans can eat only parts of some plants. Herbivores can eat some parts of plants that humans cannot eat and some plants that humans cannot eat at all. When people eat these herbivores, more of the energy stored in plants becomes available for human consumption. The most dramatic example of this is in aquatic food chains. Because people cannot digest most kinds of algae, which are the base of most aquatic food chains, they depend on eating fish that eat algae and fish that eat other fish. So if people were to become entirely herbivorous, they would be excluded from many food chains. In addition, there are major areas of Earth where crop production damages the land but graz-

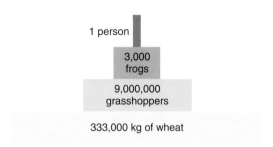

The total bulk (mass) of all the organisms making up each level of a food chain is less than that of the previous level. (*Source*: Nelson, 1970, *Fundamental Concepts of Biology*, 2nd ed., Wiley, New York, p. 299.)

ing by herbivores does not. In those cases, conservation of soil and biological diversity leads to arguments that support the use of grazing animals for human food. This creates an environmental issue: How low on the food chain should people eat?

Critical Thinking Questions

1. Why does the energy content decrease at each higher level of a food chain? What happens to the energy that is lost at each level?

2. The pyramid diagram uses mass as an indirect measure of the energy value for each level of the pyramid. Why is it appropriate to use mass to represent energy content?

3. Using the average of 21 kJ of energy to equal 1 g of completely dried vegetation (see Working It Out 9.3) and assuming that wheat is 80% water, what is the energy content of the 333,000 kg of wheat shown in the pyramid?

4. Make a list of the environmental arguments for and against an entirely vegetarian diet for people. What might be the consequences for U.S. agriculture if everyone in the country began to eat lower on the food chain?

5. How low do you eat on the food chain? Would you be willing to eat lower? Explain.

flies. Some animals capture prey from the land or air, as in the case of trout that catch flying insects.

An extreme case of a food chain based on external food input occurs in the floodplain of the Amazon River basin, in which fish feed on fruits and nuts carried into the streams during the rainy season. Here, the production of herbivorous fish exceeds what would be possible from aquatic primary production alone, yielding an abundant food supply for the human populations of the region.[11]

Energy Flow in Ocean Ecosystems

In the open ocean, the first trophic level is composed of many species of floating algae, collectively called phytoplankton.[12] These are fed on by many animals, including small, floating ones known as zooplankton, which in turn are fed on by larger organisms. A food chain is based on dead organic material that is produced at the surface and sinks toward the ocean floor, including cells of algae;

dead bodies and excrement of heterotrophs; and some material, such as logs, transported from the land. These form the energy base for two groups of carnivores and detritivores: those that swim in the depths and catch the material as it sinks and those that live on the ocean bottom and hunt for material once it has been deposited. These food chains thus depend on an external input of energy as the base of energy flow.

Chemosynthetic Energy Flow in the Ocean

Earlier, we mentioned organisms that make their own food from energy in sulfur compounds. This process creates a curious class of food chains recently discovered in the depths of the oceans. These food chains support previously unknown life-forms. The basis of the food chains is *chemosynthesis*, in which the source of energy is not sunlight but hot, inorganic sulfur compounds emitted from vents in the ocean floor.

Sulfur-laden water is emitted from hot-water vents occurring at depths of 2,500–2,700 m (8,200–8,900 ft) associated with areas where flowing lava causes seafloor spreading. A rich biological community exists in and around the vents, including large white clams up to 20 cm (7.8 in) in diameter, brown mussels, and white crabs. Clams and mussels filter chemoautotrophic bacteria and particles of dead organic matter from the water. Some vent communities contain limpets, pink fish, tube worms, and octopuses. Among the most curious creatures found in vents are giant worms, some 3.9 m (12.8 ft) long.[13]

Large areas of ocean have low productivity; combined, however, oceans account for a major portion of total energy fixed. Highly productive areas of the oceans occur in upwelling zones, which occur when deep ocean waters, rich in nutrients from dead organic material, flow upward, allowing abundant growth of algae and photosynthetic bacteria. Herbivores and carnivores move organic nutrients through the food chain. Although only 1/1000 of the oceans' surface has natural upwellings, these zones account for more than 44% of the fish eaten by the world's human population.[14]

Summary

- The study of energy flow is important to determine limits on food supply and on the production of all biological resources, such as wood and fiber.

- In every ecosystem, energy flow provides a foundation for life and thus imposes a limit on the abundance and richness of life. The amount of energy available to each trophic level in a food chain depends not only on the strength of the energy source but also on the efficiency with which the energy is transferred along the food chain.

- Energy is fixed by autotrophs, which are organisms that make their own food from energy and small inorganic compounds. The initial energy comes from two sources: light (mainly sunlight) and the energy in small sulfur compounds. Plants, algae, and some bacteria are autotrophs.

- Only autotrophs can make their own food; all other organisms are heterotrophs that must feed on other organisms.

- Biological production is the production of new organic matter, which we measure as change in biomass, change in stored energy, or change in stored carbon. Another way to think about biological production is that it is the change in biomass over time.

- Gross production is production measured before any utilization. Net production is the amount stored (not used) at the end of some time period. Respiration uses stored energy, so that net production equals gross production minus respiration.

- The laws of thermodynamics connect life to order in the universe. The second law of thermodynamics tells us that order always decreases when any real process occurs in the universe. However, life is more ordered than its environment. The ability to create order is the essence of what we get from our food.

- Energy efficiency is the ratio of output to input, or the amount of useful work obtained from some amount of available energy. Trophic-level efficiency is the ratio of production of one trophic level to the production of the next lower trophic level. This efficiency is never very high, often only about 1%.

● REEXAMINING THEMES AND ISSUES

Human Population: The ultimate limit on the human population and its use of resources is set by available energy, although many other factors can set limits well below the maximum.

Sustainability: Ecological energy flow sets an upper limit to sustainable biological production.

From a cosmic perspective, Earth is a small planet where life uses the one-way flow of energy from the sun to create life's order: organisms, species, ecosystems, and the environment of the whole earth.

One of the fundamental characteristics of life is that it uses energy to create order on a local scale. From an ecological perspective, cities are an extreme example of this process.

Until we know the upper limit on biological production, we cannot know the practical limits of our harvest and use of living resources. Understanding energy flow is fundamental to understanding the limits of our actions and what we can consider a good—that is, sustainable—set of activities.

● KEY TERMS

● STUDY QUESTIONS

1. What is the difference between ecological production, ecological productivity, and biomass? Gross production and net production? Primary and secondary production?

2. What does the statement "Any living or life-containing system is always more ordered than its nonliving environment" mean?

3. Keep track of the food you eat during one day and make a food chain linking yourself with the sources of those foods. Determine the biomass (grams) and energy (kilocalories) you have eaten? Using an average of 5 kcal/g, then using the information on food packaging or assuming that your net production is 10% efficient in terms of the energy intake, how much additional energy might you have stored during the day? What is your weight gain from the food you have eaten?

4. Referring to question 3, what amount of vegetation did you eat during one day? If vegetation was 1% efficient in converting sunlight to organic matter stored as net production, how much sunlight was required to provide the vegetation you took in during the day?

● FURTHER READING

Relative to many other topics discussed in this book, energy flow and productivity have a long history, and the most useful and interesting references tend to be classics published several decades ago. Although in most cases we try to provide the most up-to-date references, with this subject we believe some of the easiest-to-read and most important references are among the classical earlier works.

Blum, H. F. 1962. *Time's Arrow and Evolution.* New York: Harper & Row. A very readable book discussing how life is connected to the laws of thermodynamics and why this matters.

Gates, D. M. 1980. *Biophysical Ecology.* New York: Springer-Verlag. A discussion about how energy in the environment affects life.

Morowitz, H. J. 1979. *Energy Flow in Biology.* Woodbridge, Conn.: Oxbow. The most thorough and complete discussion available about the connection between energy and life, at all levels, from cells to ecosystems to the biosphere.

Morowitz, H. J. 1981. "The Six Million Dollar Man." In *The Wine of Life and Other Essays on Societies, Energy, and Living Things.* New York: Bantam Books. A fun essay about the second law of thermodynamics and life.

Peterson, R. O. 1995. *The Wolves of Isle Royale: A Broken Balance.* Minocqua, Wis.: Willow Creek Press. A first-hand account of Rolf Peterson's 25-year association with a long-running study of the wild wolves of Isle Royale National Park and their primary prey, the moose.

Schrödinger, E. 1942. *What Is Life?* Cambridge: Cambridge University Press. The original statement about how the use of energy differentiates life from other phenomena in the universe. Easy to read, and a classic.

Sherman, K. 1990. *Large Marine Ecosystems: Patterns, Processes, and Yields.* Portland, Ore.: Book News. A book based on an AAAS symposium dealing with the possible impacts of global change on ocean productivity. It discusses managing large marine ecosystems as multinational units in order to sustain biomass yields of major coastal regions.

CHAPTER 10 Ecological Restoration

"Sunset in Yosemite," a 19th-century painting by Albert Bierstadt, illustrates the idea of the balance of nature. Nature is portrayed as still (in a single state), beautiful, and without people.

● LEARNING OBJECTIVES

Restoration ecology is a new field. In this chapter, we explore the concepts of restoration ecology, with a special emphasis on how ecosystems restore themselves through the process of ecological succession. After reading this chapter, you should understand:

- What ecological restoration means.
- What kinds of goals are possible for ecological restoration.
- What basic approaches, methods, and limits apply to restoration.
- How an ecosystem restores itself following a disturbance through ecological succession.
- What role disturbances play in the persistence of ecosystems.
- How physical forces and biological processes affect the land.
- Why ecosystems do not maintain a steady-state condition.

Restoring a Ponderosa Forest

At Fort Valley Experimental Forest, about seven miles northwest of Flagstaff, Arizona, a ponderosa pine forest was in trouble—in danger of suffering a seriously damaging wildfire. The reason the forest was in trouble was ironic: It was suffering from too much care of the wrong kind.

Ponderosa forests burn naturally, and the species that live in them have evolved with and adapted to fire over millions of years. Natural fires occur frequently and as a result are light, clearing out many young trees but not burning through the bark of most of the large, thick-barked, seed-producing seeds. Nor are the fires intense enough to destroy much of the organic matter in the soil. Ponderosa seeds germinate after a fire. The young trees do well in bright light, not in the shade of other trees, and therefore require the openings that traditionally result from wildfires. A century ago, the woodlands were open, with scattered trees among which grasses grew. Most of the trees were large and mature.

But throughout most of the 20th century, forests fires were believed to be only bad for forests, and so the general practice was to suppress fires and fight them once they started. The ponderosa pine forest at Fort Valley had been protected from fires for over a century; and during that century, the forest had changed. By the 1980s, the forest was a thicket of small and large stems. Whereas in 1876, density averaged 57 trees per hectare (23 trees per acre), in 1992, after a century of fire suppression, density had increased to almost 3,000 trees per hectare (1,200 trees per acre).

A restoration project was launched at Fort Valley with the goal of returning the forest to the way it was in about 1876. But this could not be done simply by reintroducing fire into the modern, densely stocked forest. The high density of trees in 1992 made it extremely likely to burn. In addition, a large buildup of organic matter had accumulated in the soil and on the soil surface. The kind of fire that would occur under these conditions could be highly destructive. The fire could be hot enough to kill even the mature, seed-bearing trees, and it might burn through the soil, destroying the soil's organic matter. Such a fire could either permanently eliminate ponderosa pines from the burned area or so damage the trees and the soil that recovery would be extremely slow.

Thus, the restoration process required that the forest be returned to a lower fuel condition before fire could be reintroduced. Under the direction of Professor Wallace Covington of Northern Arizona University, Flagstaff, trees were carefully removed to reproduce

conditions typical in the 19th century, with trees clumped around large grassy openings. The researchers removed, one by one, 2,200 trees per hectare (900 trees per acre) a process called thinning. They also raked organic matter from the forest floor. This was a labor-intensive approach.

When these modified stands were burned, fire intensity was low; flames averaged about six inches high. Figure 10.1*a* shows part of the Coconino Natural Forest in Arizona prior to thinning. The same area is shown on Figure 10.1*b* after thinning and controlled burning. The fires did not kill the mature trees and left some of the younger trees to replace the older ones.

(a)

(b)

Figure 10.1 • (*a*) Ponderosa pine forest in Coconino National Forest after a period of fire suppression. The forest became much denser and the threat of seriously damaging wildfires was great. (*b*) In a restoration project, stands were thinned and controlled burns were conducted.

Since then, grass has returned to the openings. Now the modified forest can undergo its historic pattern of frequent but light fires. The forest has been restored, but the case study of Fort Valley illustrates that ecological restoration can be complex and can require great care and considerable effort.[1]

 Wherever people have lived, they have changed their environment. In North America, this was true of the original people to arrive on the continent, just as it has been since European settlement. While some of the changes people make are considered desirable, others are considered degradation. For lands and waters considered degraded, the question is: How can we restore them? In recent years, a new field called **restoration ecology** *has developed within the science of ecology. Its goal is to return damaged ecosystems to some set of conditions considered functional, sustainable, and "natural." Whether restoration can always be successful is still an open question. In the case of some ecosystems and species, it appears that success can be achieved, but at Fort Valley, success may require great effort.*

10.1 Restore to What?

The idea of ecological restoration raises a curious question: Restore to what?

The Balance of Nature

Until the second half of the 20th century, the predominant belief in Western civilization was that any kind of nature—a forest, a prairie, an intertidal zone—left undisturbed by human beings achieved a single condition that would persist indefinitely. This condition, as mentioned in Chapter 3, has been known as the **balance of nature**. The major tenets of a belief in the balance of nature are as follows:

Nature undisturbed achieves a permanency of form and structure that persists indefinitely.

When disturbed, and the disturbing force is removed, nature returns to exactly the same permanent state.

In this permanent state of nature, there is a "great chain of being" with a place for each creature (a habitat and a niche) and each creature is in its appropriate place.

The roots of these ideas can be traced to Greek and Roman philosophies about nature, but they have played an important role in modern environmentalism as well.

In the early 20th century, ecologists continued this belief, formalizing it. They believed that succession proceeded to a fixed, classic condition which they called **climax state**, defined as a steady-state stage that would persist indefinitely and have maximum organic matter, maximum storage of chemical elements, and maximum biological diversity. At that time, people thought that wildfires were always detrimental to wildlife, vegetation, and natural ecosystems. *Bambi*, the Walt Disney movie of the 1930s, expressed this belief; a fire only brought death to friendly animals. In the United States, Smokey Bear is a well-known symbol employed for many decades by the U.S. Forest Service to warn visitors to national forests to be careful with fire and avoid setting wildfires. The message is that wildfires are always harmful to wildlife and ecosystems. All this suggests a belief that the balance of nature does in fact exist. If that were true, the answer to the question, "Restore to what?," would be simple: restore to the original, natural, permanent condition. The method of restoration would be simple, too: Get out of the way, and let nature take its course.

Since the second half of the 20th century, though, ecologists have learned that nature is not constant and that forests, prairies—all ecosystems—undergo change. Moreover, since change has been a part of natural ecological systems for millions of years, many species have evolved and adapted to change. Indeed, many require specific kinds of change in order to survive.

Dealing with change—natural and human-induced—poses questions of human values as well as science. This is illustrated by wildfires in forests, grasslands, and shrublands, which can be extremely destructive to human life and property. Scientific understanding tells us fires are natural and some species require them. But whether we choose to allow fires to burn, or light fires ourselves, is a matter of values. In 1991, a wildfire that started in chaparral shrubland in Santa Barbara, California, burned just 83 minutes but did $500 million damage. A few years later, a wildfire in Oakland, California, did $1 billion damage. Restoration ecology depends on science to discover what used to be, what is possible, and how different goals can be achieved. The selection of goals for restoration is a matter of human values.

The Boundary Waters Canoe Area Wilderness: An Example of the Naturalness of Change

One of the best-documented examples of natural disturbance is the role of fire in the northern woods of North America. The Boundary Waters Canoe area—an area of more than 400,000 ha (a million acres) in northern Minnesota designated as wilderness under the U.S. Wilderness Act—exemplifies nature relatively undisturbed by people. The area is no longer open to logging or other direct human-induced disturbances. In the early days of European exploration and settlement of North America, French voyagers traveled through this region hunting and trading for furs. In some places, logging and farming were common in the 19th and early 20th centuries, but for the most part the land has been relatively untouched. In spite of the lack of human influence, the forests show a persistent history of fire. Fires occur somewhere in this forest almost every year, and on the average the entire area burns once in a century. Fires cover areas large enough to be visible by satellite remote sensing (Figure 10.2).

When fires occur in the forests of the Boundary Waters Canoe Area at natural rates and natural intensities, they have some beneficial effects. For example, trees in unburned forests appear more susceptible to insect outbreaks and disease. Thus, recent ecological research suggests that wilderness depends on change and that succession and disturbance are continual processes. The landscape is dynamic.[2]

Goals of Restoration: What Is "Natural"?

With the examples of the forests at Fort Valley and in the Boundary Waters Canoe Area, we can now return to the question: "Restore to what?" If an ecosystem passes naturally through many different states, and all of them are "natural," and if change itself—including certain kinds of wildfire—is natural, then what can it mean to "restore" nature? And how can restoration that involves such things as wildfires occur without undue damage to human life and property?

Can we restore an ecological system to any one of its past states and claim that this is natural and successful restoration? A frequently accepted answer is that restora-

Figure 10.2 • Forest fires can be natural or induced by human beings. This figure shows the change in a large area of the Superior National Forest in Minnesota between 1973 and 1983, as observed by the Landsat Satellite. The black boundaries show a central corridor where logging is permitted, surrounded by the Boundary Waters Canoe Area at the top and bottom. This is an area that is protected from all uses except certain kinds of recreation. The bright yellow shows areas that were clear of trees in 1973 but had regenerated to young forest by 1983. Most of this change is due to regrowth following a large fire that burned both inside and outside the wilderness. Red areas were forested in 1973 but cleared in 1983. Most of these are outside the wilderness, and some of these are due to logging (red) and some to fire or storms. Greens show areas that were forested both years.

Table 10.1   Some Possible Restoration Goals

Goal	Approach
1. Pre-Industrial	Maintain ecosystems as they were in 1500 A.D.
2. Presettlement (e.g., of North America)	Maintain ecosystems as they were about 1492 A.D.
3. Preagriculture	Maintain ecosystems as they were about 5,000 B.C.
4. Before any significant impact of human beings	Maintain ecosystems as they were about 10,000 B.C.
5. Maximum production	Independent of a specific time
6. Maximum diversity	Independent of a specific time
7. Maximum biomass	Independent of old growth
8. Preserve a specific endangered species	Whatever stage it is adapted to
9. Historic range of variation	Create the future like the known past

tion means restoring an ecosystem to its past (historic) range of variation and to an ability to sustain itself and its crucial functions, including the cycling of chemical elements (see Chapter 4), the flow of energy (Chapter 9), and the maintenance of the biological diversity that existed previously (Chapters 7 and 8). *According to this interpretation, ecological restoration means restoring processes and a set of conditions that are known to have existed for that ecosystem.* From this point of view, we examine populations that have declined and ecosystems that have been damaged and try to learn what is lacking. Then we go about trying to restore what is lacking.

Other ecologists have other suggestions. Table 10.1 gives a number of possibilities that have been proposed as goals of restoration. At the extreme are those who argue that all human impacts on nature are "unnatural" and therefore undesirable and that the only true goal of restoration is to bring nature back to a condition that existed prior to human influence. The anthropologist Paul S. Martin takes this position. He proposes that the only truly "natural" time is before any significant human influence occurs. Specifically, he suggests that restoration to conditions of 10,000 B.C.—before farming—should be our goal. He even suggests introducing the African elephant into North America to replace the extinct mastodon, whose extinction, he argues, was the result of hunting by Indians.

The new way of thinking about restoration leaves open choice. The choice is a matter of science and values. Science tells us what has been and what can be; our values determine what we want nature to be. *There is no single perfect condition.* However, for some of the goals in Table 10.1, specific conditions are especially desirable. It is possible to restore an ecosystem so that, most of the time, it supports conditions that people desire for one reason or another.

The idea that there might be human choice involved in restoration is controversial. Many people want to restore nature to what is "natural." But upon examination, the word *natural* turns out to be complicated. So when we ask "Restore to what?" we understand now that this means

restoring processes and a set of conditions and that some choice is involved. Restoration no longer means returning an ecosystem to a single, supposedly perfect state.

 ## 10.2   What Needs to Be Restored?

Specific ecosystems of all types have undergone degradation, and an argument can be made that examples of each kind of ecosystem need restoration. However, certain kinds of ecosystems have undergone especially widespread loss and degradation and are therefore a focus of current attention. In addition to forests, these include wetlands (both freshwater, Figure 10.3, and coastal); grasslands, especially the North American prairie; streams and rivers and the riparian zones alongside them; lakes; and habitats of threatened and endangered species. It also includes areas that people desire to restore for aesthetic and moral reasons, showing once again that restoration involves values. In this section, restoration of wetlands, rivers, and streams and of prairies are discussed briefly.

Wetlands, Rivers, and Streams

Both freshwater and coastal wetlands have been greatly altered over large areas, especially in North America, during the past 200 years. It is estimated that California has lost more than 90% of its wetlands, both freshwater and coastal, for example. Total wetland loss for the U.S. is about 50% (see Chapter 20).

One of the largest and most expensive single restoration projects in the United States is the restoration of the Kississimmi River in Florida. This river was channelized by the U.S. Army Corps of Engineers to provide a ship passage through Florida. However, this water transportation never developed, and the river and its adjacent ecosystems were greatly altered. Several hundred million dollars will be spent to put the Kississimmi River back as it was before channelization, including restoring the meandering flow of the river channel and replacing the

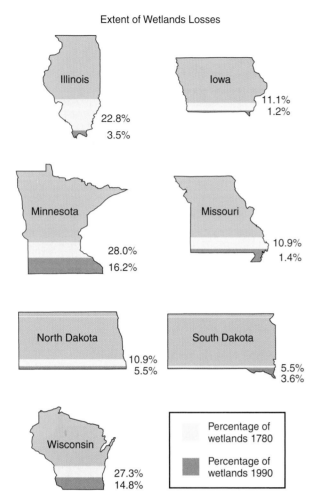

Extent of Wetlands Losses

Illinois
22.8%
3.5%

Iowa
11.1%
1.2%

Minnesota
28.0%
16.2%

Missouri
10.9%
1.4%

North Dakota
10.9%
5.5%

South Dakota
5.5%
3.6%

Wisconsin
27.3%
14.8%

| | Percentage of wetlands 1780 |
| | Percentage of wetlands 1990 |

Figure 10.3 • Loss of wetlands in the Midwestern United States. These maps illustrate the extensiveness of wetland losses. In some states, the losses are even greater. It is estimated that about 90% of the wetlands in California have been lost, for example. (*Source*: Based on Scientific Assessment and Strategy Team, *Science for Floodplain Management into the 21st Century*, p. 84. Faber, 1996.)

soil layers in the order in which they had lain on the bottom of the river prior to channelization (see Chapter 20).

The Everglades National Park is also the subject of a variety of restoration efforts. As another example, a large effort is underway to restore bottomland—that is, wetland—forests of the lower Mississippi Valley (see discussion in Chapter 14).

Prairie Restoration

As noted in Chapter 8, prairies once occupied more land area in the United States than any other kind of ecosystem. Today, only a few small remnants of prairie remain.

Prairie restoration is of two kinds. In a few places, original prairie exists that has never been plowed. Here, the soil structure is intact, and restoration is simpler. One of the best known of these areas is the Kanza Prairie near Manhattan, Kansas. In other places, where the land has

been plowed, restoration is more complicated. Nevertheless, the restoration of prairies has gained considerable attention in recent decades, and restoration of prairie on previously plowed and farmed land is occurring in many midwestern states. The Allwine Prairie, which is within the city limits of Omaha, Nebraska, has been undergoing restoration from farm to prairie for many years. Similarly, prairie restoration has been taking place near Chicago.

A peculiarity about the history of prairies is that, while most prairie land was converted to agriculture, this was not done along roads and railroads, so that long, narrow strips of unplowed native prairie remain on these rights-of-way. In Iowa, for example, prairie once covered more than 80% of the state—11 million ha (28 million acres). More than 99.9% of prairie land has been converted to other uses, primarily agriculture. However, along roadsides, there are 242,000 ha (600,000 acres) of prairie—more than in all of Iowa's county, state, and federal parks. These roadside and railway stretches of prairie provide some of the last habitats for native plants, and restoration of prairies elsewhere in Iowa are making use of these habitats as seed sources.[3]

10.3 When Nature Restores Itself: The Process of Ecological Succession

We have been discussing human attempts to restore damaged ecosystems. Often, the damage has been caused by human influence. Natural areas are subject to disturbances of many kinds, however. Storms and fires, for example, have always been a part of the environment.[4] Furthermore, recovery of disturbed ecosystems can occur naturally, through a process called **ecological succession**. This natural recovery can occur if the damage is not too great. Sometimes, though, the rate of recovery is longer than humans would like it to be.

We can classify ecological succession as either primary or secondary. **Primary succession** is the initial establishment and development of an ecosystem where one did not exist previously. **Secondary succession** is reestablishment of an ecosystem following disturbances. In secondary succession, there are remnants of a previous biological community, including such things as organic matter and seeds. Forests that develop on new lava flows (Figure 10.4*a*) and at the edges of retreating glaciers (Figure 10.4*b*) are examples of primary succession. Forests that develop on abandoned pastures or following hurricanes, floods, or fires are examples of secondary succession (see A Closer Look 10.1).

Succession is one of the most important ecological processes, and the patterns of succession have many management implications. We see examples of succession all around us. When a house lot is abandoned in a city, weeds begin to grow. After a few years, shrubs and trees

(a)

(b)

Figure 10.4 • Primary succession. (*a*) Forests developing on new lava flows in Hawaii and (*b*) at the edge of a retreating glacier.

can be found; secondary succession is taking place. A farmer weeding a crop and a homeowner weeding a lawn are both fighting against the natural processes of secondary succession.

Patterns in Succession

Succession occurs in most kinds of ecosystems, and when it occurs, it follows certain general patterns. Next, we consider succession in three classic cases involving forests: (1) on dry sand dunes along the shores of the Great Lakes in North America, (2) in a northern freshwater bog, and (3) in an abandoned farm field.

Dune Succession

Sand dunes are continually being formed along sandy shores and then breached and destroyed by storms. In Indiana, on the Shores of Lake Michigan, soon after a dune is formed, dune grass invades. This grass has special adaptations to the unstable dune. Just under the surface, it puts out runners with sharp ends (if you step on one, it will hurt). The dune grass rapidly forms a complex network of underground runners, criss-crossing almost like a coarsely woven mat. Above the ground, the green stems carry out photosynthesis, and the grasses grow.

Once the dune grass is established, its runners stabilize the sand, and seeds of other plants are less easily buried too deep or blown away. The seeds germinate and grow. An ecological community of many species begins to develop. The plants of this early stage tend to be small, to grow well in bright light, and to withstand the harshness of the environment—high temperatures in the summer, low temperatures in the winter, and intense storms.

Slowly, larger plants, such as eastern red cedar and eastern white pine, are able to grow on the dunes.

Eventually, a forest develops which may include species such as beech and maple. Such a forest can persist for many years, but at some time an extreme storm breaches even these heavily vegetated dunes, and the process begins again (Figure 10.6).

Bog Succession

A bog is an open body of water with surface inlets—usually small streams—but no surface outlet. As a result, the waters of a bog are quiet, flowing slowly if at all. Many bogs that exist today originated as lakes that filled depressions in the land, which in turn were created by actions of glaciers during the Pleistocene ice age.

Succession in a northern bog, such as Livingston Bog in Michigan (Figure 10.7), begins when a sedge (grasslike herbs) puts out floating runners (Figure 10.8 *a*, *b*). These runners form a complex matlike network, similar to the matlike network formed by dune grass. The stems of the sedge grow on the runners and carry out photosynthesis. Wind blows particles onto the mat, and a soil, of a kind, develops. Once the mat and its soil are present, seeds of other plants land on the mat and do not sink into the water. They can germinate. The floating mat becomes thicker, and small shrubs and trees, adapted to wet environments, grow. In the north, these include species of the blueberry family.

The bog also fills in from the bottom as streams carry fine particles of clay into it (Figure 10.8 *b*, *c*). At the shoreward end, the floating mat and the bottom sediments meet, forming a solid surface. But farther out, a quaking bog occurs. You can walk on this quaking bog mat; and if you jump up and down, all the plants around you bounce and shake. The mat is really floating. Eventually, as the bog fills in from the top and the bottom, trees that can withstand wetter conditions, such as

An Example of Forest Secondary Succession

Within a few years after a field is abandoned, seeds of many kinds sprout, some of short-lived weedy plants and some of trees (Figure 10.5*a*). After a few years have passed, certain species, generally referred to as pioneer species, become established. In the Poconos of Pennsylvania, red cedar is such a species (Figure 10.5*b*). In New England, white pine, pin cherry, white birch, and yellow birch are especially abundant. These trees are fast growing in bright light and have widely distributed seeds. For example, red cedar seeds are eaten by birds and dispersed; birches have very light seeds that are dispersed widely by the wind. After several decades, forests of the pioneer species are well established, forming a dense stand of trees (Figure 10.5*c*).

Once the initial forest is established, other species begin to grow and become important. Typical dominants in the Northeastern United States are sugar maple and beech. These later successions of trees are slower growing than the ones that came into the forest first, but they have other characteristics that make them well adapted to the later stages in succession. These species are what foresters call shade tolerant; they grow relatively well in the deep shade of the redeveloping forest.

After three or four decades, most of the short-lived species have matured, borne fruit, and died; these trees cannot grow in the shade of the forest and so do not regenerate in a forest that has been reestablished. For example, after five or six decades a New England forest is a rich mixture of birches, maples, beeches, and other species. The trees are a variety of sizes, although dominant trees are generally taller than those dominating earlier stages. After 1 or 2 centuries, such a forest will be composed mainly of shade-tolerant species.

(*a*)

(*b*)

(*c*)

Figure 10.5 ● A series of photographs of succession on an abandoned pasture in the Poconos: (*a*) a second-year field; (*b*) young red cedar trees, grasses, and other plants some years after a field was abandoned; (*c*) mature forests developed along old stone farm walls.

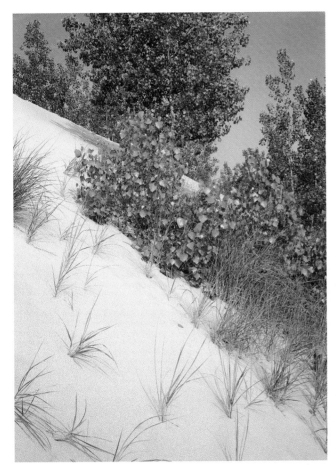

Figure 10.6 • Dune succession on the shores of Lake Michigan. Dune grass shoots appear scattered on the slope where they emerge from underground runners.

northern cedar, black spruce, and balsam fir, grow. The formerly open water bog becomes a wetland forest.

Old Field Succession

In the eastern United States, such as the states of New England, a great deal of land was cleared and farmed in the 18th and 19th centuries. Today, much of this land has been abandoned for farming and allowed to grow back to forest (see Figure 10.5). The first plants to enter the abandoned crop or pasture lands are plants of small stature adapted to the harsh and highly variable conditions of an opening—a wide range of temperatures and precipitation. As these plants become established, other, larger plants enter. Eventually, large trees such as sugar maple, beech, yellow birch, and white pine grow on the land, forming a dense forest.

General Patterns of Succession

Reviewing these three examples of ecological succession involving forests, you can see common elements in them, even though the environments are very different. Common elements of succession in such cases include:

1. An initial kind of vegetation specially adapted to the unstable conditions. These plants are typically small in stature, with adaptations that help stabilize the physical environment.

2. A second stage with plants still of small stature, rapidly growing, with seeds that spread rapidly, occupying the site.

Figure 10.7 • Livingston Bog, a famous bog in the northern part of Minnesota's lower peninsula.

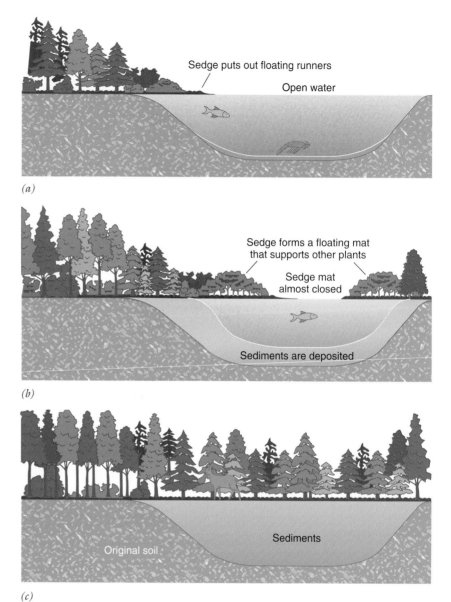

Sedge puts out floating runners

Open water

(a)

Sedge forms a floating mat
that supports other plants

Sedge mat
almost closed

Sediments are deposited

(b)

Original soil

Sediments

Figure 10.8 • Diagram of bog succession. Open water (*a*) is transformed through formation of a floating mat of sedge and deposition of sediments (*b*) into wetland forest (*c*).

(c)

3. A third stage in which larger plants—including trees—enter and begin to dominate the site.

4. A fourth stage in which mature forest develops.

Although we list four stages, it is common practice to combine the first two and to speak of early, middle, and late successional stages. These general patterns in succession can be found in most ecosystems, although the species differ. The stages in succession are described here in terms of vegetation, but similarly adapted animals and other life forms are associated with each stage. Other general properties of the process of succession are discussed later in this chapter.

Species characteristic of the early stages of succession are called pioneers, or **early successional species**. These species have evolved and are adapted to the environmen-

tal conditions in early stages in succession. Plant species that dominate late stages of succession, called **late successional species**, tend to be slower growing and longer lived. These species have evolved and are adapted to environmental conditions in the late stages. For example, they grow well in shade and have seeds that, while not as widely dispersing, can persist a rather long time.

In early stages of succession, biomass and biological diversity increase (Figure 10.9). In middle stages of succession, we find trees of many species and many sizes. Gross and net production (see Chapter 8) change during succession as well; gross production increases, and net production decreases. Chemical cycling also changes. The organic material in the soil increases, as does the amount of chemical elements stored in the soils and trees.[5] We look more closely at chemical cycling changes in the next section.

Figure 10.9 • Graphs showing changes in biomass and diversity with succession.

10.4 Succession and Chemical Cycling

One of the important effects of succession is a change in the storage of chemical elements necessary for life. On the land, the storage of chemical elements (including nitrogen, phosphorus, potassium, and calcium, essential for plant growth and function) generally increases during the progression from the earliest stages of succession to middle succession. There are two reasons for this.

First, organic matter stores chemical elements; as long as there is an increase in organic matter within the ecosystem, there will be an increase in the storage of chemical elements. This is true for live and dead organic matter. Additionally, many plants have root modules containing bacteria that can assimilate atmospheric nitrogen, which is then used by the plant in a process known as nitrogen fixation.

The second reason is indirect: The presence of live and dead organic matter helps retard erosion. Both organic and inorganic soil can be lost to erosion because of the effects of wind and water. Vegetation tends to prevent such losses and therefore causes an increase in total stored material.

Organic matter in soil contributes to the storage of chemical elements in two ways. First, it contains chemical elements itself. Second, dead organic matter functions as an ion exchange column that holds onto metallic ions

that would otherwise be transported in the groundwater as dissolved ions and lost to the ecosystem. As a general rule, the greater the volume of soil and the greater the percentage of organic matter in the soil, the more chemical elements will be retained.

The amount of chemical elements stored in a soil depends not only on the total volume of soil but also on its storage capacity for each element. Chemical storage capacity of soils varies with the average size of the soil particles. Large, coarse particles, like sand, have a smaller surface area per unit volume and can store a smaller quantity of chemical elements. Clay, which is made up of the smallest particles, stores the greatest quantity of chemical elements.

Soils contain greater quantities of chemical elements than do live organisms. However, much of what is stored in a soil may be relatively unavailable, or it may only become available slowly, because the elements are tied up in complex compounds that decay slowly. In contrast, the elements stored in living tissues are readily available to other organisms through food chains.

Rates of cycling and average storage times are system characteristics (as discussed in Chapter 3). Soils store more elements than live tissue but cycle them at a slower rate. The increase in chemical elements that occurs in the early and middle stages of succession does not continue indefi-

Changes in Chemical Cycling During a Disturbance

When an ecosystem is disturbed by fire, storms, or human actions, changes occur in chemical cycling. For example, when a forest is burned, complex organic compounds (such as wood) are converted to smaller inorganic compounds (including carbon dioxide, nitrogen oxides, and sulfur oxides). Some of the inorganic compounds from the wood are lost to the ecosystem during the fire as particles of ash that are blown away or as vapors that escape into the atmosphere and are distributed widely. Other compounds are deposited on the soil surface as ash. These compounds are comparatively highly soluble in water and readily available for vegetation uptake. Therefore, immediately after a fire, the availability of chemical elements increases. This is true even if the ecosystem as a whole has undergone a net loss in total stored chemical elements.

If sufficient live vegetation remains after a fire, then the sudden, temporary increase, or *pulse*, of newly available elements is taken up rapidly, especially if it is followed by a moderate amount of rainfall (enough for good vegetation growth but not so much as to cause excessive erosion). The pulse of inorganic nutrients can then lead to a pulse in growth of vegetation and an increase in the amount of stored chemical elements in the vegetation. This in turn provides an increase in nutritious food for herbivores, which can subsequently undergo a population increase. The pulse in chemical elements in the soil can therefore have effects that extend through the food chain.

Other disturbances on the land have effects similar to those of fire. For example, severe storms, such as hurricanes and tornadoes, knock down and kill vegetation. The vegetation decays, increasing the concentration of chemical elements in the soil, which are then available for vegetation growth. Storms have another effect in forests; when trees are uprooted,

the soil can be turned over, and chemical elements that were near the bottom of the root zone are brought to the surface, where they are more readily available.

Knowledge of the changes in chemical cycling and the availability of chemical elements from the soil during succession can be useful to us in restoring damaged lands. We know that nutrients must be available within the rooting depth of the vegetation. The soil must have sufficient organic matter to hold onto nutrients. Restoration will be more difficult on lands where the soil has lost its organic matter and has been leached than on lands where the soil retains organic matter and has been little leached. A leached soil is one that has lost nutrients as water drains through it, especially acidic water, and dissolves and carries away chemical elements. Heavily leached soils, such as those subjected to acid rain and acid mine drainage, pose special challenges to the process of land restoration.

nitely. If an ecosystem persists for a very long period with no disturbance, it will experience a slow but definite loss of stored chemical elements. Thus, the ecosystem will slowly run downhill and become *depauperate*—that is, less

able to support rapid growth, high biomass density, and high biological diversity (Figure 10.10).[6] The changes in chemical cycling during disturbance and successional recovery are discussed further in A Closer Look 10.2.

(a)

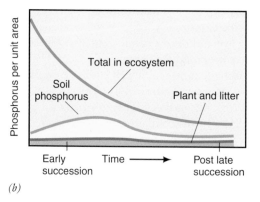

(b)

Figure 10.10 ● (*a*) Hypothesized changes in soil nitrogen during the course of soil development. (*b*) Change in total soil phosphorus over time with soil development. (*Source:* P. M. Vitousek and P. S. White, 1981, "Process Studies in Forest Succession," in D. C. West, H. H. Shugart, and D. B. Botkin, eds. *Forest Succession: Concepts and Applications* [New York: Springer-Verlag, 1981], Figure 17.1, p. 269.)

10.5 Species Change in Succession: Do Early Successional Species Prepare the Way for Later Ones?

To restore an ecosystem, it is important to understand what causes one species to replace another during the succession process. If we understand these causes and effects, then we can use them to better restore ecosystems. These processes are patterns of interaction among earlier and later species in succession: Facilitation; Interference; Life history differences; and Chronic patchiness (Table 10.2).[7,8]

Facilitation

In dune and bog succession, the first plant species, dune grass and floating sedge, prepare the way—make it possible—for other species to grow. This is called **facilitation** (early successional species facilitate the ability of later successional species to become established; Figure 10.11a).

Facilitation has been found to take place in tropical rain forests.[9] Early successional species speed the reappearance of the microclimatic conditions that occur in a mature forest. Because of the rapid growth of early successional plants, in tropical forests, temperature, relative humidity, and light intensity at the soil surface can reach conditions similar to those of a mature rain forest after only 14 years.[5] Once these conditions are established, species that are adapted to deep forest shade can germinate and persist.

Knowing the role of facilitation can be useful in the restoration of damaged areas. Plants that facilitate the presence of others should be planted first, as on sandy areas, where dune grasses can help hold the soil before attempts are made to plant larger shrubs or trees.

Interference

Facilitation does not always occur. Sometimes certain early successional species prevent for a while and so delay the entrance of other species (Figure 10.11b). For example, in the old fields of mid-Atlantic states, such as the states from Connecticut to Virginia, among the early suc-

Facilitation—pine provides shade that helps oaks

(a)

Interference—grass interferes with seeds of other species

Grass continues to remain grass

Bare ground goes to pine tree

(b)

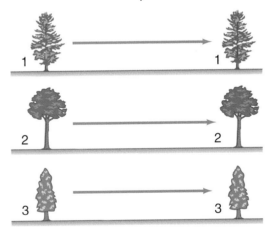

Chronic patchiness—no effect of one species on another

(c)

Figure 10.11 • Three of the four patterns of interaction among species in ecological succession. (a) *Facilitation.* As Henry David Thoreau observed in Massachusetts more than 100 years ago, pines provide shade and act as "nurse trees" for oaks. Pines do well in openings. If there were no pines, few or no oaks would survive. Thus the pines *facilitate* the entrance of oaks. (b) *Interference.* Some grasses that grow in open areas form dense mats that prevent seeds of trees from reaching the soil and germinating. The grasses *interfere* with the addition of trees. (c) *Chronic patchiness.* Earlier-entering species neither help nor interfere with others species; instead, as in a desert, the physical environment dominates.

Table 10.2 Patterns of Interaction among Earlier and Later Species in Succession

1. *Facilitation.* One species can prepare the way for the next (and may even be necessary for the occurrence of the next).

2. *Interference.* Early successional species can, for a time, prevent the entrance of later successional species.

3. *Life history differences.* One species may not affect the time of entrance of another; two species may appear at different times during succession because of differences in transport, germination, growth, and longevity of seeds.

4. *Chronic patchiness.* Succession never occurs, and the species that enters first remains until the next disturbance.

Source: J. H. Connell and R. O. Slatyer, "Mechanisms of Succession in Natural Communities and Their Role in Community Stability and Organization," *American Naturalist* III (1977): 1119–1144; S. T. A. Pickett, S. L. Collins, and J. J. Armesto, "Models, Mechanisms, and Pathways of Succession," *Botanical Review* 53 (1987): 335–371.

cessional species are grasses that form dense cover, including the prairie grass little bluestem. The living and dead stems of this grass form a mat so dense that seeds of other plants that fall onto it cannot reach the ground and therefore do not germinate. The grass interferes with the entrance of other plant species, a process called, naturally enough, **interference**.

Interference does not last forever, however. Eventually, some breaks occur in the grass mat—perhaps from physical events such as surface water erosion, the death of a patch of grass from disease, or removal by fire. Breaks in the grass mat allow seeds of trees such as red cedar to reach the ground. Red cedar is also adapted to early succession because its seeds are spread rapidly and widely by birds who feed on them and because this species can grow well in the bright light and otherwise harsh conditions of early succession. Once started, red cedar soon grows taller than the grasses, shading them so much that they cannot grow. More ground is open, and the grasses are eventually replaced.

In parts of Asia, interference occurs where bamboo grows, and, in tropical areas, where another, smaller grass, *Imperata*, grows. Like little bluestem in the United States, these grasses form stands so dense that seeds of other, later successional species cannot reach the ground, germinate, or obtain enough light, water, and nutrients to survive. *Imperata* either replaces itself or is replaced by bamboo, which then replaces itself.[9] Once established, *Imperata* and bamboo appear able to persist for a long time. Once again, when and if breaks occur in the cover of these grasses, other species can germinate and grow, and a forest eventually develops.

Life History Differences

In other cases, species do not so much affect one another. Rather, differences in the life histories of the species allow some to arrive first and grow quickly, while others arrive later and grow more slowly. An example of such a **life history difference** is seed disbursal. When early successional species produce seeds, the seeds are readily transported by wind or animals and so reach a clearing sooner and grow faster than seeds of late successional species. In many forested areas of eastern North America, for example, birds eat the fruit of cherries and red cedar, and their droppings contain the seeds, which are spread widely.

In contrast, late successional species can have life histories that bring them in much later. For example, sugar maple can grow in open areas, but its seeds take longer to travel, and its seedlings are tolerant of shade. Beech produces large nuts that store much food for a newly germinated seedling. This helps the seedling establish itself in the deep shade of a forest, providing time for it to be able to feed itself through photosynthesis. But these seeds are heavy and are moved relatively short distances by seed-eating animals.

Chronic Patchiness

A fourth possibility is that species just do not interact and that succession, as it has been described, does not take place. This is called **chronic patchiness** (Figure 10.11*c*), and it is the case in some deserts. For example, in the warm deserts of California, Arizona, and Mexico, the major shrub species grow in patches, which are often composed of mature individuals with few seedlings. These patches tend to persist for long periods until there is a disturbance.[10] Similarly, in highly polluted environments, a sequence of species replacement may not occur.

What kinds of changes occur during succession depend on the complex interplay between life and its environment. Life tends to build up, or *aggrade*, whereas nonbiological processes in the environment tend to erode and degrade. In harsh environments, where energy or chemical elements required for life are limited and where disturbances are frequent, the physical, degrading environment dominates, and succession does not occur.

 10.6 **Applying Ecological Knowledge to Restore Heavily Damaged Lands and Ecosystems**

An example of how ecological succession can aid in the restoration of heavily damaged lands is the restoration of mine tailings in Great Britain. Mining is a widespread cause of destruction and damage to land. In some areas, efforts are being made to undo mining damage and restore the landscape, if not to a natural ecosystem, then to some kind of vegetation cover. In Great Britain, where some mines have been used since medieval times, approximately 55,000 ha (136,000 acres) have been damaged by mining. Recently, programs have been initiated to restore these damaged lands to useful biological production, to remove toxic pollutants from the mines and mine tailings, and to restore the visual attractiveness of the landscape.[11]

One area damaged by a long history of mining lies within the British Peak District National Park, where lead has been mined since the Middle Ages and waste tailings are as much as 5 m (16.4 ft) deep. The first attempts to restore this area used a modern agricultural approach: heavy application of fertilizers and planting of fast-growing agricultural grasses. It was hoped that rapidly growing grasses would be able to revegetate the site quickly. These grasses quickly green on the good soil of a level farm field; the plan was that, with fertilizer, they would do the same in this situation. But after a short period of growth, the grasses died. On the poor soil, leached of its nutrients and lacking organic matter, erosion continued, and the fertilizers that had been added were soon leached away by water runoff. Without the fertilizers, the fast-growing grasses died. As a result, the areas quickly returned to a barren condition.

How Can We Evaluate Constructed Ecosystems?

What happens when restoring damaged ecosystems, as described for abandoned land mines in Great Britain in this chapter, is not an option? In such cases, those responsible for the damage may be required to establish alternative ecosystems to replace the damaged ones.

An example involved some saltwater wetlands on the coast of San Diego County, California. In 1984, construction of a flood control channel and two projects to improve interstate freeways damaged an area of saltwater marsh. The projects were of concern because California had lost 91% of its wetland area since 1943 and the few remaining coastal wetlands were badly fragmented. In addition, the damaged area provided habitat for three endangered species: the California least tern, the light-footed clapper rail, and a plant, the salt-marsh bird's beak.

The California Department of Transportation, with funding from the Army Corps of Engineers and the Federal Highway Administration, was required to compensate for the damage by constructing new areas of marsh in the Sweetwater Marsh National Wildlife Refuge. To meet these requirements, eight islands, known as the Connector Marsh, with a total area of 4.9 ha, were constructed in 1984. An additional 7 ha area, known as Marisma de Nacion, was established in 1990.

Goals for the constructed marsh, which were established by the U.S. Fish and Wildlife Service, included:

1. Establishment of tide channels with sufficient fish to serve as food for the California least tern.

2. Establishment of a stable or increasing population of salt-marsh bird's beak for three years.

3. Establishment of habitat for the light-footed clapper rail with the elements needed for successful foraging and nesting of at least two breeding pairs.

The Pacific Estuarine Research Laboratory (PERL) at San Diego State University was selected to monitor progress on the goals and conduct research on the constructed marsh. In 1997, PERL reported that goals for the least tern and bird's beak had been met, but attempts to establish a habitat suitable for the rail had been only partially successful (see table).

During the past decade, PERL scientists have conducted extensive research on the constructed marsh to determine the reasons for its limited success. They found that rails live, forage, and nest in cordgrass more than 60 cm tall. Nests are built of dead cordgrass attached to stems of living cordgrass so that the

Species	Mitigation Goals	Progress in Meeting Requirements	Status as of 1997
California least tern	Tidal channels with 75% of the fish species and 75% of the number of fish found in natural channels	Met standards	In compliance
Salt-marsh bird's beak	Through reintroduction, at least 5 patches (20 plants each) that remain stable or increase for 3 years	Did not succeed on constructed islands but an introduced population on natural Sweetwater Marsh thrived for 3 years (reached 140,000 plants); continue to monitor because plant is prone to dramatic fluctuations in population	In compliance
Light-footed clapper rail	Seven home ranges (82 ha), each having tidal channels with:	Constructed	Not in compliance
	a. Forage species equal to 75% of the invertebrate species and 75% of the number of invertebrates in natural areas	Met standards	
	b. High marsh areas for rails to find refuge during high tides	Sufficient in 1996 but two home ranges fell short in 1997	
	c. Low marsh for nesting with 50% coverage by tall cordgrass	All home ranges met low marsh acreage requirement and all but one met cordgrass requirement and six lacked sufficient tall cordgrass	
	d. Population of tall cordgrass that is self-sustaining for 3 years	Plant height can be increased with continual use of fertilizer but tall cordgrass is not self-sustaining	

nests can float above water as the level rises and falls. If the cordgrass is too short, the nests cannot float high enough to prevent being washed out during high tides. Researchers suggest that the coarse soil used to construct the marsh does not retain the amount of nitrogen needed for cordgrass to grow tall. Adding nitrogen-rich fertilizer to the soil resulted in taller plants in the constructed marsh, but only if fertilizer was added on a continuing basis.

Another problem is that diversity and numbers of large invertebrates, which are the major food source of the rails, are lower in the constructed marsh than in natural marshes. PERL researchers suspect that this, too, is linked to low nitrogen levels. Because nitrogen stimulates the growth of algae and plants, which provide food for small invertebrates, and these in turn provide food for larger invertebrates, low nitrogen can affect the entire food chain.

Critical Thinking Questions

1. Make a diagram of the food web in the marsh showing how the clapper rail, cordgrass, invertebrates, and nitrogen are related.

2. The headline of an article about the Sweetwater Marsh project in the April 17, 1998 issue of *Science* declared "Restored Wetlands Flunk Real-World Test." Based on the information you have about the project, would you agree or disagree with this judgment? Explain your answer.

3. How do you think one can decide whether a constructed ecosystem is an adequate replacement for a natural ecosystem?

4. The term *adaptive management* refers to the use of scientific research in ecosystem management. In what ways has adaptive management been used in the Sweetwater Marsh project? What lessons from the project could be used to improve similar projects in the future?

When the agriculturally oriented methods failed, an ecological approach was tried. This approach applied knowledge about ecological succession. Instead of planting fast-growing but vulnerable agricultural grasses, ecologists planted slow-growing native grasses that were known to be adapted to minerally deficient soils and to the harsh conditions that exist in cleared areas. In choosing these plants, the ecologists relied on their observations of what vegetation first appeared in areas of Great Britain that had undergone succession naturally.[11] The result of the ecological approach has been successful restoration of once-damaged lands (Figure 10.12).

Lands as heavily altered as mine tailings occur in many places. Restoration similar to that in Great Britain is done in the United States to reclaim lands where surface coal mining has taken place. In such cases, restoration could be done during the mining process rather than afterwards. Similar methods could also be used to restore areas within cities where buildings once occupied the land.

Summary

- Restoration of damaged ecosystems is a major new emphasis in environmental sciences, and is developing into a new field. The process of restoration involves a combination of natural processes of ecological succession and human activities.

- Disturbance, change, and variation in the environment are natural, and ecological systems and species have evolved and adapted to these changes.

- When ecosystems are disturbed, they undergo a process of recovery known as ecological succession, the establishment and development of an ecosystem. Knowledge of succession is important for the restoration of damaged lands.

- During succession, there is usually a clear, repeatable pattern of changes in species. Some species, called early successional species, are adapted to the first stages, when the environment is harsh and variable but when necessary resources may be available in abundance. This contrasts with late stages in succession, when biological effects have modified the environment and reduced some of the variability but also have tied up some resources. Typically, early successional species are fast growing, whereas late successional species are slow growing and long lived.

Figure 10.12 An old lead mining area in Great Britain, now undergoing restoration. Restoration involves planting of early succession native grasses adapted to low-nutrient soils with little physical structure.

- Biomass, production, diversity, and chemical cycling change during succession. Biomass and diversity peak in midsuccession, increasing at first to a maximum, and then declining and varying over time.
- The causes of changes in the species found during succession can be due to facilitation, interference, or simply life-history differences. In facilitation, one species prepares the way for others. In interference, an early successional species prevents the entrance of later successional ones. Life-history characteristics of late successional species sometimes make them slow to enter an area.

REEXAMINING THEMES AND ISSUES

Human Population: If we degrade ecosystems so that their recovery from disturbance is slowed or so that they cannot recover at all, then we reduce the local carrying capacity of an area for human beings. Therefore, an understanding of the factors that determine ecosystem restoration is important to developing a sustainable human population.

Sustainability: Life tends to build up; nonbiological forces in the environment tend to tear down. By helping ecosystems to develop, we promote sustainability. Heavily degraded land, such as that significantly affected by pollution or overgrazing, loses the capacity to recover—to undergo ecological succession. Knowledge of the causes of succession can be useful in restoring ecosystems and therefore in achieving sustainability.

Global Perspective: Each degradation of land takes place locally, but such degradation has been happening around the world since the beginnings of civilization. Ecosystem degradation is therefore now a global issue.

Urban World: In cities, we generally eliminate or damage the processes of succession and the ability of ecosystems to recover. As our world becomes more and more urban, we must learn to maintain these processes within cities as well as in the countryside. Ecological restoration is an important method for the improvement of city life.

Science and Values: Because ecological systems naturally undergo changes and exist in a variety of conditions, there is no single "natural" state for an ecosystem. Rather, there is the process of succession, with all of its stages. In addition, there are major changes in the species composition of ecosystems over time. While science can tell us what conditions are possible and have existed in the past, the ones which we choose to promote in any location is a question of values. Values and science are intimately integrated in ecological restoration.

KEY TERMS

balance of nature **171**
chronic patchiness **182**
climax state **171**

early successional species **178**
ecological succession **174**
facilitation **181**

interference **182**
late successional species **178**
life history difference **182**

successional species **178**
primary succession **174**
restoration ecology **171**
secondary succession **174**

STUDY QUESTIONS

1. Farming has been described as managing land to keep it in an early stage of succession. What does this mean, and how is it achieved?
2. Redwood trees reproduce successfully only after disturbances (including fire and floods), yet individual redwood trees may live more than 1,000 years. Is redwood an early or late successional species?
3. Why could it be said that succession does not take place in a desert shrubland (an area where rainfall is very low and the only plants are certain drought-adapted shrubs)?

4. Develop a plan to restore an abandoned field in your town to natural vegetation for use as a park. The following materials are available: bales of hay; artificial fertilizer; and seeds of annual flowers, grasses, shrubs, and trees.

5. Oil has leaked for many years from the gasoline tanks of a gas station. Some of the oil has oozed to the surface. As a result, the gas station has been abandoned and revegetation has begun to occur.

   a. What effects would you expect this oil to have on the process of succession?

   b. What might you do to restore a different sequence of succession?

6. You are put in charge of a project to grow yellow birch, a species common in midsuccession, as a commercial crop. (Yellow birch is used to make dowels and hardwood plywood panels.) Considering the process of succession, how would you go about setting up plantations of yellow birch? How often would you log these plantations? Would the logging be complete or selective; that is, would you remove all the trees at one time or only some at any one time? Explain your answer.

7. Early in the 20th century, a large meteorite collided with the earth in Siberia and destroyed boreal forests over a large area. In what ways might restoration and succession following this large-scale disturbance differ from restoration and succession when a fire burned a few hectares in the same forest? (See Chapter 8 for information about boreal forests.)

● FURTHER READING

Berger, J. J. 1990. *Environmental Restoration: Science and Strategies for Restoring the Earth.* Washington, D.C.: Island Press. An informed and lively overview of the beginning of the restoration movement that includes scientific and technical papers given at the first national conference on restoration, held in 1933.

Cairns, J., Jr., ed. 1995. *Rehabilitating Damaged Ecosystems,* 2nd ed. Lewis Publishers. Discussions of natural and human-assisted restoration of various ecosystem types after either natural or human-caused disturbance.

Foster, D. R., and J. F. O'Keefe. 2000. *New England Forests through Time: Insights from the Harvard Forest Dioramas,* Cambridge, Mass.: Harvard University Press. A beautifully illustrated short book that discusses secondary succession of forests in New England, using as the centerpiece a famous set of miniatures at the Harvard Forest.

Gorham, E., P. M. Vitousek, and W. A. Reiners, 1979. "The Regulation of Chemical Budgets over the Course of Terrestrial Ecosystem Succession," *Annual Review of Ecology and Systematics* 10:53–84. A paper that examines the biogeochemical cycles of an ecosystem from the perspective of chemical inputs and outputs to that system.

Petts, G., and P. Calow, eds. 1996. *River Restoration,* London: Blackwell Science. An overview of international efforts to restore rivers designed for undergraduates as well as other audiences. The book begins with a general introduction to rivers, and includes chapters on the control of weeds, the conservation and restoration of fish, and the relationships between disturbance and recovery.

Stevens, W. K. 1995. *Miracle Under the Oaks, the Revival of Nature in America.* Pocket Books. The story of several citizen action groups' efforts to restore damaged ecosystems.

West, D. C., H. H. Shugart, and D. B. Botkin, eds. 1981. *Forest Succession: Concepts and Application.* New York: Springer-Verlag. A volume of papers presenting succession in forests from a variety of perspectives.

CHAPTER 11 World Food Supply

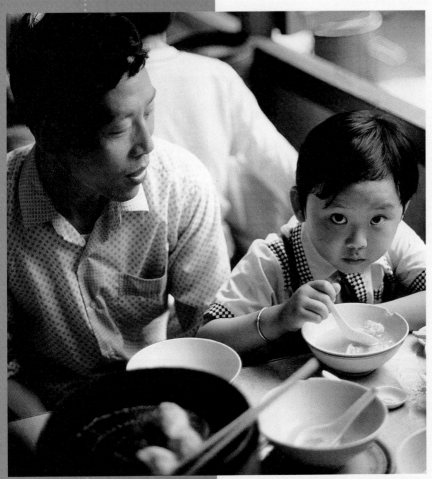

With one-fifth of the world's population, China requires a large food supply which begins simply as in this picture with a family meal.

● LEARNING OBJECTIVES

The major agricultural challenges facing us today are to achieve sustainable production of crops and domestic animals; to distribute food adequately around the world, and to decrease negative environmental effects of agriculture; and avoid creating new kinds of environmental problems as agriculture advances. After reading this chapter, you should understand:

- What it means to take an ecological perspective on agriculture.
- How agroecosystems differ from natural ecosystems.
- How the food supply depends on the environment.
- What role limiting factors play in determining crop yield and what the likelihood is that water will become the major limiting factor for crop growth in the next decade for many areas of the world.
- How the growing human population, the loss of fertile soils, and the lack of water for irrigation can affect future food shortages worldwide.
- What are the possibilities and limitations of some of the techniques of modern agriculture that may lead to increased food production.
- What is the relative importance of food distribution and food production.
- What are the potential benefits and environmental effects of genetic engineering of crops.

Food for China

China, with one-fifth of the world's people, is the most populous country on Earth. Between 1980 and 1995, China's population grew by 200 million people—about three-fourths of the population of the United States—to reach 1.2 billion. Although its growth rate is expected to slow somewhat in the coming decades, population experts predict that there will be 1.4 billion Chinese by 2025 (Figure 11.1).[1] Farming is, of course, an ancient practice in China—but can China's food production continue to keep pace with its growing population? Should China develop a food deficit, it may need to import more grain from other countries than those countries can spare from their own needs.

To give some idea of the potential impact of China on the world's food supply, consider the following examples. All of the grain produced by Norway would be needed to supply two more beers a year to each person in China. If the Chinese were to eat as much fish as the Japanese do, China would consume the entire world fish catch. At one point, it was estimated that food for all the chickens required for China to reach its goal of 200 eggs per person per year by 2000 would require all the grain exported by Canada—the world's second largest grain exporter.[2] Increased demand by China for world grain supplies could result in dramatic increases in food prices and precipitate famines in other areas of the world.

In just two years, between 1959 and 1961, 30 million Chinese starved to death during a famine brought on by a state policy of modernization that forced millions of farmers to work on large construction projects. In the wake of famine, China attempted to slow popula-

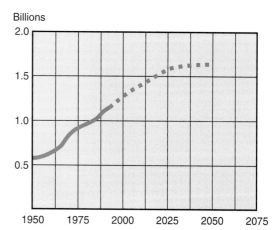

Figure 11.1 • Population of China, 1950–1994, with projections to 2050. (*Source:* L. R. Brown, *Who Will Feed China? Wakeup Call for a Small Planet.* [New York: W. W. Norton, 1995], p. 36.)

tion growth through a "one couple, one child" policy and to increase food production through extensive agricultural reform. Between 1980 and 1995, the total fertility rate in China fell from 4.8 children per woman to less than 2 children. In the same period, China achieved a remarkable increase in annual grain production from the 200 kg per person needed to maintain a minimal level of physical activity to 300 kg.

Can China continue to produce enough food for its people? China has relatively little cropland—approximately 100 million ha, or 0.08 ha per capita. By comparison, India has 0.19 hectares per capita—170 million ha of land with which to feed its population of 987 mil-

(a)

(b)

Figure 11.2 • (*a*) Meat consumption in China, 1975–1994. (*Source:* L. R. Brown, *Who Will Feed China? Wakeup Call for a Small Planet.* [New York: W. W. Norton, 1995], p. 47.) (*b*) Grain used for feed in China, 1960–1994. (*Source:* L. R. Brown, *Who Will Feed China? Wakeup Call for a Small Planet.* [New York: W. W. Norton, 1995], p. 51.)

lion.[1] And industrialization, including the development of the vast new hydroelectric "Three Gorges dam," will eliminate some of the existing farmland. As a result of industrial development, China is losing agricultural land to roads, railroads, and manufacturing plants at the rate of 1.6% per year over a 10-year period.[3]

At the same time, China, with half of its cropland under irrigation, is facing a severe water crisis because of seasonal variations in rainfall, periodic floods, and diversion of water to nonagricultural uses. Finally, with increased prosperity, the Chinese people are eating higher up the food chain; they are consuming more meat and dairy products—a diet that requires more grain production per person (Figure 11.2). Whether China's agricultural resources can sustain its population in coming decades is seen by many as a critical test of whether the world as a whole can find a sustainable balance between food production, consumer demand, and population growth.[4-9]

● *The needs of this one country, then, illustrate several of the themes of this book. We can see that the human population problem is a fundamental, underlying problem and that it is linked to a global perspective, because the desires and demands of the people of China, including an improvement in the material quality of their life, will place a stress on world food production. This stress leads to questions about the sustainability of agriculture, both within a nation and worldwide; and it ultimately leads to decisions that involve values and science.*

11.1 World Food Supply and the Environment

China's population and food production problem illustrate major aspects of the relationships between world food supply and the environment. Here is a brief overview of these relationships.

● At present, world food production is adequate in quantity and quality (nutritional value) to feed the current human population. Our modern food problem is the result of two processes: the great increase in human populations, which outstrips local food production in many areas, and an inadequate food distribution system for these growing populations.

● The distribution problem is at present the most important factor in starvation; however, in the future, if the human population continues to increase, there will be a limit on the ability of the earth to produce food.

● If the human population continues to grow at something like its present rate, it will double within the next hundred years, meaning that world food production will have to double just to provide the same amount and quality of food per person available today.

● As living standards are raised in nations like China, the desire and demand for higher-quality food will increase. As a result, per capita food demand will grow faster than the human population.

● Increases in world food production in the past have been accomplished in two ways: (1) increasing the area in food production and (2) increasing the production per unit area. The first method predominated during most of human history; but during the 20th century, the major increases in food production were due largely to increasing yields per unit area.

● Some of the best farmland in the world is being withdrawn from food production and put to other uses, including urbanization and suburbanization.

● Potential effects of future global environmental problems could decrease world food production. These include possible effects of global warming (Chapter 22) and possible transportation of pests from one part of the world to another (Chapter 8).

● Many agricultural experts believe that production per unit area can increase through the use of genetic engineered crops. However, there are concerns that these crops may cause major, novel environmental problems; and the jury is out as to whether such crops will be a net benefit or a net disbenefit.

Agriculture is an environmental issue for many reasons, not the least of which is its linkage to the human population.

Without enough food, the human population cannot continue to increase, and major social disruptions will occur if the world food supply falls behind the human population growth. Therefore, as stated in the case study, agriculture is closely linked to several major themes of this book.

11.2 Land, People, and Agriculture

The history of agriculture can be viewed as a series of human attempts to overcome environmental limitations and problems. It is also a history in which each new solution has created new environmental problems, which have in turn required their own solutions. Thus, in seeking to improve agricultural systems, we should expect some undesirable side effects and be ready to cope with them.

A surprisingly large percentage of the world's land area is in agriculture, approximately 11% of the total land area of the world, excluding Antarctica—enough to make agriculture a human-induced biome (Table 11.1). The percentage of land in agriculture varies considerably among continents, from 30% of the land in Europe to 6% in Australia. The total amount of land in agriculture covers an area approximately equal to the area of South and North America.

In the future, when the human population doubles, the production of agriculture must double just to meet the present level of per capita food consumption. If there were no increases in production per unit area, then an additional area equal to the area of the New World would have to be put into farmland. This seems unlikely, not only because it is a huge area but also because the land that is best suited for agriculture has already been put to this use. Moreover, as urbanization and suburbanization increases, some of this best agricultural land will be converted to other uses. More and more of the land in agriculture will be marginal.

In the United States, for example, about 1 million ha/year (2.5 million acres/year) are lost to urbanization and soil erosion. In 1982, 68.6 million acres of the 339 million acres of prime farmland in the United States were considered by the Census Bureau to be under urban influence.[10] By 1992, prime farmland inside urban areas had increased to 82 million acres, while total acres nationwide had declined from 340 million acres in 1982 to 330 million in 1992.[11] About 79% of the yearly destruction of wetlands is attributed to agricultural practices. The drainage of wetlands adds about 117,000 acres per year, so the net loss to agricultural lands is 2.38 million acres per year.[12]

At present, in the United States, considerable farmland is "banked." That is, farmers are paid not to grow crops, because at present there is more than enough production. As more land is needed, this banked farmland will go back into production. Since it seems unlikely that enough land area will be found to provide the necessary doubling of food production, then the productivity of crops and livestock must increase per unit area. Whether this can happen is an open and intensively debated question.

On the one side, many environmental activists argue that farmland is being degraded and that per unit area will, if anything, decline. On the other side, some agricultural scientists and agricultural corporations believe that production per area will continue to increase and that advances in genetically engineered crops (GMCs) offer one of the keys. This new methodology, however, raises some important potential environmental problems, discussed in Chapter 13. Furthermore, increased production in the past has depended on increases in the use of water and fertilizers. Water is a limiting factor in many parts of the world and will become limiting in more areas in the future.

Food production depends on the environment. When there is a drought, or a lack of any crucial resource, food production decreases. Because the environment is always varying, there are good years and bad years for agriculture, and farmers must continually adjust to these environmental variations. Agriculture, in turn, changes the environment. When

Table 11.1   Land, People, and Agriculture

Location	Total Land Area (mi²)	Total Land Area (km²)	Human Population	People per km²	Crop Land Area (km²)	Crop Land per Person (ha/person)	Crop Land as % of Total Land
Asia	17,388,686	45,036,697	3,600,000,000	79.93	4,827,520	0.13	11%
Africa	11,715,721	30,343,717	800,000,000	26.36	2,737,040	0.34	9%
N. America	9,529,129	24,680,444	300,000,000	12.16	2,247,000	0.75	9%
S. America	6,878,572	17,815,501	500,000,000	28.07	1,595,940	0.32	9%
Europe	4,065,945	10,530,798	800,000,000	75.97	3,112,050	0.39	30%
Australia	3,405,792	8,821,001	30,000,000	3.40	521,480	1.74	6%
World	52,983,845	137,228,159	6,030,000,000	43.94	15,041,030	0.25	11%

Notes: World land area excludes Antarctica; S. American includes Central America and the Caribbean; Europe includes former USSR states; Australia includes Oceania; Africa cropland includes Middle East. An additional 300 million ha (740 million acres) is in pasture and rangeland.

Source: For cropland and populations, *Information Please Almanac*, 1998, p. 487; for total land area, *Random House Atlas of the World*, 2000.

these changes are detrimental, large numbers of people can suffer, as can many other species, along with natural ecosystems and habitats. Typically, landscape beauty suffers as well.

World food supply is also greatly influenced by social disruptions and social attitudes, which affect the environment and in turn affect agriculture. In Africa, social disruptions since 1960 have included 12 wars, 70 coups, and 13 assassinations. This social instability makes sustained agricultural yields difficult.[13]

11.3 An Ecological Perspective on Agriculture

In a sense, as we mentioned earlier, agriculture can be viewed as a human-developed biome (see Chapter 8), because agriculture covers a large area of the earth's land surface, occurs on all continents except Antarctica, and involves different crops in different places, each fulfilling the same agricultural "niche."

When we farm, we create abnormal ecological conditions and novel ecosystems, known as **agro-ecosystems** (Figure 11.3). Traditionally agro-ecosystems differ from natural ecosystems in six ways.

First, in farming we try to stop ecological succession, so that we can keep an agro-ecosystem in an early successional state (see Chapter 10). Most crops are early successional species, which means that they do best when sunlight, water, and chemical nutrients in the soil are abundant. Under natural conditions, however, crop species would eventually be replaced by later successional plants. Any attempt to prevent natural successional processes from occurring requires actions, time, and effort on our part, as well as an input of energy. Most crops are planted on cleared land, which is then kept clear of other vegetation.

Pre-agricultural ecosystem

Agro ecosystem

Figure 11.3 • How farming changes an ecosystem. It converts complex ecosystems of high species diversity and structural diversity to a monoculture of uniform structure. The soil is greatly modified.

In contrast, when a clearing is created by a natural disturbance, such as a fire or a storm, the vegetation returns—first early successional species and then later ones.

The second way in which agro-ecosystems differ from natural ones is that large areas are planted to a single species or even to a single strain or subspecies, such as a single hybrid of corn. This practice is called **monoculture**. It opens the entire crop to attack by a single disease and makes it vulnerable to the same changes in environmental conditions. It also tends to reduce soil fertility, because each species has its own set of requirements for specific chemical elements and repeated planting of a single species can reduce the availability of certain essential elements. This effect on soil fertility can be counteracted to a certain extent by crop rotation and with artificial fertilizers. In *crop rotation*, different crops are planted in turn in the same field, with the field occasionally left *fallow*. A fallow field is allowed to grow with a cover crop (sometimes planted, sometimes whatever plants germinate and grow) and is not harvested for at least one season. Often, the vegetation that grows on the fallow field is plowed under to add to soil fertility.

An important tradeoff is implicit in the choice to plant a single hybrid. The tradeoff is as follows. Each year, seed companies make use of forecasts of climatic conditions for the growing season and knowledge of the most likely genetic strains of diseases and insect pests in the area to be planted. They then develop hybrids resistant to those strains and well adapted to the forecasted weather. If the predictions are correct, then crop production can be very high. However, if the predictions are incorrect, then the entire area can have very low production.

An alternative approach is to plant a mixture of crops and/or a broad range of genotypes at a particular time and place. This approach is typical of preindustrial agriculture such as still occurs in many developing world countries, and it is promoted today by organic farmers. The result of the approach is lower average production in any year; but there is the possibility that long-term production may be higher, and that years of very low production will be very rare. The first approach, then, trades off long-term stability and the possibility of a very bad year for the opportunity to have very high production this year. The second approach trades off the potential for very high production this year for long-term stability of food supply. Which approach to choose is a question of values.

A third way in which agro-ecosystems differ from natural ecosystems is that crops are planted in neat, regular rows. In most natural ecosystems, many species of plants grow mixed together in complex patterns, rarely forming regular rows. A simple regular planting of one species makes it easy for pests; a crop has no place to hide. A complex pattern of many species makes it harder for the pests to find their victims.

Fourth, biological diversity and food chains are greatly simplified in agro-ecosystems. Most pest control methods reduce the abundance and diversity of natural predators. In addition, parasites of crop plants—undesired, non-crop-plant species that compete for nutrients with the desired crop—are eliminated. This, as explained before, makes it easier for pests to find the crop and therefore makes the agro-ecosystem more susceptible to undesirable changes.

Fifth, plowing is unlike any natural disturbance of the soil. Nothing in nature repeatedly and regularly turns over the soil to a specific depth. Plowing exposes the soil to erosion and can damage its physical structure, leading to a decline in organic matter and a loss of chemical elements to erosion. This is illustrated by studies in Santa Barbara County, California, of previously plowed land that has been abandoned for agriculture. Instead of returning to California oak woodlands, which originally occupied the land, the plowed land becomes shrub-herb-vegetation. Alternatively, it undergoes succession to oak woodlands very slowly—much more slowly than in non-tilled land.

Sixth, the application of biotechnology to food production is a novel effect. According to a recent Secretary of Agriculture, Daniel Glickman, no issue facing the U.S. Department of Agriculture grew as quickly between 1995 and 2000 as the use of biotechnology.[14]

11.4 Sources of Food

Primitive societies obtained food through hunting and gathering, and some societies today still depend solely on these methods. Hunting and gathering remain the predominant way in which food is obtained from the oceans. On the land, though, only about 5 percent of the human population's protein is obtained from hunting and gathering.[15] The great majority of people obtain food from cultivated land plants and domesticated animals.

Crops

There are 270,000 species of plants, but only about 3,000 have been used as agricultural crops, and only 150 species have been cultivated on a large scale. Some crops provide food; others provide commercial products, including oils, medicines, pesticides, and fibers. In the United States, 200 plants are grown as crops.

It is useful to classify crops as cash crops or subsistence crops. **Cash crops** are grown to be sold or traded in a large market. **Subsistence crops** are used directly for food by the farmer or sold locally where the food is used directly. Some cash crops may provide nonfood products. For example, rubber plantations produce latex from the sap of rubber trees; the latex is then sold to make rubber products. Other cash crops, such as coffee and tea, do not provide primary nutrition and are not necessary for survival; but in many countries, such cash crops are a major source of international trade and foreign hard currencies. As a result, there is governmental and economic pressure to grow these cash

Figure 11.4 • Some of the world's major crops, including (*a*) wheat, (*b*) rice, (*c*) soybeans, (*d*) bananas, and (*e*) coconuts. See text for a discussion of the relative importance of these five crops.

crops instead of food crops; sometimes, this works against the provision of adequate locally grown food.

Most of the world's food is provided by only 14 crop species. These are, in approximate order of importance, wheat, rice, maize, potatoes, sweet potatoes, manioc, sugarcane, sugar beet, common beans, soybeans, barley, sorghum, coconuts, and bananas (Figure 11.4). Of these 14, six species provide more than 80% of the total calories consumed by human beings either directly or indirectly.[16] Wheat, rice, and maize alone make up the majority of food from crops (Figure 11.5).

Other crops, called forage, are grown as food for domestic animals and are not eaten directly by people; these include alfalfa, sorghum, and various species of grasses grown as hay. Alfalfa is the most important such crop in the United States, where 14 million ha (about 30 million acres) are planted in alfalfa—one-half the world's total.

There is a large world trade in small grains. Only the United States, Canada, Australia, and New Zealand are major exporters; the rest of the world's nations are net importers. In 1999, the world small grain production was 1.900 billion tons, slightly below a 1998 record crop of

Figure 11.5 • Geographic distribution of world production of a few major small grain crops.

Legend:
▲ Wheat
■ Corn (maize)
○ Rice

1.904 billion tons.[17] In spite of record-breaking production of small grains, food emergencies affected 34 countries worldwide in 1998. The effect of weather patterns in Africa, Latin America, and Asia, as well as an inadequate international trade in food, contributed to these emergencies.[18] Africa remains the continent with the most acute food shortages, due to adverse weather and civil strife.[19]

Livestock

Domesticated animals are an important food source and have a major impact on the land. Most domesticated animals used as food are **ruminants** (see Chapter 7, Figure 7.8). A ruminant has a four-chambered stomach within which bacteria convert the woody tissue of plants to proteins and fats; these, in turn, are digested by the animal. In this way, ruminants convert cellulose, indigestible by people but Earth's most abundant organic compound, to human food.

The major livestock are cattle, sheep, and goats, but other species are important as well. These include the water buffalo of Southeast Asia, the camel of North Africa, and the llama of the South American Andes. Worldwide, there are more than 1.7 billion sheep and goats, 1.3 billion cattle, 19.1 million camels, and 158 million buffalo. These animals are maintained on pasture and range. **Rangeland** is land that provides or can provide food for grazing and browsing animals without plowing and planting; *pasture* is land that is plowed and planted to provide forage for the animals.

Approaches to Agriculture

In terms of technology, we can identify three major approaches: (1) production based on highly mechanized technology, with a high demand for resources, including land, water, and fuel, and with little use of biologically based technologies; (2) production based on biological technology, with conservation of land, water, and energy; and (3) organic food production. In the first kind, called **demand-based agriculture**, production is determined by economic demand and limited by that demand and not by resources. In the second approach, **resource-based agriculture**, production is limited by environmental sustainability, and the availability of resources and economic demand usually exceeds production (Figure 11.6).

The history of agriculture can be summarized most simply as four stages:

1. Resource-based agriculture and what we now call organic agriculture were introduced about 10,000 years ago.
2. A shift to mechanized, demand-based agriculture occurred during the Industrial Revolution of the 18th and 19th centuries.
3. A return to resource-based agriculture began in the 20th century.
4. Today, we see a growing interest in organic agriculture, as well as a potentially large-scale use of genetically engineered crops.

This resource-based agriculture, however, differs from the original in that it includes many modern approaches

(a)

Mining and production of fertilizers

Mechanized groundwater irrigation

Mechanized transfer of raw materials and products

Heavy use of chemical pesticides

(b)

Introduction of natural enemies of pests

Contour plowing

Chemical herbicides to control weeds; otherwise, limited use of chemical pesticides

Drip irrigation

No–till agriculture

(c)

Reduced yields due to allowance for some competition for weeds and some effects of insect pests

No artificial chemicals or fertilizers; use of natural parasites and predators to control pests

Interplanting of crops to provide greater structural diversity

Greater mixture of crops per unit area

Emphasis on no-tilling

Figure 11.6 • Agricultural technologies: (*a*) demand-based agriculture, (*b*) resource-based agriculture, (*c*) an organic farm.

and uses many forms of technology. (See A Closer Look 11.1 and 11.2.)

Let us look more closely at the nature and extent of organic food production today. Organic agriculture has three goals: food that is "natural" in the sense of not containing human-developed artificial compounds; farming that is done in systems that are more like natural ecosystems than monocultures; and farming that minimizes

negative environmental impacts. During the 1990s, organic farming was one of the fastest-growing segments of U. S. agriculture. Organic certification programs became well known. Certified organic croplands more than doubled, and the number of farmers certifying their products grew by 40%. In the United States, 1.347 million acres were certified as organic in 1997. Organic production of livestock and animal products, especially eggs and dairy products, grew even faster. From 1992 to 1997, organic milk cows increased from 2.3 thousand to 12.9 thousand, and organic layer hens increased from 44,000 to more than 500,000.

Although adoption of organic farming systems showed strong gains between 1992 and 1997, and the adoption rate continues high, the overall adoption level is still low—only 0.2% of all U.S. cropland was certified organic by 1997. Only 0.01 percent of U.S. corn and soybean crops acreage was grown under certified organic farming systems in 1997; however, over 1% of dry peas and tomatoes were grown organically, and about 2% of apple, grape, lettuce, and carrot acreage was organic. On the high end, nearly one-third of U.S. buckwheat, herb, and mixed vegetable crops were grown under organic farming conditions in 1997.[20]

Sources of Food: Aquaculture

Aquaculture is the production of food from aquatic habitats—both marine and freshwater. Aquaculture provides only a small amount of the world's food at present; but it is important as a source of protein for many nations, especially in Asia and Europe, and offers a potentially important cash crop in other parts of the world. Although it is relatively new in the United States, aquaculture has a long history elsewhere.

Animals grown in aquaculture include oysters in many countries; shrimp and yellowtail in Japan; herbivorous crayfish in east Texas and Los Angeles; fish such as the carp, grown widely; tilapia in Africa, Israel, the United States, and many other countries; eels, Chinese carp, and minnows in China; catfish in the southern and midwestern United States; salmon in Norway and the United States; trout in the United States; plaice, sole, and the southeast Asian milkfish in Great Britain; mussels in France, Spain, and southeast Asian countries; and sturgeon in the Ukraine. A few species—trout and carp—have been subject to genetic breeding programs.[21] Sometimes aquaculture is combined with agriculture. In the Szechwan area of China, there are more than 100,000 ha (about 250,000 acres) of rice fields where fish are farmed in the flooded fields with the rice. In China, growing fish in rice fields is an ancient practice that can be traced back to a treatise on fish culture written by Fan Li in 475 B.C.[21]

Aquaculture holds promise as a contributor to the sustainability of commercial fisheries. Indeed, until recently, freshwater and marine aquaculture were seen only as beneficial. Recently, however, concern has grown about the environmental effluents from aquaculture. One of the best known environmental issues involves Atlantic salmon, the major species of salmon grown as a "crop." In 2000, Atlantic salmon were put on the endangered species list; one of the causes proposed for the problems of this species is extensive mariculture of Atlantic salmon in places such as the coast of Maine. Observers have identified two concerns. First, water pollution from excrement and excess food provided for the salmon damages the habitat of wild salmon. Second, breeding between farmed salmon not native to a locale and native salmon may have a negative effect on the genetics of wild salmon, leading to genetic strains that are less fit for a particular stream.

Freshwater Aquaculture

In China and other Asian countries, farmers often grow several species of fish in the same pond, exploiting their different ecological niches. For example, ponds developed mainly for carp, a bottom-feeding fish, also contain minnows, which feed at the surface on plant leaves added to the pond. Sometimes, fish ponds are designed to use otherwise wasted resources. Some countries make use of fertilized water available from treated sewage. Other fish ponds make use of natural hot springs (Idaho) and of the warmed water used in cooling electric power plants (Long Island, New York, and Great Britain).[21]

An interesting example of an ecosystem approach to aquaculture occurs in China, where the production of freshwater fish and silkworms is combined. The silkworms feed on mulberry leaves, and the mulberry trees are planted around fish ponds. The leaves and the droppings from the silkworms are added to the ponds to provide fertilizers for the growth of algae, which in turn provide food for the fish. The sediment from the pond is used to fertilize the trees. In this way the chemical elements are recycled (Figure 11.7).

In China, aquaculture includes a wide variety of organisms, from 50 species of fish, shrimp, crab, and shellfish to marine algae, sea turtles, sea cucumbers (a marine animal), and water chestnuts.[22] Aquaculture can be extremely productive on a per area basis, especially because flowing water brings food into the pond or enclosure from outside. Although the area of Earth that can support freshwater aquaculture is small, we can expect this kind of aquaculture to increase in the future and to become a more important source of protein.

Mariculture

Mariculture—aquaculture of ocean fish—represents a relatively small part of the total marine fish catch. For the most part, on the ocean, we are still hunters and gatherers, not farmers. Nevertheless, mariculture has grown rapidly in the last decade and can be expected to increase in the future. Although mariculture provides only a small fraction of the world's protein, it is important in certain countries and can provide valuable cash crops.

Such delicacies as abalone and oysters, whose natural production is limited, are undergoing increased production as part of mariculture. In the United States and Canada, for example, researchers are working to learn how to attract the young, swimming stages of these shellfish to areas where they can be conveniently grown and harvested.

Figure 11.7 • Diagram of the traditional aquaculture practice in China. Leaves from mulberry trees and droppings from silkworms are added to ponds to provide fertilizers for algae, which in turn provide food for fish. The pond sediment is used to fertilize the trees. In this way, ecosystem chemical recycling is used to increase the production of both silk and fish.

Oysters and mussels are grown on rafts that are lowered in the ocean, a common practice on the Atlantic Ocean in Portugal and in the Mediterranean in such nations as France. As filter feeders, these animals obtain food from water that moves past them in currents. Because a small raft is exposed to a large volume of water, thus a large volume of food, rafts can be extremely productive. For example, mussels grown on rafts in bays of Galicia, Spain, produce 300 metric tons/ha; by comparison, public harvesting grounds of wild shellfish in the United States yield about 10 kg/ha.[21] Oysters and mussels are also grown on artificial pilings in the intertidal zone, a practice in the state of Washington (Figure 11.8).

11.5 Soil and Agriculture

Ironically, agriculture depends heavily on soil quality but can lead to a decline in that quality. A high-quality agricultural soil has all the chemical elements required for plant growth and a physical structure that lets both air and water move freely through the soil, yet it retains water well. A soil with a high organic matter content has three qualities. Soils with a mixture of sediment particle sizes are also usually beneficial to agriculture. Small particles, especially fine clays, help retain moisture and chemical elements; and larger particles, including sand and pebbles, help the flow of water. Desirable physical qualities of the soil vary with the crop. Lowland rice grows in flooded ponds and requires a heavy, water-saturated soil, while watermelons grow best in very sandy soil.

When a forest or prairie is cleared for agriculture, changes take place in the soil. The original soil developed over a long period and is typically rich in organic matter, which is concentrated in its upper layers. Organic matter in soil is rich in chemical nutrients and provides a physi-

Figure 11.8 • An oyster farm in Poulsbo, Washington. Oysters are grown on artificial pilings in the intertidal zone.

Traditional Farming Methods

In industrialized countries of temperate zones, there is a long history of plowing to clear land for agriculture; but in less industrialized, tropical areas, there is a history of agricultural methods that depend on clearing the vegetation without plowing. Where the loss of nutrients from the soil occurs rapidly following clearing, as in some tropical rain forests, the traditional practice is to cut the forest in small patches but not cut it completely (Figure 11.9). Some shrubs and herbaceous plants are left. Several crops are planted together among existing vegetation. Crops are harvested for a few years. Then the land is allowed to grow back to a forest. The natural process of secondary succession—the redevelopment of the ecosystem—is allowed to occur. In fact, these farming practices promote this redevelopment and increase the conservation of chemical elements in the ecosystem. After the forest has grown back, the process is repeated.

This kind of agriculture has many names. It is sometimes called *cultivation with forest* or *bush fallow*. In Latin America, it is called *milpa* agriculture; in Great Britain, *swidden* agriculture; in west Africa, *fang* agriculture. In this type of agriculture, a mixture of crops is utilized, including root, stem, and fruit crops. For example, in western Africa, fang agriculture includes yams (a root crop) plus maize; in Southeast Asia, root crops are grown with rice and millet or with rice and maize.[23]

In theory, this method could be sustainable if human population density remained low. Erosional losses are minimized, and the soil eventually recovers its fertility. Uncut vegetation provides future seed sources. When human population pressures are low, a longer time occurs between the periods of use of any site. This is known as a *long rotation period*. Under high population pressures, such as occur in many places today, the rotation period is much shorter, and the land may not be able to recover sufficiently from previous usage. In such cases, production is not sustainable.

For many years, agricultural experts from developed nations viewed this method of agriculture as a poor process with low, short-term productivity, used only by primitive peoples. Now it is understood that this kind of agriculture is well suited to high rainfall lands where soils readily become impoverished when the land is completely cleared. The mixture of crops allows different species to contribute to soil fertility in different ways. Some perennial plants slow physical erosion; native legumes add nitrogen to the soil; and so forth.

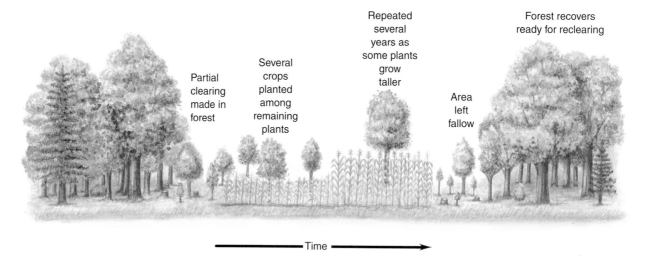

Partial clearing made in forest

Several crops planted among remaining plants

Repeated several years as some plants grow taller

Area left fallow

Forest recovers ready for reclearing

→ Time →

Figure 11.9 ● Bush fallow, also called milpa, fang, or swidden agriculture. Over time, secondary succession will take place on land partially cleared by slash and burn.

cal structure conducive to plant growth. When the original vegetation is cleared and planted in crops, most of whose organic matter is harvested and removed, there is less input of dead organic matter from the vegetation to the soil. And soil is exposed to sunlight, which warms it and increases the rate of decomposition of its organic matter. For these reasons, the amount of organic matter declines. Plowing and harvesting open the soil to erosion and lead to compaction of the soil. The soil's physical structure becomes less conducive to plant growth.

Traditionally, farmers combated the decline in soil fertility by using organic fertilizers, such as animal manure. Organic fertilizers have the advantage of improving both chemical and physical characteristics of soil. But organic fertilizers can have drawbacks, especially under intense agriculture on poor soils. The development of modern industrially produced fertilizers, commonly called "chemical" or "artificial" fertilizers, was a major factor in the great increases in crop production in the 20th century. One of the most important was the development of industrial processes to convert molecular nitrogen gas in the atmosphere to nitrate that can be used directly by plants. Phosphorus is mined, usually from a fossil source that was biological in origin, such as deposits of bird guano on islands used for nesting and uplifted marine limestones. Nitrogen, phosphorus, and other elements are combined in proportions that are appropriate for specific crops in specific locations.

11.6 Limiting Factors

In many areas of the world, soils lack one or more of the chemical elements required by crops. Such an element is called a **limiting factor**. An agricultural limiting factor is the single requirement for growth that is available in the least supply in comparison to the need of the crop. In most cases, this lack involves one of the *macronutrients*— chemical elements required in relatively large amounts by all living things. Macronutrients are sulfur, phosphorus, magnesium, calcium, potassium, nitrogen, oxygen, carbon, and hydrogen. In other cases, soils are lacking in *micronutrients*, which are specific trace elements, typically metals, required only in small amounts. Micronutrients include molybdenum, copper, zinc, manganese, iron, boron, and chlorine. (Macronutrients and micronutrients are also discussed in Chapter 4.)

The older a soil, the more likely it is to lack trace elements, because as a soil ages its supply of chemical elements tends to be leached by water from the upper layers to deeper layers. (See the Chapter 10 discussion of ecological succession.) When crucial chemical elements are leached below the reach of crop roots, the soil becomes infertile. Some of the most striking cases of soil nutrient limitations have been found in Australia, which has some of the oldest soils in the world. The Australian soils occur on land that has been above sea level for many millions of years, during which time severe leaching has taken place.

Sometimes trace elements are required in extremely small amounts. For example, crops require very small amounts of molybdenum, and the addition of very small amounts has very great effects. It is estimated that adding an ounce of molybdenum to a field increased the yield of grass in sheep pastures by 1 ton/year.

The idea that some single factor determines the growth, and therefore the presence, of a species is known as *Liebig's law of the minimum*. Justus von Liebig was a 19th-century agriculturalist. He knew that crops required a number of nutrients in the soil and that crop yields could be increased by adding these as fertilizers. However, the factor that caused an increase varied.

A general statement of Liebig's law is: *The abundance and distribution of a species is limited by the single factor in shortest supply or having the greatest impact.* Liebig wrote in 1840 that the yield of a crop was limited by the single nutrient that was in the shortest supply. Since then, ecologists have generalized Liebig's law beyond soil nutrients to include all life requirements. Although the law can be helpful, it should be used within limits and with caution.

If Liebig's law were always true, then environmental factors would always act one by one to limit the distribution of living things. For example, commercial fertilizers commonly contain nitrogen, phosphorus, and potassium. If a particular soil had nitrogen in least supply, phosphorus second, and potassium third, then, if Liebig's law were correct, the growth of plants would be increased by adding nitrogen up to some maximum amount. Adding phosphorus or potassium before adding nitrogen would have no effect. However, once the nitrogen content of the soil reached the maximum a particular plant required, then the plant's growth would be limited by phosphorus. Because phosphorus would now be the limiting factor, adding it would increase the yield. We could add phosphorus until the maximum was reached, at which point potassium would become the limiting factor.

This could continue until we had fertilized the soil to the point at which the plant had all the nutrients it could use. Its growth might then be limited by water and light. When all life requirements were available in maximum amounts, the yield of the plant would be restricted by its own genetic characteristics and by the constraints imposed by fundamental laws of energy, the role of which was discussed in Chapter 9. All this seems very reasonable, and Liebig's law seems to be true for the Australian soil that lacked molybdenum.

Liebig's law of the minimum as stated earlier is too restrictive, however, as it implies that there is no interaction between environmental factors—that each operates independently. This may not always be true. For example, nitrogen is a necessary part of every protein, and proteins are essential building blocks of cells. Enzymes, which make many cell reactions possible, contain nitrogen. A plant subject to little nitrogen and phosphorus might not make enough of the enzymes involved in taking up and using phosphorus. Increasing nitrogen to the plant might therefore increase the plant's uptake and use of phosphorus. If this were so, the two elements would have a **synergistic effect**. In a synergistic effect, a change in availability of one resource affects the response of an organism to some other resource.

It is also important to understand that there can be too much of a good thing. As a simple example, plants die

Potential Future Advances in Agriculture

New Genetic Strains and Hybrids

From the beginning, farming has affected the genetics of domesticated animals and plants, as people selected strains that were easy to grow and harvest (Figure 11.10). The act of farming makes certain species abundant where they were rare; thus, farming selectively favors certain genotypes (populations with certain genetic characteristics).

Features that make a species or a genotype a weaker competitor under natural conditions sometimes make it more desirable as a crop. For example, wild relatives of wheat lose their seeds when the seeds are ripened and gently shaken by wind or animals. If you try to cut wild wheat stems and take them home, most of the seeds will be shaken off and few will be left when you arrive. This is an adaptation that helps spread the seeds.

Some individuals of wild wheat have a mutation that makes the seeds remain on the stalk. In the wild, these mutants leave fewer offspring and do not persist in the population; in the wild, the mutants are less fit genetically (see Chapter 7). Early farmers selected these mutants because they were easier to collect, transport, and use. In this way, people changed selective pressures on wheat and hastened the evolution of wheat strains that are useful to us but that could not survive on their own. One might also say that the result has been a symbiotic relationship between people and these otherwise less competitive forms of wheat.

People also domesticated wheat by moving it to habitats to which it was not originally adapted, then developing new strains that could persist in these environments and also changing the environmental conditions of the new habitat to better suit the wheat.[24] Corn (maize) went through a similar process of domestication. Likewise, domestic animals have been bred to make them more docile and better producers of the meat and dairy

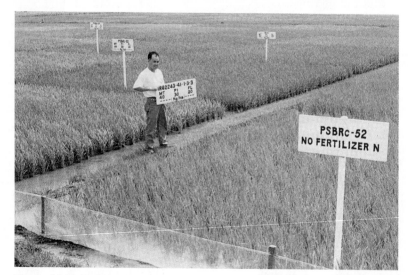

Figure 11.10 ● Experimental rice plots at the International Rice Research Institute, Philippines, showing visual crop variation based on the use of fertilizers.

products that we prefer. Modern agriculture has carried this process a step further with the frequent, intentional development of hybrids of different genotypes, bred to overcome new strains of disease and changes in climate.

New Crops

Developing new crops by domesticating species that are currently wild offers considerable potential. Although new crops are unlikely to replace current crop species as major food sources in this century, there is great interest in new crops to increase production in marginal areas and to increase the production of nonfood products, such as oils.

The development of new crops has been a continuing process in the history of agriculture. As people spread around the world, new crops were discovered and transported from one area to another. The process of introduction and increase in the production of crops continues. For example, the seed area harvested in the United States has increased from 17,000 ha in 1961 to 1.15 million ha in 1998 (42,000 acres to 2.8 million acres).[25]

Among the likely candidates for new crops are amaranth for seeds and leaves; *Leucaena*, a legume useful for animal feed; and triticale, a synthetic hybrid of wheat and rye. A promising source of new crops is the desert; none of the 14 major crops are plants of arid or semiarid regions, yet there are vast areas of desert and semidesert. The United States has 200,000 million ha (about 500,000 million acres) of arid and semiarid rangeland. In Africa, Australia, and South America, the areas are even greater. Several species of plants can be grown commercially under arid conditions, allowing us to use a biome for agriculture that has been little used in this way in the past. Examples are guayule (a source of rubber), jojoba (for oil), bladderpod (for oil from seeds), and gumweed (for resin). Jojoba, a native shrub of the American Sonoran Desert, produces an extremely fine oil, remarkably resistant to bacterial degradation, which is useful in cosmetics and as a fine lubricant. Jojoba is now grown commercially in Australia, Egypt, Ghana, Iran, Israel, Jordan, Mexico, Saudi Arabia, and the United States.[23]

when they have too little water but also when they are flooded, unless they have specific adaptations to living in water. So it is with chemical elements required for life. In this section, we have discussed the problems of too little; we discuss the problems of too much in the chapter on environmental toxicology (Chapter 15).

11.7 Sustainable Yields

As with other biological resources, sustainability is seen today as the key practice in agriculture. In the past, there was a search for a **maximum sustainable yield**. This is the maximum amount of a crop that can be produced per unit area indefinitely, year after year, without decreasing the ability of soil and that crop species to sustain that yield. An **optimum sustainable yield** of a crop is the largest yield that can be obtained without decreasing the ability of the agro-ecosystem, as well as of the crop itself, to sustain that yield in the future.

Little of modern agriculture is sustainable in the sense of retaining the major qualities of the preagricultural ecosystem. In China and other parts of Asia and in parts of Europe, agriculture has been practiced in the same areas for thousands of years. The effects of this long-term cropping are often quite apparent: a soil with little organic matter and a need for much fertilization.

A major challenge in the future will be to create sustainable agriculture. This links agriculture to major themes of this book: the human population problem, an urban world, a global perspective, and science and values. In terms of these themes, the major agricultural question is: *How many people can be fed on a sustainable basis?* An additional challenge before the world is how to produce adequate food to meet the demands for improved diets for an expanding and increasingly affluent human population.

A major obstacle to answering this question is that technology for food production, processing, and distribution changes faster than we can understand the effects of the changes, and these changes have both positive and negative impacts on the environment. With modern technology, given enough energy and materials, we can grow food in a completely artificial environment, such as a spacecraft. But this is too expensive a solution in terms of money, resources, and effects on the environment to be a large-scale solution to food production.

Optimum Environmental Conditions

The concept of sustainable yield leads to the concept of optimum environmental conditions for growth of any single crop. Because every crop has its own set of optimum environmental conditions for growth, crops differ in where they grow best. Some crops, such as rye, grow best under relatively cool, moist conditions. Other crops, such as maize, do better in warmer, dryer climates.

Regions of Earth differ greatly in their capacity for crop production. Many factors influence which crops are produced in which areas, including tradition, access to technology and supplies, and local politics. For example, even within the United States, states differ greatly in their crop production (Figure 11.11). California has the greatest value of crop production in the United States, totaling $12.4 billion in 1998, and producing 36.4 million tons of fruits, nuts, and vegetables in 1998. This accounts for about half of all U.S. production.[26] Much of California would be too dry to support this food production without irrigation.

In the high plains east of the Rocky Mountains (where shortgrass prairie originally grew), rangelands and irrigated farmlands are common. There, irrigated cropland is mainly planted in small grains (corn, wheat, etc.). Farther east in Nebraska are areas where winter wheat is most important; spring wheat is important in the Dakotas. The Midwest is the corn belt. In northern states from Minnesota to Maine, dairy and hay for cattle are important products. In the Southeast, major crops include cotton, tobacco, vegetables, and fruits.

11.8 Human Populations and World Food Supply: The Global Perspective

People "starve" in two ways: undernourishment and malnourishment. *Undernourishment* is the lack of sufficient calories in available food, so that one has little or no ability to move or work and eventually dies from the lack of energy. *Malnourishment* is the lack of specific chemical components of food, such as proteins, vitamins, or essential chemical elements. Both are global problems.

Undernourishment leads to famines that are obvious, dramatic, and fast acting. Malnourishment is long term and insidious. Although people may not die outright, they are less productive than normal and can suffer permanent brain damage. Among the major problems of malnourishment are *marasmus*, progressive emaciation caused by a lack of protein and calories; *kwashiorkor*, a lack of sufficient protein in the diet, which leads in infants to a failure of neural development and therefore to learning disabilities; and *chronic hunger*, which occurs when people have enough food to stay alive but not enough to lead satisfactory and productive lives. We must be aware that providing the world's food supply requires providing enough food to avoid both undernourishment and malnourishment.

Since the end of World War II, not a single year has passed without a famine somewhere in the world.[27] Examples include famines in Brazil (1979–1984), Ethiopia (1984–1985), and Somalia (1991–1993) and the 1998 crisis in Sudan. The distribution problem is clearly illustrated by these recent famines. Food distribution fails because poor people cannot buy food and pay

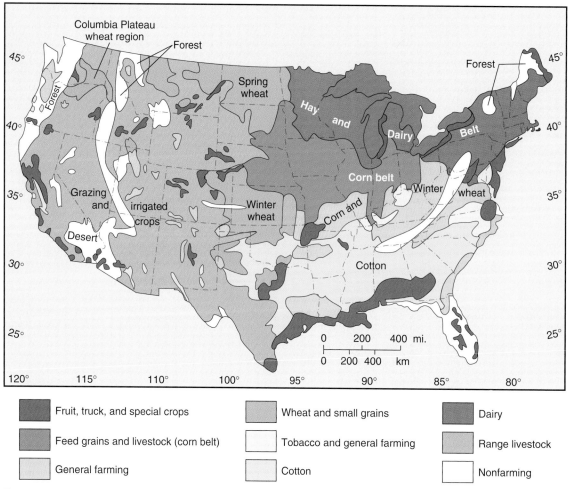

Figure 11.11 ● Major types of agricultural production in the United States.

Legend:
- Fruit, truck, and special crops
- Feed grains and livestock (corn belt)
- General farming
- Wheat and small grains
- Tobacco and general farming
- Cotton
- Dairy
- Range livestock
- Nonfarming

for its delivery; because transportation is lacking or too expensive; or because food is withheld for political or military reasons. Although there is considerable international trade in food, most of the trade is among rich nations.

Another solution, one that is commonly used, is food aid among nations, where one nation provides food to another or gives or lends money that is used to purchase food. This type of aid has become highly publicized in recent years, especially with the interest of pop music stars in the African famine. In the 1950s and 1960s, only a few industrialized countries provided food aid by using stocks of surplus food. A peak in international food aid occurred in the 1960s, when a total of 13.2 million tons per year of food was given. The world food crisis in the early 1970s raised awareness of the need for greater attention to food supply and stability. During the 1980s, donor commitments only reached a high of 7.5 million tons. Yet a record level of 15 million tons of food aid in 1992/1993 met less than 50% of the minimum caloric norms of the people fed. If food aid alone were to bring the world's malnourished people to a desired nutritional status, 55 million tons would be required by the year 2010—more than six times the level of food aid available in 1995.[28]

As long as food distribution is the primary problem, the best solution is to increase local production.

When a group of people starve, the world sorrows for them, and there has been much recent activity to provide starving people with food. Humanitarian gestures are important, but in themselves such efforts cannot provide a solution to the world food problem. Food aid is a short-term answer. In the longer run, local food production must increase. Ironically, food aid can work against an increase in the availability of locally grown food. Free food undercuts local farmers; they cannot compete with free food, and local production of food decreases. The only complete solution to famine problems is to develop long-term sustainable agriculture at local levels. The old saying, "Give a man a fish and feed him for a day; teach him to fish and feed him for life," is true. Agricultural techniques must be developed and taught that do not use up resources but that can be maintained over long periods. Availability of food grown locally also solves problems arising from disruptions in distribution and the need to transport food over long distances.

11.9 Measures of Food Availability

How do we know whether there is enough food in the world for today, for tomorrow, and over the longer term? A measure of the world's ability to deal with undernourishment and famine is the *supply of grain on hand, measured in days*. This number has fluctuated since 1960. It peaked at a 104-day supply in 1987. Recent stocks have been declining; the 1995 estimate was a 62-day supply, and the 1996 estimate was a 49-day supply, less than the previous low, a 56-day supply in 1973.[29] In the United States, where we have some of the best soils and climates for agriculture in the world, we have tended to take the availability of food for granted. But a worldwide 49-day supply will not last long.

The United States stores more than one-half the world's surplus grains and has 4 million metric tons of grain for emergency relief. Since World War II, the United States has been the leading food exporter and donor in the world. It is important to have a worldwide distribution system that includes reserves of food along with the capability to transport this food to where it is needed. For the entire world, such a system would require a standing stock of 10 million to 20 million metric tons. At a price of $137 per ton, this would be about $2 billion worth of crops; maintenance of this stock would require $200 million per year.[25]

Another measure of the availability of food is **per capita food production**, the amount of food produced per person. Today, population growth and food production are in a tug-of-war, and it is unclear in the short run which will increase faster, although we know in the long run there are limits to both. Both population and food supply have increased, but food production has varied more; on the average, food available per person has not increased. This trend supports Malthus's 1798 prediction that at some point the growth of population will outstrip the capacity of Earth to supply food (see Chapter 5's Critical Thinking Issue).

The *demand for food* is the amount of food that would be bought at a given price if it were available. The demand increases with population and with per capita income. **Per capita demand** is the economic demand per person. The per capita demand increases as the standard of living and average income increase. The per capita demand also increases with rising expectations, when people become aware that they could have more to eat and have a greater variety of food. Modern communication has led to rising expectations and to growth in per capita demand. As we saw in Chapter 5, an increase in either the environmental impact per person or the total number of people results in an increase in the total human environmental impact.

11.10 Limits to Food Production

Even given the great strides in food production that can result from the use of new crops, genetic engineering, more fertilizer, and more water, there is a limit to the production of food on Earth. For one thing, only some lands can be used for agriculture (Figures 11.12, 11.13).

Figure 11.12 • Physical and ecological considerations in watershed development such as slope, elevation, floodplain, and river delta location, limit land available for agriculture.

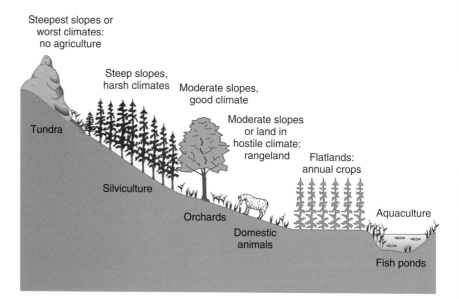

Figure 11.13 • Land that is unsuitable for crop production on a sustainable basis can be used for other purposes.

In addition, for many crops, we appear to have reached the limit of production gains realized for fertilizer applied. In the future, increases in yield per unit area must come from the development of new, higher-producing crops or new superstrains of existing ones. Such increases can cause additional environmental problems.

Another limitation is that climatic change is more likely to decrease yield than to increase it because areas with the best soils in the world also happen to have climates well suited to agriculture. Thus, most climatic changes are likely to make things worse. If global warming takes place as forecast by models of global climate, major disruptions to agriculture will result.[30] The best climates for agriculture might shift from their present locations to locations farther north. For example, the climate of the midwestern corn belt of the United States, which has been so good for crop production, could move northward, favoring Canada. This move could have a real impact on the amount of grain produced because soils in Canada are generally not as suitable for grain production as are U.S. midwestern soils. Global warming could also lead to an increase in evapotranspiration in many midlatitude areas.[31] Supplying water for irrigation would become an even greater problem than it is today. Those concerned about the future of agriculture must take the possibility of global warming into account in their planning.

Plants require water, and much modern agriculture involves irrigation. The importance of water is indicated by the fact that soil moisture in spring is a major determinant for agricultural production in the corn and wheat belts of the United States.

Like the use of fertilizers, techniques to add water to cropland are as old as agriculture (see Figure 11.14). Sources of irrigation water include groundwater; diversion of water from nearby streams, rivers, and natural lakes; and artificial reservoirs. But large-scale irrigation

(a)

(b)

Figure 11.14 • Irrigation. (a) A United Nations Food and Agricultural Organization project helps farmers in the Sudan develop irrigation canals. (b) At a modern agricultural research station in Saudi Arabia, a grated pipe irrigation system brings water to a corn field.

projects cause environmental problems (discussed in Chapter 12). Construction of reservoirs changes the local environment. Some habitats disappear. Stream patterns change, and erosion rates increase in the watershed of the reservoir (see Chapter 20).

11.11 Ways to Increase the Food Supply

As stated, most increases in agricultural production must come from the development of high-yield strains of crops. During the 20th century, rapid strides were made in the increase in production of crops per unit area. Some of these advances, which involved the development of new hybrids, became known as the green revolution. Several other methods of increasing the food supply may hold some promise as well, although always with limitations.

The Green Revolution

The **green revolution** is the name attached to post–World War II programs that have led to the development of new strains of crops with higher yields, better resistance to disease, or better ability to grow under poor conditions. An advancement of the green revolution was the development of super strains of rice at the International Rice Research Institute in the Philippines. (Figure 11.10). Although genetic hybridization of rice vastly increased rice production per acre, the new strains required a greater use of fertilizers and as much as four to seven times the water. In some cases, they produced a rice that was not considered desirable to eat. Another development in the green revolution was the formation of strains of maize with improved disease resistance at the International Maize and Wheat Improvement Center in Mexico.

Improved Irrigation

A possible means to increase yield is with better irrigation techniques, which can improve yield and reduce overall water use. **Drip irrigation**, the application of water to the soil from tubes that drip water slowly, greatly reduces the loss of water from direct evaporation and increases yield. However, it is an expensive approach, most likely to be applied in developed nations or nations with a large surplus of hard currency—in other words, in few of the countries where hunger is most severe.

Increasing the Amount of Agricultural Land

What are the opportunities to add new lands to agriculture? As we have noted, because most of the best land for agriculture is already so used, more and more marginal lands will be farmed. These lands are considered marginal because they lack water, have poor soil quality, or lie on slopes so steep that the soil readily erodes.

Some people suggest that in the future we will rely increasingly on artificial agriculture, such as **hydroponics**. Hydroponics is the growing of plants in a fertilized water solution on a completely artificial substrate in an artificial environment, such as a greenhouse. This approach is extremely expensive and is unlikely to be effective in the developing nations, where hunger is the greatest.

Eating Lower on the Food Chain

Some people believe that it is ecologically unsound to use domestic animals as food on the grounds that each step farther up a food chain leaves much less food to eat per acre. This argument is as follows (see Chapter 9). No organism is 100% efficient. Of the energy in food taken in, only a fraction is converted to new organic matter. Crop plants may convert 1–10% of sunlight to edible food, and cows may convert only 1–10% of hay and grain to meat. Thus, the same area could produce 10–100 times as much vegetation per year as meat (see the Critical Thinking Issue in Chapter 5). This holds true for the best agricultural lands, which have deep, fertile soils on level ground.

However, as with so many environmental issues, a simple generalization does not apply to all cases. Land too poor for crops that people can eat can be excellent rangeland and grow grasses and woody plants that domestic livestock can eat (Figure 11.13). These lands occur on steeper slopes, with thinner soils, or with lower rainfall. Thus, from the point of view of *sustainable* agriculture, there is value in rangeland or pasture. *The wisest approach to sustainable agriculture involves a combination of different kinds of land use:* using the best agricultural lands for crops and poorer lands for pastures and rangelands and avoiding the use of the best lands for grain production for animal feed.

Another problem with the argument that we should eat lower on the food chain is that food is more than just calories, and animals are important sources of proteins and minerals. Animals provide the major source of protein in human diets—56 million metric tons of edible protein per year worldwide. In the United States, 75% of the protein, 33% of the energy, and most of the calcium and phosphorus in human nutrition come from animal products.

A third factor to keep in mind is that domestic animals are often used for other purposes—as beasts of burden for plowing, carrying loads, and transportation and as sources of clothing material (wool and leather)—as well as for food. In addition, their excrement is an important fertilizer and, in some areas of the world, an important—sometimes the only—fire fuel. In these ways, the use of animals as part of food production represents an increase in efficiency.

Some people maintain vegetarian diets because of specific dietary problems. And others, of course, do not eat meat for moral, ethical, or religious reasons. Thus, the

Will There Be Enough Water to Produce Food for a Growing Population?

Between 2000 and 2025, scientists estimate that the world population will increase from 6.1 billion to 7.8 billion,[33] approximately double what it was in 1974. To keep pace with the growing population, the United Nations Food and Agricultural Organization predicts that food production will have to double by 2025, and so will the amount of water consumed by food crops. Will the supply of freshwater be able to meet this increased demand, or will water supply limit global food production?

Growing crops consume water through transpiration (loss of water from leaves as part of the photosynthetic process) and evaporation from plant and soil surfaces. The volume of water consumed by crops worldwide—including rainwater and irrigated water—is estimated at 3,200 billion m³ per year. An almost equal amount of water is used by other plants in and near agricultural fields; thus, it takes 7,500 billion m³ per year of water to supply crop ecosystems around the world. (See Table 11.2.) Grazing and pasture land account for another 5,800 billion m³ and evaporation from irrigated water another 500 billion m³ for a total of 13,800 billion m³ of water per year for food production, or 20% of the water evaporated and transpired worldwide. By 2025, therefore, humans will be appropriating almost half of all the water available to life on land for growing food for their own use. Where will the additional water come from?

Although the amount of rainwater cannot be increased, it can be more efficiently used through farming methods such as terracing, mulching, and contouring. Forty percent of the global food harvest now comes from irrigated land, and some scientists estimate that the volume of irrigated water available to crops will have to triple by 2025—a volume equal to that of 24 Nile Rivers or 110 Colorado Rivers.[34] A significant saving of water can, therefore, come from more efficient irrigation methods, such as improved sprinkler systems, drip irrigation, night irrigation, and surge flow.

Additional water could be diverted from other uses to irrigation. But this may not be as easy as it sounds because of competing needs for water. For example, if water were provided to the 1 billion people in the world who currently lack drinking and household water, less would be available for growing crops. And the new billions of people to be added to the world population in the next decades will also need water. Already, humans use 54% of the world's runoff. Increasing this to more than 70%, as will be required to feed the growing population, may result in loss of freshwater ecosystems, decline in world fisheries, and extinction of aquatic species.

Irrigation projects are costly to construct, and fewer sites than in the past are now available that are acceptable ecologically and socially. As a result, irrigation has been growing slower than population in recent years. If this trend continues, the irrigation gap will become even larger. Irrigation has also been responsible for

Table 11.2	Estimated Water Requirements of Food and Forage Crops
Crop	*Liters/kg*
Potatoes	500
Wheat	900
Alfalfa	900
Sorghum	1,110
Corn	1,400
Rice	1,912
Soybeans	2,000
Broiler chicken	3,500*
Beef	100,000*

* Includes water used to raise feed and forage.
Source: D. Pimentel et al., "Water Resources: Agriculture, The Environment, and Society," *Bioscience* 4, no. 2 (February 1997): 100.

salinization of agricultural areas, which makes them less suitable for growing crops and requires more water to flush the salts from the soil. In many places, groundwater and aquifers are being used faster than they are replaced—a process that is unsustainable in the long run. Many rivers are already so heavily used that they release little or no water to the ocean. These include the Ganges and most other rivers in India, the Huang He (Yellow River) in China, the Chao Phraya in Thailand, the Amu Dar'ya and Syr Dar'ya in the Aral Sea basin, and the Nile and Colorado Rivers.

Two hundred years ago, Thomas Malthus put forth the proposition that population grows more rapidly than the ability of the earth to grow food and that at some time the human population will outstrip the food supply (see A Closer Look 5.1). Malthus might be surprised to know that, by applying science and technology to agriculture, food production has so far kept pace with population growth. For example, between 1950 and 1995, world population increased by 122% while grain productivity increased 141%. Since 1995, however, grain production has slowed down (see Figure 11.15), and the question remains as to whether Malthus will be proven right in the 21st century. Will science and technology be able to solve the problem of water supply for growing food for people, or will water prove a limiting factor in agricultural production?

Critical Thinking Questions

1. How might changes in diet in developed countries affect water availability?

Figure 11.15 • (*a*) World irrigated area per thousand people, 1961–1995. (*Source:* L. R. Brown, M. Renner, and C. Flavin, *Vital Signs: 1998* [New York: W. W. Norton, 1998], p. 47.) (*b*) World grain production per person, 1950–1977. (*Source:* L. R. Brown, M. Renner, and C. Flavin, *Vital Signs: 1998:* [New York: W. W. Norton, 1998], p. 29.)

2. How might global warming affect estimates of the amount of water needed to grow crops in the 21st century?

3. Withdrawing water from aquifers faster than the replacement rate is sometimes referred to as "mining water." Why do you think this term is used?

4. Many countries in warm areas of the world are unable to raise enough food, such as wheat, to supply their popula-tions. Consequently, they import wheat and other grains. How is this equivalent to importing water?

5. Malthusians are those who believe that sooner or later, unless population growth is checked, there will not be enough food for the world's people. Anti-Malthusians believe that technology will save the human race from a Malthusian fate. Analyze the issue of water supply for agriculture from both points of view.

decision about when and whether to use animals as part of food production is an issue of both science and values.

 11.12 Biotechnology

The discovery that DNA is the universal carrier of genetic information has led to the development of a completely new technology known as genetic engineering. Now that the chemistry of inheritance is understood, scientists have been able to develop methods to transfer specific genetic characteristics from one individual to another, from one population to another, and from one species to another. In terms of agriculture, this allows several modifications of traditional agriculture involving **genetically modified organisms**.

Genetic engineering in agriculture involves several different kinds of practices, which we can group as follows: (1) faster and more efficient methods of developing new hybrids; (2) the introduction of the "terminator gene" (discussed in the following chapter); and transfer of genetic properties from widely divergent kinds of life. These three practices provide quite different potentials and problems. Of the three, the first, hybridization, is the least novel in the sense that crop hybridization has become a standard methodology in modern agriculture.

In this case, biotechnology adds a new way to create hybrids, not an entirely new process. In contrast, the second and third enable actions never previously possible.

There is considerable interest in the potential for genetic engineering to develop strains of crops with entirely new characteristics. One focus of this research is the development of new crops that fix nitrogen, creating the same symbiotic relationship found in legumes (members of the pea family). Recall that bacteria grow in the root nodules of legumes. The bacteria live on products produced by the legumes; in turn, the bacteria fix nitrogen (meaning that they convert atmospheric gaseous nitrogen to a form that can be used by green plants). Legumes are often rotated with other crops so that the soil is enriched in nitrogen. It may be possible to develop new strains of corn and other crops that, along with new strains of bacteria, can form a symbiotic nitrogen-fixing relationship. Such an accomplishment would lead to an increase in the production of these crops and a reduction in the utilization of fertilizers.

Another goal of agricultural genetic engineering is the development of strains with improved tolerance to drought, cold, heat, and toxic chemical elements. For example, one effort is aimed at developing wheat that is resistant to high levels of aluminum, an element that has negative effects on the growth of many plants.[32] Yet another goal of agricultural genetic engineering is to create crops that produce their own pesticides. This topic is discussed in the next chapter.

Although genetic modifications have proved to have great benefit, they involve both limitations and environmental concerns. These, too, are discussed in the next chapter.

Summary

- Among the major aspects of an environmental perspective on agriculture and therefore an improved relationship between agriculture and the conservation of land, water, and biological resources are the following: (1) agriculture depends on the environment, and good agricultural practices require an understanding of the relationships between crop growth and the environment; (2) agriculture changes the environment, and the more intense the agriculture the greater the changes; and (3) as human population increases, we need to conserve our agricultural resources and to increase production and at the same time to decrease the negative effects of agriculture on the environment.

- From an ecological perspective, agriculture is an attempt to keep an ecosystem in a specific stage, usually an early, highly productive successional stage. Such stages favor early successional plants and animals, which become competitors and predators of crops.

- There are three major approaches to agriculture: (1) demand-based agriculture, in which production is based on highly mechanized technology, with a high demand for resources, including land, water, and fuel, and with little use of biologically based technologies; (2) resource-based agriculture, in which production is based on biological technology, with conservation and renewal of land, water, and energy; and (3) organic farming.

- The history of agriculture can be viewed as a series of attempts to overcome environmental limitations and problems. Each new solution has created new environmental problems, which have in turn required their own solutions.

- Our modern food problem is the result of the great increase in human population growth, which outstrips local food production in many areas, and an inadequate food distribution system for this growing population. If the human population continues to grow, however, there will be limits on total food production in the future. One solution is to seek new sources of food.

- Most of the world's food is provided by only 14 crop species. Although some food is obtained from oceans and fresh waters, 95% of the protein and most of the calories humans consume are obtained from traditional land-based agriculture of crops and livestock. There is great potential for new crops from among the 270,000 species of plants.

- Farming greatly simplifies ecosystems, creating short, simple food chains, reducing species diversity, reducing the organic matter content and overall fertility of soils, and straining water supplies. Agro-ecosystems also differ from natural ecosystems in their spatial regularity. This simplicity makes these systems more vulnerable to outbreaks of diseases, undesired competitors, and herbivores.

- The major focus for the future must be on sustainable agriculture, which means that production per unit area is restricted to an amount that does not damage soils, eliminate water supplies, cause extinction of wild relatives of crops or extinction of potential new food species, or lead to permanent pollution downstream (the topic of the next chapter).

- Biotechnology makes genetic manipulation possible, creating entirely new ways to produce new crops and new strains within a crop. This provides a potential for increased production and for the production of new products from crops and domesticated animals but also poses novel environmental threats and raises many issues regarding science and values.

● REEXAMINING THEMES AND ISSUES

Human Population:

Our modern food problem is the result of the great increase in human population growth, which outstrips local food production in many areas, and an inadequate food distribution system for this growing population. Human population growth will eventually be limited by total food production.

Sustainability:

In the future, a major goal of agriculture must be to achieve sustainable food production in any location. This requires the development of farming methods that do not damage soils, eliminate water supplies, cause extinction of wild relatives of crops or of potential new food species, or lead to permanent pollution downstream.

The world as a whole is moving from demand-based to resource-based agriculture. The latter is more consistent with a sustainable approach to agriculture.

From an ecological perspective, agriculture is an attempt to keep an ecosystem in a specific stage, usually an early, highly productive, successional stage. In this way, agriculture works against natural mechanisms of sustainability, and we must compensate for this.

Global Perspective: Most of the world's food is provided by only 14 crop species. A challenge for the future is to search the world's diversity of life for additional food species. Food is a global resource in that it is traded internationally, and the availability of food in any area is the result of both local production and global agricultural markets.

Urban World: The artificiality of urban environments tends to lead us to believe that when we live in cities, we are independent of the environment. But when urban people in Brazil starved, they broke into food storage facilities and caused social disruption, thus illustrating the intimate connection between urban life and agricultural production.

Science and Values: The fundamental ethical problem we face concerning food supply is: Are we going to continue to attempt to feed more and more people? Do we attempt this at the risk of sacrificing habitats of non-crop species, natural ecosystems, and landscapes, reducing water supply for other purposes, and increasing the erosion of landscape? Or will we attempt to limit our numbers and also limit total food production?

● KEY TERMS

agro-ecosystem **191**
aquaculture **196**
cash crops **192**
demand-based agriculture **194**
drip irrigation **205**
genetically modified organisms **207**
green revolution **205**

hydroponics **205**
limiting factor **199**
mariculture **196**
maximum sustainable yield **201**
monoculture **192**
optimum sustainable yield **201**
per capita demand **203**

per capita food production **203**
rangeland **194**
resource-based agriculture **194**
ruminants **194**
subsistence crops **192**
synergistic effect **199**

● STUDY QUESTIONS

1. Explain differences between demand-based and resource-based agriculture. Which characterizes traditional swidden agriculture (see A Closer Look 11.1)?

2. A city garbage dump is filled; it is suggested that the area be turned into a farm. What factors in the dump might make it a good area to farm, and what might make it a poor area to farm?

3. Most crops are characteristic of what stages in ecological succession? How might we use our knowledge of succession to make agriculture sustainable?

4. Ranching wild animals—that is, keeping them fenced but never tamed—has been suggested as a way to increase food production in Africa, where there is a great abundance of wildlife. Based on this chapter,

what are the environmental advantages and disadvantages of such game ranching?

5. Explain what is meant by the following statement: The world food problem is one of distribution, not of production. What are the major solutions to this world food problem?

6. You are sent into the Amazon rain forest to look for new crop species. In what kinds of habitats would you look? For what kinds of plants would you look?

7. How does agriculture simplify an ecosystem? In what ways is this simplification beneficial to people? In what ways does it pose problems for a sustainable food supply?

8. A vegetable garden is planted in a vacant lot in a city. Peas and beans grow well, but tomatoes and lettuce

do poorly. What is the likely problem? How could it be corrected?

9. A second vegetable garden is planted in a vacant lot. Nothing grows well. Outside the city, in otherwise similar environments, vegetables grow vigorously. What might explain the difference?

10. Should organic farming be allowed to include genetically modified organisms? Why or why not?

● FURTHER READING

Achebe, C., G. Hyden, C. Magadza, and A. P. Okeyo, eds. 1990. *Beyond Hunger in Africa: Conventional Wisdom and an Alternative Vision.* Portsmouth, N.H.: Heinemann.

Borgstrom, G. 1965. *The Hungry Planet: The Modern World at the Edge of Famine.* New York: Macmillan. The major book by one of the leaders of agricultural change.

Fisher, G., K. Frohberg, M. L. Parry, and C. Rosenzweig. 1994. "Climate Change and World Food Supply, Demand, and Trade: Who Benefits, Who Loses," *Global Environmental Change* 4 (1):7–23. Uses computer simulation of climate change to examine implications for agriculture.

Pesek, J., ed. Board on Agriculture, National Research Council. 1989. *Alternative Agriculture.* Washington, D.C.: National Academy. A very readable report presenting the scientific and economic soundness of alternative agricultural systems.

Wittwer, S., Y. Yce, H. Sun, and L. Wong. 1987. *Feeding a Billion: Agriculture in China.* Lansing: Michigan State University Press. A look at modern and traditional agricultural technologies of China and a discussion of how China feeds almost one quarter of the world's population on 7% of the world's arable land.

Frey, K. J. 1996, *National Plant Breeding Study—1.* Iowa State University Agriculture and Horne Economics Experiment Station, Special Report 98.

CHAPTER 12

Effects of Agriculture on the Environment

Overgrazing and subsequent erosion at this site in the Central Valley of California has damaged and nearly destroyed riverbank (riparian) vegetation while increasing soil erosion.

● LEARNING OBJECTIVES

Agriculture changes the environment in many ways, both locally and globally. After reading this chapter, you should understand:

- How agriculture can lead to soil erosion, the severity of the problem, what methods are available to minimize erosion, and how application of these methods has reduced soil erosion in the United States.
- How farming can deplete soil fertility and why agriculture in most cases requires the use of fertilizers.
- Why some lands are most effectively used for grazing and how overgrazing can damage land.
- What causes desertification.
- How farming creates conditions that tend to promote pest species, the importance of effective pest (including weed) control, and the problems associated with chemical pesticides.
- How alternative agricultural methods, including integrated pest management, no-till agriculture, mixed cropping, and other methods of soil conservation, can provide major environmental benefits.
- How new methods of genetic modification of crops could improve food production and benefit the environment and how these methods could create new environmental problems.

Clean Water Farms

Steve Burr, a farmer near Salina, Kansas, has 300 acres of crops and 400 acres of grassland on which he raises cattle. In 1994, Steve became concerned about the amount of erosion on his cropland and the effects of fertilizers, pesticides, and livestock wastes on water quality. He decided to make two major changes in the management of his farm. By converting some of the crop acreage to grass, he reduced the erosion and chemical pollution in the runoff from his land. At the same time, he divided his grazing land into sections, called paddocks, and rotated his animals through them so that they had a constant supply of fresh forage. The cows now dropped their wastes evenly as they moved through the series of paddocks, and the animal wastes no longer piled up in feed lots, where they posed a major disposal problem. The rotation system prevented overgrazing as it fertilized the pastures. In addition, Steve now could spend less money on dry feed for his cows, allowing him to increase the size of his herd and generate more income. This method of raising livestock is known as intensive rotational grazing or management intensive grazing (Figure 12.1).

Steve Burr is not alone in his efforts to adopt farm management practices that reduce pollution and improve water quality while improving his balance sheet. The Burr farm is one of 36 participants of the Kansas Rural Center's Clean Water Farms Project, which began in 1995. The goal of the Clean Water Farms Project is to farm in a manner that is beneficial both to the environment and to the economics of farming.[1] Farmers in many other states are using intensive rotational grazing systems. Their numbers are increasing every year. And American farmers are not the only ones adopting this more environmentally benign and economically advantageous approach to raising livestock. In fact, the first system of rotational grazing was developed in France, and an adaptation of it was pioneered in South Africa. New Zealand ranchers are known for their expertise in intensive rotational grazing and the world-famous Argentinian beef is raised using similar methods.

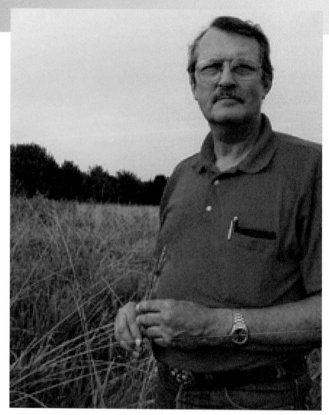

Figure 12.1 • Steve Burr rotates his cattle on this grassland to prevent overgrazing on his farmland.

According to Missouri farmer David Shafer, consumers also reap benefits from pasture-raised beef—they get higher-quality products at lower prices. Shafer has also found that his animals have fewer parasites and that soil fertility, as well as water quality of runoff, has improved. On a small scale, farmers like David Shafer and Steve Burr are recreating migration patterns similar to those of bison and elk. Shafer hopes that over time rotational grazing will rejuvenate the prairie biome that once existed in the central United States.

Intensive rotational grazing is just one of many approaches to farming that are sustainable. Other methods include crop rotation, use of cover crops to reduce fertilizer needs and erosion, composting livestock wastes, no-till farming, integrated pest and weed management, and redesigning livestock waste management and watering systems.

• *This case history shows that practices that are environmentally benign can be economically advantageous. The environmental advantages of such alternative approaches to agriculture and the environmental effects of various forms of agriculture are the subjects of this chapter.*

12.1 How Agriculture Changes the Environment

In the previous chapter, we learned how agriculture depends on the environment and how modern agricultural approaches offer benefits in producing food and therefore in feeding the growing population of the world. This chapter focuses on how agriculture affects the environment. It therefore discusses some environmental problems resulting from agriculture and some actual and potential solutions to these problems. Agriculture effects the environment locally, regionally, and globally. It is therefore part of one of the major themes of this book: a global perspective. Because agriculture occupies a large percentage of the Earth's land area (as explained in the previous chapter, approximately 11% excluding Antarctica), the environmental effects of agriculture are a global issue as well as a local and regional issue (Figure 12.2). Agriculture has long affected the environment. Nothing in nature resembles a plow, so as soon as plowing was invented, agriculture began to induce large changes in the local environment. This is an ancient lineage, extending back thousands of years.

There is, in fact, a curious puzzle about agriculture and the plow. Great differences exist between soils in unplowed forests and soils from previously forested land that have been plowed and used for crops for thousands of years. These differences were observed and written about by one of the great originators of the modern study of the environment, George Perkins Marsh. Born in Vermont in the 19th century, Marsh became the American ambassador to Italy and Egypt. While in Italy, he was so struck by the differences in the soils of the forests of his native Vermont and the soils that had been farmed for thousands of years on the Italian peninsula that he made this a major theme in his landmark book, *Man and Nature*. The farmland he observed in Italy had once been forests. But while soil in his nature Vermont was rich in organic matter, was deep, and had definite layers, the soil of Italian farmland had little organic matter and lacked definite layers.

One would expect that such changes in the soil would lead to an unsustainable condition—that is, that agriculture would eventually be impossible. But much of the farmland in Italy and France has been in continuous use since pre-Roman times and still is highly productive. How can this be? And what has been the long-term effect of such agriculture on the environment—local, regional, and global? In this chapter we explore such questions, adding to the original effect of the plow the many environmental effects of modern agriculture.

WORLD AGRICULTURE

- Dairying
- Fruit, Truck and Specialized Crops
- Mixed Livestock and Crop Farming
- Commercial Grain Farming
- Subsistence Crop and Livestock Farming
- Mediterranean Agriculture
- Diversified Tropical Agriculture –chiefly plantation
- Intensive Subsistence Farming –chiefly rice
- Intensive Subsistence Farming –chiefly wheat and other crops
- Rudimentary Sedentary Cultivation
- Shifting Cultivation
- Livestock Ranching
- Nomadic and Semi-Nomadic Herding
- Nonagricultural Areas

Figure 12.2 • World map showing major land areas in agriculture.

Agriculture has both on-site, or *primary*, effects and *off-site* or *secondary* effects. An **on-site effect** is an effect on the area where the agriculture takes place. An **off-site effect** is an effect on an environment away from the agricultural site, typically downstream and downwind (see this chapter's Critical Thinking Issue).

Local effects are those that occur at or near the site of the farming; these effects include erosion, loss of soils, and increases in sedimentation downstream in local rivers. Regional effects are those that generally result from the combined effects of farming practices in the same large region, such as a county, state, or small nation. Regional effects include the creation of deserts, large-scale pollution, increases in sedimentation in major rivers and in the estuaries at the mouths of the rivers, and changes in the chemical fertility of soils over large areas. Global effects include climatic changes as well as potentially extensive changes in global biogeochemical cycles (see Chapter 4).

Major environmental problems that result from agriculture include deforestation, desertification, soil erosion, overgrazing, degradation of water resources, salinization, accumulation of toxic metals, accumulation of toxic organic compounds, and soil, air, and water pollution.

12.2 Agriculture and Soil Erosion

Farming easily damages soils (see A Closer Look 12.1). When land is cleared of its natural vegetation, such as forest or grassland, the soil begins to lose its fertility. Some of this occurs by physical erosion. Once the protection of the vegetative cover is lost, the soil is exposed directly to water and wind, which remove the loosened soil. In addition, the introduction of heavy, earth-moving machinery after World War II has led to a considerable increase in the compaction (packing down) of soil and the loss of the proper soil structure for crop production.

Farmed soil also loses fertility when chemical elements are dissolved in water and transported away in streams and subsurface runoff. The rate of loss of fertility is sometimes measured as the time required for the soil to lose one-half of its original store of the chemical elements necessary for crops. The time over which a soil loses one-half of a chemical element varies. It is much faster in warmer and wetter climates, such as in tropical rain forests, than it is in colder or dryer climates, such as those where the natural vegetation is a temperate-zone grassland or forest.[2]

Much concern with soil erosion in the United States arose in the 1930s with the occurrence of the "dust bowl." Intense and environmentally poor plowing practices, combined with a major drought, loosened the soil over large areas, which then blew away, creating dust storms that buried automobiles and houses.

Since the end of World War II, human food production activities have seriously damaged more than 1 billion ha (2.47 billion acres) of land (about 10.5% of the world's best soil), equal in area to China and India. Overgrazing, deforestation, and destructive crop practices have damaged approximately 9 million ha (22 million acres) to the point that recovery will be difficult; restoration of the rest will require serious actions.[3]

In the United States, about one-third of the country's topsoil has been lost, resulting in 80 million ha (198 million acres) either totally ruined by soil erosion or made only marginally productive.[4] Although soil loss in the United States continues, measurements suggest that rates have been considerably reduced as a result of improvements in plowing and the use of no-till agriculture (discussed later in this chapter). For example, the drainage area of Coon Creek, Wisconsin, an area of 360 km², has been heavily farmed. This stream watershed was the subject of a detailed study in the 1930s by the United States Soil Conservation Service. Then the area was restudied in the 1970s and 1990s. Measurements at these three times indicate that present soil erosion is only 6% of what occurred in the 1930s.[5,6]

Although soil scientists seem to agree in general that soil erosion has decreased in the United States since the 1930s, there is a lack of consensus about the rate of decline and percentage of decline in soil erosion for the United States as a whole. Estimates suggest that, on average, soil erosion in the United States has declined from 17 metric tons per hectare per year (t/ha/yr) to about 13 t/ha/yr.[7]

Sediment Damage

Soil eroded from one location has to move somewhere else. Much is moved downstream by surface water flows and deposited at the mouth of a stream or river. Thus, an increase in erosion also means an increase in downstream sedimentation. Rivers carry about 3.6 billion metric tons/year (4 billion U.S. tons/year) of sediment in the United States, 75% of which is from agricultural lands. Of this total, 2.7 billion metric tons/year (3 billion U.S. tons/year) is deposited in reservoirs, rivers, and lakes.

Downstream sedimentation is a serious environmental effect of modern agriculture. Sediments fill in otherwise productive waters, destroying some fisheries. In tropical waters, sediments entering the ocean can destroy coral reefs that are near shore. Nitrates, ammonia, and other fertilizers carried by sediments can cause eutrophication in downstream waters; the resulting buildup of algae reduces fish production. Polluted sediments also can transport toxins. Sediment damage costs the United States about $500 million/year in dredging expenses and costs attributable to reduction in the useful life of reservoirs. Since the 1930s, agriculture-induced sedimentation has decreased overall, accompanying the decrease in the rate of soil erosion, as illustrated in Figure 12.3.

(a)

(b)

Figure 12.3 • Improvement in conditions near a stream as a result of reduction in soil erosion and sedimentation. (*Source:* S. W. Trimble and P. Crosson, "U.S. Soil Erosion Rates: Myth and Reality," *Science* 289 [July 14, 2000]: 248–250.)

Making Soils Sustainable

Soil is continuously formed. Ideal farming would result in soil loss no greater than the production of new soil. On good lands, soil forms at a rate that ranges from 1 mm/decade to 1 mm/40 years. Obviously, one way to counter soil erosion is to promote new soil formation. Other practices that can aid in the sustainability of soils include fall plowing, which can be less harmful to the soil; multiculture (planting several crops intermixed in the same field); strip-cropping; terracing; and proper crop rotation. Two additional practices are contour plowing and no-till agriculture.

Contour Plowing

One way to counter erosion from plowing is contour plowing. In traditional plowing, the plowed furrows make a path for water to flow, and if the furrows go downhill, then the water moves rapidly along them, increasing the erosion rate. In **contour plowing**, the land is plowed along the contours, perpendicular to the slope and as much in the horizontal plane as possible. Plowing along the contours can greatly reduce erosion loss owing to water runoff.

In the recent past, contour plowing has been the single most effective method for reducing soil erosion, as demonstrated by an experiment with sloping land planted in potatoes. Part of the land was plowed in uphill and downhill rows, and part was contour plowed. The uphill and downhill section lost 32 metric tons/ha (14.4 tons/acre) of topsoil, whereas the contour-plowed section lost only 0.22 metric ton/ha (0.1 ton/acre) (Figure 12.4*a*). In addition to drastically reducing soil erosion, contour plowing used less fuel and time. However, only a small fraction of the land receives this treatment. For example, of Minnesota's 4 million ha (10 million acres) of cropland, only 526,110 ha (1.3 million acres) are contour plowed.

(a)

(b)

Figure 12.4 • Alternative agricultural plowing and tilling methods: (*a*) contour strip crops in the midwestern United States; (*b*) no-till soybean crop planted in wheat stubble on a Kansas farm.

Soils

Soils can be defined as earth materials modified over time by physical, chemical, and biological processes such that, in addition to supporting rooted plant life, they are altered from the original *parent material* into a series of horizons that are subparallel to the surface (Figure 12.5).

As soils develop, materials such as clay along with minerals and nutrients such as iron, calcium, and magnesium are leached from the upper horizons (*A* and *E*) and may be deposited in a lower horizon (*B*). *Leaching* is accomplished as water infiltrates from the surface, dissolves soil materials as part of chemical weathering processes, and transports the dissolved materials laterally or downward. These processes change the composition of horizons as well as the ability of the soil to retain water. The upper horizons (*O* and *A*) are complex ecosystems; each cubic

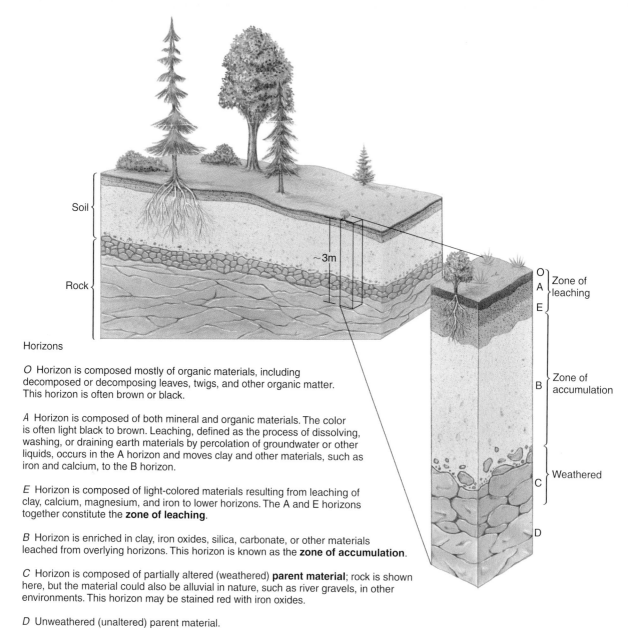

Horizons

O Horizon is composed mostly of organic materials, including decomposed or decomposing leaves, twigs, and other organic matter. This horizon is often brown or black.

A Horizon is composed of both mineral and organic materials. The color is often light black to brown. Leaching, defined as the process of dissolving, washing, or draining earth materials by percolation of groundwater or other liquids, occurs in the A horizon and moves clay and other materials, such as iron and calcium, to the B horizon.

E Horizon is composed of light-colored materials resulting from leaching of clay, calcium, magnesium, and iron to lower horizons. The A and E horizons together constitute the **zone of leaching**.

B Horizon is enriched in clay, iron oxides, silica, carbonate, or other materials leached from overlying horizons. This horizon is known as the **zone of accumulation**.

C Horizon is composed of partially altered (weathered) **parent material**; rock is shown here, but the material could also be alluvial in nature, such as river gravels, in other environments. This horizon may be stained red with iron oxides.

D Unweathered (unaltered) parent material.

Figure 12.5 ● Idealized diagram of a soil, showing soil horizons.

centimeter of these horizons contains a multitude of microorganisms as well as insects, earthworms, and plant roots.

The type of soil at a particular site depends on such factors as climate, parent material, slope or topography, biological activity, and time. The soil horizons shown in Figure 12.5 are not necessarily all present in any one soil. Very young soils may have only an upper *A* horizon over a *C* horizon, whereas older mature soils may have nearly all the horizons shown.

Soil fertility refers to the capacity of a soil to supply the nutrients necessary for plant growth when other factors, such as availability of water and climate, are also favorable. Soils that have formed on geologically young materials—for example, the glacial deposits of northern Indiana and Illinois that form the famous midwestern corn belt—are nutrient rich. Soils in semiarid regions are often nutrient rich and need only water to become very productive for agriculture. Soils in humid areas and tropics may be heavily leached and relatively nutrient poor. In such soils, nutrients may be cycled through the organic-rich upper horizons; and if forest cover is removed, reforestation may be very difficult (see Chapter 13). Soils that accumulate certain clay minerals in semiarid regions may expand (swell) and contract (shrink) on wetting and drying, cracking roads, walls, buildings, and other structures. Expansion and contraction of soils in the United States causes billions of dollars in property damage each year.

Some soils, such as those composed of very small clay particles (less than 0.004 mm in diameter), retard the movement of fluids because the inner connecting space between soil particles is very small and tenaciously holds fluids. Other soils with coarser grains (greater than 0.06 mm in diameter), such as sand or gravel, have relatively large spaces between grains (pores), and fluids may move quickly through them. Thus, the type of soil particles present is important in siting facilities such as landfills where retention of pollutants on site is an objective (see Chapter 28).

Some soils are more susceptible to erosion than are others. For example, soils consisting of unconsolidated sand particles (0.06–2.0 mm in diameter) are particularly susceptible to erosion by running water and wind. Soils composed of coarser (heavier) particles or finer particles that are usually more cohesive (held together by clay minerals) are more resistant to erosion.

In summary, soils are extremely important in many environmental considerations, such as agriculture (fertility), land use, siting of facilities, erosion potential, and shrink-swell potential. It is difficult to think of a human use of the near-surface land environment that does not involve consideration of the soils present. As a result, the study of soils continues to be an important part of environmental sciences.

No-Till Agriculture

An even more efficient technique to slow erosion is to avoid plowing altogether. **No-till agriculture**, also referred to as *conservation tillage*, is a recent term for a combination of farming practices that includes not plowing the land, using herbicides to keep down the weeds, and allowing some weeds to grow. Stems and roots that are not part of the commercial crop are left in the fields and allowed to decay in place (Figure 12.4b). In contrast to standard modern approaches, the goal in managing weeds is to suppress and control them but not to eliminate them at the expense of soil conservation. A variety of methods are used to control weeds, including integrated pest management (discussed later in this chapter) and chemical pesticides. These practices can greatly reduce soil loss, as well as the use of tractor fuel. Land in no-till agriculture increased from about 2 million ha (5 million acres) in 1962 to 30 million ha (74 million acres) in 1990 in the United States. The term is applied to land that is either never plowed or left unplowed most years.

Pest Control and Agricultural Chemicals

All agriculture suffers from pests. From an ecological point of view, *pests* are undesirable competitors, parasites, or predators. The major agricultural pests are insects (feeding mainly on the live parts of plants, especially leaves and stems); nematodes (small worms that live mainly in the soil and feed on roots and other plant tissues), bacterial and viral diseases, weeds (flowering plants that compete with the crops), and vertebrates (mainly rodents and birds that feed on grain or fruit). Even today, with modern technology, the total losses from all pests are huge; in the United States, pests account for an estimated loss of one-third of the potential harvest and about one-tenth of the harvested crop. Preharvest losses are due to competition from weeds, diseases, and herbivores; postharvest losses are largely due to herbivores.[8]

Weeds

Although we tend to think that the major agricultural pests are insects, in fact, weeds are the major problem in terms of potential crop loss. Farming produces special environmental and ecological conditions that tend to promote weeds. Remember that the process of farming is an attempt to (1) hold back the natural processes of ecological succession; (2) prevent the normal entrance of migrating organisms into an area; and (3) prevent natural interactions (including competition, predation, and parasitism) between populations of different species.

As explained in Chapter 11, a farm is maintained in a very early stage of ecological succession. With the addition of fertilizers and water, it is a very good place not only for crops but also for other early successional plants to grow. These noncrop and therefore undesirable plants are what we call "weeds." (A weed is just a plant in a place we do not want it to be.) Recall that early successional plants tend to be fast growing and to have seeds that are easily blown by the wind or spread by animals that have eaten the plant's fruit or carried its seeds on their fur or feet. These plants spread and grow rapidly in the inviting habitat of open, early successional croplands.

There are about 30,000 species of weeds, and in any year a typical farm field is infested with between 10 and 50 weed species. Weeds compete with crops for all resources: light, water, and nutrients. The more weeds, the less crop. Some weeds can have a devastating effect on crops. For example, the production of soybeans can be reduced 60% if just one individual of the weed called cocklebur per row foot is allowed to compete with the crop over the course of a growing season.[9] Agricultural losses in the United States as a result of weeds exceed $16 billion/year. About $3.6 billion is spent annually for chemical weed control, amounting to 60% of all pesticide sales in the United States.

Development of Pesticides

Before the Industrial Revolution, farmers could do little to prevent pests except remove them when they appeared or use farming methods that tended to decrease their density. For example, swidden (more familiarly known as "slash and burn") agriculture reduces the density of pests by allowing noncrop plants to grow and by maintaining a diversity of crops in a single field (see Chapter 11). Preindustrial farmers also could have planted vegetation, such as aromatic herbs, that tends to repel insects.

With the rise of modern agricultural sciences, chemical pesticides were developed. The use of pesticides has grown, reaching $31.25 billion worldwide in 1996. About 80% of the pesticides in use are applied in developing countries. Agricultural scientists work continuously to improve pesticides. The most desirable pesticide would be *narrow spectrum*—that is, lethal to a single pest species but not harmful to other forms of life. Scientists refer to such an ideal, ultimate pesticide as a *magic bullet*, a chemical that rapidly seeks out individuals of a particular species and kills them, with no effect on any other form of life. No pesticide is that perfect.

Early chemical pesticides were *broad spectrum*, meaning that they affected a wide range of organisms. One of the earliest pesticides was arsenic, a chemical element toxic to all life, including people. It was certainly effective in killing pests, but it killed beneficial organisms as well and was very dangerous to use.

The second stage in the development of pesticides, which began in the 1930s, included the development of oil sprays and the use of natural plant chemicals. Many plants produce natural chemicals as a defense against disease and herbivores. Many of these natural chemicals are effective pesticides. Nicotine, from the tobacco plant, is one of these chemicals; it is the primary agent in some insecticides still in wide use today. However, although natural plant pesticides are comparatively safe, they are not as effective as desired.

Monitoring Pesticides in the Environment

Today, about 700 million pounds of pesticides of more than 100 compounds are applied each year nationwide in the United States. Once applied, these chemicals may decompose in place or may be blown by the wind or transported by surface and subsurface waters, meanwhile continuing to undergo some amount of decomposition. Sometimes the initial breakdown products (the first, still complex chemicals produced from the original pesticides) are toxic, as is the case with DDT (see A Closer Look 12.2). Eventually, the toxic compounds are decomposed to their original inorganic or simple, nontoxic organic compounds. However, for some chemicals, this can take a very long time. Public health standards and environmental-effects standards have been established for some but not all of these compounds.

Herbicides account for about 60% of the total pesticides found in the nation's waters. Surprisingly little is known about the history of the concentrations of pesticides in the major rivers of America. For example, there is no well-established program to monitor the changes in the concentration of pesticides in the Missouri River, one of the longest rivers in the world. The Missouri drains one-sixth of the United States, much of which is from the major agricultural states of the nation.

The United States Geological Survey has established a network for monitoring 60 sample watersheds throughout the nation. These are medium-sized watersheds, not the entire flow from the nation's major rivers. One such watershed is that of the Platte River, a major tributary of the Missouri River. The most common herbicides used for growing corn, sorghum, and soybeans along the Platte River were alachlor, atrazine, cyanazine, and metolachlor, all organonitrogen herbicides. Monitoring on the Platte near Lincoln, Nebraska, suggested that, during heavy spring runoff, concentrations of some herbicides might be reaching or exceeding established public health standards. But this research is just beginning, and it is difficult to reach definitive conclusions about whether present concentrations are causing harm in public water supplies or to wildlife, fish, algae in fresh waters, or vegetation. Advances in knowledge give us much more information, on a more regular basis, about how much of many artificial compounds are in the waters; but we are still unclear about their environmental effects.

DDT

The real revolution in chemical pesticides—the development of more sophisticated pesticides—began with the end of World War II and the discovery of DDT and other chlorinated hydrocarbons, including Aldrin and Dieldrin. The pesticide DDT was developed in the 1940s. At first, DDT was thought to be the long-sought magic bullet: It appeared to have no short-term effect on people and seemed to kill only insects. At the time, scientists believed that a chemical could not be readily transported from its original site of application unless it was water soluble. DDT was not very soluble in water and therefore did not appear to pose an environmental hazard.

DDT was used very widely until three things were discovered

1. It has long-term effects on other, desirable organisms. For example, it affect the ability of birds to produce eggs and may cause increased incidence of cancer in other organisms.

2. It is stored in oils and fats and is concentrated as it is passed up food chains, so that the higher an organism is on a food chain, the higher the concentration of DDT it contains. This process is known as *food chain concentration* or *biomagnification* (discussed in detail in Chapter 15).

3. The storage of DDT in fats and oils allows the chemical to be transferred biologically even though it is not very soluble in water.

In birds, intake of DDT and the products of its chemical breakdown (known as DDD and DDE) results in thin eggshells that break easily, so that the birds cannot reproduce successfully. This problem was especially severe in birds that are high on the food chain—predators that feed on other predators, such as the bald eagle, the osprey, and the pelican, which feed on fish that may be predators of other fish.

As a result of these problems, DDT was banned in most developed nations. It was banned in the United States in 1971. Since then, a dramatic recovery has been seen in the populations of affected birds. For example, the brown pelican of the California coast, which had become rare and endangered and whose reproduction had been restricted to offshore islands where DDT had not been used, is common again. Another example, the bald eagle, has become abundant again in the north woods, where it can be seen in Voyageurs National Park and the Boundary Waters Canoe area of northern Minnesota. However, DDT is still being produced in the United States for use in developing and less developed nations, especially as a control for malaria-spreading mosquitoes.

The use of DDT has had some benefits. It was primarily responsible for eliminating malaria and yellow fever as major diseases, reducing incidence of malaria in the United States from an average of 250,000 cases a year prior to the spraying program to fewer than 10 per year in 1950. However, the fact remains that DDT is toxic to animals. Furthermore, its effectiveness has been decreased over the years because many species of insects have developed a resistance to it. Nevertheless, DDT continues to be used because it is cheap and effective and because people have become accustomed to using it. About 35,000 metric tons of DDT are produced annually in at least five countries, and it is legally imported and used in dozens, including Mexico.

Although people in developed nations believe they are free from the effects of DDT, in fact this chemical is transported back to industrial nations in agricultural products from nations still using the chemical. Also, migrating birds that spend part of the year in malarial regions are still subject to the effects of DDT on eggshells. Thus, in spite of its banning in the developed nations, DDT remains an important world issue in pest control. (The problem of developing nations' using pesticides banned in other nations is not an issue only for DDT; the situation extends to other chemicals as well.)

With the banning of DDT in developed nations, other chemicals became more prominent. Chemicals were sought that were less persistent in the environment. Among the next generation of insecticides were organophosphates: phosphorus-containing chemicals that affect the nervous system. These chemicals are more specific and decay rapidly in the soil. Therefore, they do not have the same persistence as DDT. However, they are toxic to people and must be handled extremely carefully by those who apply them.

Chemical pesticides have created a revolution in agriculture. However, in addition to the negative environmental effects of individual chemicals such as DDT, their use has other major drawbacks. One such problem is known as *secondary pest outbreaks*, which occur after extended use (and possibly because of extended use) of a pesticide. Secondary pest outbreaks can come about in two ways: (1) reduction in one target species reduces competition with a second, which then increases and becomes a pest, or (2) the pest develops resistance to the pesticides through evolution and natural selection, which favor those in the population with a greater immunity to the chemical.[10] Developed resistance has occurred with many pesticides. For example, Dasanit, an organophosphate, was first introduced in 1970 to control maggots that attack onions in Michigan; although it was originally successful, it is now so ineffective that it can no longer be used for that crop.

Modern approaches to pest control involve **integrated pest management** (IPM), an ecosystem approach to pest management that integrates a variety of techniques. These techniques include:

- The use of natural enemies of pests, including parasites, diseases, and predators.
- The planting of a greater diversity of crops to reduce the chance that pests will find a host plant.
- No-till or low-till agriculture, which helps natural enemies of some pests to build up in the soil.
- The application of a set of highly specific chemicals, used much more sparingly than in earlier approaches.

Integrated pest management recognizes ecological communities and ecosystems. It takes into account the effect of one species on others. For instance, a decrease in one species may lead to an increase in another and decreases in still others.

Four principles govern IPM.[11]

1. The goal is control, not extinction. Pests are allowed to continue to exist at low, tolerable levels; the method is considered a success if pests are kept at these levels. (In contrast, advocates of older approaches involving heavy application of chemicals sought complete elimination of the pests.)
2. The use of natural control agents is maximized.
3. The ecosystem is the management unit (instead of a single species).
4. Any control action can have unexpected and unwanted effects.

The components of IPM include chemicals, the development of genetically resistant stock, biological control, and land culture. *Land culture* refers to how the land is physically managed, including whether and how it is plowed, what kind of crop rotation is used, the dates of planting, and basic means of handling crop harvests to reduce the presence of pests in residues and products sold.

A part of IPM, **biological control** is a set of methods to control pest organisms by using natural ecological interactions, including predation, parasitism, and competition. It includes the intentional introduction of predators, diseases, or other parasites of a pest. For example, ladybugs are common predators of many plant-eating pests. In most parts of the United States, you can buy ladybugs for release in home gardens or farms.

There are many specialized and effective biological controls. One of the most effective is a bacterial disease, *Bacillus thuringiensis*, which kills larval forms of many insect pests, including many caterpillars. It is used widely. Other effective biological control agents are small wasps that are parasites of caterpillars. The wasps lay their eggs on the caterpillars; the larval wasps then feed on the caterpillars and kill them. These wasps tend to have very specific relationships (one species of wasp will be a parasite of one species of pest), and so they are both an effective and narrow spectrum (Figure 12.6).

The control of the oriental fruit moth, which attacks a number of fruit crops, is an example of IPM biological con-

Figure 12.6 • Integrated pest management, the biological control of pests.

(a)

(b)

Figure 12.7 • The wasp that preys on the oriental fruit moth and that has been used in biological control (*a*) and the strawberry field where the wasps live (*b*).

A study by the U.S. Office of Technology Assessment concluded that IPM could reduce the use of pesticides by as much as 75% and at the same time reduce preharvest pest-caused losses by 50%. This would also greatly reduce the costs to farmers for pest control.[8]

Another technique to control insects involves the use of sex pheromones. In most species of adult insects, one sex (usually the female) releases a chemical called a sex pheromone, which acts as an attractant to members of the opposite sex. In some species, it has been shown to be effective up to 4.3 km (2.7 mi) away. These chemicals have been identified, synthesized, and used in insect control as bait in traps, in insect surveys, or simply to confuse the mating patterns of the insects involved.

Current agricultural practices in the United States involve a combination of approaches, but in most cases, they are more restricted than an IPM strategy. Biological control methods are used to a comparatively small extent. They are the primary tactic for the control of vertebrate pests (mice, voles, and birds) of lettuce, tomatoes, and strawberries in California, but not a major technique for grains, cotton, potatoes, apples, or melons. Chemicals are the principal control methods for insect pests. For weed control, the principal controls are methods of land culture. The use of genetically resistant stock is important for disease control in wheat, corn, cotton, and some vegetable crops, such as lettuce and tomatoes.

Integrated pest management reduces the release of toxic chemicals into the environment while enabling economically viable production of crops. The more it can be employed, the better for the environment and for individual ecosystems.

trol. The moth was found to be a prey of a species of wasp. The introduction of the wasp into fields helped control the moth. Interestingly, the effectiveness of the wasp in peach fields was increased when nearby strawberry fields were nearby. These strawberry fields provided an alternative habitat for the wasp, especially important for overwintering (Figure 12.7).[9] As this example shows, spatial complexity and biological diversity also become part of the IPM strategy.

12.5 Genetically Modified Crops: Biotechnology Applied to Agriculture

In the previous chapter, the concept of genetically modified organisms was introduced. In this chapter we discuss the potential environmental problems that might be associated with genetic modifications. The basic methods for genetic modification of crops are illustrated in Figure 12.8.

Figure 12.8 • How crops are genetically modified: an example.

Remember from Chapter 11 that the methods of genetic modification of organisms can be grouped into three major types: (1) a faster and more efficient method of developing new hybrids, (2) the introduction of the "terminator gene," and (3) transfer of genetic properties from widely divergent kinds of life. These methods pose different kinds of potential environmental problems. Here, we need to keep in mind a general rule of environmental actions: If actions we take are similar in kind and frequency to natural changes, then the effects are likely to be benign. This is because species have had a long time to evolve and adapt to these changes. In contrast, changes that are novel—that do not exist in nature—are more likely to have negative or undesirable effects, both direct and indirect. We can apply this rule to the three categories of genetically engineered crops.

New Hybrids

The development of hybrids within a species is a natural phenomenon, and the development of hybrids of major crops, especially for small grains, has been a major factor in the great increase in productivity of 20th century agriculture. So, strictly from an environmental perspective, using genetic engineering methods to develop hybrids within a species is likely to be as benign as has been the development of agricultural hybrids using conventional methods.

There is an important caveat, however. Some are concerned that the great efficiency of genetic modification methods may allow "superhybrids" that are so productive they can grow where they are not wanted and become pests. There is also a concern that some of the new hybrid characteristics could be transferred by interbreeding with closely related weeds (Figure 12.9). If this were to happen, it might be possible to inadvertently create a "superweed" whose growth, persistence, and resistance to pesticides would make it difficult to control.

Another environmental concern is that new hybrids might be developed that could grow on more and more marginal lands; the development of crops on such marginal lands might then increase erosion and sedimentation and lead to decreases in biological diversity in specific biomes. Another potential problem is that the "superhybrids" might require much larger amounts of fertilizers, pesticides, and water. This, in turn, could lead to greater pollution and the need for more irrigation.

On the other hand, it is possible that genetic engineering could lead to hybrids that require less fertilizer, pesticide, and water. For example, at present, only legumes (peas and their relatives) have symbiotic relationships with bacteria and fungi that allow them to fix nitrogen. Attempts are underway to transfer this capability to other crops, so that more kinds of crops would enrich the soil with nitrogen and require much less external application of nitrogen fertilizer.

The Terminator Gene

The **terminator gene** makes seeds from a crop sterile after the first year. This means that the corporation that has developed and marketed the seed has a major market for its production every year; farmers cannot get around purchasing seeds by saving some of their crop and using the hybrid seeds the next year. But this poses social and political problems. Farmers in less developed nations and governments of nations that lack genetic engineering capabilities are concerned that the terminator gene will allow the United States, and a few of its major corporations, to control the world food supply. These concerned observers believe that farmers in poor nations must be able to save some of their crop and grow the next year's crops from their own seeds, because they do not have the funds to buy new seeds every year. This is not directly an environmental problem, but it can become an environmental problem indirectly by affecting total world food production, which then affects the human population and affects how land is used in those areas that have been in agriculture.

Transfer of Genes Among One Major Form of Life to Another

Most environmental concerns have to do with the third category of genetic modification of crops: the transfer of

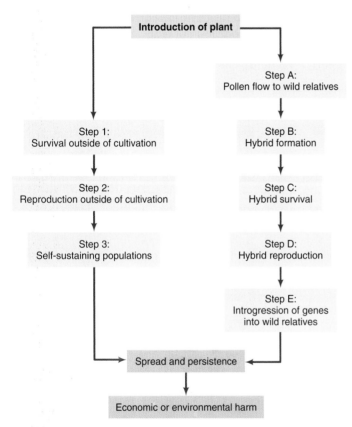

Figure 12.9 • Ways in which the genetic characteristics of a modified crop might spread.

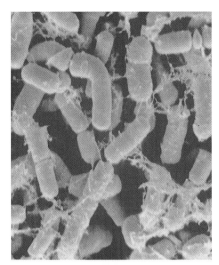

(a) *Bacillus thuringiensis bacteria (a natural pesticide) The gene that caused the pesticide (BT) was placed into corn through genetic engineering.*

(b) *BT corn contains its own pesticide in every cell of the plant.*

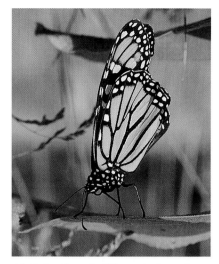

(c) *Pollen from the BT corn is also toxic and when it lands on milkweed, Monarch butterflies that eat the milkweed may die.*

Figure 12.10 • The flow of the BT toxin from bacteria (*a*) to corn through genetic engineering (*b*) and the ecological transfer of the toxic substances to Monarch butterflies (*c*).

genes from one major kind of life to another. This is a novel effect and therefore more likely to have negative and undesirable impacts. In several incidences, this type of genetic modification has led to unforeseen and undesirable environmental effects. Perhaps the best known involves potatoes and corn, caterpillars that eat these crops, a disease of caterpillars that controls these pests, and an endangered species, Monarch butterflies. Here is what happened.

Caterpillars are pests of many plants, from forest trees to many kinds of crops. These pests suffer from a disease caused by bacteria named *Bacillus thuringiensis*. Decades ago, scientists realized that the bacteria could be used as a pesticide—they could be sprayed over forests and crops. Since the disease does not affect mammals or plants, BT, as the treatment became known, was seen as a benign approach. It did not involve a novel toxic chemical but instead made use of a natural ecological interaction. BT has been a great success.

With the development of biotechnology, agricultural scientists studied the bacteria and discovered the toxic chemical and the gene that caused its production within the bacteria. This gene was then placed into potatoes and corn so that the biologically engineered plants produced their own pesticide (Figure 12.10). At first, this was believed to be a constructive step in pest control, because it was no longer necessary to spray a pesticide. However, the genetically engineered potatoes and corn produced the toxic BT substance in every cell—not just in the leaves that the caterpillars ate, but also in the potatoes and corn sold as food, in flowers, and in pollen.

Pollen released from fields of these genetically modified crops spread widely. A concern arose that the toxic pollen could be ingested by Monarch butterflies as they migrated through areas where the crops were grown. The Monarch butterfly is an endangered species protected by the U.S. Endangered Species Act (see Chapter 14). In fact, experiments showed that the butterflies inadvertently ingested the toxic pollen and died. (This pollen is not their food; they feed on milkweed, but they ingested pollen that fell on to the milkweed.) Although this effect has not yet been observed in the field, outside of experimental plots, it is a cause for concern.

The indirect, complex effect just described was not originally expected. In hindsight, it is easy to see that such complex, indirect effects are typical of ecosystems and ecological interactions. It is an example of the transfer of genes from one kind of life (bacteria) to a different type of life (green plants). This is a novel effect and therefore one that should be most carefully monitored.

On the benefit side, a strain of rice has been developed that produces beta carotene, important in human nutrition. The rice thus has added nutritional benefits, particularly valuable for the poor of the world that depend on rice as a primary source of food. The gene to allow rice to make beta carotene comes from daffodils. This modification required the introduction of four specific genes and would have been unlikely—probably impossible—without genetic engineering techniques. Genes were transferred between plants, but between plants that would not exchange genes in nature. Once again, the rule of natural change suggests that we should monitor such actions carefully.

Although the genetically modified rice appears to have beneficial effects, the government of India has refused to allow it to be grown in that country. There is much concern worldwide about the political, social, and environmental effects of genetic modification of crops. This is a story in process, one that will change rapidly in the next few years. You can check on these fast-moving events on the textbook's website.

12.6 Grazing and Rangelands

To this point in the chapter, the emphasis has been on crops that we harvest to eat directly; however, another major kind of agricultural land is that used to graze domestic animals. This is the subject of the next sections.

Rangeland is defined as land where native vegetation is mostly grasses or shrubs suitable for grazing or browsing. More than 99% of United States rangeland is west of the Mississippi River. Much of the world's rangeland is considered to be in poor condition from overgrazing (Figure 12.11). In the United States, rangeland conditions have improved since the 1930s, especially in upland areas. However, land near streams and streams themselves continue to be heavily affected by grazing. Grazing cattle with access to streams (Figure 12.12) trample stream banks and release their waste into stream water. Maintaining a high quality stream environment requires that cattle be fenced behind a buffer zone from streams.

Traditional and Industrialized Use of Grazing and Rangelands

Traditional herding practices and industrialized production of domestic animals have different effects on the environment. In modern industrialized agriculture, cattle are initially raised on open range and then transported to feedlots, where they are fattened for market. Feedlots have become widely known in recent years as sources of local pollution. The cattle are often kept at extremely high densities and fed grain or forage that is transported to the feedlot. Manure builds up in large mounds. When it rains, the manure pollutes local streams. Feedlots are popular with meat producers because they are economical for rapid production of good-quality meat. However, large feedlots require intense use of resources and have negative environmental effects.

Traditional herding practices, by comparison, chiefly affect the environment through overgrazing. Goats are especially damaging to vegetation, but all domestic herbivores can destroy rangeland. The effect of domestic herbivores on the land varies greatly with their density (relative to rainfall and soil fertility). At low to moderate densities, the animals may actually aid the growth of above-ground vegetation by fertilizing soil through their manure and stimulating plant growth by clipping off ends

Figure 12.11 • Grazing longhorn cattle in Nebraska. Note exposed soil that is susceptible to erosion.

in grazing, just as pruning stimulates plant growth. But at high densities, the vegetation is eaten faster than it can grow; some species are lost, and the growth of others is greatly reduced.

The Biogeography of Agricultural Animals

People have distributed cattle, sheep, goats, and horses, as well as other domestic animals, around the world and then promoted the growth of these animals to densities

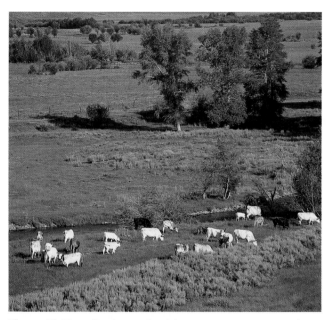

Figure 12.12 • Grazing cattle with access to a stream trample stream banks and increase bank erosion where trails cross the stream. Cattle also urinate and defecate in streams.

that have changed the landscape. Pre-industrial people made such introductions. For example, Polynesian settlers brought pigs and other domesticated animals to Hawaii and other Pacific islands. Since the age of exploration by Western civilization, starting in the 15th century, domestic animals have been introduced into Australia, New Zealand, and the Americas. The horse, cow, sheep, and goat were brought to North America after the 16th century. The spread of cattle brought new animal diseases and new weeds, which arrived on the animals' hooves and in their manure. Introductions of domestic animals into new habitats have many environmental effects:

1. Native vegetation, not adapted to the introduced grazers, may be greatly reduced and threatened with extinction.
2. The introduced animals may compete with native herbivores and may lead a reduction in their numbers to a point at which they too may be threatened with extinction.

A recent important issue in cattle production is the opening up of tropical forest areas and their conversion to rangeland—for example, in the Brazilian Amazon basin. In a typical situation, the forest is cleared by burning, and crops are planted for about four years. After that time, the soil has lost so much fertility that crops can no longer be grown economically. Ranchers then purchase the land, already cleared, and run cattle bred to survive in the hot, humid conditions. After about another four years, even grazing can no longer be supported, and the land is abandoned. In such areas, the land's capability for many uses, including forest growth, has been greatly reduced by grazing.[12] Clearly, this is a non-sustainable approach to agriculture and therefore undesirable.

The spread of domestic herbivores around the world is one of the major ways we have changed the environment through agriculture. As the human population increases, and as income and expectations rise, the demand for meat increases. As a result, we can expect an increased demand for rangeland and pastureland in the next decades. A major challenge in agriculture will be to develop ways to make the production of domestic animals sustainable.

Carrying Capacity of Grazing Lands

Carrying capacity is the maximum number per unit area of a species that can persist without decreasing the ability of that population or its ecosystem to maintain that density in the future. The carrying capacity of land for cattle varies with rainfall, topography, soil type and soil fertility.

When the carrying capacity is exceeded, the land is overgrazed. **Overgrazing** reduces the diversity of plant species, leads to reduction in the growth of vegetation and to dominance of plant species that are relatively undesirable to the cattle, increases the loss of soil by erosion as the plant cover is reduced, and results in damage from the cattle's trampling on the land (Figure 12.13). For example, paths made as the cattle travel to the same water hole or stream develop into gullies, which erode rapidly in the rain. Overgrazing leads to a decrease in the density of cattle that can be supported on the land.

In areas with moderate to high rainfall evenly distributed throughout the year, cattle can be maintained at high densities; but in arid and semiarid regions, the density drops greatly. In the United States, the carrying capacity for cows drops from $190/km^2$ ($500/mi^2$) in the east, to $40/km^2$ ($100/mi^2$) in the region that was prairie, to $4/km^2$ ($10/mi^2$) or less in semiarid and desert regions (Figure 12.14). In Arizona, for example, rainfall is low,

(a)

(b)

Figure 12.13 • (a) Soil erosion in northern Natal caused by overgrazing and other land-use practices. (b) A commercial feedlot.

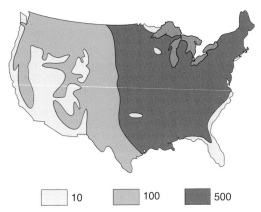

| | 10 | | 100 | | 500 |

Figure 12.14 • Carrying capacity of pasture and rangeland in the United States, in average number of cows per square kilometer. (*Source:* U.S. Department of Agriculture statistics.)

and cattle can be maintained only at low densities—one head of cattle for 7–10 ha (17–25 acres). Near Paso Robles, California, in an area where the rainfall is about 25 cm/year (9.8 in./year)—that is, in desert to semiarid conditions—a ranch where cattle are grazed without artificial irrigation or fertilization supports about one head for 6 ha (15 acres).

12.7 Desertification: Regional Effect and Global Impact

Deserts occur naturally where there is too little water for substantial plant growth; the plants that do grow are too sparsely spaced and unproductive to create a soil rich in organic matter. Desert soils are mainly inorganic—that is, coarse and typically sandy. When rain does come, it often is heavy, and erosion is severe. The principal climatic condition that leads to desert is low or undependable precipitation. The warmer the climate, the greater the rainfall required to convert an area from desert to a nondesert, such as grassland. But even in cooler climates or at higher altitudes, deserts may form if precipitation is too low to support more than sparse plant life. The crucial factor is the amount of water in the soil available for plants to use. Factors that destroy the ability of a soil to store water can create a desert.

The Earth has five natural warm desert regions, all of which lie primarily between latitudes 15° and 30° north and south of the equator. These include the deserts of the southwestern United States and Mexico; Pacific coast deserts of Chile and southern Ecuador; the Kalahari desert of southern Africa; the Australian deserts that cover most of that continent; and the greatest desert region of all—the desert that extends from the Atlantic coast of North Africa (the Sahara) eastward to deserts of Arabia, Iran, Russia, Pakistan, India, and China.[13] Only Europe lacks a major warm desert; it lies north of the desert latitudinal band.

Based on climate, about one-third of Earth's land area should be desert, but estimates suggest that 43% of the land is desert. This additional desert area is believed to be a result of human activities.[14] **Desertification** is the deterioration of land in arid, semiarid, and dry subhumid areas due to changes in climate and human activities.[15]

Desertification is a serious global problem. It affects one-sixth of the world's population (about 1 billion people), 25% of the world's total land area, and 70% of all drylands (3.6 billion ha). Land degradation caused by people has altered 73% of drier rangelands (3.3 billion ha) and the soil fertility and structure of 47% of dryland areas with marginal rainfall for crops. Land degradation also affects 30% of dryland areas with high population density and agricultural potential.

A large part of desertification occurs in the poorest countries. These regions include Asia, Africa, and South America. Six million ha of land per year are lost to this process. Over $40 billion is lost annually, and the cost of recovery of these lands worldwide could reach $10 billion per year.[16]

Causes of Desertification

Some areas of Earth are *marginal* lands, easily converted to deserts when used for light grazing and crop production. In these areas, the rainfall is just barely enough to enable the land to produce more vegetation than a desert, and then only if the soils are maintained within a certain range of conditions.

The leading causes of desertification are bad farming practices (such as failure to use contour plowing or simply too much farming; see Figure 12.15); overgrazing (Figure 12.13); the conversion of rangelands to croplands in marginal areas where rainfall is not sufficient to support crops over the long term; and poor forestry practices, including cutting all the trees in an area marginal for tree growth.

In northern China, for example, areas that were once grasslands were overgrazed, and then some of these rangelands were converted to croplands. Both practices led to the conversion of the lands to deserts. Between 1949 and 1980, an area of 65,000 km² (25,100 mi², an area larger than Denmark) became desert, and an additional 160,000 km² (61,760 mi²) is in danger of becoming deserts. As a result of this desertification, the occurrence of sandstorms increased from about 3 days/year in the early 1950s to an average of 17 days/year in the next decade and to more than 25 days/year by the early 1980s.[13]

Desertlike areas can be created anywhere by the poisoning of soils. Poisoning can result from the application of persistent pesticides or other toxic organic chemicals or from industrial processes that lead to improper disposal of toxic chemicals. Airborne pollutant acidification, excessive manuring in feedlots, and oil or chemical spills can all lead to degradation of soils, forcing abandonment or reduced agricultural use of lands. Worldwide, chemical

(a)

(b)

(c)

Figure 12.15 • (*a*) Gully soil erosion on cleared and plowed farmland in South Australia. (*b*) Agricultural runoff carrying heavy sediment load. (*c*) A farmer and his children flee the dust storm of the 1930s in Oklahoma.

degradation accounts for about 12% of all soil degradation. Irrigation of soils in arid areas can also lead to desertification. When irrigation water evaporates, a residue of salts is left behind. Although these salts may have been in very low concentrations in the irrigation water, over time the salts can build up in the soil to the point at which they become toxic. This effect can sometimes be reversed if irrigation is increased greatly; the larger volume of water then redissolves the salts and carries them with it as it percolates down into the water table.[17]

Preventing Desertification

The major symptoms of desertification are:

- Lowering of the water table (wells have to be dug deeper and deeper).
- Increase in salt content of the soil.
- Reduced surface water (streams and ponds dry up).
- Increased soil erosion (the dry soil, losing its organic matter, begins to blow away and to be washed away in heavy rains).
- Loss of natural native vegetation (not adapted to desert conditions, native vegetation can no longer survive).[17]

Preventing desertification begins with monitoring these factors. Monitoring of aquifers and soils is an important activity in marginal agricultural lands. When undesirable changes are observed, activities producing these changes can be stopped, reduced, or changed.

Proper methods of soil conservation, forest management, and irrigation can help prevent the spread of deserts. (See Chapters 11 and 20 for a background discussion of soil and farming and irrigation practices.) In addition to the practices discussed earlier, good soil conservation includes the use of windbreaks (narrow lines of trees that help slow the wind) to prevent sandstorms and wind erosion of the soil. A landscape with trees is a landscape with a good chance of avoiding desertification. Practices that lead to deforestation in marginal areas should be avoided. Reforestation, including the planting of windbreaks, should be encouraged.

12.8 Global Effects of Agriculture

Local and regional effects of agriculture on the environment are easily identified. Global effects are not as obvious, although they are potentially as serious. This section discusses a few of these global changes. First, agriculture changes land cover, resulting in changes in the reflection of light by the land surface; the evaporation of water; the roughness of the surface; and the rate of exchange of chemical compounds (such as carbon dioxide) produced and removed by living things. Each of these changes can have regional and global climatic effects.

Should Rice Be Grown in a Dry Climate?

Water is a precious resource, especially in California, where average rainfall is low (38–51 cm/year, or 15–20 in/year), population is large and growing (31 million, growing by about 2,000 a day), and agricultural water use is high (farmers use 85% of the state's water to irrigate 3.2 million ha, or 7.8 million acres, more than in any other state). (Figure 12.16 shows declining water levels in California reservoirs in the late 1980s.)

With cities and industries, not to mention fish and other wildlife, in need of water, the practice of growing crops with high water demand has come under heavy criticism. The conflict has been intensified because much of the water farmers receive is subsidized by the government, allowing many to sell it at a profit to municipal water districts. Some farmers have responded by reducing the acreage of water-intensive crops, and others have switched to crops such as fruits, vegetables, and nuts that require less water. Rice farmers have tried to change agricultural practices without changing their crop.

California produces 18% of the nation's rice, making it the second largest rice-growing state. The rice industry, concentrated in the Sacramento valley, is worth about $500 million a year and uses enough water to supply one-fourth of the state's population for a year. Although rice growers are not the biggest water users in the state, they have been particular targets of attack because the flooded fields required for rice are a very visible reminder of the amount of water used by agriculture. In addition to high water use, rice growing has had other adverse environmental effects: high pesticide and herbicide use, causing contamination of rivers and drinking supplies; and burning of stubble left after harvesting, contributing to air pollution in the valley.

Rice growers have responded to pressure to clean up their act in a number of ways. They have decreased water use by 32% and pesticide use by 98% in the past 10 years; they have con-

verted to biodegradable pesticides; and they have decreased the burning of stubble by plowing it under, harvesting it, or flooding the fields in winter and allowing it to rot. In addition, the fields provide a wetland habitat for many migrating birds and

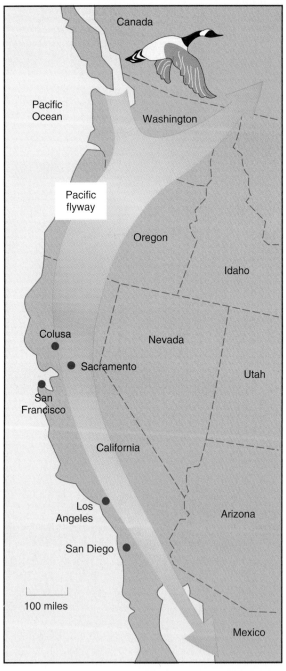

Figure 12.17 • The Pacific flyway, used by many birds that stop at agricultural wetlands.

Figure 12.16 • Dropping water levels in California reservoirs, 1986–1990: capacity, 38 million acre-feet; average, 22.3 million (1 acre-foot = 325.851 gallons). (*Source:* California Department of Water Resources, 1991).

other species. Winter flooding benefits an even wider diversity and greater number of species: waterfowl are of particular interest because their numbers declined from 10–12 million in 1967 to 4–5 million in 1990; the state lost 90% of its wetlands to development and agriculture in that period.

The Sacramento valley lies along the Pacific flyway (Fig. 12.17), a major migration route for waterfowl. The area is host to 20% of the ducks of the United States and 50% of all waterfowl in the flyway. Twenty-one wildlife species with special status (endangered, threatened, candidate species, or species of special concern) use the rice fields, attracted by the 114–136 kg (250–300 lb) of rice grains per acre left behind after harvesting and the 273–318 kg (600–700 lb) of invertebrates per acre that grow in the waters. An experimental program was begun in 1992 to test the feasibility of winter flooding as an alternative to burning stubble, as a supplement to the state's water storage capacity, and as a way to create additional wetlands. Experts in the program are also trying to find ways to protect young salmon, which run in the rivers of the Sacramento valley, from being pumped into the channels leading to the rice fields. Although drawing water out of the rivers might have a negative impact on salmon, the release of water at the end of the winter, when the rivers are low, could help the spring run of salmon.

Critical Thinking Questions

1. Most of the areas where rice is grown have alkaline, hardpan soil, unsuited to other crops. If rice were not grown on the land, it probably would be developed for housing. Each acre of rice requires 5 acre-feet of water. Less than 1 acre-foot could supply a family of four for a year. If housing lots were one-eighth acre and all housed families of four, how many acre-feet of water per acre of housing would be used in a year? Which uses more, an acre of rice or an acre of people? How would real estate development affect wildlife habitat?

2. Farmers find that the presence of waterfowl in flooded fields speeds up the rotting of stubble. What are at least two reasons for this?

3. Although birds can feed on rice grains in dry fields, they get a more balanced diet by feeding in flooded fields. Why would this be so?

4. Two of the unknowns in this system are the long-term effects of flooding on the ability of the soil to support rice growing and the long-term effects on dryland species, such as rattlesnakes and rats. What is a scientific way of investigating one of these questions?

5. The new rice-growing practices are referred to as "win-win." What is meant by the term in general, and how does this situation exemplify the term?

Second, modern agriculture increases carbon dioxide (CO_2) in two ways. As a major user of fossil fuels, it contributes to the increase in carbon dioxide concentration in the atmosphere, adding to the buildup of greenhouse gases (discussed in detail in Chapter 22). In addition, clearing land for agriculture leads to an increase in the decomposition of organic matter in the soil, transferring the carbon stored in the organic matter into carbon dioxide, which also increases the CO_2 concentration in the atmosphere.

Agriculture can also affect climate through fire. Fires associated with clearing land for agriculture, especially in tropical countries, may have significant effects on the climate because they add small particulates to the atmosphere. Another global effect of agriculture results from the artificial production of nitrogen compounds for use in fertilizer, which may be leading to significant changes in global biogeochemical cycles. Finally, agriculture affects species diversity. The loss of competing ecosystems (because of agricultural land use) reduces biological diversity and increases the number of endangered species.

Summary

- The Industrial Revolution and the rise of agricultural sciences have led to a revolution in agriculture, with many benefits and some serious drawbacks. These developments have been accompanied by an increase in soil loss, erosion, and resulting downstream sedimentation, as well as the pollution of soil and water with pesticides, fertilizers, and heavy metals that are concentrated as a result of irrigation.

- Modern fertilizers have greatly increased the yield per unit area. Modern chemistry has led to the development of a wide variety of pesticides that have reduced, but not eliminated, the loss of crop production to weeds, diseases, and herbivores.

- Most 20th century agriculture has relied on machinery and the use of abundant energy, with relatively little attention paid to the loss of soils, the limits of groundwater, and the negative effects of chemical pesticides.

- Overgrazing has caused severe damage to lands. It is important to properly manage livestock, including using appropriate lands for grazing and keeping livestock at a sustainable density.

- Desertification is a serious problem that can be caused by poor farming practices as well as by the conversion of marginal grazing lands to croplands, Additional desertification can be avoided by improving farming practices, planting trees as windbreaks, and monitoring land for symptoms of desertification.

Two revolutions are occurring in agriculture, one ecological and the other genetic. In the ecological approach to agriculture, pest control will be dominated by integrated pest management. Agriculture will be approached in terms of ecosystems and biomes, taking into account the complexity of these systems. Soil conservation through no-till agriculture and contour plowing will be emphasized along with water conservation through methods discussed in Chapter 20). The genetic revolution is already the subject of great controversy, offering both great benefits and great environmental dangers. Dangers will result if genetic modification is used without considering the ecosystem, landscape, biome, and global context within which it is done.

● REEXAMINING THEMES AND ISSUES

 Human Population: Agriculture is the world's oldest and largest industry; more than one-half of all the people in the world still live on farms. Because the production, processing, and distribution of food alter the environment, and because of the size of the industry, large effects on the environment are unavoidable.

 Sustainability: Alternative agricultural methods appear to offer the greatest hope of sustaining agricultural ecosystems and habitats over the long term, but more tests and development of methods are needed. One must separate the sustainability of a crop from the sustainability of the original ecosystem where it is grown. As the experience with European agriculture shows, crops can be produced on the same lands for thousands of years as long as sufficient fertilizers and water are available; however, the soils and other aspects of the original ecosystem are greatly changed—these are not sustained.

 Global Perspective: Agriculture has numerous global effects. It changes land cover, affecting climate at regional and global levels. It also increases carbon dioxide in the atmosphere, adding to the buildup of greenhouse gases (discussed in detail in Chapter 22). Fires associated with clearing land for agriculture may have significant effects on the climate because of the small particulates they add to the atmosphere. Genetic modification is a new global issue having not only environmental, but also political and social effects.

 Urban World: The agricultural revolution allows fewer and fewer people to produce more and more food and leads to greater productivity per acre. People, freed from dependence on farming, flood to cities. This leads to increased urban effects on the land. Thus, agricultural effects on the environment indirectly extend to the cities.

 Science and Values: Human activities have seriously damaged one-fourth of the world's total land area, impacting one-sixth of the world's population (about 1 billion people). Six million ha of land per year are lost to desertification. A large part of desertification occurs in the poorest countries. Overgrazing, deforestation, and destructive crop practices have caused so much damage that recovery in some areas will be difficult; restoration of the rest will require serious actions. A major value judgment we must make in the future is whether our societies will allocate the funds required to restore these damaged lands. Restoration requires scientific knowledge, both about present conditions and about actions required for restoration. Will we seek this knowledge and pay the costs for it?

● KEY TERMS

biological control **220**
contour plowing **215**
desertification **226**

integrated pest management **220**
no-till agriculture **217**
off-site effect **214**

on-site effect **214**
overgrazing **225**
terminator gene **222**

● STUDY QUESTIONS

1. How can an insect pest species become resistant to a pesticide?

2. What is meant by calling a pesticide a magic bullet? What would its characteristics be?

3. How can farming lead to the spread of deserts? What might be done to stop this desertification?

4. It has been said that farming can never be sustainable. What is meant by this statement? Do you agree or disagree? List your reasons.

5. What is meant by food chain concentration? How does this lead to differences in the effects of the same pesticide on different organisms?

6. Design an integrated pest management scheme for use in a small vegetable garden in a city lot behind a house. How would this scheme differ from IPM used on a large farm? What aspects of IPM could not be employed? How might the artificial structures of a city be put to use to benefit IPM?

7. Arsenic is a potent toxic chemical. Why does it make a poor pesticide?

8. What are the off-site effects of an urban vegetable garden?

9. Under what conditions might grazing cattle be sustainable when growing wheat is not? Under what conditions might a herd of bison provide a sustainable supply of meat when cattle might not?

10. How can we avoid another dust bowl in the United States?

11. Consider the genetically modified rice that produces vitamin A. Make a table of the potential benefits and potential environmental problems that such a crop could produce.

12. Should genetically modified crops be considered acceptable for "organic" farming?

● FURTHER READING

Committee on the Future Role of Pesticides in U.S. Agriculture, Board on Agriculture and Natural Resources, Board on Environmental Studies and Toxicology. 2000. *The Future Role of Pesticides in U.S. Agriculture*. Washington, D.C.: National Research Council.

Committee on Genetically Modified Pest-Protected Plants. 2000. *Genetically Modified Pest-Protected Plants: Science and Regulation*. Washington, D.C.: National Research Council.

Grainger, A. 1982. *Desertification: How People Make Deserts, How People Can Stop and Why They Don't*. 2nd ed. London: Earthscan Books. A book that provides examples and discussion of the connections between desertification and human activity, particularly in nonindustrialized countries.

Lashof, J. C., ed. 1979. *Pest Management Strategies in Crop Protection*, vol. 1. Washington, D.C.: Office of Technology Assessment, U.S. Congress. An assessment and evaluation of crop protection strategies.

Matthews, A. 1992. *Where the Buffalo Roam*. Weidenfeld, N.Y.: Grove Publishing.

Mellon, M. 1993. *Perils Amidst the Promise: Ecological Risks of Transgeni Crops in a Global Market*. Union of Concerned Scientists.

Rissler, J., and M. Mellon. 1996. *The Ecological Risks of Engineered Crops*, MIT Press, Cambridge, MA.

Sheridan, D. 1981. *Desertification of the United States*. Washington, D.C.: Council on Environmental Quality. A report that reviews scientific information on desertification in the United States, identifies policies and technologies that promote or control desertification, and provides an excellent overview of continuing conditions.

Steiner, F. 1990. *Soil Conservation in the United States: Policy and Planning*. Baltimore: Johns Hopkins University Press.

Young, M. D. 1991. *Towards a Sustainable Agricultural Development*. New York: John Wiley & Sons. A broad survey of the use of agricultural chemicals, intensive animal production, soil erosion, land-use patterns, and the impact on agriculture of pollution from other sources.

CHAPTER 13

Forests, Parks, and Landscapes

Lower Mississippi River. The Straight Channel is the Intercoastal Canal. Southwest Pass, Louisiana. (Bruce Coleman Photo.)

● LEARNING OBJECTIVES

Conservation and management of natural resources have begun to take account of a larger view in which populations, species, and ecosystems are connected across landscapes. After reading this chapter, you should understand:

- What ecological services are provided by landscapes of various kinds.
- What is the landscape context for conservation and management of forests and parks.
- What are the basic principles of forest management, including its historical context.
- What roles parks and nature preserves play in the conservation of wilderness.

Conserving Wetland Forests at the Mouth of the Mississippi

The lower valley of the Mississippi River once contained the largest area of wetland forests in the United States—more than 8 million ha (20 million acres or 31,250 mi^2), an area almost as large as Connecticut, Massachusetts, New Hampshire, and Vermont combined (Figures 13.1 and 13.2). Difficult to walk through, these bottomland forests must have seemed to the early Spanish explorers to extend forever. Biological diversity was high and there were many useful biological resources. Among the many native tree species, 70 were commercially valuable. And the rich, heavily watered soil made the land productive.

These wetland forests never existed in isolation from other landscapes. Their existence depended on the local high water table and fertile soils, which in turn derived from water flow and erosion of sediments far upriver. The high productivity and diversity resulted in part from frequent flooding and deposition of sediments of the Mississippi, which supplied a renewable source of nutrient-rich soil; the upper Missouri and Mississippi and

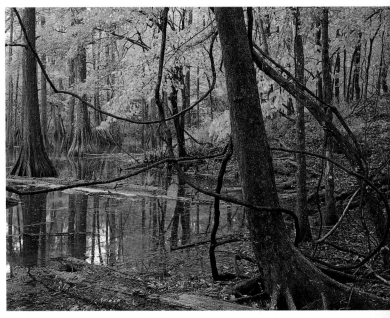

Figure 13.1 • Lower Mississippi wetland forests, Lake Chicot, Louisiana.

(a)

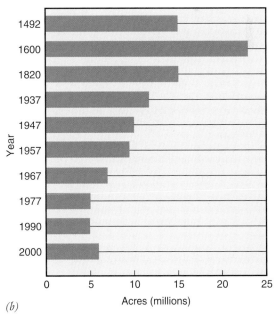

(b)

Figure 13.2 • (*a*) Original distribution of lower Mississippi wetland forests. (*b*) Changes in the area of lower Mississippi wetland forests since European settlement of North America. (*Source:* J. A. Stanturf, E. S. Gardiner, P. B. Hamil, M. S. Devall, T. D. Leininger, and M. E. Warren, Jr., "Restoring Bottomland Hardwood Ecosystems in the Lower Mississippi Alluvial Valley," *Journal of Forestry 98* no. 8 [2000]: 10–16.)

their tributaries fed this vast valley. Another reason for the high biological diversity was local, subtle variations in elevation above and below the water table, along with changes in soil type.[1] Land a meter above the water table could grow trees typical of dry climates; land frequently submerged under water was habitat to specially adapted bog plants. This locally complex matrix was important to wildlife. Thus, the characteristics of these vast wetland forests were the result of processes at many spatial scales, vastly different in area, from events that took place over a few square meters to events that took place over thousands of square kilometers.

Since European settlement, 96% of the wetland forests of the lower valley of the Mississippi have been converted to agriculture or lost to U.S. Army Corps flood control projects. About 2 million ha (5 million acres) remain, mostly in Louisiana, Mississippi, and Arkansas. But these are polluted by high levels of nitrogen—the result of agricultural fertilizers flowing off of fields both nearby and up the watersheds throughout the Mississippi–Missouri River Valley.

How do we conserve and save these wetlands? Years ago, the approach would have been simply to establish a preserve, put some of the land aside, and do nothing more. Now it is recognized that conservation and restoration of these wetland forests requires actions at many spatial scales, consistent with the variety of spatial scales that create and maintain the forests. Reducing agricultural runoff near and far away is important, as is maintaining the natural variation of water flow and the level and amount of flow into the wetlands from the Mississippi–Missouri River system. The original subtle variations in elevation of the lands need to be maintained or restored. And corridors are needed to connect conservation areas.

This new approach is referred to as a *landscape perspective*. Following this approach, various kinds of restoration and conservation projects are underway. The U.S. Fish and Wildlife Service is planting trees, taking into account the complex patterns natural to the area.

The Army Corps of Engineers is now required to mitigate the effects of its construction projects. The Natural Resources Conservation Service is purchasing uneconomical farmlands and converting them back to forests. These efforts take into account pollution from near and far and the need to allow the river to flow in natural patterns.

Because of modern land uses, much of the land currently available for conservation and restoration activities is in relatively small parcels—on the order of 120 ha (300 acres)—that lie within a complex landscape matrix of farmlands. Connectedness among these areas needs to be developed.

Much of the concern is to restore wildlife habitat. Many species of wildlife require forests with a complex structure—many tree species of many sizes, as well as patches of forests of different ages and growing on different kinds of drainage. Some of the restoration and conservation projects underway appear to be succeeding.[2]

The wetland forests of the Mississippi Delta exist first of all because the Missouri–Mississippi River system, one of the 20 largest river systems in the world, drains a huge area. The Missouri alone drains one-sixth of the United States. From this drainage area, these rivers and their tributaries transport eroded sediment, and some of it is transported all the way to the mouth of the Mississippi. This is an ancient process that has gone on as long as the Mississippi–Missouri have been the principle rivers of the central United States. What happens within a bottomland forest therefore depends in part on events and processes far away—the amount, timing, and intensity of precipitation upstream, throughout the drainage basin; the condition of the riparian zones, including the condition of their vegetation, which affects their rate of erosion. But at the same time, local events, within a stand, affect a specific bottomland forest. A local outbreak of an insect that eats the leaves of one of the trees, or an increase in the abundance of deer, have local effects on the density and regeneration of the trees.

● *The task for those who seek to manage and conserve the bottomland forest landscapes of the lower Mississippi requires that they understand events and processes on a variety of spatial scales. This chapter discusses modern approaches to conserving and managing landscapes for the production of useful products and the conservation of ecosystems and biological diversity.*

13.1 The Landscape Concept

As the case study of the Mississippi valley wetland forests illustrates, populations, species, and ecosystems are connected across landscapes. Each ecosystem affects others and is affected by others. Many species require complex habitats involving more than one type of ecosystem to complete their life cycle. Thus, life is connected spatially on surprisingly large regional scales. While in some cases it is possible to conserve a small area without considering its surrounding landscape, in most cases, management and conservation of natural living resources requires a larger view. One of the recent advances in environmental sciences is the acceptance of this larger view—the **landscape perspective**.

In the early part of the 20th century, when most of the professions that engage in natural resource management were just beginning to develop as professions in the modern sense, with a scientific base, the perspective was quite different. At the time of the founding of the U.S. Forest Service, around the turn of the 20th century, the goal of natural resource management was the maximization of production of a single species or population in a single place. The landscape of concern was typically local—what foresters call a *stand*, an informal term for a specific local area of forest being managed. Management was conceived primarily at the stand level.

Today, it is recognized that management and conservation of forests, prairies, rivers, wetlands, and so forth require knowledge on several spatial scales, from the local (the forester's "stand") to the regional, complex matrix of many kinds of ecosystems. The need for both breadth and depth of knowledge makes conservation and management of landscapes more difficult, but it also makes successful management and conservation more likely.

Several questions remain unanswered, however. One question concerns the percentage of a landscape that is appropriately included in parks and preserves and the percentage that is appropriately used to grow and harvest natural resources. The percentage will vary with the kind of biome, with the habitat conditions required by the species that live within the biome, and with the values of the people participating in the decision making. Another important question involves the topology of the landscape: What shapes of parks, preserves, and areas open to harvest work best, and how are these different land uses best connected?

This chapter discusses the landscape perspective in terms of forests and parks, because forests provide major resources and are the subject of major environmental controversies. However, the general principles apply to all biomes (see Chapter 8).

The chapter also presents basic principles of forest management and conservation. Only a few examples can be presented, but again the principles discussed apply widely. Then the chapter discusses parks and nature preserves. Finally, it considers the spatial connections among these areas.

13.2 Forests and Forestry

In recent decades, conservation of forests has become an international *cause celebre*, especially conservation of remaining old-growth forests, and especially conservation of the giant trees of the rainforests in the North American Pacific Northwest and of all tropical rainforests (Figure 13.3). In this section, some basic principles about the roles of forests are introduced.

Forests as Resources

Modern conflicts about forests center on whether a forest should be used as a source of resources for people and civilization or should function to conserve natural ecosystems and biological diversity, including specific endangered species. (See Table 13.1 for a list of some major forestry issues). Forests have always been important to people in both ways, and forests and civilization and always been closely linked.

Since the earliest civilizations—in fact, since some of the earliest human cultures—wood has provided one of the major building materials and the major source of fuel. Forests provided materials for the first boats and the first wagons. Even today, nearly half the people in the world depend on wood for cooking. In developing nations, wood remains the primary heating fuel.

At the same time, people have appreciated forests for spiritual and aesthetic qualities. There is a long history of sacred forest groves. When Julius Caesar was trying to conquer the Gauls in what is now southern France, he found the soldiers difficult to defeat on the battlefield, so he burned the society's sacred groves to demoralize

Figure 13.3 • Temperate rainforest on Vancouver Island.

Table 13.1   Major Forestry Issues

- *Sustainability:* How can we achieve sustainable forestry? (This is the fundamental question.)
- *Clearcutting:* Is clearcutting ever good? Should it ever be allowed?
- *Old-growth forests:* Should all old-growth forests be preserved, or should some logging be allowed in some old-growth stands?
- *Plantations:* Are plantations intrinsically bad because they involve intentional manipulation of land to grow trees, or are plantations one of the keys to achieving the goals of biological conservation of forests?
- *Stream protection zones:* Should all streams everywhere have a wooded buffer zone in which no logging and no other harvesting or destructive activities are permitted? If so, how wide should that buffer zone be?

- *National Forests:* Is the purpose of national forests to provide the nation with a reliable source of timber, or is it to conserve living resources? What is the role of recreation in national forests
- *Forest fires:* Are forests fires almost all bad, are they occasionally beneficial to a forest, or are they frequently important and essential to forests?
- *Certification:* Should our society certify forests as sustainable? If so, how should this be done and who should do it?
- *Scale of management:* What is the appropriate spatial and temporal scale for management of forests?
- *People's role:* Should people play any active role in the management of forests? If so, what kinds of activities, how often, and where?

them—demonstrating the spiritual importance of forests to the Gauls (and also serving as an early example of psychological warfare). In the Pacific Northwest, the great forests of Douglas fir provided many necessities of life to the Indians, from housing to boats, but also were important spiritually.

Forests benefit people and the environment in yet another way, through indirect benefits referred to as the **public service functions** of forests. Forests retard erosion and moderate the availability of water (Figure 13.4), improving the water supply from major watersheds to cities. Forests are habitats for endangered species and other

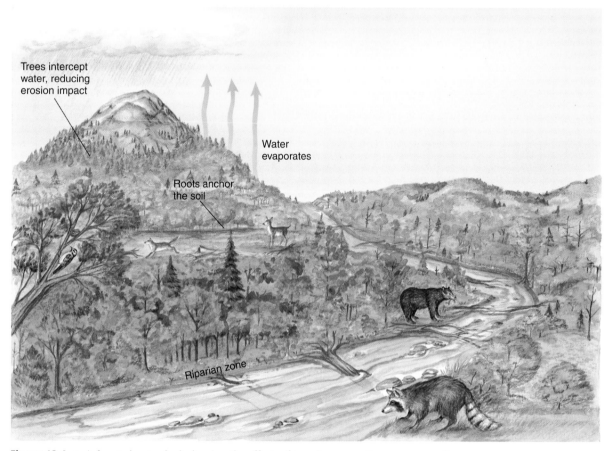

Figure 13.4 • A forested watershed, showing the effects of trees in evaporating water, retarding erosion, and providing wildlife habitat.

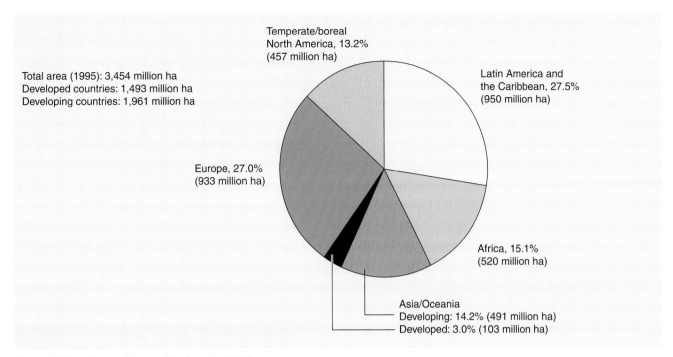

Figure 13.5 • State of the world's forest in 1995.

wildlife. They are important for recreation, hiking, hunting, and bird and wildlife viewing. At regional and global levels, forests may be significant factors in the climate.

Global Production and Consumption of Forest Resources

The world today contains approximately 3.45 billion ha (8.4 billion acres) of forested area (Figure 13.5), down from 4 billion ha in 1980, covering approximately 26.6% of the Earth's surface (Figure 13.6). Countries differ greatly in forest resources, depending on the potential of the land and climate for tree growth and on the history of land use and deforestation. About 61% of Indonesia is forest, 65% of Brazil, 22% of India, and less than 15% of China.[3] Some of the largest forest reserves are in the far north, in the boreal forests of Russia and Siberia, the United States, Canada, and the Scandinavian nations.

Developed countries account for 70% of the world's total production and consumption of industrial wood products; developing countries produce and consume about 90% of wood used as firewood. International trade in timber for construction, pulp, and paper makes up approximately 90% of world timber trade (the rest are hardwoods used for furniture, such as teak, mahogany, oak, and maple). North America is the world's dominant supplier (Table 13.2).[3] Total global production/consumption is about 1.5 billion m³ annually. (To think of this in terms that are easier to relate to, a cubic meter of timber is a block of wood 1 m thick on all sides. A million cubic meters would be a block of wood 1 m thick in a square 1 km on a side.)

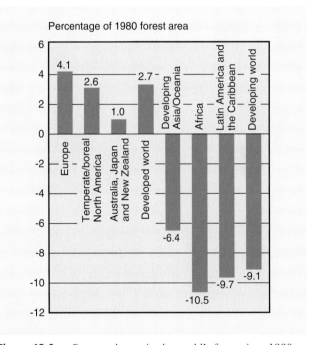

Figure 13.6 • Percent change in the world's forest since 1980.

The United States has approximately 212 million ha (524 million acres) of commercial grade forest, which is defined as forest capable of producing at least 1.4 m³/ha (20 ft³/acre) per year of wood. Commercial timberland occurs in many parts of the United States. Nearly 75% is in the eastern half of the country (about equally divided

Table 13.2 World Industrial Wood Production, 1994

Country	Volume (m³)	Percent Total
U.S.A.	399,725	25.8
Canada	181,054	11.7
Europe (excluding Nordic)	194,166	12.5
Nordic	114,301	7.4
Russian Federation	152,743	9.9
Japan	32,362	2.1
Indonesia/Malaysia	74,608	4.8
China/India	126,694	8.1
Other	273,242	17.6
World (total)	1,548,895*	100.0

Sources: R. A. Sedjo, and D. B. Botkin, "Using Forest Plantations to Spare the Natural Forest," *Environment*, 39 no. 10 (1997): 14–20; and FAO, *Forest Yearbook*, 1994.

* In addition, the world supply of fiber is supplemented by paper recycling and by nontimber fibers used for paper production.

between north and south); the rest is in the West (Oregon, Washington, California, Montana, Idaho, Colorado, and other Rocky Mountain states) and in Alaska.

In the United States, 70% of forestland is privately owned, 15% is on U.S. Forest Service lands, and 15% is on other federal lands.[4] Publicly owned forests are primarily in the Rocky Mountain and Pacific Coast states on sites of poor quality and high elevation. In addition to these forests, about 300 million ha (741 million acres) in the United States are 10% stocked with trees, and another 312 million ha (770 million acres) are rangeland, including natural grasslands, shrublands, most deserts, tundra, coastal marshes, and meadows (Figure 13.7).[5]

In recent years, the world trade in timber has been relatively flat—has not grown substantially. Thus, the amount traded annually (about 1.5 billion m³, as mentioned earlier) is a reasonable estimate of the total present world demand for the six billion people on the Earth, at their present standards of living. *The fundamental question is whether and how the Earth's forests can continue to*

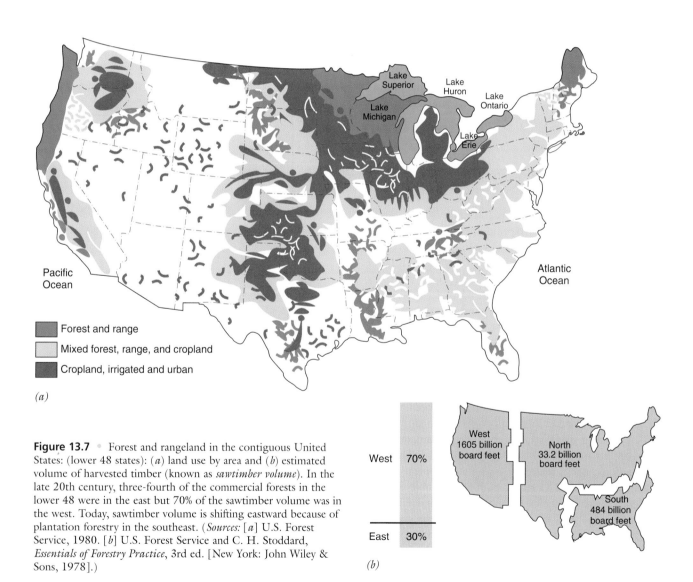

(a)

Figure 13.7 • Forest and rangeland in the contiguous United States: (lower 48 states): (*a*) land use by area and (*b*) estimated volume of harvested timber (known as *sawtimber volume*). In the late 20th century, three-fourth of the commercial forests in the lower 48 were in the east but 70% of the sawtimber volume was in the west. Today, sawtimber volume is shifting eastward because of plantation forestry in the southeast. (*Sources:* [*a*] U.S. Forest Service, 1980. [*b*] U.S. Forest Service and C. H. Stoddard, *Essentials of Forestry Practice*, 3rd ed. [New York: John Wiley & Sons, 1978].)

(b)

provide at least this amount of production, sustainably, for an indefinite period, and how they can provide even greater amounts as the world's human population continues to grow and as standards of living increase worldwide. All this has to happen while forests continue to meet their other functions, which include public service functions, biological conservation functions, and functions involving the aesthetic and spiritual needs of people.

How can this be achieved? How can we even approach thinking about the set of problems and goals associated with forests? A useful path is to begin with the life of an individual tree and to understand its needs. From that, we can expand outward.

13.3 The Life of a Tree

In order to solve the big issues about forestry, it is necessary to understand basic principles about how a tree grows. This section provides a brief summary of how an individual tree grows, as illustrated in Figure 13.8.

How a Tree Grows

Leaves of the tree take up carbon dioxide from the air and absorb sunlight. These, in combination with water that is transported up from the roots, provide the energy and chemical elements for leaves to carry out photosynthesis. Through photosynthesis, the leaves convert carbon dioxide and water to a simple sugar and molecular oxygen.

The first product of photosynthesis is sugar. Leaves and other parts of the tree combine this sugar (made up of carbon and hydrogen) with other chemical elements to form all the necessary organic compounds for life (Figure 13.8). For example, the addition of nitrogen in the form of nitrate (an oxide of nitrogen) or ammonia (nitrogen and hydrogen) results in the production of proteins, which in turn make possible enzymes, DNA, and many other large compounds necessary to life. Enzymes are chemicals that allow chemical reactions to take place at much lower temperatures and pressures than would otherwise be required.

Root hairs at the ends of roots take up water, chemical elements dissolved in the water, and small inorganic com-

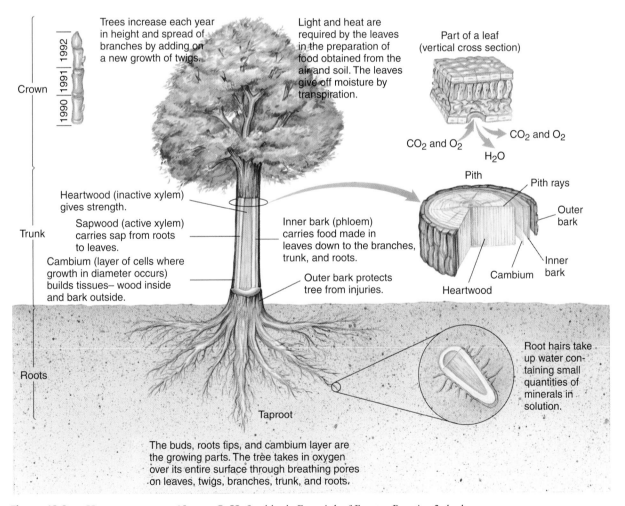

Figure 13.8 • How a tree grows. (*Source:* C. H. Stoddard, *Essentials of Forestry Practice*, 3rd ed. [New York: John Wiley & Sons, 1978].)

pounds, such as the nitrate or ammonia necessary to make proteins. Often, the process of extracting minerals and compounds from the soil is aided by symbiotic relationships between the tree roots and fungi. Tree roots release sugars and other compounds that are food for the fungi, and the fungi benefit the tree as well. Without these organisms, the tree roots either would not be able to take up elements and compounds—or would take them up at a much slower rate.

Leaves and roots are connected by two transportation systems located primarily in the inner bark. Phloem, on the outside of the living part of the bark, transports sugars and other organic compounds downward to stems and roots. Xylem, on the inside of the living bark, transports water and inorganic molecules upward to the leaves. Water is transported upward by a sun-powered pump. Sunlight heats a leaf and water from the leaf (primarily from small pores in the leaf called stomata) is lost to the atmosphere through the process of transpiration and to a lesser extent to evaporation (together called evapotranspiration). Water is pushed upwards in the tree to replace losses (Figures 13.4 and 13.8).

Tree Niches

Each species of tree has its own niche (see Chapter 6) and is therefore adapted to specific environmental conditions. For example, in boreal forests (see Chapter 8), one of the determinants of a tree niche is the water content of the soil. White birch grows well on dry soils; balsam fir grows well on well-watered sites; and northern white cedar grows well in bogs (Figure 13.9).

Another way in which trees separate their niches is through tolerance to shade. Some trees can only grow in the bright sun of open areas and are therefore found in clearings. Other species can grow in deep shade, and seedlings and saplings of these species can be found within the deep shade of an old forest. Sugar maple and beech

are typical of these trees, called "shade tolerant." Birch and cherry are examples of trees that require bright sunlight; they are called "shade intolerant."

Most of the big trees of the western United States require open, bright conditions and certain kinds of disturbances in order to germinate and survive early stages in their lives. These include coastal redwood, which only wins in competition with other species if both fires and floods occasionally occur; Douglas fir, which begins its growth in openings; and the giant Sequoia, whose seed will only germinate on bare, mineral soil. Where there is a thick layer of organic mulch, the Sequoia's seeds cannot reach the soil surface, and they die before they can germinate.

We discussed these requirements in somewhat different terms in our discussion of succession in Chapter 10. Some trees are adapted to early stages in succession, where sites are open and there is bright sunlight. Others are adapted to later stages in succession, where there is a high density of trees.

Understanding the specific requirements of individual tree species helps us to determine where they will grow, where we might best plant them as a commercial crop, and where they might do best as contributing to biological conservation or the beauty of a landscape. Clearly, there is no single answer to what is the best set of conditions for a forest. There are many kinds of forests adapted to many kinds of conditions. The same is true for all landscapes—grasslands, deserts, rivers, and lakes.

13.4 A Global Perspective on Forests

The functioning of an individual tree provides basic understanding about how forests can affect the entire biosphere. Vegetation of any kind can affect the atmosphere in four ways (Figure 13.10):

Figure 13.9 • Some characteristics of tree niches. Tree species have evolved to be adapted to different kinds of environments. In northern boreal forests, white birch grows on dry sites (and early successional sites); balsam fir grows in wetter soils, up to wetlands; and white cedar grows in even the wetter sites of northern bogs.

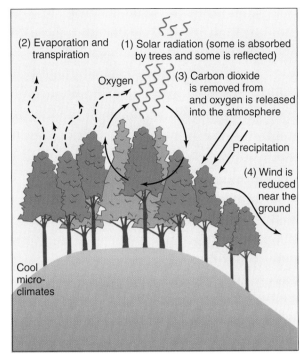

Figure 13.10 • Four ways that a forest can affect the atmosphere: (1) solar radiation, some is absorbed by trees and some is reflected, changing the local energy budget, compared to a non-forest environment; (2) evaporation and transpiration, together called evapotranspiration, from trees, transfers water to the atmosphere; (3) carbon dioxide is removed from the air and oxygen is released to the atmosphere by photosynthesis from trees (carbon dioxide is a greenhouse gas associated with climate change, and reducing the gas cools the temperature of the atmosphere, see Chapter 22); and (4) near surface wind is reduced because the trees produce a roughness near the ground that slows the wind.

1. Changing the color of the surface and therefore the amount of sunlight reflected and absorbed.
2. Increasing the amount of water transpired and evaporated from the surface to the atmosphere.
3. Changing the rate of release of greenhouse gases between the Earth's surface and the atmosphere.
4. Changing "surface roughness," which affects the wind speed at the surface.

In general, vegetation makes the surface darker so that it absorbs more sunlight and reflects less, warming the Earth. This contrast is especially strong between the dark needles of conifers and winter snow in northern forests and between the dark green of shrublands and the yellowish sandstone-derived soils in semiarid climates.

Vegetation in general, and forests in particular, tend to evaporate more water than bare surfaces. This is because of differences in surface area. There can be many leaves above any point, so that the surface area available to evaporate water is many times larger than the area of the soil surface.

Is this increased evaporation good or bad? That depends on one's goals. Increasing evaporation means

that less water runs off the surface. This reduces erosion. Tree roots hold back soil and also retard erosion. However, increased evaporation also means that less water is available for human water supply and for streams. In most situations, however, the ecological and environmental benefits of increased evaporation outweigh the disadvantages.

13.5 Sustainable Forestry

A major modern goal is to have **sustainable forests**. Stated in the most general theoretical terms, a sustainable forest is one from which a resource can be harvested at a rate that does not decrease the ability of the forest ecosystem to continue to provide that same rate of harvest indefinitely in the future. In reality, the situation is more complicated.

What Is Sustainability?

There are two basic kinds of sustainability: sustainability of the harvest of trees (logging) and sustainability of the forest ecosystem. Of course, these two kinds of sustainability apply to any biological resource that we wish to harvest, from fish to pine nuts. Although sustainability has long been discussed in forestry, we have adequate scientific data to demonstrate that sustainability of either kind has been achieved.

Certification of Forest Practices

If we cannot prove scientifically that a particular set of practices has led to sustainable forestry, then what do we do? How do we seek to achieve sustainable forestry? The general answer in use today is to develop a set of criteria—a set of practices believed to be consistent with sustainable forestry—and then compare the actual practices of specific corporations or government agencies with criteria. If the practices conform to the practices believed to be consistent with sustainability, then we certify the forest practices as *sustainable*. This is called **certification of forestry**.

The catch here is that nobody actually knows that the practices will turn out to be sustainable. And since trees take a long time to grow, and a series of harvests is necessary to prove that the set of practices is sustainable, the proof lies in the future. In spite of this limitation, certification of forestry is becoming common practice.

Worldwide concern with the need for forest sustainability has led to international activities of two kinds: attempts to impose bans on imports of wood obtained from purportedly nonsustainable forest practices and development of international programs for certification of forest practices. Some European nations have banned the import of certain tropical woods. Some environmental organizations have led demonstrations in support of such bans.

There is a gradual movement away from calling certified forest practices "sustainable" and instead referring to "well-managed forests" or "improved management."[6] A small industry has developed consisting of companies that review the management of forests and provide a certification; however, no uniform criteria have been established. The search for acceptable uniform criteria has led to an ongoing series of international meetings, including major ones in Montreal and Helsinki.

Certification programs began because of the concern of environmental organizations that forests were being cut in an unsustainable fashion. However, the certification process has evolved into a means for producers of wood and wood products (private corporations and nations that depend on wood exports) to reach new markets, keep existing ones, or increase the value of the product (because people will pay more for a "certified" wood product) and therefore increase the selling price.

Recognizing factors that should be taken into account, some scientists have begun to call for a new forestry that includes a variety of practices to increase the likelihood of sustainability, including recognition of the dynamic characteristics of forests and of the need for ecosystem and landscape contexts as part of management. In developing certification criteria, it must be recognized that any new methods create an experiment and should be treated accordingly. Therefore, *any new programs that claim to provide sustainable practices must include control areas where no cutting is done, for comparative purposes, and must include adequate scientific monitoring of the status of the forest ecosystem.*

13.6 Forestry as a Profession

Forestry has a long history as a profession. The professional growing of trees is called **silviculture** (from *silvus*, Latin for "forest," and *culture*, as in "agriculture.") Traditionally, foresters have managed trees at a local level in what are called **stands**. As mentioned earlier, a *stand* is an informal term to identify a group of trees being managed or cared for. Often, a stand consists of a set of trees of the same species composition and the same age structure composition, living in the same kind of environment; but sometimes, a stand consists of a group of such similar areas. Stands can be small (half a hectare) to medium-sized (several hundred hectares) in area.

Some forests are managed in **plantations**, which are similar to mechanized farms. A single species is planted in straight rows, and the land is fertilized, sometimes by helicopter (Figure 13.11). Modern machines make harvesting rapid; some remove the entire tree, root and all. Intensive management such as this is characteristic of Europe and parts of the northwestern United States. Other forests, such as those of New England, are man-

Figure 13.11 • A modern forest plantation in the southeastern United States.

aged less actively. In these, regeneration takes place from seeds from existing trees, and ecological succession follows (see Chapter 10). What approach is best depends on the type of forest, the environment, and the characteristics of the commercially valuable species.

Foresters classify stands on the basis of tree composition. The two major kinds of commercial stands are *even-aged stands*, where all live trees began growth from seeds and roots germinating the same year, and *uneven-aged stands*, which have at least three distinct age classes. In even-aged stands, trees are approximately the same height but differ in girth and vigor.

A forest that has never been cut is called a *virgin forest*; one that has been cut and has regrown is called a **second-growth forest**. **Old-growth forests**, a term that has gained popularity in several well-publicized disputes about forests, is not a scientific term and does not yet have an agreed-on, precise meaning. In popular usage, it often refers to virgin forest. Another important management term is **rotation time**. As mentioned earlier, this is the time between cuts of a stand. Trees are divided into the **dominants** (tallest, most common, and most vigorous), **codominants** (fairly common, sharing the canopy or top part of the forest), **intermediate** (forming a layer of growth below dominants), and **suppressed** (growing in the understory).

Productivity of a forest varies according to soil fertility, water supply, and local climate. Foresters classify sites by **site quality**, which is the maximum timber crop the land can produce in a given time. Site quality can decrease with poor management. For example, too-frequent burning of forests decreases the potential for tree growth by lowering soil fertility. Traditionally, foresters developed site indices for types of forest lands and derived yield tables to estimate future production. Today, forecasts of forest production rely more and more on computer simulation methods.

The management of forests can involve removing poorly formed and unproductive trees (or selected other trees) to permit larger trees to grow more rapidly, planting genetically controlled seedlings, and fertilizing the soil. Forest geneticists breed new strains of trees just as agricultural geneticists breed new strains of corn, wheat, tomatoes, and other crop plants. New "supertrees" are supposed to be able to maintain a high rate of growth and increase the total production of forests.

One aspect of silviculture is the control of diseases and pests. There has been relatively little success in controlling diseases in forests. Tree diseases are primarily fungal. Often, as with Dutch elm disease, an insect spreads the fungus from tree to tree. Insect outbreaks tend to occur infrequently, but when they do occur, they can have devastating results. Some insect problems are due to introductions of exotic species. For example, the gypsy moth was introduced intentionally into New England around the turn of the century as a source of silk, but it escaped and has spread through many eastern states. Other insect outbreaks appear to be naturally recurrent and to have existed for a long time. For example, a 19th century New Hampshire gazetteer referred to a "plague of loathsome worms" that removed all the leaves from large areas of forest.

Insects affect trees by defoliating them; by eating the buds at the tops of the trees and destroying the main trunk, causing forked growth; by eating fruits; and by serving as carriers of diseases. Insecticides are sometimes used to combat these insects.

Forests are complex and difficult to manage. However, trees provide many kinds of easily obtained information that can be of great assistance in management. For example, the age and growth rate of trees can be measured from tree rings (see Figure 13.8). In temperate and boreal forests, trees produce one growth ring per year, and a tree can be aged by counting the number of rings.

Silviculture developed into a modern profession with a hope for a scientific basis around the turn of the 20th century, along with the establishment of the U.S. Forest Service and the Yale School of Forestry, the first professional graduate school in the United States devoted to this profession. In the early days of the 20th century, the goal of silviculture was generally to maximize the yield in the harvest of a single resource. Ecosystem status was given a minor place, as were nontarget, noncommercial species and associated wildlife. Today, most forestry schools take a much broader view, considering the sustainability of timber harvest and of the ecosystem and considering the entire range of goals for managing forests.

Figure 13.12 • A satellite image showing clearings in the tropical rain forests in the Amazon in Brazil. The image is in false infrared. Rivers are black, and the bright red is the leaves of the living rain forest. The straight lines of other colors, mostly light blue to gray, are clearing sites of deforestation by people extending from roads. Much of the clearing is for agriculture. The distance across the image is about 100 km (63 mi).

the temperate zone, such as in the eastern and midwestern United States, were cleared in the 19th and early 20th centuries and are now regenerating. Most experts disagree. Because few forests are actually successfully managed to achieve sustainability, it seems likely that forests of the world are undergoing a net decline, perhaps rapid.

A surprising and important point is that information is lacking on which to base an accurate evaluation. Because forests cover large, often remote areas that are little visited or studied, it is difficult to assess the total amount of forest area. Only recently have programs begun to obtain accurate estimates of the distribution and abundance of forests, and these suggest that past methods overestimated forest biomass by 100% to 400%.[7] Recognizing the limits of existing information, it is useful to review standard information about forests. It is estimated that forests covered one quarter of Earth's land area in 1950 but only one-fifth in 1980.[8] Modern technologies, including remote sensing and geographic information systems, now make possible assessments of the total world land area in forests and will enable us to track future changes in this area (Figure 13.12).

 ## 13.7 Deforestation: A Global Dilemma

It is difficult to determine the worldwide net rate of change in forest resources. Some experts have argued that there is a worldwide net increase in forests because large areas in

History of Deforestation

Forests were cut in the Near East, Greece, and the Roman Empire before the modern era. Removal of forests continued northward in Europe as civilization advanced. Fossil records suggest that prehistoric farmers in Denmark cleared forests so much that early successional weeds occupied large areas. In medieval times, Great Britain was cut over, and many forested areas were removed. With colonization of the New World, much of North America was cleared.[9]

Today, much deforestation is concentrated in the developing world, which lost approximately 200 million ha (494 million acres) from 1980 to 1995.[10] Many of these forests are in the tropics, mountain regions, or high latitudes, places difficult to exploit before the advent of modern transportation and machines.[11] The problem is especially severe in the tropics because of high human population growth. For example, fire-related deforestation rose sharply in Indonesia in 1997. Estimates of forest area destroyed in 1997 range from 150,000 to 300,000 ha (370,500 to 741,000 acres).[10] Other areas where tropical deforestation is a major concern include the Amazon basin, western and central Africa, and Southeast Asia and the Pacific regions, including Borneo, the Philippines, and Malaysia. According to the World Resources Institute, Asia cleared 30%, Africa 18%, Latin America 18%, and the world as a whole cleared 20% of tropical forests between 1960 and 1990. It is important to note that estimates are, if anything, underestimates, because organizations must rely primarily on official statements from nations, and losses resulting from illegal cutting or other unstated cutting are not necessarily included.[11]

Global and Regional Effects

Forests are a global resource, so cutting forests in one country affects other countries. For example, by 1990, erosion due to deforestation had caused the loss of 562 million ha (1.4 billion acres) of soil worldwide, with an additional 5 to 6 million hectares lost annually since then.[12]

Nepal, one of the most mountainous countries in the world, lost more than half its forest cover between 1950 and 1980. Such cutting destabilizes soil, increasing the rate of landslides, amount of runoff, and sediment load in streams. Many Nepalese streams feed rivers that flow into India (Figure 13.13). Recent heavy flooding in India's Ganges Valley has caused $1 billion a year in property damage and is blamed on the loss of large forested watersheds in other countries.[13] The loss of forest cover in Nepal continues at a rate of about 100,000 ha (247,000 acres) per year. Reforestation efforts replace less than 15,000 ha (37,050 acres) per year. If present trends continue, little forestland will remain in Nepal, thus permanently impacting India's flooding problems.[14]

Causes of Deforestation

The most common reason that people cut forests is to use or sell timber for lumber and paper products or fuel. Logging by large timber companies and local cutting by villagers are both major causes of deforestation. Another important cause

is clearing of forestland for agriculture, including conversion to crops and pasture. Land clearing for agriculture is a principal cause of deforestation in Nepal and Brazil. It was one of the major reasons for clearing forests in New England during the first settlement by Europeans.

The World Firewood Shortage

In many parts of the world, wood is a major energy source. Sixty-three percent of all wood produced in the world, or 2.1 million m^3, is used for firewood. In developed countries, firewood provides 2% of total commercial

(a)

(b)

Figure 13.13 • (a) Planting pine trees on the steep slopes in Nepal to replace forests that have been cut. The dark green in the background is yet uncut forest, and the contrast between foreground and background suggests the intensity of clearing that is taking place. (b) The Indus River in Northern India carries a heavy load of sediment, as shown by the sediments deposited within and along the flowing water and by the color of the water itself. This scene, near the headwaters, shows that erosion takes place at the higher reaches of the river.

Community Forestry

In many parts of the world, people cut nearby forests to meet the needs of small communities. This is particularly true in the developing nations, where, as noted, the use of firewood for fuel constitutes an important part of energy used. In the past, most government forestry departments concentrated their efforts on government-owned forestland or acted merely to police a country's forests. Now, many realize that this approach must change.

Some countries have placed new emphasis on community forestry, in which professional foresters help villagers develop woodlots with the goal of achieving some kind of sustainable local harvest to meet local needs. The FAO and the World Bank are supporting these programs. For example, in Malawi, Africa, a World Bank and FAO project sponsors reforestation in which almost 40% of the households have planted trees. In South Korea, villagers have been reforesting the country at the rate of 40,000 ha/yr (98,840 acres/yr).

In community forestry, good management practices include limiting access; cutting the slower-growing and poorer-burning species first to promote the growth of the better firewood species; making use of plantations; and supplementing firewood with other, more easily renewable fuels. However, some of these practices are in conflict with local, traditional activities or are difficult to implement for other reasons.

Such community efforts are impressive. However, in total they have only a small effect on the problem of the worldwide shortage of firewood. It is not clear whether developing nations can implement a successful management policy in time to prevent serious damage to their forests and the land. If alternative fuels for developing nations cannot be found, the effects will be severe, not only for the land but also for the people.

---

energy, but it provides 15% of the energy in developing countries and is the major source of energy for most countries of sub-Saharan Africa, Central America, and continental Southeast Asia.[15]

As the human population grows, firewood use increases. In this situation, intentional management practices need to be introduced, including management of woodland stands to improve growth. However, well-planned management of firewood stands has been the exception rather than the rule. Some successful community-based projects are discussed in A Closer Look 13.1.

Indirect Deforestation

A more subtle cause of the loss of forests is indirect deforestation—the death of trees from pollution or disease. Acid rain and other pollutants may be killing trees in many areas in and near industrial countries. In Germany, there is talk of *Waldsterben* ("forest death"). The German government has estimated that one-third of the country's forests have suffered damage: death of standing trees, yellowing of needles, or poorly formed shoots. The causes are unclear but appear to result from a number of influences, including acid rain, ozone, and other air pollutants that tend to weaken the trees and increase their susceptibility to disease. This problem extends throughout central Europe and is especially acute in Poland, the Czech Republic, and Slovakia. In the New England area of the United States, similar, curious damage is affecting red spruce.

If global warming occurred as projected by global models of climate, indirect forest damage could occur over vast regions, with major die-offs in many areas and major shifts in the potential areas for the growth of each species of trees.[16,17] The combination of temperature and rainfall required for various tree species could be altered by global warming, and some species might not continue to grow in their current locations. Such changes would dwarf other causes of damage, affecting the habitat for many endangered species.[18]

Some suggest that global warming would merely change the location, not the total area or production, of forests. However, even if a climate conducive to forest growth were to move to new locations, trees would have to reach these areas; and a long time is required for large changes in the geographic distribution of trees, since distribution is accomplished primarily by seeds blown by the wind or carried by animals. In addition, for production to remain as high as it is now, climates that meet the needs of forest trees would have to occur on soils that also meet these needs. This combination of climate and soils occurs widely now but might decrease with large-scale climate change.

Approaches to Forest Management

Clearcutting

Clearcutting (Figure 13.14) is the practice of cutting all trees in a stand at the same time. Alternatives to clearcutting are selective cutting, strip cutting, shelterwood cutting, and seed tree cutting.

Figure 13.14 • A clearcut forest in Western Washington.

Shelterwood cutting is the practice of cutting dead and less desirable trees first, and then later cutting mature trees. As a result, there are always young trees left in the forest. **Seed tree cutting** removes all but a few seed trees (trees with good genetic characteristics and with high seed production that are mature and seed producing), so as to promote regeneration of the forest. In **selective cutting**, individual trees are marked and cut. Sometimes, smaller, poorly formed trees are selectively removed; this practice is called **thinning**. At other times, trees of specific species and sizes are selectively removed. For example, some forestry companies in Costa Rica cut only some of the largest mahogany trees, leaving other, less valuable trees to help maintain the ecosystem and permitting some of the large mahogany trees to continue to provide seeds for future generations.

In **strip cutting**, narrow rows of forest are cut, leaving wooded corridors. Strip cutting offers several advantages. The uncut strips protect regenerating trees from wind and direct sunlight, and these remaining trees provide seeds. In addition, strip cutting can minimize the negative aesthetic effects of logging by leaving buffer zones and allowing the corridors of forest that remain to be used for recreation and as wildlife habitats.

Experimental Tests of Clearcutting

Scientists have tested the effects of clearcutting.[19] For example, in the U.S. Forest Service Hubbard Brook experimental forest in New Hampshire, an entire watershed was clearcut (see Figure 6.9) and herbicides were applied to prevent regrowth for two years. The results were dramatic. Erosion increased, and the pattern of water runoff changed substantially. The exposed soil decayed more rapidly, and the concentrations of nitrates in the stream water exceeded public health standards.

In another experiment, at the U.S. Forest Service H. J. Andrews experimental forest in Oregon, clearcutting greatly increased the frequency of landslides, as did the construction of logging roads. In this forest, rainfall is high (about 240 cm, or 94 in, annually) and the trees (mainly Douglas fir, western hemlock, and Pacific silver fir) grow very tall and live a long time.[20]

These experiments show that clearcutting also changes chemical cycling in forests and results in the loss from the soil of chemical elements necessary for life. When a forest is clearcut, trees are no longer available to take up nutrients. Opened to the sun and rain, the ground becomes warmer. The process of decay is accelerated; chemical elements, such as nitrogen, are converted more rapidly to forms that are water soluble and can be readily lost in runoff during rains (Figure 13.15).[21]

The Forest Service experiments show that clearcutting can be a poor practice on steep slopes in areas of moderate to heavy rainfall. In addition, the worst effects of clearcutting were demonstrated by the logging of vast areas of North America during the 19th and early 20th centuries, when huge areas were clearcut. Clearcutting on such large scales is neither necessary nor desirable for the best timber production. However, where the ground is level or the slope slight, where rainfall is moderate, and where the desirable species require open areas for growth, clearcutting on an appropriate spatial scale may be a useful method to regenerate desirable species. The key here is that clearcutting is neither all good nor all bad for timber production or forest ecosystems. The use of clearcutting must be evaluated on a case-by-case basis, taking into account the size of cuts, the environment, and the available species of trees.

Plantation Forestry

One alternative solution to the pressure on natural forests is to emphasize forest plantations, where trees are planted as crops on land totally devoted to that use. Much commercial timber is grown in this way. If plantations were used where forest production is high, then a comparatively small percentage of the world's forest land could provide all the world's timber. For example, high-yield forests produce 15 to 20 m³/ha/yr. According to one estimate, if plantations were put on timber land that could produce at least 10 m³//ha/yr, then 10% of the world's forest land could provide the timber quantity for the world's timber trade.[22] This approach could reduce pressure on old-growth forests, on forests important for biological conservation, and on forest lands important for recreation.

 13.9 **Parks, Nature Preserves, and Wilderness**

Landscapes are often protected from harvest and other potentially destructive uses by the establishment of parks, nature preserves, and legally designated wilderness areas.

① Trees shade ground
② In cool shade, decay is slow
③ Trees take up nutrients from soil

(a)

① Branches and so on, decay rapidly in open, warm areas
② Soil is more easily eroded without tree roots
③ Runoff is greater without evaporation by trees

(b)

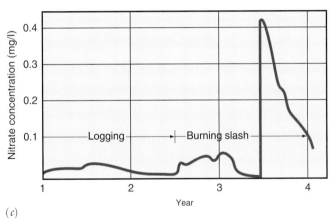

(c)

Figure 13.15 • Effects of clearcutting on forest chemical cycling. Chemical cycling (a) in an old-growth forest and (b) after clearcutting. (c) Increase in nitrate concentration in streams following logging and the burning of slash (leaves, branches, and other tree debris). (*Sources:* [a] and [b] adapted from R. L. Fredriksen, "Comparative Chemical Water Quality—Natural and Disturbed Streams Following Logging and Slash Burning," in *Forest Land Use and Stream Environment* [Corvallis: Oregon State University, 1971], pp. 125–137.)

In addition, some privately held lands have been set aside for biological conservation. Organizations such as the Nature Conservancy, the Southwest Florida Nature Conservancy, and the Land Trust of California purchase lands privately and maintain them as nature preserves. Whether government or private conservation areas succeed better in reaching the goals listed in Table 13.3 is a matter of much present controversy.

Parks, natural areas, and wildernesses provide benefits within their boundaries and can also serve as migratory corridors between other natural areas. Originally, parks were established for specific purposes related to the land within the park boundaries (see A Closer Look 13.2). In the future, the design of large landscapes composed of a

matrix of different land uses—including parks, preserves, and wilderness areas—will become more important and a greater focus for discussion.

Parks and Preserves as Islands

A park is an area set aside for use by people, whereas a nature preserve (although it may be used by people) has as its primary purpose the conservation of some resource, typically a biological one. Every park or preserve is an ecological island of one kind of landscape surrounded by a different kind of land use.

Ecological and physical islands have special ecological qualities (discussed in detail in Chapter 8), and concepts

Table 13.3   Goals of Parks, Nature Preserves, and Wilderness Areas

Parks are as old as civilization. The goals of park and nature preserve management can be summarized as follows:

1. Preservation of unique geological and scenic wonders of nature, such as Niagara Falls and the Grand Canyon.

2. Preservation of nature without human interference (preserving wilderness for its own sake).

3. Preservation of nature in a condition thought to be representative of some prior time period (e.g., the United States prior to European settlement).

4. Wildlife conservation, including conservation of the required habitat and ecosystem of the wildlife.

5. Conservation of specific endangered species and habitats.

6. Conservation of the total biological diversity of a region.

7. Maintenance of wildlife for hunting.

8. Maintenance of uniquely or unusually beautiful landscapes for aesthetic reasons.

9. Maintenance of representative natural areas for an entire country.

10. Maintenance for outdoor recreation, including a range of activities from viewing scenery to wilderness recreation (hiking, cross-country skiing, rock climbing) to tourism (car and bus tours, swimming, downhill skiing, camping).

11. Maintenance of areas set aside for scientific research, both as a basis for park management and for the pursuit of answers to fundamental scientific questions.

12. Provision of corridors and connections between separated natural areas.

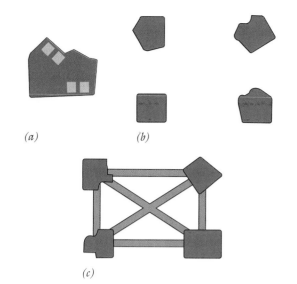

Figure 13.16 • Park shapes and island biogeography. A large park (*a*) can maintain more species, but several small parks (*b*) provide a kind of insurance against catastrophe. For example, if a storm struck one park and killed all individuals of one species in it, other populations of that species could survive in the other parks. A combination (*c*) can provide the benefits of both a single large park and several small ones. Here, the small parts are connected by migration corridors that allow occasional migration among them; the total area is equal to that of (*a*), and the distribution over the landscape provides greater insurance.

of island biogeography are used in the design and management of parks. Specifically, park and preserve planners know that the size of the park and the diversity of habitats affect the number of species that can be maintained there. Also, the farther the park is from other parks or sources of species, the fewer species are found (Figure 13.16). Even the shape of a park can determine what species can survive within it.

One of the important differences between a park and a truly natural wilderness area is that a park has definite boundaries. These boundaries are usually arbitrary from an ecological viewpoint and have been established for political, economic, or historical reasons not related to the natural ecosystem. In fact, many parks have been developed on what are otherwise considered wastelands, useless for any other purposes.

Even where parks or preserves have been set aside for the conservation of some species, the boundaries are usually arbitrary, and such arbitrariness has caused problems. For example, Lake Manyara National Park in Tanzania,

famous for its elephants, was originally established with boundaries incorrect for elephant habits. Prior to the establishment of this national park, elephants would spend part of the year feeding along a steep incline above the lake. At other times of the year, they would migrate down to the valley floor, depending on the availability of food and water. The annual movements of the elephants were necessary so that they could obtain food of sufficient nutritional quality throughout the year.

When the park was established, farms were laid out along its northern border. These farms crossed the traditional pathways of the elephants, creating two negative effects. First, elephants came into direct conflict with farmers. Elephants crashed through farm fences, eating corn and other crops and disrupting the farms. Second, whenever the farmers were successful in preventing the movement of the elephants, the animals were cut off from reaching their feeding ground near the lake. It became clear that the park boundaries were arbitrary and inappropriate, and adjustments were made to extend the boundaries to cover those traditional migratory routes. This solution reduced the conflicts between elephants and farmers.

Because parks isolate populations genetically, they may provide too small a habitat for the maintenance of a minimum safe population size. If parks are to function as biological preserves, they must be adequate in size and habitat diversity to maintain a minimum population that will not undergo the serious genetic difficulties that can occur in small populations. An alternative, if necessary, is for a manager to move individuals of one species—say, lions in African preserves—from one park to another to maintain genetic diversity.

How Much Land Should Be in Parks?

Among conservationists, there is now considerable discussion about what percentage of a landscape should be in a park or nature preserve, especially in regard to goals related to the conservation of biological diversity. One project in the United States, called "The Wildness Project," proposes putting large areas of land in some kind of conservation use, so that in some states more than half of the land area either would be in a park or a nature preserve or would have some kind of restrictions on its use in order to benefit biological diversity. At the same time, however, the growing human population and the desire for an increase in the standard of living puts pressure on existing parks in many parts of the world.

Nations differ greatly in the percentage of their total area set aside as national parks. Costa Rica, a small country with great biological diversity, has more than 12% of its land in national parks.[23] Kenya, a larger nation also with numerous biological resources, has 7.6% of its land in national parks.[24] In France, an industrialized nation in which civilization has altered the landscape for several thousand years, only 0.7% of the land is in the nation's six national parks. However, France has 38 regional parks that encompass 11% (5.9 million ha) of the nation's area.[25]

The total amount of protected natural areas in the United States is more than 104 million ha (about 240 million acres), approximately 11.2% of total U.S. land area.[26] However, the percentage of land in parks, preserves, and other conservation areas differs greatly among the states. The West, for example, has vast parks, whereas the six Great Lakes states (Michigan, Minnesota, Illinois, Indiana, Ohio, and Wisconsin), covering an area approaching that of France and Germany combined, allocate less than 0.5% to parks and less than 1% to designated wilderness.[27]

Conserving Wilderness

As a modern legal concept, **wilderness** is an area undisturbed by people. The only people in a wilderness area are visitors who do not remain. The conservation of such an area is a new idea introduced in the second half of the 20th century and one that is likely to become more important as the human population increases and the effects of civilization become more pervasive throughout the world (Figure 13.17).

The United States Wilderness Act of 1964 was a landmark piece of legislation, marking the first time anywhere that wilderness was recognized by national law as a

Figure 13.17 • Marble Mountain wilderness area, Northern California.

national treasure to be preserved. Under this law, wilderness includes "an area of undeveloped Federal land retaining its primeval character and influence, without permanent improvements or human habitation, which is protected and managed so as to preserve its natural conditions." Such lands are those in which (1) the imprint of human work is unnoticeable, (2) there are opportunities for solitude and for primitive and unconfined recreation, and (3) there are at least 5,000 acres.

The law also recognizes that these areas are valuable for ecological processes, geology, education, scenery, and history. The Wilderness Act required certain maps and descriptions of wilderness areas, resulting in the U.S. Forest Service's Roadless Area Review and Evaluation (RARE I and RARE II), which evaluated lands for inclusion as legally designated wilderness.

Countries with a significant amount of wilderness include New Zealand, Canada, Sweden, Norway, Finland, Russia, and Australia; some countries of eastern and southern Africa; many countries of South America, including parts of the Brazilian and Peruvian Amazon basin; the mountainous high-altitude areas of Chile and Argentina; some of the remaining interior tropical forests of Southeast Asia; and the Pacific Rim countries (parts of Borneo, the Philippines, Papua New Guinea, and Indonesia). In addition, wilderness can be found in the polar regions, including Antarctica, Greenland, and Iceland.

Many countries have no wilderness left to preserve. Switzerland is a country in which wilderness is not a part of preservation. In the Danish language, even the word for wilderness has disappeared, although that word was important in the ancestral languages of the Danes.[28] The historian Roderick Nash contrasts a national park in Switzerland with a wilderness preserve. The Swiss park lies in view of the awesome scenery of the Alps—the scenery that inspired the English romantic poets of the early 19th century to praise what they saw as wilderness. But the park is situated in an area that has been heavily exploited for such activities as mining and foundries since the Middle Ages. All the forests are planted.[28]

In another, perhaps deeper sense, wilderness is an idea and an ideal that can be experienced in many places, such as in Japanese gardens. Henry David Thoreau distinguished between "wilderness" and "wildness." He thought of wilderness as a physical place and wildness as a state of mind. During his travels through the wilderness of the Maine woods in the 1840s, he concluded that wilderness was an interesting place to visit but not to live. Instead, he preferred long walks through the woods and near swamps around his home in Concord, Massachusetts. There, he was able to achieve a feeling of wildness. Thus, Thoreau raised a fundamental question: Can one experience true wildness only in a huge area set aside as a wilderness and untouched by human actions, or can wildness be experienced in small, heavily modified and naturalistic landscapes, such as those around Concord in the 19th century?

As Thoreau suggests, small, local naturalistic parks may have more value as places of solitude and beauty than some more traditional wilderness areas. In Japan, for instance, there are roadless recreation areas, but they are filled with people. One two-day hiking circuit leads to a high-altitude marsh where people can stay in small cabins. Trash is removed from the area by helicopter. People taking this hike experience a sense of wildness.

In some ways, the answer to the question raised by Thoreau is highly personal. We must discover for ourselves what kind of natural or naturalistic area meets our spiritual, aesthetic, and emotional needs. This is yet another area in which one of our key themes, science and values, is evident.

Conflicts in Managing Wilderness

The legal definition of wilderness has produced several controversies. The wilderness system in the United States began in 1964 with 3.7 million ha (9.2 million acres) designated in U.S. Forest Service control. By 1979, it had grown to include nearly 8 million ha (19 million acres). *Currently, over 4% of the United States, or 104 million acres of land, has wilderness protection. Another 200 million acres could also be protected as wilderness; half of this area is in Alaska.*[29]

Those interested in developing the natural resources of an area, including mineral ores and timber, have argued that the rules are unnecessarily stringent, protecting too much land from exploitation when, they say, there is plenty of wilderness elsewhere. Those who wish to conserve additional wild areas have argued that the interpretation of the U.S. Wilderness Act is too lenient and that mining and timber-cutting is inconsistent with the wording of the Act.

The management of wilderness may seem a paradox: A true wilderness would need no management. In fact, though, in modern life, with the great numbers of people in the world, even wilderness must be defined, legally set aside, and controlled. We can view the goal of managing wilderness in two distinct ways: one in terms of the wilderness itself and the other in terms of people. In the first, the goal is to preserve nature undisturbed by human beings. In the second, the purpose is to provide people with a wilderness experience.

Legally designated wilderness can be seen as one extreme in a spectrum of environments to manage.[30] Its management should involve as little direct action as possible, so as to minimize any human influence. Ironically, one of the necessities is to control human access and, therefore, to impose a carrying capacity for visitors so that any visitor has little, if any, sense that there are other people present. Access to wilderness for various kinds of activities, such as operation of motor vehicles, snowmobiles, and offroad vehicles must also be restricted.

As an example of considerations in wilderness management, consider the Desolation Wilderness Area in California, an area of more than 24,200 ha (60,000 acres), which in one year had more than 250,000 visitors.[41] Could each visitor really have a wilderness experience in that area, or had the human carrying capacity been exceeded? This is a subjective judgment. If all visitors saw only their own companions and believed they were alone, then the actual number of visitors did not matter for each visitor's wilderness experience. If, on the other hand, every visitor found the solitude ruined by strangers, then the management failed, no matter how few people visited the area.

Wilderness must be managed with a knowledge of adjacent lands. A wilderness next to a garbage dump or a smoke-emitting power plant is a contradiction. Whether a wilderness can be adjacent to high-intensity campgrounds or near a city is a more subtle question that must be resolved by citizens.

Today, those involved in wilderness management recognize that wild areas change over time and that these changes should be allowed to occur as long as they are natural. This is different from earlier views that nature undisturbed was unchanging and should be managed so that it did not change. In addition, it is generally argued now that, in choosing what activities can be allowed in a wilderness, emphasis should be given to those activities that depend on wilderness (the experience of solitude or the observation of shy and elusive wildlife) rather than to those that can be carried out elsewhere (such as downhill skiing).

A source of conflict is that wilderness areas frequently contain resources of economic importance, including timber, mineral ores, and sources of energy. This is exempli-

A Brief History of Parks

The old French word *parc* referred to an enclosed area for keeping wildlife to be hunted. Such areas were set aside for nobility, excluding the public. An example is the area now known as Donana National Park on the southern coast of Spain. Originally a country home of nobles, today it is one of the most important natural areas of Europe, used by 80% of birds migrating between Europe and Africa.

The first major public park of the modern era was Victoria Park in Great Britain, authorized in 1842. The concept of a national park, whose purposes include protection of nature as well as public access, originated in North America in the 19th century. In the 20th century the purpose of national parks was broadened to emphasize biological conservation, and this idea was applied worldwide.[31]

The first designated national park was Yosemite National Park in California, which was made a park by passage of a bill signed by President Lincoln in 1864. The term *national park*, however, was first used with the establishment of Yellowstone in 1872.

The purpose of the earliest national parks established in the United States was to preserve the unique, awesome, and grand landscapes of the country, a purpose the historian Alford Runte refers to as *monumentalism*. In the 19th century, Americans saw their national parks as a contribution to civilization equivalent to the architectural treasures of the Old World and sought to preserve them as a basis of their pride in their country.[31]

In the second half of the 20th century, the emphasis of park management became more ecological, with parks established both for scientific research and to maintain examples of representative natural areas. For example, Zimbabwe established Sengwa National Park solely for scientific research; there are no tourist areas, and tourists are not generally allowed. Its purpose is the study of natural ecosystems with as little human interference as possible so that the principles of wildlife and wilderness management can be better formulated and understood. Other national parks in the countries of eastern and southern Africa, including those of Kenya, Uganda, Tanzania, Zimbabwe, and South Africa, have been established primarily for viewing wildlife and for biological conservation.

In recent years, the number of national parks throughout the world has increased rapidly. The first law establishing national parks in France was enacted in 1960. Taiwan had no national parks prior to 1980 but now has six.[32] In the United States, the area in national and state parks has increased from less than 12 million ha (30 million acres) in 1950 to nearly 83.6 million acres today, much of the increase resulting from the establishment of parks in Alaska.[33]

The conservation of representative natural areas of a country is an increasingly common goal of national parks. For example, the goal of New Zealand's park planning is to include in national parks at least one area representative of each major ecosystem of the nation, from seacoast to mountain peak.

---

fied by the current conflict over whether wilderness areas should be open to the extraction of oil and mineral ores.

Another controversy involves the need to study wilderness, to understand wilderness so that it can be better conserved on the one hand, and the desire to leave the wilderness undisturbed. Those in favor of scientific research in the wilderness argue that it is necessary for the conservation of the wilderness. Those opposed argue that scientific research is contradictory to the purpose of a designated wilderness as an area undisturbed by people. One solution is to establish separate research preserves.

Summary

- In the past, the primary focus on the management of land for the harvest of resources and the conservation of nature was local in scale; each parcel of land was considered independently. Today, a landscape perspective has developed, and lands used for the harvest of resources are seen as part of a matrix, along with lands for the conservation of biological diversity and for landscape beauty.

- Forests are among civilization's most important renewable resources. Forest management seeks a sustainable harvest and sustainable ecosystems.

- There are few direct demonstrations of successful sustainable forestry. As a result, a practice has developed called "certification of sustainable forestry." Certification involves determining the methods that appear most consistent with sustainability and then comparing actual practices in the management of a specific forest with those standards.

- The continued use of firewood as an important fuel in developing nations is a major threat to forests, given the rapid population growth of those areas. It is doubtful whether developing nations can implement a successful management program in time to prevent serious damage to their forests and land and severe effects on their people.

How Does Fragmentation of Tropical Forests Contribute to Habitat Destruction?

Although tropical rain forests occupy only about 7% of the land area of the world, they provide the habitat for at least one-half of the world's species of plants and animals. Approximately 100 million people live in rain forests or depend on them for their livelihood. Tropical plants provide useful products, such as chocolate, nuts, fruits, gums, coffee, wood, rubber, pesticides, fibers, and dyes. Drugs used to treat high blood pressure, Hodgkin's disease, leukemia, multiple sclerosis, and Parkinson's disease have been made from tropical plants, and medical scientists believe many more are yet to be discovered.

Most of the interest in tropical rain forests has focused on Brazil, whose forests are believed to have more species than any other habitat. Estimates of destruction in the Brazilian rain forest range from 6% to 12%. Numerous studies have shown, however, that deforested area alone does not adequately measure habitat destruction. For example, David Skole and Compton Tucker estimated that the clearing of 6% of the Brazilian rain forest by 1988 had affected 15% of the forest. To estimate the affected area, Skole and Tucker assumed an *edge effect* of 1 km—that is, they assumed that an area of forest bordering a deforested area would be affected for a distance of 1 km (0.6 miles) into the forest. The more fragmented a forest is, the more edges there are, and the greater the impact on the living organisms.

Field studies have shown that the edge effect depends on the species, characteristics of the land surrounding the forest fragment, and distance between fragments. For example, a forest surrounded by farmland is more deeply affected than one sur-

rounded by abandoned land in which secondary growth presents a more gradual transition between forest and deforested areas. Some insects, small mammals, and many birds find only 80 m (262.5 ft) a barrier to movement from one fragment to another, whereas one small marsupial has been found to cross distances of 250 m (820.2 ft). Corridors between forested areas also help to offset the negative effects of deforestation on plants and animals of the forest.

Critical Thinking Questions

1. Assuming an edge effect of 1 km, what is the approximate area affected by deforestation of 100 km^2 (38.6 mi^2) in the form of a square, 10 km (6.2 miles) on each side? If the 100-km^2 area were in the form of 10 rectangles, each 10 km long and 1 km wide, separated from each other by a distance of 5 km (3.1 miles), how large an area would be affected?

2. What environmental factors at the edge of a fragment would differ from those in the center? How might the differences affect plants and animals at the edge?

3. Why is a simple rule of thumb, such as assuming an edge effect of 1 km, too simplistic as a model of the effects of deforestation? Given that it is too simplistic, what advantages are there in using the rule?

4. Forest fragments are sometimes compared to islands. What are some ways in which this is an appropriate comparison? Some ways in which it is not?

- Clearcutting is a major controversy in forestry. Some species require clearing to reproduce and grow. As a result, clearings are sometimes useful, but the size and method of cutting must be examined carefully in terms of the needs of the species of trees and of the forest ecosystem type.

- Properly managed, plantations can offer a way to relieve pressure on forests.

- The management of parks for biological conservation is a relatively new idea that began in the 19th century. The manager of a park must be concerned with its shape and size. Parks that are too small or the wrong shape or that have too small a population of the species for which the park was established may not be able to sustain the species.

- A special extreme in conservation of natural areas is the management of wilderness. In the United States, the 1964 Wilderness Act provided a legal basis for this conservation. The management of wilderness involves a basic contradiction: Trying to preserve an area undisturbed by people requires the interference of the manager to limit user access and to maintain the wilderness in a natural state.

- Parks, nature preserves, wilderness areas, and actively harvested forests affect each other. The geographic pattern of these areas on a landscape, including corridors and connections among different types, is part of the modern approach to biological conservation and the harvest of forest resources.

Human Population:

Forests provide essential resources for civilization. As human population grows, there will be greater and greater demand for these resources. Because plantations can be highly productive, it is likely that a greater emphasis in the future will be placed on forest plantations as a source of timber. This frees more forest land for other uses.

Sustainability:

Sustainability is the key to conservation and management of wild living resources. However, in few cases has it been demonstrated that sustainable harvests have been achieved in forestry. In even fewer (if any) cases has it been shown that the ecosystem within the harvested forest has been sustained. Sustainability must be the central focus for these resources in the future.

Global Perspective:

Forests are global resources. Decline in the availability of forest products in one region affects the rate of harvest and economic value of these products in other regions. Biological diversity is also a global resource. As the human population grows, the conservation of biological diversity would seem to depend more and more on legally established parks, nature preserves, and wilderness areas.

Urban World:

We tend to think of cities as separated from living resources, but urban parks are important in making cities pleasant and livable. Urban parks, if properly designed, can also help to conserve wild living resources.

Science and Values:

Throughout history, people have shown a duality in the values they attach to forests. On the one hand, forests have provided products essential to civilization, large areas of majestic trees have been an inspiration for many people, and forested environments have been an important focus for biological conservation. On the other hand, the dark interiors of forests have often been viewed negatively. Rather than maintain forests, people have tended to clear and eliminate them. We have yet to come to terms with the various values we attach to forests.

Similarly, many conflicts over parks, nature preserves, and legally designated wilderness areas are conflicts that involve science and values. Science tells us what is possible and, at least in potential, what is required for the conservation of a specific species and for the conservation of biological diversity. But what our society desires for such areas, given that scientific knowledge, is a matter of values and experience.

● KEY TERMS

certification of forestry **241**
clearcutting **245**
codominants **242**
dominants **242**
intermediate **242**

landscape perspective **235**
old growth **242**
plantations **242**
public service functions **236**
rotation time **242**

second growth **242**
seed tree cutting **246**
selective cutting **246**
shelterwood cutting **246**
silviculture **242**
site quality **242**

stands **242**
strip cutting **246**
suppressed **242**
sustainable forests **241**
thinning **246**
wilderness **249**

● STUDY QUESTIONS

1. What environmental conflicts might arise when a forest is managed for the multiple uses of (a) commercial timber, (b) wildlife conservation, and (c) a watershed for a reservoir? In what ways would management for one use benefit another?

2. What arguments could you offer for and against the statement "Clearcutting is natural and necessary for forest management"?

3. Can a wilderness park be managed to supply water to a city? Discuss your answer.

4. A park is being planned in rugged mountains with high rainfall. What are the environmental considerations if the purpose of the park is to preserve a rare species of deer? If the purpose is recreation, including hiking and hunting?

5. What are the environmental effects of decreasing the rotation time in forests from an average of 60 years to 10 years? Compare these effects for (a) a woodland in a dry climate on a sandy soil and (b) a rain forest.

6. In a small but heavily forested nation, two plans are put forward for forest harvests. In Plan A, all the forests to be harvested are in the eastern part of the nation, while all the forests of the west are in wilderness areas, parks, and nature preserves. In Plan B, small areas of forests to be harvested are distributed throughout the country, in many cases adjacent to parks, preserves, and wilderness areas. Which plan would you choose? Note that in Plan B, wilderness areas would be smaller than in Plan A.

● FURTHER READING

Barnes, B. V., D. R. Aak, S. R. Denton, and S. H. Spurr. 1998. *Forest Ecology*, 4th ed. New York: John Wiley & Sons.

Food and Agricultural Organization. 1999. *State of the World's Forests.* Washington, D.C.: United Nations.

Grossman, D. H., D. Faber-Langendoen, A. S. Weakley, M. Anderson, P. Bourgeoron, R. Crawford, K. Goodin, S. Landaal, K. Metzler, K. D. Patterson, M. Pyne, M. Reid, and M. L. Hunter, Jr. 1990. *Wildlife, Forests and Forestry: Principles of Managing Forests for Biological Diversity.* Englewood Cliffs, N.J.: Prentice-Hall.

Kimmins, J. P. 1996. *Forest Ecology: A Foundation for Sustainable Management.* New York: Prentice-Hall. A textbook that applies recent developments in ecology to the practical problems of managing forests.

Kohm, K. A., and J. F. Franklin. 1996. *Creating a Forestry for the 21st Century: The Science of Ecosystem Management.* Washington, D.C.: Island Press. A book that applies new understanding of ecological processes, including the role of disturbances, to management of forests. The book views forest management within an economic and societal context.

Noss, R. 1987. *Protecting Natural Areas in Fragmented Landscapes. Natural Areas J.* 7:2–7.

Oliver, C. D., and B. C. Larson. 1996, *Forest Stand Dynamics.* New York: John Wiley and Sons.

Perlin, J. 1989. *A Forest Journey: The Role of Wood in the Development of Civilization.* New York: W. W. Norton. A fascinating presentation of the story of wood, forests, and people, covering a period of five thousand years and spanning five continents.

Runte, A. 1979. *National Parks: The American Experience.* Lincoln: University of Nebraska Press. A work that interprets selected people, events, and legislation to provide a history of the concept of national parks.

Sharma, N., ed. 1992. *Contemporary Issues in Forest Management: Policy Implications.* Washington, D.C.: The World Bank. A series of articles that analyze the reasons for conserving and sustaining forests.

Sneddon. 1998. *International Classification of Ecological Communities: Terrestrial Vegetation of the United States. Vol. I.: National Vegetation Classification System: Development, Status and Applications.* Arlington, VA: The Nature Conservancy.

Spurr, S. H., and B. V. Barnes. 1992. *Forest Ecology.* New York: Ronald Press. The most recent edition of a classic textbook that provides an ecological basis for the management of forestland.

Yallee, S. L., A. F. Phillips, I. C. Frentz, P. W. Hardy, S. M. Maleki, and B. E. Thorpe. 1996. *Ecosystem Management in the United States: An Assessment of Current Experience.* Washington, D.C.: Island Press. A guide to where ecosystem management is being attempted. It reviews 105 representative sites and discusses the status of management efforts.

CHAPTER 14

Wildlife, Fisheries, and Endangered Species

The nene goose or Hawaiian goose (Branta Sandwicensis). Only about 1,000 of these nonmigratory and nonaquatic birds remain in the wild today.

● LEARNING OBJECTIVES

There are many reasons for preserving wildlife and endangered species. Much can be done to improve the ways in which we go about the conservation of species. After reading this chapter, you should understand:

- Why habitat is important in the conservation of endangered species and the management of all species.
- What is the ecosystem context for wildlife management and endangered species conservation.
- What is the landscape perspective for wildlife management and endangered species conservation.
- What is the traditional role of the S-shaped curve in wildlife and fisheries management and the conservation of endangered species, and what are the limitations and problems with this method.
- What steps we can take to achieve sustainability of wildlife, fisheries, and endangered species.
- What are the major current causes of extinction.
- Why we conserve wildlife and endangered species.
- What are the major concepts and terms related to conservation, including carrying capacity, maximum sustainable yield, and minimum viable population size.
- How the goals and emphasis of modern wildlife management differ from those of traditional wildlife management.

A Story of Two Geese

Two closely related species, the Aleutian goose (Figure 14.1) and the Hawaiian goose (the state bird of Hawaii, also called the nene, opening photograph), were both threatened with extinction. The two birds share a common heritage with the Canada goose and also share food habits, both feeding on grasses. Their habitats, however, are distant from each other and very different in climate. The Aleutian goose nests in the high-latitude Aleutian islands surrounded by cold waters. The Nene lives on lava slopes in the balmy climes of Hawaii (Figure 14.2a).

Threats to the continued existence of these two species were human-induced. Both suffered from human introduction of non-native predators and from introduced exotic grasses that were not palatable to the geese or were much less nutritious for them.

Recovery programs were established. The Aleutian goose recovered, while the Hawaiian goose has not. The reasons are clear.

By the end of the 1960s, the Aleutian goose population had declined to 800 individuals. The decline began early in the 20th century when arctic foxes were introduced for fur farming to most of the goose's nesting

islands. The geese nest on the ground and have no natural defenses against the fox. In the recovery program, introduced predators were completely removed from more than half of the goose's current range and 25 to 50% of its historic range. In addition, many of the introduced grasses were removed, so that the native food grasses, more palatable and more nutritious, are abun-

(a)

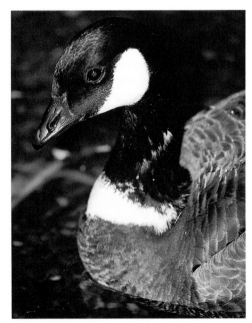

Figure 14.1 • The Aleutian goose, *Branta canadensis leucopareia*, is one of five subspecies of the Canada goose. The U.S. Endangered Species Act allows listing of subspecies as well as species as threatened or endangered.

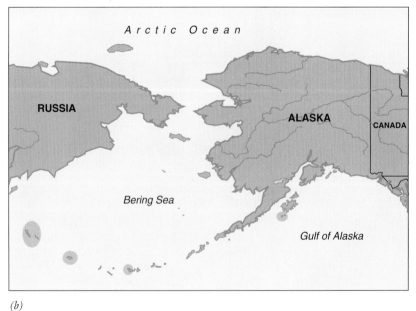

(b)

Figure 14.2 • (*a*) Range of the Hawaiian goose. (*b*) Range of the Aleutian goose (blue areas are where the goose occurs today). (*Source:* Map [*b*] from website of the Alaska Department of Fish and Game, http://www.state.ak.us/local/akpages/FISH.GAME/wildlife/geninfo/game/al_goose.htm; map [*a*] from http://www.aloha_hawaii.com/a_nene.html.)

dant once again. The Aleutian goose now numbers more than 15,000, primarily inhabiting three of the islands of its former range (Figure 14.2b).

The nene declined with human introductions of mongooses, rats, feral dogs, and pet cats, as well as with hunting (outlawed in 1911), and the introduction of exotic grasses. The population declined to a few small flocks by the 1950s outside of captivity.[1] The nene have been successfully bred in captivity, but have done poorly when released back into their native habitat. In the recovery program, only intermittent removal of introduced predators has occurred, and exotic grasses continue to dominate much of the birds' habitat, so that palatable and nutritious food is much less abundant and harder to find. At present, there are approximately 1,000 nene outside of captivity.

● *The comparison between the successful recovery of the Aleutian goose and the lack of recovery of its near relative, the nene, provides some insight into how to help endangered species. In modern times, many species became threatened with extinction because of human actions, including introduction of non-native species and habitat alteration. Restoration of the habitat is an essential part of the recovery of a species. In most cases, it is better to have a small population in a healthy habitat than to have a large population in a poor, heavily degraded habitat. In this chapter, we discuss the conservation and management of endangered species, building on the concepts introduced in the previous chapter with reference to forests, parks, and landscapes.*

14.1 Introduction

Wildlife, fisheries, and endangered species are considered together in this chapter because they have a common history of exploitation, management, and conservation and because modern attempts to manage and conserve them follow the same approaches. Although an endangered species can, of course, assume any form of life, from bacteria and fungi to flowering plants and animals, the primary issues about endangered species will be discussed in terms of wildlife. Remember that the general principles apply to all species.

Historically, attempts to understand, conserve, and manage endangered species were built primarily from principles and practices in wildlife management. This creates an order to the material in this chapter. We begin with a discussion of the basic status of endangered species, the processes that lead to extinction, and the goals of biological conservation. With that background, we then turn to the basic concepts of wildlife manage-

ment. These two discussions then allow us to example a variety of real-world examples, considering how attempts to conserve and manage these have been attempted in the past, and what we might do in the future to improve the success in saving such species.

The beginning of attempts to apply science to conserve and manage wildlife and fisheries—beginning around the turn of the 20th century—viewed each species as a single population in isolation, under the belief that it could be managed successfully this way. Assumptions included the following:

1. The population could be represented by a single number, its total size.
2. Undisturbed by human activities, a population grew to a fixed size called the *carrying capacity*.
3. Environment was treated as constant.

This perception of wildlife and fisheries was formalized in the S-shaped logistic growth equation, which we met in

Figure 14.3 • The basic logistic growth curve.

Chapter 5 (Figure 14.3). The goal for species to be harvested was a *maximum sustainable growth*. The goal for species to be conserved was to allow them to reach their carrying capacity.

This approach failed. It is being replaced today by a broader view, which includes understanding that a population exists in a naturally changing environment (including human-induced changes); that populations interact, and that it is necessary to include an ecosystem and landscape context for conservation and management (as we learned in Chapter 13). The goal for a species to be harvested is an *optimum sustainable population*. The goal for a threatened or endangered species is sometimes a *minimum viable population*, sometimes *the carrying capacity*, and sometimes an *optimum sustainable population*. (These terms will be defined later in this chapter.)

In general, when a human population enters a new territory, land or sea, four stages occur in the relationship with wildlife: discovery, exploitation, single-species management, and ecosystem- and landscape-context management. (Of course, if a species becomes extinct in any of these stages, the process stops.)

In North and South America, Australia, and New Zealand, the period of discovery extended from the 15th century (the time of Columbus's voyage to the New World) to the early 19th century (as with the Lewis and Clark expedition of 1804–1806 in the United States). Exploitation—basically, mining without expectation of renewal—began around 1820 in North America and lasted approximately to the First World War, the exact time varying with species and location. Single-species management with a desire for a scientific foundation began around the end of World War I and continued until the 1970s—and still continues for some species.

The process by which people affect the status of species has existed for thousands of years—probably tens of thousands of years (some examples of these prehistoric alterations will be discussed in this chapter). There are exceptions to this story, but this is the basic story and the primary lesson of this chapter.

Extinction is a natural event and ultimately the fate of every species. But people have greatly increased the rate of extinctions, principally through hunting, habitat alteration and destruction, habitat pollution, and introduction of exotic species.

Today, major questions about wildlife (including fisheries and endangered species) include the following:

• How can we achieve sustainability of a species of wildlife?
• When is a species truly threatened with extinction?
• How can we help an endangered species?
• What do we have to know, including what do we have to monitor, about a species and its habitat?
• How do we restore habitats?
• What are the relative effects of human actions and "natural" forces?

The Current Status of Endangered Species

The number of species of animals listed as threatened or endangered has increased, from about 1,700 in 1988, to 3,800 in 1996, to 5,400 in 2000 (Table 14.1 for data in 2000).[2] A standard list of threatened and endangered ani-

Table 14.1   Worldwide Status of Endangered and Threatened Animals

Animal Group	Critically Endangered	Endangered	Vulnerable (threatened)	Total
Mammals	180	340	610	1130
Birds	182	321	680	1183
Reptiles	56	79	161	296
Amphibians	25	38	83	146
Fish	152	126	431	709
Crustacians	56	72	280	408
Insects	45	118	392	555
Molluscs	170	209	467	846

Source: Data from IUCN (World Conservation Union) Red List 2000.

mals is maintained by the International Union for the Conservation of Nature (IUCN) in a publication known as *The Red Book*. A recent *IUCN Red Book of Threatened Animals* reported that about 25% of all known species of mammals are at risk of extinction, as well as 11% of known birds, 20% of known reptiles, 25% of amphibians, and 34% of fish, primarily freshwater fish. An especially large number of mammals and birds are endangered or vulnerable (see Table 14.1).

The *IUCN Red List of Threatened Plants* estimates that 33,798 species of vascular plants (the familiar kind of plants—trees, grasses, shrubs, flowering herbs), or 12.5% of those known, have recently become extinct or endangered.[3] Of the approximately 100,000 species of trees, more than 8,700, or approximately 9%, are listed as globally threatened by the IUCN.[4]

In the United States, about 1,000 species are listed as threatened by the IUCN. Table 14.2 shows the number of species threatened in selected countries. Note that most threatened species are plants.

The definitions of the terms "endangered" and "threatened" species are stated in the U.S. Federal Endangered Species Act of 1973. According to that Act, "The term 'endangered species' means any species which is in danger of extinction throughout all or a significant portion of its range other than a species of the Class Insecta determined by the Secretary to constitute a pest whose protection under the provisions of this Act would present an overwhelming and overriding risk to man." And "the term 'threatened species' means any species which is likely to become an endangered species within the foreseeable future throughout all or a significant portion of its range."

It is interesting to note that insect pests can be excluded by this legal definition, but there is no mention of disease-causing bacteria or other microorganisms.

The Good News:
Species Whose Status Has Improved

There is some good news about endangered species as a result of direct human activities during the past 40 years (in some cases, much longer). A number of previously endangered species have recovered, just as has the Aleutian goose of the introductory case study. Among the recovered species are:

- The elephant seal, reduced to about a dozen animals around 1900, now numbers in the hundreds of thousands.

- The sea otter, reduced in the 19th century to several hundred, now numbers approximately 10,000.

- Many species of birds became endangered because the insecticide DDT caused thinning of egg shells and failure of reproduction. With the elimination of DDT in the United States, many bird species recovered, including the bald eagle, brown pelican, white pelican, osprey, and peregrine falcon.

- The blue whale, thought to have been reduced to about 400 when whaling was still actively pursued by a number of nations. Today, 400 blue whales are sighted annually in the Santa Barbara channel along the California coast, a fraction of the total population.

- The gray whale, hunted near to extinction, has recovered and is abundant along the California coast and in its annual migration to Alaska.

Since The U.S. Endangered Species Act became law in 1973, 43 species have improved to the extent that they have been either reclassified from endangered to threatened or removed completely from the list. In addition, the U.S. Fish and Wildlife Service—which, along with the

Table 14.2 Threatened Species by Country: Selected Examples

	Mammals	Birds	Reptiles	Amphibians	Fish	Molluscs	Other Invertebrates	Plants	Total
North America									
United States	37	54	27	25	131	255	300	168	997
Canada	14	8	2	1	16	0	10	1	52
Mexico	69	38	18	4	88	4	36	161	418
South America									
Brazil	79	113	22	6	16	21	13	338	608
Columbia	36	77	13	0	6	0	0	213	345
Ecuador	31	60	10	0	3	23	0	197	324
Peru	47	71	6	1	0	0	2	269	396
Asia									
China	76	73	31	1	33	1	3	167	385
Japan	37	32	11	10	13	25	20	11	159
India	86	70	25	3	8	2	21	244	459
Indonesia	140	113	28	0	67	3	28	384	763
Malaysia	47	37	21	0	16	1	2	681	805
Philippines	50	67	8	22	28	3	16	193	387
Africa									
South Africa	41	20	19	9	30	10	101	45	275
Tanzania, United Republic of	43	33	5	0	15	41	6	236	379
Oceania									
Australia	63	32	39	25	44	175	106	38	522

Source: IUCN (World Conservation Union) Red List 2000.

National Marine Fisheries Services, administers the Endangered Species Act—lists 33 species that have the potential for reclassification to an improved condition.[5]

But Too Much of a Good Thing Can Lead to Problems

A species can change in status from endangered, to recovered, to abundant, to overabundant. At that point, it can become a pest.

All marine mammals are protected in the United States by the Federal Marine Mammal Protection Act of 1972. This act has led to improvement in the status of many marine mammals. Sea lions have become so abundant as to be local problems. For example, in San Francisco harbor and in Santa Barbara harbor, sea lions haul out and sun themselves on boats and pollute the water with their excrement near shore. In one case, so many sea lions hauled out on a sailboat in Santa Barbara harbor that they sank the boat; and some of the animals were trapped and drowned. (For more information on the history and status of marine mammals, see A Closer Look 14.1.)

In the eastern United States, white-tailed deer have increased so much that they destroy vegetation desired by people, including plantings around houses; have negative effects on woodlands; and contribute to the spread of Lyme disease. Mountain lions, too, have become locally overabundant. In the 1990s, California voters passed an initiative that protected the endangered mountain lion but contained no provisions for management of the lion, should it become abundant, except in cases where it threatened human life and property. Few people thought that the mountain lion could ever recover enough to become a problem. But in several cases, in recent years, mountain lions have attacked and even killed humans. These attacks increase as the mountain lion population grows and as the human population grows and people build houses in what was mountain lion habitat. Such cases make it evident that it is useful to approach wildlife as a whole; endangered species, abundant species, and pest or problem species.

Conservation of Whales and Other Marine Mammals

All marine mammals were originally inhabitants of the land. During the last 80 million years, several separate groups of mammals returned to the oceans and underwent adaptations to marine life. Each group of marine mammals shows a different degree of transition to ocean life. Understandably, the adaptation is greatest for those that began the transition longest ago. Some marine mammals, such as dolphins, porpoises, and great whales, complete their entire life cycle in the oceans and have organs and limbs that are highly adapted to life in the water. They cannot move on the land. Others, such as seals and sea lions, spend part of their time on shore.

Whales

Whales fit into two major categories: baleen and toothed. The sperm whale is the only great whale that is toothed; the rest of the toothed group are smaller whales, dolphins, and porpoises. The other great whales, in the baleen group, have highly modified teeth that look like giant combs and act as water filters. Baleen whales feed by filtering ocean plankton.

Drawings of whales have been dated as early as 2200 B.C.[6] Eskimos used whales for food and clothing as long ago as 1500 B.C. In the ninth century A.D., whaling by Norwegians was reported by travelers whose accounts were written down in the court of the English king Alfred.

The earliest whale hunters killed these huge mammals from the shore or from small boats near shore, but gradually whale hunters ventured farther out. In the 11th and 12th centuries, Basques hunted the Atlantic right whale from open boats in the Bay of Biscay, off the western coast of France. Whales were brought ashore for processing, and the boats returned to land once the search for a whale was finished.

Eventually, whaling became pelagic: Whalers took to the open ocean and searched for whales from ships that remained at sea for long periods. The whales were brought on board and processed there. This became possible with the invention of furnaces and boilers for extracting whale oil at sea. Thus, pelagic whaling was a product of the Industrial Revolution. With this invention, whaling grew as an industry. American fleets developed in the 18th century in New England; and by the 19th century, the United States dominated the industry, providing most of the ships and even more of the crews for whaling.[7,8]

Whales provided many 19th century products. Whale oil was used for cooking, lubrication, and lamps. Whales provided the main ingredients for the base of perfumes. The elongated teeth (whalebone, or baleen) that enable baleen whales to strain the ocean waters for food are flexible and springy and were used for corset stays and other products before the invention of inexpensive steel springs.

Whale harvests increased in the 20th century. Although the 19th century whaling ships are more famous, made popular by such novels as *Moby Dick*, more whales were killed in the 20th century than were killed in the previous century. The worldwide decline in most species of whales as a result of these human activities made this a global environmental issue.

Conservation of whales has been a concern among conservationists for many years. Attempts to control whaling began with the League of Nations in 1924. The first agreement, the Convention for the Regulation of Whaling, was signed by 21 countries in 1931. In 1946, a conference in Washington, D.C., initiated the International Whaling Commission (IWC). In 1982, the IWC established a moratorium on commercial whaling. Currently, 12 of approximately 80 species of whales are protected.[6]

The IWC has played a major role in the reduction (almost elimination) of the commercial harvest of whales. Since the formation of the IWC, no species has become extinct, the total take of whales has decreased, and harvesting of species considered endangered has ceased. Endangered species protected from hunting have had a mixed history (see Table 14.3). Blue whales appear to have recovered some but remain rare and endangered. Gray whales, have increased and are now relatively abundant, numbering about 26,000 individuals.[9] However, global climate change, pollution, and ozone depletion have become greater risks to whale populations than whaling.[10]

The establishment of the IWC was a major landmark in wildlife conservation. It was one of the first major attempts by a group of nations to agree on a reasonable harvest of a biological resource. The annual meeting of the IWC has become a forum for discussing international conservation, working out basic concepts of maximum and optimum sustainable yields, and formulating a scientific basis for commercial harvesting. The IWC demonstrates that even an informal commission whose decisions are accepted voluntarily by nations can function as a powerful force for conservation.

In the past, each marine mammal population has been treated as if it were isolated, had a constant supply of food, and was subject only to the effects of human harvesting. That is, it has been assumed that its growth followed the logistic curve. It is now realized that management policies for marine mammals must be expanded to include ecosystem concepts and the understanding that populations interact in complex ways.

The goal of marine mammal management is to prevent extinction and maintain large population sizes rather than to maximize production. For this reason, the Marine Mammal Protection Act,

Table 14.3   Recent Whale Population Estimates

Area/Region	Blue	Bowhead	Humpback	Fin	Right	Gray	Sperm	Sei	Minke
WESTERN ARCTIC		8,000					—		
PACIFIC						23,109			
California–Mexico	2,134		482						
Central Pacific			1,407						
North Pacific				UNK	UNK		UNK	UNK	
California–Washington			597	575			756		526
East Tropical Pacific	1,400						22,700		
ATLANTIC									
Western North	308		5,543	2,700	295		337	870	2,650
Northern Gulf of Mexico							213		
ANTARCTICA IWC Estimates for South 30°S	610		15,000	18,000		UNK	290,000	17,000	700,000

Source: National Marine Fisheries Service, 1994; J. Barlow et al. *U.S. Pacific Marine Mammal Stock Assessments.* NOAA Technical Memorandum NMFS (Washington, D.C.: U.S. Department of Commerce, NOAA, National Marine Fisheries Service, 1995); and, R. J. Small, and D. P. DeMaster, *Alaska Marine Mammal Stock Assessments 1995*, NOAA Technical Memorandum NMFS-AFSC-57 (Washington, D.C.: U.S. Department of Commerce, NOAA, National Marine Fisheries Service, 1995). Antarctic estimates are from the International Whaling Commission.

enacted by the United States in 1972, has as its goal an optimum sustainable population (OSP) rather than a maximum or optimum sustainable yield. An OSP means the largest population that can be sustained indefinitely without deleterious effects on the ability of the population or its ecosystem to continue to support that same level.

Dolphins and Other Small Cetaceans

Among the many species of small "whales," or cetaceans, are dolphins and porpoises, more than 40 species of which have been hunted commercially or have been killed inadvertently as a part of other fishing efforts.[11] A classic case is the inadvertent catch of the spinner, spotted, and common dolphins of the eastern Pacific. Because these carnivorous, fish-eating mammals often feed with yellow fin tuna, a major commercial fish, more than 100,000 of them have been netted and killed inadvertently in recent years.[11]

The U.S. Marine Mammal Commission and commercial fishermen have cooperated in seeking methods to reduce dolphin mortality. Research conducted on dolphin behavior helped in the design of new netting procedures that would be less likely to trap these mammals. When the new procedures were adopted, the trapping of dolphins was greatly reduced. The attempt to reduce dolphin mortality illustrates cooperation among fishermen, conservationists, and government agencies and the role of scientific research in the management of renewable resources.

14.3   The Goals of Biological Conservation

The introductory case study about the Aleutian and Hawaiian geese raises an interesting philosophical question: When we say that we want to save a species, what is it that we really want to save? The answers go beyond science and call for an integration of science and values. The possibilities include four goals:

1. A wild creature in a wild habitat, as a symbol to us of wilderness.
2. A wild creature in a managed habitat, so that the species persists, feeds, and reproduces with little interference and so that we can see it in a naturalistic habitat. (The recovery of the Aleutian goose fits this goal.)
3. Preservation of a population in a zoo so that the genetic characteristics are maintained in live individuals.
4. Conservation of genetic material only—frozen cells containing DNA from a species for future scientific research.

Although the specific goals of conservation are rarely spelled out in this fashion, policies and actions differ widely depending on which goal is chosen.

This is a list of what we want to see as a product of our work to help an endangered species. Another important list includes the *reasons* for conservation of endangered species. The major reasons for saving endangered species are: utilitarian, ecological, cultural, recreational, spiritual, inspirational, aesthetic, and moral. Some of these are discussed in detail in A Closer Look 14.2.

Reasons for the
Conservation of Endangered Species

Some important reasons for conserving endangered species can be classified as utilitarian, ecological, aesthetic, moral, and cultural.

Utilitarian Justification

Utilitarian justification is based on the consideration that many wild species might be useful to us and thus it is imprudent to destroy them before we have a chance to test their uses. Many of the arguments presented for the conservation of endangered species, and for biological diversity in general, have focused on the utilitarian justification.

One utilitarian justification is the need to conserve wild strains of grains and other crops. Modern agricultural production of crops such as wheat and corn depends on the continued introduction of fresh genetic characteristics from wild strains to create new genetic hybrids. Disease organisms that attack crops evolve continually, changing their genetic characteristics. As new disease strains develop, crops become vulnerable. By introducing fresh genetic characteristics from the wild, new hybrid strains can be developed that are disease-resistant.

Related to this justification is the possibility of finding new crops among the many species of plants. Many horticultural crops and products have come from tropical rain forests, and hopes are high that new products will be found.[12] For example, of 275 species found in 1 ha (0.4 acre) in a Peruvian tropical forest, 72 yielded products with direct economic value.

Another utilitarian justification for biological conservation is that many important chemical compounds come from wild organisms. Digitalis, an important drug in treating certain heart ailments, comes from purple foxglove. Aspirin is a derivative of willow bark. A recent example of a utilitarian was the discovery of a cancer-fighting chemical named *Taxol* in the Pacific Yew tree (genus name *Taxus*, hence the name of the chemical). Well-known medicines derived from tropical forests include anticancer drugs from rosy periwinkles, steroids from Mexican yams; antihypertensive drugs from serpent wood; and antibiotics from tropical fungi.[13] Some 25% of prescriptions dispensed in the United States contain ingredients extracted from vascular plants.[14] Only a small fraction of the estimated 270,000 existing plant species are used for drugs.[13] Other organisms may produce useful medical compounds that are as yet unknown. For example, scientists are testing marine organisms for use in pharmaceutical drugs. Coral reefs offer a promising area of study for such compounds, because many coral reef species produce toxins to defend themselves.[15]

Some species are also used directly in medical research. For example, the armadillo, one of only two animal species known to contract leprosy, is important to studies seeking a cure for disease.[16] Other animals, such as the horseshoe crab and barnacles, are important because of physiologically active compounds they make. Others may have similar uses as yet unknown to us.

Another utilitarian justification is that many species provide pollution control. Plants, fungi, and bacteria remove toxic substances from air, water, and soils. Carbon dioxide and sulfur dioxide are removed by vegetation, carbon monoxide is reduced and oxidized by soil fungi and bacteria, and nitric oxide is incorporated into the biological nitrogen cycle.[17] Because species have different capabilities for removal, a diversity of species can provide the best range of pollution control.

Tourism provides yet another utilitarian justification. Ecotourism is a growing source of income for many developing countries. Of course, ecotourism occurs because the tourists value nature, including its endangered species, and this value is usually aesthetic or spiritual; so the *result* of aesthetic and spiritual justifications can be utilitarian activity.

Ecological Justification

When we reason that organisms are necessary to maintain the functions of ecosystems and the biosphere, we are using an ecological justification for their conservation. Individual species, entire ecosystems, and the biosphere provide public service functions essential or important to the persistence of life, and as such they are indirectly necessary for our survival. When bees pollinate flowers, for example, they provide a benefit to us that would be costly to replace with human labor. Trees remove certain pollutants from the air; and some soil bacteria fix nitrogen, converting it from molecular nitrogen in the atmosphere to nitrate and ammonia that can be taken up by other living things. The fact that some such functions involve the entire biosphere reminds us of the global perspective on conserving nature and specific species.

Aesthetic Justification

An aesthetic justification asserts that biological diversity adds to the quality of life, providing some of the most beautiful and appealing aspects of our existence. Biological diversity is an important quality of landscape beauty. Many organisms—birds, large land mammals, and flowering plants, as well as many insects and ocean animals—are appreciated for their beauty. This appreciation of nature is ancient. Whatever other reasons Pleistocene people had for creating paintings in caves in France and Spain, their paintings of wildlife, done about 14,000 years ago, are beautiful. The paintings include species that have since become extinct, such as mastodons.

Poetry, novels, plays, paintings, and sculpture often celebrate the beauty of nature. It is a very human quality to appreciate nature's beauty and a strong reason for the conservation of endangered species.

Moral Justification

Moral justification is based on the belief that species have a moral right to exist, independent of our need for them. Consequently, the argument follows that in our role as global stewards, we are obligated to promote continued existence of species and to conserve biological diversity. This right to exist was stated in the U.N. General Assembly World Charter for Nature, 1982. The U.S. Endangered Species Act also includes statements concerning the rights of organisms to exist. Thus, a moral justification for the conservation of endangered species is part of the intent of the law.

Moral justification has deep roots within human culture, religion, and society. Those who focus on cost-benefit analyses tend to downplay moral justification. However, although moral justification may not seem to have economic ramifications, in fact it does. As more and more citizens of the world assert the validity of moral justification, more actions that have economic effects are taken to defend a moral position.

The moral justification has grown in popularity in recent decades, as indicated by increasing interest in the deep ecology movement, mentioned in an earlier chapter: Arnee Naess, one of its principle philosophers, explains "The right of all the forms [of life] to live is a universal right which cannot be quantified. No single species of living being has more of this particular right to live and unfold than any other species."[18,19]

Cultural Justification

Specific species, some threatened or endangered, are of great importance to many indigenous peoples, for whom diversity in forests and wildlife provides food, shelter, and tools, fuel, materials for clothing, and medicine. A reduction in biological diversity can severely increase the poverty of these people. For poor, indigenous people who depend on forests, there may be no reasonable replacement for these benefits except continual external assistance, which development projects are supposed to eliminate. Urban residents, too, share in the benefits of biological diversity, even if these benefits may not be apparent or may become apparent too late (see Chapter 27).

 14.4 **How People Cause Extinctions and Affect Biological Diversity**

People have become an important cause of threatened and endangered species. The question: What causes extinction? has a number of answers, which we discuss soon. But before we turn to that subject, it is useful to view those processes against the ways people are influencing the persistence of many species.

Human actions cause extinction of species in several ways:

1. Through intentional hunting or harvesting (for commercial purposes, for sport, or to control a species that is considered a pest).
2. Through disruption or elimination of habitats.
3. Through introduction of exotic species, including new parasites, predators, or competitors of a native species.
4. Through pollution of the environment.

People have caused extinctions over a long time, not just in recent years. The earliest people probably caused extinctions through hunting. This practice continues, especially for specific products from animals considered valuable, such as elephant ivory and rhinoceros horns. When people learned to use fire, they began to change habitats over large areas. With the development of agriculture and the rise of civilization, rapid deforestation and other habitat changes became significant factors. Later, as people explored new areas, introductions of exotic species became an important cause of extinction (see Chapter 8), especially after Columbus's voyage to the New World and the spread of European civilization and technology. In the 20th century, with the introduction of thousands of novel chemicals into the environment, pollution has become an increasingly significant cause of extinction. Pollution control has proved to be a successful method to help species endangered by pollution. *All* bird species in the United States that were threatened with extinction by a specific pollutant have recovered when the pollutant was removed from their environment.[5]

The International Union for the Conservation of Nature (IUCN) estimates that 75% of the extinctions of birds and mammals since 1600 have been caused by human beings. Hunting is estimated to have caused 42% of the extinctions of birds and 33% of the extinctions of mammals. In the United States, of 47 species of wildlife that became extinct between 1700 and 1970, 25 were lost in the last 50 years of the period.

Modern civilization has contributed greatly to the increased rate of extinction, and tropical deforestation accounts for much of the increased rate. In recent years, the number of identified extinct species has increased significantly because of efforts to catalog species being lost. The current extinction rate among most groups of mammals is estimated to be 1,000 times greater than the extinction rate at the end of the Pleistocene epoch.[20] Estimates today put the rate of extinctions worldwide as high as 50,000 species each year, resulting in a loss of 10% of all species alive today within 25 years.[21]

14.5 How a Species Becomes Endangered and Extinct

Extinction is the rule of nature. **Local extinction** occurs when a species disappears from a part of its range but persists elsewhere. **Global extinction** occurs when a species becomes extinct everywhere. Although extinction is the ultimate fate of all species, the rate of extinctions has varied greatly over geologic time and has increased rapidly since the Industrial Revolution. From 580 million years ago until the beginning of the Industrial Revolution, on average, about one species per year became extinct.

Over much of the history of life on Earth, the rate of evolution of new species equaled or slightly exceeded the rate of extinction. The average longevity of a species has been about 10 million years.[31] However, the fossil record suggests that there have been periods of catastrophic losses of species and other periods of rapid evolution of new species (Figure 14.4), which some refer to as "punctuated extinctions." About 250 million years ago a mass extinction occurred in which approximately 53% of marine animal species disappeared; about 65 million years ago, most of the dinosaurs became extinct. Interspersed with the episodes of mass extinctions, there seem to have been periods of hundreds of thousands of years with comparatively low rates of extinction.

Natural extinctions often appear to follow understandable patterns, with the replacement of one form by a more successful one through the process of competition and evolution (see Chapter 7). This was not the case, however, about 10,000 years ago, at the end of the last great continental glaciation. At that time, massive extinctions of large birds and mammals occurred: 33 genera of large mammals—those weighing 50 kg (110 lb) or more—became extinct, whereas only 13 genera had become extinct in the preceding 1 or 2 million years (Figure 14.5). Smaller mammals were not so affected, nor were marine mammals. As early as 1876, Alfred Wallace, an English biological geographer, noted that "we live in a zoologically impoverished world, from which all of the hugest, and fiercest, and strangest forms have recently disappeared."[22] It has been suggested that these sudden extinctions coincided with the arrival, on different continents, at different times, of Stone Age people and therefore may have been caused by hunting.[23] Causes of extinction are summarized in A Closer Look 14.3.

14.6 Traditional Single-Species Wildlife Management

The goal of the first attempt at a scientific-based management of wildlife was a **maximum sustainable population**, defined as the population size that has the maximum production (measured either as a net increase in the number of individuals in the population or as a net change in biomass) that will allow this population to be sustained indefinitely without decreasing the population's ability to provide the same level of production.

The Logistic Growth Curve

The concept of maximum sustainable population was made explicit in the logistic growth curve, which was first proposed in 1838. Recall from Chapter 5 that this is an S-shaped curve representing the growth of a population, from a small population, over time (see Figure 14.3 and Figure 14.6). The logistic growth curve includes the following ideas:

- A population that is small in relation to its resources grows at a nearly exponential rate.
- Competition among individuals in the population slows down the growth rate.
- The greater the number of individuals, the greater the competition and the slower the rate of growth.
- Eventually, a point is reached, called the *logistic carrying capacity*, at which the number of individuals is just sufficient for the available resources.
- At this level, the number of births in a unit time equals the number of deaths, and the population is constant.
- A population can be described simply by its total number.
- Therefore, all individuals are equal.
- The environment is assumed to be constant.

A logistic population is *stable* about its carrying capacity, meaning that it will return to that number when it is disturbed from it. Mathematically, if the population increases above the carrying capacity, deaths exceed births, and the population declines back to the carrying capacity. If the population falls below the carrying capacity, births exceed deaths, and the population increases. Only if the population is exactly at the carrying capacity do births exactly equal deaths, and then the population does not change.

Carrying capacity is an important term in wildlife management. It has three definitions. The first is the carrying capacity as defined by the logistic growth curve; we will refer to this as the **logistic carrying capacity**. The second definition contains the same idea but is not dependent on that specific equation. It states that the carrying capacity is an abundance a population can sustain without any detrimental effects that would decrease the ability of that species to maintain that abundance. The third, more recent, definition is usually referred to as the **optimum sustainable population** and is defined as the maximum population that can be sustained indefinitely without decreasing the ability of that species or its habitat or ecosystem to sustain that population level in the future.

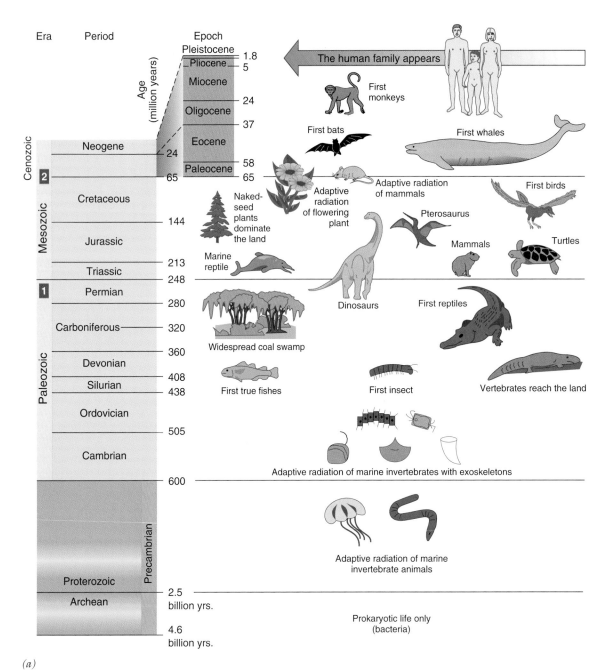

Era | Period | Epoch | Age (million years)

The human family appears

First monkeys

First bats   First whales

Adaptive radiation of mammals   First birds

Naked-seed plants dominate the land   Adaptive radiation of flowering plant   Pterosaurus

Marine reptile   Mammals   Turtles

Dinosaurs   First reptiles

Widespread coal swamp

First true fishes   First insect   Vertebrates reach the land

Adaptive radiation of marine invertebrates with exoskeletons

Adaptive radiation of marine invertebrate animals

Prokaryotic life only (bacteria)

Cenozoic — Neogene (24); 2; Mesozoic — Cretaceous (144), Jurassic (213), Triassic (248); Paleozoic — 1, Permian (280), Carboniferous (320), Devonian (360), Silurian (408, 438), Ordovician (505), Cambrian (600); Precambrian — Proterozoic, Archean (2.5 billion yrs., 4.6 billion yrs.)

Epoch: Pleistocene (1.8), Pliocene (5), Miocene (24), Oligocene (37), Eocene (58), Paleocene (65), 65

(a)

Figure 14.4 • (a) Brief diagrammatic history of evolution and extinction of life on Earth. There have been periods of rapid evolution of new species and episodes of catastrophic losses of species. Two major catastrophes were the Permian loss, which included 52% of marine animals, as well as land plants and animals, and the Cretaceous loss of dinosaurs. (b) Graph of the number of families of marine animals in the fossil records, showing long periods of overall increase in the number of families punctuated by brief periods of major declines. (c) Extinct vertebrate species and subspecies, 1760–1979. The number of species becoming extinct increases rapidly after 1860. Note that most of the increase is due to the extinction of birds. (*Sources:* [a] D. M. Raup, "Diversity Crisis in the Geological Past," in E. O. Wilson, ed., *Biodiversity* [Washington, D.C.: National Academy Press, 1988], p. 53; derived from S. M. Stanley, *Earth and Life through Time* [New York: W. H. Freeman, 1986]. Reprinted with permission. [b] D. M. Raup and J. J. Sepkoski, Jr., "Mass Extinctions in the Marine Fossil Record," *Science* 215 [1982]:1501–1502. [c] Council on Environmental Quality; additional data from B. Groombridge, England: IUCN, 1993].)

Another key concept from the logistic growth curve is the population size that provides the **maximum sustainable yield**, or **MSY** (Figure 14.6). In the logistic curve, the greatest production occurs *exactly* when the population is one-half of the carrying capacity. Even the slightest change from that abundance—whether an increase or a decrease—leads to a decline in production. Unless a logistic population is maintained at exactly one-

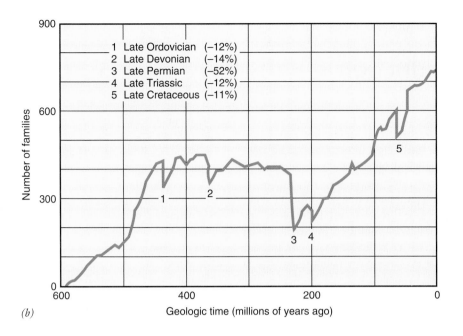

(b)

Number of families (y-axis, values: 0, 300, 600, 900)

1 Late Ordovician (−12%)
2 Late Devonian (−14%)
3 Late Permian (−52%)
4 Late Triassic (−12%)
5 Late Cretaceous (−11%)

Geologic time (millions of years ago) (x-axis values: 600, 400, 200, 0)

(c)

Cumulative number of extinct species and subspecies (y-axis values: 0, 10, 20, 30, 40, 50)

Fish
Mammals
Birds

(x-axis ranges: 1760–1779, 1780–1799, 1800–1819, 1820–1839, 1840–1859, 1860–1879, 1880–1899, 1900–1919, 1920–1939, 1940–1959, 1960–1979)

Figure 14.4 • (*continued*)

Figure 14.5 • Artist's restoration (drawing) of an extinct saber-toothed cat with prey. The cat is an example of one of the many large mammals that became extinct about 10,000 years ago.

Causes of Extinction

Causes of extinction are usually grouped into four risk categories: population risk, environmental risk, natural catastrophe, and genetic risk. Risk here means the chance that a species or population will become extinct owing to one of these causes.

Population Risk

Random variations in population rates (in birth rates and death rates) can cause a species in low abundance to become extinct. This is termed *population risk*. For example, blue whales swim over vast areas of ocean. Because whaling once reduced the total population to only several hundred individuals, the success of individual blue whales in finding mates probably varied from year to year. If in one year most whales were unsuccessful in finding mates, then births could be dangerously low. Such random variation in populations, typical among many species, can occur without any change in the environment. It is a risk especially to species that consist of only a single population in one habitat. Mathematical models of population growth can help calculate the population risk and determine the minimum viable population size.[24]

Environmental Risk

Population size can be affected by changes in the environment that occur from day to day, month to month, and year to year, even though the changes are not severe enough to be considered environmental catastrophes. *Environmental risk* involves variation in the physical or biological environment, including variations in predator, prey, symbiotic, or competitor species. In some cases, species are sufficiently rare and isolated that such normal variations can lead to their extinction.

As an example, Paul and Anne Ehrlich described the local extinction of a population of butterflies in the Colorado mountains.[25] These butterflies lay their eggs in the unopened buds of a single species of lupine (a member of the legume family), and the hatched caterpillars feed on the flowers. One year, however, a very late snow and freeze killed all the lupine buds, leaving the caterpillars without food and causing local extinction of the butterflies. The plants survived, and their roots produced new stems, leaves, and flowers the next year. Had this been the only population of that butterfly, the entire species would have become extinct.

Natural Catastrophe

A sudden change in the environment not the result of human action is a *natural catastrophe*. Fires, major storms, earthquakes, and floods are natural catastrophes on land; changes in currents and upwellings are ocean catastrophes. For example, the explosion of a volcano on the island of Krakatoa in Indonesia in 1883 caused one of recent history's worst natural catastrophes. Most of the island was blown to bits, bringing about local extinction of most life forms there.

Genetic Risk

Detrimental change in genetic characteristics not caused by external environmental changes is called *genetic risk*.[26] Genetic changes can occur in small populations from reduced genetic variation, genetic drift, and mutation (see Chapter 7). In a small population, only some of the possible inherited characteristics will be found. The species is vulnerable to extinction because it lacks variety or because a mutation can become fixed in the population.

Consider the last 20 condors in the wild in California. It stands to reason that this small number was likely to have less genetic variability than the much larger population that existed several centuries ago. This increased the condors' vulnerability. Suppose that the last 20 condors, by chance, had inherited characteristics that made them less able to withstand lack of water. If left in the wild, these condors would have been more vulnerable to extinction than a larger, more genetically variable population.

half its carrying capacity, its growth therefore must decline.

The logistic growth curve has been used for all wildlife, but it has been especially important in fisheries. Harvest levels have been set, for most of the 20th century and for some species into the 21st century, based on the logistic growth equation. The goal was to maintain the population at exactly the MSY level—that is, at one-half its carrying capacity.

There is a problem with this approach, however. The logistic growth curve is a hypothetic curve, quite unrealistic and found rarely, if at all, in nature. In addition, it is almost impossible to maintain a wild population at some exact number, such as the MSY level. Therefore, the logistic growth curve makes a poor basis for conservation and management. Its limitations have led a group of organizations to proposed specific improvements, discussed later in this chapter.

An Example of Problems with the Logistic Curve

Suppose you are in charge of managing a deer herd for recreational hunting in one of the 50 U.S. states. Your goal is to maintain the population at its MSY level, which,

(a)

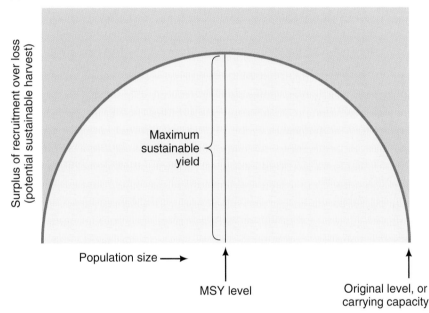

(b)

Figure 14.6 • (*a*) The logistic growth curve, showing the carrying capacity and the maximum sustainable yield (MSY) population (where the population size is one-half the carrying capacity). The figure shows what happens to a population when we assume it is at MSY and it is not. Suppose a population grows according to the logistic curve from a small number to a carrying capacity of 100,000 with an annual growth rate of 5%. The correct maximum sustainable yield would be 50,000. When the population reaches exactly that amount and we harvest at exactly the calculated maximum sustainable yield, the population continues to be constant.

But if we make a mistake in estimating the size of the population (for example, if we believe that it is 60,000 when it is only 50,000), then the harvest will always be too large, and we will drive the population to extinction. (*b*) Another view of a logistic population. The growth in the population here is graphed against population size. The growth peaks when the population is exactly at one-half the carrying capacity. This is a mathematical consequence of the equation for the curve. Hypothetical, it is rarely, if at all, observed in nature.

as you can see from Figure 14.6, occurs at exactly one-half of the carrying capacity. At this abundance, the population increases by the greatest number during any time period.

To accomplish this goal, first you have to determine the logistic carrying capacity. Immediately, you are in trouble because (1) in few cases has the carrying capacity ever been determined by legitimate scientific methods

Elephants and the Logistic Growth Curve

Let's apply the logistic growth curve to elephants. Traditionally, this curve was thought to be the correct description of the growth of elephant populations and of all wildlife if nature was allowed to take its course. The logistic growth curve shows the growth of a hypothetical population plotted against time (Figure 14.6*a*). We can also graph the relationship defined in the logistic equation between population growth and population size (Figure 14.6*b*). In these graphs, we see that when the number of animals (such as elephants) is small and when the number is very high, the increase in population in a single time period is small. At the logistic carrying capacity, as explained earlier, net growth is zero. This must be true by definition; if the net growth were not zero, the population would be changing and thus would not be at its carrying capacity.

Figure 14.6*b* shows that the rate of growth changes between the smallest population and the carrying capacity and reaches a maximum between these extremes. *For a population growing exactly according to the logistic curve, the maximum rate of growth occurs when the population is exactly one-half its carrying capacity.* This point—a population size exactly one-half the carrying capacity—is therefore known as the point of maximum sustainable yield, or the MSY level.

It is sustainable only in the narrow context of the logistic curve, however.

If an elephant population really did grow according to the logistic curve, then to achieve the traditional management goal of maximizing population growth, the strategy would be to harvest enough individuals every year to bring the population to exactly one-half its carrying capacity. Then the population growth in the second year would be the same as in the first, and it would be the maximum that could be added by that species. In the third year, the same number could be harvested as in the second and first years, and the population would grow again by exactly the same amount. The process could continue year after year. Harvests would be constant and sustainable forever (Figures 14.6*a* and *b*).

For the MSY concept to be true, and therefore for successful management to be based on it, three things must hold:

1. The population must have an exact and single carrying capacity, and its growth must be determined exactly by the classical logistic curve.

2. It must be possible to know precisely both the carrying capacity and the present population size.

3. It must be possible to obtain complete cooperation from all who harvest the animal involved (for ele-

phants, this includes any leopards, lions, or hyenas that prey on young elephants) so that individual kills sum exactly the MSY level.

None of these conditions are met in the real world of wildlife management. If a population did grow according to the logistic growth curve, and measurements were subject to error, our management could easily lead to disaster for the population. Suppose we assume that an elephant population is at the MSY size but in fact it is slightly less. We have overestimated the population. Based on our overestimate, we assume that we can harvest at the calculated MSY level, and we proceed to do that. As illustrated in Figure 14.6(*a*), this leads to a disastrous situation; the population goes into a decline. Every year we harvest at the supposed MSY level, believing the population has returned to the calculated MSY population size. But each year, the population harvests exceed births, and the population size declines.

Even the slightest mistake in overestimating the MSY will push the population to a smaller size the next year. If we continue to maintain the same harvest levels, expecting the population to be at MSY, then we will drive the population to lower and lower levels and finally to extinction unless we recognize our error.

(see Chapter 2) and (2) we know now that the carrying capacity varies with changes in the environment.

The procedure in the past has been to estimate the carrying capacity by nonscientific means and then attempt to maintain the herd at one-half that level. This method requires accurate counts of the herd each year. It also requires that the environment not vary, or if it does, that it vary in a way that does not affect the population. Since these conditions cannot be met, the logistic curve has to fail as a basis for management of the deer herd. A Closer Look 14.4 illustrates in greater detail the problems of this method.

In spite of its lack of correspondence to the real world, the logistic growth curve has remained a mainstay of wildlife management. An interesting example can be found in the Marine Mammal Protection Act of 1972. This act states that its primary goal is to conserve "the health and stability of marine ecosystems," which is part of the modern approach. It then goes on to state that second only to that goal, the goal of the act is to maintain an "optimum sustainable population" of marine mammals. What is this? The wording of the act allows two interpretations. One is the logistic carrying capacity, and the other is the MSY population level of the logistic growth curve.

Chapter 2 pointed out that studies in the environmental sciences sometimes do not seem suited to the standard scientific method. This is true of some aspects of wildlife management and conservation. Several examples illustrate the needs and problems.

The Grizzly Bear

A classic example of wildlife management is the North American grizzly bear. An endangered species, the grizzly has been the subject of efforts on the part of the U.S. Fish and Wildlife Service to meet the requirements of the U.S. Endangered Species Act and restore the population of the grizzly. The grizzly is endangered from habitat destruction and direct hunting. It is arguably the most dangerous North American mammal, famous for its unprovoked attacks on people, and has been extirpated from much its range for that reason. At the same time, however, people like to watch grizzlies for recreation and in appreciation of the beauty and complexity of nature. Restoration is also justified on the basis that the grizzly has a moral right to exist.

The goal of restoring the grizzly at first glance seems simple enough, but then the question arises: Restore to what? It turns out that there is very little historic information on the abundance of the grizzly at the time of first European contact, so it is not simple to find a number of grizzlies (or a density in grizzlies per square kilometer) to select as a "restored" population.

Adding to this difficulty is the lack of a good estimate of the present abundance. The grizzly is difficult to study. It is large and dangerous and tends to be reclusive. The U.S. Fish and Wildlife Service attempted to obtain a count of the grizzlies in Yellowstone National Park by installing automatic flash cameras that were set off when the grizzlies took a bait. This seemed a good idea, but the grizzlies did not like the flash cameras and destroyed them, and counts turned out to be impossible by this method.

How do we obtain an estimate of a population from a time when nobody thought of counting its members? Where possible, we can make use of historic records, as discussed in Chapter 2. With the grizzly, a crude estimate of its abundance at the beginning of the 19th century can be obtained from the journals of the Lewis and Clark expedition. Lewis and Clark did not list the numbers of most wildlife they saw, simply writing down that they saw "many" bison, elk, and so forth. But the grizzlies were especially dangerous and tended to travel alone, so Lewis and Clark wrote down each encounter, stating the exact number they met.

On that expedition, Lewis and Clark saw 37 grizzly bears over a distance of approximately 1,000 miles.

Estimating based on their sightings, and making use of other sightings that suggest the original geographic range of the bear, we can make a crude estimate of the possible original abundance of the grizzly.

Lewis and Clark saw grizzlies from what is now near Pierre, South Dakota, to what is today Missoula, Montana. A northern and southern geographic limit to the grizzlies' range can be obtained from other explorers (Figure 14.7). Assuming Lewis and Clark could see a half-mile to each side of their line of travel on average, the density of the bears was approximately 3.7 per 100 square miles (their records were in miles, so for this example we will use the British system). If we estimate that the bear's geographic range was an area of 320,000 square miles in the mountain and western plains states, we arrive at a total population of ($320,000 \times 0.37$), or about 12,000 bears.

Suppose we phrase this as a hypothesis: "The number of grizzly bears in 1805 in what is now the United States was 12,000." Is this a statement open to disproof? Not without time travel. Therefore, it is not a scientific statement. It can only be taken as an educated guess or, more formally, as an assumption or premise. It has some basis in historical documents, and it is better than no information; we have few alternatives but to use this kind of information to determine "what used to be." We can use this assumption and create a plan to restore the grizzly to that abundance. But is that the best approach?

Another approach is to ask what is the **minimum viable population** of grizzly bears—forget completely about what might have been in the past and make use of modern knowledge of population dynamics and genetics, along with food requirements and potential production of that food. Using this approach, we could estimate how

Figure 14.7 • Grizzly bear. Records of bear sightings by Lewis and Clark have been used to estimate their population at the beginning of the 19th century.

many bears appear "safe." More precisely, we could phrase this statement as: How many bears are necessary so that the probability that the grizzly will become extinct in the next 10 years (or some other time period that we consider reasonable for planning) is less than 1 percent (or some other percentage that we would like)? With appropriate studies, this approach could have a scientific base. Consider a statement of this kind phrased as a hypothesis: "A population of 1,000 bears (or some other number) results in a 99% chance that at least one mature male and one mature female will be alive 10 years from today." We can disprove this statement, but only by waiting for the next 10 years to go by. Although it is a scientific statement, it is a difficult one to deal with in planning for the present.

American Bison

Another classic case of wildlife management (or mismanagement) is the demise of the American bison or buffalo (Figure 14.8a). The bison was brought close to extinction in the 19th century for two reasons. Bison were hunted for their skins because coats made of bison had become fashionable in Europe. Bison were also killed as part of warfare against the Indians in an effort to eliminate the Plains peoples (Figure 14.8b). In 1865, General Mitchell, in response to Indian attacks in the fall of 1864, set fires to drive away the Indians and the buffalo, killing vast numbers of animals.[27] Colonel R. I. Dodge was quoted in 1867 to have said, "Kill every buffalo you can. Every buffalo dead is an Indian gone."[27] The rapidity with which bison were almost eliminated was surprising—even to many of those involved in hunting bison.

Once again, it is difficult to obtain an estimate of the numbers of bison, as there was no established science of wildlife management, and few in the West were attempting to monitor the abundance of this or any other species. So once again we have to rely on written accounts that were not intended to be scientific.

All the early writers tell of immense herds of buffalo, but rarely do they count the buffalo to give us an estimate of their abundance. An exception was General Isaac I. Stevens, who, on July 10, 1853, was surveying for the transcontinental railway in North Dakota. He and his men climbed a high hill and saw "for a great distance ahead every square mile" having "a herd of buffalo upon it." He wrote that "their number was variously estimated by the members of the party—some as high as half a million. I do not think it any exaggeration to set it down at 200,000."[28]

One of the better attempts to estimate the number of buffalo in a herd was made by Colonel R. I. Dodge, who took a wagon from Fort Zarah to Fort Larned on the Arkansas River in May 1871, a distance of 34 miles. For at least 25 of those miles, he found himself in a "dark blanket" of buffalo. Dodge estimated that there were

(a)

(b)

Figure 14.8 • (a) A bison ranch in the United States. In recent years, interest in growing bison on ranches has increased greatly. In part, the goal is to restore bison to a reasonable percentage of its numbers before the Civil War. In part, bison are ranched because people like them. In addition, there is a growing market for bison meat and other products, including cloth made from bison hair. (b) Painting of a buffalo hunt by George Catlin in 1832–1833 at the mouth of the Yellowstone River.

480,000 in the mass of animals he saw in one day. At one point, he and his men traveled to the top of a hill, where he estimated that he could see six to ten miles, and from that high point there appeared to be a single solid mass of buffalo extending over 25 miles. At 10 animals per acre, not a particularly high density, the herd would have numbered between 2.7 and 8 million animals.[27]

In the fall of 1868, "a train traveled one hundred twenty miles between Ellsworth and Sheridan through a continuous, browsing herd, packed so thick that the engineer had to stop several times, mostly because the buffaloes would scarcely get off the tracks for the whistle and the belching smoke."[27] That spring, a train had been delayed for eight hours while a single herd passed "in one steady, unending stream." We can use such experiences to set bounds on the possible number of animals seen. At the largest extreme, we can assume that the train bisected a circular herd with a diameter of 120 miles. (Once again, we use the British measurement system, because those were the terms used in the 19th century.) Such a herd would cover 11,310 square miles, or more than 7 million acres. Suppose people exaggerated the density of the buffalo, and there were only 10 per acre—a moderate density for a herd. This single herd would have numbered 70 million animals!

Some might say that this estimate is probably too high, because the herd would more likely form a broad, meandering, migrating line, rather than a circle. The impression remains the same—there were huge numbers of buffalo in the American West even as late as 1868—numbering in the tens of millions and probably 50 million or more. Ominously, that same year, the Kansas Pacific Railroad advertised a "Grand Railway Excursion and Buffalo Hunt."[27] Some say that many hunters believed that the buffalo could never be brought to extinction, because there were so many. This belief was common about all living resources of America throughout the 19th century.

We tend to think that environmentalism is a social and political movement of the 20th century; but to the contrary, after the Civil War there were "angry protestations in every legislature" over the slaughter of the buffalo. In 1871, the U.S. Biological Survey sent George Grinnell to survey the herds along the Platte River. He estimated that only 500,000 buffalo remained there and that at the present rate of killing, the animals would not last long. As late as the spring of 1883, a herd of an estimated 75,000 crossed the Yellowstone River near Miles City, Montana, but fewer than 5,000 reached the Canadian border.[27] By the end of that year—only 15 years after the Kansas Pacific train was delayed for eight hours by a huge herd of buffalo—only a thousand or so buffalo could be found, 256 in captivity and about 835 roaming the plains. A short time later, there were only 50 buffalo wild on the plains.

The number of bison is increasing today. More and more ranchers are finding ways to maintain bison, and there is a growing market for bison meat and other products, along with a growing interest in reestablishing bison herds for aesthetic, spiritual, and moral reasons.

Once again, we can use historical records to get a general idea about the original number of bison and then use that estimate as a premise or assumption not directly open to standard scientific proof. But the history of the bison raises another question. Even with our crude estimates of the original abundances, the numbers would have varied from year to year. So what does it mean to "restore" bison? Not to a single number independent of the ability of its habitat to support the population, but to some range of abundances. How do we approach that problem and obtain an estimate of the range? This is a problem that is not completely solved and a challenge for the next generation of conservation scientists, perhaps including you who are reading this chapter.

Fisheries

Human beings began as hunter-gatherers, and some hunter-gatherer culture still exists. In the past, wildlife on the land provided a major source of food for hunter-gatherers. It is now a minor source of food for people of developed nations, although it still serves as a major food source for some indigenous peoples, such as the Eskimo. In contrast, even developed nations are still primarily hunter-gatherers in the harvesting of fish (see the discussion of aquaculture in Chapter 11).

People catch many species of fish and shellfish for food, but fewer than 20 species provide about 66% of the annual catch and 90% of the value. The worldwide harvest of fish is more than 122 million metric tons/year. Most of this (93.3 million metric tons) is from ocean catch of wild fish; aquaculture provides 28.8 million metric tons (see Chapter 11).[29] The contribution of fish to the human diet is 6.6% in North America, 9.7% in Western Europe, 21.1% in Africa, 8.2% in Latin America, 27.8% in the Far East, and 21.7% in Central Asia; in total, fish provide about 16% of the world's protein. In developing countries, fish are an especially important part of the diet.[30]

Marine catch has changed dramatically over the last 45 years. Marine harvest peaked in 1989 at 86 million metric tons (94.8 million tons). In 1993, it dropped to 84 million metric tons (92.6 million tons), although total global fish harvest continues to climb because of increased aquaculture production. The prices of most fish species continue to rise as harvests shrink.[30] Fish vary greatly in value, from tens of dollars to a thousand dollars per ton, depending on the species and the port at which they are landed. Fishing is an international trade, but a few countries dominate. Japan, China, Russia, Chile, and the United States are among the major fisheries nations.[31]

Commercial fisheries are concentrated in relatively few areas of the world's oceans (Figure 14.9). Continental shelves, which make up only 10% of the oceans, provide more than 90% of the fishery harvest. Fish are abundant where their food is abundant and, ultimately, where there is a high production of algae at the base of the food chain. Algae are most abundant in areas with relatively high concentrations of the chemical elements necessary for life, particularly nitrogen and phosphorus. These areas occur most commonly along the continental shelf, particularly in regions of wind-induced upwellings and sometimes quite close to shore.

Figure 14.9 • The world's major fisheries. Red areas are major fisheries; the darker the red, the greater the harvest and the more important the fishery. Most major fisheries occur in areas of ocean upwellings—locations where currents rise, bringing nutrient-rich waters from the depths of the ocean. Upwellings tend to occur near continents.

During the last 200 years, one fishery after another has been developed and has failed. The problem has continued during the last 30 years in spite of the rise of modern environmentalism, and a number of major fisheries have experienced failures (see Table 14.4).[32] Perhaps best known of the recent declines in North America are those of salmon of the west coast; the failure of the Grand Banks fisheries off of Newfoundland in the Atlantic Ocean; and the decline since 1950 by 95% of the oyster catch in the Chesapeake Bay (Figure 14.10).[33] The repeated failure of fisheries around the world makes this a global environmental problem.

14.8 Improved Approaches to Wildlife Management

The U.S. Council on Environmental Quality (an office within the executive branch of the federal government), the World Wildlife Fund United States, the Ecological Society of America, the Smithsonian Institution, and the IUCN have proposed four principles of wildlife conservation:

1. A safety factor in terms of population size, to allow for limitations of knowledge and the imperfections of procedures. An interest in harvesting a population

---

Table 14.4   Problems of Some Major Fisheries

Anchovy: Reached peak in 1970 (10 million metric tons), then declined.

Atlantic herring: Exploitation so high that recruitment was decreased.

Arcto-Norwegian cod: High fishing level followed by 4 years of poor recruitment.

Downs' stock of herring in the North Sea: Managers failed to grasp stock and recruitment problems.

North Atlantic haddock: Catch averaged 50,000 metric tons for many years; increased to 155,000 in 1965 and 127,000 in 1966, then fell to 12,000 in 1971–1974. In 1973, the International Commission for the Northwest Atlantic Fisheries (ICNAF) established a quota of 6,000 metric tons. Apparently, haddock could sustain a 50,000-metric-ton catch, but when this was tripled, the population was do decreased that only a smaller catch could be sustained.

Atlantic menhaden: Peak catch was 712,000 metric tons in 1956; in 1969, it was 161,400. Fisheries experts believe the drop was due to overfishing.

Salmon: Loss has occurred worldwide wherever salmon and their relatives once used fresh waters to spawn and rear. This has been the result of construction of dams on rivers, channelization of rivers, pollution, overharvesting, and habitat alteration of many kinds.

Pacific sardines: Declined catastrophically from the 1950s through the 1970s.

Source: D. Cushing, *Fisheries Resources of the Sea and Their Management* (London: Oxford University Press, 1975).

Figure 14.10 • An oyster fishery on the Chesapeake Bay.

should not allow the population to be depleted to some theoretical minimum size.

2. Concern with the entire community of organisms and all the renewable resources, so that policies developed for one species are not wasteful of other resources.

3. Maintenance of the ecosystem of which the wildlife are a part, minimizing risk of irreversible change and long-term adverse effects as a result of use.

4. Continual monitoring, analysis, and assessment. The application of science and the pursuit of knowledge about the wildlife of interest and its ecosystem should be maintained and the results made available to the public.

These principles broaden the scope of wildlife management from the narrow focus on a single species to inclusion of the ecological community and ecosystem. They call for a safety net in terms of population size, meaning that no population should be held at exactly the MSY level or reduced to some theoretical minimum abundance. These new principles provide a starting point for an improved approach to wildlife management, to which we can add several other suggestions.

Time Series and Historic Range of Variation

As the history of the buffalo illustrates, we would like to have an estimate of population over a number of years. This set of estimates is called a **time series** and could provide us with a measure of the **historic range of variation**. Such records exist for few species. One is the American whooping crane (Figure 14.11). Because this species became so rare, and because it migrated as a single flock, people have begun counting it since the late 1930s. At that time, there were only 14 whooping cranes. Not only was the total number counted but also the number born

that year. From this, the number dying can be calculated. And from these numbers, it is possible to estimate the probability of extinction. At present, in spite of the importance, mathematical estimates of the probability of extinction have been done for just a handful of species.

Age Structure as Useful Information

An additional key to successful management of wildlife is monitoring the age structure of the population (see Chapter 5). Age structure provides many different kinds of information. For example, the age structure of the catch of salmon from the Columbia River in Washington for two different periods: 1941–1943 and 1961–1963 is quite different. In the first period, most of the catch consisted of four-year-olds (60% of the catch); three-year-olds and five-year-olds each made up about 15% of the population. Twenty years later, in 1961 and 1962, half the catch consisted of three-year-olds, and the number of five-year-olds had declined to about 8%. The total catch had declined considerably also. During the period 1941–1943, 1.9 million fish were caught; while during the second period, 1961–63, the total catch dropped to 849,000, 49% of the total in the earlier period.

The shift in catch toward younger ages, along with an overall decline in catch, suggests that the fish are exploited to a point at which they are not reaching older ages. Such a shift in the age structure of a harvested population is a good early sign of overexploitation and of a need to alter allowable catches. This applies to all wildlife, on land and in waters.

Catch Per Unit Effort

Another method of estimating animal populations is to use the number harvested. One approach uses the number harvested as an index of a minimum abundance. Records of the number of buffalo killed were neither organized nor all that well kept, but enough are available to give us some idea of the number taken. In 1870, about two million buffalo were killed. In 1872, one company in Dodge City handled 200,000 hides. Estimates based on the sum of reports from such companies, together with guesses at how many more animals would have been taken by small operators and not reported, suggest that about 1.5 million hides were shipped in 1872 and again in 1873.[28] In these years, buffalo hunting was the main economic activity in Kansas. The Indians were also killing large numbers of buffalo for their own use and for trade. Estimates range to 3.5 million buffalo killed per year during the 1870s.[27]

Another way in which harvest counts are used to estimate previous animal abundance is called the **catch per unit effort**. The assumption of this method is that the same effort is exerted by all hunters/harvesters per unit of time. So if you know the total time spent in hunting/harvesting and you know the catch per unit of effort, you can estimate the total population. The method has

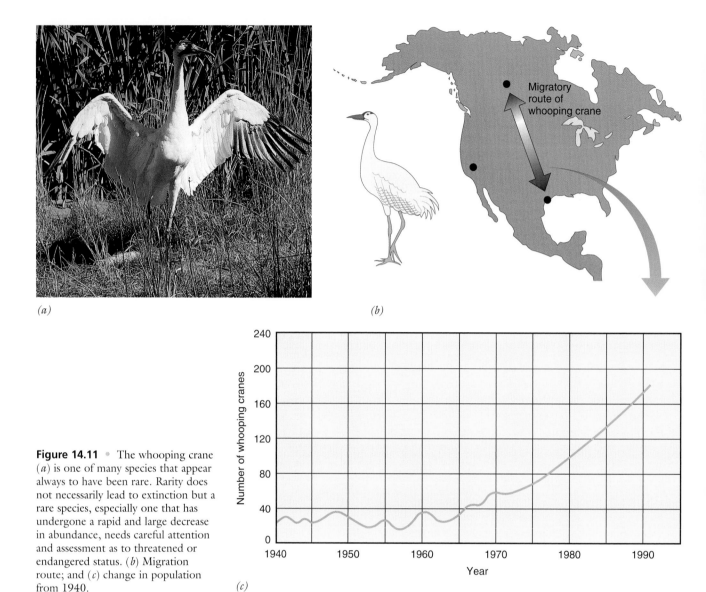

(a)

(b)

Figure 14.11 • The whooping crane (*a*) is one of many species that appear always to have been rare. Rarity does not necessarily lead to extinction but a rare species, especially one that has undergone a rapid and large decrease in abundance, needs careful attention and assessment as to threatened or endangered status. (*b*) Migration route; and (*c*) change in population from 1940.

(c)

major limitations and leads to rather crude estimates with a large observational error; but where there is no other source of information, it can provide unique insights.

An interesting application of this method is the reconstruction of the harvest of the bowhead whale (Figure 14.12*a*) and, from that, an estimate of the total bowhead population. Taken traditionally by Eskimos, the bowhead was the object of "Yankee" or American whaling (Figure 14.12*b*) from 1820 until the beginning of World War I. (See A Closer Look 14.1 for a general discussion of marine mammals.) Every ship's voyage was recorded, so we know essentially 100% of all ships that went out to catch bowheads. In addition, on each ship a daily log was kept, with records including the number of whales caught and their size in terms of barrels of oil; the sea conditions; the visibility; and the ice conditions. Of these logbooks, 20% still exist, and their entries have been computerized. Using some crude statistical techniques, it was possible to

estimate the abundance of the bowhead in 1820, which was 20,000 plus or minus 10,000. The total catch of whales and the catch for each year, and therefore the history of the entire whaling of this species, was obtained.

14.9 The Kirtlands Warbler and Environmental Change

Many endangered species are adapted to natural environmental change and require it. When human actions remove that change, a species can become threatened with extinction. This happened with the Kirtlands warbler, which nests in jack pine forests in Michigan (Figure 14.13). In 1951, a survey was made of the Kirtlands warbler, making it the first songbird in the United States to be subject to a complete census. About 400 nesting males were found. Concern about the species grew in the 1960s

(a)

(b)

Figure 14.12 • (a) A bowhead whale. (b) A 19th-century whaling ship.

and increased when only 201 nesting males were found in the third census, in 1971.[34] Conservationists and scientists began to try to understand what was causing the decline, which threatened the species with extinction.

Kirtlands warblers are known to nest only in jack pine woodlands that are between 6 and 21 years old. At these ages, the trees, 5 to 20 feet tall, retain dead branches at ground level. Jack pine is a "fire species," which persists only where there are periodic forest fires. Cones of the jack pine open only after they have been heated by fire. Furthermore, the trees are intolerant of shade, able to

grow only when their leaves can reach into full sunlight; so even if seeds were to germinate under mature trees, the seedlings could not grow in the shade and would die. Jack pine produces an abundance of dead branches that promote fires, which is interpreted by some as an evolutionary adaptation to promote those conditions most conducive to the survival of the species.

The Kirtlands warbler thus requires change at rather short intervals—forest fires approximately every 20 to 30 years, which was about the frequency of fires in jack pine woods in presettlement times.[35] At the time of the first

(a)

(b)

Figure 14.13 • (a) A Kirtlands warbler and (b) its jack pine habitat.

European settlement of North America, jack pine may have covered a large area in what is now Michigan. Even as recently as the 1950s, the pine was estimated to cover nearly 500,000 acres in the state. Jack pine, a small, poorly formed tree, was considered a trash species to the commercial loggers and was left alone; but many large fires followed the logging operations when large amounts of slash—branches, twigs, and other economically undesirable parts of the trees—were left in the woods. Elsewhere, fires were set in jack pine areas to clear them and promote the growth of blueberries. Some experts think that the population of Kirtlands warblers peaked in the late 19th century as a result of these fires. After 1927, fire suppression became the practice, and control of forest fires reduced the area burned and the size of individual fires. Where possible, it was the practice to encourage the replacement of jack pine with economically more useful species. These actions reduced the areas conducive to the nesting behavior of the warbler.[36]

Although it may seem obvious today that the warbler requires forest fire, this was not always understood. In 1926, one expert wrote that "fire might be the worst enemy of the bird."[37] Only with the introduction of controlled burning after vigorous advocacy by conservationists and ornithologists was habitat for the warbler maintained. The Kirtlands Warbler Recovery Plan, published by the Department of the Interior and the Fish and Wildlife Service in 1976 and updated in 1985, calls for the creation of 38,000 acres of new habitat for the warbler for which "prescribed fire will be the primary tool used to regenerate non-merchantable jack-pine stands on poor sites."[34,38]

(a)

(b)

14.10 Ecological Islands and Endangered Species

The history of the Kirtlands warbler illustrates that a species may inhabit "ecological islands," which the isolated jack pine stands of the right range of ages are for that bird. Recall from our discussion in Chapter 8 that an ecological island is an area that is biologically isolated so that a species occurring within the area cannot mix (or only rarely mixes) with any other population of the same species (Figure 14.14). Mountaintops and isolated ponds are ecological islands. Geographic, or real, islands are also ecological islands. Insights gained from studies of the biogeography of islands have important implications for the conservation of endangered species and for the design of parks and preserves for biological conservation.[39]

Almost every park is a biological island for some species. A small city park occupying a square between streets may be an island for trees and squirrels. At the other extreme, even a large national park is an ecological island. For example, the Masai Mara Game Reserve in the Serengeti Plain, which stretches from Tanzania to Kenya in East Africa, and other great wildlife parks of eastern and southern Africa are becoming islands of natural land-

(c)

Figure 14.14 • Ecological islands: (*a*) Central Park in New York City; (*b*) a mountaintop in Arizona where there are big-horn sheep; (*c*) an African wildlife park.

scape surrounded by human settlements. Lions and other great cats exist in these parks as isolated populations, no longer able to roam completely freely and to mix over large areas. Other examples are islands of uncut forests left by logging operations and oceanic islands, where intense fishing has isolated parts of fish populations.

What is a sufficiently large ecological island to guarantee the survival of a species? The size varies with the species but can be estimated. Some islands that seem large to us are too small for species we wish to preserve. For example, a preserve was set aside in India in an attempt to reintroduce the Indian lion into an area where it had been eliminated by hunting and changing patterns of land use. In 1957, a male and two females were introduced into a 95-km^2 (36-mi^2) preserve in the Chakia forest, known as the Chandraprabha Sanctuary.[62] The introduction was carried out carefully and the population counted annually. There were 4 lions in 1958, 5 in 1960, 7 in 1962, and 11 in 1965, after which they disappeared and were never seen again.

Why did they disappear? Although 95 km^2 seems large to us, male Indian lions have territories of 130 km^2 (50 mi^2). Within such a territory, females and young also live. A population that could persist for a long time would need a number of such territories, so an adequate preserve would require 640 to 1,300 km^2 (247 to 500 mi^2). Various other reasons were suggested for the disappearance of the lions, including poisoning and shooting by villagers; but regardless of the immediate cause, a much larger area than was set aside was required for long-term persistence of the lions.

14.11 Using Spatial Relationships to Conserve Endangered Species

The Redcockaded woodpecker (Figure 14.15a) is an endangered species of the American southeast. One of its foods is the pine bark beetle. The woodpecker makes its nests in old dead or dying pines. To conserve this species, these pines must be preserved. But the old pines also serve as a home to the beetles, which are pests to the trees and harm them for commercial logging. This presents an intriguing problem: How can we maintain the woodpecker and its food (the bark beetle, Figure 14.15b) and also maintain productive forests?

The classic 20th century way to view the relationship between the pines, the bark beetle, and the woodpecker would be to show a food chain (see Chapter 6). But this alone does not solve the problem for us.

A newer approach is to consider the habitat requirements of the pine bark beetle and the woodpecker. Their requirements are somewhat different. Maps of habitat requirements for each can be overlain on each other, and the co-occurrence of habitats can be compared. Beginning with such maps, it becomes possible to conceive of a landscape design that would allow the maintenance of all three: pines, beetles, and birds.

Summary

- The management and conservation of wildlife, whatever its status (abundant, threatened, endangered, or overabundant), are best viewed in terms of a whole, because species can change in status. Modern approaches to management and conservation of wildlife use a broad perspective that includes interactions among species and an ecosystem and landscape context.

- To successfully manage wildlife for harvest, conservation, and protection from extinction, a set of information is desirable. We would like to have measures of total abundance and of births and deaths over a long time period. We also need to know the habitat, viewed spatially and in quantitative terms. Age structure and other characteristics of a population can help in management, forecasting, and conservation. However, it is often difficult to obtain these data, especially for historic information.

- A common goal of current conservation of wildlife is to "restore" the abundance of a species to some previous condition, usually to numbers thought to have existed prior to the influence of modern technological civilization. Information about such numbers is rarely available. Sometimes, numbers can be estimated indirectly, as in the use of the Lewis and Clark journals to reconstruct the 1805 population of grizzly bears or the logbooks from ships hunting the bowhead whale. Adding to the complexity, the abundance of wildlife changes over time even in natural systems uninfluenced by modern technological civilization. And historic information often cannot be subjected to formal tests of disproof

Figure 14.15 • (*a*) Endangered redcockaded woodpecker, and (*b*) the bark beetle, the food for the woodpecker.

(*a*)

(*b*)

Should Wolves be Reestablished in the Adirondack Park?

With an area slightly over 24,000 km², the Adirondack Park in northern New York is the largest park in the lower 48 states. Unlike most parks, however, it is a mixture of private (60%) and public (40%) land and home to 130,000 people. When colonists from Europe first came to the area, it was, like much of the rest of North America, inhabited by gray wolves. By 1960, wolves had been exterminated in all of the lower 48 states except for northern Minnesota. The last official sighting of a wolf in the Adirondacks was in the 1890s.

Although the gray wolf was not endangered on a global level—there were more than 60,000 in Canada and Alaska—it was one of the first animals listed as endangered under the 1972 Endangered Species Act. As required, the United States Fish and Wildlife Service developed a plan for recovery that included protection of the existing population and reintroduction of wolves to wilderness areas. The recovery plan was to be considered a success if survival of the Minnesota wolf population was assured and at least one other population of more than 200 wolves had been established at least 320 km from the Minnesota population.

Under the plan, the wolf population in Minnesota increased, and some individuals from that population, as well as others from southern Canada, dispersed into northern Michigan and Wisconsin, both of which had populations of approximately 100 wolves in 1998. Also, 31 wolves from Canada were introduced into Yellowstone National Park in 1995, where the population increased to over 100. By the end of 1998, it seemed fairly certain that the criteria for removing the wolf from the endangered species list would soon be met.

In 1992, when the fate of the recovery was still uncertain, the Fish and Wildlife Service proposed to investigate the possibility of reintroducing wolves to northern Maine and the Adirondack Park. A survey of New York State residents in 1996 funded by Defenders of Wildlife found that 76% of those living in the park supported reintroduction. However, many residents and organizations within the park vigorously opposed reintroduction and questioned the validity of the survey. Concerns expressed focused primarily on the potential dangers to humans, livestock, and pets and the possible impact on the deer population. In response to the public outcry, Defenders of Wildlife established a citizens' advisory committee that initiated two studies by outside experts, one on the social and economic aspects of reintroduction and another on whether there was sufficient prey and suitable habitat for wolves.

The primary prey of wolves are moose, deer, and beaver. In recent years, moose have been returning to the Adirondacks and now number about 40, but this is far less than the moose population in areas where wolves are successfully reestablishing. Beaver are abundant in the Adirondacks, with an estimated population of over 50,000. Because wolves feed on beaver primari-

ly in the spring and the moose population is small, the main food source for Adirondack wolves would be deer.

Deer thrive in areas of early successional forest and edge habitats, both of which have declined in the Adirondacks as logging has decreased on private forest land and been eliminated altogether on public lands. Furthermore, the Adirondacks are at the northern limit of the range for white-tailed deer, where harsh winters can result in significant mortality. Deer density in the Adirondacks, which is estimated at 3.25/km², is less than that found in wolf habitat in Minnesota, which also has 8,500 moose. If deer were the only prey available, wolves would kill between 2.5% and 6.5% of the deer population, while hunters take approximately 13% each year. Determining whether there is a sufficient prey base to support a population of wolves is complicated by the fact that coyotes have moved into the Adirondacks and occupy the niche once filled by wolves. Whether wolves would add to the deer kill or replace coyotes, with no net impact on the deer population, is difficult to predict.

An area of 14,000 km² in various parts of the Adirondack Park meets criteria established for suitable wolf habitat, but this is about half of the area required to maintain a wolf population for the long term. Based on the average deer density and weight, as well as the food requirements of wolves, biologists estimate that this habitat could support about 155 wolves. Because human communities are scattered throughout much of the park, many residents are concerned that wolves would not remain on public land and would threaten local residents as well as backcountry hikers and hunters. Also, private lands around the edges of the park, with their greater density of deer, dairy cows, and people, could attract wolves.

Critical Thinking Questions

1. Who should make decisions about wildlife management, such as returning wolves to the Adirondacks—scientists, government officials, or the public?

2. Some people advocate leaving the decision to the wolves—that is, waiting for them to disperse from southern Canada and Maine into the Adirondacks. Study a map of the northeastern United States and southeastern Canada. What is the likelihood of natural recolonization of the Adirondacks by wolves?

3. Do you think that wolves should be reintroduced to the Adirondack Park? If you lived in the park, would that affect your opinion? How would removal of the wolf from the endangered species list affect your opinion?

4. Recently, some biologists have concluded that wolves in Yellowstone and the Great Lakes region belong to a different subspecies (Rocky Mountain timber wolf) than those

that formerly lived in the northeastern United States (eastern timber wolf). This means that the eastern timber wolf is still extinct in the lower 48 states. Would this affect your opinion on reintroduction of wolves to the Adirondacks?

and therefore does not by itself qualify as scientific. Adequate information exists for relatively few species.

- Another approach is to seek a minimum viable population, a carrying capacity, or an optimal sustainable population or harvest based on data that can be obtained and tested today. This approach abandons the goal of restoring a species to some hypothetical past abundance.

- The good news is that many species once endangered have been successfully restored to an abundance that suggests they are unlikely to become extinct. Success is achieved when the habitat is restored to conditions required by a species. The conservation and management of wildlife presents great challenges but also offers great rewards of long-standing and deep meaning to people.

REEXAMINING THEMES AND ISSUES

Human Population:

Today, human beings are a primary cause of the extinction of species. People also contributed to extinctions in the past. Nonindustrial societies have caused extinction by such methods as hunting and the introduction of exotic species into new habitats. With the age of exploration in the Renaissance and with the Industrial Revolution, the rate of extinctions increased greatly. People altered habitats more rapidly and over greater areas. Hunting efficiency increased, as did the introduction of exotic species into new habitats. As the human population grows, conflicts for habitat between people and wildlife will be increased. Once again, we find that the human population problem underlies this environmental issue.

Sustainability:

At the heart of issues concerning wild living resources is the question of sustainability of species and the ecosystems of which they are a part. One of the key questions is whether we can sustain these resources at a constant abundance. In general, it has been assumed that fisheries and wildlife hunted for recreation, such as deer, could be maintained at some constant, highly productive level. Constant production is desirable economically because it would provide a reliable, easily forecast income each year. But in spite of direct management attempts, few wild living resources have remained at constant levels. New ideas about the intrinsic variability of ecosystems and populations lead us to question the assumption that such resources can or should be maintained at constant levels.

Global Perspective:

Although the final extinction of a species takes place in one locale, the problem of biological diversity and the extinction of species is global because of the worldwide increase in the rate of extinction and because of the growth of the human population and its effects on wild living resources.

Urban World:

Although we have tended to think of wild living resources as existing outside of cities, there is a growing recognition that urban environments will be more and more important in the future in the conservation of biological diversity. This is in part because cities now occupy many sensitive habitats around the world, such as coastal and inland wetlands. It is also because appropriately designed parks and backyard plantings can provide habitats for some endangered species. As the world becomes increasingly urbanized, this function of cities will grow more important.

Science and Values:

The reasons that people want to save endangered species begin with human values, including values placed on the continuation of life and on the public service functions of ecosystems. Among the most controversial environmental issues in terms of values are the conservation of biological diversity and the protection of endangered species. Science tells us what is possible in terms of how to conserve species, which species are likely to persist, and which are not. Ultimately, therefore, our decisions about where to spend our efforts in sustaining wild living resources depend on our views of these values.

● KEY TERMS

catch per unit effort **275**
global extinction **265**
historic range of variation **275**
local extinction **265**

logistic carrying capacity **265**
maximum sustainable population **265**
maximum sustainable yield **266**

minimum viable population **271**
optimum sustainable population **265**
time series **275**

● STUDY QUESTIONS

1. Why are we so unsuccessful in making rats an endangered species?

2. What have been the major causes of extinction (a) in recent times and (b) before people existed on Earth?

3. Describe a feasible plan for restoring the grizzly bear.

4. What are some arguments for and against the use of historic records that are not necessarily scientific to establish a "restored" abundance?

5. Five justifications for preserving endangered species were discussed in this chapter. Which of them apply to the following? (You can decide that none apply.)
   a. The black rhinoceros of Africa.
   b. The furbish lousewort, a rare, small flowering plant of New England, seen by few people.
   c. An unnamed and newly discovered beetle from the Amazon rain forest.
   d. Smallpox.
   e. Wild strains of potatoes in Peru.
   f. The North American bald eagle.

6. Locate an ecological island close to where you live and visit it. Which species are most vulnerable to local extinction?

7. Visit a local zoo or botanical garden. What activities are conducted there to promote biological conservation? Are these activities likely to be of much benefit?

8. What are the primary reasons that a single-species approach to conservation and management of wildlife did not succeed?

9. Using information available in libraries, determine the minimum area required for a minimum viable population of the following.
   a. Domestic cat.
   b. The cheetah.
   c. The American bison.
   d. Swallowtail butterflies.

10. Both a ranch and a preserve will be established for the North American bison. The goal of the ranch owner is to show that bison can provide a better source of meat than introduced cattle and at the same time have a less detrimental effect on the land. The goal of the preserve is to maximize the abundance of the bison. How will the plans for the ranch and preserve differ, and how will they be similar?

● FURTHER READING

Botkin, D. B. 2001. *No Man's Garden: Thoreau and a New Vision for Civilization and Nature.* New York: Island Press. A work that discusses Deep Ecology and its implications for biological conservation, as well as reasons for the conservation of nature, both scientific and beyond science.

Cadieux, C. L. 1991. *Wildlife Extinction.* Washington, D.C.: Stone Wall Press. An engaging overview of the issues surrounding endangered wildlife. Case studies are presented, along with lists of endangered wildlife and key players in their protection.

Caughley, G., and A. R. E. Sinclair. 1994. *Wildlife Ecology and Management.* London: Blackwell Scientific. A valuable textbook based on new ideas of wildlife management.

Ehrlich, P. R., and E. O. Wilson, eds. 1988. *Biodiversity.* Washington, D.C.: National Academy Press. An excellent compilation of papers providing an overview of biological diversity and the threats that can lead to extinction.

Ferry, Luc. 1995. *The New Ecological Order.* Chicago: University of Chicago Press. Translated by Carol Volk from the original French. An important book in environmental philosophy. It includes a fascinating discussion of Deep Ecology and its implications for biological conservation, as well as views of nature from medieval times to the present.

McCullough, D. R. 1996. *Metapopulations and Wildlife Conservation.* Washington, D.C.: Island Press. Introduction to a new geographically based approach to wildlife conservation, including the idea that isolated patches, such as parks and preserves, can help stabilize wildlife populations.

Talbot, L. 1996. "Living Resource Conservation: An International Overview." *Ecological Applications.* A description of how the conservation of wild living resources is actually being done today, based on interviews and conferences involving more than 400 scientists and policy makers.

15 Environmental Health, Pollution, and Toxicology

Children playing in the South Bronx of New York City. Lead in the soil, a legacy of using lead in gasoline and lead in the paint of old buildings, remains a hazard to children.

● LEARNING OBJECTIVES

Serious environmental health problems and diseases may arise from toxic elements in water, air, soil, and even the rocks on which we build our homes. After reading this chapter, you should understand:

- How the terms *toxic, pollution, contamination, carcinogen, synergism,* and *biomagnification* are used in environmental health.
- What the classification and characteristics are of major groups of pollutants in environmental toxicology.
- Why there is controversy and concern about synthetic organic compounds such as dioxin.
- Whether we should be concerned with exposure to human-produced electromagnetic fields.
- What the dose-response concept is and how it relates to LD-50, TD-50, ED-50, ecological gradients, and tolerance.
- How the process of biomagnification works and why it is important in toxicology.
- Why the threshold effects of environmental toxins are important.
- What is the process of risk assessment in toxicology and why such processes are often difficult and controversial.

Is Lead in the Urban Environment Contributing to Antisocial Behavior?

Lead is one of the most common toxic (harmful or poisonous) metals in our inner-city environments, and it may be linked to delinquent behavior in children. Lead is found in all parts of the urban environment (air, soil, older pipes, and some paint, for example) and in biological systems, including people (Figure 15.1). There is no apparent biologic need for lead, but it is sufficiently concentrated in the blood and bones of children living in inner cities to cause health and behavior problems. In some populations, over 20% of the children have blood concentrations of lead that are higher than those believed safe.[1]

Lead affects nearly every system of the body. Acute lead toxicity may be characterized by a variety of symptoms, including anemia, mental retardation, palsy, coma, seizures, apathy, uncoordination, subtle loss of recently acquired skills, and bizarre behavior. Lead toxicity is particularly a problem for young children, who apparently are more susceptible to lead poisoning than are adults. Following acute toxic response to lead, some children manifest aggressive, difficult-to-manage behavior.[1–3]

The occurrence of lead toxicity or lead poisoning has cultural, political, and sociological implications. Over two thousand years ago, the Roman Empire produced and used tremendous amounts of lead for a period of several hundred years. Production rates were as high as 55,000 metric tons per year. Romans had a wide variety of uses for lead. Lead was used in pots in which grapes were crushed and processed into a syrup for making wine, in cups and goblets from which wine was drunk, and as a base for cosmetics and medicines. In the homes of Romans wealthy enough to have running water, lead was used to make the pipes that carried the water. It has been argued that lead poisoning among the upper class in Rome was partly responsible for Rome's decline.

Lead poisoning probably resulted in widespread stillbirths, deformities, and brain damage. Studies analyzing the lead content of bones of ancient Romans tend to support this hypothesis.[4] More recently, the occurrence of lead has been studied in glacial ice cores from Greenland. Glaciers have an annual growth layer of ice. Older layers are buried by younger layers, allowing us to identify the age of each layer. Researchers drill glaciers, taking continuous samples of the layers that look like long solid rods of glacial ice; these are called cores. Measurements of the concentration of lead from cores show that during the Roman period, from approximately 500 B.C. to A.D. 300,

lead concentrations in the glacial ice are about four times higher than before and after this period. This suggests that the mining and smelting of lead in the Roman Empire added small particles of lead to the atmosphere that eventually settled out in the glaciers of Greenland.[5]

Lead toxicity, then, seems to have been a problem for a long time. Now, an emerging, interesting, and potentially significant hypothesis is that, in children, lead concentrations below the levels known to cause physical damage may be associated with an increased potential for antisocial, delinquent behavior. This is a testable hypothesis. (See Chapter 2 for a discussion of hypotheses.) If the hypothesis is correct, then some of our urban crime may be traced to environmental pollution!

A recent study[1] in children aged 7 to 11 years old measured the amount of lead in bones and compared it with data concerning behavior over a four-year period. The study concluded that an above-average concentration of lead in children's bones was associated with an increased risk for attention deficit disorder, aggressive behavior, and delinquency. The study took into account factors such as maternal intelligence, socioeconomic status, and quality of child rearing.

Figure 15.1 • The lead in urban soils (a legacy of our past use of lead in gasoline) is still concentrated where children are likely to play. Lead-based paint in older buildings, such as these in New York, also remains a hazard to young children, who sometimes ingest flakes of paint.

The data and analysis presented in this case study can be used to argue that exposure of children to lead—a preventable occurrence—should be included among the multivariant factors that contribute to antisocial and delinquent behavior.[1] The environmental significance is that although we have often considered social unrest, antisocial behavior, and crime to be related to social and economic factors, they may in fact be due, in part, to environmental factors related to pollution of inner-city environments. Thus, we see again that many environmental issues are found in the urban environment. In this chapter, we discuss a wide variety of environmental issues, problems, and solutions linked to environmental health and pollution.

15.1 Some Basics

Disease is often due to an imbalance resulting from a poor adjustment between the individual and the environment.[6] However, disease seldom has a one-cause, one-effect relationship with the environment. Rather, the incidence of a disease depends on several factors, including physical environment, biological environment, and lifestyle. Linkages between these factors are often related to other factors, such as local customs and the level of industrialization. More primitive societies that live directly off the local environment are usually plagued by different environmental health problems than is our urban society. Industrial societies have nearly eliminated such diseases as cholera, dysentery, and typhoid. However, we are more likely to suffer from chronic and acute diseases such as respiratory problems and cancer.

People are often surprised to learn that the water we drink, the air we breathe, the soil in which we grow crops, and the rocks on which we build our homes and places of work may affect our chances of experiencing serious environmental health problems and diseases (although, as suggested, direct causative relationships between the environment and disease are difficult to establish). At the same time, the environmental factors that contribute to disease—soil, rocks, water, and air—can also influence our chances of living a longer, more productive life.

Many people believe that soil, water, or air in a so-called natural state must be good and that if human activities have changed or modified them, they have become contaminated, polluted, and therefore bad.[7] This is by no means the entire story; many natural processes, including dust storms, floods, and volcanic processes, can introduce materials harmful to humans and other living things into the soil, water, and air.

A tragic example occurred on the night of August 21, 1986, when there was a massive natural release of carbon dioxide (CO_2) gas from Lake Nyos in Cameroon, Africa. The carbon dioxide was probably initially released from volcanic vents at the bottom of the lake and accumulated there with time. Pressure of the overlying lake water normally keeps the dissolved gas at the bottom of the lake. However, the water was evidently agitated by a slide or small earthquake, and the bottom water moved upward. When the CO_2 gas reached the surface of the lake, it was released quickly into the atmosphere. The CO_2 gas, which is heavier than air, flowed downhill from the lake and settled in nearby villages, killing many animals and more than 1,700 people by asphyxiation (Figure 15.2). It was estimated that a similar event could recur within about 20 years, assuming that carbon dioxide continued to be released at the bottom of the lake.[8] Fortunately, a project funded by the U.S. Office of Foreign Disaster Assistance Hazards Reduction Project (scheduled to be completed early in the 21st century) is inserting pipes into the bottom of Lake Nyos. The gas-rich water is pumped to the surface where it is safely discharged into the atmosphere. In 2001 a warning system was installed, and one degassing pipe released a little more CO_2 than was seeping naturally into the lake. At least five more pipes will be necessary to minimize the hazard.

Terminology

What do we mean when we use the terms *pollution, contamination, toxic,* and *carcinogen*? A polluted environment is one that is impure, dirty, or otherwise unclean. The term **pollution** refers to the occurrence of an unwanted change in the environment caused by the introduction of harmful materials or the production of

(a)

(b)

Figure 15.2 • (*a*) In 1986, Lake Nyos in Cameroon, Africa, released carbon dioxide that moved down the slopes of the hills to settle in low places, asphyxiating animals and people. (*b*) Animals asphyxiated by carbon dioxide.

harmful conditions (heat, cold, sound). **Contamination** has a meaning similar to that of pollution and implies making something unfit for a particular use through the introduction of undesirable materials—for example, the contamination of water by hazardous waste. The term **toxic** refers to materials (pollutants) that are poisonous to people and other living things. **Toxicology** is the science that studies chemicals that are known to be or could be toxic, and toxicologists are scientists in this field. A **carcinogen** is a particular kind of toxin that increases the risk of cancer. Carcinogens are among the most feared and regulated toxins in our society.

An important concept in considering pollution problems is **synergism**, the interaction of different substances resulting in a total effect greater than the sum of the effects of the separate substances. For example, both sulfur dioxide and coal dust particulates are air pollutants. Either one taken separately may cause adverse health effects, but when they combine, as when sulfur dioxide (SO_2) adheres to the coal dust, the dust with SO_2 is inhaled deeper than sulfur dioxide alone, causing greater damage to lungs. Another aspect of synergistic effects is that the body may be more sensitive to a toxin if it is simultaneously subjected to other toxins.

Pollutants are commonly introduced into the environment by way of **point sources**, such as smokestacks (see A Closer Look 15.1), pipes discharging into waterways (Figure 15.3), or accidental spills. **Area sources**, also called *non-point sources*, are more diffused over the land and include urban runoff and **mobile sources** such as automobile exhaust. Area sources are difficult to isolate and correct because the problem is often widely dispersed over a region, as in agricultural runoff that contains pesticides (Chapter 21).

Measuring the Amount of Pollution

How the amount or concentration of a particular pollutant or toxin present in the environment is reported varies widely. The amount of treated wastewater entering Santa Monica Bay in the Los Angeles area is a big number reported in millions of gallons per day. Emission of nitrogen and sulfur oxides into the air is also a big number reported in millions of tons per year. Small amounts of pollutants or toxins in the environment, such as pesticides, are reported in units as parts per million (ppm) or parts per bil-

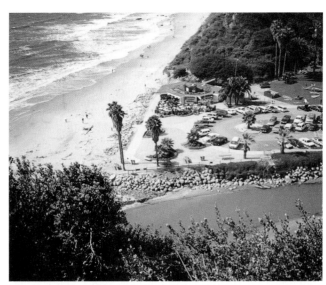

Figure 15.3 • This southern California urban stream flows into the Pacific Ocean at a coastal park. The stream water often carries high counts of fecal coliform bacteria. As a result, the stream is a point source of pollution for the beach, which is sometimes closed to swimming following runoff events.

Sudbury Smelters: A Point Source

A famous example of a point source of pollution is provided by the smelters that refine nickel and copper ores at Sudbury, Ontario. Sudbury contains one of the world's major nickel and copper ore deposits. A number of mines, smelters, and refineries lie within a small area. The smelter stacks at one time released large amounts of particulates containing nickel, copper, and other toxic metals into the atmosphere. In addition, because the areas contain a high percentage of sulfur, the emissions included large amounts of sulfur dioxide (SO_2). During its peak output in the 1960s, this complex was the largest single source of sulfur dioxide emissions in North America.

In 1969, regulations were mandated to improve local air quality, forcing a reduction in emissions. Concentrations of sulfur dioxide were reduced locally by more than 50% after 1972. However, attempts to minimize the pollution problem in the immediate vicinity of the smelting operation by increasing smokestack height spread the problem as wind carried the pollutants greater distances. In order to better control emissions from Sudbury, the Ontario government set standards to reduce emissions to less than 365,000 tons per year by 1994 (about 14% of earlier emissions of 2,560,000 tons per year). The goal was achieved by reducing production from the smelters and by treating the emissions to reduce pollution.[9]

As a result of years of pollution, nickel has been found to contaminate soils 50 km (about 31 mi) from the stacks. The forests that once surrounded Sudbury were devastated by decades of acid rain (produced from SO_2 emissions) and the deposition of particulates containing heavy metals. An area of approximately 250 km² (96 mi²) is nearly devoid of vegetation, and damage to forests in the region is visible over an area of approximately 3,500 km² (1,350 mi²; see Figure 15.4). Secondary effects, in addition to loss of vegetation, include soil erosion and drastic changes in soil chemistry resulting from the influx of the heavy metals.

Reductions in emissions from Sudbury have allowed surrounding areas to slowly begin to recover from these effects. Species of trees once eradicated from some areas have begun to grow again. Lakes that were damaged due to acid precipitation in the area are rebounding and now support populations of plankton and fish.[9] The case of the Sudbury smelters thus provides a positive example of pollution reduction, emphasizing the key theme of thinking globally but acting locally to reduce air pollution. It also illustrates the theme of science and values: Scientists and engineers can design pollution abatement equipment, but spending the money to purchase the equipment reflects what value we place on clean air.

Figure 15.4 ● Lake St. Charles, Sudbury, Ontario. Note high stacks (smelters) in the background and lack of vegetation in the foreground resulting from air pollution (acid and heavy-metal deposition).

lion (ppb). It is important to keep in mind that the concentration in ppm or ppb may be by either volume, mass, or weight. In some toxicology studies, the units used are milligrams of toxin per kilogram of body mass (1 mg/kg is equal to 1 ppm). Concentration may also be recorded as a percent. For example, 100 ppm (100 mg/kg) is equal to 0.01%. (How many ppm is equal to 1%?)

When dealing with water pollution, units of concentration for a pollutant may be milligrams per liter (mg/L) or micrograms per liter (µg/L). A milligram is one-thousandth of a gram, and a microgram is one-millionth of a gram. For water pollutants that do not cause significant change in density of water (1 g/cm³), a concentration of pollution of 1 mg/L is approximately equivalent to 1 ppm.

Air pollutants are commonly measured in units such as micrograms of pollutant per cubic meter of air ($\mu g/m^3$).

Units such as ppm, ppb, or $\mu g/m^3$ reflect very small concentrations. For example, if you were to use 3 g (one-tenth of an ounce) of salt to season popcorn in order to have salt at a concentration of 1 ppm by weight of the popcorn, you would have to pop approximately 3 metric tons of kernels!

15.2 Categories of Pollutants

A partial classification of pollutants by arbitrary categories is presented below. Examples of other pollutants are discussed in other parts of the book.

Infectious Agents

Infectious diseases, spread from the interactions between individuals and food, water, air, or soil, constitute some of the oldest health problems that humans face. Today, infectious diseases have the potential to pose rapid local to global threats by spreading in hours through airplane travelers. Terrorist activity may also spread diseases. Inhalation anthrax, caused by a bacterium, sent in a powdered form in envelopes through the mail in 2001, killed several people. New diseases are emerging, and previous ones may emerge again. Although we have cured many diseases, we have no known reliable vaccines for others, such as HIV, hantavirus, and Dengue fever.

Diseases that can be controlled by manipulating the environment, such as by improving sanitation or treating water, are classified as environmental health concerns. Although there is great concern about the toxins and carcinogens produced in industrial society today, the greatest mortality in developing countries is caused by environmentally transmitted infectious disease. In the United States, thousands of cases of water-borne illness and food poisoning occur each year. These diseases can be spread by people; mosquitoes or fleas; or contact with contaminated food, water, or soil or can be transmitted through ventilation systems in buildings.

Some examples of environmentally transmitted infectious diseases are:

- Legionellosis, or Legionnaires' disease, which often occurs where air conditioning systems have been contaminated by disease-causing organisms.
- Giardiasis, a protozoan infection of the small intestine spread via food, water, or person-to-person contact.
- Salmonella, a food-poisoning bacterial infection spread via water or food.
- Malaria, a protozoan infection transmitted by mosquitoes.
- Lyme Borreliosis, or Lyme disease, transmitted by ticks.

- Cryptosporidosis, a protozoan infection transmitted via water or person-to-person contact (see Chapter 21).[10]
- Anthrax, spread by terrorist activity.

We sometimes hear about epidemics in developing nations. An example is the highly contagious Ebola virus in Africa, which causes external and internal bleeding resulting in death of 80% of those infected. We may tend to think of such epidemics as problems for only developing nations. This belief may give us a false sense of security! Monkeys and bats spread Ebola but the origin of the virus in the tropical forest remains unknown. Developed countries, where outbreaks may occur in the future, will need to learn from the developing countries' experiences. To accomplish this and avoid potential global tragedies, more funds need to be provided for the study of infectious diseases in developing countries.

Toxic Heavy Metals

The major **heavy metals** (metals with relatively high atomic weight; see Chapter 4) that pose health hazards to people and ecosystems include mercury, lead, cadmium, nickel, gold, platinum, silver, bismuth, arsenic, selenium, vanadium, chromium, and thallium. Each of these elements may be found in soil or water not contaminated by humans. Each metal has uses in our modern industrial society, and each is also a by-product of the mining, refining, and use of other elements. Heavy metals often have direct physiological toxic effects. Some are stored or incorporated in living tissue, sometimes permanently. Heavy metals tend to be stored (accumulate with time) in fatty body tissue. A little arsenic each day may eventually result in a fatal dose, the subject of more than one murder mystery.

The content of heavy metals in our bodies is referred to as the *body burden*. The body burden of toxic heavy elements for an average human body (70 kg) is about 8 mg for antimony, 13 mg for mercury, 18 mg for arsenic, 30 mg for cadmium, and 150 mg for lead. Lead (for which we apparently have no biological need) has an average body burden of about twice that of the others combined, reflecting our heavy use of this potentially toxic metal (see the chapter-opening case study).

Mercury, thallium, and lead are very toxic to humans. They have long been mined and used, and their toxic properties are well known. Mercury, for example, is the "Mad Hatter" element. At one time, mercury was used in making felt hats stiff; because mercury damages the brain, hatters were known to act peculiarly in Victorian England. Thus, the Mad Hatter in Lewis Carroll's *Alice in Wonderland* had real antecedents in history.

Toxic Pathways

Chemical elements released from rocks or human processes can become concentrated in humans (see

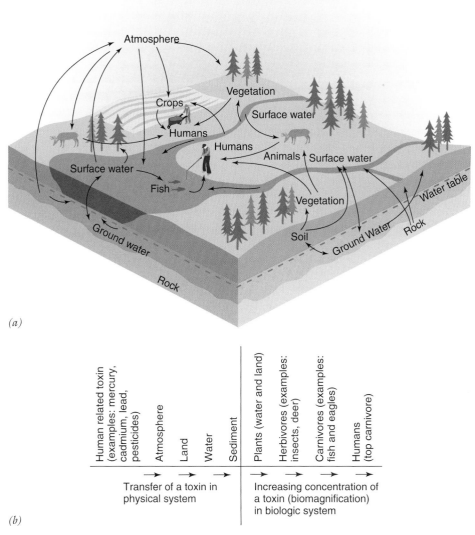

(a)

(b)

Figure 15.5 • (*a*) Potential complex pathways for toxic materials through the living and nonliving environment. Note the many arrows into humans and other animals, sometimes in increasing concentrations, as they move through the food chain (*b*).

Chapter 4) through many pathways (Figure 15.5). These pathways may involve what is known as **biomagnification**—the accumulation or increase in concentration of a substance in living tissue as it moves through a food web (also known as *bioaccumulation*). For example, cadmium, which influences the risk of heart disease, may enter the environment via ash from burning coal. The cadmium in coal exists in very low concentrations (less than 0.05 ppm). After coal is burned in a power plant, the ash is collected in a solid form and disposed of in a landfill. The landfill is covered with soil and revegetated. The low concentration of cadmium in the ash and soil is taken into the plants as they grow. But the concentration of cadmium in

the plants is three to five times greater than the concentration in the ash. As the cadmium moves through the food chain, it becomes more and more concentrated. By the time it is incorporated in the tissue of people and other carnivores, the concentration is approximately 50–60 times the original concentration in the coal.

Mercury in aquatic ecosystems offers another example of biomagnification. Mercury is a potentially serious pollutant of aquatic ecosystems such as ponds, lakes, rivers, and the ocean. Natural sources of mercury in the environment include volcanic eruption and erosion of natural mercury deposits, but we are most concerned with human input of mercury into the environment through

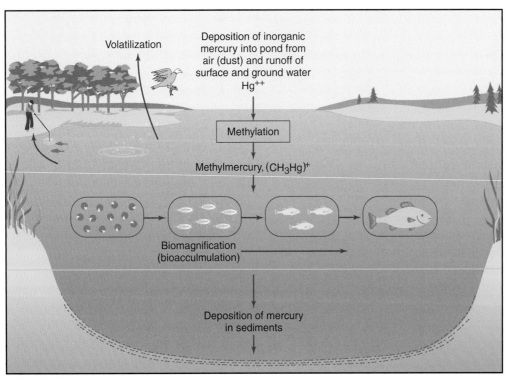

Figure 15.6 • Idealized diagram showing selected pathways for movement of mercury into and through an aquatic ecosystem. (*Source:* Modified from G. L. Waldbott, *Health Effects of Environmental Pollutants,* 2nd ed. [Saint Louis: C. V. Mosby, 1978].)

processes such as burning coal in power plants, incinerating waste, and processing metals such as gold. Rates of input of mercury into the environment through human processes is poorly understood. However, it is believed that human activities have doubled or tripled the amount of mercury in the atmosphere, and it is increasing at about 1.5% per year.[11]

A major source of mercury in many aquatic ecosystems is deposition from the atmosphere through precipitation. Most of the deposition is of inorganic mercury (Hg^{++}, ionic mercury). Once this mercury is in surface water, it enters into complex biogeochemical cycles and a process known as *methylation* may occur. Methylation changes inorganic mercury to methyl mercury $[CH_3Hg]^+$ through bacterial activity.

Methyl mercury is much more harmful (toxic) than is inorganic mercury, and it is eliminated more slowly from animals' systems. As the methyl mercury works its way through food chains, biomagnification occurs, so that higher concentrations of methyl mercury are found further up the food chain. Thus, big fish in a pond that eat little fish contain a higher concentration of mercury than do smaller fish and the aquatic insects that the fish feed upon.

Selected aspects of the mercury cycle in aquatic ecosystems are shown in Figure 15.6. The figure emphasizes the input side of the cycle, from deposition of inorganic mercury through formation of methyl mercury, biomag-

nification, and sedimentation of mercury at the bottom of a pond. On the output side of the cycle, the mercury that enters fish may be taken up by animals that eat the fish; and sediment may release mercury by a variety of processes, including resuspension in the water, where eventually the mercury enters the food chain or is released into the atmosphere through volatilization (conversion of liquid mercury to a vapor form).

During the 20th century, several significant incidents of methyl mercury poisoning have been recorded. One, in Minamata Bay, Japan, involved the industrial release of methyl mercury (see A Closer Look 15.2). Another, in Iran, involved a methyl mercury fungicide used to treat wheat seeds. In each of these cases, hundreds of people were killed and thousands were permanently damaged.[11]

Organic Compounds

Organic compounds are compounds of carbon produced naturally by living organisms or synthetically by human industrial processes. It is difficult to generalize about the environmental and health effects of artificially produced organic compounds because there are so many of them, they have so many uses, and they can produce so many different kinds of effects (see A Closer Look 15.3). **Synthetic organic compounds** are used in industrial processes, pest control, pharmaceuticals, and food additives.

Mercury and Minamata, Japan

In the Japanese coastal town of Minamata, on the island of Kyushu, a strange illness began to occur in the middle of the 20th century. It was first recognized in birds that lost their coordination and fell to the ground or flew into buildings and in cats that went mad, running in circles and foaming at the mouth.[12] The affliction, known by local fishermen as the "disease of the dancing cats," subsequently affected people, particularly families of fishermen. The first symptoms were subtle: fatigue, irritability, headaches, numbness in arms and legs, and difficulty in swallowing. More severe symptoms involved the sensory organs; vision was blurred, and the visual field was restricted. Afflicted people became hard of hearing and lost muscular coordination. Some complained of a metallic taste in their mouths; their gums became inflamed, and they suffered from diarrhea. Eventually, 43 people died and 111 were severely disabled; in addition, 19 babies were born with congenital defects. Those affected lived in a small area, and much of the protein in their diet came from fish from the Minamata Bay.

A vinyl chloride factory on the bay used mercury in an inorganic form in its production processes. The mercury was released in waste that was discharged into the bay. Mercury forms few organic compounds, and it was believed that the mercury, although poisonous, would not get into food chains. But the inorganic mercury released by the factory was converted by bacterial activity in the bay into methyl mercury, an organic compound that turned out to be much more harmful. Unlike inorganic mercury, methyl mercury readily passes through cell membranes. It is transported by the red blood cells throughout the body, and it enters and damages brain cells.[13] Fish absorb methyl mercury from water 100 times faster than they absorb inorganic mercury. (This was not known before the epidemic in Japan.) Once absorbed, methyl mercury is retained two to five times longer than is inorganic mercury.

Harmful effects of methyl mercury depend on a variety of factors, including the amount and route of intake, the duration of exposure, and the species affected. The effects of the mercury are delayed from three weeks to two months from the time of ingestion. If mercury intake ceases, some symptoms may gradually disappear, but others are difficult to reverse.[13]

The mercury episode at Minamata illustrates four major factors that must be considered in evaluating and treating toxic environmental pollutants.

1. *Individuals vary in their response to exposure to the same dose, or amount, of a pollutant.* Not everyone in Minamata responded in the same way; there were variations even among those most heavily exposed. Because we cannot predict exactly how any single individual will respond, we need to find a way to state an expected response of a particular percentage of individuals in a population.

2. *Pollutants may have a threshold*—that is, a level below which the effects are not observable and above which the effects become apparent. Symptoms appeared in individuals with concentrations of 500 ppb of mercury in their bodies; no measurable symptoms appeared in individuals with significantly lower concentrations.

3. *Some effects are reversible.* Some people recovered when the mercury-filled seafood was eliminated from their diet.

4. *The chemical form of a pollutant, its activity, and its potential to cause health problems may be changed markedly by ecological and biological processes.* In the case of mercury, its chemical form and concentration changed as the mercury moved through the food webs.

Some synthetic compounds are called "persistent organic pollutants" or POPs. Many were first produced decades ago when their harm to the environment was not known, and are now banned or restricted (see Table 15.1).

POPs have several properties that define them: [14]

- They have a carbon-based molecular structure, often containing highly reactive chlorine.
- Most are manufactured by humans. That is, they are synthetic chemicals.
- They are persistent in the environment. That is, they do not easily break down in the environment.

- They are polluting and toxic.
- They are soluble in fat and likely to accumulate in living tissue.
- They occur in forms that allow them to be transported by wind, water, and sediments for long distances.

We have produced over 20 million synthetic chemicals, and new ones are appearing at a rate of about 1 million per year! Most are not produced commercially, but up to 100,000 chemicals have now, or in the past, been used. Once used and dispersed in the environment, they may produce a hazard for decades to hundreds of years.

Dioxin: The Big Unknown

Dioxin may be one of the most toxic of the human-made chemicals in the environment. The history of the scientific study of dioxin and its regulation illustrate once again the interplay of science and values. Although science isn't entirely certain of the toxicity of dioxin to humans and ecosystems, society has made a number of value judgments involving regulation of the substance. Controversy has surrounded these judgments and will surely continue to do so.

Dioxin is a colorless crystal made up of oxygen, hydrogen, carbon, and chlorine. It is classified as an organic compound because it contains carbon. About 75 types of dioxin (and dioxin-like compounds) are known; they are distinguished from one another by the arrangement and number of chlorine atoms in the molecule.

Dioxin is not normally manufactured intentionally but is a by-product resulting from chemical reactions, including the combustion of compounds containing chlorine in the production of herbicides.[15] In the United States, there are a variety of sources for dioxin-like compounds (specifically, chlorinated dibenzo-p-dioxin, or CDD, and chlorinated dibenzofurans, or CDF). These compounds are emitted into the air through such processes as incineration of municipal waste (the major source), incineration of medical waste, burning of gasoline and diesel fuels in vehicles, burning of wood as a fuel, and refining of metals such as copper. The good news is that releases of CDDs and CDFs decreased about 75% from 1987 to 1995. However, we are only beginning to understand the many sources of emission of dioxins to the air, water, and land, and linkages and rates of transfer from dominant airborne transport to deposition in water, soil, and the biosphere. In too many cases the amounts of dioxins emitted are based more on expert opinion, rather than high-quality data or even limited data.[16] As a result of scientific uncertainty the controversy concerning dioxin is sure to continue.

Although dioxin is known to be extremely toxic to mammals, its actions in the human body are not well known. What is known is that sufficient exposure to dioxin (usually from meat or milk containing the chemical) produces a skin condition (a form of acne) that may be accompanied by loss of weight, liver disorders, and nerve damage.[17]

Studies of animals exposed to dioxin suggest that some fish, birds, and other animals are sensitive to even small amounts; as a result, it can cause widespread environmental damage to wildlife, including birth defects and death. However, the concentration necessary to cause human health hazards is still controversial. Studies suggest that workers exposed to high concentrations of dioxin for longer than a year have an increased risk of dying of cancer.[18]

The Environmental Protection Agency (EPA) has set an acceptable intake of dioxin at 0.006 pg per kilogram of body weight per day (1 pg = 10^{-12} g; see appendix for prefixes and multiplication factors). This level is deemed too low by some scientists, who argue that the acceptable intake ought to be 100–1,000 times higher, or approximately 1–10 pg/day.[18] The EPA believes that setting the level this much higher could result in health effects. However, some scientists assert that lack of data precludes establishment of a specific threshold concentration of dioxin at which health hazards begin.[19] As indicated by these uncertainties, toxicity of dioxin will remain unclear until additional studies better delineate the potential hazard.

Dioxin is a stable, long-lived chemical that is accumulating in the environment. Analysis of sediments taken from the bottom of Lake Superior suggest that the rate of deposition of dioxin increased eightfold from 1940 to 1970. However, since then, rates have slowly declined.[20] As yet, we have not been able to determine a safe, reliable, and economically feasible way to clean up areas contaminated by dioxin. Many old waste disposal sites are contaminated by dioxin; it may also be found in soil and streams several kilometers around the sites.

Figure 15.7 ● Soil samples from Times Beach, Missouri, thought to be contaminated by dioxin.

The dioxin problem became well known in 1983 when Times Beach, Missouri, a river town just west of Saint Louis with a population of 2,400, was evacuated and purchased for $36 million by the government. The evacuation and purchase occurred after the discovery that oil sprayed on the town's roads to control dust contained dioxin and the entire area had been contaminated. Times Beach was labeled a dioxin ghost town (Figure 15.7). Today, the buildings have been bulldozed, and all that is left is a grassy and woody area enclosed by a barbed-wire-topped chain-link fence. The evacuation has since been viewed by some scientists (including the person who ordered the evacuation) as an overreaction by the government to a perceived dioxin hazard.

The EPA convened a panel of scientists to reevaluate the risk to the environment and to people from exposure to dioxin. Subsequent reports concluded that:

- Dioxin is a probable human carcinogen, and exposure is known to cause disruption of the endocrine, immune, and reproductive systems. However, it is not a widespread, significant cancer threat to people at ordinary (very low) levels of exposure.[21]
- The cancer risk to workers exposed to chemicals containing high concentrations of dioxin may be higher than previously thought.[21]
- Very small levels of dioxin released into the environment can cause serious damage to wildlife sensitive to the chemical, potentially causing significant damage to ecosystems.[22]

As noted, the controversy concerning the toxicity of dioxin is not over.[23] Some environmental scientists argue that the regulation of dioxin must be tougher, whereas the industries producing the chemical argue that the dangers of exposure are overestimated.

Table 15.1 Production and Use of Selected Common Persistent Organic Pollutants (POPs)

Chemical	Example of Use	Date Introduced	Cumulative World Production (tons)
Aldrin[1]	insecticide	1949	240,000
DDT[1]	insecticide	1942	2.8–3 million
Dieldrin[1]	insecticide	1948	240,000
Endrin[2]	insecticide	1951	(3,119 tons in 1977)
PCBs[1]	liquid insulators in electric transformers	1929	1–2 million
Dioxins	by-product of herbicide production	1920s	11

Source: Data in part from Anne Platt McGinn, "Phasing Out Persistent Organic Pollutants," in Lester R. Brown et al., State of the World 2000 (New York: W.W. Norton & Company, 2000).

NOTES: [1]. Banned in the U.S. and many other countries.
[2]. Restricted or banned in many countries.

For example, consider PCBs, originally used as an insulator in electric transformers.[14]

- A factory in Alabama manufactured PCBs in the 1940s, shipping them to a General Electric factory in Massachusetts. They were put in insulators and mounted on poles in thousands of locations.
- The transformers deteriorated over time. Some were damaged by lightning, and others were damaged or destroyed during demolition. The PCBs leaked into the soil or were carried by surface runoff into streams and rivers. Others combined with dust and were transported by wind around the world.
- The dust containing PCBs was deposited in ponds, lakes, or rivers where it entered the food chain. First it entered algae along with nutrients it combined with. Insects ate the algae, which were eaten by shrimp and fish. In each stage up the food web, the concentration of PCBs increased.

- Fish are caught by fisherman and eaten. The PCBs are then passed on to people where they are concentrated in fatty tissue and mother's milk.

Radiation

Nuclear radiation is introduced here as a category of pollution. It is discussed in detail in Chapter 19 in conjunction with nuclear energy. We are concerned about nuclear radiation because excessive exposure is linked to serious health problems, including cancer. (See Chapter 24 for a discussion of radon gas as an indoor air pollutant.)

Thermal Pollution

Thermal pollution, also called heat pollution, occurs when heat released into water or air produces undesirable effects. Heat pollution can occur as a sudden, acute event or as a long-term, chronic release. Sudden heat releases

Figure 15.8 • Two types of cooling towers. (*a*) Wet cooling tower. Air circulates through tower; hot water drips down and evaporates, cooling the water. (*b*) Dry cooling tower. Heat from the water is transferred directly to the air, which rises and escapes the tower. (*c*) Cooling towers emitting steam at Didcot power plant, Oxfordshire, England. Red and white lines are vehicle lights resulting from long exposure time (photograph taken at dusk).

may result from natural events, such as brush or forest fires and volcanoes, or from human-induced events, such as agricultural burning. The major sources of chronic heat pollution are electric power plants that produce electricity in steam generators.

The release of large amounts of heated water into a river changes the average water temperature and the concentration of dissolved oxygen (warm water holds less oxygen than cooler water), thereby changing the river's species composition (see the discussion of eutrophication in Chapter 21). Every species has a range of temperature within which it can survive and an optimal temperature for living. For some species of fish, the range is small, and even a small change in water temperature is a problem. Lake fish move away when the water temperature rises more than about 1.5°C above normal; river fish can withstand a rise of about 3°C.[24] Heating river water can change its natural conditions and disturb the ecosystem in several ways. Fish spawning cycles may be disrupted, and the fish may have a heightened susceptibility to disease; warmer water causes physical stress in some fish, and they may be easier for predators to catch; and warmer water may change the type and abundance of food available for fish at various times of the year.

There are several solutions to chronic thermal discharge into bodies of water. The heat can be released into the air by cooling towers (Figure 15.8), or the heated

(*c*)

water can be temporarily stored in artificial lagoons until it is cooled to normal temperatures. Some attempts have been made to use the heated water to grow organisms of commercial value that require warmer water temperatures. Waste heat from a power plant can also be captured and used for a variety of purposes, such as warming buildings (see Chapter 16 for a discussion of cogeneration).

Particulates

Particulates are small particles of dust (including soot and asbestos fibers, discussed below), released into the atmosphere by many natural processes and human activities. Modern farming and combustion of oil and coal add

Figure 15.9 • Fires in Indonesia in 1997 caused serious air pollution problems. People here are purchasing surgical masks in an attempt to breathe cleaner air.

considerable amounts of particulates to the atmosphere, as do dust storms, fires (Figure 15.9), and volcanic eruptions. The 1991 eruptions of Mount Pinatubo in the Philippines were the largest volcanic eruptions of the 20th century, explosively delivering huge amounts of volcanic ash, sulfur dioxide, and other volcanic material and gases into the atmosphere to elevations up to 30 km (18.6 mi). Eruptions can have a significant impact on the global environment and are linked to global climate change and stratospheric ozone depletion (see Chapters 22 and 25). In addition, many chemical toxins, such as heavy metals, enter the biosphere as particulates. Sometimes, nontoxic particulates link with toxic substances, creating a synergetic threat (see discussion of particulates in Chapter 23).

Asbestos

Asbestos is a term for several minerals that take the form of small, elongated particles, or fibers. Industrial use of asbestos has contributed to fire prevention and has provided protection from the overheating of materials. Asbestos is also used as insulation for a variety of purposes. Unfortunately, however, excessive contact with asbestos has led to asbestosis (a lung disease caused by the inhalation of asbestos) and to cancer in some industrial workers. Experiments with animals have demonstrated that asbestos can cause tumors to develop if the fibers are embedded in lung tissue.[25] The hazard related to certain types of asbestos under certain conditions is thought to be so serious that extraordinary steps have been taken to reduce the presence of asbestos or to ban it outright. The expensive process of asbestos removal from old buildings (particularly schools) in the United States is one of those steps.

There are several types of asbestos, and they are not equally hazardous. The form most commonly utilized in the United States is white asbestos, which comes from the mineral chrysolite. It has been used as an insulation material around pipes, in floor and ceiling tiles, and for brake linings of automobiles and other vehicles. Approximately 95% of the asbestos that is now in place in the United States is of the chrysolite type. Most of this asbestos was mined in Canada, and environmental health studies of Canadian miners show that exposure to chrysolite asbestos is not particularly harmful. Studies involving another type of asbestos, known as crocidolite asbestos (blue asbestos), suggest that exposure to this mineral can be very hazardous and evidently does cause lung disease. Several other types of asbestos have also been shown to be harmful.[25]

A great deal of fear has been associated with nonoccupational exposure to chrysolite asbestos in the United States. Tremendous amounts of money have been expended to remove it from homes, schools, public buildings, and other sites in spite of the fact that there has been no asbestos-related disease recorded among those exposed to chrysolite in nonoccupational circumstances. It is now thought that much of the removal was unnecessary and that chrysolite asbestos doesn't pose a significant health hazard. Additional research into health risks from other varieties of asbestos is necessary to better understand the potential problem and to outline strategies to eliminate potential health problems.

Electromagnetic Fields

Electromagnetic fields (EMFs) are part of everyday urban life. Electric motors, electric transmission lines for utilities, and electrical appliances, such as toasters, electric blankets, and computers, all produce magnetic fields.

There is currently a controversy over whether these fields produce a health risk.

Early on, investigators did not believe that magnetic fields were harmful, because fields drops off quickly with distance from the source and the strength of the fields that most people come into contact with are relatively weak. For example, the magnetic fields generated by power transmission lines, or by a computer terminal are normally only about 1% of Earth's magnetic field; directly below power lines the electric field induced in the body is about what the body naturally produces within cells.

Several early studies, however, concluded that children exposed to EMFs from power lines have an increased risk of contracting leukemia, lymphomas, and nervous system cancers.[26] Investigators concluded that children so exposed are about one and a half to three times as likely to develop cancer as children with very low exposure to EMFs, but the results were questioned because of perceived problems with the research design (problems of sampling, tracking children, and estimating exposure to EMF).

A later study analyzed over 1,200 children, approximately half of whom suffered from acute leukemia. It was necessary to estimate residential exposure to magnetic fields generated by nearby power lines in the children's present and former homes. Results of that study, which is the largest such investigation to date, concluded that there is no association between childhood leukemia and measured exposure to magnetic fields.[26]

Another study compared exposure of electric utility workers to magnetic fields with incidents of brain cancer and leukemia. That study revealed a weak association between exposure to magnetic fields and both brain cancer and leukemia. However, the associations are not strong and were not statistically significant.[27]

Saying that data are not statistically significant is another way of stating that the relationship between exposure and disease cannot be reasonably established, given the data base that was analyzed. It does not mean that additional data in a future study will not find a statistically significant relationship. Statistics can predict the strength of the relationship between variables such as exposure to a toxin and the incidence of a disease, but statistics cannot prove a cause-and-effect relationship between them.

In summary, in spite of the many studies that have been completed to evaluate relationships between disease and exposure to magnetic fields in our modern urban environment, the jury is still out. There seems to be some indication that magnetic fields may cause problems, but so far the risks are relatively small and difficult to quantify.

Noise Pollution

Noise pollution is unwanted sound. Sound is a form of energy that travels as waves. We hear sound because our ears respond to sound waves through vibrations of the eardrum. The sensation of loudness is related to the

Table 15.2 Examples of Sound Levels

Sound Source	Intensity of Sound (dB)	Human Perception
Threshold of hearing	0	
Rustling of leaf	10	Very quiet
Faint whisper	20	Very quiet
Average home	45	Quiet
Light traffic (30 m away)	55	Quiet
Normal conversation	65	Quiet
Chain saw (15 m away)	80	Moderately loud
Jet aircraft flyover at 300 m	100	Very loud
Rock music concert	110	Very loud
Thunderclap (close)	120	Uncomfortably loud
Jet aircraft takeoff at 100 m	125	Uncomfortably loud
	140	Threshold of pain
Rocket engine (close)	180	Traumatic injury

intensity of the energy carried by the sound waves and is measured in units of decibels (dB). The threshold for human hearing is 0 dB; the average sound level in the interior of a home is about 45 dB; the sound of an automobile, about 70 dB; and the sound of a jet aircraft taking off, about 120 dB (see Table 15.2). A 10-fold increase in the strength of a particular sound adds 10 dB units on the scale. An increase of 100 times adds 20 units.[12] The decibel scale is logarithmic; it increases exponentially as a power of 10. For example, 50 dB is 10 times louder than 40 dB and 100 times louder than 30 dB.

Environmental effects of noise depend not only on the total energy but also on the sound's pitch, frequency, and time pattern and the length of exposure to the sound. Very loud noises (more than 140 dB) cause pain, and high levels can cause permanent hearing loss. Human ears can take sound up to about 60 dB without damage or hearing loss. Any sound above 80 dB is potentially dangerous. The noise of a lawn mower or motorcycle will begin to damage hearing after about eight hours of exposure. In recent years, there has been concern for teenagers (and older people, for that matter) who have suffered some permanent loss of hearing following extended exposure to amplified rock music (110 dB). At a noise level of 110 dB, damage to hearing can occur after an exposure time of only half an hour. Loud sounds at the workplace are another hazard. Levels of noise that are below the hearing-loss level may still interfere with human communication and may cause irritability. Noise in the range of 50–60 dB is sufficient to interfere with sleep, producing a feeling of fatigue upon awakening.

Voluntary Exposure

Voluntary exposure to toxins and potentially harmful chemicals is sometimes referred to as exposure to personal pollutants. The most common of these are tobacco, alcohol, and other drugs. Use and abuse of these substances has led to a variety of human ills, including death and chronic disease, criminal activity such as reckless driving and manslaughter, loss of careers, street crime, and the straining of human relations at all levels.

Scientific evidence has demonstrated that use of tobacco, in all of its forms, is both habit forming and dangerous to human health. Tobacco contains a variety of components that are toxic, carcinogenic, radioactive, and addictive. It has been estimated that 30% of all cancers in the United States are tied to smoking-related disorders. According to the American Cancer Society, cigarette smoking is responsible for approximately 80% of lung cancers; second-hand smoke is also a hazard, as are tobacco products such as chewing tobacco. Although the number of people who smoke in the United States, as a percentage of adults, has decreased in recent years, young people and people in developing countries are still becoming addicted to cigarettes and cigars.

Many people in our society use alcohol at social gatherings and celebrations. Approximately 70% of all American adults drink some alcohol, and moderate use of alcohol is legal and accepted in our society. However, when abused, alcohol causes very serious problems. Approximately one-half of all deaths in automobile accidents are related to alcohol use by drivers. Furthermore, significant numbers of violent crimes and other criminal activities are committed by people under the influence of alcohol. Some people believe that alcohol is the most abused drug in our society today. Young people unfamiliar with the potential toxicity of alcohol have died from alcohol overdose (for example, by drinking 21 drinks on their 21st birthdays). Chronic alcoholism has many toxic consequences, including liver and heart disease.

A variety of illegal drugs are commonly used in the United States and other parts of the world. These drugs have various effects on their users, but the end result is often the degradation of the mind and/or body. Illegal drugs are particularly dangerous, because their strength, composition, and other chemical characteristics are seldom subject to quality control. Of particular concern in recent years has been the development of synthetic (designer) drugs that are addictive and capable of causing significant health problems and even death—in some cases—in first-time users.

15.3 General Effects of Pollutants

Almost every part of the human body is affected by one pollutant or another, as shown in Figure 15.10*a*. For

Table 15.3	Effects of Pollutants on Wildlife
Effect on Population	*Pollutants*
Changes in abundance	Arsenic, asbestos, cadmium, fluoride, hydrogen sulfide, nitrogen oxides, particulates, sulfur oxides, vanadium
Changes in distribution	Fluoride, particulates, sulfur oxides
Changes in birth rates	Arsenic, lead
Changes in death rates	Arsenic, asbestos, beryllium, boron, cadmium, fluoride, hydrogen sulfide, lead, particulates, selenium, sulfur oxides
Changes in growth rates	Boron, fluoride, hydrochloric acid, lead, nitrogen oxides, sulfur oxides

Source: J. R. Newman, *Effects of Air Emissions on Wildlife*, U.S. Fish and Wildlife Service, 1980. Biological Services Program, National Power Plant Team, FWS/OBS-80/40, U.S. Fish and Wildlife Service, Washington, D.C.

example, lead and mercury (remember the Mad Hatter) affect the brain; arsenic, the skin; carbon monoxide, the heart; and fluoride, the bones. Wildlife is affected as well. Sites of effects of major pollutants in wildlife are shown in Figure 15.10*b*; effects of pollutants on wildlife populations are listed in Table 15.3.

The lists of potential toxins and affected body sites for humans and other animals in Figure 15.10 may be somewhat misleading. For example, chlorinated hydrocarbons, such as dioxin, are stored in the fat cells of animals, but they cause damage not only to fat cells but to the entire organism through disease, damaged skin, and birth defects. Similarly, a toxin that affects the brain, such as mercury, causes a wide variety of problems and symptoms, as illustrated in the Minamata, Japan, example (discussed in A Closer Look 15.2). The value of Figure 15.10 is in helping us to understand in general the adverse effects of excess exposure to chemicals.

Concept of Dose and Response

Five centuries ago, the physician and alchemist Paracelsus wrote that "everything is poisonous, yet nothing is poisonous." By this, he meant essentially that a substance in too great an amount can be dangerous yet in an extremely small amount can be relatively harmless. Every chemical element has a spectrum of possible effects on a particular organism. For example, selenium is required in small amounts by living things, but may be toxic or increase the probability of cancer in cattle and wildlife when it is present in high concentrations in the soil. Copper, chromium,

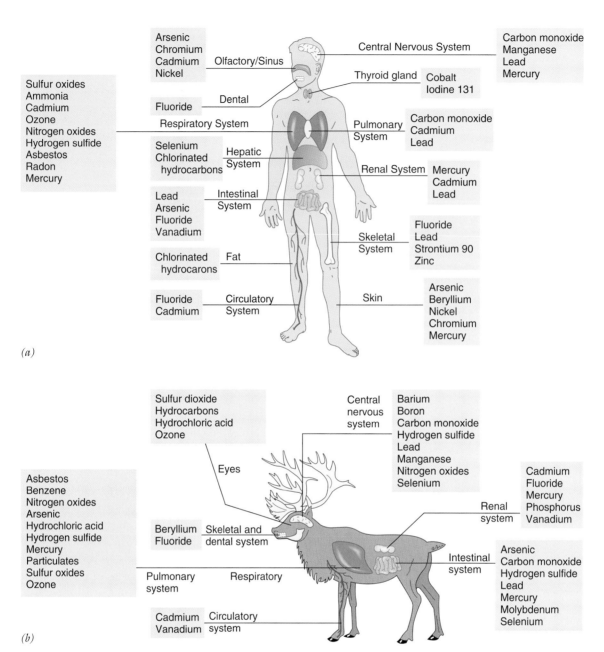

Figure 15.10 • (*a*) Effects of some major pollutants in human beings. (*b*) Known sites of effects of some major pollutants in wildlife. (*Source:* [*a*] G. L. Waldbott, *Health Effects of Environmental Pollutants*, 2nd ed. [St. Louis: Moseby, 1978].

Copyright 1978 by C. V. Moseby. [*b*] J. R. Newman, *Effects of Air Emissions on Wildlife Resources*, U.S. Fish and Wildlife Services Program, National Power Plant Team, FWS/OBS-80/40, [Washington, D.C.: U.S. Fish and Wildlife Service, 1980]).

and manganese are other chemical elements required in small amounts by animals but toxic in higher amounts.

It was recognized many years ago that the effect of a certain chemical on an individual depends on the dose. This concept is termed **dose response**. Dose dependency can be represented by a generalized dose-response curve such as that shown in Figure 15.11.

When various concentrations of a chemical present in a biological system are plotted against the effects on the organism, two things are apparent. First, relatively large

concentrations are toxic and even lethal (points *D*, *E*, and *F* in Figure 15.12). Second, trace concentrations may be beneficial for life (between points *A* and *D*); and the dose–response curve forms a plateau of optimal concentration and maximum benefit between two points (*B* and *C*). Points *A*, *B*, *C*, *D*, *E*, and *F* in Figure 15.11 are important thresholds in the dose-response curve. Unfortunately, the amounts at which points *E* and *F* occur are known only for a few substances, for a few organisms, including people; and the very important

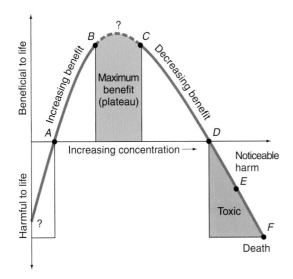

Figure 15.11 • Generalized dose-response curve. Low concentrations of a chemical may be harmful to life (below point *A*). As the concentration of the chemical increases from *A* to *B*, the benefit to life increases. The maximum concentration that is beneficial to life lies within the benefit plateau (*B–C*). Concentrations greater than this plateau provide less and less benefit (*C–D*) and will harm life (*D–F*) as toxic concentrations are reached. Increased concentrations above the toxic level may result in death.

point *D* is all but unknown. Doses that are beneficial, harmful, or lethal may differ widely for different organisms and are difficult to characterize.

Fluorine provides a good example of the general dose-response concept. Fluorine forms fluoride compounds that prevent tooth decay and promote the development of a healthy bone structure.

Relationships between the concentration of fluoride (in a compound of fluorine, such as sodium fluoride, NaF) and health show a specific dose-response curve (Figure 15.12). The plateau for an optimal concentration

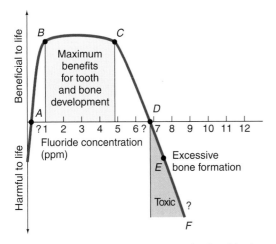

Figure 15.12 • General dose-response curve for fluoride showing the relationship between fluoride concentration and physiological benefit.

of fluoride (point *B* to point *C*) to reduce dental caries (cavities) is from about 1 ppm to just less than 5 ppm. Levels greater than 1.5 ppm do not significantly decrease tooth decay but do increase the occurrence of discoloration of teeth. Concentrations of 4 to 6 ppm reduce the prevalence of osteoporosis, a disease characterized by loss of bone mass; and toxic effects are noticed between 6 and 7 ppm (point *D* in Figure 15.12).

Dose-Response Curve (LD-50, ED-50, and TD-50)

Individuals differ in their response to chemicals, and it is difficult to predict the dose that will cause a response in a particular individual. For this reason, it is practical to predict what percentage of a population will respond to a specific dose of a chemical.

For example, the dose at which 50% of the population die is called the lethal dose 50, or **LD-50**. The LD-50 is a crude approximation of a chemical's toxicity. It is a gruesome index that does not adequately convey the sophistication of modern toxicology and is of little use in setting a standard for toxicity. However, the LD-50 determination is required for new synthetic chemicals as a way of estimating their toxic potential. Table 15.4 lists, as examples, LD-50 values in rodents for selected chemicals.

The **ED-50** (effective dose 50%) is the dose that causes an effect in 50% of the population of observed subjects. For example, the ED-50 of aspirin would be the dose that relieves headaches in 50% of the people.[29]

The **TD-50** (toxic dose 50%) is defined as the dose that is toxic to 50% of the population. TD-50 is often used to indicate responses such as reduced enzyme activity, decreased reproductive success, or onset of specific

Table 15.4	Approximate LD-50 Values (for rodents) for Selected Agents
Agent	$LD_{50}(mg/kg)^*$
Sodium chloride (table salt)	4,000
Ferrous sulfate (to treat anemia)	1,520
2,4-D (a weed killer)	368
DDT (an insecticide)	135
Caffeine (in coffee)	127
Nicotine (in tobacco)	24
Strychnine sulfate (used to kill certain pests)	3
Botulinum toxin (in spoiled food)	0.00001

*Milligrams per kilogram of body mass (termed mass weight, although it really isn't a weight) administered by mouth to rodents. Rodents are commonly used in such evaluations, in part because they are mammals (as we are), are small, have a short life expectancy and because their biology is well known.

Source: H. B. Schiefer, D. C. Irvine, and S. C. Buzik, *Understanding Toxicology* (New York: CRC Press, 1997).

symptoms, such as loss of hearing, nausea, or slurred speech.

For a particular chemical, there may be a whole family of dose-response curves, as illustrated in Figure 15.13. Which dose is of interest depends on what is being evaluated. For example, for insecticides, we may wish to know the dose that will kill 100% of the insects exposed; therefore the LD-95 (the dose that kills 95% of the insects) may be the minimum acceptable level. However, when considering human health and the exposure to a particular toxin, we often want to know the LD-0—the maximum dose that does not cause any deaths.[23] For potentially toxic compounds such as insecticides, which may form a residue on food or food additives, we want to ensure that the expected levels of human exposure will have no known toxic effects. From an environmental perspective, this is important because of concerns about increased risk of cancer associated with exposure to toxic agents.[29]

For drugs used to treat a particular disease, the efficiency of the drug as a treatment is of paramount importance. In addition to knowing what the therapeutic value (ED-50) is, it is also important to know the relative safety of the drug. For example, there may be an overlap between the therapeutic dose (ED) and the toxic dose (TD) (see Figure 15.13). That is, the dose that causes a positive therapeutic response in some individuals might be toxic to others. A quantitative measure of the relative safety of a particular drug is the *therapeutic index*, defined as the ratio of the LD-50 to the ED-50. The greater the therapeutic index, the safer the drug is believed to be.[30] In other words, a drug with a large difference between the lethal and theraputic dose is safer than one with a smaller difference.

Threshold Effects

Recall from A Closer Look 15.2 that a **threshold** is a level below which no effect occurs and above which effects begin to occur. If a threshold dose of a chemical exists, then a concentration of that chemical in the environment below the threshold is safe. If there is no threshold dose, then even the smallest amount of the chemical has some negative toxic effect (Figure 15.14).

Whether or not there is a threshold effect for environmental toxins is an important environmental issue. For example, the U.S. Federal Clean Water Act originally stated as a goal to reduce to zero the discharge of pollutants into water. The goal implies there is no such thing as a threshold effect, since no level of toxin is to be legally permitted. However, it is unrealistic to believe zero discharge of a water pollutant can be achieved or to believe that we can reduce to zero the concentration of chemicals shown to be carcinogenic.

A problem in evaluating thresholds for toxic pollutants is that it is difficult to account for synergistic effects. Little is known about if or how thresholds might change

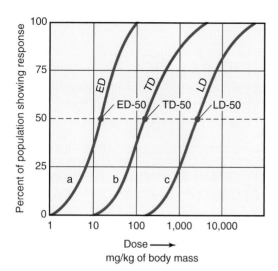

Figure 15.13 • Idealized diagram illustrating a family of dose-response curves for a specific drug: ED (effective dose), TD (toxic dose), and LD (lethal dose). Notice the overlap for some parts of the curves. For example, at ED-50, a few percent of the people exposed to that dose will suffer a toxic response, but none will die. At TD-50 dose, about 1% of the people exposed to that dose will die.

if an organism is exposed to more than one toxin at the same time or to a combination of toxins and other chemicals, some of which are beneficial. Exposures of people to chemicals in the environment are complex, and we are only beginning to understand and conduct research on the possible interactions and consequences of multiple exposures.

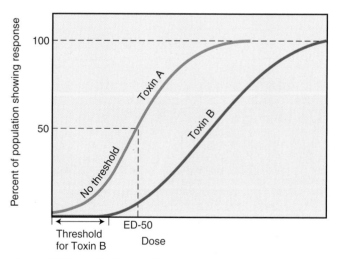

Figure 15.14 • In this hypothetical toxic dose-response curve, toxin *A* has no threshold; even the smallest amount has some measurable effect on the population. The TD-50 for toxin *A* is the dose required to produce a response in 50% of the population. Toxin *B* has a threshold (flat part of curve) where response is constant as dose increases. After the threshold dose is exceeded, the response increases.

Ecological Gradients

Dose-response effects differ among species. For example, the kinds of vegetation that can live nearest to a toxic source are often small plants with relatively short lifetimes (grasses, sedges, and weedy species usually regarded as pests) that are adapted to harsh and highly variable environments. Farther from the toxic source, trees may be able to survive. Changes in vegetation with distance from a toxic source define the **ecological gradient**.

Ecological gradients may be found around smelters and other industrial plants that discharge pollutants into the atmosphere from smokestacks. For example, ecological gradient patterns can be observed in the area around the smelters of Sudbury, Ontario, discussed earlier in this chapter (A Closer Look 15.1). Near the smelters, an area that was once forest is now a patchwork of bare rock and soil occupied by small plants.

Tolerance

The ability to resist or withstand stress resulting from exposure to a pollutant or harmful condition is referred to as **tolerance**. Tolerance can develop for some pollutants in some populations, but not for all pollutants in all populations.

Tolerance may result from behavioral, physiological, or genetic adaptation. *Behavioral tolerance* results from changes in behavior. For example, mice learn to avoid traps.

Physiological tolerance results when the body of an individual adjusts to tolerate a higher level of pollutant. For example, in studies at the University of California Environmental Stress Laboratory, students were exposed to relatively high levels of ozone (O_3), an air pollutant often present in large cities (Chapter 23). The students at first experienced symptoms that included irritation of eyes and throat and shortness of breath. However, after a few days, their bodies adapted to the ozone, and they reported that they believed they were no longer breathing ozone-contaminated air, even though the concentration of O_3 stayed the same. This phenomenon explains why some people who regularly breathe polluted air report that they do not notice the pollution. Of course, it does not mean that the ozone is doing no damage; it is, especially in people with existing respiratory problems. There are many mechanisms for physiological tolerance, including *detoxification*, in which the toxic chemical is converted to a nontoxic form, and the internal transport of the toxin to a part of the body where it is not harmful, such as fat cells.

Genetic tolerance, or adaptation, results when some individuals in a population are naturally more resistant to a toxin than others. They are less damaged by exposure and more successful in breeding. Resistant individuals pass on the resistance to future generations, who are also more successful at breeding. Adaptation has been observed among some insect pests following exposure to

some chemical pesticides. For example, certain strains of malaria-causing mosquitoes are now resistant to DDT (see the discussion in Chapter 12); and some organisms that cause deadly infectious diseases have become resistant to common antibiotic drugs, such as penicillin.

Acute and Chronic Effects

Pollutants can have acute and chronic effects. An *acute effect* is one that occurs soon after exposure, usually to large amounts of a pollutant. A *chronic effect* takes place over a long period, often as a result of exposure to low levels of a pollutant. For example, a person exposed all at once to a high dose of radiation may be killed by radiation sickness soon after exposure (an acute effect). However, that same total dose, received slowly in small amounts over an entire lifetime, may instead cause mutations and lead to disease or affect the person's DNA and offspring (a chronic effect).

15.4 Risk Assessment

Risk assessment can be defined as the process of determining potential adverse environmental health effects to people exposed to pollutants and potentially toxic materials (recall the discussion of measurements and methods of science in Chapter 2). Such an assessment generally includes four steps:[31]

1. *Identification of the hazard.* Identification consists of testing materials to determine whether exposure is likely to cause environmental health problems. One method used is to investigate populations of people who have been previously exposed. For example, to understand the toxicity of radiation produced from radon gas, researchers studied workers in uranium mines. Another method is to perform experiments to test effects on animals, such as mice, rats, or monkeys. This method has drawn increasing criticism from groups of people who believe such experiments are unethical. Another approach is to try to understand how a particular chemical works at the molecular level on cells. For example, research has been done to determine how dioxin interacts with living cells to produce an adverse response. After quantifying the response, scientists can develop mathematical models to predict or estimate dioxin's risk.[18] This relatively new approach might also be applicable to other potential toxins that work at the cellular level.

2. *Dose-response assessment.* The next step involves identifying relationships between the dose of a chemical (therapeutic drug, pollutant, or toxin) and the health effects to people. Some studies involve administering fairly high doses of a chemical to animals. The effects, such as illness or symptoms (rash, tumor development) are recorded for varying doses, and the results are used to

Are We Getting too Much Fluoride?

The drinking water of approximately 132 million Americans contains fluoride at concentrations of 0.7 mg/L or higher. More than 40 years after the introduction of fluoridation in the United States, controversy concerning its health benefits and risks remains. Two questions dominate the debate: First, are parts of the population receiving more fluoride than necessary? Second, does the amount of fluoride pose a health risk? Let us examine these issues by following three steps associated with risk assessment.

1. *Identification of the hazard.* Fluoride is harmful at concentrations above 6 ppm, and doses of 5 mg/kg of body weight can cause poisoning. Overexposure affects teeth and bones. At low doses, white areas may appear on the teeth; at moderate doses, brown discoloration may occur; and at high doses, destruction of tooth enamel may occur. Mild or moderate effects on teeth are found in 1.3% of children in the United States.

In combination with other factors, fluoride may cause bones to become opaque and brittle, leading to abnormal bone growth and crippling. However, this is rare in the United States. Absorption of fluoride in the intestines and the percentage retained in the body vary with the person's age, the type of fluoride compound, the compound's pH (a measure of acid or base content), and the presence of other ions. Of three studies to determine whether fluoride causes mutations, two were negative and one was positive.

A 1993 study by the National Research Council found no evidence that fluoride (at concentrations to which humans are exposed) caused kidney toxicity, gastrointestinal problems, reproductive problems, or genetic toxicity. However, further studies need to be conducted to determine the role of fluoride in bone fractures and cancer occurrence. In 1990, the National Toxicology Program concluded that there was no evidence relating fluoride to birth defects or to gastrointestinal, genital-urinary, or respiratory problems and that evidence was insufficient to determine whether fluoride had any effects on reproduction or development. The National Cancer Institute reviewed data on cancer from 1973 to 1987 and found no links between cancer and fluoride. Exposing animals to moderate and high doses of fluoride produced equivocal results; in four studies, results were negative, but one study found bone cancers in 5% of the rats, at least 10 times the normal rate.

2. *Dose-response assessment.* The dose-response curve in Figure 15.13 shows the relationship between concentration of fluoride and human health. Although high doses of fluoride are associated with adverse, even toxic effects, too little fluoride intake is associated with increased incidence of dental caries (cavities). When first introduced, fluoridated water decreased the incidence of caries in children by 60%. Research in the 1980s, however, showed a difference of only 20% to 40%. One interpretation of the change is that fluoride is now used in beverages, food, dental products, and dietary supplements, so the control groups are actually receiving some fluoride.

3. *Exposure assessment.* Groundwater contains trace amounts to 25 ppm of fluoride. Optimal concentration for preventing dental caries and avoiding adverse effects has been set at 0.7–1.2 ppm for drinking water. People may ingest additional fluoride in food or in substances to which fluoride has been added—for example, toothpaste. Estimates of daily consumption of fluoride appear in Table 15.5.

Critical Thinking Questions

1. Given the information provided in Table 15.5, what is the risk from fluoride to both children and adults in the general population?

Table 15.5   Daily Fluoride Consumption

Source of Fluoride	Amount in Source	Total Intake (mg)		Concentration in Body *(mg/kg)	
		Child	*Adult*	*Child*	*Adult*
Drinking water	1.20 mg/L	1.68	2.40	0.084	0.034
Toothpaste, mouthwash	0.145–0.66 mg	0.66	0.145	0.033	0.002
Diet	0.2–0.4 mg	0.20	0.40	0.010	0.006
Total intake	—	2.54	2.95	0.127	0.042
Safe intake	—	0–1 year: 0.1–1,	1.5–4.0		
		1–2 years: 0.5–1.5			

* Assumes that a child weighs 20 kg and drinks 1.4 L/day; an adult weighs 70 kg and drinks 2.0 L/day; and absorption in both cases is 100%.

2. Fluoridated toothpastes are not required to display information on the concentration of fluoride, but they average 1,000 mg/L. Some dental experts have recommended half this amount for children's toothpaste. What would your recommendation be, and why?

3. People living in areas in which groundwater levels exceed the recommended 1 ppm can remove fluoride, at considerable expense, or look for other sources of water. In an area with a fluoride concentration varying from 3 to 8 ppm, what recommendation would you make?

4. What recommendations would you make for the U.S. government concerning fluoride policy and research?

5. According to B. Walker, codirector of the Center for Epidemiologic Research and dean of public health at the University of Oklahoma, risk assessments are not science in the classical sense. What do you think he means by this? Hint: Think about what science is in terms of testing hypotheses.

predict the response in people. This is difficult, and the results are controversial for several reasons:

- The dose that results in a particular response may be very small and subject to measurement errors.
- There may be arguments over whether thresholds are present or absent.
- Experiments on animals such as rats, mice, or monkeys may not be directly applicable to humans.
- The assessment may rely on probability and statistical analysis. Although statistically significant results from experiments or observations are accepted as evidence to support an argument, statistics cannot establish that the substance tested *caused* the observed response.

3. *Exposure assessment.* Exposure assessment evaluates the intensity, duration, and frequency of human exposure to a particular chemical pollutant or toxin. The hazard to society is directly proportional to the total population exposed. The hazard to an individual is generally greater closer to the source of exposure. Like dose-response assessment, exposure assessment is difficult, and the results are often controversial, in part because of difficulties in measuring the concentration of a toxin present in doses as small as parts per million, billion, or even trillion. Some questions that exposure assessment attempts to answer are:

- How many people were exposed to concentrations of a toxin thought to be dangerous?
- How large an area was contaminated by the toxin?
- What are the ecological gradients for exposure to the toxin?
- How long were people exposed to a particular toxin?

4. *Risk characterization.* During this final step, the goal is to delineate health risk in terms of the magnitude of the potential environmental health problem that might result from exposure to a particular pollutant or toxin. To do this, it is necessary to identify the hazard, complete the dose-response assessment, and evaluate the exposure assessment as outlined above. This step involves all the uncertainties of the prior steps, and results are again likely to be controversial.

In summary, risk assessment is difficult, costly, and controversial. Each chemical is different, and there is no one method of determining responses of humans for specific EDs or TDs. Toxicologists use the scientific method of hypothesis testing with experiments (see Chapter 2) to generate predictions of how specific doses of a chemical may affect humans. Warning labels listing potential side effects of using a specific medication are required by law, and these warnings result from toxicology studies to determine a drug's safety.

Finally, risk assessment requires making scientific judgments and formulating actions to help minimize environmental health problems related to human exposure to pollutants and toxins. The process of *risk management* integrates the assessment of risk with technical, legal, political, social, and economic issues.[18] Scientific arguments concerning the toxicity of a particular material are often open to debate. For example, there is debate concerning whether the risk from dioxin is linear. That is, do effects start at minimum levels of exposure to dioxin and gradually increase, or is there a threshold exposure beyond which environmental health problems occur? (See A Closer Look 15.3.)[18,23] It is the task of people in appropriate government agencies assigned to manage risk to make judgments and decisions based on the risk assessment and then to take appropriate actions to minimize the hazard resulting from exposure to toxins.

Summary

- Pollution produces an impure, dirty, or otherwise unclean state. Contamination means making something unfit for a particular use through the introduction of undesirable materials. Toxic materials are poisonous to people and other living things; toxicology is the study of toxic materials. A concept important in studying pollution problems is synergism, whereby actions of different substances produce a combined effect greater than the sum of the effects of the individual substances.

- How we measure the amount of a particular pollutant introduced into the environment or the concentration of that pollutant varies widely, depending on the substance. Common units for expressing the concentration of pollutants are parts per million (ppm) and parts per billion (ppb). Air pollutants are commonly measured in units such as micrograms of pollutant per cubic meter of air ($\mu g/m^3$).

- Categories of environmental pollutants include toxic chemical elements (particularly heavy metals), organic compounds, radiation, heat, particulates, electromagnetic fields, and noise.

- Organic compounds of carbon are produced by living organisms or synthetically by humans. Artificially produced organic compounds may have physiological, genetic, or ecological effects when introduced into the environment. Organic compounds vary with respect to their potential hazards: Some are more readily degraded in the environment than others; some are more likely to undergo biomagnification; and some are extremely toxic even at very low concentrations. Organic compounds that cause serious concern include pesticides and dioxin.

- The effect of a chemical or toxic material on an individual depends on the dose. It is also important to determine tolerances of individuals as well as acute and chronic effects of pollutants and toxins.

- Risk assessment involves hazard identification, assessment of dose response, assessment of exposure, and risk characterization.

● REEXAMINING THEMES AND ISSUES

Human Population: As the total population and population density increase, the probability that more people will be exposed to hazardous materials increases as well. Finding acceptable sites for the disposal of toxic chemicals also becomes more difficult as populations increase and people live closer to industrial areas and waste disposal sites.

Sustainability: Ensuring that future generations inherit a relatively unpolluted, toxin-free, healthy environment remains a difficult problem. Sustainable development requires that our use of chemicals and other materials not damage the environment.

Global Perspective: Release of toxins into the environment may result in global patterns of contamination or pollution. This is particularly true when a toxin or contaminant enters the atmosphere, surface water, or oceans and becomes widely dispersed. For example, pesticides, herbicides, or heavy metals emitted into the atmosphere in the midwestern United States may be transported by winds and deposited on glaciers in polar regions.

Urban World: Processes of industrial activity in urban areas concentrate potentially toxic materials that may be inadvertently, accidentally, or deliberately released into the environment. Human exposure to a variety of toxins—including lead, asbestos, particulates, organic chemicals, radiation, and noise—are often greater in urban areas.

Science and Values: Because we value both human and nonhuman life, we are interested in expanding our knowledge concerning risk of exposure of living things to chemicals, pollutants, and toxins in the natural and human-made environment. Unfortunately, the state of our knowledge of risk assessment is often incomplete, and the dose response for many chemicals is poorly understood.

What we decide to do about exposure to toxic chemicals reflects our values. Increased control of toxic materials in homes and the work environment is expensive. Do we sufficiently value the health of workers in countries, where goods are manufactured, to pay more for those goods in order to reduce environmental hazards for the workers at their work sites?

STUDY QUESTIONS

1. Do you think the hypothesis that some crime is caused in part by environmental pollution is valid? Why? Why not? How might the hypothesis be further tested? What are the social ramifications of the tests?

2. What kinds of life-forms would most likely survive in a highly polluted world? What would be their general ecological characteristics?

3. Some environmentalists argue that there is no such thing as a threshold for pollution effects. What is meant by this statement? How would you determine whether it was true for a specific chemical and a specific species?

4. What is biomagnification, and why is it important in toxicology?

5. You are lost in Transylvania while trying to locate Dracula's castle. Your only clue is that the soil around the castle has an unusually high concentration of the heavy metal arsenic. You wander in a dense fog, able to see only the ground a few meters in front of you. What changes in vegetation warn you that you are nearing the castle?

6. Distinguish between acute and chronic effects of pollutants.

7. Design an experiment to test whether tomatoes or cucumbers are more sensitive to lead pollution.

8. Why is it difficult to establish standards for acceptable levels of pollution? In giving your answer, consider physical, climatological, biological, social, and ethical reasons.

9. A new highway is built through a pine forest. Driving along the highway, you notice that the pines nearest the road have turned brown and are dying. You stop at a rest area and walk into the woods. One hundred meters away from the highway, the trees seem undamaged. How could you make a crude dose-response curve from direct observations of the pine forest? What else would be necessary to devise a dose-response curve from direct observation of the forest? What else would be necessary to devise a dose-response curve that could be used in planning the route of another highway?

FURTHER READING

Amdur, M., J. Doull, and C. D. Klaasen, eds. 1991. *Casarett & Doull's Toxicology: The Basic Science of Poisons*, 4th ed. Tarrytown, N.Y.: Pergamon. A comprehensive and advanced work on toxicology.

Carson, R. 1962. *Silent Spring*. Boston: Houghton Mifflin. A classic book on problems associated with toxins in the environment.

Schiefer, H. B., D. G. Irvine, and S. C. Buzik. 1997. *Understanding Toxicology: Chemicals, Their Benefits and Risks*. Boca Raton, Fla.: CRC Press. A concise introduction to toxicology as it pertains to everyday life, including information about pesticides, industrial chemicals, hazardous waste, and air pollution.

Travis, C. C., and H. A. Hattemer-Frey. 1991. "Human Exposure to Dioxin," *The Science of the Total Environment* 104:97–127. An extensive technical review of dioxin accumulation and exposure.

CHAPTER 16 Energy: Some Basics*

Power plant generating electricity at Delano, California—Mark Gibson photo

● LEARNING OBJECTIVES

Understanding the basics of what energy is, as well as the sources and uses of energy, is essential for effective energy planning. After reading this chapter, you should understand:

- That energy is neither created nor destroyed but is transformed from one kind to another.
- Why in all transformations energy tends to go from a more usable to a less usable form.
- What energy efficiency is and why it is always less than 100%.
- That people in industrialized countries consume a disproportionately large share of the world's total energy and how efficiency and conservation of energy can help make better use of global energy resources.
- Why some energy planners propose a hard-path approach to energy provision and others a soft-path approach, and why both of these approaches have positive and negative points.
- Why moving toward global sustainable energy planning with integrated energy planning is an important goal.
- What elements are needed to develop integrated energy planning.

*Written with assistance from Mel S. Manalis.

California Energy Crisis: A Wake-up Call

California in 2001 was faced with "rolling blackouts"—electrical power outages that caused disruption in homes and industry. It was an "energy shock" that startled the entire country, from coast to coast, into thinking about energy policy in the United States.

A basic problem in California was that while the economic growth of the 1990s brought prosperity and more people to the state, energy demand rose, and few new sources came on-line to meet the growing demand. Utility companies were forced to buy "emergency power" from other power suppliers at very high prices (as much as 900% higher). They also had to buy natural gas, which was increasing nationwide in price. By law, the utility companies could not pass on their increasing costs, and they claimed they were driven to near bankruptcy.

Faced with potential bankruptcy of the utility companies, the state of California launched a plan to end the crisis. The governor of California started an investigation of potential price-gouging and illegal market control of energy sources. Ultimately, the cost to utility customers will be more than $20 billion. Electricity bills to homes and businesses will rise to about 20% more than before the crisis.

The utility companies argue that the solution to the crisis is to let them raise the rates to cover the increased costs. However, such simple solutions to complex problems are usually wrong. California has an aging power grid (the system of powerlines that distribute the electricity) that is in need of upgrades to meet new demands. New power plants are also needed.

The 2001 power crisis has changed some opinions in California. An opinion poll suggests that most people (about 80%) want new power plants as long as they are not nuclear and are willing to have them located near where they live. Furthermore, most people report they are taking steps to conserve energy. The answer to California's energy crisis of 2001 will involve aspects of conservation and efficiency, construction of new power plants, development and use of more alternative energy sources (solar and wind), and the upgrading of existing facilities (power plants and power grid).

The California energy crisis sparked debate in Congress that is resulting in a new national energy policy—something that hasn't happened since the 1970s. Lawmakers have recognized that we could face a nationwide crisis. Some congresspersons support the so-called hard path: build more oil, natural gas, coal, and nuclear power plants. Others want more emphasis on energy conservation and alternative energy sources such as solar and wind. As of 2002 the price of energy had decreased in California and across the United States, but the shock remains a wake-up call.

 This case history demonstrates that the United States faces serious energy problems. With this in mind, we explore in this chapter some of the basic principles associated with what energy is, how much energy we consume, and how we might manage energy for the future.

16.1 Outlook for Energy

Energy crises are nothing new. People have encountered energy problems for thousands of years, as far back as the early Greek and Roman cultures.

Energy Crises in Ancient Greece and Rome

The climate in coastal areas of Greece 2,500 years ago was characterized by warm summers and cool winters, much as it is today. To warm their homes in winter, the Greeks used small, charcoal-burning heaters that were not very efficient. Since charcoal is produced from burning wood, wood was their primary source of energy, as it is today for half the world's people.

By the fifth century B.C., fuel shortages had become common, and much of the forest in many parts of Greece was depleted of firewood. As local supplies diminished, it became necessary to import wood from farther away. Olive groves became sources of fuel, which was made into charcoal for burning, reducing a valuable resource. By the fourth century B.C., the city of Athens had banned the use of olive wood for fuel.

At about this time, the Greeks began to build their houses facing the south, designing them so that the low winter sun entered the houses, providing heat, and the higher summer sun was partially blocked, cooling the houses. Recent excavations of ancient Greek cities suggest that large areas were planned so that individual homes could make maximum use of this solar energy. The Greeks' use of solar energy in heating homes was a logical answer to their energy problem.[1]

The use of wood in ancient Rome is somewhat analogous to the use of oil and gas in the United States today. Wealthy Roman citizens about 2,000 years ago had central heating in their large homes, burning as much as 125 kg (275 lb) of wood every hour. Not surprisingly, local wood supplies were exhausted quickly, and the Romans had to import wood from outlying areas. Eventually, wood had to be imported from as far away as 1,600 km (about 1,000 mi).[1]

The Romans turned to solar energy for the same reasons as the Greeks but with much greater application and success. The Romans used glass windows to increase the effectiveness of solar heating, developed greenhouses to raise food during the winter, and oriented large public bathhouses (some of which accommodated up to 2,000 people) to use passive solar energy (Figure 16.1). The Romans believed that sunlight in bathhouses was healthy, and it also saved greatly on fuel costs. The use of solar energy in ancient Rome was widespread and resulted in laws to protect a person's right to solar energy. In some areas, it was illegal for one person to construct a building that shaded another's.[1]

The ancient Greeks and Romans experienced an energy crisis in their urban environments. In turning to solar energy, they moved toward what today we call sustainability. We are on that same path as fossil fuels become scarce.

Energy Today and Tomorrow

The energy situation facing the United States and the world today is in some ways similar to that faced by the early Greeks and Romans. The use of wood in the United States peaked in the 1880s, when the use of coal became widespread. The use of coal in turn began to decline after 1920, when oil and gas started to become available. Today, we are facing the global peak of oil production, which is expected by about 2020. Fossil fuel resources, which took millions of years to form, may be essentially exhausted in just a few hundred years.

The decisions we make today will affect energy use for generations. Should we choose complex, centralized energy production methods, or use simpler and widely dispersed energy production methods, or use a combination of the two? Which sources of energy should be

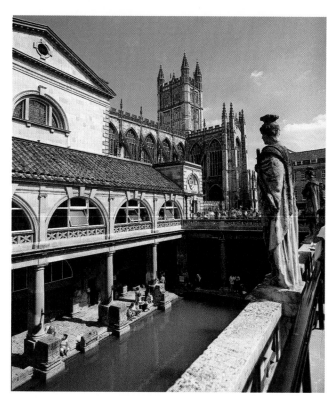

Figure 16.1 • Roman bathhouse (lower level) in the town of Bath, England. The orientation of the bathhouse and the placement of windows are designed to maximize the benefits of passive solar energy.

emphasized? Which uses of energy should be emphasized for increased efficiency? How can we rely on current energy sources and provide for developing a sustainable energy policy? There are no easy answers.

Use of fossil fuels, especially oil, has resulted in improvements in sanitation, medicine, and agriculture. These improvements have helped to make possible the global human population increase that we have discussed in other chapters. Many of us are living longer, with a higher standard of living than people before us. However, burning fossil fuels imposes growing environmental costs that are causing concerns ranging from urban pollution to a change in global climate.

One thing certain about the energy picture for tomorrow is that it will involve living with uncertainty when it comes to energy availability and cost. The sources and patterns of energy utilization will undoubtedly change. Problems with energy supply and cost can be expected to occur, as they did in California in 2000–2001, as a result of growing demand and insufficient supply. Supplies will continue to be regulated, and there exists a growing potential for the disruption of supplies. Oil embargoes could cause significant economic impact in the United States and other countries; and a war or revolution in a petroleum-producing country would cause exports of petroleum to be reduced significantly.

It is clear that we need to rethink our entire energy policy in terms of sources, supply, consumption, and environmental concerns. We can begin by understanding basic facts about what energy is.

16.2 Energy Basics

The concept of **energy** is somewhat abstract; you cannot see it or feel it, even though you have to pay for it.[2] To understand energy, it is easiest to begin with the idea of a force. We all have had the experience of exerting force by pushing or pulling. The strength of a force can be measured by how much it accelerates an object.

What if your car stalls while you are going up a hill and you get out to push it uphill to the side of the road (Figure 16.2). You apply a force against gravity, which would otherwise cause the car to roll downhill. If the brake is on, the brakes, tires, and bearings might heat up from friction. The longer the distance over which you exert force in pushing the car, the greater the change in the car's position, and the greater the amount of heat from friction in the brakes, tires, and bearings.

In physicists' terms, exerting the force over the distance moved is the work done. That is, *work is the product of a force times a distance*. Conversely, *energy is the ability to do work*. If you push hard but the car does not move, you have exerted a force, but you have not done any work on the car (according to the definition), even if you feel very tired and sweaty.[2]

In pushing your stalled car, you have moved it against gravity and caused some of its parts (brakes, tires, bearings) to be heated. These effects have something in common: They are forms of energy. You have converted chemical energy in your body to the energy of motion of the car (kinetic energy). When the car is higher on the hill, the potential energy of the car has been increased, and friction produces heat energy.

Energy can be, and often is, converted or transformed from one kind to another, but the total energy is always conserved. The principle that energy cannot be created or destroyed but is always conserved is known as the **first law of thermodynamics**. (Recall the discussion of matter and energy in Chapter 4 and the detailed discussion of energy flow in the biosphere in Chapter 9.) Thermodynamics is the science that keeps track of energy as it undergoes various transformations from one type to another. We use the first law to keep track of the quantity of energy.[3]

To illustrate the conservation and conversion of energy, think about a tire swing over a creek (Figure 16.3). When the tire swing is held in its highest position, it is not moving. It does, however, contain stored energy owing to its position. We refer to the stored energy as *potential energy*. Other examples of potential energy are the gravitational energy in water behind a dam; the chemical energy in coal, fuel oil, and gasoline, as well as in the fat in your body; and nuclear energy, which is related to the forces binding the nuclei of atoms.[2]

The tire swing, when released from its highest position, moves downward. At the bottom (straight down), the speed of the tire swing is greatest, and no potential energy remains. At this point, all the swing's energy is the energy of motion, called *kinetic energy*. As the tire swings back and forth, the energy continuously changes between the two forms, potential and kinetic. However, with each swing, the tire slows down a little more and goes a little less high because of friction created by the movement of the tire and rope through air and friction at the pivot where the rope is tied to the tree. The friction slows the swing, generating *heat energy*, which is energy from random motion of atoms and molecules. Eventually, all the energy is converted to heat and emitted to the environment, and the swing stops.[2]

The example of the swing illustrates the tendency of energy to dissipate and end up as heat. Indeed, physicists

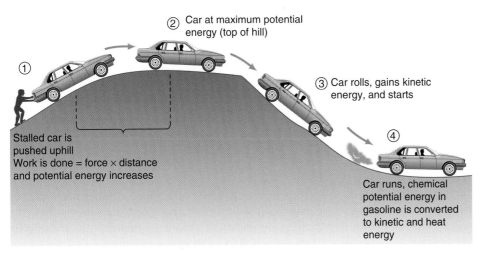

② Car at maximum potential energy (top of hill)

①

③ Car rolls, gains kinetic energy, and starts

④

Stalled car is pushed uphill
Work is done = force × distance and potential energy increases

Car runs, chemical potential energy in gasoline is converted to kinetic and heat energy

Figure 16.2 • Some basic energy concepts, including potential energy, kinetic energy, and heat energy.

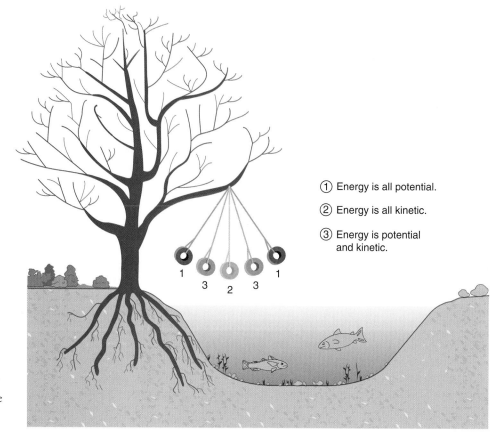

① Energy is all potential.

② Energy is all kinetic.

③ Energy is potential and kinetic.

Figure 16.3 • Diagram of a tire swing over a creek, illustrating the relation between potential and kinetic energy.

have found that it is possible to change all the gravitational energy in a tire swing (a type of pendulum) to heat. However, it is impossible to change all the heat energy thus generated back into potential energy.

Energy is conserved in the tire swing. All the initial gravitational potential energy has been transformed by way of friction to heat energy when the tire swing finally stops. If the same amount of energy, in the form of heat, were returned to the tire swing, would you expect the swing to start again? The answer is no! What, then, is used up? It is not energy, because energy is always conserved. What is used up is the energy *quality*—or the availability of the energy to perform work. The higher the quality of the energy, the more easily it can be converted to work; the lower the energy quality, the more difficult it is to convert to work.

This example illustrates another fundamental property of energy: Energy always tends to go from a more usable (higher quality) form to a less usable (lower-quality) form. This is the **second law of thermodynamics**, and it means that, when you use energy, you lower its quality.

Let us return to the example of the stalled car, which you have now pushed to the side of the road. Having pushed the car a little way uphill, you have increased its potential energy. You can convert this to kinetic energy by letting it roll back downhill. You engage the gears to restart the car. As the car idles, the potential chemical energy (from the gasoline) is converted to waste heat energy and other energy forms, including electricity to charge the battery and play the radio.

Why can we not collect the wasted heat and use it to run the engine? Again, as the second law of thermodynamics tells us, once energy is degraded to low-quality heat, it can never regain its original availability or energy grade. When we refer to low-grade heat energy, we mean that relatively little of it is available to do useful work. High-grade energy, such as that of gasoline, coal, or natural gas, has high potential to do useful work. The biosphere continuously receives high-grade energy from the sun and radiates low-grade heat to the depths of space.[2,3]

16.3 Energy Efficiency

Two fundamental types of energy efficiencies are derived from the first and second laws of thermodynamics: the first-law efficiency and the second-law efficiency. The **first-law efficiency** deals with the amount of energy without any consideration of the quality or availability of the energy. It is calculated as the ratio of the actual amount of energy delivered where it is needed to the amount of energy supplied to meet that need. Expressions for efficiencies are given as fractions; multiplying the fraction by 100 converts it to a percentage. As an example, consider a

Table 16.1   Examples of First- and Second-Law Efficiencies

Energy (End Use)	First-Law Efficiency (%)	Waste Heat (%)	Second-Law Efficiency (%)	Potential for Savings
Incandescent light bulb	5	95		
Fluorescent light	20	80		
Automobile	20–25	75–80	10	Moderate
Power plants (electric); fossil fuel and nuclear	30–40	60–70	30	Low to moderate
Burning fossil fuels (used directly for heat)	65	35		
Water heating			2	Very high
Space heating and cooling			6	Very high
All energy (U.S.)	50	50	10–15	High

furnace system that keeps a home at a desired temperature of 18°C (65°F) when the outside temperature is 0°C (32°F). The furnace, which burns natural gas, delivers 1 unit of heat energy to the house for every 1.5 units of energy extracted from burning the fuel. That means it has a first-law efficiency of 1 divided by 1.5, or 67% (see Table 16.1 for other examples).[3] The "unit" of energy for our furnace is arbitrary for the purpose of discussion; we could use British thermal units (Btu) or some other units (see A Closer Look 16.1 on units of energy).

First-law efficiencies are misleading, because a high value suggests (often incorrectly) that little can be done to save energy through additional improvements in efficiency. This problem is addressed by the use of the **second-law efficiency**, which is defined as the ratio of the minimum available work needed to perform a particular task to the actual work used to perform the task. *Available work* is energy that is accessible to do work; it is referred to simply as *availability*. Availability is higher for sources of high-quality energy. For our home-heating example, the second-law efficiency would compare the minimum availability necessary to heat the home with the energy actually used by the gas furnace. If we calculated the second-law efficiency (which is beyond the scope of this discussion), the result might be 5%—much lower than the first-law efficiency of 67%.[3] (We will see why later.) Table 16.1 also lists some second-law efficiencies for common uses of energy.

Values of second-law efficiency are important because low values indicate where improvements in energy technology and planning may save significant amounts of high-quality energy. Notice that the definition of the second-law efficiency calls for specification of the *task* needing energy. This task is referred to as the energy's *end use*. We know that the second-law efficiency monitors the availability used in performing the defined task. Because availability is higher for high-quality energy, it also monitors the energy quality and tells us whether the quality is appropriate to the task. For example, you could use a

welder's acetylene blowtorch to light a candle, but a match is much more efficient (and safer as well).

We are now in a position to understand why the second-law efficiency is so low (5%) for the house-heating example discussed earlier. This low efficiency implies that the furnace is consuming too much available work (high-quality energy) in carrying out the task of heating the house. In other words, the task of heating the house requires heat at a relatively low temperature, near 18°C (65°F), not heat with temperatures in excess of 1,000°C (1,832°F), such as is generated inside the gas furnace. Energy with less availability (i.e., lower-quality energy), such as solar energy, could do the task and yield a higher second-law efficiency, because there is a better match between the required energy quality and the house-heating end use. Through better energy planning, such as matching the quality of energy supplies to the end use, higher second-law efficiencies can be achieved, resulting in substantial savings of high-quality energy.

Examination of Table 16.1 indicates that electricity-generating plants have nearly the same first-law and second-law efficiencies. These generating plants are examples of *heat engines*. A heat engine produces work from heat. Most of the electricity generated in the world today comes from heat engines that use nuclear fuel, coal, gas, or other fuels. Our own bodies are examples of heat engines, operating with a capacity (power) of about 100 watts (W) and fueled indirectly by solar energy. (See A Closer Look 16.1 for an explanation of watts and other units of energy.) The internal combustion engine (used in automobiles) and the steam engine are additional examples of heat engines. A great deal of the world's energy is used in heat engines, with profound environmental effects such as thermal pollution, urban smog, acid rain, and global warming.

The maximum possible efficiency of a heat engine, known as *thermal efficiency*, was discovered by the French engineer Sadi Carnot in 1824, before the first law of thermodynamics was formulated.[4] Modern heat engines have thermal efficiencies that range between 60% and 80% of

Energy Units

When we buy electricity by the kilowatt-hour, what are we buying? We say we are buying energy, but what does that mean? Before we go deeper into the concepts of energy and its uses, we need to define some basic units.

The fundamental energy unit in the metric system is the *joule*; 1 joule is defined as a force of 1 newton* applied over a distance of 1 meter. To work with large quantities such as the amount of energy used in the United States in a given year, we use the unit *exajoule*, which is equivalent to 10^{18} (a billion billion) joules or roughly equivalent to 1 quadrillion, or 10^{15}, British thermal units (Btu), referred to as a *quad*. To put these big numbers in perspective, the United States today consumes approximately 86 exajoules (or quads) of energy per year, and world consumption is about 325 exajoules (quads) annually.

In many instances, we are particularly interested in the rate of energy use, or *power*, which is energy divided by time. In the metric system, power may be expressed as joules per second, or *watts*, W (1 joule per second is equal to 1 watt). When larger power units are required, we can use multipliers, such as *kilo* (thousand), *mega* (million), and *giga* (billion). For example, the rate of pro-

duction of electrical energy in a modern nuclear power plant is 1000 megawatts (MW) or 1 gigawatt (GW).

Sometimes it is useful to use a hybrid energy unit, such as the watt-hour, Wh (remember energy is power multiplied by time). Electrical energy is usually expressed and sold in *kilowatt-hours* (kWh, or 1000 Wh). This unit of energy is 1000 W applied for 1 hour (3600 seconds), the equivalent energy of 3,600,000 J (3.6 MJ).

The average estimated electrical energy in kilowatt-hours used by various

household appliances over a period of a year is shown in Table 16.2. The total annual energy used is the power rating of the appliance multiplied by the time the appliance was actually used. The appliances that use most of the electrical energy are water heaters, refrigerators, clothes dryers, and washing machines. A list of common household appliances and the amounts of energy they consume is useful in identifying those appliances that might help save energy through conservation or improved efficiency.

Table 16.2 Average Estimated Electrical Energy Use per Year for Typical Household Appliances

Appliance	Power (W)	Average Hours Used per Year	Approximate Energy Used (kWh/yr)
Clock	2	8,760	17
Clothes dryer	4,600	228	1,049
Hair dryer	1,000	60	60
Light bulb	100	1,080	108
Compact fluorescent[a]	18	1,080	19
Television	350	1,440	504
Water heater (150 L)	4,500	1,044	4,698
Energy-efficient model[a]	2,800	1,044	2,900
Toaster	1,150	48	552
Washing machine	700	144	1,008
Refrigerator	360	6,000	2,160
Energy-efficient model[a]	180	6,000	1,100

Sources: Data from U.S. Department of Energy and D. G. Kaufman and C. M. Franz, *Biosphere 2000: Protecting Our Global Environment* (New York: Harper Collins, 1993).
[a]Newer, energy-efficient model.

*A newton (N) is the force necessary to produce an acceleration of 1 m per sec per sec (m/s^2) to a mass of 1 kg.

their ideal Carnot efficiencies. Modern 1000-megawatt (MW) electrical generating plants have thermal efficiencies ranging between 30% and 40%; that means at least 60% to 70% of the energy input to the plant is rejected as waste heat. For example, assume that the electric power output from a large generating plant is 1 unit of power (typically 1000 MW). Producing that 1 unit of power requires 3 units of input (such as burning coal) at the power plant, and the entire process produces 2 units of waste heat, for a thermal efficiency of 33%. The significant number here is

the waste heat, 2 units, which amounts to twice the actual electric power produced.

Electricity may be produced by large power plants that burn coal or natural gas, by plants that use nuclear fuel, or by smaller producers, such as geothermal, solar, or wind sources (see Chapters 17, 18, and 19). Once produced, the electricity is fed into the grid, which is the network of power lines, or the distribution system. Eventually, it reaches homes, shops, farms, and factories, where it lights, heats, and drives motors and other machinery used by society. As electricity moves through the grid, there are losses.[5] The wires that transport electricity (power lines) have a natural resistance to electrical flow, known as *electrical resistivity*. This resistance converts some of the electric energy in the transmission lines to heat energy, which is radiated into the environment surrounding the lines.

16.4 Energy Sources and Consumption

People living in industrialized countries make up a relatively small percentage of the world's population, but they consume a disproportionate share of the total energy produced in the world. For example, the United States, with only 5% of the world's population, uses approximately 25% of the total energy consumed in the world. There is a direct relationship between a country's standard of living (as measured by gross national product) and energy consumption per capita. After the peak in oil production, expected in 2020–2030, supplies of oil and gasoline will be reduced and will be more expensive. Before then, use of these fuels may be curtailed to reduce potential global climate change. As a result, within the next 30 years, both developed and developing countries will need to find innovative ways to obtain energy. In the future, affluence may be as related to more efficient use of a wider variety of energy sources as to total energy consumption.

Fossil Fuels and Alternative Energy Sources

Today, approximately 90% of the energy consumed in the United States is produced by petroleum, natural gas, and coal. Because of their organic origin, these are called fossil fuels. They are produced from plant and animal material and are forms of stored solar energy that are part of our geologic resource base. They are essentially nonrenewable. Other sources of energy—which include geothermal, nuclear, hydropower, and solar, among others— are referred to as alternative energy sources. The term *alternative* designates these as sources that might replace fossil fuels in the future. Some of the alternative sources, such as solar and wind, are not depleted by consumption and are known as *renewable energy*.

The shift to alternative energy sources may be gradual, as fossil fuels continue to be utilized, or it could be accelerated as a result of concern over the potential environmental effects of burning fossil fuels. Regardless of which path we take, one thing is certain: Fossil fuel resources are finite. It has taken millions of years to form them, yet fossil fuels will be consumed in only a few hundred years of human history. Using even the most optimistic predictions, the fossil fuel epoch that started with the Industrial Revolution will represent only about 500 years of human history. Therefore, although fossil fuels have been extremely significant in the development of modern civilization, their use will be a short-lived event in the span of human history.

Energy Consumption in the United States Today

Energy consumption in the United States from 1950 to 1999 is shown in Figure 16.4. The figure dramatically illustrates our present dependence on the three major fossil fuels (coal; natural gas; and petroleum). From approximately 1950 through the mid-1970s, energy consumption increased tremendously, from approximately 30–70

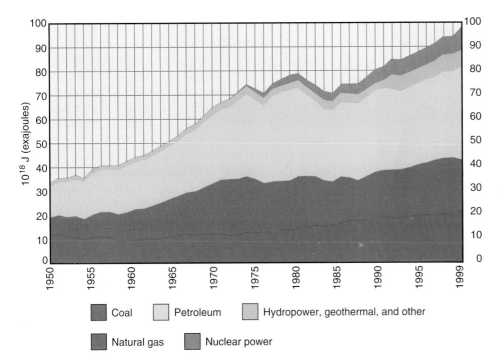

Figure 16.4 • Total energy consumption for the United States, 1950–1999. (*Source*: Energy Information Administration Annual Energy Review [Washington, D.C.: U.S. Department of Energy, 1999].)

Table 16.3 Generalized Annual Energy Flow for the United States in 1997

Energy Source	Energy Production[a] +	Net Imports (Imports – Exports) ±	Adjustments =	Energy Consumed	Consumed by Sector
Coal	23.2	–2.0			Residential/commercial
Natural gas	21.9	+2.9			34.2
Oil	13.6	+19.2			Industrial
					34.2
Nuclear	6.7	0			
Hydropower	3.5	0			
Other	0.1	+0.3			Transportation
					26.6
Total	69.0 +	20.4	± 5.6[b] =	95.0	95.0

Source: OECD, Energy Information Administration, *Annual Energy Review* (Washington, D.C.: U.S. Department of Energy, 1998).

[a]Exajoules (10^{18} J).

[b]Balancing to account for a variety of items, including unaccounted-for supply, blending components, and changes in stock.

exajoules (energy units are defined in A Closer Look 16.1). Since about 1980, energy consumption has increased by only about 20 exajoules. This situation is encouraging because it suggests that policies to improve energy conservation through efficiency improvements (such as requiring new automobiles to be more fuel efficient and buildings to be better insulated) have been at least partially successful.

What is not shown in the figure, however, is the tremendous energy loss. For example, energy consumption in the United States in 1965 was approximately 50 exajoules; and of this, about half was effectively used. Energy losses were about 50% (the number shown in Table 16.1 for all energy). In 1999, energy consumption in the United States exceeded 95 exajoules, and again about 50% was lost in conversion processes. Energy losses in 1999 were about equivalent to total energy consumption in 1965! The largest energy losses are associated with the production of electricity and with transportation.

We have already seen that energy is lost both in the generation of electricity at power plants and in the transmission of the electricity through the grid system. Major losses of energy in the transportation sector of our economy are related to the burning of fossil fuels in the engines of cars, buses, and trucks. Most of the losses related to the production of electricity and to transportation occur through the use of heat engines, which produce waste heat that enters the environment.

Another way to examine energy use is to look at the generalized energy flow of the United States for a particular year and by use user (Figure 16.5). The data on Table 16.3 show that for 1999 we imported considerably more oil than we produced and that consumption of energy is fairly evenly distributed in three sectors: residential/commercial, industrial, and transportation. It is clear that we remain dangerously vulnerable to changing world conditions affecting the production and delivery of crude oil. Evaluation of the entire spectrum of potential energy sources is necessary to ensure the availability of sufficient energy in the future while sustaining environmental quality. Energy consumption patterns have also come under scrutiny.

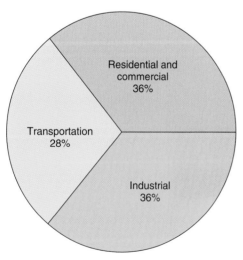

Figure 16.5 • Energy consumption in the United States by sector. (*Source*: Energy Information Administration, Annual Energy Review [Washington, D.C.: U.S. Department of Energy, 1998].)

 Energy Conservation, Increased Efficiency, and Cogeneration

There is a movement to change patterns of energy consumption in the United States through measures such as conservation, increased energy efficiency, and cogeneration. **Conservation** of energy refers simply to getting by with less demand for energy. In a pragmatic sense, this has to do with adjusting our energy needs and uses to minimize the amount of high-quality energy necessary to accomplish a given task. Increased **energy efficiency** involves designing equipment to yield more energy output from a given amount of input energy (first-law efficiency) or better matches between energy source and end use (second-law efficiency).[6] **Cogeneration** includes a number of processes designed to capture and use waste heat rather than simply to release it into the atmosphere, water, or other part of the environment as a thermal pollutant. Energy conservation is particularly attractive because it provides more than a one-to-one savings. Remember that it takes 3 units of fuel such as coal to produce 1 unit of power such as electricity (two-thirds is waste heat). Therefore, not using (conserving) 1 unit of power saves 3 units of fuel!

These three concepts—energy conservation, energy efficiency, and cogeneration—are all interlinked. For example, when electricity is produced at large, coal-burning power stations, large amounts of heat may be emitted into the atmosphere. Cogeneration, by using that waste heat, can increase the overall efficiency of a typical power plant from 33% to as much as 75%, effectively reducing losses from 67% to 25%.[6] Cogeneration also involves generating electricity as a by-product from industrial processes that produce steam as part of their regular operations. Optimistic energy forecasters estimate that eventually we may meet approximately one-half the electrical power needs of industry through cogeneration.[6] Another source has estimated that more than 10% of the power capacity of the United States could be provided through cogeneration.

The average first-law efficiency of only 50% (Table 16.1) illustrates that large amounts of energy are currently lost in producing electricity and in transporting people and goods. Innovations in how we produce energy for a particular use can help prevent this loss, raising second-law efficiencies. Of particular importance will be energy uses with applications below 100°C (212°F), because a large portion of the total U.S. energy consumption (for uses below 300°C, or 572°F) is for space heating and water heating (Figure 16.6).

In considering where to focus our efforts to develop more energy efficiency, we need to look at the total energy-use picture. In the United States, space heating and cooling of homes and offices, water heating, industrial processes (to produce steam), and automobiles account for nearly 60% of the total energy use. By comparison, transportation by train, bus, and airplane account for only about 5%. Therefore, the areas we should target for development of more energy efficiency are building design, industrial energy use, and automobile design.[6,7] We should note, however, that debate continues as to how much efficiency improvements and conservation can reduce future energy demands and the need for increased production from traditional sources, such as fossil fuel.

Building Design

A spectrum of possibilities exists for increasing energy efficiency and conservation in residential buildings. For new homes, the answer is to design and construct homes that minimize the energy consumption necessary to ensure comfortable living.[8] For example, we can design buildings to take advantage of passive solar potential, as did the early Greeks and Romans and the Native American cliff dwellers. (Passive solar energy systems collect solar heat without using moving parts.) Windows and overhanging structures can be positioned so that the overhangs shade the windows from incoming summer solar energy, thereby keeping the house cool while allowing winter sun to penetrate the windows and warm the house.

The potential for energy savings through architectural design for older buildings is extremely limited. The position of the building on the site is already established, and reconstruction and modifications are often not cost-effective. The best approach to energy conservation for these buildings is insulation, caulking, weather-stripping, installation of window coverings and storm windows, and regular maintenance.

Buildings constructed to conserve energy are more likely to develop indoor air pollution problems, as pollutants emitted in buildings become concentrated due to reduced ventilation. Indoor air pollution is emerging as one of our most serious environmental problems. Potential problems can be reduced by better designs for air circulation systems that purify indoor air and bring in fresh, clean air (see Chapter 24). Construction that incorporates environmental principles is more expensive owing to higher fees for architects and engineers as well as higher initial construction costs. Nevertheless, moving toward improved design of homes and residential buildings to conserve energy remains an important endeavor.

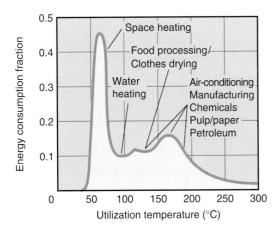

Figure 16.6 • Energy use below 300°C in the United States. (*Source*: Los Alamos Scientific Laboratory, LASL 78–24, 1978.)

Industrial Energy

The graph of total energy consumption in the United States (Figure 16.4) shows that the rate of increase in energy utilization leveled off in the early 1970s. Nevertheless, industrial production of goods (automobiles, appliances, etc.) continued to grow significantly! Today, U.S. industry consumes about one-third of the energy produced. The reason we have had higher productivity with less energy use is that more industries are using cogeneration and more energy-efficient machinery, such as motors and pumps designed to use less energy.[6,9]

Automobile Design

Steady improvements have been made in the development of fuel-efficient automobiles during the last 30 years. In the early 1970s, the average U.S. automobile burned approximately 1 gallon of gas for every 14 miles traveled. By 1996, the miles per gallon (mpg) had risen to an average of 28 for highway driving and as high as 49 for some automobiles.[10] Fuel consumption rates did not improve much from 1996 to 1999. This was due in part to lower gasoline prices in the 1990s and in part to rising concern for vehicle safety, which resulted in heavier automobiles. Today, however, the fuel consumption of some hybrid (gasoline-electric) vehicles exceeds 90 mpg on the highway and 60 mpg in the city. This improvement has several causes: increased efficiency and resulting conservation of fuel; cars that are smaller, with engines constructed of lighter materials[6]; and the combination of a fuel-burning engine with an electric motor.[11] Of course, there is a price to pay for this change. Smaller cars may be more prone to damage on impact; and as cars have gotten smaller, trucks have tended to stay the same or increased in size. As a result, the number of serious accidents between cars and trucks in the United States have increased.

Values, Choices, and Energy Conservation

A potentially effective method of conserving energy is to change our behavior by using less energy. This involves our values and the choices we make to act at a local level to address global environmental problems, such as human-induced warming caused by the burning of fossil fuels. For example, we make choices as to how far we commute to school or work and what method of transport we use to get there. Some people commute more than an hour by car to get to work, while others ride a bike, walk, or take a bus or train. Other ways of modifying behavior to conserve energy include:

- Using carpools to travel to and from work or school.
- Purchasing a hybrid car (gasoline-electric).
- Turning off lights when leaving rooms.
- Taking shorter showers (conserves hot water).
- Putting on a sweater and turning down the thermostat during winter.

What other ways of modifying your behavior would help conserve energy?

 16.6 Energy Policy

U.S. energy policy during the past half-century has not moved us closer to energy self-sufficiency. We import more oil than ever. The United States spent $110 billion on energy research and development (R&D) from 1949 to 1999 (Figure 16.7*a*). Nearly 60% of this went to nuclear energy, which supplies only 11% of our total energy. Over 20% went to fossil fuels. Renewable energy (wind, solar, biomass, and geothermal) and energy efficiency received only 11% and 7%, respectively. Investments in renewable energy are paying off slowly. Although renewable energy sources (solar and wind) provide only about 5% of the energy we use (Figure 16.7*b*), they are growing rapidly. The dollars spent on R&D for energy efficiency have yielded big returns. The average family in 1978 consumed about 27% more energy than in 1997. Savings came from such improvements as more efficient television sets, furnaces, refrigerators, and automobiles.

In the late 1990s, the United States spent about $2 billion per year on research and development for energy. By comparison, $45 billion per year went to research and development for the military. As we look to the future, we should attempt to separate politics and "pork barrel" expenditures (money politicians arrange for projects in their districts, regardless of their merit) from sound energy policy that will result in reaching sustainability with respect to energy production and use.

Hard Path versus Soft Path

Energy policy today is at a crossroads. One road leads to what is termed the **hard path**, which involves finding greater amounts of fossil fuels and building larger power plants. Taking the hard path means continuing the past emphasis on the quantity of energy we use. In this respect, the hard path is more comfortable. It requires no new thinking, no realignment of political, economic, or social conditions, and little anticipation of coming reductions in production of oil.

People heavily invested in the continued use of fossil fuels and nuclear energy favor the hard path. They argue that much environmental degradation around the world has been caused by people who have been forced to utilize local resources, such as wood, for energy. As a result, their lands have suffered from loss of plant and animal life and from soil erosion (as discussed in Chapters 10 and 11). They argue that the way to solve these environmen-

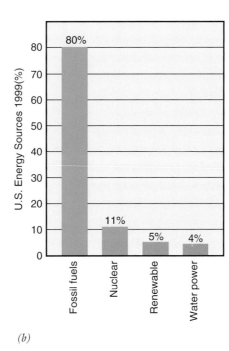

(a)

(b)

Figure 16.7 • (*a*) Expenditures for U.S. energy research and development (R&D) from 1949 to 1999 (as a percent of $110 billion). (*b*) Total U.S. energy consumed in 1999. (*Sources*: Energy Information Administration and Congressional Research Service; modified from Nona Yeats, Los Angeles Times, February 27, 2001.)

tal problems is to provide cheap, high-quality energy, such as fossil fuels or nuclear energy.

In countries like the United States, with sizable energy resources of coal, natural gas, and petroleum, people supporting the hard path argue that we should exploit those resources while finding ways to reduce the environmental impact of their use. According to these hard-path proponents, we should (1) let the energy industry develop the available energy resources; and (2) let industry, free from government regulations, provide a steady supply of energy with less total environmental damage.

The U.S. energy plan suggested by President George W. Bush in 2001 is largely a hard-path proposal. The plan is to find and use more coal, oil, and natural gas; to use more nuclear power; and to build more than 1,000 new fossil fuel plants in the next 20 years. Energy conservation and development of alternative energy sources, while encouraged, are not considered primary in importance.

The second road of energy policy is called the **soft path**.[12] Amory Lovins, the scientist who has defined and championed this soft path, states that it involves energy alternatives that emphasize energy quality, are renewable, are flexible, and are environmentally more benign than those of the hard path. As defined by Lovins, these alternatives have several characteristics:

- They rely heavily on renewable energy resources, such as sunlight, wind, and biomass (wood and other plant material).
- They are diverse and are tailored for maximum effectiveness under specific circumstances.
- They are flexible, accessible, and understandable to many people.

- They are matched in energy quality, geographic distribution, and scale, to end-use needs, increasing second-law efficiency.

Lovins points out that people are not particularly interested in having a certain amount of oil, gas, or electricity delivered to their homes; rather, they are interested in having comfortable homes, adequate lighting, food on the table, and energy for transportation.[12] According to Lovins, only about 5% of end uses require high-grade energy, such as electricity. Nevertheless, a lot of electricity is used to heat homes and water. Lovins shows that there is an imbalance in using nuclear reactions at extremely high temperatures and in burning fossil fuels at high temperatures simply to meet needs where the necessary temperature increase may be only a few 10s of degrees. Such large discrepancies are thought wasteful and a misallocation of high-quality energy.

Energy for Tomorrow

The availability of energy supplies and the future demand for energy are difficult to predict because the technical, economic, political, and social assumptions underlying predictions are constantly changing. In addition, seasonal and regional variations in energy consumption must also be considered. For example, in areas with cold winters and hot, humid summers, energy consumption peaks during the winter months, with a secondary peak in the summer (the former resulting from heating and the latter from air conditioning). Regional variations in energy sources and consumption are significant. For example, in the United States as a whole, the transportation sector uses an average of 23% of the energy consumed. However, in California, where people often commute long distances to work, 48% of the energy is used for transportation, more than double the national average. Energy sources also vary by region. For example, in the eastern and southwestern United States, the fuel of choice for power plants is often coal, but on the West Coast, power plants are more likely to burn oil or natural gas or use hydro power from dams to produce electricity.

Future changes in population densities as well as intensive conservation measures will probably change existing patterns of energy use. This might involve a shift to more reliance on alternative (particularly renewable) energy sources. One recent prediction holds that energy consumption in the United States in the year 2030 may be as high as 120 exajoules or as low as 60 exajoules (Figure 16.8). The higher value assumes little change in energy policies, whereas the low value assumes the implementation of aggressive energy conservation policies.

To bring about an environmental scenario that would also stabilize the climate in terms of global warming, use of energy from fossil fuels would need to be cut by about 50%. One reason such a large reduction in the use of fossil fuels may not come quickly is that some policy-makers are still not convinced that significant global warming will actually occur as a result of burning fossil fuels. (Global warming is discussed at length in Chapter 22.)

Over the past 30 years, energy scenarios have consistently overestimated the energy demands of the future. It seems unlikely that energy consumption in the United States in the year 2030 will be as low as 60 exajoules, and it is also unlikely that it will be as high as 120.

Low-energy scenarios for the future generally assume a moderate decrease in energy consumption accompanied by a shift from dependence on fossil fuels to the softer technologies (renewable, alternative energy resources). Reductions in energy use need not be associated with a lower quality of life. For example, Japan and some Western European nations with standards of living as high as or higher than that in the United States consume significantly less energy per person than the United States. What is needed is increased conservation of energy and more efficient use of energy, including the following:[7]

- More energy-efficient land-use planning that maximizes the accessibility of services and minimizes the need for transportation.
- Agricultural practices and personal choices that emphasize (1) eating more locally grown foods and so reducing energy use for transporting crops and (2) eating more foods such as vegetables, beans, and grains. These foods require less total energy to produce than beef, chicken, and pork, when crops are grown to feed these animals.
- Industrial guidelines for factories that promote energy conservation and minimize production of waste.

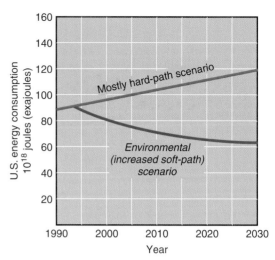

Figure 16.8 • Two possible future energy scenarios for the United States. (*Source*: Union of Concerned Scientists, 1991.)

All projections of specific sources of energy use in the future must be considered speculative. Perhaps most speculative of all is the idea that we really can obtain most of our energy needs from alternative renewable energy sources in the next several decades. From an energy viewpoint, the next 20 to 30 years, as we move through the maximum production of petroleum, will be crucial to the United States and to the rest of the industrialized world.

The transformation of energy use from a hard to a softer path has a long history, as seen in the example of the early Greek and Roman cultures. In the latter part of the 20th century, the United States experienced a shock in 1973 due to a shortage of oil. Long lines at the gas pumps caused anxiety concerning our energy supply and the lifestyle that depends on abundant oil. The 1973 oil shock spawned new research and development of alternative energy sources. It also was the impetus for the government to provide financial incentives for the utilization of sunlight, wind, and other alternative energy sources. However, with the return of abundant, inexpensive oil in the 1980s, there was much less government support of alternative energy. During this period, China, with one-fifth of the world's population, continued to develop its emerging big industry by burning huge quantities of coal. Today, the industrialized countries of the world are even more dependent on imported oil than they were in the 1970s. One way to reduce oil consumption might be to establish new taxes on energy (see A Closer Look 16.2).

The energy decisions we make in the very near future will greatly affect both our standard of living and our quality of life. From an optimistic point of view, we have the necessary information and technology to ensure a bright, warm, lighted, and moving future—but time may be running out, and we need action now. We can continue to take things as they come and live with the results of our present dependence on fossil fuels. Or we can choose

Energy Taxes

One proposal to lower demand for oil in the United States is to tax all forms of energy based on their heat content. The tax would encourage the conservation and development of alternative lower-heat-content energy sources such as solar and wind. For example, the price of gasoline in the United States (even at its peak) remains low compared with the price in other industrialized countries. This low price has reduced the incentive to conserve fuel and drive smaller automobiles. If the price of gasoline were more than $4/gal in the United States, as it is in countries such as Spain, France, Germany, and England, there certainly would be more of an incentive at the personal level to use less gasoline and to drive more fuel-efficient automobiles.

For the sake of discussion, consider the arguments for and against raising the gasoline tax by $1/gal. The taxes raised could go toward building or improving rapid-transit systems and roads, energy conservation programs, and development of alternative energy sources. Even with this increase, our gasoline prices would be moderate compared with those in the rest of the world. To minimize the shock to consumers, the tax could be put in place over a period of several years. A tax increase would be very unpopular, however, and people would be suspicious that the tax would become a revenue-generating scheme, as it is in Europe.

---

to build a sustainable energy future based on careful planning, innovative thinking, and a willingness to move from our dependency on petroleum.

Integrated, Sustainable Energy Management

The concept of **integrated energy management** recognizes that no single energy source can provide all the energy required by the various countries of the world.[13] A range of options that vary from region to region will have to be employed. Furthermore, the mix of technologies and sources of energy will involve both fossil fuels and alternative, renewable sources.

A basic goal of integrated energy management is to move toward **sustainable energy development** that is implemented at the local level. Sustainable energy development would have the following characteristics:

- It would provide reliable sources of energy.
- It would not cause destruction or serious harm to our global, regional, or local environments.
- It would help ensure that future generations inherit a quality environment with a fair share of the Earth's resources.

To implement sustainable energy development, leaders in various regions of the world will need to implement energy plans based on local and regional conditions. The plans will integrate the sources of energy most appropriate for a particular region with potential for conservation and efficiency and with desired end uses for energy. Such plans will recognize that the preservation of resources can be profitable and that degradation of the environment and poor economic conditions go hand in hand.[14] In other words, the degradation of air, water, and land resources results in a depletion of assets that ultimately will lower both the standard of living and the quality of life.

A good energy plan is part of an aggressive environmental policy with the goal of producing a quality environment for future generations. A good plan should do the following:[14]

- Provide for sustainable energy development.
- Provide for aggressive energy efficiency and conservation.
- Provide for the diversity and integration of energy sources.
- Provide for a balance between economic health and environmental quality.
- Use second-law efficiencies as an energy policy tool (that is, strive to produce a good balance between quality of energy source and end uses for that energy).

Such a plan recognizes that energy demands can be met in environmentally preferred ways. An important element of the plan involves the energy used for automobiles. This builds on policies of the past 30 years to develop hybrid engines that are part electric and part internal combustion and improve fuel technology to reduce both fuel consumption and emission of air pollutants. Finally, the plan should factor in the marketplace through pricing that reflects the economic cost of using the fuel as well as its cost to the environment. In summary, the plan should be an integrated energy management statement that moves toward sustainable development. Those who develop such plans recognize that a diversity of energy

Micropower

It is likely that sustainable energy management will include the emerging concept of **micropower**—smaller, distributed systems for production of electricity. Such systems are not new. The inventor Thomas Edison evidently anticipated that electricity-generating systems would be dispersed; and by the late 1890s, many small electrical companies were marketing and building power plants, often located in the basements of businesses and factories. These early plants evidently used cogeneration principles, since waste heat was reused for heating buildings.[16] Imagine if we had followed this early model: Homes would have their own power systems, electric power lines wouldn't snake through our neighborhoods, and we could replace older, less efficient systems as we do refrigerators.

Instead, in the 20th century in the United States, power plants producing electricity grew larger. Industrializing countries, by the 1930s, had set up utility systems based on large-scale central power plants, as diagrammed in Figure 16.9a. Today, though, we are again evaluating the merits of distributive power systems, as shown in Figure 16.9b.

Large, centralized power systems are consistent with the hard path, while the distributive power system is more aligned with the soft path. Micropower devices rely heavily on renewable energy sources such as wind and sunlight, which feed into the electric grid system shown in Figure 16.9b. Use of micropower systems in the future is being encouraged because they are reliable and are associated with

less environmental damage than are large fossil-fuel-burning power plants.[16]

Uses for micropower are emerging in both developed and developing countries. In countries that lack a centralized power generating capacity, small-scale electrical power generation from solar and wind has become the most economical option. In nations with a high degree of industrialization, micropower may emerge as a potential replacement for aging electric power plants. For micropower to be a significant factor in energy

production, a shift in policies and regulations to allow micropower devices to be more competitive with centralized generation of electrical power is required. Irrespective of the obstacles that micropower devices face, distributive power systems will probably play an important role in our goal of achieving integrated, sustainable energy management for the future.

(**a**) Central power plant, fossil fuel or nuclear

(**b**) Distributive power systems, solar, fuel cell, wind or biomass

Figure 16.9 ● Idealized diagram comparing (*a*) a centralized power system such as those used in industrial developed countries today with (*b*) a distributive power system based on generating electricity from biomass, wind, solar, and other sources, all of which feed into the transmission and distribution system. (Source: Modified from S. Dunn, "Micropower, the Next Electrical Era," Worldwatch Paper 151 [Washington, D.C.: Worldwatch Institute, 2000].)

supplies will be necessary and that the key components are (1) improvements in energy efficiency and conservation and (2) matching energy quality to end uses.[14]

The global pattern of ever-increasing energy consumption led by the United States and other nations cannot be sustained without a new energy paradigm that includes

changes in human values rather than a breakthrough in technology. Choosing to own lighter, more fuel-efficient automobiles and living in more energy-efficient homes is consistent with a sustainable energy system that focuses on how to provide and use energy to improve human welfare. A sustainable energy paradigm establishes and

Is There Enough Energy to Go Around?

The developing countries have most of the world's population (about 5 billion of 6 billion people) and are growing in population faster than developed countries. The average rate of energy use for individuals in developing countries is 1.0 kW, whereas that for persons in developed countries is 7.5 kW. If the current annual growth rate of 1.3% is maintained, the world's population will double in 54 years, to 12 billion people. More people will mean more energy use. In addition, people in developing countries will likely consume more energy per capita if they are to achieve a higher standard of living.

With a worldwide average energy use rate of 2.6 kW per person, the 6 billion people on Earth use 15.6 terrawatt (TW) years annually (a terrawatt is a trillion watts). A population of 12 billion with an average per capita energy use rate of 6.0 kW would use 72 TW years annually, or about five times as much as is presently used. Can the world support this much energy use?

John Holdren, an energy expert, believes that a realistic goal is for annual per capita energy use to reach 3 kW, with the world population peaking at 10 billion individuals by the year 2100. If this goal is to be achieved, developing nations will be able to increase their populations by no more than 60% and their energy use by no more than 100%; developed nations can increase their population by only 10% and will have to reduce their energy use by 2% each year.

Critical Thinking Questions

1. What would the energy use rate be if Holdren's goals were realized? How much total energy would be required for all people on Earth to have a standard of living supported by 7.5 kW per person? How do these totals compare with the present energy use rate worldwide?

2. Compared with today's mix of energy sources, how much greater efficiency would be required to provide an additional 45 TW years annually? How realistic is Holdren's assumption? Explain.

3. In what specific ways could energy be used more efficiently in the United States? Compare your list with those of your classmates, and compile a class list.

4. In addition to increasing efficiency, what other changes in energy consumption might be required to provide an average energy use rate of 7.5 kW per person or 3.0 kW per person in the future?

5. Would you consider Holdren's vision of the energy future an example of a hard or a soft path? Explain.

maintains multiple linkages between energy production, energy consumption, human well-being, and environmental quality.[15] It might also involve using more distributed production of energy (see A Closer Look 16.3).

Summary

- The first law of thermodynamics states that energy is neither created nor destroyed but is always conserved and is transformed from one kind to another. We use the first law to keep track of the quantity of energy.

- The second law of thermodynamics tells us that as energy is used, it always goes from a more usable (higher-quality) form to a less usable (lower-quality) form.

- Two fundamental types of energy efficiency are derived from the first and second laws of thermodynamics. In the United States today, first-law efficiencies average about 50%, which means that about 50% of the energy produced is returned to the environment as waste heat. Second-law efficiencies average 10–15%, so there is a high potential for saving energy through

better matching of the quality of energy sources with their end uses.

- Energy conservation and improvements in energy efficiency can have significant effects on energy consumption. It takes three units of a fuel such as oil to produce one unit of electricity. As a result, each electricity unit conserved or saved through improved efficiency saves three units of fuel.

- There are arguments for both the hard and soft energy paths. The former has a long history of success and has produced the highest standard of living ever experienced. However, present sources of energy are causing serious environmental degradation and are not sustainable. Soft-path alternative energy sources are renewable, decentralized, diverse, and flexible; provide a better match between energy quality and end use; and emphasize second-law efficiencies.

- Sustainable, integrated energy management is needed to make the transition from fossil fuels to other energy sources. The goal is to provide reliable sources of energy that do not cause serious harm to the environment and ensure that future generations inherit a quality environment.

REEXAMINING THEMES AND ISSUES

Human Population:

The industrialized and urbanized countries of the world produce and use most of the world's energy. As societies change from rural to urban, energy demands generally increase. Controlling the increase of human population is an important factor in reducing total demand for energy (total demand is the product of average demand per person and number of people).

Sustainability:

It will be impossible to achieve sustainability in the United States if we continue with our present energy policies. Present use of fossil fuels is not sustainable. We need to rethink the sources, uses, and management of energy. Sustainability is the central issue in our decision to continue on the hard path or change to the soft path.

Global Perspective:

Understanding trends in energy production and consumption on a global basis is important if we are to directly address the global impact of burning fossil fuels on problems of air pollution and global warming. Furthermore, the use of energy resources greatly influences global economics as these resources are transported and utilized around the world.

Urban World:

A great deal of the total energy demand is in urban regions, such as Tokyo, Beijing, London, New York, and Los Angeles. How we choose to manage energy in our urban regions greatly affects the quality of urban environments. Burning cleaner fuels results in far less air pollution. This has been observed in several urban regions, such as London. Burning of coal in London once caused deadly air pollution events. Today, natural gas and electricity heat homes, and the air is cleaner. Burning coal in Beijing continues to cause significant air pollution and health problems for millions of people living there.

Science and Values:

Public opinion polls consistently show that people value a quality environment. In response, energy planners are evaluating how to make more efficient use of our present energy resources, practice conservation of energy, and reduce adverse environmental effects of energy consumption. Science is providing options in terms of energy sources and uses; which choices we make will reflect our values.

KEY TERMS

cogeneration **315**
conservation **314**
energy **309**
energy efficiency **315**
first-law efficiency **310**

first law of thermodynamics **309**
hard path **316**
integrated energy management **319**
second-law efficiency **311**
second law of thermodynamics **310**

soft path **317**
sustainable energy development **319**
work **309**

STUDY QUESTIONS

1. What evidence supports the notion that although present energy problems are not the first in human history, they are unique in other ways?

2. How do the terms *energy, work*, and *power* differ in meaning?

3. Compare and contrast potential advantages and disadvantages of a major shift from hard-path to soft-path energy development.

4. You have just purchased a 100-ha wooded island in Puget Sound. Your house is uninsulated and built of

raw timber. Although the island receives some wind, trees over 40 m tall block most of it. You have a diesel generator for electric power, and hot water is produced by an electric heater run by the generator. Oil and gas can be brought in by ship. What steps would you take in the next five years to reduce the cost of the energy you use with the least damage to the natural environment of the island?

5. How might better matching of end uses with potential sources yield improvements in energy efficiency?

6. Complete an energy audit of the building in which you live, and develop recommendations that might lead to lower utility bills.

7. How might plans using the concept of integrated energy management differ for the Los Angeles area and the New York City area? How might both of these plans differ from an energy plan for Mexico City, which is quickly becoming one of the largest urban areas in the world?

8. A recent energy scenario for the United States suggests that in the coming decades, energy sources might be natural gas (10%), solar power (30%), hydropower (20%), wind power (20%), biomass (10%), and geothermal energy (10%). Do you think this is a likely scenario? What would be the major difficulties and points of resistance or controversy?

● FURTHER READING

Alliance to Save Energy, American Council for an Energy-Efficient Economy, Natural Resources Defense Council, and Union of Concerned Scientists. 1991. *America's Energy Choices: Investing in a Strong Economy and a Clean Environment*. Cambridge, Mass.: Union of Concerned Scientists. A detailed approach to altering America's energy use to achieve a cleaner environment and a stronger economy. Includes four scenarios that use improvements in energy efficiency and renewable energy technologies to substantially improve environmental quality and save money.

Berger, J. J. 2000. *Beating the Heat*. Berkeley, Calif.: Berkeley Hills Books. Excellent overview of all aspects of energy as they relate to global warming. Discusses how we can reduce or eliminate warming as an environmental threat.

Committee for Economic Development, Research and Policy Committee. 1993. *What Price Clean Air? A Market Approach to Energy and Environmental Policy*. New York: Committee for Economic Development. A discussion of air quality, energy use, and environmental policy in the United States, including recommendations for policy changes to improve environmental quality at a lower cost economically.

Gibbons, J. H., and P. D. Blair. 1991. "U. S. Energy Transition: On Getting from Here to There." *Physics Today* 44(7):22–32. A good overview concerning our energy supplies, including dependence on Middle East oil, advances in alternative energy systems, energy efficiency, and energy policy.

International Energy Agency. 1992. *Global Energy: The Changing Outlook*. Paris, France: Organization for Economic Cooperation and Development. A good discussion of the future shifts expected in the global energy balance, including how energy demand will change for each geographic region.

Lovins, A. B. 1979. *Soft Energy Path: Towards a Durable Peace*. New York: Harper & Row. An interesting book that presents the argument for the soft path. Its message is more important today than when it was written.

Organization for Economic Cooperation and Development (OECD), International Energy Agency (IEA). 1994. *The Economics of Climate Change*. Proceedings of an OECD/IEA Conference. Discussions of the economics of climate change, differences of opinion on the matter, methods of linking economic studies and climate change policy, and directions of international policy dealing with climate change.

CHAPTER 17 Fossil Fuels and the Environment

Slow-moving traffic during afternoon rush hour in San Francisco, California.

● LEARNING OBJECTIVES

We rely almost completely on fossil fuels (oil, natural gas, and coal) for our energy needs. However, these are nonrenewable resources, and their production and use have a variety of serious environmental impacts. After reading this chapter, you should understand:

● Why we may have serious, unprecedented supply problems with oil and gasoline within the next 20 to 50 years.
● How oil, natural gas, and coal form.
● What the environmental effects are of producing and using oil, natural gas, and coal.

Oil Shocks

During the first half of the year 2000, people in the United States were shocked when gasoline prices shot up nearly 100% in only a few months. The shock of the price increase caught the attention of a society that had grown dependent on private automobiles and gasoline, especially since the increase followed several years of a glut of inexpensive oil and gasoline.

The reasons for the sharp price increase are complex but are based on several factors. First, as demand rose in response to low fuel prices, oil-producing countries reduced production rates, thereby reducing the supply. Second, in order to reduce air pollution from burning gasoline, producers in the United States are required in many areas to reformulate gasoline with additives that produce a cleaner-burning fuel. This reformulating, which is done at a limited number of refineries, further lowered the supply. Reduced supply together with high demand resulted in a quick increase in the price of gasoline at pumps.

As dramatic as the price increases were, this was the fourth oil shock in the space of only 28 years. The first of the four was the Arab oil embargo of 1973–1974, when Arab oil-producing nations cut back production and imposed an embargo on shipments. For the first time, Americans were forced to wait in long lines to receive limited amounts of gasoline for their automobiles. During that period, the price per barrel of oil more than doubled. A second shock occurred in 1978 with the overthrow of the shah of Iran, which disrupted petroleum supplies and

again led to long lines at gasoline pumps in the United States. The third shock occurred in 1990 when Iraq invaded Kuwait, causing panic in the world financial markets and resulting in an increase of approximately 170% in oil prices in just three months.[1,2] Just when everyone thought that we would be faced with permanently high gasoline costs at the pumps, the oil glut of the mid- to late 1990s occurred; but this abruptly ended when prices rose dramatically in 2000. By late 2001 and early 2002, prices dropped once again, so we continue our roller coaster ride with gasoline prices.

It seems we have learned little since the first, second, third, or even fourth oil shortage. We are just as dependent as ever on imported oil to maintain our economic system and our livelihoods. This seems foolhardy in light of the fact that there are several viable options whereby we could decrease our dependency on foreign oil. For example, we could use more biomass energy crops, processed to produce ethanol and methanol, both of which could be used to fuel our automobiles. It has been estimated that the idle farmland in the United States today could produce sufficient fuels for our automobiles and light-duty vehicles.

Regardless of the choices we make, we must be aware that the world's fossil fuel resources, which took millions of years to form, will soon be depleted, and production will be naturally reduced. Continuing to rely on fossil fuels with the certainty of future fuel shortages is not a viable long-term option.

● *The path we take for our future energy needs will undoubtedly be a compromise between the hard and soft paths introduced in Chapter 16, and we will need to develop plans to fill the gap during the conversion from fossil fuels to other energy sources. One fuel that may assist in this transitional period appears to be natural gas, which is relatively clean-burning compared with coal. In this chapter, we discuss fossil fuels in terms of origin, estimated time to depletion, and environmental concerns.*

17.1 Fossil Fuels

Fossil fuels are forms of stored solar energy. Plants are solar energy collectors, because they can convert solar energy to chemical energy through photosynthesis (see

Chapter 4). The main fossil fuels used today were created from incomplete biological decomposition of dead organic matter (mostly land and marine plants). This occurred when buried organic matter that was not completely oxidized was converted by chemical reactions over

Figure 17.1 • World energy consumption by primary sources, 1967–1998. (*Source:* British Petroleum Company, *BP Statistical Review of World Energy* [London: British Petroleum Company, 1999].)

hundreds of millions of years to oil, natural gas, and coal. Biological and geologic processes in various parts of the geologic cycle produce the sedimentary rocks in which we find these fossil fuels.

The major fossil fuels—crude oil, natural gas, and coal—are our primary energy sources; on a worldwide basis, they provide approximately 90% of the energy consumed (Figure 17.1). In this chapter, we focus primarily on these major fossil fuels. We also briefly discuss two other fossil fuels, oil shale and tar sands, that may become increasingly important as oil, gas, and coal reserves are depleted.

17.2 Crude Oil and Natural Gas

Most geologists accept the hypothesis that **crude oil** (petroleum) and **natural gas** are derived from organic materials (mostly plants) that were buried with marine or lake sediments in what are known as depositional basins. Oil and gas are found primarily along geologically young tectonic belts at plate boundaries, where large depositional basins are more likely to occur (see Chapter 4). There are, however, exceptions, such as in Texas, the Gulf of Mexico, and the North Sea, where oil in depositional basins far from active plate boundaries has been discovered.

The source material, or *source rock*, for oil and gas is fine-grained (less than 1/16 mm, or 0.0025 in, in diameter), organic-rich sediment buried to a depth of at least 500 m (1640 ft), where it is subjected to increased heat and pressure. The elevated temperature and pressure initiate the chemical transformation of the organic material in the sediment into oil and gas. The elevated pressure causes the sediment to be compressed; this, along with the elevated temperature in the source rock, initiates the upward migration of the oil and gas, which are relatively light, to a lower-pressure environment, (known as the *reservoir rock*. The reservoir rock is coarser grained and relatively porous (has more and larger open spaces

between the grains). Sandstone and porous limestone, which have a relatively high proportion (about 30%) of empty space in which to store oil and gas, are common reservoir rocks.

As mentioned, oil and gas are light; and if their upward mobility is not blocked, they will escape to the atmosphere. This explains why oil and gas are not generally found in geologically old rocks. Oil and gas in rocks older than about 0.5 billion years have had ample time to migrate to the surface, where they have either vaporized or eroded away.[5]

The oil and gas fields from which we extract resources are places where the natural upward migration of the oil and gas to the surface is interrupted or blocked by what is known as a *trap* (Figure 17.2). The rock that helps form the trap, known as the *cap rock*, is usually a very fine-grained sedimentary rock, such as shale, which is composed of silt and clay-sized particles. A favorable rock structure, such as an anticline (arch-shaped fold) or a fault (fracture in the rock along which displacement has occurred), is necessary to form traps, as shown in Figure 17.2. The important concept is that the combination of favorable rock structure and the presence of a cap rock allow deposits of oil and gas to accumulate in the geologic environment, where they are then discovered and extracted.[5]

Petroleum Production

Production wells in an oil field recover oil through both primary and enhanced methods. *Primary production* involves simply pumping the oil from wells, but this method can recover only about 25% of the petroleum in the reservoir. To increase the amount of oil recovered to about 60%, *enhanced* methods are used. In enhanced recovery, steam, water, or chemicals, such as carbon dioxide or nitrogen gas, are injected into the oil reservoir to push the oil toward the wells, where it can be more easily recovered by pumping.

(a)

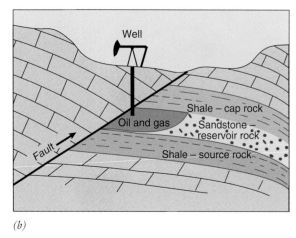

(b)

Figure 17.2 • Two types of oil and gas traps: (*a*) anticline and (*b*) fault.

Next to water, oil is the most abundant fluid in the upper part of the Earth's crust. Most of the known, proven oil reserves, however, are located in a few fields. Proven oil reserves are the part of the total resource that has been identified and can be extracted now at a profit. Of the total reserves, 65% is located in 1% of the fields, the largest of which are in the Middle East (Figure 17.3). Although new oil and gas fields have recently been and continue to be discovered in Alaska, Mexico, South America, and other areas of the world, the present known world reserves will be depleted in the next few decades.

The total resource always exceeds known reserves; it includes petroleum that cannot be extracted at a profit and petroleum that is suspected but not proved to be present. Several decades ago, the size of the proven reserves was estimated to be about 350 billion barrels (bbl). Today, that estimate is just over 1 trillion bbl.[6] The increases in proven reserves of oil in the last few decades have primarily been due to discoveries in the Middle East and Venezuela.

Because so much of the world's oil is found in the Middle East, oil revenues have flowed into that area, resulting in huge trade imbalances. Figure 17.4 shows the major routes of trade for oil. Interestingly, in recent years the United States imported more oil from Venezuela than from the Middle East. Most of the Middle Eastern oil goes to Europe, Japan, and Southeast Asia.

Oil in the 21st Century

Recent estimates of proven oil reserves suggest that, at present production rates, oil and natural gas will last only a few decades.[6,7] The important question, however, is not how long oil is likely to last at present and future production rates but when will we reach peak production? This is important because following peak production, less oil will be available, leading to shortages and price shocks. World oil production is likely to peak between the years 2020 and 2030, within the lifetime of many people living today.[8] Whichever projection is correct, there is little time left to adjust to potential changes in lifestyle and economies in a post-petroleum era.[8] We will never entirely run out of crude oil, but people of the world depend on oil for nearly 40% of their energy, and significant shortages will cause major problems.[9]

Consider the following argument that we are heading toward a potential crisis in availability of crude oil:

- We are approaching the time when approximately 50% of the total crude oil available from traditional oil fields

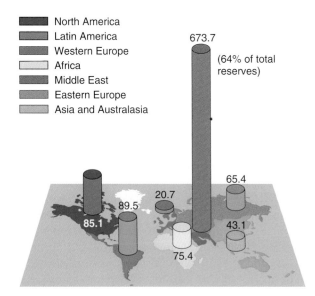

Figure 17.3 • Proven world oil reserves (billions of barrels) in 1998. The Middle East dominates with 65% of total reserves. (*Source:* British Petroleum Company, *BP Statistical Review of World Energy* [London: British Petroleum Company, 1999].)

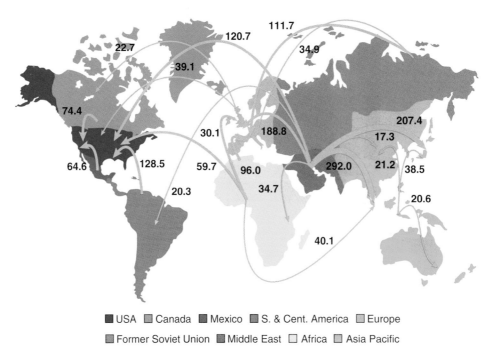

■ USA ■ Canada ■ Mexico ■ S. & Cent. America ■ Europe

■ Former Soviet Union ■ Middle East □ Africa ■ Asia Pacific

Figure 17.4 • Major trade routes for the world's oil, emphasizing the countries that use Middle Eastern oil. Units are millions of metric tons (1 metric ton = 2205 pounds) (*Source:* British Petroleum Company, *BP Statistical Review of World Energy* [London: British Petroleum Company].)

will have been consumed.[8] Recent studies suggest that about 20% more oil awaits discovery than predicted a few years ago and that there is more oil in known fields than earlier thought. However, the volumes of new oil discovered and recovered in known fields will not significantly change the date when world production will peak and a decline in production will begin.[7]

- Proven reserves are about one trillion barrels.[6] It is estimated that approximately 2 to 3 trillion barrels of crude oil may ultimately be recovered from remaining oil resources. World consumption today is about 27 billion barrels per year (74 million barrels per day). We are using what is left fast.

- Today, for every four barrels of oil we consume, we are finding only one barrel.[9] In other words, output is four times higher than input.

- Forecasts that predict a decline in production of oil are based on the estimated amount of oil that may ultimately be recoverable (2 to 3 trillion barrels, two to three times today's proven reserves), along with projections of new discoveries and rates of future consumption. As mentioned, it has been estimated that the peak in world crude oil production, about 90 million bbl/yr, will occur between the years 2020 and 2050.[7,9]

- In the United States, it is expected that the production of oil as we know it now will end by about 2090. World production of oil will be nearly exhausted by 2100.[9]

What is an appropriate response to the likelihood that production rates of oil will likely fall in the mid-21st century? First, we need an educational program early in the 21st century to inform people and governments of the potential depletion of crude oil and the consequences of shortages. Presently, many people are operating in ignorance or denial in the face of a potentially serious situation. Planning and appropriate action are necessary to avoid military confrontation (we have already had one oil war), food shortages (oil is used to make the fertilizers modern agriculture depends on), and social disruption. Before significant shortages of oil occur, we need to develop alternative energy sources such as solar energy and wind power and perhaps rely more on nuclear energy.

Natural Gas

We have only begun to seriously search for natural gas and to utilize this resource to its potential. One reason for this slow start is that natural gas is transported primarily by pipelines, and only in the last few decades have these been constructed in large numbers. In fact, until recently, natural gas found with petroleum was often simply burned off as waste; and in some cases, this practice continues.[10]

The worldwide estimate of recoverable natural gas is about 140 trillion m³, which at the current rate of world consumption will last approximately 70 years.[6] Considerable natural gas has recently been discovered in the United States, and at present U.S. consumption levels, this resource is expected to last about 30 years. Furthermore, new supplies are being found in surprisingly large amounts, particularly at depths that are deeper than those at which oil is found. Optimistic estimates of the total resource suggest that at current rates of consumption the supply may last approximately 120 years.[3]

This possibility has important implications. Natural gas is considered a clean fuel; burning it produces fewer pollutants than does burning oil or coal, so it causes fewer environmental problems than do the other fossil fuels. As a result, it is being considered as a possible transition fuel from other fossil fuels (oil and coal) to alternative energy sources, such as solar power, wind power, and hydropower.

In spite of the new discoveries and the construction of pipelines, long-term projections for a steady supply of natural gas has uncertainties. The supply is finite, and at present rates of consumption, it is only a matter of time before the resources are depleted.

The Powder River Basin in Wyoming contains a large amount of natural gas (methane) locked by groundwater in coal beds. The resource in the basin is about a trillion cubic meters and is known as "coal-bed methane." To obtain the gas from wells, huge amounts of salty groundwater are also produced. Much of the water may be reinjected at the surface possibly contaminating the semi-arid environment. Safe disposal of the waste water is a major environmental concern because it can damage or kill rangeland vegetation. With as many as 50 or more thousand wells possibly drilled to develop the gas resource in the early decades of the 21st century, the potential impact is considerable and being carefully evaluated.

Methane Hydrates

Beneath the seafloor, at depths of about 1,000 m, there exist deposits known as **methane hydrate**, a white, ice-like compound made up of molecules of methane gas (CH_4) molecular "cages" of frozen water. The methane has formed as a result of microbial digestion of organic matter in the sediments of the seafloor and has become trapped in these ice cages. Methane hydrates in the oceans were discovered over 30 years ago and are widespread in both the Pacific and Atlantic Oceans. Methane hydrates are also found on land; the first ones discovered were in permafrost areas of Siberia and North America, where they are known as marsh gas.[11]

Methane hydrates in the ocean are found in areas where deep, cold sea water provides high pressure and low temperatures. They are not stable at lower pressure and warmer temperatures. At a water depth of less than about 500 m, methane hydrates decompose rapidly, and methane gas is freed from the ice cages to move up as a flow of methane bubbles (like rising helium balloons) to the surface and the atmosphere.

Researchers from Russia in 1998 discovered the release of methane hydrates off the coast of Norway. During the release, scientists documented plumes of methane gas as tall as 500 m being emitted from methane hydrate deposits on the seafloor. It appears that there have been large emissions of methane from the sea. The physical evidence includes fields of depressions, looking something like bomb craters, that pock-mark the bottom of the sea near methane hydrate deposits. Some of the craters are as large as approximately 30 m deep and 700 m in diameter, suggesting that they were produced by rapid if not explosive eruptions of methane.

Methane hydrates in the marine environment are a potential energy resource with approximately twice as much energy as all the known natural gas, oil, and coal deposits on Earth.[11] Methane hydrates look particularly attractive to countries such as Japan that rely exclusively on foreign oil and coal for fossil fuel needs. Unfortunately, mining methane hydrates will be a difficult task, at least in the near future. The hydrates tend to be found along the lower parts of the continental slopes, where water depths are often greater than 1 km. The deposits themselves extend into the ocean floor sediments another few hundred meters. Most drilling rigs cannot operate safely at these depths, and development of a method to produce and transport the gas to land will be challenging.

Environmental Effects of Oil and Natural Gas

There is no escaping the fact that recovery, refining, and use of oil—and, to a lesser extent, gas—cause well-known, documented environmental problems, such as air and water pollution, acid rain, and global warming. Humans have gained many benefits from abundant, inexpensive energy, but at a price to the global environment and human health.

Recovery
Development of oil and gas fields involves drilling wells on land or beneath the sea floor (Figure 17.5). Possible environmental impacts on land include the following:

- Use of land to construct pads for wells, pipelines, and storage tanks and to build a network of roads and other production facilities.
- Pollution of surface waters and groundwater from: (1) leaks from broken pipes or tanks containing oil or other oil-field chemicals and (2) salty water (brine) that is brought to the surface in large volumes with the oil. The brine is toxic and may be disposed of by evaporation in lined pits, which may leak. Alternatively, brine may be disposed of by pumping it into the earth using deep disposal wells outside the oil fields. However, disposal wells may cause groundwater pollution (see Chapter 28).
- Accidental release of air pollutants, such as hydrocarbons and hydrogen sulfide (a toxic gas).
- Land subsidence (sinking) as oil and gas are withdrawn.
- Loss or disruption and damage of fragile ecosystems, such as wetlands or other unique landscapes. This is

Figure 17.5 • Drilling for oil in (*a*) the Sahara Desert of Algeria and (*b*) the Cook Inlet of southern Alaska.

(*a*)
(*b*)

the center of the controversy over the development of petroleum resources in pristine environments such as the Arctic National Wildlife Refuge in Alaska (see A Closer Look 17.1).

Environmental impacts associated with oil production in the marine environment include:

- Oil seepage into the sea from normal operations or large spills from accidents, such as blowouts or pipe ruptures.
- Release of drilling muds (heavy liquids injected into the bore hole during drilling to keep the hole open) containing heavy metals, such as barium, that may be toxic to marine life.
- Aesthetic degradation from the presence of offshore oil-drilling platforms, which some people think are unsightly.

Refining

Refining crude oil and converting it to related products also create environmental impacts. At refineries, crude oil is heated so that its components can be separated and collected (this process is called *fractional distillation*). Other industrial processes are then used to make products such as gasoline and heating oil.

Refineries may have accidental spills and slow leaks of gasoline and other products from storage tanks and pipes. Over years of operation, large amounts of liquid hydrocarbons may be released, polluting soil and groundwater resources below the site. Massive groundwater cleaning projects have been required at several West Coast refineries.

Crude oil and its distilled products are used to make fine oil, a wide variety of plastics, and organic chemicals used by society in huge amounts. The industrial processes involved in the production of these chemicals have the potential for releasing a variety of pollutants into the environment.

Delivery and Use

Some of the most extensive and significant environmental problems associated with oil and gas occur when the fuel is delivered and consumed. Crude oil is mostly transported on land in pipelines or across the ocean by tankers, and both methods present the danger of oil spills. For example, a bullet from a high-powered rifle punctured the trans-Alaskan pipeline in 2001, causing a small but damaging oil spill. Although most effects of oil spills are relatively short lived (days to years), marine spills have killed thousands of seabirds, temporarily spoiled beaches, and caused loss of tourist and fishing revenues (see Chapter 21).

Air pollution is perhaps the most familiar and serious environmental impact associated with the use (burning) of oil. Combustion of gasoline in automobiles produces pollutants that contribute to urban smog. The adverse effects of smog on vegetation and human health are well documented and are discussed in detail in Chapter 23.

 17.3 Coal

Partially decomposed vegetation, when buried in a sedimentary environment, may be slowly transformed into the solid, brittle, carbonaceous rock we call **coal**. This process is shown in Figure 17.8. Coal is by far the world's

The Arctic National Wildlife Refuge: To Drill or Not to Drill

The Arctic National Wildlife Refuge (ANWR) on the North Slope of Alaska is one of the few pristine wilderness areas remaining in the world (Figure 17.6). The U.S. Geological Survey estimates that the refuge contains about 3 billion bbl of recoverable oil. According to the oil industry, several times more oil than that can be recovered. The oil industry has long argued in favor of drilling for oil in the ANWR, but the idea was unpopular for decades among many members of the public and the U.S. government, and no drilling was permitted. When George W. Bush, who favored drilling in the ANWR, was elected president in 2000, however, controversy surrounding drilling emerged.

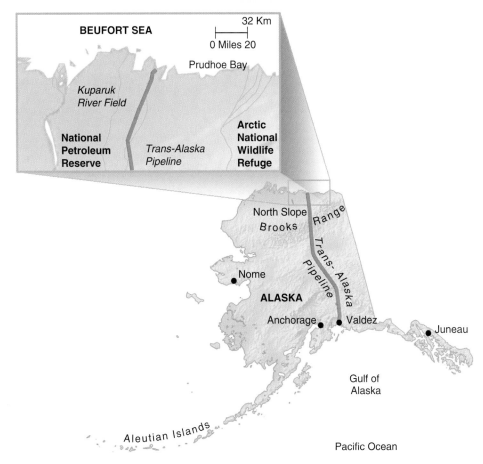

Figure 17.6 ● The National Arctic Wildlife Refuge, North Slope, Alaska, is valued for its scenery, wildlife, and oil.

Argument in Favor of Drilling in the ANWR

Those in favor of drilling for oil in the refuge make the following arguments:

- The United States needs the oil, and it will help us to be more independent of imported oil.
- New oil facilities will bring jobs and dollars to Alaska.

- New exploration tools to evaluate the subsurface for oil pools require far fewer exploratory wells.
- New drilling practices have much less impact on the environment (Figure 17.7a). These include: (1) constructing roads of ice in the winter that melt in the summer instead of constructing permanent roads (Figure 17.7b); (2) elevating pipelines to allow for animal migration (Figure 17.7c); (3) drilling in various directions from a central location, minimizing land needed for wells; and (4)

Shale

Sandstone

Conglomerate

Limestone

Oil wells- precision directional drilling (4-6 km long, deep)

Elevated pipeline- (2m) allows animal passage

Road constructed of ice in winter, in spring it melts

Oil

Pond in summer

Disposal well for oil field waste

Oil field wastes

Fault

Patterned ground with ice wedges "▼" on permafrost

Figure 17.7 • (*a*) New technology can reduce the impact of developing oil fields in the Arctic: Wells are located in a central area and use directional drilling; roads are constructed of ice in the winter, melting to become invisible in the summer; pipelines are elevated to allow animals, in this case caribou, to pass through the area, and oil field and drilling wastes are injected deep underground. (*b*) Oil wells being drilled on frozen ground (tundra), North Slope Alaska; and (*c*) caribou passing under a pipeline near the Arctic National Refuge in Alaska.

(a)

(b)

(c)

disposing of oil field wastes by putting them back into the ground to minimize surface pollution.

Argument Against Drilling in the ANWR

Those who are opposed to drilling in the ANWR argue as follows:

- Advances in technology are irrelevant to the question of whether or not the NAWR should be drilled. Some wilderness should remain wilderness! Drilling will forever change the pristine environment of the North Slope.

- Even with the best technology, oil exploration and development will impact the ANWR. Intensive activity, even in winter on roads constructed of ice, will probably disrupt wildlife.

- Ice roads are constructed from water from the tundra ponds (Figure 17.7). To build a road 1 km (0.63 mi) long requires about 3,640 m³ (1 million gallons) of water.

- Heavy vehicles used in exploration permanently scar the ground—even if the ground is frozen hard when the vehicles travel across the surface of the open tundra.

- Oil development is inherently damaging because it involves a massive industrial complex of people, vehicles, equipment, pipelines, and support facilities.

The decision we make concerning the ANWR will reflect both science and values at a basic level. New technology will produce less environmental impact from drilling oil. How we value the need for energy from oil compared with how we value preservation of a pristine wilderness area will determine our action: to drill or not to drill. Congress in 2002 voted not to drill.

(a) Coal swamps form.

(b) Rise in sea level buries swamps in sediment.

(c) Compression of peat forms coal.

Figure 17.8 • Processes by which buried plant debris (peat) is transformed into coal.

most abundant fossil fuel, with a total recoverable resource of about 1,000 billion metric tons (Figure 17.9). The annual world consumption of coal is about 4 billion metric tons, sufficient for about 250 years at the current rate of use. However, if consumption of coal increases in the coming decades, the resource will not last nearly so long.[12]

Coal is classified according to its energy and sulfur content as anthracite, bituminous, subbituminous, or lignite (see Table 17.1). Energy content is greatest in anthracite coal, and lowest in lignite coal. The distribution of coal in the contiguous United States is shown in Figure 17.10.

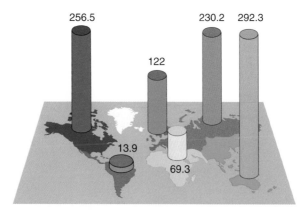

Figure 17.9 • World coal reserves (billions of tons) in 1999. The United States has about 26% of the total reserves. (*Source:* British Petroleum Company, *BP Statistical Review of World Energy* [London: British Petroleum Company].)

Table 17.1   U.S. Coal Resources

Type of Coal	Relative Rank	Energy Content (millions of joules/kg)	Sulfur Content (%)		
			Low (0–1)	Medium (1.1–3.0)	High (3 +)
Anthracite	1	30–34	97.1	2.9	—
Bituminous coal	2	23–34	29.8	26.8	43.4
Subbituminous coal	3	16–23	99.6	0.4	—
Lignite	4	13–16	90.7	9.3	—

Sources: U.S. Bureau of Mines Circular 8312, 1966; P. Averitt, "Coal," in D. A. Brobst and W. P. Pratt, eds., United States Mineral Resources, *U.S. Geological Survey, Professional Paper 820*, pp. 133–142.

The sulfur content of coal is important because low-sulfur coal emits less sulfur dioxide (SO_2) and as a result is more desirable as a fuel for power plants. Most of the low-sulfur coal in the United States is relatively low-grade, low-energy lignite and subbituminous coal found west of the Mississippi River. Power plants on the East Coast treat the high-sulfur coal mined in their own region to lower its sulfur content before, during, or after combustion and thus avoid excessive air pollution. Although it is expensive, treating coal to reduce pollution may be more economical than transporting low-sulfur coal from the western states.

Coal Mining and the Environment

In the United States, thousands of square kilometers of land have been disturbed by coal mining, and only about half this land has been reclaimed. Reclamation is the process of restoring and improving disturbed land, often by reforming the surface and replanting vegetation (see

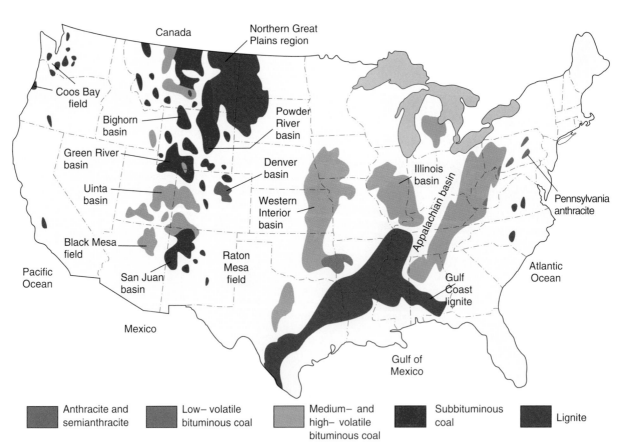

Figure 17.10 • Coal areas of the contiguous United States. This is a highly generalized map, and numerous relatively small occurrences of coal are not shown. (*Source:* S. Garbini and S. P. Schweinfurth, *U.S. Geological Survey Circular 979*, 1986.)

Figure 17.11 • Strip coal mine in Wyoming. The land in the foreground is being mined, and the green land in the background has been reclaimed following mining.

Chapter 10). Unreclaimed coal dumps from open-pit mines are numerous and continue to cause environmental problems. Because little reclamation occurred before about 1960, and mining started much earlier, abandoned mines are common in the United States. One surface mine in Wyoming, abandoned more than 40 years ago, caused a disturbance so intense that vegetation has still not been reestablished on the waste dumps. Such barren, ruined landscapes emphasize the need for reclamation.[12]

Strip Mining

Over half of the coal mining in the United States is done by *strip mining*, a surface mining process in which the overlying layer of soil and rock is stripped off to reach the coal. The practice of strip mining started in the late 19th century and has steadily increased because it tends to be cheaper and easier than underground mining. More than 40 billion metric tons of coal reserves are now accessible to surface mining techniques. In addition, approximately another 90 billion metric tons of coal within 50 m (165 ft) of the surface are potentially available for strip mining. It is likely that more and larger strip mines will be developed as the demand for coal increases.

The impact of large strip mines varies from region to region, depending on topography, climate, and reclamation practices. One serious problem in areas of the eastern United States with abundant rainfall is *acid mine drainage*—the drainage of acidic water from mine sites (see Chapter 21). Acid mine drainage occurs when surface water (H_2O) infiltrates the spoil banks (rock debris left after the coal is removed). The water reacts chemically with sulfide minerals, such as pyrite (FeS_2), a natural component of some sedimentary rocks containing coal, to pro-

duce sulfuric acid (H_2SO_4). The acid then pollutes streams and groundwater resources. Acid water also drains from underground mines and roads cut in areas where coal and pyrite are abundant, but the problem of acid mine drainage is magnified when large areas of disturbed material remain exposed to surface waters. Acid mine drainage can be minimized from active mines by channeling surface runoff or groundwater before it enters a mined area and diverting it around the potentially polluting materials.[13] However, diversion is not feasible in heavily mined regions where spoil banks from unreclaimed mines may cover hundreds of square kilometers. In these areas, acid mine drainage will remain a long-term problem.

In arid and semiarid regions, water problems associated with mining are not as pronounced as in wetter regions, but the land may be more sensitive to activities related to mining, such as exploration and road building. In some arid areas of the western and southwestern United States, the land is so sensitive that tire tracks can remain for years. (Indeed, wagon tracks from the early days of the westward migration reportedly have survived in some locations.) To complicate matters, soils are often thin, water is scarce, and reclamation work is difficult.

Strip mining has the potential to pollute or damage water, land, and biologic resources. However, good reclamation practices can minimize the damage (Figure 17.11). Reclamation practices required by law necessarily vary by site. Some of the principles of reclamation are illustrated in the case history of a modern coal mine in Colorado (see a Closer Look 17.2).

Coal mining in the Appalachian Mountains of West Virginia is a major component of the state's economy. However, there is a growing environmental concern over

The Trapper Mine

The Trapper Mine on the western slope of the Rocky Mountains in northern Colorado is a good example of a new generation of large coal strip mines. The operation, in compliance with mining laws, is designed to minimize environmental degradation during mining and to reclaim the land for dryland farming and grazing of livestock and big game.

Over a 35-year period, the mine will produce 68 million metric tons of coal from the 20–24 km^3 (4.8–5.8 mi^3) site, to be delivered to an 800-MW power plant located adjacent to the mine. Four coal seams, varying from about 1 to 4 m (3.3–13.1 ft) thick, will be mined. The seams are separated by layers of rock, called overburden (rocks without coal), and there is additional overburden above the top seam of coal. The depth of the overburden varies from 0 to about 50 m (165 ft).

A number of steps are involved in the actual mining. First, bulldozers and scrapers remove the vegetation and topsoil from an area up to 1.6 km long and 53 m wide (1 mile by 175 ft), and the soil is stockpiled for reuse. Then the overburden is removed with a 23-m^3 (800-ft^3) dragline bucket. Next, the exposed coal beds are drilled and blasted to fracture the coal, which is removed with a backhoe and loaded onto trucks (Figure 17.12). Finally, the cut is filled, the topsoil replaced, and the land either planted with a crop or returned to rangeland.

At the Trapper Mine, the land is reclaimed without artificially applying water. Precipitation (mostly snow) is about 35 cm/year (about 14 in/year), which is sufficient to reestablish vegetation provided there is adequate topsoil. The fact that reclamation is possible at this site emphasizes an important point about reclamation: It is site specific. What works at one location may not be applicable to other areas.

Water and air quality are closely monitored at the Trapper Mine. Surface water is diverted around mine pits, and groundwater is intercepted while pits are open. Settling basins, constructed downslope from the pit, allow suspended solids in the water to settle out before the water is discharged into local streams. Although air quality at the mine can be degraded by dust produced from the blasting, hauling, and grading of the coal, the dust is minimized by regular sprinkling of water on the dirt roads.

Reclamation at the Trapper Mine has been successful during the first years of operation. Although reclamation increases the cost of the coal by as much as 50%, it will pay off in the long-range productivity of the land as it is returned to farming and grazing uses. On the one hand, it might be argued that the Trapper Mine is unique in its combination of geology, hydrology, and topography, which has allowed for a successful reclamation. To some extent this is true, and perhaps The Trapper Mine presents an overly optimistic perspective on mine reclamation compared with other sites with less favorable conditions. On the other hand, the success of the mine operation demonstrates that, with careful site selection and planning, the development of energy resources can be compatible with other land uses.

(a)

(b)

Figure 17.12 ● (a) Mining an exposed coal bed at the Trapper Mine, Colorado, and (b) the land during restoration following mining. Topsoil (lower right) is spread prior to planting of vegetation.

the mining technique known as "mountaintop removal." This strip mining method is very effective in obtaining coal as it levels the tops of mountains and fills valleys with coal mining waste rock. As mountains are destroyed the flood hazard is increased as valleys are filled with mine waste and toxic waste water is stored behind coal waste sludge dams. Mining can produce voluminous amounts of coal dust that settles on towns and fields, polluting the land and causing or facilitating lung disease including asthma. Protests and complaints by communities in the path of mining that formerly were ignored are now receiving more attention by the state mining board. As people become better educated about mining laws they are more effective in confronting mining companies to have them reduce potential adverse consequences of mining—however, much more needs to be done. In May of 2002 a federal judge ordered the government to no longer allow mining companies to dump mining waste into streams and valleys. The decision upholds laws to protect our streams and rivers, but the ruling will be fought by the mining industry and those elected officials who favor mining coal at any costs to the environment.

Since the adoption of the Surface Mining Control and Reclamation Act of 1977, the U.S. government has required that mined land be restored to support its pre-mining use. The regulations also prohibit mining on prime agricultural land and give farmers and ranchers the opportunity to restrict or prohibit mining on their land, even if they do not own the mineral rights. Reclamation includes disposing of wastes, contouring the land, and replanting vegetation. Reclamation is often difficult, and it is unlikely that it will be completely successful. In fact, some environmentalists argue that success stories with reclamation are the exception and that strip mining should not be allowed in the semiarid southwestern states because reclamation is uncertain in that fragile environment.

Underground Mining

Underground mining accounts for approximately 40% of the coal mined in the United States. In addition, underground mines have been abandoned, particularly in the eastern U.S. coalfields of the Appalachian Mountains. Underground coal mining is a dangerous profession; there are always hazards of collapse, explosion, and fire. Respiratory illnesses are a risk, especially black lung disease, which is related to exposure to coal dust and has killed or disabled many miners.

Some of the environmental problems associated with underground mining include the following:

- Acid mine drainage from the mines and waste piles has polluted thousands of kilometers of streams (see Chapter 21).
- Land subsidence can occur over mines. Vertical subsidence occurs when the ground above coal mine tunnels collapses, often resulting in a crater-shaped pit at

Figure 17.13 • Subsidence below coal mines in the Appalachian coal belt.

the surface (Figure 17.13). Coal mining areas in Pennsylvania and West Virginia, for example, are well known for serious subsidence problems. In recent years, a parking lot and crane collapsed into a hole over a coal mine in Scranton, Pennsylvania; and damage from subsidence caused condemnation of many buildings in Fairmont, West Virginia.

- Coal fires in underground mines may be either naturally caused or deliberately set. The fires may belch smoke and hazardous fumes, causing people exposed to them to suffer from a variety of respiratory diseases.

Transport of Coal

Transporting coal from mining areas to large population centers where energy is needed is a significant environmental issue. Although coal can be converted at the production site to electricity, synthetic oil, or synthetic gas, these alternatives have their own problems. Power plants necessary to convert coal to electricity require water for cooling, and in semiarid coal regions of the western United States there may not be sufficient water. Furthermore, transmission of electricity over long distances is inefficient and expensive (see Chapter 16). The conversion of coal to synthetic oil or gas also requires a tremendous amount of water; and the conversion process is expensive.[14]

Freight trains and coal-slurry pipelines (designed to transport pulverized coal mixed with water) are options to transport the coal itself over long distances. Trains are typically used and will continue to be used because they provide relatively low-cost transportation compared with the cost of constructing pipelines. The economic advantages of slurry pipelines are tenuous, especially in the western United States, where large volumes of water to transport the slurry are difficult to obtain.[14]

The Future of Coal

The burning of coal produces nearly 60% of the electricity used and about 25% of the total energy consumed in the United States today.[15] Coal accounts for nearly 90% of the fossil fuel reserves in the United States, and we have enough coal to last at least several hundred years. However, serious concern has been raised about burning that coal. Giant power plants that burn coal as a fuel to produce electricity in the United States are responsible for about 70% of the total emissions of sulfur dioxide, 30% of the nitrogen oxides, and 35% of the carbon dioxide. (The effects of these pollutants are discussed in Chapter 23.)

Legislation as part of the Clean Air Amendments of 1990 mandated that sulfur dioxide emissions from coal-burning power plants be eventually cut by 70% to 90%, depending on the sulfur content of the coal, and that nitrogen oxide emissions be reduced by about 2 million metric tons per year. As a result of this legislation, utility companies are struggling with various new technologies designed to reduce emissions of sulfur dioxide and nitrogen oxides from burning coal. Options being used or developed include the following:[16]

- Chemical and/or physical cleaning of coal prior to combustion.
- New boiler designs that permit a lower temperature of combustion, which reduces emissions of nitrogen oxides.
- Injection of material rich in calcium carbonate (such as pulverized limestone or lime) into the gases following burning of the coal. This practice, known as **scrubbing**, removes sulfur dioxides. In the scrubber—a large, expensive component of a power plant—the carbonate reacts with sulfur dioxide, producing hydrated calcium sulfite ($CaSO_3 \cdot 1/2H_2O$) as a sludge. The sludge has to be collected and disposed of, which is a major problem.
- Conversion of coal at power plants into a gas (syngas, a methane-like gas) before burning. This technology is being tested at the Polk Power Station in Florida. The syngas, although cleaner-burning than coal, is still more polluting than natural gas.
- Consumer education about energy conservation and efficiency to reduce the demand for energy and thus the amount of coal burned and emissions released.

The real shortages of oil and gas are still a few years away, but when they do come, they will put pressure on the coal industry to open more and larger mines in both the eastern and western coal beds of the United States. Increased use of coal will have significant environmental impacts for several reasons.

1. More and more land will be strip mined and thus will require careful and expensive restoration.

2. Unlike oil and gas, burning coal produces large amounts of air pollutants, as already mentioned. Also produced are ash, which can be as much as 20% of the coal burned; boiler slag, a rocklike cinder produced in the furnace; and calcium sulfite sludge, produced from removing sulfur through scrubbing. Coal-burning power plants in the United States today produce about 90 million tons per year of these materials. Calcium sulfite from scrubbing can be used to make wallboard (by converting calcium sulfite to calcium sulfate, which is gypsum) and other products. Gypsum is being produced for wallboard in Japan and Germany. However, this practice is unlikely to be widely used in the United States, where wallboard can be made less expensively from abundant natural gypsum deposits. Another waste product, boiler slag can be used for fill along railroad tracks and at construction projects. Nevertheless, about 75% of the combustion products of burning coal in the United States today end up on waste piles or in landfills.[17]

3. The handling of large quantities of coal through all stages (mining, processing, shipping, combustion, and final disposal of ash) will have potentially adverse environmental effects. These include aesthetic degradation, noise, dust, and—most significant from a health standpoint—release of harmful or toxic trace elements into the water, soil, and air.

It seems unlikely that coal will be abandoned in the near future in the United States, because we have so much of it and we have spent so much time and money developing coal resources. It has been suggested that we should now promote the use of natural gas in preference to coal because it burns so much cleaner. However, there is a concern that we might then become dependent on imports of natural gas. Regardless, it remains a fact that coal is the most polluting of all the fossil fuels.

Allowance Trading

An innovative approach to managing U.S. coal resources and reducing pollution is **allowance trading**. In this system, the Environmental Protection Agency grants utility companies tradable allowances for polluting. One allowance is good for 1 ton of sulfur dioxide emissions per year. In theory, some companies would not need all their allowances because they use low-sulfur coal or have installed equipment and methods that have reduced their emissions. Their extra allowances may then be traded and sold by brokers to other utility companies that are unable to stay within their allocated emission levels. The idea is to encourage competition in the utility industry and reduce overall pollution through economic market forces.[4,16]

Some environmentalists are not comfortable with the concept of allowance trading. They argue that, although buying and selling may be profitable to both parties in the transaction, it is less acceptable from an environmental viewpoint. They believe that companies should not be able to buy their way out of pollution problems.

17.4 Oil Shale and Tar Sands

Oil shale and tar sands play a minor role in today's mix of available fossil fuels. However, they may be more significant in the future when traditional oil from wells becomes scarce.

Oil Shale

Oil shale is a fine-grained sedimentary rock containing organic matter (kerogen). When heated to 500°C (900°F) in a process known as *destructive distillation*, oil shale yields up to nearly 60 L (14 gal) of oil per ton of shale. If not for the heating process, the oil would remain in the rock. The oil from shale is one of the so-called **synfuels** (from the words *synthetic* and *fuel*), which are liquid or gaseous fuels derived from solid fossil fuels. The best known sources of oil shale in the United States are found in the Green River formation, which underlies approximately 44,000 km^2 (17,000 mi^2) of Colorado, Utah, and Wyoming.

Total identified world oil shale resources are estimated to be equivalent to about 3 trillion bbl of oil. However, evaluation of the oil grade and the feasibility of economic recovery with today's technology is not complete. Oil shale resources in the United States amount to about 2 trillion bbl of oil, or two-thirds of the world total. Of this, 90%, or 1.8 trillion bbl, is located in the Green River oil shales.[6,19]

The environmental impact of developing oil shale varies with the recovery technique used. Both surface and subsurface mining techniques have been considered.

Surface mining is attractive to developers because nearly 90% of the shale oil can be recovered, compared with less than 60% by underground mining. However, waste disposal will be a major problem with either surface or subsurface mining. Both require that oil shale be processed, or *retorted* (crushed and heated), at the surface. The volume of waste will exceed the original volume of shale mined by 20% to 30%, owing to the fact that crushed rock, because of added pore spaces, has more volume than the solid rock from which it came. (For example, pour some concrete into a milk carton. When it hardens, remove the block of concrete from the carton and break it into small pieces with a hammer. Then try to place the pieces in the carton. You will find that not all of the pieces fit.) The mines from which the shale is removed thus will not be able to accommodate all the waste, and its disposal will become a problem.[10]

In the 1970s, a tremendous rush to develop oil shale was expected. Interest in oil shale was heightened by the oil embargo in 1973 and by fear of continued shortages of crude oil. In the 1980s through the mid-1990s, though, plenty of cheap oil was available, and oil shale development was put on the back burner; it is much more expensive to extract a barrel of oil from oil shale than to pump it from a well. However, shortages of oil will occur in the future, and we may again turn to oil shale. This would result in significant environmental, social, and economic impacts in the oil shale areas resulting from rapid urbanization to house a large workforce, construction of industrial facilities, and an increased demand on water resources.

Tar Sands

Tar sands are sedimentary rocks or sands impregnated with tar oil, asphalt, or bitumen. Petroleum cannot be recovered from tar sands by pumping wells or other usual commercial methods because the oil is too viscous (thick) to flow easily. Oil in tar sands is recovered by first mining the sands (which are very difficult to remove) and then washing the oil out with hot water.

Some 75% of the world's known tar sand deposits are in the Athabasca Tar Sands near Alberta, Canada. The total Canadian resource represents about 2 trillion bbl, but it is not known how much of this will eventually be recovered. Today's production of the Athabasca Tar Sands is about 500,000 bbl of synthetic crude oil per day, about 0.7% of the world's production of oil.[8]

In Alberta, tar sand is mined in a large open-pit mine. The mining process is complicated by the fragile native vegetation, a water-saturated mat known as a muskeg swamp—a kind of wetland that is difficult to remove except when frozen. Restoration of this fragile, naturally frozen (permafrost) environment is difficult. In addition, there is a waste disposal problem because the mined sand material (as discussed for oil shale) occupies a greater volume than unmined material. The land surface after mining can be up to 20 m (66 ft) greater than the original ground surface.

Summary

- The United States has an energy problem caused by dependence on fossil fuels, especially oil. Maximum global production is expected between 2020 and 2050, followed by a decline in production. The challenge is to plan now for the decline in oil supply and shift to alternative energy sources.

- Fossil fuels are forms of stored solar energy. Most are created from the incomplete biological decomposition of dead organic material that is buried and converted by complex chemical reactions in the geologic cycle.

- Because fossil fuels are nonrenewable, we will eventually have to develop other sources to meet our energy demands. We must decide when the transition to alternative fuels will occur and what will be the impacts of the transition.

Should the Gasoline Tax Be Raised?

Petroleum, like other fossil fuels, is a limited and nonrenewable resource that must be conserved and used wisely during a period of transition to other energy sources. Because petroleum occurs naturally in liquid form and is easily portable, it is particularly well suited to transportation uses. The number of motor vehicles in the world, most of which use petroleum as fuel, have increased about nine times since 1950. There are about 750 million motor vehicles in the world; and about one third of them are in the United States. Each day, the United States uses over 10 million bbl of oil for transportation—about 70% of the country's total oil consumption.

In addition to the drain on a limited resource, the use of petroleum in transportation contributes significantly to air pollution. Approximately 60% of the air pollutants annually emitted into the atmosphere in the United States are attributed to automobiles. Gasoline accounts for 25% of the carbon dioxide emissions from burning fossil fuels in the United States. Each gallon of gasoline burned adds 8.6 kg (19 lb) of carbon dioxide to the air. A car that runs for 160,000 km (100,000 mi) with an average fuel consumption of 11.7 km/l (27.5 mi/gal) emits 35 metric tons of carbon dioxide. Vehicles also emit carbon monoxide, nitrogen oxides, and organic chemicals called hydrocarbons. In the presence of sunlight, the hydrocarbons react with the nitrogen oxides to produce ozone. Carbon monoxide competes with oxygen in the blood, nitrogen oxides form nitric acid in the presence of water, some organic molecules are carcinogenic, and ozone is an irritant to the lungs.

Proposals for reducing the dependence on petroleum for transportation, and thereby slowing the drain on oil resources and decreasing air pollution, include designing more efficient vehicles; using alternative fuels; shifting to other modes of transportation, such as mass transit and bicycles; improving roads; and encouraging consumers to conserve gasoline. These proposals have to some extent been implemented.

Some energy policy makers have also suggested that increasing the tax on gasoline would encourage Americans to purchase more fuel-efficient vehicles and discourage unnecessary use of motor transportation, thereby lowering the demand for gasoline. One of the main arguments advanced for higher gasoline taxes is that U.S. gasoline taxes are only 12% to 18% of those of other industrialized countries (Figure 17.14). Opponents, however, argue that higher taxes at the pump might have a greater impact on

poorer families and would also unfairly punish those living in western states, where driving distances are typically greater. As with many controversial issues, the facts themselves are in question. For example, do the poor spend proportionally more on transporta-

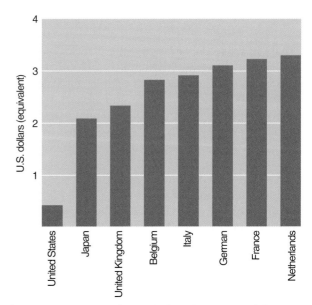

Figure 17.14 • Taxes for 1 gal of gasoline in U.S. dollars equivalent for selected countries, 1995. (*Source:* Federal Highway Administration, U.S. Department of Transportation, *Monthly Motor Fuel Reports by State*, Washington, D.C., August 1995.)

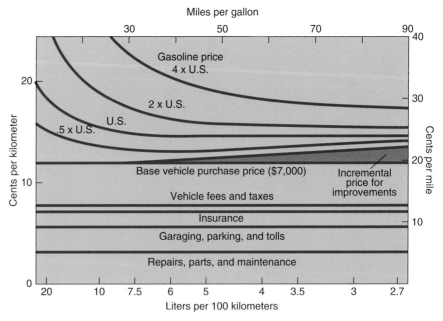

Figure 17.15 • Estimating cost of driving with increased fuel economy.

tion, and would they be disproportionately and adversely affected by a rise in gasoline taxes?

Not surprisingly, higher gas taxes have been advocated by many environmentalists. Automobile companies have also supported the proposal, hoping that it would eliminate pressure for more stringent fuel-efficiency standards. Americans are divided on the issue. In the past people have supported, by a slim majority, a modest tax increase of about 15 cents per gallon.

Critical Thinking Questions

1. Refer to Figure 17.15 to answer the following questions:
   a. What is the relationship of fuel efficiency (mpg) to gasoline price (cents per mile)?

b. What effect would increasing the price of gasoline have in terms of giving consumers an incentive to choose more fuel-efficient vehicles? What increase in the cost would motivate you to purchase a more fuel-efficient vehicle over other factors, such as size of car, comfort, and safety?

2. Make a table of the pros and cons of raising the price of gasoline through taxation in relation to environmental, economic, and social issues. What are the direct and indirect effects?

3. When you have completed the table, evaluate the arguments and decide where you stand on the issue. Write a paragraph explaining your position.

- Environmental impacts related to oil and natural gas include those associated with exploration and development (damage to fragile ecosystems, water pollution, air pollution, and waste disposal); those associated with refining and processing (soil, water, and air pollution); and those associated with burning oil and gas for energy to power automobiles, produce electricity, run industrial machinery, heat homes, and so on (air pollution).
- Coal is a source of energy particularly damaging to the

environment. The environmental impacts associated with mining, processing, transporting, and using coal are many. Problems associated with mining include fires, subsidence, acid mine drainage, and difficulties related to land reclamation. Burning coal can release air pollutants, including sulfur dioxide and carbon dioxide. Finally, burning coal produces a large volume of combustion products and by-products such as ash, slag, and calcium sulfite (from scrubbing).

● REEXAMINING THEMES AND ISSUES

Human Population:
As human population (particularly in developed countries such as the United States) has increased, so has the total impact from the use of fossil fuels. Total impact is the product of the impact per person and the total number of people. Reducing the impact will require that all countries adopt a new energy paradigm emphasizing use of the minimum energy needed to complete a task (end use) rather than the current prodigious over-use of energy.

Sustainability:
It has been argued that we cannot achieve sustainable development and maintenance of a quality environment for future generations if we continue to increase our consumption of fossil fuels. Achieving sustainability will require wider use of a variety of alternative renewable energy sources and less dependence on fossil fuels.

Global Perspective:
The global environment has been significantly affected by burning fossil fuels. This is particularly true for the atmosphere, where fast-moving processes operate (see Chapters 22 and 23). Solutions to global problems from burning fossil fuels are implemented at the local and regional levels, where the fuels are consumed.

Urban World:
Burning of fossil fuels in urban areas has a long history of problems. Not too many years ago, black soot from burning coal covered the buildings of most major cities of the world, and historic pollution events killed thousands of people. Today, we are striving to improve our urban environments and reduce urban environmental degradation from burning fossil fuels.

Science and Values: Scientific evidence for the adverse effects of burning fossil fuels is well documented. The controversy concerning their use is linked to our values. Do we value burning huge amounts of fossil fuels to increase economic growth more than living in a quality environment? Economic growth is possible without damaging the environment; developing a sustainable energy policy that doesn't harm the environment is possible with present technology. What is required are changes in values and lifestyle that are linked to energy production and use, human well-being, and environmental quality.

● KEY TERMS

allowance trading **338**	fossil fuels **325**	oil shale **339**	synfuels **339**
coal **330**	methane hydrate **329**	scrubbing **338**	tar sands **339**
crude oil **326**	natural gas **326**		

● STUDY QUESTIONS

1. Assuming that oil production will peak in about 2020 and then decline at about 3% per year, when will production be half of that in 2020? What could be the consequences? Why? How could these consequences be avoided?

2. Compare the potential environmental consequences of burning oil, burning natural gas, and burning coal.

3. What actions can you take at a personal level to reduce consumption of fossil fuels?

4. What environmental and economic problems could result from a rapid transition from fossil fuels to alternative sources?

5. What are some of the technical solutions to reducing air-pollutant emissions from burning coal. Which are best? Why?

6. What do you think about the idea of allowance trading as a potential solution to reducing pollution from burning coal?

● FURTHER READING

Brookins, D. G. 1990. *Mineral and Energy Resources.* Columbus, Ohio: Merrill. A book written by a geologist that includes a good description of how fossil fuels form and how they are being used today, along with information concerning the environmental effects of burning fossil fuels.

Fulkerson, W., R. R. Judkins, and M. K. Sanghvi. 1990. "Energy from Fossil Fuels." *Scientific American* 263(3):128–135. A good overview of fossil fuels in terms of sources, consumption, and environmental considerations.

Liu, P. I. 1993. *Introduction to Energy and the Environment.* New York: Van Nostrand Reinhold. A good summary of energy sources and issues, with discussions of the environmental effects of various energy sources.

Miller, E. W., and R. M. Miller. 1993. *Energy and American Society: A Reference Handbook.* ABC-CLIO. Follows patterns of energy use from early American history through today and includes looks at future energy sources.

Vergara, W., N. E. Hay, and C. W. Hall. eds. 1990. *Natural Gas—Its Role and Potential in Economic Development.* Boulder, Colo.: Westview. A good discussion of the role of natural gas in providing future energy, including a chapter on how it compares with other energy sources in terms of environmental effects.

CHAPTER 18 Alternative Energy and the Environment

Spain is bullish on wind power. These wind-mills are on the Spanish island of Tenerife in the Canary Islands just offshore of the west coast of north Africa. Strong winds from the Sahara Desert are often laden with desert dust particles, visible on satellite images as they move across the islands and into the Atlantic Ocean. (Edward A. Keller)

● LEARNING OBJECTIVES

Alternatives to fossil fuels include geo-thermal energy, solar energy, water power, wind power, and biomass fuels. Some of these alternatives are already being used, and efforts are underway to develop others. After reading this chapter, you should understand:

● What passive and active solar systems are, the advantages and limitations of each, and the environmental effects of developing and using the various types of active solar systems.
● What are the potential uses of photovoltaics.
● Why hydrogen may be an important fuel of the future.
● The advantages, disadvantages, and environmental impacts of developing hydropower.
● Why wind power has considerable potential, and how its development and utilization could affect the environment.
● What are the potential environmental consequences of using biomass as an energy source.
● What geothermal energy is, and how developing and using it affects the environment.
● What important policy issues will affect large-scale use of alternative energy sources.

Pamplona—Bullish on Wind Power

Renewable alternative energy sources that flow naturally and continuously through the environment are likely to make a much more significant contribution to the global supply of energy in the future beyond 2050. One of the sources most likely to make an important contribution is wind power, a rapidly maturing technology that is competitive with conventional sources of energy such as oil.[1]

A leader in demonstrating the usefulness of wind energy is the city of Pamplona, Spain. Although it is best known for its annual celebration of the "running of the bulls," Pamplona is a city with a population approaching 200,000 and an economy heavily based on manufacturing. In past years, the city's economy depended on energy imported from other parts of Spain, where it was produced from burning coal and uranium. But then the city made some changes, including tax incentives for companies that produced energy from wind.

The new policies have paid off. Several wind energy companies became established in the area. By the end of 1998, the state of Navarra, in which the city of Pamplona resides, was producing sufficient wind energy to provide approximately 23% of its total demand for electricity. Navarra's success may be a sign of the advantages in the use of renewable energy sources.[2]

The bright future of wind energy around Pamplona, Spain, is probably just the beginning of what will be an energy revolution occurring in the next 50 years. The revolution will likely be the result of two factors. (1) Oil, which is the dominant energy source today, will in the next 20 to 50 years begin a slow decline in production, precipitating shortages and price increases. (2) Much more energy will be recovered from renewable sources that constitute an extremely large potential resource—10,000 times the present global energy consumption.[1]

> • *The growth of wind power, an alternative and renewable source of energy, is a positive step toward sustainable energy development. In this chapter, we explore alternative, renewable energy in terms of sources, advantages and disadvantages, environmental concerns, and policy issues.*

18.1 Introduction to Alternative Energy Sources

As we have seen, the primary energy sources today are fossil fuels, which supply approximately 90% of the energy consumed by people. All other sources are considered **alternative energy** and are divided into **renewable energy** and **nonrenewable energy**. Nonrenewable alternative energy sources include nuclear energy (discussed in Chapter 19) and geothermal energy. Nuclear energy is nonrenewable because it requires a mineral fuel mined from Earth. Geothermal energy is considered nonrenewable for the most part because heat can be extracted from the Earth faster than it is naturally replenished from the Earth (that is, output exceeds input; see Chapter 3). The renewable sources are solar energy, water (hydro) power, wind power, hydrogen produced from water, and energy derived from biomass (crops, wood, and so forth).

Renewable energy sources—sun, water, wind, and biomass—are often discussed as a group because they are all derived from the sun's energy. In other words, solar energy, broadly defined, comprises all the renewable energy sources, as shown in Figure 18.1. These energy sources are renewable because they are regenerated (renewed) by the sun within a time period useful to humans.

Water and biomass, although renewable, are fundamentally different from solar and wind energy, which are there as long as the sun shines and the wind blows. Water power and biomass energy are not always automatically renewed by nature. They may be depleted if the environment necessary for their renewal is not maintained. For biomass to be renewable, both water and soil are necessary for plant growth. If either of these is depleted, then biomass production may decrease or even stop. Likewise, the water behind a dam depends on the climate to produce runoff. If the climate becomes arid, runoff may be reduced, and the dam will store less water and produce less electricity. If people divert river water for crops upstream of a dam, there will be less water in the reservoir and hydropower will be reduced.

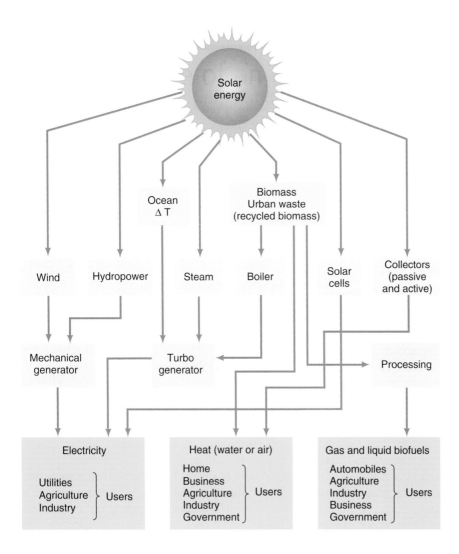

Figure 18.1 • Routes of various types of renewable solar energy.

The primary difference between renewable alternative energy sources and fossil fuels is analogous to the difference between a checking account that receives regular monthly deposits and a checking account that receives an initial large deposit but no further deposits. The account receiving regular deposits will, over the long term, not be depleted (similar to renewable energy sources). Even if you spend the present balance, more will be added later. The second account will eventually be depleted (similar to oil), although the timing will depend on the rate at which the funds are spent. Financial planning would differ for these two accounts. Similarly, energy planning based on oil differs fundamentally from planning based on renewable energy—for example, solar energy. Using solar energy requires that we live and plan within the constraints of the energy provided by the sun. Using oil requires planning based on a resource that is finite and is being depleted.[1]

The total energy we may be able to extract from alternative energy sources is enormous (Table 18.1). For example, the estimated recoverable energy from solar energy is about 1,000 times the present human global energy consumption of about 10 TW (320 exajoules) per year. The estimated recoverable energy from wind is

Table 18.1 Global Resource-Based and Recoverable Energy for Selected Alternative Energy Sources

Source	Resource Base (TW)*	Recoverable Resource (TW)
Solar	90,000	1,000
Wind	300–1,200	10
Water	10–30	2
Biomass	30	10
Geothermal	30	3

*1 TW = 10^{12} W; global energy consumption is about 10 TW (320 exajoules) per year.

Source: Modified from T. Jackson, and R. Lofstedt, "Royal Commission on Environmental Pollution, Study on Energy and the Environment," 1998. Accessed November 29, 2000, at http://www.rcep.org.uk/studies/energy/98-6061/jackson.html.

about equivalent to current global energy consumption, as is the estimated recoverable energy from biomass

It is true, however, that this energy may not necessarily always be available when we need it. Renewable energy sources, with the exception of biomass and water power (which can be stored), are intermittent, with daily and seasonal variations in supply. These variations will become a problem when renewable energy supplies make up about 40% of our total energy use. Beyond about 40%, the natural variation will periodically produce shortages in energy supply. As a result, the development of energy storage systems to smooth out the variations in supply will be necessary.[1] For example, suppose you have solar collectors on your roof to produce electricity. At night, no energy is produced, but if you store some of the energy produced during the day in batteries, you can use stored energy at night. For large-scale production of renewable energy, large energy storage systems are necessary.

It is also true that renewable energy sources are not equally available in all locations. Therefore, matching energy production to appropriate sites is important. For example, in the United States, we might build solar energy power plants in the Southwest, where sunlight is often intense, and construct wind farms in the Great Plains, Texas, the Northwest, and California, where the wind is strong and steady. Likewise, we could build biomass power plants at locations where forest and agriculture fuel resources are abundant.[3]

Renewable alternative energy sources are associated with minimal environmental degradation. In general, because no fuel is burned, these energy sources do not increase atmospheric carbon dioxide, which causes global warming. (An exception is the burning of biomass or its derivative, urban waste.) Renewable energy sources such as solar and wind power will not cause climate change or raise sea levels, increasing coastal erosion. Another advantage is that the construction lead time necessary to implement the technology for production of energy from renewable sources is often short compared with the construction of power plants that use fossil or nuclear fuels. Although in the United States most renewable energy sources are more expensive than energy from fossil fuels or nuclear energy, this is in part because in the past the federal subsidies they received were much smaller than those granted to domestic fossil fuel and nuclear energy production.[4]

Alternative energy sources, particularly solar and wind, are growing at tremendous rates. For the first time, it has become apparent that these energy sources may compete with fossil fuels. Alternative renewable energy sources offer our best chance to develop a truly sustainable energy policy that will not harm the planet.[5]

This section has supplied a brief introduction to alternative energy sources. Next, we discuss individual sources and, where appropriate, the environmental advantages and disadvantages of each.

18.2 Solar Energy

The total amount of solar energy reaching the Earth's surface is tremendous. For example, on a global scale, 10 weeks of solar energy is roughly equivalent to the energy stored in all known reserves of coal, oil, and natural gas on Earth. Solar energy is absorbed at Earth's surface at an average rate of 120,000 TW (1 TW is 10^{12} W), which is 10,000 times the total global demand for energy.[1] In the United States, on average, 13% of the sun's original energy entering the atmosphere arrives at the surface (equivalent to approximately 177 W/m², or about 16 W/ft², on a continuous basis). The estimated year-round availability of solar energy in the United States is shown on Figure 18.2. However, solar energy is site specific, and detailed observation of a potential site is necessary to evaluate the daily and seasonal variability of its solar energy potential.[6]

Solar energy may be used directly through passive solar systems or active solar systems. **Passive solar energy systems** often involve architectural designs that enhance the absorption of solar energy by using and adjusting for natural changes that occur throughout the year without requiring mechanical power (Figure 18.3). Various societies during the past few thousand years have used passive solar energy (see Chapter 16). For example, Islamic architects have traditionally used passive solar energy in hot climates to cool buildings.[1]

One simple technique is to design overhangs on buildings that block summer (high-angle) sunlight but allow winter (low-angle) sunlight to penetrate and warm rooms. Another technique is to build a wall that absorbs the solar energy and then radiates heat that warms the

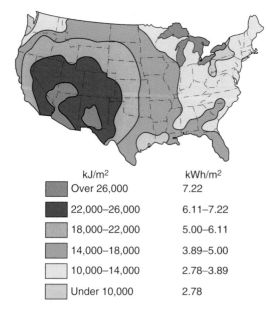

kJ/m²	kWh/m²
Over 26,000	7.22
22,000–26,000	6.11–7.22
18,000–22,000	5.00–6.11
14,000–18,000	3.89–5.00
10,000–14,000	2.78–3.89
Under 10,000	2.78

Figure 18.2 • Estimated solar energy for the contiguous United States. (*Source*: Modified from Solar Energy Research Institute, 1978.)

(a)

(b)

Figure 18.3 • (*a*) Essential elements of passive solar design. High summer sunlight is blocked by the overhang, but low winter sunlight enters the south-facing window. Other features are designed to facilitate the storage and circulation of passive solar heat. (*b*) Design of this home utilizes passive solar energy. Sunlight enters through the windows and strikes a specially designed masonry wall that is painted black. The masonry wall heats up and radiates this heat, warming the house during the day and into the evening. (*Source*: Moran, Morgan, and Wiersma, *Introduction to Environmental Science* [New York: Freeman, 1986]. Copyright by W. H. Freeman & Company. Reprinted with permission.)

room. Yet another technique makes use of trees that lose their leaves during the winter, which are common in the midwestern and eastern United States. When planted on the sunny side of a building, these trees provide summer shade that cools the building. In the winter, with the leaves gone, winter sunlight enters the house.

Many thousands of homes and other buildings in the southwestern United States, as well as other parts of the country, now use passive solar systems for at least part of their energy needs.[3] Passive solar energy also provides natural lighting to buildings through windows and skylights. A special glazing on the glass transmits light and provides insulation. Use of passive solar energy features is most cost effective when they are part of the design for a new building. However, in some cases, adding passive solar features to an existing building can also be cost effective.

Active solar energy systems require mechanical power, usually electric pumps and other apparatuses, to circulate air, water, or other fluids from solar collectors to a location where the heat is stored until used. We discuss solar collectors next and then describe several specific types of active solar energy systems.

Solar Collectors

Solar collectors to provide space heating or, more commonly, hot water are usually flat panels consisting of a glass-covered plate over a black background where an absorbing fluid (water or a volatile liquid) is circulated through tubes (Figure 18.4). Solar radiation enters the glass and is absorbed by the black background. Heat is emitted from the black material, heating the fluid circu-

Flat-plate collector

Glazing

Cool fluid in

Copper tube

Insulation

Absorber plate

Warm fluid out

Solar water heater

Insulated pipe

Cold in

Heat for use

Drain

Pump

Solar tank

Pressure relief
valve

Existing water heater

Figure 18.4 • Detail of a flat-plate solar collector and pumped solar water heater. (*Source*: Farallones Institute, *The Integral Urban House* [San Francisco: Sierra Club Books, 1979]. Copyright 1979 by Sierra Club Books. Reprinted with permission.)

lating in the tubes. If water is the absorbing fluid, it is heated to 38–93°C (100–200°F).[4] In cold climates, water is a poor absorbing fluid, because it can freeze. As a result, a liquid with a low freezing temperature is used.

A second type of solar collector is called the evacuated tube collector. Its design is similar to that of the flat-plate collector. The difference is that each tube, along with its absorbing fluid, passes through a larger tube that helps reduce heat loss.[1]

Photovoltaics

Photovoltaics is a technology that converts sunlight directly into electricity (Figure 18.5). The systems use solar cells, also called photovoltaic cells, made of thin layers of semiconductors (silicon or other materials) and solid-state electronic components with few or no moving parts. Solar cell technology is advancing rapidly, and first-law efficiency

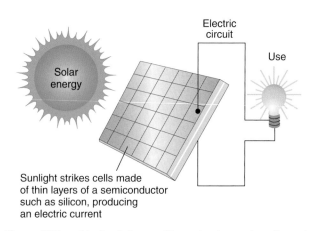

Electric circuit

Solar energy

Use

Sunlight strikes cells made of thin layers of a semiconductor such as silicon, producing an electric current

Figure 18.5 • Idealized diagram illustrating how solar cells work.

(a)

(b)

Figure 18.6 • (a) Panels of photovoltaic cells are being used here to power a small refrigerator to keep vaccines cool. The unit is designed to be carried by camels to remote areas in Chad. (b) Photovoltaics are used to power emergency telephones along a highway on the island of Tenerife in the Canary Islands.

is as high as 10%. The cells are constructed in standardized modules, encapsulated in plastic or glass, which can be combined to produce systems of various sizes so that power output can be matched to the intended use.

Electricity is produced when sunlight strikes the cell. The different electronic properties of the layers cause electrons to flow out of the cell through electrical wires. Photovoltaic cells can also work with light energy other than the sun. For example, using the light of your desk lamp, a photovoltaic cell can produce electricity to light a small bulb in a clock or to power your hand-held calculator.

An early and continuing use for photovoltaics is to supply power to satellites and space vehicles. Another commercial use is as a power source in remote areas for electric equipment such as water-level sensors, meteorological stations, and emergency telephones (Figure 18.6).

Photovoltaics is emerging as a significant contributor to developing countries that do not have the financial ability to build large central power plants that burn fossil fuels. Developing countries and their use of photovoltaics are demonstrating that solar technology can be simple, relatively inexpensive, and capable of meeting energy needs for people in many places in the world by matching the demand with an appropriate supply. For example, a solar company in the United States is providing people in a number of developing countries with photovoltaic systems that power lights and televisions at an installed cost of less than $400 per household.[7] About a half million

homes, mostly in villages not linked to a countrywide electrical grid, now receive their electricity from photovoltaic cells.

Used in developing countries and other nations as well are solar roofing tiles that utilize the roof of a building as a platform for a power plant.[8] Panels of solar cells can also be placed on walls or window glass.[5] By 1995, Germany had installed photovoltaic cells on 1,000 roofs. The governments of India and the United States have both announced that a million roofs will be installed with photovoltaic systems by 2010.

Utility companies are interested in photovoltaics and are constructing power plants on a variety of scales, ranging from solar cell panels on roofs to larger power plants. A large photovoltaic power plant is located in the Carrizo plain of central California, where about 5,000 kW of power has been produced by large panels of solar cells.

In some remote locations today, the cost of using photovoltaics is comparable with the cost of using grid-connected power. These are places where the high cost of joining the grid exceeds the cost of photovoltaics.

In the future, the alternative energy source for many people and communities might well be lightweight photovoltaics in which the solar cells are mounted together to form modules. The growth and technology changes in photovoltaics suggest that in the 21st century, solar energy is likely to become a multi-gigawatt-per-year industry that will provide a significant portion of the energy we use.[9]

Power Towers

An interesting type of solar energy system is the **solar power tower**, shown in Figure 18.7. The system works by collecting heat from solar energy and delivering this energy in the form of steam to turbines that produce electric power. An experimental 10 MW power tower near Barstow, California is approximately 100 m (330 ft) high and is surrounded by approximately 2,000 mirror modules, each with a reflective area of about 40 m² (430 ft²). The mirrors adjust continually to reflect as much sunlight into the tower as possible. In 1999, the power tower was able to produce electricity 24 hours per day by storing the heat produced during the day in a molten salt heat-storage system. The stored heat was used at night or on cloudy days to produce energy.

At the end of 1999, the power tower was shut down, in part because the plant was not economically competitive with other sources of electricity. Given the energy crisis in California in 2001, however, that decision was premature. It is expected that the cost of electricity from power towers will be reduced sufficiently by the year 2030 to be economically competitive with more traditional sources of electricity.[3]

Luz Solar Electric-Generating System

A system developed at Luz International is the most technically successful solar power experiment to date. The site, in the Mojave Desert, consists of nine solar farms. Each farm comprises a power plant surrounded by hundreds of solar collectors, and each produces 350 MW of electrical power, sufficient for about 550,000 people. The system uses solar collectors (curved mirrors) to heat a synthetic oil that flows through heat exchangers, which in turn drive steam turbine generators. The basic components of the system are shown in Figure 18.8. One advantage of this system is that it is modular, which allows for individual solar farms to be constructed relatively quickly.

The success of the system is based in part on the fact that it uses a natural gas burner backup system (75% solar and 25% natural gas), ensuring uninterrupted power generation both on cloudy days and during times of peak demand. Thus, the Luz system is really a combination of solar technology and conventional power generation.[10]

Luz international filed for bankruptcy in 1992, not because it was a technological failure but because it expanded too fast, borrowed too much money, and went too deeply into debt. Reorganization of the company has allowed the power plants to continue generating electricity.[11]

Solar Ponds

The use of shallow, salt-gradient **solar ponds** to generate relatively low-temperature water is an interesting prospect to produce heat for commercial, industrial, and agricultural uses. The ponds are designed to collect incoming solar radiation, which produces a bottom-water temperature of about 70°C (144°F). The hot water is kept on the bottom by the addition of salt, which makes it heavier. Circulation is restricted so that the dense bottom water does not mix with the water above. However, solar radiation penetrates to warm the lower water. The heat can then be extracted from the bottom and used.[12]

Ocean Thermal Conversion

A final example of the direct utilization of solar energy, **ocean thermal conversion**, involves using part of the oceanic environment as a gigantic solar collector. The surface temperature of ocean water in the tropics is often about 28°C (82°F). However, at the bottom of the ocean, at a depth of about 600 m (1,970 ft), the temperature may be 1–3°C (35–38°F). Equipment has been designed to take advantage of this temperature difference, either by using the seawater directly or by using a heat exchange system in which a fluid such as ammonia or propane is vaporized by the warm water. The expand-

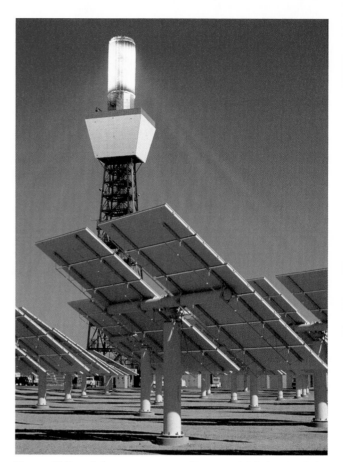

Figure 18.7 • Solar power tower at Barstow, California. Sunlight is reflected and concentrated at the central collector where the heat is used to produce steam to drive turbines and generate electric power.

(a)

Power Block Assembly:
1. solar collector assembly
2. natural gas boiler
3. turbine generator
4. steam generator and solar superheater
5. control building
6. cooling tower
7. Southern California Edison interconnect

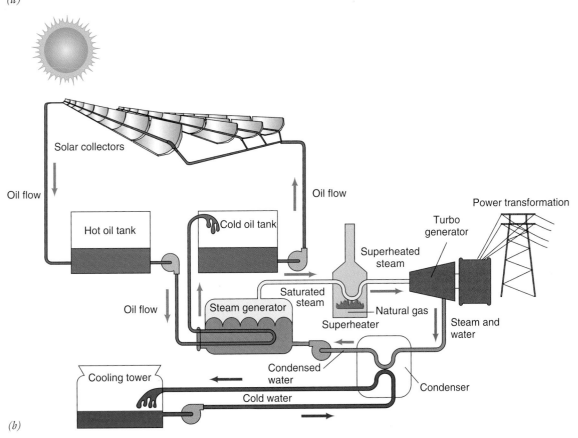

(b)

Figure 18.8 • (*a*) Luz International Solar Farm, showing the system of solar collectors, natural gas boiler, turbine generator, and other components. (*b*) Diagram illustrating how the Luz system works. (*Source*: Courtesy of Luz International.)

ing vapor then propels a turbine and generates electricity. Following the generation of electricity, the vapor is cooled and condensed by the cold water.

Experiments with ocean thermal energy conversions have been conducted on the island of Hawaii. The exper-iments, located at Keahole Point on the Kona coast, were well situated because the seafloor there drops off very steeply to deep, cold waters. Only a small amount of energy was actually produced. Part of the cost of energy production was offset by aquaculture at the site (see

Chapter 11). Cold, nutrient-rich water brought up as part of the energy production was used to grow seafoods, such as salmon, oysters, and abalone, as well as kelp. When the laboratory was fully operating, it also used the cold water for air conditioning, reducing the utility bill.

The construction of large-scale ocean thermal plants depends on whether they can be built near potential markets (the people who use the energy) and whether they are economically viable.[6] Answers to these questions remain uncertain, as the future of ocean thermal plants is speculative.

Solar Energy and the Environment

The use of solar energy generally has a relatively low impact on the environment, but there are environmental concerns. One disadvantage of solar energy is that it is relatively dispersed, arriving at Earth's surface like a fine mist, so that a large land area is required to generate a large amount of energy. This problem is negligible when solar collectors can be combined with existing structures, as with the addition of solar hot-water heaters on the roofs of existing houses. Highly centralized and high-technology solar energy units, such as solar power towers, have a greater impact on the land because they need considerable space. The impact of large centralized solar energy systems can be minimized by locating them in remote areas not used for other purposes and by making use of dispersed solar energy collectors on existing structures wherever possible.

Another concern involves the large variety of metals, glass, plastics, and fluids used in the manufacture and use of solar equipment. Some of these substances may cause environmental problems through production and by accidental release of toxic materials.

18.3 Hydrogen

Hydrogen, the fuel burned by our Sun (producing solar energy), is the lightest, most abundant element in the universe. Hydrogen gas may be an important fuel of the future.[3]

Hydrogen is a high-quality fuel that can be easily used in any of the ways in which we normally use fossil fuels, such as to power automobile and truck engines and to heat water and buildings. When hydrogen is used and stored in **fuel cells** (see A Closer Look 18.1, which are similar to batteries, it can be used to produce electricity. Hydrogen, like natural gas, can be transported in pipelines and stored in tanks; and it can be produced using solar and other renewable energy sources. It is a clean fuel; the combustion product of burning hydrogen is water, so it does not contribute to global warming, air pollution, or acid rain (see Chapters 22 and 23).

Technological improvements in producing hydrogen are certain, and the fuel price of hydrogen may be substantially reduced in the future.[13]

One way to produce hydrogen is to use an electric current to separate hydrogen (H) from water (H_2O), a process known as *electrolysis*. If the electricity for electrolysis is produced by solar or wind power, the hydrogen is produced in a clean, carbon-free process.[3] However, the most economical way to produce hydrogen is to use a thermal process in which steam is combined with natural gas (CH_4), removing the carbon (C) and leaving hydrogen (H). This process depends on a fossil fuel and produces some carbon dioxide (CO_2). Hydrogen can also be produced from gasification of biomass.

Hydrogen is an ideal medium for the storage of energy. It can help smooth out the natural variability of using renewable energy sources such as solar and wind power. As mentioned, hydrogen can be stored in fuel cells which are somewhat similar to batteries. Thus, hydrogen produced by electricity from solar or wind power and stored in fuel cells can be later used to produce electricity on cloudy days and at night or on days when the wind isn't blowing.[1]

18.4 Water Power

Water power is a form of stored solar energy, as mentioned earlier, because climate, which dictates the flow of water on Earth, is driven in part by differential solar heating of the atmosphere. Water power has been successfully harnessed since the time of the Roman Empire. Waterwheels that convert water power to mechanical energy were turning in western Europe in the 17th century; during the 18th and 19th centuries, large waterwheels provided energy to power grain mills, sawmills, and other machinery in the United States.

Today, hydroelectric power plants use the water stored behind dams. In the United States, hydroelectric plants generate about 80,000 MW of electricity—about 10% of the total electricity produced in the nation. In some countries, such as Norway and Canada, hydroelectric power plants produce most of the electricity used. Figure 18.10a shows the major components of a hydroelectric power station.

Hydropower can also be used to store energy through the process of pump storage (Figure 18.10b and c). During times when demand for power is low (for example, at night in the summer), electricity produced from oil, coal, or nuclear plants in excess of the demand is used to pump water uphill to a higher reservoir (high pool). Then, during times when demand for electricity is high (for example, on hot days in the summer), the stored water flows back down to a low pool through generators to help provide energy. The advantage of pump storage lies in the timing of energy production and use.

Fuel Cells—An Attractive Alternative

Power produced from burning fossil fuels, particularly coal and fuels used in internal combustion engines (cars, trucks, ships, and locomotives), is associated with serious environmental problems. As a result, we are searching for and developing environmentally benign technologies capable of generating power.[14] One promising technology uses fuel cells, which produce fewer pollutants, are relatively inexpensive, and have the potential to store and produce high-quality energy.

Fuel cells are highly efficient power-generating systems that produce electricity by combining fuel and oxygen in an electrochemical reaction. Hydrogen and phosphoric acid are the most common fuel types, although fuel cells that run on methanol, ethanol, and natural gas are available. Traditional generating technologies require combustion of fuel in order to convert the resultant heat into mechanical energy (to drive pistons or turbines), and this mechanical energy is then converted into electricity. With fuel cells, however, chemical energy is converted directly into electricity, thus increasing second-law efficiency (see Chapter 16) while reducing harmful emissions.

Basic components of a hydrogen-burning fuel cell are shown in Figure 18.9. Both hydrogen and oxygen are added to the fuel cell in an electrolyte

solution. The reactants remain separated from one another and, upon ionization, migrate through the electrolyte solution from one electrode to another. The flow of electrons from the negative to the positive electrode is diverted along its path into an electrical motor, supplying current to keep the motor running. In order to maintain this reaction, hydrogen and oxygen are added as needed. When hydrogen is used in a fuel cell, the only waste products are oxygen and water. Using natural gas (CH_4) in fuel cells produces some pollutants, but the amount is only about 1% of what would be produced by burning fossil fuels in an internal combustion engine or a conventional power plant.[14]

Fuel cells are efficient and clean, and they can be arranged in a series to produce the appropriate amount of energy for a particular task. Additionally, the efficiency of a fuel cell is largely independent of its size and energy output. For these reasons, fuel cells are well suited to providing

power for automobiles, homes, and large-scale power plants. They can also be used to store energy to be used as needed. Fuel cells are used in many locations. For example, they power buses at Los Angeles International Airport and provide heat and power at Vandenberg Air Force Base in California.[15]

Figure 18.9 ● Idealized diagram showing how a fuel cell works and its application to power a vehicle.

Small-Scale Systems

The total amount of electrical power produced by running water from large dams will probably not increase in the coming years in the United States, where most of the acceptable dam sites are already being utilized. However, small-scale hydropower systems, designed for individual homes, farms, or small industries, may be more common in the future. These small systems, known as *micro-hydropower* systems, have power output of less than 100 kW.[16]

Numerous sites in many areas have the potential for producing small-scale electrical power. This is particularly true in mountainous areas, where potential energy from stream water is often available. Micro-hydropower development is site specific, depending on local regulations, economic situations, and hydrologic limitations. Hydropower can be used to generate either electrical power or mechanical power to run machinery; its use may help reduce the high cost of importing energy. It may also help small operations become more independent of local utility providers.[16]

Common hydroelectric system

Water

Dam

Power line

Generator

Turbine

(a)

Pump storage system (Pump cycle)

High pool

Low pool

(b)

Pump storage system (Generating cycle)

High pool

Low pool

(c)

Figure 18.10 • (*a*) Basic components of a hydroelectric power station. (*b*) A pump storage system. During light power load, water is pumped from low pool to high pool. (*c*) During peak power load, water flows from high pool to low pool through a generator. (*Source*: Modified from Council on Environmental Quality, Energy Alternatives: A Comparative Analysis [Norman: University of Oklahoma Science and Policy Program, 1975].)

Water Power and the Environment

Water power is clean power; it requires no burning of fuel, does not pollute the atmosphere, produces no radioactive or other waste, and is efficient. However, there are environmental prices to pay (see Chapter 20):

- Large dams and reservoirs flood large tracts of land that could have had other uses. For example, towns and agricultural lands may be lost.
- Dams block the migration of some fish, such as salmon.
- Water falling over high dams may pick up nitrogen gas; if the gas enters the blood of fish, it expands and kills them. This is similar to the condition known as the bends, which occurs if divers using a compressed air source surface too rapidly from deep water. The fish respire nitrogen gas under pressure of the falling water; and when the pressure is reduced quickly, the gas bubbles in the blood expand and damage tissue. Nitrogen has killed migrating game fish in the Pacific Northwest in this manner.
- Dams trap sediment that would otherwise reach the sea and eventually replenish the sand on beaches.
- For a variety of reasons, many people do not want to turn wild rivers into a series of lakes.

For all these reasons, and because many good sites for dams already have one, the growth of large-scale water power in the future (with the exception of a few areas, including Africa, South America, and China) appears limited. Indeed, in the United States, there is an emerging social movement to remove dams. Hundreds of dams, especially those with few useful functions, are being considered for removal; and a few have already been removed (see Chapter 20).

As mentioned, there does seem to be continued interest in micro-hydropower to supply either electricity or mechanical energy. However, small dams and reservoirs tend to fill more quickly with sediment than large reservoirs, rendering their useful life much shorter. In fact, many dams likely to be removed are small ones filled with sediment. Because micro-hydropower development can adversely affect stream environments by blocking fish passage and changing downstream flow, careful consideration must be given to their construction. A few small dams cause little environmental degradation beyond the specific sites. However, if the number of dams in a region is large, the total impact may be appreciable. This principle applies to many forms of technology and development. The impact of a single development may be nearly negligible over a broad region; but as the number of such developments increases, the total impact may become significant.

18.5 **Tidal Power**

The use of **tidal power**, a type of water power derived from ocean tides, can be traced back to 10th-century Britain, where tides were used to power coastal mills.[1] However, only in a few places with favorable topography—such as the north coast of France, the Bay of Fundy in Canada, and the northeastern United States—are the tides sufficiently strong to produce commercial electricity. The

Figure 18.11 • Tidal power station on the River Rance near Saint-Malo, France.

tides in the Bay of Fundy have a maximum range of about 15 m (49 ft). A minimum range of about 8 m (26 ft) is necessary for development of tidal power to be considered.

To harness tidal power, a dam is built across the entrance to a bay or estuary, creating a reservoir. As the tide rises (flood tide), water is initially prevented from entering the bay landward of the dam. Then, when there is sufficient water (from the ocean-side high tide) to run the turbines, the dam is opened, and water flows through it into the reservoir (the bay), turning the blades of the turbines and generating electricity. When the reservoir (the bay) is filled, the dam is closed, stopping the flow and holding the water in the reservoir. When the tide falls (ebb tide), the water level in the reservoir is higher than that in the ocean. The dam is then opened to run the turbines (which are reversible), and electric power is produced as the water is let out of the reservoir.

Figure 18.11 shows the La Rance tidal power plant on the north coast of France. La Rance, constructed in the 1960s, is the first and largest modern tidal power plant. The plant at capacity produces about 240,000 kW from 24 power units spread out across the dam. At the La Rance power plant, most electricity is produced from the ebb tide, which is easier to control.

Tidal power does have environmental impacts. The change in the hydrology of a bay or estuary caused by the dam can adversely affect the vegetation and wildlife. The dam restricts upstream and downstream passage of fish. Furthermore, the periodic rapid filling and emptying of the bay as the dam opens and closes with the tides rapidly changes habitats for birds and other organisms.

18.6 Wind Power

Wind power, like solar power, has evolved over a long period of time, from early Chinese and Persian civilizations to the present. Wind has propelled ships and has driven windmills to grind grain and pump water. In the past, thousands of windmills in the western United States were used to pump water for ranches. More recently, wind has been used to generate electricity.

Basics of Wind Power

Winds are produced when differential heating of Earth's surface creates air masses with differing heat contents and densities. The potential for energy from the wind is large, and yet there are potential problems with its use because wind tends to be highly variable in time, place, and intensity.[17]

Wind prospecting has become an important endeavor. On a national scale, regions with the greatest potential for wind energy are the Pacific Northwest coastal area, the coastal region of the northeastern United States, and a belt extending from northern Texas through the Rocky Mountain states and The Dakotas. Other good sites include mountain areas in North Carolina and northern Coachella Valley in southern California. A site with sustained wind velocity of about 5 m/sec (16 ft/sec) or greater is considered a good prospect for wind energy development.[1]

Even at a particular site, the direction, velocity, and duration of wind may be quite variable, depending on local topography temperature differences in the atmosphere.[17] For example, wind velocity often increases over hilltops, and wind may be funneled through a mountain pass (Figure 18.12). The increase in wind velocity over a mountain is due to a vertical convergence of wind, whereas in a pass, the increase is partly due to a horizontal convergence. Because the shape of a mountain or a pass is often related to the local or regional geology, prospecting for wind energy is a geologic as well as a geographic and meteorologic problem. The primary task in evaluating the wind energy potential of a region or site is to place instruments that measure and monitor over time the strength, direction, and duration of the wind.[12]

Strong winds

Mountain

| Wind turbine | Convergence of wind over a ridge or mountain |

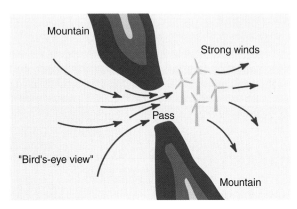

Mountain

Strong winds

Pass

"Bird's-eye view"

Mountain

Figure 18.12 • Idealized diagram showing how wind energy is concentrated by topography.

(a)

(b)

Figure 18.13 • (a) Location of major wind energy projects in the United States. (*Source*: National Wind Technology Center, 1999 [http://www.nrel.gov/wind/deploy.html].) (b) Windmills on a wind farm near Altamont, California, a mountain pass region east of San Francisco.

Significant improvements in the size of windmills and the amount of power they produce occurred from the late 1800s through approximately 1950, when many European countries and the United States became interested in large-scale generators driven by the wind. In the United States, thousands of small, wind-driven generators have been used on farms. Most small windmills generate approximately 1 kW of power, which is much too small to be considered for central power generation needs. Interest in wind power declined for several decades prior to the 1970s because of the abundance of cheap fossil fuels; but in recent years, interest in building windmills has been revived.

The cost of producing electricity from wind must be competitive with other sources to be economically viable. Such viability is nowhere better exemplified than in California, where approximately 17,000 windmills, with a generating capacity of about 1,400 MW, have been installed in the 1980s. This is about 80% of the U.S. wind power capacity of 1,800 MW (Figure 18.13a). Individual, relatively small windmills produce from 60 to 75 kW of power and are arranged in wind farms, consisting of clusters of windmills located in mountain passes (Figure 18.13b). Electricity produced at these sites is connected to the general utility lines. In 1998, wind farms produced sufficient electricity to power, for example, the city of San Francisco, making a significant contribution to the modern utility grid. Tax incentives have helped establish the wind power industry. Wind power will likely be California's second least expensive source of power by 2010, second only to hydropower. Today, wind farms in California produce about 1.5% of the state's electricity, but that percentage is increasing.

Wind energy is now used in many locations on Earth (Figure 18.14). Large utility companies are considering wind power in their long-range energy planning goals. The United States in 1996 was the world leader, with the most installed wind power systems (Figure 18.14). However, by 1998, Germany, with over 2,000 MW installed, became the leader, and the European Union leads the world in using wind power. Where land for wind farms is scarce, wind turbines can be sited offshore.

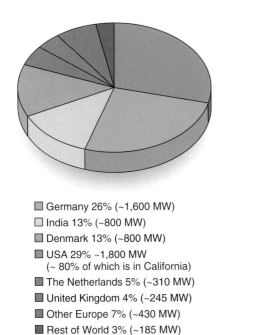

- ☐ Germany 26% (~1,600 MW)
- ☐ India 13% (~800 MW)
- ☐ Denmark 13% (~800 MW)
- ☐ USA 29% ~1,800 MW
  (~ 80% of which is in California)
- ☐ The Netherlands 5% (~310 MW)
- ☐ United Kingdom 4% (~245 MW)
- ☐ Other Europe 7% (~430 MW)
- ☐ Rest of World 3% (~185 MW)

Figure 18.14 • Worldwide installed wind power totaled 6,170 MW at the end of 1996. (*Source*: International Energy Agency, February 1998 [http://www.afm.dtu.dk/wind/iea/status.htm#cap].)

Wind Power and the Environment

The use of wind power will not solve all our energy problems, but as one of the major alternative energy sources, wind power can be used at particular sites to reduce dependence on fossil fuel and help achieve sustainability. Wind energy does have a few disadvantages:

- Windmills kill birds (birds of prey, such as hawks and falcons, are vulnerable).
- Large windmill farms use large areas of land for roads, pads for the windmills, and other equipment.
- Windmills may degrade an area's scenic resources.

However, everything considered, wind energy has a relatively low environmental impact, and its continued use is certain.

Future of Wind Power

During recent years, wind energy use has grown at approximately 25% per year, over 10 times the growth rate of oil use. It's believed that there is sufficient wind energy in Texas, South Dakota, and North Dakota to satisfy the electricity needs of the entire United States. Wind power potential in Britain is more than twice the present demand for electricity.[2] Consider the implications for nations such as China. China burns tremendous amounts of coal at a heavy environmental cost, including exposing millions of people to hazardous air pollution. In rural China, exposure to the smoke from burning coal in homes has increased the threat of lung cancer by a factor of nine or more. China could probably double its current capacity to generate electricity with wind alone![9]

As suggested earlier, power from wind is being taken very seriously. Although wind now provides less than 1% of the world's demand for electricity, its growth rate suggests that it could be a major supplier of power in the relatively near future. One scenario suggests that wind power could supply 10% of the world's electricity in the coming decades and, in the long run, could provide more power than hydropower, which today supplies approximately 20% of the electricity in the world.

The wind energy industry has created thousands of jobs in recent years; it is also becoming a major investment opportunity. Technology is producing more efficient wind turbines, thereby reducing the price of wind power. Today, a large state-of-the-art wind turbine produces about 700 kW of electricity at a cost, including installation, of about $700,000. This one wind turbine provides enough power for 200 homes.

In many regions, power generated from wind is already less expensive in some instances than electricity produced from coal-fired and natural gas power plants.[2] For example, the world's largest wind farm is on the Oregon–Washington border. The facility includes over 450 windmills sited on ridges above the Columbia River. Total power output at 300 MW is about one-third of that produced from a large fossil fuel or nuclear power plant. The cost of producing the electric energy will be about 4 cents per kilowatt hour, which is competitive with electricity from burning natural gas.

18.7 Biomass Energy

Biomass energy is energy recovered from *biomass*—organic matter, such as plant material and animal waste. Whereas the energy from most other renewable sources comes from physical processes, such as moving water and wind, the energy from biomass comes from chemical bonds formed through photosynthesis in living or once-living matter.[1] Biomass fuel is organic matter that can be burned directly or converted to a more convenient form and then burned. For example, we can burn wood in a stove or convert it into charcoal to use as a fuel. Energy from biomass is the oldest fuel used by humans. Our Pleistocene ancestors burned wood in caves to keep warm and cook food.

Biomass continued to be a major source of energy for human beings throughout most of the history of civilization. When North America was first settled, there was more wood fuel than could be used. The forests often were cleared for agriculture by girdling trees (cutting through the bark all the way around the base of a tree) to kill them and then burning the forests.

Until the end of the 19th century, wood was the major fuel source in the United States. During the mid-20th century, when coal, oil, and gas were plentiful, burning wood became old-fashioned and quaint. Burning wood was something done for pleasure in an open fireplace that conducted more heat up the chimney than it provided for space heating. Now, with other fuels reaching a limit in abundance and production, there is renewed interest in the use of natural organic materials for fuel.

More than 1 billion people in the world today still use wood as their primary source of energy for heat and cooking. Today, in developing countries, biomass provides about 35% of the total energy supply.[1] Energy from biomass can take several routes: direct burning of biomass either to produce electricity or to heat water and air; heating of biomass to form a gaseous fuel (gasification); and distillation or processing of biomass to produce biofuels such as ethanol, methanol, or methane.[18]

Sources of Biomass Energy

Firewood is the best known and most widely used biomass fuel, but there are many others. In India and other countries, cattle dung is burned for cooking. Peat, a form of compressed dead vegetation, provides heating and cooking fuel in northern countries, such as Scotland, where it is abundant. The primary sources of biomass fuels in North America are forest products, agricultural residues, energy crops, animal manure, and urban waste (Table 18.2).

In some facilities, manure from livestock and other organic waste is converted (the technical term is *digested*) by microorganisms in specially designed digestion chambers to form methane (CH_4), which is burned to produce electricity, used in fuel cells, or used as fuel for tractors or other vehicles. Energy crops such as sugarcane are grown to be fermented to produce ethanol, a fuel that can be used in automobiles. For example, Brazil produces about 12 billion liters of ethanol per year from sugarcane. This is equivalent to about 60% of the fuel used for the nation's automobiles. The wastes from ethanol distillation were originally dumped in rivers, causing water pollution. The wastes are now treated to produce biogas and liquid fertilizers (recycled to sugarcane fields).[1] In Hawaii, sugarcane waste is burned to produce electricity.

Methane and biogas can also be produced from urban waste in landfills and sewage at wastewater treatment plants. In rural China, about 5 million small facilities are used to treat sewage. The original purpose was to reduce disease, but their potential as an energy source was soon recognized and used.[1]

Today, a number of facilities in the United States process urban waste to generate electricity or to be used as a fuel. Presently, only about 1% of the nation's municipal solid wastes are being recovered for energy. However, if all the power plants were operating at full capacity and if additional plants under construction were completed and operating, about 10% of the country's waste, or 24 million metric tons per year, could be burned to extract energy. The United States has been slower than other countries to utilize urban waste as an energy source. For example, in Western Europe a number of countries utilize from one-third to one-half their municipal waste for energy production. With the end of inexpensive, available fossil fuels, additional energy recovery systems utilizing urban waste will no doubt emerge.[19]

Table 18.2   Selected Examples of Biomass Energy Sources, Uses, and Products

Sources	Examples	Uses/Products	Comment
Forest products	Wood, chips	Direct burning,[a] charcoal[b]	Major source today in developing countries
Agriculture residues	Coconut husks, sugarcane waste, corncobs, peanut shells	Direct burning	Minor source
Energy crops	Sugarcane, corn, sorghum	Ethanol (alcohol)[c] Gasification[d]	Ethanol is major source of fuel in Brazil for automobiles
Animal residues	Manure	Methane[e]	Used to run farm machinery
Urban waste	Waste paper, organic household waste	Direct burning or methane from wastewater treatment or from landfills[f]	Minor source

[a]Principal biomass conversion.

[b]Secondary product from burning wood.

[c]Ethanol is an alcohol produced by fermentation, which uses yeast to convert carbohydrates into alcohol in fermentation chambers (distillery).

[d]Biogas from gasification is a mixture of methane and carbon dioxide produced by pyroclytic technology, which is a thermochemical process that breaks down solid biomass into an oil-like liquid and almost pure carbon char.

[e]Methane is produced by anaerobic fermentation in a digester.

[f]Naturally produced in landfills by anaerobic fermentation.

Biomass Energy and the Environment

The use of biomass fuels can pollute the air and degrade the land. For most of us, the smell of smoke from a single campfire is part of a pleasant outdoor experience; but under certain weather conditions, wood smoke from many campfires or chimneys in narrow valleys can lead to air pollution.

The use of biomass as fuel places pressure on an already heavily used resource. A worldwide shortage of firewood is adversely affecting natural areas and endangered species. For example, the need for firewood has threatened the Gir Forest in India, the last remaining habitat of the Indian lion (not to be confused with the Indian tiger). The world's forests will also decrease if our need for forest products and forest biomass fuel exceeds the productivity of the forests.

If our forest and crop resources are managed properly (for sustainability), it may be possible to make biomass energy more attractive. Current estimates for the United States suggest that many millions of hectares of land that are unsuitable for food production could be used for growing biomass crops (trees and other plants) using short rotation times (time between harvests). However, forest plantations would have to be managed for sustainability, because deforestation accelerates the process of soil erosion (soils without vegetation cover erode more quickly). When fine-grained (silt and clay) particles from soil erosion enter streams and rivers, the water becomes muddy, and its quality is degraded.

Advocates of using biomass for energy point out that, assuming areas harvested for bioenergy are replanted, the burning of biofuels contributes no net carbon dioxide emissions to the atmosphere. Replacing fossil fuels with sustainably produced biomass will actually reduce net CO_2 emissions, according to some estimates.[20] Finally, combustion of biomass-derived fuels generally releases fewer pollutants such as sulfur dioxide and nitrogen oxides than combustion of coal and gasoline. This is not always the case for burning urban waste, however. Although plastics, glass, and hazardous materials are removed before burning, some inevitably slip through the sorting process and are burned, releasing air pollutants, including heavy metals. Burning urban waste to recover energy is preferable to landfill disposal. However, incineration of waste paper competes with recycling, which is preferable to disposal or burning.

Future of Biomass Energy

Biomass energy in its various forms appears to have a future as an energy source. At the global scale, recoverable biomass energy is about equivalent to present energy consumption.[1] Biomass energy is also flexible in that solid biomass can be stored, as can biogas and biofuels. This allows biomass to be used as needed to meet demands.[1] However, questions remain about the amount of energy biomass can reasonably provide and the rate of depletion. Biomass energy is renewable, but it is limited

and is dependent on favorable environmental conditions for renewal. Any use of biomass fuel must be part of the general planning for all uses of the land's resources.[21]

Biomass fuel could someday help free the United States from importing oil for gasoline. There is now sufficient idle farmland to produce enough ethanol to meet fuel needs for most, if not all, U.S. automobiles and light-duty trucks.[3] However, at present, this is not practical for economic reasons, and automobile engines would need modifications to run smoothly on pure ethanol.

It has been estimated that in the United States by 2030, biomass fuel crops could provide about 100,000 MW of electricity (12% of U.S. capacity), equivalent to 100 large coal burning or nuclear power plants. Raising the necessary crops would require only about 20 million ha (50 million acres) of the 50 million ha (128 million acres) that are expected to be idle in 2030. Today, biomass in the United States is a small but growing renewable source of electric power.[3] Sweden is a leader in biomass energy, with 17% of its primary energy consumption from biomass. The pulp and paper industry of Sweden provides much of the source for this fuel.[1]

18.8 Geothermal Energy

Geothermal energy is natural heat from the interior of the Earth that is converted to heat buildings and generate electricity. The idea of harnessing Earth's internal heat is not new. As early as 1904, geothermal power was used in Italy. Today, Earth's natural internal heat is being used to generate electricity in 21 countries, including Russia, Japan, New Zealand, Iceland, Mexico, Ethiopia, Guatemala, El Salvador, the Philippines, and the United States. Total worldwide production is approaching 9,000 MW (equivalent to nine large modern coal-burning or nuclear power plants)—double the amount in 1980. Some 40 million people today receive their electricity from geothermal energy at a cost competitive with that of other energy sources.[22] In El Salvador, geothermal energy is supplying 30% of the total electric energy used. However, at the global level, geothermal energy supplies less than 0.15% of the total energy supply.[1]

Geothermal energy may be considered a nonrenewable energy source when rates of extraction are greater than rates of natural replenishment. However, geothermal energy has its origin in the natural heat production within Earth, and only a small fraction of the vast total resource base is being utilized today. Although most geothermal energy production involves the tapping of high heat sources, people are also using the low-temperature geothermal energy of groundwater in some applications.

Geothermal Systems

The average heat flow from the interior of the Earth is very low, about 0.06 W/m². This amount is trivial com-

Figure 18.15 • Geysers Geothermal Field, located north of San Francisco, California. The Geysers is the largest geothermal power operation in the world and produces energy directly from steam.

pared with the 177 W/m² from solar heat at the surface in the United States. However, in some areas, heat flow is sufficiently high to be useful for producing energy.[1] For the most part, areas of high heat flow are associated with plate tectonic boundaries (see Chapter 4). Oceanic ridge systems (divergent plate boundaries) and areas where mountains are being uplifted and volcanic island arcs are forming (convergent plate boundaries) are areas where this natural heat flow is anomalously high. One such region is located in the western United States, where recent tectonic and volcanic activity has occurred.[22,23]

On the basis of geologic criteria, several types of hot geothermal systems (with temperatures greater than about 80°C, or 176°F) have been defined, and the resource base is larger than that of fossil fuels and nuclear energy combined. A common system for energy development is hydrothermal convection, characterized by the circulation of steam and/or hot water that transfers heat from depths to the surface. An example is the Geysers Geothermal Field, a geothermal system 145 km (90 miles) north of San Francisco, California, where about 2,000 MW of electrical energy is produced. The Geysers is the largest geothermal power operation in the world (Figure 18.15).

Yellowstone National Park is an area with a significant potential for geothermal energy. The park contains about 10,000 hot springs and geysers (including the famous Old Faithful geyser). However, development in this area presents problems. Federal legislation has been passed to protect the hot springs and geysers of Yellowstone. What is required to ensure protection is not clear. Certainly,

though, there must be an adequate buffer zone to ensure that any geothermal energy development near the park will not damage Yellowstone's geothermal features. In other words, we want to make sure that geothermal energy use outside of the park will not "shut off" Old Faithful, or other hot springs or geysers.

It may come as a surprise to learn that relatively low temperature groundwater can be considered a source of geothermal energy. It is geothermal because the normal internal heat flow from Earth keeps the temperature of groundwater at a depth of 100 m (320 ft) at about 13°C (55°F). Water at 13°C is cold for a shower; but compared with winter temperatures in much of the United States, it is warm and can help heat a house. Furthermore, compared with summer temperatures of 30–35°C, groundwater at 13°C is cool and can be used to provide air conditioning. In the summer, heat can be transferred from the warm air in a building to the cool groundwater. In the winter, when the outdoor temperature is below about 4°C (40°F), heat can be transferred from the groundwater to the air in the building, reducing the heating needed from other sources. Technology for such heat transfer is well known and available.

Geothermal Energy and the Environment

The environmental impact of geothermal energy may not be as extensive as that of other sources of energy, but it can be considerable. When geothermal energy is developed at a particular site, environmental problems include on-site noise, emissions of gas, and disturbance of the land at drilling sites, disposal sites, roads and pipelines, and power plants. Development of geothermal energy does not require large-scale transportation of raw materials or refining of chemicals, as development of fossil fuels does. Furthermore, geothermal energy does not produce the atmospheric pollutants associated with burning fossil fuels or the radioactive waste associated with nuclear energy. However, geothermal development often does produce considerable thermal pollution from hot wastewaters, which may be saline or highly corrosive, producing disposal and treatment problems.

Geothermal power is not very popular in some locations among some people. For instance, geothermal energy has been produced for years on the island of Hawaii, where active volcanic processes provide abundant near-surface heat. There is controversy, however, over further exploration and development. Native Hawaiians and others have argued that the exploration and development of geothermal energy degrade the tropical forest as developers construct roads, build facilities, and drill wells. In addition, religious and cultural issues in Hawaii relate to the use of geothermal energy. For example, some people are offended by using the "breath and water of Pele" (the volcano goddess) to make electricity. This issue points out the importance of being sensitive to the values and cultures of people where development is planned.

Future of Geothermal Energy

At present, geothermal energy supplies only a small fraction of the electrical energy produced in the United States. However, if developed, known geothermal resources in the United States could produce about 20,000 MW, which is about 10% of the electricity needed for the western states.[22] Geohydrothermal resources not yet discovered could conservatively provide four times that amount (approximately 10% of total U.S. electric capacity), about equivalent to the electricity produced from water power today.[3]

  ### 18.9 Policy Issues

Renewable energy requires a broad energy policy that takes the long-term view. Policy issues and goals for alternative renewable energy include the following:[1]

- Encouraging and funding alternative energy research and development.
- Ensuring fair access to alternative energy sources.
- Developing a fiscal framework that takes into account the external costs of fossil fuels (air pollution and health costs) and the benefits of alternative energy sources (less environmental damage).
- Developing planning initiatives that promote environ-

mentally preferred and socially acceptable alternative energy.
- Establishing industrial training programs to retrain employees in alternative energy industries, especially in rural agricultural regions and former coal mining communities.
- Encouraging funding institutions to support community investment in alternative energy.

These policy issues and goals are important in discussions concerning the future of alternative renewable energy sources. Before oil production declines, we need to rethink our entire energy planning strategy. Phasing in alternative renewable energy in time to meet our future energy demands is critical to avoiding a looming energy crisis in the years to come.

Summary

- The use of renewable alternative energy sources, such as wind and solar energy, is growing rapidly. These energy sources do not cause air pollution, create health problems, or cause climate changes. They offer our best chance to replace fossil fuels and develop a sustainable energy policy.
- Passive solar energy systems often involve architectural designs that enhance absorption of solar energy without requiring mechanical power or moving parts.
- Some active solar energy systems use solar collectors to heat water for homes. Systems to produce heat or electricity include power towers, solar farms, solar ponds, and ocean thermal conversion.
- Photovoltaics is a technology that converts sunlight directly into electricity. Photovoltaic systems use solar cells for a variety of purposes, such as powering remote equipment. Experiments are ongoing to evaluate photovoltaic power plants. This emerging technology remains expensive.
- Hydrogen gas may be an important fuel of the future, especially when used in fuel cells.
- Water power today provides about 15% of the total electricity produced in the United States. Most good sites for large dams have already been utilized. Water power is clean, but there is an environmental price to pay in terms of disturbance of ecosystems, sediment trapped in reservoirs, loss of wild rivers, and loss of productive land.
- Wind power has tremendous potential as a source of electrical energy in many parts of the world. Many utility companies are presently experimenting with or using wind power as part of energy production or as part of long-term energy planning. Environmental impacts include loss of land to wind farms and killing of birds, as well as degradation of scenic resources.

How Can We Evaluate Alternative Energy Sources?

The world is moving into a new era, one of transition from almost total dependence on fossil fuels to greater use of alternative renewable sources of energy. Although each of the alternatives offers a way out of the energy dilemma created by population growth and technological development, each has advantages and disadvantages. How can we evaluate the alternatives and select the right mix of energy sources for the coming decades? We can begin by comparing them on the basis of those characteristics most important to us: cost, jobs lost or gained, environmental impact, and potential for supplying energy.

Critical Thinking Questions

1. Using what you have learned about alternative energy in this chapter and elsewhere, evaluate the environmental impacts of the energy sources listed in the accompanying table. Complete the last column of the table. You may wish to subdivide the column into advantages and disadvantages.

2. Using the numbers 1–10, where 10 represents the best and 1 the worst, assign a rating to each value in the table. For example, in the column for carbon reduction, you might assign a rating of 10 to wind because it results in 100% reduc-

tion of carbon emissions. Solar thermal energy would then receive a rating of 8.4. In rating environmental impact, you will have to use your judgment in assigning numerical values.

3. One way to evaluate the various alternatives would be to add up the rating scores for each energy source and see which ones received the highest score. However, you may feel that some of the characteristics are more important than others and therefore should be weighted more heavily. Assign a weight to each column of the table, taking into consideration the importance you believe each should have in decision making. For example, if you believe that costs are more important than land used, you will assign a higher value to costs. In order to be able to compare your evaluation with those of your classmates, use decimal fractions for the weights, such as 0.2. The total should add up to 1.0.

4. Now, for each energy source, multiply its rating in each column by the weight you have assigned to the column. What is the total weighted score for each energy source? What are the sources in order of score, from highest to lowest?

5. Based on this analysis, what policy and research recommendations would you make to the U.S. government concerning alternative sources of energy?

Energy Source	U.S. Recoverable Resource[a] (exajoule/yr)	Costs in 1998 Cents[b] (per KWh)		Land Use[c] (m²/GWh for 30 years)	Carbon Reduction (%)	Carbon Avoidance Cost[d] ($/ton)	Number of Jobs[e] (per thousand GWh/yr)	Environmental Impact
		1988	2000					
Wind	10–40	8	5	1,355	100	95	542	
Geothermal	Small	4	4	404	99	110	112	
Photovoltaic	35	30	10	3,237	100	819	—	
Solar thermal	65	8	6	3,561	84	180	248	
Biomass	13–26	5	NA	—	100[f]	125	—	
Combined-cycle coal	—	6[g]	—	3,642	10	954	116	
Nuclear	—	15[h]	—	—	86	535	100	

[a]*Recoverable resource* is a measure of how much of the energy can be captured or exploited. From M. Brower, *Cool Energy* (Washington, D.C.: Union of Concerned Scientists, 1990), p. 19.

[b]L. R. Brown, C. Flavin, and S. Postel, *Saving the Planet* (New York: W. W. Norton, 1991), p. 27.

[c]Ibid., p. 60.

[d]Based on comparison with existing coal-fired plants. From C. Flavin, "Slowing Global Warming," in *State of the World* (New York: W.W. Norton, 1990), p. 27.

[e]Brown et al., p. 62.

[f]Assumes that the amount of carbon dioxide released in combustion will be consumed by replanted vegetation.

[g]C. Flavin, "Building a Bridge to a Sustainable Future," in *State of the World* (New York: W.W. Norton, 1992), p. 35.

[h]A. K. Reddy and J. Goldenberg, "Energy for the Developing World," *Scientific American* 263, no. 3 (1990):116.

- Biomass is the oldest human fuel. More than 1 billion people still use wood as their primary source of energy. Recently, there have been efforts to produce ethanol as a fuel from crops such as sugarcane. The environmental impacts of burning wood include deforestation, soil erosion, water pollution, and air pollution. Burning organic urban waste can release toxins and cause air pollution as well. Conversion of crops to ethanol can also produce pollution and, considering total costs, may not be economically viable for many applications.

- Geothermal energy is natural heat from Earth's interior used as an energy source. The environmental effects of developing geothermal energy relate to specific site conditions and the type of heat utilized (steam, hot water, or warm water). Geothermal energy may involve disposal of saline or corrosive waters, as well as on-site noise, emission of gas, and industrial scars.

- Significant policy issues are related to the future of alternative renewable energy. These include financing of research and development, access to energy sources, education, and community involvement.

● REEXAMINING THEMES AND ISSUES

Human Population:

As human population continues to increase, so does total world demand for and consumption of energy. Environmental problems related to increased consumption of fossil fuels could be minimized through controlling human population, increasing conservation efforts, and using those alternative renewable energy sources that do not harm the environment.

Sustainability:

Use of fossil fuels is not sustainable. In order to plan for sustainability from an energy perspective, we need to rely more on alternative energy sources that are naturally renewable and do not pollute the atmosphere or otherwise damage the environment. To do otherwise is antithetical to the concept of sustainability.

Global Perspective:

Evaluation of the potential of alternative energy sources requires us to understand global Earth systems and identify regions that are likely to produce high-quality alternative energy that could be utilized in urban regions of the world.

Urban World:

Alternative renewable energy sources have a future in our urban environment. For example, the roofs of buildings can be used for solar collectors or photovoltaic systems. Patterns of energy consumption can be regulated through use of innovative systems such as pump storage to augment production of electrical energy when demands are high in urban areas.

Science and Values:

We are seriously considering alternative energy today because we value environmental quality. Recognizing that burning fossil fuels creates many serious environmental problems, we are trying to increase our scientific knowledge and improve our technology to produce energy supplies to meet our needs for the future while minimizing environmental damage. Our present science and technology can produce a sustainable energy future, but we will need to change our values and our behavior to achieve it.

● KEY TERMS

active solar energy systems **347**
alternative energy **344**
biomass energy **357**
fuel cells **352**
geothermal energy **359**
nonrenewable energy **344**

ocean thermal conversion **350**
passive solar energy systems **346**
photovoltaics **348**
renewable energy **344**
solar collectors **347**

solar ponds **350**
solar power tower **350**
tidal power **354**
water power **352**
wind power **355**

1. What type of government incentives could be used to encourage use of alternative energy sources? Would their widespread use impact our economic and social environment?

2. Your town is near a large river that has a nearly constant water temperature of about 15°C (60°F). Could the water be used to cool buildings in the hot summers? How? What would be the environmental effects?

3. Which has greater future potential for energy production, wind or water power? Which causes more environmental problems? Why?

4. What are some of the problems associated with producing energy from biomass?

5. It is the year 2500, and natural oil and gas are rare curiosities that people see in museums. Given the technologies available today, what would be the most sensible fuel for airplanes? How would this fuel be produced to minimize adverse environmental effects?

6. When do you think the transition from fossil fuels to other energy sources will (or should) occur? Defend your answer.

● FURTHER READING

Ahmed, K. 1994. "Renewable Energy Technologies: A Review of the Status and Costs of Selected Technologies." World Bank Technical Paper No. 240. Discussion of the relative costs of energy production with biomass, solar, and photovoltaic energy sources compared with traditional sources and how these technologies have and will continue to become more competitive with oil and coal.

Brower, M. 1992. *Cool Energy: Renewable Solutions to Environmental Problems.* Cambridge, Mass.: MIT Press. Good coverage of alternative energy sources, showing fewer environmentally negative side effects than conventional sources.

Chartier, P., A. A. C. M. Beenackers, and G. Grassi, eds. 1995. *Biomass for Energy, Environment, Agriculture and Industry.* Proceedings of the 8th European Communities Conference, Vienna, Austria, October 3–5, 1994. London: Pergamon. A valuable, up-to-date summary of the status and promise of biomass use, including papers covering policy issues, assessments of the resource base, economic assessments, environmental impacts, and commercialization issues.

Flavin, C., and N. Lenssen. 1994. *Power Surge: Guide to the Coming Energy Revolution.* Worldwatch Environmental Alert Series. New York: W. W. Norton. An informative book that looks at problems of energy production and includes examples of alternative energy use and future paths for these alternative sources.

Sheffield, J. 1997. "The Role of Energy Efficiency and Renewable Energies in the Future and World Energy Market." *Renewable Energy* 10(2):315–318.

U.S. Department of Energy. 1993. *Electricity from Biomass—Renewable Energy Today and Tomorrow.* Washington, D.C.: Solar Thermal Biomass Power Division, Office of Solar Energy Conversion. A brief introduction to the potential for biomass-generated electricity in the United States.

Van Koevering, T. E., and N. J. Sell. 1986. *Energy: A Conceptual Approach.* Englewood Cliffs, N.J.: Prentice-Hall. A short book on energy that includes several chapters on alternative energy sources and related environmental concerns. It provides a good overview of a variety of alternative renewable and nonrenewable energy sources.

Weinberg, C. J., and R. H. Williams. 1990. "Energy from the Sun." *Scientific American* 263(3):146–155. A review article that discusses technologies in solar energy and argues that advances in solar alternative sources will become more important in the future.

CHAPTER 19 Nuclear Energy and the Environment

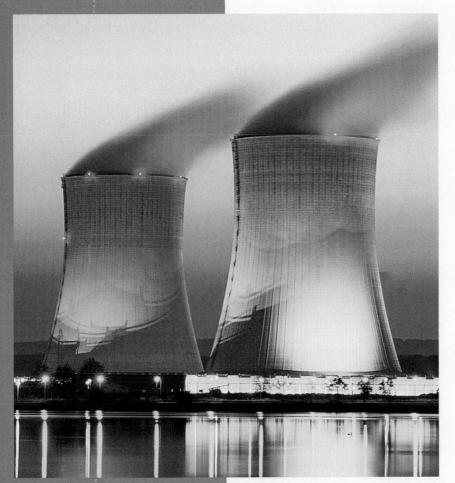

Nuclear power plant at dusk in Cattenom, France.

● LEARNING OBJECTIVES

As one of the alternatives to fossil fuels, nuclear energy generates much controversy. After reading this chapter, you should understand:

- What nuclear fission is and what are the basic components of a nuclear power plant.
- What nuclear radiation is and what are the three major types.
- Why it is important to know the type of radiation and the half-life for a particular radioisotope.
- What are the basic parts of the nuclear fuel cycle, and how each is related to our environment.
- How radioisotopes affect the environment and the major pathways of radioactive materials in the environment.
- What the breeder reactor is, and why it is important for the future of nuclear energy.
- What are the relationships between radiation doses and health.
- What we have learned from accidents at nuclear power plants.
- How we might safely dispose of high-level radioactive materials.
- What the future of nuclear power is likely to be.

Nuclear Energy and Public Opinion

In 1953, Dwight D. Eisenhower, a popular World War II general and U.S. president, described an optimistic vision of the future of atomic power, which he called "atoms for peace." The chairman of the Atomic Energy Commission, in a 1954 address to science writers, predicted that nuclear-powered generators would provide power so cheap, so nearly unlimited, and so clean that we would never need to meter it. More than a hundred nuclear power plants were constructed in the next several decades, and orders had been placed for more than a hundred additional plants by 1978.

But the public mood changed in the early 1980s. No more nuclear reactors were ordered; today, there are still only 104 commercial nuclear power plants in the United States. Polls suggest that the majority of Americans have been opposed to construction of nuclear power plants since the early 1980s. The reversal of opinion occurred three years after the Three Mile Island accident and four years before the Chernobyl accident (discussed later in this chapter). People's anxiety concerning nuclear weapons and nuclear war peaked in the early 1980s at about the time attitudes changed toward constructing more nuclear power plants, suggesting that worry over nuclear weapons had spilled over onto commercial nuclear power.[1]

Public support for building new nuclear plants is increasing today, especially in the western United States.

The California energy crisis (Case Study, Chapter 16) caused many people to rethink the value of nuclear energy. In one January 2001 survey, about 50 percent of the U.S. adults interviewed agreed that we should build more nuclear energy plants in the future. It appears that nuclear power is at a crossroads. Polls like the one just mentioned suggest that many Americans believe nuclear power should and will be an important energy source in the future. However, many people still oppose construction of new nuclear power plants in the near future and remain uncomfortable with issues of safety.

An important question is this: If nuclear energy were phased out, would it make any significant difference? The answer depends on whom you listen to. To some people, the answer is no. They argue that nuclear energy is too expensive and makes a relatively small part of the U.S. energy package; only minor substitution of other sources would be required to compensate for its loss.[2] Others point out, however, that nuclear energy does supply about 20% of our electricity and may be absolutely necessary if we are to sustain our environment.

Nuclear energy is one of several technologies that may eventually replace fossil fuels. Therefore, nuclear power, which does not contribute to global warming, may again be looked at as a major source of energy, particularly if issues of cost, availability of nuclear fuel, safety, and storage of waste can be resolved.[2]

● *The discussion of public opinion about nuclear power suggests a dilemma for the United States. Although many people remain uneasy about its use, nuclear power cannot be ignored as an alternative to fossil fuels. This chapter explores nuclear power reactors, radiation, accidents, waste management, and the future of nuclear power.*

19.1 Nuclear Energy

Nuclear energy is the energy of the atomic nucleus. Two nuclear processes can be used to release that energy to do work: fission and fusion. Nuclear **fission** is the splitting of atomic nuclei, and nuclear **fusion** is the fusing, or combining, of atomic nuclei. A by-product of both fission and fusion reactions is the release of enormous amounts of energy. (You may wish to review the discussion of matter and energy in A Closer Look 4.1.)

Nuclear energy for commercial use is produced by the splitting of atoms in **nuclear reactors**, which are devices that produce controlled nuclear fission. In the United States, almost all of these reactors use a form of uranium oxide as fuel. Nuclear fusion is not yet used commercially, although it has been accomplished in experimental fusion reactors.

Fission Reactors

The first human-controlled nuclear fission, demonstrated in 1942 by Italian physicist Enrico Fermi at the University of Chicago, led to the development of nuclear energy to produce electricity. Today, in addition to power plants to supply electricity to homes and industry, nuclear reactors power submarines, aircraft carriers, and icebreaker ships. Russia is building ships that contain reactors to provide electricity to coastal towns, and the United States is designing reactors for deep space missions.

Nuclear fission produces much more energy than other sources, such as burning fossil fuels. One kilogram (2.2 lb) of uranium oxide produces heat equivalent to approximately 16 metric tons of coal, making uranium an important source of energy in the United States and the world.

Three types, or isotopes, of uranium occur in nature: uranium-238, which accounts for approximately 99.3% of all natural uranium; uranium-235, which makes up about 0.7%; and uranium-234, which makes up about 0.005%. Uranium-235 and uranium-238 are two naturally radioactive isotopes of uranium. However, uranium-235 is the only naturally occurring fissionable (or *fissile*) material; therefore, it is essential to the production of nuclear energy. Processing uranium (called *enrichment*) to increase the concentration of uranium-235 from 0.7% to about 3% produces enriched uranium, which is used as fuel for the fission reaction. Radiation is explained and related terms are defined in A Closer Look 19.1.

Fission reactors split uranium-235 by neutron bombardment (Figure 19.1). The reaction produces neutrons, fission fragments, and heat. The released neutrons strike other uranium-235 atoms, releasing more neutrons, fission products, and heat. The neutrons released are fast moving and must be slowed down slightly, or *moderated*, to increase the probability of fission. In *light water reactors*, the kind most commonly used in the United States, ordinary water is used as the moderator. As the process continues, a chain reaction develops as more and more uranium is split, releasing more neutrons and more heat.

Most reactors now in use consume more fissionable material than they produce and are known as **burner reactors**. The reactor is part of the nuclear steam supply system, which produces steam to run the turbine generators that produce the electricity.[4] Therefore, the reactor has the same function as the boiler that produces the heat in coal-burning or oil-burning power plants (Figure 19.4).

Figure 19.5 shows the main components of a reactor: the core (consisting of fuel and moderator), control rods, coolant, and reactor vessel. The core of the reactor is enclosed in the heavy, stainless steel reactor vessel; then, for extra safety and security, the entire reactor is contained in a reinforced concrete building.

Chain reaction ⟶

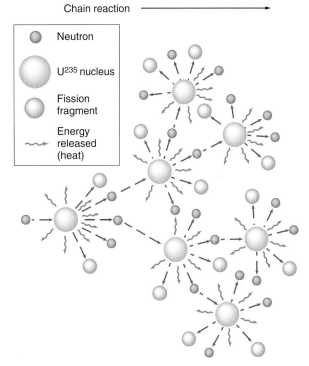

Figure 19.1 • Fission of uranium-235. A neutron strikes the U-235 nucleus, producing fission fragments and free neutrons and releasing heat. The released neutrons may then strike other U-235 atoms, releasing more neutrons, fission fragments, and energy. As the process continues, a chain reaction develops.

In the reactor core, fuel pins, consisting of enriched uranium pellets in hollow tubes (3–4 m long and less than 1 cm, or 0.4 in. in diameter), are packed together (40,000 or more in a reactor) in fuel subassemblies. A minimum fuel concentration is necessary to keep the reactor *critical*—that is, to achieve a self-sustaining chain reaction. A stable fission chain reaction in the core is maintained by controlling the number of neutrons that cause fission. The control rods, which contain materials that capture neutrons, are used to regulate the chain reaction. As the control rods are moved out of the core, the chain reaction increases; as they are moved into the core, the reaction slows. Full insertion of the control rods into the core stops the fission reaction.[7]

The function of the coolant is to remove the heat produced by the fission reaction. This is an important point—the rate of generation of heat in the fuel *must match* the rate at which heat is carried away by the coolant. All the major nuclear accidents have occurred when something went wrong with the balance and heat built up in the reactor core.[3] A **meltdown** generally refers to a nuclear accident in which the nuclear fuel becomes so hot that it forms a molten mass that breaches the containment of the reactor and contaminates the outside environment with radioactivity.

Radioactive Decay

To many people, radiation is a subject shrouded in mystery. They feel uncomfortable with it. We learn from an early age that nuclear energy may be dangerous because of radiation and that nuclear fallout from the detonation of atomic bombs can cause widespread human suffering. One thing that makes radiation scary is that we cannot see it, taste it, smell it, or feel it. In this feature, we try to demystify some aspects of radiation by discussing the natural process of radiation, or radioactivity.

First, we need to understand that radiation is a natural process that has been going on since the creation of the universe. Understanding the process of radiation involves understanding the **radioisotope**, which is a form of a chemical element that spontaneously undergoes **radioactive decay**. During the decay process, the radioisotope changes from one isotope to another and emits one or more forms of radiation.

You may recall from Chapter 4 that isotopes are atoms of an element that have the same atomic number (the number of protons in the nucleus) but that vary in atomic mass number (the number of protons plus neutrons in the nucleus). For example, two isotopes of uranium are $^{235}U_{92}$ and $^{238}U_{92}$. The atomic number for both isotopes of uranium is 92 (revisit Table 4.1); however, the atomic mass numbers are 235 and 238. The two different uranium isotopes may be written as uranium-235 and uranium-238 or U-235 and U-238.

An important characteristic of a radioisotope is its *half-life*, the time required for one-half of a given amount of the isotope to decay to another form. For example, uranium-235 has a half-life of 700 million years, a very long time indeed! Radioactive carbon-14 has a half-life of 5,570 years, which is in the intermediate range, and radon-222 has a relatively short half-life of 3.8 days. Other radioactive isotopes have even shorter half-lives; for example, polonium-218 has

a half-life of about 3 minutes. Still other radioactive isotopes have half-lives as short as a fraction of a second.

The main point here is that each radioactive isotope has its own unique and unchanging half-life. Isotopes with very short half-lives are present for only a brief time, whereas those with long half-lives remain in the environment for long periods. Table 19.1 illustrates the general pattern for decay in terms of the elapsed half-lives and the fraction remaining. For example, suppose we start with 1 g polonium-218 with a half-life of approximately 3 minutes. After an elapsed time of 3 minutes, 50% of the polonium-218 remains. After 5 elapsed half-lives, or 15 minutes, only 3% is still present; and after 10 elapsed half-lives (30 minutes), 0.1% is still present. Where has the polonium gone? It has decayed to lead-214, another radioactive isotope, which has a half-life of about 27 minutes. The progression of changes associated with the decay process is often known as a *radioactive decay chain*. Now suppose we had started with 1 g uranium-235, with a half-life of 700 million years. Following 10 elapsed half-lives, 0.1% of the uranium would be left—but this

Table 19.1	Generalized Pattern of Radioactive Decay	
Elapsed Half-life	*Fraction Remaining*	*Percent Remaining*
0	—	100
1	1/2	50
2	1/4	25
3	1/8	13
4	1/16	6
5	1/32	3
6	1/64	1.5
7	1/128	0.8
8	1/256	0.4
9	1/512	0.2
10	1/1024	0.1

process would take 7 billion years.

Radioisotopes with short half-lives initially undergo a more rapid rate of change (nuclear transformation) than do radioisotopes with long half-lives. Conversely, radioisotopes with long half-lives have a less intense and slower initial rate of nuclear transformation but may be hazardous for a much longer time.[5]

There are three major kinds of nuclear radiation: *alpha particles, beta particles*, and *gamma rays*. Alpha particles consist of two protons and two neutrons (a helium nucleus) and have the greatest mass of the three types of radiation (Figure 19.2*a*). Because alpha particles have a relatively high mass, they do not travel far. In air, alpha particles can travel approximately 5–8 cm (about 2–3 in) before they stop. However, in living tissue, which is much denser than air, they can travel only about 0.005–0.008 cm (0.002–0.003 in). Because this is a very short distance, to cause damage to living cells, alpha particles must originate very close to the cells.[6] Alpha particles can be stopped by a sheet or so of paper.

Beta particles are electrons and have a mass of 1/1,840 of a proton. Beta decay occurs when one of the protons or neutrons in the nucleus of an isotope spontaneously changes. What happens is that a proton turns into a neutron or a neutron is transformed into a proton (Figure 19.2*b*). As a result of this process, another particle, known as a *neutrino*, is also ejected. A neutrino is a particle with no rest mass (the mass when the particle is at rest with respect to an observer).[6] Beta particles travel farther through air than the more massive alpha particles but are blocked by even moderate shielding, such as a thin sheet of metal (aluminum foil) or a block of wood.

The third and most penetrating type of radiation comes from *gamma decay*. When gamma decay occurs, a gamma ray, a type of electromagnetic radiation, is emitted from the isotope. Gamma rays are similar to X rays but are more energetic

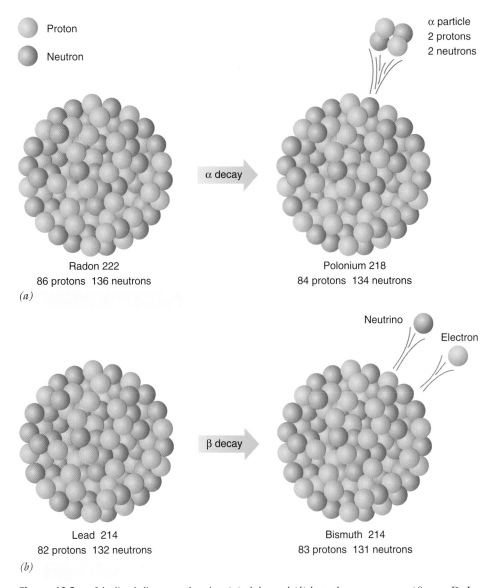

Proton

Neutron

α particle
2 protons
2 neutrons

α decay

Radon 222
86 protons 136 neutrons

Polonium 218
84 protons 134 neutrons

(a)

Neutrino

Electron

β decay

Lead 214
82 protons 132 neutrons

Bismuth 214
83 protons 131 neutrons

(b)

Figure 19.2 • Idealized diagrams showing (*a*) alpha and (*b*) beta decay processes. (*Source*: D. J. Brenner, Radon: Risk and Remedy [New York: Freeman, 1989]. Copyright 1989 by W. H. Freeman and Company. Reprinted with permission.)

and penetrating; they travel the longest average distance of all types of radiation. Protection from gamma rays requires thick shielding such as about a meter of concrete or several centimeters of lead.

Each radioisotope has its own characteristic emissions; some isotopes emit only one type of radiation, whereas others emit a mixture. In addition, the different types of radiation have different toxicities (degrees or intensities of potential to harm or poison). In terms of human health, and the health of other organisms, alpha radiation is most toxic or dangerous when inhaled or ingested.

Because alpha radiation is stopped within a very short distance by living tissue, much of the damaging radiation is absorbed by the tissue. When alpha-emitting isotopes are stored in a container, however, they are relatively harmless. Beta radiation is intermediate in its toxicity, although most beta radiation is absorbed by the body when a beta emitter is ingested. Gamma emitters are toxic and dangerous inside or outside the body; but when they are ingested, some of the radiation passes outside the body.

Some radioisotopes, particularly those of very heavy elements such as ura-

nium, undergo a series of radioactive decay steps (a decay chain) before finally reaching a stable nonradioactive isotope. For example, uranium decays through a series of steps to the stable nonradioactive isotope of lead. A decay chain for uranium-238 (with a half-life of 4.5 billion years) to stable lead-206 is shown in Figure 19.3. Also listed are the half-lives and types of radiation that occur during the transformations. Note that the simplified radioactive decay chain shown in Figure 19.3 involves 14 separate transformations and includes several environmentally important radioisotopes,

Radioactive Elements	Radiation Emitted			Half-life		
	Alpha	Beta	Gamma	Minutes	Days	Years
← Uranium–238 ↓	☢		☢			4.5 billion
← Thorium–234 ↓		☢	☢		24.1	
← Protactinium–234 ↓		☢	☢	1.2		
← Uranium–234 ↓	☢		☢			247,000
← Thorium–230 ↓	☢		☢			80,000
← Radium–226 ↓	☢		☢			1,622
← Radon–222 ↓	☢				3.8	
← Polonium–218 ↓	☢	☢		3.0		
← Lead–214 ↓		☢	☢	26.8		
← Bismuth–214 ↓		☢	☢	19.7		
← Polonium–214 ↓	☢			0.00016 (sec)		
← Lead–210 ↓		☢	☢			22
← Bismuth–210 ↓		☢			5.0	
← Polonium–210 ↓	☢		☢		138.3	
← Lead–206	None			Stable		

Figure 19.3 • Uranium-238 decay chain. (*Source*: F. Schroyer, ed., Radioactive Waste, 2nd printing [American Institute of Professional Geologists, 1985].)

including, in addition to uranium-238, radon-222, polonium-218, and lead-210. The decay from one radioisotope to another is often stated in terms of parent and daughter products. For example, uranium-238, with a half-life of 4.5 billion years, is the parent of daughter product thorium-234.

To sum up, when considering radioactive decay, two important facts to remember are: (1) the half-life and (2) the type of radiation emitted.

Other parts of the nuclear steam supply system are the primary coolant loops and pumps, which circulate the coolant through the reactor, extracting heat produced by fission, and heat exchangers or steam generators, which use the fission-heated coolant to make steam (Figure 19.4*b*). In light water reactors, water is used as the coolant as well as the moderator.

A design philosophy has emerged in the nuclear industry to build less complex, smaller reactors that are safer.

Large nuclear power plants, which produce about 1,000 MW of electricity, require an extensive set of pumps and backup equipment to ensure that adequate cooling is available to the reactor. Smaller reactors can be designed with cooling systems that work by gravity and as a result are not as vulnerable to pump failure caused by power loss. Such cooling systems are said to have *passive stability*, and the reactors are said to be *passively safe*.[8] Another approach is the use of helium gas to cool reactors that

(a)

(b)

Figure 19.4 • Comparison of (*a*) a fossil fuel power plant and (*b*) a nuclear power plant with a boiling water reactor. Notice that the nuclear reactor has exactly the same function as the boiler in the fossil fuel power plant. The coal-burning plant (*a*) is Ratcliffe-on-Saw, located in Nottinghamshire, England; and the nuclear power station (*b*) is located in Leibstadt, Switzerland. (*Source*: American Nuclear Society, Nuclear Power and the Environment, 1973.)

have specially designed fuel capsules capable of withstanding temperatures as high as 1,800°C (about 3,300°F). The idea is to design the fuel assembly so that it cannot hold sufficient fuel to reach this temperature and thus cannot experience a core meltdown.

Nuclear Energy and Sustainability

Energy from reactors that use uranium-235 as fuel will not be sustainable, particularly if many more such reactors are built. The total known amount of uranium oxide that can be mined and used in burner reactors is not sufficient to provide a sustainable source of energy. Today, about 440 nuclear power plants supply about 16% of the world's electricity. If use of nuclear energy based on uranium-235 increased enough to provide about 40% of the energy presently supplied by fossil fuels, all known uranium deposits would be mined and used in about 30 years. Clearly, this is not sustainable.[3]

Our present light water reactors use uranium very inefficiently. Only about 1% of the uranium is used in the reactor; the other 99% ends up as waste. Therefore, our present reactors are part of the nuclear waste problem and not a long-term solution to the energy problem. One way for nuclear power to be sustainable for at least hundreds of years would be the use of a process known as *breeding*. **Breeder reactors** are designed to produce new nuclear fuel. They do this through a process in which they transform waste or lower-grade uranium into fissionable material. Breeder reactors, if constructed in sufficient numbers (several thousand), could supply about half the energy presently produced by fossil fuels for about 2,000 years.[3]

Fuel and control rods

Heat

Reactor vessel
(stainless steel)

Core

Coolant

(a)

Figure 19.5 • *(a)* Main components of a nuclear reactor. *(b)* Glowing spent fuel elements being stored in water at a nuclear power plant.

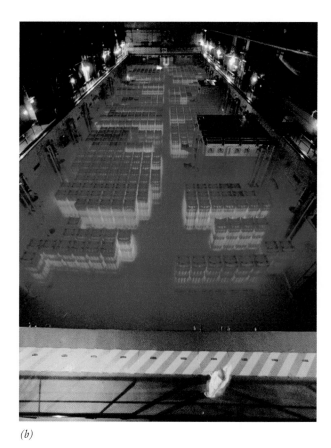

(b)

Breeding is apparently the future of nuclear power, if sustainability in terms of nuclear fuel is the objective. Bringing breeder reactors "on line" to produce safe nuclear power will take planning, research, and advanced reactor development. Also, fuel for the breeder reactors will have to be recycled, as reactor fuel must be replaced every few years. What is needed is a new type of breeder reactor comprising an entire system that includes reactor, fuel cycle (especially recycling and reprocessing of fuel), and less production of waste. Such a new reactor is possible but will require redefining our national energy policy and turning energy production in new directions. It remains to be seen whether this will happen.

Fusion Reactors

In contrast to fission, which involves splitting heavy nuclei (such as of uranium), fusion involves combining the nuclei of light elements (such as hydrogen) to form heavier ones (such as helium). As fusion occurs, heat energy is released (Figure 19.6). Nuclear fusion is the source of energy in our sun and other stars.

Figure 19.6 • Deuterium-tritium (DT) fusion reaction. (*Source*: Modified from U.S. Department of Energy, 1980.)

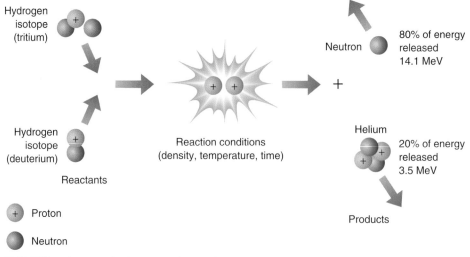

Hydrogen isotope (tritium)

Hydrogen isotope (deuterium)

Reactants

Reaction conditions (density, temperature, time)

Neutron

Helium

80% of energy released 14.1 MeV

+

20% of energy released 3.5 MeV

Products

Proton

Neutron

MeV Million electron volts (measure of energy)

Figure 19.7 • Experimental fusion nuclear reactor that magnetically confines plasma at very high temperatures.

In a hypothetical fusion reactor, two isotopes of hydrogen—deuterium and tritium—are injected into the reactor chamber, where the necessary conditions for fusion are maintained. Products of the deuterium-tritium (DT) fusion include helium, producing 20% of the energy released, and neutrons, producing 80% of the energy released (Figure 19.7).[9]

Several conditions are necessary for fusion to take place. First, the temperature must be extremely high (approximately 100 million degrees Celsius for DT fusion). Second, the density of the fuel elements must be sufficiently high. At the temperature necessary for fusion, nearly all atoms are stripped of their electrons, forming a plasma. Plasma is an electrically neutral material consisting of positively charged nuclei, ions, and negatively charged electrons. Third, the plasma must be confined for a sufficient time to ensure that the energy released by the fusion reactions exceeds the energy supplied to maintain the plasma.[9,10]

The potential energy available when and if fusion reactor power plants are developed is nearly inexhaustible. One gram of DT fuel (from a water and lithium fuel supply) has the energy equivalent of 45 barrels of oil. Deuterium can be extracted economically from ocean water, and tritium can be produced in a reaction with lithium in a fusion reactor. Lithium can be extracted economically from abundant mineral supplies.

Many problems remain to be solved before nuclear fusion can be used on a large scale. Research is still in the first stage, which involves basic physics, testing of possible fuels (mostly DT), and magnetic confinement of plasma. Progress in fusion research has been steady in recent years, so there is optimism that useful power will eventually be produced from controlled fusion.[11]

19.2 Nuclear Energy and the Environment

The **nuclear fuel cycle** includes the processes involved in producing nuclear power from the mining and processing of uranium to controlled fission, the reprocessing of spent nuclear fuel, the decommissioning of power plants, and the disposal of radioactive waste. Throughout the cycle, radiation can enter and affect the environment (Figure 19.8). In order to understand the environmental effects of radiation, it is useful to be acquainted with the units used to measure radiation and the amount or dose of radiation that may cause a health problem. These are explained in A Closer Look 19.2.

Problems with Nuclear Power

Let's look a little closer at the environmental effects of the nuclear fuel cycle (Figure 19.8):

- Uranium mines and mills produce radioactive waste material that can pollute the environment. There have been instances in which radioactive mine tailings have been used for foundation and building materials and have contaminated dwellings. (Tailings are materials that are removed by mining activity but are not processed and generally remain at the site.)

Figure 19.8 • Idealized diagram showing the nuclear fuel cycle for the U.S. nuclear energy industry. Disposal of tailings, which because of their large volume may be more toxic than high-level waste, was treated casually in the past. (Source: Office of Industry Relations, The Nuclear Industry, 1974.)

- Uranium-235 enrichment and fabrication of fuel assemblies also produce waste materials that must be carefully handled and disposed of.

- Site selection and construction of nuclear power plants in the United States have been extremely controversial. The environmental review process is extensive and expensive, often centering on hazards related to the probability of such events as earthquakes.

- The power plant or reactor is the site most people are concerned about because it is the most visible part of the cycle. It is also the site of past accidents, including partial meltdowns that have released harmful radiation into the environment.

- The United States does not reprocess spent fuel from reactors to recover uranium and plutonium at this time, so there currently is no environmental impact from them. However, many problems are associated with the handling and disposal of nuclear waste, as discussed later in this chapter.

- Waste disposal is a controversial part of the nuclear cycle because no one wants a nuclear waste disposal facility nearby. There is general public concern that hazardous nuclear waste cannot be adequately isolated from the environment for the long period of time (millions of years) that it remains hazardous.

- Nuclear power plants have a limited lifetime of between 20 and 30 years. Decommissioning (removal from service) or modernization of a plant is a controversial part of the cycle with which we have little experience. Contaminated machinery needs to be disposed of or stored so that environmental damage will not occur. Decommissioning or refitting will be very expensive (perhaps $200 million to $500 million) and is an important aspect of planning for the use of nuclear power. It is possible that the dismantling of decommissioned reactors may become one of the highest costs for the nuclear industry.[12]

In addition to the hazards of transporting and disposing of nuclear material, potential hazards are associated with supplying other nations with reactors. Terrorist activity and the possibility of irresponsible persons in governments add risks that are present in no other form of ener-

Radiation Units and Doses

The units used to measure radioactivity are complex and somewhat confusing. Nevertheless, a modest acquaintance with them is useful in understanding and talking about the effects of radiation on the environment.

A commonly used unit for radioactive decay is the *curie* (Ci), a unit of radioactivity equal to 37 billion nuclear transformations per second.[5]

The curie is named for Marie Curie and her husband, Pierre, who discovered radium in the 1890s. They also discovered polonium, which they named after Marie's homeland, Poland. The harmful effects of radiation were not known at that time, and both Marie Curie and her daughter died of radiation-induced cancer.[6] Her laboratory (Figure 19.9) is still contaminated today.

In the International System (SI) of measurement, the unit commonly used for radioactive decay is the *becquerel* (Bq), which is one radioactive decay per second. Units of measurement often used in discussions of radioactive isotopes, such as radon-222, are becquerels per cubic meter and picocuries per liter (pC/l). A picocurie is one-trillionth (10^{-12}) of a curie. Becquerels per cubic meter or picocuries per liter are therefore measures of the number of radioactive decays that occur each second in a cubic meter or liter of air.

When dealing with the environmental effects of radiation, we are most interested in the actual dose of radiation delivered by radioactivity. That dose is commonly measured in terms of *rads* (rd) and *rems*. In the International System (SI), the corresponding units are *grays* (Gy) and *sieverts* (Sv). Rads and grays are the units of the absorbed dose of radiation; 1 gray is equivalent to 100 rads. Rems and sieverts are units of equivalent dose, or effective equivalent dose, where 1 sievert is 100 rems.[5] The energy retained by living tissue that has been exposed to radiation is called the *radiation absorbed dose*, which is where the term *rad* comes from. Because different types of radiation have

different penetrations and so result in different degrees of damage to living tissue, the rad is multiplied by a factor known as the *relative biological effectiveness* to produce the rem or sievert units. When very small doses of radioactivity are being considered, the millirem (mrem) or millisievert (mSv)—that is, one-thousandth (0.001) of a rem or sievert—is used.[5,13,14] For gamma rays, the unit commonly used is the roentgen (R), or, in SI units, coulombs per kilogram (C/kg).

People are exposed to a variety of sources of radiation from the sky, the air, and the food we eat (Figure 19.10). What is the background radiation received by people? This question is commonly asked by people concerned with radiation. The average American receives about 2 to 4 mSv/yr. Of this, about 1 to 3 mSv/yr, or 50% to 75%, is natural. The differences are primarily due to elevation and geology. More cosmic radiation from outer space (which delivers about 0.3 to 1.3 mSv/yr) is received at higher elevations. Radiation from rocks and soils (such as granite and

organic shales) containing radioactive minerals delivers about 0.3 to 1.2 mSv/yr. The amount of radiation delivered from rocks, soils, and water may be much larger in areas where radon gas (a naturally occurring radioactive gas) seeps into homes. As a result, mountain states that also have an abundance of granitic rocks, such as Colorado, have a greater background radiation than do states that have a lot of limestone bedrock and are low in elevation, such as Florida.[13] In spite of this general pattern, in locations in Florida where phosphate deposits occur, background radiation is above average because of a relatively high uranium concentration found in the phosphate rocks.[15]

The amount of radiation received by people from human sources is about 1.35 mSv/yr. Two sources include naturally occurring radioactive potassium-40 and carbon-14, which are present in our bodies and produce about 0.35 mSv/yr. Potassium is an important electrolyte in our blood, and one isotope of potassium (potassium-40) has a very long half-life.

Figure 19.9 • Marie Curie in her laboratory.

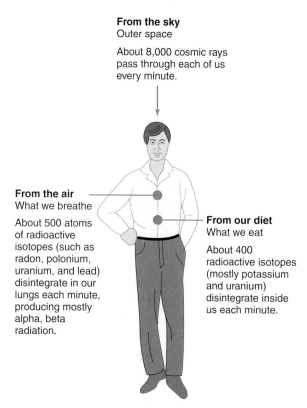

From the sky
Outer space

About 8,000 cosmic rays pass through each of us every minute.

From the air
What we breathe

About 500 atoms of radioactive isotopes (such as radon, polonium, uranium, and lead) disintegrate in our lungs each minute, producing mostly alpha, beta radiation.

From our diet
What we eat

About 400 radioactive isotopes (mostly potassium and uranium) disintegrate inside us each minute.

Figure 19.10 • The major sources of natural radiation are the sky, the air we breathe, and the food we eat. (*Source*: Modified from National Radiological Protection Board, *Living with Radiation*, 3rd ed. [Reading, England: National Radiological Protection Board, 1986].)

Although potassium-40 makes up only a very small percentage of the total potassium in our bodies, it is present in all of us, so we are all slightly radioactive. As a result, if you choose to share your life with another person, you are also exposed to a little bit more radiation.

Sources of low-level radiation from human processes include X rays for medical and dental purposes, which may deliver an average of 0.8 to 0.9 mSv/yr; nuclear weapons testing, approximately 0.04 mSv/yr; the burning of fossil fuels such as coal, oil, and natural gas, 0.03 mSv/yr; and nuclear power plants (under normal operating conditions), 0.002 mSv/yr.[5,13,14]

A person's occupation and lifestyle can also affect the annual dose of radiation received. If you fly at high altitudes in jet aircraft, you receive an additional small dose of radiation—about 0.05 mSv for each flight across the United States. If you work at a nuclear power plant, you can receive up to about 3 mSv/yr. Living next door to a nuclear power plant adds 0.01 mSv/year, and sitting on a bench watching a truck carrying nuclear waste pass by would add 0.001 mSv to your annual exposure. Sources of radiation are summarized in Figure 19.11a, assuming an annual total of 3 mSv/yr.[15,16] The amount of radiation received at certain job sites, such as nuclear power plants and laboratories where X rays are produced, is closely monitored. At such locations, personnel wear badges that indicate the dose of radiation received.

Figure 19.11b lists some of the common sources of radiation to which we are exposed. Notice that exposure to radon gas can equal what people were exposed to as a result of the Chernobyl nuclear power accident, which occurred in the Soviet Union in 1986. In other words, in some homes, people are exposed to about the same radiation as that experienced by the people evacuated from the Chernobyl area. (Radon gas is discussed in detail in Chapter 24.)

(a)

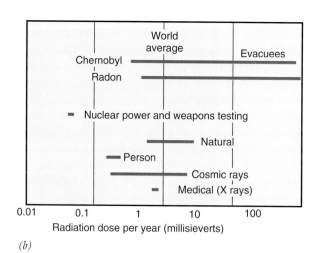

(b)

Figure 19.11 • (a) Sources of radiation received by people; assumes annual dose of 3.0 mSv/yr, with 66% natural and 33% medical and other (occupational, nuclear weapons testing, television, air travel, smoke detectors, etc.). (*Sources*: U.S. Department of Energy, 1999; *New Encyclopedia Britannica*, S.v. "radiation." (b) Range in annual radiation dose to people from major sources. (*Source*: Data in part from A. V. Nero, Jr., "Controlling Indoor Air Pollution," *Scientific American*, 258, no. 5 [1998]:42–48.

gy production. Nuclear energy may indeed be an answer to our energy problems, and perhaps someday it will provide unlimited cheap energy. However, with nuclear power comes responsibility.

Effects of Radioisotopes

As explained in A Closer Look 19.1, **radioisotope** is an isotope of a chemical element that spontaneously undergoes radioactive decay. Radioisotopes affect the environment in two ways: by emitting radiation that affects other materials and by entering the normal pathways of mineral cycling and ecological food chains.

The explosion of a nuclear atomic weapon does damage in both ways. At the time of the explosion, intense radiation of many kinds and energies is sent out, killing organisms directly. The explosion generates large amounts of radioactive isotopes, which are dispersed into the environment. Nuclear bombs exploded in the atmosphere produce a huge cloud that sends radioisotopes directly into the stratosphere,

where the radioactive particles may be widely dispersed by winds. Atomic *fallout*—the deposit of these radioactive materials around the world—was an environmental problem in the 1950s and 1960s, when the United States, the former Soviet Union, China, France, and Great Britain were testing and exploding nuclear weapons in the atmosphere.

The pathways (Figure 19.12) of some of these isotopes illustrate the second way in which radioactive materials can be dangerous in the environment. They can enter ecological food chains. Let's consider an example. One of the radioisotopes emitted and sent into the stratosphere by atomic explosions was cesium-137. This radioisotope was deposited in relatively small concentrations but was widely dispersed in the Arctic region of North America. It fell on reindeer moss, a lichen that is a primary winter food of the caribou. A strong seasonal trend in the levels of cesium-137 in caribou was discovered; the level was highest in the winter, when reindeer moss was the principle food, and lowest in the summer. Eskimos who obtained a high percentage of their protein from caribou ingested the

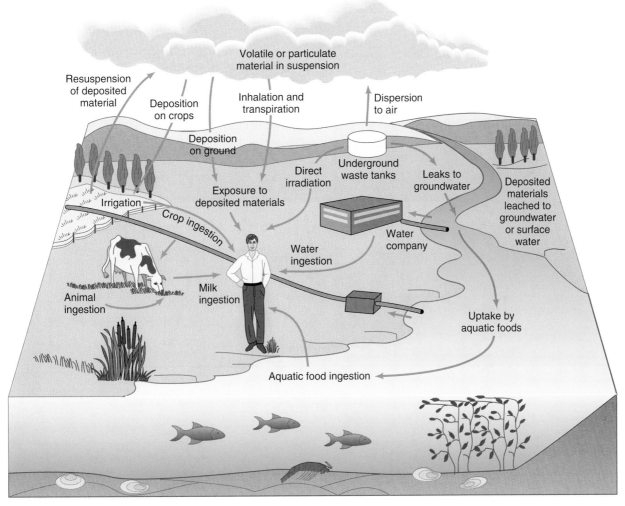

Figure 19.12 • How radioactive substances reach people. (*Source:* F. Schroyer, ed., *Radioactive Waste*, 2nd printing [American Institute of Professional Geologists, 1985].)

radioisotope by eating the meat, and their bodies concentrated the cesium. The more that members of a group depended on caribou as their primary source of food, the higher was the level of the isotope in their bodies.

What had happened? After entering specific ecosystems through the vegetation, the cesium-137 had undergone biomagnification, or ecological food chain concentration. That is, at each level in the food chain (moss to caribou to people), the concentration of the toxic material relative to concentrations of other materials in the bodies of organisms increased. Cesium-137 concentrations approximately doubled with each higher level in the food chain. This biomagnification was unknown until

radioisotopes and toxic organic compounds were found to be occurring in higher and higher concentrations at higher and higher levels in the food chain. Biomagnification is one of the major ecological factors in environmental toxicology, as we saw in Chapter 15.

The actual body burden of cesium varied within the Eskimo population. Adult males between the ages of 20 and 50 had the highest levels, apparently because their diet contained the most caribou meat. Concentration varied seasonally and increased over the years of intensive atmospheric bomb testing (Figure 19.13).[17]

Another important factor in the toxicity of cesium-137 from fallout is the length of time the isotope remains in

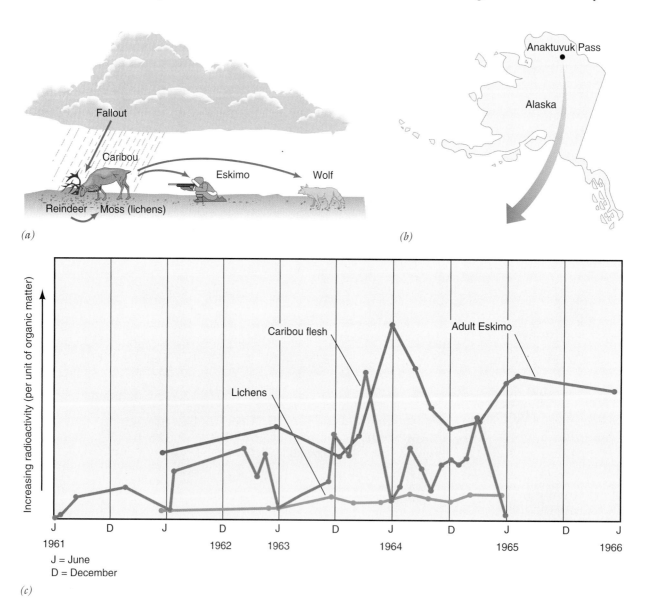

Figure 19.13 • Cesium-137 released into the atmosphere by atomic bomb tests was part of the fallout deposited onto soil and plants. (*a*) The cesium fell on lichens, which were eaten by caribou. The caribou were in turn eaten by the Eskimo. (*b*) Measurements of cesium were taken in the lichens, caribou, and Eskimo in the Anaktuvuk Pass of Alaska. (*c*) The cesium was concentrated by the food chain. Peaks in concentrations occurred first in the lichens, then in the caribou, and last in the Eskimo. (*Source:* [*c*] W. G. Hanson, "Cesium-137 in Alaskan Lichens, Caribou, and Eskimos," *Health Physics*, 13 [1967]: 383–389. Copyright 1967 by Pergamon Press. Reprinted with permission.)

the body. This is measured by the biological half-life, the time for one-half the concentration to be lost. In the caribou, the average biological half-life was 15 days. The biological half-life in the Eskimo, however, was approximately 65 days.

It is possible to predict the environmental pathways that radioisotopes will follow because we know the normal pathways of nonradioactive isotopes with the same chemical characteristics. Our knowledge of biomagnification and of large-scale air and oceanic movements that transport radioisotopes throughout the biosphere will also help us to understand the effects of radioisotopes.

Radiation Doses and Health

The most important question in studying radiation exposure in people involves determining the point at which the exposure or dose becomes a hazard to health (see A Closer Look 19.2). Unfortunately, there are no simple answers to this seemingly simple question. We do know that a dose of about 5,000 mSv (5 sieverts) is considered lethal to 50% of people exposed to it. Exposure to 1,000–2,000 mSv is sufficient to cause health problems, including vomiting, fatigue, potential abortion of pregnancies of less than 2 months' duration, and temporary sterility in males. At 500 mSv, physiological damage is recorded. The maximum allowed dose of radiation per year for workers in industry is 50 mSv, which is approximately 30 times the average natural background radiation received by people.[5,13] For the general public, the maximum permissible annual dose (for infrequent exposure) is set in the United States at 5 mSv, which is about three times the annual natural background.[5] For continuous or frequent exposure, the limit for the general public is 1 mSv.

Most information concerning the effects of high doses of radiation on people comes from studies of the people who survived the atomic bomb detonations in Japan at the end of World War II. We also have information concerning people exposed to high levels of radiation in uranium mines, workers who painted watch dials with luminous paint containing radium, and people treated with radiation therapy for disease.[18] Workers in uranium mines who were exposed to high levels of radiation have been shown to suffer a significantly higher rate of lung cancer than the general population. Studies have shown that there is a delay of 10 to 25 years between the time of exposure and the onset of disease. Starting in about 1917 in New Jersey, approximately 2,000 young women were employed painting watch dials with luminous paint. To maintain a sharp point on their brushes, they licked them and as a result were swallowing radium, which was in the paint. By 1924, dentists in New Jersey were reporting cases of jaw rot; and within five years radium was known to be the cause. Many of the women died of anemia or bone cancer.[6]

Although there is a vigorous and ongoing debate about the nature and extent of the relationship between radiation exposure and cancer mortality, most scientists agree that radiation can cause cancer. Some scientists believe that there is a linear relationship, such that any increase in radiation beyond the background level will produce an additional hazard. Others believe that the body is able to successfully handle and recover from low levels of exposure to radiation but that health effects (toxicity) become apparent beyond some threshold. The verdict is still out on this subject, but it seems prudent to take a conservative viewpoint and accept that there may be a linear relationship. Unfortunately, long-term chronic health problems related to low-level exposure to radiation are neither well known nor well understood.

Radiation has a long history in the field of medicine. Drinking waters that contain radioactive materials goes back to Roman times. By 1899, the adverse effects of radiation had been studied and were well known; and in that year, the first lawsuit for malpractice in using X rays was filed. Because science had shown that radiation could destroy human cells, however, it was a logical step to conclude that drinking water containing radioactive material such as radon might help fight diseases such as stomach cancer. In the early 1900s, it became popular to drink water containing radon, and the practice was supported by doctors, who stated that there were no known toxic effects. Although we now know that statement to be incorrect, radiotherapy, which uses radiation to kill cancer cells in humans, has been widely and successfully used for a number of years.[6]

19.3 Nuclear Power Plant Accidents

Although the chance of a disastrous nuclear accident is estimated to be very low, the probability that an accident will occur increases with every reactor put into operation. For example, according to the U.S. Nuclear Regulatory Commission's performance goal for a single reactor, the probability of a large-scale core meltdown in any given year should be no greater than 0.01% (one chance in 10,000). However, if there were 1,500 nuclear reactors (about four times the present world total), a meltdown could, at the low annual probability of 0.01%, be expected every seven years. This is clearly an unacceptable risk.[3] Increasing safety by about 10 times would result in lower, more manageable risk; but the risk would still be appreciable, because the potential consequences remain large.

Next, we discuss the two most well-known nuclear accidents, which occurred at the Three Mile Island and Chernobyl reactors. It is important to understand that these serious accidents resulted in part from human error.

Three Mile Island

One of the most dramatic events in the history of U.S. radiation pollution occurred on March 28, 1979, at the Three Mile Island nuclear power plant near Harrisburg,

Pennsylvania. The malfunction of a valve, along with human errors (thought to be the major problem), resulted in a partial core meltdown. Intense radiation was released to the interior of the containment structure. Fortunately, the containment structure functioned as designed, and only a relatively small amount of radiation was released into the environment. Exposure from the radiation emitted into the atmosphere has been estimated at 1 mSv, which is low in terms of the amount of radiation required to cause acute toxic effects. Average exposure to radiation in the surrounding area is estimated to have been approximately 0.012 mSv, which is only about 1% of the natural background radiation received by people. However, radiation levels were much higher near the site. On the third day after the accident, 12 mSv/hour was measured at ground level near the site. By comparison, the average American receives about 2 mSv/year from natural radiation.

Because the long-term chronic effects of exposure to low levels of radiation are not well understood, the effects of Three Mile Island exposure, although apparently small, are difficult to estimate. However, the incident revealed many potential problems with the way in which U.S. society dealt with nuclear power. Historically, nuclear power had been relatively safe, and the state of Pennsylvania was unprepared to deal with the accident. For example, there was no state bureau for radiation help, and the state Department of Health did not have a single book on radiation medicine (the medical library had been dismantled two years earlier for budgetary reasons). One of the major impacts of the incident was fear; yet there was no state office of mental health, and no staff member from the Department of Health was allowed to sit in on briefing sessions.[19]

Chernobyl

Lack of preparedness to deal with a serious nuclear power plant accident was dramatically illustrated by events that began unfolding on the morning of Monday, April 28, 1986 (Figure 19.14). Workers at a nuclear power plant in Sweden, frantically searching for the source of elevated levels of radiation near their plant, concluded that it was not their installation that was leaking radiation. Rather, the radioactivity was coming from the Soviet Union by way of prevailing winds. Confronted, the Soviets announced that an accident had occurred at a nuclear power plant at Chernobyl two days earlier, on April 26. This was the first notice to the world of the worst accident in the history of nuclear power generation.

It is speculated that the system that supplies cooling waters for the Chernobyl reactor failed as a result of human error, causing the temperature of the reactor core to rise to over 3,000°C (about 5,400°F), melting the uranium fuel. Explosions removed the top of the building over the reactor, and the graphite surrounding the fuel rods used to moderate the nuclear reactions in the core ignited. The fires produced a cloud of radioactive particles that rose high into the atmosphere. There were 237 confirmed cases of acute radiation sickness, and 31 people died of radiation sickness.[20]

In the days following the accident, nearly 3 billion people in the Northern Hemisphere received varying amounts of radiation from Chernobyl. With the exception of the 30-km (19-mi) zone surrounding Chernobyl, the world human exposure was relatively small. Even in Europe, where exposure was highest, it was considerably less than the natural radiation received during one year.[20]

In that 30-km zone, approximately 115,000 people were evacuated; and as many as 24,000 people were estimated to have received an average radiation dosage of 0.43 Sv (430 mSv). This group of people is being studied carefully.

It was expected, based on results from Japanese bomb survivors, that approximately 122 spontaneous leukemias would likely occur during the period from 1986 through 1998.[20] Surprisingly, as of late 1995, there was no significant increase in the incidence of leukemia, even among the most highly exposed people, but leukemia could still be expected in the future.[21]

Studies have found that the number of childhood thyroid cancer cases per year has risen steadily in the three countries of Belarus, Ukraine, and the Russian Federation (those most affected by Chernobyl) since the accident. In 1994, a combined rate of 132 new thyroid cancer cases was identified. Since the accident, a total of 653 thyroid cancer cases have been diagnosed in children and adolescents. These cancer cases are believed to be linked to the released radiation from the accident, although other factors, such as environmental pollution, may also play a role. It is predicted that a few percent of the roughly 1 million children exposed to the radiation eventually will contract thyroid cancer.[22]

Outside the 30-km zone, the increased risk of contracting cancer is very small and not likely to be detected from an ecological evaluation.[20] Nevertheless, according to one estimate, Chernobyl will ultimately be responsible for approximately 16,000 deaths worldwide.[8]

Vegetation within 7 km of the power plant was either killed or severely damaged following the accident. Pine trees examined in 1990 around Chernobyl showed extensive tissue damage and still contained radioactivity. The distance between annual rings (a measure of tree growth) had decreased since 1986.[23]

Scientists returning to the evacuated zone in the mid-1990s found, to their surprise, thriving and expanding animal populations. Species such as wild boar, moose, otters, waterfowl, and rodents seemed to be enjoying a population boom in the absence of humans. The wild boar population had increased 10-fold since the evacuation. However, these animals may be paying a genetic price for living within the contaminated zone. Study of gene mutations in voles (small mammals resembling rats

Figure 19.14 • Guard halting entry of people into the forbidden zone evacuated in 1986 as a result of the Chernobyl nuclear accident.

or mice) within the contaminated zone found gene mutation rates of over 5 mutations per animal, compared with a rate of only 0.4 per animal outside the zone. It is puzzling to scientists that the high mutation rate has not crippled the animal populations, but it appears so far that the benefit of excluding humans outweighs the negative factors associated with radioactive contamination.[24]

In the areas surrounding Chernobyl, radioactive materials continue to contaminate soils, vegetation, surface water, and groundwater, presenting a hazard to plants and animals. The evacuation zone may be uninhabitable for a very long time unless some way is found to remove the radioactivity.[20] For example, the city of Prypyat, 5 km from Chernobyl, is a "ghost city." At the time of the accident, the population of the city was 48,000. Today, it is abandoned, with blocks of vacant apartment buildings and rusting vehicles. Roads are cracking and trees are growing as new vegetation transforms the urban land back to green fields. Cases of thyroid cancer are still increasing, and the number of cases is many times higher for people who lived as children in Prypyat at the time of the accident. The final story of the world's most serious nuclear accident is yet to completely unfold.[25] Estimates of the total cost of the Chernobyl accident vary widely, but the cost will probably exceed $200 billion.

Although the Soviets were accused of not giving attention to reactor safety and of using outdated equipment, people are still wondering if such an accident could happen again elsewhere. Because there are several hundred reactors producing power in the world today, the answer has to be yes. About 10 accidents have released radioactive particles during the past 34 years. Therefore, although Chernobyl is the most serious nuclear accident to date, it certainly was not the first, and it is not likely to be the last. Although the probability of a serious accident is very small at a particular site, the consequences may be great, perhaps resulting in an unacceptable risk to society. This is really not so much a scientific issue as a political one involving a question of values.

Advocates of nuclear power have argued that nuclear power is safer than other sources of energy. They say that the number of additional deaths caused by air pollution resulting from burning fossil fuels is much greater than the number of lives lost through nuclear accidents. For example, the 16,000 deaths that might eventually be attributed to Chernobyl are fewer than the number of deaths caused each year by air pollution from burning coal.[8] Those arguing against nuclear power state that as long as people build nuclear power plants and manage them, there will be the possibility of accidents. We can build nuclear reactors that are safer, but people will continue to make mistakes, and accidents will continue to occur.

19.4 Radioactive Waste Management

Examination of the nuclear fuel cycle (Figure 19.8) illustrates some of the sources of waste that must be disposed of as a result of using nuclear energy to produce electricity. Radioactive wastes are byproducts that must be expected when electricity is produced at nuclear reactors; they may be grouped into three general categories: low-level waste, transuranic waste, and high-level waste. In addition, the tailings from uranium mines and mills must also be considered hazardous. In the western United States more than 20 million metric tons of abandoned tailings will continue to produce radiation for at least 100,000 years.

Low-level Radioactive Waste

Low-level radioactive waste contains sufficiently low concentrations or quantities of radioactivity that it does not present a significant environmental hazard if properly handled. Low-level waste includes a wide variety of items, such as residuals or solutions from chemical processing; solid or liquid plant waste, sludges, and acids; and slightly contaminated equipment, tools, plastic, glass, wood, and other materials.[26]

Low-level waste has been buried in near-surface burial areas in which the hydrologic and geologic conditions were thought to severely limit the migration of radioactivity.[26] However, monitoring has shown that several of the U.S. sites for disposal of low-level radiation have not provided adequate protection for the environment, and leaks of liquid waste have polluted groundwater. Of the original six burial sites, three had closed prematurely by 1979 due to unexpected leaks, financial problems, or loss of license. As of 1995, only two remaining low-level nuclear waste repositories remained in operation in the United States, one in Washington and the other in South Carolina. Construction of new burial sites, such as the Ward Valley site in southeastern California, has been met with strong public opposition, and controversy remains as to whether low-level radioactive waste can be disposed of safely.[27]

Transuranic Waste

Transuranic waste is composed of human-made radioactive elements heavier than uranium. It is produced in part by neutron bombardment of uranium in reactors and includes plutonium, americum, and einsteineum. Most transuranic waste is industrial trash, such as clothing, rags, tools, and equipment, that has been contaminated. The waste is low-level in terms of intensity of radioactivity, but plutonium has a long half-life and requires isolation from the environment for about 250,000 years. Most transuranic waste is generated from the production of nuclear weapons and, more recently, from cleanup of former nuclear weapons facilities.

Some transuranic wastes, as of 1999, are being transported to a disposal site near Carlsbad, New Mexico. The waste is isolated at a depth of 655 m (2,150 ft) in salt beds (rock salt) that are several hundred meters thick (Figure 19.15). Rock salt at the New Mexico site has several advantages:[28,29]

- The salt is about 225 million years old, and the area is geologically stable, with very little earthquake activity.
- The salt has no flowing groundwater and is easy to mine. Excavated rooms in the salt about 10 m wide and 4 m high will be used for disposal.
- Rock salt flows slowly into mined openings. The waste-filled spaces in the storage facility will be naturally closed by the slow-flowing salt in 75 to 200 years, sealing the waste.

The New Mexico disposal site is important because it is the first geologic disposal site for radioactive waste in the United States. It is a pilot project that will be evaluated very carefully. Safety is the primary concern. Procedures to transport the waste to the disposal site as safely as possible and place it underground in the disposal facility have been established. Because the waste will be hazardous for many thousands of years, and there are uncertainties con-

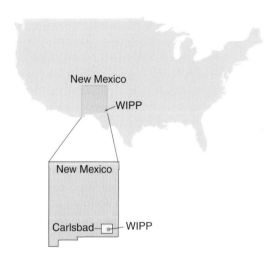

Figure 19.15 • Waste isolation pilot plant (WIPP) in New Mexico for disposal of transuranic waste. (*Source:* U.S. Department of Energy.)

cerning future cultures and languages, clear warnings above and below ground have been created. The site is clearly marked to help ensure that human intrusion in the future is avoided.[29]

High-level Radioactive Waste

High-level radioactive waste consists of commercial and military spent nuclear fuel; uranium and plutonium derived from military reprocessing; and other radioactive nuclear weapons materials. It is extremely toxic, and a sense of urgency surrounds its disposal as the total volume of spent fuel accumulates. At present, in the United States, tens of thousands of metric tons of high-level waste are being stored at more than a hundred sites in 40 states. Seventy-two of the sites are commercial nuclear reactors.[30,31]

Storage of high-level waste is at best a temporary solution, and serious problems with radioactive waste have occurred where it is being stored. Although improvements in storage tanks and other facilities will help, eventually some sort of disposal program must be initiated. Some scientists believe the geologic environment can best provide safe containment of high-level radioactive waste. Others disagree and have criticized proposals for long-term disposal of high-level radioactive waste underground. A comprehensive geologic disposal development program should have the following objectives:[32]

- Identification of sites that meet broad geologic criteria of ground stability and slow movement of groundwater with long flow paths to the surface.
- Intensive subsurface exploration of possible sites to positively determine geologic and hydrologic characteristics.
- Predictions of behavior of potential sites based on present geologic and hydrologic situations and assumptions for future changes in variables such as climate, groundwater flow, erosion, and earth movements.
- Evaluation of risk associated with various predictions.
- Political decision making based on risks acceptable to society.

The Nuclear Waste Policy Act of 1982 initiated a high-level nuclear waste disposal program. The Department of Energy was given the responsibility to investigate several potential sites and make a recommendation. The 1982 act was amended in 1987; and the amendment, along with the Energy Power Act of 1992, specified that high-level waste was to be disposed of underground in a deep, geologic waste repository. It was also specified that the Yucca Mountain site in Nevada was to be the only site evaluated. This does not mean that Yucca Mountain had been selected for actual disposal of nuclear waste, only

that it is the only site being completely evaluated at this time. If the site is suitable, then it could accept high-level waste as early as 2010. Following are some of the key issues being addressed by the Department of Energy at the Yucca Mountain site:[31,33]

- Assessment of the probability and consequences of volcanic eruptions.
- Evaluation of the earthquake hazard.
- Estimation of changes in the storage environment over long periods of time.
- Estimation of the time the waste may be contained and the types and rates of radiation that may escape from deteriorated waste containers.
- Evaluation of how heat generated by the waste may affect moisture in and around the repository and the design of the repository.
- Characterization of ground water flow near the repository.
- Identification and understanding of major geochemical processes that control the transport of radioactive materials.

Extensive scientific evaluations of the Yucca Mountain site are nearly completed.[31] However, use of this site is controversial and is generating considerable resistance from the state and people of Nevada as well as those scientists not confident in the plan. Some of the scientific questions at Yucca Mountain have concerned natural processes and hazards that might allow radioactive materials to escape, such as surface erosion, groundwater movement, earthquakes, and volcanic eruptions.

One of the major questions concerning the disposal of high-level radioactive waste is this: How credible are long-range geologic predictions—those covering several thousand to a few million years?[32] Unfortunately, there is no easy answer to this question, because geologic processes vary over both time and space. Climates change over long periods of time, as do areas of erosion, deposition, and groundwater activity. For example, large earthquakes even thousands of kilometers from a site may permanently change groundwater levels. The earthquake record for most of the United States extends back in time for only a few hundred years; therefore, estimates of future earthquake activity are tenuous at best.

The bottom line is that geologists can suggest sites that have been relatively stable in the geologic past, but they cannot absolutely guarantee future stability. Therefore, decision makers (not geologists) need to evaluate the uncertainty of predictions in light of pressing political, economic, and social concerns.[32] In the end, the geologic environment may be deemed suitable for safe containment of high-level radioactive waste, but care must be taken to ensure that the best possible decisions are made on this important and controversial issue.

Nuclear energy as a power source for electricity is now being seriously evaluated. Advocates for nuclear power have argued that nuclear power is good for the environment for these reasons:

- It does not produce potential global warming through release of carbon dioxide (see Chapter 22).

- It does not cause air pollution or emit precursors (sulfates and nitrates) that cause acid rain (see Chapter 23).

- If breeder reactors are developed for commercial use, the amount of fuel available will be greatly increased.

Those in favor of nuclear power argue that it is safer than other means of generating power and that we should build many more nuclear power plants in the future. This argument is predicated on the understanding that such power plants would be considerably safer than those being used today. That is, if we standardize nuclear reactors and make them safer and smaller, nuclear power could provide much of our electricity in the future.[8] The safety of nuclear power under normal operating procedures is not disputed, although the possibility of accidents and the disposal of spent fuel are concerns.

The argument against reviving nuclear power is based on political and economic considerations as well as scientific uncertainty concerning safety issues. Those opposed to expanding nuclear power argue that converting from coal-burning plants to nuclear power plants for the purpose of reducing carbon dioxide emissions would require an enormous investment in nuclear power to make a real impact. This is true. Furthermore, critics say, given the fact that safer nuclear reactors are only just being developed, there will be a time lag. As a result, nuclear power is not likely to have a real impact on environmental problems, such as air pollution, acid rain, and potential global warming, before at least the year 2050.[34] This point is debatable; the lag time could be shortened if research and development in advanced reactors become a higher priority and are better funded.

Until 2001, the politics of nuclear energy was losing ground. Nearly all energy scenarios were based on the expectation that nuclear power will continue to grow slowly or perhaps even decline in the coming years. Since the Chernobyl accident, many countries in Europe have been reevaluating the use of nuclear power; and in most instances, the number of nuclear power plants being built has been significantly reduced. Indeed, in Germany, where about one-third of the country's electricity is produced by nuclear power, the decision has been made to shut down all nuclear power plants in the next 25 years as they become obsolete.

Nuclear power produces about 20% of electricity in the United States today. As mentioned earlier, there have been no new orders for nuclear power plants in the United States. Nevertheless, research and development into the smaller, safer nuclear power plants and breeder reactors are going forward; and the Bush administration's 2001 energy plan suggests that nuclear power use should be increased in the future. The nuclear option is again being evaluated in light of the environmental problems associated with fossil fuels. However, the benefits of nuclear power must be balanced with the safety and waste disposal issues that have made nuclear energy an uncertain option for many people. The full impact of what began in 1942, when the atom was first split, is still to be determined.

Summary

- Nuclear fission is the process of splitting an atomic nucleus into smaller fragments. As fission occurs, energy is released. The major components of a fission reactor are the core, control rods, coolant, and reactor vessel.

- Nuclear radiation occurs when a radioisotope spontaneously undergoes radioactive decay and changes into another isotope.

- The three major types of nuclear radiation are alpha, beta, and gamma.

- Each radioisotope has its own characteristic emissions. Different types of radiation have different toxicities; and in terms of the health of humans and other organisms, it is important to know the type of radiation emitted and the half-life.

- The nuclear fuel cycle consists of the mining and processing of uranium, the generation of nuclear power through controlled fission, the reprocessing of spent fuel, the disposal of nuclear waste, and the decommissioning of power plants. Each part of the cycle is associated with characteristic processes, all with different potential environmental problems.

- The present burner reactors (mostly light water reactors) use uranium-235 as a fuel. Uranium is a nonrenewable resource mined from the Earth. If many more burner reactors were constructed, we would face fuel shortages. Nuclear energy based on burning uranium-235 in light water reactors is thus not sustainable. For nuclear energy to be sustainable, safe, and economical breeder reactors will need to be developed.

- Radioisotopes affect the environment in two major ways: by emitting radiation that affects other materials and by entering ecological food chains. Major environmental pathways by which radiation reaches people include uptake by fish ingested by people, uptake by crops ingested by people, inhalation from air, and exposure to nuclear waste and the natural environment.

- The dose response for radiation is fairly well established. We know the dose response for higher exposures, when illness or death occurs. However, there are vigorous debates concerning the health effects of low-

Does Nuclear Power Have a Future in the United States?

We return now to the important question of the future of nuclear power in the United States. In 1996, the United States had 109 nuclear power plants in operation in 32 states, producing 670 billion kWh of electricity. This is almost twice as much as the next leading country, France, which generates 77% of its total electricity by nuclear power. Why, then, is there still a question as to the future of nuclear power in the United States? One reason is that, despite the impressive output of nuclear-generated electricity, the United States ranks 19th out of 33 countries with nuclear power plants in the percentage of its electricity generated by nuclear power (only 20%). Other reasons are that no new reactors have been ordered since 1978 and that 61 plants were 20 years old (more than half the expected lifetime for nuclear plants) by 2000. Finally, we are a long way from predictions made in the 1970s that there would be 1,000 large nuclear facilities in the United States by the beginning of the next century.

Why has the nuclear power industry been stalled in the United States? One factor has been that growth in energy demand has slowed since the 1970s in reaction to the oil scares in that decade. Conservation and the use of alternative energy sources since then have exceeded expectations, so that predictions of growth in energy demand have had to be continuously revised downward (see Figure 19.16) and may yet be inaccurate. Prior to 1970, energy use in this country was increasing parallel with the gross domestic product (GDP); but since then, growth in energy use has slowed while GDP has continued to increase.

At the same time, the cost of construction of nuclear power plants and the time required to complete them have increased at least two to three times, discouraging power utilities from becoming involved in new projects. Several factors may have contributed to these conditions: concern over safety in the wake of the accident at Three Mile Island; new safety regulations established by the Nuclear Regulatory Commission (NRC); public anger over the higher prices for electricity that would have to be charged to cover construction costs; poor management; and industry response to the decline in energy demand.

Public opinion about nuclear power in the United States is perceived as negative by many, primarily because of concerns about safety (low-level exposures as well as catastrophic accidents), disposal of radioactive wastes, and proliferation of raw materials that could be used to manufacture nuclear weapons. Certainly there have been outspoken critics of nuclear power and local opposition to plants and disposal sites. A 1992 poll found that 65% of the public opposed construction of new reactors but went on to suggest that "opinion leaders" may exaggerate the opposition of the general public. A 2001 poll during the California energy crisis indicated that 50% of the people at that time believed we should build more nuclear energy plants.

Critical Thinking Questions

1. If the United States is only 19th in the world in percentage of electrical energy generated by nuclear energy, how is it that it produces almost twice as much energy from nuclear power plants as its nearest competitor?

2. In the United States, about 45,000 deaths per year result from automobile use, but no deaths have been attributed to the nuclear industry. Why are Americans more willing to accept one risk than another?

3. What are some reasons for the decline in the U.S. nuclear power industry? How would you rank these reasons in order of importance from the point of view of the utility companies? From the point of view of the public? Explain your rankings and the differences, if any, between the two points of view.

4. Under what conditions could nuclear power make a comeback in the United States? How likely do you think it is that these conditions will occur?

Figure 19.16 • Growth in electric generating capacity in the United States compared with annual forecasts by the National Electricity Reliability Council (based on utilities' construction plants). (*Source:* J. F. Ahearne, "The Future of Nuclear Power," *American Scientist* 81, no. 1 [1993]:24–35.)

level exposure to radiation and what relationships exist between exposure and cancer mortality. Most scientists believe that radiation can cause cancer. Ironically, radiation can be used to kill cancer cells, as in radiotherapy treatments.

- We have learned from accidents at nuclear power plants that it is difficult to plan for the human factor. People make mistakes. We have also learned that we are not as prepared for accidents as we would like to think. Some people believe that humans are not ready for the responsibility of nuclear power. Others believe that we can design much safer power plants where serious accidents are impossible.

- Transuranic nuclear waste is now being disposed of in salt beds—the first disposal of radioactive waste in the geologic environment in the United States.

- There is consensus that high-level nuclear waste may be safely disposed of in the geologic environment. The problem has been to locate a site that is safe and not objectionable to the people who make the decisions and to those who live in the region.

- Nuclear power, as a result of the California energy shortage of 2000–2001, is being seriously evaluated as an alternative to fossil fuels. On the one hand, it has advantages in that it emits no carbon dioxide and will not contribute to global warming or cause acid rain. On the other hand, people are uncomfortable with nuclear power because of possible accidents and waste disposal problems. For nuclear power to make a serious contribution to our future energy needs, we will need breeder reactors designed to be at least 10 times safer than any today.

● REEXAMINING THEMES AND ISSUES

Human Population:

As human population has increased, so has demand for electrical power; and as a result, a number of countries have turned to nuclear energy. The California energy crisis has caused many people in the United States to rethink the value of nuclear energy. Although relatively rare, accidents at nuclear power plants such as Chernobyl have exposed people to increased radiation. There is considerable debate over potential adverse effects of that radiation. Nevertheless, the fact remains that as world population increases, and if the number of nuclear power plants increases, the total number of people exposed to a potential hazardous release of toxic radiation will increase as well.

Sustainability:

It has been argued that sustainable energy development will require a return to nuclear energy, because nuclear energy does not contribute to a variety of environmental problems related to burning of fossil fuels. For nuclear energy to significantly contribute to sustainable energy development, however, we cannot depend on burner reactors that will quickly use Earth's uranium resources; rather, development of safer breeder reactors will be necessary.

Global Perspective:

Use of nuclear energy fits into our global management of the entire spectrum of energy sources. In addition, testing of nuclear weapons has spread radioactive isotopes around the entire planet, as have nuclear accidents. Radioactive isotopes that enter rivers and other waterways may eventually enter the oceans of the world, where oceanic circulation may further disperse and spread them.

Urban World:

Development of nuclear energy is a product of our technology and our urban world. In some respects, it is near the pinnacle of our accomplishments in terms of technology.

Science and Values:

We have a good deal of knowledge concerning nuclear energy and nuclear processes. However, people remain suspicious and, in some cases, frightened by nuclear power—in part because of the value they place on a quality environment and their perception that nuclear radiation is toxic to that environment. As a result, the future of nuclear energy will be related to political decisions based in part on the risk acceptable to society. It will also depend on research and development to produce much safer nuclear reactors.

KEY TERMS

STUDY QUESTIONS

1. If exposure to radiation is a natural phenomenon, why are we worried about it?

2. What is a radioisotope, and why is knowing its half-life important?

3. What is the normal background radiation that people receive? Why is it variable?

4. What are the possible relationships between exposure to radiation and adverse health effects?

5. What processes in our environment may result in radioactive substances reaching people?

6. Suppose it is recommended that high-level nuclear waste be disposed of in the geologic environment of the region in which you live. How would you go about evaluating potential sites?

7. Are there good environmental reasons to develop and build new nuclear power plants? Discuss both sides of the issue.

FURTHER READING

Cohen, B. L. 1990. *The Nuclear Energy Option*. New York: Plenum. A book that presents the case for nuclear energy and provides a complete discussion of the science concerning radiation and the potential health effects from exposure to radiation.

Murray, R. L. 1993. *Nuclear Energy: An Introduction to the Concepts, Systems, and Applications of Nuclear Processes*, 4th ed. New York: Pergamon Press.

Nuclear Energy Agency (NEA), Organization for Economic Co-Operation and Development (OECD). 1993. *The Cost of High-Level Waste Disposal in Geological Repositories: An Analysis of Factors Affecting Cost Estimates*. NEA, OECD. Analyses of management of spent fuel, methods of geological disposal, and costs associated with disposal of high-level radioactive waste.

Nuclear Energy Agency (NEA), Organization for Economic Co-Operation and Development (OECD). 1994. *Power Generation Choices: Costs, Risks, and Externalities*. Proceedings of an international symposium, Washington, D.C., September 23–24, 1993. NEA, OECD. Includes discussion of the economics of nuclear power versus other energy sources.

Nuclear Energy Agency (NEA), Organization for Economic Co-Operation and Development (OECD). 1995. *Environmental and Ethical Aspects of Long-Lived Radioactive Waste Disposal*. Proceedings of an international workshop, Paris, September 1–2, 1994. NEA, OECD. Essays covering topics of environmental policies, ethical and environmental considerations, cost-benefit analysis, and disposal issues of long-lived radioactive waste.

Thomas, S., and F. Berkhout. 1992. "The First 50 Years of Nuclear Power: Legacy and Lessons—Parts 1 and 2." *Energy Policy* 20(7,8). A special issue of *Energy Policy* containing more than 20 reviews, overviews, and case studies of the current state of nuclear energy supply, technology, and waste management, as well as the political situation.

U.S. Department of Energy, Office of Environmental Management. 1995. *Closing the Circle on the Splitting of the Atom*. Washington, D.C.: U.S. Department of Energy. Descriptions of environmental, safety, and health problems associated with production of nuclear weapons and how the U.S. Department of Energy plans to deal with the problem.

World Health Organization. 1995. *Health Consequences of the Chernobyl Accident*. Geneva, Switzerland: World Health Organization. A short book covering the accident, response, health consequences, findings, and proposed future work.

Young, J. P. and R. S. Yalow, eds. 1995. *Radiation and Public Perception: Benefits and Risks*. Washington, D.C.: American Chemical Society. A comprehensive look at public perception of radiation risks and health effects of radiation through experimentation, occupational exposure, atomic detonation, and nuclear reactor accidents.

CHAPTER 20

Water Supply, Use, and Management

Edwards Dam on the Kennebec River, near Augusta, Maine in June of 1999, just before it was removed.

● LEARNING OBJECTIVES

Although water is one of the most abundant resources on Earth, many important issues and problems are involved in water management. After reading this chapter, you should understand:

- Why the total abundance of water on Earth is not a problem, but making it available where and when it is needed *is* a problem.
- Why the residence times of water in various parts of the hydrologic cycle are important to water use and pollution potential.
- What a water budget is, and why it is useful in analyzing water supply problems and potential solutions.
- What groundwater is, and what environmental problems are associated with its use.
- How water can be conserved at home and in industrial and agricultural practice.
- Why sustainable water management will become more difficult as the demand for water increases.
- What the environmental impacts are of water projects such as dams, reservoirs, canals, and channelization.
- What a wetland is, how wetlands function, and why they are important.
- What hazards are presented by river flooding.

Removal of a Dam: A Maine Experience

The idea that dams, unlike the pyramids of Egypt, are not a permanent feature of the human-made landscape is somewhat new to Americans. The idea of actually removing a dam was, until recently, almost unheard of. Nevertheless, in the past few years, more than 200 dams have been removed. The rate of removal has been accelerating since the Edwards Dam on the lower Kennebec River in southern Maine was removed in 1999 (Figure 20.1a).

The Edwards Dam was constructed in 1837 following three years of protest that the structure would damage salmon and other fisheries by blocking the fishes' access to upriver spawning grounds. A fish ladder was constructed along with the dam, but it washed away in 1838 and was not replaced. From 1842 to 1847, the dam provided mechanical power for sawmills. Although the dam was overtopped by floods in 1839, 1846, 1855, and 1870, it survived; and by 1882, it was providing mechanical power for a textile mill employing over 700 workers. The dam first generated electric power in 1913. Flood water in 1974 removed about 46 m

(150 ft) of the dam; but it was repaired in 1975, in spite of the arguments of biologists who favored removing it. The textile mill closed in the 1980s; and in 1991, the company that owned the dam applied to increase electric power generation from 3.5 to 11.5 MW. At that time, the governor of Maine recommended that the dam be removed, as did the Maine legislature. In 1995, the owners withdrew their proposal to increase the amount of electric power produced at the dam, and an Environmental Impact Statement recommended that the dam be removed. The company gave the dam to the state of Maine, and it was removed in the summer of 1999 (Figure 20.1b).

Removal of the Edwards Dam opened about 29 km (18 mi) of river habitat to migrating fish including Atlantic salmon, striped bass, shad, alewives, and Atlantic sturgeon. Following removal of the dam, the Kennebec River came back to life as millions of fish migrated upstream for the first time in 160 years. The 29 km of free flowing river has sparked renewed interest in drift boat fishing, contributing to some $48 million in economic benefits generated from sport fishing. Water quality improved during the first year after the dam was removed; and the project spawned interest in removing other dams, including others upstream on the Kennebec. For the lower Kennebec River, the removal is a new beginning of river restoration.[1]

(a)

(b)

Figure 20.1 • (a) Location of Edwards Dam near Augusta, Maine. (b) Dam being removed in June of 1999. (AP Photo Robert Butakn)

Water is a critical, limited, renewable resource in many regions on Earth. This chapter discusses our water resources in terms of supply, use, management, and sustainability. It also addresses important environmental concerns related to water: wetlands, dams and reservoirs, channelization, and flooding. The chapter ends with a capstone discussion of water use, management, and environmental concerns for the Colorado River.

20.1 Water

To understand water as a necessity, as a resource, and as a factor in the pollution problem, we must understand its characteristics, its role in the biosphere, and its role in living things. Water is a unique liquid; without it, life as we know it is impossible. Consider the following:

- Compared with most other common liquids, water has a high capacity to absorb and store heat. The capacity of water to hold heat has important climatic significance. Solar energy warms the oceans of the world, storing huge amounts of heat. The heat can be transferred to the atmosphere to develop hurricanes and other storms. The heat in warm oceanic currents such as the Gulf Stream warms Great Britain and Western Europe, making these areas much more hospitable for humans then would otherwise be possible at such high latitudes.

- Water is the universal solvent. Because many natural waters are slightly acidic, they can dissolve a great variety of compounds, from simple salts to minerals, including sodium chloride (common table salt) and calcium carbonate (calcite) in limestone rock. Water also reacts with complex organic compounds, including many amino acids found in the human body.

- Compared with other common liquids, water has a high surface tension, a property that is extremely important in many physical and biological processes that involve moving water through, or storing water in, small openings or pore spaces.

- Among the common compounds, water is the only one whose solid form is lighter than its liquid form (it expands by about 8% when it freezes, becoming less dense). That is why ice floats. If ice were heavier than liquid water, it would sink to the bottom of the oceans, lakes, and rivers. If water froze from the bottom up, shallow seas, lakes, and rivers would freeze solid. All life in the water would die, because cells of living organisms are mostly water, and as water freezes and expands, cell membranes and walls rupture. If ice were heavier than water, the biosphere would be vastly different from what it is, and life, if it existed at all, would be greatly altered.[2]

- Sunlight penetrates water to variable depths, permitting photosynthetic organisms to live below the surface.

A Brief Global Perspective

A review of the global hydrologic cycle, introduced in Chapter 4, is important here. The main process in the cycle is the global transfer of water from the atmosphere to the land and oceans and back to the atmosphere (Figure 20.2). Table 20.1 lists the relative amounts of water in the major storage compartments of the cycle. Notice that more than 97% of Earth's water is in the oceans; the next largest storage compartment, the ice caps and glaciers, accounts for another 2%. Together, these sources account for more than 99% of the total water, and both are generally unsuitable for human use because of salinity (seawater) and location (ice caps and glaciers). Only about 0.001% of the total water on Earth is in the atmosphere at any one time. However, this relatively small amount of water in the global water cycle, with an average atmosphere residence time of only about 9 days, produces all our freshwater resources through the process of precipitation.

On a global scale, then, total water abundance is not the problem; the problem is water's availability in the right place at the right time in the right form. Water can be found in either liquid, solid, or gaseous form at a number of locations at or near Earth's surface. Depending on the specific location, the residence time may vary from a few days to many thousands of years (see Table 20.1). However, as mentioned, more than 99% of Earth's water in its natural state is unavailable or unsuitable for beneficial human use. Thus, the amount of water for which all the people, plants, and animals on Earth compete is much less than 1% of the total.

As the world's population and industrial production of goods increase, the use of water will also accelerate. The world per capita use of water in 1975 was about 700 m^3/year, or 2,000/day (185,000 gal/yr), and the total human use of water was about 3,850 km^3/year (about 10^{15} gal/yr). Today, world use of water is about 6,000 km^3/yr (about 1.58×10^{15} gal/yr), which is a significant fraction of the naturally available fresh water.

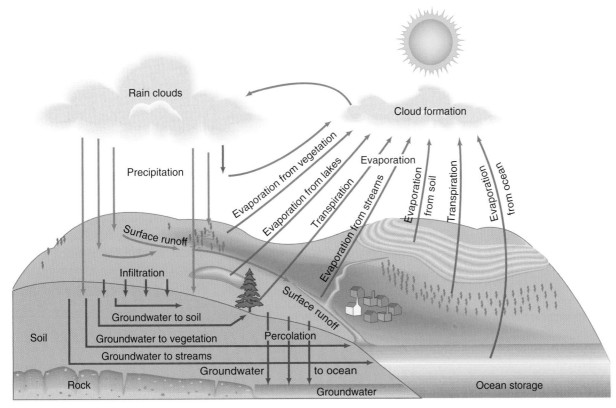

Figure 20.2 • The hydrologic cycle, showing important processes and transfer of water. (*Source:* Modified from Council on Environment Quality and Department of State, *The Global 2000 Report to the President*, vol. 2 [Washington, D.C.].)

Compared with other resources, water is used in very large quantities. In recent years, the total mass (or weight) of water used on Earth per year has been approximately 1,000 times the world's total production of minerals, including petroleum, coal, metal ores, and nonmetals.[3] Because of its great abundance, water is generally a very inexpensive resource. However, in the southwestern United States, the cost of water has been kept artificially low as a result of government subsidies and programs.

Because the quantity and quality of water available at any particular time are highly variable, shortages of water have occurred and will probably continue to occur with increasing frequency. Such shortages can lead to serious economic disruption and human suffering.[4] In the Middle East and Northern Africa, scarce water has resulted in harsh words and threats between countries. War over water is a possibility. The U.S. Water Resources Council estimates that water use in the United States by the year 2020 may exceed surface water resources by

Table 20.1	The World's Water Supply (Selected Examples)			
Location	Surface Area (km²)	Water Volume (km³)	Percentage of Total Water	Estimated Average Residence Time of Water
Oceans	361,000,000	1,230,000,000	97.2	Thousands of years
Atmosphere	510,000,000	12,700	0.001	9 days
Rivers and streams	—	1,200	0.0001	2 weeks
Groundwater (shallow to depth of 0.8 km)	130,000,000	4,000,000	0.31	Hundreds to many thousands of years
Lakes (fresh water)	855,000	123,000	0.0009	Tens of years
Ice caps and glaciers	28,200,000	28,600,000	2.15	Tens of thousands of years and longer

Source: U.S. Geological Survey.

13%.[4] Therefore, an important question is, how can we best manage our water resources, use, and treatment to maintain adequate supplies?

Groundwater and Streams

Next, before moving on to issues of water supply and management, we introduce groundwater and surface water and the terms used in discussing them. An acquaintance with this terminology is important in understanding many environmental issues, problems, and solutions.

The term **groundwater** usually refers to the water below the water table, where saturated conditions exist. The upper surface of the groundwater is called the *water table*.

Rain that falls on the land evaporates, runs off the surface, or moves below the surface and is transported underground. Locations where surface waters move into, or infiltrate, the ground are known as *recharge zones*. Places where groundwater flows or seeps out at the surface, such as springs, are known as *discharge zones* or *discharge points*.

Water that moves into the ground from the surface first seeps through pore spaces (empty spaces between soil particles or rock fractures) in the soil and rock known as the *vadose zone*. This area is seldom saturated (not all pore spaces are filled with water). The water then enters the groundwater system, which is saturated (all pore spaces are filled with water).

An *aquifer* is an underground zone or body of earth material from which groundwater can be obtained (from a well) at a useful rate. Loose gravel and sand with lots of pore space between grains and rocks or many open fractures generally make good aquifers. Groundwater in aquifers usually moves slowly at rates of centimeters or meters per day. When water is pumped from an aquifer, the water table is depressed around the well, forming a *cone of depression*. Figure 20.3 shows the major features of a groundwater and surface water system.

Streams may be classified as effluent or influent. In an **effluent stream**, the flow is maintained during the dry season by groundwater seepage into the stream channel from the subsurface. A stream that flows all year is called a perennial stream. Most perennial streams flow all year because they constantly receive groundwater to sustain flow. An **influent stream** is entirely above the water table and flows only in direct response to precipitation. Water from an influent stream seeps down into the subsurface. An influent stream is called an ephemeral stream, because it doesn't flow all year.

A given stream may have reaches (unspecified lengths of stream) that are perennial and other reaches that are ephemeral. It may also have reaches, known as intermittent, that have a combination of influent and effluent flow varying with the time of year. For example, streams flowing from the mountains to the sea in southern California often have reaches in the mountain that are perennial, supporting populations of trout or endangered southern steelhead, and lower intermittent reaches that transition to ephemeral reaches. At the coast, these streams may receive fresh or salty groundwater and tidal flow from the ocean to become a perennial lagoon.

Interactions Between Surface Water and Groundwater

Surface water and groundwater interact in many ways and should be considered part of the same resource. Nearly all natural surface water environments, such as rivers and

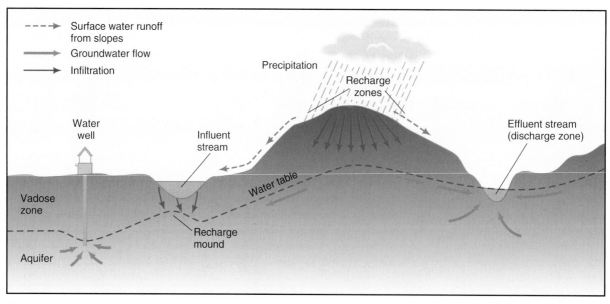

Figure 20.3 · Groundwater and surface water flow system.

Figure 20.4 • Idealized diagram illustrating some interactions between surface water and groundwater for a city in a semiarid environment with adjacent agricultural land and reservoir. (1) Water pumped from wells lowers the groundwater level. (2) Urbanization increases runoff to streams. (3) Sewage treatment discharges nutrient-rich waters to stream, groundwater, and reservoir. (4) Agriculture uses irrigation waters from wells, and runoff to stream from fields contains nutrients from fertilizers.

lakes, as well as human-constructed water environments, such as reservoirs, have strong linkages with groundwater. For example, withdrawal of groundwater by pumping from wells may reduce stream flow, lower lake levels, or change the quality of surface water. Reduction of effluent stream flow by lowering of the groundwater level may change a perennial stream that flows all year to an intermittent influent stream. Similarly, withdrawal of surface water by diversion from streams and rivers can deplete groundwater resources or change the quality of groundwater. Diversion of surface waters that recharge groundwaters may result in an increase in concentrations of dissolved chemicals in the groundwater. This happens because dissolved chemicals present in the groundwater are not diluted by mixing with infiltrated surface water. Finally, pollution of groundwater may result in pollution of surface water, and vice versa.[5]

Selected interactions between surface water and groundwater in a semiarid urban and agricultural environment are shown in Figure 20.4. Urban and agricultural runoff increase the volume of water in the reservoir.

Pumping of groundwater for agricultural and urban uses lowers the groundwater level. Quality of surface water and groundwater is reduced by urban and agricultural runoff, which adds nutrients from fertilizers, oil from roads, and nutrients from treated waste waters to streams and groundwater.

20.2 Water Supply: A U.S. Example

The water supply at any particular point on the land surface depends on several factors in the hydrologic cycle, including the rates of precipitation, evaporation, transpiration (water in vapor form that directly enters the atmosphere from plants through pores in leaves and stems), stream flow, and subsurface flow. A concept useful in understanding water supply is the **water budget**, which is a model that balances the inputs, outputs, and storage of water in a system. Simple annual water budgets (precipitation – evaporation = runoff) for North America and other continents are shown on Table 20.2. The total

Table 20.2   Annual Water Budgets for the Continents[a]

| Continental | Precipitation | | Evaporation | | Runoff |
	mm/yr	km³	mm/yr	km³	km³/yr
North America	756	18,300	418	10,000	8,180
South America	1,600	28,400	910	16,200	12,200
Europe	790	8,290	507	5,320	2,970
Asia	740	32,200	416	18,100	14,100
Africa	740	22,300	587	17,700	4,600
Australia and Oceania	791	7,080	511	4,570	2,510
Antarctica	165	2,310	0	0	2,310
Earth (entire land area)	800	119,000	485	72,000	47,000[b]

[a] Precipitation – evaporation = runoff.

[b] Surface runoff is 44,800; groundwater runoff is 2,200.

Source: I. A. Shiklomanov, "World Fresh Water Resources," in P. H. Gleick, ed., *Water in Crisis* (New York: Oxford University Press, 1993), pp. 3–12.

average annual water yield (runoff) from Earth's rivers is approximately 47,000 km³ (1.2×10^{16} gal), but its distribution is far from uniform (see Table 20.2). Some runoff occurs in relatively uninhabited regions, such as Antarctica, which produces about 5% of Earth's total runoff. South America, which includes the relatively uninhabited Amazon basin, provides about one-fourth of Earth's total runoff. Total runoff in North America is about two-thirds that of South America. Unfortunately, much of the North American runoff occurs in sparsely settled or uninhabited regions, particularly in the northern parts of Canada and Alaska.

The daily water budget for the contiguous United States is shown in Figure 20.5. The amount of water vapor passing over the United States every day is approximately 152,000 million m³ (40 trillion gal); and of this, approximately 10% falls as precipitation in the form of rain, snow, hail, or sleet. Approximately 66% of the precipitation evaporates quickly or is transpired by vegetation. The remaining 34% enters the surface water or

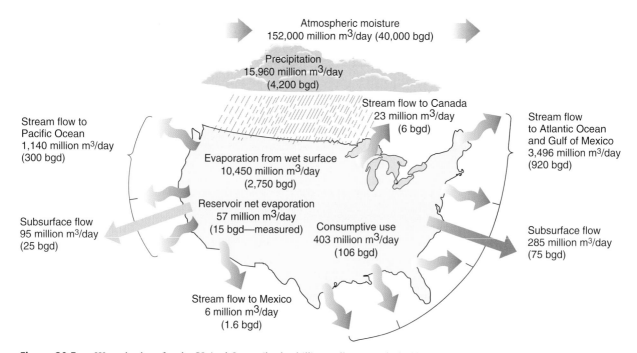

Figure 20.5 • Water budget for the United States (bgd = billion gallons per day). (*Source:* Water Resources Council, *The Nation's Water Resources 1975–2000* [Washington, D.C.: Water Resources Council, 1978].)

groundwater storage systems, flows to the oceans or across the nation's boundaries, is used by people, or evaporates from reservoirs. Owing to natural variations in precipitation that cause either floods or droughts, only a portion of this water can be developed for intensive uses (only about 50% is considered available 95% of the time).[4]

Precipitation and Runoff Patterns

In developing water budgets for water resources management, it is useful to consider annual precipitation and runoff patterns. Potential problems with water supply can be predicted in areas where average precipitation and runoff are relatively low, such as in the arid and semiarid parts of the southwestern and Great Plains regions of the United States. Surface water supply can never be as high as the average annual runoff, because not all runoff can be successfully stored. Total storage of runoff is not possible because of evaporative losses from river channels, ponds, lakes, and reservoirs. As a result, shortages in the water supply are common in areas with natural low precipitation and runoff coupled with strong evaporation. In these areas, strong conservation practices are necessary to help ensure an adequate supply of water.[4]

Droughts

Because there are large annual and regional variations in stream flow, even areas with high precipitation and runoff may periodically suffer from droughts. For example, the dry years near the end of the 20th century in the western United States produced serious water shortages. Fortunately for the more humid eastern United States, stream flow there tends to vary less than in other regions, and drought is less likely.[4] Nevertheless, droughts in the summers of some years in the southeastern United States will continue to cause hardships and billions of dollars of damage.

Groundwater Use and Problems

Nearly half the people in the United States use groundwater as a primary source for drinking water. It accounts for approximately 20% of all water used. Fortunately, the total amount of groundwater available in the United States is enormous. In the contiguous United States, the amount of shallow groundwater within 0.8 km (about 0.5 mi) of the surface of the land is estimated to be between 125,000 and 224,000 km^3 (3.3 x 10^{16} to 05.9 x 10^{16} gal). To put this in perspective, the lower estimate of the amount of shallow groundwater is about equal to the total discharge of the Mississippi River during the last 200 years. However, the high cost of pumping limits the total amount of groundwater that can be economically recovered.[4]

In many parts of the country, groundwater withdrawal from wells exceeds natural inflow. In such cases of **over-draft**, we can think of water as a nonrenewable resource that is being *mined*. This can lead to a variety of problems, including damage to river ecosystems and land subsidence (see A Closer Look 20.1). Groundwater overdraft is a serious problem in the Texas–Oklahoma–High Plains area (which includes much of Kansas and Nebraska and parts of other states), as well as in California, Arizona, Nevada, New Mexico, and isolated areas of Louisiana, Mississippi, Arkansas, and the south Atlantic region.

In the Texas–Oklahoma–High Plains area, the overdraft amount per year is approximately equal to the natural flow of the Colorado River for the same period.[4] The Ogallala aquifer, which is composed of water-bearing sands and gravels that underlie an area of about 400,000 km^2 from South Dakota into Texas, is the main groundwater resource in this area. Although the aquifer holds a tremendous amount of groundwater, it is being used in some areas at a rate up to 20 times higher than the rate at which it is being naturally replaced. The water table in many parts of the aquifer has declined in recent years (Figure 20.7), causing yields from wells to decrease and energy costs for pumping the water to increase. The most severe water depletion problems in the Ogallala aquifer today are in locations where irrigation was first used in the 1940s. There is concern that eventually a significant portion of land now being irrigated will be returned to dry-land farming as the resource is used up.

Desalination as a Water Source

Seawater is about 3.5% salt; that means each cubic meter of seawater contains about 40 kg (88 lb) of salt. **Desalination**, a technology to remove salt from water, is being used at several hundred production plants around the world to produce water with reduced salt. The salt content of the water must be reduced to about 0.05% for the water to be used as a freshwater resource. Large desalination plants produce 20,000–30,000 m^3 (about 5–8 million gal) of water per day.

The cost of desalinated water is about 10 times that paid for traditional water supplies in the United States. Desalinated water has a *place value*, which means that the price increases quickly with the transport distance and the cost of moving water from the plant. Because the various processes that remove the salt require large amounts of energy, the cost of the water is also tied to ever-increasing energy costs. For these reasons, desalination will remain an expensive process used only when alternative water sources are not available.

Desalination also has environmental impacts. Discharge of very salty water from a desalination plant into another body of water, such as a bay, may locally increase salinity and kill some plants and animals intolerant to salt. The discharge from desalination plants may also cause wide fluctuations in salt content of local environments, which may damage ecosystems.

Loss of Riparian Vegetation and Land Subsidence

In many areas, the pumping of groundwater has forever changed the character of the land. For instance, Tucson, Arizona, gets most of its water supply from groundwater sources, and overdraft has caused the water table to drop. Before the water table was lowered through pumping, rivers in the Tucson area were perennial, with healthy populations of trout, beavers, and other animals. Today, the rivers are dry much of the year, and the native *riparian* trees (those on or adjacent to riverbanks) have died. Ironically, these processes have also increased the flood hazard in Tucson. Loss of riparian trees and the root strength they provided to stream banks have rendered the banks of the channels much more vulnerable to erosion. During the 1983 and 1993 floods in Tucson, the effects of loss of riparian trees and resulting bank erosion became apparent as roads, bridges, and buildings were damaged by the shifting channels. Tree-lined channels are much more stable, but riparian trees need a groundwater table close to the surface. Unfortunately, restoration of trees along the streams is not possible, because the water table is now too low.

When groundwater is pumped at a rate faster than it is naturally replaced, the water table or fluid pressure in an aquifer may be lowered. If this occurs, effects may be seen at the surface. For example, withdrawal of groundwater in Mexico City over many years has resulted in surface subsidence—(sinking or settling of the land)—of about 5 m (16 ft) in some locations. The Palace of Fine Arts, constructed in 1939, originally had steps going up to the first floor. It now has steps going down to that floor. Subsidence occurs in such instances because when the water is pumped out of the aquifer, the natural fluid pressure that helps support the mineral particles in the aquifer in a more open framework (similar to a house built from playing cards) is reduced, and the space between grains partially collapses or flattens (similar to when the house of cards collapses). The cumulative effect is subsidence at the surface. Similar problems have been reported in Phoenix, Las Vegas, Houston, and the San Joaquin Valley in California.

Areas with limestone bedrock and aquifers may experience a sudden and spectacular form of subsidence owing in part to withdrawal of groundwater. Limestone is subject to chemical weathering that dissolves the stone, making a series of interconnected caves with large vertical or dome-shaped rooms that may come close to the surface. If a room close to the surface is filled with groundwater, then the water helps support the rocks above. The combination of a high water table and the strength of the overlying rocks helps prevent collapse. If the water table drops and a collapse does occur, a feature known as a collapsed sinkhole forms at the surface. The collapse can be considerable, and it can happen quickly. Figure 20.6 shows a sinkhole that formed on May 8, 1981, in Winter Park, Florida. The sinkhole grew rapidly for three days, swallowing a house, two businesses, part of a community swimming pool, a street, and several automobiles. The sinkhole developed during a drought, when groundwater levels were at a record low. Similar events have occurred in Alabama, Pennsylvania, Puerto Rico, the Yucatan Peninsula in Mexico, and many other areas around the world where extensive limestone cave systems are part of groundwater systems.

Figure 20.6 ● Sinkhole that formed in Winter Park, Florida, on May 8, 1981. Note damage to the municipal swimming pool, roads, and vehicles trapped in the depression.

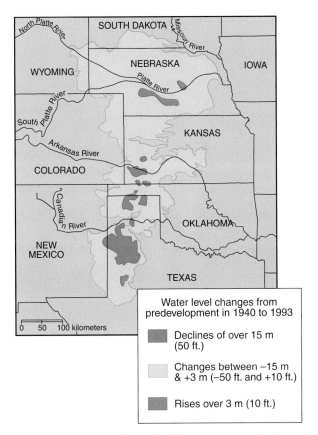

Figure 20.7 • Groundwater level changes as a result of pumping in the Texas–Oklahoma-High Plains region. (*Source:* U.S. Geological Survey.)

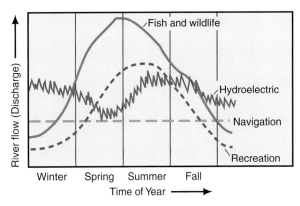

Figure 20.8 • In-stream water uses and optimal discharges (volume of water flowing per second) for each use. Discharge is the amount of water passing by a particular location and is measured in cubic meters per second.

20.3 Water Use

In discussing water use, it is important to distinguish between off-stream and in-stream uses. **Off-stream use** refers to water removed from its source (such as a river or reservoir) for use. Much of this water is returned to the source after use; for example, the water used to cool industrial processes may go to cooling ponds and then be discharged to a river, lake, or reservoir. **Consumptive use** is an off-stream use in which water is consumed by plants and animals or used in industrial processes. The water enters human tissue or products or evaporates during use and is not returned to its source.[4]

In-stream use includes the use of rivers for navigation, hydroelectric power generation, fish and wildlife habitats, and recreation. These multiple uses usually create controversy, because each requires different conditions to prevent damage or detrimental effects. For example, fish and wildlife require certain water levels and flow rates for maximum biological productivity; these levels and rates will differ from those needed for hydroelectric power generation, which requires large fluctuations in dis-

charges to match power needs. Similarly, in-stream uses of water for fish and wildlife will likely conflict with requirements for shipping and boating. Figure 20.8 demonstrates some of these conflicting demands on a graph that shows optimal discharge for various uses throughout the year. In-stream water use for navigation is optimal at a constant fairly high discharge. Some fish, however, prefer higher flows in the spring for spawning.

Another problem for in-stream use is how much water can be removed from a stream or river without damaging the stream's ecosystem. This is an issue in the Pacific Northwest, where fish, such as steelhead trout and salmon, are on the decline partly because diversions (removal of water for agricultural, urban, and other uses) have reduced stream flow to the extent that fish habitats are damaged.

The Aral Sea in Kazakhstan and Ubekistan provides a wake-up call of environmental damage that can be caused by diversion of water for agricultural purposes. Diversion of water from the two rivers that flow into the Aral Sea has transformed one of the largest bodies of inland water in the world from a vibrant ecosystem into a dying sea. The present shoreline is surrounded by thousands of square kilometers of salt flats that formed as the surface area of the sea was reduced by about 40% in the past 40 years (Figure 20.9). The volume of the lake was reduced by more than 50%. The salt content of the water increased fish kills, such as sturgeon that are an important component of the economy. Dust raised by winds from the dry salt flats is producing a regional air pollution problem, and the climate in the region has changed as the moderating effect of the sea has been reduced. Winters have grown colder and summers warmer. Fishing centers such as Muynak in the south and Aralsk to the north that were once on the shore of the sea are now many kilometers inland. Loss of fishing with decline of tourism has damaged the local economy.

Figure 20.9 ● The Aral Sea is drying up and dying as a result of diversion of water for agriculture (Courtesy of Philip P. Micklin.)

Transport of Water

In many parts of the world, demands are being made on rivers to supply water to agricultural and urban areas. This is not a new trend. Ancient civilizations, including the Romans and Native Americans, constructed canals and aqueducts to transport water from distant rivers to where it was needed. In our modern civilization, as in the past, water is often moved long distances from areas with abundant rainfall or snow to areas of high usage (usually agricultural areas). For instance, in California, two-thirds of the state's runoff occurs north of San Francisco, where there is a surplus of water. However, two-thirds of the water use in California occurs south of San Francisco, where there is a deficit. In recent years, canals of the California Water Project have moved tremendous amounts of water from the northern to the southern part of the state, mostly for agricultural uses but increasingly for urban uses as well.

On the opposite coast, New York City has imported water from nearby areas for more than 100 years. Water use and supply in New York City show a repeating pattern. Originally, local groundwater, streams, and the Hudson River itself were used. However, as population increased and the land was paved over, surface waters were diverted to the sea rather than percolating into the soil to replenish groundwater. Furthermore, what water did infiltrate the soil was polluted by urban runoff. Water needs in New York exceeded local supply, and in 1842 the first large dam was built.

As the city rapidly expanded from Manhattan to Long Island, water needs increased. The shallow aquifers of Long Island were at first a source of drinking water, but this water was used faster than the infiltration of rainfall could replenish it. At the same time, the groundwater became contaminated with urban and agricultural pollutants and from salt water seeping in underground from the ocean. (The pollution of Long Island groundwater is explored in more depth in the next chapter.) A larger dam was built at Croton in 1900. Further expansion of the population created the same pattern: initial use of groundwater; pollution, salinization, and overuse of the resource; and subsequent building of new, larger dams farther and farther upstate in forested areas.

In a broader perspective, the cost of obtaining water for large urban centers from long distances, along with competition for available water from other sources and users, will eventually place an upper limit on the water supply of the city. As shortages develop, stronger conservation measures are implemented, and the cost of water increases. As with other resources, as the water supply is reduced and demand for water increases, so does its price. If the price becomes high enough, more expensive sources may be developed—for example, pumping from deeper wells or using desalination.

Some Trends in Water Use

Trends in freshwater withdrawals and human population for the United States from 1950 to 1995 are shown in Figure 20.10. You can see that withdrawal of surface water far exceeds withdrawal of groundwater. In addition, withdrawals of both surface water for human uses and groundwater increased between 1950 and 1980, reaching a total maximum of approximately 375 thousand million gal/day. However, since 1980, water withdrawals decreased and leveled off. It is encouraging that water withdrawals have decreased after 1980 while the population of the United States has continued to increase. This suggests that improvements have been made in water management and water conservation.[6, 7]

Trends in freshwater withdrawals by water-use categories for the United States from 1960 to 1995 are shown in Figure 20.11. Examination of this graph suggests that:

1. The major uses of water are for irrigation and the thermoelectric industry.

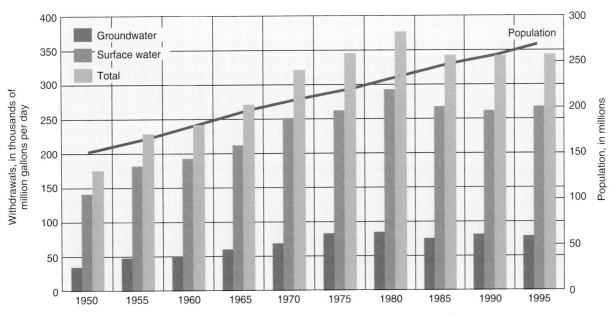

Figure 20.10 • Trends in United States fresh groundwater and surface-water withdrawals and human population (1950–1995). (*Source:* W. B. Solley, R. P. Pierce, and H. A. Perlman. 1998. "Estimated Use of Water in the United States in 1995," *U.S. Geological Survey Circular 1200*, 1998.)

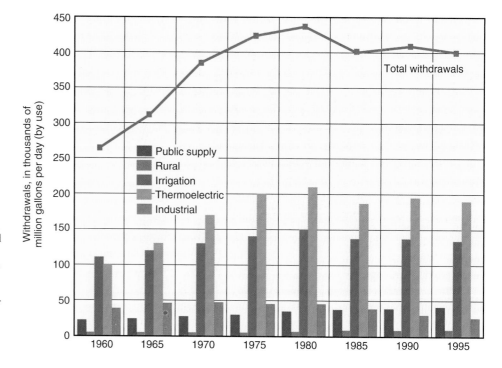

Figure 20.11 • Trends in United States water withdrawals (fresh and saline) by water-use category and total (fresh and saline) withdrawals (1960–1995). (*Source:* W. B. Solley, R. P. Pierce, and H. A. Perlman. 1998. "Estimated Use of Water in the United States in 1995," *U.S. Geological Survey Circular 1200*, 1998.)

2. The use of water for irrigation by agriculture leveled off starting in about 1980.

3. Water use by the thermoelectric industry and other industries decreased slightly beginning in 1980.

4. Use of water for public and rural supplies continued to increase through the period from 1950 to 1995, presumably related to the increase in human population.[7]

20.4 Water Conservation

Water conservation is the careful use and protection of water resources. It involves both the quantity of water used and its quality. Conservation is an important component of sustainable water use. Because the field of water conservation is changing rapidly, it is expected that a number of innovations will reduce the total withdrawals of water for various purposes, even though consumption will continue to increase.[3]

Agricultural Use

Improved irrigation (Figure 20.12) could reduce agricultural withdrawals by between 20 and 30%. Because agriculture is the biggest water user, this would be a tremendous savings. Suggestions for agricultural conservation include the following:

- Price agricultural water to encourage conservation (subsidizing water will encourage overuse).
- Use lined or covered canals that reduce seepage and evaporation.

- Use computer monitoring and schedule release of water for maximum efficiency.
- Integrate the use of surface water and groundwater to more effectively use the total resource. That is, irrigate with surplus surface water when it is abundant, and also use surplus surface water to recharge groundwater aquifers by applying the surface water to specially designed infiltration ponds or injection wells. When surface water is in short supply, use more groundwater.
- Irrigate at times when evaporation is minimal, such as at night or in the early morning.
- Use improved irrigation systems, such as sprinklers or drip irrigation, that more effectively apply water to crops.
- Improve land preparation for water application; that is, improve the soil to increase infiltration and minimize runoff. Where applicable, use mulch to help retain water around plants.
- Encourage the development of crops that require less water or are more salt tolerant so that less periodic flooding of irrigated land is necessary to remove accumulated salts in the soil.

Domestic Use

Domestic use of water accounts for only about 10% of total national water withdrawals. However, because domestic water use is concentrated in urban areas, it may pose major local problems in areas where water is periodically or often in short supply. Most water in homes is used in the bathroom and for washing clothing and dishes. Water use for domestic purposes can be substantially

Agriculture, 1990

Groundwater pump (unmetered)

Flood irrigation

Fertilizer and
pesticide storage

Agricultural
runoff

Drip
irrigation

To river or groundwater

Agriculture, 2020

Groundwater pump
(metered)

Habitat for
wildlife

Biological pest
control shed

Restored
stream

Ground station
for weather data

Precision
sprinklers

All drip
irrigation

Moisture sensors

To river or groundwater

Figure 20.12 • Comparison of agricultural practices in 1990 with what they might be by 2020. The improvements call for a variety of agricultural procedures, from biological pest control to more efficient application of irrigation water to restoration of water resources and wildlife habitat. (*Source:* P. H. Gleick, P. Loh, S. V. Gomez, and J. Morrison, *California Water 2020, a Sustainable Vision* [Oakland, Calif.: Pacific Institute for Studies in Development, Environment and Security, 1995].)

reduced at a relatively small cost by implementing the following measures:

- In semiarid regions, replace lawns with decorative gravels and native plants.
- Use more efficient bathroom fixtures, such as low-flow toilets that use 1.6 gallons or less per flush rather than the standard 5 gallons and low-flow shower heads that deliver less but sufficient water.
- Turn off water when not absolutely needed for washing, brushing teeth, shaving, and so on.
- Flush the toilet only when really necessary.
- Fix all leaks quickly. Dripping pipes, faucets, toilets, or garden hoses waste water. A small drip can waste sev-

eral liters per day; multiply this by millions of homes with a leak, and a large volume of water is lost.

- Purchase dishwashers and washing machines that minimize water consumption.
- Take a long bath rather than a long shower.
- Don't wash sidewalks and driveways with water (sweep them).
- Consider using gray water (from showers, bathtubs, sinks, and washing machines) to water vegetation. The gray water from washing machines is easiest to use, as it can be easily diverted before entering a drain.
- Water lawns and plants in the early morning, late afternoon, or at night to reduce evaporation.
- Use drip irrigation and place water-holding mulch around garden plants.
- Plant drought-resistant vegetation that requires less water.
- Learn how to read the water meter to monitor for unobserved leaks and record your conservation successes.

In addition, local water districts should encourage water pricing policies in which water is more expensive beyond some baseline amount determined by the number of people in a home and the size of the property.

Industry and Manufacturing Use

Water conservation measures taken by industry can be improved. For instance, water removal for steam generation of electricity could be reduced 25% to 30% by using cooling towers that use less or no water. Manufacturing and industry could curb water withdrawals by increasing in-plant treatment and recycling of water and by developing new equipment and processes that require less water.[4]

Perception and Water Use

How the water supply is perceived by people is important in determining how much water is used. Perception of water is based partly on its price and availability. If water is abundant and inexpensive, we don't think much about it. If water is scarce or expensive, it is another matter. For example, people in Tucson, Arizona, perceive the area as a desert (which it is) and use a lot of native plants (cactus and other desert plants) in yards and gardens around homes and buildings there. Tucson's water supply is mostly from groundwater, which is being mined (used faster than it is being naturally replenished). Tucson also receives some Colorado River water, used for irrigation and industrial purposes. Water use in Tucson is about 409 (108 gal) per person per day. Not far from Tucson, the people of Phoenix use about 60% more at 662 (175 gal) per person per day. In parts of Phoenix, as much as 3,780 (1,000 gal) of water per person per day is used to water mulberry trees and high hedges.

Phoenix has been accused of having an oasis mentality concerning water use, reflected in part by water rates. The people in Tucson often pay about 100% more for water than the people in Phoenix, where the rates are among the lowest in the western United States. Water rates in Tucson are structured to encourage conservation, and some industries consider water conservation a cost control measure.[8] For example, for residential water use in Tucson (for a family of four), the price increases from about $2.99 to $5.83 per 100 ft^3 as water use per month increases from 1,600 ft^3 to more than 4,500 ft^3. In Phoenix, the low price of water ($1.12 to $1.68 per 100 ft^3) encourages people to use more, and there is little incentive for water conservation. The message is that we could all do with a little of Tucson's desert mentality. This is particularly true for those in large urban areas, such as Los Angeles and San Diego, in Southern California.

20.5 Sustainability and Water Management

Water is essential to sustain life and to maintain ecological systems necessary for the survival of humans. As a result, water plays important roles in ecosystem support, economic development, cultural values, and community well-being. Managing water use for sustainability is thus important in many ways.

Sustainable Water Use

From a water supply use and management perspective, **sustainable water use** can be defined as use of water resources by people in a way that allows society to develop and flourish into an indefinite future without degrading the various components of the hydrologic cycle or the ecological systems that depend on it.[9] Some general criteria for water use sustainability are as follows.[9]

- Develop water resources in sufficient volume to maintain human health and well-being.
- Provide sufficient water resources to guarantee the health and maintenance of ecosystems.
- Ensure minimum standards of water quality for the various users of water resources.
- Ensure that actions of humans do not damage or reduce long-term renewability of water resources.
- Promote the use of water-efficient technology and practice.
- Gradually eliminate water pricing policies that subsidize the inefficient use of water.

Groundwater Sustainability

The concept of sustainability by its very nature involves a long-term perspective. With groundwater resources, the length of time for effective management for sustainability is even longer than for other renewable resources. Surface waters, for example, may be replaced over a relatively short time. In contrast, groundwater development may take place over many years, at relatively slow rates. Effects of pumping groundwater at rates greater than natural replenishment rates may take years to be recognized. Similarly, effects of withdrawal of groundwater, such as drying up of springs or reduction of stream flow, may not be recognized until years after pumping begins. The long-term approach to sustainability with respect to groundwater often involves balancing withdrawals of groundwater resources with recharge of those resources, which is an important component of water management.[10]

Water Management

Management of water resources for water supply is a complex issue that will become more difficult as demand for water increases in the coming years. This difficulty will be especially apparent in the southwestern United States and other semiarid and arid parts of the world where water is or soon will be in short supply. Options for minimizing potential water supply problems include locating alternative water supplies and managing existing supplies better. In some areas, location of new supplies is unlikely, and serious consideration is being given to ideas as original as towing icebergs to coastal regions where fresh

water is needed. It seems apparent that water will become much more expensive in the future; and if the price is right, many innovative programs are possible.

A method of water management utilized by a number of municipalities is known as the *variable-water-source approach*. For example, the city of Santa Barbara, California, has developed a variable-water-source approach that uses several interrelated measures to meet present and future water demands. Details of the plan (shown in Figure 20.13) include importing state water, developing new sources, using reclaimed water, and instituting a permanent conservation program. In essence, this seaside community has developed a master water plan.

A Master Plan for Water Management

Luna Leopold, a famous U.S. hydrologist, suggests that a new philosophy of water management is needed, one based on geologic, geographic, and climatic factors as well as on the traditional economic, social, and political factors. He argues that the management of water resources cannot be successful as long as it is naively perceived from an economic and political standpoint.

The essence of Leopold's water management philosophy is that surface water and groundwater are both subject to natural flux with time. In wet years, there is plenty of surface water, and the near-surface groundwater resources are replenished. During dry years, which must be expected

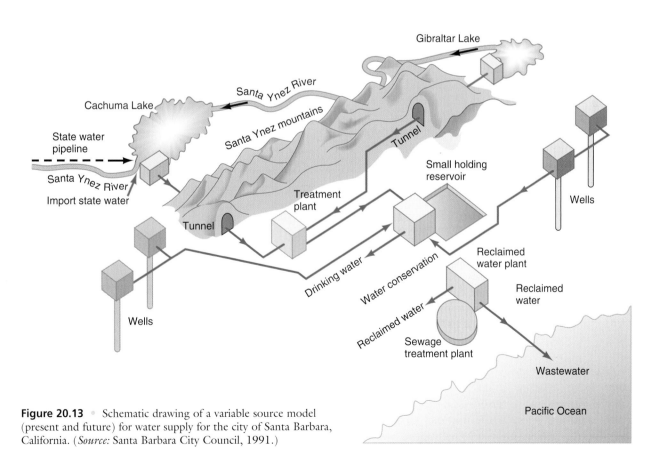

Figure 20.13 • Schematic drawing of a variable source model (present and future) for water supply for the city of Santa Barbara, California. (*Source:* Santa Barbara City Council, 1991.)

even though they may not be accurately predicted, specific plans to supply water on an emergency basis to minimize hardships must be in place and ready to use.

For example, there are subsurface waters in various locations in the western United States that are too deep to be economically pumped from wells or that have marginal water quality. These waters may be isolated from the present hydrologic cycle and therefore not subject to natural recharge. Such water might be used when the need is great. However, advance planning to drill the wells and connect them to existing water lines is necessary if they are to be ready when the need arises.

Another possible emergency plan might involve the treatment of wastewater. Reuse of such water on a regular basis might be too expensive, but advance planning to reuse treated water during emergencies could be a wise decision.

Finally, we should develop plans to use surface water when available, and we should not be afraid to use groundwater as needed in dry years. During wet years, natural recharge as well as artificial recharge (pumping excess surface water into the ground) will replenish the groundwater resources. This water management plan recognizes that excesses and deficiencies in water are natural and can be planned for.[11]

Water Management and the Environment

Many agricultural and urban areas require water to be delivered from nearby (and, in some cases, not-so-nearby) sources. To deliver the water, a system is needed for water storage and routing by way of canals and aqueducts from reservoirs. As a result, dams are built, wetlands may be modified, and rivers may be channelized to help control flooding. Often, a good deal of controversy surrounds water development.

The days of developing large projects in the United States without environmental and public review have passed. The resolution of development issues now involves input from a variety of government and public groups, which may have very different needs and concerns. These range from agricultural groups that see water development as critical for their livelihood to groups primarily concerned with wildlife and wilderness preservation. It is a positive sign that the various parties with interests in water issues are encouraged—and, in some cases, required—to meet and communicate their desires and concerns. Next, we address more directly the subjects of some of these concerns: wetlands, dams, channelization, and flooding.

20.6 Wetlands

Wetlands is a comprehensive term for landforms such as salt marshes, swamps, bogs, prairie potholes, and vernal pools (shallow depressions that seasonally hold water).

Their common feature is that they are wet at least part of the year and as a result have a particular type of vegetation and soil. Figure 20.14 shows several types of wetlands.

Wetlands may be defined as areas that are inundated by water or where the land is saturated to a depth of a few centimeters for at least a few days per year. Three major components used to determine the presence of wetlands are: hydrology, or wetness; type of vegetation; and type of soil. Of these, hydrology is often the most difficult to define, because some freshwater wetlands may be wet for only a few days a year. The duration of inundation or saturation must be sufficient for the development of wetland soils, which are characterized by poor drainage and lack of oxygen, and for the growth of specially adapted vegetation.[12]

Functions of Wetlands

Wetland ecosystems may serve a variety of natural service functions for other ecosystems and for people, including the following:

- Freshwater wetlands are a natural sponge for water. During high river flow they store water, reducing downstream flooding. Following a flood, they slowly release the stored water, nourishing low flows.
- Many freshwater wetlands are important as areas of groundwater recharge (water seeps into the ground from a prairie pothole, for instance) or discharge (water seeps out of the ground in a marsh that is fed by springs).
- Wetlands are one of the primary nursery grounds for fish, shellfish, aquatic birds, and other animals. It has been estimated that as many as 45% of endangered animals and 26% of endangered plants either live in wetlands or depend on them for their continued existence.[12]
- Wetlands are natural filters that help purify water; plants in wetlands trap sediment and toxins.
- Wetlands are often highly productive and are places where many nutrients and chemicals are naturally cycled.
- Coastal wetlands provide a buffer for inland areas from storms and high waves.
- Wetlands are an important storage site for organic carbon; carbon is stored in living plants, animals, and rich organic soils.
- Wetlands are aesthetically pleasing to people.

Freshwater wetlands are threatened in many areas. One percent of the nation's total wetlands are lost every two years, and freshwater wetlands account for 95% of this loss. Wetlands such as prairie potholes in the midwestern United States and vernal pools in Southern California are

(a)

(b)

(c)

(d)

Figure 20.14 • Several types of wetlands: (*a*) aerial view of part of the Florida Everglades at a coastal site; (*b*) marsh environment at the edge of Chesapeake Bay; (*c*) cypress swamp, water surface covered with a floating mat of duckweed, northeast Texas; and (*d*) aerial view of farmlands encroaching on prairie potholes, North Dakota.

particularly vulnerable because their hydrology is poorly understood and establishing their wetland status is more difficult.[13] Over the past 200 years, over 50% of the wetlands in the United States have disappeared because they have been diked or drained for agricultural purposes or filled for urban or industrial development. Perhaps as much as 90% of the freshwater wetlands have disappeared.

Although most coastal marshes are now protected in the United States, the extensive salt marshes at many of the nation's major estuaries, where rivers entering the ocean widen and are influenced by tides, have been modified or lost. These include deltas and estuaries of major rivers such as the Mississippi, Potomac, Susquehanna (Chesapeake Bay), Delaware, and Hudson.[14] The San Francisco Bay estuary, considered the estuary most mod-

ified by human activity in the United States today, has lost nearly all its marshlands to leveeing and filling (Figure 20.15).[14] Modifications result not only from filling and diking but also from loss of water. The freshwater inflow has been reduced by more than 50%, dramatically changing the hydrology of the bay in terms of flow characteristics and water quality. As a result of the modifications, the plants and animals in the bay have changed as habitats for fish and wildfowl have been eliminated.[14]

The delta of the Mississippi River includes some of the major coastal wetlands of the United States and the world. Historically, coastal wetlands of southern Louisiana were maintained by flooding of the Mississippi River, which delivered water, mineral sediments, and nutrients to the coastal environment. The mineral sediments contributed to the vertical accretion (building up)

Figure 20.15 • Loss of marshlands in the San Francisco Bay and estuary from about 1850 to the present. (*Source:* T. J. Conomos, ed., *San Francisco, the Urbanized Estuary* [San Francisco: American Association for the Advancement of Science, 1979]; F. H. Nichols, J. E. Cloern, S. N. Luoma, and D. H. Peterson, "The Modification of an Estuary," *Science* 231 [1986]: 567–573. Copyright 1986 by the American Association for the Advancement of Science.)

of wetlands. The nutrients enhanced growth of wetland plants, whose coarse organic components (leaves, stems, roots) also accreted. These accretion processes counter processes that naturally submerge the wetlands, including a slow rise in sea level, and subsidence (sinking) due to compaction. If the rates of submergence of wetlands exceed the rates of accretion, then the area of open water increases, and the wetlands are reduced.

Today, human-constructed levees line the lower Mississippi River, confining the river and directing flood-waters, mineral sediments, and nutrients into the Gulf of Mexico rather than into the coastal wetlands. Deprived of water, sediments, and nutrients in a coastal environment where the sea level is rising, the coastal wetlands are being lost. Global sea level is rising 1–2 mm/yr as a result of human-induced and natural global warming. Regional and local subsidence in the Mississippi Delta region combine with the global rise in sea level to produce a relative sea level rise of about 12 mm/yr. To keep the coastal wetlands from declining, the rate of vertical accretion would thus need to be about 13 mm/yr. Currently, natural vertical accretion is only about 5–8 mm/yr.[15]

Most people agree that wetlands are valuable and productive lands for fish and wildlife. But wetlands are also valued as potential lands for agricultural activity, mineral exploitation, and building sites. Wetland management is drastically in need of new incentives for private land owners (who own the majority of several types of wetlands in the United States) to preserve wetlands rather than fill them in and develop the land.[13] Management strategies must also include careful planning to maintain the water quantity and quality necessary for wetlands to flourish or at least survive. Unfortunately, although laws govern the filling and draining of wetlands, no national wetland policy for the United States is in place. Debate continues as to what constitutes a wetland and how property owners should be compensated for preserving wetlands.[12, 16]

Restoration of Wetlands

A related management issue is restoration of wetlands. A number of projects have attempted to restore wetlands, with varied success. The most important factor to be considered in most freshwater marsh restoration projects is the availability of water. If water is present, wetland soils and vegetation will likely develop. The restoration of salt marshes is more difficult because of the complex interactions among the hydrology, sediment supply, and vegetation that allow salt marshes to develop. Careful studies of relationships between the movement of sediment and the flow of water in salt marshes is providing information crucial to restoration, which makes successful reestablishment of salt marsh vegetation more likely. The restoration of wetlands has become an important topic in the United States because of the mitigation requirement related to environmental impact analysis, as set forth in the National Environmental Policy Act of 1969 (see Chapter 30). According to this requirement, if wetlands are destroyed or damaged by a particular project, the developer must obtain or create additional wetlands at another site to compensate.[12] Unfortunately, the state of the art of restoration is not adequate to ensure that specific restoration projects will be successful.

Constructing wetlands for the purpose of cleaning up agricultural runoff is an idea being attempted in areas with extensive agricultural runoff. Wetlands have the natural ability to remove excess nutrients, break down pollutants, and cleanse water. A series of wetlands are being created in Florida to remove nutrients (especially phosphorus) from agricultural runoff and thus help restore the Everglades to more natural functioning. The Everglades are a huge wetland ecosystem that functions as a wide, shallow river flowing south through southern Florida to the ocean. Fertilizers applied to farm fields north of the Everglades make their way directly into the Everglades by way of agricultural runoff, disrupting the ecosystem. (Phosphorus enrichment causes undesired changes in water quality and aquatic vegetation; see the discussion of eutrophication in the next chapter.) The human-made wetlands are designed to intercept and hold the nutrients so they do not enter and damage the Everglades.[17]

In southern Louisiana, restoration of coastal wetlands in southern Louisiana has recently included the application of treated wastewater, which adds nutrients, nitrogen, and phosphorous to accelerate plant growth. As plants grow, organic debris (stems, leaves, and so forth) build up on the bottom of the wetland and cause the wetland to grow vertically. This growth helps offset wetland submergence resulting from relative rise in sea level, maintaining and restoring the wetland.[15]

20.7 Dams and the Environment

Dams and their accompanying reservoirs generally are designed to be multifunctional structures. People who propose the construction of dams and reservoirs point out that reservoirs may be used for recreational activities and generating electricity as well providing flood control and ensuring a more stable water supply. However, it is often difficult to reconcile these various uses at a given site. For example, water demands for agriculture might be high during the summer, resulting in a drawdown of the reservoir and the production of extensive mudflats or an exposed bank area subject to erosion (Figure 20.16). Recreational users find the low water level and the mudflats aesthetically displeasing. Also, high water demand may cause quick changes in lake levels, which may interfere with wildlife (particularly fish) by damaging or limiting spawning opportunities. Another consideration is that dams and reservoirs tend to give a false sense of security to those living below these water retention structures. Dams may fail. Flooding may originate from tributary rivers that enter the main river below a dam; and dams cannot be guaranteed to protect people against floods larger than those for which they have been designed.

The environmental effects of dams are considerable and include the following:

- Loss of land, cultural resources, and biological resources in the reservoir area.
- Storage behind the dam of sediment that would otherwise move downstream to coastal areas, where it would supply sand to beaches. The trapped sediment also reduces water storage capacity, limiting the life of the reservoir.
- Downstream changes in hydrology and in sediment transport that change the entire river environment and the organisms that live there.

For a variety of reasons that include displacement of people, loss of land, loss of wildlife, and permanent, adverse changes to river ecology and hydrology, many people today are vehemently against turning remaining rivers into a series of reservoirs with dams. In the United States, several dams have recently been removed, and others are being considered for removal as a result of the environmental damages they are causing. In contrast, China is currently constructing the world's largest dam, as described in A Closer Look 20.2.

There is little doubt that if our present practices of water use continue we will need additional dams and reservoirs, and some existing dams will be heightened to increase water storage. However, there are few acceptable sites for new dams. Conflicts over the construction of additional dams and reservoirs are bound to occur. Water developers may view a canyon dam site as a resource for water storage, whereas others may view it as a wilderness area and recreation site for future generations. The conflict is common, because good dam sites are often sites of high-quality scenic landscape.

There is also an economic aspect to dams: They are expensive to build and operate. They are often constructed with federal tax dollars in the western United States, where they provide inexpensive subsidized water for agriculture. This has been a point of concern to some taxpayers in the eastern United States, who do not have the benefit of federally subsidized water. Perhaps a different pricing structure for water would encourage conservation, and fewer new dams and reservoirs would be needed.

Canals

Water from upstream reservoirs may be routed downstream by way of natural watercourses or canals and aqueducts. Canals are not hydrologically the same as creeks or rivers. They often have smooth, steep banks; and the water moves deceptively fast. Canals are a hazard attracting children to swim in them and animals to swim across

Figure 20.16 • Bank erosion along the shoreline of a reservoir in Central California following release of water, exposing bare banks.

20.7 • DAMS AND THE ENVIRONMENT   **407**

Three Gorges Dam

In spite of known adverse effects of large dams, the world's largest is now being constructed in China. Three Gorges Dam on the Yangtze River (Figure 20.17) will drown cities, farm fields, important archeological sites, and highly scenic gorges while displacing approximately two million people from their homes. In the river, habitat for endangered dolphins will likely be damaged.

The dam, which will be approximately 185 m high and more than 1.6 km wide, will produce a reservoir nearly 600 km long. Raw sewage and industrial pollutants that are discharged into the river will be discharged into the reservoir, and there is concern that the reservoir will become seriously polluted. In addition, the Yangtze River has a high sediment load, and it is feared that the upstream end of the reservoir, where sediments will likely be deposited, will fill with sediment, damaging deep-water shipping harbors.

The dam may produce a false sense of security in the people living in downstream cities. The presence of the dam may encourage further development in flood-prone areas, which will be damaged or lost if the dam and reservoir are unable to hold back

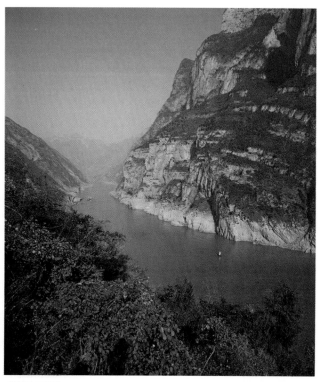

Figure 20.17 ● Three Gorges on the Yangtze River is a landscape of high scenic value. Shown here is the Wu Gorge, near Wushan, one of the gorges that will be flooded by the water in the reservoir.

floods in the future. If this happens, loss of property and life from flooding may be greater than if the dam had not been constructed. Contributing to this problem is the dam's location in a seismically active region where earthquakes and large landslides have been common in the past. Should the dam fail, then downstream cities such as Wushan, with a population of several million people, might be submerged with catastrophic loss of life.[18]

A positive attribute of the giant dam and reservoir will be the capacity to produce about 18,000 MW of electricity, which is equivalent to about 18 large coal-burning power plants. As has been pointed out in earlier discussions, pollution from coal burning is a serious problem in China. Some opponents to the dam have pointed out, however, that a series of dams on tributaries to the Yangtze River could have produced similar electric power while not causing the anticipated environmental damage to the main river.[19]

them. Where they flow, drownings of people and other animals are an ever-present threat.

The construction of canal systems, especially in developing countries, has led to serious, unanticipated environmental problems. For example, when the High Dam on the Nile River at Aswan, Egypt, was completed in 1964, a system of canals was built to convey the water to agricultural sites. The canals became infested with snails that carry the disease schistosomiasis (snail fever). This disease has always been a problem in Egypt, but the swift currents of the Nile floodwaters flushed the snails out each year. The tremendous expanse of waters in irrigation canals now pro-

vides happy homes for these snails. The disease is debilitating and so prevalent in parts of Egypt that virtually the entire population of some areas may be affected by it.

Removal of Dams

We opened this chapter with the story of the removal of the Edwards Dam in Maine. As mentioned, a number of U.S. dams have been removed; and the largest project to date is the removal of the Matilija Dam on the Ventura River in Southern California. The dam, completed by Ventura County in 1942, is about 190 m (620 ft) wide and 60 m (200 ft) high. The structure is in poor condition, with leaking, cracked concrete and a reservoir nearly filled with sediment. The dam serves no useful purpose and blocks endangered southern steelhead trout from their historic spawning grounds. The sediment trapped in the dam also reduces the natural nourishment of sand on beaches, increasing coastal erosion.

The trapped sediment presents a problem in removing the dam. If released quickly, it could damage the downstream river environment, filling pools and killing river organisms such as fish, frogs, and salamanders. If the sediment can be slowly, more naturally released, the downstream damage can be minimized.

The removal process began with much fanfare in October 2000, when a 27 m (90 ft) section was removed from the top of the dam. The entire removal process may take years, after scientists have determined how to safely remove the sediment stored behind the dam. The cost of the dam in 1948 was about $300,000. The cost to remove the dam and sediment will be more than 10 times that amount.[20]

The perception concerning dams as permanent structures has clearly changed. What is learned from studying the removal of the Edwards Dam in Maine and the Matilija Dam in California will be useful in planning other dam removal projects. The studies will also provide important case histories to evaluate ecologic restoration of rivers following removal of dams.

20.8 Channelization and the Environment

Channelization of streams consists of straightening, deepening, widening, clearing, or lining existing stream channels. It is an engineering technique that has been used to control floods, improve drainage, control erosion, and improve navigation.[21] Of these objectives, flood control and drainage improvement are cited most often in channel improvement projects.

Thousands of kilometers of streams in the United States have been modified by channelization. The practice all too often has produced adverse environmental effects, including the following:

- Degradation of the stream's hydrologic qualities, turning a meandering stream with pools (deep, slow flow) and riffles (faster, shallow flow) into straight channels that are nearly all riffle flow, resulting in loss of important fish habitats.
- Removal of vegetation along the watercourse, which removes wildlife habitats and shading of the water.
- Downstream flooding where the channelized flow ends, because the channelized section has a larger channel area and carries a greater amount of floodwater than the downstream natural channel can carry without overbank flow or flooding.

- Damage or loss of wetlands (because their source of water is removed by the channelization, which often lowers the groundwater table and thus drains the wetlands).
- Aesthetic degradation (channelized streams are much less attractive than natural streams).

Figure 20.18 compares selected environmental features of natural and channelized streams.

A case study in problems with channelization involves the Kissimmee River in Florida. Channelization of the river started in 1962. After nine years and $24 million of construction, the meandering river with many bends had been converted into a straight ditch 83 km (about 50 mile) long (Figure 20.19). Unfortunately, not only did the channelization fail to provide the expected flood protection, but it also damaged a valuable wildlife habitat, contributed to water quality problems associated with the land drainage, and caused aesthetic degradation. Then, in the 1990s, efforts began to return the river to its original meandering channel. The restoration of the Kissimmee River may become the most ambitious restoration project ever attempted in the United States, and the cost may exceed the original cost of channelization. Work began on a 16-km (10-mi) stretch of river. By 2001, 12 km (7.5 miles) of nearly straight flood control channel had been restored to a 24-km (15 miles) meandering channel with wetlands, returning ecosystems to a more natural state.

Not all channelization causes serious environmental degradation, and in many cases, drainage projects are beneficial. Moreover, channel design is being improved on the basis of past experience. Currently, more consideration is being given to the environmental aspects of channelization, and some projects are being designed with modified channels that behave more like natural streams than do straight ditches.

20.9 Flooding

River flooding is the most universally experienced natural hazard in the world today. A river and the flatland adjacent to it, known as the **floodplain** (Figure 20.20), together constitute a natural system. In most natural rivers, the water flows over the riverbanks and onto the floodplain every year or so. This natural process has many benefits for the environment:

- Water and nutrients are stored on the floodplain.
- Deposits on the floodplain contribute to the formation of nutrient-rich soils.
- Wetlands on the floodplain provide an important habitat for many birds, animals, plants, and other living things.
- The floodplain functions as a natural greenbelt that is distinctly different from adjacent environments and provides environmental diversity.

General characteristics

Natural stream

Suitable water temperatures:
adequate shading; good cover for fish life; minimal
temperature variation; abundant leaf material output.

Channelized stream

Increased water temperatures:
no shading; no cover for fish life; rapid daily
and seasonal temperature fluctuations;
reduced leaf material output.

Low flow

Pool—riffle sequence

Pool—deep, slow flow,
sand, and fine gravel
on stream bed.

Riffle—shallow,
fast flow, coarse
gravel on stream bed.

Pools and riffles provide diversified habitats
for many stream organisms.

Low flow

Mostly riffle

Lack of pools causes reduction in habitats;
few organisms.

Figure 20.18 • A natural stream compared with a channelized stream in terms of general characteristics and pool environments. (*Source:* Modified from R. V. Corning, *Virginia Wildlife*, February 1975, pp. 6–8.)

Natural flooding is not a problem until people choose to build homes and other structures on floodplains. These structures are damaged when inundated by floodwaters. In the United States, about 100 persons lose their lives in floods every year; and accompanying damages exceed $3 billion. The 1993 flood of the Mississippi River took about 50 lives and caused over $10 billion in damages (Figure 20.21). This loss of life, although terrible, is rel-

(a)

(b)

Figure 20.19 • (*a*) Channelized Kissimmee River and (*b*) restoration work in progress on the Kissimmee River near Cornwell Florida. (AP Photo Tom Eran)

Figure 20.20 • Floodplain of the Owens River in California.

Urbanization has caused significant environmental impacts on these basins and streams. With urbanization, much of the land is paved over or covered with buildings and rendered impervious to water (pavement and buildings do not allow for infiltration). As a result, rainwater quickly runs off the artificial surfaces to drainage systems (storm sewers) and then to streams. Because both the amount of runoff and the speed at which it reaches streams increases as a result of urbanization, there is also an increase in the flood hazard. Thus, urban areas experience more frequent and larger floods than do natural systems of similar size.

The flooding hazards associated with urbanization can be minimized in several ways. For example, runoff from urban areas during storms can be delayed by storing water in retention ponds, parking lots, and other creative means.

Adjustments to the Flood Hazard

By far the most appropriate adjustment to flooding is land-use planning or floodplain zoning that restricts development and building on floodplains. This option is also the most desirable from an environmental perspec-

atively low compared with that in areas of the world that lack the sophisticated monitoring and warning systems the United States has for its rivers. For example, flooding associated with two cyclones that struck Bangladesh in 1970 and 1991 killed more than half a million people.

Urbanization and Flooding

A drainage basin is the land area that contributes water to a particular stream system. Many small drainage basins with areas less than 20 km² or so have been intensively urbanized. In these drainage basins, it may be hard to even find the streams, which have been forced to flow through underground pipes or concrete lined channels.

(a)

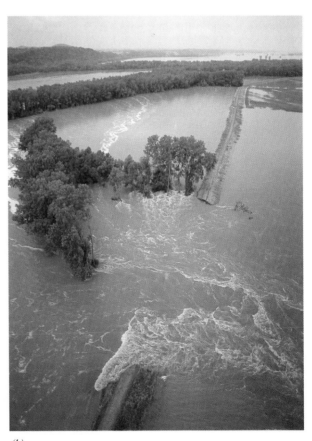

(b)

Figure 20.21 • (*a*) Flood damage to farmlands and homes as a result of the 1993 Mississippi River floods. (*b*) Failure of a levee in Illinois caused flooding in the town of Valmeyer.

tive. The federal flood insurance program in the United States has in a sense had the opposite effect; the program has been criticized as encouraging people to live on floodplains by allowing repeated insurance coverage in hazardous areas. The insurance program is being reevaluated, and repeated coverage may be denied in favor of buyout programs for floodplain homes.

As we pointed out earlier, reliance on upstream dams, levees, and floodwalls to prevent flooding fails to recognize that these structures can fail. During the 1993 flood of the Mississippi, for example, about 70% of the levees failed (Figure 20.21). They simply were not designed to withstand a flood that lasted over two months. Following the flood, some communities relocated rather than rebuilding on the floodplain. Urban land presently developed on floodplains will in most cases continue to be protected, but we should not repeat past mistakes with future development. Not allowing homes to be built on floodplains in the future would be a good start.

Getting the message across to people wishing to build on floodplains is not easy. For example, in Great Britain, thousands of homes and millions of people are threatened by flooding; and it is expected that by 2020, thousands of additional homes will be constructed on the nation's floodplains. The Environment Agency in Great Britain has often recommended that new buildings not be allowed on floodplains, but it does not have the authority to deny applications for development.

That may change as a result of floods in the fall of 2000, which occurred following record rainfall. During the six-month period before the floods, which inundated more than 5,000 buildings (Figure 20.22), the agency advised that 200 applications for new buildings not be allowed. Local authorities in over 40% of the cases allowed the development to proceed. After the flood, the national government acknowledged that the Environment Agency

and local authorities must now say no to future development on floodplains. It will be interesting to see the final policy decisions concerning floodplain development, as housing is expensive in Britain and there is pressure to build new homes, even in hazardous areas.

The Colorado River: Water Resources Management and Environment

We present the history of Colorado River water as a capstone to this chapter. It emphasizes linkages among physical, biological, and social systems that are at the heart of environmental science.

The Colorado is the major river of the southwestern United States and extends into Mexico, where it ends in the Gulf of California (Figure 20.23). The river basin occupies approximately 632,000 km² (244,000 mi²). Considering its size, the river has only a modest flow. However, it is one of the most regulated and controversial bodies of water in the world. The total flow of water in the river was apportioned among various users, including seven U. S. states and Mexico, by the Colorado River Compact of 1922. The compact allocated no water for environmental purposes, as the concept of sustainable water management was not considered.

Today, Colorado River water only occasionally flows into the Gulf of California—it's all stored in dams and used upstream. As a result, ecosystems of the lower river and delta, deprived of water and nutrients, have been damaged. The size of the delta has been reduced from about 7,500 km² (3,000 mi²) to less than half of that, damaging fish populations and forcing native people who depended on fishing to move away.

The complex issues of water management for the Colorado River illustrate major problems that are likely to

Figure 20.22 • Floods in Great Britain in 2000. Shown here are people being rescued in Lewes, England from floodwaters of the River Ouse, which inundated their homes. (AP Photo, Alastair Grant)

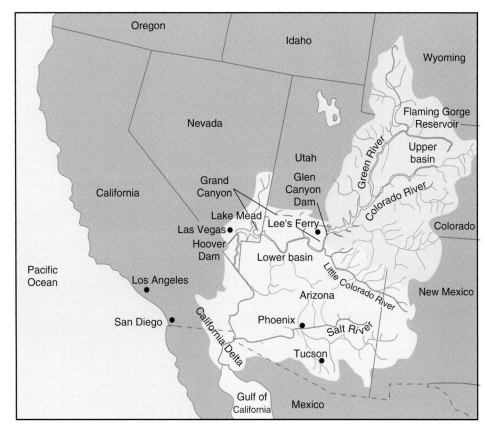

Figure 20.23 • The Colorado River basin.

be faced by other semiarid regions of the world in the coming years: How are scarce water resources to be allocated? How can we best control water quality? How can we protect river ecosystems? There are no easy answers to these questions.

The Colorado River originates in the Wind River Mountains of Wyoming and, in its 2,300-km (1,400-mile) journey to the Gulf of California, flows through some of the most spectacular scenery in the world. Eight hundred years ago, Native Americans living in the Colorado River basin constructed a sophisticated water distribution system. In the 1860s, settlers cleared the debris from these early canals and used them once again for irrigation.[22]

Today, the water of the Colorado River basin is completely spoken for and distributed by canals and aqueducts to millions of people in urban areas as well as to agricultural areas. The basin itself remains sparsely populated; and the largest city on the river is Yuma, Arizona, with a population of 42,000. The historian Rod Nash, writing about his wilderness experience on the river, states that at the junction of the Green and Colorado Rivers, in the heart of a national park, it is 80 km (50 mi) to the nearest video game.[23]

Managing the waters of the Colorado River has been frustrating in part because of inherent uncertainties concerning how much water can be expected in a given year.[22] Water resources in the basin include snowmelt, long-term winter precipitation, and short-term summer thunderstorms; as a result, the total water available on a year-to-year basis is tremendously variable. Table 20.3 compares the legal entitlements to water with the actual distribution to the users. Interestingly, the annual legal entitlements of 21.5 km³ (17.5 million acre-feet) exceed the actual distribution of 17.8 km³ (14.5 million acre-feet). (One acre-foot of water is equal to 1,238,000, or 325,829 gal. It is the volume of water necessary to cover one acre to a depth of one foot.[24])

Table 20.3 shows that not all users take the full legal entitlement of water. If all did, serious problems would result, because the entitlements exceed the natural annual flow. The reason the distribution can be maintained is that the Colorado River is one of the most managed rivers in the world. Figure 20.24 shows the profile of the river and some of the major dams and reservoirs in the system, which store approximately 86.3 km³ (70 million acre-feet) of water. The two largest reservoirs—Hoover Dam and Glen Canyon Dams—store about 80% of the total in the basin. Total storage represents (with careful management) a buffer of several years' water supply. However, if there are several drought years in a row, maintaining a sufficient water supply for all users may not be possible.

Table 20.3	Legal and Actual Distribution of Colorado River Water	
Area	Legal Entitlements (km³/yr)	Actual Distribution[a] (km³/yr)
Lower basin		
California	5.461[b]	5.461[c]
Arizona	4.674[b]	2.522[c]
Nevada	0.369[b]	0.369
Subtotal	10.455[d]	8.303
Upper basin		
Colorado	4.774[e]	2.959
Utah	2.122[e]	1.316
Wyoming	1.292[e]	0.801
New Mexico	1.038[e]	0.643
Subtotal	9.225[d]	5.720
Mexico	1.845[f]	1.845
Total	21.525	17.835

Source: W. L. Graf, *The Colorado River* (Washington, D. C.: Association of American Geographers, 1985).

[a] Includes losses to evaporation of 0.6 million acre-feet/year in the upper basin and 0.9 million acre-feet/year in the lower basin and 0.9 million acre-feet/year inflow to the lower basin from local streams.

[b] 1928 Boulder Canyon Project Act.

[c] Agreement at time of Central Arizona Project authorization by Congress. Upper basin amounts agreed to by states as percentages.

[d] 1922 Colorado River Compact.

[e] 1948 Upper Colorado River Basin Compact.

[f] 1944 Mexico–U.S. Treaty.

The Glen Canyon Dam was completed in 1963 (Figure 20.25). From a hydrologic viewpoint, the Colorado River has been changed by the dam. The river has been tamed. The higher flows have been reduced, the average flow has increased, and the flow changes often

because of fluctuating needs to generate electrical power. Changing the hydrology of the river has also changed other aspects, including the rapids; the distribution of sediments that form sandbars, called beaches by rafters (Figure 20.26); and the vegetation near the water's edge.[25] The sandbars, which are valuable wildlife habitats, shrank in size and number following construction of the dam because sediment that would have moved downstream to nourish them was trapped in the reservoir. All these changes affect the Grand Canyon, which is downstream from the dam.

A record snowmelt in the Rocky Mountains in June 1983 forced the release of about 2,500 m³ (88,000 ft³) of water per second from the Glen Canyon Dam—about three times the amount normally released and similar to the amount in a spring flood before the dam was built. Resulting floods scoured the river bed and banks, releasing stored sediment that replenished sandbars and breaking off some vegetation that had taken root.[26] This release of water was beneficial to the river environment and highlighted the importance of floods in maintaining the system in a more natural state. Natural disturbances are a necessary part of the river ecosystem if it is to function on a sustainable basis.

As an experiment, a flood with discharge about half the size of that in 1983 was deliberately released for a period of a week in 1996. Between March 26 and April 2, water was allowed to flow at full flood; then the flow was reduced for the last two days in order to redistribute the sand supply. The flood resulted in 55 new sandbars and increased the size of 75% of the existing sandbars. It also helped rejuvenate marshes and backwaters, which are critical habitats to the native fish and some endangered species.[26]

This experimental release of high flows of water marked a turning point in river management—the first time the U.S. government had opened the floodgates of a dam to improve a river's ecosystem. Some scientists were concerned that floodwaters would not be of sufficient volume and that the duration of the flooding would

Figure 20.24 • Longitudinal profiles of the Colorado River and Green River, showing the major dams, reservoirs, and canyons. (*Source*: W. L. Graf, *The Colorado River*, Resource Publications in Geography [Washington, D.C.: Association of American Geographers, 1985].)

Figure 20.25 • Glen Canyon Dam on the Colorado River in Arizona.

be too short. The flood was hailed as a success, but it will take some time to see how long the results will last. It is hoped that what was learned will be used to restore river environments and improve ecosystems of other rivers impacted by dams. Today, the Colorado River in the Grand Canyon is more accessible to recreational river rafting and recreation is better managed than ever before.

Before 1950, fewer than 100 explorers and river runners had made a trip through the Grand Canyon. Today, the number of people rafting through the canyon is limited to 15,000 per year; and even with this limitation, long-range impacts on the canyon resources are appreciable. Even though all waste is removed by boat, increased traffic and camping are placing stress on canyon ecosystems.

We must concede that the Colorado River is changed and will remain changed as long as the Glen Canyon dam is present. Although human-made floods have counteracted some of the changes, river restoration by itself cannot be expected to return the river to what it was before construction of the dam. However, better management of the flows and sediment transport will improve and better maintain river ecosystems.[25, 26, 27]

Conflicts over Colorado River water use and quality have spanned decades and extend beyond the river basin to urban centers and agricultural areas in California, Colorado, New Mexico, and Arizona. The crucial need for water in these semiarid regions has resulted in overuse of limited supplies of water as well as deterioration of water quality. This is a problem in the lower river particularly. By the time the water reaches Mexico, it is so salty that at times Mexican farmers are unable to use it. But Mexico is entitled to receive 1.8 billion m^3 (1.5 million acre-feet) of Colorado River water per year according to a treaty with the United States.

Figure 20.26 • Large rubber rafts docked on a beach (sandbar) on the Colorado River in the Grand Canyon, Arizona.

The world's largest desalination plant went on-line in May 1992 on the Colorado River near Yuma, Arizona (Figure 20.27). The cost of the plant was about $260 million, and its purpose was to reduce the salinity of the water taken into the plant from about 1,200 ppm salt to 800 ppm. Plant capacity is about 97 million m³/yr (700 million gal/per day). The upstream salt causing the problem comes mostly from agricultural runoff from the Welton-Mohawk Irrigation and Drainage District. The desalination plant had operated only a few months when floods destroyed canals that carried drainage water from the irrigation district to the plant. Since 1993, the plant has been on stand-by status (that is, it is closed but can be restarted). The annual cost of maintaining the stand-by status is about $6 million. There is some question whether the plant should resume operation. Some years, the Gila River has sufficient natural flows to dilute the salty agricultural waters, and desalination is not needed. In addition, salty waters from the irrigation district are diverted to the desert wetland of Cienga de Santa Clara, which has grown from about 200 to 20,000 ha, becoming an important desert wetland. Using the desalination plant rather than diverting the salty water would cause the wetland to disappear. Finally, Mexico is receiving better-quality water from the Coachella Canal, which flows into the Colorado River. The canal was concrete lined to keep water from infiltrating the soil. The concrete lining saves more water than the desalination plant can produce. Mexico thus now receives its water as required by treaty without need for the desalination plant.

International and interstate agreements and court settlements have periodically intensified or eased tension among water users of the Colorado River. These court decisions and laws, along with changes in water use, continue to significantly influence the lives of millions of people in the United States and Mexico.[28]

Summary

- Water is a liquid with unique characteristics that has made life on Earth possible.
- Although it is one of the most abundant and important renewable resources on Earth, more than 99% of Earth's water is unavailable or unsuitable for beneficial human use because of its salinity or location.
- The pattern of water supply and use on Earth at any particular point on the land surface involves interactions and linkages between the biological, hydrological, and rock cycles. To evaluate a region's water resources and use patterns, a water budget is developed to define the natural variability and availability of water.
- During the next several decades, it is expected that the total water withdrawn from streams and groundwater in the United States will decrease slightly, but the consumptive use will increase because of greater demands from a growing population and industry.
- Water withdrawn from streams competes with in-stream needs, such as maintaining fish and wildlife habitats and navigation, and may therefore cause conflicts.

Figure 20.27 • Location of the Yuma desalination plant, Welton Mohawk Irrigation and Drainage District, and Coachella Canal. The

photograph ia a view of the Little Colorado River which transports sand to the Colorado River in the Grand Canyon, see Figure 20.23.

How Wet Is a Wetland?

Areas where land meets water, whether fresh or salt, are places where *wetlands* are found. Characteristically, wetlands are covered by surface water or have saturated soils. Landscape features such as swamps, bogs, marshes, potholes, and sloughs are wetlands. For most of our history, wetlands were considered wastelands and were destroyed by filling, draining, or polluting. Only 230 of the 530 million ha (568.3 of the 1,309.6 million acres) of wetlands that existed in the lower 48 states at the time of European colonization remain, and 40% of them suffer from some degree of pollution.

Today, however, it is generally recognized that wetlands have many values. They provide food, water, and cover for fish, shellfish, waterfowl, game animals, and many amphibians and reptiles. In addition, soil and plants in wetlands purify water by absorbing or destroying pollutants. Wetlands also help to recharge groundwater stores and, by holding water, control flooding and erosion. One-third of endangered and threatened species, two-thirds of the commercial saltwater fish and shellfish, one-third of the birds, and almost all the amphibians in our country depend on wetlands. Wetlands are among the most productive ecological communities in the world, many times more productive than a heavily fertilized cornfield.

Protection of these unique communities began with the goal of preserving wetlands used by wildlife, particularly ducks, but federal protection has since been broadened to include most of the remaining wetlands. Still, 81,000–162,000 ha (200,151–400,302 acres) of wetlands are lost in the United States each year. Of particular concern are the freshwater wetlands, many of which are on private land. A recent policy of no net loss of wetlands has been praised by some environmental scientists but attacked by farmers and developers.

Particularly controversial are the small, seasonal potholes in the agricultural areas of the Midwest and northern Great Plains and other seasonal wet areas of the West that may not appear wet to the casual observer. They do, however, provide habitats for many species, including about half the 10 to 31 million waterfowl that nest in the lower 48 states. Critics of applying strict regulations to potholes say that if an area is not wet enough for a duck to land in and splash, it is not wet enough to qualify as a wetland.

Critical Thinking Questions

1. Results of a study comparing wildlife use of seasonal wetlands with wildlife use of wetlands that are permanently inundated or saturated in the San Francisco estuary are shown in the accompanying table. What conclusions can you draw from the data about the importance of seasonal wetlands to wildlife in the estuary? What additional data would you need to extend your conclusions to potholes in the Midwest?

2. Some people have proposed defining wetlands so as to exclude many of the seasonal ones, some 4.5 million ha (11.1 million acres). What position would you take on such a proposal, and why? How would you reconcile the conflicting needs of the farmers and developers with the need to preserve wildlife habitats?

3. In what ways would you expect a substantial decrease in wetlands to affect populations of migratory birds?

Area and Wildlife Use of Wetlands in the San Francisco Estuary

	Area		Wildlife Use	
Type of Wetland	*ha*	*%*	*Number of Species*	*%[a]*
Permanent	58,765	23		
Mudflat	25,949	10.2	49	11.9
Marsh	17,964	7.1		
Tidal Salt			95	23.1
Brackish			192	46.6
Tidal Fresh			174	42.2
Salt Pond	14,852	5.8	82	19.9
Seasonal	195,709	5.8	82	19.9
Diked marsh and other	34,467	13.5		
Diked Marsh			72	17.5
Other			—	
Farmed wetlands	156,176	61.3	92	22.3
Riparian woodlands	5066	1.9	207	50.2
Total	**254,474**	**100[b]**	**412**	**100**

[a] Percent of total number of species found in all wetlands
[b] Permanent + seasonal = 100%

- Groundwater use has resulted in a variety of environmental problems, including overdraft, loss of vegetation along watercourses, and land subsidence.

- Because agriculture is the biggest user of water, conservation of water in agriculture has the most significant effect on sustainable water use. However, it is also important to practice water conservation at the personal level in our homes and to price water to encourage conservation and sustainability.

- There is a need for a new philosophy in water resource management that considers sustainability and uses creative alternatives and variable sources. Development of a master plan involves inclusion of normal sources of surface water and groundwater, conservation programs, and use of reclaimed water.

- Development of water supplies and facilities to more efficiently move water may cause considerable environmental degradation; construction of reservoirs and canals and channelization of rivers should be considered carefully in light of potential environmental impacts.

- Wetlands serve a variety of functions at the ecosystem level that benefit other ecosystems and people.

- Flooding is perhaps the most universal natural hazard in the world; both the frequency and the severity of flooding of small streams are increased by urbanization. The preferable and most environmentally sound adjustment to flooding is land-use planning that avoids building on floodplains.

- The Colorado River in the southwestern United States and northern Mexico is one of the most regulated rivers in the world. Understanding links among physical, biological, and social systems of the Colorado River is necessary to manage its water resources and ecosystems.

● REEXAMINING THEMES AND ISSUES

Human Population:

As human population has increased, so has the demand for water resources. As a result, we must be more careful in managing Earth's water resources, particularly near urban centers.

Sustainability:

The water resources of the planet are sustainable provided we manage them properly and they are not overused, polluted, or wasted. This requires good water management strategies. We believe that the move toward sustainable water use must be facilitated now to avoid conflicts in the future. Principles of water management presented in this chapter help delineate what needs to be done.

Global Perspective:

The water cycle is one of the major global geochemical cycles. It is responsible for the transfer and storage of water on a global scale. Fortunately, on a global scale, the total abundance of water on Earth is not a problem. However, ensuring that it is available when and where it is needed in a sustainable way *is* a problem.

Urban World:

Although urban areas consume only a small portion of the water resources used by people, it is urban areas where shortages are often most apparent. Thus, the concepts of water management and water conservation are critical in urban areas.

Science and Values:

Conflicts result from varying values related to water resources. We value natural areas such as wetlands and free-running rivers, but we also want water resources and protection from hazards such as flooding. As a result, we must learn to plan more effectively with nature to minimize natural hazards, maintain a high-quality water resource, and provide the water necessary for the ecosystems of our planet. The test floods of the Colorado River discussed in this chapter are examples of new river management practices based on the science of understanding river processes linked to values that recognize the wish to sustain the Colorado as a vibrant, living river.

● KEY TERMS

channelization **409**	floodplain **409**	off-stream use **397**	water budget **393**
consumptive use **397**	groundwater **392**	overdraft **395**	water conservation **400**
desalination **395**	influent stream **392**	sustainable water use **402**	wetlands **404**
effluent stream **392**	in-stream use **397**		

STUDY QUESTIONS

1. If water is one of our most abundant resources, why are we concerned about its availability in the future?

2. Which is more important from a national point of view, conservation of water use in agriculture or in urban areas? Why?

3. Distinguish between in-stream and off-stream uses of water. Why is in-stream use controversial?

4. What are some important environmental problems related to groundwater use?

5. How might your community better manage its water resources?

6. What are some of the major environmental impacts associated with the construction of dams and canals? How might these be minimized?

7. What are some of the environmental problems associated with channelizing rivers? How might the potential adverse effects of channelization be minimized?

8. How does urbanization affect the flood hazard, and how can this hazard best be managed?

FURTHER READING

Brookes, A. 1988. *Channelized Rivers: Perspectives for Environmental Management*. New York: Wiley. Covers consequences of river channelization and methods to improve river modification.

Graf, W. L. 1985. *The Colorado River*. Resource Publications in Geography. Washington, D.C.: Association of American Geographers. A good summary of the Colorado River water situation.

James, W., and J. Neimczynowicz. eds. 1992. *Water, Development and the Environment*. Florida: CRC Press. Covers problems with water supplies imposed by a growing population, including urban runoff, pollution and water quality, and management of water resources.

La Riviere, J. W. M. 1989. "Threats to the World's Water." *Scientific American* 261(3):80–84. Summary of supply and demand for water and threats to continued supply.

Leopold, L. B. 1974. *Water: A Primer*. San Francisco: W. H. Freeman. An excellent introduction to water processes.

Reisner, M. 1990. *Overtapped Oasis: Reform or Revolution for Western Water*. Washington, D.C.: Island Press. An overarching review of water supply, water resource development, and government water policy in the American West.

Spulber, N., and A. Sabbaghi. 1994. *Economics of Water Resources: From Regulation to Privatization*. Kluwer Academic. Discussions of water supply and demand, pollution and its ecological consequences, and water on the open market.

Twort, A. C., F. M. Law, F. W. Crowley, and D. D. Ratnayaka. 1994. *Water Supply*, 4th ed. Edward Arnold. Good coverage of water topics from basic hydrology to water chemistry, its use, management, and treatment.

Van der Leeden, F. 1990. *The Water Encyclopedia*, 2nd ed. Lewis Publishers. A great source for all sorts of water facts.

Water Resources Council. 1978. *The Nation's Water Resources, 1975–2000*, vol. 1. Washington, D.C.: Water Resources Council. A good summary of water use in the United States.

Wheeler, B. D., S. C. Shaw, W. J. Fojt, and R. A. Robertson. 1995. *Restoration of Temperate Wetlands*. New York: Wiley. Discussions of wetland restoration around the world.

CHAPTER 21

Water Pollution and Treatment

Pig stranded on a vehicle east of Burgaw, North Carolina, during a flood caused by Hurricane Floyd in 1999. Flood waters inundated pig farms, killing many thousands of the animals. Their carcasses, feces, and urine flowed through homes, churches, and schools, causing a major pollution event. (Photo by John Althonse.)

● **LEARNING OBJECTIVES**

Degradation of our surface water and groundwater resources is a serious problem, the effects of which are not fully known. There are a number of steps we can take to treat water and to minimize pollution. After reading this chapter, you should understand:

- What constitutes water pollution and what are the major categories of pollutants.
- Why the lack of disease-free drinking water is the primary water pollution problem in many locations around the world.
- How point and nonpoint sources of water pollution differ.
- What biochemical oxygen demand is, and why it is important.
- What eutrophication is, why it is an ecosystem effect, and how human activity can cause cultural eutrophication.
- Why sediment pollution is a serious problem.
- What acid mine drainage is, and why it is a problem.
- How urban processes can cause shallow aquifer pollution.
- What the various methods of wastewater treatment are, and why some are more environmentally preferable than others.
- What the environmental laws are that protect water resources and ecosystems.

North Carolina's Bay of Pigs

Hurricane Floyd struck the Piedmont area of North Carolina in September 1999. The killer storm took a number of lives while flooding many homes and forcing some 48,000 people into emergency shelters. The storm had another, more unusual effect as well. Floodwaters containing thousands of dead pigs along with their feces and urine flowed through schools, churches, homes, and businesses. The stench was reported to be overwhelming, and the count of pig carcasses may have been as high as 30,000. The storm waters had overlapped and washed out over 38 pig lagoons with as much as 950 million liters (250 million gal) of liquid pig waste, which ended up in flooded creeks, rivers, and wetlands. In all, something like 250 large commercial pig farms flooded out, drowning hogs whose floating carcasses had to be collected and disposed of (Figure 21.1).

Prior to the catastrophe brought on by Hurricane Floyd, the pig farm industry in North Carolina had been involved in a scandal reported by newspapers and television—and even *60 Minutes*. North Carolina has a long history of hog production, and the population of pigs swelled from about 2 million in 1990 to nearly 10 million in 1997. At that time, North Carolina became the second largest pig-farming state in the nation.[1] As the number of large commercial pig farms grew, the state allowed the hog farmers to build automated and very confining farms housing hundreds or thousands of pigs. There were no restrictions on farm location, and many farms were constructed on floodplains.

Each pig produces approximately two tons of waste per year. The North Carolina herd was producing approximately 20 million tons of waste a year, mostly manure and urine, which was flushed out of the pig

(a)

(b)

(c)

Figure 21.1 • North Carolina's Bay of Pigs. (*a*) Map of areas flooded by Hurricane Floyd in 1999 with relative abundance of pig farms. (*b*) Pig farm near Seven Springs, North Carolina remains flooded six weeks after Hurricane Floyd in 1999. (*c*) Collecting dead pigs near Boulaville, North Carolina. The animals were drowned when flood waters from the Cape Fear River inundated commercial pig farms.

barns and into open, unlined lagoons about the size of football fields. Favorable regulations, along with the availability of inexpensive waste disposal systems (the lagoons) were responsible for the tremendous growth of the pig population in North Carolina during the 1990s.

Following the hurricane, mobile incinerators were moved into the hog region to burn the carcasses; but there were so many dead pigs that hog farmers had to bury some animals in shallow pits. The pits were supposed to be at least 1 m deep and dry, but there wasn't always time to find dry ground; and for the most part, the pits were dug and filled on floodplains. As these pig carcasses rot, bacteria will leak into the groundwater and surface water for some appreciable time.

The pig farmers blamed the hurricane for the environmental catastrophe. However, it was clearly a human-induced disaster that had been expected. An early warning occurred in 1995, when a pig-waste-holding lagoon failed and sent approximately 95 million liters (250 million gal) of concentrated pig feces down the New River past the city of Jacksonville and into the New River Estuary. Adverse environmental effects of that spill on marine life lasted for approximately three months.

The lesson to be learned from North Carolina's "Bay of Pigs" is that we are vulnerable to environmental catastrophes caused by large-scale industrial agriculture. Economic growth and production of livestock must be carefully planned to anticipate problems, and waste management facilities must be designed so as not to pollute local streams, rivers, and estuaries.

Was the lesson learned in North Carolina? The pig farmers had powerful friends in government and big money. Incredible as it may seem, following the hurricane, the farmers asked for $1 billion in grants to help repair and replace the pig facilities, including waste lagoons, destroyed by the hurricane. Furthermore, they asked for exemptions from the Clean Water Act for a period of six months so that waste from the pig lagoons could be discharged directly into streams. This was not allowed.[2] With regard to future management, considering that North Carolina frequently is struck by hurricanes, not allowing pig operations to be located on floodplains seems obvious. However, this is only the initial step. The whole concept of waste lagoons needs to be rethought and alternative waste management practices put into effect if pollution of surface waters and groundwaters is to be avoided.

● *North Carolina's Bay of Pigs disaster produced a particularly visible and serious water pollution episode in a beautiful state with abundant natural resources. Other types of water pollution, such as water-borne disease in surface waters and pesticides in groundwater, are often much more difficult to identify without careful sampling and testing. This chapter discusses major categories of water pollution and traditional, as well as innovative, options for wastewater treatment.*

21.1 Water Pollution

Water pollution refers to degradation of water quality. In defining pollution, we generally look at the intended use of the water, how far the water departs from the norm, its effects on public health, or its ecological impacts. From a public health or ecological view, a pollutant is any biological, physical, or chemical substance that, in an identifiable excess, is known to be harmful to other desirable living organisms. Water pollutants include heavy metals, sediment, certain radioactive isotopes, heat, fecal coliform bacteria, phosphorus, nitrogen, sodium, and other useful (even necessary) elements, as well as certain pathogenic bacteria and viruses. In some instances, a material may be considered a pollutant to a particular segment of the population although it is not harmful to other segments. For example, excessive sodium as a salt is not generally harmful, but it may be harmful to people who must restrict salt intake for medical reasons.

Today, the primary water pollution problem in the world is the lack of clean, disease-free drinking water. In the past, epidemics (outbreaks) of waterborne diseases such as cholera have been responsible for the deaths of thousands of people in the United States. Fortunately, epidemics of such diseases have been largely eliminated in the United States as a result of treating drinking water prior to consumption. This certainly is not the case worldwide, however. Every year, several billion people are exposed to waterborne diseases. For example, an epidemic of cholera occurred in South America in the early 1990s, and outbreaks of waterborne diseases continue to be a threat, even in developed countries.

It is a fundamental principle that the quality of water determines its potential uses. The major uses for water today are agriculture, industrial processes, and domestic (household) supply. Water for domestic supply must be free from constituents harmful to health, such as insecticides, pesticides, pathogens, and heavy-metal concentrations; it should taste good, should be odorless, and should not damage plumbing or household appliances. The quality of water required for industrial purposes varies widely depending on the process involved. Some processes may require distilled water; others need water that is not highly corrosive or that is free of particles that could clog or otherwise damage equipment. Because most vegetation is tolerant of a wide range of water quality, agricultural waters may vary widely in physical, chemical, and biological properties.[3]

Many different processes and materials may pollute surface water or groundwater. Some of these are listed in Table 21.1. All segments of society (urban, rural, industrial, agricultural, and military) may contribute to the problem of water pollution. Most of the sources result from runoff and leaks or seepage of pollutants into surface water or groundwater. Pollutants are also transported by air and deposited in water bodies.

Increasing population often results in the introduction of more pollutants into the environment as well as demands on finite water resources.[4] As a result, it can be expected that sources of drinking water in some locations will be degraded in the near future. More than one-quarter of drinking water systems in the United States have reported at least one violation of federal health standards.[5] Approximately 36 million people in the United States were recently supplied with water from systems that violated (at least once) federal drinking water standards.[6]

The U.S. Environmental Protection Agency has set thresholds, or limits, on water pollution levels for some (but not all) pollutants. As a result of difficulties in determining effects of exposure to low levels of pollutants, maximum concentration standards have been set for only a small fraction of the more than 700 identified drinking water contaminants. If the pollutant is present in a concentration greater than an established threshold, then the water is unsatisfactory for a particular use. A list of selected pollutants (contaminants) included in the national drinking water standards for the United States can be found in Table 21.2.

The following sections focus on several water pollutants to emphasize principles that apply to pollutants in general. (See Table 21.3 for categories and examples of water pollutants.) Other water pollutants are discussed elsewhere in this book (for example, heavy metals, organic chemicals, and thermal pollution in Chapter 15 and radioactive materials in Chapter 19). Before proceeding to our discussion of pollutants, however, we first consider biochemical oxygen demand and dissolved oxygen; dissolved oxygen is not a pollutant but is rather needed for healthy aquatic ecosystems.

Table 21.1   Some Sources and Processes of Water Pollution

Surface Water	*Groundwater*
Urban runoff (oil, chemicals, organic matter, etc.) (U, I, M)	Leaks from waste disposal sites (chemicals, radioactive materials, etc.) (I, M)
Agricultural runoff (oil, metals, fertilizers, pesticides, etc.) (A)	Leaks from buried tanks and pipes (gasoline, oil, etc.) (I, A, M)
Accidental spills of chemicals including oil (U, R, I, A, M)	Seepage from agricultural activities (nitrates, heavy metals, pesticides, herbicides, etc.) (A)
Radioactive materials (often involving truck or train accidents) (I, M)	Saltwater intrusion into coastal aquifers (U, R, I, M)
Runoff (solvents, chemicals, etc.) from industrial sites (factories, refineries, mines, etc.) (I, M)	Seepage from cesspools and septic systems (R)
Leaks from surface storage tanks or pipelines (gasoline, oil, etc.) (I, A, M)	Seepage from acid-rich water from mines (I)
Sediment from a variety of sources, including agricultural lands and construction sites (U, R, I, A, M)	Seepage from mine waste piles (I)
Air fallout (particles, pesticides, metals, etc.) into rivers, lakes, oceans (U, R, I, A, M)	Seepage of pesticides, herbicide nutrients, and so on from urban areas (U)
	Seepage from accidental spills (e.g., train or truck accidents) (I, M)
	Inadvertent seepage of solvents and other chemicals including radioactive materials from industrial sites or small businesses (I, M)

Key: U = urban; R = rural; I = industrial; A = agricultural; M = military.

Table 21.2   National Drinking Water Standards

Contaminant	Maximum Contaminant Level (mg/l)
Inorganics	
Arsenic	0.05
Cadmium	0.01
Lead	0.015 action level[a]
Mercury	0.002
Selenium	0.01
Organic chemicals	
Pesticides	
Endrin	0.0002
Lindane	0.004
Methoxychlor	0.1
Herbicides	
2,4-D	0.1
2,4,S-TP	0.01
Silvex	0.01
Volatile organic chemicals	
Benzene	0.005
Carbon tetrachloride	0.005
Trichloroethylene	0.005
Vinyl chloride	0.002
Microbiological organisms	
Fecal coliform bacteria	1 cell/100 ml

Source: U.S. Environmental Protection Agency.

[a] Action level is related to the treatment of water to reduce lead to a safe level. There is no maximum contaminant level for lead.

Figure 21.2 • Pollution control officer measuring oxygen content of the River Severn near Shrewsbury, England.

21.2 Biochemical Oxygen Demand (BOD)

Dead organic matter in streams decays. Bacteria carrying out this decay use oxygen. If there is enough bacterial activity, the oxygen in the water available to fish and other organisms can be reduced to levels so low that they may die. A stream with low oxygen content is a poor environment for fish and most other organisms. A stream with an inadequate oxygen level is considered polluted for those organisms that require dissolved oxygen above the existing level.

The amount of oxygen required for biochemical decomposition processes is called the **biochemical oxygen demand (BOD)**. BOD is commonly used in water quality management (Figure 21.2). It measures the amount of oxygen consumed by microorganisms as they break down organic matter within small water samples, which are analyzed in a laboratory. The BOD is routinely measured as part of water quality testing, particularly at discharge points into surface water, such as at wastewater treatment plants. At treatment plants, the BOD of the incoming sewage water from sewer lines is measured, as is water from locations both upstream and downstream of the plant. This practice allows comparison of the upstream, or background, BOD with the BOD of the water being discharged by the plant.

Dead organic matter—which produces BOD—is added to streams and rivers from natural sources (such as dead leaves from a forest) as well as from agricultural runoff and urban sewage. Approximately 33% of all BOD in streams results from agricultural activities. However, urban areas, particularly those with older, combined sewer systems (in which stormwater runoff and urban sewage share the same line), also considerably increase the BOD in streams. This results because during times of high flow, when sewage treatment plants are unable to handle the total volume of water, raw sewage mixed with storm runoff overflows and is discharged untreated into streams and rivers.

When the BOD is high, as suggested earlier, the *dissolved oxygen content* of the water may become too low to support the life in the water. The U.S. Environmental Protection

Table 21.3   Categories of Water Pollutants

Pollutant Category	Examples of Sources	Comments
Dead organic matter	Raw sewage, agricultural waste, urban garbage	Produces biochemical oxygen demand and diseases.
Pathogens	Human and animal excrement and urine	Examples: Recent cholera epidemics in South America and Africa; 1993 epidemic of cryptosporidiosis in Milwaukee, Wisconsin. See discussion of fecal coliform bacteria under "Waterborne Disease."
Organic chemicals	Agricultural use of pesticides and herbicides (see Chapter 12); industrial processes that produce dioxin (Chapter 15)	Potential to cause significant ecological damage and human health problems. Many of these chemicals pose hazardous waste problems (see Chapter 28).
Nutrients	Phosphorous and nitrogen from agricultural and urban lead use (fertilizers) and wastewater from sewage treatment	Major cause of artificial eutrophication. Nitrates in groundwater and surface waters can cause pollution and damage to ecosystems and people.
Heavy metals	Agricultural, urban, and industrial use of mercury, lead, selenium, cadmium, and so on (Chapter 15)	Example: Mercury from industrial processes that is discharged into water (see Chapter 15). Heavy metals can cause significant ecosystem damage and human health problems
Acids	Sulfuric acid (H_2SO_4) from coal and some metal mines; industrial processes that dispose of acids improperly	Acid mine drainage is a major water pollution problem in many coal mining areas, damaging ecosystems and spoiling water resources.
Sediment	Runoff from construction sites, agricultural runoff, and natural erosion	Reduces water quality and results in loss of soil resources.
Heat (thermal pollution)	Warm to hot water from power plants and other industrial facilities	Causes ecosystem disruption (see Chapter 15).
Radioactivity	Contamination by nuclear power industry, military, and natural sources (see Chapter 19)	Often related to storage of radioactive waste. Health effects vigorously debated (see Chapters 15 and 19).

Agency defines the threshold for a water pollution alert as a dissolved oxygen content of less than 5 mg/l of water. Figure 21.3 illustrates the effect of high BOD on dissolved oxygen content in a stream when raw sewage is introduced as a result of an accidental spill. Three zones are identified:

1. *A pollution zone*, where a high BOD exists. As decomposition of the waste occurs, oxygen is used by microorganisms, and dissolved oxygen content of the water decreases.

2. *An active decomposition zone*, where the dissolved oxygen content reaches a minimum owing to rapid biochemical decomposition by microorganisms as the organic waste is transported downstream.

3. *A recovery zone*, where the dissolved oxygen increases

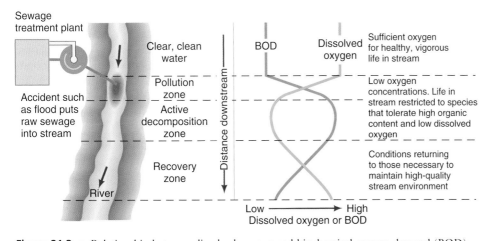

Figure 21.3 • Relationship between dissolved oxygen and biochemical oxygen demand (BOD) for a stream following the input of sewage.

and the BOD is reduced, because most oxygen-demanding organic waste from the input of sewage has decomposed and the natural stream processes are replenishing the water with dissolved oxygen. For example, in quickly moving water, the water at the surface mixes with air, and oxygen enters the water.

All streams have some capability to degrade organic waste. Problems result when the stream is overloaded with biochemical oxygen-demanding waste, overpowering the stream's natural cleansing function.

21.3 Waterborne Disease

As mentioned earlier, the primary water pollution problem in the world today is the lack of clean, disease-free drinking water. Each year, particularly in less-developed countries, several billion people are exposed to waterborne diseases whose effects vary in severity from an upset stomach to death. As recently as the early 1990s, epidemics of cholera, a serious waterborne disease, caused widespread suffering and death in South America.

In the United States, we tend not to think much about waterborne illness. Although historically, epidemics of waterborne disease killed thousands of people in U.S. cities such as Chicago, public health programs largely eliminated such epidemics by treating drinking water to remove disease-carrying microorganisms and not allowing sewage to contaminate drinking water supplies. As we will see, however, North America is not immune to **outbreaks**—or sudden occurrences—of waterborne disease.

Outbreak in Milwaukee, Wisconsin

The largest outbreak of waterborne disease outbreak in the history of the United States occurred in April 1993 in Milwaukee, Wisconsin. The disease, which causes flulike symptoms, is a gastrointestinal illness carried by a microorganism (a parasite) known as *Cryptosporidium* (the disease is known as *cryptosporidiosis*). Between March 11 and April 9, approximately 400,000 people of a total of 1.6 million people in a five-county area became ill following exposure to *Cryptosporidium* in the drinking water. Most people who contracted the illness were sick for approximately nine days; but the disease can be fatal to people with depressed immune systems, such as cancer patients and AIDS patients. Approximately 100 people died. The parasite is resistant to chlorination and evidently passed through one of the city's water treatment plants. The source of the parasite remains unknown, but possible sources include cattle grazing along rivers that flow into Milwaukee Harbor, slaughterhouses, and human sewage. It is possible that runoff from spring rains and snowmelt transported the parasites to Lake Michigan, where they entered the intake to water treatment plants.[7,8]

The outbreak in Milwaukee was a "wake-up call" concerning the quality of U.S. drinking water. Many other cities in the United States that utilize surface water resources are just as vulnerable as Milwaukee.[7] In fact, recent tests suggest that *Cryptosporidium* is present in 65% to 97% of the surface waters in the United States.[9] In May 1994, another outbreak of the same disease, caused by the same parasite, killed 19 people in Las Vegas.[6]

The outbreak in Milwaukee happened even though the water treatment plants met all existing federal and state quality standards. Federal guidelines are even stricter now, but there still is concern as to whether waterborne parasites such as *Cryptosporidium* are being effectively removed. Although *Cryptosporidium* is very resistant to disinfectants, it may be removed by filtration. Upgrading water treatment plants is a cost-effective mechanism for reducing the threat of waterborne disease. The price of inaction is very high. Considering the high costs of illness and death associated with drinking contaminated water, future investments in technology and facilities to treat water is an important government service that should be considered a bargain.[6]

Fecal Coliform Bacteria

Because it is difficult to monitor disease-carrying organisms directly, we use the count of **fecal coliform bacteria** as a standard measure and indicator of disease potential. The presence of fecal coliform bacteria in water indicates that fecal material from mammals or birds is present, so organisms that cause waterborne diseases may be present as well. Fecal coliform bacteria are usually (but not always) harmless bacteria that are normal constituents of human and animal intestines. They are present in all human and animal waste. The threshold used by the U.S. Environmental Protection Agency for swimming water is not more than 200 cells of fecal coliform bacteria per 100 ml of water; if fecal coliform is above the threshold level, the water is considered unfit for swimming. Water with any fecal coliform bacteria is unsuitable for drinking.

One type of fecal coliform bacteria, *Escherichia coli*, or *E. coli*, has been responsible for causing human illness and death. Outbreaks have resulted from eating contaminated meat and drinking contaminated juices or water. There was an outbreak caused by contaminated meat at a popular fast food chain in 1993. In 1998, 26 children became ill and one died after visiting a Georgia water park. As recently as July 1998, the community of Alpine, Wyoming, suffered a major outbreak of illness due to the presence of *E. coli* in the town's drinking water supply.[10] It is clear that *E. coli* bacteria can be a real threat to human health and must be carefully regulated.

Threat of waterborne diseases in nearshore coastal waters as well as lakes and rivers is responsible for thousands of warnings and beach closings per year in the United States. Advisories are often posted that swimming may be hazardous to health (Figure 21.4). In most cases,

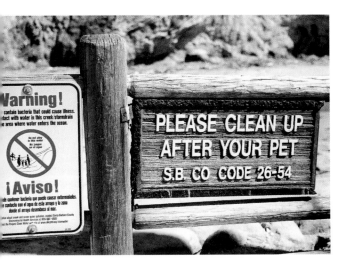

Figure 21.4 ● This beach in southern California is occasionally closed as a result of contamination by bacteria.

the identified pollutant is fecal coliform bacteria, which may indicate the presence of a specific disease-causing virus, such as hepatitis. Pollutants to coastal waters have a variety of sources, including urban storm runoff, leaking sewer lines, sewage treatment plant overflow or failure, and leaking sewage treatment facilities from individual homes (septic tanks; see Section 21.10). Coastal communities in many areas are facing potential loss of income from tourism resulting from beach closures. As a result, increased research and testing of nearshore coastal waters is becoming more routine. As sources of pollutants are identified, management plans are being designed and implemented to reduce the threat of pollution and future beach closures.

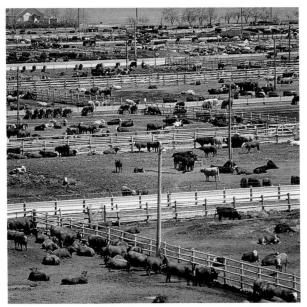

Figure 21.5 ● Cattle feedlot in Colorado. High numbers of cattle in small areas have the potential to create both surface-water and groundwater pollution because of runoff and infiltration of urine.

Outbreak in Walkerton, Ontario

One of the most serious outbreaks of *E. coli* bacterial infection in Canadian history developed in May 2000 in Walkerton, Ontario, a town with about 5,000 inhabitants. The strain of *E. coli* involved was a dangerous one found in the digestive system of cows. The likely cause of the contamination was cow manure washed into the water supply during heavy rains and flooding on May 12, 2000. The local Public Utility Commission knew as early as May 18 that water from wells serving the town was contaminated by the *E. coli*, but the commission did not report it immediately to health authorities. People were not advised to boil water (thus killing the bacteria) until it was too late to avoid the outbreak. By May 26, five people had died, more than twenty were in intensive care units, and over five hundred were ill with severe symptoms that included cramps, vomiting, and diarrhea. The old and very young are most vulnerable to the ravages of the disease, which can damage the kidneys. Two of the first deaths were a 2-year-old baby and an 82-year-old woman. Finally, officials took over management of the water supply, and bottled water was distributed.

The town of Walkerton experienced a full-blown outbreak of a waterborne disease. Doctors warned that additional deaths were likely and that modern medicine could do little to treat the disease. The best advice doctors could give was to consume large amounts of safe water and avoid dehydration as the infection ran its course. Soon, people were asking why the outbreak had occurred. An investigation was launched by authorities, and one of the questions asked was why there had been a delay between identifying the potential problem and issuing a warning. If people had been notified on May 18, much of the sickness might have been avoided. The outbreak might also have been detected earlier if the government of Ontario had not cut back on the level of testing of the public water supply (earlier regulations had required more testing). A major lesson learned from Walkerton is that we remain vigilant in testing our water supplies and raising the alarm quickly if potential problems arise.

21.4 Nutrients

Two important nutrients that cause water pollution problems are phosphorus and nitrogen, and both are released from sources related to land use. Forested land has the lowest concentrations of phosphorus and nitrogen in stream waters. In urban streams, concentrations of these nutrients are greater because of fertilizers, detergents, and products of sewage treatment plants. Often, however, the highest concentrations of phosphorus and nitrogen are found in agricultural areas, where the sources are fertilized farm fields and feedlots (Figure 21.5). Over 90% of total nitrogen added to the environment by human activity comes from agriculture.

Medical Lake: An Example

An example will illustrate the effect nutrients can have on a body of water. During the summer of 1971, the waters of Medical Lake in Washington became clogged with algae and bacteria and turned a dark, turbid green. Fish, algae, and bacteria died, and the wind blew them into stinking masses onto the shore.[12] It was apparent that the Medical Lake ecosystem was dying.

Fish in the lake did not die from a direct effect of pollution, from disease organisms in the water, or from something in the water that was directly toxic to them. They died as the result of a complex effect produced by a community of interacting organisms linked to the local environment (recall the discussion of ecosystems and the community effect in Chapter 6).

This is how the process worked. Phosphorus, a chemical element necessary for all living things and one that is often in short supply, became available in large concentrations in treated sewage water that entered the lake. Just as when phosphorus is added as a lawn fertilizer and causes increased growth of grass, the phosphorus added to the lake caused a population explosion of photosynthetic blue-green bacteria and algae. Mats of the algae and bacteria became so thick that those at the top shaded those at the bottom; not receiving enough light, those at the bottom died. The dead algae and bacteria became food for other bacteria, and as these bacteria increased, they used more and more of the oxygen dissolved in the lake water. (Bacteria require oxygen for respiration, just as we do.) The supply of oxygen in the water was soon depleted. Algae, bacteria, and fish began to die. Because fish require a higher concentration of oxygen in water than do bacteria, most of the fish died, whereas some bacteria continued to live.

The fish did not die in Medical Lake from phosphorus poisoning. If you added phosphorus to water in an aquarium where there were only fish and no algae or bacteria, so that the phosphorus concentration was the same as that in Medical Lake, the phosphorus would not affect the fish. The fish died from a lack of oxygen resulting from a chain of events that started with the input of phosphorus and affected the whole ecosystem. The unpleasant effects resulted from the interactions among different species, the effects of the species on chemical elements in their environment, and the condition of the environment (the lake and the air above it). This is what we call an **ecosystem effect**.

Eutrophication

What happened at Medical Lake is called **eutrophication**—the process by which a body of water develops a high concentration of nutrients, such as nitrogen and phosphorus (in the forms of nitrates and phosphates). The nutrients cause an increase in the growth of aquatic plants in general, as well as production of photosynthetic blue-green bacteria and algae. Algae may form surface mats (Figure 21.6), shading the water and reducing light to algae below the surface, greatly reducing photosynthesis. The bacteria and algae die; as they decompose, the BOD increases, oxygen in the water is consumed, and oxygen content is reduced. If the oxygen content is sufficiently lowered, other organisms, such as fish, will die.

The process of eutrophication of a lake is shown in Figure 21.7. A lake that has a natural high concentration of the chemical elements required for life is called a *eutrophic lake*. A lake with a relatively low concentration of chemical elements required by life is called an *olig-*

Figure 21.6 • Mats of dying green algae in a pond undergoing eutrophication.

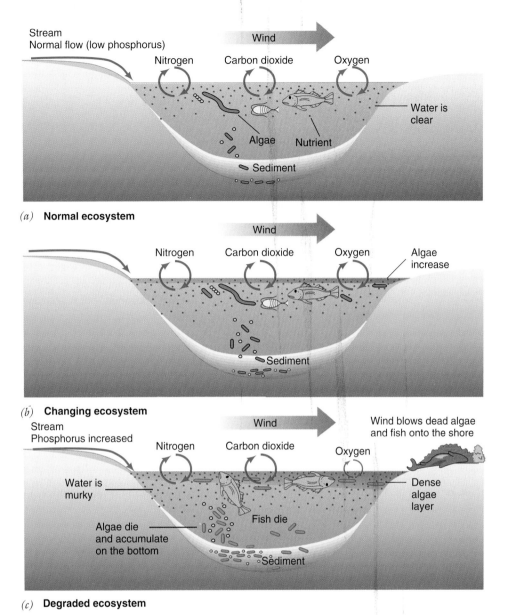

Figure 21.7 • The eutrophication of a lake. (*a*) In an oligotrophic, or low-nutrient, lake, the abundance of green algae is low, the water clear. (*b*) Phosphorus is added to streams and enters the lake. Algae growth is stimulated, and a dense layer is formed.

(*c*) The algae layer becomes so dense that the algae at the bottom die. bacteria feed on the dead algae and use up the oxygen. Finally, fish die from lack of oxygen.

otrophic lake. Oligotrophic lakes have clear water that is pleasant for swimmers and boaters, with relatively low abundance of life. Eutrophic lakes have an abundance of life, often with mats of algae and bacteria and murky, unpleasant water.

When eutrophication is accelerated by human processes that add nutrients to a body of water, we say that **cultural eutrophication** is occurring. Problems associated with the artificial eutrophication of bodies of water are not restricted to lakes (See A Closer Look 21.1). In recent years concern has grown about the outflow of sewage from urban areas into tropical coastal waters and

cultural eutrophication on coral reefs.[13,14] For example, parts of the famous Great Barrier Reef of Australia, as well as some reefs fringing the Hawaiian Islands (Figure 21.8), are being damaged by eutrophication.[15,16] The damage to corals occurs as nutrient input stimulates algal growth on the reef, which covers and smothers the coral.

The solution to artificial eutrophication is fairly straightforward and involves ensuring that high concentrations of nutrients from human sources do not enter lakes and other water bodies. This goal can be accomplished by reducing pollutants by using phosphate-free detergents, controlling nitrogen-rich runoff from agricul-

Figure 21.8 • The Great Barrier Reef of Australia.

tural and urban lands, disposing of or reusing treated wastewater, and using more advanced water treatment methods, such as special filters and chemical treatments that remove more of the nutrients.

21.5 Oil

Oil discharged into surface water, usually in the ocean but also on land and in rivers, has caused major pollution problems. Several large oil spills from underwater oil drilling have occurred in recent years. However, although spills make headlines, normal shipping activities probably release more oil over a period of years than is released by the occasional spill. The cumulative impacts of these releases are not well known.

Exxon Valdez: Prince William Sound, Alaska

The best known oil spills are caused by tanker accidents. On March 24, 1989, the supertanker *Exxon Valdez* ran aground on Bligh Reef south of Valdez in Prince William Sound, Alaska. Alaskan crude oil that had been delivered to Valdez through The Trans-Alaska Pipeline poured out of ruptured tanks of the vessel at about 20,000 barrels per hour. The tanker was loaded with about 1.2 million barrels of oil, and about 250,000 barrels (11 million gal) entered the sound (Figure 21.9). The spill could have been larger than it was, but fortunately some of the oil in the tanker was off-loaded (pumped out) into another vessel. The *Exxon Valdez* spill produced an environmental shock that resulted in passage of the Oil Pollution Act of 1990 and in a renewed evaluation of cleanup technology.[17,18]

An environmental shock resulted from the fact that the oil was spilled into what is considered one of the most pristine and ecologically rich marine environments of the world.[17] Many species of fish, birds, and marine mammals are present in the sound. The effects of the spill included the death of 13% of harbor seals, 28% of sea otters, and 100,000–645,000 seabirds.[18]

Within three days of the spill, winds began blowing the slick beyond any hope of containment. Of the 11 million gal of spilled oil, about 20% evaporated, and 50% was

Figure 21.9 • *Exxon Valdez* tanker accident in Prince William Sound (1989). Oil is being offloaded from the leaking *Exxon Valdez* (left) to the smaller *Exxon Baton Rouge* (right).

Cultural Eutrophication in the Gulf of Mexico

Each summer, a so-called "dead zone" develops off the nearshore environment of the Gulf of Mexico, south of Louisiana. The zone varies in size from about 13,000 to 18,000 km² (5,000 to 7,000 mi²), an area about the size of the small country of Kuwait or the state of New Jersey. Within the zone, bottom water generally has low concentrations of dissolved oxygen (less than 2 mg/l; a water pollution alert occurs if dissolved oxygen has a concentration of less than 5 mg/l). Shrimp and fish can swim away from the zone, but bottom dwellers such as shellfish, crabs, and snails are killed. Nitrogen is believed to be the most significant cause of the dead zone.

The low concentration of oxygen occurs because the nitrogen causes cultural eutrophication. Algae bloom; and as the algae die, sink, and decompose, the oxygen in the water is depleted. The source of nitrogen is believed to be in one of the richest, most productive agricultural regions of the world—the Mississippi River drainage basin.

The Mississippi River drains about 3 million km², which is about 40% of the land area of the lower 48 states. The use of nitrogen fertilizers in that area greatly increased beginning in the mid-20th century but leveled off in the 1980s and 1990s. The level of nitrogen in the river water has also leveled off, suggesting that the dead zone may have reached its max-

imum size. This provides the time necessary to study the cultural eutrophication problem carefully and make sound decisions to reduce or eliminate it.

A partial reduction of nitrogen (nitrates) reaching the Gulf of Mexico via the Mississippi River may be accomplished by the following actions:[11]

- Modify agricultural practices to reduce the nitrogen that enters the river from fertilizers by using fertilizers more effectively and efficiently.
- Restore and create river wetlands between farm fields and streams and rivers, particularly in areas known to contribute high amounts of nitrogen. The wetland plants use the nitrogen, lowering the amount that enters the river.
- Implement nitrogen reduction processes at wastewater treatment plants for towns, cities, and industrial facilities.
- Implement better flood control in the upper Mississippi River to retain floodwaters on floodplains, where nitrogen can be used by riparian vegetation.
- Divert floodwater from the Mississippi to backwaters and coastal wetlands of the Mississippi River Delta. At present, levees in the delta push river waters directly into the

gulf. Plants in coastal wetlands will use the nitrogen, reducing the concentration of the nutrient that reaches the Gulf of Mexico.

Improving agricultural practices could result in a reduction of up to 20% in the nitrogen reaching the Mississippi. This would require a reduction of about 20% in fertilizers used, something farmers say would harm productivity. Restoration and creation of river wetlands and riparian forests holds the promise of reducing nitrogen input to the river by up to 40%. This would require some combination of wetlands and forest of about 10 million ha (24 million acres), which is about 3.4% of the Mississippi River Basin.[11] That is a lot of land!

There is no easy solution to cultural eutrophication in the Gulf of Mexico. Clearly, though, a reduction in the amount of nitrogen entering the Gulf is needed. Also needed is a better understanding of details of the nitrogen cycle within the Mississippi River basin and delta. Developing this understanding will require monitoring of nitrogen and development of mathematical models of sources, sinks, and rates of transfer of nitrogen. With an improved understanding of the nitrogen cycle, a management strategy to reduce or eliminate the dead zone can be created and implemented.

deposited on the shoreline. Only 14% was collected by skimming and waste recovery. The extent of oil sheens, tar balls (formed by sticky, less volatile components in the oil), and mousse (a thick, weathered patch of oil with the consistency of soft pudding) is shown in Figure 21.10.

Before the *Exxon Valdez* spill, it was generally believed that the oil industry was capable of dealing with oil spills. However, more than $3 billion has been spent to clean the spill, and few people are satisfied with the results. Some scientists argue that the recovery might have been faster if

some of the cleanup methods, such as spraying rocks and beaches with high-pressure hot water, had not been used. They argue that the coastal organisms that live under the rocks and that had survived the initial impact of the spill were killed by the high pressure and heat.[18] There is no doubt that the cleanup work posed enormous problems (Figure 21.11). Photographs and videotapes of workers attempting to clean individual pebbles on beaches are a vivid reminder of the difficulty and virtual futility of achieving an effective cleanup after an event of this magnitude. In

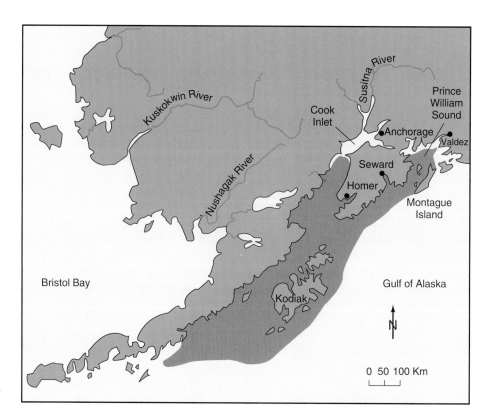

Figure 21.10 • Extent of Alaskan oil spill of 1989. (*Source:* Alaska Department of Fish and Game *21[4],* 1989.)

addition, the oil spill disrupted the lives of the people who live and work in the vicinity of Prince William Sound.

The long-term effects of large oil spills are uncertain. We know that the effects can last several decades; toxic levels of oil have been identified in salt marshes 20 years following a spill.[18] Fortunately, the ocean abounds with natural process-es capable of eventually consuming the oil. Therefore, compared with some other water pollutants (such as pathogens, heavy metals, and organic chemicals), oil is relatively low in toxicity, and recovery can occur fairly quickly. Nevertheless, spills have significant short-term effects, including disruption of ecosystems and commercial fisheries, curtailment of tourism, and loss of seabirds and mammals.

The *Exxon Valdez* spill demonstrated that the technology for dealing with oil spills is inadequate. The first and most important step is to avoid large spills; a primary method for doing so is to use supertankers with double hulls designed to minimize the release of oil on collision and rupture of tanks. The second most important step is to pump the oil out of a tanker as soon as an accident occurs, thereby preventing further spillage into the sea. Once a spill occurs, the collection of oil at sea using floating barriers and skimmers (oil is lighter than water and so floats on water) is a worthwhile endeavor; but if conditions include high winds and rough seas, it is nearly impossible. Cleaning oil from birds and mammals is also a worthwhile endeavor, but because the animals ingest oil and are difficult to clean, mortality is high. Oil on beaches may be collected by spreading absorbent material, such as straw, allowing the oil to soak in, and then collecting and disposing of the oily straw.

Jessica: Galapagos Islands

Figure 21.11 • Attempt to clean oil from the shoreline of Eleanor Island, Alaska, four months after the oil spill from the *Exxon Valdez.*

Another shock occurred on January 22, 2001, when a small tanker, the *Jessica*, made a navigational error off the coast of Ecuador near the Galapagos Islands and ran aground,

Figure 21.12 • Ecuadorean tanker *Jessica* ran aground in January of 2001 near Cristobal Island of the Galapagos. Over 100,000 gallons of diesel fuel spilled into the ocean near the fragile island environment. Shown here are park rangers attempting to collect the fuel.

spilling light diesel oil into the ocean (Figure 21.12). Although the spill (over 100,000 gallons) was small compared with the *Exxon Valdez* spill in Alaska, there was grave concern, and Ecuador declared a state of emergency. The Galapagos are environmental treasures and icons of the environment, where Charles Darwin worked in developing his theory of evolution of species. The United States responded quickly with a Coast Guard ship designed to pump oil from a damaged tanker. Some of the oil was washed ashore onto a small island, injuring birds, seals, and other marine life. The oil slick spread over 3,000 km² in the first week but was fortunately carried by currents and trade winds away from the nearby Galapagos. The spill was yet another warning regarding the use of better-built tankers with double hulls and, where possible, the routing of tankers away from ecological areas of particular concern.

21.6 Sediment

Sediment consisting of rock and mineral fragments ranging from gravel particles greater than 2 mm in diameter to finer sand, silt, clay, and even finer colloidal particles can produce a *sediment pollution* problem. In fact, by volume and mass, sediment is our greatest water pollutant. In many areas, it chokes streams; fills lakes, reservoirs, ponds, canals, drainage ditches, and harbors; buries vegetation; and generally creates a nuisance that is difficult to remove. Sediment pollution is a twofold problem: It results from erosion, which depletes a land resource (soil) at its site of origin (Figure 21.13), and it reduces the quality of the water resource it enters.[19]

Many human activities affect the pattern, amount, and intensity of surface-water runoff, erosion, and sedimentation. Streams in naturally forested or wooded areas may be nearly stable; that is, there is relatively little excessive erosion or sedimentation. However, converting forested land to agriculture generally increases the runoff and sediment yield or erosion of the land. Application of soil conservation procedures to farmland can minimize but not eliminate soil loss. The change from agricultural, forested, or rural land to highly urbanized land has even more dramatic effects. Large quantities of sediment may be produced during the construction phase of urbanization. Fortunately, sediment production and soil erosion can be minimized by on-site erosion control measures.

Figure 21.13 • Erosion of a highway embankment in Georgia. The erosion produces gullies and removes vegetation. The sediment may be transported offsite to degrade water resources (streams, rivers, ponds, and lakes).

Reduction of sediment pollution in an urbanizing area through control measures is demonstrated by a study in Maryland.[20] The suspended sediment transported by the northwest branch of the Anacostia River near Colesville, Maryland, with a drainage area of 54.6 km² (21 mi²), was measured over a 10-year period. During that time, urban construction within the basin involved about 3% of the area each year, and the total urban land area in the basin was about 20% at the end of the 10-year study. Sediment pollution was a problem because of the amount of rainfall and the type of soil, which was highly susceptible to erosion when not protected, during spring and summer rainstorms, by a vegetative cover.

A sediment control program reduced the sediment yield by an estimated 35%. The program's basic principles were to tailor the development to the natural topography, expose a minimum amount of land, provide temporary protection for exposed soil, minimize surface runoff from critical areas, and trap eroded sediment on the construction site.[20]

Figure 21.14 • Acid drainage from an abandoned mine site is entering a small stream channel and polluting the surface waters. This site is located in the mountains of southwestern Colorado.

21.7 Acid Mine Drainage

The term **acid mine drainage** refers to water with a high concentration of sulfuric acid (H_2SO_4) that drains from mines—mostly coal mines but also metal mines (copper, lead, and zinc). Coal and the rocks containing coal are often associated with a mineral known as fool's gold or pyrite (FeS_2), which is iron sulfide. When the pyrite, which may be finely disseminated in the rock and coal, comes into contact with oxygen and water, it weathers. A product of the chemical weathering is sulfuric acid. In addition, pyrite is associated with metallic sulfide deposits, which, when weathered, also produce sulfuric acid. The acid is produced when surface water or shallow groundwater runs through or moves into and out of mines or tailings (Figure 21.14). If the acid-rich water runs off to a natural stream, pond, or lake, significant pollution and ecological damage may result. The acidic water is toxic to the plants and animals of an aquatic ecosystem; it damages biological productivity, and fish and other aquatic life may die. Acid-rich water can also seep into and pollute groundwater.

Acid mine drainage is a significant water pollution problem in Wyoming, Indiana, Illinois, Kentucky, Tennessee, Missouri, Kansas, and Oklahoma and is probably the most significant water pollution problem in West Virginia, Maryland, Pennsylvania, Ohio, and Colorado. The total impact is significant, because thousands of kilometers of streams have been damaged.

Even abandoned mines can cause serious problems. For example, subsurface mining in the tristate area of Kansas, Oklahoma, and Missouri for sulfide deposits containing lead and zinc began in the late 19th century and ended in some areas in the 1960s. When the mines were operating, they were kept dry by pumping out the groundwater that seeped in. However, since the termination of mining, some

of them have flooded and overflowed into nearby creeks, polluting the creeks with acid-rich water. The problem was so severe in the Tar Creek area of Oklahoma that it was at one time designated by the U.S. Environmental Protection Agency as the nation's worst hazardous waste site.

21.8 Surface-Water Pollution

Pollution of surface waters occurs when too much of an undesirable or harmful substance flows into a body of water, exceeding the natural ability of that water body to remove the undesirable material, dilute it to a harmless concentration, or convert it to a harmless form.

Water pollutants, like other pollutants, are categorized as being emitted from point or nonpoint sources (see Chapter 15). **Point sources** are distinct and confined, such as pipes from industrial or municipal sites that empty into streams or rivers (Figure 21.15). In general, point source pollutants from industries are controlled through on-site treatment or disposal and are regulated by permit. Municipal point sources are also regulated by permit. In older cities in the northeastern and Great Lakes areas of the United States, most point sources are outflows from combined sewer systems. As mentioned earlier, such systems combine stormwater flow with municipal wastewater. During heavy rains, urban storm runoff may exceed the capacity of the sewer system, causing it to overflow and deliver pollutants to nearby surface waters.

Nonpoint sources, such as runoff, are diffused and intermittent and are influenced by factors such as land use, climate, hydrology, topography, native vegetation, and geology. Common urban nonpoint sources include runoff from streets or fields; such runoff contains all sorts of pollutants, from heavy metals to chemicals and sediment. Rural sources of nonpoint pollution are generally associated with agriculture, mining, or forestry. Nonpoint sources are difficult to monitor and control.

From an environmental view, two approaches to dealing with surface-water pollution are (1) to reduce the sources and (2) to treat the water to remove pollutants or convert them to forms that can be disposed of safely. Which of the options is used depends on the specific circumstances of the pollution problem. Reduction at the source is the most environmentally preferable way of dealing with pollutants. For example, air-cooling towers, rather than water-cooling towers, may be utilized to dispose of waste heat from power plants, thereby avoiding thermal pollution of water. The second method—water treatment—is used for a variety of pollution problems. Water treatments include chlorination to kill microorganisms such as harmful bacteria and filtering to remove heavy metals.

There is a growing list of success stories in the treatment of water pollution. One of the most notable is the cleanup of the Thames River in Great Britain. For centuries, London's sewage had been dumped into that river, and there were few fish to be found downstream in the estuary. In recent decades, however, improvement in water treatment has led to the return of a number of species of fish, some not seen in the river in centuries.

Many large cities in the United States, such as Boston, Miami, Cleveland, Detroit, Chicago, Portland, and Los Angeles, grew on the banks of rivers; but the rivers were often nearly destroyed with pollution and concrete. Today, there are grassroots movements all around the country to restore urban rivers and adjacent lands as greenbelts, parks, or other environmentally sensitive developments. For example, the Cuyahoga River in Cleveland, Ohio, was so polluted by 1969 that sparks from a train ignited oil-soaked wood in the river, setting the surface of the river on fire! The burning of an American river became a symbol for a growing environmental consciousness. The Cuyahoga River today is cleaner and no longer flammable. From Cleveland to Akron, the river is a beautiful greenbelt (Figure 21.16). The greenbelt changed part of the river from a sewer into a valuable public resource and focal point for economic and environmental renewal.[21] However, in downtown Cleveland and Akron, the river remains an industrial stream, and parts remain polluted.

Figure 21.16 The Cuyahoga River (lower left) flows toward Cleveland, Ohio, and Lake Erie. A trail along the Ohio and Erie Canal (lower right) is in the Cuyahoga National Park. Industrial Cleveland is on the skyline.

Figure 21.15 This pipe is a point source of chemical pollution from an industrial site entering a river in England (photo by David Woodfall).

Approximately half of all people in the United States today depend on groundwater as their source of drinking water. (Water for domestic use in the United States is discussed in A Closer Look 21.2). People have long believed that groundwater is in general pure and safe to drink. In fact, however, groundwater can be easily polluted by any one of several sources (see Table 21.1); and the pollutants, even though they are very toxic, may be difficult to recognize. (Groundwater processes were discussed in Section 20.1, and you may wish to review them.)

In the United States today, only a small portion of the groundwater is known to be seriously contaminated; but as mentioned earlier, the problem may become worse as human population pressure on water resources increase. Already, the extent of the problem is growing as the testing of groundwater becomes more common. For example, Atlantic City and Miami are two eastern cities threatened by polluted groundwater that is slowly migrating toward their wells.

It is estimated that 75% of the 175,000 known waste disposal sites in the United States may be producing plumes of hazardous chemicals that are migrating into groundwater resources. Because many of the chemicals are toxic or are suspected carcinogens, it appears that we have inadvertently been conducting a large-scale experiment on the effects on people of chronic low-level exposure to potentially harmful chemicals. The final results of the experiment will not be known for many years.[22] Preliminary results suggest that we had better act now before a hidden time bomb of health problems explodes.

The hazard presented by a particular groundwater pollutant depends on several factors, including the concentration or toxicity of the pollutant in the environment and the degree of exposure of people or other organisms to the pollutants.[23] (See the section on risk assessment in Chapter 15.)

Principles of Groundwater Pollution: An Example

Some general principles of groundwater pollution are illustrated by an example. Pollution from leaking buried gasoline tanks belonging to automobile service stations is a widespread environmental problem that no one thought very much about until only a few years ago. Underground tanks are now strictly regulated. Many thousands of old, leaking tanks have been removed, and the surrounding soil and groundwater have been treated to remove the gasoline. Cleanup can be a very expensive process, involving removal and disposal of soil (as a hazardous waste) and treatment of the water using a process known as vapor extraction (Figure 21.17). Treatment may also be accomplished under the ground by micoorganisms that consume the gasoline. This is known as **bioremediation** and is much less expensive than removal, disposal, and vapor extraction.

Pollution from leaking buried gasoline tanks emphasizes some important points about groundwater pollutants:

- Some pollutants, such as gasoline, are lighter than water and thus float on the groundwater.
- Some pollutants have multiple phases: liquid, vapor, and dissolved. Dissolved phases chemically combine with the groundwater (e.g., salt dissolves into water).

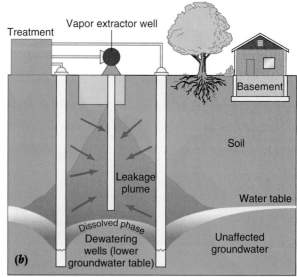

Figure 21.17 • Diagram illustrating (a) leak from a buried gasoline tank and (b) possible remediation using vapor extractor system. Notice that the liquid gasoline and the vapor from the gasoline are above the water table; a small amount dissolves into the water. All three phases of the pollutant (liquid, vapor, and dissolved) float on the denser groundwater. The extraction well takes advantage of this situation. The function of the dewatering wells is to pull the pollutants in where the extraction is most effective. (*Source:* Courtesy of University of California Santa Barbara Vadose Zone Laboratory and David Springer.)

Water for Domestic Use: How Safe Is It?

Water for domestic use in the United States is drawn from surface waters and groundwater. Although some groundwater sources have high water quality and need little or no treatment, most sources are treated to conform to national drinking water standards (revisit Table 21.3).

Before treatment, water is usually stored in reservoirs or special ponds. Storage allows for solids, such as fine sediment and organic matter, to settle out, improving the clarity of water. The water is then run through a water plant, where it is filtered and chlorinated before it is distributed to individual homes. Once in people's homes, water may be further treated. For example, many people run their tap water through readily available charcoal filters before using it for drinking and cooking.

A growing number of people prefer not to drink tap water. Instead, they use bottled water for personal consumption; as a result, production of bottled water has become a multi-billion-dollar industry.[6] Some people prefer not to drink water that contains chlorine or that runs through metal pipes. Furthermore, water supplies vary in clarity, hardness (concentration of calcium and magnesium), and taste; and the water available locally may not be to some people's liking. A common complaint about tap water is a chlorine taste, which may occur with chlorine concentrations as low as 0.2–0.4 mg/l. People may also fear contamination by minute concentrations of pollutants. The drinking water in the United States is some of the safest in the world; there is no doubt that treatment of water with chlorine has nearly eliminated waterborne diseases, such as typhoid and cholera, which previously caused widespread suffering and death in the developed world and still do in many parts of the world. However, we need to know much more about the long-term effects of exposure to low concentrations of toxins in our drinking water. How safe is the water in the United States? It's much safer than it was 100 years ago, but low-level contamination (below what is thought dangerous) of organic chemicals and heavy metals is a concern that requires continued research and evaluation.

- Some pollutants are heavier than water and sink or move downward through groundwater. Examples of sinkers include some particulates and cleaning solvents. Pollutants that sink may become concentrated deep in groundwater aquifers.
- The method used to treat or eliminate a water pollutant must take into account the physical and chemical properties of the pollutant and how these interact with surface water or groundwater. For example, the extraction well for removing gasoline from a groundwater resource (Figure 21.17) takes advantage of the fact that gasoline floats on water.
- Because cleanup or treatment of water pollutants in groundwater is very expensive, and undetected or untreated pollutants may cause environmental damage, the emphasis should be on preventing pollutants from entering groundwater in the first place.

Groundwater pollution differs in several ways from surface-water pollution. Groundwater often lacks oxygen, a situation that kills aerobic types of microorganisms (which require oxygen-rich environments) but may provide a happy home for anaerobic varieties (which live in oxygen-deficient environments). The breakdown of pollutants that occurs in the soil and in material a meter or so below the surface does not occur readily in groundwater. Furthermore, the channels through which groundwater moves are often very small and variable. Thus, the rate of movement is low in most cases, and the opportunity for dispersion and dilution of pollutants is limited.

Long Island, New York

Another example—that of Long Island, New York—illustrates several groundwater pollution problems and how they affect people's water supply. Two counties on Long Island, New York (Nassau and Suffolk), with a population of several million people, depend entirely on groundwater. Two major problems associated with the groundwater in Nassau County are intrusion of salt water and shallow-aquifer contamination.[24] Saltwater intrusion is a problem in many coastal areas of the world (see Figure 21.18).

The general movement of groundwater under natural conditions for Nassau County is illustrated in Figure 21.19. Salty groundwater is restricted from migrating inland by the large wedge of fresh water moving beneath the island. Notice also that the aquifers are layered, with those closest to the surface being the most salty.

In spite of the huge quantities of water in Nassau County's groundwater system, intensive pumping in recent years has caused water levels to decline as much as 15 m (50 ft) in some areas. As groundwater is removed near coastal areas, the subsurface outflow to the ocean

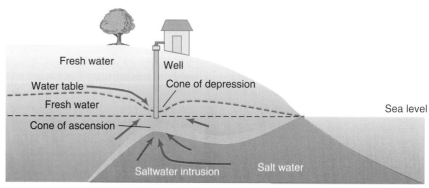

Figure 21.18 • How saltwater intrusion might occur. The upper drawing (*a*) shows the groundwater system near the coast under natural conditions, and the lower drawing (*b*) shows a well with both a cone of depression and a cone of ascension. If pumping is intensive, the cone of ascension may be drawn upward, delivering salt water to the well.

decreases, allowing salt water to migrate inland. Saltwater intrusion has become a problem for south shore communities, which now must pump groundwater from a deeper aquifer, below and isolated from the shallow aquifers where saltwater intrusion problems exist.

The most serious groundwater problem on Long Island is shallow-aquifer pollution associated with urbanization. Sources of pollution in Nassau County include urban runoff, household sewage from cesspools and septic tanks, salt used to de-ice highways, and industrial and solid waste. These pollutants enter surface waters and then migrate downward, especially in areas of intensive pumping and declining groundwater levels.[24] Landfills for municipal solid waste have been a significant source of shallow-aquifer pollution on Long Island because pollutants (garbage) placed on sandy soil over shallow groundwater quickly enter the water. For this reason, most Long Island landfills were closed in the last decade.

21.10 Wastewater Treatment

Water used for industrial and municipal purposes is often degraded during use by the addition of suspended solids, salts, nutrients, bacteria, and oxygen-demanding material. In the United States, by law, these waters must be treated before being released back into the environment. **Wastewater treatment**, or sewage treatment costs over

$15 billion per year in the United States, and the cost continues to increase. Wastewater treatment will continue to be big business.

Conventional methods of wastewater treatment include septic-tank disposal systems in rural areas and centralized wastewater treatment plants in cities. Recent, innovative approaches include the application of wastewater to the land and wastewater renovation and reuse. We discuss the conventional methods in this section and some newer methods in later sections.

Septic-Tank Disposal Systems

In many rural areas, no central sewage systems or wastewater treatment facilities are available. As a result, individual septic-tank disposal systems, not connected to sewer systems, continue to be an important method of sewage disposal in rural areas, as well as outlying areas of cities. Because not all land is suitable for the installation of a septic-tank disposal system, an evaluation of each site is required by law before a permit can be issued. An alert buyer should make sure that the site is satisfactory for septic-tank disposal before purchasing property in a rural setting or on the fringe of an urban area where such a system is necessary.

The basic parts of a septic-tank disposal system are shown in Figure 21.20. The sewer line from the house leads to an underground septic tank in the yard. The tank

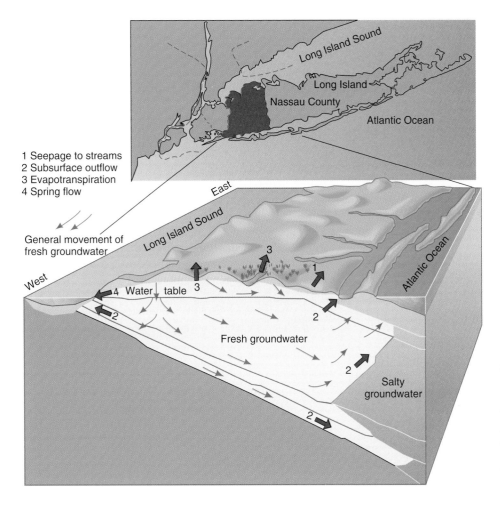

1 Seepage to streams
2 Subsurface outflow
3 Evapotranspiration
4 Spring flow

General movement of
fresh groundwater

Figure 21.19 • The general movement of fresh groundwater for Nassau County, Long Island. (*Source:* G. L. Foxworth, Nassau County, Long Island, New York—Water Problems in Humid Country, in G. D. Robinson and A. M. Spieke, eds., *Nature to be Commanded*, U.S. Geological Survey Professional Paper 950, 1978, pp. 55–68.)

is designed to separate solids from liquid, digest (biochemically change) and store organic matter through a period of detention, and allow the clarified liquid to discharge into the drain field (absorption field) from a system of piping through which the treated sewage seeps into the surrounding soil. As the wastewater moves through the soil, it is further treated by the natural processes of oxidation and filtering. By the time the water reaches any freshwater supply, it should be safe for other uses.

Sewage absorption fields may fail for several reasons. The most common causes are failure to pump out the septic tank when it is full of solids, and poor soil drainage,

Figure 21.20 • Septic-tank sewage disposal system and location of the absorption field with respect to the house and well. (*Source:* Based on Indiana State Board of Health.)

which allows the effluent to rise to the surface in wet weather. When a septic-tank absorption field does fail, pollution of groundwater and surface water may result. Solutions to septic-tank system problems include siting septic tanks on well-drained soils, making sure systems are of adequate size, and practicing proper maintenance.

Wastewater Treatment Plants

In urban areas, wastewater treatment occurs at specially designed plants that accept municipal sewage from homes, businesses, and industrial sites. The raw sewage is delivered to the plant through a network of sewer pipes. Following treatment, the wastewater is discharged into the surface-water environment (river, lake, or ocean) or, in some limited cases, used for another purpose, such as crop irrigation. The main purpose of standard treatment plants is to break down and reduce the BOD and kill bacteria with chlorine. A simplified diagram of a wastewater treatment plant is shown in Figure 21.21.

Wastewater treatment methods are usually divided into three categories: **primary treatment, secondary treatment**, and **advanced treatment**. Primary and secondary treatment are required by federal law for all municipal plants in the United States. However, treatment plants may qualify for a waiver to be exempt from secondary treatment if the installment of secondary treatment facilities poses an excessive financial burden. Where secondary treatment is not sufficient to protect the quality of the surface water into which the treated water is discharged—for example, a river with endangered fish species that must be protected—advanced treatment may be required.[25]

Primary Treatment

Incoming raw sewage enters the plant from the municipal sewer line and is first passed through a series of screens, the purpose of which is to remove large floating organic material. The sewage next enters the grit chamber, where sand, small stones, and grit are removed and disposed of. The sewage then enters the primary sedimentation tank, where particulate matter settles out to form a sludge. Sometimes, chemicals are used to help the settling process. The sludge is removed and transported to the digester for further processing. Primary treatment removes approximately 30% to 40% of the BOD by volume from the wastewater, mainly in the form of suspended solids and organic matter.[25]

Secondary Treatment

There are several methods of secondary treatment. What is described here is known as *activated sludge*, the most common treatment. In this procedure, the wastewater from the primary sedimentation tank enters the aeration tank (Figure 21.21), where the wastewater is mixed

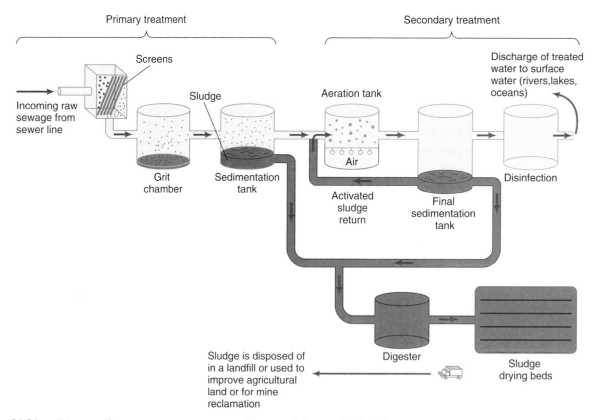

Figure 21.21 • Diagram of sewage treatment processes. The use of digesters is relatively new, and many older treatment plants do not have them.

with air (which is pumped in) and some of the sludge from the final sedimentation tank. The sludge contains aerobic bacteria that consume organic material (BOD) in the waste. The wastewater then enters the final sedimentation tank, where sludge settles out. Some of this activated sludge, which is rich in bacteria, is recycled and mixed again in the aeration tank with air and new, incoming wastewater, acting as a starter; and the bacteria are used again and again. Most of the sludge from the final sedimentation tank, however, is transported to the sludge digester. There, along with sludge from the primary sedimentation tank, it is treated by anaerobic bacteria, which further degrade the sludge by microbial digestion.

Methane gas (CH_4) is a product of the anaerobic digestion and may be used at the plant as a fuel to run equipment or heat and cool buildings. In some cases it is burned off. Wastewater from the final sedimentation tank is then disinfected, usually by chlorination, to eliminate disease-causing organisms. The treated wastewater is then discharged into a river, lake, or ocean or, in some limited cases, used to irrigate farmland. Secondary treatment removes about 90% of the BOD that entered the plant in the sewage.[25]

The sludge from the digester is dried and disposed of in a landfill or applied to improve soil. In some instances, treatment plants in urban and industrial areas contain many pollutants, such as heavy metals, that are not removed in the treatment process. Sludge from these plants is too polluted to use to improve the soil, and the sludge must be disposed of. Some communities, however, require industries to pre-treat sewage to remove heavy metals before the sewage is sent to the treatment plant; in these instances, the sludge can be more safely used for soil improvement.

Advanced Wastewater Treatment

Primary and secondary treatment do not remove all pollutants from incoming sewage. Some additional pollutants can be removed by adding more steps of treatment. For example, nutrients such as phosphates and nitrates; organic chemicals; and heavy metals can be removed by specifically designed treatments such as sand filters, carbon filters, and chemicals applied to assist in the removal process.[25] Treated water is then discharged into surface water or may be used for irrigation of agricultural lands or municipal properties, such as golf courses, city parks, and grounds surrounding wastewater treatment plants.

Advanced wastewater treatment is used when it is particularly important to maintain good water quality. For example, if a treatment plant discharges treated wastewater into a river and there is concern that nutrients remaining after secondary treatment may cause damage to the river ecosystem (eutrophication), advanced treatment may be used to reduce the nutrients.

Chlorine Treatment

As mentioned chlorine is frequently used to disinfect water as part of wastewater treatment. Chlorine treatment is very effective in killing the pathogens that historically caused outbreaks of serious waterborne diseases, which killed many thousands of people. However, a recently discovered potential is that chlorine treatment produces minute quantities of chemical byproducts, some of which have been identified as potentially hazardous to humans and other animals. For example, a recent study in Britain revealed that in some rivers, male fish sampled downstream from wastewater treatment plants had testes containing both eggs and sperm. This is likely related to the concentration of sewage effluent and the treatment method used.[26] Evidence also suggests that these byproducts in the water may pose risk of cancer and other human health effects. The degree of risks is controversial and is currently being debated.[27]

21.11 Land Application of Wastewater

The practice of applying wastewater to the land results from the fundamental belief that waste is simply a resource out of place. Land application of untreated human waste was practiced for hundreds if not thousands of years before the development of wastewater treatment plants, which have sanitized the process through the reduction of BOD and the use of chlorination.

The Wastewater Renovation and Conservation Cycle

The ideal land application system is sometimes called the **wastewater renovation and conservation cycle** and is shown schematically in Figure 21.22. The major steps in the cycle are the following:

1. Return of treated wastewater (following primary treatment) to crops via a sprinkler or other irrigation system.
2. *Renovation*, or natural purification by slow percolation of the wastewater into the soil, to eventually recharge the groundwater resource with clean water (an advanced form of treatment).
3. Reuse of the treated water, which is pumped out of the ground for municipal, industrial, institutional, or agricultural purposes.

Recycling of wastewater is now being practiced at many sites around the United States. In a large-scale wastewater recycling program near Muskegon and Whitehall, Michigan, raw sewage from homes and industry is transported by sewers to the treatment plant, where it receives primary and secondary treatment. The wastewater (over 32 million gallons per day) is then chlorinated and pumped into a network that transports the effluent to a series of spray irrigation rigs, which apply the treated water to about 2000 ha (5000 ac) of corn, soybeans, and

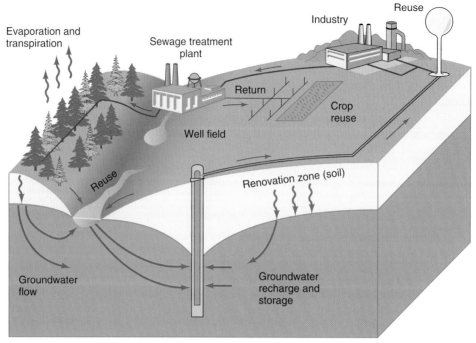

Figure 21.22 • The wastewater renovation and conservation cycle. (*Source:* R. R. Parizek, L. T. Kardos, W. E. Sopper, E. A. Myers, D. E. Davis, M. A. Farrel, and J. B. Nesbitt, "Pennsylvania State Studies: Waste Water Renovation and Conservation,"

University Studies 23. Copyright 1967 by the Pennsylvania State University. Reproduced by permission of the Pennsylvania State University Press.

alfalfa. After the wastewater trickles down through the soil, it is collected in a network of tile drains and transported to the Muskegon River for final disposal. This last step is an indirect advanced treatment that uses the natural physical and biological environment as a filter. This system removes most potential pollutants, meeting state water quality standards.

Technology for wastewater treatment is rapidly evolving. An important question being asked is: Can we develop environmentally preferred, economically viable wastewater treatment plants that are fundamentally different from those in use today? An idea for such a plant, called a resource recovery wastewater treatment plant, is shown in Figure 21.23. The term *resource recovery* here refers to the production of resources, including methane gas (which can be burned as a fuel) as well as ornamental plants and flowers that have commercial value.

The processes in the resource recovery treatment plant are as follows: First, the wastewater is run through filters to remove large objects. Second, the water undergoes anaerobic processing (this process produces the methane gas). Third, the nutrient-rich water flows over an incline surface containing plants (the plants use the nutrients and further purify the water). This process is thought to clean the water to the same standards obtained from secondary treatment in conventional wastewater treatment plants. If further purification is necessary, then the water may be processed by other living plants before being discharged into the environment.

Wastewater treatment that utilizes the resource recovery concept is in the experimental stage in small pilot plants. This technology must overcome several problems before it is likely to be used more widely. First, there has been a tremendous investment in traditional wastewater treatment plants, and engineers and other technicians are familiar with how to build and operate them. Second, economic incentives to provide for new technologies are not sufficient. Third (and perhaps most significant), there are not sufficient personnel trained to design and operate new types of wastewater treatment plants. This may be changing, however, because more universities are developing environmental engineering programs that take a broader view of technological development and applications.[28]

Wastewater and Wetlands

Wastewater is being applied successfully to natural and constructed wetlands at a variety of locations.[29,30,31] Natural or human constructed wetlands are potentially effective in treating the following water quality problems:

- Municipal wastewater from primary or secondary treatment plants (BOD, pathogens, phosphorus, nitrate, suspended solids, metals).
- Stormwater runoff (metals, nitrate, BOD, pesticides, oils).
- Industrial wastewater (metals, acids, oils, solvents).

Grasses, roses, trees

Hardy plants such as cattails

Outdoor nutrient-flow beds useful seasonally in cold climates

Secondary anaerobic bed

Primary anaerobic bed

Initial filtration for large objects

Discharged into environment

Bypass

Figure 21.23 • Components of a resource recovery wastewater treatment plant. For this model, two resources are recovered: methane, which can be burned to produce energy from the anaer-obic beds, and ornamental plants, which can be sold. Based on W. J. Jewell, "Resource-Recovery Wastewater Treatment," *American Scientist* [1994] 82:366–375.

- Agricultural wastewater and runoff (BOD, nitrate, pesticides, suspended solids).
- Mining waters (metals, acidic water, sulfates).
- Groundwater seeping from landfills (BOD, metals, oils, pesticides).

Treatment of wastewater through wetland systems is particularly attractive to communities that find it difficult to purchase traditional wastewater treatment plants. For example, the city of Arcata, in northern California, makes use of a wetland as part of its wastewater treatment system. The wastewater comes mostly from homes, with minor inputs from the numerous lumber and plywood plants in Arcata. It is treated by standard primary and secondary methods, then chlorinated and dechlorinated before being discharged into Humboldt Bay.[29]

Louisiana Coastal Wetlands

The state of Louisiana, with its abundant coastal wetlands, is a leader in the development of advanced treatment using wetlands following secondary treatment (Figure 21.24). Nitrogen- and phosphorus-rich wastewater, when applied to coastal wetlands, increases the production of wetland plants, thereby improving water quality as these nutrients are used by the plants. When the

plants die, their organic material (stems, leaves, roots) causes the wetland to grow vertically (or accrete), partially offsetting wetland loss due to sea level rise.[32] There are also significant economic savings in the application of treated wastewater to wetlands, because the financial investment needed is small compared with the cost of advanced treatment at conventional treatment plants. Over a 25-year period, a savings of about $40,000 per year is likely.[31,33] In conclusion, the utilization of isolated wetlands, such as those in coastal Louisiana, is a practical solution to improving water quality in small, widely dispersed communities in the coastal zone. As water quality standards are tightened, wetland wastewater treatment will become a viable, effective alternative that is cost effective compared with traditional treatment.[32,33]

Phoenix, Arizona: Constructed Wetlands

Wetlands can be constructed in arid regions to treat poor quality water. For example, at Avondale, Arizona, near Phoenix, a wetland treatment facility for agricultural wastewater is sited in a residential community (Figure 21.25). The facility is designed to eventually treat about 17,000 m^3/day (4.5 million gal/day) of water. The water entering the facility has nitrate (NO_3) concentrations as high as 20mg/l. The artificial wetlands contain naturally

(a)

(b)

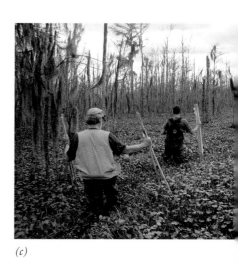

(c)

Figure 21.24 • (*a*) Wetland Pointe au Chene Swamp, three miles south of Thibodaux, Louisiana, that receives wastewater; (*b*) one of the outfall pipes delivering wastewater; and (*c*) ecologists doing field work at the Point au Chene Swamp to evaluate the wetland. (Courtesy of John Day, Louisiana State University).

Figure 21.25 • (*a*) Map of artificial wetlands project for treatment of agricultural wastewater at Avondale (near Phoenix), Arizona; and (*b*) photograph of the site with wetlands integrated with housing development (lower left). (*Source:* Integrated Water Resources, Inc., Santa Barbara, Calif.).

occurring bacteria that reduce the nitrate to below the maximum contaminant level of 10mg/l. Following treatment, the water flows by pipe to a recharge basin on the nearby Agua Fria River, where it seeps into the ground to become a groundwater resource. The wetland treatment facility cost about $11 million, about half the cost of a more traditional treatment facility.

21.12 Water Reuse

Water reuse can be inadvertent, indirect, or direct. *Inadvertent water reuse* results when water is withdrawn, treated, used, treated, and returned to the environment, followed by further withdrawals and use. Inadvertent water use is very common and a fact of life for millions of people who live along large rivers. Many sewage treatment plants are located along rivers and discharge treated water into the rivers. Downstream, other communities withdraw, treat, and consume the water.

Several risks are associated with inadvertent reuse:

1. Inadequate treatment facilities may deliver contaminated or poor-quality water to downstream users.
2. Because the fate of all disease-causing viruses during and after treatment is not completely known, the environmental health hazards of treated water remain uncertain.
3. Every year, new potentially hazardous chemicals are introduced into the environment. Harmful chemicals are often difficult to detect in the water; and if they are ingested in low concentrations over many years, their effects on humans may be difficult to evaluate.[29]

Indirect water reuse is a planned endeavor. An example is the wastewater renovation and conservation cycle previously discussed and illustrated in Figure 21.22. Similar plans have been used in many places in the southwestern United States, where several thousand cubic meters of treated wastewater per day have been applied to surface recharge areas. The treated water eventually enters groundwater storage to be reused for agricultural and municipal purposes.

Direct water reuse refers to use of treated wastewater that is piped directly from a treatment plant to the next user. In most cases, the water is used in industry, in agricultural activity, or for irrigation of golf courses, institutional grounds (such as university campuses), and parks. Direct water reuse is growing rapidly. Direct reuse of water by factories for industrial processes is the norm. In Las Vegas, Nevada, new resort hotels that use a great deal of water for fountains, rivers, canals, and lakes are required to treat wastewater and reuse it (Figure 21.26). Very little direct reuse of water is planned for human consumption (except in emergencies) because of perceived risks and negative cultural attitudes toward using treated wastewater.

Figure 21.26 • Water reuse at a Las Vegas, Nevada, resort hotel.

21.13 Water Pollution and Environmental Law

Environmental law, the branch of law dealing with conservation and use of natural resources and control of pollution, is very important as we debate environmental issues and make decisions about how best to protect our environment (see Chapter 30). In the United States, laws at the federal, state, and local levels address these issues.

Federal laws to protect water resources go back to the Refuse Act of 1899, which was enacted to protect navigable streams, rivers, and lakes from pollution. Table 21.4 lists major federal laws that have a strong water resource/pollution component. Each of these major pieces of legislation had a significant impact on water quality issues. Many federal laws have been passed with the purpose of cleaning up or treating pollution problems or treating wastewater. However, there has also been a focus on preventing pollutants from entering water. Prevention has the advantage of avoiding environmental damage and costly cleanup and treatment.

From the standpoint of water pollution, the mid-1990s in the United States was a time of debate and controversy. Congress in 1994 attempted to rewrite major environmental laws, including the Clean Water Act (1972, amended 1977). The purpose was to give industry greater flexibility in choosing how to comply with environmental regulations concerning water pollution. Industry interests favored proposed new regulations that, in their estimation, would be more cost effective without causing an increase in environmental degradation. Environmentalists, on the other hand, viewed the attempts to rewrite the Clean Water Act as a giant step backward in the nation's 20-year fight to clean up our

Table 21.4   Federal Water Legislation

Date	Law	Comment
1899	Refuse Act	Protects navigable water from pollution.
1956	Federal Water and Pollution Control Act	Enhances the quality of water resources and prevents, controls, and abates water pollution.
1958	Fish and Wildlife Coordination Act	Water resources projects such as dams, power plants, and flood control must coordinate with U.S. Fish and Wildlife Service to enact wildlife conservation measures.
1969	National Environmental Policy Act	Requires environmental impact statement prior to federal actions (development) that significantly affect the quality of the environment. Included are dams and reservoirs, channelization, power plants, bridges, and so on.
1970	Water Quality Improvement Act	Expands power of 1956 Act through control of oil pollution and hazardous pollutants and provides for research and development to eliminate pollution in Great Lakes and acid mine drainage.
1972 (amended in 1977)	Federal Water Pollution Control Act (Clean Water Act)	Purpose is to clean up nation's water. Provides billions of dollars in federal grants for sewage treatment plants. Encourages innovative technology, including alternative water treatment methods and aquifer recharge of wastewater.
1974	Federal Safe Drinking Water Act	Aims to provide all Americans with safe drinking water. Sets contaminant levels for dangerous substances and pathogens.
1980	Comprehensive Environmental Response, Compensation, and Liability Act	Established revolving fund (Superfund) to clean up hazardous-waste disposal sites, reducing groundwater pollution.
1984	Hazardous and Solid Waste Amendments to the Resource Conservation and Recovery Act	Regulates underground gasoline storage tanks. Reduces potential for gasoline to pollute groundwater.
1987	Water Quality Act	Established national policy to control nonpoint sources of water pollution. Important in development of state management plants to control nonpoint water pollution sources.

water resources. Apparently, Congress had incorrectly read the public's values on this issue. Survey after survey established that there is strong support for a clean environment in the United States and that people are willing to pay to have clean air and clean water. Congress has continued to debate changes in environmental laws, but little has been resolved.[34]

The president of the United States, in July 2000, imposed new water pollution regulations aimed at protecting thousands of streams and lakes from nonpoint sources to agricultural, industrial, and urban pollution sources. The regulations are to be administered by the Environmental Protection Agency, which will work with local communities and states to develop detailed plans with the objective of reducing pollution to streams, rivers, lakes, and estuaries that do not now meet the minimum standards of water quality. The new regulations acknowledged that nonpoint sources of water pollution are a serious problem that is difficult to regulate. The plan, which will take at least 15 years to implement completely, has been opposed for years by Congress as well as some agricultural groups, the utility industry, and even the U. S. Chamber of Commerce. The main objections are that the requirements would be costly (billions of dollars) and that local and state governments are better suited to implement their own water pollution regulations. These regulations mark a new phase in water pollution control measures in the United States.

Summary

- The primary water pollution problem in the world today is the lack of disease-free drinking water.
- Water pollution is degradation of quality that renders water unusable for its intended purpose.
- Major categories of water pollutants include disease-causing organisms, dead organic material, heavy metals, organic chemicals, acids, sediment, heat, and radioactivity.
- Sources of pollutants may be point sources, such as pipes that discharge into a body of water, or nonpoint sources such as runoff, which are diffused and intermittent.
- Eutrophication is a natural or human-induced increase

CRITICAL THINKING ISSUE

How Can Polluted Waters Be Restored?

The Illinois River begins in the northeast part of the state and flows west and south, draining parts of Indiana and Wisconsin

Figure 21.27 • Illinois River Watershed.

(see Figure 21.27). From Lake Michigan, which is connected to the river by a canal at Chicago, to the river's confluence with the Mississippi is a distance of 526 km (327 mi). The surrounding floodplains, once a mixture of prairie and oak-hickory forest, are now primarily used for raising crops. Formerly, the river was highly productive, especially in the lower 320 km (200 mi); it produced 10% of the U.S. freshwater fish catch in 1908 (11 million kg, or 24 million lb; 200 kg/ha, or 178 lb/acre). By the 1970s, the same stretch of river produced a mere 0.32% of the total freshwater fish harvest (4.5 kg/ha or 4 lb/acre). Two major factors are responsible for the change in the productivity of the Illinois River: diversion of Chicago's sewage from Lake Michigan to the river, and agriculture. A brief history of events related to water quality in the Illinois River is given in Table 21.5.

Critical Thinking Questions

1. Develop a hypothesis to explain why the fish population peaked in 1908, after the construction of the Chicago Sanitary and Ship Canal, and declined after that. Your hypothesis should also be able to explain the recovery of the fish in the 1920s and 1930s and the causes of the environmental problems of the 1940s and 1950s. Design a controlled experiment to test your hypothesis.

2. Why did water quality show some improvement by 1990, although the Tunnel and Reservoir Plan (TARP) was not yet completed? (*Hint:* See Table 21.5.)

3. The most important variables that affect the life of a river or stream are energy source (the amount of organic material entering the stream from sources outside it), water quality, habitat quality, water flow, and interactions among living things. In the case of the Illinois River, which variables are

Year	Critical Event	Environmental Impact
1854–1855	After heavy rains, untreated sewage from Chicago entered Lake Michigan and then the city's drinking water	Cholera and typhoid epidemic in Chicago
1900	Chicago Sanitary and Ship Canal built to convey sewage away from Lake Michigan and into Illinois River	Waste entered Illinois River; commercial fish yield from river reached peak in 1908; by 1920, fish populations in river had declined
1920–1940	Most cities along river built sewage treatment plants	Some recovery in fish population
1940–1960	Rapid population growth in Chicago and other cities along the river; increase in agricultural acreage	Lower oxygen levels in river; further declines in fish populations; sport fish and ducks declined in backwaters and lakes of the river
1977	Construction of Chicago Tunnel and Reservoir Plan (TARP) to capture and treat sewage overflows initiated	Some improvement in water quality by 1990, but no change in turbidity or total phosphorus; sodium increased

Table 21.5   Illinois River Water Quality History

affected by human activities? For each variable, cite examples of specific activities, their environmental effects, and what could be done to further improve water quality in the river.

4. There is a conflict between managing the Illinois River for waterfowl and managing it for fish. Why is this so? How could the conflict be resolved?

in the concentration in water of nutrients required for living things, such as phosphorous and nitrogen. A high concentration of such nutrients may cause a population explosion of photosynthetic bacteria. As the bacteria die and decay, the concentration of dissolved oxygen in the water is lowered, leading to the death of fish.

- Sediment pollution is a twofold problem: soil where sediment is lost through erosion, and water quality is reduced when sediment enters a body of water.

- Acid mine drainage is a serious water pollution problem that results when water and oxygen react with sulfide minerals, often associated with coal or metal sulfide deposits, forming sulfuric acid. Acidic water draining from mines or tailings pollutes streams and other bodies of water, damaging aquatic ecosystems and degrading water quality.

- Urban processes—for example, waste disposal in landfills, application of fertilizers, and dumping of chemicals such as motor oil and paint—can contribute to shallow-aquifer contamination. Overpumping of aquifers near the ocean may cause salt water, found below the fresh water, to rise closer to the surface, thereby contaminating the water resource by a process called saltwater intrusion.

- Wastewater treatment at conventional treatment plants includes primary, secondary, and perhaps advanced treatment. In some locations, natural ecosystems, such as wetlands and soils, are being used as part of the treatment process.

- Water reuse is the norm for millions of people living along rivers where series of sewage treatment plants discharge treated wastewater back into the river water. People who withdraw river water downstream are reusing some of the treated wastewater.

- Industrial reuse of water is the norm for many factories.

- Deliberate use of treated wastewater for irrigating agricultural lands, parks, golf courses, and the like is growing rapidly as demand for water increases.

- Cleanup and treatment of water pollution for both surface-water and groundwater resources is expensive and may not be completely successful. Furthermore, environmental damage may result before a pollution problem is identified and treated. Therefore, we should continue to focus on preventing pollutants from entering water, which is a goal of much water quality legislation.

● REEXAMINING THEMES AND ISSUES

Human Population: We state in this chapter that the number one water pollution problem in the world today is the lack of disease-free drinking water. This problem is likely to get worse in the future as the number of people, particularly in developing countries, continues to increase. As population increases, so does the possibility of continued water pollution from a variety of sources relating to agricultural, industrial, and urban activities.

Sustainability: Any human activity that leads to water pollution—such as the building of pig farms and their waste facilities on floodplains, as discussed in the opening case—is antithetical to sustainability. Groundwater resources are fairly easy to pollute, and once degraded, these waters may remain polluted for long periods of time. Therefore, if we wish to leave a fair share of groundwater resources to future generations, we must ensure that these resources are not polluted, degraded, or made unacceptable for use by people and other living organisms on Earth.

Global Perspective: Several aspects of water pollution have global implications. For example, some pollutants may enter the atmosphere and be transported long distances around the globe, where they may be deposited and degrade water quality.

Examples include radioactive fallout from nuclear reactor accidents or experimental detonation of nuclear devices. Waterborne pollutants from rivers and streams may enter the ocean and circulate with marine waters around the ocean basins of the world.

Urban World:

Urban areas are centers of activities that may result in serious water pollution. A broad range of chemicals and disease-causing organisms are present in large urban areas, and these may enter surface waters and groundwaters to pollute them. An example is bacterial contamination of coastal waters, resulting in beach closures. Many large cities have grown along the banks of streams and rivers, and the water quality of those streams and rivers is often degraded as a result. There are positive signs that some U.S. cities are viewing their rivers as valuable resources, with a focus on environmental and economic renewal. Thus, rivers flowing through some cities are designated as greenbelts, with parks and trail systems along river corridors. Examples include New York City; Cleveland, Ohio; San Antonio, Texas; Corvallis, Oregon; and Sacramento and Los Angeles, California.

Science and Values:

It is clear that the people of the United States place a high value on the environment and in particular on protection of critical resources such as water. Attempts to weaken water quality standards are viewed negatively by the public. There is also considerable concern for protection of water resources necessary for the variety of ecosystems found on Earth. This concern has led to research and development to find new technologies to reduce, control, and treat water pollution. Examples include development of new methods of wastewater treatment and support of laws and regulations that protect water resources.

● KEY TERMS

acid mine drainage **434**
advanced wastewater treatment **440**
biochemical oxygen demand (BOD) **424**
bioremediation **436**
cultural eutrophication **429**
ecosystem effect **428**

environmental law **445**
eutrophication **428**
fecal coliform bacteria **426**
nonpoint sources **434**
outbreaks **426**
point sources **434**
primary treatment **440**

secondary treatment **440**
wastewater renovation and
    conservation cycle **441**
wastewater treatment **438**
water reuse **445**

● STUDY QUESTIONS

1. Do you think that outbreaks of waterborne diseases will be more common or less common in the future? Why? Where are outbreaks most likely to occur?

2. What was learned from the *Exxon Valdez* oil spill that might help reduce the number of future spills and their environmental impact?

3. What is meant by the term *water pollution*, and what are several major processes that contribute to water pollution?

4. Compare and contrast point and nonpoint sources of water pollution. Which is easier to treat, and why?

5. What is the twofold effect of sediment pollution?

6. In the summer, you buy a house with a septic system that appears to function properly. In the winter, effluent discharges at the surface. What could be the envi-

ronmental cause of the problem? How could the problem be alleviated?

7. Describe the major steps in wastewater treatment (primary, secondary, advanced). Can natural ecosystems perform any of these functions? Which ones?

8. In a city along an ocean coast, rare waterbirds inhabit a pond that is part of a sewage treatment plant. How could this have happened? Is the water in the sewage pond polluted? Consider this question from the birds' and from your point of view.

9. How does water that drains from coal mines become contaminated with sulfuric acid? Why is this an important environmental problem?

10. What is eutrophication, and why is it an ecosystem effect?

11. How safe do you believe the drinking water is in your home? How did you reach your conclusion? Are you worried about low-level contamination by toxins in your water? What could the sources of contamination be?

12. Do you think our water supply is vulnerable to terrorist attacks? Why? Why not? How could potential threats be minimized?

● FURTHER READING

Borner, H., ed. 1994. *Pesticides in Ground and Surface Water.* Vol. 9 of *Chemistry of Plant Protection.* New York: Springer-Verlag. Essays on the fate and effects of pesticides in surface water and groundwater, including methods to minimize water pollution from pesticides.

Dinar, A., and E. T. Loehman, eds. 1995. *Water Quantity/Quality Management and Conflict Resolution.* Praeger. Includes essays and case studies on water pollution problems and resolutions.

Dunne, T., and L. B. Leopold. 1978. *Water and Environmental Planning.* San Francisco: W. H. Freeman. A great summary and detailed examination of water resources and problems.

Hester, R. E., and R. M. Harrison, eds. 1996. *Agricultural Chemicals and the Environment.* Cambridge: Royal Society of Chemistry, Information Services. A good source for information about the impact of agriculture on the environment, including eutrophication and the impact of chemicals on water quality.

Manahan, S. E. 1991. *Environmental Chemistry.* Chelsea, Mich.: Lewis. A detailed primer on the chemical processes pertinent to a broad array of environmental problems, including water pollution and treatment.

Newman, M. C. 1995. *Quantitative Methods in Aquatic Ecotoxicology.* Chelsea, Mich.: Lewis. Up-to-date text on fate, effects, and measurement of pollutants in aquatic ecosystems.

Nichols, C. 1989. "Trouble at the Waterworks," *The Progressive* 53:33–35. A concise report on the problem of tainted water supplies in the United States.

Rao, S. S., ed. 1993. *Particulate Matter and Aquatic Contaminants.* Chelsea, Mich.: Lewis. Coverage of the biological, microbiological, and ecotoxicological principles associated with interaction between suspended particulate matter and contaminants in aquatic environments.

22 The Atmosphere, Climate, and Global Warming

Polar bear moving through thin ice while hunting for seals in Hudson Bay. (Photo by Johnny Johnson.)

LEARNING OBJECTIVES

Earth's atmosphere is a dynamic system that is changing continuously while undergoing complex physical and chemical processes. After reading this chapter, you should understand:

- What are the basic composition and structure of the atmosphere.
- How the processes of atmospheric circulation, climate, and microclimate work.
- What are the four major processes that remove materials from the atmosphere.
- How the climate has changed during the last million years.
- What is the science behind human-induced global warming.
- How human activity has resulted in increased emissions of greenhouse gases.
- How positive- and negative-feedback cycles in the atmosphere might affect global temperature change.
- What effects global warming might have, and how we can adjust to those changes.

Global Warming and the Polar Bears of Hudson Bay

The mean surface temperature of Earth will likely increase from 1.5°C to 6.0°C by 1990–2100. Although scientific uncertainties remain, it is apparent that human-induced warming is now occurring. The questions have been reduced to how much, how fast, and where. Adverse effects of global warming will impact the goal of attaining sustainable development. The good news is that something can be done to minimize these effects if we act now.[1] The story of the plight of polar bears in the western Hudson Bay begins our exploration of climate change and global warming.

Polar bears are the largest carnivore in North America; they can reach 2.5 m in length and weigh over 700 kg (about as much as a three-quarter-ton truck). Polar bears are in trouble in western Hudson Bay, and the early breaking up of sea ice each spring is thought to be the problem. During the past 40 years, sea ice, on a global basis, has shrunk by about 5%, presumably in response to global warming. Some of the most significant impacts of global warming on wildlife are occurring in the Arctic, because temperature changes are more dramatic there than at lower latitudes.

Polar bears in the western Hudson Bay use sea ice as critical habitat for hunting seals. In the spring, polar bears prey on seals and especially on very young seals, providing the opportunity for the bears to fatten up before the annual melting of the sea ice. After the ice melts the bears move to land, where they go on a fast that may last for months. In particular, pregnant female bears fast for up to eight months and need a significant reserve of fat to carry, care for, and feed their cubs until they can again return to the ice to feed.

Since 1981, when studies on polar bears in the western Hudson Bay began, the bears there have weighed less than average and have given birth to fewer cubs. Biologists have established a link between the decline in polar bears and the earlier breakup of sea ice. If the trend continues, then, there will be a continued drop in the population of bears. Another consequence is that bears will be forced on land earlier, where dangerous contact with people is more likely.[2,3]

The situation in the eastern Hudson Bay at present is different, because sea ice is more permanent there, and some populations of polar bears live their entire lives on sea ice. These bears are less likely to be adversely affected by global warming. However, massive melting of sea ice would be a serious problem for all polar bears, including those of eastern Hudson Bay.[2]

• *Our story of polar bears in the Hudson Bay suggests that global change with warming can cause serious problems in the biosphere. This chapter addresses global issues related to weather, climate change, and global warming, with an emphasis on the role of humans in climate change.*

22.1 The Atmosphere

The **atmosphere** is the thin layer of gases that envelops Earth. We begin the chapter by examining basic features of atmospheric composition, structure, and processes.

Composition of the Atmosphere

The atmosphere is composed of gas molecules held close to Earth's surface by a balance between gravitation and thermal movement of air molecules (90% of the weight of the atmosphere is in the first 12 km above the surface of the Earth). Major gases in the atmosphere are nitrogen (78%), oxygen (21%), argon (0.9%), and carbon dioxide (0.03%). The atmosphere also contains trace amounts of numerous elements and compounds, including methane, ozone, hydrogen sulfide, carbon monoxide, oxides of nitrogen and sulfur, hydrocarbons, chlorofluorocarbons (CFCs), and various particulates or aerosols (small particles). Water vapor is also present in the lower few kilometers of atmosphere.

The atmosphere is a dynamic system, changing continuously. Physical movement of air masses, each with a different temperature, pressure, moisture, and aerosol content, produces weather and climate. A vast, chemically active system, the atmosphere is fueled by sunlight, high-energy compounds (for example, oxygen, methane, and carbon dioxide) emitted by living things, and human industrial and agricultural activities. Many complex chemical reactions take place in the atmosphere, changing from day to night and with the chemical elements available.

The main chemical components in Earth's atmosphere either are produced primarily by biological activity or are greatly affected by it. Although the atmosphere has been greatly modulated by life during the last 3.5 billion years, most of these alterations have been natural—that is, they have produced an atmosphere whose makeup is relatively constant and essential for our survival.

Ever since life began on Earth, the atmosphere has been an important resource for chemical elements and a medium for the deposition of wastes. The earliest bacteria that carried out photosynthesis released oxygen as a waste product into the atmosphere. The long-term increase in atmospheric oxygen, in turn, made possible the development and survival of life forms that required high rates of metabolism and rapid use of energy. For our biological ancestors and for ourselves, oxygen became a necessary resource for respiration, the cellular process by which we burn our internal biological fuels and provide the energy to sustain our life processes.

Structure of the Atmosphere

The atmosphere is made up of several layers. The structure of the atmosphere and the relationship between altitude and air temperature for the lower atmosphere is shown as an idealized diagram in Figure 22.1.

The lower part of the atmosphere (lower 10 to 12 km, or about 6 to 8 mi) is known as the *troposphere*, and it is here that weather occurs. In the troposphere, the temperature of the atmosphere decreases systematically with elevation (from about 17°C at the surface to –60°C at 12 km above Earth's surface, the top of the troposphere) at a global average rate of decrease of approximately 6.5°C/km (7.3°F/mi). At the top of the troposphere, the *tropopause* (about 12 to 20 km above sea level), with a constant temperature of about –60°C, produces a lid, or cold trap, on the troposphere. The cold trap causes condensation of atmospheric water vapor. Therefore, there is very little water vapor in the *stratosphere*, which lies above the troposphere and where the atmosphere warms with increasing altitude. Condensation of water in the troposphere produces clouds. The role of clouds, including how they develop and move, is an important area of research to better understand the global processes that operate in the atmosphere.

Figure 22.1 also shows the *stratospheric ozone layer*, which extends from the tropopause to an elevation of approximately 40 km (25 mi), with a maximum concentration of ozone above the equator at about 25 to 30 km (16 to 19 mi). Stratospheric ozone (O_3) protects life in the lower atmosphere from receiving harmful doses of ultraviolet radiation and is discussed in detail in Chapter 25.

Atmospheric Processes: Temperature, Pressure, and Global Zones of High and Low Pressure

Two important measurable quantities in the atmosphere are pressure and temperature. *Pressure* is force per unit area. Atmospheric pressure is the weight of overlying atmosphere (air) per unit area; it decreases as altitude increases because

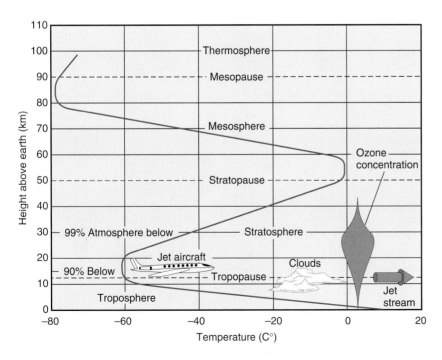

Figure 22.1 • The structure of the atmosphere showing temperature profile and ozone layer of the atmosphere to an altitude of 110 km. Note that 99% of the atmosphere (by weight) is below 30 km, the ozone layer is thickest at about 25–30 km, and the weather occurs below about 11 km—about the elevation of the jet stream. (*Source:* A. C. Duxbury and A. B. Duxbury. *An Introduction to the World's Oceans,* © 1997. Wm. C. Brown Publishers, 5th ed.).

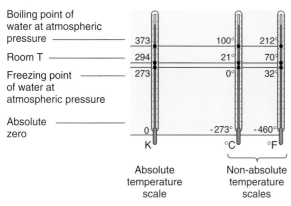

Boiling point of water at atmospheric pressure	373	100°	212°
Room T	294	21°	70°
Freezing point of water at atmospheric pressure	273	0°	32°
Absolute zero	0	-273°	-460°
	K	°C	°F

Absolute temperature scale — Non-absolute temperature scales

Figure 22.2 • Temperature scales. °C = Celsius; °F = Fahrenheit; K = kelvin. °C = 5/9 (°F −32); °C = K −273.

there is less weight from overlying air. At sea level, atmospheric pressure is 10^5 N/m² (newtons per square meter), which is equivalent to 14.7 lb/in.² *Temperature* refers to relative hotness or coldness of materials, such as air, water, soil, and living organisms, and is measured with a thermometer. In a quantitative sense, temperature is a measure of thermal energy, which is the kinetic energy of the motion of atoms and molecules in a substance. Figure 22.2 shows the three common temperature scales. This book uses the Kelvin (K) and Celsius (°C) scales.

In addition to pressure and temperature, air is characterized by its water vapor content. In the lower atmosphere, water vapor content varies from approximately 1% to 4% by volume. The amount of water vapor present in the atmosphere at a particular location depends on many

factors, including air temperature, air pressure, and availability of water vapor from processes such as evaporation from water bodies and soil and transpiration from vegetation (loss of water from plants to the air). When air holds the maximum amount of water that it can, given its particular temperature, it is said to be *saturated*, which means that no more water vapor can be added to the air. The term *relative humidity* is a measure of how close the air is to saturation. For example, a relative humidity of 100% means that the air is completely saturated; a relative humidity of 5% means that the air is very dry. At high relative humidity, people and other animals may feel uncomfortable and perceive the air as sticky. In desert regions with low relative humidity, the air is dry; even with a warm temperature, the air to many people feels more comfortable than air with a lower temperature but a higher relative humidity.[4]

Winds, the movement of clouds, and transitions from stormy to clear skies show us that the atmosphere changes rapidly and continuously. Heat energy is transferred to the air from the land and from bodies of water by the evaporation of water. As warm air rises, it cools by expansion; winds are produced as cooler surface air is drawn in to replace rising warm air. The lower atmosphere is said to be unstable because it tends to circulate and mix, particularly in the lowest 4 km (2.5 mi) or so.

On a global scale, atmospheric circulation results primarily from Earth's rotation and the differential heating of Earth's surface and atmosphere. These processes produce global patterns that include prevailing winds and latitudinal belts of low and high air pressure from the equator to the poles (Figure 22.3). In general, belts of low air

Figure 22.3 • Generalized circulation of the atmosphere. (*Source:* Williamson, *Fundamentals of Air Pollution,* © 1973. Figure 5.5. Reprinted with permission of Addison-Wesley, Reading, Mass.)

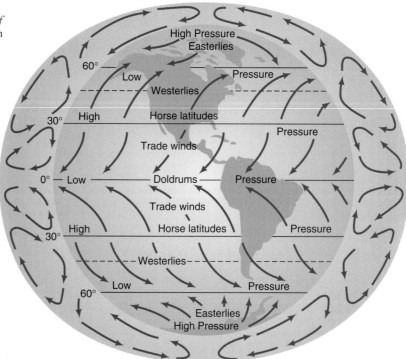

pressure develop at the equator and at 50° to 60° north and south latitude as a result of rising columns of air, producing precipitation. Belts of high pressure resulting from descending air develop at 25° to 30° north and south latitude, producing arid conditions.

The latitudinal belts have names, such as the "doldrums," regions at the equator with little air movement; "trade winds," northeast and southeast winds important in the early days of international trade, when clipper sailing ships moved the world's goods; and "horse latitudes," two belts centered about 30° north and south of the equator with descending air and high pressure. Earth's major deserts occur in the horse latitudes as a result of pervasive high pressure and low precipitation sandwiched between the equatorial and midlatitudinal zones of low pressure with higher precipitation.

Processes that Remove Materials from the Atmosphere

Understanding processes that remove materials from the atmosphere is important in solving atmospheric pollution problems. Four processes are responsible for removing human-induced particles and chemicals from the atmosphere:

- *Sedimentation.* Particles heavier than air settle out as a result of gravitational attraction to Earth. For example, particulates from volcanic eruptions or burning coal will settle out over time as a dry deposition.

- *Rain Out.* Precipitation (rain, ice, or snow) can physically and chemically flush material from the atmosphere. For example, rain drops form by condensation of water on small particles in the atmosphere, bringing the particles to Earth with the raindrops. Carbon dioxide combines with water in the atmosphere to form weak carbonic acid by the following chemical reaction:

$$CO_2 + H_2O \rightarrow H_2CO_3$$

This process effectively removes some carbon dioxide from the atmosphere and explains why natural rainfall is slightly acidic (see Chapter 23).

- *Oxidation.* Oxidation is a reaction in which oxygen is chemically combined with another substance. For example, sulfur dioxide in the atmosphere oxidizes easily to sulfur trioxide (SO_3), which may dissolve in water, forming sulfuric acid and producing acid rain (see Chapter 23).

- *Photodissociation.* Solar radiation (light) can break down chemical bonds in a chemical process known as photodissociation. For example, ozone (O_3) in the atmosphere may break down to O_2 as a result of photodissociation.

With our basic discussion of the composition of the atmosphere and some of the important atmospheric processes behind us, we next discuss climate and climatic change.

22.2 Climate

Climate refers to the representative or characteristic atmospheric conditions for a region on Earth. The term *climate* refers to these conditions over long time periods, such as seasons, years, or decades, whereas the term *weather conditions* refers to shorter periods of time, such as hours, days, or weeks. Climate depends in part on precipitation and temperature, both of which show tremendous variability on a global scale. Because the climate of a particular location may depend on extreme or infrequent conditions, however, climate is more than just the average temperature and precipitation of a region.

The simplest classification of climate is by latitude—tropical, subtropical, midlatitudinal (continental), subarctic (continental), and arctic. Several other categories are used as well, including humid continental, Mediterranean, monsoon, desert, and tropical wet–dry (Figure 22.4). Although detailed discussion of climatic types is beyond the scope of this book, it is important to recognize the significance of potential climatic variability in determining what kinds of organisms live where. Recall from the discussion of biogeography in Chapter 7 that similar climates produce similar kinds of ecosystems. This concept is important and useful for environmental science. Knowing the climate, we can predict a great deal about what kinds of life we will find in an area and what kinds could survive there if introduced.

On a regional scale, air masses that cross oceans and continents can have a profound influence on seasonal patterns of precipitation and temperature. On a local scale, climatic conditions can also vary considerably and produce a local effect referred to as a **microclimate**. Microclimate can vary even from one side of a small rock to another or from one side of a tree to another. Organisms often take advantage of these different conditions.

Urban areas produce a characteristic microclimate with important environmental consequences. The very presence of a city affects the local climate, and as the city changes, so does its climate. For example, in the middle of the 18th century, according to the writer Peter Kam, who was traveling in North America, Manhattan Island was "generally reckoned very healthy," perhaps because of its nearness to the ocean and its relatively unobstructed ocean breezes. Today, air pollution and the effects of tall buildings on air flow lead the average visitor to Manhattan to a quite different conclusion.

Although air quality in urban areas is in part a function of the amount of pollutants present or produced, it is affected also by the city's ability to ventilate and flush out pollutants. The amount of ventilation depends on several aspects of the urban microclimate. In terms of climate, cities in midlatitude regions tend to differ from surrounding rural areas in several ways:

1. They are warmer (0.5–2°C) because of heat from burning fuel for transportation, industry, and homes.

Figure 22.4 • The climates of the world and some of the major climate types in terms of characteristic precipitation and temperature conditions. (*Source:* Modified from W. M. Marsh and J. Dozer, *Landscape,* © 1981, John Wiley & Sons, reprinted with permission of John Wiley & Sons, Inc.)

2. They are less humid (6–10%) because of less evaporation of water from standing water and soil.

3. They have up to 10 times as many dust particles, due to pollution, forming the **urban dust dome**.

4. They have 30–100% more fog and 5–10% more precipitation because of the presence of small particles in the city air.

The abundance of particulates above a city reduces incoming solar radiation by as much as 30% and thus cools the city, but the cooling effects of particles are relatively small compared with the processes that heat the city.[6] The combination of lingering air and abundant particles and other pollutants produces a heat island effect (see Chapter 26). The effect causes a circulation pattern in which cooler, often cleaner air moves from rural or suburban areas toward the inner city, where heat rises and then flows out to rural areas, completing the pattern.

Climatic Change

Another important aspect of climate is **climatic change**. The mean annual temperature of the Earth has swung up and down by several degrees Celsius over the past million years (Figure 22.5). Times of high temperature reflect relatively ice-free periods (interglacial periods) over much of the planet; times of low temperature reflect the glacial events (Figure 22.5*a, b*). It is not yet clear whether our current warm climate marks the end of the ice ages or is merely an interglacial period with another glacial age due.

Global climate also changes in shorter time frames. For example, continental glaciation ended about 12,500 years ago with very rapid warming, perhaps over a period as brief as a few decades.[7] This was followed by a short global cooling about 11,500 years ago (Figure 22.5*c*). Climatic change over the last 18,000 years reflects several warming and cooling trends that have greatly affected people. For

(a)

(b)

(c)

Figure 22.5 • Changes in Earth's temperature over varying time periods during the past million years. (*Sources:* UCAR/DIES, "Science Capsule, Changes in the Temperature of the Earth," *Earth Quest* 5, no. 1 [Spring 1991]; Houghton, J. T., G. L. Jenkins, and J. J. Ephranns, eds. *Climate Change, the Science of Climate Change* [Cambridge: Cambridge University Press, 1996]; U.K. Meteorological Office, 1997, *Climate Change and Its Impacts: A Global Perspective,* 1997).

(d)

(e)

example, during a major warming trend from 1100 to 1300 (medieval warm period, Figure 22.5*c,d*), the Vikings colonized Iceland, Greenland, and North America. When glaciers advanced during the cold period starting around 1400 (Little Ice Age), the Viking settlements in North America and parts of Greenland were abandoned.

Starting in approximately 1850, a warming trend became apparent. It lasted until the 1940s, when temperatures began again to cool. Figure 22.5*e* shows change over the last 140 years. On this scale, the 1940s warming event is clearer. As you can see, it is followed by a leveling off of temperature in the 1950s and then a further drop during the 1960s. After that time, temperature increased steadily through the 1990s. What is evident from the record of the last 100 years is that global mean annual temperature has increased by approximately 0.6°

C. Furthermore, the last two decades have been the warmest in the 140 years that global temperatures have been monitored (Figure 22.5*e*). This period includes the warmest years of the 20th century.[8,9] This very recent increase in warming is known as the "late-20th century increase in global temperature."[10]

The question that begs to be asked is: Why does climate change occur? Examination of Figure 22.5a suggests there are cycles of change about 100,000 years long separated by shorter cycles of 20,000 to 40,000 years in duration. These cycles were first identified by Milutin Milankovitch in the 1920s as a hypothesis to explain climate change. Milankovitch realized that the spinning Earth is like a wobbling top unable to keep a constant position in relation to the sun, and that position determines (in part) the amount of sunlight reaching and warming Earth. He discovered that variations in Earth's orbit around the sun follow a cycle of approximately 100,000 years, which correlates with the major glacial and interglacial periods of Figure 22.5a. Cycles of approximately 40,000 and 20,000 years are the result of changes in the tilt and wobble of the Earth's axis. Milankovitch cycles are consistent with most of the long-term cycles we see in the climate. However, the cycles are not sufficient, by themselves, to produce the large-scale climatic variations in the geologic record. Therefore, the Milankovitch cycles can be looked at as natural mechanisms that, along with other processes, may produce climatic change.

Shorter cycles have also been suggested. In fact, one study suggests that during the past 4,000 years there has been a cycle of about 1,500 years that may help explain the medieval warm period followed by the Little Ice Age, as well as the present warming trend, which is predicted to continue naturally until approximately 2400 a.d.[11] If this is correct, then any warming caused by human activity would be superimposed on a system that is already slowly warming.

Our climate system may be inherently unstable and capable of changing quickly from one state (cold) to another (hot) in as short a time as a few decades.[12] Part of what may drive the climate system and its changes is the "ocean conveyor belt"—a global circulation of ocean waters characterized by strong northward movement of upper warm waters of the Gulf Stream in the Atlantic Ocean. These waters are approximately 12–13°C when they arrive near Greenland and are cooled in the North Atlantic to a temperature of 2–4°C (Figure 22.6).[12] As the water is cooled, it becomes more salty and so increases in density, causing it to sink to the bottom. The cold, deep current flows southward then eastward and finally northward in the Pacific Ocean. Upwelling in the north Pacific starts the warm shallow current again. The flow in this conveyor belt current is huge (20 million m³/sec), about equal to 100 Amazon rivers. The amount of warm water and heat released at the surface is sufficient to keep northern Europe 5–10°C warmer than it would be if the conveyor belt were not present. If the system were to shut down, global cooling could result, and northern Europe would become much less habitable. If this were to happen in the future when there are a few more billion people to feed, a global catastrophe might result![12]

Some scientific uncertainties remain regarding the human role in the observed warming of Earth's climate. However, according to a 2002 report from the Intergovernmental Panel on Climate Change (IPCC), the bulk of the scientific evidence suggests that the warming is due in part to human activities. As a consequence of projected emissions of carbon dioxide (CO_2), mean surface temperatures are expected to increase by 1.5–6.0°C by 2100.[1] The science of global warming is discussed in detail following consideration of the tools used to study global change.

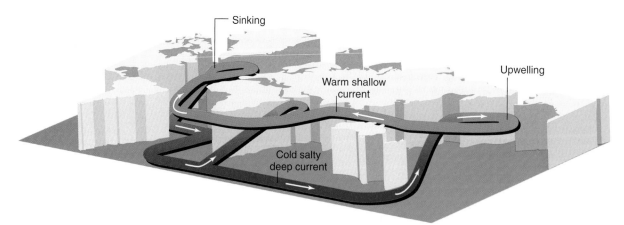

Figure 22.6 • Idealized diagram of the oceanic conveyor belt. The actual system is more complex; but in general, warm surface water (red) is transported westward and northward (increasing in salinity because of evaporation) to near Greenland, where it cools from contact with cold Canadian air. As the water increases in density, it sinks to the bottom and flows south, then east to the Pacific, then north where upwelling occurs in the north Pacific.

The masses of sinking and upwelling waters balance, and the total flow rate is about 20 million m³/sec. The heat released to the atmosphere from the warm water keeps northern Europe 5°C to 10°C warmer than if the oceanic conveyor belt were not present. (*Source*: Modified from W. Broker, "Will Our Ride into the Greenhouse Future Be a Smooth One?" *Geology Today* 7 no. 5 [1997]: 2–6.)

22.3 Earth System Science and Global Change

Until very recently, it was generally thought that human activity could only cause local or, at most, regional environmental changes. We now know otherwise! The main goal of the emerging science known as **Earth system science** is to obtain a fundamental understanding of how our planet works as a system. From a pragmatic point of view, the research priorities of Earth system science and global change can be summarized as follows:[13]

- Establishment of worldwide measurement stations to better understand physical, hydrologic, chemical, and biological processes that are significant in the evolution of Earth on a variety of time scales.
- Documentation of global changes, especially those that occur in a time period of several decades, that are of particular interest to the human environment.
- Development of quantitative models useful in the prediction of future global change.
- Provide information needed by decision makers at regional and global levels.

The major tools for studying global change are:

- Evaluation of the geologic record
- Monitoring
- Mathematical models

Geologic Record

The sediments deposited on floodplains, in bogs, and on lake and ocean bottoms can be read just as the pages of a history book can. Organic material, such as skeletal material, shells, pollen, and bits of wood, leaves, and other plant parts, is often deposited with sediments and can yield valuable information concerning Earth's history. In addition, organic material can be dated to provide the necessary chronology to establish past changes. Finally, both sediments and organic material can be used to evaluate the past climate—what lived where, what kinds of changes have occurred, and how extensive the changes were.

One interesting use of the geologic record has been the examination of glacial ice. The process of transformation of snow to glacial ice involves recrystallization and an increase in the density of the ice. The process also traps air bubbles, which can be analyzed to provide information concerning the concentration of carbon dioxide in the atmosphere at the time the ice formed. Thus, glacial ice can be thought of as a time capsule that stores information about the atmosphere in the past. To study the ice, scientists extract long cores of glacial ice (Figure 22.7) and carefully sample trapped air bubbles. This method has been used to analyze the carbon dioxide con-

Figure 22.7 • Scientists collecting a glacial ice core.

tent of the atmosphere up to 160,000 years ago. Figure 22.8 shows the record from 1500 to 2000.[14] Ice cores have also yielded data on variability of solar radiation through measurement of accumulation of cosmogenic isotopes in the ice.[9]

Real-Time Monitoring

Monitoring can be defined as the regular collection of data for specific purposes. For example, we monitor rainfall and the flow of water in rivers to evaluate water resources or flood hazards. We collect the data to provide baseline conditions from which to evaluate changes in the future. Similarly, samples of atmospheric gases, particu-

Figure 22.8 • Average concentration of atmospheric carbon dioxide, 1500–2000. (*Source:* Modified from W. M. Post, T. Peng, W. R. Emanuel, A. W. King, V. H. Dale, and D. L. De Angelis, *American Scientist* 78, no. 4 [1990]: 310–326. By permission of *American Scientist*, Journal of Sigma Xi, The Scientific Research Society.)

Monitoring of Atmospheric Carbon Dioxide Concentrations

The activities of humans and other living things affect characteristics of Earth's surface, waters, and atmosphere, even in areas removed from such activities. For example, air pollutants and other artifacts of human society are found in glacial ice, and pesticides are found in lakes far from agriculture.

Air samples taken near the summit of Mauna Loa, Hawaii, have helped show another dimension of how life and human activity affect the atmosphere. Mauna Loa is one of the largest active volcanoes and the highest mountain in the world, based on elevation change from base to top. Air samples are taken there because it is located far away from direct effects of human life and other biological activity. One substance of interest in the Mauna Loa samples is carbon dioxide. Carbon dioxide is taken up by green plants during photosynthesis and released in respiration from all oxygen-breathing organisms; thus, a measure of carbon dioxide in the atmosphere is analogous to a measure of the breathing in and out of all life on Earth.

The Mauna Loa measurements of carbon dioxide are truly remarkable. Two important aspects of the record are shown in Figure 22.9: a strong upward trend over the years and an annual cycle, which is obvious and regular. Carbon dioxide concentration reaches a peak in winter and a trough in summer. The annual curve is a measure of life activities in the entire Northern Hemisphere. In summer, green plants are most active, and the total amount of photosynthesis exceeds the total amount of respiration. As a result, carbon dioxide is removed from the atmosphere in summer. In winter, photosynthesis decreases and becomes less than total respiration, so the carbon dioxide concentration in the atmosphere increases.

Similar observations have been made for the Southern Hemisphere in Antarctica. Because Antarctica is less accessible, measurements are more sporadic. Nevertheless, the same trends are observed: a strong upward trend in concentration of carbon dioxide in the atmosphere and an annual cycle. The annual cycle from Antarctica is smaller in amplitude than the Mauna Loa cycle because of the relatively smaller land area in the Southern Hemisphere and the smaller amount of vegetation.

The upward trend, or increase, in carbon dioxide concentration in the atmosphere shown in data from both Mauna Loa and Antarctica is believed to be due to the addition of carbon dioxide from burning of fossil fuels and other human activities, such as cutting of forests and burning of wood (burning releases the stored carbon in wood, which combines with oxygen, producing carbon dioxide). The Mauna Loa data provided some of the first evidence that directly indicated that life touches the entire Earth and that human activities have begun to affect the atmosphere of our planet.

A third, less obvious global trend is shown in Figure 22.9—an apparent increase in the amplitude of the annual cycle from 1960 to present. This evidently reflects the increase in photosynthetic biomass as a result of the increase in total atmospheric carbon dioxide.

The Mauna Loa data clearly demonstrate the benefits of long-term collection of information. Funding for long-term projects is often difficult to maintain because funding agencies may prefer to sponsor new projects rather than long-term monitoring. Nevertheless, understanding global change depends on collection and maintenance of supportive data. To that end, the Mauna Loa CO_2 measurement project is unique, and its effectiveness is a tribute to the people who initiated it and nurtured it for so many years.

Figure 22.9 ● Monthly average carbon dioxide concentration and long-term trend. (*Source:* Mauna Loa Observatory, http://mloserv.mlo.hawaii.gov/mloinfo/programs/gases/CO2graph.htm, accessed October 14, 1998).

lates, and chemicals are helping establish trends in the composition of the atmosphere (see A Closer Look 22.1). Finally, measurements of the temperature, composition, and chemistry of ocean waters can be used to help evaluate changes in the marine environment.

Mathematical Models

Mathematical models are attempts to represent, through numerical means, real-world phenomena, linkages, and interactions among physical, chemical, and biological processes. The models use equations to describe the phenomena and linkages being considered. Models have been developed to predict the flow of surface water and groundwater, ocean circulation, and atmospheric circulation. In the area of global change, the models that have gained the most attention are the **global circulation models (GCMs)**. The objective of GCMs is to predict atmospheric changes on a global scale.[15]

The variables utilized in GCMs include temperature, relative humidity, and wind conditions. Values for many of these variables are estimated for the past based on data derived from such sources as tree-ring records. Annual growth rings of trees can be evaluated to provide a time scale (called *dendrochronology*). Growth rings can also be helpful in inferring climatic information, such as precipitation and runoff, which are useful in calibrating results from mathematical models. To organize the data necessary for the calculations in GCMs, the surface of Earth is divided into large cells measuring several degrees of latitude and longitude. The typical cell is about the size of Oregon or of Indiana and Ohio together. Several layers of data are necessary. Most GCMs use 6 to 20 levels of vertical data collected throughout the lower atmosphere (Figure 22.10). Mathematical relationships are used in the calculations to predict future atmospheric circulation. The GCMs are complex and require supercomputers for their operation.

Results from these models are relatively crude and so may not accurately predict future conditions.[15,16] Furthermore, it is difficult to estimate interactions of other factors, such as cloud cover, which may significantly affect atmospheric energy relations. As a result, GCMs can only be considered a first approach to solving complex atmospheric problems. In spite of their limitations, mathematical models are providing information necessary for evaluating global change and Earth as a system. The GCMs are also helping to pinpoint what additional data are necessary to better predict future change. The models do predict, in a relative sense, areas or regions that are likely to be wetter or drier if certain changes in the atmosphere occur. Thus, predictions of future global change from mathematical models are being taken seriously, and their importance as a tool for studying global change will continue to increase.

Our discussion of the tools used to study global change leads us to consider human-induced climate change. The major question is: How much has human activity contributed to global warming?

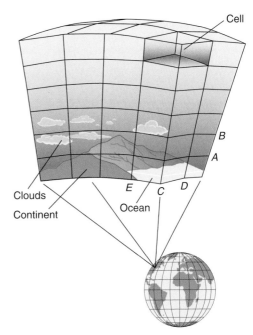

Figure 22.10 • Idealized diagram showing how cells are arranged for a global circulation model. *A–B* is a few kilometers in elevation; *E–C* is a few hundred kilometers in longitude; *C–D* is a few hundred kilometers in latitude.

22.4 Global Warming: Earth's Energy Balance and the Greenhouse Effect

Global warming is defined as a natural or human-induced increase in the average global temperature of the atmosphere near the Earth's surface. The temperature at or near the surface of the Earth is determined by four main factors:[15]

- The amount of sunlight Earth receives.
- The amount of sunlight Earth reflects.
- Retention of heat by the atmosphere.
- Evaporation and condensation of water vapor.

Electromagnetic Radiation and Earth's Energy Balance

In order to better understand global warming, it is necessary to have a modest acquaintance with electromagnetic radiation and Earth's energy balance. Our Earth is part of a planetary system receiving energy from the sun. This energy undergoes changes, affects life, oceans, atmosphere, and climate; and is eventually emitted as heat back into the depths of space. In this system, Earth is an intermediate between the *source* (the sun) and the *sink* (space).

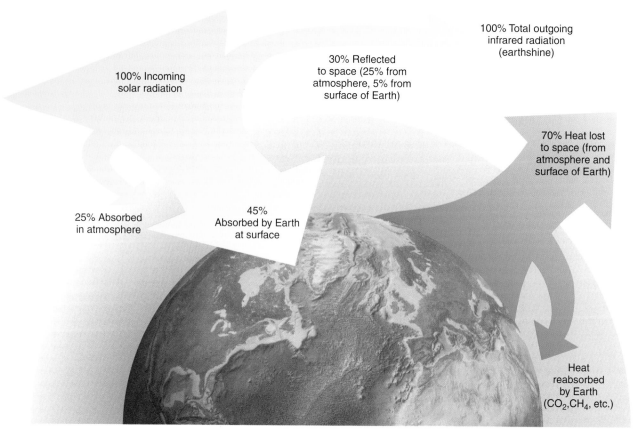

100% Incoming solar radiation

30% Reflected to space (25% from atmosphere, 5% from surface of Earth)

100% Total outgoing infrared radiation (earthshine)

70% Heat lost to space (from atmosphere and surface of Earth)

25% Absorbed in atmosphere

45% Absorbed by Earth at surface

Heat reabsorbed by Earth (CO_2, CH_4, etc.)

Figure 22.11 • Idealized diagram showing Earth's energy balance. Incoming solar radiation amounts to approximately 5.5 million exajoules, and that is balanced by a similar amount leaving Earth. A very small component of heat (1,000 exajoules) entering the system is generated within the Earth; it represents only about 0.002% of the budget and so is neglected in the values presented here. Approximately 30% of the total incoming solar radiation (1.7 million exajoules) is reflected immediately in the atmosphere, mostly by clouds and at the surface of the Earth. Of the remaining 70% of incoming solar radiation, approximately 25% is absorbed in the upper atmosphere, and 45% reaches the Earth and is absorbed at the surface. As the energy from the sun is absorbed by plants, soils, rocks, water, and other surface materials, it is transformed into heat, radiates as infrared radiation, and is eventually lost to space from Earth's upper atmosphere. Part of the infrared radiation thus released is reabsorbed by compounds, including carbon dioxide and methane, known as greenhouse gases (discussed later in the chapter). This energy too is ultimately lost to space as heat, completing the energy balance of Earth. About two-thirds of the incoming solar radiation is available to drive processes at or near the surface of the Earth, including the hydrologic cycle, wind, and so forth. (*Source:* Modified from N. L. Pruitt, L. S. Underwood, and W. Surver. *Bio Inquiry* [New York: John Wiley & Sons, 2002].)

Nearly all the energy available at Earth's surface comes from the sun (Figure 22.11), with small additional amounts coming from human activities, geothermal energy (from the interior of Earth), and tides. Because energy is conserved, the amount emitted to space matches that received from the sun plus the other small contributions. The matching of input of energy from the sun with output from the Earth defines Earth's energy balance. Although Earth intercepts only a very tiny fraction of the total energy emitted by the sun, solar energy sustains life on Earth and greatly influences climate and weather.

Energy is emitted from the sun in the form of electromagnetic radiation (EMR). Different forms of electromagnetic energy can be distinguished by their wavelengths. (See the special feature in Appendix A on EMR for a discussion of wavelength and two important laws concerning EMR.) The collection of all possible wavelengths of electromagnetic energy, considered a continuous range, is known as the **electromagnetic spectrum** (Figure 22.12). The electromagnetic spectrum is one of the most important phenomena in the physical sciences and is fundamental to understanding many environmental topics. Gamma rays, X rays, ultraviolet light, visible light, infrared radiation, television waves, radio waves, and radar are all different types of electromagnetic radiation. The relatively long wavelengths (greater than 1 m in the electromagnetic spectrum) include radio waves, and the shortest wavelengths are those of gamma rays and X rays. The electromagnetic radiation to which our eyes are sensitive, *visible electromagnetic radiation*, is a very small fraction of the total spectrum, as you can see in Figure 22.12. Other types of electromagnetic radiation with environmental significance include radar, microwaves, and infrared radiation.

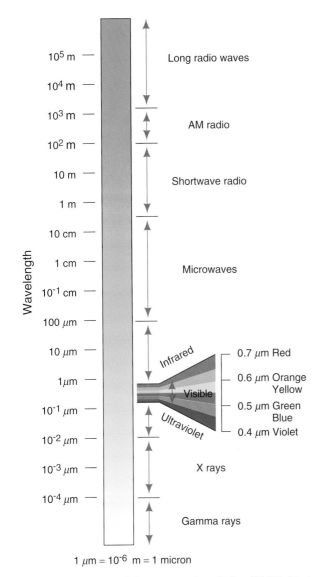

1 μm = 10^{-6} m = 1 micron

Figure 22.12 • Types of electromagnetic radiation (EMR). Notice that the spectrum of wavelengths from radio waves to gamma rays is over nine orders of magnitude (1 μm = 10^{-6} m = 1 micron).

The amount of energy per unit time radiated from a body such as the Sun or Earth varies with the fourth power of the absolute temperature of the body. Thus, if a body's temperature doubles, the energy radiated increases by 2^4, or 16 times. This phenomenon explains why the sun, with a temperature of 5,800°C (10,500°F), radiates a tremendously greater amount of energy than does the surface of Earth, which radiates at an average temperature of 15°C (59°F). Figure 22.13 illustrates this and another important point: The sun emits strongly in the visible region (relatively short-wave radiation of about 0.4–0.7 µm), whereas Earth emits in the infrared region (relatively long-wave radiation of about 10 µm). The hotter an object is, the more rapidly it radiates heat, and the shorter the wavelength of its predominant radiation. Earth's surface and the surfaces of animals, plants, clouds, water,

and rocks are cool enough to radiate heat predominantly in the infrared wavelength, which is invisible to us.

Energy from the sun travels to Earth at the speed of light through the vacuum of space. Less than half the solar energy that reaches our atmosphere is absorbed by the Earth and lower atmosphere. The rest is either absorbed by the upper atmosphere or reflected back into space. Much radiation that is harmful to living organisms, such as X rays and ultraviolet radiation, is filtered out by the upper atmosphere. The stratospheric ozone layer, extending from about 15 to 40 km (9 to 25 mi) above Earth's surface (Figure 22.1), is particularly important in absorbing ultraviolet radiation from the sun and protecting living organisms at Earth's surface. The depletion of the ozone layer is discussed in Chapter 25.

The Greenhouse Effect

Sunlight that reaches Earth warms both the atmosphere and the surface. Earth's surface and atmospheric system then re-radiate heat as infrared radiation.[15] Certain gases in Earth's atmosphere absorb and re-emit this radiation. Some of it returns to the Earth's surface, making Earth warmer than it otherwise would be. In trapping heat, the gases act a little like the panes of glass in a greenhouse (although the process by which the heat is trapped is not the same as in a greenhouse). Accordingly, the effect is called the **greenhouse effect**, and the gases—which include water vapor, carbon dioxide, methane, and chlorofluorocarbons (CFCs)—are called **greenhouse gases**.

It is important to understand that the greenhouse effect is a natural phenomenon that has been occurring for millions of years on Earth as well as on other planets in our solar system. After learning how the greenhouse effect works, you will be on your way to understanding how the effect moderates the temperature of the lower atmosphere and helps keep Earth habitable for life. You will also understand how human activity is interacting with the greenhouse effect to cause human-induced global warming.

Most natural greenhouse warming is due to water in the atmosphere. On a global level, water vapor and small particles of water in the atmosphere produce about 85% and 12%, respectively, of our total greenhouse warming. The greenhouse gases we are concerned with here are those that result in part from *anthropogenic* processes—that is, from human activities. These include carbon dioxide, CFCs, methane, nitrous oxides, and ozone, all of which have increased significantly in the atmosphere in recent years. It has been hypothesized that Earth is warming because of the increases in these anthropogenic greenhouse gases.

The following discussion focuses on the anthropogenic greenhouse effect as it relates to three factors:

• The burning of fossil fuels, which in recent years added about 5.5 GtC (gigatons, or billions of metric tons, of

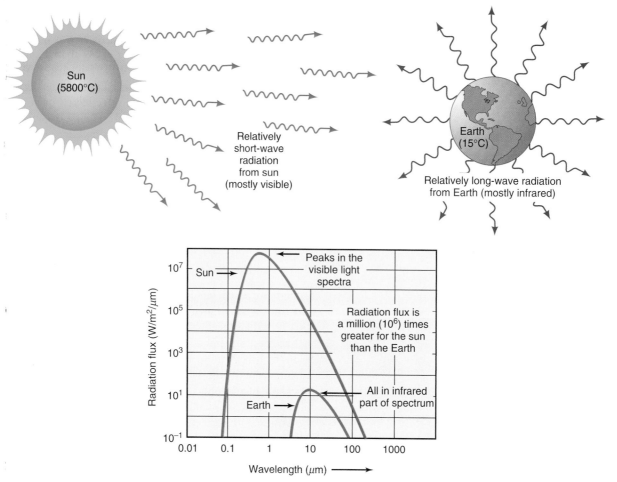

Figure 22.13 • Emission of energy from the sun compared with that from Earth. Notice that solar emissions have a relatively short wavelength, whereas those from Earth have a relatively long wavelength. (*Sources:* Modified from W. M. Marsh and J. Dozier, *Landscape* [Reading, Mass.: 1981]; and L. R. Kump, J. F. Kasting, and R. G. Crane, *The Earth System*. [Upper Saddle River, N.J.: Prentice Hall, 1999].)

carbon) (GtC) per year to the atmosphere. The carbon combines with oxygen to form CO_2. The Intergovernmental Panel on Climate Change (IPCC) reports present emissions of carbon to be 13% higher than this average at about 6.3 GtC per year.[1]

- Deforestation, by burning trees, increases the concentration of atmospheric CO_2, adding 1.6 GtC per year. Burning trees releases carbon stored in the wood that combines with oxygen to form CO_2.

- Human activities that emit other greenhouse gases, such as CFCs, ozone, methane, and nitrous oxides.

How the Greenhouse Effect Works

A highly idealized diagram showing some important aspects of the greenhouse effect is presented in Figure 22.14. The arrows labeled "energy input" represent the energy from the sun absorbed at or near the surface of Earth. The arrows labeled "energy output" represent energy emitted from the upper atmosphere and the surface of the Earth, which balances the input, consistent with Earth's energy balance. The highly contorted lines near the surface of Earth represent the absorption of infrared radiation (IR) occurring there and producing the 15°C (59°F) near-surface temperature. Following many scatterings and absorptions and reemissions, the infrared radiation emitted from levels near the top of the atmosphere (troposphere) corresponds to a temperature of approximately –18°C (0°F).

The one output arrow that goes directly through Earth's atmosphere represents radiation emitted through what is called the *atmospheric window* (Figure 22.15). The atmospheric window, centered on a wavelength of 10 μm, represents a region of wavelengths (8–12 μm) where outgoing radiation from Earth is not absorbed well by natural greenhouse gases (water vapor and carbon dioxide). Anthropogenic CFCs do absorb in this region, however; and CFCs significantly contribute to the greenhouse effect in this way.

Energy input
Close to a third of the energy that descends on the earth from the sun is reflected (scattered) back into space. The bulk of the remaining incoming visible solar radiation is absorbed by the Earth's surface.

Energy output
The atmosphere transmits outgoing infrared radiation from the surface (about 8% of the total outgoing radiation) at wavelengths between 8 and 13 microns and corresponds to a surface temperature of 15° C. This radiation appears in the atmospheric window where the natural greenhouse gases do not absorb very well. However, the anthropogenic chlorofluorocarbons do absorb well in this wavelength region.

Most of the outgoing radiation after many scatterings, absorptions, and re-emissions (about 92% of the total outgoing radiation) is emitted from levels near the top of the atmosphere (troposphere) and corresponds to a temperature of −18° C. Most of this radiation originates at the earth's surface, and the bulk of it is absorbed by greenhouse gases at heights on the order of 100m. By various atmospheric energy exchange mechanisms, this radiation diffuses to the top of the troposphere, where it is finally emitted to outer space.

Earth's surface (not to scale)

Atmosphere (troposphere)

Figure 22.14 • Idealized diagram showing the greenhouse effect. Incoming visible solar radiation is absorbed by the Earth's surface, to be re-emitted in the infrared region of the electromagnetic spectrum. Most of this re-emitted infrared radiation is absorbed by the atmosphere, maintaining the greenhouse effect. (*Source:* Developed by M. S. Manalis and E. A. Keller, 1990.)

Let us look more closely at the relation of the greenhouse effect to the Earth's energy balance, which was introduced in Figure 22.11 in a simplistic way. The figure showed that, of the incoming solar radiation, approximately 30% is reflected back to space from the atmosphere as short-wave solar radiation, while 70% is absorbed by the Earth's surface and atmosphere. The 70% that is absorbed is eventually re-emitted as infrared radiation (IR) into space. Thus, the sum of the reflected solar radiation and the outgoing infrared radiation balances with the energy arriving from the sun.

This simple balance becomes much more complicated when we consider exchanges of IR within the atmospheric and Earth surface systems. In some instances, these internal radiation fluxes may have magnitudes greater than the amount of energy entering the Earth's atmospheric system from the sun, as shown in Figure 22.16. A major contributor to the fluxes is the greenhouse effect.

At first glance, you might think it would be impossible to have internal radiation fluxes greater than the total amount of incoming solar radiation (shown as 100 units in Figure 22.16). It is possible because the infrared radiation bounces around many times in the atmosphere, resulting in high internal fluxes. For example, in terms of the figure, the amount of IR absorbed at the surface of the Earth from the greenhouse effect is approximately 88 units, which is about twice the amount of short-wave solar radiation (45 units) absorbed by the Earth's surface. In spite of the large internal fluxes, the overall energy balance remains the same. At the top of the atmosphere, the net downward solar radiation (70 units) balances the outgoing IR from the top of the atmosphere (70 units).

The important point in taking a more detailed look at Earth's energy balance is recognizing the strength of the greenhouse effect. For example, notice in the figure that, of the 104 units of IR emitted by the surface of the Earth, only 4 go directly to the upper atmosphere and are emitted. The remainder are reabsorbed and re-emitted by greenhouse gases. Of these, 88 units are directed downward to Earth and 66 units upward to the upper atmosphere.

All this may sound somewhat complicated, but if you read and study carefully the points mentioned in Figure 22.16 and work through the balances of the various parts of the energy fluxes, you will gain a deeper understanding of why the greenhouse effect is so important. The greenhouse effect keeps the lower atmosphere of the Earth approximately 33°C warmer than it would otherwise be. In addition, the greenhouse effect provides other important service functions. For example, the strong downward

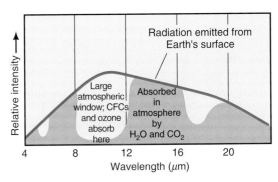

Figure 22.15 • Absorption spectra of greenhouse gases, water vapor, and carbon dioxide in the long-wave infrared radiation region. The atmospheric window is a region where neither water vapor nor carbon dioxide absorbs, but where CFCs do absorb. (*Source:* Modified from T. G. Spiro and W. M. Stigliani, *Environmental Science in Perspective* [Albany: State University of New York Press, 1980].)

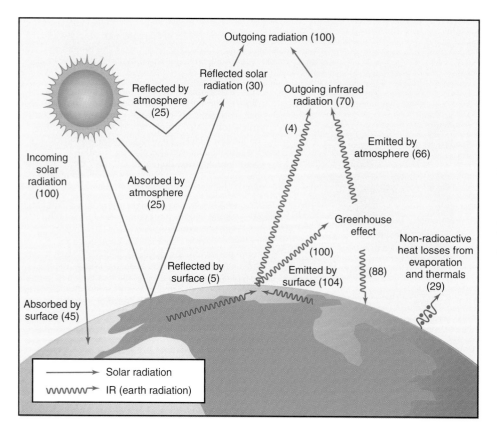

Figure 22.16 ● Idealized diagram showing Earth's energy balance and the greenhouse effect. Incoming solar radiation is arbitrarily set at 100 units, and this is balanced by outgoing radiation of 100 units. Notice that some of the fluxes (rates of transfer) of infrared radiation (IR) are greater than 100, reflecting the role of the greenhouse effect. Some of these fluxes are explained in the diagram. (*Source:* Modified from D. L. Hartmann, *Global Physical Climatology*, International Geophysics Series, vol. 56 [New York: Academic Press, 1994]; and S. Schneider, "Climate Modeling," *Scientific American* 256, no. 5 [1987]:72–80.)

•Total incoming solar radiation = 100 units

•Total absorbed by surface = 133 units
    45 from solar radiation (short wave)
    88 from greenhouse effect IR (infrared)

•Total emitted by surface = 133 units
    104 IR (of this, only 4 units pass directly to space without being absorbed or re-emitted in greenhouse effect)
    29 from evaporation and thermals (non-radioactive heat loss)

•Total IR emitted by upper atmosphere to space = 70 units
    66 units emitted by atmosphere
    4 units emitted by surface

•The 25 units of solar radiation absorbed by atmosphere is eventually emitted as IR (part of the 66 units)

•Total outgoing radiation = 100 units
    70 IR
    30 reflected solar radiation

emission of IR from the atmosphere that results from the greenhouse effect keeps variations in surface temperature from day to night relatively small. Without this effect, the land surface would cool much more rapidly at night and warm much more quickly during the day. Thus, the greenhouse effect not only maintains our relatively comfortable warm surface temperatures but also helps to limit temperature swings from day to night over the land.[3] It is, then, not the greenhouse effect itself that causes concern. Rather, it is the contribution of global warming from anthropogenic greenhouse gases.

Changes in Greenhouse Gases

The major anthropogenic greenhouse gases are listed in Table 22.1. The table also lists the recent rate of increase for each gas and its relative contribution to the anthropogenic greenhouse effect.

Carbon Dioxide

Approximately 200 billion metric tons of carbon in the form of carbon dioxide enters and leaves Earth's atmosphere each year as a result of a number of biological and physical processes. Not surprisingly, carbon dioxide has received a lot of attention with respect to global warming; 50 to 60% of the anthropogenic greenhouse effect is attributed to this gas.

In order to evaluate recent changes in atmospheric carbon dioxide, we need to take a broad view of Earth's history. As explained earlier, we can find out about ancient air by sampling the air bubbles trapped in glacial ice. Measurements of carbon dioxide trapped in air bubbles in the Antarctic ice sheet suggest that during the 160,000 years prior to the Industrial Revolution, the atmospheric concentration of carbon dioxide varied from approximately 200 to 300 ppm.[14] The highest level or concentration of carbon dioxide in the atmosphere other than at present occurred during the major interglacial period about 125,000 years ago.

Table 22.1	Relative Contribution of Trace Gases to the Anthropogenic Greenhouse Effect	
Trace Gases	Relative Contribution (%)	Growth Rate (%/yr)
CFC	15[a]–25[b]	5
CH_4	12[a]–20[b]	0.4[c]
O_3 (troposphere)	8[d]	0.5
N_2O	5[d]	0.2
Total	40–50	
Contribution of CO_2	50–60	0.3[e]–0.5[d,f]

[a] W. A. Nierenberg, "Atmospheric CO_2: Causes, Effects, and Options," *Chemical Engineering Progress* 85, no. 8 (August 1989): 27.

[b] J. Hansen, A. Lacis, and M. Prather, "Greenhouse Effect of Chlorofluorocarbons and Other Trace Gases," *Journal of Geophysical Research* 94 (November 20, 1989): 16, 417.

[c] Over the past 200 yrs.

[d] H. Rodhe, "A Comparison of the Contribution of Various Gases to the Greenhouse Effect," *Science* 248 (1990):1218, Table 2.

[e] W. W. Kellogg, "Economic and Political Implications of Climate Change," paper presented at Conference on Technology-based Confidence Building: Energy and Environment, University of California, Los Alamos National Laboratory, July 9–14, 1989.

[f] H. Abelson, "Uncertainties about Global Warming," *Science* 247 (March 30, 1990):1529.

About 140 years ago, at the beginning of the Industrial Revolution, the atmospheric concentration of carbon dioxide was approximately 280 ppm, a level that had apparently remained constant over the previous 700 years.[15] Since the beginning of the Industrial Revolution, the concentration of carbon dioxide in the atmosphere has grown exponentially. Currently, the rate of increase is about 0.5% per year. If growth continues at this rate, we will see a doubling of the concentration in approximately 140 years (or before the end of the 22nd century; recall the general rule for exponential growth in Chapter 3). Today, the concentration of carbon dioxide in the atmosphere is approaching 370 ppm, and it is predicted that the level may rise to approximately 450 ppm by the year 2050, more than 1.5 times the preindustrial level.[17] (Data prior to the mid-20th century are from measurements of air trapped in glacial ice. The remaining data are from direct measurement from the monitoring station at Mauna Loa, Hawaii, discussed in A Closer Look 22.1.) It is interesting to note that the rate of increase of carbon emissions (not carbon dioxide) from anthropogenic processes has been approximately 4.3% per year since the Industrial Revolution began, more than 8 times the 0.5% per year rate of increase in the concentration of carbon dioxide in the atmosphere.

The high rate of carbon dioxide emissions and the high growth rate in emissions would seem to suggest that the increase in carbon dioxide in the atmosphere is a direct result of the anthropogenic input of carbon. Establishing this seemingly simple relationship as a fact has been difficult, however, because the global carbon cycle is complex; all the linkages and flows of carbon from various sources to sinks are not yet well understood. What is apparent is that if all the carbon dioxide produced by human activities remained in the atmosphere, the concentration of that gas should be even higher than it is today. Therefore, we must hypothesize that there are sinks for carbon dioxide in the oceans or on the land that are not identified or well understood (see Chapter 4).

In spite of all these cautions, it is clear that carbon dioxide concentrations in the atmosphere have increased significantly since the Industrial Revolution. Furthermore, it is a reasonable assumption that these increases will continue to contribute to global warming via the greenhouse effect. Assuming that approximately 50%–60% of the anthropogenic greenhouse effect is due to carbon dioxide, we can conclude that the remaining greenhouse gases must account for approximately 40%–50% of the effect.

Methane

The concentration of methane (CH_4) in the atmosphere more than doubled in the past 200 years, and it is thought to contribute approximately 12%–20% of the anthropogenic greenhouse effect.[18] As with carbon dioxide, there are important uncertainties in our understanding of the sources and sinks of methane in the atmosphere.

Natural environments release methane into the atmosphere. Major contributors are termites, which produce methane as they process wood, freshwater wetlands, where decomposing plants in oxygen-poor environments produce and release methane as a decay product, seepage from oil fields, and seepage from methane hydrates (see Chapter 17). The several anthropogenic sources of methane include emissions from landfills (the major source in the U.S.), burning of biomass, production of coal and natural gas, and agricultural activities, such as the cultivation of rice and the raising of cattle. (Methane is released by anaerobic activity in flooded lands where rice is grown, and cattle expel methane gas as part of their digestive processes.)

Rates of increase in emissions of methane in recent years have been variable. In the 1980s to 1992, leaks in production and transport of methane pushed the global rate of increase to about 1% per year.[19,20] In the last decade of the 20th century the rate of increase decreased, but variability of emissions continued. For example, emissions of methane were anomalously high in 1998. The increase most likely resulted from unusually warm, wet weather in 1998 that caused an increase in methane released from warming wetlands along with large forest fires that also released methane, as the trees burned.

Chlorofluorocarbons

Chlorofluorocarbons, or CFCs, are inert, stable compounds that have been or are being used in spray cans as aerosol propellants and in refrigeration units. Use of CFCs as propellants was banned in the United States in 1978. Although CFCs are no longer used in spray cans in many countries, however, they have not yet been banned worldwide.

Deliberate release and accidental leaks of CFCs into the atmosphere have been considerable. The rate of increase of CFCs in the atmosphere in the recent past was about 5% per year. It has been estimated that approximately 15 to 25% of the anthropogenic greenhouse effect may be related to CFCs.[21]

In 1987, 24 countries signed a treaty, the Montreal Protocol, to reduce and eventually eliminate the production of CFCs and to accelerate the development of alternative chemicals. Because of the treaty, production of CFCs was nearly phased out by 2000. However, not all countries signed the treaty, and illegal production and use of CFCs continues in some countries. If CFCs had not been regulated by the Montreal Protocol, by the early 1990s, they would have become the major contributor to the anthropogenic greenhouse effect.[15] Reduced emissions are evidently responsible for the recent decrease in the concentration of atmospheric CFCs.

The potential global warming from CFCs is considerable, because they absorb in the atmospheric window, as explained earlier; and each CFC molecule may absorb hundreds or even thousands of times more infrared radiation emitted from Earth than is absorbed by a molecule of carbon dioxide. Furthermore, because CFCs are highly stable, their residence time in the atmosphere is long. Even if production of these chemicals is drastically reduced or eliminated within the next few years, their concentrations in the atmosphere will remain significant (although reduced from today's concentrations) for many years, perhaps for as long as a century.[16,21] (CFCs are discussed in more detail in Chapter 25, which examines stratospheric ozone depletion.)

Nitrous Oxide

Nitrous oxide (N_2O) is increasing in the atmosphere and is probably contributing as much as 5% of the anthropogenic greenhouse effect.[17] Anthropogenic sources of nitrous oxide include agricultural activities (application of fertilizers) and the burning of fossil fuels. Reductions in the use of fertilizers and the burning of fossil fuels would reduce emissions of nitrous oxide. However, this gas also has a long residence time; even if emissions were stabilized or reduced, elevated concentrations of nitrous oxide would persist for at least several decades.[18]

In summary, as indicated in Table 22.1, carbon dioxide contributes between 50 and 60% of the anthropogenic greenhouse effect. The rest of the human-made effect comes from trace gases, the most important of which are the CFCs and methane. These trace gases contribute between 27% and 45% of the anthropogenic

greenhouse effect, and they have accumulated in the atmosphere at much faster rates than carbon dioxide.

22.5 Science of Global Warming

As we suggested earlier, there is good reason to argue that increases in carbon dioxide and other greenhouse gases are related to an increase in the mean global temperature of Earth.[11,21] Over the past 160,000 years, a strong correlation has existed between the concentration of atmospheric CO_2 and global temperature (Figure 22.17). When CO_2 has been high, temperature has also been high; conversely, low concentrations of CO_2 have correlated with a low global temperature.

Global models of the climate suggest that warming as a result of anthropogenic increases in greenhouse gases will occur (some scientists, however, find faults with the models). According to the model, the average global temperature will rise 1.5–6.0°C by 2100. In the most optimistic case, if there are large reductions in emissions of greenhouse gases, the warming may be closer to 1°C.[1]

In order to better understand global warming, we need to consider both the positive and negative feedbacks that occur on Earth, on the Sun, and in the atmospheric system. We also need to consider the major variables that may cause (force) a global change in temperature, including solar emissions, volcanic eruption, El Niño warming, and anthropogenic input of greenhouse gases. The process of changing global temperature is called **forcing**.

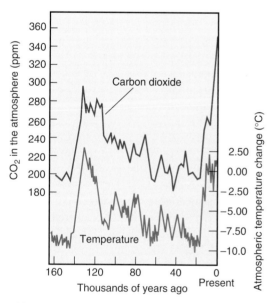

Figure 22.17 • Inferred concentration of atmospheric carbon dioxide and temperature change for the past 160,000 years. The relation is based on evidence from Antarctica and indicates a high correlation between temperature and CO_2. (*Source:* Adapted from S. H. Schneider "The Changing Climate," © September 1989, *Scientific American, Inc.* All rights reserved. Vol. 261, p. 74.)

Negative and Positive Feedbacks

Greenhouse warming is very complex. The warming effect initiates both negative and positive feedback loops that can offset any temperature increase or enhance it. Negative feedback loops are self-regulating and help stabilize global temperature change in response to a warming circumstance. Positive feedbacks are self-enhancing; thus, a process that causes an increase in global temperature leads to further increases in temperature. We first discussed positive and negative feedbacks with respect to Earth systems and changes in Chapter 3; you may wish to review those concepts.

Several of the potential negative and positive feedbacks concerning global warming are shown in Figure 22.18.[22] It is important to remember that if the negative feedbacks are strong and persistent, global warming may not occur.

In contrast, if the negative feedback systems are weak relative to positive feedback, warming is likely to occur more readily.

As shown in Figure 22.18a, negative feedbacks can be fairly complex. For example, negative feedback cycle 1 is based on the hypothesis that as global warming occurs, there will be an increase in the algae populations in the warming ocean. Algae will absorb more carbon dioxide, reducing the concentration of CO_2 in the atmosphere and cooling it. Negative feedback cycle 2 is related to terrestrial vegetation. Here, it is hypothesized that an increase in carbon dioxide concentration will stimulate plant growth (as it does in laboratory experiments), and the increased amount of vegetation will absorb more carbon dioxide from the atmosphere, facilitating cooling of the atmosphere. Negative feedback cycle 3 may result if polar regions receive more precipitation from warmer air

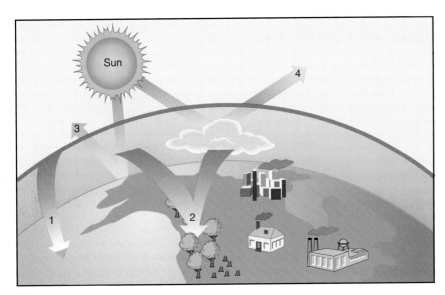

(a)

Figure 22.18 • (a) Negative feedback cycles and (b) positive feedback cycles associated with the greenhouse effect. (*Source:* J. R. Luoma, "Gazing into Our Greenhouse Future," *Audubon* 93, no. 2 [1991]:57.)

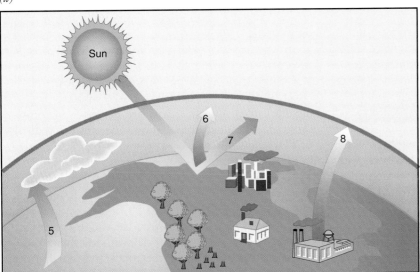

(b)

carrying more moisture. Increasing snow pack and ice buildup will cause solar energy to be reflected from Earth's surface, causing cooling. Finally, negative feedback cycle 4 is related to cloud cover, which is poorly understood but extremely important. The idea is that as the global temperature increases, more water will evaporate from the ocean, leading to more water vapor in the atmosphere and thus more clouds. Because clouds tend to reflect incoming solar radiation, Earth will then be cooled by the increased cloud cover.

Positive feedback processes are shown in Figure 22.18b. In positive feedback cycle 5, the warming Earth causes an increase in the evaporation of water from the oceans, which adds additional water vapor to the atmosphere. But here, the water vapor causes additional warming (water vapor is an important greenhouse gas). The warming Earth causes increased melting of permafrost at high latitudes, which may result in additional release of the greenhouse gas methane (positive feedback cycle 6). The methane is a by-product of decomposition of organic material in the melted permafrost layer. Another effect of the warming trend would be a reduction in the summer snowpack, which would reduce the amount of solar energy reflected from Earth (positive feedback cycle 7). Collectively, these processes would result in additional warming and thus are part of a positive feedback loop. In positive feedback cycle 8, the warming of Earth is felt by people in urban areas, who use additional air conditioning, thus increasing the burning of fossil fuels, which in turn releases additional carbon dioxide into the environment and results in additional global warming.

Both negative and positive feedback processes occur simultaneously in the atmosphere. Which are more important? The nod goes to positive feedbacks, because the mean global temperature is increasing. A great deal of research is currently being carried out to better understand negative-feedback processes associated with clouds and their water vapor. Many discussions on the greenhouse effect state that if Earth's atmosphere did not trap heat, our planet would be approximately 33°C (60°F) cooler at the surface; and as a result, all water would be frozen. However, since water vapor is the major greenhouse gas in the atmosphere, no greenhouse effect implies no (or very little) water vapor in the atmosphere. This implies no clouds, which would lead to a substantial reduction in the atmospheric reflection of incoming sunlight, which would result in warmer surface temperatures on Earth. The dual role of atmospheric water vapor as both a negative and a positive feedback with respect to global warming is extremely important to understanding possible climatic modifications created by an anthropogenic greenhouse effect.

Solar Forcing

The Sun is responsible for heating the Earth; therefore, in evaluating climatic change, we must consider solar varia-tion a possible cause. As mentioned earlier, the Milankovich cycles are consistent with most of the long-term variability of climate that has occurred during the past several hundred thousand years of Earth's history. However, the Milankovich cycles by themselves are not sufficient to produce the magnitude of observed climate changes.

When we examine the history of climate during the past thousand years, the variability of solar energy plays a role. Evaluation of isotopes from the sun found in ice cores from glaciers reveals that during the medieval warm period, from approximately 1100 to 1300, the amount of solar energy reaching Earth was relatively high, comparable to the warming we see today. Evaluation of the data also suggests that minimum solar activity occurred during the 14th century, coincident with the beginning of the Little Ice Age (Figure 22.5). Thus, it appears that the variability of solar input of energy to the Earth can indeed explain some of the climatic variability we have experienced in the past thousand years. The effect, however, is relatively small.[10]

Aerosols and Volcanic Forcing

An aerosol is a particle with a diameter less than 10 μm. Because the effects of collisions with air molecules dominate over gravity, these particles tend to remain in the atmosphere for a long time. Bigger particles (with diameters greater than 10 μm) drop out of the atmosphere faster because of gravity. Emissions of aerosols to the atmosphere by human processes have increased significantly since the Industrial Revolution, and volcanic eruptions have periodically released aerosols as well. Recent research has indicated that aerosols emitted from coal (sulfates) and volcanic eruptions may contribute to global cooling, because sulfates act as seeding for clouds. The aerosol particles provide surfaces for water to condense on; clouds form and reflect incoming solar energy. The aerosol particles also reflect a significant amount of sunlight directly.

In the early 1990s, net cooling owing to sulfate aerosols from burning fossil fuels offset some of the global warming expected from the anthropogenic greenhouse effect.[23,24] However, emissions of the sulfate SO_2 (sulfur dioxide) from human activity are decreasing because of programs to reduce air pollution. The effect of reduction of SO_2 emissions will be a strengthening of the greenhouse effect from CO_2.

Naturally released aerosols have also produced some net cooling. Tremendous explosions from Mount Pinatubo in the Philippines in 1991 sent volcanic ash to elevations of 30 km (19 mi) into the stratosphere (Figure 22.19). The aerosol cloud of ash, with 20 million tons of sulfur dioxide, remained in the atmosphere circling Earth for several years. The particles of ash and sulfur dioxide scattered incoming solar radiation, resulting in a slight

Figure 22.19 • Eruption column from Mount Pinatubo in the Philippines during 1991 eruptions. Such eruptions injected vast amounts of dust and sulfur dioxide as high as 30 km into the atmosphere.

cooling of the global climate during 1991 and 1992. Calculations suggest that aerosol additions to the atmosphere from the Mount Pinatubo eruption counterbalanced the warming effects of greenhouse gas additions through 1992. However, by 1994, most aerosols from the eruption had fallen out of the atmosphere, and global temperatures had returned to previous higher levels. Volcanic climate forcing from pulses of volcanic eruptions is also believed to have significantly contributed to the cooling associated with the Little Ice Age, from about 1450 to 1850 (see Figure 22.6).[10]

El Niño

Another natural perturbation that affects global climate is the occurrence of El Niño events (see A Closer Look 22.2). During an El Niño, the normal conditions of equatorial upwelling of deep oceanic waters are diminished or eliminated. Upwelling releases carbon dioxide to the atmosphere as carbon-dioxide-rich deep water reaches the surface. El Niño events thus reduce the amount of oceanic carbon dioxide outgassing, influencing the global carbon dioxide cycle. Climatic models generally predict that as the Earth warms, El Niño events will become more common.[1]

The 1982–1983 El Niño event was particularly strong. It is thought to have produced climatic events such as floods and droughts that killed several thousand people and caused billions of dollars in damage to crops, structures, and other facilities. El Niño events may change the patterns of the upper troposphere (jet streams) to produce wetter winters and larger storms in parts of the United States. El Niño conditions developed in 1991–1993 and again in 1997–1998; although they were not as strong as the 1982–1983 event, they contributed to a slight global warming.

Recognition of El Niño events is important for understanding processes that could affect the global climate. The events also provide data for global models that predict change. Because volcanic eruptions and El Niño events are relatively short term, they provide important information from which models can be calibrated and tested.

Methane Forcing

Methane (CH_4), or natural gas, is a strong greenhouse gas contributing as much as 20% of the anthropogenic greenhouse effect. There may also be a periodic natural methane forcing that causes rapid warming to end a glacial period. (See the discussion of methane hydrates in Chapter 17.)

One mechanism that releases large amounts of methane is linked to lowering of sea level. Remember from Chapter 17 that methane hydrates remain stable in deep (high pressure) cold water. During glacial times, sea level drops because of the huge volume of water stored in glaciers. The lower sea level causes water pressure at the bottom of the sea to be reduced. Under these conditions, methane hydrates could become unstable and be released into the water. The gas, once released from the ocean bottom, would rise and enter the atmosphere. A second process favoring the release of methane from hydrates is the warming of ocean water. Warming of deep water occurs at the beginning of interglacial periods when sea level is low.[26]

If sufficient methane is released, either by lowering of the sea level or warming of the water, the gas may cause global warming, contributing to ending the glacial period. For example, release of methane may have been responsible for a rapid increase in the global temperature that occurred over a relatively short period at the end of the last ice age about 15,000 years ago.

Anthropogenic Forcing

Until recently, there was controversy concerning whether the anthropogenic component of global warming was significant or not. The question now seems to have been answered, and we can state that there is a significant human footprint on observed global warming. A recent study evaluated climate change over the past thousand years, allowing late-20th-century warming to be placed within a historical context. The study used a mathematical model to remove major natural forcing mechanisms, including solar radiation and volcanic eruptions, so that climatic forcing by greenhouse gases could be directly estimated. The research established that the natural variability in the climate system in the past thousand years is far less than what occurred at the end of the 20th centu-

El Niño

"El Niño" became a household word during the winter of 1997–1998, when it was blamed for everything from tornadoes and thunderstorms in Florida to catastrophic fires in Indonesia (see Chapter 23). The term *El Niño* means "little boy" in Spanish and refers in particular to the Christ Child, because the event often begins off the coast of South America near Christmas time.

El Niño events disrupt the ocean-atmosphere system in the tropical Pacific. They are in part responsible for weather phenomena that can cause billions of dollars in property damage and the loss of thousands of human lives. Occurring at intervals of two to seven years, El Niño events typically last for 12 to 18 months. They start with the weakening of east to west trade winds and the warming of eastern Pacific Ocean waters. As a result, tropical rainfall shifts from Indonesia to South America, as shown in Figure 22.20.

Let's look at this process in more detail. Under non–El Niño conditions, trade winds blow west across the tropical Pacific. The warm surface water in the western Pacific tends to pile up, so that the sea surface can be as much as 0.5 m higher at Indonesia than at Peru. In contrast, during El Niño, the trade winds weaken and may even reverse. As a result, the eastern equatorial Pacific Ocean becomes unusually warm, and the westward moving equatorial ocean current weakens or reverses. The rise in temperature of sea surface waters off the South American coast inhibits the upwelling of nutrient-rich cold water from deeper levels; the upwelling normally supports a diverse marine ecosystem and major fisheries. Because rainfall follows warm water eastward during El Niño years, there are high rates of precipitation and flooding in Peru, while droughts and fires are commonly observed in Australia and

Indonesia. Because warm ocean water provides an atmospheric heat source, El Niño changes global atmospheric circulation, which causes changes in weather in regions that are far removed from the tropical Pacific.

It is important to remember that El Niño events are natural phenomena and part of a dynamic system involving the coupling of Earth's atmosphere and ocean. El Niño events can alternate with contrasting phenomena called La Niña events. La Niña events are characterized by unusually cool ocean water temperatures, and the effects are opposite those of El Niño. For example, during a 1998–1999 La Niña, winter temperatures were cooler than normal in the U.S. Northwest. A La Niña event does not necessarily follow every El Niño event; but together, they constitute a natural cycle of ocean-atmospheric disruption with global consequences.[25]

Figure 22.20 ● Idealized diagram comparing selected normal and El Niño conditions. (*Source:* NOAA.)

ry. That is, present warming far exceeds natural variability and closely agrees with the warming predicted by global circulation models that take account of increasing greenhouse gases.[10] Results of the modeling are shown in Figure 22.21. Notice that the late-20th-century rise in temperature far exceeds the zone of maximum variability

that could result from forcing mechanisms such as solar variability and volcanic activity.

It appears that the major scientific issues concerning global warming have been resolved. Significant global warming as a result of human activity is occurring. We now discuss some potential effects of and possible adjust-

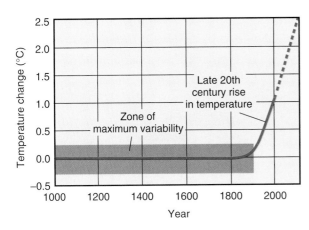

Figure 22.21 • Global temperature change during the past 1,000 years. Solar and volcanic forcing have been removed, leaving anthropogenic forcing from greenhouse gases. (*Source:* Modified from T. J. Crowley, 2000. "Causes of Climate Change over the Past 1000 Years," *Science* 289 [2000]:270–277.)

ments to global warming. What we ultimately decide to do about global warming will reflect our values.

22.6 Potential Effects Of Global Warming

If we continue emitting large amounts of carbon dioxide into the atmosphere, it is estimated that by 2030 the concentration of carbon dioxide in the atmosphere will have doubled from pre–Industrial Revolution concentrations. The average global temperature (according to mathematical models) will have risen approximately 1°–2°C (2°–4°F), with significantly greater temperature change at the polar regions.[17] Global warming causes greater temperature increases at polar regions in part because of a positive feedback mechanism; sea ice melts from warming, and water reflects much less light than white ice, resulting in enhanced warming. Solar energy that would have been reflected by sea ice is absorbed by the ice-free water. This mechanism is part of what is termed **polar amplification**.[3]

Specific effects of global warming are difficult to predict from global models. However, it is expected that warming will likely have the following consequences:[1]

- As a results of regional changes in climate, semiarid land areas will become drier, while other regions will become wetter.
- World distribution of biomes will likely change. Some, such as alpine tundra, may be lost, while others, such as midlatitude deserts, may expand.
- Agricultural production in some regions will decrease.
- The incidence of diseases such as malaria and dengue fever will increase in tropical countries.
- Rising sea levels will displace tens of millions of people.

- The biosphere will change as a result of damage to ecosystems, from tropical coral reefs to birds and bears in the Arctic.

Next, we briefly discuss selected aspects of these consequences.

Changes in Climate

Various estimates have been made of what changes in annual temperature and precipitation are likely to occur as a result of global warming. Figure 22.22 shows changes expected to occur by 2050 assuming concentrations of greenhouse gases increasing at about 1% per year.[1,27] In central North America (the grain-growing region), warming is expected to vary from approximately 2° to 4°C (4° to 7°F), with a small increase in precipitation. As a result, soil moisture may decrease in the summer by as much as 20%. Clearly, this could have a significant effect on the grain-growing areas of the United States.

Changes in relative runoff of water linked to changes in temperature and precipitation for 2080 are shown in Figure 22.23. Lower runoff is projected for much of Mexico, South America, Southern Europe, India, Southern Africa, and Australia. It is important to keep in mind that these projections are based on global circulation models that are controversial and subject to variability. Nevertheless, most of the models predict changes in the directions indicated and as a result are being taken seriously by both scientists and policy makers.

As already suggested, global rise in temperature is expected to significantly change patterns of rainfall, soil moisture, and other climatic factors related to agricultural productivity. Studies using global circulation models to predict patterns for the Northern Hemisphere suggest that some of the more northern areas, such as Canada and Russia, may become more productive. Although global warming might move North America's prime farming climate north from the midwestern United States to the region of Saskatchewan, Canada, the U.S. loss would not simply be translated into a gain for Canada. Saskatchewan would have the optimum climate for growing, but the Canadian soils are thinner and less fertile than the prairie-formed soils of the U. S. Midwest. Therefore, a climate shift could have serious negative effects on midlatitude food production. In addition, lands in the southern part of the Northern Hemisphere may become more arid; and as a result, soil moisture relationships will change.

People are anxious when uncertainty is present, and they are particularly anxious when that uncertainty involves their food supply. There is real concern that hydrologic changes associated with climatic change resulting from global warming may seriously affect the global food supply. Today, 800 million people are malnourished; and as population increases, so will demand

(a)

Predicted annual mean temperature change (°C) by the 2050s

(b)

Predicted change in precipitation by the 2050s (mm day⁻¹)

Figure 22.22 • (*a*) Projected changes in annual temperatures from today for the 2050s. Notice that changes are greatest at the polar regions. (*b*) Projected changes in annual precipitation from today for the 2050s. Highest increases are near the equator. Modeling that predicted these changes assumes an increase in greenhouse gas of about 1% per year. (*Source:* Met Office, Hadley Center for Climate Prediction and Research, in R. T. Watson, presentation at the Sixth Conference of the Parties of the United Nations Framework Convention on Climate Change, Intergovernment Panel on Climate Change, November 13, 2000. Accessed December 1, 2000, www.ipcc.ch/press/sp-COPG.htm.)

for food. During the 21st century, agricultural production in many subtropical and tropical regions (especially in South America and Africa) will likely decrease (Figure 22.24) if the mean annual temperature increases by more than 2°C. Unfortunately, millions of the poorest people on Earth live in the tropics and subtropics, where risk of hunger is the greatest.[1]

There is also concern that global warming will alter normal weather and climatic patterns, including the frequency or intensity of violent storms. This possibility may be more important than changes in climate. The hypoth-esis is that warming ocean waters could feed more energy into high-magnitude storms, such as cyclones and hurricanes, causing a significant increase in their frequency or intensity. Approximately half of Earth's human population live in coastal areas. Potential problems are exacerbated by the fact that many of these areas are low lying and are experiencing rapid population growth. In addition, greenhouse warming is expected to result in wetter winters, hotter and drier summers, an increased frequency of large storm events, and an increased possibility of droughts in the northern temperate latitudes.[28]

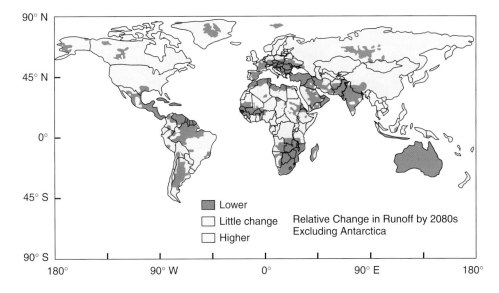

Figure 22.23   Predicted relative change in runoff of water by the 2080s linked to changes in temperature and precipitation (Figures 22.22a,b). (*Source:* University of Southampton, in R. T. Watson, presentation at the Sixth Conference of the Parties of the United Nations Framework Convention on Climate Change, Intergovernment Panel on Climate Change, November 13, 2000. Accessed December 1, 2000, at www.ipcc.ch/press/sp-COPG.htm.)

In summary, global warming could affect people and ecosystems on Earth in various ways, often for the worse. These changes may affect hydrology, crop production, forestry, and human health. Midlatitude climate zones could shift northward by as much as 550 km (330 mi) over the next century. At this rapid rate, some tree species may be nearly eliminated. Furthermore, the expected expansion of tropical climate zones will lead to an increase in tropical diseases such as malaria, dengue fever, yellow fever, and viral encephalitis.[29]

Rise in Sea Level

A rise in sea level is a potentially serious problem related to global warming. As mentioned, about half of the people on Earth live in the coastal zone, and about 50 million people per year experience flooding due to storm surges. As sea level rises and population increases, the number of people vulnerable to coastal flooding will increase. For example, two cyclones that hit highly populated Bangladesh in the last 25 years killed more than 400,000 people and caused over $1.6 billion in property damage (see Chapter 5). The double impact of rising sea level and more frequent and powerful cyclones and other tropical disturbances (owing to warmer oceans, as discussed earlier) would have devastating effects on people in developing countries.

Although a precise estimate of the total potential rise in sea level is not possible at this time, there is a consensus that the level of the sea will in fact rise. In fact, sea level along much of the U.S. coast is already rising at a rate of 2.5–3.0 mm/yr, or about 10–12 in/century.[30]

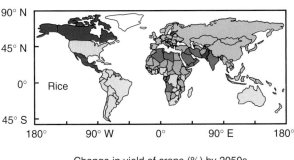

Figure 22.24   Predicted changes in average yield of wheat, corn, and rice in the 2050s linked to changes in temperature and precipitation (Figures 22.22a,b). (*Sources:* Jackson Institute, University College, London; Goddard Institute for Space Studies; International Institute for Applied Systems Analysis; and in R. T. Watson, presentation at the Sixth Conference of the Parties of the United Nations Framework Convention on Climate Change, Intergovernment Panel on Climate Change, November 13, 2000. Accessed December 1, 2000, at www.ipcc.ch/press/sp-COPG.htm.)

The cause for the rise is thought to be twofold: thermal expansion of warming ocean water (the primary cause) and melting of glacial ice (a secondary cause).

Various models predict that sea level may rise anywhere from 20 cm to approximately 2 m (8–80 in) in the next century; the most likely rise is probably 20–40 cm (8–16 in). One estimate is that sea level will likely rise 15 cm by 2050 and 34 cm by 2100. When other factors, such as land subsidence and compaction, groundwater depletion, and natural climate variation, are considered, some coastal regions could experience a sea level rise of 45–55 cm by 2100.[30]

Such a change would have significant environmental impacts. It could easily cause increased coastal erosion on open beaches of 50–100 m (165–230 ft), making buildings and other structures in the coastal zone more vulnerable to damage from waves generated by high-magnitude storms. It could also cause a landward migration of estuaries and salt marshes, lead to loss of coastal wetlands (see Chapter 20), and put additional pressure on human structures in the coastal zone.[17] The people and wetlands at risk from a 44-cm rise in sea level are shown by region on Figure 22.25. Finally, groundwater supplies for coastal communities may be threatened by saltwater intrusion should sea levels rise (see Chapter 21).

A rise in sea level of approximately 1 m (3.3 ft) would have even more serious consequences. People would have to significantly alter the coastal environment to protect investments, and communities would be forced to choose either to make very heavy expenditures to control coastal erosion or to allow considerable loss of property.[17]

It seems inevitable that a rise in sea level will increase coastal erosion and lead to further investments to protect cities in the coastal zone. Construction of seawalls, dikes, and other erosion-controlling structures will become more common as coastal erosion threatens urban property. In more rural areas, where development is set well back from the coastal zone, the most likely response to rising sea level will be to adjust to the erosion that occurs. Coastal erosion is a difficult problem that is very expensive to deal with. In many cases, it is best to allow erosion to take place naturally where feasible and only defend against coastal erosion where absolutely necessary.

Changes in Biosphere

A growing body of evidence indicates that global warming is probably initiating a number of changes in the biosphere, sometimes threatening ecological systems and people. These changes include shifts in the ranges of plants and animals. Some such shifts are summarized in Figure 22.26.

One change mentioned on Figure 22.26 is a shift in the range of mosquitoes that carry diseases including malaria and dengue fever. Malaria, which causes chills and fever, is particularly worrisome, as it kills about 3,000 people per day, mostly children. Projected global warming will increase the land area where the disease can be transmitted. Today, that area contains 45% of the world's population. With global warming, malaria could threaten 60% of the world's population.[32]

The possible relationship between global warming and the emergence of the West Nile virus in New York City in 1999 is an example of linkages between physical, biological, and social systems—the essence of environmental science. How the West Nile virus arrived in North America is not known, but its spread from mosquitoes to birds to people was enhanced by a warm winter followed by a dry spring and a hot, wet summer; these conditions are symptoms of global warming. This is how it works:[32]

- During a mild winter, more mosquitoes survive in sewers, as well as in still water in a variety of locations, including ponds, waste cans, and abandoned tires.

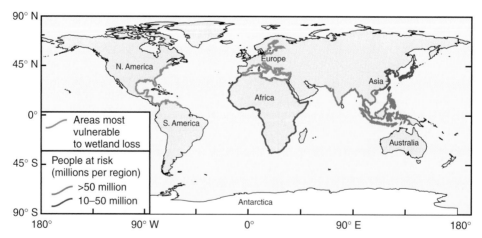

Figure 22.25 • People and wetlands at risk from a 44-cm rise in sea level by the 2080s. It is assumed that coastal flood protection remains as it is today. (*Source:* Modified from R. Nicholls, Middlesex University in the U.K. Met Office, in R. T. Watson presentation at the Sixth Conference of the Parties of the United Nations Framework Convention on Climate Change, Intergovernment Panel on Climate Change, November 13, 2000. Accessed December 1, 2000, at www.ipcc.ch/press/sp-COPG.htm.)

1. Black guillemots, sea birds living in the Arctic environment, have been declining in numbers on Copper Island, Alaska. As sea ice receded earlier each spring during the 1990s, the birds' prey, the Arctic cod, moved with the ice, causing the decline in numbers of the sea birds.

2. Edith's checkerspot butterflies are moving north and to higher elevations. This migration has been going on over the last century and represents a shift that matches global warming in the butterflies' western North American habitat.

3. Sachem skipper butterflies have been expanding their range from northern California into Oregon and southeastern Washington. This has been going on over a 50-year period, coinciding with gradual global warming. Each move northward has occurred during an unusually warm summer.

4. Snails, sea stars, and other intertidal organisms have been shifting northward in Monterey Bay as a result of increases in shoreline water temperatures over the past 60 years.

5. Mexican jays are breeding earlier in the Chiricahua Mountains, according to a 30-year study. It is hypothesized that the birds have adjusted to earlier warm spring temperatures by breeding earlier to match emergence of food sources, including insects.

6. Mexican voles (small rodents with short tails and stout bodies) have expanded their range during the past 100 years into northeastern Arizona and Colorado from the south and west. Studies suggest that a number of other mammal species are also shifting northward to adjust to changes in climate and habitat.

7. Zone-tailed hawks and other birds have gradually moved northward over a 50-year period. The trend is observed for a number of bird species of the southwest and Mexico, including common blackhawks, whippoorwills, and brown-crested flycatchers.

8. Prairie grasses are on the decline in the central plains. In some cases, exotic weeds are replacing the native grasses. This phenomenon correlates with global warming since 1964.

9. Fire ants are expanding their range in all directions from the Deep South as far north as North Carolina. The ants' spread is limited by cold temperatures and correlates with warming trends in recent years.

10. Palm trees along Florida's west coast are dying as a result of exposure to salt water. This has resulted in part from a rise in global sea level associated with a trend over several thousand years. However, more recently, sea levels have been rising faster as a result of human-induced climate change, increasing the threat to low-lying palm trees.

11. Polar bears in western Hudson Bay are losing weight and bearing fewer cubs. Warmer temperatures are melting sea ice earlier in the spring. The sea ice is critical habitat for the bears, who use the ice surface as a hunting platform for capturing seals.

12. Peary caribou populations are declining in the high Arctic environment of Canada. This may be related to a series of unusually warm, wet, and snowy winters, which have created vast ice cover. The animals are unable to forage below the solid crust of the ice.

13. Mosquitoes carrying dengue fever (a severe flu-like viral illness that may cause fatal internal bleeding) are moving to higher elevations in Mexico.

14. Coral reefs in the Florida Keys, Bermuda, and the Caribbean are declining as a result of bleaching thought to be caused by warming marine waters.

15. Subalpine forests are moving to higher elevations in the Olympic Mountains of Washington.

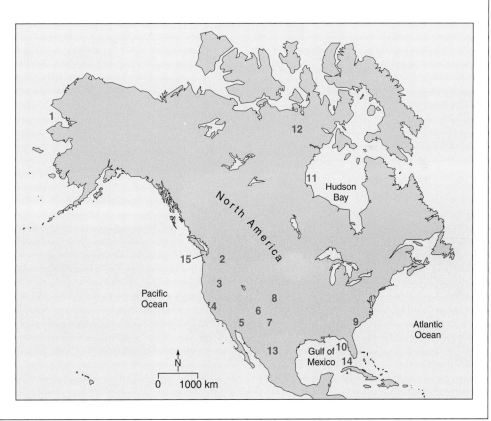

Figure 22.26 • Selected examples of how rising global temperatures are shifting the range of plants and animals in North America. (*Sources:* Modified from S. Levy, "Wildlife on the Hot Seat," *National Wildlife* 38, no. 5[2000]:20–27; and N. Holmes, "Has Anyone Checked the Weather [Map]?" *Amicus Journal* 21, no. 4[2000]:50–51.)

- During a dry spring, surface water sites, such as small ponds, decrease in size, concentrating both birds and mosquitoes at the water sites.
- Mosquitoes infected with the virus bite uninfected birds, passing the virus on to them.
- Infected birds are bit by uninfected mosquitoes, passing the virus to more mosquitoes.
- Hot summer months, with their warm air and heavy rains, cause the mosquito population to mature and grow rapidly. More mosquitoes become infected and pass on the virus both to birds and, eventually, to people.

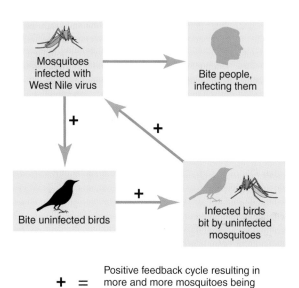

Positive feedback cycle resulting in more and more mosquitoes being infected with virus

Figure 22.27 As a result of a positive feedback cycle, more and more mosquitoes are infected with West Nile virus, eventually infecting people.

Thus, the unusual weather, possibly from global warming, results in an increase in the population of mosquitoes. Once the virus appears, a strong positive feedback cycle between birds and mosquitoes (Figure 22.27) increases the number of infected mosquitoes, eventually producing a health problem for people.

What is happening to some butterfly species is also of concern. For example, as early as 1989, Edith's checkerspot butterflies (Figure 22.28) in some high meadows of the Sequoia National Forest in California died, not from poisoning or disease, but because the snowpack melted early, and the warm temperatures caused the butterflies to emerge before the nectar-rich plants they feed on were available.

As a final example of ecological effects of global warming, consider the declining numbers of black guillemots on Copper Island, Alaska (Figure 22.29). Increases in temperature in the 1990s caused the sea ice to recede farther from Copper Island each spring. The recession of sea ice occurred before the black guillemot chicks were mature enough to survive on their own. Parent birds feed on Arctic cod found under the sea ice, then return to the nest to feed their chicks. The distance from feeding grounds to nest must be less than about 30 km; but in recent years, the ice in the spring has been receding as much as 250 km from the island before the chicks are able to leave the nests. As a result, the black guillemots on the island have lost an important source of food, and their numbers have been reduced. The future of black guillemots on Copper Island depends on future springtime weather. Too warm, and the birds may disappear. Too cold, and there may be too few snow-free days for breeding; in this case, too, they will disappear. Thus, the survival of black guillemots on Copper Island is precarious, given our changing global climate.[2]

What is apparent from our examples is that the ecological consequences of global warming are not the con-

Figure 22.28 Edith's checkerspot butterflies in some locations are emerging before the plants they feed on bloom and produce the nectar necessary for their survival.

Figure 22.29 Black guillemots on Copper Island, Alaska, experienced a reduction in population in the 1990s as a result of receding sea ice in the spring.

cern of a distant future. Changes in the biosphere and ecological communities as a result of warming are happening now. As individual species in an ecosystem are affected, changes will occur in other species. This is the principle of environmental unity discussed in Chapter 3.

22.7 Adjustments To Potential Global Warming

Debate continues over some details of climatic change related to human activities that release greenhouse gases into the atmosphere. Scientists do agree that we need to develop better models and gain a better understanding of both climatic change and variability and short-term trends in weather. At this point, two important statements can be made concerning global warming and climate change:

- Global temperature in the last few decades has increased approximately 0.5°C (0.9°F).
- The bulk of the evidence suggests that massive emissions of carbon dioxide into the atmosphere from use of fossil fuels are causing some significant portion of the observed global warming. Furthermore, computer models predict that global warming will continue to occur as a result of increased atmospheric levels of greenhouse gases.

Given these statements, should we spend large sums of money to reduce CO_2 and other greenhouse gas emissions? There is no easy answer to this question, and the answer will not come from scientific evaluation. In terms of science, the question of whether human-induced warming is occurring is solved: A significant part of global warming is human-induced. We now must evaluate potential consequences and risks to society and the biosphere. Policy makers must take action to address risks posed by global warming. They must do this even though significant scientific uncertainties remain concerning whether global warming and associated climate-induced changes can be reversed.[1]

What should we do about potential global warming? There are two basic adjustments:

- Do nothing to combat it, and live with future global climatic change.
- Work to mitigate the situation (reduce its severity) by reducing emissions of greenhouse gases and enhancing biospheric sinks for CO_2.

Living with Global Change

Considering our present understanding of global climate and recent trends in energy use and deforestation, a likely adjustment will be learning to live with the changes. As discussed earlier, these include a warmer climate, more variability in weather patterns, changes in the biosphere, and a higher sea level. If the changes are relatively slow and total mean global warming is less than 2°C, then learning to live with new conditions may be feasible. In some cases, the changes will offer opportunities. However, overall, this may not be the best adjustment. Surprises and unexpected problems will emerge. Furthermore, a 2°C change is at the low end of predicted global warming; at increases greater than 2°C, potential consequences to humans increase significantly. Finally, although yields of crops will vary on a regional and local level, problems in food production will occur in tropical and subtropical regions, where many millions of people live.[1]

In summary, one possible adjustment to global warming will be to live with it. However, if we are prudent, we will plan instead to reduce emissions of carbon dioxide and other greenhouse gases. By doing so, we may be able to limit warming to less than 2°C.

Mitigating Global Warming

The attempt to control emissions of greenhouse gases, particularly carbon dioxide, was initiated when a major scientific conference on the issue of global warming was held in 1988 in Toronto, Canada. At that meeting, scientists recommended a 20% reduction in carbon dioxide emissions by 2005. The meeting was a catalyst for scientists and other concerned people to work with politicians to initiate international agreements for reductions in emissions of greenhouse gases. Although at that time many uncertainties remained concerning global warming, the prevailing attitude was that it was advisable to be conservative and reduce emissions before problems became apparent.

In 1992, at the Earth Summit in Rio de Janeiro, Brazil, a general blueprint for the reduction of global emissions was suggested. Some in the United States, however, objected that the reductions in CO_2 emissions would be too costly. Furthermore, agreements from the Earth Summit did not include legally binding limits. Following the meetings in Rio de Janeiro, governments worked to strengthen a climate control treaty that included specific limits on the amounts of greenhouse gases that could be admitted into the atmosphere by each industrial country.

Legally binding emission limits were discussed in Kyoto, Japan, in December 1997, but specific aspects of the agreement divided the delegates. The United States eventually agreed to cut emissions to about 7% below 1990 levels. However, that was far below the reductions suggested by scientists, who recommended reductions of 60–80% below 1990 levels. In fact, following the conference, it was realized that emissions of carbon dioxide in 2010 would likely be about 30% above the 1990 emissions.

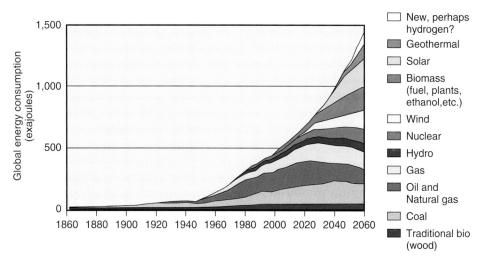

Figure 22.30 • Global energy consumption, 1860–2000, predicted to 2060, based on a model of movement toward sustainable energy development. (*Source:* Modified from Shell International, in R. T. Watson, presentation at the Sixth Conference of the Parties of the United Nations Framework Convention on Climate Change, Intergovernment Panel on Climate Change, November 13, 2000. Accessed December 1, 2000, at www.ipcc.ch/press/sp-COPG.htm.)

Since that time, the science question has been answered, and the human component of global warming has increased, as evidenced by a series of very warm years and biosphere changes. In spite of this, some sectors of the fossil fuels industries continue to resist lowering emissions, arguing that the economic cost would be too great.

Every time an environmental solution to a problem has been suggested, there have been those who said it couldn't be done, yet it has been proved time and time again that it can be done. It appears that advances in alternative renewable energy are sufficient that society could in fact meet goals for reductions of carbon dioxide emissions without jeopardizing our economic future. An energy scenario that emphasizes sustainable alternative renewable energy sources is shown in Figure 22.30. This scenario (prepared by a major oil company) predicts that total energy consumed will triple by 2060. All of the growth in energy use after about 2020 is assumed to be from alternative sources, which under this scenario would supply about 60% of the total global demand. The energy scenario of Figure 22.30 continues the present growth path and is not sustainable. Nevertheless, the scenario is significant in recognizing a path toward more sustainable energy development. To have alternative energy resources supplying twice the total energy consumed today in the next 60 years seems optimistic but possible, with significant energy policy reforms, education, and enhanced research and development in alternative energy.

The United States, with 5% of the world's population, emits about 20% of the atmospheric carbon dioxide. It is encouraging to note that the U.S. government has acknowledged that reduction of emissions of carbon dioxide into the atmosphere is an important goal. Providing economic incentives to improve and design new energy-efficient technologies and alternative energy sources would be a positive step. However, the U.S. Congress has been slow to act or to fund necessary studies. It is discouraging that the United States, in the Hague meetings of late 2000, refused to honor reductions in emissions of CO_2 agreed to at Kyoto in 1997. Japan and Russia were also unwilling to cut emissions that they had earlier agreed to. In March 2001, the new administration in Washington disappointed European allies by announcing that the United States would not abide by the Kyoto protocol. More than 180 countries meeting in Bonn, Germany, in July 2001 adopted the Kyoto protocol in spite of the U.S. rejection of the agreement. Perhaps the positive action of so many nations will spur the United States to rethink its position.

How can carbon dioxide emissions be reduced? Energy planning that relies heavily on energy conservation and efficiency, along with use of alternative energy sources, would reduce emissions of carbon dioxide. Increased use of nuclear power would also reduce CO_2 emissions. Changing the balance of fossil fuels to burn more natural gas would also be helpful, because natural gas releases less carbon (Table 22.2).[33] Other ideas include increasing tax for using fossil fuels; providing economic incentives to increase use of mass transit and decrease use of automobiles; providing greater economic incentives to improve development of energy-efficient technology; requiring higher fuel economy standards for cars, trucks, and buses; and requiring higher standards of energy efficiency. Because burning forests to convert lands to agricultural purposes accounts for about 20% of anthropogenic carbon loading into the atmosphere, management plans that seek to minimize burning and protect the world's forests would help reduce the threat of global warming, as would

Table 22.2 Comparison of Common Fossil Fuels in Terms of Share of World Energy Supply and Percentage of Total Carbon Released

Fossil Fuel	World Energy Supply (%)*	Carbon Emitted as % of Total Carbon from Fossil Fuels	Comparison
Oil	32	43	Releases 23% less carbon per unit of energy than coal
Natural gas	22	21	Releases 28% less carbon per unit of energy than oil or gas
Coal	21	36	Releases much more carbon per unit of energy than oil or gas

* In total, fossil fuels comprise 75% of world energy supply.

Source: Data from J. Dunn, "Decarbonizing the Energy Economy," in *World Watch Institute State of the World 2001* (New York: W. W. Norton, 2001), pp. 83–102.

plans to plant trees (reforestation). The latter is an attempt to increase the biospheric sinks for CO_2. Other sinks may be enhanced through better management of soils, forest, and grasslands to sequester more carbon dioxide.[34]

Summary

- The atmosphere, a layer of gases that envelops Earth, is a dynamic system that is constantly changing. A great number of complex chemical reactions take place in the atmosphere, and atmospheric circulation takes place on a variety of scales, producing the world's weather and climates.

- Nearly all the compounds found in the atmosphere either are produced primarily by biological activity or are greatly affected by life.

- The four main processes that remove particles and pollutants from the atmosphere are sedimentation, rain out, oxidation, and photodissociation.

- On a local scale, climatic conditions vary considerably and may produce microclimates. For example, the presence of a city significantly affects local climate, producing an urban microclimate. Thus, some midlatitude cities are cloudier, warmer, rainier, and less humid than the surrounding areas.

- Major climatic changes have occurred during the past 2 million years, with periodic appearances and retreats of glaciers. During the past 1,500 years, several warming and cooling trends have affected people. During the past 100 years, the mean global annual temperature has apparently increased by about 0.5°C.

- Water vapor and several other gases, including carbon dioxide, tend to warm Earth's atmosphere through the greenhouse effect. The vast majority of the greenhouse effect is produced by water vapor, which is a natural constituent of the atmosphere. Carbon dioxide and other greenhouse gases also occur naturally in the atmosphere. However, especially since the Industrial Revolution, human activity has added substantial amounts of carbon dioxide to the atmosphere, along with such greenhouse gases as methane and CFCs.

- The mean global temperature of Earth is rising. Climatic models suggest that when carbon dioxide has doubled from its preindustrial levels in the next few decades, the mean global temperature may rise by 1–2°C. Total warming during the 21st century may range from 1.5°C to 6.0°C.

- Many complex positive feedback and negative feedback cycles affect the atmosphere. Nevertheless, it appears that increases in the emission of greenhouse gases from human activity is a significant cause of late-20th-century and present global warming. Natural cycles, solar forcing, aerosols, volcanic eruptions, and El Niño events also affect the temperature of Earth.

- Major effects of global warming include: (1) changes in climatic patterns and frequency and intensity of storms, (2) a rise in sea level, and (3) changes in the biosphere.

- Changes in climatic patterns and storms are worrisome because they may adversely affect agriculture and thus food supply. A rise in sea level is a potentially serious problem because so many people live in or near coastal areas and coastal erosion is a difficult problem. As sea level rises and more heat energy is fed into the atmosphere, cyclones and tropical disturbances will pose a greater hazard for people living in vulnerable areas.

- Changes in the biosphere include shifts in where specific plants and animals live, bleaching of coral, and spreading of diseases such as malaria to higher elevations.

- Adjustments to global warming include learning to live with the changes and attempting to mitigate warming through reduction of emissions of green-

Will Planting Trees Offset Global Warming?

Some scientists, foresters, environmentalists, and government officials have suggested that planting trees could ease the threat of global warming. Trees take up carbon dioxide during photosynthesis and can store carbon for long periods of time. It is an enticing proposal, since it does not require that we stop doing anything, such as driving large cars. Furthermore, it requires only a simple biological solution and would have other advantages, such as shading buildings, improving the aesthetics of urban areas, and providing more wood—a renewable resource.

Who could argue against planting a tree? Or even millions or billions of trees? But if we are to ask millions of people to plant billions of trees, we should have some confidence that they will have a substantial effect on reducing carbon dioxide in the atmosphere. (We should also be careful of what trees are planted where. Planting trees that are not native to a region—for example, eastern deciduous trees in Colorado—could cause other environmental problems, such as increase in water demand, that could offset any contribution to carbon dioxide removal.)

According to current estimates, humans are adding about 7.1 Gt (1 Gt, or gigaton, equals 1 billion metric tons, or 1.1 billion U.S. tons) of carbon, as carbon dioxide, to the environment each year, most as a result of burning fossil fuels. About 5.2 Gt remains in the atmosphere or is absorbed in the oceans. What happens to the remaining 1.9 Gt is controversial. Some scientists think it is absorbed by the oceans; others think it is absorbed by the forests; still others think that both the oceans and the forests are involved.

Living trees contain about 500 Gt of carbon; the atmosphere, 750 Gt (as carbon dioxide); and soils, an estimated 1,000 to 2,000 Gt. Each year, the trees of the world remove about 5 Gt of carbon through photosynthesis and return an equal amount to the atmosphere through respiration and decay. The forests of the world occupy an area of 4 Gha (1 Gha, or giga-hectare, equals 1 billion ha, or 2.5 billion acres). Annual rates of deforestation are estimated at 0.2% (0.008 Gha, or 8 million ha, or 19.8 billion acres), a rate of 1 ha every 4 seconds.

Critical Thinking Questions

1. If we planted trees equal to 1% of our present forests, how much carbon dioxide would be removed from the atmosphere, and why?

2. How many hectares of trees would we need to plant to offset the present rate of deforestation? Using an atlas, compare this area with some specific area of the world, such as Peru, the United States, or Australia.

3. Estimates of how much carbon could be taken up by trees are based on maximum net growth, but trees do not always grow at their maximum rate. What factors could interfere with maximum growth? How would such interference affect your earlier calculations?

4. Would you recommend a program of tree planting? If so, why, and what type of program? If not, why not?

house gases. It seems likely that the adjustment chosen will be to learn to live with change. A danger of this path is that climatic change may be rapid, resulting in problems that will be difficult to address.

- Because a significant component of global warming appears to result from anthropogenic processes, it is

prudent to reduce emissions of greenhouse gases. With control of CO_2 emissions, global warming may be limited to less than 2°C, which would cause far less harm than would a larger increase. Carbon dioxide emissions can be reduced through energy conservation and the use of alternative energy sources.

REEXAMINING THEMES AND ISSUES

Human Population: Much of this chapter centers on human-induced climatic change. Burning of fossil fuels and trees, among other processes, have resulted in increased emissions of carbon dioxide into the atmosphere, forcing climate change. Continued increases in the world's human population will place further demands on resources, which will result in continued increases in emissions of greenhouse gases. Thus, as stated many times in this book, control of

Sustainability:

increases in human population is needed to maximize the chances of solving many environmental problems—such as global warming.

Through our emissions of greenhouse gases, we are conducting global experiments, the results of which are difficult to predict. Because of this, achieving sustainability in the future will be more difficult. If we do not know what the consequences or magnitude of human-induced climatic change will be, then it is difficult to predict how we might achieve sustainable development for future generations. Some of the uncertainties concerning global climatic change will be resolved in the coming years, and we will be better prepared to develop plans that will work in addressing variables such as providing a sustainable food supply for the world's population.

Global Perspective:

One of the major topics of this chapter is global warming. In particular, we are interested in the potential effects of human activities. Our research on global warming centers on developing ways to look at what Earth is doing as a system with respect to climate; this is the essence of having a global perspective.

Urban World:

In this chapter, we explore the effects of urbanization on the local climate. As large urban areas continue to expand, we can expect that urban microclimate effects will become more apparent. As a result, it is important in urban planning to incorporate the development of green zones and other such areas to partially offset potentially adverse urban microclimate effects.

Science and Values:

It is becoming increasingly clear that people in industrialized countries, as well as other areas of the globe, are placing more value on the quality of the environment. Numerous research programs have been established to try to understand what factors affect global climate change and in particular what impact human activities are having. It is expected that people will continue to place a high priority on understanding the global climate system and how it is likely to change in the future. The tough questions will be the choices we make. Will we choose to significantly reduce emissions of carbon dioxide by greatly reducing our use of the fossil fuels for the food of the environment, or will we continue our economically successful use of fossil fuels?

● KEY TERMS

atmosphere **452**
climate **455**
climatic change **456**
Earth system science **459**
electromagnetic spectrum **462**

forcing **468**
global circulation models (GCMs) **461**
global warming **461**
greenhouse effect **463**

greenhouse gases **463**
microclimate **455**
polar amplification **473**
urban dust dome **456**

● STUDY QUESTIONS

1. What is the composition of Earth's atmosphere and how has life affected the atmosphere during the past several billion years?
2. How do midlatitude cities differ from surrounding rural areas in climate?
3. What is the greenhouse effect? What is its importance to global climate?
4. What is an anthropogenic greenhouse gas? Discuss the various anthropogenic greenhouse gases in terms of their potential to cause global warming.
5. What are some of the major negative feedback cycles

and positive feedback cycles that might increase or decrease global warming?
6. In terms of the effects of global warming, do you think that a change in climate patterns and storm frequency and intensity is likely to be more serious than a global rise in sea level? Illustrate your answer with specific problems and areas where the problems are likely to occur.
7. How would you refute or defend the statement that the most likely adjustment to global warming is to do little or nothing and learn to live with change?

● FURTHER READING

Anthes, A. R. 1992. *Meteorology*, 6th ed. New York: Macmillan. A short text providing a good overview of basic meteorology and atmospheric processes.

Goodess, C. M., J. P. Palutikof, and T. D. Davies. 1992. *The Nature and Causes of Climate Change: Assessing the Long-Term Future*. Lewis. A good text discussing short- and long-term climate changes and anthropogenic versus natural forcing in global warming.

Houghton, J. 1994. *Global Warming: The Complete Briefing*. Lion. A review for the nonscientist of knowledge regarding human-induced climate change, including discussions of ancient climates, weather forecasting, global freshwater use, and technological advances that could reduce the use of fossil fuels.

IPCC. 1991. *The Intergovernmental Panel on Climate Change Scientific Assessment*. New York: Oxford University Press. A detailed scientific review and assessment of global warming.

Mohnen, V. A., W. Goldstein, and C. W. Wang. 1991. "The Conflict over Global Warming," *Global and Environmental Change* 1(2):109–123. A paper providing detailed information on the global warming controversy that existed at the time.

Organisation for Economic Co-operation and Development, International Energy Agency. 1994. *The Economics of Climate Change*. Proceedings of an OECD/IEA Conference, OECD, IEA. These proceedings discuss the economics of climate change, differences of opinion on the matter, methods to link economic studies and climate change policy, and directions of international policy to deal with climate change.

Titus, J. G., and V. K. Narayanan. 1995. *The Probability of Sea Level Rise*. Washington, D.C.: U.S. Environmental Protection Agency. A close look at future sea-level rise due to global warming using mathematical models to predict changes.

CHAPTER 23 Air Pollution

A person guides a London bus at midday through thick fog on December 8, 1952.

LEARNING OBJECTIVES

The atmosphere has always been a sink—a place for deposition and storage—for gaseous and particulate wastes. When the amount of waste entering the atmosphere in an area exceeds the ability of the atmosphere to disperse or degrade the pollutants, problems result. After reading this chapter, you should understand:

- Why human activities that pollute the air, combined with meteorological conditions, may exceed the natural abilities of the atmosphere to remove wastes.
- What are the major categories and sources of air pollutants.
- Why air pollution problems are different in different regions.
- What acid rain is, how it is produced, what its environmental impacts are, and how they might be minimized.
- What methods are useful in the collection, capture, and retention of pollutants before they enter the atmosphere.
- What air quality standards are, and why they are important.
- Why determining the economics of air pollution is controversial and difficult.

London Smog and Indonesian Fires

In December 1952, air in London became stagnant, and cloud cover blocked incoming solar radiation. The temperature dropped rapidly. At noontime it was about −1°C (30°F), and humidity climbed to 80%. Thick fog developed; cold and dampness increased demand for home heating. Because the primary fuel used in homes was coal, emissions of ash, sulfur oxides, and soot increased quickly. Stagnant air was filled with pollutants from home heating fuels and automobile exhaust. At the height of the crisis, visibility was so reduced that automobiles used headlights at midday. Between December 4 and 10, about 4,000 Londoners died from the pollution. The siege of smog ended when weather changed and air pollution was dispersed. Environment, not human activities, finally solved this problem. What had gone wrong?

During the London smog crisis, stagnant weather conditions, combined with emissions from homes burning coal and from cars burning gasoline, exceeded the atmosphere's ability to remove or transform pollutants. As a result, sulfur dioxide remained in the air, and fog became acidic, adversely affecting people and other organisms. Human health effects were especially destructive because small acid droplets became fixed on larger particulates, which were easily drawn into the lungs.

The 1952 London smog crisis was a landmark event. Finally, the natural abilities of the atmosphere to serve as a sink for removal of wastes had been exceeded by human activities. The crisis was due, in part, to a reinforcing feedback situation. Burning fossil fuels added particulates to the air, increasing the formation of fog and decreasing visibility and light transmission. The dense, smoggy layer increased the dampness and cold and accelerated use of home heating fuels. As the weather and pollution worsened, more people burned fuel, which further worsened weather and pollution. London had been known for its fogs; long before 1952, what had been little known was the role of coal burning in exacerbating the fog conditions. The good news is that, since 1952, London fogs associated with air pollution have been greatly reduced because coal has been replaced by cleaner natural gas as the primary home heating fuel.

Forty-five years after the London smog crisis and following the most severe drought in 50 years, a serious air pollution event in Indonesia dwarfed the London event. As a result of a strong El Niño (see Chapter 22), huge

(a)

(b)

Figure 23.1 • *(a)* Image of air pollution (smoke from forest fires) over Indonesia and Indian Ocean on Oct. 22, 1997. White is smoke near fires. Red, yellow, and green represent decreasing

amounts of lower atmosphere pollution being transported westward. *(b)* Indonesian family members in Sumatra use air filters to help protect them from hazardous smoke on Oct. 6, 1997.

fires ravaged Indonesia for months (Figure 23.1), producing a thick, toxic haze of smoke. What had gone wrong in this instance?

Slash-and-burn practices have been part of farming in tropical rain forests for centuries. Each family burns a few hectares of rainforest, plants it, and after harvest moves on to a new area. Fire is the preferred method of land clearing because clearing land with bulldozers is too expensive. It is generally thought that slash-and-burn farming has not been particularly disturbing to the ecology of rain forests, as long as it is not too widespread and fires are carefully controlled.[1] Unfortunately, in 1997 a combination of events related to El Niño and out-of-control burns resulted in what some consider one of the world's greatest environmental disasters.

Monsoon rains were late that year, and people took advantage of dry weather to burn more forest. Although much of the blame was placed on small farmers, agriculture on an industrial scale also was responsible for clear-cut logging and burning. During the 1997 dry season, at least 20,000 hectares of land burned. Vast amounts of smoke and particulate matter entered the atmosphere,

causing a severe pollution problem. At one point, people were so desperate for relief from the dense, blinding haze that they attempted to use surgical masks as filters, but this provided little protection. Some 20 million Indonesians were treated for a variety of illnesses directly caused or aggravated by the fires. Smoke was so dense that a passenger airline crashed in Sumatra, killing 234 people. The Air Quality Index (AQI)—a measure of air quality—was at about 800 because of the fire. An AQI of 500 is considered very hazardous, and it is recommended that all people remain indoors (the AQI is discussed later in this chapter).

Near the end of summer, haze began to drift across the South China Sea to Malaysia and Singapore. Many of the fires were extinguished with the arrival of monsoons, but when the rains stopped in February 1998, more fires followed. There is worry that annual cycles of fire have begun that will have cumulative adverse environmental impacts. The damage from the fires is not limited to humans. Wildlife—including endangered species, such as rhinoceros—and wilderness reserves have also suffered damage.[1]

 The London smog of 1952 and the Indonesian fires of 1997–1998 are examples of serious, human-induced air pollution. This chapter discusses the major air pollutants, urban air, acid rain, and control of air pollution.

23.1 A Brief History of Air Pollution

As the fastest-moving fluid medium in the environment, the atmosphere has always been one of the most convenient places to dispose of unwanted materials. Ever since fire was first used by people, the atmosphere has been a sink for waste disposal.

People have long recognized the existence of atmospheric pollutants, both natural and human-induced. Leonardo da Vinci wrote in 1550 that a blue haze formed from materials emitted into the atmosphere from trees. He had observed a natural photochemical smog resulting from hydrocarbons given off by living trees whose cause is still not completely understood. This haze gave rise to the name Smoky Mountains for the range in the southeastern United States. The phenomenon of acid rain was first described in the 17th century; and by the 18th century, it was known that smog and acid rain damaged plants in London. Beginning with the Industrial Revolution in the 18th century, air pollution became more noticeable. The word *smog* was introduced by a physician at a public health

conference in 1905 to denote poor air quality resulting from a mixture of smoke and fog.

A major event in Donora, Pennsylvania, in 1948, was responsible for increasing research on air pollution in the United States. The Donora event remains the worst industrial air pollution incident in U.S. history, causing 20 deaths and 5,000 illnesses. What was called the "Donora Fog" involved pollutants from the Donora Zinc Works metal smelting plant and other sources. Pollutants, including sulfur dioxide, carbon monoxide, and heavy metals, were trapped by weather conditions in a narrow valley. The event lasted about three days until pollutants were washed out and dispersed by rainstorms.

The Donora event was followed in 1952 by the London smog crisis described in the opening case study. Following the Donora and London events, regulations to control air quality began. Today, in the United States and other countries, legislation to reduce emission of air pollutants has been successful, but more needs to be done. Chronic exposure to high levels of air pollutants continues to contribute to illnesses that kill people around the world.

What are the chances that another killing smog will occur somewhere in the world? Unfortunately, the chances are all too good, given the tremendous amount of air pollution in some large cities. For example, Beijing might be a candidate; the city uses an immense amount of coal, and coughing is so pervasive among its residents that they often refer to it as the "Beijing cough." Another likely candidate is Mexico City, which has one of the worst air pollution problems anywhere in the world today.

23.2 Stationary and Mobile Sources of Air Pollution

What are the sources of air pollution? The two major categories are stationary sources and mobile sources. **Stationary sources** are those that have a relatively fixed location. These include point sources, fugitive sources, and area sources.

- *Point sources*, as discussed in Chapter 15, emit pollutants from one or more controllable sites, such as smokestacks of power plants (Figure 23.2).
- *Fugitive sources* generate air pollutants from open areas exposed to wind processes. Examples include burning for agricultural purposes (Figure 23.3), as well as dirt roads, construction sites, farmlands, storage piles, surface mines, and other exposed areas from which particulates may be removed by wind.
- *Area sources*, also discussed in Chapter 15, are well-defined areas within which are several sources of air pollutants—for example, small urban communities, areas of intense industrialization within urban complexes, and agricultural areas sprayed with herbicides and pesticides.

Mobile sources of air pollutants move from place to place while emitting pollutants. These include automobiles, trucks, buses, aircraft, ships, and trains (Figure 23.4).[2]

23.3 General Effects of Air Pollution

Air pollution affects many aspects of our environment: its visual qualities, vegetation, animals, soils, water quality, natural and artificial structures, and human health. Air pollutants affect visual resources by discoloring the atmosphere and by reducing visual range and atmospheric clarity so that the visual contrast of distant objects is decreased. We cannot see as far in polluted air, and what we do see has less color contrast. These effects were once limited to cities, but they now extend to some wide-open spaces of the United States. For example, near the area where the borders of New Mexico, Arizona, Colorado, and Utah meet, emissions from the Four Corners fossil fuel–burning power plant are altering the visibility in a region where in the past, visibility was normally 80 km (50 mi) from a mountaintop on a clear day.[2]

Figure 23.2 • This steel mill in Beijing, China is a major source of air pollution.

Figure 23.3 • Burning sugar cane fields, Maui, Hawaii—an example of a fugitive source of air pollution.

Figure 23.4 • Exhaust from older diesel buses such as this one in Thailand are a major source of air pollution.

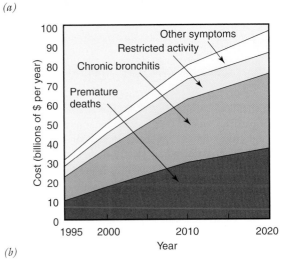

(a)

(b)

Figure 23.5 • *(a)* Air pollution in Beijing, China, in February 2001 is causing discomfort to this woman who is covering her nose and mouth. *(b)* Annual cost to human health in China. (*Source:* Modified from World Bank, Clear Water, Blue Skies, China's Environment in the New Century [World Bank, 1997].)

The effects of air pollution on vegetation are numerous. They include damage to leaf tissue, needles, and fruit; reduction in growth rates or suppression of growth; increased susceptibility to a variety of diseases, pests, and adverse weather; and disruption of reproductive processes.[2]

Air pollution is a significant factor in the human death rate for many large cities. For example, it has been estimated that in Athens, Greece, the number of deaths is several times higher on days when the air is heavily polluted; and in Hungary, where air pollution has been a serious problem in recent years, it may contribute to as many as 1 in 17 deaths. The United States is certainly not immune to health problems related to air pollution. The most polluted air in the country is found in the Los Angeles urban area, where millions of people are exposed to unhealthy air. It is estimated that as many as 150 million people live in areas of the United States where exposure to air pollution contributes to lung disease, which causes more than 300,000 deaths per year. Air pollution in the United States is directly responsible for annual health costs of about $50 billion.[3] In China, whose large cities have serious air pollution problems, mostly from burning coal, the health cost is now about $50 billion per year and may increase to about $100 billion per year by 2020 (Figure 23.5).

Air pollutants can affect human health in several ways (Figure 23.6). The effects on an individual depend on the dose or concentration (see the discussion of dose response in Chapter 15) and other factors, including indi-

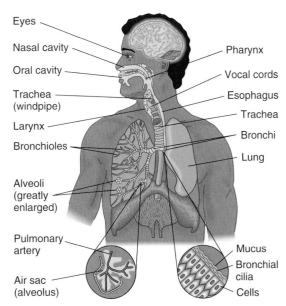

Figure 23.6 • Idealized diagram showing some of the parts of the human body (brain, cardiovascular system, and pulmonary system) that can be damaged by common air pollutants. The most severe health risks from normal exposures are related to particulates. Other substances of concern include carbon monoxide, photochemical oxidants, sulfur dioxide, and nitrogen oxides. Toxic chemicals and tobacco smoke also can cause chronic or acute health problems.

vidual susceptibility. Some of the primary effects of air pollutants include toxic poisoning, cancer, birth defects, eye irritation, and irritation of the respiratory system; increased susceptibility to viral infections, causing pneumonia and bronchitis; an increased susceptibility to heart disease; and aggravation of chronic diseases, such as asthma and emphysema. People suffering from respiratory diseases are the most likely to be affected by air pollutants. Healthy people tend to acclimate to pollutants in a relatively short period of time. However, this is a physiological tolerance; as explained in Chapter 15, it does not mean that the pollutants are doing no harm.

Many air pollutants have *synergistic effects* (in which the combined effects are greater than the sum of the sep-

---

Table 23.1   Effects of Major Air Pollutants on People, Plants, and Materials

Pollutant	Effects on People [a,b]	Effects on Plants [a,c]	Effects on Materials [a,d]
Ozone (O_3)	Strong irritant, aggravates asthma; causes injury to cells in respiratory system, decreased elasticity of lung tissue, coughing, chest discomfort; eye irritation	Flecking, stippling, spotting, and/or bleaching of plant tissue (leaves, stems, etc.); oldest leaves are most sensitive; tips of needles of conifers become brown and die; reduction of yields and damage to crops including lettuce, grapes, and corn	Cracks rubbers; reduces durability and appearance of paint, causes fabric dyes to fade
Sulfur dioxide (SO_2)	Increase in chronic respiratory disease; shortness of breath; narrowing of airways for people with asthma	Bleaching of leaves; decay and death of tissue; younger leaves are more sensitive than older; sensitive crops and trees include alfalfa, barley, cotton, spinach, beets, white pine, white birch, and trembling aspen; if oxidized to sulfuric acid, causes damage associated with acid rain	If oxidized to sulfuric acid, damages buildings and monuments, corrodes metal; causes paper to become brittle; turns leather to red-brown dust; SO_2 fades dyes of fabrics, damages paint
Nitrogen oxides (NO_x)[e]	A mostly nonirritating gas; may aggravate respiratory infections and symptoms (some throat, cough, nasal congestion, fever) and increase risk of chest cold, bronchitis, and pneumonia in children	No perceptible effects on many plants, but may suppress plant growth for some, and may be beneficial at low concentrations; oxidized to nitric acid, causes damage associated with acid rain	Causes fading of textile dyes; if oxidized to nitric acid, may damage buildings and monuments
Carbon monoxide (CO)	Reduces the ability of the circulatory system to transport oxygen; causes headache, fatigue, nausea, impairs performance of tasks that require concentration; reduces endurance; may be lethal, causing asphyxiation	None perceptible	None perceptible
Particulate matter (PM 2.5, PM 10)	Increased chronic and acute respiratory diseases; depending on chemical composition of particulates, may irritate tissue of throat, nose, lungs, and eyes	Depending on chemical composition of particles, may damage trees and crops; dry deposition of SO_2, when oxidized, is a form of acid rain	Contributes to and may accelerate corrosion of metal; may contaminate electrical contacts; damages paint appearance and durability; fades textile dyes

[a] Effects depend on dose (concentration of pollutant and time of exposure) and susceptibility of people, plants, and materials to a particular pollutant. For example, older people, children, and those with chronic lung diseases are more susceptible to O_3, SO_2, and NO_x

[b] Annual U.S. losses exceed $50 billion.

[c] Annual U.S. losses to crops is $1 billion to $5 billion.

[d] Annual U.S. losses exceed $5 billion.

[e] In NO_x, x refers to the number of oxygen atoms in the gas molecules, as in NO (nitric oxide) and NO_2 (nitrogen dioxide).

Sources: Modified from U.S. Environmental Protection Agency; R. W. Bunbel, D. L. Fox, D. B. Turner, and A. C. Stern. *Fundamentals of Air Pollution*, 3rd ed. (San Diego: Academic Press, 1994); and T. Godish *Air Quality*, 3rd ed. (Boca Raton: Lewis Publishers, 1997).

arate effects). For example, sulfate and nitrate may attach to small particles in the air, facilitating their inhalation deep into lung tissue. There, they may do greater damage to the lungs than a combination of the two pollutants would be expected to do based on their separate effects. This phenomenon has obvious health consequences; consider joggers breathing deeply of particulates as they run along the streets of a city.

The effects of air pollutants on vertebrate animals in general include impairment of the respiratory system; damage to eyes, teeth, and bones; increased susceptibility to disease, parasites, and other stress-related environmental hazards; decreased availability of food sources (such as vegetation affected by air pollutants); and reduced ability for successful reproduction.[2]

Air pollution can also degrade soil and water resources when pollutants from the air are deposited. Soils and water may become toxic from the deposition of various pollutants. Soils may also be leached of nutrients by pollutants that form acids. The effects of air pollution on human-made structures include discoloration, erosion, and decomposition of building materials. These effects are described when we explore the topic of acid rain later in this chapter. Effects of major air pollutants discussed in this chapter are listed in Table 23.1.

23.4 Primary and Secondary Pollutants, Natural and Human

The major air pollutants occur either in gaseous forms or as particulate matter (PM). Particulate matter pollutants (PM 10) are particles of solid or liquid substances less than 10 μm in diameter and may be organic or inorganic. The gaseous pollutants include sulfur dioxide (SO_2), nitrogen oxides (NO_x), carbon monoxide (CO), ozone (O_3); volatile organic compounds (VOC) such as hydrocarbons, hydrogen sulfide (H_2S), and hydrogen fluoride (HF).

Air pollutants can be classified as primary or secondary. **Primary pollutants** are those emitted directly into the air. They include particulates, sulfur dioxide, carbon monoxide, nitrogen oxides, and hydrocarbons. **Secondary pollutants** are produced through reactions between primary pollutants and normal atmospheric compounds. For example, ozone forms over urban areas through reactions of primary pollutants, sunlight, and natural atmospheric gases. Thus, ozone is a secondary pollutant.

The primary pollutants that account for nearly all air pollution problems are carbon monoxide (58%), volatile organic compounds (11%), nitrogen oxides (15%), sulfur oxides (13%), and particulates (3%). In the United States today, about 140 million metric tons of these materials enter the atmosphere from human-related processes. If these pollutants were uniformly distributed in the atmos-

phere, the concentration would be only a few parts per million by weight. Unfortunately, pollutants are not uniformly distributed but tend to be released, produced, and concentrated locally or regionally—for example, in large cities.

In addition to pollutants from human sources, our atmosphere contains many pollutants of natural origin. Examples of natural emissions of air pollutants include the following:

- Release of sulfur dioxide from volcanic eruptions. For example, volcanic activity on the island of Hawaii emits SO_2 and other pollutants, which react in the atmosphere to produce volcanic smog called "vog." The smog can present a health hazard to people and cause local acid rain.
- Release of hydrogen sulfide from geysers and hot springs and from biological decay in bogs and marshes.
- Release of ozone in the lower atmosphere as a result of unstable meteorological conditions, such as violent thunderstorms.
- Emission of a variety of particles from wildfires and windstorms.[2]

The data in Table 23.2 suggest that, with the exception of sulfur and nitrogen oxides, natural emissions of air pollutants exceed human-produced emissions. Nevertheless, it is the human component that is most abundant in urban areas and that leads to the most severe problems for human health.

23.5 Major Air Pollutants: Some Details

We now consider some major air pollutants in more detail.

Sulfur Dioxide

Sulfur dioxide (SO_2) is a colorless and odorless gas normally present at Earth's surface at low concentrations. A significant feature of SO_2 is that once it is emitted into the atmosphere it can be converted through complex oxidation reactions into fine particulate sulfate (SO_4) and removed from the atmosphere by wet or dry deposition. The major anthropogenic source of sulfur dioxide is the burning of fossil fuels, mostly coal in power plants (Table 23.2). Another major source comprises a variety of industrial processes, ranging from petroleum refining to the production of paper, cement, and aluminum.[2,4]

Adverse effects associated with sulfur dioxide depend on the dose or concentration present (see Chapter 15) and include injury or death to animals and plants, as well as corrosion of paint and metals. Crops such as alfalfa, cotton, and barley are especially susceptible. Sulfur dioxide is capable of causing severe damage to the lungs of human and other ani-

Table 23.2 • Major Natural and Human-Produced Components of Air Pollutants

Air Pollutants	Emissions (% of total)		Major Sources of Human-Produced Components	Percent
	Natural	Human-Produced		
Particulates	89	11	Industrial processes	51
			Combustion of fuels (stationary sources)	26
Sulfur oxides (SO$_x$)	50	50	Combustion of fuels (stationary sources, mostly coal)	78
			Industrial processes	18
Carbon monoxide (CO)	91	9	Transportation (mostly automobiles)	75
			Agricultural burning	9
Nitrogen dioxide (NO$_2$)		Nearly all	Transportation (mostly automobiles)	52
			Combustion of fuels (stationary sources, mostly natural gas and coal)	44
Ozone (O$_3$)	A secondary pollutant derived from reactions with sunlight, NO$_2$, and oxygen (O$_2$)		Concentration present depends on reactions in lower atmosphere involving hydrocarbons and thus automobile exhaust	
Hydrocarbons (HC)	84	16	Transportation (mostly automobiles)	56
			Industrial processes	16
			Evaporation or organic solvents	9
			Agricultural burning	8

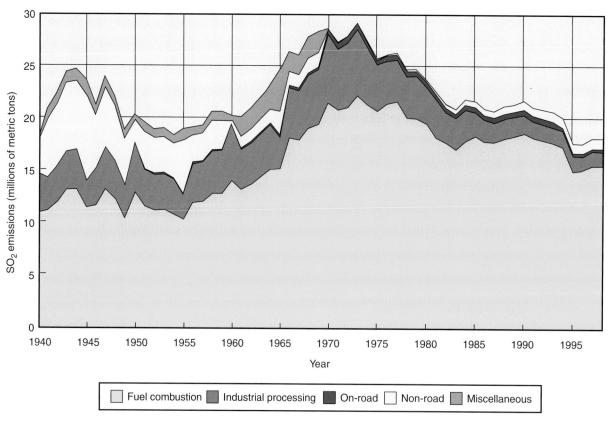

Figure 23.7 • Trend in emissions of sulfur dioxide from 1940 to 1998. On-road emissions are from light-duty gasoline vehicles and motorcycles, while non-road emissions are from burning gasoline in lawn mowers and industrial and recreational marine engines.

(*Source*: U.S. Environmental Protection Agency, National Air Pollution Emission Trends, 1990–1998, accessed August 19, 2000, at http://www.epa.gov.)

mals, particularly in the sulfate form. It is also an important precursor to **acid rain**.[2,4] (See A Closer Look 23.1.)

U.S. emission rates of SO_2 from 1940 to 1998 are shown in Figure 23.7. Emissions peaked at about 30 million metric tons in the early 1970s and since have been reduced to about 18 million metric tons, a reduction of about 40%, as a result of effective emission controls.

Nitrogen Oxides

Although nitrogen oxides (NO_x) occur in many forms in the atmosphere, they are emitted largely in two forms: nitric oxide (NO) and nitrogen dioxide (NO_2); and only these two forms are subject to emission regulations. The more important of the two is NO_2, a yellow-brown to reddish brown gas. A major concern with nitrogen dioxide is that it may be converted by complex reactions in the atmosphere to an ion, NO_3^{2-}, within small water particulates, impairing visibility. Both NO and NO_2 are major contributors to the development of smog, and NO_2 is a major contributor to acid rain (see A Closer Look 23.1). Nearly all NO_2 is emitted from anthropogenic sources. The two main sources are automobiles and power plants that burn fossil fuels.[2]

The environmental effects of nitrogen oxides on humans are variable but include irritation of eyes, nose, throat, and lungs and increased susceptibility to viral infections, including influenza (which can cause bronchitis and pneumonia).[2] Nitrogen oxides may suppress plant growth. When the oxides are converted to their nitrate form in the atmosphere, they impair visibility. However, when nitrate is deposited on the soil, it can promote plant growth through nitrogen fertilization.

U.S. emission rates of NO_x from 1940 to 1998 are shown in Figure 23.8. Emissions are primarily from combustion of fuels in power plants and vehicles. They increased from 1940 to the early 1970s and then leveled to about 22 million metric tons.

Carbon Monoxide

Carbon monoxide (CO) is a colorless, odorless gas that even at very low concentrations is extremely toxic to humans and other animals. The high toxicity results from a physiological effect—carbon monoxide and hemoglobin in blood have a strong natural attraction for one another. The hemoglobin in our blood will take up carbon monoxide nearly 250 times more rapidly than it will oxygen. Therefore, if there is any carbon monoxide in the vicinity, a person will take it in very readily, with potentially dire effects. Many people have been accidentally

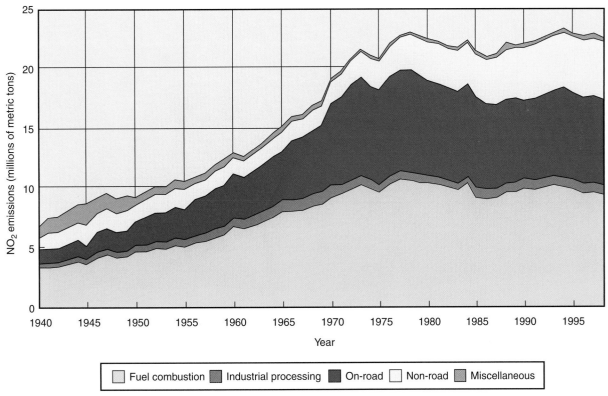

Figure 23.8 • Trend in emissions of nitrogen oxide from 1940 to 1998. On-road sources are light-duty gasoline vehicles and motorcycles, and non-road sources are gasoline-burning lawn mowers and industrial and recreational marine engines. (*Source*: U.S. Environmental Protection Agency, National Air Pollution Emission Trends, 1990–1998, accessed August 19, 2000, at http://www.epa.gov.)

asphyxiated by carbon monoxide produced from incomplete combustion of fuels in campers, tents, and houses. The effects depend on the dose or concentration of exposure and range from dizziness and headaches to death. Carbon monoxide is particularly hazardous to people with known heart disease, anemia, or respiratory disease. In addition, it may cause birth defects, including mental retardation and impairment of growth of the fetus.[2] Finally, the effects of carbon monoxide tend to be worse at higher altitudes, where oxygen levels are naturally lower. Detectors (similar to smoke detectors) are now in common use to warn people if CO in a building becomes concentrated at a potentially harmful level.

Approximately 90% of the carbon monoxide in the atmosphere comes from natural sources. The other 10% comes mainly from fires, automobiles, and other sources of incomplete burning of organic compounds. Concentrations of carbon monoxide can build up and cause serious health effects in a localized area.

Anthropogenic emissions of CO in the United States (1940–1998) are shown in Figure 23.9. Most emissions are through tailpipes of vehicles. Emissions peaked in the early 1970s at about 120 million metric tons. The rate in 1998 was about 80 million metric tons, a very significant reduction considering the increase in vehicles.

Reductions have resulted largely from cleaner-burning automobile engines.

Ozone and Other Photochemical Oxidants

Photochemical oxidants result from atmospheric interactions of nitrogen dioxide and sunlight. The most common photochemical oxidant is ozone (O_3), a colorless gas with a slightly sweet odor. In addition to ozone, a number of photochemical oxidants known as PANs (peroxyacyl nitrates) occur with photochemical smog (discussed later in the chapter).

Ozone is a form of oxygen in which three atoms of oxygen occur together rather than the normal two. Ozone is relatively unstable and releases its third oxygen atom readily, so it oxidizes or burns things more readily and at lower concentrations than does normal oxygen. Ozone is sometimes used to sterilize; for example, bubbling ozone gas through water is a method used to purify water. The ozone is toxic to and kills bacteria and other organisms in the water. When it is released into the air or produced in the air, ozone may injure living things.

Ozone is very active chemically, and it has a short average lifetime in the air. Because of the effect of sunlight on normal oxygen, ozone forms a natural layer high in the

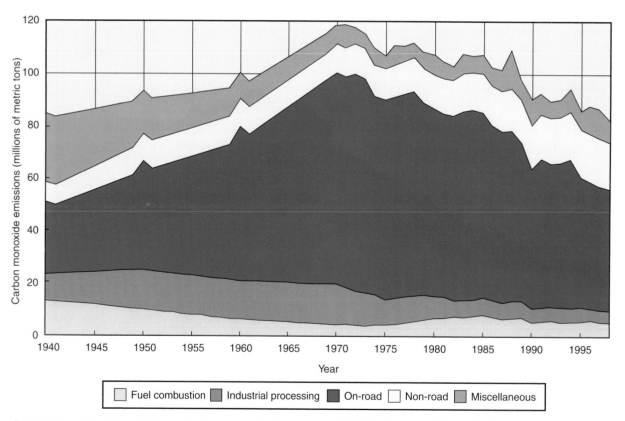

Figure 23.9 • Trend in emissions of carbon monoxide from 1940 to 1998. On-road sources are light-duty gasoline vehicles and motorcycles, and non-road sources are gasoline-burning lawn mowers and industrial, and recreational marine engines. (*Source*: U.S. Environmental Protection Agency, National Air Pollution Emission Trends, 1990–1998, accessed August 19, 2000, at http://www.epa.gov.)

atmosphere (stratosphere). This ozone layer protects us from harmful ultraviolet radiation from the sun. Thus, although ozone is considered a pollutant in the lower atmosphere when concentrations are above the National Ambient Air Quality Standard threshold, it is beneficial in the stratosphere. The important topic of stratospheric ozone depletion is discussed in Chapter 25.

Ozone is a secondary pollutant produced on bright, sunny days in areas where there is much primary pollution. The major sources of the chemicals that produce ozone, as well as other oxidants, are automobiles, fossil fuel burning, and industrial processes that produce nitrogen dioxide. Because of the nature of its formation, ozone is difficult to regulate. It is the pollutant whose health standard is most frequently exceeded in urban areas of the United States.[6] The adverse environmental effects of ozone and other oxidants, like those of other pollutants, depend in part on the dose or concentration of exposure and include damage to plants and animals as well as to materials such as rubber, paint, and textiles.

Effects of ozone on plants can be subtle. At very low concentrations, ozone can reduce growth rates while not producing any visible injury. At higher concentrations, ozone kills leaf tissue and, if pollutant levels remain high, whole plants. The death of white pine trees along highways in New England is believed due in part to ozone pollution. Ozone's effect on animals, including people, involves various kinds of damage, especially to the eyes and the respiratory system. At ozone levels to which many millions of Americans are often exposed, the exposure damages cell walls in lungs and airways. This causes tissue to redden and swell, inducing cellular fluids to seep into the lungs. Eventually, the lungs decrease in elasticity and are more susceptible to bacterial infection; and scars and lesions may form in the airways. Even young, healthy people may not be able to breathe normally, and on especially polluted days, breathing may be shallow and painful.[6]

Volatile Organic Compounds

Volatile organic compounds (VOCs) include a variety of organic compounds used as solvents in industrial processes such as dry cleaning, degreasing, and graphic arts. Hydrocarbons—compounds composed of hydrogen and carbon—comprise one group of VOCs. Thousands of hydrocarbon compounds exist, including natural gas, or methane (CH_4); butane (C_4H_{10}); and propane (C_3H_8). Analysis of urban air has identified many hydrocarbons, some of which react with sunlight to produce photochemical smog. Potential adverse effects of hydrocarbons are numerous. Many are toxic to plants and animals, and some may be converted to harmful compounds through complex chemical changes that occur in the atmosphere.

On a global basis, only about 15 percent of hydrocarbon emissions (primary pollutants) are anthropogenic. However, nearly half of hydrocarbons entering the atmosphere in the United States are emitted from anthropogenic sources. The largest human source of hydrocarbons in the United States is automobiles. Anthropogenic sources are particularly abundant in urban regions. However, in some southeastern U.S. cities such as Atlanta, Georgia, natural emissions probably exceed those from automobiles and other human sources.[3]

Emission trends for VOCs from 1940 to 1998 are shown in Figure 23.10. Like CO emissions, VOCs peaked in the early 1970s. As noted, a major source of hydrocarbons (VOC) is automobiles. Effective government-mandated emission controls for automobiles are thus responsible for the lowered emissions.

Particulate Matter (PM 10 and PM 2.5)

Particulate matter (PM 10) is made up of particles less than 10 μm in diameter. The term is used for varying mixtures of particles suspended in the air we breathe. Particles are present everywhere, but high concentrations and/or specific types of particles have been found to present a serious danger to human health. Asbestos, for example, is particularly dangerous.

Farming adds considerable particulate matter to the atmosphere, as do wind storms in areas with little vegetation and volcanic eruptions. Nearly all industrial processes, as well as the burning of fossil fuels, release particulates into the atmosphere. Much particulate matter is easily visible as smoke, soot, or dust; other particulate matter is not easily visible. Particulates include materials such as airborne asbestos particles and small particles of heavy metals, such as arsenic, copper, lead, and zinc, which are usually emitted from industrial facilities such as smelters.

Of particular concern are very fine particle pollutants (PM 2.5) less than 2.5 μm in diameter (2.5 millionths of a meter; Figure 23.11). For comparison, the diameter of human hair is about 60 μm to 150 μm. Fine particles are easily inhaled into the lungs, where they can be absorbed into the bloodstream or remain embedded for a long period of time. Among the most significant fine particulate pollutants are sulfates and nitrates. These are mostly secondary pollutants produced in the atmosphere through chemical reactions between normal atmospheric constituents and sulfur dioxide and nitrogen oxides. These reactions are important in the formation of sulfuric and nitric acids in the atmosphere and are further discussed when we consider acid rain.[2]

When measured, particulate matter is often referred to as *total suspended particulates* (TSPs). Values for TSPs tend to be much higher in large cities in developing countries such as Mexico, China, and India than in developed countries such as Japan and the United States (Figure 23.12).

Particulates affect human health, ecosystems, and the biosphere. In the United States, particulate air pollution contributes to the death of 60,000 people annually.[7] Recent studies estimate that 2–9% of human mortality in cities is associated with particulate pollution; risk of mortality is

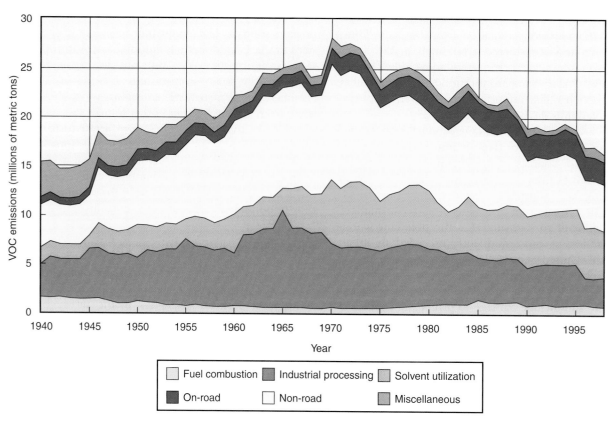

Figure 23.10 • Trend in emissions of volatile organic compounds from 1940 to 1998. On-road sources are light-duty gasoline vehicles and motorcycles, and non-road sources are gasoline-powered lawn mowers and industrial and recreational marine engines. (*Source*: U.S. Environmental Protection Agency, National Air Pollution Emission Trends, 1990–1998, accessed August 19, 2000, at http://www.epa.gov.)

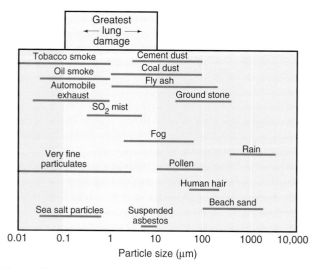

Figure 23.11 • Size of selected particulates. Shaded area shows size range that produces the greatest lung damage. (*Source*: Modified from Fig. 7-8, p. 244 from Chemistry, Man and Environmental Change: An Integrated Approach, by J. Calvin Giddings. Copyright © 1973 by J. Calvin Giddings. Reprinted by permission of HarperCollins Publishers, Inc.)

about 15–25% higher in cities with the highest levels of fine particulate pollution.[8] As mentioned, particulates that enter lungs may lodge there, with chronic effects on respiration. Particulates are linked to both lung cancer and bronchitis. Particulate matter is especially hazardous to the elderly and to individuals with respiratory problems, such as asthma. There is a direct relationship between particulate pollution and increased hospital admissions for respiratory distress.

Dust raised by road building and plowing not only makes breathing more difficult for animals (including humans) but also can be deposited on surfaces of green plants, where it may interfere with absorption of carbon dioxide and oxygen and release of water (transpiration). On a larger scale, particulates associated with large construction projects, such as housing developments, shopping centers, and industrial parks, may injure or kill plants and animals and damage surrounding areas, changing species composition, altering food chains, and thereby affecting ecosystems. Modern industrial processes have greatly increased total suspended particulates in Earth's atmosphere. Particulates block sunlight and may cause changes in climate. Such changes have lasting effects on the biosphere.

Anthropogenic emissions of PM 10 particles in the United States are shown on Figure 23.13. Emissions peaked in the early 1950s and today are low by comparison (75% lower).

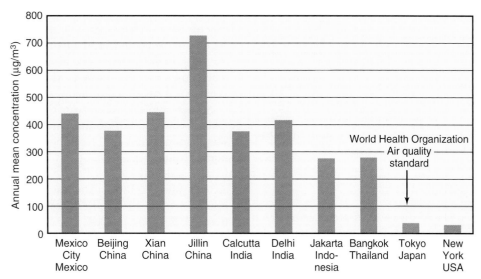

Figure 23.12 • Total suspended particulates (TSP) for several large cities in developing countries (red) and developed countries (yellow). The value of 100 μg/m3 is the air quality standard set by the World Health Organization. (*Source*: Modified from R. T. Watson, Intergovernment Panel on Climate Change, presentation at the Sixth Conference of Parties to the United Nations Framework Convention on Climate Change, November 13, 2000, Figure 20.)

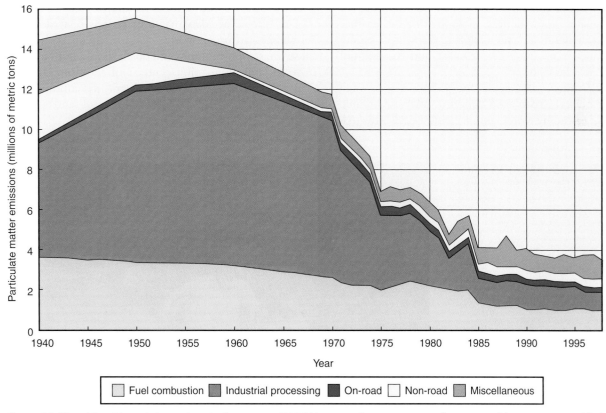

Figure 23.13 • Trend in emissions of particulate matter (PM 10) from 1940 to 1998. These emissions do not include fugitive dust sources, which are those stationary sources that generate particulate pollutants from open areas exposed to wind processes. On-road sources are light-duty vehicles and motorcycles, and non-road sources are gasoline-powered lawn mowers and industrial and recreational marine engines. (*Source*: U.S. Environmental Protection Agency, National Air Pollution Emission Trends, 1990–1998, accessed August 19, 2000, at http://www.epa.gov.)

Hydrogen Sulfide

Hydrogen sulfide (H_2S) is a highly toxic corrosive gas easily identified by its rotten egg odor. Hydrogen sulfide is produced from natural sources such as geysers, swamps, and bogs and from human sources such as industrial plants that produce petroleum or that smelt metals. Potential effects of hydrogen sulfide include functional damage to plants and health problems ranging from toxicity to death for humans and other animals.[4]

Hydrogen Fluoride

Hydrogen fluoride (HF) is a gaseous pollutant released by some industrial activities such as production of aluminum, coal gasification, and burning of coal in power plants. Hydrogen fluoride is extremely toxic. Even a small concentration (as low as 1 ppb) of HF may cause problems for plants and animals. HF is potentially dangerous to grazing animals because some forage plants can become toxic when exposed to this gas.[2]

Other Hazardous Gases

It is a rare month when the newspapers do not carry a story of a truck or train accident that releases toxic chemicals in a gaseous form into the atmosphere. People are often evacuated until the leak is stopped or the gas has dispersed to a nontoxic level. Chlorine gases and a variety of other materials used in chemical and agricultural processes may be involved.

Another source of gaseous air pollution is sewage treatment plants. Urban areas deliver a variety of organic chemicals, including paint thinner, industrial solvents, chloroform, and methyl chloride, to treatment plants by way of sewers. These materials are not usually removed in treatment plants; in fact, the treatment processes facilitate the evaporation of the chemicals into the atmosphere, where they may be inhaled. Many of the chemicals are toxic or are suspected of being carcinogens.

Some chemicals are so toxic that extreme care must be taken to ensure they do not enter the environment. This was demonstrated on December 3, 1984, when a toxic liquid from a pesticide plant leaked, vaporized, and formed a deadly cloud of gas that settled over a 64 km^2 area of Bhopal, India. The gas leak lasted less than an hour; yet more than 2,000 people were killed and more than 15,000 were injured. The colorless gas that resulted from the leak was methyl isocyanate, which causes severe irritation (burns on contact) to eyes, nose, throat, and lungs (Figure 23.14). Breathing the gas in concentrations of only a few parts per million (ppm) causes violent coughing, swelling of the lungs, bleeding, and death. Less exposure can cause a variety of problems, including loss of sight.

Methyl isocyanate is an ingredient of a common pesticide known in the United States as Sevin, as well as two other insecticides used in India. An industrial plant in West Virginia also makes the chemical, and small leaks not leading to major accidents occurred there before and after the catastrophic accident in Bhopal.

The Bhopal accident has implications related to our key theme "an urban world." Clearly, chemicals that can cause catastrophic injuries and death should not be stored close to large population centers. In addition, chemical plants need to have reliable accident-prevention equipment and personnel trained to control and prevent potential problems.

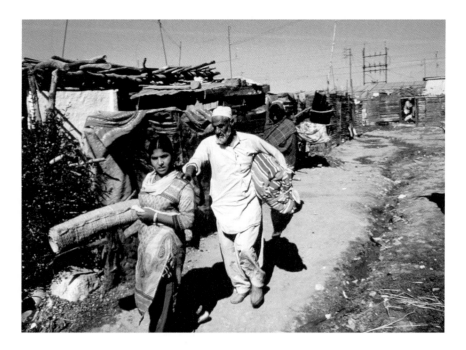

Figure 23.14 • Evacuation from Bhopal, India, in 1984, following a leak of toxic gas from a pesticide plant.

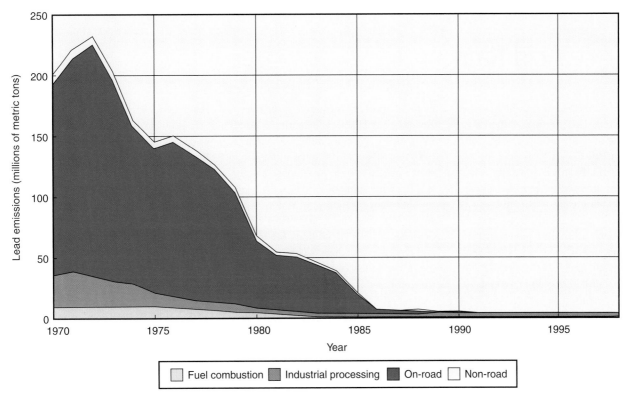

Figure 23.15 • Trend in emissions of lead from 1970 to 1998. On-road sources are light-duty gasoline vehicles and motorcycles, and non-road sources are gasoline-powered lawn mowers and industrial and recreational marine engines. (*Source*: U.S. Environmental Protection Agency, National Air Pollution Emission Trends, 1990–1998, accessed August 19, 2000, at http://www.epa.gov.)

Lead

Lead is an important constituent of automobile batteries and many other industrial products. When lead is added to gasolines, it helps protect engines and promotes more even fuel consumption. Lead in gasoline (which is still used in some countries) is emitted into the air with the exhaust. By this process, lead has been spread widely around the world and has reached high levels in soils and waters along roadways.

Once released, lead can be transported through the air as particulates to be taken up by plants through the soil or deposited directly on plant leaves. Thus, it enters terrestrial food chains. When lead is carried by streams and rivers, deposited in quiet waters, or transported to oceans or lakes, it is taken up by aquatic organisms and enters aquatic food chains.

Lead reaches Greenland as airborne particulates and via seawater and is stored in glacial ice. The concentration of lead in Greenland glaciers was essentially zero in A.D. 800 and reached measurable levels with the beginning of the Industrial Revolution in the mid-18th century. The lead content of the glacial ice increased steadily from 1750 until about 1950, when the rate of lead accumulation began to increase rapidly. This sudden upsurge reflects rapid growth in the use of lead additives in gaso-

line. The accumulation of lead in the Greenland ice illustrates that our use of heavy metals in the 20th century reached the point of affecting the entire biosphere.

Lead has now been removed from nearly all gasoline in the United States, Canada, and much of Europe. In the United States, lead emissions have been reduced by about 98% since the early 1970s (Figure 23.15). The reduction and eventual elimination of lead in gasoline is a good start in reducing levels of anthropogenic lead in the biosphere.

23.6 Variability of Air Pollution

Pollution problems vary in different regions of the world. There is great variance even within the United States. For example, s we will see, in the Los Angeles basin and many U.S. cities, nitrogen oxides and hydrocarbons are particularly troublesome because they combine in the presence of sunlight to form photochemical smog. Most of the nitrogen oxides and hydrocarbons are emitted from automobiles, a collection of mobile sources.[18] In other U.S. regions, such as in Ohio and the Great Lakes region, air quality problems also result from emissions of sulfur dioxide and particulates from industry and from coal-burning power plants, which are point sources.

Acid Rain

Acid rain is precipitation in which the pH is below 5.6. The pH of a solution is an expression of relative acidity and alkalinity. It is the negative logarithm of the concentration of the hydrogen ion (H^+). Many people are surprised to learn that all rainfall is slightly acidic; water reacts with atmospheric carbon dioxide to produce weak carbonic acid. Thus, pure rainfall has a pH of about 5.6, where 1 is highly acidic and 7 is neutral (see Figure 23.16). (Natural rainfall in tropical rain forests has been observed in some instances to have a pH of less than 5.6; this is probably related to acid precursors emitted by the trees.) Because the pH scale is logarithmic, a pH value of 3 is 10 times more acidic than a pH value of 4 and 100 times more acidic than a pH value of 5. Automobile battery acid has a pH value of 1.

Acid rain includes both wet (rain, snow, fog) and dry (particulate) acidic depositions. The depositions occur near and downwind of areas where the burning of fossil fuels creates major emissions of sulfur dioxide (SO_2) and nitrogen oxides (NO_x). Although these oxides are the primary contributors to acid rain, other acids are also involved. An example is hydrochloric acid, which is emitted from coal-fired power plants.

Acid rain has likely been a problem at least since the beginning of the Industrial Revolution. In recent decades, however, acid rain has gained more and more attention; and today, it is a major global environmental problem affecting all industrial countries. In the United States, nearly all of the eastern states are affected, as well as West Coast urban centers such as Seattle, San Francisco, and Los Angeles. The problem is also of great concern in Canada, Germany, Scandinavia, and Great Britain. Developing countries that rely heavily on coal, such as China, are facing serious acid rain problems as well.

Causes of Acid Rain

As noted, sulfur dioxide (SO_2) and nitrogen oxides (NO_x) are the major contributors to acid rain. Amounts of these substances emitted into the environment in the United States are shown in Figures 23.7 and 23.8. Emissions of SO_2 peaked in the 1970s at about 30 million metric tons per year and had declined to about 18 million metric tons per year by 1998. Nitrogen oxides leveled off at about 23 million metric tons per year in the mid-1980s.

In the atmosphere, sulfur dioxide and nitrogen oxides are transformed by reactions with oxygen and water vapor to sulfuric and nitric acids. These acids may travel long distances with prevailing winds to be deposited as acid precipitation (Figure 23.17). As mentioned, such precipitation may take the form of rainfall, snow, or fog. Sulfate and nitrate particles may also be deposited directly on the surface of the land as dry deposition. These

Figure 23.16 ● The pH scale. (Source: Modified from U.S. Environmental Protection Agency, 1980.)

Figure 23.17 ● Idealized diagram showing selected aspects of acid rain formation and paths.

particles may later be activated by moisture to become sulfuric and nitric acids.

Sulfur dioxide is emitted primarily from stationary sources, such as power plants that burn fossil fuels, whereas nitrogen oxides are emitted from both stationary sources and transport-related sources, such as automobiles. Approximately 80% of sulfur dioxide and 65% of nitrogen oxides in the United States come from states east of the Mississippi River.

In some areas, stationary sources have attempted to reduce the local effects of emissions by constructing taller emission stacks. Taller stacks reduced local concentrations of air pollutants but increased regional effects by spreading pollution more widely. Tall stacks increase average residence time of pollutants emitted into the atmosphere from 1–2 days to 10–14 days, because pollutants enter the atmosphere at a greater altitude, where mixing and transport by wind are more effective. Thus, this practice simply created more widespread problems. For example, problems associated with acid precipitation in Canada can be traced to emissions of sulfur dioxide and other pollutants in the Ohio Valley.

Analysis of the distances over which sulfur compounds can be transported before deposition suggests that approximately one-third of the total amount deposited over the eastern United States originates from sources farther away than 500 km (300 mi). Another one-third comes from sources between 200 and 500 km (about 125–300 mi) away, and the remainder comes from sources less than 200 km away.[9]

Sensitivity to Acid Rain

Geology and climatic patterns as well as types of vegetation and soil composition affect the potential impact of acid rain. Figure 23.18, showing areas of the United States and Canada sensitive to acid rain, is based on some of these factors. Sensitive areas are those in which bedrock or soil cannot buffer acid input. Materials (chemicals) that have the ability to neutralize acids are called *buffers*. Calcium carbonate ($CaCO_3$), the mineral calcite, which is present in many soils and rock (limestone), is an important natural buffer to acid rain. Hydrogen in

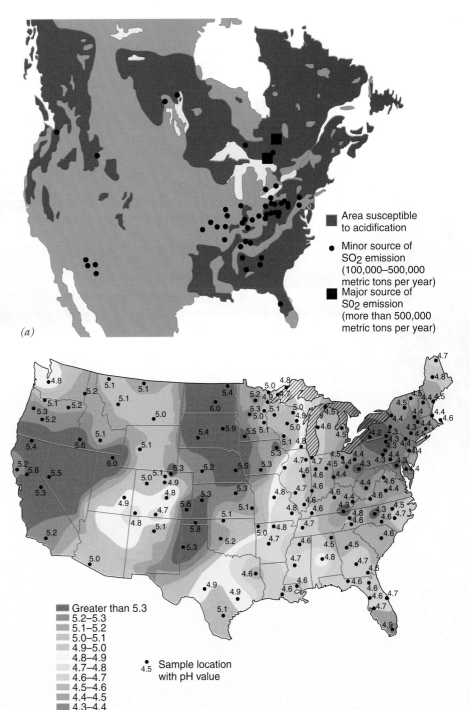

Figure 23.18 • (a) Areas in Canada and the United States that are sensitive to acid rain. (*Source*: "How Many More Lakes Have to Die?" *Canada Today* 12, no. 2 [1981] (b) pH of precipitation over the United States in 2000. Notice the relationship between the numerous sources of SO_2 in the eastern U. S. [in (a)] and low (more acidic) pH values. (*Source*: National Atmospheric Deposition Program/National Trends Network, 2000.)

acid reacts with calcium carbonate, and the reaction neutralizes acid. Thus, areas less likely to suffer damage from acid rain are those in which bedrock contains limestone or other carbonate material or where soils contain calcium carbonate, which buffers the acid. In contrast, areas with abundant granitic rocks and areas in which soils have little buffering action are sensitive to acid rain. Soils may lose their fertility when exposed to acid rain, either because nutrients are leached out by acid water moving through the soil or because the acid in the soil releases elements that are toxic to plants.

Forest Ecosystems

It has long been suspected that acid precipitation, whether snow, rain, fog, or dry deposition, adversely affects trees. Studies in Germany led scientists to cite acid rain and other air pollution as the cause of death for thousands of acres of evergreen trees in Bavaria. Similar studies in the Appalachian Mountains of Vermont (where many soils are naturally acidic) suggest that in some locations half the red spruce trees have died in recent years. Damage is attributed in part to acid rain and fog with pH levels of 4.1 and 3.1, respectively.

We have already seen that acid rain can cause loss of nutrients in the soil.

This loss of nutrients results in weakening of trees, and they become more susceptible to disease, drought, and consumption by herbivores. As trees weaken and die, there is less habitat and food for birds and animals. Trees stripped of leaves and needles allow more light through to the forest floor, changing the temperature and water content of soil at the surface. This in turn affects what grows on the forest floor and what lives in the soil. Thus, the entire forest ecosystem is affected by acid rain. This is an example of the principle of environmental unity introduced in Chapter 3.

Lake Ecosystems

In recent years, fish have disappeared from lakes in Scandinavia. Records from Scandinavian lakes show an increase in acidity accompanied by a decrease in fish (Figure 23.19). The increased acidity has been traced to acid rain, the result of

industrial processes in other countries, particularly Germany and Great Britain.

Acid rain affects lake ecosystems in two ways. First, it damages aquatic species (fish, amphibians, and crayfish) directly by disrupting their life processes in ways that limit growth or causes death. For example, crayfish produce fewer eggs in acidic water, and the eggs produced often grow into malformed larvae. Second, acid rain dissolves chemical elements necessary for life in the lake. Once in solution, the necessary elements leave the lake with water outflow. Thus, elements that once cycled in the lake are lost. Without these nutrients, algae do not grow, animals that feed on the algae have little to eat, and animals that feed on these animals also have less food.

To better study the effects of acidification on lakes, scientists in Canada added sulfuric acid to a lake in northwest Ontario over a period of years and observed the

Figure 23.19 ● *(a)* In Norway, many lakes in the south have severe problems with acid rain. *(b)* The rain has become more acidic over the last 20 years, as measured at Oslo and at five other sites in southern Norway. *(c)* The catch of fish, as illustrated by the catch of salmon in the Tovdalselva River of southern Norway, has decreased dramatically. (*Sources*: I. P. Muniz, H. Leivestad, E. Gjessing, E. Joranger, and D. Svalastog, Acid Precipitation: Effects on Forest and Fish, SNSF Project, IR 13/75 [As, Norway: Government of Norway, 1976]; S. Oden, "The Acidity

Problem—An Outline of Concepts," *Water, Air, and Soil Pollution* 6 [1976]:137–166; E. Snekvik, Norwegian Directorate for Game and Freshwater Fish, unpublished report, 1970; and R. F. Wright, T. Dale, E. G. Gjessing, G. R. Hendrey, A. Henriksen, M. Johannessen, and I. P. Muniz, "Impact of Acid Precipitation on Freshwater Ecosystems in Norway," *Water, Air, and Soil Pollution* 6 [1976]:438–499 [reprinted by permission of Kluwer Academic Publishers].)

effects. When the experiment started, the pH of the lake was 6.8. The following year, owing to addition of the acid, the pH dropped to 6.1. The initial drop in pH was not harmful to the lake; but as more and more acid was added, the pH dropped first to 5.8, then to 5.6, then to 5.4, and finally, five years after the project started, to 5.1. The problems started when the pH was lowered to 5.8. Some species disappeared, and others experienced reproductive failure. At a pH of 5.6, the death rate among lake trout embryos increased. When the pH was lowered to 5.4, lake trout reproduction failed.[10]

These experiments have proved valuable in pointing out what we can expect in thousands of other lakes that are now becoming increasingly acidified. The precise processes involved in the toxicity and damage to the lake are poorly understood. However, it is known that acid rain leaches metals, such as aluminum, lead, mercury, and calcium, from the soils and rocks in a drainage basin and discharges them into rivers and lakes. The elevated concentrations of aluminum are particularly damaging to fish, because the metal can clog the gills and cause suffocation. The heavy metals may pose health hazards to humans, because they may become concentrated in fish and then passed on to people, mammals, and birds when the fish are eaten. Drinking water taken from acidic lakes may also have high concentrations of toxic metals.

Not all lakes are as vulnerable to acidification as was the lake in the Ontario experiment. Acid is neutralized in waters with a high carbonate content (in the form of the ion HCO_3^-). Lakes on limestone or other rocks rich in calcium or magnesium carbonates can therefore readily buffer river and lake water against the addition of acids. Lakes with high concentrations of such elements are called hard-water lakes. Lakes on sand or igneous rocks, such as granite, tend to lack sufficient buffering to neutralize acids and are more susceptible to acidification.[11]

Thousands of rivers and lakes in the United States and Canada located in areas sensitive to acid rain are currently in various stages of acidification. In Nova Scotia, at least a dozen rivers have water so acidic part of the year that they no longer support healthy populations of Atlantic salmon. In the northeastern United States, about 200 lakes in the Adirondacks are no longer able to support fish; and thousands more are slowly losing the battle with acid rain.

Human Society

Acid rain damages not only forests and lakes but also many building materials, including steel, galvanized steel, paint, plastics, cement, masonry, and several types of rock, especially limestone, sandstone, and marble (Figure 23.20). Classical buildings on the Acropolis in Athens and in other cities show considerable decay (chemical weathering) that accelerated in the 20th century as a result of air pollution. The problem has grown to such an extent that buildings require restoration, and statues and other monuments must have protective coatings replaced quite frequently, resulting in costs of billions of dollars a year. Particularly important statues in Greece and other areas have been removed and placed in protective glass containers, with replicas standing in their former outdoor locations for tourists to view.[10]

In the United States, cities along the eastern seaboard are more susceptible to acid rain today because emissions of sulfur dioxide and nitrogen oxide are more abundant there. However, as noted, the problem is moving westward; acid precipitation has been recorded in California. Acid fog in Los Angeles may have a pH as low as 3—over 10 times as acidic as the average acid rain in the eastern United States. In contrast to acid rain, which may form relatively high in the atmosphere and travel long distances, acid fog forms near the ground. Water vapor mixes with pollutants and turns into an acid, which evidently condenses around very fine particles of smog; if the air is sufficiently humid, a fog may form. When the fog eventually dissipates, nearly pure drops of sulfuric acid may be left behind. Tiny particles containing the acid may be inhaled deeply into people's lungs—a considerable health hazard.

Stone decay occurs about twice as rapidly in cities as it does in less urban areas. The damage comes mainly from acid rain and humidity in the atmosphere, as well as from corrosive groundwater.[12] This implies that measuring

Figure 23.20 • Damage to a statue in Chicago resulting from an acid deposition (left) and the same statue following restoration (right).

rates of stone decay will tell us something about changes in the acidity of rain and groundwater in different regions and ages. It is now possible, where the ages of stone buildings and other structures are known, to determine if the acid rain problem has changed through time.

Control of Acid Rain

The cause of acid precipitation is known. It is the solution we are struggling with.

One solution to lake acidification is rehabilitation by the periodic addition of lime, as has been done in New York state, Sweden, and Ontario. This solution is not satisfactory over a long period, however, because it is expensive and requires a continuing effort. This solution to the acid rain problem is to ensure that the production of acid-forming components in the atmosphere is minimized. The only long-term solution involves decreasing emissions of sulfur dioxide and nitrogen oxides. From an environmental point of view, the best strategy is increasing energy efficiency and conservation measures that result in burning less coal and using nonpolluting alternative energy sources. Another strategy is to utilize pollution abatement technology at power plants to lower emissions of air pollutants.[13] Such technology is expensive and an additional cost in producing energy.

Air pollution also varies with the time of year. For example, smog is usually a problem mostly in the summer months when there is a lot of sunlight; and particulates are a problem in dry months when wildfires are likely and during months when the wind blows across the desert.

Las Vegas: Particulates

Pollution from particulates is a problem in arid regions where little vegetation is present and wind can easily pick up and transport fine dust. For example, the brown haze over Las Vegas, Nevada, is mostly due to naturally occurring particles (PM 10) from the desert environment. Las Vegas in the 1990s was the most rapidly growing urban area in the United States. Population in Clark County, which includes Las Vegas, increased from less than 300,000 in 1970 to nearly 1.5 million in 2000. Las Vegas also has some of the most polluted air in the southwestern United States (Figure 23.21). As mentioned, the main problem is the nearly 80,000 metric tons of PM 10 particles that enter the air in the Las Vegas region. About 60% of the dust comes from new construction sites, dirt roads, and vacant land. The remainder is natural windblown dust. Las Vegas also has a carbon monoxide pollution problem from vehicles; but it is the particulates that are causing concern, possibly leading to future Environmental Protection Agency sanctions and growth restrictions.

Haze from Afar

Air quality concerns are not restricted to urban areas. For example, the North Slope of Alaska is a vast strip of land approximately 200 km (125 mi) wide that is considered by many to be one of the last unspoiled wilderness areas left on Earth. It seems logical to assume that air in the Arctic environments of Alaska would be pristine in quality, except perhaps near areas where petroleum is being vigorously developed. However, ongoing studies suggest that the North Slope has an air pollution problem that originates from sources in Eastern Europe and Eurasia.

It is suspected that pollutants from the burning of fossil fuels in Eurasia are transported via the jet stream, moving at speeds that may exceed 400 km/hr (250 mi/hr), northeast from Eurasia over the North Pole and eventually to the North Slope of Alaska. There, the air mass slows, stagnates, and produces what is known as the Arctic haze. Concentrations of air pollutants, which include oxides of sulfur and nitrogen, are high enough that the air quality is being compared with that of some eastern U.S. cities, such as Boston. Air quality problems in remote areas such as Alaska have significance as we try to understand air pollution at the global level.[19]

Another global event occurred in the spring of 2001, when a white haze consisting of dust from Mongolia and industrial particulate pollutants arrived in North America.

Figure 23.21 • Las Vegas haze resulting from particulate (PM 10) pollution. Sources of particulates include construction sites and dirt roads (60% of total particulates) and natural and other sources (40%). (*Source*: Bob Carey, Los Angeles Times, February 2, 2001.)

The haze affected one fourth of the United States and could be seen from Canada to Mexico. The particulates were close enough to the ground to cause respiratory problems for people. In the United States, pollution levels from the haze were as high as two-thirds of federal health limits. The haze demonstrates what was formerly believed—that pollution from Asia is carried by winds across the Pacific Ocean.

23.7 Urban Air Pollution

Wherever many sources emit air pollutants over a wide area (whether automobile emissions in Los Angeles or smoke from wood-burning stoves in Vermont), air pollution can develop. Whether air pollution does develop depends on topography and meteorological conditions, because these factors determine the rate at which pollutants are transported away from their sources and converted to harmless compounds in the air. When the rate of production exceeds the rate of degradation and of transport, dangerous conditions can develop, as illustrated in the case study that opened this chapter.

Influences of Meteorology and Topography

Meteorological conditions can determine whether air pollution is a nuisance or a major health problem. The primary adverse effects of air pollution are damage to green plants and aggravation of chronic illnesses in people; most of these effects are due to relatively low-level concentrations of pollutants over a long period of time.

Periods of pollution generally do not directly cause large numbers of deaths. However, as with the London and Pennsylvania cases described earlier, serious pollution events (disasters) can develop over a period of days and lead to increases in illnesses and deaths.

In the lower atmosphere, restricted circulation associated with inversion layers may lead to pollution events. An **atmospheric inversion** occurs when warmer air is found above cooler air, and it poses a particular problem when there is a stagnant air mass. Figure 23.22 shows two types of atmospheric inversion that may contribute to air pollution problems. In the upper diagram, which is somewhat analogous to the situation in the Los Angeles area, descending warm air forms a semipermanent inversion layer. Because the mountains act as a barrier to the pollution, polluted air moving in response to the sea breeze and other processes tends to move up canyons, where it is trapped. The air pollution that develops occurs primarily during the summer and fall.

The lower part of Figure 23.22 shows a valley with relatively cool air overlain by warm air. This type of inversion can occur when cloud cover associated with a stagnant air mass develops over an urban area. Incoming solar radiation is blocked by the clouds, which reflect and absorb some of the solar energy, and are warmed. On the ground, or near Earth's surface, the air cools. If there is moisture in the air (humidity), then as the air cools, the dew point (the temperature at which water vapor condenses) is reached, and a fog may form. Because the air is cold, people burn more fuel to heat their homes and factories, and more pollutants are delivered into the atmosphere. As long as the stagnant conditions exist, the pol-

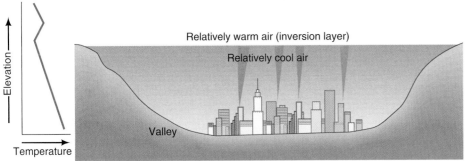

Figure 23.22 • Two causes for the development of atmospheric inversion, which may aggravate air pollution problems.

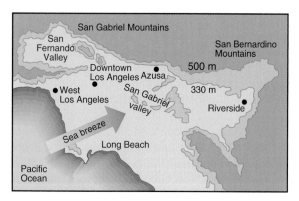

Figure 23.23 • Part of southern California showing the Los Angeles basin (south coast air basin). (*Source*: Modified from S. J. Williamson, *Fundamentals of Air Pollution*, © 1973, by Addison-Wesley, Reading, Mass.)

lutants will build up. It was this mechanism that caused the deadly 1952 London smog.

Cities situated in a valley or topographic bowl surrounded by mountains are more susceptible to smog problems than are cities in open plains. Surrounding mountains and the occurrence of temperature inversions prevent the pollutants from being transported by winds and weather systems. The production of air pollution is particularly well documented for Los Angeles, which has mountains surrounding part of the urban area and lies within a region where the air lingers, allowing pollutants to build up (Figure 23.23).

Potential for Urban Air Pollution

We have seen that topographical and meteorological conditions are important in the development of air pollution.

More specifically, the potential for air pollution in urban areas is determined by the following factors:

1. The rate of emission of pollutants per unit area.
2. The downwind distance that a mass of air moves through an urban area.
3. The average speed of the wind.
4. The elevation to which potential pollutants can be thoroughly mixed by naturally moving air in the lower atmosphere (Figure 23.24).[14]

The concentration of pollutants in the air is directly proportional to the first two factors. That is, as either the emission rate or downwind travel distance increases, so will the concentration of pollutants in the air. A good example is provided by the Los Angeles basin (Figure 23.23). If there is a wind from the ocean, as is generally the case, coastal areas such as West Los Angeles will experience much less air pollution than inland areas such as Riverside.

Assuming a constant rate of emission of air pollutants, the air mass will collect more and more pollutants as it moves through the urban area. The inversion layer acts as a lid for the pollutants; however, near a geological barrier, such as a mountain, there may be a chimney effect, in which the pollutants spill over the top of the mountain (see Figures 23.23 and 23.24). This effect has been noticed in the Los Angeles basin, where pollutants may climb several thousand meters, damaging mountain pine trees and other vegetation and spoiling the air of mountain valleys.

City air pollution decreases with increases in the third and fourth factors, which are meteorological: the wind velocity and the height of mixing. The stronger the wind and the higher the mixing layer, the lower the pollution.

Figure 23.24 • The higher the wind velocity and the thicker the mixing layer (shown here as H), the less the air pollution. The greater the emission rate and the longer the downwind length of the city, the greater the air pollution. The chimney effect allows polluted air to move over a mountain and down into an adjacent valley.

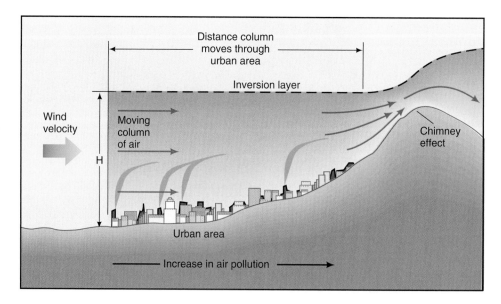

Smog

Smog is a general term first used in 1905 for a mixture of smoke and fog that produced unhealthy urban air. It is the most recognized term for urban air pollution. There are two major types of smog: **photochemical smog**, which is sometimes called *L.A.–type smog* or *brown air*; and **sulfurous smog**, which is sometimes referred to as *London-type smog, gray air*, or *industrial smog*.

Solar radiation is particularly important in the formation of photochemical smog (Figure 23.25). The reactions that occur in the development of photochemical smog are complex and involve both nitrogen oxides (NO_x) and organic compounds (hydrocarbons).

The development of photochemical smog is directly related to automobile use. Figure 23.26 shows a characteristic pattern in terms of how the nitrogen oxides, hydrocarbons, and oxidants (mostly ozone) vary through a typically smoggy day in southern California. Early in the morning, when commuter traffic begins to build up, the concentrations of nitrogen oxide (NO) and hydrocarbons begin to increase. At the same time, the amount of nitrogen dioxide NO_2 may decrease, because sunlight breaks it down to produce NO plus atomic oxygen (NO + O). The atomic oxygen (O) is then free to combine with molecular oxygen (O_2) to form ozone (O_3). As a result, the concentration of ozone also increases after sunrise. Shortly thereafter, oxidized hydrocarbons react with NO to increase the concentration of NO_2 by midmorning. This reaction causes the NO concentration to decrease and allows ozone to build up, producing a midday peak in ozone and a minimum in NO. As the smog develops, visibility may be greatly reduced (Figure 23.27) as light is scattered by the pollutants.

Sulfurous smog is produced primarily by the burning of coal or oil at large power plants. Sulfur oxides and particulates combine under certain meteorological conditions to produce a concentrated sulfurous smog (Figure 23.28).

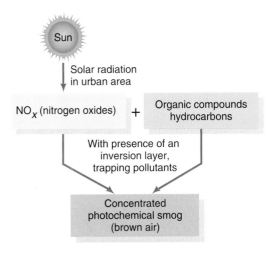

Figure 23.25 • How photochemical smog is produced.

Figure 23.26 • Development of photochemical smog over the Los Angeles area on a typical warm day.

(a)

(b)

Figure 23.27 • The city of Los Angeles, California, on (*a*) a clear day and (*b*) a smoggy day.

Figure 23.28 • How concentrated sulfurous smog and smoke might develop.

Future Trends for Urban Areas

What does the future hold for urban areas with respect to air pollution? The optimistic view is that urban air quality will continue to improve as it has in the past 35 years because we know so much about the sources of air pollution and have developed effective ways to reduce it. The pessimistic view is that in spite of this knowledge, population pressures and economics will dictate what happens in many parts of the world, and the result will be poorer air quality (more air pollution) in many locations.

The actual situation in the 21st century is likely to be a mixture of the optimistic and pessimistic points of view. Large urban areas in developing countries may experience a reduction in air quality even as they attempt to improve the situation, because the population and economic factors will likely outweigh pollution abatement. Large urban areas in developed and more affluent countries, however, may well experience improved air quality in the coming years (see Figures 23.7 to 23.10).

The United States

As an example of trends in developed countries, consider the Los Angeles urban area. This region, which has the worst air quality in the United States, is coming to grips with the problem. The people studying air pollution in the Los Angeles region now understand that pollution abatement will take massive efforts much different from past strategies, which have taken a limited approach. Los Angeles is considering a new, controversial, multifaceted air quality plan involving the entire urban region and including the following features:[15]

- Strategies to discourage automobile use and reduce the number of cars.
- Stricter emission controls for automobiles.
- A requirement for a certain number of zero-pollutant automobiles (electric cars) and hybrid cars with fuel cell and gasoline engines.
- A requirement for more gasoline to be reformulated to burn cleaner.
- Improvements in public transportation and incentives for people to use it.
- Mandatory carpooling.
- Increased controls on industrial and household activities known to contribute to air pollution.

At the household level, for example, common materials such as paints and solvents will be reformulated so that their fumes will cause less air pollution. Eventually, certain equipment, such as gasoline-powered lawn mowers that contribute to air pollution, may be banned.

There are encouraging signs of improvement in southern California's air quality. For example, from the 1950s to the present, the peak level of ozone (considered one of the best indicators of air pollution) has declined from about 0.68 to 0.3 ppm. The reduction has occurred in spite of the fact that during this period the population nearly tripled and the number of motor vehicles quadrupled.[16] Nevertheless, air quality in southern California remains the nation's worst. Even if all the aforementioned controls in urban areas are implemented, air quality will continue to be a significant problem in coming decades, particularly if the urban population continues to increase.

We have focused on air pollution in southern California because air quality is often poor there. However, many large and not-so-large U.S. cities have poor air quality a significant part of the year. Based on the criteria of 30 days per year of unhealthy air resulting from ozone pollution, many millions of Americans live in cities where hazardous air pollution exists. Table 23.3 lists the 25 most polluted counties and metropolitan areas with respect to ozone in the United States These include Riverside, California; Houston, Texas; Baltimore, Maryland, Charlotte, North Carolina; and Atlanta, Georgia. In contrast, some of the cities with the cleanest air in the United States include Bellingham, Washington; Cedar Rapids, Iowa; Colorado Springs, Colorado; and Des Moines, Iowa.[6] However, with the exception of the Pacific Northwest, no U.S. region is free from ozone pollution and associated adverse health.

Developing Countries

As mentioned earlier, cities in less developed countries with burgeoning populations are particularly susceptible to air pollution now and will be in the future (see Figures 23.5 and 23.12). They often do not have the financial base necessary to fight air pollution; they are more concerned with finding ways to house and feed their growing populations.

A good example is Mexico City. With a population of about 25 million, Mexico City is one of the four largest

Table 23.3 Twenty-five U.S. Counties with Greatest Ozone Pollution (1996–1998)

| County | State | Metropolitan Area | Number of High Ozone Days in the Unhealthy Ranges (AQI exceeds 101)* | | |
			Orange	Red	Purple
1. San Bernardino	CA	Riverside	138	88	81
2. Harris	TX	Houston	65	34	19
3. Anne Arundel	MD	Baltimore	69	21	2
4. Mecklenburg	NC	Charlotte	76	15	0
5. Fulton	GA	Atlanta	66	14	5
6. Knox	TN	Knoxville	62	14	0
7. Camden	NJ	Philadelphia	63	7	1
8. Maricopa	AZ	Phoenix	63	1	0
9. Washington	PA	Pittsburgh	47	8	0
10. Fairfield	CT	Bridgeport	38	11	1
11. Fairfax	VA	Washington, DC	41	8	0
12. Dist. of Columbia	DC	Washington, DC	44	4	1
13. Iberville	LA		42	6	0
14. New Castle	DE	Wilmington	40	5	0
15. Saint Charles	MO	St. Louis	39	4	1
16. Butler	OH	Hamilton	43	2	0
17. New York	NY	New York	35	7	0
18. Warrick	IN	Evansville	40	3	0
19. Jefferson	AL	Birmingham	33	10	0
20. Jefferson	KY	Louisville	36	4	0
21. Berrien	MI	Benton Harbor	35	4	0
22. Manitowoc	WI		24	4	2
23. Bristol	MA	Providence	27	4	0
24. Aiken	SC	Augusta	33	0	0
25. Tulsa	OK	Tulsa	28	2	0

*AQI: Orange, 101–150, unhealthy for people sensitive to pollution (young, old, ill); red, 151–199, unhealthy; purple, 200–300, very unhealthy.

Source: Based on American Lung Association, *State of the Air 2000*.

urban areas in the world. Cars, buses, industry, and power plants in the city emit hundreds of thousands of metric tons of pollutants into the atmosphere each year. The city is at an elevation of about 2,255 m (7,400 ft) in a natural basin surrounded by mountains, a perfect situation for a severe air pollution problem. It is becoming a rare day in Mexico City when the mountains can be seen, and physicians report a steady increase in respiratory diseases. Headaches, irritated eyes, and sore throats are common when the pollution settles in. Doctors advise parents to take their children out of the city permanently. The people in Mexico City do not need to be told they have an air pollution problem; it is all too apparent. However, developing a successful strategy to improve the quality of the air is difficult.

A major source of pollutants in Mexico City is motor vehicles. There are some 50,000 buses and taxis and several million automobiles in the city. Most are old and in poor running condition, and they pump immense amounts of pollutants into the atmosphere. Another major source of air pollution is leaks of liquefied petroleum gas (LPG, a hydrocarbon) used in homes for cooking and heating water. The leaking LPGs produce atmospheric precursors to the formation of ozone, a major component of urban photochemical smog, and LPG leaks in Mexico City may be responsible for a significant portion of the city's ozone pollution.[17]

In an attempt to reduce air pollution in the urban area, officials shut down a large oil refinery. For almost 60 years, the refinery had emitted nearly 90,000 metric tons of air pollutants into the atmosphere annually. Thousands of other industrial plants have been ordered to relocate. Although these measures will help the air quality of the urban area, industrial facilities are not the primary source of pollutants. If the city is unable to control increases in population; the use of buses, taxis, and automobiles; and leaks of liquefied petroleum gas, air pollution will continue to be a serious problem for many years. Indeed, Mexico City may eventually experience a pollution event of catastrophic proportions.

In summary, future trends in urban air problems and solutions, will include a mixture of success stories and potential or actual tragedies. What is apparent is that urban air pollution is important to people, and ambitious air pollution control plans are being drawn up in many urban areas. Whether these plans are put into action will depend on numerous factors: global, regional, and local economies (reducing air pollution is expensive); population growth (more people mean more air pollution); international cooperation (air pollutants travel across international borders); and the priority given to pollution abatement relative to other environmental concerns such as sanitation and clean water. With these thoughts in mind, we turn to a discussion of how to reduce air pollution.

23.8 Pollution Control

For both stationary and mobile sources of air pollutants, the most reasonable strategies for control have been to reduce, collect, capture, or retain the pollutants before they enter the atmosphere. From an environmental viewpoint, the reduction of emissions through energy efficiency and conservation measures (for example, burning less fuel) is the preferred strategy, with clear advantages over all other approaches (see Chapters 16–18). Here, we discuss pollution control for selected air pollutants.

Pollution Control: Particulates

Particulates emitted from fugitive, point, or area stationary sources are much easier to control than are the very small particulates of primary or secondary origin released from mobile sources, such as automobiles. As we learn more about these very small particles, we will have to devise new methods to control them.

A variety of settling chambers or collectors are used to control emissions of coarse particulates from power plants and industrial sites (point or area sources) by providing a mechanism that causes particles in gases to settle out in a location where they can be collected for disposal in landfills. In recent decades, tremendous gains have been made in the control of particulates, such as ash, from power plants and industry. Cities in the eastern United States, where buildings used to turn black from soot and ash, are now much cleaner. Particulate pollutants no longer pose

serious health risks in these cities. In contrast, such risks have plagued parts of Eastern Europe in recent years.

Particulates from fugitive sources (such as waste piles) must be controlled on-site so that the wind does not blow them into the atmosphere. Methods include protecting open areas, controlling dust, and reducing the effects of wind. For example, waste piles can be covered by plastic or other material and soil piles can be vegetated to inhibit wind erosion; water or a combination of water and chemicals can be spread to hold dust down; and structures or vegetation can be positioned to lessen wind velocity near the ground, thus retarding wind erosion of particles.

Pollution Control: Automobiles

Control of pollutants such as carbon monoxide, nitrogen oxides, and hydrocarbons in urban areas is best achieved through pollution control measures for automobiles. Control of these materials will also limit ozone formation in the lower atmosphere, since ozone forms through reactions with nitrogen oxides and hydrocarbons in the presence of sunlight.

Nitrogen oxides from automobile exhausts are controlled by recirculating exhaust gas, diluting the air-to-fuel mixture being burned in the engine. Dilution reduces the temperature of combustion and decreases the oxygen concentration in the burning mixture, resulting in the production of fewer nitrogen oxides. Unfortunately, the same process increases hydrocarbon emissions. Nevertheless, exhaust recirculation to reduce nitrogen oxide emissions has been common practice in the United States for more than 20 years.[20]

The most common device to reduce carbon monoxide and hydrocarbon emissions from automobiles is the exhaust system's catalytic converter. In the converter, oxygen from outside air is introduced and exhaust gases from the engine are passed over a catalyst, typically platinum or palladium. Two important chemical reactions occur: (1) carbon monoxide is converted to carbon dioxide; and (2) hydrocarbons are converted to carbon dioxide and water.

As government regulations controlling emissions became stronger, it became difficult to meet new standards without the aid of computer-controlled engine systems. Computer-controlled fuel injection began to replace carburetors in the 1980s and has resulted in lower fuel consumption and lower exhaust emissions.[20]

It has been argued that the automobile emission regulation plan in the United States has not been effective in reducing pollutants. Pollutants may be relatively low when a car is new, but many people do not take care of their automobiles well enough to ensure that the emission control devices continue to work. Some people even disconnect smog control devices. Evidence suggests that these devices tend to become less efficient with each year following purchase.

It has been suggested that effluent fees replace emission controls as the primary method of regulating air pollution from automobiles in the United States.[21] Under this scheme, vehicles would be tested each year for emission control, and fees would be assessed on the basis of test results. Fees would provide an incentive to purchase automobiles that pollute less, and annual inspections would ensure that pollution control devices were properly maintained. Although there is considerable controversy regarding enforced pollution inspections, such inspections are common in a number of areas and are expected to increase as air pollution abatement becomes essential.

Another approach to reducing urban air pollution from vehicles involves various measures aimed at reducing the number and type of cars on roads. Some of these methods were mentioned earlier and are being tried or discussed in Los Angeles and other areas. Other measures include developing cleaner automobile fuels through use of fuel additives and reformulation; requiring new cars to use less fuel; and encouraging the use of cars with electric engines and hybrid cars that have both an electric engine and an internal combustion engine.

Pollution Control: Sulfur Dioxide

Sulfur dioxide emissions can be reduced through abatement measures performed before, during, or after combustion. Technology to clean up coal so it will burn cleanly is already available. Although the cost of removing sulfur makes fuel more expensive, the expense must be balanced against the long-term consequences of burning sulfur-rich coal.

Changing from high-sulfur coal to low-sulfur coal seems an obvious solution to reducing emissions of sulfur dioxide. In some regions, this change will work. Unfortunately, most low-sulfur coal in the United States is located in the western part of the country, whereas most coal is burned in the east. Thus, transportation is an issue; and use of low-sulfur coal is a solution only in cases where it is economically feasible.

Another possibility is cleaning up relatively high-sulfur coal by washing it to remove sulfur. In this process, finely ground coal is washed with water. Iron sulfide (mineral pyrite) settles out because of its relatively high density. Although the washing process is effective in removing nonorganic sulfur from minerals such as pyrite (FeS_2), it is ineffective for removing organic sulfur bound up with carbonaceous material. Cleanup by washing is therefore limited, and it is also expensive.

Another option is **coal gasification**, which converts coal that is relatively high in sulfur to a gas in order to remove the sulfur. The gas obtained from coal is quite clean and can be transported relatively easily, augmenting supplies of natural gas. The synthetic gas produced from coal is now fairly expensive compared with gas from other sources, but its price may become more competitive in the future.

Sulfur oxide emissions from stationary sources, such as power plants, can be reduced by removing the oxides from the gases in the stack before they reach the atmosphere. Perhaps the most highly developed technology for the cleaning of gases in tall stacks is flue gas desulfurization, or **scrubbing** (Figure 23.29). The technology to scrub sulfur dioxide and other pollutants at power plants was developed in the 1970s in the United States in response to the passage of the Clean Air Act. However, the technology was not initially implemented in the United States because regulators chose to allow plants to disperse pollutants using tall smokestacks rather than use scrubbing to remove them. This increased the regional acid rain problem.

Scrubbing occurs after coal is burned. The SO_2-rich gases are treated with a slurry (a watery mixture) of lime

Figure 23.29 • Scrubber used to remove sulfur oxides from the gases emitted by tall stacks.

Clean gases

SO_2-rich gases

Stack

Furnace (burns coal)

Precipitator (removes ash)

Scrubber (slurry of water and lime reacts with SO_2 to remove it from the gases)

Drain

Tank (collects calcium sulfite, $CaSO_3 \cdot 1/2\ H_2O$)

(calcium oxide, CaO) or limestone (calcium carbonate, $CaCO_3$). The sulfur oxides react with the calcium to form calcium sulfite ($CaSO_3 \cdot 1/2\ H_2O$), which is collected and then disposed of, usually in a landfill.

A German company in 1980 purchased coal-scrubbing technology and improved on it as part of efforts to reduce air pollution and acid rain. Rather than disposing of the calcium sulfite-rich sludge formed during the scrubbing process, the company further processes it to produce building materials (gypsum, $CaSO_4 \cdot 2H_2O$, which is sheet rock or wallboard), which are sold worldwide.

Another innovative approach to removing sulfur has been taken at a large coal-burning power plant near Mannheim, Germany. The smoke from combustion is cooled, then treated with liquid ammonia (NH_3), which reacts with the sulfur to produce ammonium sulfate. In this process, the sulfur-contaminated smoke is cooled in a heat-exchange process by outgoing clean smoke to a temperature that allows the chemical reaction between the sulfur-rich smoke and ammonia to take place. The cooled, cleaned, outgoing smoke is then heated by dirty smoke in the same sort of heat-exchange process to force it out the vent. Waste heat from the cooling towers is used to heat nearby buildings, and the plant sells the ammonium sulfate in a solid granular form to farmers to use as fertilizer.

Thus, Germany, in response to tough pollution control regulations, has substantially reduced its sulfur dioxide emissions (as well as many other pollutant emissions). In the process it has boosted its economy.[22]

 23.9 **Air Pollution Legislation and Standards**

The killer smog events in Donora, Pennsylvania, in 1948 and London, England, in 1952 were the impetus for legislation to control air pollution in both England and the United States.

Clean Air Act Amendments of 1990

The **Clean Air Act Amendments of 1990** are comprehensive regulations enacted by the U.S. Congress that address acid rain, toxic emissions, ozone depletion, and automobile exhaust. In dealing with acid deposition (acid rain), the amendments establish limits on the maximum permissible emissions of sulfur dioxide from utility companies burning coal. The goal of the legislation was to reduce such emissions by about 50%, to 10 million tons a year by 2000. This was not achieved, but emissions have been greatly reduced (see Figure 23.7).

An innovative aspect of the legislation is to provide incentives to utility companies to reduce emissions of sulfur dioxide by providing marketable permits that allow companies to buy and sell the right to pollute.[23] The total amount of pollution allowed is divided into a given number of permits. Utilities with clean power plants that don't need their permits sell the permits to those that do need them. Environmentalists can also purchase these permits to keep them from being bought by utility companies, forcing the utility companies to use more vigorous pollution abatement technology. Buying of permits by environmentalists, however, has not been a major factor. As the permits are bought and sold, they take on an economic value, and polluters begin to view polluting as an expensive way to do business.[23]

The 1990 amendments also call for a reduction in emissions of nitrogen dioxides by approximately 2 million tons from the 1980 level. Greater reductions would be difficult because large amounts of nitrogen oxide emissions are related to automobiles rather than to coal-burning power plants.[24]

Emissions of toxins into the atmosphere are targeted by the legislation to be reduced by as much as 90%. The toxins chosen are those thought to have the most potential for damaging human health, including causing cancer. Abatement will depend heavily on pollution control equipment, which will be required for large manufacturers and small businesses alike. Although this requirement will undoubtedly result in an increase in the cost of many goods and services, there should be a compensating improvement in the health of people.

The Clean Air Amendments also deal with ozone depletion in the stratosphere (see Chapter 25). The goal is to end the production of all chlorofluorocarbons (CFCs) and other chlorine chemicals in steps by 2030.[24]

As we have seen, air pollution in urban areas is commonly associated with automobile exhaust. Strategies outlined in the legislation include more stringent emission controls on automobiles and requirements for cleaner-burning fuels. The aim is to reduce the occurrence of urban smog. Expected impacts of the legislation include increases in the cost of automobile fuels and in the price of new automobiles.

Ambient Air Quality Standards

Air quality standards are important because they are tied to emission standards that attempt to control air pollution. Many countries have developed air quality standards, including France, Japan, Israel, Italy, Canada, Germany, Norway, and the United States. National Ambient Air Quality Standards (NAAQS) for the United States, defined to comply with the Clean Air Act, are shown in Table 23.4.

Tougher Standards for PM 2.5 and Ozone

In 1997, the U.S. Environmental Protection Agency set new standards for PM 2.5 and ozone levels in an attempt to reduce adverse health effects for children and elderly people. The new standards (NAAQS) for these substances are shown in Table 23.4. The agency estimat-

Table 23.4   U.S. National Ambient Air Quality Standards (NAAQS)

Pollutant	Standard Value[a]		Standard Type
Carbon monoxide (CO)			
8-hour average	9 ppm	(10 mg/m³)	Primary [c]
1-hour average	35 ppm	(40 mg/m³)	Primary
Nitrogen dioxide (NO₂)			
Annual arithmetic mean	0.053 ppm	(100 μg/m³)	Primary and secondary
Ozone (O₃)			
1-hour average	0.12 ppm	(235 μg/m³)	Primary and secondary
8-hour average [b]	0.08 ppm	(157 μg/m³)	Primary and secondary
Lead (Pb)			
Quarterly average	1.5 μg/m³		Primary and secondary
Particulate (PM 10) *Particles with diameters of 10 micrometers or less*			
Annual arithmetic mean	50 μg/m³		Primary and secondary
24-hour average	150 μg/m³		Primary and secondary
Particulate (PM 2.5)[b] *Particles with diameters of 2.5 micrometers or less*			
Annual arithmetic mean	15 μg/m³		Primary and secondary
24-hour average	65 μg/m³		Primary and secondary
Sulfur Dioxide (SO₂)			
Annual arithmetic mean	0.03 ppm	(80 μg/m³)	Primary
24-hour average	0.14 ppm	(365 μg/m³)	Primary
3-hour average	0.50 ppm	(1300 μg/m³)	Secondary

[a] Parenthetical value is an approximately equivalent concentration.

[b] The ozone 8-hour standard and the PM 2.5 standards are included for information only. A 1999 federal court ruling blocked implementation of these standards, which EPA proposed in 1997. EPA has asked the U.S. Supreme Court to reconsider that decision. (*Note:* In March 2001, the Court ruled in favor of EPA, and the new standards are expected to take effect within a few years.)

[c] Primary standards set limits to protect public health, including the health of sensitive populations such as asthmatics, children, and the elderly.

[d] Secondary standards set limits to protect public welfare, including protection against decreased visibility and damage to animals, crops, vegetation, and buildings.

Source: U.S. Environmental Protection Agency.

ed that the new standards could save the lives of 15,000 children per year and reduce the incidence of respiratory disease in children by 250,000 cases per year. However, the new regulations were opposed by U.S. business leaders, who argued that implementing them would cost hundreds of billions of dollars and up to a million jobs. In 1999, a lower federal court blocked the implementation of the new standards and the EPA requested that the U.S. Supreme Court hear the case. In early March of 2001, the Supreme Court, in a unanimous decision, upheld the new, tougher standards. The justices held that the EPA's responsibility is to consider benefits to public health from pollution abatement—not financial costs.

Implementing the new standards will take years; ways will have to be found to enforce them, and the EPA will have to defend the standards to the lower court. Nevertheless, it is expected that the new standards will eventually be implemented. The decision of the Supreme Court is a turning point in the fight to reduce air pollution and its known adverse health effects. Standards from now on will be based on improving human health rather than on the economic costs of implementing standards.

Air Quality Index

In the United States, the Air Quality Index (AQI) (Table 23.5) is used to describe air pollution on a given day. For example, air quality in urban areas is often reported as good, moderate, unhealthy for sensitive groups, unhealthy, very unhealthy, or hazardous, corresponding to a color code of the Air Quality Index. The AQI is determined from measurements of the concentration of five major pollutants: particulate matter, sulfur dioxide, carbon monoxide, ozone, and nitrogen dioxide. An AQI value of greater than 100 is unhealthy. In most U.S. cities, values of AQI range between 0 and 100. AQI values greater than 100 generally are recorded for a particular city only a few times per year. However, some cities with serious air pollution problems may exceed an AQI of 100 many times per year (see Table 23.3). In a typical year, AQI values above 200 (for all U.S. sites) are

Table 23.5   Air Quality Index (AQI) and Health Conditions

Index Values	Descriptor	Cautionary Statement	General Adverse Health Effects	Action Level (AQI)[a]
0–50	Good	None	None	None
51–100	Moderate	Unusually sensitive people should consider limiting prolonged outdoor exertion.	Very few symptoms[b] for the most susceptible people[c]	None
101–150	Unhealthy for sensitive groups	Active children and adults and people with respiratory disease, such as asthma, should limit prolonged outdoor exertion.	Mild aggravation of symptoms in susceptible people, few symptoms for healthy people	None
151–199	Unhealthy	Active children and adults and people with respiratory disease, such as asthma, should avoid prolonged outdoor exertion; everyone else, especially children, should limit prolonged outdoor exertion.	Mild aggravation of symptoms in susceptible people, irritation symptoms for healthy people	None
200–300	Very unhealthy	Active children and adults and people with respiratory disease, such as asthma, should avoid outdoor exertion; everyone else, especially children, should limit outdoor exertion.	Significant aggravation of symptoms in susceptible people, widespread symptoms in healthy people	Alert (200+)
Over 300	Hazardous	*Everyone* should avoid outdoor exertion.	300–400: Widespread symptoms in healthy people 400–500: Premature onset of some diseases Over 500: Premature death of ill and elderly people; healthy people experience symptoms that affect normal activity	Warning (300+) Emergency (400+)

[a] Triggers preventative action by state or local officials.

[b] Symptoms include eye, nose, and throat irritation; chest pain; breathing difficulty.

[c] Susceptible people are young, old, and ill people and people with lung or heart disease.

AQI 51–100	Health advisories for susceptible individuals.
AQI 101–150	Health advisories for all.
AQI 151–200	Health advisories for all.
AQI 200+	Health advisories for all; triggers an alert; activities that cause pollution might be restricted.
AQI 300	Health advisories to all; triggers a warning; probably would require power plant operations to be reduced and carpooling be used.
AQI 400+	Health advisories for all; triggers an emergency; cessation of most industrial and commercial activities, including power plants; nearly all private use of vehicles prohibited.

Source: U.S. Environmental Protection Agency.

rare, and those above 300 are very rare. In comparison, for large urban cities outside of the United States with dense human populations and numerous uncontrolled sources of pollution, AQIs greater than 200 are frequent.

During a pollution episode, hourly ozone levels are reported; and a smog episode begins if the primary National Ambient Air Quality Standard (NAAQS) of 0.12 ppm (equivalent to 235 µg/m^3 in Table 23.3) is exceeded. This measure corresponds to unhealthy air with an AQI between 100 and 300 (Table 23.5). An air

pollution alert is issued if the AQI exceeds 200. An air pollution warning is issued if the AQI exceeds 300, a point at which air quality is hazardous to all people. If the AQI exceeds 400, an air pollution emergency is declared, and people are requested to remain indoors and minimize physical exertion. Driving private automobiles may be prohibited, and industry may be required to reduce emissions to a minimum during the episode. Recall that the AQI was 800, twice as high as the level that constitutes an emergency, during the Indonesian fires of 1997–1998!

The cost of air pollution control varies tremendously from one industry to another. For example, consider the incremental control costs (costs to remove an additional unit of pollution) for utilities burning fossil fuels and for an aluminum plant. The cost for incremental control in a fossil fuel–burning utility is a few hundred dollars per additional ton of particulates removed. For the aluminum plant, the cost to remove an additional ton of particulates may be as much as several thousand dollars.[24] Some economists would argue that it is wise to increase the standards for utilities and relax or at least not increase them for aluminum plants. This practice would lead to more cost-efficient pollution control while maintaining good air quality. However, the geographic distribution of various facilities will obviously determine the trade-offs possible.[25]

Another economic consideration is that, as the degree of control of a pollutant increases, a point is reached at which the cost of incremental control is very high in relation to the additional benefits of the increased control. Because of this and other economic factors, it has been argued that enforcing fees or taxes for emitting pollutants might make more economic sense than attempting to evaluate the uncertain costs and benefits associated with enforcement of standards. Another approach is to issue vouchers to allow businesses to emit a certain total amount of pollution in a region. These vouchers are bought and sold on the open market. All these economic alternatives are controversial and may be objectionable to people who believe that polluters should not be allowed to buy their way out of doing what is socially responsible (that is, not polluting our atmosphere).

Economic analysis of air pollution is not simple. There are many variables, some of which are hard to quantify. We do know the following:

- With increasing air pollution controls, the capital cost for technology to control air pollution increases.
- As the controls for air pollution increase, the loss from pollution damages decreases.
- The total cost of air pollution is the cost of pollution control plus the environmental damages of the pollution.

Although the cost of pollution abatement technology is fairly well known, it is difficult to adequately determine the loss from pollution damages, particularly when considering health problems and damage to vegetation, including food crops. For example, exposure to air pollution may cause or aggravate chronic respiratory diseases in human beings, with a very high cost. A recent study of the health benefits of cleaning up the air quality in the Los Angeles basin estimated that the annual cost associated with air pollution in the basin is 1,600 lives and about $10 billion.[26] Air pollution also leads to loss of revenue from people who choose not to visit areas such as Los Angeles and Mexico City because of known air pollution problems.

How do we determine the real and total benefits and costs of controlling or reducing air pollution? As we have seen, there are no easy answers to this question. In spite of our inability to determine all benefits and costs, it seems worthwhile to reduce the air pollution level below some particular standard. Thus, in the United States, the National Ambient Air Quality Standards have been developed as a minimum acceptable air quality level. However, as discussed, it is also a good idea to consider alternatives, such as charging fees or taxes for emissions. If such charges are determined carefully and emissions are closely monitored, the charges should provide an incentive for the installation of control measures. The end result would be better air quality.[27,28]

Summary

- Every year, approximately 150 million metric tons of primary pollutants enter the atmosphere above the United States from processes related to human activity. Considering the enormous volume of the atmosphere, this is a relatively small amount of material. If it were distributed uniformly, there would be little problem with air pollution. Unfortunately, the pollutants generally are not evenly distributed but are concentrated in urban areas or in other areas where the air naturally lingers.

- The two major types of pollution sources are stationary and mobile. Stationary sources have a relatively fixed position and include point sources, area sources, and fugitive sources.

- There are two main groups of air pollutants: primary and secondary. Primary pollutants are those emitted directly into the air: particulates, sulfur dioxide, carbon monoxide, nitrogen oxides, and hydrocarbons. Secondary pollutants are those produced through reactions between primary pollutants and other atmospheric compounds. A good example of a secondary pollutant is ozone, which forms over urban areas through photochemical reactions between primary pollutants and natural atmospheric gases.

- The effects of the major air pollutants are considerable. They include effects on the visual quality of the environment, vegetation, animals, soil, water quality, natural and artificial structures, and human health.

- The combustion of large quantities of fossil fuels results in the emission of sulfur and nitrogen oxides into the atmosphere, creating acid rain. Environmental degradations associated with acid rain include loss of fish and other life in lakes, damage to trees and other plants, leaching of nutrients from soils, and damage to stone statues and buildings in urban areas.

CRITICAL THINKING ISSUE

Where Does Arctic Haze Come From, and How Does It Affect the Environment?

A dark gray haze, full of industrial pollutants, hovers over the ground and extends to an altitude of 8 km (5 mi). It is not Los Angeles but the frozen Arctic, thousands of miles from heavy industry. The polluted air mass includes all of the atmosphere above the Arctic Circle, as well as lobes extending into Eurasia and North America. The total area of polluted air mass, about as large as the African continent, is present during winter months and disappears in summer.

Scientists describe this haze as an *aerosol*—that is, microscopic particles dispersed in a gas, smoke, or fog. They have found dust from Mongolia and sea salt in the haze; but of greater concern are the pollutants, mainly sulfates, carbon soot (dark carbon), organic compounds, and toxic metals, including mercury, lead, and vanadium. The gaseous atmosphere itself contains elevated levels of carbon dioxide, methane, and carbon monoxide, as well as chemicals destructive of the ozone layer.

Like detectives searching for fingerprints, scientists used ratios of six elements (arsenic, antimony, zinc, indium, manganese, and vanadium) to a seventh, selenium, to track down sources of Arctic haze. Emissions from burning fossil fuels differ in amounts of these elements, depending on the type of fuel used. The ratios are so characteristic that scientists can tell whether pollution is from hard or soft coal. The ratios found in the haze were compared with ratios characteristic of types of fuel known to be used in various regions of the world. For example, manganese was present in greater amounts and vanadium in lesser amounts in the haze than in emissions typical of North America.

Based on the ratios and on knowledge of air circulation patterns, scientists identified Eastern Europe and Russia as major sources of Arctic air pollution, with a significant but lesser contribution from the United

Kingdom and Western Europe. Two major routes (see Figure 23.30) are (1) from Eastern Europe and Russia across the Taimyr Peninsula and the North Pole to the Alaskan Arctic and (2) from the United Kingdom and Western Europe across Scandinavia to the Norwegian Arctic. (A different study argued that none of the pollution comes from Western Europe or from China. Rather, it comes from Eastern Europe and Russia, with some possibly originating within the Arctic Circle itself.)

Major atmospheric forces driving pollution along these routes begin with large temperature differences between the equator and the poles in winter, creating strong air currents from 0° to 90° latitude. The air currents are also propelled by seasonal lows in the North Atlantic and highs on the Eurasian continent. Once masses of pollutant-laden air reach the dry, sta-

Figure 23.30 • Two airstreams out of the former Soviet Union (top arrows) carry most of the pollution that becomes Arctic haze. According to some findings, Europe is also an important source of smog (arrow at right). But the northeastern United States (bottom arrows) is not a big contributor to the haze, because pollutants are often washed out by storms before they reach the Arctic. The dots represent air-sampling stations.

ble air of Arctic winter, they form layers, which relatively intact. In spring, when the northward flow of air diminishes, the haze disperses and is carried to higher levels in the atmosphere and back to the midlatitudes.

Now that scientists have a better idea of the sources of Arctic haze, they are beginning to speculate on the effects it has on the ecology of the Arctic and on global climate. Furthermore, they are calling for research into environmental impacts of the haze and for cooperation among countries in the Northern Hemisphere to reduce the amount of haze. In the meantime, humans are producing enough toxic emissions to carry as far as 10,000 km (6,215 mi) and produce pollution levels at the North Pole as great as those found in medium-sized industrial cities.

Critical Thinking Questions

1. Although laden with pollutants, Arctic haze drops less of its aerosol on the ground than would happen in other areas of the world. What is it in the Arctic environment that could account for this?

2. Elements such as mercury and lead are present in small amounts in Arctic haze (the levels are often comparable to those near industrial sites and 10–20 times those in Antarctica). If the amounts are so low, why are scientists concerned about their effects? (You may wish to review Chapter 15.)

3. Sulfates in haze react with water to form acids. Scientists predict that the Arctic will be sensitive to additional acids. Why might this be so?

4. What are possible effects of dark carbon particles such as soot on the Arctic climate? Make a diagram. Put the term *dark carbon particles* in the center of a sheet of paper. Write as many direct consequences of dark carbon as you can around the term, and connect the consequences to the term with arrows. Now, what would be the secondary consequences of these direct effects? Place them around the edge of the diagram, connecting them by arrows to the relevant direct effect. If you can think of third- and fourth-order effects, place these on the diagram in a similar manner.

- There are two major types of smog: photochemical and sulfurous. Each type of smog brings particular environmental problems that vary with geographic region, time of year, and local urban conditions.

- Meteorological conditions greatly affect whether polluted air is a problem in an urban area. In particular, restricted circulation in the lower atmosphere associated with temperature inversion layers may lead to pollution events.

- There are two major types of smog: photochemical and sulfurous. Each type of smog brings particular environmental problems that vary with geographic region, time of year, and local urban conditions.

- From an environmental viewpoint, the preferred method of reducing the emission of air pollutants produced from burning fossil fuels is to practice energy efficiency and conservation so that smaller amounts of fossil fuels are burned. Another option is to increase the use of alternative energy sources, such as solar and wind power, that do not emit air pollutants.

- Methods to control air pollution are tailored to specific sources and types of pollutants. These methods vary from settling chambers for particulates to scrubbers that use lime to remove sulfur before it enters the atmosphere.

- Efforts to reduce air pollution in urban regions center on automobiles, buses, and other vehicles, because they account for most of the pollutants that enter the urban atmosphere.

- Emissions of air pollutants in the United States are decreasing. In developing countries, air pollution in large urban centers is often a serious problem.

- Air quality in U.S. urban areas is usually reported in terms of whether the quality is good, moderate, unhealthy for sensitive groups, unhealthy, very unhealthy, or hazardous. These levels are defined in terms of the Air Quality Index (AQI).

- The relationships between emission control and environmental cost are complex. The minimum total cost is a compromise between capital costs to control pollutants and losses or damages resulting from pollution. If additional controls are necessary to lower the pollution to a more acceptable level, additional costs are incurred. Beyond a certain level of pollution abatement, these costs can increase rapidly.

Human Population:

Increases in human population are expected to continue to have a significant adverse effect on air pollution problems. As the human population increases, so does the total use of resources, many of which are related to emissions of air pollutants. This may be partially offset in developed countries, where the per capita emissions of air pollutants have been reduced in recent years.

Sustainability:

Ensuring that future generations inherit a quality environment with minimal air pollution is an important objective of sustainability. As a result, it remains an important objective to find and develop technology that minimizes air pollution.

Global Perspective:

Atmospheric processes and pollution of the atmosphere occur by their very nature on regional and global scales. Pollutants emitted into the atmosphere at a particular site may join the global circulation pattern and spread pollutants throughout the world. Air pollutants emitted from urban or agricultural areas may be dispersed to pristine areas far removed from human activities. Therefore, an understanding of global atmospheric processes is critical to finding solutions to many air pollution problems, including acid deposition.

Urban World:

Cities and urban corridors are sites of intense human activity, and many of these activities are associated with emission of air pollutants. Some of the most significant adverse effects of air pollution are found in our urban areas. Some large cities have air pollution problems so severe in extent that the health and very lives of people are being affected.

Science and Values:

The science and technology necessary to reduce air pollution is well known; what we do with these tools involves a value judgment. It is clear that people value a high-quality environment, and clean air is at the top of the list. The developed countries have an obligation to take a leadership role in finding ways to utilize resources while minimizing air pollution. Of particular importance is finding methods and technologies that will allow for reduction of air pollution while stimulating economies. What is considered waste in one part of the urban-industrial complex may be used as resources in another part. This idea is at the heart of what is sometimes called industrial ecology.

● KEY TERMS

acid rain **493**
air quality standards **512**
atmospheric inversion **505**
Clean Air Act Amendments of 1990 **512**

coal gasification **511**
lead **499**
mobile sources **488**
photochemical smog **507**
primary pollutants **491**

scrubbing **511**
secondary pollutants **491**
smog **507**
stationary sources **488**
sulfurous smog **507**

● STUDY QUESTIONS

1. Compare and contrast the London 1952 fog event with smog problems in the Los Angeles basin.

2. Why do we have air pollution problems when the amount of pollution emitted into the air is a very small fraction of the total material in the atmosphere?

3. What is the difference between point and nonpoint sources of air pollution? Which type is easier to manage?

4. What are the differences between primary and secondary pollutants?

5. Examine carefully Figure 22.24, which shows a column of air moving through an urban area, and Figure 22.26, which shows relative concentrations of pollutants that develop on a typical warm day in Los Angeles. What linkages between the information in

these two figures might be important in trying to identify and learn more about potential air pollution in an area?

6. Why is acid deposition a major environmental problem, and how can it be minimized?

7. Why will air pollution abatement strategies in developed countries probably be much different in terms of methods, process, and results from air pollution abatement strategies in developing countries?

8. Why is it so difficult to establish national air quality standards?

9. In a highly technological society, is it possible to have 100% clean air? Is it feasible or likely?

10. How good are the air quality standards being used by the United States? How might their usefulness be evaluated? Do you think the standards will change in the future? If so, what are the likely changes?

● FURTHER READING

Boubel, R. W., D. L. Fox, D. B. Turner, and A. C. Stern. 1994. *Fundamentals of Air Pollution*, 3rd ed. New York: Academic. A thorough book covering the sources, mechanisms, effects, and control of air pollution.

Bryner, G. C. 1995. *Blue Skies, Green Politics: The Clean Air Act of 1990 and Its Implementation*, 2nd ed. Washington, DC: CQ Press. A good text on the Clean Air Act and air pollution in the United States.

Hesketh, H. E. 1991. *Air Pollution Control: Traditional and Hazardous Pollutants*. Lancaster, Pa.: Technomic. An in-depth text covering the methodology of controlling air pollution.

Hewitt, D. N., W. T. Sturges, and NOAA, eds. 1995. *Global Atmospheric Chemical Change*. New York: Chapman & Hall. A book describing aspects of global air pollution, including chemical changes in the atmosphere, climate change, acid deposition, and other anthropogenic pollutants.

Krupnick, A. J., and P. R. Portney. 1991. "Controlling Urban Air Pollution: A Benefit-Cost Assessment." *Science* 252: 522–528. An evaluation of air pollution control proposals for Los Angeles and the United States as a whole. The authors determine that the measures would cost less than the pollution damage they would alleviate.

Rose, J., ed. 1994. *Acid Rain: Current Situation and Remedies*. Philadelphia: Gordon and Breach Science. Essays covering issues and consequences of acid rain in Europe and the United States.

Shaw, R. W. 1987. "Air Pollution by Particles," *Scientific American* 257:96–103. An article that discusses the sources, effects, and mitigation of particulate pollution.

CHAPTER 24 Indoor Air Pollution

This modern art museum which opened in 1998 in Stockholm, Sweden, closed for repairs in 2002 due to building-related illness. Employees in the kitchen and bookstore complained of a variety of symptoms, including persistent coughs, headaches, and difficulty in breathing. The indoor air pollution is thought to be in part the result of toxic mold related to the problems with the ventilation system in the building. Similar problems have occurred in American buildings including the Massachusetts Registry of Motor Vehicles Building.

● LEARNING OBJECTIVES

Indoor air pollution from human fires for cooking and heating has affected human health for thousands of years. Today, lack of adequate ventilation in many energy-efficient homes and offices has increased the risk from pollutants. After reading this chapter, you should understand:

- Why indoor air pollutants cause some of our most serious environmental health problems.
- What the major indoor air pollutants are, and where they come from.
- Why concentrations of pollutants found in the indoor environment may be much greater than concentrations of the same pollutants generally found outdoors.
- Why environmental tobacco smoke (ETS) is a serious indoor air pollutant.
- What radon gas is, and why it can be considered one of our most serious environmental health problems.
- How radon gas enters homes and other buildings, and how its indoor concentration may be minimized.
- What the major strategies are to control and minimize indoor air pollution.

Massachusetts Registry of Motor Vehicles Building: Sick Building Syndrome

Indoor air pollution is a serious environmental problem. Buildings such as the modern art museum in Stockholm (opening photograph) have been closed for repair or even abandoned due to pollution. Consider the Massachusetts Registry of Motor Vehicles Building.

Employees of the Massachusetts Registry of Motor Vehicles moved into their newly constructed building on April 19, 1994. First hints of problems were reported in June 1994, shortly after the building was fully occupied. Early reports concerned a variety of air quality problems, including unpleasant odors and a multitude of symptoms that were experienced by a large number of employees in the building. Some of the symptoms reported were respiratory problems; irritation of the eyes, nose, and throat; skin rashes; and effects on the central nervous system. The symptom most often reported was fatigue, followed closely by headaches and problems with mucous membranes. Some type of respiratory problem was reported by 52% of the staff in the building, compared with 17% having similar problems before moving to the new building.[1]

Studies by the Massachusetts Department of Public Health and private consultants suggested that the problem in the building was related to contaminated air caused by a poorly constructed ventilation system. The air circulation system in the building drew outside air into a space at the top of the structure (called the plenum). There, the air was cooled and pumped through ventilation ducts throughout the building. This produced a problem, because when the warmer outside air was cooled, water condensed onto ceiling tiles. A major component of the ceiling tiles was a starch that fermented when wet, producing an acid that smells like vomit. More serious was the discovery that the fireproofing material sprayed on all surfaces in the ventilation system was also wet and was falling apart. Mineral and other fibers from the fireproofing material were then released and spread throughout the building in the ventilation system. This exposed people in the building to potentially dangerous particulate matter. After occupying the building only 15 months, the staff of the Registry of Motor Vehicles moved out, and the building was closed.[2]

●*The museum of art in Stockholm and the Massachusetts Registry of Motor Vehicles buildings are recent examples of indoor air pollution problems that have been with us for many thousands of years. Indoor air pollution has been present for as long as people have constructed buildings for protection from the elements and burned fuels to heat human-made environments. A detailed autopsy of a fourth-century Native American woman, frozen shortly after death, revealed that she suffered from* **black lung disease** *from breathing very polluted air over many years. The pollutants included hazardous particles from lamps that burned seal and whale blubber.[3] This same disease has long been recognized as a major health hazard for underground coal miners and has been called "coal miners' disease" (Figure 24.1). As recently as the mid-1970s, black lung disease was estimated to be responsible for about 4,000 deaths per year in the United States.[4]*

People today spend between 70% and 90% of their time indoors or in other enclosed places (homes, workplaces, automobiles, restaurants, and so

Figure 24.1 • Coal miners in Eastern Europe covered with lung-damaging coal dust.

forth). But only recently have we begun to fully study the indoor environment and how pollution of that environment affects our health. The World Health Organization has estimated that as many as one in three workers may be working in a building that causes them to become sick. As many as 20% of public schools in the United States have problems related to indoor air quality. The U.S. Environmental Protection Agency considers indoor air pollution to be one of the most significant environmental health hazards people face in the modern workplace.[5]

24.1 Sources of Indoor Air Pollution

The potential sources of indoor air pollution are incredibly varied (Figure 24.2). Two common pollutants are shown in Figure 24.3. Indoor air pollutants can arise from both human activities and natural processes. In recent years the public has been made aware of several of these sources, described in the following list:

- Environmental tobacco smoke (second-hand smoke) is the most hazardous common indoor air pollutant, associated with over 40,000 deaths per year (mostly heart disease and lung cancer) in the United States.

- *Legionella pneumophila*, a bacterium that normally lives in pond water, causes a type of pneumonia called Legionnaire's disease when inhaled. About 20 species of the disease-causing bacteria have been identified. Most commonly, this disease is spread by way of air-conditioning equipment, which harbors the disease-causing bacteria in pools of stagnant water in air ducts and filters. Bacteria are transported through a building as a bacterial aerosol when heating or cooling units are in use. However, spread of the disease is not limited to this pathway. One epidemic occurred in a hospital as a result of contamination from an adjacent construction site. A recent outbreak of Legionnaire's disease caused by bacteria in the air-conditioning system of an aquarium killed four people.[5]

- Some molds (fungal growths) in buildings release toxic spores. When inhaled over a period of time, the spores can cause chronic inflammation and scarring of lungs, as well as hypersensitivity pneumonitis and pulmonary fibrosis. These medical conditions are painful, disabling, and can even cause death.[5] It is believed that molds may be responsible for up to half of all health complaints resulting from indoor air environments.

- Radon gas seeps up naturally from soils and rocks below buildings and is thought to be the second most common cause of lung cancer.

- Pesticides that are deliberately or inadvertently applied in buildings to control ants, flies, fleas, moths, and rodents are toxic to people as well.

- Some varieties of asbestos, used as an insulating material and fireproofing material in homes, schools, and offices, are known to cause a particular type of lung cancer (see Chapter 15).

1. Heating, ventilation, and air-conditioning systems of buildings may be sources of indoor air pollutants, including molds and bacteria, if filters and equipment are not maintained properly. Gas and oil furnaces release carbon monoxide, nitrogen dioxide, and particles.

2. Rest rooms may be sources of a variety of indoor air pollutants, including second-hand smoke, molds, and fungi resulting from humid conditions.

3. Furniture and carpets in buildings often contain toxic chemicals (formaldehyde, organic solvents, asbestos), which may be released over time in buildings.

4. Coffee machines, fax machines, computers, and printers can release particles and chemicals, including ozone (O_3), which is highly oxidizing.

5. Pesticides can contaminate buildings with cancer-causing chemicals.

6. Fresh air intake that is poorly located—as, for example, above a loading dock or first-floor restaurant exhaust fan—can bring in air pollutants.

7. People who smoke inside buildings, perhaps in restaurants, and people who smoke outside buildings, particularly near open or revolving doors, may cause pollution as the environmental tobacco smoke (second-hand smoke) is drawn into and up through the building by the chimney effect.

8. Remodeling, painting, and other such activities often bring a variety of chemicals and materials into a building. Fumes from such activities may enter the building's heating, ventilation, and air-conditioning system, causing widespread pollution.

9. A variety of cleaning products and solvents used in offices and other parts of buildings contain harmful chemicals, whose fumes may circulate throughout a building.

10. People can increase carbon dioxide levels; they can emit bioeffluents, and spread bacterial and viral contaminants.

11. Loading docks can be sources of organics from garbage containers, of particulates, and of carbon monoxide from vehicles.

12. Radon gas can seep into a building from soil; rising damp (water), which facilitates the growth of molds, can enter foundations and rise up walls.

13. Dust mites and molds can live in carpets and other indoor places.

14. Pollen can come from inside and outside sources.

Figure 24.2 • Some potential sources of indoor air pollution.

• Formaldehyde (a volatile organic compound, VOC, with a chemical formula of CH_2O) is used in some foam insulation materials, as a binder in particleboard and wood paneling, and in many other materials found in homes and offices. These materials can emit formaldehyde as a gas into buildings. Some mobile homes have been found

(a)

(b)

Figure 24.3 • (a) This dust mite (magnified about 140 times (is an eight legged relative to spiders. It feeds on human skin in household dust. It lives in materials such as fabrics on furniture. Dead dust mites and their excrement can produce allergic reactions and asthma attacks in some people. (b) Microscopic pollen grains that in large amounts may be visible as a brown or yellow powder. The pollen here are dandelion and horse chestnut.

to have high concentrations of formaldehyde because products containing the chemical are used in their construction (wood paneling, for example).

- Dust mites and pollen irritate the respiratory system, nose, eyes, and skin of people who are sensitive to them.

Common indoor air pollutants and guidelines for allowable exposure are listed in Table 24.1. Many of the products and processes used in our homes and workplaces are sources of pollution. Furthermore, common indoor air pollutants are often highly concentrated compared with outdoor levels. For example, carbon monoxide, particulates, nitrogen dioxide, radon, and carbon dioxide are generally found in much higher concentrations indoors than outdoors. This important concept is shown in more detail in Figure 24.4, which provides a comparison of indoor with outdoor pollutants.

Why are concentrations of indoor air pollutants generally greater than those found outdoors? One obvious reason is that there are so many potential indoor sources of pollutants. Another reason is somewhat ironic: The effectiveness

Table 24.1 Sources, Concentrations, Occurrences, and Possible Health Effects of Indoor Air Pollutants

Pollutant	Source	Guidelines (Dose or Concentrations)	Possible Health Effects
Asbestos	Fireproofing; insulation, vinyl floor, and cement products; vehicle brake linings	0.2 fibers/mL for fibers larger than 5 μm	Skin irritation, lung cancer
Biological aerosols/ microorganisms	Infectious agents, bacteria in heating, ventilation, and air-conditioning systems; allergens	None available	Diseases, weakened immunity
Carbon dioxide	Motor vehicles, gas appliances, smoking	1,000 ppm	Dizziness, headaches, nausea
Carbon monoxide	Motor vehicles, kerosene and gas space heaters, gas and wood stoves, fireplaces; smoking	10,000 μg/m^3 for 8 hours; 40,000 μg/m^3 for 1 hour	Dizziness, headaches, nausea, death
Formaldehyde	Foam insulation; plywood, particleboard, ceiling tile, paneling, and other construction materials	120 μg/m^3	Skin irritant, carcinogen
Inhalable particulates	Smoking, fireplaces, dust, combustion sources (wildfire, burning trash, etc.)	55–110 μg/m^3 annual; 350 μg/m^3 for 1 hour	Respiratory and mucous irritant, carcinogen
Inorganic particulates			
Nitrates	Outdoor air	None available	
Sulfates	Outdoor air	4 μg/m^3 annual; 12 μg/m^3 for 24 hours	
Metal particulates			Toxic, carcinogen
Arsenic	Smoking, pesticides, rodent poisons	None available	
Cadmium	Smoking, fungicides	2 μg/m^3 for 24 hours	
Lead	Automobile exhaust	1.5 μg/m^3 for 3 months	
Mercury	Old fungicides; fossil fuel combustion	2 μg/m^3 for 24 hours	
Nitrogen dioxide	Gas and kerosene space heaters, gas stoves, vehicular exhaust	100 μg/m^3 annual	Respiratory and mucous irritant
Ozone	Photocopying machines, electrostatic air cleaners, outdoor air	235 μg/m^3 for 1 hour	Respiratory irritant, causes fatigue
Pesticides and other semivolatile organics	Sprays and strips, outdoor air	5 μg/m^3 for chlordane	Possible carcinogens
Radon	Soil gas that enters buildings, construction materials, groundwater	4 pCi/L	Lung cancer
Sulfur dioxide	Coal and oil combustion, kerosene space heaters, outside air	80μg/m^3 annual; 365 μg/m^3 for 24 hours	Respiratory and mucous irritant
Volatile organics	Smoking, cooking, solvents, paints, varnishes, cleaning sprays, carpets, furniture, draperies, clothing	None available	Possible carcinogens

Sources: N. L. Nagda, H. E. Rector, and M. D. Koontz, 1987; M. C. Baechler et al., 1991; E. J. Bardana Jr. and A. Montaro (eds.), 1997; M. Meeker, 1996; D. W. Moffatt, 1997.

Figure 24.4 • Concentrations of common indoor air pollutants compared with outdoor concentrations. (*Source:* A. V. Nero, Jr., "Controlling Indoor Air Pollution," *Scientific American*, 258, no. 5 [1998]: 42–48.)

of the steps we have taken to conserve energy in homes and other buildings has led to the trapping of pollutants inside.

Two of the best ways to conserve energy in homes and other buildings are to increase the insulation and to decrease the infiltration of outside air. Constructing our buildings with windows that do not open and applying extensive caulking and weather stripping does reduce energy consumption, but it also tends to affect the air quality of the building by reducing natural ventilation. An important function of ventilation is that it replaces the indoor air with outdoor air in which the concentrations of pollutants are generally much lower. With less natural ventilation, we must depend more on the ventilation systems that are part of heating and air-conditioning systems.

24.2 Heating, Ventilation, and Air-Conditioning Systems

Heating, ventilation, and air-conditioning systems are designed to provide a comfortable indoor environment for people. Design of these systems depends on a number of variables, including the activity of people in the building, air temperature and humidity, and air quality. The interaction among these factors determines whether people are comfortable indoors. If the heating, ventilation, and air-conditioning system is designed correctly and functions properly, it will provide thermal comfort for people inhabiting the building. It will also provide the

necessary ventilation (utilizing outdoor air) and remove common air pollutants via exhaust fans and filters.[6]

Personal comfort levels in terms of temperature and humidity vary depending on age, physiology, and level of activity. Furthermore, different portions of buildings may have different temperatures and air quality because of their location in relation to heat sources, cold surfaces, and large windows. Humidity should be carefully controlled. High humidity may facilitate the growth of adverse mildews or molds, whereas low humidity may be a source of discomfort to some people.[6]

Regardless of the type of heating, ventilation, and air-conditioning system used in a home or other building, the effectiveness of that unit depends on the proper design of the equipment relative to the building, on proper installation, and on correct maintenance and operating procedures.[6] Indoor air pollution may result if any one of these factors concentrates pollutants from the many possible sources. If filters become plugged or contaminated with fungi, bacteria, or other potentially infectious agents, serious problems can result. In addition, as we see later in this chapter, ventilation systems are not generally designed to reduce some types of indoor pollution.[6,7]

24.3 Pathways, Processes, and Driving Forces

Many air pollutants originate within buildings and may be concentrated there because of lack of proper ventilation with the outside atmosphere. Other air pollutants may enter a building by infiltration, either through cracks and other openings in the foundations and walls or by way of ventilation systems.

The driving forces that control or modify the flow of air in buildings result from a variety of processes related to both natural forces and human activity. Both natural and human processes in buildings create differential pressures that move air and contaminants from one area to another. Areas of high pressure may develop on the windward side of a building, whereas pressure is lower on the leeward, or protected, side. As a result, air is drawn into a building from the windward side. Opening and closing doors produces pressure differentials that induce air to move within buildings, and wind can affect the movement of air in a building, particularly if the structure is leaky.[6]

A **chimney effect** (or **stack effect**) occurs when there is a temperature differential between the indoor and outdoor environments. Warm air rises within a building. If the indoor air is warmer than that found outdoors, as the warmer air rises in the building to the upper levels, it is replaced in the lower portion of the building by outdoor air drawn in through a variety of openings, such as windows, doors, or cracks in the foundations and walls. Facilities such as elevator shafts and stairwells provide corridors through which air can move from one floor to another.[6] Environmental tobacco smoke, or second-hand smoke, may also be drawn into a building by the chimney effect if smokers move outside near revolving or open doors to smoke.[5]

Because air is such a fluid medium, the possible interactions between the driving forces and the building are complex, and the distribution of potential air contaminants and pollutants is extensive. One outcome of this situation is that people in various parts of a building may complain about the air quality even if they are separated by considerable distances from each other and from potential sources of pollution.[6]

24.4 Building Occupants

Typically, the people living or working in particular indoor environments react in different ways to pollutants:

- Some groups of people are particularly susceptible to indoor air pollution problems.
- The symptoms reported by people in a particular environment vary.
- In some cases, the symptoms reported result from factors other than air pollution.

Particularly Susceptible People

People have varying sensitivity to air pollutants. One person may be adversely affected by a particular pollutant, whereas others in nearby areas may seem to be unaffect-ed. Sometimes, the problem is a matter of concentration rather than sensitivity; the person most affected by a particular indoor air pollutant might be experiencing the greatest exposure. A person's susceptibility to a particular air pollutant also depends on genetic factors, lifestyle, and age (see Chapter 15).

As a result of these individual differences, the response of one person bothered by a particular pollutant may differ from the response of another affected individual, making it difficult to evaluate indoor pollution problems.

Nevertheless, older people with impaired health, and children (because of their activity level and developing lungs) are generally more sensitive to air pollutants.[7] People suffering from chronic lung or respiratory diseases, such as chronic bronchitis, allergies, or asthma, are especially likely to be affected adversely by poor indoor air quality. Another group more strongly affected comprises individuals who have suppressed immune systems owing to disease or medical treatment, such as chemotherapy or radiation therapy.[6] Some people, when exposed to chemicals, develop what is known as multiple chemical sensitivity (MCS), a controversial disease in which people are allergic to any material or product containing human-produced chemicals. Some sufferers of MCS are so sensitive that they effectively have to live in a "plastic bubble" that keeps out chemicals.[5]

Symptoms of Indoor Air Pollution

A great variety of symptoms can result from exposure to indoor air pollutants (Table 24.2). Some chemical pollutants can cause nosebleeds, chronic sinus infections, headaches, and irritation of the skin or eyes, nose, and throat. More serious problems include loss of balance and memory, chronic fatigue, difficulty in speaking, and allergic reactions, including asthma. For example, chlorine tablets, which are often used in swimming pools and hot tubs, are a tremendous irritant if dust from the tablets is inhaled; shortness of breath and coughing result. Other pollutants cause dizziness or nausea. Exposure to carbon monoxide results in shortness of breath at low concentrations. At high concentrations, extreme toxicity and death can result. Tissues sensitive to carbon monoxide include the brain, heart, and muscles.[8]

The symptoms just described may have a quick onset after exposure. Other pollutants, including radon, asbestos, and chemicals such as benzene, may have long-term chronic health effects, including diseases such as cancer. Because of long lag times between exposure and disease, it may be difficult to establish relationships between a particular indoor air environment and disease in an individual.

Sick Buildings

As we saw in the opening case study, an entire building can be considered sick because of environmental problems. There are two types of sick buildings:

Table 24.2 Some Symptoms of Indoor Air Pollution

Symptoms	ETS [a]	Combustion Products [b]	Biologic Pollutants [c]	VOCs [d]	Heavy Metals [e]	SBS [f]
Respiratory						
Inflammation of mucous membranes of the nose, nasal congestion	Yes	Yes	Yes	Yes	No	Yes
Nosebleed	No	No	No	Yes	No	Yes
Cough	Yes	Yes	Yes	Yes	No	Yes
Wheezing, worsening asthma	Yes	Yes	No	Yes	No	Yes
Labored breathing	Yes	No	Yes	No	No	Yes
Severe lung disease	Yes	Yes	Yes	No	No	Yes
Other						
Irritation of mucous membranes of eyes	Yes	Yes	Yes	Yes	No	Yes
Headache or dizziness	Yes	Yes	Yes	Yes	Yes	Yes
Lethargy, fatigue, malaise	No	Yes	Yes	Yes	Yes	Yes
Nausea, vomiting, anorexia	No	Yes	Yes	Yes	Yes	No
Cognitive impairment, personality change	No	Yes	No	Yes	Yes	Yes
Rashes	No	No	Yes	Yes	Yes	No
Fever, chills	No	No	Yes	No	Yes	No
Abnormal heartbeat	Yes	Yes	No	No	Yes	No
Retinal hemorrhage	No	Yes	No	No	No	No
Muscle pain, cramps	No	No	No	Yes	No	Yes
Hearing loss	No	No	No	Yes	No	No

[a] Environmental tobacco smoke.

[b] Combustion products include particles, NO_x, CO, and CO_2.

[c] Biologic pollutants include molds, dust mites, pollen, bacteria, and viruses.

[d] Volatile organic compounds, including formaldehyde and solvents.

[e] Heavy metals include lead and mercury.

[f] Sick building syndrome.

Source: Modified from American Lung Association, Environmental Protection Agency, and American Medical Association, "Indoor Air Pollution—An Introduction for Health Professionals," 523-217/81322 (Washington, D.C.: GPO, 1994).

- Buildings with identifiable problems, such as occurrences of toxic molds or bacteria known to cause disease. The diseases are known as *building-related illnesses* (BRI).
- Buildings with **sick building syndrome (SBS)**, where the symptoms people report cannot be traced to any one known cause.

Sick building syndrome is a condition associated with an indoor environment that appears to be unhealthy. A number of people in such a building report adverse health effects that they believe are related to the amount of time they spend in the building. The range of complaints may vary from funny odors to more serious symptoms, such as headaches, dizziness, nausea, and so forth. In addition, a number of people in the building may be sick; and a group of people may have contracted a disease, such as cancer.

In many cases, it is difficult to establish what may be causing a particular sick building syndrome. Sometimes, the problem has been found to be related to poor management practices and worker morale rather than exposure to toxins in the building. When the occupants of a building report adverse health effects and a study follows, often the cause is not detected. A number of things may be happening:[6]

- The complaints result from the combined effects of a number of contaminants present in the building.
- Environmental stress from a source other than air quality—such as noise, high or low humidity, poor lighting, or overheating—is responsible.
- Employment-related stress—such as stress resulting from poor relations between labor and management, poor morale, or overcrowding—may be leading to the symptoms reported.
- Other unknown factors may be responsible. For example, pollutants or toxins may be present but not identified.

Of course, sick building syndrome may be the combined effect of various aspects of some or all of these factors. As noted, one common aspect of sick building syndrome is that often no one specific disease or cause is easily identified.[6]

We have reviewed some basics concerning sources, processes, and effects of indoor air pollution. Next, we consider two selected pollutants: environmental tobacco smoke and radon gas.

24.5 Environmental Tobacco Smoke

Environmental tobacco smoke (ETS), also known as *second-hand smoke*, comes from two sources: smoke exhaled by smokers and smoke emitted from burning tobacco in cigarettes, cigars, or pipes. People who are exposed to ETS are referred to as *passive smokers*.[9]

ETS is the best known of the hazardous indoor air pollutants. It is hazardous for the following reasons:[9,10]

- Tobacco smoke contains several thousand chemicals, many of which are irritants. Examples include NO_x, CO, hydrogen cyanide, and about 40 carcinogenic chemicals.
- Studies of nonsmoking workers exposed to ETS found that they have reduced airway functions comparable to what would be caused by smoking up to 10 cigarettes per day. They suffer more illnesses, such as coughs, eye irritation, and colds, and lose more work time than those not exposed to ETS.
- In the United States, about 3,000 deaths from lung cancer and 40,000 deaths from heart disease are thought to be associated with ETS.

ETS exposure depends on a variety of factors, including the number of people in a room smoking, the size of the room, and the rate of ventilation. Separating smokers from nonsmokers reduces but does not eliminate exposure to ETS. Thus, some U.S. state and local governments have banned smoking in restaurants, bars, and public buildings to protect citizens from ETS.

Smokers must realize that they are harming not only themselves but others, including their families, their friends, and the general public, by polluting the air that others breathe. Indeed, many people consider being exposed to ETS as a transgression of their right to breathe clean air in their homes, their workplaces, and public buildings.

The number of smokers in the United States has declined, but there are still about 40 million. The rate is higher in the developing world, where health warnings are few or nonexistent. Smoking is extremely addictive, because tobacco contains nicotine, a highly addictive substance. Nevertheless, education and social pressure have worked to some extent to influence some thoughtful people to quit smoking and encourage others to quit.

24.6 Radon Gas

It has become apparent within the past few decades that radon gas may constitute a significant environmental health problem in the United States (see A Closer Look 24.1).[11,12] An interesting aspect of the radon gas hazard is that it comes from natural processes rather than human activities.

Radon is a naturally occurring radioactive gas that is colorless, odorless, and tasteless. It is a member of the naturally occurring radioactive decay chain from radiogenic uranium to stable lead (Figure 24.5). Radon-222,

Figure 24.5 • Simplified diagram showing the radioactive decay chain for radon. Not all isotopes are shown. Half-lives and types of decay are shown for some.

Is Radon Gas Dangerous?

Many people today are anxious and worried about radon gas in homes because studies have shown that exposure to elevated concentrations of radon is associated with increased risk of lung cancer. It is believed that the risk increases with the level of exposure, the length of exposure, and certain habits, such as smoking.[12] The Environmental Protection Agency (EPA) estimates that 14,000 lung cancer deaths per year in the United States are related to exposure to radon gas and its daughter products (products that result from its radioactive decay), primarily polonium-218. (The estimate actually ranges from 7,000 to 30,000.) By comparison, there are approximately 140,000 total lung cancer deaths in the United States each year. If these estimates are correct—and they are controversial—approximately 10% of the lung cancer deaths in the United States can be attributed to radon gas. Exposure to radon gas has also been linked to other forms of cancer, such as melanoma (a deadly form of skin cancer) and leukemia, but such linkages are highly controversial.[19]

It is also believed that exposure to radon gas combined with smoking produces a synergistic effect that is particularly hazardous. One estimate is that the combination of exposure to radon gas and tobacco smoke is 10–20 times as hazardous as exposure to either pollutant by itself.[12]

Few comprehensive studies directly link radon gas exposure in houses to increased incidence of lung cancer. One study in Sweden did conclude that exposure to radon in homes is an important cause of lung cancer in the general population.[18] However, the linkage between radon and cancer is mostly based on studies of uranium miners, a group of people exposed to high concentrations of radon in mines.

It is believed that the health risk from radon gas is primarily related to its daughter products, such as polonium-218, which is a particle and thus adheres to dust. It is hypothesized that the dust is then inhaled into lungs, where cell-damaging alpha radiation can occur when polonium-218 decays (it has a half-life of approximately 3 minutes) (Figure 24.6).

The risk related to radon gas has been estimated by the EPA. Table 24.3 relates levels of exposure to radon gas to the estimated number of lung cancer deaths for smokers and nonsmokers.[12] The average concentration of radon gas in the outdoor environment is approximately 0.4 pCi/L; the average indoor level is approximately 1 pCi/L. The EPA has set the action level for radon at 4 pCi/L. This level is the concentration below which exposure is thought by the EPA to be an acceptable risk. In its risk charts (see Table 24.3), the EPA equates the risk associated with exposure to 4 pCi/L for a nonsmoker to about the risk of drowning. For a smoker, the risk increases to about 100 times the risk of dying in an airplane crash. These risks are calculated in terms of long-term (lifetime) exposure. Prediction of the number of deaths from lung cancer resulting from radon exposure is controversial. It has been argued that there is insufficient evidence to support direct correlation of smoking habits and exposure to radon gas with deaths from lung cancer. Furthermore, there are difficulties with measurements.

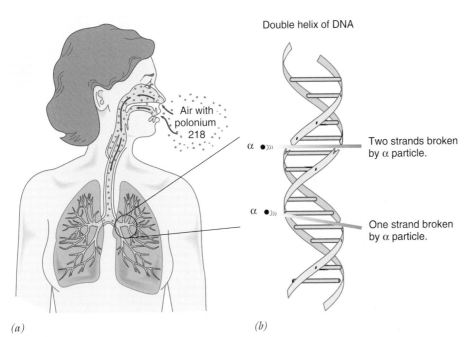

(a)　　　　　　(b)

Figure 24.6 ● (*a*) Deposition of polonium-218 in the lungs. (*b*) Alpha radiation (particle) breaking one or both DNA strands at the cell level. (*Source:* Modified from D. J. Brenner, *Radon: Risk and Remedy* [New York: W. H. Freeman & Co., 1989]. Reprinted with permission.)

Table 24.3   Estimated Risk Associated with Radon

Radon Level (pCi/L)	If 1,000 People Were Exposed to This Level Over a Lifetime	Risk of Cancer from Radon Exposure Compares to...	What to Do
Radon Risk if You Smoke[a]			Stop smoking and...
20	About 135 people could get lung cancer	←100 times the risk of drowning	Fix your home
10	About 71 people could get lung cancer	←100 times the risk of dying in a home fire	Fix your home
8	About 57 people could get lung cancer		Fix your home
4	About 29 people could get lung cancer	←100 times the risk of dying in an airplane crash	Fix your home
2	About 15 people could get lung cancer	←2 times the risk of dying in a car crash	Consider fixing between 2 and 4 pCi/L
1.3	About 9 people could get lung cancer	Average indoor radon level	
0.4	About 3 people could get lung cancer	Average outdoor radon level	Reducing radon levels below 2 pCi/L is difficult
Radon Risk if You Never Smoke[b]			
20	About 8 people could get lung cancer	←The risk of being killed in a violent crime	Fix your home
10	About 4 people could get lung cancer		Fix your home
8	About 3 people could get lung cancer	←10 times the risk of dying in an airplane crash	Fix your home
4	About 2 people could get lung cancer	←The risk of drowning	Fix your home
2	About 1 person could get lung cancer	←The risk of dying in a home fire	Consider fixing between 2 and 4 pCi/L
1.3	Less than 1 person could get lung cancer	Average indoor radon level	
0.4	Less than 1 person could get lung cancer	Average outdoor radon level	Reducing radon levels below 2 pCi/L is difficult

[a]If you are a former smoker, your risk may be lower.

[b]If you are a former smoker, your risk may be higher.

Source: U.S. Environmental Protection Agency, *A Citizen's Guide to Radon,* 2nd ed., ANR-464, 1992.

Radon concentrations in one part of a home may be much different from those in another. If homes have basements, concentrations tend to be higher there than on upper level floors. Nevertheless, all scientists agree that exposure to high levels of radon can cause cancer.

If the estimated risks from radon gas are anywhere close to the actual risk, then the hazard is a large one. The Surgeon General of the United States has stated that "indoor radon gas is a national health problem." The risks posed by exposure are thought to be hundreds of times higher than risks resulting from outdoor pollutants present in air and water. Such pollutants are generally regulated to reduce the risk of premature death and disease to less than 0.001%. Risks from some indoor pollutants, such as organic chemicals, may be as high as 0.1%.[20] These risks still are very small compared with the risk for radon. For example, people who live in homes for about 20 years with an average concentration of radon of about 25 pCi/L are estimated to have a 1% to 2% chance of contracting lung cancer.[12,20]

which has a half-life of 3.8 days, is the product of radioactive decay of radium-226. Radon decays with emission of an alpha particle to polonium-218, which has a half-life of approximately 3 minutes. (The discussion of radiation, radiation units, radiation doses, and health problems related to radiation in Chapter 19 will help you in understanding the following materials.)

Radon was discovered in 1900 by Ernest Dorn, a German chemist. The use and misuse of radon has an interesting history. In the early 1900s, bathing in radon water became a health fad. During this period, when radon was thought to be beneficial to health, many products containing radium, and thus radon, hit the market. These included chocolate candies, bread, and toothpaste.

As recently as 1953, a contraceptive jelly containing radium was marketed in the United States.[11]

Geology and Radon Gas

The concentration of radon gas that reaches the surface of the Earth and thus can enter our dwellings is related to the concentration of radon in the rocks and soil, as well as the efficiency of the transfer processes from the rocks or soil to the surface. Some regions in the United States contain bedrock with an above-average natural concentration of uranium. A large area that includes parts of Pennsylvania, New Jersey, and New York is now famous for elevated concentrations of radon gas. This area, known as the Reading Prong, contains many homes with elevated radon gas concentrations.[11] Areas with elevated concentrations of radon have also been identified in a number of other states, including Florida, Illinois, New Mexico, South Dakota, North Dakota, Washington, and California.

How Does Radon Gas Enter Homes and Other Buildings?

A radon legend was born in Boyer Town, Pennsylvania, in 1984, when Stanley Watres, who had a job as a technical advisor to the Limerick Nuclear Power Station, set off radiation alarms on the way into the plant. The reactor in the power plant had not yet been turned on when the alarms went off, and extensive testing of Watres's clothing suggested that the contamination came not from where he worked but from where he lived. Investigators were astounded to find that the radiation level in his home was 3,200 pCi/L, 800 times higher than the action level of 4 pCi/L set by the Environmental Protection Agency! Until that time, scientists had not believed that radon could occur naturally in concentrations high enough to be hazardous to anyone.[11,13–15] The Watres home held the record until the latter part of the 1980s, when a home in Whispering Hills, New Jersey, was found to have a radiation level of 3,500 pCi/L.[14]

Radon gas enters homes and other buildings in three main ways (Figure 24.7):

1. It migrates up from soil and rock into basements and lower floors.
2. Dissolved in groundwater, it is pumped into wells and then into homes.
3. Radon-contaminated materials, such as building blocks, are used in construction.

It is very difficult to estimate the number of homes in the United States that may have elevated concentrations of radon gas. The EPA estimates that about 7% of U.S. homes have elevated radon levels and recommends that

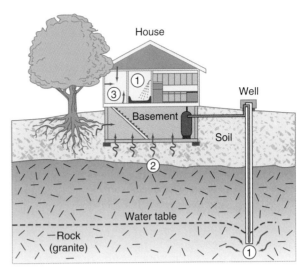

Figure 24.7 • How radon can enter homes. (1) Radon in groundwater enters a well and goes to a house, where it is used for water supply, dishwashing, showers, and other purposes. (2) Radon gas in rocks and soil migrates into a basement through cracks in the foundation and pores in construction. (3) Radon gas is emitted from construction materials used in building a house. (*Source:* Environmental Protection Agency.)

all homes and schools be tested.[12] The test is simple and inexpensive.

How Can Radon Be Reduced in Homes and Other Buildings?

The good news concerning the radon gas hazard is that, once identified, it is often easily fixed. There are well-known ways to reduce the concentration of radon gas in buildings. The simplest step is to locate the entry points of radon and seal them. This action, however, is often not sufficient; so additional ventilation, using fans and other devices, may be necessary. Increased ventilation is the primary remedy for radon problems. If these methods are not successful, a venting system can be constructed.[16,17]

24.7 Control of Indoor Air Pollution

In terms of the workplace, there are strong financial incentives to provide workers with a clean air environment. As much as $250 billion per year might be saved by decreasing illnesses and increasing productivity through improving the work environment.[5] A good starting point would be passing environmental legislation requiring minimum indoor air quality standards. This should include increasing the inflow of fresh air through ventilation. In Europe, systems of filters and pumps in many office buildings circulate air three times as frequently as is typical in buildings in the United States.

Table 24.4   Strategies to Control Indoor Air Pollution

Ventilation: General ventilation system; spot (zone or localized) ventilation (exhaust fans, etc.)

Source removal: Material or product substitution; restrictions on source use (e.g., establishment of smoking areas; restrictions on sale of particular items and on activities that cause indoor air pollution)

Source modification: Change in combustion design (e.g., maximize efficiency of a gas stove); material substitution (use materials that don't cause air pollution); reduction in emission rates by intervention of barriers (e.g., apply coatings over lead paint or asbestos)

Air cleaning (pollutant removal): Particle filtering; gas and vapor removal; passive scavenging or absorption

Education: Consumer information on products and materials; public information on health, productivity, and nuisance effects; resolution of legal rights and liabilities of consumer, tenant, manufacturer, and so on, related to indoor air quality

Source: Modified from Committee on Indoor Pollutants, *Indoor Pollutants* (Washington, D.C.: 1981), p. 489.

Many building codes in Europe require that workers have access to fresh air (windows) and natural light. Unfortunately for U. S. workers, no similar codes exist in the United States, and many buildings use central air conditioning with windows permanently sealed.[5]

You might think that heating, ventilating, and air-conditioning systems, when operating properly and well maintained, will ensure good indoor air quality. Unfortunately these systems are not designed to maintain all aspects of air quality. For example, commonly used ventilation systems do not generally reduce radon gas. Ventilation is one control strategy when faced with high concentrations of any indoor air pollutant, including radon. Other strategies, shown in Table 24.4, include source removal, source modification, and air cleaning.[8] These strategies do not constitute a complete list, and some combination of them may be the best approach.

One of the principal means for controlling the quality of indoor air is by dilution with fresh outdoor air via a ventilating air-conditioning system and windows that can be opened. Outside air is brought in and mixed with air in the return flow system from the building; the air is filtered, heated or cooled, and supplied to the building. Various types of air-cleaning systems for residential and nonresidential buildings are available to reduce potential pollutants, such as particles, vapors, and gases. These systems can be installed as part of the heating, ventilation, and air-conditioning system or as stand-alone appliances.[8]

Education also plays an important role in understanding and developing strategies for reducing indoor air pollution problems; it empowers people with knowledge necessary to make intelligent decisions. At one level, this may involve deciding not to install unvented or poorly vented appliances. A surprising (and tragic) number of people are killed each year by carbon monoxide poisoning resulting from poor ventilation in homes, campers, and tents. At other levels, educated people are more aware of their legal rights with respect to product liability and safety. Furthermore, education provides people with the information necessary to make decisions concerning exposure to chemicals, such as paints and sol-

vents, and strategies to avoid potentially hazardous conditions in the home and workplace.[8]

Summary

- Indoor air pollution has been with us for thousands of years, since people first built structures and burned fuel indoors. It is one of our most serious environmental health problems.

- Sources of indoor air pollution are extremely varied. They may be associated with the materials with which we build our buildings, the furnishings we put in them, and the types of equipment we use for heating and cooling, as well as natural processes that allow gases to seep into buildings.

- Concentrations of indoor air pollutants are generally greater than concentrations of the same pollutants found outdoors. Part of the reason is that in recent years we have attempted to conserve energy through better insulation of buildings. The common method for controlling indoor air pollution is ventilation. However, natural ventilation has been reduced through tighter construction of buildings, and commonly used ventilation systems are not generally designed to reduce certain types of indoor air pollutants. In addition, these systems require careful maintenance.

- A variety of pathways, processes, and driving forces affect the air quality of a building. The most common natural processes involve the differential pressure produced by wind and the chimney, or stack, effect, which occurs when there is a temperature differential between indoor and outdoor environments.

- Indoor air pollution has different effects on different people, and some groups of people are particularly susceptible to air pollution problems. Often, the symptoms reported by people working in a building vary; and some symptoms may result from factors other than air pollution.

- There are two basic types of sick buildings: those with

Are Airplanes Adequately Ventilated?

The environment at 10,667 m (35,000 ft) above Earth, a typical cruising altitude for jet airplanes, is not fit for humans. The temperature is –54°C (–65°F), the humidity is close to zero, and the air pressure is low. To supply air for the passengers and crew of a jet airplane, outside air is compressed and heated by the jet engines. Because the air emerges from the engines at a higher pressure and temperature than would be comfortable, the air is cooled by mixture with outside air before entering the cabin, and the pressure is adjusted to that of an altitude of 1,524 m (5,000 ft). Humidity, between 5% and 20% at cruising altitudes, is supplied primarily by the breath and perspiration of passengers and crew.

Until the mid-1980s, commercial planes used 100% fresh air, which was recirculated every three minutes. But using fresh air taken in by the jet engines to provide air inside the plane reduces fuel efficiency (Figure 24.8). Providing fresh air in an airplane costs 22–37 times as much as supplying the same amount inside a building in Washington, D.C., in January. Airlines found they could save money by recirculating air already inside the cabin in combination with fresh air (usually in a 1:1 proportion). This practice reduced the flow rate, so that there was a complete change of air only every seven minutes or even less often.

Air flow is measured in cubic feet per minute (cfm); ventilation rate is the air flow per person (cfm/number of people). Ventilation rates vary in commercial planes from 150 cfm/person in the cockpit to 50 cfm/person in first class to 7 cfm/person in economy class. The latter is comparable to the rates found in trains and subways (5–7 cfm/person). Airline officials point out that the rate of air change in planes is much better than that found in office buildings (10–12 times an hour as opposed to 1–2 times).

Many flight attendants and passengers have complained about symptoms, such as headaches, dry eyes, fatigue, nausea, and upper respiratory problems, that could be associated with the air quality on planes. The carbon dioxide level, which is often used as an indicator of indoor air quality, averages 1,500 ppm in planes but has been measured as high as 2,000 ppm on 25% of flights studied. Even with ventilation rates as high as 35 cfm/person, carbon dioxide levels have been found at 1,200 ppm or more. The American Society of Heating, Refrigerating, and Air-Conditioning Engineers (ASHRAE) recommends a maximum of 1,000 ppm for indoor air, and the Occupational Safety and Health Administration (OSHA) has a standard of 5,000 ppm for industrial buildings. The Federal Aviation Administration (FAA) standard allows carbon dioxide levels up to 30,000 ppm.

Some experts are concerned that high carbon dioxide levels may be associated with other undesirable constituents, such as chemicals from cleaning fluids, pesticides, and fuels. Airborne contamination from sick passengers and animals below deck is another source of concern. Outbreaks of influenza have been tied to airplane air. Three passengers on one flight in 1992 became infected with tuberculosis, although it could not be

Figure 24.8 • Fuel required for ventilation with outside air on a jet airplane. (*Source:* Data from National Research Council, *The Airliner Cabin Environment* [Washington, D.C.: 1986], National Academy Press, p. 61.)

clearly determined that infection occurred onboard. This occurrence and the fact that the incidence of tuberculosis has been on the rise worldwide since the 1980s sparked an inquiry by the U.S. Congress. Although medical experts concluded that long periods of close contact were required for transmission of tuberculosis, concern was great enough for the Centers for Disease Control (now called the Centers for Disease Control and Prevention) in Atlanta to undertake a study of the subject in 1993. The results of this study confirmed that although there is a possibility of tuberculosis transmission aboard airplanes, the possibility is extremely small. Thus, it is unlikely that airplane cabin air circulation standards will be affected.

Critical Thinking Questions

1. Airlines contend that higher ventilation rates are necessary in the cockpit. What reasons can you think of to support their argument? Flight attendants claim that their requirements for air are different from those of passengers. What do you think is the basis for this claim?

2. If air exchange in airliners is as good or better than that in trains and subways, is there any reason to question the air quality? Develop a rationale for a position on this question.

3. What criteria would you accept for establishing a standard for the level of carbon dioxide in airplanes?

4. Three options exist for reducing carbon dioxide levels in airliner cabins: reducing emissions, increasing ventilation rates, and absorbing carbon dioxide from the ambient air. What are the advantages and disadvantages of each option?

5. Very little research has been done on air quality and the health of employees and passengers. Why, even if research is conducted, might it be hard to prove a connection?

an identifiable problem, such as mold or bacteria, and those with sick building syndrome, in which people's symptoms can't be traced to any one cause.

- Environmental tobacco smoke, or second-hand smoke, is the most hazardous indoor air pollutant.
- Radon gas that seeps into homes is thought to be a serious environmental health hazard in the United States today. Studies have suggested that exposure to elevated concentrations of radon is associated with an increased risk of lung cancer.
- Control of indoor air pollution involves several strategies, including ventilation, source removal, source modification, and installation of air-cleaning equipment, as well as education.

● REEXAMINING THEMES AND ISSUES

Human Population: Indoor air pollution has been with us since we made the decision to build homes and move inside. As human populations grew and people constructed a broader variety of homes, they were exposed to additional pollutants. As the number of people on Earth continues to increase, people will use more and different resources to build homes, and more people are likely to be living in smaller spaces. As a result, effects of indoor air pollution are expected to increase in the future and to continue to present serious environmental health problems.

Sustainability: If we accept the premise that sustainability begins at home, then air quality in our homes and our workplaces is an important part of a sustainable future. Our health and that of future generations depend on breathing pollution-free air where we spend most of our time—indoors. Conflicts may arise, however, because we also wish to conserve resources and build energy-efficient homes as part of sustainability. As we have seen, that may lead to restricted circulation of air and resulting indoor air pollution problems. This conflict may be solved through technological advances that minimize air pollution while maximizing energy efficiency.

Global Perspective: Indoor air pollution occurs everywhere in the world where people live inside or work in buildings or other human-constructed enclosures such as mines. Strictly speaking, indoor air pollution is not a global problem; but it *is* a significant issue if we wish to address globally experienced human health concerns. Environmental tobacco smoke is a serious health problem, as is exposure to high levels of radon gas. These are examples of toxicity and health problems found almost everywhere people live.

Urban World: Indoor air pollution problems may occur anywhere people live inside. However, problems are more likely in urban areas that have significant outdoor air pollution problems. Therefore, in cities, if we wish to help eliminate indoor air pollution, we must also pay attention to outdoor air quality.

Science and Values: For too long, we have not paid sufficient attention to our indoor environment and the consequences of poor air quality there. This issue is now moving to the forefront, however; and we believe that, in the future, technological advances will allow for better design of homes, buildings, and other structures to maximize air quality for the people that spend their time there. People value clean air, as can be seen in the attitude of many people toward second-hand smoke. Second-hand smoke presents a serious health risk to nonsmokers, and there is considerable social pressure to encourage people to quit smoking.

● KEY TERMS

black lung disease **521**
chimney effect (stack effect) **526**

environmental tobacco smoke (ETS) **528**

radon **528**
sick building syndrome (SBS) **527**

STUDY QUESTIONS

1. What are some of the common sources of air pollutants where you live, work, or attend classes?

2. Develop a research plan to complete an audit of the indoor air quality in your local library. How might that research plan differ from a similar audit for the science buildings on your campus?

3. What do you think about the concept of the sick building syndrome? If you were working for a large corporation and a number of your employees stated that they were becoming sick and listed a series of symptoms and problems, how would you react? What could you do? Play the role of the administrator, and develop a plan to look at the potential problem.

4. Some people argue that the potential hazard from radon gas in homes is much less than suggested by the Environmental Protection Agency. Do you agree or disagree? How might potential differences of opinion ultimately be answered?

5. Develop a plan to study the potential radon hazard in your community. Where would you start? How would you gather data, and so on? If your community has undergone extensive testing already, review the results and decide if further testing is necessary.

FURTHER READING

Brenner, D. J. 1989. *Radon: Risk and Remedy.* New York: W. H. Freeman. A wonderful book concerning the hazard of radon gas. It covers everything from the history of the problem to what was happening in 1989, as well as solutions, and is highly recommended.

Brooks, B. O., and W. F. Davis. 1992. *Understanding Indoor Air Quality.* Ann Arbor, Mich.: CRC Press. A comprehensive evaluation of indoor air pollution. It discusses most of the sources of indoor air pollutants, as well as health effects and controls.

Kay, J. G., G. E. Keller, and J. F. Miller. 1991. *Indoor Air Pollution.* Chelsea, Mich.: Lewis. Essays on problems with biological and non-biological air pollution in the indoor environment.

Marconi, M., B. Seifert, and T. Lindvall. 1995. *Indoor Air Quality: A Comprehensive Reference Book.* New York: Elsevier. A thorough text covering all aspects of indoor air pollution and air quality.

Nagda, N. L., H. E. Rector, and M. D. Koontz. 1987. *Guidelines for Monitoring Indoor Air Quality.* New York: Hemisphere. A good overview of indoor air pollution, including some of the factors affecting indoor air pollutants and how they might be measured.

U.S. Environmental Protection Agency. 1991. *Building Air Quality.* EPA/400/1-91/033, DHHS (NIOSH). Publication No. 91-114. A guide to air quality issues for people building structures and managing them. It has a good section on factors affecting indoor air quality and another on resolving problems.

U.S. Environmental Protection Agency. 1995. *The Inside Story: A Guide to Indoor Air Quality.* A brief introduction to the concepts of indoor air pollution.

CHAPTER 25 Ozone Depletion

Sunscreen lotion being applied to a child's skin. Sunscreen reduces exposure to UVB, which is known to cause skin cancer.

● LEARNING OBJECTIVES

Ozone depletion in the stratosphere is recognized as a major environmental problem with serious effects. After reading this chapter, you should understand:

- What ozone is and how ozone is naturally formed and destroyed in the stratosphere.
- What the so-called ozone shield is, and why it is important.
- How chemical and physical processes and reactions link emissions of chlorofluorocarbons (CFCs) to stratospheric ozone depletion.
- What role polar stratospheric clouds play in ozone depletion.
- Why the problem of ozone depletion is a long-term problem.
- What the environmental effects of ozone depletion are, and what options are available to minimize ozone depletion.
- Why international cooperation, including significant economic aid from wealthy to less wealthy nations, is necessary to encourage future reduction or elimination of emissions of ozone-depleting chemicals into the atmosphere.

Epidemic of Skin Cancers in the United States

People in the United States are experiencing an epidemic of skin cancer. A worldwide increase in skin cancer has occurred since early 1970, and the increase in the United States over that period has been about 90%. It is suspected that the increase in skin cancer rates is linked to ozone depletion.[1] Ozone depletion and, in particular, development of the "ozone hole" (discussed later in this chapter) has been most dramatic since about 1975, and this correlates with the increased incidence of skin cancers.

As ozone depletion occurs, ultraviolet B (UVB) radiation, a normal part of the solar radiation that reaches Earth's surface (see Figure 22.12), increases. UVB causes damage and mutation to DNA in skin cells, which may initiate cancer. It is believed that a 1% decrease in ozone causes an increase of UVB radiation of about 1–2%. For each 1% increase in UVB radiation, it is projected that skin cancer will increase about 2%.

Since 1970, ozone depletion in the atmosphere above the United States has been about 10%. Under a worst-case scenario, this 10% decrease in ozone could cause an increase in skin cancer of 20–40% over the same period. However, the observed increase has been 90%! Thus, either we are incorrectly evaluating effects of ozone depletion and increased UVB radiation on skin cancers; or diagnosis of skin cancer has improved, so more are found; or other factors are affecting the incidence of skin cancers. Indeed, because disease seldom involves a one cause–one effect relationship, the epidemic of skin cancers is probably due to multiple interrelated causes, including the following:

- People are living longer, and cancers are a disease of aging.
- Cancer seems to be more a problem of urbanized and industrial societies.
- People in the United States today are more affluent than in the past and spend more time outdoors exposed to cancer-inducing UVB radiation.

Older people who for years worked outside as fishermen, farmers, lifeguards, or in construction, as well as those who sunbathed regularly, are at greater risk of contracting skin cancer. Sunburns that produce blistering and severe sunburns in childhood are thought to be especially hazardous, increasing the risk of melanoma in later life.[1] The risk to an individual depends on two factors: the extent of exposure to UVB and the amount of protective melanin pigment in the skin. Melanin pigment absorbs UVB and provides a natural protection. Thus, fair-skinned people who have little melanin compared with darker-skinned people have a higher incidence of skin cancer.

> • *Ozone depletion is a serious global environmental problem. We understand what causes ozone depletion and have implemented appropriate policies to solve the depletion problem.[2] The science of ozone depletion and the policies to reduce it are the two main subjects of this chapter.*

25.1 Ozone

The air we breathe at sea level is composed of approximately 21% diatomic oxygen (O_2), which is two oxygen atoms bonded together. **Ozone** (O_3) is a triatomic form of oxygen, in which three atoms of oxygen are bonded. Ozone is a strong oxidant and chemically reacts with many materials in the atmosphere. In the lower atmosphere, ozone is a pollutant produced by photochemical reactions involving sunlight, nitrogen oxides, hydrocarbons, and diatomic oxygen.

Figure 25.1 shows the structure of the atmosphere and concentrations of ozone. The highest concentrations are in the stratosphere, ranging from about 15 km to 40 km (9 to 25 mi) in altitude. Approximately 90% of the ozone in the atmosphere is found in the stratosphere, where peak concentrations are about 400 ppb. The altitude of peak concentration varies from about 30 km (19 miles) near the equator to about 15 km (9 mi) in polar regions.[3]

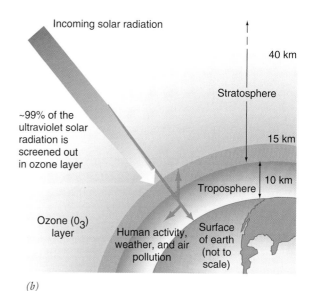

(a)

(b)

Figure 25.1 • (a) Structure of the atmosphere and ozone concentration. (b) Reduction of the most potentially biologically damaging ultraviolet radiation by ozone in the stratosphere. (*Source:* Ozone concentrations modified from R. T. Watson, "Atmospheric Ozone," in J. G. Titus, ed., *Effects of Change in Stratospheric Ozone and Global Climate*, vol. 1, *Overview*, p. 70 [U.S. Environmental Protection Agency].)

Ultraviolet Radiation and Ozone

The ozone layer in the stratosphere is often called the **ozone shield**, because it absorbs most of the potentially hazardous ultraviolet radiation that enters Earth's atmosphere from the sun. Figure 25.2 shows part of the electromagnetic spectrum, discussed in Chapter 22. Ultraviolet radiation consists of wavelengths between 0.1 and 0.4 μm and is subdivided into ultraviolet A (UVA), ultraviolet B (UVB), and ultraviolet C (UVC). Ultraviolet radiation with a wavelength of less than about 0.3 μm is potentially very hazardous to life. If much of this radiation reached Earth's surface, it would injure or kill most living things.

Ultraviolet C (UVC) has the shortest wavelength and is the most energetic of the types of ultraviolet radiation. It has sufficient energy to break down diatomic oxygen (O_2) in the stratosphere into two oxygen atoms. Each of these oxygen atoms may combine with an O_2 molecule to create ozone. Ultraviolet C is strongly absorbed in the stratosphere, and negligible amounts reach the surface of Earth.[3,4]

Figure 25.2 • Part of the electromagnetic spectrum showing ultraviolet radiation with wavelengths between 0.01 and 0.4 μm.

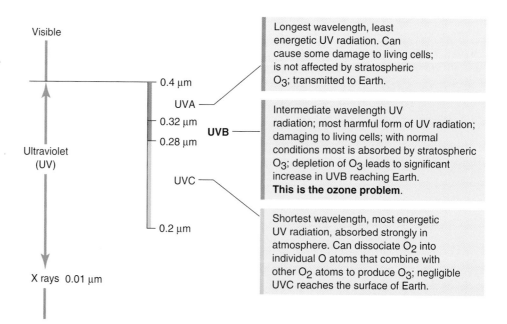

Ultraviolet A (UVA) radiation has the longest wavelength and the least energy of the three types of ultraviolet radiation. UVA can cause some damage to living cells, is not affected by stratospheric ozone, and is transmitted to the surface of Earth.[3]

Ultraviolet B (UVB) radiation is energetic and is strongly absorbed by stratospheric ozone. Ozone is the only known gas that absorbs UVB. As a result, depletion of ozone in the stratosphere results in an increase in the UVB that reaches the surface of the Earth. Because UVB radiation is known to be hazardous to living things,[3,4] this increase in UVB is the hazard we are talking about when we discuss the ozone problem.

Processes that produce ozone in the stratosphere are illustrated in Figure 25.3. The first process, or step, in ozone production occurs when intense ultraviolet radiation (UVC) breaks apart an oxygen molecule (O_2) through the process of photodissociation into two oxygen atoms. These atoms then react with another oxygen molecule to form two ozone molecules. Ozone, once produced, may absorb UVC radiation, which breaks the ozone molecule into an oxygen molecule and an oxygen atom. This is followed by the recombination of the oxygen atom with another oxygen molecule to reform into ozone. As part of this process, UVC radiation is converted to heat energy in the stratosphere.[5] Natural conditions that prevail in the stratosphere result in a dynamic balance between the creation and destruction of ozone.

In summary, approximately 99% of all ultraviolet solar radiation (all UVC and most UVB) is absorbed or screened out in the ozone layer. The absorption of ultraviolet radiation by ozone is a natural service function of the ozone shield and protects us from the potentially harmful effects of ultraviolet radiation.

Measurement of Stratospheric Ozone

Scientists first measured the concentration of atmospheric ozone in the 1920s, from the ground, using an instrument known as a Dobson ultraviolet spectrometer. The **Dobson unit (DH)** is still commonly used to measure the concentration of ozone; 1 DU is equivalent to a concentration of 1 ppb O_3. Today, we have a record of ozone concentrations from more than 30 locations around the world over about 30 years. Most of the measurement stations are in the midlatitudes, and the accuracy of the data varies with different levels of quality control.[3] Satellite measurements of concentrations of atmospheric ozone began in 1970 and continue today.

Ground-based measurements first identified ozone depletion over the Antarctic. Members of the British Antarctic Survey began to measure ozone in 1957 and in 1985 published the first data that suggested significant ozone depletion over Antarctica. Figure 25.4 shows the data from the British Antarctic Survey from the Halley Bay site. The data were taken during October of each

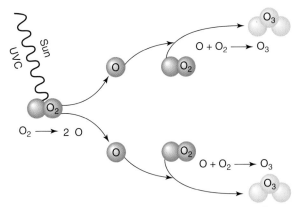

Figure 25.3 • Production of ozone (O_3) in the stratosphere. Photo dissociation of the oxygen molecule (O_2) yields two atoms of oxygen. Each combines with an oxygen molecule to form ozone (O_3). (*Source:* Modified from NASA-GSFC, "Stratospheric Ozone," accessed August 22, 2000, at http://see.gsfc.nasa.gov.)

year—the Antarctic spring—and show that the concentration of ozone hovered around 300 DU from 1957 to about 1970 and sharply dropped to approximately 200 DU in 1984. Since 1984, the variability of the ozone concentration has been considerable, with a high of about 250 DU in 1988 and a low of about 120 DU in 1993 and 1994 (Figure 25.4). In spite of the variations, the direction of change, with minor exceptions, is clear; ozone concentrations in the stratosphere during the Antarctic spring have been decreasing since the mid-1970s.[6–9]

Satellite measurements of ozone recorded prior to 1985 had also indicated a significant reduction in ozone concentration; however, the values were so low that they were not believed. After the 1985 announcement of the decrease in ozone over Antarctica, the satellite measure-

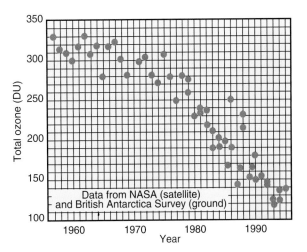

Figure 25.4 • Average Antarctic minimum ozone concentration during October 1957–1995. (*Source:* Modified from NASA-GSFC, "Stratospheric Ozone," accessed August 22, 2000, at http://see.gsfc.nasa.gov.)

ments were reevaluated and found to confirm the observations reported by the British Antarctic Survey. This depletion in ozone was dubbed the *ozone hole*. However, there is no actual hole in the ozone shield where all the ozone is depleted; rather, the term describes a relative depletion in the concentration of ozone that occurs during the Antarctic spring.

25.2 Ozone Depletion and CFCs

The hypothesis that ozone in the stratosphere is being depleted by the presence of **chlorofluorocarbons (CFCs)** was first suggested in 1974 by Mario Molina and F. Sherwood Rowland.[10] This hypothesis, based for the most part on physical and chemical properties of CFCs and knowledge about atmospheric conditions, was immediately controversial. The idea received a tremendous amount of exposure both in newspapers and on television and was vigorously debated by scientists, companies producing CFCs, and other interested parties.

The public became concerned because everyday products, such as shaving cream, hair spray, deodorants, paints, and insecticides, were packaged in spray cans that carried CFCs as a propellant. The idea that these products could be responsible for threatening their health and the well-being of the environment captured the imagination of the American people, many of whom responded by writing to their senators and representatives and making individual decisions to purchase fewer products containing CFCs.[11]

The major features of the Molina and Roland hypothesis are as follows[3]:

- The CFCs emitted in the lower atmosphere by human activity are extremely stable. They are unreactive in the lower atmosphere and therefore have a very long residence time (about 100 years). Another way of stating this is to say that no significant tropospheric sinks for CFCs are known. A possible exception is soils, which evidently do remove an unknown amount of CFCs from the atmosphere at Earth's surface.[12]
- Because CFCs have a long residence time in the lower atmosphere and because the lower atmosphere is very fluid, with abundant mixing, the CFCs eventually (by the process of dispersion) wander upward and enter the stratosphere. Once they have reached altitudes above most of the stratospheric ozone, they may be destroyed by the highly energetic solar ultraviolet radiation. This process releases chlorine, a highly reactive atom.
- The reactive chlorine released may then enter into reactions that deplete ozone in the stratosphere.
- The result of the depletion of ozone is an increase in the amount of UVB radiation that reaches Earth's surface. Ultraviolet B is a cause of human skin cancers and is also thought to be harmful to the human immune system.

Emissions and Uses of Ozone-Depleting Chemicals

Emissions of the commonly used chemicals related to ozone depletion are shown in Table 25.1. Emissions of chemicals thought to destroy stratospheric ozone amounted to

Table 25.1 Emissions of Some Chemicals Associated with Stratospheric Ozone Depletion in 1989, with Growth Rate of CFCs for 1998 and Substitute for CFCs

Chemical	Emissions (thousands of tons)	Atmospheric Lifetime[a] (years)	Applications	Annual Growth Rate to 1989 (%)	Share of Contribution to Depletion (%)	Annual Growth Rate, 1998 (%)
CFC-12	454	139	Air conditioning, refrigeration, aerosols, foams	5	45	0.5
CFC-11	262	76	Foams, aerosols, refrigeration	5	26	−0.1
CFC-113	152	92	Solvents	10	12	0.0
Carbon tetrachloride	73	35	Solvents	1	8	
Methyl chloroform	522	5	Solvents	7	5	
Halon 1301	3	65	Fire extinguishers	n.a.	4	
Halon 1211	79	11	Refrigeration, foams	11	0	
HCFC-22		12	Substitute for CFC			
HCFC-123		2	Substitute for CFC			
HCFC-124		6	Substitute for CFC			

[a.] Sources for atmospheric lifetimes of substances include C. P. Shea, "Mending the Earth's Shield," *World Watch* (January–February 1989):27–34; NASA-GSFC, "Stratospheric Ozone," accessed August 22, 2000, at http://see.gsfc.nasa.gov; and U.S. Environmental Protection Agency, "Ozone-Depleting Substances," accessed March 1, 2001, at http://www.epa.gov/ozone/ods/html.

approximately 1.5 million metric tons per year, with CFCs accounting for approximately 60% of the total emissions.

Table 25.1 also shows the approximate atmospheric lifetimes of these chemicals, which varies from about 5 years to more than 100 years. Because CFCs, in particular, have such long lifetimes in the atmosphere, they will be with us for many years.

Various uses of the chemicals are also shown in the table. CFCs have been used as aerosol propellants in spray cans, as a working gas in refrigeration and air-conditioning units, and in the foam-blowing process for the production of Styrofoam. A variety of cleaning solvents, such as carbon tetrachloride and methyl chloroform, contain chlorine and thus destroy ozone, as does halon, which contains bromine (another chemical like chlorine) and is used in fire extinguishers.[3,11]

One of the first restrictions on CFCs included its use as a propellant gas for spray cans. This practice was banned in the late 1970s in a number of countries, setting a trend that has continued, with the result that CFCs as aerosol propellants are no longer a problem.[3] In contrast, the use of CFCs as a refrigerant has increased dramatically in recent years (Figure 25.5), especially in developing countries, such as China.

Simplified Stratospheric Chlorine Chemistry

CFCs are considered responsible for most of the ozone depletion observed by scientists. Let us look more closely at how this effect occurs.

Earlier, we noted that there are no tropospheric sinks for CFCs. That is, the processes that remove most chemicals in the lower atmosphere—destruction by sunlight, rain-out, and oxidation—do not break down CFCs, because CFCs are transparent to sunlight, are essentially insoluble, and are nonreactive in the oxygen-rich lower atmosphere.[13] Indeed, the fact that CFCs are nonreactive in the lower atmosphere was one reason they were attractive for use as propellants.

When CFCs wander to the upper part of the stratosphere, however, reactions do occur. Highly energetic ultraviolet radiation (UVC) splits up the CFC, releasing chlorine. When this happens, the following two reactions can take place:[13]

$$(1)\ Cl + O_3 \rightarrow ClO + O_2$$

$$(2)\ ClO + O \rightarrow Cl + O_2$$

Figure 25.5 • Uses of CFC and amounts emitted in (*a*) 1974; (*b*) 1986; and (*c*) 1991. Use of CFCs as propellants in spray cans decreased dramatically, while use of CFCs as refrigerants increased. Total emissions in 1991 are far below those in 1986 and less than those in 1974. (*Source:* Modified from NASA-GSFC, "Stratospheric Ozone," accessed August 22, 2000, at http://see.gsfc.nasa.gov.)

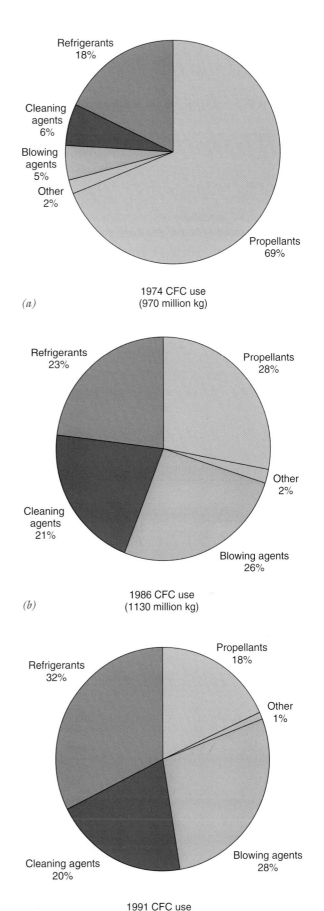

(*a*) 1974 CFC use (970 million kg)

Refrigerants 18%
Cleaning agents 6%
Blowing agents 5%
Other 2%
Propellants 69%

(*b*) 1986 CFC use (1130 million kg)

Refrigerants 23%
Propellants 28%
Other 2%
Blowing agents 26%
Cleaning agents 21%

(*c*) 1991 CFC use (680 million kg)

Refrigerants 32%
Propellants 18%
Other 1%
Blowing agents 28%
Cleaning agents 20%

These two equations define a chemical cycle that can deplete ozone (Figure 25.6). In the first reaction, chlorine combines with ozone to produce chlorine monoxide, which in the second reaction combines with monatomic oxygen to produce chlorine again. The chlorine can then enter another reaction with ozone and cause additional ozone depletion. This series of reactions is what is known as a *catalytic chain reaction*. Because the chlorine is not removed but reappears as a product of the second reaction, the process may be repeated over and over again. It has been estimated that each chlorine atom may destroy approximately 100,000 molecules of ozone over a period of one or two years before the chlorine is finally removed from the stratosphere through other chemical reactions and rain-out.[13] The significance of these reactions is apparent when we realize how many metric tons of CFCs have been emitted into the atmosphere (see Table 25.1).

It should be noted that what actually happens chemically in the stratosphere is considerably more complex than the two equations shown here. The atmosphere is essentially a chemical soup in which a variety of processes related to aerosols and clouds take place (some of these are addressed in the discussion of the ozone hole). Nevertheless, these equations show us the basic chemical chain reaction that occurs in the stratosphere to deplete ozone.

The catalytic chain reaction just described can be interrupted through storage of chlorine in other compounds in the stratosphere. Two possibilities are as follows:

1. Ultraviolet light breaks down CFCs to release chlorine, which combines with ozone to form chlorine monoxide (ClO), as already described. This is the first reaction discussed. The chlorine monoxide may then react with nitrogen dioxide (NO_2) to form a chlorine nitrate ($ClONO_2$). If this reaction occurs, ozone depletion is minimal. The chlorine nitrate, however, is only a temporary reservoir for chlorine. The compound may be destroyed, and the chlorine released again.

2. Chlorine released from CFCs may combine with methane (CH_4) to form hydrochloric acid (HCl). The hydrochloric acid may then diffuse downward. If it enters the troposphere, rain may remove it, thus removing the chlorine from the ozone-destroying chain reaction. This is the ultimate end for most chlorine atoms in the stratosphere. However, while the hydrochloric acid molecule is in the stratosphere, it

Figure 25.6 • Processes of natural formation of ozone and destruction by CFCs, N_2O, and methylbromide. (*Source:* Modified from NASA-GSFC, "Stratospheric Ozone," accessed August 22, 2000, at http://see.gsfc.nasa.gov.)

Figure 25.7 • Average change in concentration of stratospheric ozone, 1970–1994 (October mean total), by latitude. Notice that ozone reduction varies most in the Antarctic, whereas concentrations at the equator are nearly constant. (*Source:* Modified from NASA-GSFC, "Stratospheric Ozone," accessed August 22, 2000 at http://see.gsfc.nasa.gov.)

may be destroyed by incoming solar radiation, releasing the chlorine for additional ozone depletion.

It has been estimated that the chlorine chain reaction that destroys ozone may be interrupted by the processes just described as many as 200 times while a chlorine atom is in the stratosphere.[3,14]

In part as a result of the ozone depletion reactions, concentrations of ozone have declined in both northern and southern temperate latitudes. Figure 25.7 shows the estimated decline in stratospheric ozone from 1970–1994 by latitude in the Southern Hemisphere. Similar declines are thought to be occurring in the Northern Hemisphere. Massive destruction of ozone identified in the Antarctic constitutes the ozone hole, which continues to be a source of concern.[3]

In discussing the global distribution of ozone, it is important to remember that, for the southern hemisphere, under natural conditions, the highest concentration of ozone is found in the polar regions (about 60° south latitude) and the lowest near the equator. (See Figure 25.7 for 1970, before massive ozone destruction occurred.) At first, this may seem strange, because ozone is produced in the stratosphere by solar energy, and more solar energy is found near the equator. Much of the world's ozone is produced near the equator, but the ozone in the stratosphere moves from the equator toward the poles with global air circulation patterns, which are not well understood.[8]

25.3 The Antarctic Ozone Hole

Since the Antarctic ozone hole was first reported in 1985, it has captured the interest of many people around the world. Every year since then, ozone depletion has been observed in the Antarctic in October, the spring season there.

The thickness of the ozone layer above the Antarctic during the springtime has been declining since the mid-1970s, and the geographic area covered by the ozone hole continues to increase.[15] The ozone hole has increased in size from a million or so square kilometers in the late 1970s and early 1980s to about 28 million square kilometers today, exceeding the area of North America (Figure 25.8).

Development of the ozone hole from the 1970s to the early to middle 1990s is shown in Figure 25.9. Also shown are the percent differences between the two time periods, emphasizing the development of the massive ozone hole in the early to middle 1990s.[5]

Polar Stratospheric Clouds

The amount of ozone depletion in the Antarctic has varied from about 15 to 60% (refer to Figure 25.4). In 1988,

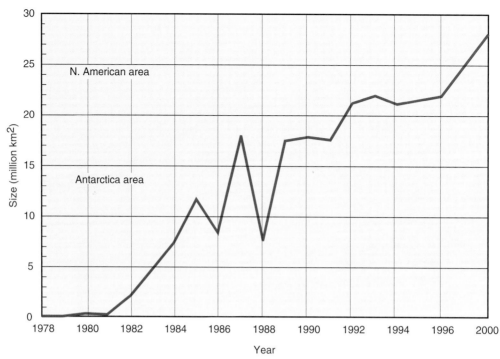

Figure 25.8 ● Size of the Antarctic ozone hole, 1979–2000 (September 7–October 13 average). The size of the hole started increasing rapidly in the 1980s, leveled off in the early to middle 1990s, and then increased again. (*Source:* Modified from NASA-GSFC, "Stratospheric Ozone," accessed August 22, 2000, at http://see.gsfc.nasa.gov; and "Record-setting Ozone Hole," Science at NASA September 8, 2000, accessed March 8, 2001, at http://www.science.nasa.gov/headlines/y2000/ast08sep_1.htm.)

ozone depletion was 15%. The lesser amount is thought to be related to the fact that during 1988 there were fewer polar stratospheric clouds over the Antarctic. By contrast, in years when the ozone depletion was greater, the regions with the most ozone depletion were in the lower stratosphere at an altitude between 14 and 24 km (9 and 15 miles), where polar stratospheric clouds are present. What is the significance of these clouds?

Polar stratospheric clouds have been observed for at least the past hundred years at altitudes of approximately 20 km (about 12 mi) above the polar regions. The clouds are approximately 10–100 km (6–60 mi) in length and several kilometers thick.[14] They have an eerie beauty and an iridescent glow, with a color reminiscent of mother-of-pearl (Figure 25.10).[14]

Polar stratospheric clouds form during the polar winter (called the polar night because of the lack of sunlight, which results from the tilt of Earth's axis). During the polar winter, the Antarctic air mass is isolated from the rest of the atmosphere and circulates about the pole in what is known as the Antarctic **polar vortex**. The vortex, which rotates counterclockwise because of the rotation of Earth in the Southern Hemisphere, forms as the isolated air mass cools, condenses, and descends.[7] The cooling occurs because the isolated air mass continues to lose heat through radiation, and no more heat is supplied because of the lack of sunlight.

Clouds are formed in the vortex when the air mass reaches a temperature between 195 and 190 K (−78° to −83°C; −108° to 117°F). At these very low temperatures, small sulfuric acid particles (approximately 0.1 μm) are frozen and serve as seed particles for nitric acid (HNO_3). These clouds are called Type I polar stratospheric clouds.

If temperatures drop below 190 K (−83°C; −117°F), water vapor condenses around some of the earlier-formed Type I cloud particles, forming Type II polar stratospheric clouds, which contain larger particles. Type II polar stratospheric clouds have a mother-of-pearl color visible in polar areas.

During the formation of polar stratospheric clouds, nearly all the nitrogen oxides in the air mass are held in the clouds as nitric acid. The nitric acid particles grow large enough to fall out by gravitational settling from the stratosphere. This phenomenon has the important result of leaving very little nitrogen oxide in the atmosphere in the vicinity of the clouds.[3,7,14] This process facilitates ozone-depleting reactions, which ultimately may reduce stratospheric ozone in the polar vortex by as much as 1% to 2% per day in the early spring, when sunlight returns to the polar region.

An idealized diagram of the polar vortex that forms over Antarctica is shown in Figure 25.11*a*. Ozone-depleting reactions that occur within the vortex are illustrated in Figure 25.11*b*. As this figure shows, in the dark Antarctic winter, almost all the available nitrogen oxides are tied up on the edges of particles in the polar stratospheric clouds or have settled out. Hydrochloric acid and chlorine nitrate

Total DU

100.   500.

Percent difference

.60.0   0.0

Figure 25.9 • Average development of the ozone hole in the 1970s compared with the early to middle 1990s, with percent differences between those two periods. (*Source:* Modified from NASA-GSFC, "Stratospheric Ozone," accessed August 22, 2000, at http://see.gsfc.nasa.gov.)

Figure 25.10 • Polar stratospheric clouds, February 12, 1989, photographed from an aircraft cruising at an altitude of approximately 12 km in the polar region north of Stavanger, Norway. The red haze and thin orange or brown layers at lower altitudes are Type I clouds. The red coloring is probably due to scattering from nitric acid particles. The higher white clouds are Type II polar stratospheric clouds, which consist mostly of water molecules frozen as ice.

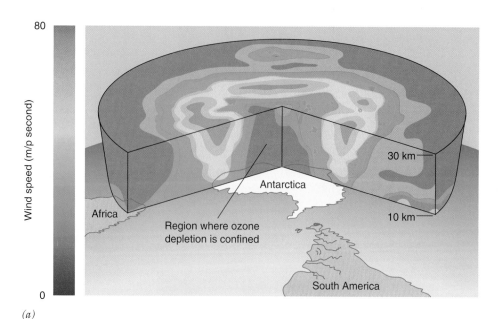

Figure 25.11 • (*a*) Idealized diagram of the Antarctic polar vortex and (*b*) the role of polar stratospheric clouds in the ozone depletion chain reaction. (*Source:* Based on O. B. Toon and R. P. Turco, "Polar Stratospheric Clouds and Ozone Depletion," *Scientific American*, 264, no. 6, [1991]:68–74.)

(the two important sinks of chlorine) act on particles of polar stratospheric clouds to form dimolecular chlorine (Cl₂) and nitric acid through the following reaction:[16]

$$HCl + ClONO_2 \rightarrow Cl_2 + HNO_3$$

In the spring, when sunlight returns and breaks apart chlorine (Cl₂), the ozone-depleting reactions discussed earlier occur. Nitrogen oxides are absent from the Antarctic stratosphere in the spring; thus, the chlorine cannot be sequestered to form chlorine nitrate, one of its major sinks. Therefore chlorine is free to destroy ozone. In the early spring of the Antarctic, these ozone-depleting reactions can be rapid, producing the 80% reduction in ozone observed in 1994. Ozone depletion in the Antarctic vortex ceases later in spring as the environment warms and the polar stratospheric clouds disappear, releasing nitrogen back into the atmosphere, where it can combine with chlorine and thus be removed from ozone-depleting reactions. Stratospheric ozone concentrations then increase as ozone-rich air masses again migrate to the polar region.

An Arctic Ozone Hole?

A polar vortex also forms over the North Pole area, but it is generally weaker than the Antarctic polar vortex and does not last as long. Nevertheless, ozone depletion occurs over the North Pole and it is speculated that if the vortex persists for a month or more, ozone losses in the affected air mass may be as high as 30% to 40%.

A major concern regarding the northern polar vortex is that, as it breaks up, it sends ozone-deficient air masses southward, where they may drift over populated areas of Europe and North America. In January 1992, for example, satellite data indicated a large air mass containing high levels of chlorine monoxide (ClO) stretching from Great Britain eastward over northern Europe.[17] (ClO is sometimes referred to as the "smoking gun" in the ozone problem because it plays a major part in ozone depletion—recall the earlier discussion of ozone reactions.) In contrast, the Antarctic polar vortex tends to remain more stationary, although in 1987 an ozone-depleted air mass that formed over Antarctica in October drifted northward over Australia and New Zealand in December, resulting in record low concentrations of stratospheric ozone in that region.[3]

In summary, although not as severe as the ozone depletion over the Antarctic, ozone depletion over the Arctic each winter is troublesome. Since the Arctic polar vortex is relatively weak, warmer air from midlatitudes is usually able to dissipate the vortex before ozone depletion becomes severe. However, in 1995, ozone levels were as much as 40% below normal. Scientists studying the developing ozone hole speculated that the unusually cold Arctic winter of 1995 triggered the record losses of ozone, leading to an ozone hole similar to the one that forms over the Antarctic.[18]

25.4 Tropical and Midlatitude Ozone Depletion

It has been firmly established that ozone depletion in polar regions occurs as a result of reactions that take place on particles in polar stratospheric clouds. Ice particles also occur in the stratosphere over the tropics; and at times, sulfuric acid aerosols are abundant in the stratosphere because of injection of sulfur by volcanic eruptions. That these particles may cause ozone depletion is only a hypothesis; there is no substantial evidence. (See A Closer Look 25.1.) Nevertheless, following the 1982 eruption of the volcano El Chichón in Mexico, a 10% reduction in ozone in the Northern Hemisphere was measured.[19]

Stratospheric ozone from the 1970s to the 1990s was lost at a rate of about 35% per year, at the south pole (Figure 25.4).[20] However, evidence also suggests an increase in ozone depletion at midlatitude over areas including the United States and Europe. During the unusually cold winter of 1996, ozone depletion over the contiguous United States was reported by NOAA scientists. In summary, although we know the most information about ozone depletion in polar regions (particularly Antarctica), depletion of ozone is a global concern, from the poles to the tropics.

25.5 The Future of Ozone Depletion

A troubling aspect of ozone depletion is that if the manufacture, use, and emission of all ozone-depleting chemicals were to stop today, the problem would not go away, because millions of metric tons of those chemicals are now in the lower atmosphere, working their way up to the stratosphere. As shown in Table 25.1, several CFCs have atmospheric lifetimes of 75 to 140 years. Thus, about 35% of the CFC-12 molecules in the atmosphere are expected to be there still in 2100, and approximately 15% will be there in 2200.[3] In addition, approximately 10% to 15% of the CFC molecules manufactured in recent years have not yet been admitted to the atmosphere, because they remain tied up in foam insulation, air-conditioning units, and refrigerators.[3] Nevertheless, indicators suggest that growth in the concentrations of CFCs has been reduced and in some cases reversed.

Figure 25.13 shows concentrations of CFC-11 and CFC-12 from 1977 to 1998. CFC-11 concentrations reached a peak in about 1992 and are now decreasing, while CFC-12 concentrations have leveled off. The growth in concentrations of CFC-12, which accounts for nearly 50% of ozone depletion (see Table 25.1), was about 5% per year from 1978 to 1990. It was less than 0.5% per year by 1998. CFC-11 concentrations were in decline by 1992 (see Figure 25.13).

Environmental Effects

Ozone depletion has several serious potential environmental effects, including damage to Earth's food chains on land and in the oceans and damage to human health, including increases in all types of skin cancers and cataracts and suppression of immune systems.[21,22] As mentioned earlier, a 1% decrease in ozone can cause an increase of 1–2% in UVB radiation and an increase of 2% in the incidence of skin cancer.[22]

There has been speculation for some time that ozone depletion might lead to a reduction of primary productivity in the world's oceans. A loss of productivity of the phytoplankton (microscopic marine algae and photosynthetic bacteria that float near the surface of the ocean) would have a negative impact on a variety of other marine organisms, because phytoplankton are at the base of the food chain. Because ozone over the Antarctic area has been depleted by as much as 80% in recent years, more UVB radiation is reaching the surface of the ocean there. A study of the Antarctic waters beneath the mass of ozone-depleted air suggests that a reduction of primary productivity of at least 6–12% is associated with ozone depletion.[4] This disruption of the food chain might eventually affect people who consume marine resources, because fewer resources would be available for harvest. Also, because plankton are a sink for atmospheric carbon dioxide, their disruption might increase the concentration of CO_2 in the atmosphere, thereby increasing global warming.

If ozone depletion becomes more widespread and affects major food crops (such as beans, wheat, rice, and corn), serious social disruption could occur. Even a small decrease in food production might have large social and political consequences around the world. A loss of 10%–15% production could be catastrophic.

The range of human health effects of ozone depletion is being vigorously researched and debated. There is general agreement that the effects will be negative and will result in increases in a variety of diseases, perhaps at epidemic levels compared with what would otherwise be expected. As mentioned, one of the most serious hazards

How Did Mount Pinatubo's Eruption Affect Ozone Levels?

The June 1991 eruption of Mount Pinatubo in the Philippines was the largest volcanic eruption of the 20th century. The eruption injected approximately 20 million metric tons of sulfur dioxide into the atmosphere, creating a stratospheric aerosol cloud larger than all the polar stratospheric clouds combined.[19]

It was hypothesized that the sulfur-based aerosol cloud contained particles that facilitated ozone-depleting reactions. Figure 25.12a shows the sulfur-dioxide-rich cloud (orange band) that circled Earth in the tropics after the Mount Pinatubo eruption. Figure 25.12b shows a corresponding band of air relatively low in ozone (violet band). The correlation between the two suggests that they might be related. However, such a correlation cannot be used to establish a cause-and-effect relationship without further supporting evidence.[19] In fact, additional data and analysis from the Pinatubo event suggest that anomalously low values of tropical ozone (Figure 25.12b) were probably due to effects of the eruptions on physical atmospheric circulation rather than chemical reactions that produce loss of ozone (J. W. Waters, 1994, person communication). That is, the eruptions may have caused air at lower elevations and low ozone concentrations to heat and rise to higher elevations, where higher values of ozone are normally present.

(a)

(b)

Figure 25.12 ● (a) Tropical belt of sulfur-dioxide-rich aerosol cloud (orange) and (b) corresponding belt of air with lower concentration of ozone (violet). (*Source:* J. B. Waters, Jet Propulsion Laboratory, in J. Horgan, "Volcanic Disruption," *Scientific American* 266, no. 3 [1992]:28, 29.)

anticipated is an increase in skin cancers of all types, including the often fatal melanoma.

In 1987, as noted earlier, record low ozone measurements over Antarctica in October were followed by record low readings over Australia and New Zealand in December[23] when a mass of ozone-depleted air separated from the Antarctic vortex as it broke up in late spring. There is fear that large ozone holes will increasingly threaten populated areas in mid-latitude regions.

Already, around the world, the incidence of skin cancers has increased (see the case study at the beginning of this chapter). For many years, a suntan was considered a

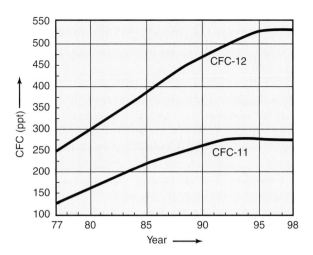

Figure 25.13 • Concentrations of CFC-11 and CFC-12 from 1977 to 1998. (*Source:* Data from Mauna Loa, Hawaii, modified from NOAA Climate Monitoring and Diagnostic Lab, accessed August 22, 2000, at http://www.cmdl.noaa.gov.)

healthy look, and people deliberately exposed their bodies to sunlight. Today, sun-block lotions and hats are replacing the tanning oils of health-conscious people. Newspapers in 58 experimental regions of the United States are providing their readers with the Ultraviolet (UV) Index (Table 25.2), developed by the National Weather Service and the Environmental Protection Agency. The index predicts UV intensity levels on a scale from 0 to 10+. Some news agencies also use the index to recommend the level of sunblock. It is speculated that the incidence of skin cancer caused by ozone depletion will increase until about 2060 and will then decline as the ozone shield recovers as a result of controls on CFC emissions.[20]

Individuals can reduce their risk of skin cancer and other skin damage from UV exposure by following a few simple precautions:

- Limit exposure to the sun between the hours of 10 A.M. and 4 P.M. That is, avoid the hours of intense solar radiation.
- When possible, remain in the shade.
- Use a sunscreen with an SPF of at least 15. (Remember that protection from sunscreen with any SPF is diminished with time of exposure.)
- Wear a wide-brimmed hat and, where possible, tightly woven full-length clothing (for example, loose-fitting light cotton).
- Wear UV-protective sunglasses.
- Avoid tanning salons and sunlamps.
- Consult the UV Index before going out (Table 25.2).

Following these precautions may help ensure that your skin will remain healthy, with fewer wrinkles into old age.

In addition to harming skin, ultraviolet radiation can damage eyes, causing cataracts, an eye disease in which the lens becomes opaque and vision is impaired. People are now more often choosing eyeglasses that block ultraviolet radiation. It is believed that an increased exposure to ultraviolet radiation may also damage or reduce the efficiency of the human immune system.[21] In turn, decreases in the effectiveness of the human immune system would result in higher numbers of a variety of diseases. Finally, a variety of environmental pollutants in the air and water could have synergistic effects, increasing potential health risks of exposure to ultraviolet radiation.

Management Issues

A key issue in management of ozone depletion is whether the depletion is natural or human-induced. If stratospheric ozone depletion were a natural process, then we would not expect to see the continuous, dramatic reductions since the mid-1970s, when the real impact of production of CFCs began. World production of CFC-11 increased sevenfold from 1970 to 1994. During this period, the most dramatic declines in ozone in Antarctica occurred. Thus, it appears that chlorine from CFCs is the "smoking gun" mentioned earlier. Supporting this hypothesis, a recent investigation determined that the concentration of stratospheric chlorine (which is responsible for ozone destruction) is more than five times what could be expected from natural emissions from oceans or other natural processes. The authors concluded that, beyond a reasonable doubt, CFCs are responsible for ozone depletion in the stratosphere.[24] We turn next to a discussion of management issues and strategies to deal with CFC-induced ozone depletion.

The Montreal Protocol

A diplomatic achievement of monumental proportions was completed with the signing of the Montreal Protocol in September 1987; 27 nations signed the agreement originally, and 119 additional nations signed later. The protocol outlined a plan for the eventual reduction of global emissions of CFCs to 50% of 1986 emissions. The protocol originally called for elimination of the production of CFCs by 1999.

Because of scientific evidence that stratospheric ozone depletion was occurring faster than predicted, the timetable for elimination of CFC production was shortened. Most industrialized countries, including the United States, had stopped production by the end of 1995; the deadline for developing countries is the end of 2005. An eventual phase-out of all CFC consumption is part of the Montreal Protocol.

Assessment of the effects of the protocol, along with effects of other agreements and amendments made in London (1990) and Copenhagen (1992), suggests that stratospheric concentrations of ozone-depleting substances (CFCs) will return to pre-1980 levels by about 2050. Already, the rate of increase of CFC emissions has been reduced (see Table 25.1 and Figure 25.13). Nevertheless, as already mentioned, because of the long residence times of CFCs in the stratosphere, ozone depletion from CFCs is expected to continue for many years to come.[25,26] That's the bad news. The good news is that ozone levels in the stratosphere will slowly increase in the next few decades.

Unfortunately, not all industrial nations are responding to the urgency of this issue, partly because of the economic gap between wealthy and poorer nations. For example, China and India chose not to participate in the protocol. Their refusal was probably related to the substantial investments they are making in refrigeration and the fact that replacement chemicals currently considered for CFCs are approximately six times as expensive. The poorer nations probably will not participate in the reduction of CFCs unless they are assisted by the wealthier countries.[27] The Montreal Protocol stipulates a transfer of both technology and funds from industrial to developing countries to speed up CFC phase-out. We return to this issue in the discussion of substitutes for CFCs.

A black-market trade in CFCs has hindered the attempt to eliminate their production and consumption. Illegal smuggling of CFCs into Europe from Russia in recent years has equaled as much as 10% of the total CFCs legally allowed in those countries. The United States has also received illegal imports of CFCs.[28]

Collection and Reuse of CFCs

One way to lower emissions of CFCs into the atmosphere is to develop ways to collect and reuse them. Every year, approximately 50 million refrigerators are discarded worldwide, and each one contains approximately 1.2 kg (2.6 lb) of CFCs, mostly tied up in the foam plastic insulation. Methods have been developed to liberate and collect these CFCs when refrigerators are recycled. The CFCs used as coolant gases can also be collected. One company in Germany recycles approximately 6,000 refrigerators a month.[29] The same techniques can be used to recover the CFCs in air conditioners used in automobiles and homes.

Substitutes for CFCs

Of primary importance in developing substitutes for CFCs is finding substitutes that are both safe and effective. Two substitutes for CFCs being experimented with today are **hydrofluorocarbons (HFCs)** and **hydrochlorofluorocarbons (HCFCs)**. These chemicals are controversial but do have advantages.

The advantage of HFCs is that they do not contain chlorine. They do contain fluorine, though; and when fluorine atoms are released into the stratosphere, they participate in reactions similar to those of chlorine. Thus, they can cause ozone depletion. However, ozone depletion is not thought to be a significant problem with fluorine, because it is approximately 1,000 times less efficient in those reactions.[3,27]

HCFCs contain an atom of hydrogen in place of a chlorine atom. They can be broken down in the lower atmosphere, in which case they do not inject chlorine into the stratosphere; however, they can cause ozone depletion if they do reach the stratosphere before being broken down. Although their atmospheric lifetime is much shorter than those of the CFCs (Table 25.1), when HCFCs are used in tremendous quantities, they do cause ozone depletion.[3] HCFCs are at best a transitional chemical to be used until substitutes that do not cause ozone depletion are readily available. HCFCs will be phased out by 2030.

The new HCFCs are also expensive, about three to five times as costly as the CFCs now used in the refrigeration industry. A major manufacturer of CFCs has estimated that the worldwide cost to replace CFCs with HCFCs is as follows:[30]

- $4 billion to the chemical industry over a 10-year period (this cost would be passed on to those using the chemicals).
- $385 billion for replacement of older refrigeration equipment that cannot be economically modified to use the new HCFCs.

Another option being considered is the use of helium as the working gas in refrigerators. At present, however, the world supply of helium is very limited. Furthermore, helium is expected to be too expensive for home refrigeration because of the cost of making the equipment. Nevertheless, research in this area continues, and the price may come down, making its use more feasible.

The hydrocarbon propane, a colorless gas easily separated from crude oil and natural gas, is a common fuel that can be easily liquefied and stored in containers. It is also an inexpensive substitute for CFCs that has not received nearly as much attention as have HFCs and HCFCs. The cost of propane is about 10% that of CFCs and less than 2% that of HFCs and HCFCs. A refrigerator in a London laboratory was inexpensively converted to use propane as the coolant, and oil refineries have used propane as a coolant in industrial processes.[30] Why, then, hasn't propane been more seriously considered as a replacement for CFCs? There are two main reasons:[30]

- The chemical companies that manufacture CFCs and

have patented HFCs and HCFCs are not in the refrigeration industry, and their focus is on their products rather than on finding an inexpensive substitute.

- Propane is thought to be dangerous in refrigerators and air conditioners because it is potentially explosive.

For most uses in household refrigerators and air conditioners, the concern about the danger of propane is overstated. The modern refrigerator would use about 100 g (3.5 oz) of propane in a sealed system, the equivalent of the amount of propane in a hand-held cigarette lighter. Because the cooling system is sealed, it is unlikely that a house fire could generate sufficient heat to rupture the system. If it did, the resulting risk should be weighed against the risk of CFCs in the same situation; combustion products associated with a fire burning CFCs includes toxic nerve gas. In contrast, in commercial refrigeration systems that use large amounts of coolant, the use of propane would be a potential hazard, and careful engineering would be required to ensure safety.

The use of propane as a refrigeration coolant is being tested in the United States and elsewhere. The results will be important, especially to developing countries, where most CFC use is for refrigeration and air conditioning. Developing countries tend not to have large industrial uses for CFCs. For example, in India, about 75% of the CFCs are for refrigerators and air conditioners, a pattern characteristic of developing countries. To reduce or eliminate CFC use by replacement with HCFCs and HFCs, developing countries will need massive financial aid. They could do it on their own if an inexpensive, safe alternative were available. Research is being conducted to determine if propane or some other chemical could be that alternative.[30]

Short-Term Adaptation to Ozone Depletion

As we have seen, there is good news in the ozone depletion story. Concentrations of CFCs in the upper atmosphere where ozone depletion occurs have apparently peaked. Depletion of stratospheric ozone will be a story of gradual recovery by the mid-21st century.[20,28] Recovery will take place as a result of restrictions in the production of ozone-depleting chemicals, especially CFCs.

Given the nature of the ozone depletion problem and the atmospheric lifetimes of the chemicals that produce the depletion, the major short-term adaptation by people will be learning to live with higher levels of exposure to ultraviolet radiation. In the long term, achievement of sustainability with respect to stratospheric ozone will require management of human-produced ozone-depleting chemicals.

Summary

- Concentration of atmospheric ozone has been measured for more than 70 years. In the last decade, measurements have been taken from instruments mounted on satellites. Evaluation of the available data shows a clear trend: Ozone concentrations in the stratosphere have been decreasing since the mid-1970s.

- In 1974, Mario Molina and F. Sherwood Rowland advanced the hypothesis that stratospheric ozone might be depleted as a result of emissions of chlorofluorocarbons (CFCs) into the lower atmosphere. Major features of the hypothesis are: CFCs are extremely stable and have a long residence time in the atmosphere; eventually, the CFCs reach the stratosphere, where they may be destroyed by highly energetic solar ultraviolet radiation, releasing chlorine; the chlorine may then enter into a catalytic chain reaction that depletes ozone in the stratosphere. An environmentally significant result of the depletion is that more ultraviolet radiation reaches the lower atmosphere, where it can damage living cells.

- The Antarctic ozone hole was first reported in 1985 and since has captured the imagination of people around the world. Of particular importance to understanding the ozone hole are the complex reactions that occur in the polar vortex and the development of polar stratospheric clouds. Reactions in the clouds tend to denitrify the air mass in the vortex; and during the polar spring, chlorine is released to react in the catalytic ozone depletion cycle. The reactions can be very rapid, producing the observed 80% reduction in stratospheric ozone in only a few weeks.

- Tropical and midlatitude ozone depletion is also hypothesized. Polar stratospheric clouds may be present above the tropics; in addition, belts of sulfur-dioxide–rich aerosol clouds may result from volcanic eruptions. These clouds may be related to processes that denitrify the atmosphere and facilitate ozone depletion.

- Millions of tons of chemicals with the potential to deplete stratospheric ozone are now in the lower atmosphere and working their way to the stratosphere. As a result, if all production, use, and emissions of these chemicals were stopped today, the problem would continue. The good news is that the concentrations of CFCs in the atmosphere have apparently peaked and are now static or in slow decline.

- Potential environmental effects related to ozone depletion include damage to Earth's food chain, both on land and in the ocean, and human health effects, including increases in skin cancers, cataracts, and suppression of the immune system.

- Many nations around the world have agreed to the Montreal Protocol, which will reduce global emissions of CFCs to 50% of the 1986 levels. The agreement called for elimination of production of the chemicals by 1996 for industrialized nations and by 2006 for developing nations. A serious hurdle to compliance for some nations relates to the economic fact that most chemical replacements for CFCs are more expensive than CFCs.

Human-Made Chemicals and the Ozone Hole: Why Was There Controversy?

By 1993, scientists had accumulated enough evidence to support earlier predictions that stratospheric ozone was being depleted over the Antarctic. Most of them blamed the damage on organic chlorine compounds (those containing both carbon and chlorine) manufactured by humans, such as CFCs. But consensus among most of the scientists in the field did not prevent a continuing storm of controversy over these findings. Critics included meteorologists, scientists from other fields, amateur scientists, journalists, talk-show hosts, and authors of nontechnical books on the environment. The critics charged that natural sources of chlorine, not those generated by humans, were responsible for ozone depletion and that the environmental and health threats of ozone depletion were greatly exaggerated.

Accusations leveled against scientists, NASA officials, and some industrialists included claims that the ozone scare was a sham, a scam, a hoax, or a conspiracy. Scientific uncertainties, such as those concerning the causes of ozone depletion in the Northern Hemisphere and whether depletion was allowing more ultraviolet radiation to reach Earth's surface, were emphasized.

In his 1993 presidential address to the American Association for the Advancement of Science, F. Sherwood Rowland, principal architect of the hypothesis that CFCs damage the ozone layer, cited poor communication between scientists and nonscientists as the cause of the controversy. His colleague, Mario Molina, stated that the arguments put forth by critics about natural causes of ozone depletion had been tested; no results had been found to support those arguments in the 20 years since the hypothesis that CFCs were responsible for ozone depletion had been proposed.

Others felt that the controversy fed on lack of understanding of the nature of science, particularly discomfort with uncertainty, on the part of most nonscientists. In making this point, science fiction writer Frederic Pohl quoted the late Nobel Prize physicist Richard Feynman: "Scientific knowledge is a body of statements of varying degrees of certainty—some most unsure, some nearly sure, but none absolutely certain."

Critical Thinking Questions

1. Table 25.3 summarizes the major points of argument about whether ozone depletion is primarily a result of natural or of human-made chemicals. Evaluate the arguments in the table as best you can in light of what you have learned about ozone in this chapter. Identify each statement as (a) a hypothesis, (b) an inference, (c) confirmed by observation or experiment (fact), (d) pseudoscience, or (e) absolutely certain. If you feel the information is incomplete, what addi-

Table 25.3	Major Arguments Regarding Sources of Ozone Depletion

Natural sources of chlorine are so large that CFCs are insignificant by comparison.

1. CFCs are heavier than air and would not reach the stratosphere in significant amounts.

2. No measurements have shown CFCs to be present in the stratosphere.

3. Volcanoes, which produce 20 times as much chlorine as do CFCs, caused the Antarctic ozone hole. Mount Erebus in Antarctica has been erupting since 1973 and emits more than 1,000 tons of chlorine a day.

4. Evaporation of sodium chloride from ocean water is a source of stratospheric chlorine.

5. Volcanoes release hydrogen chloride, which has increased in the stratosphere in the last 10 years.

6. Burning releases methyl chloride.

Humans are producing chemicals that significantly deplete stratospheric ozone in the Antarctic and possibly in the Northern Hemisphere.

1. Within the atmosphere, large masses of air are constantly mixing.

2. CFCs are found in stratospheric samples.

3. Volcanic activity has been around for a long time, and recent eruptions have not been unusual. Mount Erebus does not erupt forcefully enough to propel chlorine into the stratosphere in significant amounts. Furthermore, it produces only 15,000 tons of chorine a year.

4. Sodium chloride, unlike CFCs, is water soluble and is washed out of the atmosphere when it rains. No sodium is found in the lower stratosphere.

5. Both hydrogen chloride and hydrogen fluoride, also water soluble, are increasing, which is what would be predicted if CFCs were reaching the stratosphere. There are almost no natural sources of hydrogen fluoride, and increases in it would not be consistent with a volcanic source for chlorine. Sulfuric acid aerosols from volcanoes that do eject into the stratosphere promote destruction of ozone.

6. Twenty percent of the chlorine in the stratosphere occurs as methyl chloride, but only 25% of it comes from burning.

tional observations or experiments would you suggest? Based on your analysis, what is your position on the question of whether natural or human-made chemicals are depleting the ozone layer?

2. Some critics have pointed out that ultraviolet radiation increases by 50 times (5000%) in going from pole to equator. The expected increase in ultraviolet radiation in the mid latitudes, they say, is like moving from New York City to Philadelphia or from sea level to an elevation of 1,500 ft. What is your reaction to this criticism?

3. In February 1992, NASA scientists reported unusually high levels of chlorine over the Northern Hemisphere and warned that this might lead to significant ozone loss over heavily populated areas. Congress reacted quickly to enact legislation to speed up discontinuing use of CFCs. When the ozone loss over the Northern Hemisphere was much less than the scientists had expected, they were attacked by some people as being unnecessarily alarmist and using scare tactics to obtain additional funding. Do you think these attacks were justified? Would it have been better if the scientists had presented a much more conservative scenario?

4. F. Sherwood Rowland identified two sources of public misunderstanding about scientific issues: poor communication about science and widespread scientific illiteracy. Do you agree or disagree with Rowland? Explain your position. If you agree, what remedies do you suggest?

Therefore, it appears that financial aid will be required if the less wealthy nations are to eliminate CFC use.

- Potential management strategies for the ozone depletion problem, along with eliminating production of CFCs: include (1) collecting and reusing CFCs; and (2) using substitutes for CFCs.

- Given the lifetimes of the ozone-depleting chemicals, people will have to continue to live with higher levels of exposure to ultraviolet radiation over the next few decades. Banning chemicals that can deplete stratospheric ozone is a step in the right direction and will ultimately result in the reduction of atmospheric ozone depletion.

REEXAMINING THEMES AND ISSUES

Human Population:

As the number of humans on Earth has increased, so has our use of the chemicals that are currently causing ozone depletion. Of particular significance are developing countries wishing to increase their industrial output. Coupled with a rapid rise in human population, industrialization in these countries could dramatically increase emissions of the chemicals that are causing global changes in the atmosphere. Thus, control of human population is an important objective in addressing ozone depletion in the future.

Sustainability:

Controlling emissions of chemicals that cause ozone depletion is at the heart of sustainability, which has the objective of providing a quality environment for future generations. Potential adverse environmental effects of ozone depletion related to biological productivity and disease are significant, and reduction of emissions of ozone-depleting chemicals has been a high priority. Results of reductions will become apparent during the first half of the 21st century, as stratospheric ozone concentrations recover to preindustrial levels.

Global Perspective:

Stratospheric ozone depletion is a significant global environmental problem conclusively shown to result from human activity. Understanding the causes of and potential solutions to the problem has required careful evaluation of Earth and its atmosphere on a global scale. Interactions between emissions of chemicals and chemical reactions occurring in the atmosphere, as well as general circulation of the atmosphere, have all been involved in understanding the processes that have resulted in ozone depletion in the stratosphere. Recommended management strategies to solve this problem have come directly through our understanding of earth system science and its applications.

Urban World:

Significant quantities of CFCs have been used for refrigeration and air conditioning in urban environments. Urban areas are the places where the chemicals are most often produced, used, and emitted into the environment.

The fact that human emission of chemicals is altering the stratosphere to the detriment of people and the environment in general has come as a surprise to many people. The science of ozone depletion is well understood, however, and we are moving through management decisions, based on our values, to ensure the problem will slowly go away. Solutions to ozone depletion are costly and will require sacrifices on the part of most people. Most of the cost will fall on the wealthiest countries. It is to be hoped that these nations will continue to have the moral and ethical courage to exercise leadership and pay the costs necessary for an improved global environment.

● KEY TERMS

chlorofluorocarbons (CFCs) **540**
Dobson unit (DU) **539**
hydrochlorofluorocarbons (HCFCs)
   550

hydrofluorocarbons (HFCs) **550**
ozone **537**
ozone shield **538**
polar stratospheric clouds **544**

polar vortex **544**
ultraviolet A (UVA) **539**
ultraviolet B (UVB) **539**
ultraviolet C (UVC) **538**

● STUDY QUESTIONS

1. Do you think that a gap ever existed between people's understanding of the ozone problem and their taking action to find a solution to the problem? Compare the situation with the global warming problem?

2. Given that primary productivity is reduced in the Antarctic by ozone depletion, how could you test the hypothesis that penguins might be adversely affected?

3. Consider the possibility of building a small chamber to test the hypothesis that polar stratospheric clouds facilitate ozone depletion. What might be some of the problems in attempting to do this? What are alternative approaches to studying the role of stratospheric clouds?

4. Suppose that next year all our understanding of ozone depletion is changed by the discovery that

concentrations of stratospheric ozone have natural cycles and the lower concentrations in recent years have resulted from natural rather than human-induced processes. How would you put all the information in this chapter into perspective? Would you think that science had let you down?

5. What types of economic and political changes will be necessary to encourage less developed countries to support plans to eliminate chemicals responsible for ozone depletion? Do you believe that richer countries have an obligation to help poorer ones?

6. Do you agree or disagree that most of the people in the world will adjust to ozone depletion by doing little to decrease their personal exposure to ultraviolet radiation?

● FURTHER READING

Environmental Protection Agency. 1995. *Stratospheric Ozone Depletion: A Focus on EPA's Research*. Washington, D.C.: U.S. EPA Office of Research and Development. A brief but good overview of ozone depletion, current status of ozone holes, and consequences of ozone depletion.

Hamill, P., and O. B. Toon. 1991. "Polar Stratospheric Clouds and the Ozone Hole," *Physics Today* 44(12):34–42. A good review of the ozone problem and important chemical and physical processes related to ozone depletion.

Jones, R. R., and T. Wigley. 1989. *Ozone Depletion: Health and Environmental Consequences*. New York: Wiley. Proceedings of the International Conference on the Health and Environmental Consequences of Stratospheric Ozone Depletion, held at the Royal Institute of British Architects, London, November 28–29, 1988.

Litfin, K. T. 1994. *Ozone Discourses: Science and Politics in Global Environmental Cooperation*. New York: Columbia University Press. Discussions on the importance of science and scientific discourse in shaping world politics and policy with regard to ozone depletion.

Makhijani, A., and K. R. Gurney. 1995. *Mending the Ozone Hole*. Cambridge, Mass.: MIT Press. A good overview of causes and consequences of stratospheric ozone depletion, sources of and alternatives to ozone-depleting chemicals, and policy recommendations to reverse ozone damage.

Rowland, F. S. 1990. "Stratospheric Ozone Depletion by Chlorofluorocarbons." *AMBIO* 19(6–7):281–292. An excellent summary of stratospheric ozone depletion that discusses some of the major issues.

Shea, C. P. 1989. "Mending the Earth's Shield," *World Watch* 2(1):28–34. An article focusing on solutions to the ozone problem. It is primarily concerned with control strategies to stop ozone-depleting chemicals from being emitted into the atmosphere.

Toon, O. B., and R. P. Turco. 1991. "Polar Stratospheric Clouds and Ozone Depletion," *Scientific American* 246(6):68–74. An article that provides valuable information concerning polar stratospheric clouds and their importance in ozone depletion. It offers a good explanation of the formation of polar stratospheric clouds and the chemistry that occurs there.

26 Environmental Economics

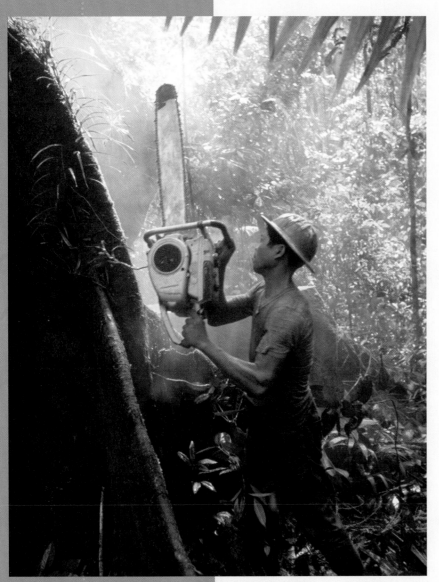

Logging of Asian mahogany in the Philippines.

● LEARNING OBJECTIVES

Other chapters in this text have explained the causes of environmental problems and discussed technical solutions. The scientific solutions, however, are only part of the answer. This chapter introduces some basic concepts of environmental economics and shows how these concepts have been applied in the analysis of environmental issues. After reading this chapter, you should understand:

- What the "tragedy of the commons" is, and how it leads to an overexploitation of resources.
- How the perceived future value of an environmental benefit affects our willingness to pay for it now.
- What externalities are, and why it is important to evaluate them in determining the costs of actions that affect the environment.
- What factors may be involved in determining a level of acceptable environmental risk and risk to human life.
- Why it is difficult, yet important, to evaluate environmental intangibles, such as aesthetics.
- What issues are involved in determining who pays the direct and indirect costs of controlling pollution and minimizing environmental damage.
- What kinds of policy methods are available to control pollution and harvesting of resources.

The Economics of Mahogany

The debate over how to harvest mahogany trees without serious damage to tropical forests illustrates the importance of economics in resolving environmental issues. Mahogany, which occurs naturally in Latin America, is valued for its beautiful grain and color, hardness, and durability. Since the Spanish landed in the Americas in the 16th century, mahogany has been used for shipbuilding, furniture, musical instruments, and coffins. As West Indian mahogany, which occurs in Florida and the West Indies, became scarce, commercial logging shifted to the big-leaf mahogany found from southern Mexico to Brazil. Big-leaf mahogany (Figure 26.1), which can grow as tall as 45 m (150 ft) with leaves up to 0.6 m (2 ft) long, wholesales for approximately $1,500 a tree in the United States.

Mahogany seedlings cannot grow in the shade of other trees. Under natural conditions, mahogany trees regenerate in clearings created by fires, hurricanes, or farmers. Traditional logging of mahogany is selective; that is, other species are left standing as long as they do not interfere with the mahogany operations. However, the size of the clearings left by selective logging is not large enough to promote regeneration, so the practice is not sustainable. Mahogany takes 120 years or more to reach marketable size, so maintaining a sustainable population in the face of extensive logging is difficult. Experts estimate that the rate at which mahogany is being logged—140,000 trees a year—is three times the rate that would be sustainable (Figure 26.2).

Because mahogany has a naturally low density (1–6 trees/ha or 2.5 acres), logging requires building extensive systems of roads through the forest. The greater access the roads provide to the forest interior makes it easier for people to clear the forest and convert the land to agriculture. The Rainforest Action Network and some other environmental groups argue that mahogany logging is, therefore, a threat to the entire rainforest ecosystem. They advocate a boycott of all mahogany products and restrictions on international trade in mahogany. Critics contend that, by lowering the economic value of the rainforest, a boycott will encourage conversion of the forest to agricultural land or other, more profitable, uses.

An alternative approach, advocated by some forest ecologists, is to maintain the economic value of the rainforest by continuing to log mahogany and by developing commercial uses for other rainforest tree species. Some scientists are experimenting in the Rio Bravo

Figure 26.1 • Big-leaf mahogany tree.

Figure 26.2 • Logging mahogany in Belize.

Conservation and Management Area of Belize with harvest methods that remove more trees in order to provide clearings in which mahogany seedlings can grow. They are monitoring the ecological impact of each method by studying bird and butterfly populations that depend on an intact forest ecosystem. At the same time, they are analyzing the economic costs and benefits of each method. Because all the methods involve removing more trees than would be done through traditional selective logging, this approach is highly controversial. Based on results of a study in the Bolivian Amazon, scientists at Conservation International estimate that the traditional method would damage the rainforest only one-third as much as management methods aimed at achieving sustainability.

A third approach, advocated by the International Institute of Tropical Forestry in Puerto Rico, is to grow mahogany on plantations and leave the old-growth tropical forest intact. Indonesia, for example, is raising mahogany on 120,000 ha (300,000 acres). Institute scientists suggest that second-growth forests and abandoned agricultural lands would be good sites for mahogany plantations and a better solution than disturbing old growth. However, this approach might not help the people in Latin America whose livelihood depends on the mahogany market.[1-6]

- *The case study of mahogany illustrates some economic aspects of environmental renewable resources.*

26.1 The Importance of Environmental Economics

In the mid-1990s, the United States spent about $115 billion a year to deal with pollution, about 2% of the nation's gross national product. The defense budget is only two and a half times larger. Present costs are closer to $170 billion.[7] Included in these figures are the amounts spent by consumers and by private companies as well as by the government. Clearly, the total is much greater than the $6 billion budget of the Environmental Protection Agency (EPA).[8] And this 2% of the nation's budget is related to pollution control only; it does not include expenses for other environmental programs, such as those for conservation of endangered species, wildlife management, and conservation of forests.

Populations subject to high levels of certain pollutants (people in inner cities, for example) have lower average life expectancies and higher incidences of certain diseases. Studies in the United States have indicated that particulate air pollution in cities alone is at least a significant contributor to 60,000 deaths annually.[9] Other studies in the United States have estimated that from 2 to 9% of total mortality in cities is associated with particulate air pollution.[10]

Thus, the economics of environment affects all of us, and a field called **environmental economics** has developed to study it. According to environmental economist Charles Kolstad, the primary areas of concern in environmental economics are:[11]

- The importance of the environment to the economy.
- The impact to the environment as a result of economic activity.
- The regulation of the economy and economic processes in an attempt to balance environmental and economic goals of society.
- The proper intervention of government policy on the economy to help minimize environmental degradation and help solve environmental problems.

Ecological economics is an area of study that has emerged from the study of ecologic systems. Environmental economics is a field of study by economists who have extended their discipline to the area of the environment. By comparison, ecological economics is a field of inquiry by ecologists who are extending their discipline of ecology to the study of relations between humans and the economy, with an emphasis on the long-term health of ecosystems, or sustainability. The important point is that the two are not the same in their approach to solving environmental problems. Environmental economics, the main topic of this chapter, compared to ecological economics, focuses on relative short-term conditions that tend to discount the future. As a result, consequences of economic decisions today that effect what happens a century from now are not generally considered by conventional economics. Those effects are left to future generations to deal with. This is morally unacceptable to many people including ecological economists who argue for sustainability. Discounting the future is one reason some environmentalist believe that economics is one of the causes of our environmental problems. However, how we evaluate costs and benefits for the future remains a complex problem.[11]

Ecologic economists would argue that we should not make decisions today that threatens future sustainability. This is a philosophy that we agree with, and a position that is becoming more appreciated in the field of environmental economics. However, conventional economics continues to have problems dealing with economic costs and benefits to society over long periods of time. There will be a continuing evolution of ideas for both environmental economists and ecological economists, and perhaps one day the two fields will appropriately merge. We will return to the important discussion of how the future is valued in Section 26.4 after we consider the environment as a commons, and economic externalities, both of which have important implications to how we value the future.

Economic analysis involves several kinds of environmental issues including: the use of desirable resources (such as fish in the ocean, oil in the ground, or forests on the land) and the minimization of undesirable pollution. An understanding of basic concepts of environmental economics will provide insight into why environmental resources have been poorly conserved in the past. This insight may help us determine how best to modify our policies so that, in the future, conservation becomes a part of economic policy and development.

Environmental decision making often involves analysis of tangible and intangible factors. A mudslide that results from altering the slope of land is an example of a tangible factor; the aesthetics of the altered slope is an example of an intangible factor. Of the two, the intangibles are obviously more difficult to deal with because they are hard to measure and to value. Nonetheless, evaluation of the intangibles is becoming more important in local, regional, and national land-use planning and environmental analysis. Therefore, one task of environmental economics is to develop a method of aesthetic evaluation that provides good guidelines, is easy to understand, and is quantitatively credible.

 26.2 Use of Desirable Resources

It may seem curious that sometimes those who use a natural resource or benefit in some way from it do not act in a way that maintains that resource and its environment in a renewable state. When individuals benefit from the resource, it would seem to be in their best interests not to damage or destroy it. However, economics shows that the profit motive, by itself, will not always lead a person to act in the best interests of the environment. Here, we give two reasons why this may be so.

The first reason has to do with what the ecologist Garrett Hardin called "the tragedy of the commons."[12] When a resource is shared, an individual's personal share of profit from exploitation of the resource is usually greater than the individual's share of the resulting loss. The second has to do with the low growth rate, and therefore the low productivity, of a resource.

The Environment as a Commons

A **commons** is land owned publicly with public access for private uses. Historically, a commons was a part of old English and New England towns where all the farmers could graze their cattle. The practice of sharing the grazing area works as long as the number of cattle is low enough to prevent overgrazing.

As Hardin points out, in such a situation, each herder is trying to maximize personal gain and must periodically consider whether to add more cattle to the herd. The addition of cattle has both a positive and a negative *utility* (in economic terms, value or worth). The positive utility is the benefit the herder receives by selling the extra cattle; the negative utility is the overgrazing caused by the addition of too many animals. Hardin argues that because the benefit to the individual of selling a cow for personal profit is greater than that individual's share of the loss in the degradation of the commons, freedom of action in a commons inevitably brings ruin to all. According to Hardin, without some management or control, all natural resources treated like commons will inevitably be destroyed.

Some biological resources, such as many of the forests in the United States, are on privately owned lands. Many other resources, including 38% of our nation's forests, are on publicly owned lands that are accessible to everyone. In addition, resources in international regions, such as ocean fisheries away from coastlines, are not controlled by any single nation.

A society that controls resources such as public lands and waters has a number of social mechanisms to achieve its environmental goals. At one extreme, a society may simply rely on individual motivation, on the assumption that what people find best for themselves will also be best for society. However, a society can also enact laws that directly regulate use, such as the U.S. Marine Mammal Protection Act, which allows no taking of marine mammals without a permit. It can set total harvest quotas while allowing open access to the resource, such as occurs with some fisheries; issue licenses, as is typical with sport fishing, so that control of harvest may be achieved by limiting the number of licenses issued; or limit the individual harvest while not directly limiting the total number harvested.

When biological resources occur in international or national areas with open access, they are like a commons. Thus, they are threatened by the tragedy of the commons. There are many examples of commons, both past and present. An important one today is the deep oceanic seabed, where valuable mineral deposits may be found. At present, these areas are a true commons, since the high seas have always been considered areas open to all and not the property of any single nation. Another example is Antarctica. Although there are some national territorial claims on this continent, most of the continent is a commons. Negotiations have continued for a number of years concerning the conservation of Antarctica and possible use of its resources.

At a national level, the overcrowding of national parks, wilderness areas, and other nature-recreation areas, as well as the pollution of the atmosphere, are also problems of the commons. An example of the first is Voyageurs National Park in northern Minnesota. For a number of years, there has been a debate in Minnesota about the status of Voyageurs National Park. The park, located within the boreal forest biome (see Chapter 8) of North America, contains many lakes and islands and is an excellent place for fishing, hiking, canoeing, and viewing wildlife. Before the area became a national park, it was used for motorboating, skimobiling, and hunting; a number of people in the region made their living from tourism based on these kinds of recreation. Some environmental groups argue that the Voyageurs National Park is ecologically fragile and needs to be legally designated a U.S. wilderness area, to protect it from overuse and from adverse effects of motorized vehicles. Others argue that the nearby million acre Boundary Waters Canoe Area (see Chapter 13) provides ample wilderness, that Voyageurs can withstand a moderate level of hunting and motorized transportation, and that these uses should be allowed. At the heart of the issue is the problem of the commons, which in this case can be phrased as: What is the appropriate public use of public lands? Should all public lands be open to all public uses? Should some public lands be protected from the people? At present, the United States has a policy of different uses for different lands. In general, national parks are open to the public and many kinds of recreation, whereas designated wildernesses have restricted visitorship and kinds of uses.

The atmosphere, too, functions as a commons. Consider the possibility of global warming. Individuals, corporations, public utilities, motor vehicles, and nations add carbon dioxide to the air through the burning of fossil fuels. The tragedy of the commons has seemed a possibility for the atmosphere. Just as Garrett Hardin suggested, people tend to respond by choosing to benefit themselves (by burning more fossil fuel) than by benefiting the commons (burning less fossil fuel). The picture here is quite mixed, however, with much effort ongoing to bring cooperation to this commons.

As another example, in the 19th century, burning wood in fireplaces was the major source of heating in the United States (and fuelwood is still the major source of heat in many nations.) Until the 1980s, a wood fire in a fireplace or woodstove was considered a simple good, providing warmth and beauty; people enjoyed sitting around a fire and watching the flames. This activity must have a long history in human societies. But in the 1980s, with increases in populations and recreational houses in states such as Vermont and Colorado, home burning of wood began to result in local air pollution. Especially in valley towns surrounded by mountains, the air became fouled, visibility declined, and there was a potential for effects on human health and environmental conditions. As a result, some communities restrict or prohibit the use of fireplaces and wood stoves. The local air was a commons, and its overuse required a societal change.

Low Growth Rate and Therefore Low Profit as a Factor in Exploitation

Recall that the second reason individuals tend to overexploit natural resources held in common is the low growth rate of the resource.[13] For example, one way to view whales economically is to consider them solely in terms of whale oil. Whale oil, a marketable product, can be thought of as the capital investment of the industry. How can whalers get the best return on their capital? (Here we need to remember that whale populations, like other populations, increase only if there are more births than deaths.) We will examine two approaches: resource sustainability and maximum profit. If whalers adopt a simple, one-factor resource sustainability policy, they will harvest only the net biological productivity each year and thus maintain the total abundance of whales at its current level (see Chapter 14); that is, they will stay in the whaling business indefinitely. In contrast, if they adopt a simple approach to maximizing immediate profit, they will harvest all the whales now, sell the oil, get out of the whaling business, and invest the profits.

Suppose they adopt the first policy. What is the maximum gain they can expect? Whales, like other large, long-lived creatures, reproduce slowly; a calf born every three or four years per female is typical. The total net growth of a whale population is unlikely to be more than 5% per year, where the net growth is defined as the net increase in biomass (see the explanation of biological growth in Chapter 9). If all the oil in the whales in the ocean today represented a value of $10 million, then the most the whalers could expect to take in each year would be 5% of this amount, or $500,000. Meanwhile, they must pay the cost of upkeep on ships and other equipment, interest on loans, and salaries of employees, all of which decrease profit.

If they adopt the second policy and harvest all the whales, then they can invest the money from the oil. Although investment income varies, even a conservative investment of $10 million would very likely yield more than 5%, especially when this income could be received without the cost of paying a crew, maintaining the ships, buying fuel, marketing the oil, and so on.

It is quite reasonable and practical, if one considers only direct profit, to adopt the second policy: Harvest all the whales, invest the money, and relax. Whales simply are not a highly profitable long-term investment under the resource sustainability policy. It is no wonder that there are fewer and fewer whaling companies and that companies have left the whaling business when their ships became old and inefficient. Few nations support whaling; those that do have stayed with whaling for cultural reasons. For example, whaling is important to the Eskimo culture, and some harvest of bowheads takes place in Alaska; whale meat is a traditional Japanese food, and whale harvest is maintained for this reason.

Other factors to consider in resource use are the relative scarcity of a necessary resource and its price. For example, if a whaler lived on an isolated island where whales were the only food, and he had no communication with other people, then the whaler's primary interest in whales would be in his remaining alive. He could not choose to sell off all whales to maximize profit, as he would have no one to sell them to. It would seem to make sense that he would harvest in a way that would maintain the population of whales. Or the whaler might estimate that his own life expectancy was only 10 years and that he had to consume the whales beyond their ability to reproduce. He might try to harvest the whales so that they would become extinct at the same time he would die anyway. "You can't take it with you" would be his attitude. However, if ships began to land regularly at this island, he could trade and begin to see the value of whales in other terms. Causing the extinction of whales would not be necessary.

If ocean property rights existed so that he could "own" the whales that lived within a certain distance of his island, then he might consider the economic value of owning this right to the whales. He could sell rights to future whalers, or mortgage against them, and thus reap the benefits during his lifetime for whales that could be caught after his death.

From these examples, we learn that if we want to conserve whales, we must think beyond the immediate, direct economic advantages of whaling. Policies that seem ethically good may not be the most profitable for an individual. Finally, in the example of whaling, the economic analysis clarifies how an environmental resource is used, and this brings us to the question of externalities.

26.3 Externalities

The gap in our thinking about the whales, an environmental economist would say, is that we must be concerned with externalities in whaling. An **externality**, also called an **indirect cost**, is an effect not normally account-

ed for in the cost-revenue analysis of producers.[13] In this case, externalities include the loss of revenue to tourist boats used to view whales and the loss of an ecological role played by whales in marine ecosystems.

Air and water pollution provide other good examples of externalities. Consider production of nickel from ore at the Sudbury, Ontario, smelters, which has serious environmental effects, as discussed in Chapter 15. Traditionally, the economic costs associated with the production of commercially usable nickel from an ore are the **direct costs**—that is, those borne by the producer and passed directly on to the user or purchaser. In this case, direct costs include the costs of purchasing the ore, of energy to run the smelter, of building the plant, and of paying employees.

Traditionally, costs associated with degradation of the environment from emissions from the plant have been considered externalities. For example, prior to implementation of pollution control, the Sudbury smelter destroyed vegetation over a wide area, which led to an increase in erosion. Although air emissions from smelters have been substantially reduced and restoration efforts have begun a slow recovery of the area, pollution remains a problem, and total recovery of the local ecosystem may take a century or more.[14] These are costs associated with the value of trees and soil and with restoration of vegetation and land to a productive state.

An important societal question is: Who should bear the burden of these costs? Some environmental advocates suggest that costs should be included in costs of production through taxation or fees. In this way, expense would be borne by the corporation that benefits directly from the sale of nickel or would be passed on, in increased sale prices, to users (purchasers) of nickel. Others suggest that these costs should be shared by the entire society and paid for by general taxation. Stated simply, the question is whether it is better to finance pollution control using tax dollars or a "polluter pays" approach. Today, economists generally agree that the "polluter pays" approach provides much stronger incentives for cost-effective pollution reduction.

26.4 How Is the Future Valued?

The preceding example reminds us of the old saying "A bird in the hand is worth two in the bush." That is, a profit now is worth much more than a profit in the future. This economic concept—the future value compared with the present value—is another important idea for environmental science.

The value of some elements of the environment may increase, decrease, or remain the same (Figure 26.3). Economists observe that it is an open question whether something promised in the future has less value than something today. The value is the result of the interaction of consumers' preferences for present and future con-

Figure 26.3 • Economic value as a function of time—a way of comparing the value of having something now with the value of having it in the future. A negative value means that there is more value attached to having something in the present than having it in the future. A positive value means that there is more value attached to having something in the future than having it today.

sumption, the technology for transferring present consumption to future consumption, and how future consumers view consumption.

As an example, suppose you are dying of thirst in a desert and meet two people; one offers to sell you a glass of water now, and the other offers to sell you a glass of water if you can be at the well tomorrow. How much is each glass worth? If you believe you will die today without water, the glass of water today is worth all your money, and the glass tomorrow is worth nothing.

In practice, things are rarely so simple and distinct. We all know we are mortal, so we tend to value personal wealth and goods more if they are available now than if they are promised in the future. This evaluation is made more complex, however, because we are accustomed to thinking of the future—to planning a nest egg for retirement or for our children. Indeed, many people today argue that we have a debt to future generations and must leave the environment in at least as good condition as we found it; these people would argue that the future environment is not to be valued less than the present.

Our discussion of future value is obviously sensitive to time. The future times associated with some important global environmental topics, such as stratospheric ozone depletion and global warming, extend nearly a century. This is so because chlorofluorocarbons (CFCs) have such a long residence time in the atmosphere (see Chapter 25) and because of the time necessary to realize potential benefits from changing energy policy to offset global cli-

mate change. How to deal with future value in an analysis of environmental issues remains a complex, contentious problem.

Another aspect of future versus present value is that spending on the environment can be viewed as diverting resources from alternative forms of productive investment that will be of benefit to future generations. (This assumes that spending on the environment is not itself a productive investment.) For example, in the 20th century, asbestos was used as an insulator in buildings, around hot water pipes and similar devices. Then it was discovered that asbestos can cause cancer. Insulation containing asbestos was found in some schools. As long as the asbestos is well contained within an external covering, it does not pose an immediate threat. In some cases, people decided to spend the money now to remove the asbestos, to reduce future risk to the students. Another approach is to leave the well-protected asbestos intact and spend the money that it would have cost to remove the asbestos on improving other educational facilities in a school, only acting to deal with the asbestos when it begins to enter the air in the school.

A further complicating issue is that as we get wealthier, the value we place on many environmental assets (like wilderness areas) increases dramatically. Thus, if society continues to grow in wealth over the next century as it has over the past century, the environment will be worth far more to our great-grandchildren than our great-grandparents, at least in terms of willingness to pay to protect it. It might be argued, however, that conserving resources and other parts of the environment for the future in that case is tantamount to taking from the poor at present and giving to the rich in the future.

If history is a guide, Americans in 2100 will be far better off than Americans in 2000. To what extent should we ask the average American today to sacrifice now for much richer great-great grandchildren? How can we know the future usefulness of today's sacrifices? Put another way, what would you have liked your ancestors in 1900 to have sacrificed for our benefit today? Should they have increased research and development on electric transportation? Should they have saved more tall-grass prairie or restricted whaling?

The two issues are these: (1) We are so much richer and better off than our ancestors that their sacrificing for us would have been inappropriate. (2) Even if they had wanted to sacrifice, how would they have known what sacrifices would be important to us? As a general rule, one answer to these thorny questions about future value is: Do not throw away or destroy something that cannot be replaced if you are not sure of its future value. For example, if we do not fully understand the value of the wild relatives of potatoes that grow in Peru but do know that their genetic diversity might be helpful in developing future strains of potatoes, then we ought to preserve those wild strains.

Death is the fate of all individuals, and every activity in life involves some risk of injury or death. What, then, does it mean to save a life by reducing the level of a pollutant? This is a question raised in another important area of environmental economics—**risk-benefit analysis**, in which the riskiness of a present action in terms of its possible outcomes is weighed against the benefit, or value, of the action. Here, too, difficulties arise.

How Is Risk Determined?

With some activities, the relative risk is clear. It is much more dangerous to stand in the middle of a busy highway than to stand on the sidewalk. Hang gliding has a much higher mortality rate than hiking. The effects of pollutants are often more subtle, so the risks are harder to pinpoint and quantify. Table 26.1 gives the risk associated with a variety of activities.

The degree of risk is an important concept in our legal processes. For example, the U.S. Toxic Substances Control Act states that no one may manufacture a new chemical substance or process a chemical substance for a new use without obtaining a clearance from the EPA. The act establishes procedures to estimate the hazard to the environment and to human health of any new chemical before it becomes widespread. The EPA examines the data provided and judges the degree of risk associated with all aspects of the production of the new chemical or process, including extraction of raw materials, manufacturing, distribution, processing, use, and disposal. The chemical can be banned or restricted in either manufacturing or use if the evidence suggests that it will pose an unreasonable risk of injury to human health or to the environment. But what is unreasonable?[15]

The level of acceptable pollution (and thus risk) is a social-economic-environmental trade-off. The level of acceptable risk changes over time in society, depending on changes in scientific knowledge, comparison with risks from other causes, the expense of decreasing the risk, and the social and psychological acceptability of the risk.

When DDT was first used, no one understood the subtle environmental transport and ecological effects of the chemical; scientific observations revealed these effects later (see A Closer Look 26.1). At that time, there was relatively little concern with environmental issues, and society was not yet willing to pay for many of these costs. Now, people widely agree that the environment is a major concern, and there is less willingness to accept indirect environmental effects. What had been considered externalities to the use of DDT have become internal cost factors.

As seen in Chapter 23, the total cost of pollution is the sum of the costs to control pollution and the loss from pollution damages. In some cases, these two factors have opposite trends in terms of economic cost (as one goes down, the other goes up), and their intersection point is

Table 26.1 Risk of Death for Various Activities

Activity	Risk of Death/Year
Sports	
Auto racing	1.2 in 1,000
Biking	1 in 100,000
Football	4 in 100,000
Water sports	1.9 in 100,000
Travel	
Air travel (1 transcontinental trip, United States)	3 in 1 million
Motor vehicles	2.4 in 10,000*
Natural hazards	
Hurricanes	4 in 10 million
Lightning	4 in 10 million
Tornadoes	5 in 10 million
Pollution	
Air pollution	2 in 10,000*
Drinking water (EPA limit, chloroform)	6 in 10 million*
Food and Smoking	
Peanut butter (4 tablespoons per day)	8 in one million*
Cigarette smoking	3.6 in 1,000 *
Work-Related	
Police killed on duty by felons	1.3 in 10,000

Sources: Modified from R. Wilson, "Risk/Benefit Analysis for Toxic Chemicals," *Ecotoxicology and Environmental Safety* 4 (1980): 370–383. Copyright 1980. Reprinted by permission of Academic Press, Orlando, Fla.: Risk figures identified by asterisks from R. Wilson and E. A. C. Crouch "Risk Assessment and Comparisons: An Introduction," *Science* 236 (1987): 267–270.

the minimum total cost, as shown by point A in Figure 26.4. In other cases, total costs may stabilize or even decline as companies utilizing pollution control become more efficient and external costs in the form of environmental damage are minimized. If the minimum total cost involves a pollution level that is too high a risk, then additional control may add considerable expense.

The risk associated with a pollutant can be determined by the present levels of exposure and predicted future trends. These trends depend on the production and origin of the pollutant, pathways it follows through the environment, and changes it undergoes along these pathways. Dose-response curves establish risk to a population from a particular level of pollutant (see Chapter 15). Relative risks of different pollutants can be determined by comparing their current levels and their dose-response curves.

It is possible to take scientific and technological steps to estimate risk and, once the level of risk is established,

Risk-Benefit Analysis and DDT

A review of the history of the use of DDT illustrates the difficulty of completely eliminating a pollution risk. As noted in Chapter 12, DDT was first applied widely in the 1940s to control the spread of diseases such as malaria via insects and was subsequently used to control crop pests. At first, tests of the safety of DDT focused on human health effects, which were believed to be small. In the late 1950s, however, DDT was discovered in the livers of sharks. Because DDT had not been used in the ocean, at first it was believed that this observation was simply a measurement error or the result of a unknown dumping of DDT directly in the ocean.

Gradually, and to their surprise, scientists began to understand that DDT had spread from farmland through surface-water runoff and the air into the ocean and that the chemical had become a worldwide contaminant. DDT was found in the tissues of penguins in the Antarctic and seals in the Pribilof Islands of the Bering Sea. By 1970, it had become clear that DDT was everywhere and was a global environmental problem.[21] Although DDT is banned for use in the United States, U.S. corporations continue to be among the largest producers of this chemical, which is shipped to other nations. DDT is still used widely elsewhere, especially in developing nations of the tropics.

Complete elimination of DDT residues from all the environments of the world—and from all the sharks, penguins, seals, and birds—no longer seems feasible. Aside from the difficulty of eliminating pollutants, another lesson of this example is that what had seemed to be an economic externality of DDT (its indirect ecological effects on birds and ocean animals) became a major societal issue.

estimate the cost of reducing risk and then compare the cost with the benefit. However, what constitutes an acceptable risk is more than a scientific or technical issue. Acceptability of a risk involves ethical and psychological attitudes of individuals and society.

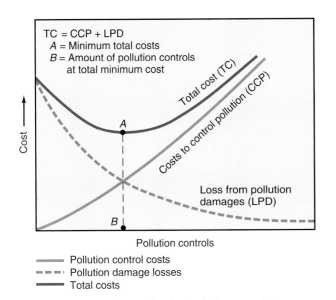

Figure 26.4 ● Total cost of pollution is the sum of the costs to control pollution and the loss from pollution damages. These two factors have opposite trends in terms of economic cost, and their intersection point forms a minimum total cost, as shown by point A. If the minimum total cost involves a pollution level that poses too high a risk, then additional control may add considerable expense.

We must therefore ask several questions. What risk from a particular pollutant is acceptable? How much is a given reduction in risk from that pollutant worth to us? How much will each of us, as individuals or collectively as a society, be willing to pay for a given reduction in that risk? The answers depend not only on facts but on societal and personal values. What also must be factored into the equation is that costs associated with cleanup of pollutants and polluted areas and with restoration programs can be minimized or even eliminated if a recognized pollutant is controlled initially. The total cost of pollution control need not increase indefinitely.

Acceptability of Risks and Costs

Societies differ in socially, psychologically, and ethically acceptable levels of risk for any cause of death or injury. It is commonly believed that future discoveries will help to decrease the risk, perhaps eventually allowing us to approach a zero-risk environment. But complete elimination of risk is generally either technologically impossible or prohibitively expensive.

We can make some generalizations about the acceptability of various risks. One factor is the number of people affected. Risks that affect a small population (such as employees at nuclear power plants) are usually more acceptable than those that involve all members of a society (such as risk from radioactive fallout).

In addition, novel, or new, risks appear to be less acceptable than long-established or natural risks, and society tends to be willing to pay more to reduce novel risks. For example, France spends approximately $1 million to reduce the likelihood of one air-traffic death but only $30,000 for the same reduction in automobile deaths.[16] Some argue that the greater safety of commercial air travel compared with automobile travel is in part a function of the relatively novel fear of flying compared with the more ordinary fear of death from a road accident. That is, because the risk is newer to us and thus less acceptable, we are willing to spend more per life to reduce the risk from flying than to reduce the risk from driving.

People's willingness to pay for reducing a risk also varies with how essential and desirable the activity is. For example, many people accept much higher risks for athletic or recreational activities than they would for transportation or employment related activities (see Table 26.1). The risks associated with playing a sport or utilizing transportation are assumed to be inherent in the activity. The risks to human health from pollution may be widespread and linked to a large number of deaths. Although risks from pollution are often unavoidable and unseen, people want a lesser risk for pollution than, say, driving a car or playing a sport.

In an ethical sense, it is impossible to put a value on a human life. However, it is possible to determine how much people are willing to pay for a certain reduction in risk or a certain probability of an increase in longevity. For example, a study by the Rand Corporation considered measures that would save the lives of heart-attack victims, including increasing ambulance services and initiating pretreatment screening programs. According to the study, which identified the likely cost per life saved and the willingness of people to pay, people were willing to pay approximately $32,000 per life saved, or $1,600 per year of longevity.[16]

Although information is incomplete, it is possible to estimate the cost of extending lives in terms of the dollars per person per year for various actions (Figure 26.5). For example, on the basis of direct effects on human health, it costs more to increase longevity through a reduction in air pollution than to directly reduce deaths through the addition of a coronary ambulance system. Such a comparison is useful as a basis for decision making. Clearly, though, when a society chooses to reduce air pollution, many factors beyond the direct, measurable health benefits are considered. Pollution directly affects more than

34 g/mi = 21 g/km

Figure 26.5 • One way to rank the effectiveness of various efforts to reduce pollutants is to estimate the cost of extending a life in dollars per year. This graph shows that reducing sulfur emissions from power plants to the Clean Air Act level (*A*) would extend a human life 1 year at a cost of about $10,000. Similar restrictions applied to automobile emissions (*B*, *C*) would increase lifetimes by 1 day. More stringent automobile controls would be much more expensive (*D*), and mobile units and screening programs for heart problems would be much cheaper (*E*). This graph represents only one step in an environmental analysis. (*Source:* Based on R. Wilson, "Risk/Benefit Analysis for Toxic Chemicals," *Ecotoxicology and Environmental Safety* 4 [1980]:370–383.)

just our health, and ecological and aesthetic damage can also indirectly affect human health (see Section 26.4). We might want to choose a slightly higher risk of death in a more pleasant environment (spend money to clean up the air instead of to increase ambulance services) rather than increase the chances of living longer in a poor environment (spend the money on reducing heart attacks).

Such comparisons may make you feel uncomfortable. But like it or not, we cannot avoid making choices of this kind. The issue boils down to whether we should improve the quality of life for the living or extend life expectancy regardless of the quality of life.[17]

Although pollution control may involve many dollars, the average cost per family in the United States is low, especially compared with other costs. It has been estimated that the per-family pollution control cost is between $30 and $60 per year for a family with a median income. In addition to the low cost per family, pollution control has many benefits whose quantitative value can be estimated. For example, federal air quality standards are estimated to reduce the risk of asthma by 3% and the risk to locally exposed adults of chronic bronchitis and emphysema by 10% to 15%. Estimates of the total cost of the direct and indirect effects on human health from stationary sources of air pollution are $250 per family per year. Air pollution contributes to inflation by reducing the number of productive workdays, reducing work efficiency, adding to direct expenditures for health treatments, and necessitating repair of nonhuman environmental damage. On this basis, air pollution control appears to be cost-effective; in fact, it has economic benefits.[18]

26.6 Environmental Intangibles

Environmental intangibles include the visual and other sensual pleasures of experiences, as well as the public service functions of living things. Intangibles are extremely difficult to evaluate economically. This is the Achilles' heel of environmental economics.

Aesthetic Qualities

A nature photographer finding an elusive subject on a lonely mountaintop, a hunter in a blind on a crisp autumn morning, a beachcomber searching misty coastal tidal pools, picnickers relaxing in an urban park, a motorist out for a drive in the country, an office worker in a city daydreaming about a canoe trip in a wilderness area—all are enjoying different aspects of the environment. Their experiences and memories cannot be readily assigned an economic value; to an individual, such experiences and memories may be priceless.

One of the perplexing problems associated with aesthetic evaluation is personal preference. One person may appreciate a high mountain meadow far removed from civilization, whereas a second person prefers visiting with others on a patio at a trailhead lodge. A third person may prefer to visit a city park; another may prefer the austere beauty of a desert. If we are going to consider aesthetic factors in environmental analysis, we must develop a method of aesthetic evaluation that allows for individual differences.

The criteria necessary to judge aesthetic qualities include unity, vividness, and variety.[19] *Unity* refers to the quality or wholeness of the perceived landscape, not as an assemblage but as a single, harmonious unit. *Vividness* refers to that quality of landscape that makes a scene visually striking; it is related to intensity, novelty, and clarity. *Variety* refers to how different one landscape is from another. Variety includes ideas of diversity and uniqueness; however, it is not always the case that the greater the diversity, the greater the aesthetic value.

As mentioned, the aesthetic aspects of resources are intangible and thus difficult to evaluate quantitatively. In contrast, quantitative evaluation of tangible natural resources, such as air, water, forests, and minerals, prior to development or management of a particular area, is standard procedure. Water resources for power or other uses may be evaluated by the amount of flow of the rivers and the quantity of water storage in rivers and lakes; forest resources may be evaluated by the number, type, and sizes of trees and their subsequent yield of lumber; and mineral resources may be evaluated by the estimated number of metric tons of economically valuable mineral material at particular locations. We can make a statement of the quality and quantity of each tangible resource compared with some known quality or quantity.

Ideally, we would like to make similar statements about the more intangible resources, such as scenery. That is, we would like to compare scenery to specific standards and compare one scene with another. Or we might want to compare the tangible economic value of a resource with the aesthetic value of the same resource. This is a very difficult task for which few standards are available.

Public Service Functions of Living Things

Another kind of environmental intangible, already discussed briefly in Chapter 13, is the public service function of the biota and the environment. For example, it is estimated that bees pollinate $20 billion worth of crops in the United States (Figure 26.6). The cost of pollinating these crops by hand would be exorbitant, and a pollutant that eliminated bees would have large, indirect economic consequences. We rarely think of this benefit of bees. Recently, however, an outbreak of parasites of bees in the United States has reduced the abundance of bees, bringing this once intangible factor to public attention.

Figure 26.6 • Public service functions of living things. Wild creatures and natural ecosystems perform public service functions for us—carrying out tasks important for our survival that would be extremely expensive for us to accomplish by ourselves. For example, bees pollinate millions of flowers important for food production, timber supply, and aesthetics.

As another example, bacteria fix nitrogen in the oceans, lakes, rivers, and soils. The cost of replacing this function in terms of production and transport of artificially produced nitrogen fertilizers would be immense, but again we rarely think about this activity of bacteria. Bacteria also clean water in the soil by decomposing toxic chemicals.

The atmosphere performs a public service by acting as a large disposal for toxic gases. For instance, carbon monoxide is eventually converted to nontoxic carbon dioxide, either by inorganic chemical reactions or by bacteria.

Only when our environment loses a public service function do we usually begin to recognize its economic benefits. Then, what had been accepted as an economic externality (indirect cost) suddenly may become a direct cost.

Estimates have been attempted to determine the dollar value of public service functions. At this time, we have to consider these estimates only rough approximations, as the value is difficult to measure. Public service functions of living things that benefit human beings and other forms of life have been estimated to provide between $3 trillion and $33 trillion per year.[20,21]

Economists have also attempted to estimate the costs of conserving biodiversity worldwide. As with estimates of the dollar value of the public service functions of nature, these estimates must be taken as rough estimates at this time. One current global estimate is presented in Table 26.2. The estimated total annual cost for maintaining and expanding to an area of 10% of each country or ecosystem, protected areas such as national parks and nature reserves for the conservation of biodiversity globally is estimated to be about $18 billion. This cost is less than 1% of the estimated benefits from the public service functions of nature.

26.7 Global Issues: Who Bears the Costs?

Global environmental problems make us more aware of the public service functions of the environment of our planet and of life around us as well as raising new economic questions. An important case in point is the possibility of global warming. The problem is that our technological society is adding carbon dioxide and other greenhouse gases to the atmosphere, which have the potential to warm the climate. (See the discussion of global warming in Chapter 21.)[22,23] The direct solution is to decrease the release of these gases, but to do so would require a worldwide decrease in the burning of fossil fuels. Although most of the production of greenhouse gases is from the industrial nations, in the future, developing nations, especially China and India, will contribute large quantities of these gases.

The economist Ralph D'Arge points out an economic problem arising from this global issue: The less developed countries did not share in the economic benefits of the burning of fossil fuels during the first two centuries of the Industrial Revolution, but they are sharing in the disadvantages of this activity.[24] Now the industrialized nations are suggesting that all nations, including the less developed ones, restrict their use of fossil fuels and therefore participate in future disadvantages without obtaining the benefits of cheap energy. Developing nations tend to think that industrial nations, which enjoyed the past benefits, should accept most of the future costs. At the same time, why shouldn't the developing nations proceed to develop and burn fossil fuels?

This perspective, limited to what benefits individual nations, may be too restricted in light of the global environmental effects of our technological civilization. It may be necessary to reduce the total production of greenhouse gases. If so, then the economic question is: Who pays, and how? At present, this is an unresolved issue in environmental economics. One suggestion is that the developed nations pay for the reduction in greenhouse emissions of the less developed nations. Another suggestion is that the developed countries share their technology with developing countries, thus helping the developing countries to reduce both local and global pollution. These issues were a major concern at the 1992 Earth Summit in Rio de Janeiro.[25]

26.8 Society and the Protection of Public Goods

A *public good*, such as clean air, is generally defined by economists as a good that is enjoyed in common, and for which it is hard to exclude people who don't pay (Figure 26.7). Suppose you decided to go into the business of cleaning up the air. Because anyone could use the clean air, people would not voluntarily pay a fee for your cleanup. It is unlikely that you could make a successful business of this.

Table 26.2   Estimated Global Cost for Current Management and Future Expansion of Protected Areas[1] Such as Wilderness Areas and Natural Reserves to Conserve Biodiversity

	Present Protection		Expanded Protection[4]		
Regions	Current Spending (billions of dollars)[2]	Budget Shortfall (billions of dollars)[3]	Cost to Purchase (billions of dollars)	Management Costs (billions of dollars)	Compensation Costs (billions of dollars)[5]
Developed					
North America	3.35	0.34	0	0	0
Europe	1.17	0.51	0	0	0
Australia/New Zealand	0.30	0.03	0.03	0.01	0
East Asia (developed)	0.45	0.05	0.60	0.19	0
Developing Regions					
Latin America	0.22	0.30	0.03	0.01	1.24
Russia	0.05	0.06	0.81	0.25	0.27
North Africa/Middle East	0.04	0.20	0.18	0.06	0.23
Africa	0.25	0.25	0.72	0.23	1.90
Asia	0.11	0.57	0.65	0.20	1.28
Pacific	0.04	0.01	0.43	0.14	0.03
Total World [6]	**5.98**	**2.32**	**3.45**	**1.09**	**4.95**

Source: Modified after A. James, K.J. Gaston, and A. Balmford 2001, "Can We Afford to Conserve Biodiversity?" *BioScience* 51 (1):43–52.

[1]Categories for protected area include national parks, national monuments, nature reserves, wilderness areas, protect marine areas, habitat or species management areas, or areas managed for sustainable use of resources.

[2]1996 dollars

[3]resulting because current spending is not adequate

[4]assumes that protected areas will expand to 10% of Earth's land area. The goal set by IUCN (World Conservation Union) would require an addition of about 3 million square kilometers across all categories.

[5]Payment to compensate people for new land for protected areas and loss of income.

[6]Total cost today about 6 billion dollars per year. Total cost to cover shortfall and expand about 18 billion dollars per year.

Figure 26.7 • Housing development in Alexandria, Virginia, near Washington, D.C. The rural and wooded Virginia countryside is considered beautiful, and this beauty is lost when dense housing tracts are built. It is difficult to place an economic value on such scenic quality, as it is on other public goods. This raises several economic questions: How much is the quality of the environment worth? How much would it cost to prevent the loss of public goods? How much economic activity is lost by not developing an area?

As discussed earlier, to some extent, natural ecosystems provide cleaning services that are public goods. Forests absorb particulates, salt marshes convert toxic compounds to nontoxic forms, and biological activity in streams removes sewage. Globally, we all profit from these public service functions of natural ecosystems. As we have seen, we have no simple way to put a value on them or even to estimate the amount of pollution removal that takes place. Nevertheless, it is in our interest to pursue environmental goals related to these functions.

Policy Instruments

How does a society achieve an environmental goal, such as preservation and use of a resource or reduction of a pollutant? Any society has several methods to achieve such goals. Means to implement a society's policies are known among economists as **policy instruments** (see Tables 26.3 and 26.4). These include moral suasion (which politicians call "jawboning," i.e., persuading people by talk, publicity, and social pressure); direct controls, which include regulations; market processes, which affect the price of goods and include taxation of various kinds, subsidies, licenses, and deposits; and government investments, which includes research and education. Society also has administrative mechanisms to ensure that the policy instruments chosen actually function.

Control of Pollutants

In discussing the control of pollutants, the concept of the marginal cost is useful. **Marginal cost** is the cost to reduce one additional unit of pollutant. In some cases the mar-

---

Table 26.3   Approaches to Environmental Policy

Policy Instruments

1. Moral suasion (publicity, social pressure, etc.)
2. Direct controls
   a. Regulations limiting the permissible levels of emissions
   b. Specification of mandatory processes or equipment
3. Market processes[a]
   a. Taxation of environmental damage
      i. Tax rates based on evaluation of social damage
      ii. Tax rates designed to achieve preset standards of environmental quality
   b. Subsidies
      i. Specified payments per unit of reduction of waste emissions
      ii. Subsidies to defray costs of damage control equipment
   c. Issue of limited quantities of pollution licenses
      i. Sale of licenses to the highest bidders
      ii. Equal distribution of licenses with legalized resale
   d. Refundable deposits against environmental damage
   e. Allocation of property rights to give individuals a proprietary interest in improved environmental quality
4. Government investment
   a. Damage prevention facilities (e.g., municipal treatment plants)
   b. Regenerative activities (e.g., reforestation, slum clearance)
   c. Dissemination of information (e.g., pollution control techniques, opportunities for profitable recycling)
   d. Research
   e. Education
      i. Of the general public
      ii. Of professional specialists (ecologists, urban planners, etc.)

Administrative Mechanisms

1. Administrative unit
   a. National agency
   b. Local agency
2. Financing
   a. Payment by those who cause the damage
   b. Payment by those who benefit from improvements
   c. General revenues
3. Enforcement mechanism
   a. Regulatory organization or police
   b. Citizen suits (with or without sharing of fines)

---

[a] Subsidies and taxes can also be distinguished by use of a property-rights framework. Per-unit subsidies implicitly confer ownership of the right to pollute on the polluter, and these rights are then purchased by the government via the subsidy. Taxes essentially say that there is public ownership of usage rights, which can be purchased from the public through its agent, the government, or by private parties on payment of the tax (price).

Source: W. J. Baumol and W. E. Oates, *Economics, Environmental Policy and the Quality of Life*, (Englewood Cliffs, N.J.: Prentice-Hall, 1979).

ginal cost increases rapidly as the percentage of reduction increases. For example, the marginal cost of reducing the biological oxygen demand in wastewater from petroleum

Table 26.4 Performance of Various Policy Instruments

Policy Instrument	Reliability	Permanence	Adaptability to Growth	Resistance to Inflation	Incentive for Improved Effort	Economy	Feasibility without Metering	Noninterference in Private Decisions	Political Attraction Actual	Political Attraction Potential
Moral suasion	Good[a]	Poor	Good[b]	Good[b]	Fair	Poor[c]	Excellent	Excellent	Excellent	—
Direct controls										
By quota	Fair	Poor	Fair	Excellent	Poor	Poor	Poor	Poor	Excellent	—
By specification of technique	Fair	Poor	Good[b]	Good[b]	Poor	Poor	Excellent	Poor	Excellent	—
Fees	Excellent	Excellent	Fair	Fair	Excellent	Excellent	Poor	Excellent	Poor	Good
Sale of permits or licenses	Excellent	Excellent	Excellent	Excellent	Excellent	Excellent[d]	Poor	Excellent	Poor	Good
Subsidies										
Per-unit reduction	Fair[e]	Good	Fair	Fair	Excellent	Good	Poor	Excellent	Good	—
For equipment purchase	Fair	Good	Fair	Fair	Excellent	Good	Poor	Excellent	Good	—
Government investment	Good	?	?	?	—	?	Excellent	—	Good	—

[a]For short periods of time when urgency of appeal is made very clear.

[b]Baumol and Oates's judgment.

[c]Induces contributions from decision makers who are most cooperative, not necessarily from those able to do the job most effectively.

[d]Tends to allocate reduction quotas among firms in a cost-minimizing manner, but if the number of emissions permitted is too small it will force the community to devote an excessive quantity of resources to environmental protection.

[e]Tends to allocate reduction quotas among firms in cost-minimizing manner but introduces inefficiency into the environmental protection process by attracting more polluting firms into the subsidized industry, so that aggregate response is questionable.

Source: W. J. Baumol and W. E. Oates, *Economics, Environmental Policy and the Quality of Life* (Englewood Cliffs, N.J.: Prentice-Hall, 1979).

Fishing Resources and Policy Instruments

The example of ocean fishing can be used to look at different policy instruments. The oceans outside of national territorial waters are commons, and thus the fish and mammals that live in them are common resources. What is a common resource may change over time, however. The move by many nations to define international waters as beginning 325 km (200 mi) from their coasts has turned some fisheries from completely open common resources to national resources open only to domestic fishermen.

In fisheries, there are four main management options:[27]

1. Establish total catch quotas for the entire fishery and allow anybody to fish until the total is reached.
2. Issue a restricted number of licenses but allow each licensed fisherman to catch many fish.
3. Tax the catch (the fish brought in) or the effort (the cost of ships, fuel, and other essential items).
4. Allocate fishing rights—that is, assign each fisherman a transferable and salable quota.

With total-catch quotas, the fishery is closed when the quota is reached. Whales, Pacific halibut, tropical tuna, and anchovies have been regulated in this way. When the practice was used in Alaska, all of the halibut were caught in a few days, with the result that restaurants no longer had halibut available for most of the year. This undesirable result led to a change in policy; the total-catch approach was replaced by the sale of licenses.

Although regulating the total catch can be done in a way that helps the fish, it tends to increase the number of fishermen and the capacity of vessels, and the end result is a hardship on fishermen. Recent economic analysis suggests that taxes that take into account the cost of externalities can work to the best advantage of fishermen and fish. Similar results are achieved by allocating a transferable and salable quota to each fisherman.

Determining which management method achieves the best use of a desirable environmental resource is not simple. The answer varies with the specific attributes of both the resource and the users. The tools of economics can be used to determine the methods that will work best within a given social framework.

refining increases exponentially. When 20% of the pollutants have been removed, the cost of removing an additional kilogram is 5 cents. When 80% of the pollutants have been removed, it costs 49 cents to remove an additional kilogram. Extrapolating from these results, it would cost an infinite amount to remove all the pollution.

Three common methods of direct control of pollution were mentioned in Tables 26.3 and 26.4: (1) setting maximum levels of pollution emission and (2) requiring specific procedures and processes that reduce pollution; and (3) charging fees for pollution emission. In the first case, a political body could set a maximum level for the amount of sulfur emitted from the smokestack of an industry. In the second, it could restrict the kind of fuel the industry could use. Many areas have chosen the latter method by prohibiting the burning of high-sulfur coal.

The problem with the first approach—controlling emissions—is that careful monitoring is required indefinitely to make certain the allowable levels are not exceeded. Such monitoring may be costly and difficult to carry out. The disadvantages of the second approach—requiring specific procedures—are that the required methodology may impose a severe financial burden on the producer of the pollutant, restrict the kinds of production methods open to an industry, and become technologically obsolete. (See the discussion of air pollution and laws regulating it in Chapter 23.) However, after an initial expenditure on procedures to reduce pollution, production efficiency may be increased and other costs such as monitoring and waste disposal can be reduced. For example, Japan, whose economy is one of the most energy efficient, has reduced its air pollution more than any other industrial nation. In recent years, Japanese industry has used only 5 megajoules to produce $1 gross domestic product (GDP), while U.S. industry required 12 megajoules. A large international chemical company, 3M, initiated a pollution prevention program and was able to stop the release of a billion pounds of toxic chemicals and save the company $500 million in the process.

Although the United States has emphasized the use of direct regulation to control pollution, other countries have been successful in controlling pollution by charging effluent fees. For example, charges for effluents into the Ruhr River in Germany are assessed on the basis of both the concentration of pollutant and the total quantity of polluted water emitted into the river. In response, plants have introduced water recirculation and internal treatment to reduce emissions.[26]

Studies of the uses of different policy instruments for environmental matters have resulted in some ability to evaluate their relative success (see Table 26.4). For example, moral suasion is reliable but not very permanent in its effect. Sale of licenses or permits has been found to be among the more successful recourses.

U.S. Fisheries: How Can They Be Made Sustainable?

Both overfishing and pollution have been blamed for the alarming decline in groundfish (cod, haddock, flounder, redfish, pollack, hakes) off the northeastern coast of the United States. Most scientists and fisheries managers have focused on overfishing, but attempts to regulate fishing have generated bitter disputes with fishermen, many of whom contend that restrictions on fishing make them scapegoats for pollution problems. The controversy has become a classic battle between short-term economic interests and long-term environmental concerns.

The Massachusetts Division of Marine Fisheries conducted a study in 1992 to settle the question of overfishing versus pollution. They concluded that key characteristics in declining flounder populations were consistent with overfishing rather than pollution. Furthermore, if pollution had been the cause, the skate and dogfish populations would have been expected to decline as well. However, S. J. Correia, the author of the study, cautioned that the issue should not be seen as an either/or question and that "overfishing may currently mask any pollution-induced population growth constraints."

Issues related to the use of U.S. fisheries were hardly new in 1992. In 1977, in response to concerns about overfishing in U.S. waters by foreign factory ships, the U.S. government extended the nation's coastal waters from 12 to 200 mi (19–322 km). To encourage domestic fishermen, the National Marine Fisheries Service provided loan guarantees for replacing older vessels and equipment with newer boats with high-tech equipment for locating fish. During this same period, demand for fish increased as Americans became more concerned about cholesterol levels in red meat. Consequently, the number of fishing boats, the number of days at sea, and fishing efficiency increased sharply. As a result, 50–60% of the populations of some species were landed each year.

A decision over Canadian and American fishing rights in the Georges Bank, the most prolific area in the North Atlantic, in favor of Canada in 1984 by the International Court of Justice in The Hague intensified competition for remaining fishing waters among U.S. fishermen. Overfishing knows no boundaries, however; and in 1992, Canada was forced to suspend all cod fishing to save the stock from complete annihilation.

In 1982, the New England Fisheries Management Council attempted to enforce harvest quotas but rescinded the order under pressure from the commercial fishing industry. In the wake of a bitter controversy that resulted from this decision, as well as further declines in fish populations, the council later instituted a series of measures prohibiting fishing at certain times and in certain areas, mandating minimum net sizes, and enacting quotas on the catch.

In 1992, the council adopted a more ambitious plan intended to cut fishing effort in half by 1997, but by means other than quotas and removal of current fishermen. The council decided to issue a limited number of fishing permits, limit the number of days at sea, use high-tech monitoring equipment to ensure compliance, and establish trip limits on some fish. Even more recently, portions of the Georges Bank were closed indefinitely to fishing for some fish species, including yellowtail, cod, and haddock. Landings of yellowtail in 1993 were 3,800 metric tons, the lowest on record and a mere 6% of the historical maximum in 1969.

The National Marine Fisheries Service advocates instead a system of Individual Transferable Quotas (ITQs) by which permits are issued to boat owners to allow them to harvest a fixed amount of fish each year. They can lease, sell, or bequeath the permits to others. Although ITQs have been successful in several U.S. fisheries and in some other countries, some small operators in New England fear that large corporations will buy up permits and dominate the industry.

Critical Thinking Questions

1. Using the example of the New England fishery, what are some arguments for and against the proposition that all natural resources treated like commons will inevitably be destroyed unless controls are instituted?

2. Which measures described attempt to convert the fishing industry from a commons system to private ownership? How might these measures help prevent overfishing? Is it right to institute private ownership of public resources?

3. What approach to future value (approximately) does each of the following people assume for fish?

    Fisherman: If you don't get it now, someone else will.

    Fisheries manager: By sacrificing now, we can do something to protect fish stocks.

4. Develop a list of the environmental and economic advantages and disadvantages of Individual Transferable Quotas (ITQs). Would you support instituting ITQs in New England? Explain.

5. Do you think it possible to reconcile economic and environmental interests in the case of the New England fishing industry? If so, how? If not, why not?

In every environmental matter, there is a desire on the one hand to maintain individual freedom of choice and on the other to achieve a specific social goal. In ocean fishing, for example, how does a society allow every individual to choose whether or not to fish and yet prevent everyone from fishing at the same time and bringing the fish species to extinction? This interplay between private good and public good is at the heart of environmental issues.

Some argue that the market itself will provide the proper control. For example, it can be argued that people will stop fishing when there is no longer a profit to be made. We have already seen, however, that this may not be so. Two factors interfere with this argument: (1) By the time the reduced fish population results in no economic gain for fishermen, it may be too late to avoid eventual extinction. (2) Even when it is not possible to make a sustained annual profit, there may be an advantage in harvesting the entire resource and getting out of the business (see A Closer Look 26.2).

Summary

- An economic analysis can help us understand why environmental resources have been poorly conserved in the past and how we might more effectively achieve conservation in the future.
- Economic analysis is applied to two different kinds of environmental issues: the use of desirable resources (fish in the ocean, oil in the ground, forests on the

land) and the minimization of undesirable pollution.

- Resources may be common property or privately controlled. The kind of ownership affects the methods available to achieve an environmental goal. There is a tendency to overexploit a common property resource and to harvest to extinction nonessential resources whose innate growth rate is low, as suggested in Hardin's "tragedy of the commons."
- Future worth compared with present worth can be an important determinant of the level of exploitation.
- The relation between risk and benefit affects our willingness to pay for an environmental good.
- Evaluation of environmental intangibles, such as landscape aesthetics and scenic resources, is becoming more common in environmental analysis. When quantitative, such evaluation balances the more traditional economic evaluation and helps separate facts from emotion in complex environmental problems.
- Societal methods to achieve an environmental goal include the categories of moral suasion, direct controls, market processes, and government investment. Many kinds of controls have been applied to the use of desirable resources and the control of pollution.

● REEXAMINING THEMES AND ISSUES

Human Population: The tragedy of the commons will increase as human population density increases, because there will be more and more individuals to seek gain at the expense of community values. For example, more and more individuals will try to make a living from harvesting natural resources. How people can use resources while at the same time conserving them requires an understanding of environmental economics.

Sustainability: From this chapter, we learn why people sometimes are not interested, economically, in sustaining an environmental resource from which they are making a living. When the goal is simply to maximize profits, it is sometimes a rational decision to liquidate an environmental resource and put the money gained in a bank or another investment. To avoid such liquidation, we need to understand economic externalities and intangible values.

Global Perspective: Solutions to global environmental issues, such as global warming, require that we understand the different economic interests of developed and developing nations. These can lead to different economic policies and different valuation of global environmental issues.

Urban World: The tragedy of the commons began with grazing rights in small villages. As the world becomes increasingly urbanized, the pressure to use public lands for private economic gain is likely to increase. An understanding of environmental economics can help us find solutions to urban environmental problems.

Science and Values: One of the central questions of environmental economics concerns how to develop equivalent economic valuation for tangible and intangible factors. For example, how can we compare the value of timber that could be harvested with the beauty people attach to the scenery, trees intact? How can we compare the value of a dam that provides irrigation water and electrical power on the Columbia River with the scenery without the dam and the salmon that could inhabit that river?

● KEY TERMS

commons **558**

direct costs **560**

ecological economics **557**

environmental economics **557**

externality **560**

indirect costs **560**

marginal cost **568**

policy instruments **568**

risk-benefit analysis **562**

● STUDY QUESTIONS

1. What is meant by the term "the tragedy of the commons"? Which of the following are the result of this tragedy: (a) the fate of the California condor, (b) the fate of the right whale, or (c) the high price of walnut wood used in furniture?

2. What is meant by risk-benefit analysis?

3. Cherry and walnut are valuable woods used to make fine furniture. Basing your decision on the information in the following table, would you invest in (a) a cherry plantation, (b) a walnut plantation, (c) a mixed stand of both species, or (d) an unmanaged woodland where you see some cherry and walnut growing? (*Hint*: Refer to the discussion in this chapter concerning whales.)

Species	Longevity	Maximum Size	Maximum Value
Walnut	400 years	1 m	$15,000/tree
Cherry	100 years	1 m	$10,000/tree

4. Flying over Los Angeles, you see smog below. Your neighbor in the next seat says, "That smog looks bad, but eliminating it would save only a few lives. Doing that isn't worth the cost. We should spend the money on other things, like new hospitals." Do you agree or disagree? Give your reasons.

5. Which of the following are intangible resources? Which are tangible?
   a. The view of Mount Wilson in California.
   b. Owning property with a view of Mount Wilson.
   c. Porpoises in the ocean.
   d. Tuna fish in the ocean.
   e. Clean air.
   f. Owning property outside the smog area of Los Angeles.

6. What kind of future value is implied by the statement "Extinction is forever"? Discuss how we might approach providing an economic analysis for extinction. (See Chapter 14 for additional information about extinction.)

7. Which of the following can be thought of as commons in the sense meant by Garrett Hardin? Explain.
   a. Tuna fisheries in the open ocean.
   b. Catfish in artificial freshwater ponds.
   c. Grizzly bears in Yellowstone National Park.
   d. A view of Central Park in New York City.
   e. Air over Central Park in New York City.

● FURTHER READING

Block, W., ed. 1990. *Economics and the Environment: A Reconciliation.* Vancouver: Fraser Institute.

Bromley, D. W. 1995. *Handbook of Environmental Economics.* Cambridge, Mass.: Blackwell.

Daly, H. E. and K. N. Townsend. eds. 1993. *Valuing the Earth: Economics, Ecology, Ethics.* Cambridge, Mass.: MIT Press. Essays that take a look at economic growth and come to the conclusion that sooner or later it is both physically and economically unsustainable as well as morally undesirable.

Goodstein, E. S. 2000. *Economics and the Environment.* 3rd ed. New York: John Wiley & Sons.

Hardin, G. 1968. "Tragedy of the Commons." *Science* 162:1243–1248. One of the most cited papers in both science and social science, this classic work outlines the differences between individual interest and the common good.

Hodge, I. 1995. *Environmental Economics: Individual Incentives and Public Choice.* New York: St. Martin's Press.

Jordan, C. F. 1995. *Conservation.* New York: John Wiley & Sons.

Mercuro, N., F. A. Lopez, and K. P. Preston. 1994. *Ecology, Law, and Economics: The Simple Analytics of Natural Resources and Environmental Economics.* Lanham, Md.: University Press of America. Presents a conceptual model that describes the interrelations among the economy, the ecology of the natural world, and the state (government and law), including conventional remedies to environmental problems.

National Academy of Public Administration. 1994. *The Environment*

Goes to Market: The Implementation of Economic Incentives for Pollution Control. Washington, D.C.: National Academy Press. Description and analysis of a number of implementation issues as evidenced in four case studies: air pollution trading, pollution charges, solid waste recycling, and a deposit-refund program. Washington, D.C.: National Academy Press.

Schnaiberg, A. and K. A. Gould. 1994. *Environment and Society: The Enduring Conflict.* New York: St. Martin's Press. An examination of several myths related to economics and environmental problems.

Tietenberg, T. H. 1992. *Environmental and Natural Resource Economics.* 3rd ed. New York: HarperCollins. A book that provides a broad range of national and international examples of the application of economic principles as they relate to environmental issues and natural resource management.

Tisdell, C. A. 1991. *Economics of Environmental Conservation: Economics of Environmental and Ecological Management.* Amsterdam and New York: Elsevier Science. A book focusing on the ecological dimensions of environmental economics, concentrating on living or biological resources and their life support systems.

Tisdell, C. A. 1993. *Environmental Economics: Policies for Environmental Management and Sustainable Development.* Aldershot, Hants, England: Edward Elgar. Includes discussions of externalities, cost-benefit analysis, sustainable development and economic activity, ecological economics, and global resource conservation.

27 Urban Environments

Tubular bus station in the city of Curitiba, Brazil with a population of over 2 million people efficiently moves people on and off buses, avoiding long delays. (Panos Pictures–Paul Smith)

LEARNING OBJECTIVES

Because the world is becoming increasingly urbanized, it is important to learn how to improve urban environments, to make cities more pleasant and healthier places in which to live, and to reduce undesirable effects on the environment. After reading this chapter, you should understand:

- How to view a city from an ecosystem perspective.
- What a city's site and situation are, and how they determine the location and qualities of the city.
- How cities have changed with changes in technology and with ideas about city planning.
- How a city changes its own environment and affects the environment of the surrounding areas, and how we can plan cities to minimize some of these effects.
- Why trees and other vegetation in cities are important as pleasing elements and as habitats for animals, and how we can alter the urban environment to encourage wildlife and to discourage pests.
- How cities can be designed to promote biological conservation and become pleasant environments for people.
- What are the two paths available for the future of the human population with regard to habitation of cities, and what are the requirements of the second path.
- What fundamental choices face us in deciding what kind of future we want and what will be the role of cities in that future.

The Ecological Capital of Brazil

In 1950, the city of Curitiba in Brazil had 300,000 inhabitants; but, by 1990 the population had grown to 2.3 million (Figure 27.1), making it the tenth largest city in Brazil (Figure 27.2). The growth of Curitiba resulted primarily from migration of rural people displaced by mechanization of agriculture. The newcomers lived in squatter huts at the edge of the city in conditions of great poverty, poor sanitation, and frequent flooding caused by conversion of rivers and streams into artificial canals. By 1970, Curitiba was well on the way to becoming an example of environmental degradation and social decay. The story of how Curitiba turned itself from an urban disaster into a model of planning and sustainability by 1995 illustrates that cities can be designed in harmony with people and the environment.[1,2]

Much of the credit for the transformation of Curitiba goes to its three-time mayor, Jaime Lerner, who believed a workable transportation system was the key to making Curitiba an integrated city where people could live as well as work. Rather than a more expensive underground rail system, Lerner spearheaded development of a bus system with five major axes, each containing lanes dedicated to express buses (see Figure 27.3), with others carrying local traffic and high-speed automobile traffic. Forty-nine blocks of the historic center of Curitiba were reserved for pedestrians. Tubular bus stations were built in which passengers paid fares before boarding, an arrangement that avoids long delays caused by collecting fares after boarding. Circular routes and smaller feeder routes between the major axes maintain vital connections between the central city and outlying areas. As a result, more than 1.3 million passengers ride buses each day. Although Curitiba has the second highest per capita number of cars in Brazil, it uses 30% less gas than eight comparable Brazilian cities, and its air pollution is among the lowest in the country.[2,3]

To solve its serious garbage problem, Curitiba required each household to sort recyclables from garbage. As a result, two-thirds of the garbage, more than 100 tons a day, is recycled, with 70% of the population participating. Where streets are too narrow to allow access by garbage trucks, residents are encouraged to bring garbage bags to the trucks. They are reimbursed with bus tokens, surplus food, or school notebooks.

Through a low-cost housing program, 40,000 new homes were built, many placed so that residents have easy access to job sites. The city also embarked on a

Figure 27.1 • City streets of Curitiba, Brazil. The quality of life for the people in this busy city has improved since the 1970s and has become a positive model for urban planning and sustainability.

program to increase the amount of green space. Artificial drainage channels were replaced with natural drainage, reducing the need for expensive flood control. Some areas, including those around the river basins,

Figure 27.2 • Located in southern Brazil, Curitiba provides an example of how a city can transform itself from an urban disaster into a model of planning and sustainability. (*Source:* Courtesy of Karl Gude.)

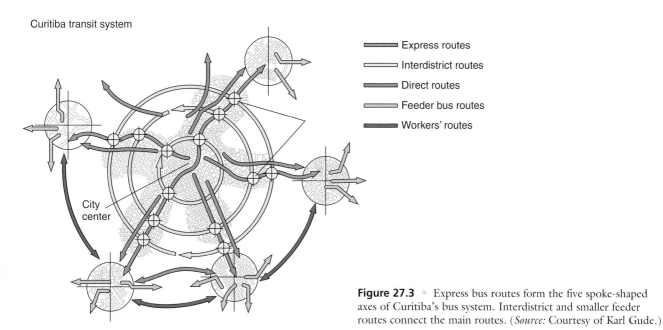

Curitiba transit system

Express routes
Interdistrict routes
Direct routes
Feeder bus routes
Workers' routes

City center

Figure 27.3 • Express bus routes form the five spoke-shaped axes of Curitiba's bus system. Interdistrict and smaller feeder routes connect the main routes. (*Source:* Courtesy of Karl Gude.)

were set aside for parks. In 1970, Curitiba had only 0.5 m² of green area per capita; by 1990, the area had increased to 50 m² for each inhabitant. The accomplish-

ments of Curitiba have led some to call it the "ecological capital of Brazil" and to hope that it is also the "city of the future."[2,4]

● *This case history illustrates that the global trend toward urbanization poses difficult challenges but that innovative ideas and strong leadership can create cities that work and are pleasant to inhabit. The solutions to urban problems need not involve high technology to improve the quality of urban life, but they must take into account the environment and the needs of the people.*

 City Life

In the past, the emphasis of environmental action has most often been on wilderness, wildlife, endangered species, and the impact of pollution on natural landscapes outside cities. Now it is time to turn more of our attention to city environments. In the development of the modern environmental movement in the 1960s and 1970s, it was fashionable to consider everything about cities bad and everything about wilderness good. Cities were thought of as polluted, lacking in wildlife and native plants, dirty, and artificial—and therefore bad. Wilderness was considered as unpolluted, clean, full of wildlife and

native plants, and natural—and therefore good. Although it was fashionable to disdain cities, the majority of people live in urban environments and have suffered directly from their decline.

Comparatively little public concern has focused on urban ecology; as a result, many urban people see environmental issues as outside their realm. However, city dwellers are at the center of some of the most important environmental issues. Today, there is a rebirth of interest in urban environments and in the development of urban ecology. People are realizing that city and wilderness are inextricably connected. We cannot fiddle in the wilderness while our Romes burn from sulfur dioxide and nitrogen oxide pollution. Environmental scientists are realizing the importance of analyses of the urban environment. The National Science Foundation has added two urban areas, Baltimore and Phoenix, to its Long-Term Ecological Research Program, a program that supports long-term monitoring as well as research on specific ecosystems and regions.

Worldwide, as we have seen, we are becoming an increasingly urbanized species. In the United States, about 75% of the population live in urban areas, and about 25% live in rural areas. Today, approximately 45% of the world's population, 2.75 billion people, live in cities.[5] It is projected that 62% of the population, 6.5 billion people, will live in cities by the year 2025.[6] Economic development leads to urbanization; 75% of people in developed countries live in cities, but only 38% of the people in the poorest developing countries are city dwellers.[5]

Not only is human population increasingly urbanized, but there is rapid growth of huge metropolitan areas with more than 8 million residents. In 1950, the world had only two such areas: New York City with its nearby New Jersey areas (12.2 million residents) and greater London (12.4 million). By 1975, Mexico City, Los Angeles, Tokyo, Shanghai, and Sao Paulo, Brazil, had joined this list. *In 1995, there were 23 such areas, 17 of them in the developing world. By 2015, there will be 36 megacities; and 23 of them will be located in Asia. Los Angeles and New York City are among the 10 largest megacities.*[7,8]

It is clear that in the future most people will live in cities. In fact, in most nations, most urban residents will live in the country's single largest city. In the future, for most people, living in an environment of good quality will mean living in a city that is managed carefully to maintain that environmental quality.

27.2 The City as a System

One of the ways in which we can improve the management is to analyze the city as an ecological system. Like any other life-supporting system, a city must maintain a flow of energy, provide necessary material resources, and have ways of removing wastes. These ecosystem functions are maintained in a city by transportation and communication with outlying areas. A city is not a self-contained ecosystem; it depends on other cities and rural areas. A city takes in raw materials from the surrounding countryside: food, water, wood, energy, mineral ores, everything that a human society uses. In turn, the city produces and exports material goods and, if it is a truly great city, exports ideas, innovations, inventions, arts, and the spirit of civilization. A city cannot exist without a countryside to support it. As was said half a century ago, city and country, urban and rural, are one thing—one connected system of energy and material flows—not two things (Figure 27.4).[9]

As a consequence, if the environment of a city declines, almost certainly the environment of its surroundings will also decline. The reverse is also true: If the environment around a city declines, the city itself will be threatened. Some people suggest, for example, that the ancient Native American settlement in Chaco Canyon, Arizona, declined after the environment surrounding that settlement was adversely affected.[10]

Cities also export waste products to the countryside, including polluted water, air, and solids. It has been estimated that the average city resident in an industrial nation annually uses (directly or indirectly) 208,000 kg (229 tons)

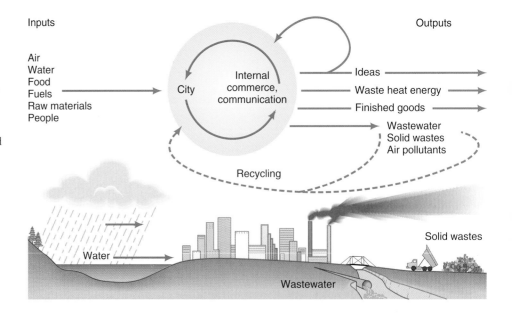

Figure 27.4 • The city as a system with flows of energy and materials. A city must function as part of a city-countryside ecosystem, with an input of energy and materials, internal cycling, and an output of waste heat energy and material wastes. As in any natural ecosystem, recycling of materials can reduce the need for input and the net output of wastes.

of water, 660 kg (0.8 tons) of food, and 3,146 kg (3.5 tons) of fossil fuels and produces 1,660,000 kg (1826 tons) of sewage, 660 kg (0.8 tons) of solid wastes, and 200 kg (440 lb) of air pollutants. If these are exported without care, they pollute the countryside, reducing the countryside's ability to provide necessary resources for the city and making life in the surroundings less healthy and less pleasant.

With such dependencies and interactions between city and surroundings, it is no wonder that relationships between people in cities and in the countryside have often been strained. Why, country dwellers want to know, should they have to deal with the wastes of those in the city? The answer is that many of our serious environmental problems occur at the interface between urban and rural areas. People who live outside but near a city have a vested interest in maintaining a good environment for that city and maintaining a good system for managing the city's resources.

As we will see in this chapter, the more concentrated the human population, the more land is available for other uses, including wilderness, recreation, conservation of biological diversity, and production of renewable resources. At the same time, if people live in densely populated cities, ways must be found to make urban life healthy and pleasant and to keep the cities from polluting the very environment that their human population in theory frees for other uses.

There are many ways to make cities pleasing environments, including the development of parks, connecting cities in an environmentally and aesthetically sound way to major landscape features such as rivers and nearby mountains. As we will also see, there is a long tradition of city planning with the goal of making cities a pleasing environment. By using both the long experience in city planning and modern knowledge from environmental sciences, we can make cities of the future healthier and more satisfying to people and better integrated within the environment. An argument has been made frequently that beautiful cities are not only healthy but attract people, thereby relieving pressures on the countryside.

With the growing human population, we can imagine two futures: In one, cities are pleasing and livable; use resources from outside the city in such a way that those resources are sustainable; minimize pollution of the surrounding country; and allow room for wilderness, agriculture, and forestry. In the other future, cities continue to be seen as environmental negatives and allowed to decay from the inside. People flee them to grander and more expansive suburbs that occupy much land; and the poor who remain in the city live in an unhealthy and unpleasant environment; without care for the city, its technological structure declines and it pollutes even more than in the past. Trends in both directions appear to be occurring.

In light of all these concerns, this chapter describes:

1. How a city fits within, uses, and avoids destroying the ecological systems on which it depends.

2. How the city itself can serve human needs and desires as well as environmental functions.

With this information, you, the student, will have the foundations for making decisions, based on science and on what you value, about what kind of urban-rural landscape you believe will provide the most benefits for people and nature.

27.3 Site and Situation: The Location of Cities

Here is an idea that our modern life, with its rapid transportation and its many electronic tools, obscures: Cities are not located at random but develop mainly because of local conditions and regional benefits. In most cases, they grow up at crucial transportation locations (an aspect of what is called the city's situation) and can be readily defended with good building locations, water supplies, and access to resources (qualities related to what is called site). The primary exceptions are cities that have been located primarily for political reasons. For example, Washington, D.C. was located to be near the geographic center of the area of the original 13 states; but the site was primarily swampland, and nearby Baltimore provided the major harbor of the region.

Importance of Site and Situation

As the case study environmental history of Venice, Italy, illustrates (see A Closer Look 27.1), the location of a city is influenced by the two factors just mentioned: **site**, which is the summation of all the environmental features of that location; and **situation**, which is the placement of the city with respect to other areas. A good site includes a good geologic substrate suitable for building construction, such as a firm rock base and well-drained soils that are above the water table; good nearby supplies of drinkable water; and good nearby lands suitable for agriculture, abundant timber, and other natural resources. It is also easier to build a city where the climate is benign—meaning that it does not suffer extremes of temperature and rainfall and is not subject to frequent storms. However, many important cities have been built in difficult climates. For example, Minneapolis–St. Paul is a city with a cold winter and hot summer; Houston, Texas, experiences hot, moist summers; and Miami, Florida, is among many cities that lie in the path of hurricanes. In these cases, one negative aspect of site has been overcome with modern engineering technology.

The environmental situation strongly affects the development and importance of a city, particularly with regard to transportation and defense. Waterways are important for transportation. Especially in early times, before railroads, automobiles, and airplanes, cities depended on

Venice Sinking

The city of Venice, Italy, has been known to be slowly sinking; but for a long time, no one knew the cause or a solution. Floods were becoming more and more common, especially during winter storms, when the winds drove waters from the Adriatic Sea into the city's streets. Famous for its canals and architectural beauty, Venice was in danger of being destroyed by the very lagoon that had sustained its commerce for more than a thousand years. Finally, the reason that the city was sinking was discovered. Groundwater in the region was being pumped out and used; and the depletion of the water table, over time, had caused the soil to compress under the weight of the city above it. The wells from which the groundwater was being pumped were located on the Italian mainland as well as on the islands that make up Venice; the wells supplied water to nearby industrial and domestic users. Once the cause was discovered, the wells were capped, and other sources of water were found. As a result, the city has stopped sinking.

Venice was founded by people escaping from the hordes that pillaged cities at the end of the era of the Roman Empire. Its location in the marshes along the Adriatic Sea was easily defended, and the seaside was also a good location for transportation and trade. The location of Venice also presented environmental problems, however, which had to be solved so that the city could become a major center. The shifting muds along the flat coast were a poor foundation for buildings. To improve the site, the early Venetians drove poles (made from saplings from forests on the mainland) into the mud and built on these. Venice still stands on this foundation, established more than 1,000 years ago.

Venice's success as a city depended on its situation—that is, its location relative to its surroundings. Nearby areas were a source of raw materials and resources for the city and provided a market for its products. Today, as in the past, Venice is affected by its surroundings in many ways, both positive and negative. It receives most raw goods and necessities from outside—food (except fish), fuel, and so forth. On the negative side, the highly decorated buildings and monuments for which the city is famous are deteriorating from acid rain generated by industries both nearby and far away. Its waters have high concentrations of heavy metals from industrial sources on the mainland and from fertilizers from mainland farms.[12]

Venice is not the only city to experience problems with subsidence, or sinking. This important problem can result from the removal not only of water but also of oil and gas or minerals and ore. The city of Long Beach, California, experienced subsidence as the result of the removal of oil and gas from wells, as well as the removal of groundwater.[13] Mexico City has also suffered from subsidence; removal of groundwater there has led to subsidence of as much as 7 m (30 ft).

As the people of Venice, Long Beach, and Mexico City have learned, cities influence and are influenced by their environment. The environment of a city affects its growth, success, and importance and can also provide the seeds of its destruction. All cities are so influenced; and those who plan, manage, and live in cities must be aware of all aspects of the urban environment.

---

water for transportation. Most early cities were located on or near waterways. In the ancient Roman Empire, for example, all important cities were located near waterways. Waterways have continued to influence the locations of cities; most major cities of the eastern United States are situated either at major ocean harbors or at the fall line on major rivers. See A Closer Look 27.2, for the definition of the term **fall line** and more information that illustrates how much environment affects the location and success of cities (Figure 27.5).

Cities are often founded at other kinds of crucial transportation points, growing up around a market, a river crossing, or a fort. Newcastle, England, and Budapest, Hungary, are located at the lowest bridging points on their rivers; other cities, such as Geneva, are located where a river enters or leaves a major lake. Some well-known cities are located at the confluence of major rivers: Saint Louis lies at the confluence of the Missouri and Mississippi rivers; Manaus, Brazil; Pittsburgh, Pennsylvania; Koblenz, Germany; and Khartoum, Sudan, are located at the confluence of several rivers. Many famous cities are located at crucial defensive locations, such as on or adjacent to easily defended rock outcrops. Examples include Edinburgh (Figure 27.6), Athens, and Salzburg, Austria. Other cities are situated on peninsulas—for example, Monaco and Istanbul.

Cities are often founded close to a mineral resource, such as salt (Salzburg, Austria); metals (Kalgoorlie, Australia); or medicated waters and thermal springs (Spa, Belgium; Bath, Great Britain; Vichy, France; and Saratoga Springs, New York).

A successful city can grow and spread over surrounding terrain, so its original purpose may be obscured to a resident. Its original market or fort may have evolved into

Cities and the Fall Line

The fall line on a river occurs where there is an abrupt drop in elevation of the land, creating waterfalls (Figure 27.5). The fall line typically is located where streams pass from harder, more erosion-resistant rocks to softer rocks. Cities have frequently been established at fall lines, for a number of reasons. Fall lines provide good sites for water power—an important source of energy in the 18th and 19th centuries, when the major eastern cities of the United States were established or rose to importance. At that time, the fall line was the farthest inland that larger ships could navigate; and just above the fall line was the farthest downstream that the river could be easily bridged. Not until the development of steel bridges in the late 19th century did it become practical to span the wider regions of the rivers below the fall line.[14] The proximity of a city to a river has another advantage, because river valleys have rich, water-deposited soils that are good for agriculture. In early times, rivers also provided an important means of waste disposal, which today has become a serious problem.

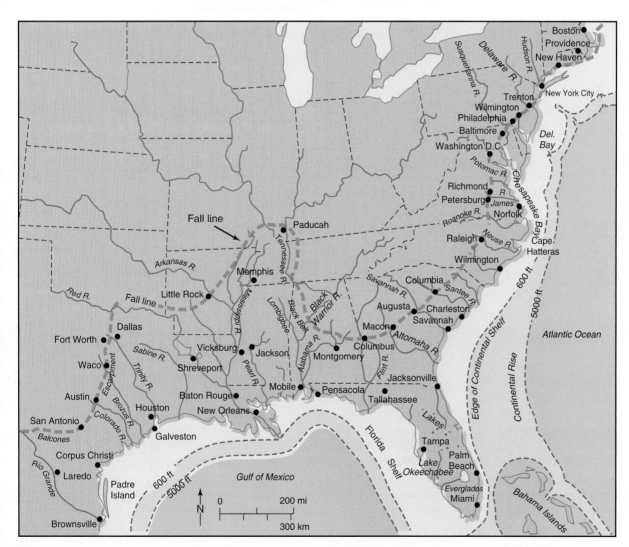

Figure 27.5 • Most major cities of the eastern and southern United States lie either at the sites of harbors or along the fall line (shown by the dashed line in the figure), which marks locations of waterfalls and rapids on major rivers. This is one way in which the location of cities is influenced by the characteristics of the environment. (*Source:* C. B. Hunt, *Natural Regions of the United States and Canada* [San Francisco: Freeman, 1974]. Copyright 1974 by W. H. Freeman & Co.)

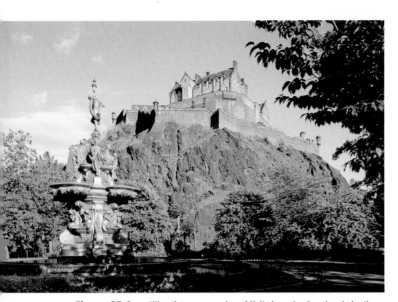

Figure 27.6 • The famous castle of Edinburgh, Scotland, built on top of bedrock overlooking the city in a readily defended location.

a square or historical curiosity. In most cases, though, cities originated where the situation led to a natural meeting point for people.

An ideal location for a city has both a good site and good situation, but such a place is difficult to find. Paris is perhaps one of the best examples of a perfect location for a city—one with both a good site and good situation. Paris began on an island more than 2,000 years ago, the situation providing a natural moat for defense and waterways for transportation. Surrounding countryside, a fertile lowland called the Paris Basin, affords good local agricultural land and other natural resources.

Site Modification

Site is provided by the environment, but technology and environmental change can alter a site for better or worse. People can improve the site of a city and have done so when the situation of the city made it important and when its citizens could afford large projects. An excellent situation can sometimes compensate for a poor site. However, improvements are almost always required to the site so the city can persist.

For example, New Orleans, at the mouth of the Mississippi, has a good situation but a poor site. An important transportation center at the mouth of the Mississippi River, it lies on low mud flats of the delta, which are unstable, are subject to frequent floods, and provide poor substrate for construction. Backwaters and swamps offer little as a local resource for agriculture but provide breeding habitats for mosquitoes. Fishing in the Gulf of Mexico is a plus, however. The situation of New Orleans was especially important at the time of the founding of the United States and throughout the 19th century, both for trans-

portation and as a defense of the interior, as illustrated by the importance of the Battle of New Orleans at the end of the War of 1812. Modern construction methods, including levees to prevent flooding, have improved the site.

Changes in a site over time can have adverse effects on a city. For example, Bruges, Belgium, developed as an important center for commerce in the 13th century because its harbor on the English Channel permitted trade with England and other European nations. By the 15th century, however, the harbor had seriously silted in, and the limited technology of the time did not make dredging possible. This problem, combined with political events, led to a decline in the importance of Bruges—a decline from which it never recovered. Today, Bruges still lives, a beautiful city with many fine examples of medieval architecture. Ironically, the fact that these buildings were never replaced with modern ones makes Bruges a modern tourist destination. Ghent, Belgium, and Ravenna, Italy, are examples of other cities whose harbors silted.[11] As human effects on the environment extend to global change, there may be rapid, serious changes in the sites of many cities. For example, if global warming occurs and sea levels rise, many coastal cities will be subject to flooding.

27.4 City Planning and the Environment

A city can never be free of environmental constraints, even though its human constructions give us a false sense of security. Lewis Mumford, the historian of cities, wrote, "Cities give us the illusion of self-sufficiency and independence and of the possibility of physical continuity without conscious renewal."[9] But this security is only an illusion. (A brief history of the relation of cities to their environments is presented in A Closer Look 27.3.)

A danger in city planning is the tendency to transform a city center from natural to artificial features and to replace grass and soil with pavement, gravel, houses, and temples, creating an impression that civilization has dominated the environment. Ironically, the very artificial aspects of the city that make it seem so independent of the rest of the world actually make it more dependent on its rural surroundings for all resources. Although such a city appears to its inhabitants to grow stronger and more independent, it actually becomes more fragile.[9]

In this way, a city grows at the expense of surrounding countryside, destroying surrounding landscape on which it depends. As nearby areas are ruined for agriculture and as the transportation network extends, the use, misuse, and destruction of the environment increase. Many cities of history were built, prospered, and declined in this way.

City Planning for Defense and Beauty

Many cities in history grew without any conscious plan. **City planning**—formal, conscious planning for new cities—can be traced back as far as the 15th century. Sometimes cities have been designed for specific social purposes, with little consideration of the environment, in other cases, the environment and its effect on city residents have been major planning considerations.

Two dominant themes in formal city planning have been defense and beauty (see A Closer Look 27.4). We can think of these two types of cities as fortress cities and park cities. The ideas of the fortress city and the park city influenced the planning of cities in North America. The importance of aesthetic considerations is illustrated in the plan of Washington, D.C., designed by Pierre Charles L'Enfant. L'Enfant mixed a traditional rectangular grid pattern of streets (which can be traced back to the Romans) with broad avenues set at angles. The intention was to design a city of beauty, with many parks, including small ones at the intersections of avenues and streets. This design has made Washington one of the most pleasant cities in the United States.

The City Park

Over time, parks have become more and more important in cities. A significant advance for U.S. cities was the 19th century planning and construction of Central Park in New York City, the first large public park in the United States. The park's creator, Frederick Law Olmsted, was one of the most important modern experts on city planning. He took site and situation into account and attempted to blend improvements to a site with the aesthetic qualities of the city.

For Olmsted, the goal of a city park was to provide relief from the city, both psychological and physiological, and to provide access to nature. A means to provide relief from city life was the creation of beauty in the park, and vegetation was one of the keys to this beauty. In his plan, Olmsted carefully considered the opportunities and limitations of the topography, geology, hydrology, and vegetation.

In contrast to the approach of a preservationist of nature, who would keep a park in as close to its original state as possible, Olmsted sought to create specific aspects that we would call "naturalistic" so that his goals for the park could be best met. He kept the rugged, rocky terrain that was present but made ponds where he thought they were desirable. He constructed "rambles," which were densely planted and followed circuitous patterns to add variety and relief. He created a "sheep meadow" by using explosives. As part of fitting the park within the other needs of the city, he built transverse roads through the park. He placed recreational areas in the southern part of the park, where there were flat meadows. He created depressed roadways that allowed traffic to cross the park without detracting from the vistas seen by park visitors.

Central Park is an example of "design with nature," a term coined much later. The design of Central Park influ-

Figure 27.7 • Boston has been modified over time to improve the environment and provide more building locations. This map of Boston shows land filled in to provide new building sites as of 1982. Although such landfill allows for expansion of the city, it can also create environmental problems, which then must be solved. (*Source:* A. W. Spirn, *The Granite Garden: Urban Nature and Human Design* [New York: Basic Books, 1984].)

enced other U.S. city parks, and Olmsted remained a major figure in American city planning throughout the 19th century. The firm he founded continued to be important in city planning into the 20th century.[15,16]

Olmsted's skill in creating designs that addressed both the physical and aesthetic needs of a city is illustrated by his work in Boston. The original site of Boston had had certain advantages: a narrow peninsula with several hills that could be easily defended, a good harbor, and a good water supply. As Boston had grown, however, demand had increased for more land for buildings, a larger area for docking ships, and a better water supply; and the need to control ocean floods and to dispose of solid and liquid wastes had grown as well. Much of the original tidal flats area, which had been too wet to build on and too shallow to navigate, had been filled in (Figure 27.7). Hills had been leveled and the marshes filled with soil. The largest project had been the filling of Back Bay, which began in 1858 and continued for decades. Once filled, however, the area had suffered from flooding and water pollution.[17]

Olmsted's solution to these problems was a water control project called the fens. His goal was to "abate existing nuisances" by keeping sewage out of the streams and ponds and providing artificial banks for the streams to prevent flooding—and to do this in a natural-looking way. His solution included creating artificial watercourses by digging shallow depressions in the tidal flats following meandering patterns like natural streams; setting aside other artificial depressions as holding ponds for tidal flooding; restoring a natural salt marsh planted with vegetation tolerant of the brackish water; and planting the entire area to serve as a recreation park when not in flood. He put a tidal gate on the Charles River—Boston's major

An Environmental History of Cities

The Rise of Towns

The first cities emerged on the landscape thousands of years ago during the New Stone Age with the development of agriculture, which provided the excess of food resources necessary for the maintenance of a city.[9] In this first stage, the density of people per square kilometer in cities was much higher than in the surrounding countryside; but the density was still too low to cause rapid, serious disturbance to the land. In fact, the waste provided by the city dwellers and their animals was an important fertilizer for the surrounding farmlands. In this stage, the city's size was restricted by the primitive transportation methods for bringing food and necessary resources into the city and removing waste. Because of such limitations, no European medieval town served only by land transportation had a population greater than 15,000.[11]

The Urban Center

In the second stage, more efficient transportation made possible the development of much larger urban centers, with a totally urban social core. Boats, barges, canals, and wharves, as well as roads, horses, carriages, and carts, made it possible for cities to be located farther from agricultural areas. Rome, originally dependent on local produce, became a city fed by granaries in Africa and the Near East.

The population of a city is limited by how far a person can travel in one day to and from work and by the density (number of people per unit area) with which people can be packed into an area. In the second stage, the internal size of a city was limited by pedestrian travel. The city could be no larger in area than the distance a worker could walk to work, do a day's work, and walk home. The density of people per square kilometer was limited by architectural techniques and primitive waste disposal. These cities never exceeded a population of 1 million; and there were few cities near this size, most notably Rome and some cities in China.

The Industrial Metropolis

The Industrial Revolution allowed greater modification of the environment than had been possible before. Two technological advances that had significant effects on the city environment were improved sanitation methods, which led to the control of many diseases, and modern transportation methods (Figure 27.8).

Improved transportation makes a larger city possible. Workers can live farther from their place of work and commerce, and communication can extend over larger areas. Air travel has freed cities even more from the traditional limitation of situation. We now have thriving urban areas where previously transportation was poor: in the far north (Fairbanks, Alaska) and on islands (Honolulu). These changes increase urban dwellers' sense of separateness from their natural environment.

Rapid transportation, such as subways and commuter trains, has also allowed the development of suburbs. In some cities, however, the negative effects of urban sprawl have led many people back to the urban centers or to smaller, satellite cities surrounding the central city. The drawbacks of suburban commuting and the destruction of the landscape in suburbs have brought new appeal to the city center.

The Center of Civilization

We are at the beginning of a new stage in the development of cities. With modern telecommunications, people can work at home or at distant locations. Perhaps, as telecommunication frees us from the necessities of certain kinds of commercial travel and related activities, the city can become a cleaner, more pleasing center of civilization.

An optimistic future for cities requires a continued abundance of energy and material resources, which are certainly not guaranteed, and wise use of these resources. If energy resources are rapidly depleted, modern mass transit may fail. Fewer people will be able to live in suburbs, and the cities will become more crowded. Reliance on coal and wood will increase air pollution. The continued destruction of the land within and near cities could compound transportation problems, making local production of food impossible. The future of our cities depends on our ability to plan and to conserve and use our resources wisely.

Figure 27.8 ● The New York City aqueduct, built in the 1860s, brought clean, clear water from the Croton Reservoir, miles away, into the heart of the city. It was one of the most important factors in creating a livable environment in that city.

A Brief History of City Planning

Two dominant themes in formal city planning have been planning for defense and planning for beauty. Ancient Roman cities were typically designed along simple geometric patterns that had both practical and aesthetic benefits. The symmetry of the design was considered beautiful but also provided a useful layout for streets.

During the height of Islamic culture, in the first millennium A.D., Islamic cities typically contained beautiful gardens, often within the grounds of royalty. One of the most famous urban gardens in the world is the Alhambra of the Islamic Alhambra Palace in Granada, Spain (Figure 27.9). The garden was created when this city was a Moorish capital and was maintained after the fall of Islamic control of Granada in 1492. Today, as a tourist attraction that receives 2 million visitors a year, the Alhambra garden demonstrates the economic benefits of aesthetic considerations in city planning. It also illustrates the use of a beautiful park as a specific focus in a city. This benefits the city environment indirectly, by providing relief from the city itself.

After the fall of the Roman Empire, the earliest planned towns and cities in Europe were walled fortress cities designed for defense. But even in these, city planners considered the aesthetics of the town. In the 15th century, one such planner, Leon Battista Alberti, argued that large and important towns should have broad and straight streets; smaller, less fortified towns should have winding streets to increase their beauty. He also advocated the inclusion of town squares and recreational areas, which continue to be important considerations in city planning.[18] One of the most successful of these walled cities is Carcasone, in southern France, now the third most visited tourist site in that country. Today, walled cities have become major tourist attractions, again illustrating the economic benefits of good aesthetic planning in urban development.

The usefulness of walled cities essentially ended with the invention of gunpowder. The Renaissance sparked an interest in the ideal city, which in turn led to the development of the park city. A preference for gardens and parks, emphasizing recreation, developed in Western civilization in the 17th and 18th centuries, characterized in the plan of Versailles, France, with its famous formal parks of many sizes and tree-lined walks, and by the work of the Englishman Capability Brown, who designed parks in England and was one of the founders of the English School of landscape design, which emphasized naturalistic gardens.

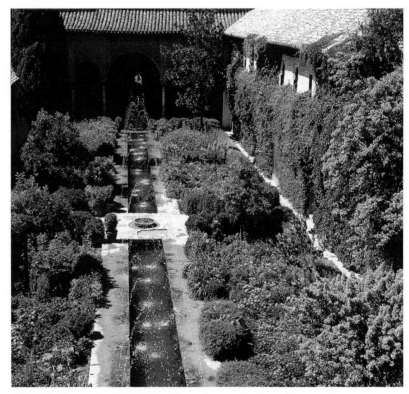

Figure 27.9 ● The Alhambra gardens of Granada, Spain, illustrate how vegetation can be used to create beauty within a city.

river—and had two major streams diverted directly through culverts into the Charles so that they flooded the fens only during flood periods. He reconstructed the Muddy River primarily to create new, accessible landscape.

The result of Olmsted's vision was that control of water became an aesthetic addition to the city. The blending of several goals made the development of the fens a landmark in city planning. Although it appears to the casual stroller to be simply a park for recreation, the area serves an important environmental function in flood control.

An important extension of the park idea was the **garden city**, a term coined in 1902 by Ebenezer Howard.

Howard believed that city and countryside should be planned together. A garden city was one that was surrounded by a **greenbelt**. The idea was to locate garden cities in a set connected by greenbelts, forming a system of countryside and urban landscapes. The garden city idea caught on, and garden cities were planned and developed in Great Britain and the United States. Greenbelt, Maryland, just outside Washington, D.C., is one of these cities, as is Lecheworth, England. Howard's garden city concept, like Olmsted's use of the natural landscape in designing city parks, continues to influence city planning today.

27.5 The City as an Environment

A city changes the landscape; and because it does, it also changes the relationship between biological and physical aspects of the environment. Many of these changes were discussed in earlier chapters as aspects of pollution, water management, or climate. They are mentioned again as appropriate in the following sections, generally with a focus on how effective city planning can reduce the problems.

The Energy Budget of a City

A city, like any ecological and environmental system, has an energy budget. The city exchanges energy with its environment in the following ways: (1) absorption and reflection of solar energy; (2) evaporation of water; (3) conduction of air; (4) winds (air convection); (5) transport into the city and burning of fuels by people within the city; and (6) convection of water (subsurface and surface stream flow). These in turn affect the climate within the city, and the city may affect the climate in the nearby surroundings, a possible landscape effect.

The Urban Atmosphere and Climate

Cities affect the local climate; as the city changes, so does its climate (see Chapter 22). Cities are generally less windy than nonurban areas, because buildings and other structures obstruct the flow of air. But city buildings also channel the wind, sometimes creating local wind tunnels with high wind speeds. The actual flow of wind around one building is influenced by nearby buildings. The total wind flow through a city is the result of the relationships among all the buildings. Thus, in planning a new building, its location among other buildings as well as its shape must be taken into account. In some cases, when this has not been done, dangerous winds around tall buildings have resulted in blown-out windows. A famous example involved the John Hancock building in Boston.

Recall that a city can receive less sunlight than the countryside because of the particulates in the atmosphere over cities. Often, urban areas have 10 or more times more particulates than surrounding areas.[19] In spite of the reduced sunlight, cities are warmer than surrounding areas (a city is a **heat island**), for two reasons. One is increased heat production (the burning of fossil fuels and other industrial and residential activities). The other is a decreased rate of heat loss, partly owing to the abundance of building and paving materials, which act as solar collectors (Figure 27.10).[20]

Sketch of an urban heat island profile

Late afternoon temperature (°F)

92°

85°

Rural | Suburban residential | Commercial | Downtown | Urban residential | Park | Suburban residential | Rural farmland

Figure 27.10 • A typical urban heat island profile. The graph shows temperature changes correlated with the density of development and trees. (*Source:* Andrasko and Huang, in H. Akbari et al., *Cooling Our Communities: A Guidebook on Tree Planting and Light-Colored Surfacing* [Washington, D.C.: U.S. EPA Office of Policy Analysis, 1992].)

Solar Energy in Cities

Until modern times, it was common to make use of solar energy to heat city houses. Our century is a major exception to this approach, because cheap and easily accessible fossil fuels have led us to forget certain fundamental lessons. Cities in ancient Greece, Rome, and China were designed so that houses and patios faced south and passive solar energy applications were accessible to each household.[20] Today, we are beginning to appreciate the importance of solar energy once again. Some cities have enacted solar energy ordinances that make it illegal to shade another property owner's building in such a way that it loses solar heating capability. (See Chapter 18 for a discussion of solar energy.)

Water in the Urban Environment

The construction of modern cities affects the water cycle greatly, in turn affecting soils and, consequently, plants and animals in the city. Paved city streets and city buildings prevent water infiltration. As a result, most rain runs off directly and is channeled into storm sewer systems. Hard city surfaces also prevent water in the soil from evaporating to the atmosphere. In natural ecosystems, evaporation is an important way of cooling the surface. City pavement increases the chances of local flooding within the city, and the increased runoff from the city to the countryside can increase the chances of flooding downstream. New methods of managing storm water can alleviate this problem by controlling and reducing the speed and quality of water running off pavements and into streams. As an example, a plan for the Alexandria, Virginia, central library parking lot includes wetland vegetation and soils. These temporarily absorb runoff from the parking lot. The vegetation removes some of the pollutants, and the plantings slow down the speed of water flow. (Figure 27.11).

Because of reduced evaporations, midlatitude cities generally record a lower relative humidity (2% lower in winter to 8% lower in summer) than the surrounding countryside. At the same time, cities can have higher local rainfall than their surroundings, because dust above a city provides particles for condensation of raindrops. Some urban areas have 5% to 10% more precipitation and considerably more cloud cover and fog than do surrounding areas. Fog is particularly troublesome in the winter and may impede ground and air traffic.

Most cities have a single underground sewage system. During times of no rain or light rain, this system handles only sewage. But during periods of heavy rain, the runoff is mixed with the sewage and can exceed the capacity of sewage treatment plants. During heavy rains, sewage is emitted downstream without sufficient treatment. In most cities that already have such systems, the expense of building a completely new and separate runoff system is prohibitive, so other solutions must be found. One city that avoids this problem is the city of Woodlands, Texas. It was designed by the famous landscape architect Ian McHarg. McHarg originated the phrase "design with nature," the subject of A Closer Look 27.5.[21]

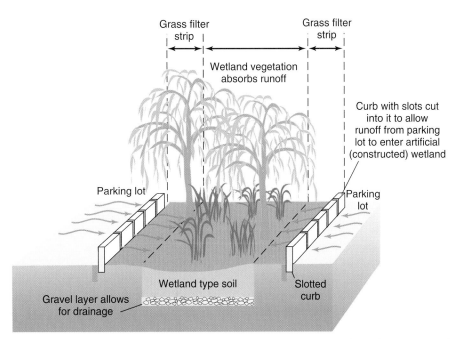

Figure 27.11 • A plan for the Alexandria, Virginia, central library parking lot includes wetland vegetation and soils, which temporarily absorb (see arrows) runoff from the parking lot. The landscape architecture firm of Rhodeside and Harwell planned the project. (*Source:* Modified after F. Harwell, Rhodeside and Harwell.)

Problems with flooding and overtaxing of storm sewage systems are made worse in many cities built on floodplains. Floodplains are often chosen as sites for cities because the land is flat and easy to build on and river transportation is available nearby. More often than not, to protect against flooding, the river is channeled and levees built along the shores. But as the citizens of St. Louis, Missouri, learned in the heavy rains and floods of the summer of 1993, these levees have their limits. They concentrate the waters in a narrower channel, which increases the speed of water flow and the destructive force of the waters when they do rise above the levee. The levee is an example of the idea expressed earlier that the artificial structures of cities give us a sense of independence from the environment, whereas in fact these structures make us more dependent on the environment. An alternative to artificial structures to control flooding is to allow for more areas that have multiple uses and can flood without severe damage during times of heavy rains, along the lines of the Olmsted design for Back Bay in Boston, discussed earlier.

In urban environments, long stretches of major rivers are channelized. Recall from Chapter 20 that channelization has two negative effects. First, when the river is maintained in an artificial channel, its sediment load is not deposited on the land, and the land's fertility is not renewed. Second, this sediment load passes downriver and is lost at the mouth, where it causes siltation and may fill in important harbors and damage cities at the ocean side. It is important to note that these negative effects are an indirect result of building cities on floodplains without proper planning and then trying to prevent the inevitable floods.

Soils in the City

A modern city has a great impact on soils. Most soil is covered by cement, asphalt, or stone; the soil no longer has its natural cover of vegetation, and the natural exchange of gases between the soil and air is greatly reduced. Such soils lose organic matter, because they are no longer replenished by vegetation growth. Soil organisms die from lack of food and oxygen. The process of construction and the weight of the buildings compact the soil, which restricts water flow. City soils, then, are more likely to be compacted, waterlogged, impervious to water flow, and lacking in organic matter.

A kind of soil important in modern cities is the soil that occurs on **made lands**, which are lands created from fill, sometimes as waste dumps of all kinds, sometimes directly to create more land for construction. The soils of made lands are different from those of the original landscape. They may be made of all kinds of trash, from newspapers to bathtubs, and may contain some toxic materials. The fill material is unconsolidated, meaning that it is loose material without rock structure. Thus, it is not well suited to be a foundation for buildings. Fill material is particularly vulnerable to the shaking from earthquakes. In such events,

the fill can act somewhat like a liquid and amplify the effects of the earthquake on buildings. Some made lands have developed into well-used parks, however. For example, a marina park in Berkeley, California, is build on a solid waste landfill. It extends into San Francisco Bay, providing public access to beautiful scenery. It is a windy location, popular for kite flying and family strolls. (See Chapter 28 for more information about solid waste disposal.)

Pollution in the City

Everything is concentrated in a city, including pollutants. City dwellers are exposed to more kinds of toxic chemicals in higher concentrations and to more human-produced noise, heat, and particulates than are their rural neighbors (see Chapter 15). This environment makes life riskier. Lives are shortened by an average of one to two years in the most polluted cities in the United States. The city with the greatest number of early deaths is Los Angeles, with an estimated 5,973 early deaths per year, followed by New York with 4,024, Chicago with 3,479, Philadelphia with 2,590, and Detroit with 2,123.[17]

Some urban pollution comes from motor vehicles, which have contributed lead in gasoline (where it is still used), nitrogen oxides, ozone, carbon monoxide, and other air pollutants from exhaust. Stationary power sources also produce air pollutants. Home heating is a third source, contributing particulates, sulfur oxides, nitrogen oxides, and other toxic gases. Industries are a fourth source, contributing a wide variety of chemicals. The primary sources of particulate air pollution, which consists of smoke and soot and tiny particles formed from emissions of sulfur dioxide and volatile organic compounds, are older, coal-burning power plants, industrial boilers, and gas- and diesel-powered vehicles.[22]

Although it is impossible to eliminate exposure to pollutants in a city, it is possible to reduce the exposure through careful design, planning, and development. For example, when lead was used in gasoline, exposure to lead was greater near a road than away from it. Exposure to lead could be reduced by placing houses and recreational areas away from roadways and by developing a buffer zone that made use of trees that were resistant to the pollutant and that absorbed pollutants and slowed the rate of spread.

27.6 Bringing Nature to the City

A practical problem for planners and managers of cities is how to bring nature to the city—that is, how to make plants and animals a part of a city landscape. This activity has evolved into several specialized professions, including urban forestry (whose professionals are often called tree wardens), landscape architecture, city planning, and civil engineering specializing in urban development. Most cities have an urban forester on the payroll who deter-

Design with Nature

When a new city is developed, our knowledge of the problems of flooding and water runoff can be used to plan a better flow of water. The new town of Woodlands, a suburb of Houston, Texas, illustrates such planning. Woodlands was designed so that most houses and roads were on ridges and the lowlands were left as natural open space. The lowlands provide areas for temporary storage of floodwater and, because the land is unpaved, allow the rain to penetrate the soil and recharge the aquifer for Houston. Preserving the natural lowlands has other environmental benefits as well. In this region of Texas, low-lying wetlands are habitats for native wildlife, such as deer. Large, pleasant trees, such as magnolias, grow there, providing food and a habitat for birds. The innovative city plan has economic as well as aesthetic and conservational benefits. It is estimated that a conventional drainage system would have cost $14 million more than the amount spent to maintain and develop the wetlands.[17]

mines the best sites for planting trees and the best species of trees to suit the environment. These professionals take into account climate, soils, and the general influences of the urban setting, such as the shading imposed by tall buildings and the pollution from motor vehicles.

Cities and Their Rivers

For cities on rivers, one way to bring nature to the city is to connect the city to its river. But traditionally, rivers have been valued for their utility in transportation and waste disposal rather than their ability to make city life more pleasant or to help in the conservation of nature. Rivers have been viewed as places to dump wastes. The old story was that a river renewed and cleaned itself every mile or every three miles (depending on who said it). That might have been relatively correct when there was one person or one family per linear river mile, but it is not true with the high density of cities nor with modern chemicals.

Kansas City, Missouri, which is located at the confluence of the Kansas and Missouri Rivers, illustrates the traditional treatment (Figure 27.12). The Missouri River's floodplain provided a convenient transportation corridor, so the south shore is dominated by railroads, while downtown the north shore forms the southern boundary of the

Figure 27.12 ● (*a*) Idealized general land use in Kansas City, Missouri. (*Source:* Modified after S. Faber, *On Borrowed Land: Public Policies for Floodplains* [Cambridge, Mass.: Lincoln Institute of Land Policy, 1998].) (*b*) A view of Kansas City's Missouri River waterfront.

Figure 27.13 • The Willamette River in Portland, Oregon, is an important urban resource with riverside parks, shops, and walkways.

city's airport. There are few parks or public spaces near the river. A small riverfront park has been established; but even so, the river has little place in this city as a source of recreation and relief for its citizens or in the conservation of nature.

Contrast Kansas City with Portland, Oregon. In Portland, interstate highway I-5 was originally built on the city's side of the Willamette River, separating the city from its river. The city raised funds and moved Interstate Highway I-5 across the river, replacing the interstate corridor with Riverfront Park, a marina, housing and shops, and pleasant walkways and developing pocket parks that lead into the city's downtown (Figure 27.13).

Vegetation in Cities

Planting of trees, shrubs, and flowers improves the beauty of a city.[23] Plants provide for different needs in different locations.[24] Trees provide shade, which reduces the need for air conditioning and makes travel much more pleasant in hot weather. In parks, vegetation provides places for quiet contemplation, trees and shrubs can block some of the city sounds, and the complex shapes and structures create a sense of solitude. Plants also provide habitats for wildlife such as birds and squirrels, which many urban residents consider pleasant additions to a city.

The use of trees in cities has expanded since the time of the European Renaissance. In earlier times, trees and shrubs were set apart in gardens, where they were viewed as scenery but not experienced as part of ordinary activities. Street trees were first used in Europe in the 18th century; among the first tree-lined streets were in Paris and London (Figure 27.14). In many cities, trees are now considered an essential element of the urban visual

scene, and major cities have large tree-planting programs. For example, in New York City, 11,000 trees are planted each year; in Vancouver, Canada, 4,000 are planted per year.[25]

There is growing use of trees to soften the effects of climate near houses as well. In colder climates, rows of conifers planted to the north of a house can protect the house from winter winds. Deciduous trees to the south can provide shade in the summer, reducing requirements for air conditioning, and yet allow sunlight to warm the house in the winter (Figure 27.15).[26]

Cities can even provide habitat for endangered plants. For example, in Lakeland, Florida, endangered plants are used in local landscaping with considerable success.

Vegetation in cities is under special kinds of stress, however. Trees along city streets are often surrounded by cement, which prevents normal access to water and air. The root systems are more likely to experience extremes of dryness on the one hand and soil saturation (immediately following or during a rainstorm) on the other. Because city soils tend to be compacted and to drain poorly, trees planted in a city sidewalk are often overwatered, and the roots may die from lack of oxygen. The solution is to prepare streets and sidewalks specially for tree growth. A tree-planting project was completed for the World Bank building in Washington, D.C., in 1996. Special care was taken to provide growing conditions for trees, including aeration, irrigation, and adequate drainage so that the soils did not become waterlogged. Five years later, the trees continued to grow and remained healthy.[26]

Many species of trees and plants are very sensitive to air pollution. For example, the Eastern white pine of North America is extremely sensitive to ozone pollution

Figure 27.14 • Paris was one of the first modern cities to use trees along streets to provide beauty and shade, as shown in this picture of the famous Champs Elysées.

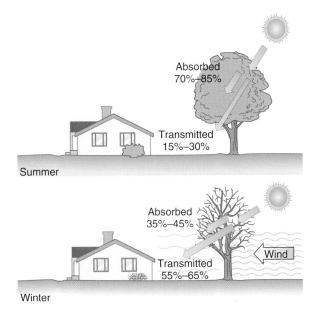

Figure 27.15 • Trees can improve the microclimate near a house, protecting the house from winter winds and providing shade in the summer while allowing sunlight through in the winter. (*Source:* J. Huang and S. Winnett, in H. Akbari et al., *Cooling Our Communities: A Guidebook on Tree Planting and Light-Colored Surfacing* [Washington, D.C.: U.S. EPA Office of Policy Analysis, 1992].)

and does not do well in cities with heavy motor vehicle traffic or along highways. Dust can interfere with the exchange of oxygen and carbon dioxide necessary for photosynthesis and respiration of the trees. City trees also suffer direct damage from physical impact of bicycles, cars, and trucks and from vandalism. Trees subject to such stresses are more susceptible to attacks by fungus diseases and insects. The lifetime of trees in a city is generally shorter than in their natural woodland habitats unless they are given considerable care.

Some species of trees are more useful and successful in cities than are others. An ideal urban tree would be resistant to all forms of urban stress; have a beautiful form and foliage; and produce no messy fruit, flowers, or leaf litter that requires cleaning. In most cities, in part because of these requirements, only a few species of trees are used for street planting. However, the reliance on one or a few species results in an ecologically fragile urban planting, as we learned when Dutch elm disease spread throughout the eastern United States, destroying urban elms. It is prudent to use a greater diversity of trees to avoid outbreaks of insect pests and tree diseases.[27]

Cities, of course, have many recently disturbed areas, including abandoned lots and areas recently made into corridors between lanes in boulevards and highways. Disturbed areas provide habitat for early successional plants, including many that we call "weeds," and these often include introduced (exotic) plants, such as European mustard. Therefore, wild plants that do partic-

ularly well in cities are those characteristic of disturbed areas and of early stages in ecological succession (see Chapter 10). City roadsides in Europe and North America have wild mustards, asters, and other early successional plants.

Wildlife in Cities

With the exception of some birds and small, docile mammals such as squirrels, most forms of wildlife in cities are considered pests. But there is much more wildlife in cities, a great deal of it unnoticed. In addition, there is growing recognition that urban areas can be modified to provide more habitats for wildlife that people can enjoy; this can be an important method of biological conservation.[28,29]

We can divide city wildlife into the following categories: (1) those species that cannot persist in an urban environment and disappear; (2) those that tolerate an urban environment but do better elsewhere; (3) those that have adapted to urban environments, are abundant there, and are either neutral or beneficial to human beings; and (4) those that are so successful they become pests.

Urban "Wilds":
The City as Wildlife and Endangered Species Habitat

We do not associate wildlife with cities, but cities provide homes to many forms of wildlife. Consider, for example, how the fox has integrated itself into the urban landscape of London. Foxes in London feed on garbage and roadkill (animals run over by motor vehicle); shy and nocturnal, they are seen by few Londoners.[30] Their food chain includes pets and other animals, fruits and vegetables, earthworms and insects, and scavenged items. Automobile run-overs account for about half of the deaths of urban foxes.

Cities are a habitat, albeit artificial. They can provide all the needs—physical structures and necessary resources such as food, minerals, and water—for many plants and animals. We can also identify ecological food chains in cities, as shown in Figure 27.16 for insect-eating birds and fox. For some species, cities' artificial structures are sufficiently like their original habitat to be home. For example, chimney swifts once lived in hollow trees but are now common in chimneys and other vertical shafts. Their nests are glued to walls with saliva. A city can easily have more chimneys per square kilometer than a forest has hollow trees. Cities also include natural habitats in parks and preserves. Modern parks provide some of the world's best wildlife habitats. In New York City's Central Park, approximately 260 species of birds have been observed—100 in a single day. Urban zoos play an important role in the conservation of endangered species, and the importance of parks and zoos will increase as truly wild areas shrink. Finally, cities that are harbors often have many species of marine wildlife at their doorsteps. New York City's waters include sharks, bluefish, mackerel, tuna,

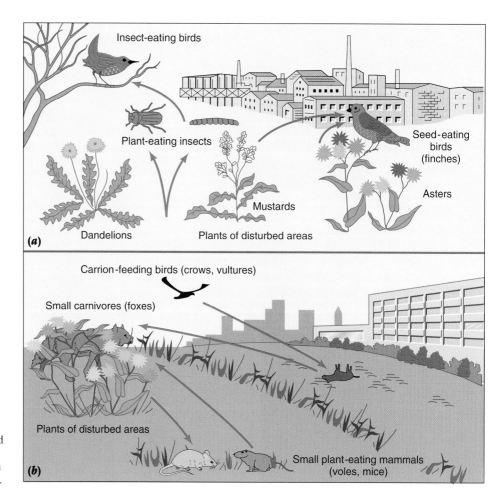

Figure 27.16 • (*a*) An urban food chain based on plants of disturbed places and insect herbivores. (*b*) An urban food chain based on roadkill.

striped bass, and nearly 250 other species of fish.[30]

Cities can even be home to rare or endangered species. Peregrine falcons once hunted pigeons above the streets of Manhattan. Unknown to most New Yorkers, the falcons nested on the ledges of skyscrapers and dived on their prey in an impressive display of predation. The falcons disappeared when DDT and other organic pollutants caused a thinning of their eggshells and a failure in reproduction, but they have been reintroduced recently into the city. The reintroduction of peregrine falcons illustrates an important recent trend: the growing understanding that city environments can assist in the conservation of nature, including the conservation of endangered species.

City environments can contribute to conservation of wildlife in a number of ways. Urban kitchen gardens—backyard gardens that provide table vegetables and decorative plants—can be designed to provide habitats. For example, flowers that provide nectar for threatened or endangered hummingbirds can be planted in such gardens. Rivers and their riparian zones, ocean shorelines, and wooded parks can provide habitat for endangered species and ecosystems. For example, prairies, which once occupied more land area than any other vegetation type

in the United States, are rare. One restored prairie exists within the city limits of Omaha, Nebraska. (Some urban nature preserves are not accessible to the public or offer only limited access, as is the case with the prairie preserve in Omaha.)

Urban drainage structures can also be designed as wildlife habitats. A typical urban runoff design depends on concrete-lined ditches that speed the flow of water from city streets to lakes, rivers, or the ocean. However, as with Boston's Back Bay design, discussed earlier, these features can be planned so that stream and marsh habitats are maintained or created, with meandering waterways and storage areas that do not interfere with city processes. Such areas can become habitats for fish and mammals (Figure 27.17). Modified to promote wildlife, cities can provide urban corridors that allow wildlife to migrate, following natural routes of movement in spite of the imposition of the city.[31] Urban corridors also serve to prevent some of the effects of ecological islands and are increasingly important to biological conservation.

Animal Pests

Pests are familiar to urban dwellers. City pests include cockroaches, fleas, termites, rats, and pigeons, but there

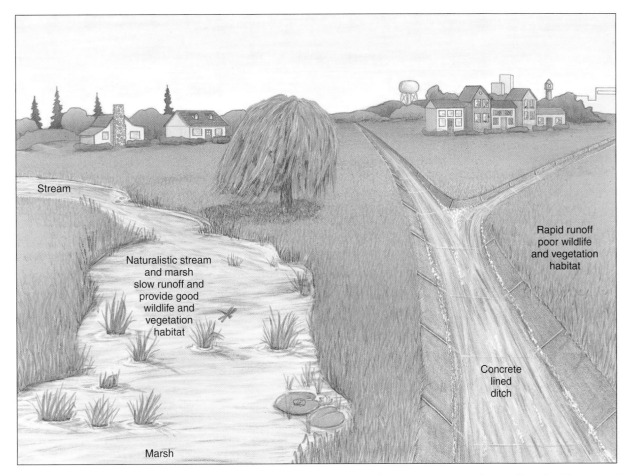

Figure 27.17 • How water drainage systems in a city can be modified to provide wildlife habitat. In the community on the right, concrete-lined ditches result in rapid runoff and have little value to fish and wildlife. In the community on the left, the natural stream and marsh were preserved; water is retained between rains and an excellent habitat is provided. (*Source:* D. L. Leedly and L. W. Adams, 1984, *A Guide to Urban Wildlife Management,* National Institute for Urban Wildlife, Columbia, MD, pp. 20–21.)

are many more, especially species of insects. In gardens and parks, pests include insects, birds, and mammals that feed on fruit and vegetables and destroy foliage of shade trees and plants. Pests compete with people for food and spread diseases. Indeed, before modern sanitation and medicine, such diseases played a major role in limiting human population density in cities. Bubonic plague is spread by fleas found on rodents; mice and rats in cities promoted the spread of the Black Death (see Chapter 1).

An animal is a pest to people when it is in an undesired place at an undesirable time doing an unwanted thing. For example, a termite in a woodland helps in the natural regeneration of wood by hastening decay and speeding return of chemical elements to the soil, where they are available to living plants. But termites in a house are pests because they threaten the physical structure of the house.

Animals that survive best in cities have certain common characteristics. Animals that are urban pests usually are generalists in their food choice, so they can share the diet of people. They have a high reproductive rate and a

short average lifetime.

Controlling Pests

We can best control pests by recognizing how they fit their natural ecosystem and identifying their natural controlling factors. It is often assumed the only way to control animal pests is poisons, but there are limitations to this approach. Early poisons used in pest control were generally toxic to people. Another problem arises with reliance on one toxic compound; through evolution, a species can develop resistance, which can lead to rebound—that is, an increase in the pest once again. If a pesticide is used once and spread widely, it will greatly reduce the population of the pest. However, when the pesticide's effectiveness is lost, the population can increase rapidly as long as habitat is suitable and food plentiful. This situation occurred when an attempt was made to control Norway rats in Baltimore.

The best way to control rats is to make the city a bad habitat for them by reducing the amount of open garbage and eliminating areas to hide and nest. Common access

How Can Urban Sprawl Be Controlled?

As the world becomes increasingly urbanized, individual cities are growing in area as well as population. Residential areas and shopping centers move into undeveloped land near cities, impinging on natural areas and creating a chaotic, unplanned human environment. More than 7.6 million ha of rural land were developed in the United States between 1970 and 1990, a process that is continuing at the rate of 160,000 ha per year. Urban sprawl has become a serious concern in communities all across the United States. In 1998, voters said yes to more than 200 state and local ballot initiatives designed to control urban sprawl.

The city of Boulder, Colorado, has been in the forefront of this effort since 1959, when it created the "blue line"—a line at an elevation of 1,761 m (the city itself is at 1,606 m) above which it would not extend city water or sewer services. Boulder's citizens felt, however, that the blue line was insufficient to control development and maintain its scenic beauty in the face of rapid population growth. (In the decade before 1959, Boulder had grown from a population of 29,000 to 66,000 and would reach 96,000 by 1998.) To prevent uncontrolled development in the area between the city and the blue line, Boulder began in 1967 to use a portion of the city sales tax to purchase land, creating a 10,800 ha greenbelt around the city proper.

In 1976, Boulder went one step further and limited new residences to an increase of 2% a year. Two years later, recognizing that planned development requires a regional approach, the city and surrounding Boulder County adopted a coordinated development plan. By the early 1990s, it had become apparent that further growth control would need to come in the area of nonresidential building. The plan finally adopted by the city reduced the allowable density of many commercial and industrial properties, in effect limiting jobs rather than building space.

The benefits of Boulder's controlled growth initiatives have been a defined urban/rural edge; rational, planned development; protection of sensitive environmental areas and scenic vistas; and large areas of open space within and around the city for recreation. And in spite of its growth control measures, Boulder's economy has remained strong. But with restraints on residential growth, many people who found jobs in Boulder were forced to find affordable housing in adjoining communities. As a result, for example, the population of Superior, Colorado, grew from 225 in 1990 to 6,500 in 1998.

As commuting workers—40,000 a day—tried to get to and from their jobs in Boulder, traffic congestion and air pollution increased. In addition, because developers had not built stores in the outlying areas, shoppers flocked into Boulder's downtown mall. When plans for a competing mall in the suburbs were announced, though, Boulder officials worried about the loss of revenue if the new mall drew shoppers away from the city. At the same time, sprawl from Denver (only 48 km from Boulder), as well as its infamous "brown cloud" of polluted air, began to spill out along the highway connecting the two communities.

Critical Thinking Questions

1. Is a city an open or a closed system (Chapter 3)? Use examples from the case of Boulder to support your answer.
2. As Boulder takes steps to limit growth, it becomes an even more desirable place to live, which subjects it to even greater growth pressures. What ways can you suggest to avoid such a positive feedback loop?
3. Some people in Boulder think that the next step is to increase residential density within the city. How do you think people living there will accept this plan? What advantages and disadvantages are there to increasing density?
4. To some, the story of Boulder is the saga of a heroic battle against commercial interests that would destroy environmental resources and a unique quality of life. To others, it is the story of an elite group building an island of prosperity and the good life for themselves while shifting the more unpleasant aspects of modern life elsewhere. How do you view Boulder's story, and why?

areas used by rats are the spaces between walls and openings between buildings where pipes and cables enter. Houses can be constructed to restrict access by rats. In older buildings, areas of access can be sealed.

Summary

- As an urban society, we must recognize the city's relation to the environment. A city influences and is influenced by its environment and is an environment itself.
- Like any other life-supporting system, a city must maintain a flow of energy, provide necessary material resources, and have ways of removing wastes. These functions are accomplished through transportation and communication with outlying areas.
- Because cities depend on outside resources, they developed only when human ingenuity resulted in

modern agriculture and thus excess food production. The history of cities divides into four stages: (1) rise of towns; (2) era of classic urban centers; (3) industrial metropolises; and (4) age of mass telecommunication, computers, and new forms of travel.

- Locations of cities are strongly influenced by environment. It is clear that cities are not located at random but in places of particular importance and environmental advantage. A city's site and situation are both important.

- A city creates an environment that is different from surrounding areas. Cities change local climate; they are commonly cloudier, warmer, rainier, and less humid than surrounding areas. In general, life in a city is riskier because of higher concentrations of pollutants and pollutant-related diseases.

- Cities favor certain animals and plants. Natural habitats in city parks and preserves will become more important as wilderness decreases.

- Trees are an important part of urban environments. Cities, however, create stresses on trees, and attention must be paid to the condition of urban soils and the supply of water for trees.

- Cities can support the conservation of biological diversity, providing habitat for some rare and endangered species.

- As the human population continues to increase, we can envision two futures: one in which people are dispersed widely throughout the countryside and cities are abandoned except by the poor; and another in which cities attract most of the human population, freeing much landscape for conservation of nature, production of natural resources, and public service functions of ecosystems.

REEXAMINING THEMES AND ISSUES

Human Population:

As the world human population increases, we are becoming an increasingly urbanized species. Present trends indicate that in the future, most citizens of most nations will live in their country's single largest city. Thus, a concern with urban environments will become increasingly important.

Sustainability:

Cities contain the seeds of their own destruction: The very artificiality of a city gives its inhabitants the sense that they are independent of their environment. But the opposite is the case: The more artificial a city, the more it depends on its surrounding environment for resources, and the more susceptible it becomes to major disasters, unless this susceptibility is recognized and planned for. The keys to sustainable cities are an ecosystem approach to urban planning and a concern with the aesthetics of urban environments.

Cities are involved with the sustainability of all renewable resources. Cities greatly affect their surrounding environments. Urban pollution of rivers that flow into an ocean can affect the sustainability of fish and fisheries. Urban sprawl can have destructive effects on endangered habitats and ecosystems, including wetlands. At the same time, cities that are designed to support vegetation and some wildlife can contribute to the sustainability of nature.

Global Perspective:

The great urban centers of the world are having global effects. As an example, because people are concentrated in cities and because many cities are located at the mouths of rivers, most major river estuaries of the world are severely polluted.

Urban World:

The primary message of this chapter is that Earth is becoming urbanized, and environmental science must deal more and more with urban issues.

Science and Values:

Modern environmental sciences tell us much that we can do to improve the environments of cities and the effects of cities on the surrounding environments. What we choose to do with this knowledge depends on our values. Scientific information can suggest new options and choices, and we can select among these for the future of our cities, depending on our values.

A tendency in the 20th century has been to focus environmental conservation efforts on wilderness, large parks, and preserves outside of cities. Meanwhile, city environments have been allowed to decay. As the world becomes increasingly urbanized, a change in values is necessary. The conservation of biological diversity requires that increased value be placed on urban environments. Indirectly, more pleasant urban environments can benefit natural areas outside of cities: The more pleasant city environments are, the more recreation people can find in them, and the less pressure there will be on the countryside.

● KEY TERMS

city planning **582**

fall line **580**

garden city **585**

greenbelt **586**

heat island **586**

made lands **588**

site **579**

situation **579**

● STUDY QUESTIONS

1. How does the environment influence the location of cities?

2. What types of cities are most likely to become ghost towns in the next 30 years? In answering this question, make use of your knowledge of changes in resources, transportation, and communication.

3. Some futurists picture a world that is one giant biospheric city. Is this possible? Under what conditions?

4. Among ancient Greeks, it was said that a city should have no more people than the number that can hear the sound of a single voice. Would you apply this rule today? If not, how would you decide how to plan the size of a city?

5. You are the manager of Central Park in New York City and receive the following offers and requests. Which would you approve? Explain your reasons.

   a. A gift of $1 million to plant trees from all the eastern states.

   b. A gift of $1 million to set aside half the park to be forever untouched, thus producing an urban wilderness.

   c. A gift of the construction of an asphalt jogging track and a gym for physical fitness. The donor says that lack of physical fitness is a major urban health problem.

   d. A request to install an ice skating rink with artificially made ice. Facilities include an elegant restaurant with many views of the park.

6. Your state asks you to locate and plan a new town. The purpose of the town is to house people who will work at a wind farm—a large area of many windmills, all linked to produce electricity. You must first locate the site for the wind farm and then plan the town. How would you proceed? What factors would you take into account?

7. Visit your town center. What changes, if any, would make better use of the environmental location? How could the area be made more livable?

8. In what ways does air travel alter the location of cities? The value of land within a city?

9. You are put in charge of ridding your city's parks of slugs, which eat up the vegetable gardens rented to residents. How would you approach controlling this pest?

10. It is popular to suggest that in the information age people can work at home and live in the suburbs and the countryside, so that cities are no longer necessary. List five arguments for and five arguments against this point of view.

● FURTHER READING

Akbari, H., S. Davis, S. Dorsano, S., J. Huang, and S. Winnett. 1992. *Cooling Our Cities: A Guidebook on Tree Planting and Light-Colored Surfacing.* Washington D.C.: U.S. EPA Office of Policy Analysis. One of the most complete reviews of research on the uses of trees in cities.

Beveridge, C. E., and P. Rocheleau. 1995. *Frederick Law Olmsted: Designing the American Landscape.* New York: Rizzoli International. The most important analysis of the work of the father of landscape architecture.

Bornkamm, R., J. A. Lee, and M. R. D. Seaward. eds. 1982. *Urban Ecology.* Oxford: Blackwell Scientific Publications. A collection of papers presented at the Second European Ecological Symposium,

held in Berlin in 1980. The papers cover a broad range of subjects, from the constitution of urban ecosystems to the impact of human activity in urban areas to the application of ecological knowledge in urban environments.

Burton, J. A. 1977. *Worlds Apart: Nature in the City.* Garden City, N.Y.: Doubleday. An intriguing description of wildlife, including foxes, inhabiting temperate-zone cities, little known to the people in the cities.

Butti, K., and J. Perlin. 1980. *The Golden Thread: 2500 Years of Solar Architecture and Technology.* New York: Cheshire. A history and outline of the use of solar energy from ancient Greece to modern times.

Cronon, W. 1991. *Nature's Metropolis: Chicago and the Great West.* New York: W. W. Norton. A book that discusses the interdependency of a city and its countryside, as exemplified by Chicago.

Duryea, M. L., E. Kampf Binelli, and L. V. Korhnak, eds. 2000. *Restoring the Urban Forest Ecosystem.* Gainesville: School of Forest Resources and Conservation, Florida Cooperative Extension Service, University of Florida.

Gilbert, O. L. 1991. *The Ecology of Urban Habitats.* London: Chapman and Hall.

Hengeveld, H., and C. De Vocht. eds. 1982. *Role of Water in Urban Ecology.* Volume 1 of *Developments in Landscape Management and Urban Planning.* Amsterdam, Oxford, New York: Elsevier Scientific. A publication that addresses the role of water in urban ecology. The contents are from the proceedings of the Second International Environmental Symposium, held in Amsterdam in 1979.

Hise, G., and W. Deverell. 2000. *Eden by Design: The 1930 Olmsted-Bartholomew Plan for the Los Angeles Region.* Berkeley: University of California Press. A reprint, with an introductory analysis of a work first printed in 1898 under the title *To-morrow.* The plan described might have made Los Angeles into a more environmentally sound and humanely pleasant environment than it is today.

Howard, Ebenezer. 1965 (reprint). *Garden Cities of Tomorrow,* Cambridge, Mass.: MIT Press. A classic work of the 19th century that has influenced modern city design, as in Garden City, New Jersey, and Greenbelt, Maryland. It presents a methodology for designing cities with the inclusion of parks, parkways, and private gardens.

Leedly, D. L., and L. W. Adams. 1984. *A Guide to Urban Wildlife Management.* Columbia, Md.: National Institute for Urban Wildlife. The proceedings of the National Symposium on Urban Wildlife, held in Cedar Rapids, Iowa, in 1990. The book presents papers by many authors, covering wildlife conservation, ecology, and management.

Lynch, Kevin. 1968. *The Image of the City.* Cambridge, Mass.: M.I.T. Press. A book explaining the deliberate processes of designing and locating urban activities for maximum ease of use by city dwellers.

Moll, G., and S. Ebernreck. 1992. *Shading Our Cities: A Resource Guide for Urban and Community Forests.* Washington, D.C.: Island Press. An overview of urban forests, including discussions of the history and value of urban trees, followed by examples of programs designed to encourage urban tree planting and types of trees for various requirements.

McHarg, I. L. 1971. *Design with Nature.* Garden City, N.Y.: Doubleday. Examines case studies and presents a methodology for suburban regions based on an understanding of ecological processes.

Nilon, C. H., A. R. Berkowitz, and K. S. Hollweg. 1999. "Understanding Urban Ecosystems: A New Frontier for Science and Education." *Urban Ecosystems* 3(1):3–5.

Spirn, A. W. 1984. *The Granite Garden: Urban Nature and Human Design.* New York: Basic Books. Comprehensive strategies for designing cities in concert with natural processes, focusing on the look and shape of the city rather than on economics or public policy.

CHAPTER 28 Waste Management

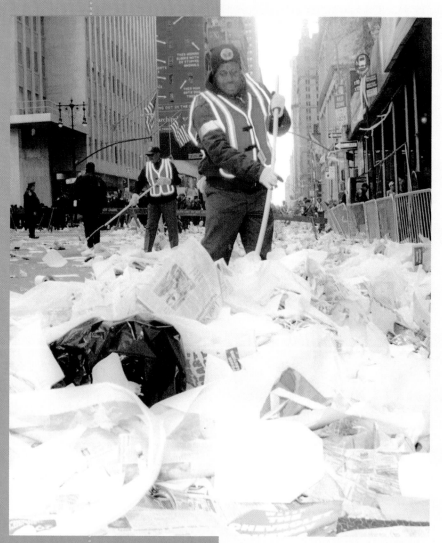

New York City Sanitation crew cleaning up after Yankees' victory parade in 2000.

● LEARNING OBJECTIVES

The waste management concept of "dilute and disperse" (for example, dumping waste into a river) is a holdover from our frontier days, when we mistakenly believed land and water to be limitless resources. We next attempted to "concentrate and contain" waste in disposal sites—a practice that also proved to pollute land, air, and water resources. We are now focusing on managing materials to eliminate waste. Finally, we are getting it right! After reading this chapter, you should understand:

- What are the advantages and disadvantages of each of the major methods that constitute integrated waste management.
- How the physical and hydrologic conditions at a site affect its suitability for a landfill.
- What multiple barriers for landfills are, and how landfill sites can be monitored.
- Why management of hazardous chemical waste is one of our most serious environmental concerns.
- What are the various methods of managing hazardous chemical waste.
- What are the major pathways by which hazardous waste from a disposal site can enter the environment.
- What problems are related to ocean dumping, and why these problems are likely to persist for some time.

Fresh Kills Landfill, New York City

Located on a mixture of wetlands (salt marsh), woodlands, and grasslands on the western shore of Staten Island, the Fresh Kills landfill (Figure 28.1) opened in 1948 and closed in March of 2001. Part of Fresh Kills was reopened following the disastrous terrorist attacks of September 11, 2001 that destroyed the towers of the World Trade Center (WTC). The site may eventually receive several million tons of material from the disaster. The location of the materials taken to the Fresh Kills is sacred ground to some people, and it has been suggested that the site be designated as a memorial to the victims of the attack.

Owned and operated by the New York City Department of Sanitation, the Fresh Kills landfill site comprises approximately 7,500 ha (3,000 acres). At its peak in 1986, the facility received more than 21,000 tons of waste per day. During the 1990s, the flow of waste began to slow down as the city eliminated commercial deliveries and New Yorkers began to recycle returnable bottles, plastic containers, and newspapers. In its final few years, Fresh Kills received between 12,000 and 14,000 tons of solid waste per day at a cost of approximately $44 per ton.[1]

The city negotiated deals to truck garbage out of New York at a cost of $50 to $70 per ton as the landfill entered its final closure stages. Closing the landfill will eventually cost the city of New York more than $1 billion, including the 30 years of monitoring after final closure. Transforming one of the world's largest landfills into an environmentally sound and aesthetically pleasing park or other public place will be a huge task.

A slurry wall (underground concrete barrier) containment system has been constructed to prevent the migration of untreated leachate outside the landfill. (Leachate is noxious, polluted liquid produced when water infiltrates through organic and other waste material.) A leachate collection trench averaging five meters in depth has been placed one meter inside the containment wall, and 32 wells collect leachate for treatment. A leachate treatment center will neutralize over 1 million gallons (3,800 m^3) of leachate per day. Furthermore, stormwater runoff is diverted into retention ponds around the perimeter of each landfill section. Currently, about 150 wells collect 10 million cu ft (283,000 m^3) of methane daily from the two active sections of the landfill. The gas is purified and sold to Union Gas in Brooklyn.

Figure 28.1 • Fresh Kills landfill, closed in 2001, was one of the largest waste disposal sites in the world.

These measures are meant to ensure that the Fresh Kills site following closure will be environmentally safe; however, another (perhaps bigger) challenge is transforming the site into an area that is aesthetically pleasing. Maintaining a "lawn" of grass the size of the entire landfill site was estimated to be too costly at more than $20 million over the 30-year monitoring period, so the city Sanitation department developed a series of test plots and experiments. To re-establish native woodland communities, three sites were planted with about 3,000 shrubs and 523 trees, many rescued from sites on Staten Island scheduled to be developed. A 40 ha meadow was planted next to the woodland. The trees require minimal care, and the plan is to let nature take over. The trees have grown moderately well, and the shrubs are doing well. Many birds came to perch in the trees, further dispersing seeds and thereby adding new species of plants to the site. It was feared that the tree roots might reach down and break the clay cap used to contain water-borne pollutants; however, so far, this has not been a problem at the Fresh Kills site.[2]

Reclamation of landfill sites to their original conditions is an action being attempted at other closed landfills in the United States today. However, preservation of the original land would obviously have been environmentally preferred. Today, at the Staten Island site, for example, a large preserve would be a treasure, as is Central Park in New York City. The need to reclaim sites today emphasizes the failure of our past waste management policy, based on a throwaway mentality.

- *Fresh Kills represents a tremendous financial burden to society and reminds us that we have failed in the past 50 years to move from a throwaway, waste-oriented society to a society that sustains natural resources through improved materials management. We are now moving in that direction by producing less waste and recycling more. With this in mind, we introduce in this chapter concepts of waste management applied to urban waste, hazardous chemical waste, and waste in the marine environment.*

28.1 Early Concepts of Waste Disposal

During the first century of the Industrial Revolution, the volume of waste produced in the United States was relatively small, and waste could be managed through the concept of "dilute and disperse." Factories were located near rivers because the water provided a number of benefits, including easy transport of materials by boat, sufficient water for processing and cooling, and easy disposal of waste into the river. With few factories and a sparse population, dilute and disperse was sufficient to remove the waste from the immediate environment.[3]

As industrial and urban areas expanded, the concept of dilute and disperse became inadequate, and a new concept, known as "concentrate and contain," came into use. It has become apparent, however, that containment was, and is, not always achieved. Containers, whether simple trenches excavated in the earth or metal drums and tanks, may leak or break and allow waste to escape. Health hazards resulting from past waste disposal practices have led to the present situation, in which many people have little confidence in government or industry to preserve and protect public health.[4]

The problem of waste disposal in New York, presented in the opening case, is not unique. In the United States, as well as in many other parts of the world, people are facing a serious solid-waste disposal problem. The problem exists because we are producing a great deal of waste and the acceptable space for permanent disposal is limited. It has been estimated that within the next few years approximately half the cities in the United States may run out of landfill space. Philadelphia, for example, is essentially out of landfill space now and is bargaining with other states on a monthly or yearly basis to dispose of its trash; the Los Angeles area has landfill space for only about 10 years.

To say we are actually running out of space for landfills isn't altogether accurate. Land used for landfills is minute compared to the land area of the United States. Rather, existing sites are being filled, and it is difficult to site new landfills. After all, no one wants to live near a waste disposal site, be it a sanitary landfill for municipal waste, an incinerator that burns urban waste, or a hazardous-waste disposal operation for chemical materials. This attitude is widely known as NIMBY ("not in my backyard").

Another major limiting factor is the cost of disposal. A decade or so ago, the cost of disposal of 1 metric ton of urban refuse was approximately $5 to $10. Today, the average cost is about $32; and some cities, such as Philadelphia, pay as much as $75 per metric ton for waste disposal.[4,5] These costs are only a small part of waste management expenditures. Disposal or treatment of liquid and solid waste costs about $20 billion every year. It is one of our most costly environmental expenditures.[6]

28.2 Modern Trends

The environmentally correct concept with respect to waste management is to consider wastes as resources out of place. Although we may not soon be able to reuse and recycle all waste, it seems apparent that the increasing cost of raw materials, energy, transportation, and land will make it financially feasible to reuse and recycle more resources. Moving toward this objective is moving toward an environmental view that there is no such thing as waste. Under this concept, waste would not exist because it would not be produced or, if produced, would be a resource to be used again. This concept is referred to as the "zero waste" movement.

Zero waste is the essence of what is known as **industrial ecology**, the study of relationships among industrial systems and their links to natural systems. Under the principles of industrial ecology, our industrial society would function much as a natural ecosystem functions. Waste from one part of the system would be a resource for another part (see A Closer Look 28.1).

The concept of zero waste production was until recently considered unreasonable in the waste management arena. The concept, however, is catching on. The city of Canberra, Australia, may be the first community to propose a plan to have zero waste, a goal it hopes to reach by 2010. Thousands of kilometers away, in the Netherlands, a national waste reduction goal of 70–90% has been set. How this is to be accomplished is not entirely clear; but a large part of the planning involves taxation of waste in all its various forms, from emissions from smokestacks to solids delivered to landfills. Already in the Netherlands, discharges of heavy metal to waterways have been nearly eliminated by implementation of pollution taxes. At the household level, the government is considering programs—known as "pay as you throw"—in which people are charged by the volume of waste they produce. As disposal of waste, including household waste, is taxed, people produce less waste.[8]

Of particular importance to waste management is the growing awareness that many of our waste management programs involve moving waste from one site to another and not really managing it. For example, waste from urban areas may be placed in landfills; but eventually, these landfills may cause new problems by producing methane gas or noxious liquids that leak from the site and contaminate the surrounding areas. Managed properly, however, methane produced from landfills is a resource that can be burned as a fuel (an example of industrial ecology).

Concepts involving our use of materials and the waste we produce are changing. Previous notions of waste disposal are no longer acceptable, and we are rethinking how we deal with materials, with the objective of eliminating the concept of waste entirely. In this way, we can reduce the consumption of virgin materials, which depletes our environment, and can live within our environment more sustainably.[8]

28.3 Integrated Waste Management

The dominant concept today in managing waste is known as **integrated waste management (IWM)**, which is best defined as a set of management alternatives that includes *reuse, source reduction, recycling, composting, landfill*, and *incineration*.[4]

Reduce, Reuse, Recycle

The three R's of IWM are **reduce, reuse**, and **recycle**. The ultimate objective of the three R's is to reduce the amount of urban and other waste that must be disposed of in landfills, incinerators, and other waste management facilities. Study of the *waste stream* (the waste produced) in areas that utilize IWM technology suggests that the amount (by weight) of urban refuse disposed of in landfills or incinerated can be reduced by at least 50% and perhaps as much as 70%. A 50% reduction by weight of urban waste could be facilitated by:[4]

- Better design of packaging to reduce waste, an element of source reduction (10% reduction).
- Establishment of recycling programs (30% reduction).
- Large-scale composting programs (10% reduction).

As this list indicates, recycling is a major player in the reduction of the urban waste stream. Can recycling in fact reduce the waste stream by 50%? Recent work suggests that the 50% goal is reasonable. In fact, it has been reached in some parts of the United States. The potential upper limit for recycling is considerably higher. It is estimated that as much as 80–90% of the U.S. waste stream might be recovered through what is known as intensive recycling.[9] A pilot study involving 100 families in East Hampton, New York, achieved a level of 84%. More realistic for many communities is partial recycling, which targets a specified number of materials, such as glass, aluminum cans, plastic,

Industrial Ecology

As noted in the text, industrial ecology can be defined as the study of relationships among industrial systems and their linkages to natural systems. Industrial ecology and sustainable development are often linked. In part, this results from the analogy drawn between industrial ecology and natural ecology, with its focus on ecosystems. Just as sustaining life on Earth is a function of ecosystems, sustainability in our industrial society is a function of industrial ecology.

To understand the concept of waste according to industrial ecology, consider waste in natural ecosystems. What might be thought of as waste in one part of an ecosystem is often a resource for another part. For example, the elephant eats and produces waste that becomes a resource for the dung beetle.

Now consider an example of industrial ecology. Suppose that there exists a coal-burning power plant that produces electricity for a town. Waste from producing the power includes ash from the coal, exhaust heat, and products from combustion that might go out a smokestack, including carbon dioxide, sulfur dioxide, and heat. For our hypothetical power plant and surrounding town, let's assume that the waste heat is used to warm homes and provide heat for industrial activities. Sulfur dioxide is removed from the system by scrubbing to produce gypsum, which is the major constituent of wallboard used in construction. Carbon dioxide is sequestered and used in local greenhouses along with waste heat to force and prolong the growing cycle of plants. We are left with the ash from the coal, which is used for road surfacing.

This hypothetical scenario is being partly played out at several power plants today. It is expected that, as principles of industrial ecology are better understood and applied, we will see more applications in a variety of industries, as well as in agriculture.

As we attempt to produce sustainable economic development on local, regional, and global scales, however, we need to acknowledge that industrial ecology based on application of science and technology alone is not adequate to achieve sustainability of global systems. We also need to clarify our values. Science can provide potential solutions to problems we face; however, which solutions we choose will reflect our values. That is one of the key themes of this book.

organic material, and newsprint. Partial recycling can provide a significant reduction, and in many places it is approaching or even exceeding 50%.[10,11]

Public Support for Recycling

An encouraging sign associated with public support for the environment is an increase in the willingness of industry and business to support recycling on a variety of scales. For example, fast-food restaurants are using less packaging for their products and providing on-site bins for recycling used paper and plastic. Grocery stores are encouraging the recycling of plastic and paper bags by providing bins for their collection and recycling. Some food stores offer inexpensive canvas shopping bags to people who prefer them over disposable plastic and paper bags. Companies are redesigning products so that they can be more easily disassembled after use and the various parts recycled. As this concept catches on, small appliances, such as electric frying pans and toasters, may be recycled rather than ending up in landfills. The automobile industry is also responding by designing automobiles with coded parts so that they can be more easily disassembled (by professional recyclers) and recycled, rather than left to become rusting eyesores in junkyards.

On the consumer front, people are now more likely to purchase products that can be recycled or that come in containers that are more easily recycled or composted. Many consumers have purchased small home appliances that crush bottles and aluminum cans, reducing their volume and facilitating recycling. The entire area is rapidly changing, and innovations and opportunities will undoubtedly continue.

Markets for Recycled Products

As with many other environmental solutions, implementing the IWM concept successfully can be a complex undertaking. In some communities where recycling has been successfully implemented, it has resulted in glutted markets for the recycled products, which has sometimes required temporary stockpiling or suspension of recycling of some items. It is apparent that if recycling is to be successful, markets and processing facilities will also have to be developed to ensure that recycling is a sound financial venture as well as an important part of IWM.

Recycling of Human Waste

The use of human waste or "night soil" on crop lands is an ancient practice. In Asia, recycling of human waste has a long history. Chinese agriculture was sustained for thousands of years through collection of human waste, which was spread over agricultural fields. The practice also spread, and by the early 20th century, land application of sewage was one of the primary methods of disposal in many metropolitan areas in countries including Mexico, Australia, and the United States.[12] These early uses of human waste for agriculture occasionally spread infectious diseases through agents, including bacteria, viruses, and parasites, applied to crops along with the waste. Today, with the globalization of agriculture, we still see occasional warnings of fruits and vegetables contaminated in this way.

One of the major problems of recycling human waste is that, along with human waste, thousands of chemicals and metals flow through our modern waste stream. Even garden waste that is composted may contain harmful chemicals such as pesticides.

Heavy metals, petroleum products, industrial solvents, and pesticides may end up in our wastewater collection systems and sewage treatment plants. Because many toxic materials are likely to be present with the waste, we must be very skeptical of utilizing sewage sludge for land application. Of course, the contents of sewage sludge vary from place to place and even from day to day. Nevertheless, studies have shown that high levels of toxic chemicals may be present in the sludge of cities or towns with industries that use toxic materials.[12] The good news is that fewer toxic materials end up at sewage treatment plants than several decades ago because many industries are now pretreating their waste to remove materials that previously contaminated wastewaters.

Discussions involving various federal, local, and other government agencies, as well as industries, have addressed the question of how much toxic material in the waste stream constitutes a problem. This is really not the correct question to ask, however. The goal should be that sewage sludge contain no toxic materials at all. A problem is that the sewer lines from urban homes are the same ones utilized by industry. As a result, conventional waste disposal technology is unlikely to produce a form of sludge that is safe for us and other living things. A possible solution is to separate urban waste from industrial waste. A second possibility is to pretreat waste from industrial sources to remove hazardous components before they enter the wastewater stream. Many industries today pretreat waste, as mentioned earlier. Finally, some communities are considering smaller wastewater treatment facilities intended to treat waste from homes; the recycled waste would be used locally by farms. In the future, as the cost of oil, which is necessary to produce fertilizers, continues to rise, the age-old practice of recycling human waste may again be financially viable and necessary in many more places than today.[12]

28.4 Materials Management

The recycling option of IWM has been seriously attempted for over two decades. Recycling has been responsible for generating entire systems of waste management that have produced tens of thousands of jobs in the United States and reduced the amount of urban waste sent from homes to landfills from 90% in the 1980s to about 65% today. Many firms have combined waste reduction with recycling to reduce by 50% to 90% the waste they deliver to landfills. However, in spite of this success, IWM has been criticized for overemphasizing recycling and failing to effectively advance policies to prevent waste production.

Futuristic waste management has the goal of zero production of waste, consistent with the ideals of industrial ecology. This visionary goal will require more sustainable use of materials combined with resource conservation in what is being termed **materials management**. It is believed that the goal could be pursued in the following ways:[13]

- Eliminate subsidies for extraction of virgin materials such as minerals, oil, and timber.
- Establish "green building" incentives that encourage the use of recycled-content materials and products in new construction.
- Assess financial penalties for production that uses negative materials management practices.
- Provide financial incentives for industrial practices and products that benefit the environment by enhancing sustainability (for example, reducing waste production and using recycled materials).
- Increase the number of new jobs in the technology of reuse and recycling of resources.

Materials management in the United States today is beginning to have effects on where industries are located. For example, approximately 50% of the steel produced in the country now comes from scrap. As a result, new steel mills are no longer located where resources such as coal and iron ore are close by. New steel mills are located in a variety of places, from California to North Carolina and Nebraska; and their resource is the local supply of scrap steel. Because they are starting with scrap metal, the new industrial facilities use far less energy and cause much less pollution than older steel mills that produce their product from virgin iron ore.[14] Similarly, recycling of paper is changing where new paper mills are constructed. In the past, mills were built near forested areas where timber necessary for paper production was being logged. As a result of the recycling of voluminous amounts of paper, mills are now being built near cities where supplies of recycled paper exist. For example, in the state of New Jersey there are 13 paper mills utilizing recycled paper and eight steel "mini-mills" producing steel from scrap metal. What is remarkable is that New Jersey has little forested land and no iron mines! Resources for the paper and steel mills come from materials already in use and exemplify the power of materials management.[14]

Although materials management is providing alternatives to waste disposal, we still need to deal with the solid waste we produce in both urban and rural areas. We now discuss solid-waste management in more detail.

28.5 Solid-Waste Management

Solid-waste management continues to be a problem in the United States as well as other parts of the world. In many areas, particularly in developing countries, waste management practices are inadequate. These practices, which include poorly controlled open dumps and illegal roadside dumping, can spoil scenic resources, pollute soil and water resources, and produce potential health hazards.

Illegal dumping is a social problem as much as a physical one, because many people are simply disposing of their waste as inexpensively and as quickly as possible. They may not see dumping their garbage or trash as an environmental problem. If nothing else, this is a tremendous waste of resources; much of what is dumped could be recycled or reused. In areas where illegal dumping has been reduced, the keys have been awareness, education, and alternatives. Environmental problems of unsafe, unsanitary dumping of waste are made known to people through education programs, and funds are provided for cleanup and inexpensive collection and recycling of trash at sites of origin.

We look next at the composition of solid waste in the United States. Then, in the remainder of this section we describe specific disposal methods: on-site disposal, composting, conservation, open dumps, and sanitary landfills.

Composition of Solid Waste

The general composition of solid waste likely to end up at a disposal site in the United States is shown in Figure 28.2. It is no surprise that paper is by far the most abundant of these solid wastes; however, this is only an average content, and considerable variation can be expected based on factors such as land use, economic base, industrial activity, climate, and season of the year.

Infectious wastes from hospitals and clinics sometimes make their way to disposal sites, where they can create potential health problems if they have not been properly sterilized before disposal. Some hospitals have facilities to incinerate such wastes, and incineration is probably the surest way to manage infectious medical waste. In addition, in urban areas, a large amount of toxic material may end up at disposal sites; as a result, many older urban landfills are now being considered hazardous waste sites that will require costly cleanup.

People have many misconceptions about our waste stream.[15] There has been much publicity concerning urban waste associated with fast-food packaging, polystyrene foam, and disposable diapers. Therefore, many

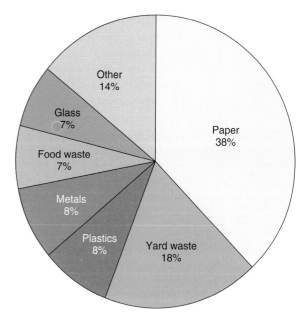

Figure 28.2 • Composition of U.S. urban solid waste (by weight), 1998. (*Source:* U.S. Environmental Protection Agency, Office of Solid Waste. Accessed October 9, 1998, at http://www.epa.gov/.)

people assume that these products make up a large percentage of the total waste stream and are responsible for the rapid filling of landfills. However, excavations into modern landfills using archeological tools have cleared up some misconceptions concerning these items. We now know that fast-food packaging accounts for only about 0.25% of the average landfill; disposable diapers, approximately 0.8%; and polystyrene products, about 0.9%.[16] Paper, as mentioned, is the major constituent in landfills (perhaps as much as 50% by volume and 40% by weight). The largest single item is newspaper, which accounts for as much as 18% by volume.[15] Newspaper is one of the major items targeted for recycling because big environmental dividends can be expected. However (and this is a value judgment), the need to deal with the major contributors does not mean that we need not reduce our use of disposable diapers, polystyrene, and other paper products. In addition to creating a need for disposal, these products are made from resources that might be better managed.

On-Site Disposal

We turn now to specific methods of solid-waste disposal. A common on-site disposal method in urban areas is the mechanical grinding of kitchen food waste. Garbage-disposal devices are installed in the wastewater pipe system at the kitchen sink, and the garbage is ground and flushed into the sewer system. This effectively reduces the amount of handling and quickly removes food waste. Final disposal is transferred to sewage treatment plants, where solids remaining as sewage sludge still must be disposed of.[17]

Composting

Composting is a biochemical process in which organic materials such as lawn clippings and kitchen scraps decompose to a rich, soil-like material. The process involves rapid partial decomposition of moist solid organic waste by aerobic organisms. Although simple backyard compost piles may come to mind, as a waste management option, large-scale composting is generally carried out in the controlled environment of mechanical digesters.[17] This technique is popular in Europe and Asia, where intense farming creates a demand for the compost.[17] A major drawback of composting is the necessity to separate organic material from other waste. Therefore, it is probably economically advantageous only where organic material is collected separately from other waste. Composting plant debris previously treated with herbicides may produce a compost toxic to some plants. Nevertheless, composting is an important component of IWM, and its contribution will undoubtedly grow in the future.

Incineration

In **incineration**, combustible waste is burned at temperatures high enough ($900-1,000°C$, or $1,650-1,830°F$) to consume all combustible material, leaving only ash and noncombustibles to dispose of in a landfill. Under ideal conditions, incineration may reduce the volume of waste by 75% to 95%.[17] In practice, however, the actual decrease in volume is closer to 50% because of maintenance problems as well as waste supply problems. Besides reducing a large volume of combustible waste to a much smaller volume of ash, incineration has another advantage in that the process of incineration can be used to supplement other fuels and generate electrical power.

Incineration of urban waste is not necessarily a clean process. Incineration may produce air pollution and toxic ash. For example, incineration in the United States apparently is a significant source of environmental dioxin, a carcinogenic toxin (see Chapter 15).[18] Smokestacks from incinerators may emit oxides of nitrogen and sulfur, which lead to acid rain; heavy metals such as lead, cadmium, and mercury; and carbon dioxide, which is related to global warming.

In modern incineration facilities, smokestacks are fitted with special devices to trap pollutants; but the process of pollutant abatement is expensive. Furthermore, the plants themselves are expensive, and government subsidization may be needed to aid in their establishment. Recent evaluation of the urban waste stream suggests that with an investment of $8 billion, a sufficient number of incinerators could be constructed in the United States to burn approximately 25% of the solid waste that is generated. However, a similar investment in source reduction, recycling, and composting could result in diversion from landfills of as much as 75% of the nation's urban waste stream.[9]

The economic viability of incinerators depends on revenue from the sale of the energy produced by burning the waste. As recycling and composting are increased, they will compete with incineration for their portion of the waste stream, and sufficient waste (fuel) to generate a profit from incineration may not be available. The main conclusion that can be drawn based on IWM principles is that a combination of reusing, recycling, and composting could reduce the volume of waste requiring disposal at a landfill by at least as much as incineration.[9]

Open Dumps

In the past, solid waste was often disposed of in open dumps, where the refuse was piled up without being covered or otherwise protected. Thousands of open dumps have been closed in recent years, and new open dumps are banned in the United States and many other countries. Nevertheless, many are still being used worldwide (Figure 28.3).

Dumps have been located wherever land is available, without regard to safety, health hazards, or aesthetic degradation. Common sites are abandoned mines and quarries, where gravel and stone have been removed (sometimes by ancient civilizations); natural low areas, such as swamps or floodplains; and hillside areas above or below towns. The waste is often piled as high as equipment allows. In some instances, the refuse is ignited and allowed to burn. In others, the refuse is periodically leveled and compacted.[17]

Figure 28.3 • Urban garbage dump in Rio de Janeiro, Brazil. At this site, people are going through the waste and recycling materials that can be reused or resold.

As a general rule, open dumps create a nuisance by being unsightly, providing breeding grounds for pests, creating a health hazard, polluting the air, and sometimes polluting groundwater and surface water. Fortunately, open dumps are giving way to the better planned and managed sanitary landfills.

Sanitary Landfills

A **sanitary landfill** is designed to concentrate and contain refuse without creating a nuisance or hazard to public health or safety. The idea is to confine the waste to the smallest practical area, reduce it to the smallest practical volume, and cover it with a layer of compacted soil at the end of each day of operation or more frequently if necessary. Covering the waste is what makes the landfill sanitary. The compacted layer restricts (but does not eliminate) continued access to the waste by insects, rodents, and other animals, such as seagulls. It also isolates the refuse, minimizing the amount of surface water seeping into and gas escaping from the waste.[19]

Leachate

The most significant hazard from a sanitary landfill is pollution of groundwater or surface water. If waste buried in a landfill comes into contact with water percolating down from the surface or with groundwater moving laterally through the refuse, **leachate**—noxious, mineralized liquid capable of transporting bacterial pollutants—is produced.[20] For example, two landfills dating from the 1930s and 1940s in Long Island, New York, have produced subsurface leachate trails (plumes) several hundred meters wide that have migrated kilometers from the disposal site. The nature and strength of the leachate produced at a disposal site depends on the composition of the waste, the amount of water that infiltrates or moves through the waste, and the length of time that infiltrated water is in contact with the refuse.[17]

Site Selection

The siting of a sanitary landfill is very important. A number of factors must be taken into consideration, including topography, location of the groundwater table, amount of precipitation, type of soil and rock, and location of the disposal zone in the surface-water and groundwater flow system. A favorable combination of climatic, hydrologic, and geologic conditions helps to ensure reasonable safety in containing the waste and its leachate.[21]

The best sites are in arid regions, where disposal conditions are relatively safe because little leachate is produced. In a humid environment, some leachate is always produced; therefore, an acceptable level of leachate production must be established to determine the most favorable sites in such environments. What is acceptable varies with local water use, regulations, and the ability of the

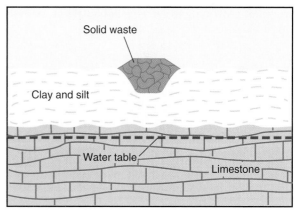

Figure 28.4 • The most desirable landfill site in a humid environment. Waste is buried above the water table in a relatively impermeable environment. (*Source:* W. J. Schneider, *Hydraulic Implications of Solid-Waste Disposal*, U.S. Geological Survey Circular 601F, 1970.)

natural hydrologic system to disperse, dilute, and otherwise degrade the leachate to harmless levels.

Elements of the most desirable site in a humid climate with moderate to abundant precipitation are shown in Figure 28.4. The waste is buried above the water table in relatively impermeable clay and silt soils, material through which water cannot easily move. Any leachate produced remains in the vicinity of the site and degrades by natural filtering action and chemical reactions between clay and leachate.[22,23]

The siting of waste disposal facilities also involves important social considerations. Often, planners choose sites where they expect local resistance to be minimal or where they perceive land to have little value. Waste disposal facilities are frequently located in areas where residents tend to have low socioeconomic status or belong to a particular racial or ethnic group. The study of social issues in siting waste facilities, chemical plants, and other such facilities is an emerging field known as **environmental justice**.[24] (See A Closer Look 28.2.)

Monitoring Pollution in Sanitary Landfills

Once a site is chosen for a sanitary landfill and before filling starts, monitoring the movement of groundwater should begin. **Monitoring** is accomplished by periodically taking samples of water and gas from specially designed monitoring wells. Monitoring the movement of leachate and gases should continue as long as there is any possibility of pollution. This procedure is particularly important after the site is completely filled and a final, permanent cover material is in place. Continued monitoring is necessary because a certain amount of settlement always occurs after a landfill is completed; and if small depressions form, surface water may collect, infiltrate, and produce leachate. Monitoring and proper maintenance of an abandoned landfill reduce its pollution potential.[19]

How Pollutants Can Enter the Environment from Sanitary Landfills

How Pollutants Can Enter
the Environment from Sanitary Landfills

Pollutants from a solid-waste disposal site can enter the environment by as many as eight paths (Figure 28.5):[26]

1. Methane, ammonia, hydrogen sulfide, and nitrogen gases can be produced from compounds in the soil and the waste and can enter the atmosphere.

2. Heavy metals, such as lead, chromium, and iron, can be retained in the soil.

3. Soluble materials, such as chloride, nitrate, and sulfate, can readily pass through the waste and soil to the groundwater system.

4. Overland runoff can pick up leachate and transport it into streams and rivers.

5. Some plants (including crops) growing in the disposal area can selectively take up heavy metals and other toxic materials. These materials are passed up the food chain as people and animals eat the plants.

6. If plant residue from crops left in fields contains toxic substances, these substances return to the soil.

7. Streams and rivers may become contaminated by waste from groundwater seeping into the channel (3) or by surface runoff (4).

8. Toxic materials can be transported to other areas by the wind.

① Atmosphere	⑤ Removal in crops
② Retention in soil	⑥ Plant residue
③ Groundwater	⑦ Stream flow
④ Surface runoff from land	⑧ Wind

Compacted cells of waste

Leachate

Groundwater

Hazardous waste pollutant

Figure 28.5 • Idealized diagram showing eight paths that pollutants from a sanitary landfill site may follow to enter the environment.

Environmental Justice: Demographics of Hazardous Waste

An emerging field in the social sciences known as environmental justice focuses on investigating social issues related to the siting of facilities that many people find objectionable on environmental grounds. These facilities include sanitary landfills and other waste facilities, as well as chemical plants; and people's concerns include risk to nearby residents in case of accidental spills, fires, explosions, or illegal discharge of waste or chemicals. In particular, environmental justice addresses the potential for inequitable impacts from the environmental hazards related to the waste facility and chemical industries.[24]

A study of 82 permitted hazardous-waste treatment, storage, and disposal facilities in Los Angeles County, California, concluded that the communities most likely to contain such facilities are those located near industrial activity. These communities generally have a large residential concentration of working-class people of color. It appears that sites are chosen where local resistance is expected to be minimal or where it is perceived that the land has little value.

By and large, people living near hazardous waste facilities in Los Angeles County are members of racial and ethnic minorities below the county average in terms of income, education, employment, and voting participation. These communities also generally lag behind the county mean in monetary values of residential homes and rental properties. In addition, industrial land use in these areas is often four to ten times the county average.

In summary, the study identified two dominant factors in location of waste facilities: the percentage of minority people living in the area and industrial land use. Of secondary importance was per capita income. Race, then, appears to be a more important factor than income in locations of waste facilities. These findings in Los Angeles County suggest that people fighting for environmental justice have real cause for concern.[25]

Modern sanitary landfills are engineered to include *multiple barriers*: clay and plastic liners to limit the movement of leachate; surface and subsurface drainage to collect leachate; systems to collect methane gas produced as waste decomposes; and groundwater monitoring to detect leaks of leachate below and adjacent to the landfill. A thorough monitoring program considers all eight possible paths by which pollutants enter the environment. In practice, however, seldom does monitoring include all pathways. It is particularly important to monitor the zone above the water table to identify potential pollution problems before they reach and contaminate groundwater resources, where correction would be very expensive. Figure 28.6 shows an idealized diagram of a landfill that utilizes the multiple-barrier approach and a photograph of a landfill site under construction.

Federal Legislation for Sanitary Landfills

New landfills opened in the United States after 1993 must comply with stricter requirements under the Resource Conservation and Recovery Act of 1980. The legislation is intended to strengthen and standardize design, operation, and monitoring of sanitary landfills. Landfills that cannot comply with regulations face closure.

States may choose between two options:

1. Comply with federal standards.
2. Seek EPA approval of solid-waste management plans, which allows greater flexibility.

Provisions of federal standards include the following:

- Landfills may not be sited on floodplains, wetlands, earthquake zones, unstable land, or airports (birds at sites are a hazard to aircraft).
- Landfills must have liners.
- Landfills must have a leachate collection system.
- Landfill operators must monitor groundwater for many specified toxic chemicals.
- Landfill operators must meet financial assurance criteria to ensure that monitoring continues for 30 years after closure of the landfill.

As mentioned, a state with EPA approval of its landfill program is allowed more flexibility:

- Groundwater monitoring may be suspended if the landfill operator can demonstrate that hazardous constituents are not migrating from the landfill.
- Alternative types of daily cover over the waste may be used.
- Alternative groundwater protection standards are allowed.
- Alternative schedules for documentation of groundwater monitoring are allowed.
- Under certain circumstances, landfills in wetlands and fault zones are allowed.
- Alternative financial assurance mechanisms are allowed.

(a)

(b)

Figure 28.6 • (*a*) Idealized diagram of a solid-waste facility (sanitary landfill) illustrating multiple-barrier design, monitoring system, and leachate collection system. (*b*) Rock Creek landfill, Calaveras County, California, under construction. This municipal solid-waste landfill is underlain by a compacted clay liner (exposed light brown slope in the center left portion of the photograph). The black slopes, covered with gravel piles, overlie the compacted clay layer. These form a vapor barrier designed to keep moisture in the clay and help avoid cracking of the clay liner. The sinuous gray trench is lined with plastic and is part of the leachate collection system for the landfill. The excavated squared pond (upper part of photograph) is a leachate evaporation pond under construction. The landfill is also equipped with a system to monitor the vadose zone below the leachate collection system. (*Source:* Photograph courtesy of John Kramer.)

Given the added flexibility, it appears advantageous for states to develop EPA-approved waste management plans.

28.6 Hazardous Waste

So far in this chapter, we have discussed integrated waste management and materials management for the everyday waste stream from homes and businesses. We now consider the important topic of hazardous waste.

Creation of new chemical compounds has proliferated in recent years. In the United States, approximately 1,000 new chemicals are marketed each year, and about 70,000 chemicals are currently on the market. Although many of these chemicals have been beneficial to people, approximately 35,000 chemicals used in the United States are classified as definitely or potentially hazardous to the health of people or ecosystems (see Table 28.1).

The United States currently produces about 700 million metric tons of hazardous chemical waste per year, referred to more commonly as **hazardous waste**. About 70% of the total is generated east of the Mississippi River, and about half of the total by weight is generated by chemical products industries, with the electronics indus-

Love Canal

The story of Love Canal has become a well-known hazardous-waste horror story. In 1976, in a residential area near Niagara Falls, New York, trees and gardens began to die (Figure 28.7). Rubber on tennis shoes and bicycle tires disintegrated. Puddles of toxic substances began to ooze through the soil. A swimming pool popped from its foundation and floated in a bath of chemicals.

The story of Love Canal started in 1892 when a canal 8 km (5 mi) long was excavated by William Love as part of the development of an industrial park. The development didn't need the canal when inexpensive electricity arrived, and it was never completed. The canal remained unused for decades and became a dump for wastes. From 1920 to 1952, 20,000 tons of more than 80 chemicals were dumped in the canal. In 1953, the Hooker Chemical Company—which produced the insecticide DDT as well as an herbicide and chlorinated solvents and which had dumped chemicals into the canal—was pressured to donate the land to the city of Niagara Falls for $1. The city knew that chemical wastes were buried there, but no one expected any problems.[27] Eventually, several hundred homes and an elementary school were built on and near the site. For years, everything seemed fine. Then, in 1976–1977, heavy rains and snows triggered the events listed in the opening of this story, making Love Canal a household word.

A study of the site identified many substances suspected of being carcinogens, including benzene, dioxin, dichlorethylene, and chloroform. Although officials admitted that little was known about the impact of these chemicals, there was grave concern for people living in the area. Eventually, concern centered on alleged high rates of miscarriages, blood and liver abnormalities, birth defects, and chromosome damage. Although a study by New York health authorities suggested that no chemically caused health effects have

been absolutely established,[29–31] the decision was made to clean up the site.

Cleanup of Love Canal demonstrates the technology for hazardous-waste treatment. The objective was to contain waste, stop migration of wastes through the groundwater flow system, and remove and treat contaminated soil and sediment in streambeds and sewers.[32] To minimize further contamination, the area was covered with 1 m (3.3 ft) of compacted clay and a polyethylene plastic cover to reduce infiltration of surface water. Water is inhibited from entering the site by specially designed barriers. These procedures greatly reduce subsurface seepage of water, and water that does seep out is collected and treated.[29–32]

By 1990, $275 million had been spent for cleanup and relocation projects at Love Canal. Homes in an adjacent area

were purchased, and about 200 homes and a school had to be destroyed by the government. The U.S. Environmental Protection Agency eventually declared the area clean, and about 280 remaining homes were marketed for resale.

As the result of a court-mediated settlement in 1994, the company responsible for the chemical waste paid the state of New York $98 million and took responsibility for $25 million for treatment operations.[33] In 1995, the company agreed to pay an additional $129 million to the federal government as reimbursement for its costs.[34]

The story of Love Canal raises many questions. What went wrong? How can we avoid such disasters in the future? Any how many other Love Canals are there across the country, each a potential time bomb waiting to explode?[29,30]

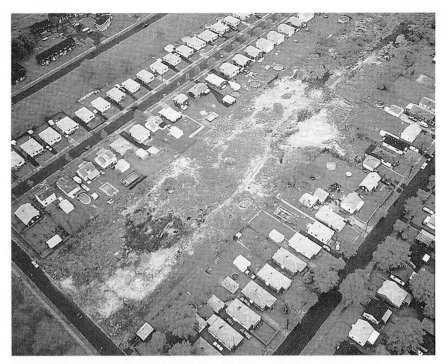

Figure 28.7 ● Aerial infrared photograph of the Love Canal area in New York state. Healthy vegetation is bright red. The canal, running from the upper left to the lower right, is a scar on the landscape where hazardous chemical waste rose to the surface, making *Love Canal* a household name for hazardous-waste problems. (*Source:* Photograph courtesy of New York State Department of Environmental Conservation.)

Table 28.1   Products and the Potentially Hazardous Waste They Generate

Products We Use	Potentially Hazardous Waste
Leather	Heavy metals, organic solvents
Medicines	Organic solvents and residues, heavy metals (e.g., mercury and zinc)
Metals	Heavy metals, fluorides, cyanides, acid and alkaline cleaners, solvents, pigments
Oil, gasoline, and other petroleum products	Oil, phenols and other organic compounds, heavy metals, ammonia salts, acids
Paints	Heavy metals, pigments, solvents, organic residues
Pesticides	Organic chlorine compounds, organic phosphate compounds
Plastics	Organic chlorine compounds
Textiles	Heavy metals, dyes, organic chlorine compounds, solvents

Source: U.S. Environmental Protection Agency, SW-826, 1980.

try and petroleum and coal products industries each contributing about 10%.[27]

Another source of hazardous chemicals is buildings destroyed by events such as fires and hurricanes. Chemicals, such as paints, solvents, and pesticides, that were stored in the buildings may be released into the environment when debris from damaged buildings is burned or buried. As a result, collection of potentially hazardous chemicals following natural disasters is an important goal in managing hazardous materials.

In the recent past, as much as half the total volume of hazardous waste produced in the United States was indiscriminately dumped.[28] Some hazardous waste was illegally dumped on public or private lands, a practice called "midnight dumping." Buried drums of hazardous waste from illegal dumping have been discovered at hundreds of sites by contractors constructing buildings and roads. Cleanup of the waste has been costly and has delayed projects (see A Closer Look 28.3).[27]

In the United States, there are an estimated 32,000–50,000, often now abandoned, waste disposal sites where past dumping was totally unregulated. Of these, probably 1,200–2,000 contain sufficient hazardous waste to be a threat to public health and the environment. For this reason, many scientists believe management of hazardous chemical materials may be the most serious environmental problem in the United States.

Uncontrolled dumping of chemical waste has polluted soil and groundwater in several ways:

- Chemical waste may be stored in barrels, either stacked on the ground or buried. The barrels eventually corrode and leak, polluting surface water, soil, and groundwater.
- When liquid chemical waste is dumped into an unlined lagoon, contaminated water may percolate through soil and rock to the groundwater table.
- Liquid chemical waste may be illegally dumped in deserted fields or even along roads.

Some sites pose particular dangers. The floodplain of a river, for example, is not an acceptable site to store hazardous chemical waste. Yet that is exactly what occurred at a site on the floodplain of the River Severn near a village in one of the most scenic areas of England. Several fires at the site in 1999 were followed by a large fire of unknown origin on October 30, 2000. In that fire, approximately 200 tons of chemicals, including industrial solvents (xylene and toluene), cleaning solvents (methylenechloride), and various insecticides and pesticides, produced a fireball that rose into the night sky (Figure 28.8). The fire occurred during a rainstorm with wind gusts of hurricane strength. Toxic smoke and ash spread to nearby farmlands and villages, necessitating evacuation. People exposed to the smoke complained of a variety of symptoms, including headaches,

Figure 28.8 • On October 30, 2000, fire ravaged a site on the floodplain of the River Severn in England where hazardous waste was being stored. Approximately 200 tons of chemicals burned.

Figure 28.9 • Flooding on November 3, 2000, followed a large fire at a hazardous-waste storage site on the floodplain of the River Severn in England (see Figure 28.8).

stomachaches and vomiting, sore throats, coughs, and difficulty in breathing. Then, a few days later, on November 3, the site flooded (Figure 28.9). The floodwaters interfered with cleanup of the site after the fire and increased the risk of contamination of downstream areas by way of waterborne transport of hazardous chemical wastes. In one small village, contaminated floodwaters apparently inundated farm fields, gardens, and even homes.[35] Of course, the solution to this problem is to clean up the site and move waste storage facilities to a safer location.

28.7 Hazardous-Waste Legislation

The recognition in the 1970s that hazardous waste was a danger to people and the environment and that the waste was not being properly managed led to important federal legislation in the United States.

Resource Conservation and Recovery Act

Management of hazardous waste in the United States began in 1976 with passage of the Resource Conservation and Recovery Act (RCRA). At the heart of the act is identification of hazardous wastes and their life cycles. The idea was to issue guidelines and responsibilities to those who manufactured, transported, and disposed of hazardous waste. This is known as "cradle to grave" management. Regulations require stringent record keeping and reporting to verify that wastes do not present a public nuisance or a health problem.

RCRA applies to solid, semisolid, liquid, and gaseous hazardous wastes. A waste is hazardous if its concentration, volume, or infectious nature may contribute to serious disease or death or if it poses a significant hazard to people and the environment as a result of improper management (storage, transport, or disposal).[27] The act classifies hazardous wastes in several categories: materials highly toxic to people and other living things; wastes that may ignite when exposed to air; extremely corrosive wastes; and reactive unstable wastes that are explosive or generate toxic gases or fumes when mixed with water.

Comprehensive Environmental Response, Compensation, and Liability Act

In 1980, Congress passed the Comprehensive Environmental Response, Compensation, and Liability Act (CERCLA). The act defined policies and procedures for release of hazardous substances into the environment (for example, landfill regulations). It also mandated development of a list of the sites where hazardous substances were likely to or already had produced the most serious environmental problems and established a revolving fund (*Superfund*) to clean up the worst abandoned hazardous-waste sites. In 1984 and 1986, CERCLA was strengthened by amendments that made the following changes:

• Improved and tightened standards for disposal and cleanup of hazardous waste (for example, requiring double liners, monitoring landfills).

• Banned land disposal of certain hazardous chemicals, including dioxins, polychlorinated biphenyls (PCBs), and most solvents.

• Initiated a timetable for phasing out disposal of all untreated liquid hazardous waste in landfills or surface impoundments.

• Increased the size of the Superfund. The fund was allocated about $8 billion in 1993. Congress approved another $7.4 billion for fiscal year 1998, which almost doubled the Superfund budget.[36]

Superfund has experienced significant management problems, and cleanup efforts are far behind schedule. Unfortunately, the funds available are not sufficient to pay for decontamination of all targeted sites. Furthermore, there is concern that present technology is not sufficient to treat all abandoned waste disposal sites; it may be necessary to simply try to confine waste to those sites until better disposal methods are developed. It seems apparent that abandoned disposal sites are likely to persist as problems for some time to come.

Other Legislation

Federal legislation has changed the ways in which real estate business is conducted. For example, there are provisions by which property owners may be held liable for costly cleanup of hazardous waste present on their property even if they did not directly cause the problem. As a result, banks and other lending institutions might be held liable for release of hazardous materials by their tenants.

The Superfund Amendment and Reauthorization Act (SARA) of 1986 provides a possible defense against such liability, provided the property owner has completed an **environmental audit** prior to the purchase of the property. Such an audit involves studying past land use at the site, usually determined from analyzing old maps, aerial photographs, and reports. It may also involve drilling and sampling groundwater and soil to determine if hazardous materials are present. Environmental audits are now completed on a routine basis prior to purchase of property for development.

SARA legislation also required that certain industries report all releases of hazardous materials, and a list of companies releasing hazardous substances—known as the "Toxic 500"—became public. No property owner or industry wants to be on such a list, and the list is thought to have placed some pressure on industries identified as polluters to develop safer ways of handling hazardous materials.[37]

In 1990, the U.S. Congress reauthorized hazardous-waste control legislation. Priorities include:

- Establishing who is responsible (liable) for existing hazardous-waste problems.
- When necessary, assisting in or providing funding for cleanup at sites identified as having a hazardous-waste problem.
- Providing measures whereby people who suffer damages from the release of hazardous materials are compensated.
- Improving the required standards for disposal and cleanup of hazardous waste.

 28.8   Hazardous-Waste Management: Land Disposal

Management of hazardous chemical waste involves several options, including recycling, on-site processing to recover byproducts with commercial value, microbial breakdown, chemical stabilization, high-temperature decomposition, incineration, and disposal by secure landfill or deep-well injection. A number of technological advances have been made in toxic-waste management; and as land disposal becomes more expensive, a recent trend toward on-site treatment is likely to continue. However, on-site treatment will not eliminate all hazardous chemical waste; disposal of some waste will remain necessary.

Table 28.2 compares hazardous-waste reduction technologies for treatment and disposal. Notice that all available technologies cause some environmental disruption. There is no simple solution for all waste management issues. In this section, we consider land disposal of hazardous waste; and in the following section we discuss alternative approaches.

Secure Landfill

A **secure landfill** for hazardous waste is designed to confine the waste to a particular location, control the leachate that drains from the waste, collect and treat the leachate, and detect possible leaks. This type of landfill is similar to the modern sanitary landfill; it is an extension of the sanitary landfill for urban waste. Because in recent years it has become apparent that urban waste contains much hazardous material, the design of sanitary landfills and that of secure landfills for hazardous waste have converged to some extent.

Design of a secure landfill is shown in Figure 28.10. A dike and liner (made of clay or other impervious material, such as plastic) confine waste, and a system of internal drains concentrates leachate in a collection basin, from which it is pumped and transported to a wastewater treatment plant. Designs of modern facilities include multiple barriers consisting of several impermeable layers and filters. The function of impervious liners is to ensure that leachate does not contaminate soil and groundwater resources. However, this type of waste disposal procedure, like the sanitary landfill from which it evolved, must have several monitoring wells to alert personnel if leachates leak out of the system and threaten water resources.

It has recently been argued that there is no such thing as a really secure landfill, implying that they all leak to some extent. Impervious plastic liners, filters, and clay layers can fail, even with several backups; and drains can become clogged and cause overflow. Animals, such as gophers, ground squirrels, woodchucks, and muskrats, can chew through plastic liners, and some may burrow through clay liners, thus promoting or accelerating leaks. Yet careful siting and engineering can minimize problems. As with sanitary landfills, preferable sites are those with good natural barriers to migration of leachate: thick clay deposits, an arid climate, or a deep water table. Nevertheless, land disposal should be used only for specific chemicals compatible with and suitable for the method.

Land Application: Microbial Breakdown

Intentional application of waste materials to the surface soil is referred to as **land application**, land spreading, or land farming. We discussed land application of human waste earlier in the chapter. Land application of waste may be a desirable method of treatment for certain biodegradable industrial waste, such as oily petroleum waste and some organic chemical-plant wastes. A good indicator of the usefulness of land application for a par-

Table 28.2 Comparison of Hazard Reduction Technologies

	Disposal			Treatment		
Parameter Compared	Landfills and Impoundments	Injection Wells	Incineration and Other Thermal Destruction	High-temperature Decomposition[a]	Chemical Stabilization	Microbial Breakdown
Effectiveness: how well it contains or destroys hazardous characteristics	Low for volatiles, high for unsoluble solids	High, for waste compatible with the disposal environment	High	High for many chemicals	High for many metals	High for many metals and some organic waste such as oil
Reliability issues	Siting, construction, and operation Uncertainties: long-term integrity and cover	Site history and geology, well depth, construction, and operation	Monitoring uncertainties with respect to high degree of DRE: surrogate measures, PICs, incinerability[b]	Mobile units; on-site treatment avoids hauling risks Operational simplicity	Some inorganics still soluble Uncertain leachate production	Monitoring uncertainties during construction and operation
Environment media most affected	Surface water and groundwater	Surface water and groundwater	Air	Air	Groundwater	Soil, groundwater
Least compatible wastes[c]	Highly toxic, persistent chemicals	Reactive; corrosive; highly toxic, mobile, and persistent	Highly toxic organics, high heavy-metal concentration	Some inorganics	Organics	Highly toxic persistent chemicals
Relative costs	Low to moderate	Low	Moderate to high	Moderate to high	Moderate	Moderate
Resource recovery potential	None	None	Energy and some acids	Energy and some metals	Possible building material	Some metals

[a] Molten salt, high-temperature fluid well, and plasma arc treatments.

[b] DRE = destruction and removal efficiency; PIC = product of incomplete combustion.

[c] Wastes for which this method may be less effective for reducing exposure, relative to other technologies. Wastes listed do not necessarily denote common usage.

Source: Modified after Council on Environmental Quality, 1983.

Figure 28.10 • A secure landfill for hazardous chemical waste. The impervious liners, systems of drains, and leak detectors are integral parts of the system to ensure that leachate does not escape from the disposal site. Monitoring in the unsaturated zone is important and involves periodic collection of soil water.

ticular waste is the *biopersistence* of the waste—how long the material remains in the biosphere. The greater the biopersistence, the less suitable the waste for land application procedures. Land application is not an effective treatment or disposal method for inorganic substances such as salts and heavy metals.[37]

Land application of biodegradable waste works because, when such materials are added to the soil, they are attacked by microflora (bacteria, molds, yeasts, and other organisms) that decompose the waste material in a process known as *microbial breakdown*. The soil thus can be thought of as a microbial farm that constantly recycles organic and inorganic matter by breaking it down into more fundamental forms useful to other living things in the soil. Because the upper soil zone contains the largest microbial populations, land application is restricted to the uppermost 15–20 cm (6–8 in.) of the soil.[38]

Surface Impoundment

Both natural topographic depressions and human-made excavations have been used to hold hazardous liquid waste in a method known as **surface impoundment**. The depressions or excavations are primarily formed of soil or other surface materials but may be lined with manufactured materials such as plastic. Examples include aeration pits and lagoons at hazardous-waste facilities.

Surface impoundments have been criticized because they are especially prone to seepage, resulting in pollution of soil and groundwater. Evaporation from surface impoundments can also produce an air pollution problem. This type of storage or disposal system for hazardous waste is controversial, and many sites have been closed.

Deep-Well Disposal

Deep-well disposal, another controversial method of waste disposal, involves injection of waste into deep wells. A deep well must penetrate to a depth below and completely isolated from all freshwater aquifers, thereby assuring that injection of waste will not contaminate or pollute existing or potential water supplies. Typically, the waste is injected into a permeable rock layer several thousand meters below the surface in geologic basins capped by relatively impervious, fracture-resistant rock such as shale or salt deposits.[3]

Deep-well injection of oil-field brine (salt water) has been important in the control of water pollution in oil fields for many years. Huge quantities of liquid waste (brine) pumped up with oil have been injected back into the rock.[39]

Deep-well disposal of industrial wastes should not be viewed as a quick and easy solution to industrial waste problems.[40] Even where geologic conditions are favorable for deep-well disposal, there are a limited number of suitable sites; and within these sites, there is limited space for disposal of waste. Finally, disposal wells must be carefully monitored by additional wells, known as monitoring wells, to determine if the waste is remaining in the disposal site.

Summary of Land Disposal Methods

Direct land disposal of hazardous waste is often not the best initial alternative. There is consensus that even with extensive safeguards and state-of-the-art designs, land disposal alternatives cannot guarantee that the waste is contained and will not cause environmental disruption in the future. This concern holds true for all land disposal

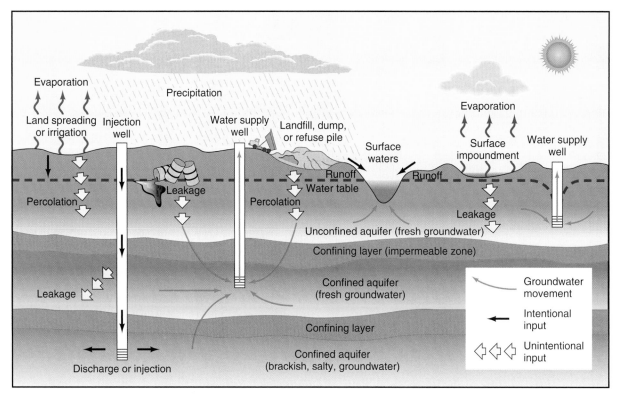

Figure 28.11 • Examples of how land disposal and treatment of hazardous waste may contaminate the environment. (*Source:* Modified from C. B. Cox, *The Buried Threat*, California Senate Office of Research, No. 115-5, 1985.)

facilities, including landfills, surface impoundments, land application, and injection wells. Pollution of air, land, surface water, and groundwater may result from failure of a land disposal site to contain hazardous waste. Pollution of groundwater is perhaps the most significant risk, because groundwater provides a convenient route for pollutants to reach humans and other living things. Figure 28.11 shows some of the paths that pollutants may take from land disposal sites to contaminate the environment. These paths include leakage and runoff to surface water or groundwater from improperly designed or maintained landfills; seepage, runoff, or air emissions from unlined lagoons; percolation and seepage from failure of surface land application of waste to soils; leaks in pipes or other equipment associated with deep-well injection; and leaks from buried drums, tanks, or other containers.

28.9 Alternatives to Land Disposal of Hazardous Waste

Our methods for handling hazardous chemical waste should be multifaceted. In addition to the disposal methods just discussed, chemical waste management should include such processes as source reduction, recycling and resource recovery, treatment, and incineration.[41] Recently, it has been argued that these alternatives to land disposal

are not being utilized to their full potential; that is, the volume of waste could be reduced and the remaining waste could be recycled or treated in some form prior to land disposal of the residues of the treatment processes.[41] Advantages to source reduction, recycling, treatment, and incineration include the following:

- Useful chemicals can be reclaimed and reused.
- Treatment of wastes may make them less toxic and therefore less likely to cause problems in landfills.
- The actual waste that must eventually be disposed of is reduced to a much smaller volume.
- Because a reduced volume of waste is finally disposed of, there is less stress on the dwindling capacity of waste disposal sites.

Although some of the following techniques have been discussed as part of IWM, the techniques have special implications and complications where hazardous wastes are concerned.

Source Reduction

The object of source reduction in hazardous waste management is to reduce the amount of hazardous waste generated by manufacturing or other processes. For example, changes in the chemical processes involved, equipment

616   CHAPTER 28   WASTE MANAGEMENT

used, raw materials used, or maintenance measures may successfully reduce the amount or toxicity of the hazardous waste produced.[41]

Recycling and Resource Recovery

Hazardous chemical waste may contain materials that can be recovered for future use. For example, acids and solvents collect contaminants when they are used in manufacturing processes. These acids and solvents can be processed to remove the contaminants and can then be reused in the same or other different manufacturing processes.[41]

Treatment

Hazardous chemical waste can be treated by a variety of processes to change the physical or chemical composition of the waste and so to reduce its toxic or hazardous characteristics. For example, acids can be neutralized, heavy metals can be separated from liquid waste, and hazardous chemical compounds can be broken up through oxidation.[41]

Incineration

Hazardous chemical waste can be successfully destroyed by high-temperature incineration. Incineration is considered a waste treatment rather than a disposal method because the process produces an ash residue, which must then be disposed of in a landfill operation. Hazardous

waste has also been incinerated offshore on ships, creating potential air pollution and ash disposal problems for the marine environment.

The technology used in incineration and other high-temperature decomposition or destruction is changing rapidly. Figure 28.12 shows a generalized diagram of one type of high-temperature incineration system that can be used to burn toxic waste. Waste (as liquid, solid, or sludge) enters the rotating combustion chamber, where it is rolled and burned. Ash from this burning process is collected in a water tank; the remaining gaseous materials move into a secondary combustion chamber, where the process is repeated. Finally, the remaining gases and particulates move through a scrubber system that eliminates surviving particulates and acid-forming components. Carbon dioxide, water, and air are then emitted from the stack. As shown in Figure 28.12, ash particulates and wastewater are produced at various stages of the incineration process; and these must be treated or disposed of in a landfill.

More advanced techniques for the incineration and thermal decomposition of waste are being developed. One of these utilizes a molten salt bed that should be useful in destroying certain organic materials. Other incineration techniques include liquid-injection incineration on land or sea and multiple-hearth furnaces. Which incineration method is used for a particular waste depends on the nature and composition of the waste and the temperature necessary to destroy the hazardous components. For example, the generalized incineration system shown in Figure 28.12 could be used to destroy PCBs.

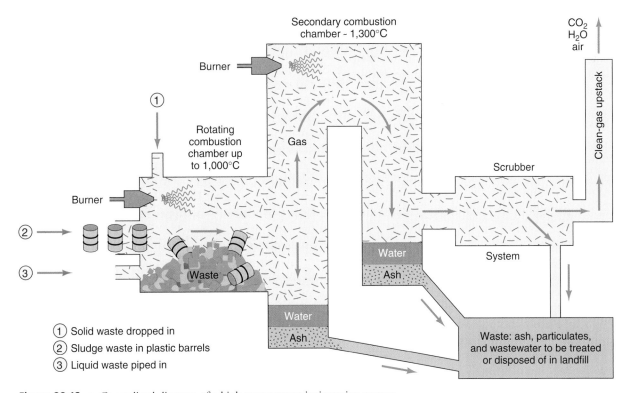

Figure 28.12 • Generalized diagram of a high-temperature incineration system.

Waste management issues are not restricted to the land. Some waste ends up in the oceans of the world.

Oceans cover more than 70% of Earth. They play a part in maintaining our global environment and are of major importance in the cycling of carbon dioxide, which helps regulate the global climate. Oceans are also important in cycling many chemical elements important to life, such as nitrogen and phosphorus, and are a valuable resource to people because they provide necessities such as foods and minerals.

It seems reasonable that such an important resource would receive preferential treatment, and yet oceans have long been dumping grounds for waste. In 1972, the Ocean Dumping Act was passed in the United States to provide for meeting U.S. commitments under the 1972 Convention on the Prevention of Marine Pollution by Dumping of Wastes and Other Matters, an international treaty signed by 80 countries. The law bans ocean dumping of radiological, chemical, and biological warfare agents and high-level radioactive waste. Amendments in 1988 extended this ban to sewage sludge, industrial waste, and medical wastes. In 1992, provisions were added to establish a national coastal water quality monitoring program. Although federal legislation has reduced the number of ocean dumping sites, ocean dumping continues.[6] Furthermore, if population growth in coastal regions continues as expected, increased amounts of waste will end up in the oceans.

The types of wastes dumped in the oceans include the following:[42]

- Dredge spoils—solid materials, such as sand, silt, clay, rocks, and pollutants, deposited in the ocean from industrial and municipal discharges and removed from water bodies, generally to improve navigation.
- Industrial waste—acids, refinery wastes, paper mill wastes, pesticide wastes, and assorted liquid wastes.
- Sewage sludge—solid material (sludge) that remains after municipal wastewater treatment.
- Construction and demolition debris—cinder block, plaster, dirt, stone, and tile.
- Solid waste—refuse, garbage, explosives, radioactive waste, and untreated urban sewage.

Ocean dumping contributes to the larger problem of ocean pollution, which has seriously damaged the marine environment and caused a health hazard. Locations in the oceans of the world that are accumulating pollution continuously, that have intermittent pollution problems, or that have potential for pollution from ships in the major shipping lanes are shown on Figure 28.13. Notice that the areas with continual or intermittent pollution are located near shore. Unfortunately, these are also areas of high productivity with valuable fisheries. Shellfish have been found to contain organisms that produce diseases, such as polio and hepatitis. In the United States, at least 20% of the nation's commercial shellfish beds have been closed (mostly temporarily) because of pollution. Beaches and bays have been closed (again, mostly temporarily) to recreational uses. Lifeless zones in the marine environment have been created. Heavy kills of fish and other organisms have occurred, and profound changes in marine ecosystems have taken place (see Chapter 21).[42,43]

Marine pollution has a variety of specific effects on oceanic life, including the following:

- Death or retarded growth, vitality, and reproductivity of marine organisms.
- Reduction in the dissolved oxygen content necessary for marine life because of increased biochemical oxygen demand.
- Eutrophication caused by nutrient-rich waste in shallow waters of estuaries, bays, and parts of the continental shelf, resulting in depletion of oxygen and subsequent killing of algae, which may wash up and pollute coastal areas. (See Chapter 21 for a discussion of eutrophication in the Gulf of Mexico.)
- Habitat change caused by waste disposal practices that subtly or drastically change entire marine ecosystems.[42]

Marine waters of Europe are in particular trouble, in part as a result of urban and agricultural pollutants that have raised concentrations of nutrients in seawater. Blooms (heavy, sudden growth) of toxic algae are becoming more common. In 1988 in the waterway connecting the North Sea to the Baltic Sea, for example, a bloom was responsible for killing nearly all marine life to a depth of about 15 m (50 ft). It is believed that urban waste and agricultural runoff contributed to the toxic bloom.

Although oceans are vast, they are basically giant sinks for materials from continents; and parts of the marine environment are extremely fragile.[43] One area of concern is the *microlayer*, the upper 3 mm of ocean water. The base of the marine food chain consists of planktonic life abundant in the microlayer. The young of certain fish and shellfish also reside in the upper few millimeters of water in the early stages of their life. Unfortunately, the upper few millimeters of the ocean also tend to concentrate pollutants, such as toxic chemicals and heavy metals. One study reported that the concentrations of heavy metals including zinc, lead, and copper, in the microlayer are from 10 to 1,000 times higher than in the deeper waters. There is fear that disproportionate pollution of the microlayer will have especially serious effects on marine organisms.[43] There is also concern that some ecosystems in the oceans, such as coral reefs, estuaries and salt marshes, and mangrove swamps, are threatened by ocean pollution.

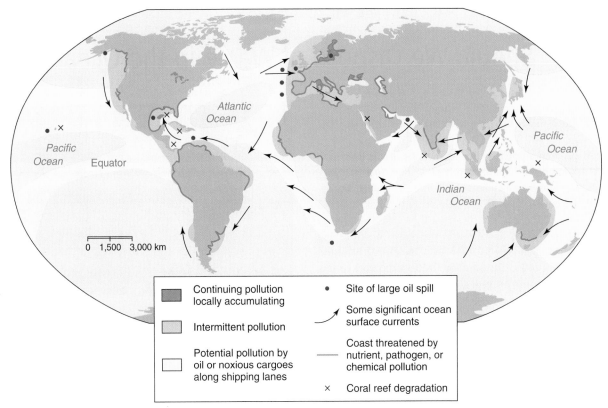

Figure 28.13 • Ocean pollution of the world. Notice that the areas of continuing and locally accumulating pollution, as well as the areas with intermittent pollution, are in nearshore environments. (*Source:* Modified from Council on Environmental Quality, *Environmental Trends*, 1981, with additional data from A. P. McGinn, "Safeguarding the Health of the Oceans," WorldWatch Paper 145 [Washington, D.C.: WorldWatch Institute, 1999], pp. 22–23.)

Marine pollution can have major impacts on people and society. Contaminated marine organisms may transmit toxic elements or diseases to people who eat them. When beaches and harbors become polluted by solid waste, oil, and other materials, there is a loss of visual appeal and other amenities. Economic loss is considerable. Loss of shellfish from pollution in the United States, for example, amounts to many millions of dollars per year. In addition, a great deal of money is spent cleaning up solid waste, liquid waste, and other pollutants in coastal areas.[42]

 Pollution Prevention

As we saw at the beginning of the chapter, approaches to waste management are changing. During the first several decades of environmental concern and management (the 1970s and 1980s), the United States approached waste management through government regulations and control of waste. Control was accomplished through chemical, physical, or biological treatment and collection (for eventual disposal), transformation, or destruction of pollutants after they had been generated. This was thought to be the most cost-effective approach to waste management.

With the 1990s came a growing emphasis on **pollution prevention**, which involves identifying ways to prevent the generation of waste rather than finding ways to dispose of it. This approach, which is part of materials management, reduces the need for management of waste, because less waste is produced. Approaches to pollution prevention include the following:[44]

- Purchasing the proper amount of raw materials so that no excess remains to be disposed of.
- Exercising better control of materials used in manufacturing processes so that less waste is produced.
- Substituting nontoxic chemicals for hazardous or toxic materials currently in use.
- Improving engineering and design of manufacturing processes so that less waste is produced.

These approaches are often called P-2 approaches, for "pollution prevention." Probably the best way to illustrate the P-2 process is through a case history.[44]

A Wisconsin firm that produced cheese was faced with disposal of about 2,000 gallons a day of a salty solution generated as part of the cheese-making process. Initially, the firm spread the salty solution on nearby agricultural lands—common practice for firms that could not dis-

Should We Dispose of Waste in the Ocean?

The New York coastal waters off the Long Island and New Jersey shores includes Newark Bay, an important commercial shipping port (Figure 28.14). To keep this $20 billion industry open to large vessels, accumulated sediments must be dredged periodically to maintain a minimum depth of 12.8 m (42 ft). Since 1976, 5.3 million m^3 (7 million yd^3) a year of dredged material has been dumped at the Mud Dump site, an area of about 5.2 k^2 (2 mi^2) about 9.6 km (6.0 mi) from shore, with an average depth of 22 m (72 ft). The site is in the midst of rich fishing grounds; only 1.6 km (1 mile) away is the famous 17 Fathoms, a prime fishing spot for bluefish, bonitos, false albacore, blackfish, and flukes and the focus of a $100 million-a-year industry.

Much of the sediment originates as runoff from the three rivers feeding the estuary in the bight: the Hudson, Passaic, and Raritan. But with the sediment come many metallic and organic toxic wastes from industries along the rivers, among them dioxin, one of the most toxic substances known. Primarily a by-product in the manufacture of Agent Orange (used as a defoliant in the Vietnam War), dioxin continues to be produced in the manufacture of paper and herbicides and in many other industrial processes (see Chapter 15 for a discussion of dioxin). Because dioxin has been associated with cancer and immune deficiencies in animals and is suspected to have many other adverse health effects, dredging the dioxin-contaminated sediments became highly controversial in the early 1990s.

A request by the Port Authority of New York to dredge and dump 383,000 m^3 (500,000 yd^3) of contaminated sediment at the Mud Dump site was held up while questions regarding the impact on the marine environment and human health were debated. Disposal of moderately contaminated sediments in the ocean requires that they be capped, or covered with noncontaminated materials, to attempt to isolate the toxic materials from the marine environment (Figure 28.15). Highly contaminated materials, which pose a significant health and environmental risk, are not allowed to be dumped in the ocean. Differing opinions as to the lower and upper limits of moderate dioxin contamination complicated the decision on disposal of the dredged materials. Concentrations greater than 4 parts per trillion (pptr) would have required capping, and ocean disposal was prohibited for concentrations above 25 pptr. Some experts felt these limits were too liberal, but others pointed out that Europe and Canada accepted even higher levels.

By 1991, scientists had reached a consensus on the mechanism by which dioxin acts on living cells and agreed on the need to question existing guidelines for safe exposure. While some scientists worked to translate the new understanding of how dioxin functions into revised guidelines, others investigated its effects on marine organisms.

The average dioxin concentration varies for different parts of Newark Bay, from 39.4 to 110.6 pptr. Sediments at the Mud

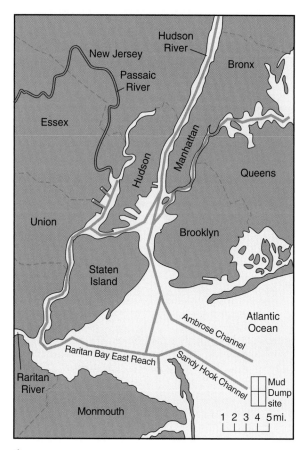

Figure 28.14 • Major federally maintained navigation channels in the port of New York and New Jersey. (*Source:* U.S. Army Corps of Engineers, New York District, *Managing Dredged Materials*, December 1989, p. 22.)

Figure 28.15 • Capping of dredged sediments. (*Source:* U.S. Army Corps of Engineers, New York District, *Managing Dredged Materials*, December 1989, p. 7A.)

Dump site average 24.1 pptr, and the average level in sandworms there is 27 pptr of dry weight. Dioxin is highly fat soluble and accumulates in fatty tissue. Predators of the sandworms and other marine organisms can be expected to accumulate concentrations as much as 1,000 times greater than those in their prey. As dioxin is passed up a food chain, concentrations in fatty tissues of fish can be as much as 10,000 times those in the surrounding water.

Table 28.3   Disposal Alternatives

Alternative	Estimated Cost per Cubic Yard	Estimated Capacity (MCY)ᵃ	Ecological Effects
Ocean disposal: Mud Dump site	$5.00	Large volumes	
Land disposal	$5.60–$13.20	2.6–8.5	
Creation of special 500-acre containment island	$10.90–$36.10	6.6–28.2	

ᵃMCY = million cubic yards.
Source: U.S. Army Corps. of Engineers, New York District, *Managing Dredged Materials,* December 1989, p. 95.

Critical Thinking Questions

1. As part of an attempt to study the effectiveness of capping, 900,000 m³ (1.2×10^6 yd³) of clean sediment was used to cap 383,000 m³ (500,000 yd³) of contaminated material in an area of the Mud Dump site in 1980. Six years later, the cap was estimated to contain only 612,000 m³ (800,000 yd³), and 99,000 m³ (130,000 yd³) of material was found outside the site. Assuming that all the measurements were accurate, how can you explain the discrepancy? What ecological effects might occur as a result of your explanation?

2. Complete Table 28.3 by filling in the Ecological Effects column. Evaluate each of the alternatives. Which alternative do you favor? Why?

3. How does the idea of the tragedy of the commons apply to the problem of ocean disposal of contaminated materials? (See Chapter 26 for a discussion of the tragedy of the commons.)

charge wastewater into publicly owned treatment plants. This method of waste disposal, if the solution was applied incorrectly, caused the level of salts in the soil to rise so much that crops were damaged. As a result, the Department of Natural Resources in Wisconsin placed limitations on the discharge of salt to the land.

The cheese manufacturing firm decided to modify its cheese-making processes to recover salt from the solution and reuse it in production. This involved developing a recovery process using an evaporator. The recovery process reduced the salty waste by about 75% and at the same time reduced the amount of the salt the company needed to purchase by 50%. The operating and maintenance costs for recovery were approximately three cents per pound of salt recovered, and the time necessary to reclaim the extra cost of the new equipment was only two months. The firm saved thousands of dollars a year by purchasing less salt.

The case history of the cheese firm suggests that often rather minor changes can result in large reductions of waste produced. And this case history is not an isolated example. Thousands of similar cases exist today as we move from the era of recognizing environmental problems and regulating them at a national level to providing economic incentives and new technology to better manage materials.[44]

Summary

- The history of waste disposal practices since the Industrial Revolution has progressed from the practice of dilution and dispersion to the concept of integrated waste management (IWM), which emphasizes the three R's of reducing waste, reusing materials, and recycling.

- The emerging concept of industrial ecology has as a goal a system in which the concept of waste doesn't exist, because waste from one part of the system would be a resource for another part.

- The most common method for disposal of solid waste is the sanitary landfill. However, around many large cities, space for landfills is hard to find; and few people wish to live near any waste disposal operation.

- Physical and hydrologic conditions of a site greatly affect its suitability as a landfill. These include landform, topography, rock and soil type, depth to groundwater, and amount of precipitation.

- Hazardous chemical waste is one of the most serious environmental problems in the United States. Hundreds or even thousands of uncontrolled disposal sites could be time bombs that will eventually cause serious public health problems. We know that we will continue to produce some hazardous chemical waste. Therefore, it is imperative to develop and use safe disposal methods.

- Management of hazardous chemical wastes involves several options, including on-site processing to recover by-products with commercial value, microbial breakdown, chemical stabilization, incineration, and disposal by secure landfill or deep-well injection.

- Ocean dumping is a significant source of marine pollution. The most seriously affected areas are near shore, where valuable fisheries often exist.

- Pollution prevention (P-2)—identifying and using ways to prevent the generation of waste—is an important emerging area of materials management.

REEXAMINING THEMES AND ISSUES

Human Population:

Waste management strategies are inextricably linked to human population. As population increases, so does the waste generated, because the total amount of waste is the product of per-capita waste generation times population. Furthermore, in developing countries, where population increase is the most dramatic, increases in industrial output, when linked to poor environmental control, produce or aggravate waste management problems.

Sustainability:

The objective of providing for a quality environment for future generations is closely linked to waste management. Of particular importance here are the concepts of integrated waste management, materials management, and industrial ecology. Carried to their natural conclusion, the ideas behind these concepts would lead to a system in which the issue would no longer be waste management but instead would be resource management. Pollution prevention (P-2) is a step in this direction.

Global Perspective:

Management of waste is becoming a global problem. Inappropriate management of waste contributes to air and water pollution, with the potential to cause environmental disruption on a regional or global scale. For example, waste generated by large inland cities and disposed of in river systems may eventually enter the oceans, where it may become dispersed by the global circulation patterns of ocean currents. Similarly, soils polluted by hazardous materials may erode, and the particles may enter the atmosphere or water system, to be dispersed widely.

Urban World:

Because so much of our waste is generated in the urban environment, it is a focus of special attention for waste management. Where population densities are high, it is easier for the principles behind "reduce, reuse, and recycle" to be implemented. There are greater financial incentives for management of waste where waste is more concentrated.

Science and Values:

People today value a quality, pollution-free environment. The way in which waste has been managed has contributed to—and continues to contribute to—health and other environmental problems. An understanding of these problems has resulted in a considerable amount of work and research aimed at reducing or eliminating the impact of waste. How a society manages its waste is a sign of the maturity of the society and its ethical framework. Accordingly, we have become more conscious of environmental justice issues related to waste management.

KEY TERMS

composting **605**	incineration **605**	leachate **606**	reduce **601**
deep-well disposal **615**	industrial ecology **601**	materials management **603**	reuse **601**
environmental audit **613**	integrated waste management (IWM) **601**	monitoring **606**	sanitary landfill **606**
environmental justice **606**		pollution prevention **619**	secure landfill **613**
hazardous waste **609**	land application **613**	recycling **601**	surface impoundment **615**

1. Have you ever contributed to the hazardous-waste problem through disposal methods used in your home, school laboratory, or other location? How big a problem do you think such events are? For example, how bad is it to dump paint thinner down a drain?

2. Why is it so difficult to ensure safe land disposal of hazardous waste?

3. Would you approve the siting of a waste disposal facility in your part of town? If not, why not, and where do you think such facilities should be sited?

4. Why might there be a trend toward on-site disposal rather than land disposal of hazardous waste? Consider physical, biological, social, legal, and economic aspects of the question.

5. Is government doing enough to clean up abandoned hazardous waste dumps? Do private citizens have a role in choosing where cleanup funds should be allocated?

6. Considering how much waste has been dumped in the nearshore marine environment, how safe is it to swim in bays and estuaries near large cities?

7. Do you think we should collect household waste and burn it in special incinerators to make electrical energy? What problems and what advantages do you see for this method compared with other disposal options?

8. Many jobs will be available in the next few years in the field of hazardous-waste monitoring and disposal. Would you take such a job? If not, why not? If so, do you feel secure that your health would not be jeopardized?

9. Should companies that dumped hazardous waste years ago when the problem was not understood or recognized be held liable today for health-related problems to which their dumping may have contributed?

10. Suppose you found that the home you had been living in for 15 years was located over a buried waste disposal site. What would you do? What kinds of studies could be done to evaluate the potential problem?

FURTHER READING

Allenby, B. R. 1999. *Industrial Ecology: Policy Framework and Implementation.* Upper Saddle River, N.J. Prentice Hall. Primer on industrial ecology.

Kreith, F. ed. 1994. *Handbook of Solid Waste Management.* New York: McGraw-Hill. Thorough coverage of municipal waste management, including waste characteristics, federal and state legislation, source reduction, recycling, and landfilling.

Rhyner, C. R., L. J. Schwartz, R. B. Wenger, and M. G. Kohrell. 1995. *Waste Management and Resource Recovery.* Boca Raton, Fla.: CRC, Lewis. Discussions of the archeology of waste, waste generation, source reduction and recycling, wastewater treatment, incineration and energy recovery, hazardous waste, and costs of waste systems and facilities.

Watts, R. J. 1998. *Hazardous Wastes.* New York: John Wiley & Sons. A to Z of hazardous wastes.

CHAPTER 29 Minerals and the Environment

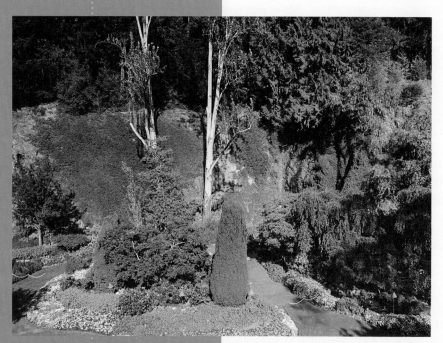

Butchart Gardens, Vancouver Island, Canada. The site of the gardens was once a limestone quarry.

● LEARNING OBJECTIVES

Modern society depends on the availability of mineral resources, which can be considered a nonrenewable heritage from the geologic past. After reading this chapter, you should understand:

- The standard of living in modern society is related in part to the availability of natural resources.
- Why minerals are not uniformly distributed throughout the Earth's crust.
- What processes are responsible for the distribution of mineral deposits.
- What the differences are between mineral resources and reserves.
- What factors control the environmental impact of mineral exploitation.
- How wastes generated from the use of mineral resources affect the environment.
- What the social impacts are from mineral exploitation.

Canada's Butchart Gardens: From Eyesore to Eden

It's often been stated that one person at the right time and in the right place has the opportunity to make a real difference. Jenny Butchart in the early 1900s had a vision to transform her husband's exhausted mine pit into a garden. Today, we would use the term *mine reclamation* for such a project. The story that followed was one of transformation. Gardens which began as a limestone quarry that was an eyesore became a tourist attraction visited by about a million people each year—Butchart Gardens, Vancouver Island, British Columbia. The Butchart Garden story offers an early illustration of the principles of environmental restoration and sequential land use.

The story is closely related to the cement industry. The process of producing Portland cement was developed in 1824 in England. The procedure includes fine grinding of an exact mixture of limestone and shale which has been heated in a rotary kiln to near-fusion temperature. This material is then finely ground again, placed in sacks, and sold to consumers as Portland cement. Adding aggregate such as gravel to the cement makes concrete, which is an integral part of construction in our urban environment.

Robert Butchart was educated in the process of making Portland cement while on honeymoon in England in 1884. In 1902, he learned of the availability of limestone located approximately 20 km north of the city of Victoria on Vancouver Island. He determined that the site was ideal for the establishment of a cement plant and started the Cod Inlet Cement Plant there in 1904. The limestone used to produce the Portland cement was exhausted by 1908, and what was left behind was a large, unsightly large open excavation about 20 m deep. The walls of the quarry were nearly vertical gray limestone.

That's when Jenny Butchart entered the picture. She had little experience with gardens but was fascinated with the site and the idea of transforming the quarry into a sunken garden. Her husband supported this endeavor, and with the help of many workmen, she brought in massive amounts of topsoil by horse and cart to form garden beds on the floor of the quarry. The garden, which was completed in 1921, soon became a tourist attraction. Today, the garden is much larger and includes fountains, rose gardens, lakes, Japanese gardens, and other attractions in addition to the original sunken garden. The Butchart Gardens case history illustrates the power of the vision of one person to transform the landscape from a degraded site to a beautiful and esthetically pleasing place.[1]

● *Butchart Gardens is a unique instance of mine reclamation; but each renovation is somewhat unique, based on local physical, hydrological, and biological conditions. This chapter discusses the origin of mineral deposits as well as environmental consequences of mineral development.*

29.1 The Importance of Minerals to Society

Modern society depends on the availability of mineral resources.[2] Many mineral products are found in a typical American home (see Table 29.1). Consider your breakfast this morning. You probably drank from a glass made primarily of sand, ate food from dishes made from clay, flavored your food with salt mined from Earth, ate fruit grown with the aid of fertilizers such as potassium carbonate (potash) and phosphorus, and used utensils made from stainless steel, which comes from processing iron ore and other minerals.

Minerals are so important to people that the standard of living increases with the availability of minerals in useful forms. The availability of mineral resources is one measure of the wealth of a society. Those who have been successful in locating and extracting or importing and using minerals have grown and prospered. Without mineral resources to grow food, construct buildings and roads, and manufacture everything from computers to televisions to automobiles, modern technological civilization as we know it would not be possible. In maintaining our standard of living in the United States, every person requires about 10 tons of nonfuel minerals per year.[3]

Minerals can be considered our nonrenewable heritage from the geologic past. Although new deposits are still forming from present Earth processes, these processes are producing new mineral deposits too slowly to be of use

Table 29.1   Mineral Products in a Typical U.S. Home

Building materials: sand, gravel, stone, brick (clay), cement, steel, aluminum, asphalt, glass

Plumbing and wiring materials: iron and steel, copper, brass, lead, cement, asbestos, glass, tile, plastic

Insulating materials: rock wool, fiberglass, gypsum (plaster and wallboard)

Paint and wallpaper: mineral pigments (such as iron, zinc, and titanium) and fillers (such as talc and asbestos)

Plastic floor tiles, other plastics: mineral fillers and pigments, petroleum products

Appliances: iron, copper, and many rare metals

Furniture: synthetic fibers made from minerals (principally coal and petroleum products); steel springs; wood finished with rotten-stone polish and mineral varnish

Clothing: natural fibers grown with mineral fertilizers; synthetic fibers made from minerals (principally coal and petroleum products)

Food: grown with mineral fertilizers; processed and packaged by machines made of metals

Drugs and cosmetics: mineral chemicals

Other items: windows, screens, light bulbs, porcelain fixtures, china utensils, jewelry made from mineral products

Source: U.S. Geological Survey, Professional Paper 940, 1975.

---

to us today. Because mineral deposits are generally located in small, hidden areas, they must be discovered. Unfortunately, most of the easy-to-find deposits have been exploited; if modern civilization were to vanish, our descendants would have a harder time discovering rich mineral deposits than we did. It is interesting to speculate that they might mine landfills for metals thrown away by our civilization. Unlike biological resources, minerals cannot be easily managed to produce a sustained yield; the supply is finite. Recycling and conservation will help, but eventually the supply will be exhausted.

 29.2   How Mineral Deposits Are Formed

Metals in mineral form are generally extracted from naturally occurring, anomalously high concentrations of Earth materials. When metals are concentrated in anomalously high amounts by geologic processes, **ore deposits** are formed. The discovery of natural ore deposits allowed early peoples to exploit copper, tin, gold, silver, and other metals while slowly developing skills in working with metals.

The origin and distribution of mineral resources is intimately related to the history of the biosphere and to the entire geologic cycle (see Chapter 4). Nearly all aspects and processes of the geologic cycle are involved to some extent in producing local concentrations of useful materials. In this section, we look first at the distribution of mineral resources on Earth and then describe processes that form mineral deposits.

Distribution of Mineral Resources

Earth's outer layer, or crust, is silica rich, made up mostly of rock-forming minerals containing silica, oxygen, and a few other elements. The elements are not evenly distributed in the crust: Nine elements account for about 99% of the crust by weight (oxygen, 45.2%; silicon, 27.2%; aluminum, 8.0%; iron, 5.8%; calcium, 5.1%; magnesium 2.8%; sodium, 2.3%; potassium, 1.7%; and titanium, 0.9%). In general, remaining elements are found in trace concentrations.

The ocean, covering nearly 71% of Earth, is another reservoir for many chemicals other than water. Most elements in the ocean have been weathered from crustal rocks on the land and transported to the oceans by rivers. Other elements are transported to the ocean by wind or glaciers. Ocean water contains about 3.5% dissolved solids, mostly chlorine (55.1% by weight). Each cubic kilometer of ocean water contains about 2.0 metric tons of zinc, 2.0 metric tons of copper, 0.8 metric ton of tin, 0.3 metric ton of silver, and 0.01 metric ton of gold. These concentrations are low compared with those in the crust, where corresponding values (in metric tons/km³) are: zinc, 170,000; copper, 86,000; tin, 5,700; silver, 160; and gold, 5. After rich crustal ore deposits are depleted, we will be more likely to extract metals from lower-grade ore deposits or even from common rock than from ocean water; however, if mineral extraction technology becomes more efficient, this prognosis could change.

We have mentioned that minerals we mine occur in deposits—with anomalously high local concentrations. Why is this so? Planetary scientists now believe that Earth, like the other planets in the solar system, formed by condensation of matter surrounding the sun. Gravitational attraction brought together the matter dispersed around the forming sun. As the mass of the proto-Earth increased, the material condensed and was heated by the process. The heat was sufficient to produce a molten liquid core, consisting primarily of iron and other heavy metals, which sank toward the center. The crust formed from generally lighter elements and is a mixture of many different elements. The elements in the crust are not uniformly distributed because geologic processes and some biological processes selectively dissolve, transport, and deposit elements and minerals. We discuss these processes next.

Plate Boundaries

Plate tectonics is responsible for the formation of some mineral deposits. According to the theory of plate tectonics (see Chapter 4), the continents (which are crustal rocks and part of the lithosphere) are composed mostly of relatively light rocks. As the tectonic plates of the lithosphere slowly move across Earth's surface, so do the continents. Metallic ores are thought to be deposited in the crust both where the tectonic plates separate, or diverge, and where they come together, or converge.

At divergent plate boundaries, cold ocean water comes in contact with hot molten rock. The heated water is lighter and more active chemically. It rises through fractured rocks and leaches metals from them. The metals are carried in solution and deposited as metal sulfides when the water cools.

At convergent plate boundaries, rocks saturated with seawater are forced together, heated, and subjected to intense pressure, which causes partial melting. The combination of heat, pressure, and partial melting mobilizes metals in the molten rocks, or magma. Most major mercury deposits, for example, are associated with the volcanic regions that occur close to convergent plate boundaries. Geologists believe that the mercury is distilled out of the tectonic plate as the plate moves downward; as the plate cools, the mercury migrates upward and is deposited at shallower depths, where the temperature is lower.

Igneous Processes

Igneous processes are related to molten rock material known as magma. Ore deposits may form when magma cools. As the molten rock cools, heavier minerals that crystallize (solidify) early may slowly sink or settle toward the bottom of the magma, whereas lighter minerals that crystallize later are left at the top. Deposits of an ore of chromium, called chromite, are thought to be formed in this way. When magma containing small amounts of carbon is deeply buried and subjected to very high pressure during slow cooling (crystallization), diamonds (which are pure carbon) may be produced (Figure 29.1).[4]

Hot waters moving within the crust are perhaps the source of most ore deposits. It is speculated that circulating groundwater is heated and enriched with minerals on contact with deeply buried rocks. This water then moves up or laterally to other, cooler rocks, where the cooled water deposits the dissolved minerals.[5]

Figure 29.1 • Diamond mine near Kimberley, South Africa. This is the largest hand-dug excavation in the world.

Sedimentary Processes

Sedimentary processes relate to the transport of sediments by wind, water, and glaciers. These processes often concentrate materials in amounts sufficient for extraction.

As sediments are transported, running water and wind help segregate the sediments by size, shape, and density. This sorting function is useful to people. The best sand or sand and gravel deposits for construction purposes, for example, are those in which the finer materials have been removed by water or wind. Sand dunes, beach deposits, and deposits in stream channels are good examples. The sand and gravel industry amounts to about $3 billion annually, and in terms of the total volume of materials mined, it is one of the largest nonfuel mineral industries in the United States.[3]

Stream processes transport and sort all types of materials according to size and density. Therefore, if the bedrock in a river basin contains heavy metals, such as gold, streams draining the basin may concentrate the metals in areas where there is less water turbulence or velocity. These concentrations, called placer deposits, are often found in open crevices or fractures at the bottoms of pools, on the inside curves of bends, or on riffles, where shallow water flows over rocks. Placer mining of gold (which was known as a poor man's method because a miner needed only a shovel, a pan, and a strong back to work the streamside claim) played an important role in the settling of California, Alaska, and other areas of the United States.

Rivers and streams that empty into oceans and lakes carry tremendous quantities of dissolved material derived from the weathering of rocks. Over geologic time, a shallow marine basin may be isolated by tectonic activity that uplifts its boundaries. In other cases, climatic variations, such as the ice ages, produce large inland lakes with no outlets. These basins and lakes eventually dry up. As evaporation progresses, the dissolved materials precipitate (drop out of solution), forming a wide variety of compounds, minerals, and rocks that have important commercial value. Most of these *evaporites* (deposits originating by evaporation) can be grouped into one of three types:[6]

- **Marine evaporites** (solids)—potassium and sodium salts, gypsum, and anhydrite.
- **Nonmarine evaporites** (solids)—sodium and calcium carbonate, sulfate, borate, nitrate, and limited iodine and strontium compounds.
- **Brines** (liquids derived from wells, thermal springs, inland salt lakes, and seawaters)—bromine, iodine, calcium chloride, and magnesium.

Heavy metals (such as copper, lead, and zinc) associated with brines and sediments in the Red Sea, Salton Sea, and other areas are important resources that may be exploited in the future. Evaporite minerals are widely used in industrial and agricultural activities, and their annual value is about $1 billion.[6] Evaporite and brine resources in the United States are substantial, ensuring an ample supply for many years.

Biological Processes

Some mineral deposits are formed by biological processes, and many are formed under conditions of the biosphere that have been greatly altered by life. Examples include phosphates (discussed in Chapter 4) and iron ore deposits.

The major iron ore deposits exist in sedimentary rocks that were formed more than 2 billion years ago.[7] There are several types of iron deposits. Gray beds, an important type, contain unoxidized iron. Red beds contain oxidized iron (the red color is the color of iron oxide). Gray beds formed when there was little oxygen in the atmosphere, and red beds formed when there was relatively more oxygen. Although the processes are not completely understood, it appears that major deposits of iron stopped forming when the atmospheric concentration of oxygen reached its present level.[8]

Organisms are able to form many kinds of minerals, such as the calcium minerals in shells and bones. Some of these minerals cannot be formed inorganically in the biosphere. Thirty-one different biologically produced minerals have been identified. Minerals of biological origin contribute significantly to sedimentary deposits.[9]

Weathering Processes

Weathering, the chemical and mechanical decomposition of rock, concentrates some minerals in the soil. When insoluble ore deposits, such as native gold, are weathered from rocks, they may accumulate in the soil unless removed by erosion. Accumulation occurs most readily when the parent rock is relatively soluble, as is limestone. Intensive weathering of certain soils derived from aluminum-rich igneous rocks may concentrate oxides of aluminum and iron. (The more soluble elements, such as silica, calcium, and sodium, are selectively removed by soil and biological processes.) If sufficiently concentrated, residual aluminum oxide forms an ore of aluminum known as bauxite. Important nickel and cobalt deposits are also found in soils developed from iron- and magnesium-rich igneous rocks.

Weathering produces sulfide ore deposits from low-grade primary ore through **secondary enrichment** processes. Near the surface, primary ore containing minerals such as iron, copper, and silver sulfides is in contact with slightly acidic soil water in an oxygen-rich environment. As the sulfides are oxidized, they dissolve, forming solutions that are rich in sulfuric acid as well as silver and copper sulfate. These solutions migrate downward, producing a leached zone devoid of ore minerals (Figure 29.2). Below the leached zone and above the groundwater table, oxida-

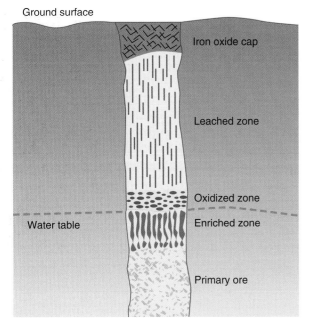

Ground surface

Iron oxide cap

Leached zone

Oxidized zone

Water table

Enriched zone

Primary ore

Figure 29.2 • Typical zones that form during secondary enrichment processes. Sulfide ore minerals in the primary ore vein are oxidized and altered and then are leached from the oxidized zone by descending groundwater and redeposited in the enriched zone. The iron oxide cap is generally a reddish color and may be helpful in locating ore deposits that have been enriched. (*Source:* R. J. Foster, *General Geology*, 4th ed. [Columbus, Ohio: Charles E. Merrill, 1983].)

tion continues, and sulfate solutions continue their downward migration. Below the water table, if oxygen is no longer available, the solutions are deposited as sulfides, enriching the metal content of the primary ore by as much as 10 times. In this way, low-grade primary ore is rendered more valuable, and high-grade primary ore is made even more attractive.[10,11] For example, secondary enrichment of a disseminated copper deposit at Miami, Arizona, increased the grade of the ore from less than 1% copper in the primary ore to as much as 5% in some areas.[10]

Minerals from the Sea

Mineral resources in seawater and on the bottom of the ocean are vast. Indeed, in some cases, they are nearly unlimited; this is true of magnesium, which is extracted from seawater. On the deep-ocean floor, two types of deposits have been identified: massive sulfide deposits associated with hydrothermal vents and manganese oxide nodules. The seabed may be the site of the next mineral rush. However, mining the seabed is controversial (see A Closer Look 29.1).

Magnesium

In the United States, magnesium was first extracted from seawater in 1940. By 1972, one company in Texas, utilizing seawater as its raw material source, produced

80% of domestic magnesium. Now companies in Alabama, California, Florida, Mississippi, and New Jersey are also extracting magnesium from seawater.

Sulfide Deposits

Massive sulfide deposits containing zinc, copper, iron, and trace amounts of silver are produced on the sea floor at divergent plate boundaries by the processes of plate tectonics. Cold seawater enters rock fractures at mid-ocean ridges, where it is heated by upwelling magma and emerges as hot springs. The circulating water leaches the rocks, removing metals that are deposited where the hot, mineral-rich water is ejected at temperatures up to 350°C (660°F) into the cold sea. Sulfide minerals precipitate, or drop out of solution, near hydrothermal vents known as *black smokers* and form massive towerlike formations rich in metals (Figure 29.3).

Hydrothermal vents are also of particular biological significance because they support a unique assemblage of animals (discussed in Chapter 9), including giant clams, tube worms, and white crabs. Communities of these animals are based on sulfide compounds extruded from black smokers and exist through a process called chemosynthesis, as opposed to photosynthesis, which supports all other known ecosystems on Earth.

The extent of sulfide mineral deposits along oceanic ridges is poorly known, and although mining is a possibility and leases to some possible deposits are being considered, it seems unlikely that such deposits will be extracted at a profit in the near future. The potential environmental effects, including serious physical and biological impacts associated with water quality and sediment pollution, will have to be carefully evaluated prior to any mining activity.

Manganese Oxide Nodules

Manganese oxide nodules cover vast areas of the deep-ocean floor. Some are found in the Atlantic Ocean off Florida, but the richest and most extensive accumulations occur in large areas of the northeastern, central, and southern Pacific, where the nodules cover 20% to 50% of the ocean floor.[12] The nodules are most abundant in those parts of the ocean where sediment accumulation is at a minimum, generally at depths of 2,500 to 6,000 m (8,200 to 19,700 ft).[13] The average size of the manganese nodules varies from a few millimeters to a few tens of centimeters in diameter, and their estimated rate of growth is 1–5 mm (about 0.04–0.2 in) per million years. The nodules are composed primarily of concentric layers of manganese in iron oxides mixed with a variety of other materials. They contain manganese (24%) and iron (14%) with secondary copper (1%), nickel (1%), and cobalt (0.25%) and form around a nucleus of broken nodules, fragments of volcanic rock, and sometimes fossils.

The origin of the metals in the nodules is not well understood. Most likely, material weathers from the con-

The Law of the Seabed

Mining the seabed generates a great deal of controversy. For one thing, there are environmental concerns about disrupting and degrading the bottom and near-bottom environments. In addition, the potential exploitation of seabed resources raises questions about who has the right to do the mining. Developing countries argue that the mineral wealth of the seabed is a common heritage of all people and should not benefit only the wealthy nations that can exploit the resource; but international consortia of companies that have spent hundreds of millions of dollars investigating seabed resources want a return on their investments.

The United Nations has sponsored Law of the Seabed conferences aimed at passing international laws (treaties) concerning ocean mining. The International Law of the Sea became a reality in 1994. The law states that seabed resources outside national limits are a common heritage of humankind. Seabed resources may be exploited (with International Seabed Authority) either privately by mining companies or publicly by international enterprises.[15]

Industrial nations have threatened to mine the seabed without an international agreement. The U.S. Congress, for example, has passed legislation authorizing U.S. companies to mine the ocean floor. Some developing countries have responded by threatening to withhold other resources, such as oil and metals, from countries that mine the seabed without approval of the United Nations. Threats aside, mining of the seabed has raised important ethical questions that must be addressed by all countries interested in the oceans' resources.

---

tinents and is transported in solution by rivers to the oceans, where it drops out of solution and forms nodules. Both inorganic precipitation and bacteria-induced precipitation are important in the formation of nodules.

Expenditures for mining and metallurgical research to recover manganese oxide nodules have surpassed several hundred million dollars. At least 20 corporations in several countries have examined metallurgical systems to process the nodules. Some would produce cobalt, copper, nickel, and manganese; and others would produce combinations of only copper and nickel.[14] However, commercial exploitation of the nodules is not expected to be economically and technically viable until well into the 21st century.[15]

 29.3 Resources and Reserves

We can classify minerals as resources or reserves. **Mineral resources** are broadly defined as elements, chemical compounds, minerals, or rocks concentrated in a form that can be extracted to obtain a usable commodity—something that can be bought and sold. It is assumed that a resource can be extracted economically or at least has the potential for economic extraction. A **reserve** is that portion of a resource that is identified and from which usable materials can be legally and economically extracted *at the time of evaluation* (Figure 29.4).

Whether a mineral deposit is classified as part of the resource base or as a reserve may be a question of economics. For example, if an important metal becomes scarce, the price may rise, which would encourage exploration and extraction (mining). As a result of the price increase, previously uneconomic deposits (part of the resource base before the scarcity and price rise) may become profitable, and those deposits would be reclassified as reserves.

The main point here is that *resources are not reserves.* An analogy from a student's personal finances may help clarify this point. A student's reserves are liquid assets, such as money in the bank, whereas the student's resources include the total income the student can expect to earn during his or her lifetime. This distinction is often critical to the student in school, because resources that may become available in the future cannot be used to pay this month's bills.[16]

Regardless of potential problems, it is important for planning purposes to estimate future resources. This task requires continual reassessment of all components of a total resource through consideration of new technology, the probability of geologic discovery, and shifts in economic and political conditions.[2]

Silver provides an illustration of some important points about resources and reserves. Based on geochemical estimates of the concentration of silver in rocks, Earth's crust (to a depth of 1 km, or 0.6 mi) contains almost 2×10^{12} metric tons of silver, an amount much larger than annual world use, which is approximately 10,000 metric tons. If this silver existed as pure metal concentrated in one large mine, it would represent a supply sufficient for several hundred million years at current levels of use. Most of the silver, however, exists in extremely low concentrations, too low to be extracted economically with current technology. The known reserves of silver, reflecting the

(a)

(b)

(c)

Figure 29.3 • (*a*) Oceanic ridge hydrothermal environment. (*b*) Detail of black smokers, where massive sulfide deposits form. (*c*) Black smoker photographed from the minisubmarine *Alvin* during a 1991 cruise to the East Pacific Rise.

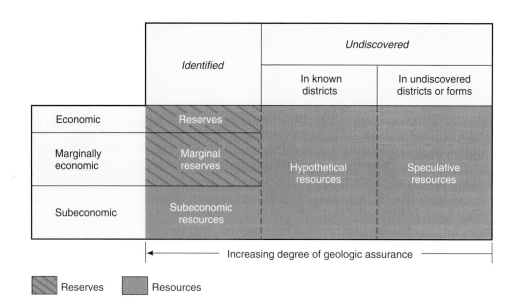

	Identified	Undiscovered	
		In known districts	In undiscovered districts or forms
Economic	Reserves		
Marginally economic	Marginal reserves	Hypothetical resources	Speculative resources
Subeconomic	Subeconomic resources		

⟵ Increasing degree of geologic assurance ⟶

 Reserves ▢ Resources

Figure 29.4 • Classification of mineral resources used by the U.S. Geological Survey and the U.S. Bureau of Mines. (*Source: Principles of a Resource Preserve Classification for Minerals*, U.S. Geological Survey Circular 831, 1980.)

amount we could obtain immediately with known techniques, is about 200,000 metric tons, or a 20-year supply at current use levels.

The problem with silver, as with all mineral resources, is not with its total abundance but with its concentration and relative ease of extraction. When an atom of silver is used, it is not destroyed but remains an atom of silver. It is simply dispersed and may become unavailable. In theory, given enough energy, all mineral resources could be recycled, but this is not possible in practice. Consider lead, which is mined from minerals in which it is concentrated. The lead that was used in gasoline for many years is now scattered along highways across the world and deposited in low concentrations in forests, fields, and salt marshes close to these highways. Recovery of this lead is, for all practical purposes, impossible.

29.4 Classification, Availability, and Use of Mineral Resources

Earth's mineral resources can be divided into several broad categories, depending on our use of them: elements for metal production and technology; building materials; minerals for the chemical industry; and minerals for agriculture. Metallic minerals can be further classified according to their abundance. The abundant metals include iron, aluminum, chromium, manganese, titanium, and magnesium. Scarce metals include copper, lead, zinc, tin, gold, silver, platinum, uranium, mercury, and molybdenum.

Some mineral resources, such as salt (sodium chloride), are necessary for life. Primitive peoples traveled long distances to obtain salt when it was not locally available. Other mineral resources are desired or considered necessary to maintain a particular level of technology.

When we think about mineral resources we usually think of the metals; but, with the exception of iron, the predominant mineral resources are not metallic. Consider the annual world consumption of a few selected elements. Sodium and iron are used at a rate of approximately 100–1,000 million metric tons per year. Nitrogen, sulfur, potassium, and calcium are used at a rate of approximately 10–100 million metric tons per year, primarily as soil conditioners or fertilizers. Elements such as zinc, copper, aluminum, and lead have annual world consumption rates of about 3–10 million metric tons, and gold and silver have annual consumption rates of 10,000 metric tons or less. Of the metallic minerals, iron makes up 95% of all the metals consumed; and nickel, chromium, cobalt, and manganese are used mainly in alloys of iron (as in stainless steel). Thus, with the exception of iron, the nonmetallic minerals are consumed at much greater rates than are elements used for their metallic properties.

Availability of Mineral Resources

The basic issue with mineral resources is not actual exhaustion or extinction but the cost of maintaining an adequate stock within an economy through mining and recycling. At some point, the costs of mining exceed the worth of material. When the availability of a particular mineral becomes a limitation, there are four possible solutions:

1. Find more sources.
2. Recycle what has already been obtained.
3. Find a substitute.
4. Do without.

Which choice or combination of choices is made depends on social, economic, and environmental factors.

The availability of a mineral resource in a certain form, in a certain concentration, and in a certain amount is a geologic issue determined by Earth's history. What is considered a resource and at what point a resource becomes limited are ultimately social questions. Before metals were discovered, they could not be considered resources. Before smelting was invented, the only metal ores were those in which the metals appeared in their pure form. For example, originally gold was obtained as a pure, or native, metal. Now gold mines are deep beneath the surface, and the recovery process involves reducing tons of rock to ounces of gold.

Mineral resources are limited, which raises important questions. How long will a particular resource last? How much short-term or long-term environmental deterioration are we willing to accept to ensure that resources are developed in a particular area? How can we make the best use of available resources? These questions have no easy answers. We are now struggling with ways to better estimate the quality and quantity of resources.

Mineral Resource Use Rates

We can use a particular mineral resource in several ways: rapid consumption, consumption with conservation, or consumption and conservation with recycling. Which option is selected depends in part on economic, political, and social criteria. Figure 29.5 shows the hypothetical depletion curves corresponding to these three options. Historically, with the exception of precious metals, rapid consumption has dominated most resource utilization. However, as the supply of resources becomes short, increased conservation and recycling are expected. Certainly, the trend toward recycling is well established for metals such as copper, lead, and aluminum.

From a global viewpoint, limits on our mineral resources and reserves threaten our affluence. As the world population and the desire for a higher standard of living increase, the demand for mineral resources expands at a faster and faster rate. Today the more developed countries, with about 16% of Earth's population, consume a disproportionate amount of the mineral resources extracted. For example, the United States, western Europe, and Japan collectively use about 70% of all aluminum, copper, and nickel that is extracted from the earth.[5] Predicted increases in the world use of iron, copper, and lead, when linked with expected population increases, suggest that the rate of production of these metals will have to increase by several times if the world per capita consumption rate is to rise to the level of consumption in developed countries today. Such an increase is very unlikely; affluent countries will thus have to find substitutes for some minerals or use a smaller proportion of the world's annual production. This situation parallels the Malthusian predictions discussed in Chapter 5: It is impossible in the long run to support an ever-increasing population on a finite resource base.

U.S. Supply of Mineral Resources

Domestic supplies of many mineral resources in the United States are insufficient for current use and must be supplemented by imports from other nations. For example, the United States imports many of the minerals needed for its complex military and industrial system, called strategic minerals (examples include bauxite, manganese, graphite, cobalt, strontium, and asbestos). Of particular concern is the possibility that the supply of a much desired or much-needed mineral may be interrupted by political, economic, or military instability in the supplying nation. For example, in recent years, a civil war in Zaire caused such concern that the price in cobalt temporarily increased by 800%.[17] Cobalt, a metal used to strengthen steel, is in great demand in many industrial countries.

The fact that the United States—along with many other countries—depends on a steady supply of imports to meet the mineral demand of industries does not necessarily mean that the imported minerals do not exist within the country in quantities that could be mined. Rather, it suggests that there are economic, political, or environmental reasons that make it easier, more practical, or more desirable to import the material. This situation has resulted in political alliances that otherwise would be unlikely. Industrial countries often need minerals from countries with whose policies they do not necessarily agree; as a result, they make political concessions on human rights and other issues that they would not otherwise make.

29.5 Impacts of Mineral Development

The impact of mineral exploitation on the environment depends on such factors as ore quality, mining procedures, local hydrologic conditions, climate, rock types, size of operation, topography, and many more interrelated factors. The impact varies with the stage of develop-

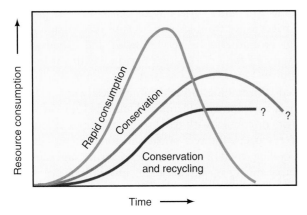

Figure 29.5 • Several hypothetical depletion curves.

ment of the resource. For example, the exploration and testing stages involve considerably less impact than do the mining and processing stages. In addition, our use of mineral resources has a significant social impact.

Environmental Impacts

Exploration activities for mineral deposits vary from collecting and analyzing remote-sensing data gathered from airplanes or satellites, to fieldwork involving surface mapping and drilling. Generally, exploration has a minimal impact on the environment, provided that care is taken in sensitive areas, such as arid lands, marshes, and areas underlain by permafrost. Some arid lands are covered by a thin layer of pebbles over fine silt several centimeters thick. The layer of pebbles, called desert pavement, protects the finer material from wind erosion. When the desert pavement is disturbed by road building or other activities, the fine silts may be eroded, impairing physical, chemical, and biological properties of the soil and possibly scarring the land for many years. Similarly, marshes and other wetlands, such as the northern tundra, are very sensitive to even seemingly small disturbances such as vehicular traffic.

The mining and processing of mineral resources generally have a considerable impact on land, water, air, and biological resources. Furthermore, as it becomes necessary to use ores of lower and lower grades, negative effects on the environment tend to become greater problems. For example, there is concern about asbestos fibers in the drinking water (from Lake Superior) of Duluth, Minnesota, as a result of the disposal of iron mining waste (tailings) from mining low-grade iron ore.

A major practical issue is whether surface or subsurface mines should be developed in an area. Surface mining is cheaper but has more direct environmental effects. The trend in recent years has been away from subsurface mining and toward large, open-pit (surface) mines, such as the Bingham Canyon copper mine in Utah (Figure 29.6). The Bingham Canyon mine is one of the world's largest human-made excavations, covering nearly 8 km² (3 mi²) to a maximum depth of nearly 800 m (2,600 ft).

Surface mines and quarries today cover less than 0.5% of the total area of the United States. Even though the impact of these operations is a local phenomenon, however, numerous local occurrences will eventually constitute a larger problem. Environmental degradation tends to extend beyond the land actually being mined. Large mining operations disturb the land by directly removing material in some areas and dumping waste in others, thus changing topography. At the very least, these actions produce severe aesthetic degradation. In addition, dust at mines may affect air resources, even though care is often taken to reduce dust production by sprinkling water on roads and on other sites that generate dust.

A potential problem associated with mineral resource development is the possible release of harmful trace ele-

Figure 29.6 • Aerial photograph of Bingham Canyon Copper Pit, Utah. It is one of the largest artificial excavations in the world.

ments to the environment. Water resources are particularly vulnerable to such degradation, even if drainage is controlled and sediment pollution is reduced. Surface drainage is often altered at mine sites, and runoff from precipitation may infiltrate waste material, leaching out trace elements and minerals. Trace elements (cadmium, cobalt, copper, lead, molybdenum, and others), when leached from mining wastes and concentrated in water, soil, or plants, may be toxic or may cause diseases in people and other animals who drink the water, eat the plants, or use the soil. Specially constructed ponds to collect such runoff help but cannot be expected to eliminate all problems. The white streaks in Figure 29.7 are mineral deposits apparently leached from tailings from a zinc mine in Colorado. Similar-looking deposits may cover rocks in rivers for many kilometers downstream from some mining areas.

Groundwater may also be polluted by mining operations when waste comes into contact with slow-moving subsurface waters. Surface water infiltration or groundwater movement causes leaching of sulfide minerals that may pollute groundwater and eventually seep into streams to pollute surface water. Groundwater problems are particularly troublesome because reclamation of polluted groundwater is very difficult and expensive (see Chapter 21 for a discussion of acid mine drainage.)

Physical changes in the land, soil, water, and air associated with mining directly and indirectly affect the biological environment. Plants and animals killed by mining activity or contact with toxic soil or water are examples of direct impacts. Indirect impacts include changes in nutrient cycling, total biomass, species diversity, and ecosystem stability owing to alterations in groundwater or surface water availability or quality. Periodic or accidental discharge of low-grade pollutants through failure of barriers, ponds, or water diversions or through breach of barriers

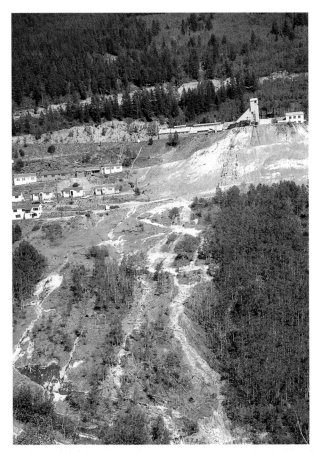

Figure 29.7 • Tailings from a lead, zinc, and silver mine in Colorado. White streaks on the slope are mineral deposits apparently leached from the tailings.

during floods, earthquakes, or volcanic eruptions also may damage local ecological systems to some extent.

Social Impacts

Social impacts associated with large-scale mining result from the rapid influx of workers into areas unprepared for growth. Stress is placed on local services such as water supplies, sewage and solid-waste disposal systems, schools, and housing. Land use shifts from open range, forest, and agriculture to urban patterns. The additional people also increase the stress on nearby recreation and wilderness areas, some of which may be in a fragile ecological balance. Construction activity and urbanization affect local streams through sediment pollution, reduced water quality, and increased runoff. Air quality is reduced as a result of more vehicles, dust from construction, and generation of power.

Adverse social impacts also occur when mines are closed; towns surrounding large mines come to depend on the income of employed miners. Closures of mines produced the well-known ghost towns in the old American West. Today, the price of coal and other minerals directly affects the livelihood of many small towns.

This relationship is especially evident in the Appalachian Mountain region of the United States, where closures of coal mines are taking their toll. These mine closings are partly the result of lower prices for coal and partly the result of rising mining costs.

One of the reasons mining costs are rising is the increased level of environmental regulation of the mining industry. Of course, regulations have also helped make mining safer and have facilitated land reclamation. Some miners, however, believe the regulations are not flexible enough, and there is some truth to their arguments. For example, some mined areas might be reclaimed for use as farmland now that the original hills have been leveled. Regulations, however, may require the restoration of the land to its original hilly state, even though hills make inferior farmland.

29.6 Minimizing Environmental Impact of Mineral Development

Minimization of environmental impacts of mineral development requires consideration of the entire cycle of mineral resources shown in Figure 29.8. Inspection of this diagram reveals that many components of the cycle are related to the generation of waste material. In fact, the major environmental impacts of mineral resource utilization are related to waste products. Waste produces pollution that may be toxic to humans, may harm natural ecosystems and the biosphere, and may be aesthetically undesirable. Waste may attack and degrade resources such as air, water, soil, and living things. Waste also depletes nonrenewable mineral resources and, when simply disposed of, provides no offsetting benefits for human society.

Minimization of environmental effects associated with mineral development can take several paths:

- *Environmental regulations at the federal, state, and local levels.* The regulations address topics such as sediment, air, and water pollution resulting from all aspects of the mineral cycle. Regulations may also address reclamation of land used for mining. Today in the United States, approximately 50% of the land utilized by the mining industry has been reclaimed.

- *On-site and off-site treatment of waste.* Minimizing on-site and off-site problems by controlling sediment, water, and air pollution through good engineering and conservation practices is an important goal. Of particular interest is the development of biotechnological processes such as biooxidation, bioleaching, and biosorption, as well as genetic engineering of microbes. These practices have enormous potential for both extracting metals and minimizing environmental degradation. For example, engineered (constructed) wetlands are being used at several sites; acid-tolerant plants in the wetlands remove metals from mine waste-

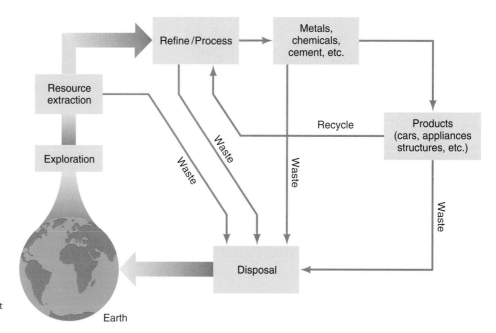

Figure 29.8 • Simplified flowchart of the resource cycle.

waters and neutralize acids by biological activity.[18] At the Homestake Gold Mine in South Dakota, biooxidation is being used to convert contaminated water from the mining operation to substances that are environmentally safe; the process uses bacteria with a natural capacity to oxidize cyanide to harmless nitrates.[19]

• *Practicing the three R's of waste management.* That is, reduce the amount of waste produced; reuse materials in the waste stream as much as feasible; and maximize recycling opportunities.

Let's look at this third option in greater detail. Wastes from some parts of the mineral cycle may themselves be referred to as ores, because they contain materials that might be recycled and used again to provide energy or useful products.[20,21] The notion of reusing waste materials is not new. Such metals as iron, aluminum, copper, and lead have been recycled for many years and are still being recycled today. For example, the metal from almost all of the millions of automobiles discarded annually in the United States is recycled.[21,23]

In 1998, the total value of recycled metals was $43 billion. Of this, iron and steel amounted to approximately 90% by weight and 40% by total value of recycled metals. Iron and steel are recycled in such large volumes for three reasons:[22] First, the market for iron and steel is huge; and as a result, there is a large scrap collection and processing industry. Second, an enormous economic burden would result from failure to recycle. Third, failing to recycle would create significant environmental impacts related to disposal of over 50 million tons of iron and steel.

It is estimated that each ton of recycled steel saves 1,136 kg (2,500 pounds) of iron ore, 455 kg (1,000 pounds) of coal, and 18 kg (40 pounds) of limestone. In addition, only one-third as much energy is required to produce steel from recycled scrap as from native ore.[23]

Other metals that are recycled in large quantities include lead (63%), aluminum (38%), and copper (36%).[3] Recycling aluminum reduces our need to import raw aluminum ore and saves approximately 95% of the energy required to produce new aluminum from bauxite.[23]

Summary

• Mineral resources are usually extracted from naturally occurring, anomalously high concentrations of earth materials. Such natural deposits allowed early peoples to exploit minerals while slowly developing technological skills.

• The origin and distribution of mineral resources is intimately related to the history of the biosphere and the geologic cycle. Nearly all aspects and processes of the geologic cycle are involved to some extent in producing local concentrations of useful materials.

• Mineral resources are not mineral reserves. Unless discovered and developed, resources cannot be used to address present shortages.

• The availability of mineral resources is one measure of the wealth of a society. Modern technological civilization would not be possible without the exploitation of mineral resources. However, it is important to recognize that mineral deposits are not infinite and that we cannot maintain exponential population growth on a finite resource base.

• The United States and many other affluent nations rely on imports for their supplies of many mineral

Will Mining with Microbes Help the Environment?

Mining is an ancient technology first practiced at least 6,500 years ago. Modern mining methods are more technologically sophisticated but use the same basic processes (digging and smelting) to isolate valuable metals. To be economic, these methods have traditionally required high-grade ore and cheap sources of energy as well as an acceptance that mining to a lesser or greater extent would damage the environment. Although these conditions have prevailed for most of human history, they are changing. Earlier exploitation of mineral resources is pushing the mining industry toward lower-grade ores; nonrenewable energy sources are expensive and disappearing; and concern over degradation of the environment and health threats to humans and other species is growing. Demand for minerals, however, is increasing, because of both population growth and technological development.

As an example, the average grade of copper ore has dropped from 6 to 0.6% over the last century, making copper mining more energy intensive and more wasteful. Five metric tons of coal are required to produce 1 metric ton of copper, and every kilogram of copper produced represents 89 kg (198 lb) of waste. Open-pit mining of copper causes acids and heavy metals, such as arsenic, to contaminate surface water and groundwater. Smelting produces sulfur dioxide and other gaseous compounds, as well as particles, which contribute to air pollution.

Microscopic organisms produced by biotechnology offer an entirely new approach to mining. By 1989, more than 30% of copper mined in the United States depended on a biochemical process that begins with a microbe, *Thiobacillus ferrooxidans*. Biological processes have also been used in mining uranium and gold. Research is underway to use microbes to remove sulfur from coal and cyanide from mining waste. The union of biological processes and mining is called *biohydrometallurgy*.

In the future, it may be possible to use microbes on ores without removing them from Earth. Metallurgists envision

drilling wells into the ore and fracturing it, then injecting bacteria into the wells and fractures. The ore could be removed by flooding the wells with water, removing the ore, and recycling the water. Biotechnologists hope to use genetic engineering to develop bacteria to mine specific metals when no naturally occurring bacteria exist to extract them.

The disadvantages of biohydrometallurgy are that it is slow, requiring decades rather than years, and that methods for breaking ores into small enough particles for efficient extraction are not yet available. Already, however, biological methods are economically feasible for low-grade ores that elude conventional methods. Further technological innovations may make them competitive in more situations.

Critical Thinking Questions

1. What are the environmental advantages of biohydrometallurgy over conventional methods? What are the possible disadvantages?

2. How would you assess the possible dangers from genetically engineered organisms developed for mining compared with the dangers from organisms engineered for use in agriculture and medicine?

3. Some experts believe that without economic pressure (for example, the decline in high-grade ore, increased energy requirements for extracting metal, governmental regulations on clean air and water, economic recessions), the mining industry would not continue to explore biochemical methods. How could the government encourage the industry to devote more research and development efforts to expanding biochemical mining? Develop a proposal for a government policy.

resources. As other nations industrialize and develop, such imports may be more difficult to obtain, and affluent countries may have to find substitutes for some minerals or use a smaller portion of the world's annual production.

- The environmental impact of mineral exploitation depends on many factors, including mining procedures, local hydrologic conditions, climate, rock types, size of operation, topography, and other factors.

- The mining and processing of mineral resources greatly affect the land, water, air, and biological resources

and create social impacts as a result of the increased demand for housing and services in mining areas.

- Because the demand for mineral resources will increase in the future, we must strive to minimize both on-site and off-site problems by controlling sediment, water, and air pollution through good engineering and conservation practices.

- Reducing consumption, reusing, and recycling mineral resources are environmentally preferable ways to delay or partly alleviate possible crises caused by the convergence of a rapidly rising population and a limited resource base.

Human Population: Increasing human population is a major issue in the availability and use of mineral resources. We would be experiencing a major mineral crisis today if the consumption rate for all the people of the world were anywhere near that of people in developed countries. Therefore, as human population increases and developing countries increase their own demands, we must find ways to reduce our per capita mineral consumption. As we have emphasized before, satisfying the mineral needs of an ever-expanding population on a finite resource base at current levels is impossible in the long term.

Sustainability: Many sustainability issues arise from our use of mineral resources, because so many parts of the mineral cycle are related to waste management problems involving pollution of soil, air, and water. If we are to provide a quality environment for future generations, we must find ways to minimize adverse environmental effects of mineral exploitation, development, and use. Of particular importance will be conservation measures aimed at reducing our use of minerals, reusing materials whenever possible, and recycling as much as is feasible.

Global Perspective: Mineral resources are scattered over the Earth at various locations, where they were concentrated by geologic processes. Because the origin of many mineral deposits is directly related to global tectonics, understanding how our world works gives us a better perspective on our mineral resources.

Urban World: The use of mineral resources is concentrated in urban areas, where most people live. Therefore, urban areas are the places where recycling of minerals such as aluminum, iron, and copper is most economically viable.

Science and Values: Because we value a quality environment and because mining and other activities related to the mineral cycle can cause environmental disruption, we have a great interest in developing the knowledge necessary to minimize adverse effects of mineral utilization. A variety of new technologies in this area have arisen in recent years, including biotechnology, which utilizes the biological environment to assist in solving environmental problems related to the mineral cycle, particularly as it relates to waste management.

● KEY TERMS

brines **628**

marine evaporites **628**

mineral resources **630**

nonmarine evaporites **628**

ore deposits **626**

reserve **630**

secondary enrichment **628**

● STUDY QUESTIONS

1. What is the difference between a resource and a reserve?

2. Under what circumstances might sewage sludge be considered a mineral resource?

3. If surface mines and quarries cover less than 0.5% of the land surface of the United States, why is there so much environmental concern about them?

4. When is recycling a mineral a viable option?

5. Which biological processes can influence mineral deposits?

6. A deep-sea diver claims that the oceans can provide all our mineral resources with no negative environmental effects. Do you agree or disagree?

7. What factors determine the availability of a mineral resource?

8. Utilizing a mineral resource involves four general phases: (a) exploration, (b) recovery, (c) consumption, and (d) disposal of waste. Which phase do you think has the greatest environmental effect?

FURTHER READING

Brookins, D. G. 1990. *Mineral and Energy Resources.* Columbus, Ohio: Charles E. Merrill. A good summary of mineral resources.

Earney, F. C. F. 1990. *Marine Mineral Resources.* Columbus, Ohio: Charles E. Merrill. A summary of marine mineral resources.

Park, C. F., Jr., and MacDiarmid, R. A. 1975. *Ore Deposits*, 3rd ed. San Francisco: W. H. Freeman. A good, basic book on geologic aspects of ore deposits.

U.S. Bureau of Mines. 1995. *FY Annual Research Report.* Washington, D.C.: U.S. Department of the Interior. A report that describes the bureau's efforts to meet U.S. mineral needs and to mitigate the associated economic, human, and environmental costs of mining and mineral processing.

30 Planning for a Sustainable Future

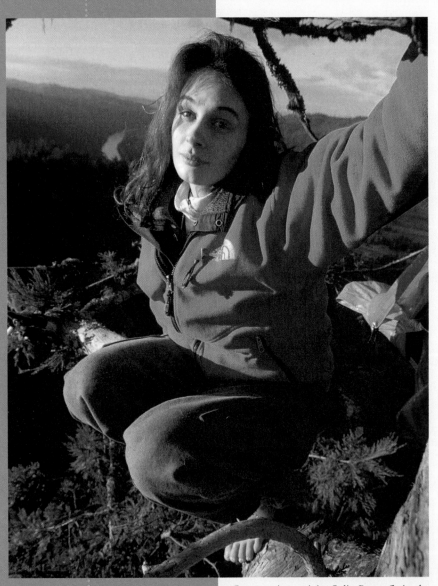

Conservation activist, Julia Butterfly in the Redwood tree named luna. She lived in the tree for about two years to protect it from being cut down by loggers.

● LEARNING OBJECTIVES

Environmental law can contribute to sustainability, as can evaluating the landscape for environmental impact and land-use. After reading this chapter, you should understand:

- How we might move toward achieving sustainability.
- How mediation is used as a tool in environmental law.
- How international environmental agreements and laws are helping solve some environmental problems.
- What are the major components of an environmental impact statement (EIS).
- What the processes of scoping and mitigation in environmental impact assessment include.
- What are the steps in land-use planning.
- Why increase in human population linked to changes in land use is increasing the occurrence of catastrophes resulting from natural hazards.
- That creating a more secure world has a significant environmental component.

Butterfly in a Tree

Butterfly is the name given to Julia Hill, a young woman who spent two years in a giant redwood to draw attention to the harvesting of these magnificent trees, many of which are more than 1,000 years old. Imagine living in a tree 60 m (180 ft) above the ground through intense rainstorms, snow, and ice. Imagine further living in a tree in an area that is being actively logged. This is nothing like sitting in a pristine forest environment; it is more like being on an industrial site with the noises of chainsaws, trucks, and helicopters. Added to this were the loggers, some of whom were extremely critical of her activities. Sometimes, helicopters would fly rapidly around the tree, the winds created by the blades ripping off branches, in hopes of intimidating Butterfly and forcing her to return to the ground.[1]

Butterfly had arrived in the redwoods only a few days prior to climbing the tree, and she had only intended to stay for a short time. She had not previously been part of any environmental protest or act of civil disobedience. The land being logged was owned by the Pacific Lumber Company. Eventually, through negotiations, the tree Butterfly was living in, which was nicknamed Luna, was protected with a 65-m (200-ft) buffer zone. Butterfly then descended from the tree and began to spread her message of the importance of forest conservation to the people of the world. In late 2000, someone using a chainsaw cut into Luna, causing damage that required reinforcing with cables and steel plates to preserve the tree.

The issues surrounding Butterfly's occupation of the giant redwood tree touch on all five key themes we have emphasized in this book. As population increases, there is increased demand for natural resources—such as timber—and this demand is leading to the worldwide problem of deforestation. The forests of the world provide many natural service functions for the entire biosphere, such as cleansing air, releasing water that produces precipitation, and sending out roots that reduce soil erosion. If we are to sustain our environment for future generations, we thus need to sustain the world's forests. This will require us to take a global perspective while recognizing that we often act at a local or regional level.

The demand for timber products is most intense in our urban centers, where wood is used to build homes, furniture, and so forth. Although much environmental activism is centered in settings such as the redwood forest, we now recognize the importance of cities as livable environments and are attempting to develop harmonious relationships between industrial and commercial activities. Finally, we recognize that while science plays an important role in our understanding of natural systems such as the redwood forest and its importance to the larger global environment, it's our values that determine the choices we make concerning these systems.

● *The story of Butterfly in the tree is a story of an individual—a story with far-reaching implications. In this chapter, we first explore some of these implications by discussing how we might move toward sustainability. We then examine environmental law, environmental impact analysis, land-use planning, and global forecasting as a means to help us move in this direction.*

30.1 Developing a Sustainable Future

Stating that we wish to develop a sustainable future acknowledges that our present practices are not sustainable. Indeed, continuing in our present paths of overpopulation, resource consumption, and pollution will not lead to sustainability. What is needed is development of new concepts that will mold industrial, social, and environmental interests into an integrated, harmonious system. In other words, we need to develop a new paradigm—a new pattern—as an alternative to our present model for running society and creating wealth.[2] The new paradigm might be described in the following ways:[3]

● *Evolutionary rather than revolutionary.* Development of a sustainable future will require an evolution in our values that involves our lifestyles as well as social, economic, and environmental justice.

- *Inclusive, not exclusive.* All peoples of Earth must be included. This means bringing the people of the world to a higher standard of living in a sustainable way that will not compromise our environment.

- *Proactive, not reactive.* We must plan for change and for events such as human population problems, resource shortages, and natural hazards rather than waiting for them to surprise us and then reacting.

- *Attracting, not attacking.* People must be attracted to the new paradigm because it is right and just. Those who speak for our environment should not take a hostile stand but, through sound, scientific argument and appropriate values, should attract people to the path of sustainability.

- *Assisting the disadvantaged, not taking advantage.* This involves issues of environmental justice. All people have the right to live and work in a safe, clean environment. Working people around the globe need to receive a living wage sufficient to support their families. Exploitation of workers to reduce costs of manufacturing goods or growing food diminishes us all.

People Value Environmental Protection

The path to a sustainable future requires careful planning as well as the diligent support of people. In the United States, a poll at the national level in 2001 found high levels of support for environmental goals:[4]

- Nearly 90% of the people polled said that protecting the environment should be a top priority of the present administration.

- Approximately 60% of the respondents believe that the United States should abide by the Kyoto agreements to control emissions of carbon dioxide, which contribute to global warming.

- Approximately 60% believe it is important to protect endangered species from extinction, even if it means people are not allowed to develop the land as they wish.

- Just over 50% believe that improving the environment does not have to conflict with economic growth.

- Half stated that if there is a conflict between the environment and economic growth, the environment should take priority; 36% stated that the economy should take priority, and 14% did not have an opinion.

- About 65% hold the view that many businesses will cut corners and damage the environment unless strong government rules and regulations are in place to prevent them from doing so.

- When asked to state the most important environmental problem facing the United States today, people selected the following as their top responses: pollution, air quality, global warming and ozone depletion, and energy production. By a large margin, pollution was the top response; and there are obviously overlaps among the categories.

These results are consistent with those of other national polls taken in recent years. Polls have consistently demonstrated that people in the United States value a quality environment and will support programs to protect the environment. Thus, it appears that there is a popular mandate for environmental protection. This mandate is necessary if we are to develop a path toward sustainability. Of particular importance will be cooperation among the industrial, environmental, and social spheres of our economy. Many examples suggest that such cooperation can be found in various parts of the world today.[2]

Curitiba, Brazil: A Positive Example

You may recall the story of Curitiba, Brazil, from Chapter 27. With a population of over 2 million and growing, Curitiba could well be a center for unemployment, poverty, pollution, and despair. However, through careful planning, responsible government, and appropriate entrepreneurship, Curitiba is achieving a far different result. The lessons to be learned from the transformation of Curitiba hold promise for urban areas throughout the world.

In Curitiba, 95% of the people are literate, nearly 100% of households have drinking water and electricity, 98% of households have trash collection, 75% of homes are owner-occupied, and people have a life expectancy of 72 years. Some other specific accomplishments include the following:[2]

- As described in the Case Study for Chapter 27, Curitiba's public transportation system moves more than a million people each day. The system uses large, fuel-efficient buses, resulting in lower fuel use and less air pollution than many other Brazilian cities.

- Green space is highly valued in the city, and new buildings outside the urban core must preserve 50% of the site's area as green space. Permanent protection is provided for green areas, and tax relief rewards the preservation of woodlands and gardens. Some visitors have reported that the city is as much like a wildlife reserve as an urban area. Many species of birds, mammals, and reptiles make their home in the greenbelts of the city.

- Recycling is important in Curitiba, and everything from trash to buildings to buses is recycled. A quarry was recycled to become an amphitheatre; a gunpowder factory was recycled to become a theatre; and an old garbage dump was recycled to become a botanical garden. Buses, following several years of service, are recycled as training centers, medical clinics, food distribution centers, and other services for which a mobile temporary building is needed.

- Education, jobs, and day care are given high priority.

The success of Curitiba has resulted from wise leadership that integrated environmental design processes with participation by the public and business sectors of the community, who widely share a common vision for the future. It seems evident that underlying the success are simple principles that have profound implications. The primary principles are respect for people and public assets, along with provision of services to the community. The people feel respected and, as a result, feel responsibility to solve common problems that otherwise would degrade urban life. So far, the results suggest that it is possible to have a healthy environment as well as a successful, just economy in a society that values humanity.[2]

Florida House

We are also learning to integrate principles of sound environmental development at the level of individual buildings. Investigations of ancient architecture suggest that earlier societies paid careful attention to the environment in constructing buildings. For example, approximately 1,000 years ago in what is now Mesa Verde, Colorado, Native Americans built their houses with solar energy in mind. Homes were set in shallow recesses along the steep sides of the mesas so that they were shaded from summer sun by the rocks above and warmed directly by winter sun. The Native Americans built their homes using the concept that today we call "design with nature."[5]

In contrast, many homes built in the United States during the 20th century, when central heating and air conditioning were common and fossil fuels were cheap, were designed with little concern for their location relative to sunlight and other environmental factors. Fortunately, more environmentally sound development is now being fostered in a number of locations around the world. One good example of this is Florida House, completed and opened to the public in May 1994.

Southwestern Florida is a rapidly growing area. Sarasota, Florida, is no exception; the city has experienced growth pains related in particular to water problems. Pollution from urban runoff and drought in the late 1980s was thought to be responsible for degradation of Sarasota Bay. In 1991, a two-year building moratorium appeared on the ballot; and although the initiative failed, it brought attention to the environmental problems of the area. Environmentalists, developers, and others joined together in an attempt to develop a new type of home that would be a model of resource efficiency (Figure 30.1). The objective was not necessarily to develop the most energy-efficient home possible and use the least possible water resources but rather to develop a home using technology that would be widely accessible to the community.

Florida House incorporates a number of interesting environmental features, including both passive and active solar systems (see Chapter 18) and the use of many recy-

Figure 30.1 • Florida House, Sarasota, Florida. This normal looking home incorporates many special design features generally available today to conserve energy and water resources. For example, the banana trees (center) use gray water from laundry and showers.

cled materials for construction. In addition, conservation of water is an important objective. For example:

- Irrigation systems use "gray water" rich in nutrients from home laundries, sinks, and baths to supply water for vegetation, including banana trees.
- Rain water from the roof is collected and stored in cisterns to augment the irrigation system.
- Use of porous concrete and other materials around the home allow for more infiltration of water into the soil, reducing surface runoff.
- Surface water is collected and routed to a small on-site artificial wetland that serves as a sediment trap and further facilitates infiltration of water into the soil.

In summary, Florida House is a successful venture promoted by both environmentalists and developers to provide readily accessible technology for resource conservation at the community level. The home also plays an important role in environmental education by providing information to other communities interested in what are sometimes termed "green choices" with respect to development. Thus, Florida House has become a learning center. Demonstration projects such as Florida House are helping to develop ideas and concepts necessary in establishing sustainable development for future generations. In fact, some 21st-century Florida construction is imitating Florida House. One development in Sarasota has deed restrictions that promote conservation of water and energy, replacing lawns with preservation of mature trees and shrubs. Innovative energy features provide much-improved insulation for the warm climate.

Environmental Law

We have briefly discussed why and how we might move toward sustainability. Both in the United States and internationally, people have used laws to help achieve this goal. The field of environmental law has grown rapidly since the 1970s. Today, environmental law is an important part of U.S. jurisprudence.

Origins of Environmental Law

The legal system of the United States has historic origins in the British system. When our legal system was formed, it preserved and strengthened the tendency of British law to protect the individual from society rather than the reverse. Individual freedom and nearly unlimited discretion to use property as the owner pleased were given high priority, and the powers of the federal government were strictly limited. However, when behavior infringed on the property or the well-being of others, the **common law** (law derived from custom, judgment, and decrees of the courts, as distinguished from legislation) provided protection through doctrines prohibiting trespass and nuisance. For example, an individual who suffered damage from erosion or flooding caused by improper management of adjacent land had recourse under common law. If the harm was more widely spread through the community, then only the government had the authority to take action—for instance, to limit certain air and water pollution to abate a public nuisance.

The common law provides another doctrine, that of **public trust**, which both grants and limits the authority of government over certain natural areas of special character. Beginning with Roman law, navigable and tidal waters were entrusted to the government to hold in trust for public use. More generally, "the Public Trust Doctrine makes the government the public guardian of those valuable natural resources which are not capable of self-regeneration and for which substitutes cannot be made by man."[6] For such resources, the government has the strong responsibility of a trustee to provide protection and is not permitted to transfer such properties into private ownership. This doctrine was considerably weakened by the exaltation of private property rights and strong development pressures in the United States but in more recent times has shown increased vitality, especially concerning the preservation of coastal areas.

Thus, although the common law provides a few examples of environmental protection possibilities, the vast majority of the activities we regulate today in the name of the environment would proceed unrestricted if left to common law. The need for regulation of these activities has been filled by legislation enacted at all levels of government. In the late 1960s, public perception in the United States that our environment was deteriorating reached a high level of awareness and concern. Congress

Year	Legislation
1969	National Environmental Policy Act (NEPA)
1970	Clean Air Act
1972	Clean Water Act, Coastal Zone Management Act Federal Insecticide, Fungicide and Rodenticide Act
1973	Endangered Species Act
1974	Safe Drinking Water Act
1976	Resource Conservation and Recovery Act
1980	Superfund (CERCLA)
1984	Hazardous and Solid Waste Amendment Act (SARA)
1986	Superfund Amendments and Reauthorization
1987	Water Quality Act
1990	Clean Air Act Amendments; Reauthorization of Superfund
1996	Water Quality Standards and Pollution Act

Figure 30.2 • Major federal environmental legislation and the year enacted. Most important environmental legislation was adopted from 1969 to 1996. Some laws were enacted earlier in a much less comprehensive form (e.g., the Clean Air Act in 1963), and most were amended subsequently.

responded with NEPA in 1969, and in the following years major environmental legislation was passed at the national level (Figure 30.2).

International Environmental Law and Diplomacy

International law is different in basic concept from domestic law because there is no world government with

(a)

(b)

Figure 30.3 • (*a*) Satellite image of Antarctica and surrounding southern oceans. (*b*) Emperor penguins and chicks, Antarctica.

enforcement authority over nations. As a result, international law must depend on the agreement of the parties concerned to bind themselves to behavior that many residents of a particular nation may oppose. Certain issues of multinational concern are addressed by a collection of policies, agreements, and treaties that are loosely called international environmental law. There have been encouraging developments in this area, such as agreements to reduce air pollutants that destroy stratospheric ozone (the Montreal Protocol of 1987).

Antarctica provides a positive example of environmental law to protect the environment. Antarctica is a continent of 14 million km² that was first visited by a Russian ship in 1820. It was soon recognized that the continent contained unique landscapes and life-forms (Figure 30.3). By 1960, a number of countries had claimed parts of the continent to exploit mineral and fossil fuel resources. Then, in 1961, an international treaty was passed that designated Antarctica as a scientific sanctuary. Thirty years later, in 1991, the Protocol of Madrid resulted in a major environmental agreement that protected Antarctica, including islands and sea south of 60° latitude. The continent was designated as "nuclear-free," and access to resources was restricted. The protocol was the first step in conserving Antarctica from territorial claims and establishing the "White Continent" as a heritage for all people of Earth.

Persistent organic pollutants (POPs) such as dioxins, DDT, and other pesticides have also been addressed at the international level. Following several years of negotiations in South Africa and Sweden, 127 nations in May 2001 adopted a treaty to greatly reduce or eliminate the use of toxic chemicals known to contribute to cancer and harm the environment.

Summit Meetings

The United Nations has sponsored several summit meeting on environmental issues, as discussed in Chapter 22. In 1997, a summit on global warming was held in Kyoto, Japan. The purpose was to negotiate a treaty (the Kyoto Protocol) to reduce greenhouse gases, especially carbon dioxide, to 5% below 1990 emissions by the year 2010. The primary long-term objective was to prevent emissions from human sources from interfering with the natural climate system. Some now consider the Kyoto conference to have been a failure, because the United States in late 2000 refused to honor reductions in CO_2 agreed to at Kyoto. Nevertheless, the summit meetings have been a step in the right direction. If we are going to move toward a sustainable society on a global basis, we need to start by developing equitable policies to ensure that all people on Earth have opportunities to work and live in a quality environment.

Environmental Litigation

Environmental groups working through the courts have been a powerful force in shaping the direction of environmental quality control since the early 1970s. Their influence arose in part because the courts, appearing to respond to the national sense of environmental crisis of that time, took a more activist stance and were less willing to defer to the judgment of agencies. At the same time, citizens were granted unprecedented access to the courts, and through them, to environmental policy.

Citizen Actions

Even without specific legislative authorization for citizens' suits, courts have allowed citizen actions in environmental cases as part of a trend to liberalize standing requirements. (*Standing* is a legal term relating here to the right to bring suit.) One of the earliest examples, and the most significant, was the willingness of the courts to entertain lawsuits to enforce NEPA. The statute itself and its legislative history do not mention either judicial review or standing, and the courts were left to decide whether a judicial remedy should be applied. Without such a remedy, and in the face of very substantial reluctance on the part of administrative agencies to change their habitual ways of doing business, it is unlikely that the EIS requirement would ever have had a significant influence on agency actions. With the courts' involvement, administrators were no longer able to take an action without explaining the reasoning that led them to conclude that the action was desirable.[8]

In the 1980s, a new type of environmentalism (which some people would label radical) arose, based in part on the premise that when it comes to the defense of wilderness there can be no compromise. Methods used by these new environmentalists have included holding sit-ins to block roads that provide access to forest areas where mining or timber-harvesting activities are scheduled to take place; climbing trees and sitting in them to block timber harvesting (see the case study that opened this chapter); implanting large steel spikes in trees to discourage timber harvesting; and sabotaging equipment, such as bulldozers (a practice known as *ecotage*).

Civil disobedience and ecotage have undoubtedly been responsible for millions of dollars of damage to a variety of industrial activities related to the use of natural resources in wilderness areas. One result of civil disobedience by some environmental groups is that other environmental groups, such as the Sierra Club, are now considered moderate in their approach to protecting the environment.

There is no doubt that civil disobedience has been successful in defending the environment in some instances. For example, members of the group Earth First successfully halted the construction of a road being built to allow access for timber harvesting in an area of southwestern Oregon. Tactics used included blockading the road by sitting down or standing in front of the bulldozer blades. Owing to the persistence of the protesters, the pace of road building was slowed considerably. In conjunction with this incident, a lawsuit was filed against the U.S. Forest Service.

Environmentalists are now relying more on the law when arguing for ecosystem protection. The Endangered Species Act has been used as a tool in attempts to halt activities such as timber harvesting and development. Although rarely is the presence of an endangered species responsible for stopping a proposed development, those species are increasingly being used as a weapon in attempts to save remaining portions of relatively undisturbed ecosystems.

Mediation

The expense and delay of litigation have led to a search for other methods to resolve disputes. An alternative that has recently received considerable attention in environmental conflicts is environmental mediation, a negotiation process between the adversaries guided by a neutral facilitator. The task of the mediator is to clarify the issues in contention, help each party understand the position and the needs of other parties, and attempt to arrive at a conclusion where each party gains enough from a compromise to prefer a settlement to the risks and costs of litigation. Often, citizens' suits or the possibility that a suit might be filed gives an environmental group a place at the table in such a mediation. Litigation, which may delay a project for years, becomes something that can be bargained away in order to gain concessions of decreased environmental impact (mitigation) from a developer. There have been some successes with mediation where bargaining positions are approximately equal and where participants who truly represent the conflicting interests can be identified and persuaded to devote the considerable effort that the process demands. In some states, mediation is required by legislation as an alternative or a precedent to litigation in the highly contentious siting of waste treatment facilities. For example, in Rhode Island, a developer who wishes to construct a hazardous-waste treatment facility must negotiate with representatives of the host community and submit to arbitration of any issues not resolved by negotiation. Costs of the negotiation process are borne by the developer.

The Storm King Mountain case is a classic example of a situation in which mediation could have saved millions of dollars in legal expenses and years of litigation. The case involved a conflict between a utility company and conservationists. In 1962, the Consolidated Edison Company of New York announced plans for a new hydroelectric project in the Hudson River highlands, an area considered to have many unique aesthetic values as well as thriving fisheries (Figure 30.4). The utility company argued that it needed the new facility, and the environmentalists fought to preserve the beautiful landscape and fisheries resource. The first lawsuit was filed in 1965, and following 16 years of intense courtroom battles, the litigation ended in 1981. Incredibly, the paper trail exceeded 20,000 pages. After millions of dollars had been spent, the various parties finally managed to get together and settle the case with the assistance of an outside mediator. If the parties had been able to sit down and talk about the issues at an early stage, mediation might have settled the issue much sooner and at much less cost to the individual parties and to society.[9] The Storm King Mountain case is

Figure 30.4 • Storm King Mountain and the Hudson River Highlands in the state of New York, the focus of environmental conflict between a utility company and conservationists for nearly 20 years, before an issue of building a power plant was finally resolved by mediation.

often cited as a major victory for environmentalists, but the cost was great to both sides.

 30.3 Environmental Impact Analysis

Environmental impact analysis is the process of determining the probable effects of human use of (and interest in) the land. Such analysis became common in the United States following the passage of the National Environmental Policy Act (NEPA) in 1969. The legislation required that each major federal project that might affect the quality of the human environment be preceded by an evaluation of the project and its potential impact on the environment. The evaluation would be presented in an **environmental impact statement (EIS)**. Guidelines to assist in the preparation of EISs were prepared by the Council on Environmental Quality.

The major components of the EIS (as revised in 1979) are as follows:[10]

- A summary of the EIS.
- A statement concerning the purpose of and need for the project.
- A rigorous comparison of the reasonable alternatives.
- A succinct description of the environment affected by the proposed project.
- A discussion of the environmental consequences of the proposed project and of the possible alternatives.

The discussion must include both direct and indirect effects (applying the principle of environmental unity, as described in Chapter 3); energy requirements; conserva-

tion potential; possible resource depletion; impact on cultural or historical resources and urban quality; possible and potential conflicts with local or state land-use plans, policies, and controls; and mitigation measures that might reduce or eliminate potential environmental degradation.

During the first 10 years under NEPA, the process of environmental impact analysis was criticized because it produced a tremendous volume of paperwork that tended to obscure the important issues. In response to that criticism, the revised regulations of 1979 introduced two important changes: scoping and the record of decision.

Scoping and the Record of Decision

Scoping involves early identification of important environmental issues that require detailed evaluation. Citizen groups as well as federal, state, and local agencies are asked to participate in the scoping process by identifying issues and alternatives that they believe should be addressed as part of the environmental analysis.

The **record of decision** is a concise statement prepared by the agency planning the proposed project that outlines the alternatives considered and discusses which alternatives are environmentally preferable. The record of decision also addresses the rationale for a particular decision and describes how environmental degradation might be minimized or avoided for the alternative chosen. Finally, the agency has the responsibility of monitoring the project to ensure that the decisions are in fact carried out as the project is initiated and completed; this responsibility puts teeth into the environmental impact analysis process. For example, if an EIS for a proposed pipeline commits an agency to a particular right-of-way and a specific design to minimize environmental degradation, it is the agency's responsibility to see that the firms doing the project incorporate those designs and plans. For this reason, special consultants are often hired to monitor the work to ensure that the procedures outlined in the record of decision are executed.

Mitigation

Mitigation identifies actions that will avoid, lessen, or compensate for anticipated adverse environmental impacts resulting from a particular project. For example, if a project in the coastal zone will damage wetlands, a possible mitigation might be the creation or enhancement of wetlands at another site. Mitigation is fast becoming a common requirement of many environmental impact reports and statements. However, there is a danger that the process may be overused. Sometimes no mitigation is possible for a particular environmental problem. Requiring mitigation to reduce or compensate for environmental degradation is a useful endeavor in many instances, but it must not be considered a standard operating procedure to circumvent the adverse environmental impacts associated with a particular project.

Methodology for Assessing Environmental Impact

There is no one methodology for assessing the environmental impact of a particular action or project. No single method of impact assessment is appropriate for the broad spectrum of projects undertaken, which may range from construction of large reservoirs or highways to urban renewal, to the hunting of migratory birds. The important principle is that those responsible for preparing the EIS produce a report that minimizes personal bias and maximizes objectivity. The analysis must be prepared according to highly objective standards and must be technically and legally defensible.[11]

Geographic Information System

A technology that has proved useful in handling the voluminous amounts of data associated with environmental impact analysis and other environmental work is the **Geographic Information System (GIS).** The GIS is a computer system capable of storing, retrieving, transferring, and displaying environmental data.[12,13] The data used in a GIS may be points, lines, areas, or numbers. Intersections of roads and locations of particular facilities, such as hospitals and fire stations, are points; roads, rivers, and landscape features such as the shoreline are lines; and land use, rock type, and soil type are areas.

The GIS is able not only to categorize and store data but also to manipulate data and create products. The creation of products is important in environmental impact analysis. For example, the product may be a land-use map or an analysis of important environmental variables, such as relationships between topography, geology, hydrology, and population density.

Figure 30.5 outlines the process used in a GIS, using an example in which the product is a land capability map. The process involves an inventory of data sources, such as topography, geology, and hydrology, and an analysis of issues such as slope stability and flooding along with production of the computer-generated map. Depending on how data are analyzed, a variety of maps may be produced as part of the environmental analysis process. This basic idea can be extended to various types of projects in which a number of environmental data sources and factors must be analyzed and interpreted.

State Environmental Impact Analysis Requirements

After the passage of NEPA in 1969, the need for state and local governments to have more active control over activities not covered by federal legislation became apparent. Today, approximately half of the states have enacted legislation requiring environmental review prior to initiation of projects that may adversely affect the environment. Nineteen states and the District of Columbia have enacted state environmental policy acts (SEPAs) patterned after the federal legislation.[14,15]

An example is the California Environmental Quality Act (CEQA), passed in 1970. CEQA requires the completion of an **environmental impact report (EIR)** for a wide spectrum of projects that may affect the environment. Following a court decision, additional legislation now requires that all private as well as public projects in California undergo environmental review. (Examples of private projects might include hotel convention centers, shopping centers, and even single family estates; examples of public projects include dams and reservoirs, as well as implementation of regulations concerning timber harvesting.)

The California law potentially provides more protection for the environment than does the federal legislation. It specifically addresses environmental quality, whereas the federal legislation is concerned with environmental policy. The EIRs required by CEQA contain information similar to that required by the federal legislation but in addition require evaluation of growth-inducing impacts and mitigation measures proposed to minimize significant adverse environmental effects.[16]

CEQA is similar to NEPA in that it allows for the filing of a negative declaration or a mitigated negative declaration. A **negative declaration** may be filed if an agency has determined that a particular project will not have significant adverse impacts on the environment. The declaration may be filed only following the completion of an environmental inventory and assessment. The negative declaration includes a description of the project, along with specific information supporting the argument that the project will not have significant environmental effects. Negative declarations need not consider alternatives to the project but must provide a comprehensive statement concerning potential environmental problems. A **mitigated negative declaration** may be filed if initial study of the project suggests that, although significant environmental problems may occur, the project could be modified in such a way as to reduce the impacts to near insignificance.[16]

Both the negative declaration and the mitigated negative declaration are important additions to the environmental review process. They provide a means whereby projects that will not cause environmental disruption can avoid the preparation of a lengthy environmental impact report. However, they are not intended to avoid the important process of public review and comment prior to approval of a project.

In general, the environmental review process at both the state and federal level has accomplished the following:

- Focused attention on potential environmental problems related to proposed projects.
- Provided a framework for evaluation of the environmental consequences of a project.
- Significantly increased environmental protection efforts in the United States.

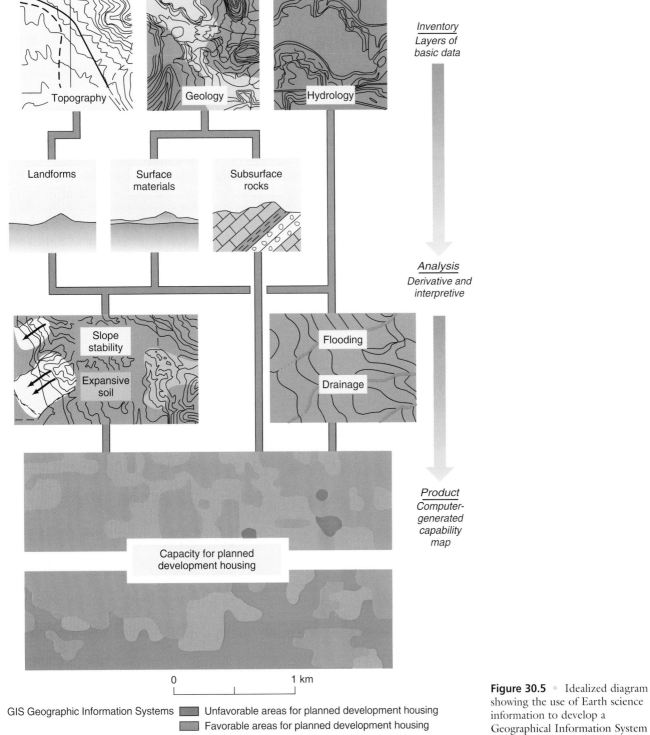

Figure 30.5 • Idealized diagram showing the use of Earth science information to develop a Geographical Information System (GIS) for land use planning.

Inventory
Layers of basic data

Analysis
Derivative and interpretive

Product
Computer-generated capability map

Topography

Geology

Hydrology

Landforms

Surface materials

Subsurface rocks

Slope stability

Expansive soil

Flooding

Drainage

Capacity for planned development housing

0   1 km

GIS Geographic Information Systems
▇ Unfavorable areas for planned development housing
▨ Favorable areas for planned development housing
▨ Developed areas

30.4 Environmental Review Cases

The environmental review process just described was developed in response to the recognized need to consider the environmental consequences of a particular action prior to its implementation. The overriding objective of environmental review is to recognize potential impacts, conflicts, and problems as early as possible so that they can be evaluated and environmental degradation minimized. The cases of the Cape Hatteras National Seashore in North Carolina and the San Joaquin Valley in California are considered next as examples.

Case History: Cape Hatteras National Seashore

The Outer Banks of North Carolina consists of long, thin barrier islands that appear as ribbons of sand on satellite images (Figure 30.6). These islands have for generations been inhabited by people living and working in a marine-dominated environment. Until recently, their way of life depended on raising livestock and on fishing, hunting, and boat building and other marine pursuits.[17]

Characteristically, the landscape of the Outer Banks can change in a very short time in response to major storms, such as hurricanes (Figure 30.7) and northeasters, which strike the islands periodically. Of the two types of storms, the more frequent northeasters probably cause the most erosion. However, hurricanes, although less frequent, can cause major changes, including extensive dune overwash and formation of new inlets.[18]

Historically, the people of the Outer Banks lived with and adjusted to a changing landscape. In recent times, however, their philosophy has changed because of economic pressure to develop coastal real estate. Because a stable landscape is necessary to encourage commercial development (people are not likely to buy property that could be eroded by the action of ocean waves), attempts have been made to stabilize the coastal environment through erosion controls.

An act establishing the Cape Hatteras National Seashore as the first national seashore was approved by Congress in 1937 and amended in 1940. The park consists of 115 km^2 (about 45 mi^2) along a 120-km (75-mi) portion of the Outer Banks; it includes portions of three islands of the more than 240 km (150 mi) of the barrier island system. The barrier islands bound and protect the largely undeveloped coastal plains lowland of North Carolina.[17]

Eight unincorporated villages are bounded by the Cape Hatteras National Seashore and are spaced along nearly its entire length. The legislation authorizing the park provided for the continued existence of these vil-

(a)

(b)

Figure 30.6 • The Outer Banks of North Carolina appear in this image from the Apollo-9 mission as a thin white ribbon of sand. The large cape in the center of the image is Cape Hatteras. The lower, hooked cape is Cape Lookout. The body of water to the left (west) of Cape Hatteras is Pamlico Sound. The distance from Cape Hatteras to Cape Lookout is about 100 km. The light color in the water of the sound and seaward of the barrier islands is sediment suspended in the water column.

Figure 30.7 • (a) Hurricane Bertha striking the coast of North Carolina on July 12, 1996. (b) Damage to homes from Hurricane Fran, which hit the coast of North Carolina in September 1996.

lages, including the beach in front of each village facing the Atlantic Ocean. This legislation was interpreted by many as creating an obligation on the part of the federal government to maintain and stabilize the beach access and frontage of these villages. Unfortunately, stabilization is difficult. In some villages, less than 60 m (200 ft) of beach remains because of recent coastal erosion.

The philosophy of protection brought changes in building patterns. The early development of the islands occurred mainly on the sound (inland) side, away from the ocean. However, with the construction of the first dune systems in the 1930s and the opening of a road link in the 1950s, people began to count on protection from the elements. The basic configuration of the villages started to change as communities began to spread toward the ocean.[17]

It is assumed that at one time much of the Outer Banks was heavily forested and that logging and overgrazing had destroyed those forests, although evidence for this is still inconclusive. The wooded areas were located in an area protected by a natural dune line, so it seemed logical to construct an artificial barrier dune system to help the area return to a natural state by inhibiting beach erosion. Of course, the dune line would also protect the roads and communities.

An artificial dune system by the 1970s was constructed along parts of the Cape Hatteras National Seashore, with an average dune height of 5 m (about 16 ft) at an average distance of 100 m (about 330 ft) from the ocean.[17] This artificial erosion control measure, along with the programs to keep sand on the beach artificially, were expected to cost several million dollars per year. The expense of this measure, along with questions as to whether the artificial dune lines would jeopardize the future of the barrier islands, led to a controversy on how best to manage the park.

The controversy about how dynamic Earth processes create and maintain barrier islands was at the philosophical heart of the U.S. Park Service's dilemma in developing a long-range program to maintain the natural environment for the use and enjoyment of present and future generations. Although there are several ideas and known ways in which coastal processes can develop barrier islands, there is heated debate on the processes needed to maintain them.

Debate on the nature and extent of the geologic processes that maintain the naturally changing barrier islands centers on whether the periodic overwash of the frontal dune system by storm waves is essential to maintaining the islands. If the overwash is essential, then the building of a dune line is contrary to natural processes, and the final result over a period of years could be the deterioration of the islands as a natural system. This would be contrary to the Park Service's objectives to preserve the islands in a natural state. Regardless of the historic role of overwash, it is apparent that overwash is a natural process and as such is probably significant in the geologic history of the barrier islands.[18] Nevertheless, it is not necessarily always bad to attempt to selectively control rapid coastal erosion by maintaining a protective frontal dune system. Proper placement of dunes and sound conservation practices for maintaining them are a feasible alternative to some short-range coastal erosion problems, particularly near settlements and critical communication lines.

Faced with the problem of selecting a management policy for Cape Hatteras National Seashore, in 1974, the Park Service presented five alternatives, ranging from attempting essentially no control of natural processes to attempting complete protection. Each alternative was analyzed to determine the spectrum of possible impacts. Few people wanted either no control or complete control of the seashore.[17] The position of the Park Service was to attempt to reach a compromise in which the barrier islands could be preserved in a nearly natural state while the need to maintain a transportation link with the mainland was recognized. People in these communities would have to live with and adjust to the dynamic high-risk environment in which they chose to reside, much as people of the Outer Banks historically had done. This approach was contrary to prevailing trends in coastal development, which assume a more stable environment. The approach acknowledged that natural processes play a significant role in preserving the natural environment.[17]

A general management plan for Cape Hatteras National Seashore was completed after several years of careful environmental impact work and public review. The plan proposed several actions:[19]

- Allowing natural seashore processes and dynamics to occur except in instances when life, health, or significant cultural resources on the major transportation link are jeopardized.
- Controlling the use of off-road vehicles.
- Expanding and allowing easier access to recreation sites on the beach and sound.
- Controlling the spread of exotic vegetation.
- Ensuring that significant natural and cultural resources are preserved and maintained.
- Encouraging state and local governments to participate in mutually beneficial planning.

For planning purposes, the national seashore was divided into four environmental resource units (ERUs): ocean beach, vegetated sand flats, interior dunes/maritime forest, and marsh/sound. Table 30.1 summarizes the planning objectives for each ERU, and Figure 30.8 illustrates the principle of management zoning for Hatteras Island.

Implications of the management plan were threefold. First, Cape Hatteras National Seashore would be preserved in a natural state. Second, residents of villages would have to live with and adjust to effects of natural

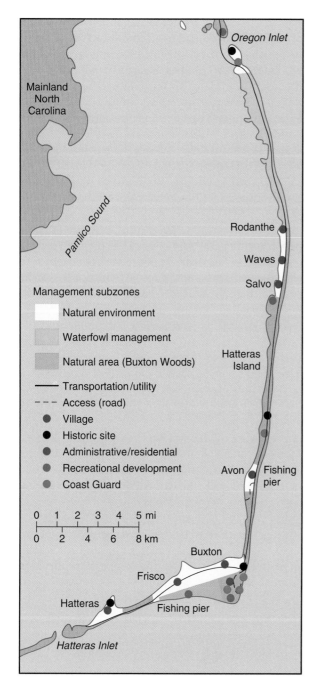

Figure 30.8 • Proposed management subzones for part of the Cape Hatteras National Seashore. (*Source:* U.S. Department of the Interior, *Cape Hatteras Shoreline Erosion Policy Statement.*)

Figure 30.9 • The famous Cape Hatteras lighthouse, originally constructed about 0.5 km from the sea in the late 19th century, was threatened by coastal erosion. Beach groins (the black structures in the surf) were designed to trap sand and retard erosion by widening the beach. Note also the light yellow seawall constructed to retard erosion in front of the lighthouse. If the lighthouse had not been moved (or greatly defended), it would eventually have been destroyed by the sea.

goal of the shoreline stabilization that has continued has been to maintain the highway that extends along the seashore. A five-year multiagency undertaking aimed at investigating and evaluating alternatives to long-term maintenance of the highway began in 1995.

One controversy has involved protection of private property and public facilities from results of natural coastal processes. A focal point has been the Cape Hatteras Lighthouse, the tallest brick lighthouse in the United States (Figure 30.9), which was constructed in 1870 approximately 0.5 km from the shoreline. By the late 1990s, it was nearly in the surf zone, and there was fear that it might be destroyed by coastal erosion. Several solutions to the lighthouse dilemma were considered:

- Do nothing and let nature take its course. In this case, coastal erosion would destroy the lighthouse.
- Artificially control coastal erosion by constructing a seawall around the lighthouse.
- Move the lighthouse out of harm's way.[20]

The decision was made to move the lighthouse inland about 500 m from the eastern shore of Hatteras Island at a cost of about $12 million. Given the present rate of sea level rise, the lighthouse should be safe in its new location from coastal erosion until the middle or end of the 21st century. This decision is consistent with the philosophy of flexible coastal planning; it avoids hazardous zones rather than trying to control natural processes. The decision is also consistent with the policy of preserving historical

events to a greater extent. Third, villages might change; although the road would be maintained, it would be subject to more periodic damage. For example, the Park Service would not perform routine dune maintenance or beach nourishment (artificially adding sand to the beach). Only in rare cases would dunes be rebuilt after overwash.

The management plan has been controversial. Coastal erosion continued to be a significant problem in several areas of the national seashore in the 1990s. The major

Table 30.1   Planning Objectives for Cape Hatteras National Seashore Environmental Resource Units (ERU)

ERU	Characteristics	Planning Objectives
Ocean beach	Shifting sands, frequent overwash, limited vegetation on dunes	Allow natural processes to continue unhampered; allow for wide range of unstructured recreational activities by visitors; no construction allowed
Vegetated sand flats	Located between dune line and edge of saltwater marsh	Continue use as transportation corridor; allow development necessary to support visitor activities and resource protection
Interior dunes/ maritime forest	Found in relatively few locations; variable topography, remoteness, dense vegetation	Maintain in natural state; allow passive recreation; design any construction to minimize impact on natural processes and systems
Marsh/sound	Includes the sound, sound shore, and associated marshes	Maintain in natural state; provide limited access to the sound; allow limited development of passive recreational activities

Source: National Park Service

Figure 30.10 • Cape Hatteras lighthouse in November of 1999 at it's new location about 1000m (3000 ft.) inland, at Buxton North Carolina.

objects (such as the lighthouse) for the enjoyment of future generations. The lighthouse was successfully moved during the summer of 1999 (Figure 30.10). Hurricane Dennis struck the island in 1999 after the lighthouse had been moved, and the historic structure sustained only superficial damage.

Case History: Selenium Toxicity in the San Joaquin Valley, California

The San Joaquin Valley in California is one of the richest farm valleys in the world. The west side of the valley (Figure 30.11), with annual precipitation of less than 25 cm (10 in), requires extensive irrigation for agriculture.

When irrigation waters are applied, however, the crop-root zone may become saturated and drown plants.[21] As a result, subsurface drainage systems are installed to remove excess water (Figure 30.12). Tile drains, pipes with holes to let water in, are placed beneath fields and connected to larger drains (ditches or canals), which are tied into a master drainage system, usually a canal.

The San Luis drain, constructed between 1968 and 1975, is a concrete-lined canal, 135 km (about 85 mi) long. Its initial purpose was to carry agricultural drainage water north to San Francisco Bay. However, the drain to the bay was not completed because of limitations in funding and uncertainty over potential environmental degradation associated with discharging water into the bay. As a result, the terminal point for drainage water became Kesterson Reservoir, then part of the much larger Kesterson Wildlife Refuge, which consisted of 12 ponds with an average depth of approximately 1.3 m (about 4.3 ft) and total surface area of about 486 ha (1,200 acres).

When the Kesterson Reservoir was first completed, it was filled primarily with fresh water purchased from water agencies to provide water for the Kesterson Wildlife Refuge. In 1978, the reservoir began receiving agricultural drainage; and by 1981, that was its main source of water. Initially, the canal and reservoir were viewed as a beneficial system that provided a sink for wastewater from farming activities. Unfortunately, however, the water infiltrated the soil, where it picked up the heavy metal selenium, as well as other salts. Selenium is a natural constituent of the soils derived from the weathering of sedimentary rocks on the west side of the San Joaquin Valley.

In 1982, unusually high selenium levels in fish from Kesterson Reservoir were identified. From 1983 to 1985, dead and deformed wildfowl chicks were reported, and analysis of the water revealed that the selenium in the reservoir ponds ranged from 60 to 390 ppb. Further study revealed that selenium concentrations in the agricultural drain waters were as high as 4,000 ppb, many times the

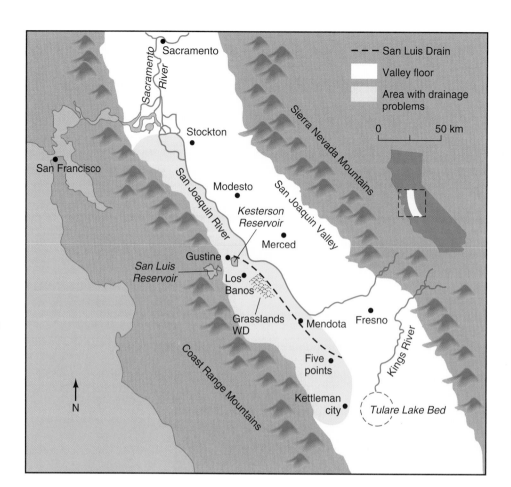

Figure 30.11 The San Joaquin Valley and the San Luis drain, which terminates at Kesterson Reservoir. (*Source:* Modified from R. Tanji, A. Lauchli, and J. Meyer, "Selenium in the San Joaquin Valley," *Environment* 28, no. 6 [1986]:6–11. Copyright 1986 by Heldref Publications. Reprinted with permission of the Helen Dwight Reid Educational Foundation.)

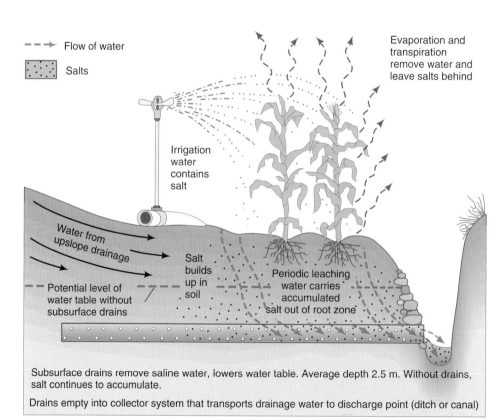

Figure 30.12 Diagram indicating the need for subsurface drains on the west side of the San Joaquin Valley. (*Source:* Modified from U.S. Department of the Interior, Bureau of Reclamation, San Luis Unit, Central Valley Project, Calif., *Drainage and Sale Disposal*, 1984.)

drinking water standard of 10 ppb set by the EPA. Subsequent work suggested that selenium toxicity was the probable cause of the death and deformities in the wildfowl.

Finally, in 1985, the Kesterson Reservoir and the water in the San Luis drain were classified as hazardous and a threat to public health. The California State Water Resources Board then ordered the U.S. Bureau of Reclamation, which was managing the wildlife refuge and reservoir, to alleviate the hazardous conditions.[22] The bureau was faced with developing a cleanup plan and implementing it.

Environmental planning and review at Kesterson illustrates the process of environmental analysis. Scoping was an important process, as was evaluation of alternatives and mitigation to minimize environmental degradation. The scoping process identified the following issues:

- The disposition of Kesterson Reservoir and cleanup of the San Luis drain.
- The environmental, social, legal, and economic impact of not allowing Kesterson to receive agricultural drain water.
- The potential impacts of selenium and other contaminants in the drain water on public health.
- The development of alternative methods to clean up Kesterson and the San Luis drain.
- The potential for contaminated groundwater to migrate away from the Kesterson area.

The Bureau of Reclamation of the U.S. Department of Interior issued its final EIS and record of decision in October 1986. The document addressed the potential impacts of alternative methods to clean up the Kesterson Reservoir as well as the San Luis ditch. The no-action alternative was considered but not accepted, and several alternative plans were developed. These varied from flooding reservoir ponds with fresh water to constructing an on-site waste disposal facility. In March 1987, the Water Resources Control Board ordered the Bureau of Reclamation to dispose of contaminated soil and vegetation in an on-site, monitored, lined, and capped landfill. That is, the site was treated as a toxic waste disposal facility. The board also required the Bureau of Reclamation to mitigate loss of Kesterson Wildlife Refuge by providing alternative wetlands for habitats.

Capping the ponds at Kesterson has been successful in reducing impacts on wildfowl. Monitoring of the site has been ongoing since the ponds were filled. Wetlands have provided valuable wildlife habitat in an area where wetlands had been lost. Selenium-rich agricultural drain water from 17,000 ha of irrigated agricultural land has been eliminated. However, the selenium problem in the broader valley hasn't disappeared, as selenium-contaminated water from agricultural drains in other areas continues to cause concern.

30.5 Land-Use Planning

Issues of environmental planning and review are closely related to how land is used. Land use in the United States is dominated by agriculture and forestry; only a small portion of land (about 3%) is used for urban purposes. However, rural lands are being converted to nonagricultural uses at about 9,000 km^2 (about 3,500 mi^2) per year. About half the conversion is for wilderness areas, parks, recreational areas, and wildlife refuges. The other half is for urban development, transportation networks, and other facilities. On a national scale, conversion of rural lands to urban uses is relatively small. In rapidly growing urban areas, however, increasing urbanization may be viewed as destroying agricultural land and exaggerating urban environmental problems. Urbanization in remote areas with high scenic and recreational values may be viewed as potentially damaging to important ecosystems.

Land-use planning offers a way to help manage land. It is a complex process aimed at identifying and studying issues of land use and determining what development is appropriate. Such planning involves important and controversial environmental issues. Although people are concerned about environmental problems, rights of property owners are also an issue, and herein lies the controversy. Nevertheless, people are beginning to realize that good land-use planning is essential for sound economic development. People at community, regional, and state levels are recognizing the role of land-use planning in avoiding conflicts and maintaining a high quality of life. An analogy comes from the business world. When a firm manages its resources efficiently, we call it good business; when a city, county, or state efficiently manages its land resources, we call it good planning.[23]

The philosophy of good land-use planning is to conserve natural resources, avoid natural hazards, and protect the environment through use of proven ecological principles. The process of land-use planning is illustrated in Figure 30.13 and includes the following steps:[24]

- Identification and definition of objectives, goals, issues, and problems.
- Collection, analysis, and interpretation of data.
- Development and testing of alternatives.
- Formulation of land-use plans.
- Review and adoption of plans.
- Implementation of plans.
- Revision and amendment of plans.

Comprehensive Planning

A **comprehensive plan** is an official land-use plan adopted by local government. The plan formally states general and long-range policies concerning future development.[23] The planning process includes development of an

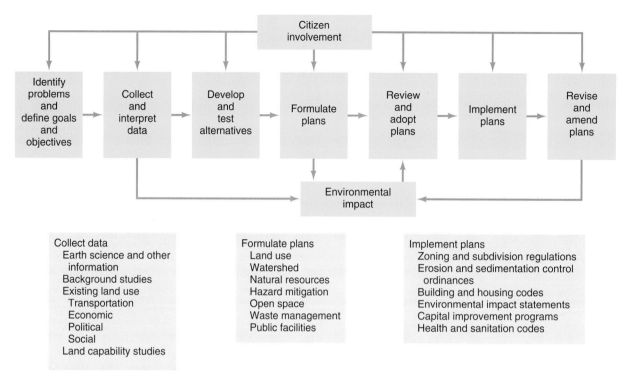

Figure 30.13 • The land-use planning process. (*Source:* Modified from William Spangle and Associates, F. Beach Leighton and Associates, and Baxter McDonald and Company, *Earth-Science Information in Land Use Planning—Guidelines for Park Scientists and Planners,* U.S. Geological Survey Circular 721, 1976.)

environmental inventory, forming the basis for planning and zoning by local governments.

Only a few states in the United States have developed statewide planning programs that require development and approval of a comprehensive plan. Oregon, a leader in the development of rigorous statewide planning, requires cities and counties to develop comprehensive plans that reflect and implement statewide goals. Comprehensive plans in Oregon address 19 goals, including:

- Citizen involvement in the planning.
- Air and water quality.
- Land resources.
- Protection from natural hazards.

Oregon state law requires that all land be planned and zoned on the basis of environmental inventories of resources and hazards. However, it is the responsibility of cities and counties to prepare plans according to state goals and guidelines and to issue permits for development.[23] Comprehensive planning in Oregon is innovative but controversial as it relates to individual property rights. Nevertheless, it appears to be working; and other states have developed similar programs.

California has also developed a system of comprehensive planning. Plans in California must address several elements identified by state law, including land use, housing, conservation, open space, and safety (natural hazards).[24] Other elements may be added that are relevant to specific needs. Because of California's state environmental quality act (CEQA, discussed earlier), environmental review is an important part of the planning process. Local governments often combine the comprehensive plan itself and the environmental impact report to save time and money.[24]

The land-use element of a comprehensive plan includes maps that designate uses of the land—for example, for housing, industry, open space, recreation, and public facilities—as well as intensity of use.[23] Ideally, the land-use maps reflect the goals and objectives of the comprehensive plan. They are the most visible part of this plan and are also the part most often used.[24]

Conservation and safety elements are important in comprehensive planning. The conservation element has the objective of setting goals and policies concerning the development and use of natural resources, such as air, water, wildlife, forests, and mineral deposits. The safety element is designed to protect people from natural hazards, such as earthquakes, landslides, floods, and wildfires (see A Closer Look 30.1).[24] Included in this element are maps that identify hazardous areas, such as active faults and floodplains. The safety element may also cover such issues as disposal of hazardous wastes, transport of hazardous materials, and failure of utility services.[24]

Natural Hazards

Earth is a dynamic, evolving system; and people on Earth have always faced hazards from natural processes such as floods, earthquakes, and violent storms. Today, however, increasing population and changing land use, particularly in developing countries, is transforming what were hazardous events to disasters and catastrophes.[25] On both personal and institutional levels, our goal should be to recognize the processes that produce disasters and catastrophes, avoid them when possible, and minimize their threat to human life and property.

At this point, it's important to define *catastrophe*, which is any situation in which the damages to people and property or society are sufficiently grave that recovery and rehabilitation are long and difficult. A catastrophe causes much more damage than what is normally termed a *disaster*, which may be very serious at the local level but is of less consequence to society on a larger scale. Table 30.2 shows natural hazards in the United States as well as their catastrophe potential. Those events in the United States most likely to produce a catastrophe are hurricanes, floods, tornadoes, earthquakes, volcanic eruptions, and large wildfires. Hazards with a low potential for catastrophe include lightning and coastal erosion; however, the former is one of the most serious in terms of lives lost.[26]

A look at mean annual losses in the United States from several natural hazards provides insight into the magnitude of the problem. As shown in Table 30.2, the largest loss of life every year results from tornadoes and windstorms, but other processes, such as lightning strikes, floods, and hurricanes, also take a heavy toll in human life. During the past 50 years, the loss of life from natural hazards in the United States has declined sharply as a result of better warning systems, forecasting, and prediction of some hazards, such as hurricanes and floods. The property damage resulting from natural hazards has increased dramatically, how-

ever, as a result of the larger numbers of people living in potentially hazardous zones, such as floodplains, steep slopes prone to landslides, low-lying coastal areas vulnerable to hurricanes, and lands likely to experience periodic wildfire.

Although loss of life from natural hazards in the United States has declined, the number of catastrophes and the associated damages worldwide have increased greatly over the past 50 years. During the years 1985–1999, about 40,000 people per year died as a result of natural hazards; fewer than 5% of the deaths were in industrialized countries. About half of all deaths were caused by floods; earthquakes and volcanoes claimed another 30%.[25] Financial losses amounted to more than $60 billion per year during this period. If social losses such as unemployment, reduced productivity, and mental anguish are considered, the losses are much larger.[25,27] The 1990s set new records, with economic losses of over $608 billion—more than the total of the previous four decades.[25] During the 20th century as a

whole, more than 10 million people on Earth were killed by natural hazards.

Linkages Between Natural Hazards

Natural processes hazardous to people involve several significant linkages. Many of the hazards themselves are linked. Hurricanes are often linked with flooding, coastal erosion, and landslides, for example. Natural hazards are also linked to a variety of other hazards, including spread of disease as a result of water pollution following hazardous events. For example, following hurricanes, polluted waters may run off, entering streams, rivers, and groundwater systems used by people. If these are contaminated with human waste, diseases such as cholera may break out.

Adjustments to Hazards

Major adjustments to natural hazards include land-use planning, structures to control natural processes, insurance, evacuation, disaster preparedness, and bearing the loss (doing nothing). Which option an individual or society chooses depends on a number of factors, includ-

Table 30.2   Natural Hazards in the United States

Hazard	Deaths per Year	Occurrence Influenced by Human Use	Catastrophe Potential[a]
Flood	86	Yes	H
Earthquake[b]	50+?	Yes	H
Landslide	25	Yes	M
Volcano[b]	<1	No	H
Coastal erosion	0	Yes	L
Expansive soils	0	No	L
Hurricane	55	Perhaps	H
Tornado and windstorm	218	Perhaps	H
Lightning	120	Perhaps	L
Drought	0	Perhaps	M
Frost and freeze	0	Yes	L

[a]Catastrophe potential: H = high; M = medium; L = low.

[b]Estimate based on recent or predicted loss over 150-year period. Actual loss of life and/or property could be much greater.

Source: Modified from G. F. White and J. E. Haas, *Assessment of Research on Natural Hazards* (Cambridge, Mass.: MIT Press, 1975).

ing risk, available funding, and hazard perception.

One of the most environmentally sound adjustments to hazards involves land-use planning. For example, people can avoid building on floodplains, in areas where there are active landslides or active fault traces, and in areas where coastal erosion is likely to occur.

In recent years, a good deal of effort has been directed toward understanding how people perceive various natural hazards. This is obviously an important endeavor, because the success of hazard reduction programs depends on the attitudes of the people likely to be affected by the hazard. For example, it may be difficult to develop an effective earthquake hazard reduction program where strong earthquakes occur only once every few generations. Similarly, it is difficult to tell an individual who has lived many years in a particular home on a floodplain that the area is dangerous. Because flooding at a particular site may occur infrequently, the individual may not perceive flooding as a serious hazard. Although there may be an adequate perception of hazards at the institutional level, this perception may not filter down to the general population. This situation is particularly true for hazards that occur infrequently.

Population Increase, Land Use, and Natural Hazards

We have said that population increase is one of our most serious environmental problems. As the population increases and puts greater demands on our land and resources,

Figure 30.14 • Damage from Hurricane Mitch (1998) at the city of Tegucigalpa, the capital of Honduras. Thousands of buildings were destroyed or damaged as was much of the country's infrastructure, including over 150 major bridges and all major highways. About one-third of the total population of the country was left homeless.

the need for planning to minimize losses from natural hazards and processes also increases. This need is dramatically illustrated by two catastrophes that occurred in 1998: flooding on the Yangtze River in China and devastation in Central America caused by Hurricane Mitch.

The flooding of the Yangtze River in central China is in part a story of land-use change and population density. Approximately 85% of the forests of the Yangtze River basin have been lost as a result of timber harvesting and conversion of land to agriculture in recent time. As a result of these land-use changes,

runoff and flooding have become much more severe and much more common.[25]

The story from Hurricane Mitch, which devastated Honduras and other areas in Central America, is much the same. As human population has increased, Honduras has lost nearly half of its forests, and a huge wildfire had occurred in the region prior to the hurricane. As a result of the deforestation and the fire, hillsides stripped of vegetation quickly washed away under the torrential rains of the hurricane. Along with the slopes went farms, homes, roads, and bridges (Figure 30.14).[25]

Urban Renewal

As part of land-use planning, many cities today are attempting to revitalize the city center, producing an "old town" or building homes and other structures to replace abandoned factories and run-down buildings. This is happening from New Jersey on the Hudson River; to Charlotte, North Carolina; to Denver, Colorado; to San Francisco, California. Sometimes, the transformations are remarkable.

For example, consider Granville Island on the delta of the Frazier River in the city of Vancouver, British

Columbia, Canada. Some think the history of Granville Island predates that of the city of Vancouver itself, going back to Native American fishing villages. By the late 1800s, in any case, such activities as timber harvesting and clearing of land for settlement had brought tremendous changes to the landscape and the island. The city of Vancouver was expanding, and in 1915 the government gave its approval to create an industrial area on Granville Island that included sawmills, iron works, slaughter houses, and other industries. In the 1920s, Granville Island became one of the largest manufacturing operations in the city of Vancouver.

The industrial boom, however, didn't last long. Following the Second World War, changes to the island were of a much different type. A series of fires caused extensive damage to factory buildings, and waterfront space became less desirable for industrial sites because of the development of highways and other infrastructure in the Vancouver area. As a result, the industrial zone of Granville Island became a very depressed area.

Then another transformation began, and by the early 1970s, Granville Island had been resurrected as a "people place" consisting of a mixture of restaurants, markets, art studios, marinas, and theaters, as well as office and industrial space. Careful environmental planning preserved the industrial heritage by keeping the best of the old look in the form of interesting buildings, walkways, and other structures. A bustling urban square contrasts with quiet places, including waterfront parks and trails. Today, Granville Island is a success story of urban environmental planning; an area given up as a "dead loss" by many has been transformed into a thriving community visited by thousands of people daily.

Urban lands degraded by pollution from factories along rivers may be attractive locations for urban renewal, but careful planning and studies are necessary to ensure that the land is safe. For example, the most expensive high school in California history was built for the Los Angeles School District in the 1990s. This state-of-the-art school will probably never be used, however. It was constructed on a site where oil development had occurred, and hazardous gases are seeping up through the soil, precluding use by students.

Another concern in urban renewal is sensitivity to issues of social and environmental justice. Displacing poor people from their homes and not allowing them to benefit from and be part of urban renewal are potential problems that must be addressed.

Regional Planning

Land-use planning occurs at a variety of levels, from local to county to state. Less frequently, it occurs at the regional (such as multistate) level. Probably one of the earliest examples of regional planning in the United States goes back to 1933, when President Franklin D. Roosevelt introduced his New Deal. The New Deal included a number of programs, and one was the Tennessee Valley Authority (TVA). This regional plan for the Tennessee River established a semi-independent authority whose responsibility was to promote economic growth and social well-being for the people of the region. The area, which includes parts of seven states, was economically depressed at the time the authority was established. There had been rampant exploitation of timber and petroleum resources in the region, and the people living there were among the poorest in the country.[28] Today, the TVA is reported to be one of the best examples of regional plan-

ning in the world. It is characterized by multidimensional and multilevel planning to manage land and water resources. The authority is involved in the production and regulation of electrical power, flood control, navigation, and outdoor recreation.

Green Plans

People around the world are becoming concerned with global environmental problems, such as ozone depletion, global warming, and acid rain. On local and regional scales, attention is focused on air and water pollution, management of toxic materials, and management of resources. Governments around the world are spending billions of dollars in attempts to solve pressing environmental problems.

Green plans are long-term strategies for identifying and solving global and regional environmental problems. At the philosophical heart of green plans is sustainability, the notion that we can live on this planet and meet our present needs and at the same time ensure that future generations will inherit a quality environment. The development of national policies for environmental protection and restoration will involve cooperative relationships between various groups, such as government agencies, industry, and labor. Those in favor of developing green plans point out that if the plans are to be successful, nations will have to place a high priority on environmental issues.

Green plans are being implemented today in several countries, including Canada and the Netherlands; and planning is underway in a number of other nations around the world. Nevertheless, environmental plans at the national level are often fragmental rather than comprehensive. The United States, for example, has no comprehensive land-use planning legislation, although it has implemented plans for certain land uses, particularly those that affect very sensitive areas, such as wilderness areas and coastal zones.

Coastal Zone Planning

Although coastal areas vary in topography, climate, and vegetation, they are all generally dynamic environments. As we saw in the Cape Hatteras example, continental and oceanic processes converge along coasts to produce a landscape that is characteristically capable of rapid change. The impact of hazardous coastal processes is considerable, because many populated areas are located near the coast. The largest cities in the United States lie in the coastal zone (including the Great Lakes), and approximately 75% of the population live in coastal states. Because of existing and potential conflicts in the coastal zone, the U.S. Congress in 1972 passed the Coastal Zone Management Act, which was amended in 1976.

The purpose of the Act was to establish a federal program to assist states in developing land-use plans for coastal areas. In particular, the Act mandated that individual states define the boundaries of the coastal zone, specify permissible land uses within the zone, and address issues of particular public concern (such as public access to the coastal zone and protection of fragile environments or endangered species). All coastal states are now involved to a lesser or greater degree in coastal zone management.

The people of North Carolina, recognizing the importance of planning for their fragile coastal environments along barrier islands, passed the Coastal Area Management Act of 1974. This Act recognizes the people's need to enjoy the aesthetic, cultural, and recreational qualities of the natural coastline. One major objective of the Act is to provide a management system that allows preservation of estuaries, barrier islands, sand dunes, and beaches so that their natural productivity and biological, economic, and aesthetic values are safeguarded. The second objective is to ensure that development in the coastal areas does not exceed the capabilities of the land and water resources. The North Carolina Act establishes a cooperative program of coastal management between local and state governments in which local government has the initiative and does the planning. The role of state government is to set standards and to review and support local government in the planning program.

An interesting case in coastal management occurs in Alaska, where some of the boundaries for the coastal zone are set many kilometers inland, at boundaries of watersheds. The rationale is that the streams that flow to the ocean are natural migratory routes for fish, such as salmon; and thus, there is a natural tie between the rivers and the coastal zone. Development, such as urbanization or timber harvesting within river basins must be reviewed by the coastal zone management system as well. Although the boundaries in Alaska seem rather remote from the ocean in some instances, the program appears to be effective in protecting marine and coastal resources (in this case, fish).

Today a focus of intense interest in coastal zone planning is centered on the Gaviota Coast west and north of Los Angeles and Santa Barbara, California. The coast contains about 50% of the last undeveloped coastline in Southern California. The 122 km (76 mi) coast is a combination of pocket beaches and rocky shorelines in an area with abundant oil and gas resources. Whether this coastline is developed like Los Angeles or preserved will reflect our values now and in the future.

Planning for Public Lands

Planning for recreational activities on U.S. government lands (including national forests and national parks) is becoming a controversial issue. Some regulation seems necessary, however, if environmental degradation resulting from recreational activities is to be minimized.

In some popular areas, such as Yosemite National Park and the Grand Canyon, protecting the environment may simply require limiting the number of permits to use trails or rivers. For example, the number of people allowed to raft in the Grand Canyon is limited to protect the river environment from overzealous recreational use. Limiting people in national parks is fairly straightforward, and most visitors understand the necessity to protect parklands. In fact, it is a congressional mandate that park environments be maintained for future generations.

On national forest lands, there is more controversy concerning controlling land use for recreational purposes. Managers for national forests tend to accommodate various interest groups through a concept known as *multiple use*. This practice has had variable success because some uses are basically incompatible, as illustrated in A Closer Look 30.2, "Off-Road Vehicles."

The management of public lands for recreational activities requires planning at a variety of levels, with considerable public input. For example, when a national forest is developing management plans, public meetings are often held to inform people about the process by which the plan is developed and to ask for ideas and suggestions. Maximizing public input provides better communication between those managing resources and those using them for recreational purposes.

Government officials and scientists involved in developing use plans for public lands are faced with complex land-use problems at a variety of levels. Often, the complexity of a problem is such that no easy answers can be found. Nonetheless, because action or inaction today can have serious consequences tomorrow, it seems best to have conservative plans to protect and preserve a quality environment for future generations.

Plans for many of the national forest and national park lands in the United States have been or are being developed. Developers of these plans tend to consider a spectrum of recreational activities and attempt to balance the desires of several user groups. At the extremes, certain areas have been set aside for intensive off-road vehicle use, and other areas have been closed entirely. For example, severe 1996–1997 winter floods in Yosemite National Park damaged roads, bridges, campgrounds, and other human structures. The flood precipitated rethinking of goals and objectives of park management. Some land claimed by the floods was returned to natural ecosystems. This eliminated private vehicles in parts of the park and further limited the number of visitors to the park.

A number of new policies have been implemented in U.S. forest and park lands. For example:

1. In wilderness areas, limits are placed on the number of people admitted.
2. In coastal areas, regulations may limit such activities as jet skiing and surfing in swimming areas.
3. In regions where endangered species exist, there may be more stringent regulations to govern the activities of people. In Yellowstone National Park in Wyoming

Off-Road Vehicles

The widespread use of off-road vehicles (ORVs) has had a significant impact on the environment. There are now more than 12 million ORVs, and many of them are invading the deserts, coastal dunes, and forested lands of the United States. Intensive use of ORVs causes soil erosion, changes in hydrology, and damage to plants and animals. The problem is severe. A single motorcycle need travel only 7.9 km (4.9 mi) to affect 1,000 m² (about 11,000 ft²); a 4-wheel-drive vehicle affects the same area by traveling only 2.4 km (1.5 mi). In some desert areas, the tracks of such vehicles produce scars that remain part of the landscape for hundreds of years.[29,30] Figure 30.15 shows damages to three ORV sites.

ORVs cause mechanical erosion and facilitate wind and water erosion of materials loosened by their passing. Runoff from ORV sites is as much as eight times greater than that from adjacent unused areas, and sediment yields are comparable to those found on construction sites in urbanizing areas.[29] Hydrologic changes from ORV activity are primarily the result of soil compaction, which reduces the soil's ability to absorb water, makes the water less available to plants and animals, and changes the variability of soil temperature. The result of intensive ORV activity is a combination of soil erosion, compaction, temperature change, and moisture content change.[30]

Environmental planning that encompasses the use of ORVs is a difficult task. There is little doubt that some land must be set aside for ORV use. The problems are how much land should be involved and how to minimize environmental damage. The possible effects of airborne soil removal (by wind) must be evaluated carefully, as must the sacrifice of nonrenewable cultural, biological, and geologic resources.[29] Intensive ORV use is incompatible with nearly all other land uses, and it is very difficult to restrict ORV damages to a specific site. Material

(a)

(b)

Figure 30.15 ● (*a*) Trails and gullies at this site in California are caused by extensive use of off-road vehicles (ORVS), mostly motorcycles. (*b*) Trails eroded into coastal bluffs by all-terrain bicycles.

removed by mechanical, water, and wind erosion will always have an impact on other areas and activities.[29,30]

In recent years, there has been a trend in some areas toward the use of all-terrain bicycles (ATBs, or mountain bikes). Controversy is growing over how to manage ATBs, particularly in areas frequently used by pedestrians and people on horseback. Riders of ATBs argue that they cause little environmental disrup-

tion and that they damage trails less than horses do. As long as the number of mountain bikes remains small, this may be true. However, because bicycles are not nearly as expensive as horses and do not have to be fed, we can expect many more ATBs than horses on trails. There is little doubt that if mountain biking continues to grow, intensive use in some areas will be associated with environmental degradation.

Chankanab Lagoon National Park

The Yucatan Peninsula is a low-lying lime-stone plateau at the southern tip of Mexico, bordered on the west by the Gulf of Mexico and on the east by the Caribbean Sea (Figure 30.16). Most of the rain that falls on the limestone quickly infiltrates into the rock, so there are few surface streams or rivers. Movement of groundwater through the limestone dissolves the rock along preferred directions, producing systems of caverns and caves. These occasionally collapse, forming sink-holes, which the Mayan people used as natural wells. Because limestone is also a relatively soft stone, it is easily cut and carved. It provided an abundant building material for the large pyramids and buildings constructed by the Maya between the years A.D. 500 and 1000. The island of Cozumel is a small island approximately 50 km long by 15 km wide (30 by 9 miles) located approximately 25 km (about 15 mi) off the coast of the peninsula.

Cozumel Island, like the nearby mainland, is composed of limestone and has a very recent geologic origin, probably emerging from the sea during the last million years or so. The island also has an interesting record of human history. The Maya who inhabited the island prior to the arrival of the Spanish in 1518 believed the land was a sacred shrine and referred to it as the island of the swallows. It is believed that some Mayan women journeyed to the island by boat to worship the goddess of fertility, and numerous archaeological sites on the island contained small dolls sacrificed during fertility ceremonies. The island was razed in 1519, when many of the Mayan temples were destroyed by the invading Spanish. Outbreaks of smallpox and other diseases reduced the Mayan population from about 40,000 in 1519 to only about 300 in 1570. The people completely abandoned the island. It was used in the 17th century by pirates and was finally resettled in about 1848.

From an environmental viewpoint, the island faces many challenges today. The

(a)

(b)

Figure 30.16 ● The location of (*a*) Cozumel Island and (*b*) Chankanab Lagoon in the Caribbean Sea, near Cancún, Mexico.

water supply, for example, is groundwater that is extremely vulnerable to pollution; and waste disposal practices are not up to high standards. Although Cozumel does not have the extensive high-rise development that is seen nearby on the Mexican mainland at Cancún, there is some evidence that high-rise development is likely to occur south of San Miguel, the island's only city. This situation may intensify in the future, and there is some concern that the island may go down the path of Cancún—intensive development and alteration of the natural environment to the extent that it scarcely resembles the natural state. However, the people of Cozumel are environmentally aware and are attempting to develop long-range comprehensive land-use planning to better protect this beautiful island.

As one element of long-term planning for the environment, national parks have been set aside. One such park is

Chankanab Lagoon, located approximately 10 km (6 mi) south of San Miguel. The government manages the site as a nature reserve, and no fishing or collecting is allowed. Chankanab Lagoon is interesting because the government is trying to restore the lagoon and, at the same time, provide a tourist attraction. Several decades ago, the lagoon was in bad shape; trash and other materials had accumulated in it. Originally, a series of caves connected the lagoon with the ocean and allowed ocean water to circulate into the lagoon. When a road was built, the caves collapsed, causing the circulation to cease. As a result, the lagoon died.

Approximately 15 years ago, a restoration project began that opened the cave system again, allowing circulation of seawater. Much of the trash was removed. At that time, a beachfront was developed, consisting of restaurants and other facilities for tourists (Figure

30.16c). Because the natural coastline at the site is rocky and contains little sand, small seawalls were constructed and the land side filled with sand to produce a beach approximately 30 m (100 ft) wide. People are allowed to swim in the ocean side of the lagoon, and the fish and marine life are completely protected.

The creation of the sandy beach turned out to be a problem. In 1989, Hurricane Gilbert struck the island. The high winds and waters moved the beach sand into the shallow lagoon, nearly filling it and burying the coral, once again killing the lagoon. Restoration work began in 1991 to dredge out part of the lagoon, and seawater again circulates from the ocean to the lagoon and back with the tides. Unfortunately, the restoration plan is likely to fail in the long term because hurricanes can be expected again, and they will again wash sand into the lagoon.

The Chankanab Lagoon story emphasizes problems that may arise when conflicts of management occur. It may be difficult to both satisfy the desires of the tourist industry and protect the natural environment. Nevertheless, a management plan could be developed to work harmoniously with the natural system and preserve the lagoon. A better design might have been to build wooden walkways with seats from which people could observe the reef and ocean, rather than to import the sand. If true restoration of the lagoon is to be successful, the management plan will have to change.

The situation at Xelha Lagoon on the Yucatan's east coast is quite different. That lagoon is much larger, and little sand has been imported. Thus, the lagoon has been preserved and remains a healthy ecosystem. That site also includes restaurants and other facilities, but they do not intrude directly on much of the total lagoon environment.

It is evident that the people of Mexico and the Yucatan Peninsula have a strong and growing consciousness of environmental concerns. However, as in many other places, additional expertise concerning natural processes is needed in the development of management plans.

and Montana, for example, special consideration is given to grizzly bear habitats through controls on where humans may venture.

Other recreational activities that are, or may become, more subject to regulation include hiking, camping, fishing, boating, skiing, snowmobiling, and such recently popularized activities as treasure hunting, which includes panning for gold.

Activities that occur on government lands can be more easily regulated than those occurring elsewhere. However, park management may be difficult if goals are not clear and natural processes not understood (see A Closer Look 30.3).

30.6 Global Forecasting

Global effects of our modern technological civilization have been discussed in earlier chapters. The burning of fossil fuels is adding a significant amount of carbon dioxide to the atmosphere; lead used in gasoline fuels has spread to the glaciers of Greenland; and DDT and other pesticides and artificial organic compounds are found in marine organisms inhabiting the Antarctic. We can imagine other effects, such as those of a nuclear war or of a large-scale release of a toxic compound. Few have yet tried to plan on a global level. However, in the last several decades, procedures for **global forecasting** and evaluation have been developed.

Important in the development of global forecasting methods was the Club of Rome. An international group of business executives, intellectuals, and government officials founded this organization, whose objectives are to promote a better understanding of humanity's predicament as it relates to global concerns; to disseminate information about this predicament; and to stimulate the development of new attitudes, policies, and institutions to redress the present situation.

From the concerns of the Club of Rome came the development of computer simulation models—models of global phenomena (such as human population growth), the use of the world's resources, and human impacts on the biosphere. The predictions of these models were reviewed in *The Limits to Growth*, a controversial book published in 1972.[31] The book forecast some dire consequences of human activities. Supporters of *The Limits to Growth* hailed its methodology as a new approach that could help save us from ourselves by forcing us to recognize the true limits to our uses of the biosphere. Opponents found fault with the models and claimed that the book said nothing more than Malthus had said hundreds of years before (see Chapter 5).

Since the publication of *The Limits to Growth*, global forecasting has become an important aspect of modern civilization's attempt to deal with its own global effects. In 1977, for example, the U.S. Council on Environmental Quality was directed by President Jimmy Carter to conduct a study of probable changes in the world's population, natural resources, and environment through the end of the century. This study resulted in *The Global 2000 Report to the President*, which used several methods to predict the status of population resources and environment by the end of the century, including a review of several computer models for global forecasting.[32]

Global forecasting remains a controversial activity. It is always difficult, if not impossible, to predict the future. The more we understand about the processes that govern change in the environment, however, the more likely it is that our forecasts will be helpful. Such forecasts are perhaps most valuable in showing us the general effects of what we assume these processes to be. In this way, the forecasts force us to recognize the meaning of what we think we know and suggest to us where our knowledge is lacking.

30.7 Global Security and Environment

The terrorist attacks on the city of New York on September 11, 2001 brought a realization that the United States and in fact the world is not as safe as we had previously assumed. The attacks launched a war on terrorists and their financial and political network around the world. However, for every terrorist removed another will fill the void unless the root causes are recognized and eliminated. This is what we mean when we state that to achieve sustainability and security we need to be proactive as well as reactive.

Achieving sustainability in the world today has strong political and economic components, but it also has an environmental component. Terrorism comes in part from poverty, overcrowding, disease, and conflicts that have environmental significance.

Over one billion people on earth today live in poverty with little hope for the future. Large urban regions with tens of millions of people are crowding people into marginal, unsanitary living conditions where safe drinking water and sewage disposal is not adequate. In the countryside, rural people in many developing countries are being terrorized and displaced by armed conflicts that have the objective of controling valuable resources such as oil, diamonds, and timber. Examples include: oil in Nigeria, Sudan, and Columbia; diamonds in Sierra Leone, Angola, and DR. Congo; and timber in Cambodia, Indonesia, and Borneo.[33]

The conflicts in developing countries over abundant natural resources are causing serious environmental degradation. For example, widespread pollution from mining of copper and gold in Papua New Guinea has damaged crops, polluted rivers, and reduced wildlife including fish that feed people. The financial benefits of resource exploitation too often pay for continued armed conflicts and terrorism. The rewards of resource extraction may go to a few while burdening the majority of people with social problems and environmental degradation.[33]

The Rio Earth Summit on Sustainable Development in 1992 had an objective of addressing global environmental problems of both developed and developing countries, with emphasis on solving conflicts between economic issues and environmental concerns. Today, in many countries the gap between the rich and poor is even wider than it was in the early 1990s. As a result, political, social, and economic security remains threatened and serious environmental damage from overpopulation and resource exploitation continues. The environment remains inade-quately funded. World Watch Institute reported in 2002 that the annual funding by the United Nations on the environment is about $100 million, while the governments of the world are spending $2 billion per day for military purposes.[33]

30.8 Charge to Students of the Environment

To end this book on an optimistic note—and there are reasons to be optimistic—the Earth Summit on Sustainable Development in the summer of 2002 in Johannesburg, South Africa, has the objectives to: continue to work toward environmental and social justice for all the people in the world; enhance the development of sustainability; minimize local, regional, and global environmental degradation resulting from overpopulation, deforestation, mining, agriculture, and pollution of the land, water, and air; and develop and support international agreements to control global warming and pollutants, and foster environmental and social justice.

Solving our environmental problems will help build a more secure and sustainable future. As students of the environment and our future leaders, this is becoming your charge and responsibility as you graduate from colleges and universities. This transfer of knowledge and leadership is a major reason why we wrote this book! Get out there and do it! In the words of a famous song—make the world a better place.

Summary

- The path toward sustainable development involves an evolution of our values to include all people within a framework that is economically, socially, and environmentally just.

- Mediation is a method of conflict or dispute resolution that seeks to avoid expensive and lengthy litigation by arriving at a conclusion whereby each party gains enough from a compromise to prefer a settlement to litigation.

- International environmental law is proving useful in addressing several important environmental problems including preservation of resources and pollution abatement.

- The major components of an environmental impact statement (EIS) are (1) summary, (2) statement of purpose and need of project, (3) rigorous comparison of alternatives, (4) succinct description of the environment affected, and (5) discussion of the environmental consequences of the proposed project and possible alternatives.

- Scoping is the process of early identification of important environmental issues requiring detailed evalua-

Euro Tunnel: England to France: Tunnel Rubble to a Nature Reserve

The tunnel across the Dover Straits of the North Sea from England to France—or the "Chunnel," as it is sometimes called—is considered one of humankind's most outstanding engineering accomplishments. The tunnel, through which trains travel between London and Paris, is approximately 50 km long; and of this, about 38 km is under the sea (Figure 30.17). A tunnel under the channel had been considered for almost 200 years, and a false start had been made in 1974. Serious construction of the tunnel started in 1984, and it was completed in 1994. Nearly 14,000 technicians, engineers, and other workers were involved in the project, which eventually cost approximately $15 billion. The project actually involved building two tunnels that are close together: an eastbound and a westbound tunnel, each with a diameter of approximately 7.6 m.

Extensive surface and subsurface geologic and hydrologic evaluations completed before construction determined that the best route for the tunnel lay within a clay-rich chalk (a type of limestone) sandwiched between an overlying chalk, which forms the famous White Cliffs of Dover, and underlying clay stones (Figure 30.18). The undersea sections of the tunnel are contained within the clay-rich chalk beds for over 90% of their total length. The primary reason the clay-rich chalk was chosen was that the clay allows very little seepage of water. The tunnel, on completion, was concrete-lined, and the train ride under the sea takes about 20 minutes, with the train moving at speeds up to 160 km/h.

As might be expected, such a large undertaking moves a great deal of rock. The British share of the tunnel rubble, or spoil, was nearly 9 million m³ mostly clay-rich chalk. Disposal of such a large volume of material was a problem. The solution finally reached was to dump the material at the base of the White Cliffs of Dover at a location known as Samphire Hoe, located approximately 3 km west of Dover. About 5 million m³ of the material was placed at Samphire Hoe. The project

Figure 30.17 • Location of the tunnel across the English Channel from England to France.

Figure 30.18 • Generalized geology along the tunnel.

involved construction of a seawall nearly 2 km long, behind which the tunnel spoil was dumped. This activity increased the size of England by an area of approximately 36 ha (90 acres). Buildings were removed, and the spoil was leveled, creating a nature reserve seaward of the cliffs of Dover. The nature reserve has walking trails; and plants have colonized the new ground, providing habitat for butterflies, birds, and other wildlife. The project has produced a truly unique area.

The natural landscape in the area consists of tall white chalk cliffs, locally known as the Shakespeare Cliffs because the famous playwright mentioned them in King Lear. The cliffs comprise a classic seacliff and beach environment that are eroding at a rate of approximately 0.1 m/yr. In the area of Samphire Hoe, this dynamic environment will be changed as seacliff erosion at the base ceases.

On the one hand, those who constructed and use the nature reserve believe it as a very positive feature, for these reasons:

1. It has added new land and new environments to England.

2. It has provided new habitats for a variety of wildlife.

3. It provides an offshore view of the White Cliffs of Dover.

4. It provides a quiet area for people to visit in our busy world.

On the other hand, one may question the advisability of dumping the tunnel spoil in this location for the following reasons:

1. The White Cliffs of Dover comprise a unique environment that will be changed as a result of the addition of the tunnel rubble.

2. There is loss of beach environment, which is a valued resource.

3. Construction of a seawall and protrusion of the land into the sea will change local coastal processes as the adjacent areas continue to erode. Eventually, the site may be turned into a point, with increased wave activity and potential for erosion and destruction of the wall.

4. Some people find seawalls offensive from an aesthetic viewpoint; and while a nature reserve is a contribution, other sites for a reserve might have been preferable.

Critical Thinking Questions

1. Consider the project of creating the nature reserve carefully, and write a few statements concerning potential environmental impact analyses that could have been done. These include: (1) statement of the purpose of the project; (2) description of the environment before the project; (3) possible alternatives to the project, including the tunnel itself; and (4) environmental consequences of each of the alternatives. In terms of the environmental consequences, do not forget the principle of environmental unity, which means that the proposed project and alternatives may have primary, secondary, and tertiary consequences. Finally, choose and defend either the project or one of your alternatives.

2. Do you think environmental analyses would have been improved if a scoping process had been completed prior to the initiation of the project? Explain your view.

tion. Mitigation is the process that identifies actions that will avoid, lessen, or compensate for anticipated adverse environmental impacts.

- Land-use planning is a controversial topic with important environmental implications. The planning process includes definition of goals and objectives; collection and analysis of data; development of alternatives; formulation of plans; and review, implementation, and revision of plans.

- The most preferable adjustment to natural hazards is land-use planning to avoid the hazard. Other adjustments include structures to control hazards, insurance, evacuation, disaster preparedness, and bearing the loss (doing nothing).

- As the human population increases, more people are forced to live in areas likely to be frequented by hazards, such as flooding and storms. Therefore, greater loss of life and property damage can be expected, especially in developing countries, where population increase and land-use change are greatest.

- Global forecasting is controversial because models that predict the future make controversial assumptions. Nevertheless, global forecasting provides an opportunity to better understand general effects of our actions and to suggest where more knowledge is needed.

- Global security, sustainability, and environment are linked in complex ways. Solving environmental problems will improve both sustainability and security.

REEXAMINING THEMES AND ISSUES

Human Population:

As human population continues to increase on a regional and global level, careful planning will become even more important to ensure that we make wise decisions concerning where we live, how we conserve and preserve ecosystems, and how we construct our urban environments.

Sustainability:

Environmental planning is at the heart of sustainable development. Planning decisions we make today and in the future will greatly affect the quality of the environment we leave for future generations. Concepts of environmental impact analysis, mitigation of adverse environmental impact, and development of environmentally preferable adjustments to natural hazards are integral land-use planning methods for obtaining sustainability.

Global Perspective:

We are becoming a global community, and our planning needs must include a significant global perspective. Development of international environmental agreements is becoming increasingly important, as are continuing attempts at global forecasting. We are only beginning to plan on a global level.

Urban World:

As we focus our attention on environmental issues in our increasingly urban world, one of our most pressing needs is to ensure that urban environments are carefully planned to maximize human well-being and minimize effects of natural hazards and pollution of land, air, and water.

Science and Values:

Because we value our environment—from the local level to global systems—we must continue to develop knowledge, technology, and tools to assist in the environmental planning process. Of particular importance is that we recognize where the knowledge needed to solve specific environmental problems is lacking so that we can focus on those areas. We must also recognize that people's values with respect to resource use and land use are variable, resulting in periodic conflicts between those who wish to use the environment and those who wish to preserve it. Environmental mediation can help resolve conflicts when values collide.

KEY TERMS

common law **644**
comprehensive plan **655**
environmental impact report (EIR) **648**
environmental impact statement (EIS) **647**

Geographic Information System (GIS) **648**
global forecasting **663**
green plans **659**
land-use planning **655**
mitigated negative declaration **648**

mitigation **647**
negative declaration **648**
public trust **644**
record of decision **647**
scoping **647**

STUDY QUESTIONS

1. Do you agree that if a person believes that he or she has a right to breathe clean air, drink clean water, and live on pollution-free land, that person is an environmentalist? Why or why not?

2. Do you believe that civil disobedience is justified to protect the environment? Why or why not? What about practices such as ecotage, which damage property?

3. Make a map of your neighborhood. Does it seem planned with the environment in mind?

4. What is meant by mitigation of an environmental impact?

5. An expert on impact analysis tells you, "Mitigation is the impact. Once you have determined the mitigating factors, you have determined the impact." Do you agree? Explain.

6. What major steps must be taken in the assessment of any environmental impact?

7. Discuss the elements of a land-use plan. Which element is likely to be most controversial? Why?

8. Why is planning immediately following disasters loaded with potential problems?

9. Discuss the major components of an environmental impact statement. Which are the most necessary? Why?

10. What are the major advantages of the scoping process in environmental impact analysis?

11. What are some potential conflicts in the management of recreational activities?

12. What are the advantages and disadvantages of regional national planning and what are some potential problems?

13. Discuss the role of mediation in environmental law. What is necessary for it to work?

14. How has population increase resulted in a greater toll in human life from natural hazards than would be expected based on the simple increase in population?

● FURTHER READING

Arnold, F. S. 1995. *Economic Analysis of Environmental Policy and Regulation*. New York: Wiley. A wide variety of practical applications of economics to environmental regulatory and policy analysis.

Bass, R. E., and A. I. Herson. 1993. *Mastering NEPA: A Step-by-Step Approach*. Point Arena, Calif.: Solano. Background and implementation of NEPA, written by environmental consultants with a thorough understanding of the act.

Bregman, J. I., and K. M. Mackenthum. 1992. *Environmental Impact Statements*. Boca Raton, Fla.: Lewis. An up-to-date examination of the many scientific and technical aspects of EIS preparation.

Jain, R. K., L. V. Urban, G. S. Stacey, and H. E. Balbach. 1993. *Environmental Assessment*. New York: McGraw-Hill. A good book covering the importance of and steps involved in investigating environmental considerations concerning a proposed action.

Kubasek, N. K., and G. S. Silverman. 1994. *Environmental Law*. Englewood Cliffs, N.J.: Prentice-Hall. A look at how our legal system is used to assist in environmental policy and preservation.

McHarg, I. L. 1971. *Design with Nature*. Garden City, N.Y.: Doubleday. A classic book on environmental planning.

Orlof, N. 1979. *The Environmental Impact Statement Process: A Guide to Citizen Action*. Washington, D.C.: Information Resources. A simple-to-understand explanation of the entire EIS process, including both political and scientific aspects. It emphasizes the role that local citizens can play in the process.

Rohse, M. 1987. *Land-Use Planning in Oregon*. Corvallis: Oregon State University Press. A good discussion of basic land-use planning.

Teubner, G., L. Farmer, and D. Murphy. eds. 1994. *Environmental Law and Ecological Responsibility: The Concept and Practice of Ecological Self-organization*. New York: Wiley. Argues that accepting that the old-style command-and-control regulation of business has reached its limits, a new, externally induced, internal self-organizing process may render business more sensitive to environmental concerns.

Appendix

A Special Feature:
Electromagnetic Radiation (EMR) Laws

Properties of Waves

- Direction of wave propagation

- W = wavelength (distance from one wave crest to the next)
- An EMR wave travels at the speed of light (C) in a vacuum, or about 300,000 km/s (3×10^8 m/s).
- The period T of a wave is the time it takes for a wave to travel a distance of one wavelength W. Then since distance is the product of speed (velocity) and time, W = CT.
- The frequency f of a wave is the number of cycles (each wavelength that passes a point is a cycle) of a wave that pass a particular point per unit time. Frequency, f is measured in cycles per second (hertz). The frequency f is the inverse of T: f = 1/T and W = C/f. For example, the period T of a 6000 hertz EMF wave is: 6000 hertz = 1/T and T = 1/6000 S, or 1.7×10^{-4} 5. The wavelength W = CT is 3×10^8 m/s times 1.7×10^{-4} 5, or 5.1×10^4 m, which according to Figure 22.12 is a long radio wave.

Absolute Temperature Scale (kelvin, K)

- Zero is really zero; there are no negative values of K
- Temperature in K = temperature in °C + 273

$$K = °C + 273$$

- Example: water freezes at °C = 0 = 273 K
  water boils at °C = 100 = 373 K

Stefan–Boltzmann Law

- All bodies with a temperature greater than absolute zero radiate EMR. These bodies are called thermal radiators. The amount of energy per second radiated from thermal radiators is called *intensity* and is given by the Stefan–Boltzmann law

$$E = aT^4$$

where E is the energy per second (intensity); T is the absolute temperature; and a is a constant (the nature of this constant involves physical ideas beyond the scope of this text).

- The Stefan–Boltzmann law states that the intensity of EMR coming from a thermal radiator is directly proportional to the fourth power of its absolute temperature.

Wien's Law

$$W_P = a/T$$

- where W_P is the wavelength of the peak intensity of a thermal radiator; T is temperature in K; and a is a constant. For example, Figure 22.13 shows that W_P for the earth is about 10 μm. Wien's law states in a general way that the hotter a substance is, the shorter the wavelength of the emitted predominant electromagnetic radiation. That is, wavelength is inversely proportional to temperature.

Prefixes and Multiplication Factors

Number	10^x, Power of 10	Prefix	Symbol
1,000,000,000,000,000	10^{18}	exa	E
1,000,000,000,000,000	10^{15}	peta	P
1,000,000,000,000	10^{12}	tera	T
1,000,000,000	10^{9}	giga	G
1,000,000	10^{6}	mega	M
10,000	10^{4}	myria	
1,000	10^{3}	kilo	k
100	10^{2}	hecto	h
10	10^{1}	deca	da
0.1	10^{-1}	deci	d
0.01	10^{-2}	centi	c
0.001	10^{-3}	milli	m
0.000 001	10^{-6}	micro	m
0.000 000 001	10^{-9}	nano	n
0.000 000 000 001	10^{-12}	pico	p
0.000 000 000 000 001	10^{-15}	femto	f
0.000 000 000 000 000 001	10^{-18}	atto	a

 ## Common Conversion Factors

Length

1 yard = 3 ft, 1 fathom = 6 ft

	in.	ft	mi	cm	m	km
1 inch (in.) =	1	0.083	1.58×10^{-5}	2.54	0.0254	2.54×10^{-5}
1 foot (ft) =	12	1	1.89×10^{-4}	30.48	0.3048	—
1 mile (mi) =	63,360	5,280	1	160,934	1,609	1.609
1 centimeter (cm) =	0.394	0.0328	6.2×10^{-6}	1	0.01	1.0×10^{-5}
1 meter (m) =	39.37	3.281	6.2×10^{-4}	100	1	0.001
1 kilometer (km) =	39,370	3,281	0.6214	100,000	1,000	1

Area

1 square mi = 640 acres, 1 acre = 43,560 ft^2 = 4046.86 m^2 = 0.4047 ha
1 ha = 10,000 m^2 = 2.471 acres

	in.^{2}	ft^2	mi^2	cm^2	m^2	km^2
1 in.^{2} =	1	—	—	6.4516	—	—
1 ft^2 =	144	1	—	929	0.0929	—
1 mi^2 =	—	27,878,400	1	—	—	2.590
1 cm^2 =	0.155	—	—	1	—	—
1 m^2 =	1,550	10.764	—	10,000	1	—
1 km^2 =	—	—	0.3861	—	1,000,000	1

Volume

	in.³	ft³	yd³	m³	qt	liter	barrel	gal (U.S.)
1 in.³ =	1	—	—	—	—	0.02	—	—
1 ft³ =	1,728	1	—	0.0283	—	28.3	—	7.480
1 yd³ =	—	27	1	0.76	—	—	—	—
1 m³ =	61,020	35.315	1.307	1	—	1,000	—	—
1 quart (qt) =	—	—	—	—	1	0.95	—	0.25
1 liter (l) =	61.02	—	—	—	1.06	1	—	0.2642
1 barrel (oil) =	—	—	—	—	168	159.6	1	42
1 gallon (U.S.) =	231	0.13	—	—	4	3.785	0.02	1

Mass and Weight

1 pound = 453.6 grams = 0.4536 kilogram = 16 ounces

1 gram = 0.0353 ounce = 0.0022 pound

1 short ton = 2000 pounds = 907.2 kilograms

1 long ton = 2240 pounds = 1008 kilograms

1 metric ton = 2205 pounds = 1000 kilograms

1 kilogram = 2.205 pounds

Energy and Power[a]

1 kilowatt-hour = 3413 Btus = 860,421 calories

2 Btu = 0.000293 kilowatt-hour = 252 calories = 1055 joules

1 watt = 3.413 Btu/hr = 14.34 calorie/min

1 calorie = the amount of heat necessary to raise the temperature of 1 gram (1 cm³) of water 1 degree Celsius

1 quadrillion Btu = (approximately) 1 exajoule

1 horsepower = 7.457×10^2 watts

1 joule = 9.481×10^{-4} Btu = 0.239 cal = 2.778×10^{-7} kilowatt-hour

[a]Values from Lange, N. A., 1967, *Handbook of Chemistry,* McGraw Hill: New York.

Temperature

$F = \frac{9}{5}C + 32$ F is degrees Fahrenheit.
C is degrees Celsius (centigrade).

Fahrenheit		Celsius
32	Freezing of H_2O (Atmospheric Pressure)	0
50	———	10
68	———	20
86	———	30
104	———	40
122	———	50
140	———	60
158	———	70
176	———	80
194	———	90
212	Boiling of H_2O (Atmospheric Pressure)	100

Other Conversion Factors

1 ft³/sec = 0.0283 m³/sec = 7.48 gal/sec = 28.32 liter/sec

1 acre-foot = 43,560 ft³ = 1233 m³ = 325,829 gal

1 m³/sec = 35.32 ft³/sec

1 ft³/sec for one day = 1.98 acre-feet

1 m/sec = 3.6 km/hr = 2.24 mi/hr

1 ft/sec = 0.682 mi/hr = 1.097 km/hr

1 atmosphere = 14.7 lb(in.$^{-2}$) = 2116 lb(ft^{-2}) = 1.013×10^5 N(m^{-2})

Era	Approximate Age in Millions of Years Before Present	Period	Epoch	Life Form
	Less than 0.01		Recent (Holocene)	
	0.01–2	Quaternary	Pleistocene	Humans
	2			
Cenozoic	2–5		Pliocene	
	5–24		Miocene	
	24–38	Tertiary	Oligocene	
	38–55		Eocene	Mammals
	55–63		Paleocene	
	63			
Mesozoic	63–138	Cretaceous		
	138–205	Jurassic		Flying reptiles, birds
	205–240	Triassic		Dinosaurs
	240			
Paleozoic	240–290	Permian		Reptiles
	290–360	Carboniferous		Insects
	360–410	Devonian		Amphibians
	410–435	Silurian		Land plants
	435–500	Ordovician		Fish
	500–570	Cambrian		
	570			
	700			Multicelled organisms
	3,400			One-celled organisms
	4,000	Approximate age of oldest rocks discovered on Earth		
Precambrian				
	4,500	Approximate age of the Earth and meteorites		

Glossary

Abortion rate The estimated number of abortions per 1000 women aged 15 to 44 in a given year. (Ages 15 to 44 are taken to be the limits of ages during which women can have babies. This, of course, is an approximation, made for convenience.)

Abortion ratio The estimated number of abortions per 1000 live births in a given year.

Acid mine drainage Does not refer to an acid mine but to acidic water that drains from mining areas (mostly coal but also metal mines). The acidic water may enter surface water resources, causing environmental damage.

Acid rain Rain made artificially acid by pollutants, particularly oxides of sulfur and nitrogen. (Natural rainwater is slightly acid owing to the effect of carbon dioxide dissolved in the water.)

Active solar energy systems Direct use of solar energy that requires mechanical power; usually consists of pumps and other machinery to circulate air, water, or other fluids from solar collectors to a heat sink where the heat may be stored.

Adaptive radiation The process that occurs when a species enters a new habitat that has unoccupied niches and evolves into a group of new species, each adapted to one of these niches.

Advanced wastewater treatment Treatment of wastewater beyond primary and secondary procedures. May include sand filters, carbon filters, or application of chemicals to assist in removing potential pollutants such as nutrients from the wastewater stream.

Aerobic Characterized by the presence of free oxygen.

Aesthetic arguments Reasons for preserving the beauty of nature and natural wilderness areas.

Aesthetic justification for the conservation of nature An argument for the conservation of nature on the grounds that nature is beautiful and that beauty is important and valuable to people.

Age dependency ratio The ratio of dependent-age people (those unable to work) to working-age people. It is customary to define working-age people as those aged 15 to 65.

Age structure (of a population) Structure of a population divided into groups by age. Sometimes the groups represent the actual number of each age in the population; sometimes the group represents the percentage or proportion of the population of each age.

Agroecosystem An ecosystem created by agriculture. It has low genetic, species, and habitat diversity.

Air Quality Standards Levels of air pollutants that delineate acceptable levels of pollution over a particular time period. Valuable because they are often tied to emission standards that attempt to control air pollution.

Allowance trading Approach to managing coal resources and reducing pollution through buying, selling, and trading of allowances to emit pollutants from burning coal. The idea is to control pollution by controlling the number of allowances issued.

Alpha particles One of the major types of nuclear radiation, consisting of two protons and two neutrons (a helium nucleus).

Alternative energy Renewable and nonrenewable energy resources that are alternatives to the fossil fuels.

Anaerobic Characterized by the absence of free oxygen.

Aquaculture Production of food from aquatic habitats.

Aquifer An underground zone or body of earth material from which groundwater can be obtained from a well at a useful rate.

Area sources Sometimes also called nonpoint sources. These are diffused sources of pollution such as urban runoff or automobile exhaust. These sources include emissions that may be over a broad area or even over an entire region. They are often difficult to isolate and correct because of the widely dispersed nature of the emissions.

Asbestos A term for several minerals that have the form of small elongated particles. Some types of particles are believed to be carcinogenic or to carry with them carcinogenic materials.

Atmosphere Layer of gases surrounding Earth.

Atmospheric inversion A condition in which warmer air is found above cooler air, restricting air circulation; often associated in urban areas with a pollution event.

Autotroph An organism that produces its own food from inorganic compounds and a source of energy. There are photoautotrophs (photosynthetic plants) and chemical autotrophs.

Average residence time A measure of the time it takes for a given part of the total pool or reservoir of a particular material in a system to be cycled through the system. When the size of the pool and rate of throughput are constant, average residence time is the ratio of the total size of the pool or reservoir to the average rate of transfer through the pool.

Balance of Nature An environmental myth that states that the natural environment, when not influenced by human activity, will reach some sort of balance or equilibrium state.

Barrier island An island separated from the mainland by a salt marsh. It generally consists of a multiple system of beach ridges and is separated from other barrier islands by inlets that allow the exchange of seawater with lagoon water.

Becquerel The unit commonly used for radioactive decay in the International System (SI) of measurement.

Beta particles One of the three major kinds of nuclear radiation; electrons that are emitted when one of the protons or neutrons in the nucleus of an isotope spontaneously changes.

Biochemical oxygen demand (BOD) A measure of the amount of oxygen necessary to decompose organic material in a unit volume of water. As the amount of organic waste in water increases, more oxygen is used, resulting in a higher BOD.

Biogeochemical cycle The cycling of a chemical element through the biosphere; its pathways, storage locations, and chemical forms in the atmosphere, oceans, sediments, and lithosphere.

Biogeography The large-scale geographic pattern in the distribution of species, and the causes and history of this distribution.

Biohydrometallurgy Combining biological and mining processes, usually involving microbes to help extract valuable metals such as gold from the ground. May also be used to remove pollutants from mining waste.

Biological control A set of methods to control pest organisms by using natural ecological interactions, including predation, parasitism, and competition. Part of integrated pest management.

Biological diversity Used loosely to mean the variety of life on the Earth, but technically this concept consists of three components: (1) genetic diversity—the total number of genetic characteristics; (2) species diversity; and (3) habitat or ecosystem diversity—the number of kinds of habitats or ecosystems in a given unit area. Species diversity in turn includes three concepts: *species richness, evenness,* and *dominance.*

Biological evolution The change in inherited characteristics of a population from generation to generation, which can result in new species.

Biomagnification Also called *biological concentration.* The tendency for some substances to concentrate with each trophic level. Organisms preferentially store certain chemicals and excrete others. When this occurs consistently among organisms, the stored chemicals increase as a percentage of the body weight as the material is transferred along a food chain or trophic level. For example, the concentration of DDT is greater in herbivores than in plants and greater in plants than in the nonliving environment.

Biomass The amount of living material, or the amount of organic material contained in living organisms, both as live and dead material, as in the leaves (live) and stem wood (dead) of trees.

Biomass energy The energy that may be recovered from biomass, which is organic material such as plants and animal waste.

Biomass fuel A new name for the oldest fuel used by humans. Organic matter, such as plant material and animal waste, that can be used as a fuel.

Biome A kind of ecosystem. The rain forest is an example of a biome; rain forests occur in many parts of the world but are not all connected with each other.

Bioremediation A method of treating groundwater pollution problems that utilizes microorganisms in the ground to consume or break down pollutants.

Biosphere That part of a planet where life exists. On Earth it extends from the depths of the oceans to the summit of mountains, but most life exists within a few meters of the surface.

Biota A general term for all the organisms of all species living in an area or region up to and including the biosphere, as in "the biota of the Mojave Desert" or "the biota in that aquarium."

Biotic province A region inhabited by a characteristic set of taxa (species, families, orders), bound by barriers that prevent the spread of the distinctive kinds of life to other regions and the immigration of foreign species.

Birth rate The rate at which births occur in a population, measured either as the number of individuals born per unit of time or as the percentage of births per unit of time compared with the total population.

Black lung disease Often called coal miner disease because it is caused by years of inhaling coal dust, resulting in damage to lungs.

Body burden The amount of concentration of a toxic chemical, especially radionuclides, in an individual.

Breeder reactor A type of nuclear reactor that utilizes between 40% and 70% of its nuclear fuel and converts fertile nuclei to fissile nuclei faster than the rate of fission. Thus breeder reactors actually produce nuclear fuels.

Brines With respect to mineral resources, refers to waters with a high salinity that contain useful materials such as bromine, iodine, calcium chloride, and magnesium.

Buffers Materials (chemicals) that have the ability to neutralize acids. Examples include the calcium carbonate that is present in many soils and rocks. These materials may lessen potential adverse effects of acid rain.

Burner reactors A type of nuclear reactor that consumes more fissionable material than it produces.

Capillary action The rise of water along narrow passages, facilitated and caused by surface tension.

Carbon cycle Biogeochemical cycle of carbon. Carbon combines with and is chemically and biologically linked with the cycles of oxygen and hydrogen that form the major compounds of life.

Carbon dioxide (CO_2) Molecule of carbon and oxygen present in the atmosphere at approximately 350 ppm. Emissions of carbon dioxide resulting from burning of fossil fuels are thought to be contributing to potential global warming through an enhanced greenhouse effect.

Carbon monoxide (CO) Colorless, odorless gas that at very low concentrations is extremely toxic to humans and animals.

Carbonate-silicate cycle A complex biogeochemical cycle over time scales as long as one-half billion years. Included in this cycle are major geologic processes, such as weathering, transport by ground and surface waters, erosion, and deposition of crustal rocks. The carbonate-silicate cycle is believed to provide important negative feedback mechanisms that control the temperature of the atmosphere.

Carcinogen Any material that is known to produce cancer in humans or other animals.

Carnivores Organisms that feed on other live organisms; usually applied to animals that eat other animals.

Carrying capacity The maximum abundance of a population or species that can be maintained by a habitat or ecosystem without degrading the ability of that habitat or ecosystem to maintain that abundance in the future.

Cash crops Crops grown to be traded in a market.

Catastrophe A situation or event that causes significant damage to people and property, such that recovery and/or rehabilitation is a long and involved process. Examples of natural catastrophes include hurricanes, volcanic eruptions, large wildfires, and floods.

Channelization An engineering technique that consists of straightening, deepening, widening, clearing, or lining existing stream channels. The purpose is to control floods, improve drainage, control erosion, or improve navigation. It is a very controversial practice that may have significant environmental impacts.

Chaparral A dense scrubland found in areas with Mediterranean climate (a long warm, dry season and a cooler rainy season).

Chemical reaction The process in which compounds and elements undergo a chemical change to become a new substance or substances.

Chemoautotrophs Autotrophic bacteria that can derive energy from chemical reactions of simple inorganic compounds. The most common use for inorganic sulfur compounds.

Chemosynthesis Synthesis of organic compounds by energy derived from chemical reactions.

Chimney (or stack) effect Process whereby warmer air rises in buildings to upper levels and is replaced in the lower portion of the building by outdoor air drawn through a variety of openings, such as windows, doors, or cracks in the foundations and walls.

Chlorofluorocarbons (CFCs) Highly stable compounds that have been or are being used in spray cans as aerosol propellants and in refrigeration units (the gas that is compressed and expanded in a cooling unit). Emissions of chlorofluorocarbons have been associated with potential global warming and stratospheric ozone depletion.

Chronic disease A disease that is usually present in a population, typically occurring in a relatively small but constant percentage of the population.

Chronic hunger Availability of sufficient food to stay alive, but not enough to lead a satisfactory and productive life.

Chronic patchiness A situation where ecological succession does not occur. Characteristic of harsh environments such as deserts. In this case, one species may replace another, or an individual of the first species may replace it, and no overall general temporal pattern is established.

City planning Conscious design of the growth and development of an urban area.

Clay May refer to a mineral family or to a very fine-grained sediment. It is associated with many environmental problems, such as shrinking and swelling of soils and sediment pollution.

Clean Air Act Amendments of 1990 Comprehensive regulations (federal statute) that address acid rain, toxic emissions, ozone depletion, and automobile exhaust.

Clear-cutting In timber harvesting, the practice of cutting all trees in a stand at the same time.

Climate The representative or characteristic conditions of the atmosphere at particular places on Earth. *Climate* refers to the average or expected conditions over long periods; *weather* refers to the particular conditions at one time in one place.

Climatic change Change in mean annual temperature and other aspects of climate over periods of time ranging from decades to hundreds of years to several million years.

Climax stage (or ecological succession) The final stage of ecological succession and therefore an ecological community that continues to reproduce itself over time, or a stage in ecological succession during which an ecological community achieves the greatest biomass or diversity. (The first definition is the classical definition.)

Climax state A hypothetical steady-state stage at the end of ecological succession. Traditionally, it was believed to be self-sustaining and also to have maximum organic matter, maximum storage of chemical elements, and maximum biological diversity. These ideas are now largely rejected among scientists doing basic ecological research, but they still have many adherents among practitioners.

Closed system A type of system in which there are definite boundaries to factors such as mass and energy such that exchange of these factors with other systems does not occur.

Closed-canopy forest Forests in which the leaves of adjacent trees overlap or touch, so that the trees form essentially continuous cover.

Coal Solid, brittle carbonaceous rock that is one of the world's most abundant fossil fuels. It is classified according to energy content as well as carbon and sulfur content.

Coal gasification Process that converts coal that is relatively high in sulfur to a gas in order to remove the sulfur.

Cogeneration The capture and use of waste heat; for example, using waste heat from a power plant to heat adjacent factories and other buildings.

Cohort All the individuals in a population born during the same time period. Thus all the people born during the year 1980 represent the world human cohort for that year.

Commensalism A relationship between two kinds of organisms in which one benefits from the relationship and the other is neither helped nor hurt.

Commercial-grade forest Forestland believed or known to be an economically profitable source of timber, often defined as capable of producing at least 1.4 m^3/ha (200 ft^3/acre) per year of wood.

Common law Law derived from custom, judgment, or decrees of courts.

Commons Land that belongs to the public, not to individuals. Historically a part of old English and New England towns where all the farmers could graze their cattle.

Community, ecological A group of populations of different species living in the same local area and interacting with one another. A community is the living portion of an ecosystem.

Community effect (community-level effect) When the interaction between two species leads to changes in the presence or absence of other species or in a large change in abundance of other species, then a community effect is said to have occurred. Here, a two-species interaction affects the entire community.

Competition The situation that exists when different individuals, populations, or species compete for the same resource(s) and the presence of one has a detrimental effect on the other. Sheep and cows eating grass in the same field are competitors.

Competitive exclusion principle The idea that two populations of different species with exactly the same requirements cannot persist indefinitely in the same habitat—one will always win out and the other will become extinct. Which one wins depends on the exact environmental conditions. Referred to as a principle, the idea has some basis in observation and experimentation.

Composting Biochemical process in which organic materials, such as lawn clippings and kitchen scraps, are decomposed to a rich, soil-like material.

Comprehensive plan Official plan adopted by local government formally stating general and long-range policies concerning future development.

Cone of depression A cone-shaped depression in the water table around a well caused by withdrawal by pumping of water at rates greater than the rates at which the water can be replenished by natural groundwater flow.

Conservation With respect to resources such as energy, refers to changing our patterns of use or simply getting by with less. In a pragmatic sense the term means adjusting our needs to minimize the use of a particular resource, such as energy.

Consumptive use A type of off-stream water use. This water is consumed by plants and animals or in industrial processes or evaporates during use. It is not returned to its source.

Contamination Presence of undesirable material that makes something unfit for a particular use.

Continental drift The movement of continents in response to seafloor spreading. The most recent episode of continental drift started about 200 million years ago with the breakup of the supercontinent Pangaea.

Continental shelf Relatively shallow ocean area between the shoreline and the continental slope that extends to approximately a 600-foot (~200 m) water depth surrounding a continent.

Contour plowing Plowing land along topographic contours, as much in a horizontal plane as possible, thereby decreasing the erosion rate.

Controlled burning Using prescribed fire to reduce the risk from wildfires, control tree diseases, increase food and habitat for wildlife, and manage forests for greater production of desirable tree species.

Controlled experiment A controlled experiment is designed to test the effects of independent variables on a dependent variable by changing only one independent variable at a time. For each variable tested, there are two set-ups (an experiment and a control) that are identical except for the independent variable being tested. Any difference in the outcome (dependent variable) between the experiment and the control can then be attributed to the effects of the independent variable tested.

Convection The transfer of heat involving the movement of particles; for example, the boiling water in which hot water rises to the surface and displaces cooler water, which moves toward the bottom.

Convergent evolution The process by which species evolve in different places or different times and, although they have different genetic heritages, develop similar external forms and structures as a result of adaptation to similar environments. The similarity in the shapes of sharks and porpoises is an example of convergent evolution.

Convergent plate boundary Boundary between two lithosphere plates in which one plate descends below the other (subduction).

Cosmopolitan species A species with a broad distribution, occurring wherever in the world the environment is appropriate.

Costs, direct In economics, costs borne by the producer and passed directly on to the user or purchaser.

Costs, indirect In environmental economics, costs associated with the degradation of the environment.

Crop rotation A series of different crops planted successively in the same field, with the field occasionally left fallow, or grown with a cover crop.

Crude oil Naturally occurring petroleum, normally pumped from wells in oil fields. Refinement of crude oil produces most of the petroleum products we use today.

Cultural Eutrophication Human-induced eutrophication that involves nutrients such as nitrates or phosphates that cause a rapid increase in the rate of plant growth in ponds, lakes, rivers or the ocean.

Curie Commonly used unit to measure radioactive decay; the amount of radioactivity from 1 gram of radium 226 that undergoes about 37 billion nuclear transformations per second.

Death rate The rate at which deaths occur in a population, measured either as the number of individuals dying per unit time or as the percentage of a population dying per unit time.

Decomposer An organism that feeds on dead organic matter and in this way reduces complex organic compounds to simpler organic compounds or smaller particles or to inorganic compounds.

Deductive reasoning Drawing a conclusion from initial definitions and assumptions by means of logical reasoning.

Deep-well disposal Method of disposal of hazardous liquid waste that involves pumping the waste deep into the ground below and completely isolated from all freshwater aquifers. A controversial method of waste disposal that is being carefully evaluated.

Deforestation Harvesting trees for commercial and other uses and burning forest to convert lands to agricultural purposes.

Demand for food The amount of food that would be bought at a given price if it were available.

Demand-based agriculture Agriculture with production determined by economic demand and limited by that demand rather than by resources.

Demographic transition The pattern of change in birth and death rates as a country is transformed from undeveloped to developed. There are three stages: (1) in an undeveloped country, birth and death rates are high and the growth rate low; (2) the death rate decreases, but the birthrate remains high and the growth rate is high; (3) the birthrate drops toward the death rate and the growth rate therefore also decreases.

Demography The study of populations, especially their patterns in space and time.

Dependent variable See **Variable, dependent**

Denitrification The conversion of nitrate to molecular nitrogen by the action of bacteria—an important step in the nitrogen cycle.

Density-dependent population effects Factors whose effects on a population change with population density. The term is usually restricted to population growth, reproduction, and mortality. For example, during a famine the mortality rate increases. In this case the food supply can be said to have a density-dependent population effect.

Density-independent population effects Changes in the size of a population due to factors that are independent of the population size. For example, certain climatic factors that are not affected by the size of a specific population can affect the entire population. A storm that knocks down all trees in a forest, no matter how many there are, is a density-independent population effect.

Desalination The removal of salts from seawater or brackish water so that the water can be used for purposes such as agriculture, industrial processes, or human consumption.

Desertification The process of creating a desert where there was not one before. Farming in marginal grasslands, which destroys the soil and prevents the future recovery of natural vegetation, is an example of desertification.

Dioxin An organic compound composed of oxygen, hydrogen, carbon, and chlorine. About 75 types are known. Dioxin is not normally manufactured intentionally but is a by-product resulting from chemical reactions in the production of other materials, such as herbicides. Known to be extremely toxic to mammals, its effects on the human body are being intensively studied and evaluated.

Direct costs See **Costs, direct**

Divergent evolution Organisms with the same ancestral genetic heritage migrate to different habitats and evolve into species with different external forms and structures, but typically continue to use the same kind of habitats. The ostrich and the emu are believed to be examples of divergent evolution.

Divergent plate boundary Boundary between lithospheric plates characterized by the production of new lithosphere; found along oceanic ridges.

Diversity, genetic The total number of genetic characteristics, sometimes of a specific species, subspecies, or group of species.

Diversity, habitat The number of kinds of habitats in a given unit area.

Diversity, species Used loosely to mean the variety of species in an area or on the Earth. Technically, it is composed of three components: species richness—the total number of species; species evenness—the relative abundance of species; and species dominance—the most abundant species.

Dobson unit Commonly used to measure the concentration of ozone. One Dobson unit is equivalent to a concentration of 1 ppb ozone.

Dominant species Generally, the species that are most abundant in an area, ecological community, or ecosystem.

Dominants Species that are most abundant or otherwise most important within the community.

Dose dependency Dependence on the dose or concentration of a substance for its effects on a particular organism.

Dose response The principle that the effect of a certain chemical on an individual depends on the dose or concentration of that chemical.

Doubling time The time necessary for a quantity of whatever is being measured to double.

Drainage basin The area that contributes surface water to a particular stream network.

Drip irrigation Irrigation by the application of water to the soil from tubes that drip water slowly, greatly reducing the loss of water from direct evaporation and increasing yield.

Early successional species Species that occur only or primarily during early stages of succession. With vegetation, these are typically rapidly growing and short-lived with high reproductive rates.

Earth system science Science of studying the Earth as a system. Includes understanding of processes and linkages between the lithosphere, hydrosphere, biosphere, and atmosphere.

Ecological community This term has two meanings. (1) A conceptual or functional meaning: a set of interacting species that occur in the same place (sometimes extended to mean a set that interacts in a way to sustain life). (2) An operational meaning: a set of species found in an area, whether or not they are interacting.

Ecological economics Study and evaluation of relations between humans and the economy with emphasis on long-term health of ecosystems and sustainability.

Ecological gradient A change in the relative abundance of a species or group of species along a line or over an area.

Ecological island An area that is biologically isolated so that a species occurring within the area cannot mix (or only rarely mixes) with any other population of the same species.

Ecological justification for the conservation of nature An argument for the conservation of nature on the grounds that a species, an ecological community, an ecosystem, or the Earth's biosphere provides specific functions necessary to the persistence of our life or of benefit to life. The ability of trees in forests to remove carbon dioxide produced in burning fossil fuels is such a public benefit and an argument for maintaining large areas of forests.

Ecological niche The general concept is that the niche is a species' "profession"—what it does to make a living. The term is also used to refer to a set of environmental conditions within which a species is able to persist.

Ecological succession The process of the development of an ecological community or ecosystem, usually viewed as a series of stages—early, middle, late, mature (or climax), and sometimes postclimax. Primary succession is an original establishment; secondary succession is a reestablishment.

Ecology The science of the study of the relationships between living things and their environment.

Ecosystem An ecological community and its local, nonbiological community. An ecosystem is the minimum system that includes and sustains life. It must include at least an autotroph, a decomposer, a liquid medium, a source and sink of energy, and all the chemical elements required by the autotroph and the decomposer.

Ecosystem effect Effects that result from interactions among different species, effects of species on chemical elements in their environment, and conditions of the environment.

Ecosystem energy flow The flow of energy through an ecosystem—from the external environment through a series of organisms and back to the external environment.

Ecotourism Tourism based on an interest in observation of nature.

ED-50 The effective dose, or dose that causes an effect in 50% of the population on exposure to a particular toxicant. It is related to the onset of specific symptoms, such as loss of hearing, nausea, or slurred speech.

Edge effect An effect that occurs following the forming of a forest island; in the early phases the species diversity along the edge is greater than in the interior. Species escape from the cut area and seek refuge in the border of the forest, where some may last only a short time.

Efficiency The ratio of output to input. With machines, usually the ratio of work or power produced to the energy or power used to operate or fuel them. With living things, efficiency may be defined as either the useful work done or the energy stored in a useful form compared with the energy taken in.

Efficiency improvements With respect to energy, refers to designing equipment that will yield more energy output from a given amount of energy input.

Effluent Any material that flows outward from something. Examples include wastewater from hydroelectric plants and water discharged into streams from waste-disposal sites.

Effluent stream Type of stream where flow is maintained during the dry season by groundwater seepage into the channel.

El Niño Natural perturbation of the physical earth system that effects global climate. Characterized by development of warm oceanic waters in the eastern part of the tropical Pacific Ocean, a weakening or reversal of the trade winds, and a weakening or even reversal of the equatorial ocean currents. Reoccurs periodically and effects the atmosphere and global temperature by pumping heat into the atmosphere.

Electromagnetic fields (EMF) Magnetic and electrical fields produced naturally by our planet and also by appliances such as toasters, electric blankets, and computers. There currently is controversy concerning potential adverse health effects related to exposure to EMF in the workplace and home from such artificial sources as power lines and appliances.

Electromagnetic spectrum All the possible wavelengths of electromagnetic energy, considered as a continuous range. The spectrum includes long wavelength (used in radio transmission), infrared, visible, ultraviolet, X rays, and gamma rays.

Endangered species A species that faces threats that might lead to its extinction in a short time.

Endemic Referring to all factors confined to a given region, such as an island or a country; the whooping crane is endemic to North America.

Endemic species A species that is native to a particular area.

Energy An abstract concept referring to the ability or capacity to do work.

Energy efficiency Refers to both first-law efficiency and second-law efficiency, where first-law efficiency is the ratio of the actual amount of energy delivered to the amount of energy supplied to meet a particular need, and the second-law efficiency is the ratio of the maximum available work needed to perform a particular task to the actual work used to perform that task.

Energy flow The movement of energy through an ecosystem from the external environment through a series of organisms and back to the external environment. It is one of the fundamental processes common to all ecosystems.

Entropy A measure in a system of the amount of energy that is unavailable for useful work. As the disorder of a system increases, the entropy in a system also increases.

Environment All factors (living and nonliving) that actually affect an individual organism or population at any point in the life cycle. Environment is also sometimes used to denote a certain set of circumstances surrounding a particular occurrence (environments of deposition, for example).

Environmental audit Process of determining the past history of a particular site, with special reference to the existence of toxic materials or waste.

Environmental economics The study of relationships of the importance of the environment to the economy, including the impact of environment as a result of economic activity, regulation of the economy and economic processes with the objective of balancing environmental and economic goals of society, and development of economic policy to minimize environmental degradation and finding solutions to environmental problems.

Environmental ethics A school, or theory, in philosophy that deals with the ethical value of the environment, including especially the rights of nonhuman objects and systems in the environment, for example, trees and ecosystems.

Environmental geology The application of geologic information to environmental problems.

Environmental impact The effects of some action on the environment, particularly action by human beings.

Environmental impact report (EIR) Similar to the environmental impact statement (EIS), a report describing potential environmental impacts resulting from a particular project, often at the state level.

Environmental impact statement A written statement that assesses and explores possible impacts associated with a particular project that may affect the human environment. The statement is required in the United States by the National Environmental Policy Act of 1969.

Environmental justice The principle of dealing with environmental problems in such a way as to not discriminate against people based upon socioeconomic status, race, or ethnic group.

Environmental law A field of law concerning the conservation and use of natural resources and the control of pollution.

Environmental risk Used in discussions of endangered species to mean variation in the physical or biological environment, including variations in predator, prey, symbiotic, or competitor species that can threaten a species with extinction.

Environmental tobacco smoke Commonly called secondhand smoke from people smoking tobacco.

Environmental unity A principle of environmental sciences that states that everything affects everything else, meaning that a particular course of action leads to an entire potential string of events. Another way of stating this idea is that you can't only do one thing.

Environmentalism A social, political, and ethical movement concerned with protecting the environment and using its resources wisely.

Epidemic disease A disease that appears occasionally in the population, affects a large percentage of it, and declines or almost disappears for a while only to reappear later.

Equilibrium A point of rest. A system that does not tend to undergo any change of its own accord but remains in a single, fixed condition is said to be in equilibrium. Compare with **steady state.**

Eukaryote Organism whose cells have nuclei and certain other characteristics that separate it from the prokaryotes. The eukaryotes include flowering plants, animals, and many single-cell organisms.

Eutrophic Referring to bodies of water having an abundance of the chemical elements required for life.

Eutrophication Increase in the concentration of chemical elements required for living things (for example, phosphorus). Increased nutrient loading may lead to a population explosion of photosynthetic algae and blue-green bacteria that become so thick that light cannot penetrate the water. Bacteria deprived of light beneath the surface die; as they decompose, dissolved oxygen in the lake is lowered and eventually a fish kill may result. Eutrophication of lakes caused by human-induced processes, such as nutrient-rich sewage water entering a body of water, is called cultural eutrophication.

Even-aged stands Forest area where all live trees began growth from seeds and roots planted in about the same year.

Evolution biological The change in inherited characteristics of a population from generation to generation, sometimes resulting in a new species or populations.

Evolution, nonbiological Outside the realm of biology, the term *evolution* is used broadly to mean the history and development of something.

Exotic species Species introduced into a new area by human action.

Experimental errors There are two kinds of experimental errors, random and systematic. Random errors are those due to chance events, such as air currents pushing on a scale and altering a measurement of weight. In contrast, a miscalibration of an instrument would lead to a systematic error. Human errors can be either random or systematic.

Exponential growth Growth in which the rate of increase is a constant percentage of the current size; that is, the growth occurs at a constant rate per time period.

Externality In economics, an effect not normally accounted for in the cost-revenue analysis of producers.

Extinction Disappearance of a life-form from existence; usually applied to a species.

Facilitation During succession, one species prepares the way for the next (and may even be necessary for the occurrence of the next).

Fact Something that is known based on actual experience and observation.

Fall line The point on a river where there is an abrupt drop in elevation of the land and where numerous waterfalls occur. The line in the eastern U.S. is located where streams pass from harder to softer rocks.

Fallow A field allowed to grow with a cover crop without harvesting for at least one season.

Fecal coliform bacteria Standard measure of microbial pollution and an indicator of disease potential for a water source.

Feedback A kind of system response that occurs when output of the system also serves as input leading to changes in the system.

First law of thermodynamics The principle that energy may not be created or destroyed but is always conserved.

First-law efficiency The ratio of the actual amount of energy delivered where it is needed to the amount of energy supplied in order to meet that need; expressed as a percentage.

Fission The splitting of an atom into smaller fragments with the release of energy.

Flooding The natural process whereby waters emerge from their stream channel to cover part of the floodplain. Natural flooding is not a problem until people choose to build homes and other structures on floodplains.

Floodplain Flat topography adjacent to a stream in a river valley that has been produced by the combination of overbank flow and lateral migration of meander bends.

Fluidized-bed combustion A process used during the combustion of coal to eliminate sulfur oxides. Involves mixing finely ground limestone with coal and burning it in suspension.

Food chain The linkage of who feeds on whom.

Food-chain concentration See **Biomagnification.**

Food web A network of who feeds on whom or a diagram showing who feeds on whom.

Force A push or pull that affects motion. The product of mass and acceleration of a material.

Forcing With respect to global change, processes capable of changing global temperature, such as changes in solar energy emitted from the sun, or volcanic activity.

Fortress city Earliest planned cities in Europe designed for defense; influenced the development of cities in North America.

Fossil fuels Forms of stored solar energy created from incomplete biological decomposition of dead organic matter. Includes coal, crude oil, and natural gas.

Fuel cell A device that produces electricity directly from a chemical reaction in a specially designed cell. In the simplest case the cell uses hydrogen as a fuel, to which an oxidant is supplied. The hydrogen is combined with oxygen as if the hydrogen were burned, but the reactants are separated by an electrolyte solution that facilitates the migration of ions and the release of electrons (which may be tapped as electricity, energy source).

Fugitive sources Type of stationary air pollution sources that generate pollutants from open areas exposed to wind processes.

Fusion, nuclear Combining of light elements to form heavier elements with the release of energy.

Gaia hypothesis The Gaia hypothesis states that the surface environment of the Earth, with respect to such factors as the atmospheric composition of reactive gases (for example, oxygen, carbon dioxide, and methane), the acidity-alkalinity of waters, and the surface temperature are actively regulated by the sensing, growth, metabolism and other activities of the biota. Interaction between the physical and biological system on the Earth's surface has led to a planetwide physiology which began more than 3 billion years ago and the evolution of which can be detected in the fossil record.

Game ranching Practice of maintaining wild herbivores in their native habitat to be harvested for meat, leather, and other products.

Gamma rays One of the three major kinds of nuclear radiation. A type of electromagnetic radiation emitted from the isotope similar to X rays but more energetic and penetrating.

Garden city Term for planning a city and countryside together, coined by Ebenezer Howard (1902).

Genetic drift Changes in the frequency of a gene in a population as a result of chance rather than of mutation, selection, or migration.

Genetic risk Used in discussions of endangered species to mean detrimental change in genetic characteristics not caused by external environmental changes. Genetic changes can occur in small populations from such causes as reduced genetic variation, genetic drift, and mutation.

Genetically modified organisms Also known as genetic engineering, refers to the altering of genes or genetic material with the objectives of producing new organisms or organisms with desired characteristics, or to eliminate undesirable characteristics in organisms. The process is produced through what is called gene-splicing, which transfers genes from one organism to another.

Geochemical cycles The pathways of chemical elements in geologic processes, including the chemistry of the lithosphere, atmosphere, and hydrosphere.

Geographic Information System (GIS) Technology capable of storing, retrieving, transferring, and displaying environmental data.

Geologic cycle The formation and destruction of earth materials and the processes responsible for these events. The geologic cycle includes the following subcycles: hydrologic, tectonic, rock, and geochemical.

Geometric growth See **Exponential growth.**

Geopressurized systems Geothermal system that exists when the normal heat flow from the Earth is trapped by impermeable clay layers that act as an effective insulator.

Geothermal energy The useful conversion of natural heat from the interior of the Earth.

Global circulation models (GCM) A type of mathematical model used to evaluate global change, particularly related to climatic change. GCMs are very complex and require supercomputers for their operation.

Global extinction Disappearance or extinction of a species everywhere.

Global forecasting Process of predicting or forecasting future change in environmental areas such as world population, natural resource utilization, and environmental degradation.

Global warming Natural or human-induced increase in the average global temperature of the atmosphere near the Earth's surface.

Gravel Unconsolidated, generally rounded fragment of rocks and minerals greater than 2 mm in diameter.

Green plans Long-term strategies for identifying and solving global and regional environmental problems. The philosophical heart of green plants is sustainability.

Green revolution Name attached to post–World War II agricultural programs that have led to the development of new strains of crops with higher yield, better resistance to disease, or better ability to grow under poor conditions.

Greenbelt The idea of locating garden cities in a set connected by undeveloped areas, forming a system of countryside and urban landscapes.

Greenhouse effect Process of trapping heat in the atmosphere. Water vapor and several other gases warm the Earth's atmosphere because they absorb and remit radiation; that is, they trap some of the heat radiating from the Earth's atmospheric system.

Greenhouse gases The suite of gases that have a greenhouse effect, such as carbon dioxide, methane, and water vapor.

Gross production (biology) Production before respiration losses are subtracted.

Groundwater Water found beneath the Earth's surface within the zone of saturation, below the water table.

Growth efficiency Gross production efficiency (P/C), or ratio of the material produced (P = net production) by an organism or population to the material ingested or consumed (C).

Growth rate The net increase in some factor per unit time. In ecology, the growth rate of a population is sometimes measured as the increase in numbers of individuals or biomass per unit time and sometimes as a percentage increase in numbers or biomass per unit time.

Habitat Where an individual, population, or species exists or can exist. For example, the habitat of the Joshua tree is the Mojave Desert of North America.

Half-life The time required for half the amount of a substance to disappear; the average time required for one-half of a radioisotope to be transformed to some other isotope; the time required for one-half of a toxic chemical to be converted to some other form.

Hard path Energy policy based on the emphasis of energy quantity generally produced from large, centralized power plants.

Hazardous waste Waste that is classified as definitely or potentially hazardous to the health of people. Examples include toxic or flammable liquids and a variety of heavy metals, pesticides, and solvents.

Heat energy Energy of the random motion of atoms and molecules.

Heat island Warmer air of a city than surrounding areas as a result of increased heat production and decreased rate of heat loss caused by the abundance of building and paving materials, which act as solar collectors.

Heat island effect Urban areas are several degrees warmer than their surrounding areas. During relatively calm periods there is an upward flow of air over heavily developed areas accompanied by a downward flow over nearby greenbelts. This produces an air-temperature profile that delineates the heat island.

Heat pumps Devices that transfer heat from one material to another, such as from groundwater to the air in a building.

Heavy metal Refers to a number of metals, including lead, mercury, arsenic, and silver (among others) that have a relatively high atomic number (the number of protons in the nucleus of an atom). They are often toxic at relatively low concentrations, causing a variety of environmental problems.

Herbivore An organism that feeds on an autotroph.

Heterotroph Organisms that cannot make their own food from inorganic chemicals and a source of energy and therefore live by feeding on other organisms.

High-level nuclear waste Extremely toxic nuclear waste, such as spent fuel elements from commercial reactors. A sense of urgency surrounds determining how we may eventually dispose of this waste material.

Homeostasis The ability of a cell or organism to maintain a constant environment. Results from negative feedback, resulting in a state of dynamic equilibrium.

Hot igneous systems A geothermal system that involves hot, dry rocks with or without the presence of near-surface molten rock.

Human demography The study of human population characteristics, such as age structure, demographic transition, total fertility, human population and environment relationships, death-rate factors, and standard of living.

Hutchinsonian niche The idea of a measured niche, a set of environmental conditions within which a species is able to persist.

Hydrocarbons Compounds containing only carbon and hydrogen are a large group of organic compounds, including petroleum products, such as crude oil and natural gas.

Hydrochlorofluorocarbons Also known as HCFCs, which are a group of chemicals containing hydrogen, chlorine, fluorine, and carbon, produced as a potential substitute for chlorofluorocarbons (CFCs).

Hydrofluorocarbons Chemicals containing hydrogen, fluorine, and carbons, produced as potential substitutes for chlorofluorocarbons (CFCs).

Hydrologic cycle Circulation of water from the oceans to the atmosphere and back to the oceans by way of evaporation, runoff from streams and rivers, and groundwater flow.

Hydrology The study of surface and subsurface water.

Hydroponics The practice of growing plants in a fertilized water solution on a completely artificial substrate in an artificial environment such as a greenhouse.

Hydrothermal convection systems A type of geothermal energy characterized by circulation of steam and/or hot water that transfers to the surface.

Hypothesis In science, an explanation set forth in a manner that can be tested and is capable of being disproved. A tested hypothesis is accepted until and unless it has been disproved.

Igneous rocks Rocks formed from the solidification of magma. They are extrusive if they crystallize on the surface of the Earth and intrusive if they crystallize beneath the surface.

Incineration Combustion of waste at high temperature, consuming materials and leaving only ash and noncombustibles to dispose of in a landfill.

Independent variable See **Variable, independent**

Indirect costs See **Costs, indirect**

Indirect deforestation Death of trees from pollution whose source is outside the forest.

Inductive reasoning Drawing a general conclusion from a limited set of specific observations.

Industrial ecology Process of designing industrial systems to behave more like ecosystems where waste from one part of the system is a resource for another part.

Industrial melanism A form of natural selection where the color of a living thing helps it to blend in with an urban, industrial environment.

Inference (1) A conclusion derived by logical reasoning from premises and/or evidence (observations or facts), or (2) a conclusion, based on evidence, arrived at by insight or analogy, rather than derived solely by logical processes.

Influent stream Type of stream that is everywhere above the groundwater table and flows in direct response to precipitation. Water from the channel moves down to the water table, forming a recharge mound.

In-stream use A type of water use that includes navigation, generation of hydroelectric power, fish and wildlife habitat, and recreation.

Integrated energy management Use of a range of energy options that vary from region to region, including a mix of technology and sources of energy.

Integrated pest management A term applied to a variety of practices whose overall goal is to contain biological pests through minimum use of artificial chemicals and minimum disruption of natural ecological processes. Biological control typically augments or replaces artificial pesticides.

Integrated waste management (IWM) Set of management alternatives including reuse, source reduction, recycling, composting, landfill, and incineration.

Interference During succession, one species prevents the entrance of later successional species into an ecosystem. For example, some grasses produce such dense and thick mats that seeds of trees cannot reach the soil to germinate. As long as these grasses persist, the trees that characterize later stages of succession cannot enter the ecosystem.

Island arc A curved group of volcanic islands associated with a deep oceanic trench and subduction zone (convergent plate boundary).

Isotope Atoms of an element that have the same atomic number (the number of protons in the nucleus of the atom) but vary in atomic mass number (the number of protons plus neutrons in the nucleus of an atom).

Keystone species Loosely speaking, a species, such as the sea otter, that has a large effect on its community or ecosystem so that its removal or addition to the community leads to major changes in the abundances of many or all other species.

Kinetic energy The energy of motion. For example, the energy in a moving car that results from the mass of the car traveling at a particular velocity.

Kwashiorkor Lack of sufficient protein in the diet, which leads to a failure of neural development in infants and therefore to learning disabilities.

Land application Method of disposal of hazardous waste that involves intentional application of waste material to surface soil. Useful for certain biodegradable industrial waste, such as oil and petroleum waste, and some organic chemical waste.

Land ethic A set of ethical principles that affirm the right of all resources, including plants, animals, and earth materials, to continued existence and, at least in some locations, to continued existence in a natural state.

Land-use planning Complex process involving development of a land-use plan to include a statement of land-use issues, goals, and objectives; a summary of data collection and analysis; a land-classification map; and a report that describes and indicates appropriate development in areas of special environmental concern.

Late successional species Species that occur only or primarily in, or are dominant in late stages in succession. With plants, these are typically slower growing and longer-lived species.

Law of the minimum (Liebig's law of the minimum) The concept that the growth or survival of a population is directly related to the life requirement that is in least supply and not to a combination of factors.

LD-50 A crude approximation of a chemical toxicity defined as the dose at which 50% of the population dies on exposure.

Leachate Noxious, mineralized liquid capable of transporting bacterial pollutants. Produced when water infiltrates through waste material and becomes contaminated and polluted.

Leaching Water infiltration from the surface, dissolving soil materials as part of chemical weathering processes and transporting the dissolved materials laterally or downward.

Lead A heavy metal that is an important constituent of automobile batteries and other industrial products. A toxic metal capable of causing environmental disruption and producing a health problem to people and other living organisms.

Liebig's law of the minimum See **Law of the minimum.**

Life expectancy The estimated average number of years (or other time period used as a measure) that an individual of a specific age can expect to live.

Limiting factor The single requirement for growth available in the least supply in comparison to the need of an organism. Originally applied to crops but now often applied to any species.

Lithosphere Outer layer of Earth, approximately 100 km thick, of which the plates that contain the ocean basins and the continents are composed.

Littoral drift Movement caused by wave motion in nearshore and beach environment.

Local extinction The disappearance of a species from part of its range, but continued persistence elsewhere.

Logistic carrying capacity In terms of the logistic curve, the population size at which births equals deaths and there is no net change in the population.

Logistic equation The equation that results in a logistic growth curve, that is, the growth rate $dN/dt = rN[(K-N)/N]$, where r is the intrinsic rate of increase, K is the carrying capacity, and N is the population size.

Logistic growth curve The S-shaped growth curve that is generated by the logistic growth equation. In the logistic, a small population grows rapidly, but the growth rate slows down, and the population eventually reaches a constant size.

Low-level nuclear waste Waste materials that contain sufficiently low concentrations or quantities of radioactivity so as not to present a significant environment hazard if properly handled.

Luz solar electric generating system Solar energy farms comprising a power plant surrounded by hundreds of solar collectors (curved mirrors) that heat a synthetic oil, which flows through heat exchangers to drive steam turbine generators.

Macronutrients Elements required in large amounts by living things. These include the big six—carbon, hydrogen, oxygen, nitrogen, phosphorus, and sulfur.

Magma A naturally occurring silica melt, a good deal of which is a liquid state.

Malnourishment The lack of specific components of food, such as proteins, vitamins, or essential chemical elements.

Manipulated variable See **Variable, independent**

Marasmus Progressive emaciation caused by a lack of protein and calories.

Marginal cost In environmental economics, the cost to reduce one additional unit of a type of degradation, for example, pollution.

Marginal land An area of the Earth with minimal rainfall or otherwise limited severely by some necessary factor, so that it is a poor place for agriculture and easily degraded by agriculture. Typically, these lands are easily converted to deserts when used for light grazing and crop production.

Mariculture Production of food from marine habitats.

Marine evaporites With respect to mineral resources, refers to materials such as potassium and sodium salts resulting from the evaporation of marine waters.

Materials management Term related to waste management consistent with the ideals of industrial ecology that makes better use of materials, leading to more sustainable use of resources.

Matter Anything that occupies space and has mass. It is the substance of which physical objects are composed.

Maximum lifetime Genetically determined maximum possible age to which an individual of a species can live.

Maximum sustainable population The population size that has the maximum production (net increase in number of individuals or net change in biomass) that allows the population to be sustained indefinitely without decreasing the ability of the population to provide the same level of production.

Maximum sustainable yield (MSY) The maximum usable production of a biological resource that can be obtained in a specified time period. The MSY level is the population size that results on maximum sustainable yield.

Mediation Negotiation process between adversaries, guided by a neutral facilitator.

Meltdown Refers to a nuclear accident in which the nuclear fuel forms a molten mass that breaches the containment of the reactor, contaminating the outside environment with radioactivity.

Methane (CH_4) Molecule of carbon and hydrogen, which is a naturally occurring gas in the atmosphere. One of the so-called greenhouse gases.

Methane hydrate A white ice-like compound made up of molecules of methane gas trapped in "cages" of frozen water in the sediments of the deep seafloor.

Microclimate The climate of a very small local area. For example, the climate under a tree, near the ground within a forest, or near the surface of streets in a city.

Micronutrients Chemical elements required in very small amounts by at least some forms of life. Boron, copper, and molybdenum are examples of micronutrients.

Migration The movement of an individual, population, or species from one habitat to another or more simply from one geographic area to another.

Migration corridor Designated passageways among parks or preserves for allowing in-migration among several of these areas.

Mineral Naturally occurring inorganic material with a definite internal structure and physical and chemical properties that vary within prescribed limits.

Mineral resources Elements, chemical compounds, minerals, or rocks concentrated in a form that can be extracted to obtain a usable commodity.

Minimum viable population The minimum number of individuals that have a reasonable chance of persisting for a specified time period.

Mitigated negative declaration Special type of negative declaration that suggests that the adverse environmental aspects of a particular action may be mitigated through modification of the project in such a way as to reduce the impacts to near insignificance.

Mitigation Process that identifies actions to avoid, lessen, or compensate for anticipated adverse environmental impacts.

Mobile sources Sources of air pollutants that move from place to place, for example, automobiles, trucks, buses, and trains.

Monitoring Process of collecting data on a regular basis at specific sites to provide a database from which to evaluate change. For example, collection of water samples from beneath a landfill to provide early warning should a pollution problem arise.

Monoculture The planting of large areas with a single species or even a single strain or subspecies in farming.

Moral justification for the conservation of nature An argument for the conservation of nature on the grounds that aspects of the environment have a right to exist, independent of human desires, and that it is our moral obligation to allow them to continue or to help them persist.

Multiple use Literally, using the land for more than one purpose at the same time. For example, forestland can be used to produce commercial timber but at the same time serve as wildlife habitat and land for recreation. Usually multiple use requires compromises and trade-offs, such as striking a balance between cutting timber for the most efficient production of trees at a level that facilitates other uses.

Mutation Stated most simply, a chemical change in a DNA molecule. It means that the DNA carries a different message than it did before, and this change can affect the expressed characteristics when cells or individual organisms reproduce.

Mutualism See **Symbiosis.**

Natural catastrophe Sudden catastrophic change in the environment, not the result of human actions.

Natural gas Naturally occurring gaseous hydrocarbon (predominantly methane) generally produced in association with crude oil or from gas wells; an important efficient and clean-burning fuel commonly used in homes and industry.

Natural selection A process by which organisms whose biological characteristics better fit them to the environment are better represented by descendants in future generations than those whose characteristics are less fit for the environment.

Nature preserve An area set aside with the primary purpose of conserving some biological resource.

Negative declaration Document that may be filed if an agency has determined that a particular project will not have a significant adverse effect on the environment.

Negative feedback A type of feedback that occurs when the system's response is in the opposite direction of the output. Thus negative feedback is self-regulating.

Net growth efficiency Net production efficiency (P/A), or ratio of the material produced (P) to the material assimilated (A) by an organism. The material assimilated is less than the material consumed, because some food taken is egested as waste (discharged) and never used by an organism.

Net production (biology) The production that remains after utilization. In a population, net production is sometimes measured as the net change in the numbers of individuals. It is also measured as the net change in biomass or in stored energy. In terms of energy, it is equal to the gross production minus the energy used in respiration.

New forestry The name for a new variety of timber harvesting practices to increase the likelihood of sustainability, including recognition of the dynamic characteristics of forests and of the need for management within an ecosystem context.

Niche (1) The "profession," or role, of an organism or species; or (2) all the environmental conditions under which the individual or species can persist. The *fundamental niche* is all the conditions under which a species can persist in the absence of competition; the *realized niche* is the set of conditions as they occur in the real world with competitors.

Nitrogen cycle A complex biogeochemical cycle responsible for moving important nitrogen components through the biosphere and other Earth systems. This is an extremely important cycle because nitrogen is required by all living things.

Nitrogen fixation The process by which atmospheric nitrogen is converted to ammonia, nitrate ion, or amino acids. Microorganisms perform most of the conversion but a small amount is also converted by lightning.

Nitrogen oxides Occur in several forms (NO, NO_2, and NO_3). Most important as an air pollutant is nitrogen dioxide, which is a visible yellow brown to reddish brown gas. It is a precursor of acid rain and produced through the burning of fossil fuels.

Noise pollution A type of pollution characterized by unwanted or potentially damaging sound.

Nonmarine evaporites With respect to mineral resources, refers to useful deposits of materials such as sodium and calcium bicarbonate, sulfate, borate, or nitrate produced by evaporation of surficial waters on the land, as differentiated from marine waters in the oceans.

Nonpoint sources Sources of pollutants that are diffused and intermittent and are influenced by factors such as land use, climate, hydrology, topography, native vegetation, and geology.

Nonrenewable energy Alternative energy sources, including nuclear and geothermal, that are dependent on fuels or a resource that may be used up much faster than it is replenished by natural processes.

Nonrenewable resource A resource that is cycled so slowly by natural Earth processes that once used, it is essentially not going to be made available within any useful time framework.

No-till agriculture Combination of farming practices that include *not* plowing the land, using herbicides to keep down the weeds, and allowing some weeds to grow.

Nuclear cycle The series of processes that begins with the mining of uranium to be processed and used in nuclear reactors and ends with the disposal of radioactive waste.

Nuclear energy The energy of the atomic nucleus that, when released, may be used to do work. Controlled nuclear fission reactions take place within commercial nuclear reactors to produce energy.

Nuclear fuel cycle Processes involved with producing nuclear power from the mining and processing of uranium to control fission, reprocessing of spent nuclear fuel, decommissioning of power plants, and disposal of radioactive waste.

Nuclear reactors Devices that produce controlled nuclear fission, generally for the production of electric energy.

Nutrients Chemicals such as phosphorus and nitrogen that, when released into water sources, may cause pollution events such as eutrophication.

Obligate symbionts A symbiotic relationship between two organisms in which neither by themselves can exist without the other.

Observations Information obtained through one or more of the five senses or through instruments that extend the senses. For example, some remote sensing instruments measure infrared intensity, which we do not see, and convert the measurement into colors, which we do see.

Ocean thermal conversion Direct utilization of solar energy using part of a natural oceanic environment as a gigantic solar collector.

Off-site effect An environmental effect occurring away from the location of the causal factors.

Off-stream use Type of water use where water is removed from its source for a particular use.

Oil shale A fine-grained sedimentary rock containing organic material known as kerogen. On distillation, yields significant amounts of hydrocarbons including oil.

Old growth A nontechnical term often used to mean a virgin forest (one never cut) but also used to mean a forest that has been undisturbed for a long, but usually unspecified, time.

Oligotrophic Referring to bodies of waters having a low concentration of the chemical elements required for life.

Omnivores Organisms that eat both plants and animals.

On-site effect An environmental effect occurring at the location of the causal factors.

Open dumps Area where solid waste is disposed of by simply dumping it. Often causes severe environmental problems, such as water pollution, and creates a health hazard. Illegal in the United States and in many other countries around the world.

Open system A type of system in which exchanges of mass or energy occur with other systems.

Open woodlands Areas in which trees are a dominant vegetation form but the leaves of adjacent trees generally do not touch or overlap, so that there are gaps in the canopy. Typically, grasses or shrubs grow in these gaps among the trees.

Operational definitions Definitions that tell you what you need to look for or do in order to carry out an operation, such as measuring, constructing, or manipulating.

Optimal carrying capacity This term has several meanings, but the major idea is the maximum abundance of a population or species that can persist in an ecosystem without degrading the ability of the ecosystem to maintain: (1) that population or species; (2) all necessary ecosystem processes; and (3) the other species found in that ecosystem.

Optimum sustainable population (OSP) The population level that results in an optimum sustainable yield; or the population level that is in some way best for the population, its ecological community, its ecosystem, or the biosphere.

Optimum sustainable yield (OSY) The largest yield of a renewable resource achievable over a long time period without decreasing the ability of the population or its environment to support the continuation of this level of yield.

Ore deposits Earth materials in which metals are concentrated in high concentrations, sufficient to be mined.

Organic compound A compound of carbon; originally used to refer to the compounds found in and formed by living things.

Outbreak The occurrence of a disease that may quickly and surprisingly occur.

Overdraft Groundwater withdrawal when the amount pumped from wells exceeds the natural rate of replenishment.

Overgrazing When the carrying capacity of land for an herbivore, such as cattle or deer, is exceeded.

Ozone (O$_3$) Form of oxygen in which three atoms of oxygen occur together. Is chemically active and has a short average lifetime in the atmosphere. Forms a natural layer high in the atmosphere (stratosphere) that protects us from harmful ultraviolet radiation from the sun. Is an air pollutant when present in the lower atmosphere above the National Air Quality Standards.

Ozone shield Stratospheric ozone layer that absorbs ultraviolet radiation.

Park city Earliest planned cities in Europe designed for beauty and recreation; influenced the development of cities in North America.

Particulate matter Small particles of solid or liquid substances that are released into the atmosphere by many activities, including farming, volcanic eruption, and burning fossil fuels. Particulates affect human health, ecosystems, and the biosphere.

Passive solar energy system Direct use of solar energy through architectural design to enhance or take advantage of natural changes in solar energy that occur throughout the year without requiring mechanical power.

Pasture Land plowed and planted to provide forage for domestic herbivorous animals.

Pebble A rock fragment between 4 and 64 mm in diameter.

Pedology The study of soils.

Pelagic whaling Practice of whalers taking to the open seas and searching for whales from ships that remained at sea for long periods.

Per-capita availability The amount of a resource available per person.

Per-capita demand The economic demand per person.

Per-capita food production The amount of food produced per person.

Permafrost Permanently frozen ground.

Pesticides, broad spectrum Pesticides that kill a wide variety of organisms. Arsenic, one of the first elements used as a pesticide, is toxic to many life-forms, including people.

Phosphorus cycle Major biogeochemical cycle involving the movement of phosphorus throughout the biosphere and lithosphere. This cycle is important because phosphorus is an essential element for life and often is a limiting nutrient for plant growth.

Photochemical oxidants Result from atmospheric interactions of nitrogen dioxide and sunlight. Most common is ozone (O$_3$).

Photochemical smog Sometimes called L.A.-type smog or brown air. Directly related to automobile use and solar radiation. Reactions that occur in the development of the smog are complex and involve both nitrogen oxides and hydrocarbons in the presence of sunlight.

Photosynthesis Synthesis of sugars from carbon dioxide and water by living organisms using light as energy. Oxygen is given off as a by-product.

Photovoltaics Technology that converts sunlight directly into electricity using a solid semiconductor material.

Physiographic province A region characterized by a particular assemblage of landforms, climate, and geomorphic history.

Pioneer species Species found in early stages of succession.

Placer deposit A type of ore deposit found in material transported and deposited by agents such as running water, ice, or wind; for example, gold and diamonds found in stream deposits.

Plantations Managed forest similar to mechanized farming, in which a single species is planted in straight rows and harvested in a specified rotation period of time.

Plate tectonics A model of global tectonics that suggests that the outer layer of Earth, known as the *lithosphere*, is composed of several large plates that move relative to one another. Continents and oceans basins are passive riders on these plates.

Point sources Sources of pollution such as smokestacks, pipes, or accidental spills that are readily identified and stationary. They are often thought to be easier to recognize and control than are area sources. This is true only in a general sense, as some very large point sources emit tremendous amounts of pollutants into the environment.

Polar amplification Processes in which global warming causes greater temperature increases at polar regions.

Polar stratospheric clouds Clouds that form in the stratosphere during the polar winter.

Polar vortex Arctic air masses that in the winter become isolated from the rest of the atmosphere and circulate about the pole. The vortex rotates counterclockwise because of the rotation of the Earth in the Southern Hemisphere.

Policy instruments The means to implement a society's policies. Includes moral suasion (jawboning—persuading people by talk, publicity, and social pressure); direct controls, including regulations; and market processes affecting the price of goods and processes, such as subsidies, licenses, and deposits.

Pollutant In general terms, any factor that has a harmful effect on living things or their environment.

Pollution The process by which something becomes impure, defiled, dirty, or otherwise unclean.

Pool (geology) Common bed form produced by scour in meandering and straight channels.

Population A group of individuals of the same species living in the same area or interbreeding and sharing genetic information.

Population age structure The number of individuals or proportion of the population in each age class.

Population dynamics The study of changes in population sizes and the causes of these changes.

Population momentum or lag effect The continued growth of a population that occurs after achievement of replacement level fertility is reached.

Population regulation See **Density-dependent population effects** and **Density-independent population effects.**

Population risk Used in discussions of endangered species to mean random variation in population rates—birthrates and death rates—possibly causing species in low abundance to become extinct.

Positive feedback A type of feedback that occurs when an increase in output leads to a further increase in output. This is sometimes known as a vicious cycle, since the more you have the more you get.

Potential energy Energy that is stored. Examples include the gravitational energy of water behind a dam; chemical energy in coal, fuel oil, and gasoline; and nuclear energy (in the forces that hold atoms together).

Power The time rate of doing work.

Predation-parasitism Interaction between individuals of two species in which the outcome benefits one and is detrimental to the other.

Predator An organism that feeds on other, live organisms, usually of other species. The term is usually applied to animals that feed on other animals, but sometimes it is used to mean herbivore.

Premises In science, initial definitions and assumptions.

Primary pollutants Air pollutants emitted directly into the atmosphere. Included are particulates, sulfur oxides, carbon monoxide, nitrogen oxides, and hydrocarbons.

Primary production See **Production, primary**

Primary succession The initial establishment and development of an ecosystem.

Primary treatment (of wastewater) Removal of large particles and organic materials from wastewater through screening.

Probability The relative probability that an event will occur.

Production, ecological The amount of increase in organic matter, usually measured per unit area of land surface or unit volume of water, as in grams per square meter (g/m^2). Production is divided into *primary* (that of autotrophs) and *secondary* (that of heterotrophs). It is also divided into *net* (that which remains stored after use) and *gross* (that added before any use).

Production, primary The production by autotrophs.

Production, secondary The production by heterotrophs.

Productivity, ecological The *rate* of production; that is, the amount of increase in organic matter per unit time (for example, grams per meter squared *per year*).

Prokaryote A kind of organism that lacks a true cell nucleus and has other cellular characteristics that distinguish it from the *eukaryotes*. Bacteria and blue-green algae are prokaryotes.

Protocooperation A symbiotic relationship that is beneficial, but not obligatory, to both species.

Pseudoscientific Ideas which are claimed to have scientific validity but are inherently untestable and/or lack empirical support and/or were arrived at through faulty reasoning or poor scientific methodology.

Public service functions Functions performed by ecosystems for the betterment of life and human existence. For examples, the cleansing of the air by trees and removal of pollutants from water by infiltration through the soil.

Public trust Grants and limits the authority of government over certain natural areas of special character.

Qualitative data Data that are distinguished by qualities or attributes that cannot or are not expressed as quantities. For example, blue and red are qualitative data about the electromagnetic spectrum.

Quantitative data Data that are expressed as numbers or numerical measurements. For example, the wavelengths of specific colors of blue and red light (460 and 650 nanometers, respectively) are quantitative data about the electromagnetic spectrum.

Radiation absorbed dose Energy retained by living tissue that has been exposed to radiation.

Radioactive decay A process of decay of radioisotopes that change from one isotope to another and emit one or more forms of radiation.

Radioactive waste Type of waste produced in the nuclear fuel cycle; generally classified as high level or low level.

Radioisotope A form of a chemical element that spontaneously undergoes radioactive decay.

Radon Naturally occurring radioactive gas. Radon is colorless, odorless, and tasteless and must be identified through proper testing.

Range Land used for grazing.

Rare species Species with a small total population size, or restricted to a small area, but not necessarily declining or in danger of extinction.

Realms Major biogeographic regions of Earth that are based upon fundamental features of the plants and animals found in those regions.

Record of decision Concise statement prepared by the agency planning a proposed project that outlines the alternatives considered and discusses which alternatives are environmentally preferable.

Recycle Integral part of waste management that attempts to identify resources in the waste stream that may be collected and reused.

Reduce With respect to waste management, refers to practices that will reduce the amount of waste we produce.

Reduce, reuse, and recycle The three Rs of integrated waste management.

Renewable energy Alternative energy sources, such as solar, water, wind, and biomass, that are more or less continuously made available in a time framework useful to people.

Renewable resource A resource, such as timber, water, or air, that is naturally recycled or recycled by artificial processes within a time framwork useful for people.

Replacement level fertility Fertility rate required for the population to remain a constant size.

Representative natural areas Parks or preserves set aside to represent presettlement conditions of a specific ecosystem type.

Reserves Known and identified deposits of earth materials from which useful materials can be extracted profitably with existing technology and under present economic and legal conditions.

Resource-based agriculture Agriculture with production limited by the availability of resources.

Resources Reserves plus other deposits of useful earth materials that may eventually become available.

Respiration The complex series of chemical reactions in organisms that make energy available for use. Water, carbon dioxide, and energy are the products of respiration.

Responding variable See **Variable, dependent**

Restoration ecology The field within the science of ecology with the goal to return damaged ecosystems to ones that are functional, sustainable, and more natural.

Reuse With respect to waste management, refers to processes whereby materials may be reused rather than disposed of.

Reuse With respect to waste mangement, refers to finding ways to reuse production and materials so they need not be disposed of.

Riffle A section of stream channel characterized at low flow by fast, shallow flow. Generally contains relatively coarse bed-load particles.

Risk assessment The process of determining potential adverse environmental health effects to people following exposure to pollutants and other toxic materials. Generally includes the four steps of identification of the hazard, dose-response assessment, exposure assessment, and risk characterization.

Risk–benefit analysis In environmental economics, the riskiness of the future that influences the value we place on things in the present.

Rock (engineering) Any earth material that has to be blasted in order to be removed.

Rock (geologic) An aggregate of a mineral or minerals.

Rock cycle A group of processes that produce igneous, metamorphic, and sedimentary rocks.

Rotation time Time between cuts of a stand or area of forest.

Rule of climatic similarity Similar environments lead to the evolution of organisms similar in form and function (but not necessarily in genetic heritage or internal makeup) and to similar ecosystems.

Ruminants Animals having a four-chambered stomach within which bacteria convert the woody tissue of plants to proteins and fats that, in turn, are digested by the animal. Cows, camels, and giraffes are ruminants; horses, pigs, and elephants are not.

Sand Grains of sediment between 1/16 and 2 mm in diameter; often sediment composed of quartz particles of this size.

Sand dune A ridge or hill of sand formed by wind action.

Sanitary landfill A method of disposal of solid waste without creating a nuisance or hazard to public health or safety. Sanitary landfills are highly engineered structures with multiple barriers and collection systems to minimize environmental problems.

Savanna An area with trees scattered widely among dense grasses.

Scientific method The systematic methods by which scientists investigate natural phenomena, including gathering data, formulating and testing hypotheses, and developing scientific theories and laws.

Scoping The process of early identification of important environmental issues that require detailed evaluation.

Scrubbing A process of removing sulfur from gases emitted from power plants burning coal. The gases are treated with a slurry of lime and limestone, and the sulfur oxides react with the calcium to form insoluble calcium sulfides and sulfates that are collected and disposed of.

Second growth Forest that has been clear-cut and regrown.

Second law of thermodynamics A fundamental principle of energy that states that energy always tends to go from a more usable (higher quality) form to a less usable (lower quality) form. When we say that energy is converted to a less useful form we mean that entropy (a measure of the energy unavailable to do useful work) of the system has increased.

Secondary enrichment A weathering process of sulfide ore deposits that may concentrate the desired minerals.

Secondary pollutants Air pollutants produced through reactions between primary pollutants and normal atmospheric compounds. An example is ozone that forms over urban areas through reactions of primary pollutants, sunlight, and natural atmospheric gases.

Secondary production See **Production, secondary**

Secondary succession The reestablishment of an ecosystem where there are remnants of a previous biological community.

Secondary treatment (of wastewater) Use of biological processes to degrade wastewater in a treatment facility.

Second-law efficiency The ratio of the minimum available work needed to perform a particular task to the actual work used to perform that task. Reported as a percentage.

Secure landfill A type of landfill designed specifically for hazardous waste. Similar to a modern sanitary landfill in that it includes multiple barriers and collection systems to ensure that leachate does not contaminate soil and other resources.

Sediment pollution By volume and mass, sediment is our greatest water pollutant. It may choke streams, fill reservoirs, bury vegetation, and generally create a nuisance that is difficult to remove.

Seed tree cutting A logging method in which mature trees with good genetic characteristics and high seed production are preserved to promote regeneration of the forest. It is an alternative to clear-cutting.

Seismic Referring to vibrations in the Earth produced by earthquakes.

Selective cutting In timber harvesting, the practice of cutting some, but not all, trees, leaving some on the site. There are many kinds of selective cutting. Sometimes the biggest trees with the largest market value are cut, and smaller trees are left to be cut later. Sometimes the best trees are left to provide seed for future generations. Sometimes trees are left for wildlife habitat and recreation.

Shelterwood cutting A logging method in which dead and less desirable trees are cut first; mature trees are cut later. This ensures that young, vigorous trees will always be left in the forest. It is an alternative to clear-cutting.

Sick building syndrome Condition associated with a particular indoor environment that appears to be unhealthy to the human occupants.

Silicate minerals The most important group of rock-forming minerals.

Silt Sediment between 1/16 and 1/256 mm in diameter.

Silviculture The practice of growing trees and managing forests, traditionally with an emphasis on the production of timber for commercial sale.

Sinkhole A surface depression formed by the solution of limestone or the collapse over a subterranean void such as a cave.

Site A factor considering the summation of all environmental features of a location that influences the placement of a city. For example, New Orleans is built on low-lying muds, which form a poor site, while New York City's Manhattan is built on an island of strong bedrock, an excellent site.

Site quality Used by foresters to mean an estimator of the maximum timber crop the land can produce in a given time.

Situation The relative geographic location of a site that makes it a good location for a city. For example, New Orleans has a good situation because it is located at the mouth of the Mississippi River and is therefore a natural transportation junction.

Smog A term first used in 1905 for a mixture of smoke and fog that produced unhealthy urban air. There are several types of fog, including photochemical smog and sulfurous smog.

Soft path Energy policy that relies heavily on renewable energy resources as well as other sources that are diverse, flexible, and matched to the end-use needs.

Soil The top layer of a land surface where the rocks have been weathered to small particles. Soils are made up of inorganic particles of many sizes, from small clay particles to large sand grains. Many soils also include dead organic material.

Soil (in engineering) Earth material that can be removed without blasting.

Soil (in soil science) Earth material modified by biological, chemical, and physical processes such that the material will support rooted plants.

Soil fertility The capacity of a soil to supply the nutrients and physical properties necessary for plant growth.

Soil horizon A layer in soil (A,B,C) that differs from another layer in chemical, physical, and biological properties.

Solar cell (photovoltaic) Device that directly converts light into electricity.

Solar collector Device for collecting and storing solar energy. For example, home water heating is done by flat panels consisting of a glass cover plate over a black background on which water is circulated through tubes. Short-wave solar radiation enters the glass and is absorbed by the black background. As long-wave radiation is emitted from the black material, it cannot escape through the glass, so the water in the circulating tubes is heated, typically to temperatures of 38° to 93°C.

Solar energy Collecting and using energy from the sun directly.

Solar pond Shallow pond filled with water and used to generate relatively low-temperature water.

Solar power tower A system of collecting solar energy that delivers the energy to a central location where the energy is used to produce electric power.

Source reduction Process of waste management, the object of which is to reduce the amounts of materials that must be handled in the waste stream.

Species A group of individuals capable of interbreeding.

Stable equilibrium A condition in which a system will remain if undisturbed and to which it will return when displaced.

Stationary sources Air pollution sources that have a relatively fixed location, including point sources, fugitive sources, and area sources.

Steady state When input equals output in a system, there is no net change and the system is said to be in a steady state. A bathtub with water flowing in and out at the same rate maintains the same water level and is in a steady state. Compare with *equilibrium*.

Stress Force per unit area. May be compression, tension, or shear.

Strip cutting In timber harvesting, the practice of cutting narrow rows of forest, leaving wooded corridors.

Strip mining Surface mining in which the overlying layer of rock and soil is stripped off to reach the resource. Large strip mines are some of the largest excavations caused by people in the world.

Subduction A process in which one lithospheric plate descends beneath another.

Subsidence A sinking, settling, or otherwise lowering of parts of the crust of the Earth.

Subsistence crops Crops used directly for food by a farmer or sold locally where the food is used directly.

Sulfur dioxide (SD_2) Colorless and odorless gas normally present at the Earth's surface in low concentrations. An important precursor to acid rain. Major anthropogenic source is burning fossil fuels.

Sulfurous smog Produced primarily by burning coal or oil at large power plants. Sulfur oxides and particulates combine under certain meteorological conditions to produce a concentrated form of this smog.

Surface impoundment Method of disposal of some liquid hazardous waste. This method is controversial and many sites have been closed.

Sustainability A concept that is emerging in the environmental sciences. With respect to resources it involves management that has the objective of ensuring that future generations will have the opportunity to use their fair share of resources and will inherit a quality environment. In an economic sense the concept means development that will not cause irreparable damage to the environment while ensuring that future generations will inherit their fair share of all Earth's resources.

Sustainable ecosystem An ecosystem that is subject to some human use, but at a level that leads to no loss of species or of necessary ecosystem functions.

Sustainable energy development A type of energy management that provides for reliable sources of energy while not causing environmental degradation and ensuring that future generations will have a fair share of the Earth's resources.

Sustainable forest ecosystem An ecosystem in which all properties of the forest are maintained.

Sustainable harvest An amount of a resource that can be harvested at regular intervals indefinitely. There are two concepts here. The first is the maximum abundance of a population or species that can be harvested without degrading the ability of that population or species to withstand the same level of harvest in the future. The second concept extends sustainability to the ecosystem. It is maximum abundance of a population or species that can be harvested without degrading (1) the ability of that population or species and (2) its ecosystem to withstand the same level of harvest in the future.

Sustainable timber harvest Amount of timber that can be removed periodically from a forest without decreasing the capacity of the forest ecosystem to sustain that level of harvest in the future.

Sustainable water use Use of water resources that does not harm the environment and provides for the existence of high-quality water for future generations.

Symbiont Each partner in symbiosis.

Symbiosis An interaction between individuals of two different species that benefits both. For example, lichens contain an alga and a fungus that require each other to persist. Sometimes this term is used broadly, so that domestic corn and people could be said to have a symbiotic relationship—domestic corn cannot reproduce without the aid of people, and some peoples survive because they have corn to eat.

Symbiotic Relationships that exist between different organisms that are mutually beneficial.

Synergism Cooperative action of different substances such that the combined effect is greater than the sum of the effects taken separately.

Synergistic effect When the change in availability of one resource affects the response of an organism to some other resource.

Synfuels Synthetic fuels, which may be liquid or gaseous, derived from solid fuels, such as oil from kerogen in oil shale or oil and gas from coal.

Synthetic organic compounds Compounds of carbon produced synthetically by human industrial processes, as for example pesticides and herbicides.

System A set of components that are linked and interact to produce a whole. For example the river as a system is composed of sediment, water, bank, vegetation, fish and other living things that all together produce the river.

Taiga Forest of cold climates of high latitudes and high altitudes, also known as boreal forest.

Tar sands Sedimentary rocks or sands impregnated with tar oil, asphalt, or bitumen.

Taxa Categories that identify groups of living organisms based upon evolutionary relationships or similarity of characters.

Taxon A grouping of organisms according to evolutionary relationships.

TD-50 The toxic dose defined as the dose that is toxic to 50% of a population exposed to the toxin.

Tectonic cycle The processes that change Earth's crust, producing external forms such as ocean basins, continents, and mountains.

Terminator gene A genetically modified crop which has a gene to cause the plant to become sterile after the first year.

Tertiary treatment (of wastewater) Advanced form of wastewater treatment involving chemical treatment or advanced filtration. An example is chlorination of water.

Theories Scientific models that offer broad, fundamental explanations of related phenomena and are supported by consistent and extensive evidence.

Thermal (heat energy) The energy of the random motion of atoms and molecules.

Thermal pollution A type of pollution that occurs when heat is released into water or air and produces undesirable effects on the environment.

Thermodynamics, first law of See **First Law of Thermodynamics.**

Thermodynamics, second law of See **Second Law of Thermodynamics.**

Thinning The timber harvesting practice of selectively removing only smaller or poorly formed trees.

Threatened species Species experiencing a decline in the number of individuals to the degree that a concern is raised about the possibility of extinction of that species.

Threshold A point in the operation of a system at which a change occurs. With respect to toxicology, it is a level below which effects are not observable and above which effects become apparent.

Tidal power Form of water utilizing ocean tides in places where favorable topography allows for construction of a power plant.

Time series The set of estimates of a population over a number of years.

Tolerance The ability to withstand stress resulting from exposure to a pollutant or harmful condition.

Total fertility Average number of children expected to be born to a woman during her lifetime. (Usually defined as the number born to a woman between the ages 15 to 44, taken conventionally as the lower and upper limit of reproductive ages for women.)

Toxic Harmful, deadly, or poisonous.

Toxicology The science concerned with study of poisons (or toxins) and their effects on living organisms. The subject also includes the clinical, industrial, economic, and legal problems associated with toxic materials.

Transuranic waste Radioactive waste consisting of human-made radioactive elements heavier than uranium. Includes clothing, rags, tools, and equipment that has been contaminated.

Trophic level In an ecological community, all the organisms that are the same number of food-chain steps from the primary source of energy. For example, in a grassland the green grasses are on the first trophic level, grasshoppers are on the second, birds that feed on grasshoppers are on the third, and so forth.

Trophic level efficiency The ratio of the biological production of one trophic level to the biological production of the next lower trophic level.

Tundra The treeless land area in alpine and arctic areas characterized by plants of low stature and including bare areas without any plants and covered areas with lichens, mosses, grasses, sedges, and small flowering plants, including low shrubs.

Ubiquitous species Species that are found almost anywhere on the Earth.

Ultraviolet A The longest wavelength of ultraviolet radiation (0.32-0.4 micrometers), not affected by stratospheric ozone, and is transmitted to the surface of the earth.

Ultraviolet B Intermediate-wavelength radiation which is the ozone problem. Wavelengths are from approximately 0.28-0.32 micrometers and is the most harmful of the ultraviolet radiation types. Most of this radiation is absorbed by stratospheric ozone, and depletion of ozone has led to increase in ultraviolet B radiation reaching the earth.

Ultraviolet C Shortest wavelength of the ultraviolet radiation with wavelengths from approximately 0.2-0.28 micrometers. Is the most energetic of the ultraviolet radiation and is absorbed strongly in the atmosphere. Only a negligible amount of Ultraviolet C reaches the surface of the earth.

Undernourishment The lack of sufficient calories in available food, so that one has little or no ability to move or work.

Uneven-aged stands Forest area with at least three distinct age classes.

Unified soil classification system A classification of soils, widely used in engineering practice, based on the amount of coarse particles, fine particles, or organic material.

Uniformitarianism The principle that processes that operate today operated in the past. Therefore, observations of processes today can explain events that occurred in the past and leave evidence, for example, in the fossil record or in geologic formations.

Urban dust dome Polluted urban air produced by the combination of lingering air and abundance of particulates and other pollutants in the urban air mass.

Urban forestry The practice and profession of planting and maintaining trees in cities, including trees in parks and other public areas. Involves determining the best species and sites for urban tree planting, taking into account climate, soil, shading from tall buildings, and motor vehicle pollution.

Utilitarian justification for the conservation of nature An argument for the conservation of nature on the grounds that the environment, an ecosystem, habitat, or species, provides individuals with direct economic benefit or is directly necessary to their survival.

Utility Value or worth in economic terms.

UVA Least energetic form of ultraviolet radiation. It is capable of causing some damage to living cells, is not affected by stratospheric ozone, and is therefore transmitted to the Earth.

UVB Intermediate-wavelength ultraviolet radiation, damaging to living cells. Most is absorbed by stratospheric ozone, and therefore depletion of ozone leads to significant increase of this radiation. This is the ozone problem.

UVC Shortest wavelength and most energetic of the ultraviolet radiation. It is strongly absorbed in the atmosphere and negligible amounts reach the surface of the Earth.

Vadose zone Zone or layer above the water table where water may be stored as it moves laterally or down to the zone of saturation. Part of the vadose zone may be saturated part of the time.

Variable, dependent A variable that changes in response to changes in an independent variable; a variable taken as the outcome of one or more other variables.

Variable, independent In an experiment, the variable that is manipulated by the investigator. In an observational study, the variable that is believed by the investigator to affect an outcome, or dependent, variable.

Variable, manipulated See **Variable Independent.**

Variable, responding See **Variable Dependent.**

Virgin forest Forest that has never been cut.

Vulnerable species Another term for *threatened species*—species experiencing a decline in the number of individuals.

Waldsterben German phenomenon of forest death as the result of acid rain, ozone, and other air pollutants.

Wallace's realms Six biotic provinces, or biogeographic regions, divided on the basis of fundamental inherited features of the animals found in those areas, suggested by A. R. Wallace (1876). His realms are Nearctic (North America), Neotropical (Central and South America), Palearctic (Europe, northern Asia, and northern Africa), Ethiopian (central and southern Africa), Oriental (the Indian subcontinent and Malaysia), and Australian.

Wastewater renovation and conservation cycle Practice of applying wastewater to the land. In some systems, treated wastewater is applied to agricultural crops, and as the water infiltrates through the soil layer it is naturally purified. Reuse of the water is by pumping it out of the ground for municipal or agricultural uses.

Wastewater treatment Process of treating wastewater (primarily sewage) in specially designed plants that accept municipal wastewater. Generally divided into three categories: pri-

mary treatment, secondary treatment, and advanced wastewater treatment.

Water budget Inputs and outputs of water for a particular system (a drainage basin, region, continent, or the entire Earth).

Water conservation Practices designed to reduce the amount of water we use.

Water power Alternative energy source derived from flowing water. One of the world's oldest and most common energy sources. Sources vary in size from microhydropower systems to large reservoirs and dams.

Water reuse The use of wastewater following some sort of treatment. Water reuse may be inadvertent, indirect, or direct.

Water table The surface that divides the zone of aeration from the zone of saturation, the surface below which all the pore space in rocks is saturated with water.

Watershed An area of land that forms the drainage of a stream or river. If a drop of rain falls anywhere within a watershed to become surface runoff it can flow out only through the same stream.

Weathering Changes that take place in rocks and minerals at or near the surface of Earth in response to physical, chemical, and biological changes; the physical, chemical, and biological breakdown of rocks and minerals.

Wetlands Comprehensive term for landforms such as salt marshes, swamps, bogs, prairie potholes, and vernal pools. Their common feature is that they are wet at least part of the year and as a result have a particular type of vegetation and soil. Wetlands form important habitats for many species of plants and animals, while serving a variety of natural service functions for other ecosystems and people.

Whole-tree harvesting Timber harvesting practice of removing all aboveground parts of a tree, most of which are chipped into small fragments for making paper.

Wilderness An area unaffected now or in the past by human activities and without a noticeable presence of human beings.

Wind power Alternative energy source that has been used by people for centuries. More recently, thousands of windmills have been installed to produce electric energy.

Work (physics) Force times the distance through which it acts when work is done we say energy is expended..

Zero population growth A population in which the number of births equals the number of deaths so that there is no net change in the size of the population.

Zone of aeration The zone or layer above the water table in which some water may be suspended or moving in a downward migration toward the water table or laterally toward a discharge point is often called the vadose zone.

Zone of saturation Zone or layer below the water table in which all the pore space of rock or soil is saturated.

Zooplankton Small aquatic invertebrates that live in the sunlit waters of streams, lakes, and oceans and feed on algae and other invertebrate animals.

Notes

Chapter 1 Notes

1. Gordon, B. B. 1993. "Pampering Our Coastlines," *Sea Frontiers* 39(2):5.
2. Mydans, S. 1996 (April 28). "Thai Shrimp Farmers Facing Ecologist's Fury," *The New York Times*.
3. "Pollution Wiping Out Shrimp Farms on Main Indonesian Island of Java." 1996. *Quick Frozen Foods International* 37(3):50.
4. Quarto, A. 1994. "Rainforests of the Sea: Mangrove Forests Threatened by Prawn Aquaculture," *E*, 5(1):16–19.
5. Wickramayanake, S. D. 1995. "East Coast Shrimp Farms Face Trouble from Both Nature and Protest Groups." *Quick Frozen Foods International* 37(1):108–109.
6. Haub, Carl, and Diana Cornelius. 2000. Nine million world population by 2050. Washington, D.C.: Population Reference Bureau.
7. Everett, G. D. 1961. One man's family. *Population Bulletin* 17:153–169.
8. Ehrlich, P. R., A. H. Ehrlich, and P. H. Holdren. 1977. *Ecoscience: Population, resources, environment.* 3d ed. San Francisco: W. H. Freeman.
9. Deevey, E. S. 1960. The human population. *Scientific American* 203:194–204.
10. Keyfitz, N. 1989. The growing human population. *Scientific American* 261:118–126.
11. Field, J. O. ed. 1993. *The challenge of famine: Recent experience, lessons learned.* Hartford, Conn.: Kumarian Press, Inc.
12. Glantz, M. H., ed. 1987. *Drought and hunger in Africa: Denying famine a future.* Cambridge: Cambridge University Press.
13. Seavoy, R. E. 1989. *Famine in East Africa: Food production and food politics.* Westport: Greenwood Press.
14. Gottfried, R. S. 1983. *The black death: Natural and human disaster in medieval Europe.* New York: Free Press.
15. Gower, B. S. 1992. What do we owe future generations? In D. E. Cooper and J. A. Palmer, eds., *The environment in question: Ethics and global issues.* Pp. 1–12. New York: Routledge.
16. Van Hyning, J. M. 1973. Factors affecting the abundance of fall chinook salmon in the Columbia River. *Research Reports of the Fish Commission of Oregon* 4:1–87.
17. Margulis, L., and J. E. Lovelock. 1989. Gaia and geognosy. In M. B. Rambler, L. Margulis, and R. Fester, eds., *Global ecology: Towards a science of the biosphere.* Pp. 1–30. Boston: Academic Press.
18. Botkin, D. B., M. Caswell, J. E. Estes, and A. Orio. eds. 1989. *Changing the global environment: Perspectives on human involvement.* New York: Academic Press.
19. Brown, L. R., and J. L. Jacobson. 1987. The future of urbanization: Facing the ecological and economic constraints. In K. Davis, M. S. Bernstam, and H. M. Sellers, eds., *Population and resources in a changing world.* Stanford, Calif.: The Morrison Institute for Population and Resource Studies, Stanford University.
20. Haub, Carl, and Diana Cornelius. 2000. *World population data sheet.* Washington, D.C.: Population Reference Bureau.
21. World Resources Institute. 1998. *Teacher's guide to world resources: Exploring sustainable communities.* Washington, D.C.: World Resources Institute. (http://www.igc.org/wri/wr-98-99/citygrow.htm)
22. World Resources Institute. 1999. *Urban growth.* Washington, D.C.: World Resources Institute.
23. Beveridge, C. E. 1990. Introduction to the landscape design reports: The California origins of Olmsted's landscape design principles for the semiarid American west. *The Papers of Frederick Law Olmsted*, vol. 5. Baltimore: Johns Hopkins University Press.
24. Rolston, H., III. 1992. Challenges in environmental ethics. In D. E. Cooper and J. A. Palmer, eds., *The environment in question: Ethics and global issues.* Pp. 135–146. New York: Routledge.
25. Nash, R. F. 1988. *The rights of nature: A history of environmental ethics.* Madison: University of Wisconsin Press.

Chapter 1 Critical Thinking Issue Notes

Bryant, Dirk, Auretta Burke, John McManus, and Mark Spalding. 1998. *Reefs at risk: A map-based indicator of threats to the world's coral reefs.* Washington, D.C.: World Resources Institute.

Coles, S. L., and L. Ruddy. 1995. Comparison of water quality and reef coral mortality and growth in southeastern Kaneohe Bay, Oahu, Hawaii, 1990 to 1992, with conditions before sewage diversion, *Pacific Science* 49(3): 247–265.

Hinrichsen, D. 1997 (October). "Coral Reefs in Crisis," *Bioscience* 47(9): 554–558.

Jameson, S. C., J. W. McManus, and M. D. Spalding. 1995 (May). State of the reefs: Regional and global perspectives. International Coral Reef Initiative Executive Secretariat (Background Paper).

Chapter 2 Notes

1. Schmidt, W. E. 1991. "Jovial con men" take credit (?) for crop circles. *New York Times.* Sept. 10, p. B1.
2. Tuohy, W. 1991. "Crop circles" their prank, 2 Britons say. *Los Angeles Times*, Sept. 10, p. A14.
3. Taylor, F. S. 1949. *Science and scientific thought.* New York: W. W. Norton.
4. Reif, F., and J. R. Larkin. 1991. Cognition in scientific and everyday domains: Comparison and learning implications. *Journal of Research in Science Teaching* 28:733–760.
5. Gibbs, A., and A. E. Lawson. 1992. The nature of scientific thinking as reflected by the work of biologists and by biology textbooks. *The American Biology Teacher* 54:137–152.
6. Pease, C. M., and J. J. Bull. 1992. Is science logical? *Bioscience* 42:293–298.
7. Lerner, L. S., and W. J. Bennetta. 1988 (April). The treatment of theory in textbooks. *The Science Teacher*, pp. 37–41.
8. Kuhn, T. S. 1970. *The structure of scientific revolutions.* Chicago: University of Chicago Press.
9. Trefil, J. S. 1978. A consumer's guide to pseudoscience. *Saturday Review* 4:16–21.
10. Hastings and Hastings, 1992. *Telepathy.* American Institute of Public Opinion poll from June 1990.
11. Feynman, Richard P. 1998. *The Meaning of it all: Thoughts of a citizen-scientist.* Reading, Mass.: Addison-Wesley, Perseus Books, p. 8. From lectures at the University of Washington in 1963.

12. Tyser, R. W., and Cerbin, W. H. 1991. Critical thinking exercises for introductory biology courses. *Bioscience* 41:41–46.

Chapter 2 Critical Thinking Issue References

Ford, R. 1998 (March). "Critically Evaluating Scientific Claims in the Popular Press." *The American Biology Teacher*, vol. 60, no. 3, pp. 174–180.

Marshall, E. 1998 (May 15). "The Power of the Front Page of *The New York Times*." *Science*, vol. 280, pp. 996–997.

Chapter 3 Notes

1. Western, D., and C. Van Prat. 1973. Cyclical changes in habitat and climate of an East African ecosystem. *Nature* 241(549):104–106.

2. Dunne, T., and Leopold, L. B. 1978. *Water in environmental planning*. San Francisco: W. H. Freeman.

3. Bartlett, A. A. 1980. Forgotten fundamentals of the energy crisis. *Journal of Geological Education* 28:4–35.

4. Meadows, D. H., D. L. Meadows, and J. Randers. 1992. *Beyond the limits*. Post Mills, Vt.: Chelsea Green Publishers.

5. Wootton, J. T., M. S. Parker, and M. E. Power. 1996. Effects of disturbances on river food webs. *Science* 273:1558–1561.

6. Leach, M. K., and T. J. Givnich. 1996. Ecological determinants of species loss in remnant prairies. *Science* 273:1555–1558.

7. Lovelock, J. 1995. *The ages of Gaia: A biography of our living earth*, revised. New York. W. W. Norton.

8. Abramovitz, J. N., and S. Dunn. 1998. Record year for weather-related disasters. *World Watch Institute*. Vital Signs Brief 98-5.

Chapter 3 Critical Thinking Issue References

Barlow, C. 1993. *From Gaia to selfish genes*. Cambridge: MIT Press.

Kirchner, J. W. 1989. The Gaia hypothesis: Can it be tested? *Reviews of Geophysics* 27:223–235.

Lovelock, J. E. 1995. *Gaia: A new look at life on Earth*. New York: Oxford University Press.

Lyman, F. 1989 (July). What hath Gaia wrought? *Technology Review* 92(5):55–61.

Resnik, D. B. 1992 (Summer). Gaia: From fanciful notion to research program. *Perspectives in Biology and Medicine* 35(4):572–582.

Schneider, S. H. 1990 (May). Debating Gaia. *Environment* 32(4):4–9, 29–32.

Chapter 3 Working It Out Reference

Bartlett, A. A. 1993. The arithmetic of growth: Methods of calculation. *Population and Environment* 14(4):359–387.

Chapter 4 Notes

1. Lehman, J. T. 1986. Control of eutrophication in Lake Washington. In *Ecological knowledge and environmental problem solving*, ed. G. H. Orians, pp. 302–316. Washington, D.C.: National Academy of Science.

2. Henderson. [1913] 1966. The fitness of the environment. Reprint. Boston: Beacon Press.

3. Van Koevering, T. E., and N. J. Sell, 1986. *Energy: A conceptual approach*. Englewood Cliffs, N.J. Prentice Hall, p. 271.

4. Isacks, B., J. Oliver, and L. Sykes. 1968. Seismology and the new global tectonics. *Journal of Geophysical Research* 73:5855–5899.

5. Dewey, J. F. 1972. Plate tectonics. *Scientific American* 22:56–68.

6. Botkin, D. B. 1990. *Discordant harmonies: A new ecology for the 21st century*. New York: Oxford University Press.

7. Ehrlich, P. R., A. H. Ehrlich, and J. P. Holdren. 1970. *Ecoscience: Population, resources, environment*. San Francisco: W. H. Freeman, p. 1051.

8. Post, W. M., T. Peng, W. R. Emanuel, A. W. King, V. H. Dale, and D. L. De Angelis. 1990. The global carbon cycle. *American Scientist* 78:310–326.

9. Keeling, C. D., T. P. Whorf, M. Wahlen, and J. van der Plicht. 1995. Interannual extremes in the rate of rise of atmospheric carbon dioxide since 1980. *Nature* 375:666–670.

10. Hudson, R. J. M., S. A. Gherini, and R. A. Goldstein. 1994. Modeling the global carbon cycle: Nitrogen fertilization of the terrestrial biosphere and the "missing" CO_2 sink. *Global Biogeochemical Cycles* 8:307–333.

11. Woods Hole. 2000. The missing carbon sink. http://www.whrc.org/science/carbon/missingc.htm. Accessed July 5, 2000.

12. Herring, D., and R. Kannenberg. 2000. The mystery of the missing carbon. http://earthobservatory.nasa.gov/cgi-bin/printall?/study/BOREAS/missing_carbon.html. Accessed July 5, 2000.

13. Chameides, W. L., and E. M. Perdue. 1997. *Biogeochemical cycles*. New York: Oxford University Press.

14. Agren, G. I., and E. Bosatta. 1996. *Theoretical ecosystem ecology*. New York: Cambridge University Press.

15. Kasting, J. F., O. B. Toon, and J. B. Pollack. 1988. How climate evolved on the terrestrial planets. *Scientific American* 258:90–97.

16. Berner, R. A. 1999. A new look at the long-term carbon cycle. *GSA Today* 9(11):2–6.

17. Carter, L. J. 1980. Phosphate: Debate over an essential resource. *Science* 209:4454.

Chapter 4 Critical Thinking Issue References

Asner, G. P., T. R. Seastedt, and A. R. Townsend. 1997 (April). The decoupling of terrestrial carbon and nitrogen cycles. *Bioscience* 47 (4):226–234.

Hellemans, A. 1998 (February 13). Global nitrogen overload problem grows critical. *Science*, 279:988–989.

Smil, V. 1997 (July). Global populations and the nitrogen cycle. *Scientific American*, 76–81.

Vitousek, P. M., J. Aber, R. W. Howarth, G. E. Likens, P. A. Matson, D. W. Schindler, W. H. Schlesinger, and G. D. Tilman. 1997. Human alteration of the global nitrogen cycle: Causes and consequences. *Issues in Ecology*. http://esa.sdsc.edu/

Chapter 5 Notes

1. Cornelius, D., and C. Haub. 1999. *World population data sheet*. Washington, D.C.: Population Reference Bureau.

2. Fazi, A. 1991 (January–February). Bangladesh bears the brunt. *World Health*. Pp. 26–27.

3. Kent, M. M. and K. A. Crews. 1998. *World population: Fundamentals of growth*. Washington, D.C.: Population Reference Bureau.

4. Hardin, G. 1971. Nobody ever dies of overpopulation. *Science* 171:527.

5. World Bank 1998. *Partnership in development: The World Bank's Assistance to Bangladesh*. Washington, D.C.: World Bank.

6. World Bank. 2000. World development indicators 2000. Washington, D.C.: World Bank.

7. World Bank. *Bangladesh 2020: A long-run perspective study*. Washington, D.C.: World Bank.

8. World Resources Institute. 2000. Nutrition, 1971–96. Washington, D.C.: World Resources Institute.

9. Erhlich, P. R. 1971. *The population bomb*. Rev. ed. New York: Ballantine Books.

10. Malthus, T. R. 1803. *An essay on the principle of population*. Selected and introduced by Donald Winch. Cambridge: Cambridge University Press. 1992.

11. Keyfitz, N. 1992. Completing the worldwide demographic transition: The relevance of past experience. 21:26–30.

12. Graunt, J. [1662] 1973. *Natural and political observations made upon the bill of mortality*. Reprint, Germany: Gregg International Publishers.

13. Dumond, D. E. 1975. The limitation of human population: A natural history. *Science* 187:713–721.

14. Zero Population Growth. 2000. *U.S. population*. Washington, D.C.: Zero Population Growth.

15. World Bank. 1984. *World development report 1984*. New York: Oxford University Press.

16. World Bank. 1992. *World development report*. The relevance of past experience.
17. World Bank. 1984.
18. Bureau of the Census. 1990. *Statistical abstract of the United States 1990*. National Data Book. Washington, D.C.: U.S. Dept. of Commerce.
19. Joint United Nations Programme on HIV/AIDS. 1999. *AIDS epidemic update*. Geneva, Switzerland.
20. Population Division. 1998. *World population growing despite AIDS spread*. United Nations Department of Economic and Social Affairs.
21. World Bank. 1985. *World development report*. New York: Oxford University Press.
22. Haupt, A., and T. T. Kane. 1978. *The Population Reference Bureau's population handbook*. Washington, D.C.: Population Reference Bureau.
23. Central Intelligence Agency. 1999. *The world factbook*. Washington, D.C.: Central Intelligence Agency.
24. Dumond, D. E. 1975. The limitation of human population: A natural history. *Science: 187*:713–721.
25. Wu, L. S., and D. B. Botkin. 1980. Of elephants and men. *American Naturalist 116*:831–849.
26. Guz, D., and J. Hobcraft. 1991. Breastfeeding and fertility: A comparative analysis. *Population Studies 45*:91–108.
27. Fathalla, M. F. 1992. Family planning: Future needs. *Ambio 21*:84–87.
28. The Alan Guttmacher Institute. Sharing responsibility: Women, society and abortion worldwide. New York: AGI, 1999.
29. World Bank. 2000. *Health, nutrition, and population*. World Bank.
30. Cohen, Susan A. 2000. Proponents say women's health, core American values at stake in global gag rule debate. *The Guttmacher Report on Public Policy 3*(2):5–7.
31. Planned Parenthood Federation of America, Public Policy Division. 1997 (June). *International family planning: The need for services*. Planned Parenthood Federation of America, Inc.
32. Singer, M. 1999. The population surprise. *The Atlantic Monthly*. August pp.22–25.

Chapter 5 Critical Thinking Issue References

Cornelius, D., and C. Haub. 1999. World population data sheet. Washington, D.C.: Population Reference Bureau.
Ehrlich, P. R., and H. A. Ehrlich. 1990. *The population explosion*. New York: Simon & Schuster.
Environment. New York: Prentice-Hall.
Environmental almanac. Boston: Houghton Mifflin.

Chapter 5 Working It Out Reference

Population Reference Bureau. 1998. World population data sheet. Washington, D.C.: Population Reference Bureau.

Chapter 6 Notes

1. Line, L. 1996 (April 16). Ticks and moths, not just oaks, linked to acorns. *The New York Times*.
2. Ostfield, R. S., C. G. Jones, and J. O. Wolff. 1996 (May). Of mice and mast: Ecological connections in eastern deciduous forests. *BioScience 46*(5):323–330.
3. Morowitz, H. J. 1979. *Energy flow in biology*. Woodbridge: Conn.: Oxbow Press.
4. Brock, T. D. 1967. Life at high temperatures. *Science 158*:1012–1019.
5. Brock, T. D. 1971. Life in the geysers basin. Washington, D.C.: National Park Service.
6. Lavigne, D. M., W. Barchard, S. Innes, and N. A. Oritsland. 1976. *Pinniped bioenergetics*. ACMRR/MM/SC/12. Rome: United Nations Food and Agriculture Organization.
7. Estes, J. A., and J. F. Palmisano, 1974. Sea otters: Their role in structuring nearshore communities. *Science 185*: 1058–1060.
8. The Marine Mammal Center, Sausalito, Calif.
9. Kenyon, K. W. 1969. The sea otter in the eastern Pacific Ocean. *North American Fauna*, no. 68. Washington, D.C.: Bureau of Sports Fisheries and Wildlife, U.S. Department of the Interior.
10. Duggins, D. O. 1980. Kelp beds and sea otters: An experimental approach. *Ecology 61*:447–453.
11. Kvitek, R. G., J. S. Oliver, A. R. DeGange, and B. S. Anderson. 1992. Changes in Alaskan soft-bottom prey communities along a gradient in sea otter predation. *Ecology 73*:413–428.
12. Paine, R. T. 1969. A note on trophic complexity and community stability. *American Naturalist 100*:65–75.

Chapter 6 Critical Thinking Issue References

Bilby, R. E., B. R. Fransen, and P. A. Bisson. 1996. Incorporation of nitrogen and carbon from spawning coho salmon into the trophic system of small streams: Evidence from stable isotopes. *Canadian Journal of Fisheries and Aquatic Sciences* 53: 164–173.
Morowitz, H. J. 1979. *Energy flow in biology*. Woodbridge, Conn.: Oxbow Press.
Spencer, C. N., R. R. McClelland, and J. A. Stanford. 1991 (January). Shrimp stacking, salmon collapse, and eagle displacement. *BioScience 41*(1):14–21.
Willson, M. F., S. M. Gende, and B. H. Marston. 1998 (June). Fishes and the forest. *BioScience 48*(6) 455–462.
Willson, M. F., and K. C. Halupka. 1995 (June). Anadromous fish as keystone species in vertebrate communities. *Conservation Biology 9*(3):489–497.

Chapter 7 Notes

1. Bourne, J. 2000. Louisiana's vanishing wetlands: Going, going. . . *Science 289*:1860–1863.
2. World Health Organization. 1998. Malaria. *Fact Sheet No. 94*. WHO.
3. World Health Organization. 2000. *Overcoming antimicrobial resistance: World health report on infectious diseases 2000*. WHO.
4. James, A. A. 1992. Mosquito molecular genetics: The hands that feed bite back. *Science 257*:37–38.
5. Kolata, G. 1984. The search for a malaria vaccine. *Science 226*:679–682.
6. Miller, L. H. 1992. The challenge of malaria. *Science 257*:36–37.
7. World Health Organization TDR. 1999 (June). Using malaria information. *News Release No. 59*. WHO TDR.
8. World Health Organization TDR. 1999 (July). Sequencing the *Anopheles gambiae* genome. *News Release No. 60*. WHO TDR.
9. Woese, C. R., O. Kandler, and M. L. Wheelis. 1990. Towards a natural system of organisms: Proposals for the domains Archaea, Bacteria, and Eucarya, *Proc. Natl. Acad. Sci.*, USA 87:4576–4579.
10. Hutchinson, G. E., 1965. *The ecological theater and the evolutionary play*. New Haven: Yale University Press.
11. Woese, Kandler, and Wheelis. Ibid.
12. Pace, N. R. 1996. New perspective on the natural microbial world: Molecular microbial ecology, 1996, *American Society for Microbiology News* 62:463–470.
13. Hardin, G. 1960. The competitive exclusion principle. *Science 131*:1292–1297.
14. Rogers, C. 1996. *Red squirrel: Sciurus vulgaris*. The Wild Screen Trust.
15. Schoener, T. W. 1983. Field experiments in interspecific competition. *American Naturalist 1222*:240–285.
16. Elton, C. S. 1927. *Animal ecology*. New York: Macmillan.
17. Hutchinson, G. E. 1958. Concluding remarks. *Cold Spring Harbor Symposium in Quantitative Biology* 22:415–427.
18. Miller, R. S. 1967. Pattern and process in competition. *Advances in Ecol. Research* 4:1–74.
19. Botkin, D. B. 1985. The need for a science of the biosphere. *Interdisciplinary Science Reviews*. 10:267–278.
20. Forman, Richard T. T. 1995. *Land mosaics: The ecology of landscapes and regions*. New York: Cambridge University Press.
21. Valentine, J. W. 1973. Plates and provinces, a theoretical history of

environmental discontinuity. In N. F. Hughes, ed., *Organisms and continents through time*, pp. 79–92. Special Papers in Paleontology 12.

Chapter 7 Critical Thinking Issue Reference

McNeely, J. A., K. R. Miller, W. V. Reid, R. A. Mittermeier, and T. B. Werner. 1990. *Conserving the world's biological diversity*. Gland, Switzerland, and Washington, D.C.: International Union for Conservation of Nature and Natural Resources, World Resources Institute, Conservation International, World Wildlife Fund–U.S., World Bank.

Chapter 8 Notes

1. Beck, K. 1994. Natural control for the purple loosestrife. *North Coast Newsletter*, Ohio Lake Erie Commission, 1–2.
2. Carroll, D. 1994. Subduing purple loosestrife. *The Conservationist* 49:6–9.
3. Malecki, R. A., B. Blossey, S. D. Hight, D. Schroeder, L. T. Kok, and J. R. Coulson, 1993. Biological control of purple loosestrife. *BioScience* 43:680–686.
4. Wallace, A. R. 1896. *The geographical distribution of animals*. Vol. 1. Reprint. New York: Hafner. 1962.
5. Pielou, E. C. 1979. *Biogeography*. New York: John Wiley & Sons.
6. Good, R. 1974. *The geography of the flowering plants*. 4th ed. London: Longman Group Ltd.
7. Takhtadzhian, A. L. 1986. *Floristic regions of the world*. Berkeley: University of California Press.
8. Udvardy, M. 1975. A classification of the biogeographical provinces of the world. *IUCN Occasional Paper* 18. Morges, Switzerland: IUCN.
9. Lentine, J. W. 1973. Plates and provinces, a theoretical history of environmental discontinuity. In N. F. Hughes, ed., *Organisms and continents through time*, pp. 79–92. Special Papers in Paleontology 12.
10. Hallam, A. 1975. Alfred Wegener and the hypothesis of continental drift. *Scientific American* 232:88–97.
11. Hurley, P. M. 1968. The confirmation of continental drift. *Scientific American* 218:52–64.
12. Mather, J. R., and G. A. Yoshioka, 1968. The role of climate in the distribution of vegetation. *Ann. Ass. American Geography* 58:29–41.
13. Prentice, I. C., W. Cramer, S. P. Harrison, R. Leemans, R. A. Monserud, and A. M. Solomon, 1992. A global biome model based on plant physiology and dominance, soil properties and climate. *Journal of Biogeography* 19:117–134.
14. Botkin, D. B. 1977. The vegetation of the west. In H. R. Lamar, ed., *The reader's encyclopedia of the American west*, pp. 1216–1234. New York: Thomas Y. Crowell.
15. Darwin, C. R. 1859. *The origin of species by means of natural selection or the preservation of favored races in the struggle for life*. London: Murray.
16. Grant, P. R. 1986. *Ecology and evolution of Darwin's finches*. Princeton, N.J.: Princeton University Press.
17. Cox, C. B., I. N. Healey, and P. D. Moore, 1973. *Biogeography*. New York: Halsted.
18. MacArthur, R. H., and E. O. Wilson, 1967. *The theory of island biogeography*. Princeton, N.J.: Princeton University Press.
19. Tallis, J. H. 1991. *Plant community history*. London: Chapman and Hall.
20. Missouri Botanic Garden, "Botany in North America: In Honor of the XVI International Botanical Congress, August 1–7, 1999, St. Louis, MO.
21. Botkin, D. B. 2001. *No Man's Garden: Thoreau and a New Vision for Civilization and Nature*, Island Press, Washington, D.C.

Chapter 8 Critical Thinking Issue Notes

The New York Times (16 August 1997), p. 1. The seaweed is *Caulerpa taxifolia*.

Chapter 9 Notes

1. Rackham, O., 1986. *The history of the countryside*. London: Dent & Sons, p. 63.

2. Perlin, J., 1989. *A forest journey: the role of wood in the development of civilization*. New York: W. W. Norton.
3. Schrödinger, E., 1942. *What is life?* Cambridge: Cambridge University Press.
4. Morowitz, H. J. 1979. *Energy flow in biology*. Oxbow Press Woodbridge, Conn.
5. Slobodkin, L. B. 1960. Ecological energy relations at the population level. *American Naturalist* 95:213–236.
6. Peterson, R.O. 1995. *The wolves of Isle Royale: A broken balance*. Minocqua, WI: Willow Creek Press.
7. Jordan, J. D., D. B. Botkin, and M. I. Wolf, 1971. Biomass dynamics in a moose population. *Ecology* 52:147–152.
8. Kozlovsky, D. G. 1968. A critical evaluation of the trophic level concept: I. Ecological efficiencies. *Ecology* 49:147–160.
9. Schaefer, M., 1991. Secondary production and decomposition. In E. Rohrig and B. Ulrich, eds., *Temperate deciduous forests. Ecosystems of the world*, vol. 7. Amsterdam: Elsevier.
10. Golley, F. B. 1989. Energy dynamics of a food chain of an old-field community. *Ecol. Monographs* 30:187–291.
11. Bagley, P. B. 1989. Aquatic environments in the Amazon basin, with an analysis of carbon sources, fish production and yield. In D. P. Dodge, ed., *Proc. Int. Large Rivers Symp. Can. Spec. Publ. Fish. Aquat. Sci.* 106:385–398.
12. Fasham, M. J. R., ed. 1984. *Flows of energy and materials in marine ecosystems: Theory and practice*. New York and London: Plenum Press.
13. Gaill, F., B. Shillito, F. Menard, G. Goffinet, and J. Childress, 1997. Rate and process of tube production by the deep sea hydrothermal vent tubeworm *Riftia pachyptila*. *Marine Ecology Progress Series* 148: 135-143.
14. Wills, J., 1996. Upwelling. *FMF Glossary*. First Millennial Foundation.

Chapter 10 Notes

1. Covington, W. W., P. Z. Ful, M. Moore, S. C. Hart, T. E. Kolb, J. N. Mast, S. S. Sackett, and M. R. Wagner. 1997. Restoring ecosystem health in ponderosa pine forests of the Southwest, *Journal of Forestry* 95:23–29.
2. Forman, R. T. T., 1995. *Landscape mosaics*. Cambridge: Harvard University Press.
3. Houseal, G., and D. Smith. 2000. Source-identified seed: The Iowa roadside experience. *Ecological Restoration* 18(3):173–183.
4. Hall, F. G., D. B. Botkin, D. E. Strebel, K. D. Woods, and S. J. Goetz, 1991. Large-scale patterns in forest succession as determined by remote sensing. *Ecology* 72:628–640.
5. Gorham, E., P. M. Vitousek, and W. A. Reiners, 1979. The regulation of chemical budgets over the course of terrestrial ecosystem succession. *Ann. Rev. Ecol. Syst.* 10:53–84.
6. Vitousek, P. M., and L. R. Walker, 1987. Colonization, succession and resource availability: Ecosystem-level interactions. In A. J. Gray, M. J. Crawley, and P. J. Edwards, eds., Colonization, succession and stability, pp. 207–223. *British Ecol. Soc. 26th Symp.* Oxford: Blackwell Scientific Publications.
7. Connell, J. H., and R. O. Slatyer, 1977. Mechanisms of succession in natural communities and their role in community stability and organization. *American Naturalist* 111:1119–1144.
8. Pickett, S. T. A., S. L. Collins, and J. J. Armesto, 1987. Models, mechanisms and pathways of succession. *Botanical Review* 53:335–371.
9. Gomez-Pompa, A., and C. Vazquez-Yanes, 1981. Successional studies of a rain forest in Mexico. In D. C. West, H. H. Shugart, and D. B. Botkin, eds., *Forest succession: Concepts and application*, pp. 246–266. New York: Springer-Verlag.
10. MacMahon, J. A. 1981. Successional processes: comparison among biomes with special reference to probable roles of and influences on animals. In D. C. West, H. H. Shugart, and D. B. Botkin, eds., *Forest succession: Concepts and application*, pp. 277–304. New York: Springer-Verlag.

11. Walthern, P. 1986. Restoring derelict lands in Great Britain. In G. Orians, ed., *Ecological knowledge and environmental problem-solving: Concepts and case studies*, pp. 248–274. Washington, D.C.: National Academy Press.

Chapter 10 Critical Thinking Issue References

Malakoff, D. (April 17) 1998. Restored wetlands flunk real-world test. *Science* 280: 371–372.

Pacific Estuarine Research Laboratory (PERL). 1997 (November). The status of constructed wetlands at Sweetwater Marsh National Wildlife Refuge. Annual Report to the California Department of Transportation.

Zedler, J. B. 1997. Adaptive management of coastal ecosystems designed to support endangered species. *Ecology Law Quarterly* 24:735–743.

Zedler, J. B., and A. Powell, 1993. Problems in managing coastal wetlands: Complexities, compromises, and concerns. *Oceanus* 36(2):19–28.

Chapter 11 Notes

1. Haub, Carl, and Diana Cornelius. 2000. World population data sheet. Washington, D.C.: Population Reference Bureau.
2. Brown, L. R., 1995. *Who will feed China? Wake-up call for a small planet*. New York: W. W. Norton.
3. World Resources Institute. 2000. Facts and figures: Country environmental data. Washington, D.C.: World Resources Institute.
4. A food crisis—or a blip? 1996 (February). *World Press Review* 43(2):34.
5. Cohen, J. E. 1995 (November/December). How many people can the earth support? *The Sciences* 18–23.
6. Holmes, B. 1993 (February 8). Feeding a world of 10 billion. *U. S. News & World Report* 114(5):55.
7. Livernash, R. 1995 (July/August). The future of populous economies: China and India shape their destinies. *Environment* 37(6):6–11, 25–34.
8. Ryan, M., and C. Flavin. 1995. Facing China's limits. *State of the world, 1995*. New York: W. W. Norton.
9. Tyler, P. E. 1996 (May 23). China's fickle rivers: Dry farms, needy industry bring a water crisis. *The New York Times*.
10. United Nations Food and Agriculture Organization. 1998. FAO-STAT database. Rome: UNFAO.
11. Biological Resources Division USGS. 1998. Historical interrelationships between population settlement and farmland in the conterminous United States, 1790–1992. *Land use history of North America*. USGS.
12. National Wildlife Federation. 1997. *Wetlands status and trends, 1985–1995: Annual rate of wetlands destruction down 60%*. National Wildlife Federation.
13. Hawthorne, P. 1998 (April 3). *Rebirth*. Time 100/Africa, 151 (15), *Time 100/Leaders and Revolutionaries*.
14. Daniel Glickman, U.S. Secretary of Agriculture. 2000 (November 29). Speech to the USDA Advisory Committee on Agricultural Biotechnology.
15. Johnson G. L., and S. H. Wittwer. 1984. *Agriculture technology until 2030: Prospects, priorities, and policies*. Michigan State University Agricultural Experiment Station, Special Report 12.
16. Raven, P. H., R. F. Evert, and S. E. Eichhorn. 1999. *Biology of plants*. New York: W. H. Freeman/Worth.
17. United States Food and Agriculture Organization. 2000 (April). Cereals: Supply/demand roundup. *Food Outlook 2*, p. 4. Rome: UNFAO.
18. United Nations Food and Agriculture Organization. 2000 (April). Food emergencies persist in 34 countries throughout the world. *Food Outlook 2*, p. 4. Rome: UNFAO.
19. United Nations Food and Agriculture Organization. 1998 (September). *Global information and early warning system on food and agriculture*. Global Watch, *Food Outlook* Rome: UNFAO.
20. USDA Magriet Caswell communication to ACAB July 2000.
21. Bardach, J. E. 1968. Aquaculture. *Science* 161:1098–106.
22. Smil, V. 1984. *The bad Earth: Environmental degradation in China*. Armonk, N.Y.: M.E. Sharpe.
23. Himnan, W. 1984. New crops for arid lands. *Science* 225:1445–1256.
24. Flannery, K. V. 1965. The ecology of early food production in Mesopotamia. *Science* 147:1247–1256.
25. Murdock, W. M. 1980. *The poverty of nations: The political economy of hunger and population*. Baltimore: Johns Hopkins University Press.
26. California Agricultural Statistics Service. 1999. *California agricultural statistics*. Sacramento, Calif.: CASS.
27. Field, J. O., ed. 1993. *The challenge of famine: Recent experience, lessons learned*. Hartford, Conn.: Kumarian Press.
28. World Food Programme. 1998. *Tackling hunger in a world full of food: Tasks ahead for food aid*. Prepared for the World Food Summit in 1996, revised March 19, 1998. Rome: UNFAO.
29. World Watch Institute. 1995. Rising food prices threaten political stability. *World Watch Magazine*. Washington, D.C.: World Watch Institute.
30. Smith, J. B., and D. Tirpak. 1989. *The potential effects of global climate change on the United States*. Report to Congress. Washington, D.C.: U.S. Environmental Protection Agency, EPA-230-05-89-050.
31. Botkin, D. B., R. A. Nisbet, and T. E. Reynolds. 1989. Effects of climate change on forests of the Great Lake states. In J. B. Smith and D. A. Tirpka, eds., *The potential effects of climate change of the United States*, pp. 2.1–2.31. Washington, D.C.: U.S. Environmental Protection Agency, EPA-302-05-89-0.
32. Borland, N. E. 1983. Contributions of conventional plant breeding to food production. *Science* 147:689–693.

Chapter 11 Critical Thinking Issue References

Brown, L. R., M. Renner, and C. Flavin. 1998. *Vital signs: 1998*. New York: W. W. Norton.

Food and Agricultural Organization of the United Nations (FAO). 1996. *Food for all*. Rome: FAO.

Pimentel, D., J. Houser, E. Preiss, O. White, H. Fang, L. Mesnick, T. Barsky, S. Tariche, J. Schreck, and S. Alpert. 1997 (February). "Water resources: agriculture, the environment, and society." *Bioscience* 4 (2), 92–106.

Population Reference Bureau. 1998. *1998 world population data sheet*. Washington, D.C.: Population Reference Bureau.

Postel, S. L. 1998 (August). "Water for food production: Will there be enough in 2025?" *Bioscience* 48(8), 629–637.

Chapter 12 Notes

1. Kansas Rural Center, 1996. *More clean water farm demonstrations selected*. March 1996, p. 5.
2. World Resources Institute. *Annual report 1992–93*. Washington, D.C.
3. Vitousek, P. M. 1987. Personal communication.
4. Pimentel, D. E. C. Terhune, R. Dyson-Hudson, S. Rochereau, R. Samis, E. A. Smith, D. Deanma, D. Reifschneider, and M. Shepard, 1976. Land degradation: Effects on food and energy resources. *Science* 194:149–155.
5. Trimble, S. W., and P. Crosson 2000, U.S. soil erosion rates—Myth and reality, *Science* 289 (14 July 2000): 248–250.
6. Trimble, S. W. 2000. Soil conservation and soil erosion in the upper midwest, *Environmental Review*, 7 (1):3–9.
7. Pimentel, D., and E. L. Skidmore. 1999, Rates of soil erosion, *Science* 286:1477–1478.
8. Lashof, J. C., ed. 1979. *Pest management strategies in crop protection*. Vol. 1. Washington, D.C.: Office of Technology Assessment, U. S. Congress.
9. Baldwin, F. L., and P. W. Santelmann, 1980. Weed science in integrated pest management. *BioScience* 30:675–678.

10. May, R. M. 1985. Evolution of pesticide resistance. *Nature* 315:12–13.

11. Barfield, C. S., and J. L. Stimac, 1980. Pest management: An entomological perspective. *BioScience* 30:683–688.

12. Buschbacher, R. J. 1986. Tropical deforestation and pasture development.

13. Grainger, A. 1982. *Desertification: How people make deserts, how people can stop and why they don't.* Earthscan Books, 2d ed. London: Russell Press, Ltd.

14. United Nations. 1978. *United Nations conference on desertification: Roundup plan of action and resolutions.* New York: United Nations.

15. U.N. Food and Agricultural Organization, 1998. *The United Nations convention to combat desertification: An explanatory leaflet.* Food and Agricultural Organization of the United Nations.

16. U.N. Food and Agricultural Organization, 1998. *What is desertification?* Food and Agricultural Organization of the United Nations.

17. Sheridan, D. 1981. *Desertification of the United States.* Washington, D.C.: Council on Environmental Quality.

Chapter 12 Critical Thinking Issue References

Martin, G. 1992 (June 29). Rice grower proud of his bird habitat. *San Francisco Chronicle*, pp. A1, A6.

Vogel, N. 1992 (December 6). Rice farmers change ways, reap good will. *Sacramento Bee*, pp. A1, A26.

Walker, S. L. 1992 (August 1). Rice growers sow good will. *San Diego Union-Tribune*, pp. A1, A15.

Wood, D. B. 1992 (September 3). California rice land does double duty. *Christian Science Monitor*, p. 10.

World Resources Institute. 1992. *The 1992 information please environmental almanac.* Boston: Houghton Mifflin.

Mechanized irrigation of potato crops in Idaho illustrates the effect of modern agriculture on the environment.

Chapter 13 Notes

1. Stanturf, J. A., E. S. Gardiner, P. B. Hamil, M. S. Devall, T. D. Leininger, and M. E. Warren, Jr. 2000. Restoring bottomland hardwood ecosystems in the lower Mississippi alluvial valley. *J. Forestry* 98 (8):10–16.

2. The information for the case study of Mississippi River Valley bottomland forests is based primarily on: Stanturf, J. A., E. S. Gardiner, P. B. Hamil, M. S. Devall, T. D. Leininger, and M. E. Warren, Jr. 2000. Restoring bottomland hardwood ecosystems in the Lower Mississippi alluvial valley. *J. Forestry* 98 (8):1–16.

3. United Nations Food and Agriculture Organization. 1999. *State of the world's forests 1999*, Annex 3: Data Tables, Table 2: Forest Cover, 1995. Rome: UNFAO.

4. Dombeck, M. 1997. (April). *Towards sustainable forest management.* Speech to USDA Forest Service.

5. Busby, F. E., et. al. 1994. *Rangeland health: New methods to classify inventory and monitor rangelands.* Washington, D.C.: National Academy Press.

6. Kimmins, H. 1995. Proceedings of the conference on certification of sustainable forestry practices. Malaysia (in press).

7. Council on Environmental Quality and U.S. Department of State. 1981. *The global 2000 report to the president: Entering the twenty-first century.* Washington, D.C.: Council on Environmental Quality.

8. Botkin, D. B., and L. Simpson. 1990. The first statistically valid estimate of biomass for a large region. *Biogeochemistry* 9:161–174.

9. Perlin, J. 1989. *A forest journey: The role of wood in the development of civilization.* New York: W. W. Norton.

10. World Resources Institute. 1999. *Deforestation: The global assault continues.* Global trends, resources at risk, world resources 1998–1999. Washington, D.C.: WRI.

11. Bryant, D., D. Nielson, and L. Tangley. 1997. *The last frontier forests: Ecosystems and economies on the edge.* Washington, D.C.: WRI.

12. World Resources Institute. *Disappearing land: Soil degradation.* Washington, D.C.: WRI.

13. World Resources Institute. 1993. *World Resources 1992–93.* New York: Oxford University Press.

14. Manandhar, A. 1997. *Solar cookers as a means for reducing deforestation in Nepal.* Nepal: Center for Rural Technology.

15. United Nations Food and Agriculture Organization. 1999. *State of the world's forests 1999.* Rome: UNFAO.

16. Botkin, D. B., R. A. Nisbet, and T. E. Reynales. 1989. Effects of climate change on forests of the Great Lake states. In J. B. Smith and D. A. Tirpak, eds., *The potential effects of global climate change on the United States*, pp. 2.1–2.31. Washington, D.C.: United States Environmental Protection Agency, EPA-203-05-89-0.

17. Zabinski, C., and M. B. Davis. 1989. Hard times ahead for Great Lakes forests: A climate threshold model predicts responses to CO_2-induced climate change. In J. B. Smith and D. A. Tirpak, eds., *The potential effects of global climate change on the United States*, pp. 5.1–5.10. Washington, D.C.: United States Environmental Protection Agency, EPA-203-05-89-0.

18. Botkin, D. B., D. A. Woodby, and R. A. Nisbet. 1991. Kirtland's warbler habitats: A possible early indicator of climatic warming. *Biological Conservation* 56:63–78.

19. Likens, G. E., F. H. Borman, R. S. Pierce, J. S. Eaton, and N. M. Johnson. 1977. *The biogeochemistry of a forested ecosystem.* New York: Springer-Verlag.

20. Swanson, F. J., and C. T. Dyrness. 1975. Impact of clearcutting and road construction on soil erosion by landslides in the western Cascade Range, Oregon. *Geology* 3:393–396.

21. Fredriksen, R. L. 1971. Comparative chemical water quality—natural and disturbed streams following logging and slash burning. In *Forest land use and stream environment*, pp. 125–137. Corvallis: Oregon State University.

22. Sedjo, R. A. 1983. The comparative economics of plantation forestry: A global assessment. Research paper. Baltimore: Johns Hopkins University Press.

23. Costa Rica's TravelNet. *National parks of Costa Rica.* 1999. Costa Rica: Costa Rica's TravelNet.

24. Kenyaweb. 1998. *National parks and reserves.* Kenya: Kenyaweb.

25. Federation des Parcs naturels regionaux de France. 1999. *List of regional nature parks in France.* Paris: Federation des Parcs naturels regionaux de France.

26. World Conservation Monitoring Centre. 1996. *Protected areas information: 1996 global protected areas summary statistics.* Cambridge, England: WCMC.

27. Botkin, D. B. 1992. Global warming and forests of the Great Lakes states. In J. Schmandt, ed., *The regions and global warming: Impacts and response strategies.* New York: Oxford University Press.

28. Nash, R. 1978. International concepts of wilderness preservation. In J. C. Hendee, G. H. Stankey, and R. C. Lucas, eds., *Wilderness management*, pp. 43–59. United States Forest Service Misc. Pub. No. 1365.

29. The Wilderness Society. 2000. *The future of wilderness: The national wilderness preservation system should triple in size.* Washington, D.C.: The Wilderness Society.

30. Hendee, J. C., G. H. Stankey, and R. C. Lucas. 1978. *Wilderness management.* United States Forest Service Misc. Pub. No. 1365.

31. Runte, A. 1979. *National parks, the American experience.* Lincoln: University of Nebraska Press.

32. Ministry of Interior, Republic of China. 1996. *National parks.* Taiwan: Ministry of Interior, ROC.

33. National Park Service. 2000. *1999 summary of acreage calendar year report.* Washington, D.C.: NPS.

Chapter 13 Critical Thinking Issue References

Bierregaard, R. O., Jr., T. E. Lovejoy, V. Kapos, A. Angusto dos Santos, and R. W. Hutchings. 1992. The biological dynamics of tropical rainforest fragments. *BioScience* 42(11):859–866.

Holloway, M. 1993. Sustaining the Amazon. *Scientific American* 269(1):90–299.

Hosmer, E. 1987 (June). Paradise lost: The ravaged rainforest. *Multinational Monitor*, pp. 6–8, 13.

Skole, D., and C. Tucker. 1993. Tropical deforestation and habitat fragmentation in the Amazon: Satellite data from 1978 to 1988. *Science* 260(5116):1905–1910.

Stevens, W. K. 1993 (June 29). Loss of species is worse than thought in Brazil's Amazon. *New York Times*, p. C4.

Chapter 14 Notes

1. Data on the Hawaiian goose are from the following website: http://www.aloha_hawaii.com/a_nene.shtml.
2. Wilcox, B., ed. 1988. *1988 IUCN red list of threatened animals*. Gland, Switzerland: IUCN.
3. International Union for the Conservation of Nature 1997. *1997 IUCN Red List of Threatened Plants*. Cambridge, England.
4. International Union for the Conservation of Nature 1994. *The World List of Threatened Trees*. Cambridge, England.
5. Abbitt, R. J. F., and J. M. Scott. In Prep. Endangered species recovery management: Are there differences between recovered and declining species?"
6. Friends of the Earth. 1979. *The whaling question: The inquiry by Sir Sidney Frost of Australia*. San Francisco: Friends of the Earth.
7. Bockstoce, J. R., and D. B. Botkin. 1980. *The historical status and reduction of the western Arctic bowhead whale* (Balaena mysticetus) *population by the pelagic whaling industry, 1848–1914*. New Bedford, Conn.: Old Dartmouth Historical Society.
8. United Nations Food and Agriculture Organization. 1978. *Mammals in the seas*. Report of the FAO Advisory Committee on Marine Resources Research, Working Party on Marine Mammals. FAO Fisheries Series 5, vol. 1. Rome: UNFAO.
9. World Wildlife Fund. 2000. *Gray whales*. Washington, D.C.: WWF.
10. Perry, M. 1996. *Climate change biggest risk to whales, says IWC*. Sydney, Australia: Reuter.
11. United Nations Food and Agriculture Organization, 1978.
12. Botkin, D. B., and L. M. Talbot, 1992. Biological diversity and forests. In N. Sharma, ed., *Contemporary issues in forest management: Policy implications*. Washington, D.C.: World Bank.
13. World Resources Institute 1993. A Short List of Plant-Based Medicinal Drugs. Washington, DC. World Resources Institute.
14. Principe, P. P. 1989. The economic significance of plants and their constituents as drugs. In H. Wagner, H. H. Hikino, and N. R. Farnsworth, eds., *Economic and medicinal plant research*. Vol. 3, pp. 1–17. New York: Academic Press.
15. Myers, N. 1983. *A wealth of wild species*. Boulder, Colo: Westview Press.
16. U.S. Congress Office of Technology Assessment. 1987. *Technologies to maintain biological diversity*. Washington, D.C.: U.S. Government Printing Office, OTA-330, p. 45.
17. Lettau, H. L., and L. C. B. Molion, 1979. Amazonia's hydrologic cycle and the role of atmospheric recycling in assessing deforestation effects. *Monthly Weather Review* 107:227–238.
18. Naess A. Ecology, Community, and Lifestyle. Cambridge University Press, Cambridge.
19. Naess does admit to the need to eat. He writes "it is against my intuition of unity to say 'I can kill you because I am more valuable' but not against the intuition to say 'I will kill you because I am hungry.' In the latter case, there would be an implicit regret: 'Sorry, I am now going to kill you because I am hungry.' In short I find obviously right, but often difficult to justify, different sorts of behavior with different sorts of living beings. But this does not imply that we classify some as intrinsically more valuable than others." Naess, p. 168.
20. Ehrenfeld, D. W. 1972. *Conserving life on Earth*. New York: Oxford University Press.
21. Rauber, P. 1996. An end to evolution. *Sierra* 81:28.
22. Wallace, A. R. 1876. *The geographical distribution of animals*. New York: Hafner.
23. Martin, P. S. 1963. *The last 10,000 years*. Tucson: University of Arizona Press.
24. Dennis, B., P. L. Munholland, and J. M. Scott, 1991. Estimation of growth and extinction parameters for endangered species. *Ecol. Monogr.* 61:115–143.
25. Ehrlich, P., and A. Ehrlich, 1981. *Extinction: The causes and consequences of the disappearance of species*. New York: Random House.
26. Cobhentz, B. E. 1990. Exotic organisms: a dilemma for conservation biology. *Conservation Biology* 4:261–265.
27. Sandoz, M. 1954. *The buffalo hunters*. University of Nebraska Press.
28. Haines, F. *The Buffalo*. Thomas Y. Crowell Co.
29. Bell, F. W. 1978. *Food from the sea: The economics and politics of ocean fisheries*. Boulder, Colorado: Westview Press.
30. Word Resources Institute. 1997. *Water and fisheries*. World Resources: A Guide to the Global Environment. Washington, D.C.: WRI.
31. United Nations Food and Agriculture Organization. Website data.
32. Cushing, D. 1975. *Fisheries resources of the sea and their management*. London: Oxford University Press.
33. Rosenthal, D. 1996. *Environmental case studies: Northeastern regions*, p. 32. New York: John Wiley & Sons, Inc.
34. Byelich, J., et al. 1985. *Kirtland's warbler recovery plan*, p. 1. U.S. Dept. of Interior and Fish and Wildlife Service.
35. Heinselman, M. F. 1985. Fire and succession in the conifer forests of northern North America. In West, Shugart, and Botkin, eds., *Forest succession*.
36. Information about the Kirtland's warbler and its habitat is from: Byelich et al., p. 12. Mayfield, H. 1969. *The Kirtland's warbler*. Bloomfield Hills, Mich.: Cranbrook Institute of Science, pp. 24–25.
37. Norman A. Wood, quoted in Mayfield, p. 23.
38. The discussion of the Kirtland's warbler is taken directly from: Botkin, D. B. 1990. *Discordant harmonies: A new ecology for the 21st Century*. N.Y.: Oxford University Press. Permission may be required.
39. Botkin, D. B. 1977. Strategies for the reintroduction of species into damaged ecosystems. In J. Cairns, ed., *Recovery and restoration of damaged ecosystems*, pp. 241–260. Charlottesville: University of Virginia Press.

Chapter 14 References

Hodgson, A. 1997 (July). Wolf restoration in the Adirondacks? Wildlife Conservation Society, Working Paper No. 8.

Hopsack, D. A. 1996. Biological potential for eastern timber wolf reestablishment in the Adirondack Park. In *Wolves of America Conference Proceedings*, November 14–16, 1996, Albany, N.Y. Washington, D.C.: Defenders of Wildlife.

Stevens, W. K. 1997 (March 3). Wolves may reintroduce themselves to East. *New York Times*.

Chapter 15 Notes

1. Needleman, H. L., J. A. Riess, M. J. Tobin, G. E. Biesecker, and J. B. Greenhouse. 1996. Bone lead levels and delinquent behavior. *Journal of American Medical Association* 275:363–369.
2. Centers for Disease Control. 1991. *Preventing lead poisoning in young children*. U.S. Department of Health and Human Services. Atlanta: Public Health Service, Centers for Disease Control.
3. Goyer, R. A. 1991. Toxic effects of metals. In M. O. Amdur, J. Doull, and C. D. Klaassen, eds., *Toxicology*, pp. 623–680. New York: Pergamon Press.
4. Bylinsky, G. 1972. Metallic nemesis. In B. Hafen, ed., *Man, Health and Environment*, pp. 174–185. Minneapolis: Burgess.
5. Hong, S., J. Candelone, C. C. Patterson, and C. F. Boutron. 1994. Greenland ice evidence of hemispheric lead pollution two millennia ago by Greek and Roman civilizations. *Science* 265:1841–1843.
6. Hopps, H. C. 1971. Geographic pathology and the medical implications of environmental geochemistry. In H. L. Cannon and H. C. Hopps, eds., *Environmental geochemistry in health*, pp. 1–11. Geological Society of America Memoir 123. Boulder, Colo.: Geological Society of America.
7. Warren, H. V., and R. E. DeLavault. 1967. A geologist looks at pollution: Mineral variety. *Western Mines* 40:23–32.

8. Evans, W. 1996. Lake Nyos. Knowledge of the fount and the cause of disaster. *Science* 379:21.

9. Gunn, J., ed. 1995. *Restoration and recovery of an industrial region: Progress in restoring the smelter-damaged landscape near Sudbury, Canada.* New York: Springer-Verlag.

10. Blumenthal, D. S., and Ruttenber. 1995. *Introduction to environmental health*, 2nd ed. New York: Springer.

11. U.S. Geological Survey. 1995. *Mercury contamination of aquatic ecosystems.* USGS FS 216-95.

12. Ehrlich, P. R., A. H. Ehrlich, and J. P. Holdren. 1970. *Ecoscience: Population, resources, environment.* San Francisco: W. H. Freeman.

13. Waldbott, G. L. 1978. *Health effects of environmental pollutants.* 2nd ed. Saint Louis: C. V. Mosby.

14. McGinn, A. P. 2000. POPs Culture. *World Watch.* April, pp. 26–36.

15. Carlson, E. A. 1983. International symposium on herbicides in the Vietnam War: An appraisal. *BioScience* 33:507–512.

16. Cleverly, D., J. Schaum, D. Winters, and G. Schweer. 1999. Inventory of sources and releases of dioxin-like compounds in the United States. Paper presented at the 19th International Symposium on Halogentated Environmental Organic Pollutants and POPs, September 12–17, Venice, Italy. Short paper in *Organohalogen Compounds* 41:467–472.

17. Grady, D. 1983 (May). The dioxin dilemma. *Discover*, pp. 78–83.

18. Roberts, L. 1991. Dioxin risks revisited. *Science* 251:624–626.

19. Johnson, J. 1995. SAB advisory panel rejects dioxin risk characterization. *Environmental Science & Technology* 29:302A.

20. Thomas, V. M., and T. G. Spiro. 1996. The U.S. dioxin inventory: Are there missing sources? *Environmental Science & Technology* 30:82A–85A.

21. U. S. Environmental Protection Agency. 1994 (June). Estimating exposure to dioxin-like compounds. Review draft. Office of Research and Development, EPA/600/6-88/005Ca-c.

22. U.S. Environmental Protection Agency. 1994. *Health assessment document for 2,3,7,8-tetrachlorodibenzo-p-dioxin (TCDD) and related compounds.* Vols. I–III. External review draft. Washington, D.C.: EPA.

23. Johnson, J. 1995. Dioxin risk assessment stalls: EPA to create new review panel. *Environmental Science & Technology* 29:492A.

24. Chanlett, E. T. 1979. *Environmental protection.* 2nd ed. New York: McGraw-Hill.

25. Ross, M. 1990. *Hazards associated with asbestos minerals.* In B. R. Doe, ed., *Proceedings of a U.S. Geological Survey Workshop on Environmental Geochemistry*, pp. 175–176. U.S. Geological Survey Circular 1033.

26. Pool, R. 1990. Is there an EMF-cancer connection? *Science* 249:1096–1098.

27. Linet, M. S., E. E. Hatch, R. A. Kleinerman, L. L. Robison, W. T. Kaune, D. R. Friedman, R. K. Severson, C. M. Haines, C. T. Hartsock, S. Niwa, S. Wacholder, and R. E. Tarone. 1997. Residential exposure to magnetic fields and acute lymphoblastic leukemia in children. *New England Journal of Medicine* 337(1):1–7.

28. Kheifets, L.I., E. S. Gilbert, S. S. Sussman, P. Guaenel, S. D. Sahl, D. A. Savitz, and G. Thaeriault, G. 1999. Comparative analyses of the studies of magnetic fields and cancer in electric utility workers: Studies from France, Canada, and the United States. *Occupational and Environmental Medicine* 56(8):567–574.

29. Francis, B. M., 1994. *Toxic substances in the environment.* New York: John Wiley & Sons.

30. Poisons and poisoning. 1997. *Encyclopedia Britannica.* Vol. 25, p. 913. Chicago: Encyclopedia Britannica.

31. Air Risk Information Support Center (Air RISC), U.S. Environmental Protection Agency. 1989. *Glossary of terms related to health exposure and risk assessment.* EPA/450/3-88/016. Research Triangle Park, N.C.

Chapter 15 Critical Thinking Issue References

1992 (December). Fluoride for kids: Too much, too soon?" *American Health* 9(10):15.

Centers for Disease Control. 1991. Public Health Service report on fluoride benefits and risks. *Journal of the American Medical Association* 266(8):1061–1067.

Hamilton, M. 1992. Water fluoridation: A risk assessment perspective. *Journal of Environmental Health* 54(6):27–32.

Marshall, E. 1990. The fluoride debate: One more time. *Science* 247:276–277.

National Research Council. 1993. *Health effects of ingested fluoride.* Washington, D.C.: National Academy Press.

Walker, B. W., Jr. 1992. Perspectives on quantitative risk assessment. *Journal of Environmental Health* 55(1):15–20.

Chapter 16 Notes

1. Butti, K., and J. Perlin. 1980. *A golden thread: 2500 years of solar architecture and technology.* Palo Alto, Calif.: Cheshire Books.

2. Morowitz, H. J. 1979. *Energy flow in biology.* New Haven, Conn.: Oxbow Press.

3. Ehrlich, P. R., A. H. Ehrlich, and J. P. Holdren. 1970. *Ecoscience: Population, resources, environment.* San Francisco: W. H. Freeman.

4. Feynman, R. P., R. B. Leighton, and M. Sands. 1964. *The Feynman lectures on physics.* Reading, Mass.: Addison-Wesley.

5. Cuff, D. J., and W. J. Young. 1986. *The United States energy atlas.* 2nd ed. New York: Macmillan.

6. Darmstadter, J., H. H. Landsberg, H. C. Morton, and M. J. Coda. 1983. *Energy today and tomorrow: Living with uncertainty.* Englewood Cliffs, N.J.: Prentice-Hall.

7. Steinhart, J. S., M. E. Hanson, R. W. Gates, C. C. Dewinkel, K. Briody, M. Thornsjo, and S. Kambala. 1978. A low energy scenario for the United States: 1975–2000. In L. C. Ruedisili and M. W. Firebaugh, eds., *Perspectives on energy*, 2nd ed., pp. 553–588. New York: Oxford University Press.

8. Olkowski, H., B. Olkowski, and T. Javits. (Farallones Institute). 1979. *The integral urban house: Self reliant living in the city.* San Francisco: Sierra Club Books.

9. Flavin, C. 1984. *Electricity's future: The shift to efficiency and small-scale power.* Worldwatch Paper 61. Washington, D.C.: Worldwatch Institute.

10. Consumers' Research. 1995. Fuel economy rating: 1996 mileage estimates. *Consumers' Research* 78:22–26.

11. Berger, J. J. 2000. *Beating the heat.* Berkeley, Calif.: Berkeley Hills Books.

12. Lovins, A. B. 1979. *Soft energy paths: Towards a durable peace.* New York: Harper & Row.

13. Brown, L. R., C. Flavin, and S. Postel (Worldwatch Institute). 1991. *Saving the planet: How to shape an environmentally sustainable global economy.* New York: W. W. Norton.

14. California Energy Commission. 1991. *California's energy plan: Biennial report.* Sacramento, Calif.

15. Flavin, C., and S. Dunn. 1999. Reinventing the energy system. In L. R. Brown et al., eds., *State of the world 1999: A Worldwatch Institute report on progress toward a sustainable society.* New York: W. W. Norton.

16. Dunn, S. 2000. *Micropower, the next electrical era.* Worldwatch Paper 151. Washington, D.C.: Worldwatch Institute.

Chapter 16 Critical Thinking Issue References

Ehrlich, P. R., and A. H. Ehrlich, 1991. *Healing the planet.* Reading, Mass.: Addison-Wesley.

Fickett, A. P. 1990. Efficient use of electricity. *Scientific American* 263(3):65–74.

Holdren, J. P. 1990. Energy in transition. *Scientific American* 263(3):157–163.

Lean, G. 1990. *Atlas of the environment.* New York: Prentice-Hall.

U.S. Census Bureau. 1998. World population and growth rates. http://www.census.gov/ipc/www/world.html.

Chapter 17 Notes

1. Flavin, C., and N. Lenssen. 1991. Designing a sustainable energy system. In L. R. Brown et al., eds., *State of the World, 1991: A*

Worldwatch Institute Report on Progress Toward a Sustainable Society, Chapter 2, pp. 21–38. New York. W. W. Norton and Company.

2. Flavin, C., and F. Dunn, 1999. Reinventing the energy system. In L. R. Brown et al., eds., *State of the World, 1999: A Worldwatch Institute report on progress toward a sustainable society.* New York. W. W. Norton and Company.

3. Van Koevering, T. E., and N. J. Sell. 1986. *Energy: A conceptual approach.* Englewood Cliffs, N.J.: Prentice-Hall.

4. Berger, J. J. 2000. *Beating the heat.* Berkeley, Calif.: Berkeley Hills Books.

5. McCulloh, T. H. 1973. In D. A. Brobst and W. P. Pratt, eds., *Oil and gas in United States mineral resources,* pp. 477–496. U.S. Geological Survey Professional Paper 820.

6. British Petroleum Company. 1999. *B.P. statistical review of world energy.* London: British Petroleum Company.

7. Kerr, R. A. 2000. USGS optimistic on world oil prospects. *Science* 289:237.

8. Youngquist, W. 1998. Spending our great inheritance. Then what? *Geotimes* 43(7):24–27.

9. Edwards, J. D. 1997. Crude oil and alternative energy production forecast for the twenty-first century: The end of the hydrocarbon era. *American Association of Petroleum Geologists Bulletin* 81(8):1292–1305.

10. Darmstadter, J., H. H. Landsberg, H. C. Morton, and M. J. Coda. 1983. *Energy today and tomorrow: Living with uncertainty.* Englewood Cliffs, N.J.: Prentice-Hall.

11. Suess, E., G. Bohrmann, J. Greinert, and E. Lauch. 1999. Flammable ice. *Scientific American* 28(5):76–83.

12. Rahn, P. H. 1996. *Engineering geology: An environmental approach,* 2nd ed. New York: Elsevier.

13. U.S. Environmental Protection Agency. 1973. *Processes, procedures and methods to control pollution from mining activities.* EPA-430/9-73-001. Washington, D.C.: U.S. Environmental Protection Agency.

14. Council on Environmental Quality. 1978. *Progress in environmental quality.* Washington, D.C.: Council on Environmental Quality.

15. Miller, E. W. 1993. *Energy and American society, a reference handbook.* Santa Barbara, Calif.: ABC-CLIO.

16. Corcoran, E. 1991. Cleaning up coal. *Scientific American* 264:106–116.

17. Energy Information Administration. 1995 (February). *Coal data: A reference.* Washington, D.C.: U.S. Department of Energy.

18. Committee on Environmental and Public Planning. 1974. Environmental impact on conversion from gas or oil to coal for fuel. *The Geologist,* supplement to vol. 9, no. 4.

19. Knapp, D. H. 1995. Non-OPEC oil supply continues to grow. *Oil & Gas Journal* 93:35–45. Paris: International Energy Agency.

Chapter 17 Critical Thinking Issue References

Bleviss, D. L. 1988. *The new oil crisis and fuel economy technologies.* New York: Quorum Books.

Bleviss, D. L., and P. Walzer. 1990. Energy for motor vehicles. *Scientific American* 263(3):103–109.

Corson, W. H., ed. 1990. *The global ecology handbook.* Boston: Beacon. 1993 (January 18). Driving down the deficit. *U.S. News and World Report* 114(2):58–60.

Energy Information Administration. 1996 (January). *Monthly energy review.* Washington, D.C.: U.S. Department of Energy.

Greenwald, J. 1993. Why not a gas tax? *Time* 141(7):25–27.

Miller, E. W. and R. M. Miller. 1993. *Energy and American society. A reference handbook.* Santa Barbara, Calif.: ABC-CLIO.

Nadis, S., and J. J. MacKenzie. 1993. *Car trouble.* Boston: Beacon.

Chapter 18 Notes

1. Jackson, T., and R. Lofstedt. 1998. Royal commission on environmental pollution. Study on energy and the environment. Accessed November 29, 2000, at http://www.rcep.org.uk/studies/energy/98-6061/jackson.html.

2. Flavin, C. 1999 (March–April). The bull market in energy. *Worldwatch.* Pp. 24–27.

3. Berger, J. J. 2000. *Beating the heat.* Berkeley, Calif.: Berkeley Hills Books.

4. Miller, E. W. 1993. *Energy and American society: A reference handbook.* Santa Barbara, Calif.: ABC-CLIO.

5. Flavin, C., and S. Dunn. 1999. Reinventing the energy system. In L. R. Browne et al., *State of the world 1999: A Worldwatch Institute report on progress toward a sustainable society.* New York: W. W. Norton.

6. Eaton, W. W. 1978. Solar energy. In L. C. Ruedisili and M. W. Firebaugh, eds., *Perspectives on energy,* 2d ed., pp. 418–436. New York: Oxford University Press.

7. Mayur, R., and B. Daviss. 1998 (October). The technology of hope. *The Futurist.* Pp. 46–51.

8. Brown, L. R. 1999 (March–April). Crossing the threshold. *Worldwatch.* Pp. 12–22.

9. Quinn, R. 1997 (March). Sunlight brightens our energy future. *The World and I.* Pp. 156–163.

10. Johnson, J. T. 1990 (May). The hot path to solar electricity. *Popular Science.* Pp. 82–85.

11. Becker, N. D. 1992 (January–February). The demise of Luz: A case study. *Solar Today.* Pp. 24–26.

12. Demeo, E. M., and P. Steitz. 1990. The U.S. electric utility industry's activities in solar and wind energy. In K. W. Böer, ed., *Advances in solar energy,* Vol. 6, pp. 1–218. New York: American Solar Energy Society.

13. Schatz Solar Hydrogen Project. N.D. Pamphlet. Arcata, Calif.: Humboldt State University.

14. Kartha, S., and P. Grimes. 1994. Fuel cells: energy conversion for the next century. *Physics Today* 47:54–61.

15. Haggin, J. 1995. Fuel-cell development reaches demonstration stage. *Chemical & Engineering News* 73:28–30.

16. Alward, R., S. Eisenbart, and J. Volkman. 1979. *Micro-hydro power: Reviewing an old concept.* Butte, Mont.: National Center for Appropriate Technology, U.S. Department of Energy.

17. Nova Scotia Department of Mines and Energy. 1981. *Wind power.*

18. U.S. Congress, Office of Technology Assessment. 1993. *Potential environmental impacts of bioenergy crop production.* Background paper. Washington, D.C.: U.S. Government Printing Office.

19. Council on Environmental Quality. 1979. *Environmental quality.*

20. Wihersaari, M. 1996. Energy consumption and greenhouse gas emissions from biomass production chains. *Energy Conversion and Management* 37:1217.

21. Sterzinger, G. 1995. Making biomass energy a contender. *Technology Review* 98:34–40.

22. Wright, P. 2000. Geothermal energy. *Geotimes* 45(7):16–18.

23. Duffield, W. A., J. H. Sass, and M. L. Sorey. 1994. *Tapping the Earth's natural heat.* U.S. Geological Survey Circular 1125.

Chapter 18 Critical Thinking Issue References

Brown, L. R., C. Flavin, and S. Postel. 1991. *Saving the planet.* New York: W. W. Norton.

Brower, M. 1990. *Cool energy.* Washington, D.C.: Union of Concerned Scientists.

Flavin, C. 1990. Slowing global warming. In L. R. Brown et al., eds., *State of the world.* New York: W. W. Norton.

Reddy, A. K. N., and J. Goldemberg. 1990. Energy for the developing world. *Scientific American* 263(3):110–118.

Chapter 19 Notes

1. Rosa, E. A. and R. E. Dunlap, 1994. Nuclear power: three decades of public opinion. *Public Opinion Quarterly* 58(2):295–324.

2. Worldwatch Institute. 1994. Nuclear power levels peak. Worldwatch News Brief, March 5, 1994. Accessed February 9, 2001, at www.worldwatch.org/alerts/990304-html.V.16no6.

3. Till, C. E. 1989. Advanced reactor development. *Ann. Nucl. Energy* 16(6):301–305.

4. Churchill, A. A. 1993 (July). Review of WEC Commission: Energy for tomorrow's world. *World Energy Council Journal*. Pp. 19–22.

5. Ehrlich, P. R., A. H. Ehrlich, and J. P. Holdren. 1970. *Ecoscience: Population, resources, environment.* San Francisco: W. H. Freeman.

6. Brenner, D. J. 1989. *Radon: Risk and remedy.* New York: W. H. Freeman.

7. Duderstadt, J. J. 1978. Nuclear power generation. In L. C. Ruedisili and M. W. Firebaugh, eds., *Perspectives on energy*, 2d ed., pp. 249–273. New York: Oxford University Press.

8. Cohen, B. L. 1990. *The nuclear energy option: An alternative for the 90s.* New York: Plenum.

9. U.S. Department of Energy. 1980. *Magnetic fusion energy.* DOE/ER-0059. Washington, D.C.: U.S. Department of Energy.

10. U.S. Department of Energy. 1979. *Environmental development plan, magnetic fusion.* DOE/EDP-0052. Washington, D.C.: U.S. Department of Energy.

11. Cordey, J. G., R. J. Goldston, and R. R. Parker. 1992. Progress toward a Tokamak fusion reactor. *Physics Today* 45:22–30.

12. Greenberg, P. A. 1993. Dreams die hard. *Sierra* 78:78.

13. Van Koevering, T. E., and N. J. Sell. 1986. *Energy: A conceptual approach.* Englewood Cliffs, N.J.: Prentice-Hall.

14. Waldbott, G. L. 1978. *Health effects of environmental pollutants.* 2nd ed. Saint Louis: C. V. Mosby.

15. *New Encyclopedia Britannica.* 1997. S.v. "radiation.".

16. U.S. Department of Energy. 1999. Radiation (in) Waste Isolation Pilot Plant. 1999. Carlsbad, New Mexico. Accessed at www.wipp.carlsbad.nm.us.

17. Hanson, W. G. 1967. Cesium-137 in Alaskan lichens, caribou, and Eskimos. *Health Physics* 13:383–389.

18. University of Maine and Maine Department of Human Services. 1983 (February). Radon in water and air. *Resource Highlights.*

19. MacLeod, G. K. 1981. Some public health lessons from Three Mile Island: A case study in chaos. *Ambio* 10:18–23.

20. Anspaugh, L. R., R. J. Catlin, and M. Goldman. 1988. The global impact of the Chernobyl reactor accident. *Science* 242:1513–1518.

21. Nuclear News. 1996. Chernobyl: WHO conference confirms excess of thyroid cancers. *Nuclear News* 39:39.

22. Balter, M. 1995. Chernobyl's thyroid cancer toll. *Science* 270:1758.

23. Skuterud, L., N. I. Goltsova, R. Naeumann, T. Sikkeland, and T. Lindmo. 1994. Histological changes in *Pinus sylvestris* L. in the proximal-zone around the Chernobyl power plant. *The Science of the Total Environment* 157:387–397.

24. Williams, N. 1995. Chernobyl: Life abounds without people. *Science* 269:304.

25. Fletcher, M. 2000 (November 14). The last days of Chernobyl. *Times 2* (London), pp. 3–5.

26. Office of Industry Relations. 1974. *Development, growth and state of the nuclear industry.* Washington, D.C.: U.S. Congress, Joint Committee on Atomic Energy.

27. Weisman, J. 1996. Study inflames Ward Valley controversy. *Science* 271:1488.

28. Weart, W. D., M. T. Rempe, and D. W. Powers. 1998 (October). The waste isolation plant. *Geotimes.*

29. U.S. Department of Energy. 1999. Waste Isolation Pilot Plant, Carlsbad, New Mexico. Accessed at www.wipp.carlsbad.nm.us.

30. Roush, W. 1995. Can nuclear waste keep Yucca Mountain dry—and safe? *Science* 270:1761.

31. Hanks, T. C., I. J. Winograd, R. E. Anderson, T. E. Reilly, and E. P. Weeks. 1999. Yucca Mountain as a radioactive-waste repository. U.S. Geological Survey Circular 1184.

32. Bredehoeft, J. D., A. W. England, D. B. Stewart, J. J. Trask, and I. J. Winograd. 1978. *Geologic disposal of high-level radioactive wastes—Earth science perspectives.* U.S. Geological Survey Circular 779. Arlington, Va.: U.S. Department of the Interior.

33. Nuclear Regulatory Commission. 2000. NRC's high level waste program. Accessed July 18, 2000, at http://www.nrc.gov/NMSS/DWM/hlw.htm.

34. Flavin, C. 1991. The case against reviving nuclear power. In L. R. Brown, ed., *The Worldwatch Reader*, pp. 205–220. New York: W. W. Norton.

Chapter 19 Critical Thinking Issue References

Ahearne, J. F. 1993. The future of nuclear power. *American Scientist* 81(1):24–35.

Fox, M. R. 1987. Perspectives in risk: Compared to what? *Vital Speeches of the Day* 53(23):730–732.

Greenberg, P. A. 1993. Dreams die hard. *Sierra* 78(6):78.

Lanouette, W. 1990 (April). Greenhouse scare reheats nuclear debate. *Bulletin of the Atomic Scientists* 46(3):34–37.

Miller, A., and I. Mintzer. 1990. Global warming: No nuclear fix. *Bulletin of the Atomic Scientists* 46(5):31–33.

United States Nuclear Regulatory Commission, 1997. *NRC Information Digest*, NUREG-1350, Vol. 9.

Chapter 20 Notes

1. State of Maine. 2001. A brief history of the Edwards Dam. Accessed 1/15/00 at http://janus.state.me.us/spo/edwards/timeline.htm.

2. Henderson, L. J. 1913. *The fitness of the environment: An inquiry into the biological significance of the properties of matter.* New York: Macmillan.

3. Council on Environmental Quality and U.S. Department of State. 1980. *The global 2000 report to the president: Entering the twenty-first century.* Vol. 2. Washington, D.C.

4. Water Resources Council. 1978. *The nation's water resources. 1975–2000.* Vol. 1. Washington, D.C.

5. Winter, T. C., J. W. Harvey, O. L. Franke, and W. M. Alley. 1998. *Groundwater and surface water: A single resource.* U.S. Geological Survey Circular 1139.

6. Solley, W. B., R. R. Pierce, and H. A. Perlman. 1993. *Estimated use of water in the United States in 1990.* U.S. Geological Survey Circular 1081.

7. Solley, W. B., R. R. Pierce, and H. A. Perlman. 1998. *Estimated use of water in the United States in 1995.* U.S. Geological Survey Circular 1200.

8. Alexander, G. 1984. (February/March) Making do with less. *National Wildlife*, special report, pp. 11–13.

9. Gleick, P. H., P. Loh, S. V. Gomez,, and J. Morrison. 1995. *California water 2020, a sustainable vision.* Oakland, Calif.: Pacific Institute for Studies in Development, Environment and Security.

10. Alley, W. M., T. E. Reilly, and O. L. Franke. 1999. Sustainability of ground-water resources. U.S. Geological Survey Circular 1186.

11. Leopold, L. B. 1977. A reverence for rivers. *Geology* 5:429–430.

12. Holloway, M. 1991. High and dry. *Scientific American* 265:16–20.

13. Levinson, M. 1984 (February/March). Nurseries of life. *National Wildlife*, special report, pp. 18–21.

14. Nichols, F. H., J. E. Cloern, S. N. Luoma, and D. H. Peterson. 1986. The modification of an estuary. *Science* 231:567–573.

15. Day, J. W., Jr., J. M. Rybczyk, L. Carboch, W H. Conner, P. Delgado-Sanchez, R. I. Pratt, and A. Westphal. 1998. A review of recent studies of the ecology and economic aspects of the application of secondary treated municipal effluent to wetlands in southern Louisiana. From L. P. Rozas et al., eds., Symposium on Recent Research in Coastal Louisiana, February 3–5, 1998, Lousiana Sea Grant College Program, pp. 1–12.

16. Hileman, B. 1995. Rewrite of Clean Water Act draws praise, fire. *Chemical & Engineering News* 73:8.

17. Gurardo, D., M. L. Fink, T. D. Fontaine, S. Newman, M. Chinmey, R. Bearzotti, and G. Goforth. 1995. Large-scale constructed wetlands for nutrient removal from stormwater runoff: An Everglades restoration project. *Environmental Management* 19:879–889.

18. Pearce, M. 1995 (January). The biggest dam in the world. *New Scientist*, pp. 25–29.

19. Zich, R. 1997. China's three gorges: Before the flood. *National Geographic* 192(3):2–33.

20. Booth, W. 2000 (December 12). Restoring rivers—at a high price. *Washington Post*, p. A3.

21. U.S. Congress. 1973. *Stream channelization: What federally financed draglines and bulldozers do to our nation's streams.* House Committee Report No. 93-530. Washington, D.C.: U.S. Government Printing Office.

22. Graf, W. L. 1985. *The Colorado River: Instability and basin management.* Resource Publications in Geography. Washington, D.C.: Association of American Geographers.

23. Nash, R. 1986. Wilderness values and the Colorado River. In G. D. Weatherford and F. L. Brown, eds., *New courses for the Colorado River*, pp. 201–214. Albuquerque: University of New Mexico Press.

24. Hundley, N., Jr. 1986. The West against itself: The Colorado River—an institutional history. In G. D. Weatherford and F. L. Brown, eds., *New courses for the Colorado River*, pp. 9–49. Albuquerque: University of New Mexico Press.

25. Dolan, R., A. Howard, and A. Gallenson. 1974. Man's impact on the Colorado River and the Grand Canyon. *American Scientist*, 62:392–401.

26. Hecht, J. 1996. Grand Canyon flood a roaring success. *New Scientist*, 151 (2045):8.

27. Lucchitta, I., and L. B. Leopold. 1999. Floods and sand bars in the Grand Canyon. *Geology Today* 9:1–7.

28. Ballard, S. C., D. D. Michael, M. A. Chartook, M. R. Clines, C. E. Dunn, C. M. Hock, G. D. Miller, L. B. Parker, D. A. Penn, and G. W. Tauxe. 1982. *Water and western energy: Impacts, issues, and choices.* Boulder, Colo.: Westview Press.

Chapter 20 Critical Thinking Issues References

Baldwin, M. F. 1987. Wetlands: Fortifying federal and regional cooperation. *Environment* 29(7):16–20, 39.

Leidy, R. A., P. L. Fiedler, and E. R. Micheli. 1992. Is wetter better? *Bioscience* 40(9):58–61, 65.

National Institute for Urban Wildlife. *Wetlands conservation and use* (Issue Pak). Columbia, Md.: National Institute for Urban Wildlife.

Stevens, W. K. 1990 (March 13). Efforts to halt wetland loss are shifting to inland areas. *New York Times*, p. C1.

World Resources Institute. 1992. *The 1992 information please environmental almanac.* Boston: Houghton Mifflin.

Chapter 21 Notes

1. Mallin, M. A. 2000. Impacts of industrial animal production on rivers and estuaries. *American Scientist* 88(1): 26–37.

2. Bowie, P. 2000. No act of God. *The Amichs Journal* 21(4):16–21.

3. Groundwater: Issues and answers 1985. Arvada, Colo.: American Institute of Professional Geologists.

4. Gleick, P. H. 1993. An introduction to global fresh water issues. In P. H. Gleick, ed., *Water in crisis*, pp. 3–12. New York. Oxford University Press.

5. Hileman, B. 1995. Pollution tracked in surface- and groundwater. *Chemical & Engineering News* 73:5.

6. Lewis, S. A. 1995. Trouble on tap. *Sierra* 80:54–58.

7. Smith, R. A. 1994. Water quality and health. *Geotimes* 39:19–21.

8. MacKenzie, W. R., et al. 1994. A massive outbreak in Milwaukee of Cryptosporidium infection transmitted through the public water supply. *The New England Journal of Medicine* 331:161–167.

9. Centers for Disease Control and Environmental Protection Agency. 1995. Assessing the public health threat associated with waterborne Cryptosporidiosis: Report of a workshop. *Journal of Environmental Health* 58:31.

10. Kluger, J. 1998. Anatomy of an outbreak. *Time* 152(5):56–62.

11. Mitch, W. J., J. W. Day, Jr., J. W. Gilliam, P. M. Groffman, D. L. Hey, G. W. Randall, and N. Wang. 2001. The Gulf of Mexico hypoxia—approaches to reducing nitrate in the Mississippi River or reducing a persistent large-scale ecological problem. *BioScience* (in press).

12. Maugh, T. H. 1979. Restoring damaged lakes. *Science* 203:425–427.

13. Hinga, K. R. 1989. Alteration of phosphorus dynamics during experimental eutrophication of enclosed marine ecosystems. *Marine Pollution Bulletin* 20:624–628.

14. Richmond, R. H. 1993. Coral reefs: Present problems and future concerns resulting from anthopogenic disturbance. *American Zoologist* 33:524–536.

15. Bell, P. R. 1991. Status of eutrophication in the Great Barrier Reef Lagoon. *Marine Pollution Bulletin* 23:89–93.

16. Hunter, C. L., and C. W. Evans. 1995. Coral reefs in Kaneohe Bay, Hawaii: Two centuries of western influence and two decades of data. *Bulletin of Marine Science* 57:499.

17. Alaska Fish and Game. 1989. *Alaska Fish and Game*, Special Issue, Vol. 21, no. 4.

18. Holway, M. 1991. Soiled shores. *Scientific American* 265:102–106

19. Robinson, A. R. 1973. Sediment, our greatest pollutant? In R. W. Tank, ed., *Focus on environmental geology*, pp. 186–192. New York: Oxford University Press.

20. Yorke, T. H. 1975. Effects of sediment control on sediment transport in the northwest branch, Anacostia River basin, Montgomery County, Maryland. *Journal of Research* 3:487–494.

21. Poole, W. 1996. Rivers run through them. *Land and People* 8:16–21.

22. Carey, J. 1984. (February/March). Is it safe to drink? *National Wildlife*, Special Report, pp. 19–21.

23. Pye, U. I., and R. Patrick. 1983. Groundwater contamination in the United States. *Science* 221:713–718.

24. Foxworthy, G. L. 1978. Nassau County, Long Island, New York—water problems in humid county. In G. D. Robinson and A. M. Spieker, eds., *Nature to be commanded*, pp. 55–68. U.S. Geological Survey Professional Paper 950. Washington, D.C.: U.S. Government Printing Office.

25. Van der Leeden, F., F. L. Troise, and D. K. Todd. 1990. *The water encyclopedia*, 2nd ed. Chelsea, Mich.: Lewis Publishers.

26. Jobling, S., M. Nolan, C. R. Tyler, G. Brighty, and J. P. Sumpter. 1998. Widespread sexual disturbance in wild fish. *Environmental Science and Technology* 32(17):2498–2506.

27. Environmental Protection Agency, Drinking Water Committee of the Science Advisory Board. 1995. *An SAB report: Safe drinking water. Future trends and challenges.* Washington, D.C.: Environmental Protection Agency.

28. Jewell, W. J. 1994. Resource-recovery wastewater treatment. *American Scientist* 82:366–375.

29. Task Force on Water Reuse. 1989. *Water reuse: Manual of practice SM-3.* Alexandria, Va.: Water Pollution Control Federation.

30. Kadlec, R. H. and R. L. Knight. 1996. *Treatment wetlands.* New York: Lewis Publishers.

31. Breaux, A. M., and J. W. Day, Jr. 1994. Policy considerations for wetland wastewater treatment in the Coastal Zone: A case study for Louisiana. *Coastal Management* (22):285–307.

32. Day, J. W., Jr., J. M. Rybczyk, L. Carboch, W. H. Conner, P. Delgado-Sanchez, R. I. Pratt, and W. Westphal. 1998. A review of recent studies of the ecology and economic aspects of the application of secondary treated municipal effluent to wetlands in Southern Louisiana. From L. P. Rozas, et al., eds., Symposium on Recent Research in Coastal Louisiana, February 1998, Louisiana Sea Grant College Program, pp. 1–12.

33. Breaux, A., S. Fuber, and J. Day, 1995. Using natural coastal wetland systems: An economic benefit analysis. *Journal of Environmental Management* (44):285–291.

34. Hileman, B. 1995. Rewrite of Clean Water Act draws praise, fire. *Chemical & Engineering News* 73:8.

Chapter 21 Critical Thinking Issue References

Allan, J. D., and A. S. Flecker. 1993. Biodiversity conservation in running waters. *BioScience* 43(1):32–43.

Armour, C. L., D. A. Duff, and W. Elmore. 1991. The effects of livestock grazing on riparian and stream ecosystems. *Fisheries* 16(1):7–11.

Karr, J. R., L. A. Toth, and D. R. Dudley. 1985. Fish communities of midwestern rivers: A history of degradation. *BioScience* 35(2):90–95.

Sparks, R. 1992. The Illinois River floodplain ecosystem. In National Research Council, *Restoration of Aquatic Ecosystems*, pp. 412–432. Washington, D.C.: National Academy Press.

Stevens, W. K. 1993 (January 26). River life through U.S. broadly degraded. *New York Times*. Pp. B5, B8.

Chapter 22 Notes

1. Watson, R. T. 2000. Presentation at the Sixth Conference of the Parties of the United Nations Framework Convention on Climate Change, Intergovernment Panel on Climate Change, November 13, 2000. Accessed December 1, 2000, at www.IPCC.ch/press/sp-COPG.htm.

2. Levy, S. 2000. Wildlife on the hot seat. *National Wildlife* 38(5):20–27.

3. Hartmann, D. L. 1994. *Global physical climatology.* International Geophysics Series, vol. 56. New York: Academic Press.

4. Anthes, R. A. 1992. *Meteorology.* 6th ed. New York: Macmillan.

5. Marsh, W. M., and J. Dozier. 1981. *Landscape.* New York: John Wiley & Sons.

6. Detwyler, T. R., and M. G. Marcus, eds. 1972. *Urbanization and the environment: The physical geography of the city.* Belmont, Calif.: Duxbury Press.

7. Steig, E. J., E. J. Brook, J. W. C. White, C. M. Sucher, M. L. Bender, S. J. Lehman, D. L. Morse, E. D. Waddington, and G. D. Clow. 1998. Synchronous climate changes in Antarctica and the North Atlantic. *Science* 282:92–94.

8. Union of Concerned Scientists. 1989. *The greenhouse effect.* Cambridge, Mass.

9. Kerr, R. A. 1996. 1995 the warmest year? Yes and no. *Science* 271:137–138.

10. Crowley, T. J. 2000. Causes of climate change over the past 1000 years. *Science* 289:270–277.

11. Campbell, I. D., C. Campbell, N. J. Apps, N. W. Rutter, and A. B. G. Bush. 1998. Late Holocene approximately 1500 year climatic periodicities and their implications. *Geology*, 26(5):471–473.

12. Broecker, W. 1997. Will our ride into the greenhouse future be a smooth one? *GSA Today*, 7(5):2–6.

13. Earth System Sciences Committee. 1988. *Earth system science: A preview.* Boulder, Colo.: University Corporation for Atmospheric Research.

14. Post, W. M., T. Peng, W. R. Emanuel, A. W. King, V. H. Dale, and D. L. De Angelis. 1990. The global carbon cycle. *American Scientist* 78:310–326.

15. Moss, M. E., and H. F. Lins. 1989. *Water resources in the twenty-first century: A study of the implications of climate uncertainty.* U.S. Geological Survey Circular 1030. Washington, D.C.: U.S. Department of the Interior.

16. Hansen, J., A. Lacis, and M. Prather. 1989. Greenhouse effect of chlorofluorocarbons and other trace gases. *Journal of Geophysical Research* 94(D13):16417–16421.

17. Titus, J. G., S. P. Leatherman, C. H. Everts, Moffatt and Nichol Engineers, D. L. Kriebel, and R. G. Dean. 1985. *Potential impacts of sea level rise on the beach at Ocean City, Maryland.* Washington, D.C.: U.S. Environmental Protection Agency, Office of Policy Planning and Evaluation.

18. Rodhe, H. 1990. A comparison of the contribution of various gases to the greenhouse effect. *Science* 248:1217–1219.

19. Kerr, R. A. 1994. Methane increases put on pause. *Science* 263:751.

20. Dlugokencky, E. J., L. P. Steele, P. M. Lang, and K. A. Masarie. 1994. The growth rate and distribution of atmospheric methane. *Journal of Geophysical Research* 99(D8):17021–17043.

21. Council on Environmental Quality. 1990. *Environmental trends 1989.* Washington, D.C.

22. Luoma, J. R., and D. Hiser. 1991. Gazing into our greenhouse future. *Audubon* 93(2).

23. Charlson, R. J., S. E. Schwartz, J. M. Hales, R. D. Cess, J. A. J. Coakley, J. E. Hansen, and D. J. Hofmann. 1992. Climate forcing by anthropogenic aerosols. *Science* 255:423–430.

24. Kerr, R. A. 1995. Study unveils climate cooling caused by pollutant haze. *Science* 268:802.

25. NOAA. 1998. *What is an El Niño?* Accessed October 2, 1998 at http://www.elnino.noaa.gov.

26. Kennett, J. P., K. G. Cannariato, I. L. Hendy, and R. J. Behl. 2000. Carbon isotopic evidence for methane hydrate instability during Quaternary interstadials. *Science* 288:128–133.

27. Mohnen, V. A., W. Goldstein, and W. Wang. 1991. The conflict over global warming. *Global Environmental Change* 1:109–123.

28. Kerr, R. A. 1995. U.S. climate tilts toward the greenhouse. *Science* 268:363–364.

29. Kerr, R. A. 1995. Greenhouse report foresees growing global stress. *Science* 270:731.

30. Titus, J. G., and V. K. Narayanan. 1995. *The probability of sea level rise.* Washington, D.C.: U.S. Environmental Protection Agency.

31. Holmes, N. 2000. Has anyone checked the weather (map)? *Amicus Journal* 21(4):50–51.

32. Epstein, P. R. 2000. Is global warming harmful to health? *Scientific American* 283(2):50–57.

33. Dunn, S. 2001. Decarbonizing the energy economy. In *WorldWatch Institute state of the world 2001.* New York: W. W. Norton. Pp. 83–102.

34. Rice, C. W. 2002. Storing carbon in soil: why and how. Geotimes 47 (1): 14–17.

A Closer Look 22.1 References

Edahl, C. A., and C. D. Keeling. 1973. *Carbon and the biosphere.* Oak Ridge, Tenn.: Technical Information Service.

Keeling, C. D., T. P. Worf, and J. Van der Plicht. 1995. *Nature* 375:666–670.

Chapter 22 Critical Thinking References

Botkin, D. B. 1989. Can we plant enough trees to absorb all the greenhouse gases? Paper delivered at the University of California Workshop on Energy Policies to Address Global Warming, September 6–8, 1989, Davis, Calif.

Rosenfeld, A. H., and D. B. Botkin. 1989 (October 27). Trees can sequester carbon, or die, decay, and amplify the threat of global warming: A primer on forestation, deforestation, and possible feedback between rising temperature, stressed forests, and CO_2. *Physics and Society* 19(4).

Stevens, W. K. 1989 (July 18). To halt climate change, scientists try trees. *New York Times*, pp. C1, C4.

Chapter 23 Notes

1. Simons, L. M. 1998. Plague of fire. *National Geographic* 194(2):100–119.

2. National Park Service. 1984. *Air resources management manual.*

3. American Lung Association. 1998. *American Lung Association Outdoor Fact Sheet.* Accessed September 18, 1998 at http://www.lungusa.org.

4. Godish, T. 1991. *Air quality*, 2nd ed. Chelsea, Mich.: Lewis Publishers.

5. Seitz, F., and C. Plepys. 1995. Monitoring air quality in healthy people 2000. *Healthy people 2000: Statistical notes no. 9.* Atlanta: Centers for Disease Control and Prevention, National Center for Health Statistics.

6. American Lung Association, 2001. *State of the Air 2000.*

7. Moore, C. 1995. Poisons in the air. *International Wildlife* 25:38–45.

8. Pope, C. A. III, D. V. Bates, and M. E. Raizenne. 1995. Health effects of particulate air pollution: Time for reassessment? *Environmental Health Perspectives* 103:472–480.

9. Office of Technology Assessment. 1984. Balancing the risks. *Weatherwise* 37:241–249.

10. Canadian Department of the Environment. 1984. *The acid rain story*. Ottawa: Minister of Supply and Services.

11. Lippmann, M., and R. B. Schlesinger. 1979. *Chemical contamination in the human environment*. New York: Oxford University Press.

12. Winkler, E. M. 1998 (September). The complexity of urban stone decay. *Geotimes*. Pp. 25–29.

13. How many more lakes have to die? 1981. *Canada Today* 12, no. 2.

14. Pittock, A. B., L. A. Frakes, D. Jenssen, J. A. Peterson, and J. W. Zillman. eds. 1978. *Climatic change and variability: A southern perspective*. New York: Cambridge University Press.

15. Brown, L. R., ed. 1991. *The Worldwatch reader on global environmental issues*. New York, W. W. Norton.

16. Lents, J. M., and W. J. Kelly. 1993. Clearing the air in Los Angeles. *Scientific American* 269:32–39.

17. Blake, D. R., and F. S. Rowland. 1995. Urban leakage of liquefied petroleum gas and its impact on Mexico City air quality. *Science* 269:953.

18. Gates, D. M. 1972. *Man and his environment: Climate*. New York: Harper & Row.

19. Tyson, P. 1990. Hazing the Arctic. *Earthwatch* 10:23–29.

20. Pountain, D. 1993 (May). Complexity on wheels. *Byte*. Pp. 213–220.

21. Stern, A. C., R. T. Boubel, D. B. Turner, and D. L. Fox. 1984. *Fundamentals of air pollution*, 2nd ed. Orlando, Fla.: Academic Press.

22. Moore, C. 1995. Green revolution in the making. *Sierra* 80:50.

23. Kolstad, C. D. 2000. *Environmental economics*. New York: Oxford University Press.

24. Molnia, B. F. 1991. Washington report. *GSA Today* 1:33.

25. Crandall, R. W. 1983. *Controlling industrial pollution: The economics and politics of clean air*. Washington, D.C.: Brookings Institution.

26. Hall, J. V., A. M. Winer, M. T. Kleinman, F. W. Lurmann, V. Brajer, and S. D. Colome. 1992. Valuing the health benefits of clean air. *Science* 255:812–816.

27. Krupnick, A. J., and P. R. Portney. 1991. Controlling urban air pollution: A benefits-cost assessment. *Science* 252:522–528.

28. Lipfert, F. W., S. C. Morris, R. M. Friedman, and J. M. Lents. 1991. Air pollution benefit-cost assessment. *Science* 253:606.

Chapter 23 Critical Thinking Issue References

Khalil, M., and R. A. Rasmussen 1993. Arctic haze—patterns and relationships to regional signatures of trace gases. *Global Biogeochemical Cycles* 7(1):27–36.

Rahn, K. A. 1984. Who's polluting the Arctic? *Natural History* 93(5):31–38.

Shaw, G. E. 1995. The Arctic haze phenomenon. *Bulletin of the American Meteorological Society* 76(12):2403–2413.

Soroos, M. S. 1992. The odyssey of Arctic haze. *Environment* 34(10):6–27.

Young, O. R. 1990. Global commons: The Arctic in world affairs. *Technology Review* 93:52–61.

Chapter 24 Notes

1. Massachusetts Department of Public Health, Bureau of Environmental Health Assessments. 1995. *Symptom prevalence survey related to indoor air concerns at the Registry of Motor Vehicles building, Ruggles Station.*

2. Horton, W. G. B. 1995. *NOVA: Can buildings make you sick?* Video production. Boston: WGBH.

3. Zimmerman, M. R. 1985. Pathology in Alaskan mummies. *American Scientist* 73:20–25.

4. Ehrlich, P. R., A. H. Ehrlich, and J. P. Holdren. 1970. *Ecoscience: Population, resources, environment*. San Francisco: W. H. Freeman.

5. Conlin, M. 2000 (June 5). Is your office killing you? *Business Week*, pp. 114–124.

6. U.S. Environmental Protection Agency. 1991. *Building air quality: A guide for building owners and facility managers*. EPA/400/1-91/033, DHHS (NIOSH) Pub. No. 91–114. Washington, D.C.:Environmental Protection Agency.

7. Zummo, S. M., and M. H. Karol. 1996. Indoor air pollution: Acute adverse health effects and host susceptibility. *Environmental Health* 58:25–29.

8. Committee on Indoor Air Pollution. 1981. *Indoor pollutants*. Washington, D.C.: National Academy Press.

9. Godish, T. 1997. *Air quality*, 3rd. ed. Boca Raton, Fla.: Lewis Publishers.

10. O'Reilly, J. T., P. Hagan, R. Gots, and A. Hedge. 1998. *Keeping buildings healthy*. New York: John Wiley and Sons.

11. Brenner, D. J. 1989. *Radon: Risk and remedy*. New York: W. H. Freeman.

12. U.S. Environmental Protection Agency. 1992. *A citizen's guide to radon: The guide to protecting yourself and your family from radon*, 2nd ed. ANR-464. Washington, D.C.: Environmental Protection Agency.

13. Hurlburt, S. 1989 (June). Radon: A real killer or just an unsolved mystery? *Water Well Journal*, pp. 34–41.

14. Egginton, J. 1989. Menace of Whispering Hills. *Audubon* 91:28–35.

15. University of Maine and Maine Department of Human Services. 1983 (February). Radon in water and air. *Resource Highlights*.

16. U.S. Environmental Protection Agency. 1986. *Radon reduction techniques for detached houses: Technical guidance*. EPA 625/5-86-019. Research Triangle Park, N.C.: Air and Energy Engineering Research Laboratory, Office of Research and Development, U.S. Environmental Protection Agency.

17. Osborne, M. C. 1988. *Radon-resistant residential new construction*. EPA 600/8-88/087. Research Triangle Park, N.C.: Air and Energy Engineering Research Laboratory, Office of Research and Development, U.S. Environmental Protection Agency.

18. Pershagen, G., G. Akerblom, O. Axelson, B. Clavensjo, L. Damber, G. Desai, A. Enflo, F. Lagarde, H. Mellander, M. Svartengren, and G. A. Swedjemark. 1994. Residential radon exposure and lung cancer in Sweden. *New England Journal of Medicine* 330:159–164.

19. Henshaw, D. L., J. P. Eatough, and R. B. Richardson. 1990. Radon as a causative factor in induction of myeloid leukaemia and other cancers. *The Lancet* 335:1008–1012.

20. Nero, A. V., Jr. 1988. Controlling indoor air pollution. *Scientific American* 258:42–48.

Chapter 24 Critical Thinking Issue References

Castleman, M. 1995. Clean air up there. *Sierra* 80(3):16.

Douglass, W. C. 1992 (September). If you fly, don't breathe. *Second Opinion*, pp. 1–5.

Kenyon, T. A. 1996. Transmission of multidrug-resistant *Myobacterium tuberculosis* during a long airplane flight. *The New England Journal of Medicine* 334(15):933.

Manning, A. 1993 (June 22). Airborne ailments: Are diseases transmitted in flight? *USA Today*, pp. 1A, 2A.

Nagda, N. L. 1993 (July 29). Testimony before Committee on Science and Technology, U.S. House of Representatives Subcommittee on Technology, Environment and Aviation, Washington, D.C.

Nagda, N. L., M. D. Koontz, and A. G. Konheim. 1991 (August). Carbon dioxide levels in commercial airliner cabins. *ASHRAE Journal* 33:35–38.

National Research Council. 1986. *The airliner cabin environment: Air quality and safety*. Washington, D.C.: National Academy Press.

Tolchin, M. 1993 (June 21). Exposures to tuberculosis on planes are investigated. *New York Times*, p. A7.

Tolchin, M. 1993 (June 25). Inquiry will check air quality on airplanes. *New York Times*, p. A16.

Chapter 25 Notes

1. Kane, R. P. 1998. Ozone depletion, UV-B changes, and increased cancer incidence. *International Journal*.

2. Brown, L. R., C. Flavin, and S. Postel. 1991. *Saving the planet: How to shape an environmentally sustainable global economy*. New York: W. W. Norton.

3. Rowland, F. S. 1990. Stratospheric ozone depletion of chlorofluorocarbons. *AMBIO* 19:281–292.

4. Smith, R. C., B. B. Prezelin, K. S. Baker, R. R. Bidigare, N. P. Boucher, T. Coley, D. Karentz, S. Macintyre, H. A. Matlick, D. Menzies, M. Ondrusek, Z. Wan, and K. J. Waters. 1992. Ozone depletion: ultraviolet saturation and phytoplankton biology in Antarctic waters. *Science* 255:952–959.

5. NASA-GSFC. 2000. Stratospheric ozone. Accessed August 22, 2000, at http://see.gsfc.nasa.gov.

6. Environmental Protection Agency. 1995. *Protection of the ozone layer*. EPA 230-N-95-00. Washington, D.C.: U.S. EPA Office of Policy, Planning, and Evaluation and Office of Air and Radiation.

7. Cutter Information Corp. 1996. Ozone hole begins to shrink. *Global Environmental Change Report* V, VIII, no. 21. Dunster, B.C., Canada: Cutter Information Corp.

8. Hamill, P., and O. B. Toon. 1991. Polar stratospheric clouds and the ozone hole. *Physics Today* 44:34–42.

9. Stolarski, R. S. 1988. The Antarctic ozone hole. *Scientific American* 258:30–36.

10. Molina, M. J., and F. S. Rowland. 1974. Stratospheric sink for chlorofluoromethanes: Chlorine-atom catalyzed distribution of ozone. *Nature* 249:810–812.

11. Brouder, P. 1986 (June). Annals of chemistry in the face of doubt. *New Yorker*, pp. 20–87.

12. Khalil, M. A. K., and R. A. Rasmussen. 1989. The potential of soils as a sink of chlorofluorocarbons and other man-made chlorocarbons. *Geophysical Research Letters* 16:679–682.

13. Rowland, F. S. 1989. Chlorofluorocarbons and the depletion of stratospheric ozone. *American Scientist* 77:36–45.

14. Toon, O. B., and R. P. Turco. 1991. Polar stratospheric clouds and ozone depletion. *Scientific American* 264:68–74.

15. Kerr, R. A. 1994. Antarctic ozone hole fails to recover. *Science* 266:217.

16. Webster, C. R., R. D. May, D. W. Toohey, L. M. Avallone, J. G. Anderson, P. Newman, L. Lait, M. Schoeberl, J. W. Elkins, and K. R. Chay. 1993. Chlorine chemistry on polar stratospheric cloud particles in the Arctic winter. *Science* 261:1130–1134.

17. Kerr, R. A. 1992. New assaults seen on Earth's ozone shield. *Science* 255:797–798.

18. Zurer, P. 1995. Record low ozone levels observed over Arctic. *Chemical & Engineering News* 73:8.

19. Horgan, J. 1992. Volcanic disruption. *Scientific American* 266:28–29.

20. Cutter Information Corp. 1996. Reports discuss present and future state of ozone layer. *Global Environmental Change Report* V, VIII, 21, no. 22, pp. 1–3. Dunster, B.C., Canada: Cutter Information Corp.

21. Shea, C. P. 1989. Mending the Earth's shield. *World Watch* 2:28–34.

22. Kerr, J. B., and C. T. McElroy. 1993. Evidence for large upward trends of ultraviolet-B radiation linked to ozone depletion. *Science* 262:1032–1034.

23. Atkinson, R. J., W. A. Matthews, P. A. Newman, and R. A. Plumb. 1989. Evidence of the mid-latitude impact of Antarctic ozone depletion. *Nature* 340:290–294.

24. Russell, J. M., M. Luo, R. J. Cicerone, and L. E. Deaver. 1996. Satellite confirmation of dominance of chlorofluorocarbons in the global stratospheric chlorine budget. *Nature* 379:526.

25. Showstack, R. 1998. Ozone layer is on slow road to recovery, new science assessment indicates. *Eos* 79(27):317–318.

26. Spurgeon, D. 1998. Surprising success of the Montreal protocol. *Nature* 389(6648):219.

27. Makhijani, A., and A. Bickel, 1990. Still working on the ozone hole. *Technology Review* 93:52–59.

28. Brown, L. R., N. Lenssen, and H. Kane. 1995. CFC production plummeting. In Worldwatch Institute, *Vital Signs 1995*. New York: W. W. Norton.

29. Shea, C. P. 1991. Disarming refrigerators. *World Watch* 4:36.

30. MacKenzie, D. 1990. Cheaper alternatives for CFCs. *New Scientist* 126:39–40.

Chapter 25 Critical Thinking Issue References

Brasseur, G., and C. Branier. 1992. Mount Pinatubo:. Aerosols, chlorofluorocarbons, and ozone depletion. *Science* 257:1239–1241.

Kerr, R. A. 1993. Ozone takes a nose dive after the eruption of Mt. Pinatubo. *Science* 260:490–491.

Pohl, F., and J. P. Hogan. 1993. Ozone politics: They call this science? *Omni* 15(8):34–42, 91.

Rowland, F. S. 1993. President's lecture: The need for scientific communication with the public. *Science* 260:1571–1576.

Russel, J. M., M. Luo, R. J. Cicerone, and L. E. Deaver. 1996. Satellite confirmation of dominance of chlorofluorocarbons in the global stratospheric chlorine budget. *Nature* 379(6565):526.

Tabazadeh, A., and R. P. Turco. 1993. Stratospheric chlorine injection by volcanic eruptions: HCl scavenging and implications for ozone. *Science* 260:1082–1085.

Taubes, G. 1993 (June 11). The ozone backlash. *Science* 260:1580–83.

Chapter 26 Notes

1. Brokaw, N. V. L., R. Wilson, A. A. Whitman, J. M. Hagan, N. Bird, P. J. Martins, L. K. Snook, E. P. Mallory, D. Novelo, D. White, and E. Losos. *Toward sustainable forestry in Belize*.

2. Line, L. 1996 (June 4). Advocates of sustainable mahogany harvests counter boycott. *New York Times* p. C-4.

Richards, E. M. 1991. The forest ejidos of southeast Mexico. *Commonwealth Forestry Review* 70:290–311. Commonwealth Forestry Foundation.

3. Schneider, D. 1996. Good wood: Can timber certification save the rain forest? *Scientific American* pp. 36–37.

4. Snook, L. K. 1991. Opportunities and constraints for sustainable tropical forestry: Lessons from the Plan Piloto Forestal, Quintana Roo, Mexico. Proceedings of the Humid Tropical Lowland Conference on Development Strategies and Natural Resource Management.

5. Snook, L. K. 1993. From circumstantial to managed conservation: The mahogany forests of Quintana Roo, Mexico. In S. Landis, ed. *Conservation by design*. Rhode Island School of Design and Woodworkers Alliance for Forest Protection.

6. Snook, L. K. 1995. Sustaining harvest of mahogany (*Swientenia macrophylla* King) from Mexico's Yucatan forests: Past, present and future. Proceedings of the Conference on Conservation and Community Development in the Mayan Forest of Belize, Guatemala and Mexico, Chetumal, Quintana Roo, Mexico.

7. Roberts, L. 1991. Costs of a clean environment. *Science* 251:1182.

8. Fairley, P. 1995. Compromise limits EPA budget cut, removes House riders. *Chemical Week* 157:17.

9. Moore, C. A. 1995. Poisons in the Air. *International Wildlife* 25:38–45.

10. Pope, C. A. III, D. V. Bates, and M. E. Raizenne, 1995. Health effects of particulate air pollution: time for reassessment? *Environmental Health Perspectives* 103:472–480.

11. Kolstad, C. D. 2000. *Environmental Economics*. New York, Oxford University Press.

12. Hardin, G. 1968. The tragedy of the commons. *Science* 162:1243–1248.

13. Clark, C. W. 1973. The economics of overexploitation. *Science* 181:630–634.

14. Gunn, J. M.. ed. 1995. *Restoration and recovery of an industrial region. Progress in restoring the smelter-damaged landscape near Sudbury, Canada*. New York Springer-Verlag.

15. Cairns, J. Jr. 1980. Estimating hazard. *BioScience* 20:101–107.

16. Schwing, R. C. 1979. Longevity and benefits and costs of reducing various risks: *Technological Forecasting and Social Change* 13:333–345.

17. Gori, G. B. 1980. The regulation of carcinogenic hazards. *Science* 208:256–261.

18. Ostro, B. D. 1980. Air pollution, public health, and inflation. *Environmental Health Perspectives* 345:185–189.

19. Litton, R. B. 1972. Aesthetic dimensions of the landscape. In J. V. Krutilla, ed., *Natural environments.* Baltimore: Johns Hopkins University Press.

20. Costanza, R. et. al., 1997. "The value of the world's ecosystem services and natural capital," *Nature:* 387:253–260.

21. James, A., Gaston, K. T. and Blamford, A. 2001. Can we afford to conserve biodiversity? *BioScience* 51(1) 43–52.

22. Office of Technology Assessment. 1991. *Changing by degrees: steps to reduce greenhouses gases.* Washington, D.C.: U.S. Superintendent of Documents.

23. Schniedere, S. H. 1989. *Global warming: Are we entering the greenhouse century?* San Francisco: Sierra Club Books.

24. D'Arge, R. 1989. Ethical and economic systems for managing the global commons. In D. B. Botkin, M. Caswell, J. E. Estes, and A. Orio, eds., *Changing the global environment: Perspectives on human involvement,* pp. 327–337. New York: Academic Press.

25. Rogers, A. 1993. *The Earth summit: A planetary reckoning.* Global View Press.

26. Baumol, W. J., and W. E. Oates, 1979. *Economics, environmental policy, and the quality of life.* Englewood Cliffs, N.J.:Prentice-Hall.

27. Clark, C. W. 1981. Economics of fishery management. In T. L. Vincent, and J. M. Skowronski, eds., *Renewable resource management: Lecture notes in biomathematics,* pp. 95–111. New York: Springer-Verlag.

Chapter 26 Critical Thinking Issue References

Anthony, V. C. 1993. The state of groundfish resources off the northeastern United States. *Fisheries* 18(3):12–17.

Correia, S. J. 1992 (3rd quarter). Flounder population declines: Overfishing or pollution? *Division of Marine Fisheries News.* Boston: Massachusetts Division of Marine Fisheries.

How to Fish. 1988 (December 10). *The Economist* 309 (7580):93–96.

Keen, E. A. 1991. Ownership and productivity of marine fisheries resources. *Fisheries* 16:18–22.

Lawren, B. 1992. Net loss, *National Wildlife* 30(6):47–52.

Leal, D. R. 1992 (July 30). Using property rights to regulate fish harvest. *Christian Science Monitor* 84:18.

National Marine Fisheries Service. 1995. *Status of the fishery resources off the northeastern United States for 1994.* Woods Hole, Mass.: U.S. Department of Commerce, NOAA, NMFS Northeast Fisheries Science Center.

Pierce, D. 1992 (2nd quarter). New England council to cut fishing effort in half over next 5 years. *Division of Marine Fisheries News.* Boston: Massachusetts Division of Marine Fisheries.

Satchell, M. 1992. The rape of the oceans. *U.S. News and World Report* 112(24):64–75.

Chapter 27 Notes

1. Dobbs, F. 1995. *Curitiba: City of the future?* Video. World Bank.

2. Rabinovitch, J. 1997. *Integrated transportation and land use planning channel Curitiba's growth.* Washington, D.C.: World Resources Institute.

3. Hunt, J. 1994 (April). Curitiba. *Metropolis Magazine.*

4. Rabinovitch, J., and J. Leitman. 1996 (March). Urban planning in Curitiba. *Scientific American,* pp. 45–53.

5. Haub, C., and D. Cornelius. 2000. *World population data sheet.* Washington, D.C.: Population Reference Bureau.

6. Population Reference Bureau. 1998. *World and regional population.* Washington, D.C.: Population Reference Bureau.

7. World Resources Institute. 1998. *Teacher's guide to world resources: Exploring sustainable communities.* Washington, D.C.: World Resources Institute.

8. World Resources Institute. 1999. *Urban growth.* Washington, D.C.: World Resources Institute.

9. Mumford, L. 1972. The natural history of urbanization. In R. L. Smith, ed., *The ecology of man: An ecosystem approach,* pp. 140–152. New York: Harper & Row.

10. Cronon, W. 1991. *Nature's metropolis: Chicago and the great west.* New York: W. W. Norton.

11. Leibbrand, K. 1970. *Transportation and town planning.* Translated by N. Seymer. Cambridge, Mass.: MIT Press.

12. Pavoni, B., A. Sfriso, D. Donazzolo, and A. A. Orio. 1990. *Influence of wastewaters on the city of Venice and the hinterland on the eutrophication of the lagoon.* The Science of the Total Environment 96:325–252.

13. Poland, J. F., and G. H. Davis. 1969. *Land subsidence due to withdrawal of fluids.* In D. J. Varnes and G. Kiersch, eds., *Reviews in engineering geology,* pp. 187–269. Boulder, Colo.: The Geological Society of America.

14. Hunt, C. B. 1974. *Natural regions of the United States and Canada.* San Francisco: W. H. Freeman.

15. McLaughlin, C. C., ed., 1977. *The formative years: 1822–1852.* Volume 1 of The papers of Frederick Law Olmsted. Baltimore: Johns Hopkins University Press.

16. Miller, L. B. 1987. Miracle on 104th Street. *American Horticulturalist* 66:14–17.

17. Spirn, A. W. 1984. *The granite garden: Urban nature and human design.* New York: Basic Books.

18. Reps, J. W. 1965. *The making of urban America: A history of city planning in the United States.* 2nd ed. Princeton, N.J.: Princeton University Press.

19. Detwyler, T. R., and M. G. Marcus, eds. 1972. *Urbanization and the environment: The physical geography of the city.* North Scituate, Mass.: Duxbury Press.

20. Butti, K., and J. Perlin. 1980. *A golden thread: 2500 years of solar architecture and technology.* New York: Cheshire.

21. McHarg, I. L. 1971. *Design with nature.* Garden City, N.Y.: Doubleday.

22. Ford, A. B., and O. Bialik. 1980. Air pollution and urban factors in relation to cancer mortality. *Archives of Environmental Health* 35:350–359.

23. Akbari, H., S. Davis, S. Dorsano, J. Huang, and S. Winnett. 1992. *Cooling our communities: A guidebook on tree planting and light-colored surfacing.* Washington, D.C.: U.S. Environmental Protection Agency Office of Policy Analysis.

24. Moll, G. 1989. Designing the ecological city. *American Forests* 85:61–64.

25. Nadel, I. B., C. H. Oberlander, and L. R. Bohm. 1977. *Trees in the city.* New York: Pergamon Press.

26. Moll, G., P. Rodbell, B. Skiera, J. Urban, G. Mann, and R. Harris. 1991 (April/May). Planting new life in the city. *Urban Forests,* pp. 10–20. Washington, D.C.: American Forestry Association.

27. Dreistadt, S. H., D. L. Dahlsten, and G. W. Frankie. 1990. Urban forests and insect ecology. *BioScience* 40:192–198.

28. Leedly, D. L., and L. W. Adams. 1984. *A guide to urban wildlife management.* Columbia, Md.: National Institute for Urban Wildlife.

29. Tylka, D. 1987. Critters in the city. *American Forests* 93:61–64.

30. Burton, J. A. 1977. *Worlds apart: Nature in the city.* Garden City, N.Y.: Doubleday.

31. Adams, L. W., and L. E. Dove. 1989. *Wildlife reserves and corridors in the urban environment.* Columbia, Md.: National Institute for Urban Wildlife.

Chapter 27 Critical Thinking Issue References

Egan, T. 1996 (December 30). Drawing a hard line against urban sprawl. *New York Times.*

Egan, T. 1996 (December 29). Urban sprawl strains western states. *New York Times.*

Pollock, P. 1998 (January). Controlling sprawl in Boulder: Benefits and pitfalls. *Land Lines,* Lincoln Institute of Land Policy, vol. 10, no. 1.

Pryne, E. 1998 (August 23). How curbing development shaped a city. *Seattle Times*.

Chapter 28 Notes

1. Wright, A. G. 1998. Big Apple is dumping its last active landfill. *ENR* 240 (7):28–31.
2. Young, W. 1995 (Autumn). A dump no more. *American Forests*, pp. 59–62.
3. Galley, J. E. 1968. Economic and industrial potential of geologic basins and reservoir strata. In J. E. Galley, ed., *Subsurface disposal in geologic basins: A study of reservoir strata*, pp. 1–19. American Association of Petroleum Geologists Memoir 10. Tulsa, Okla.: American Association of Petroleum Geologists.
4. Relis, P., and A. Dominski. 1987. *Beyond the crisis: Integrated waste management*. Santa Barbara, Calif.: Community Environmental Council.
5. Repa, D. W., and A. Blakey. 1996. Municipal solid waste disposal trends: 1996 update. *Waste Age* 27:42–54.
6. Council on Environmental Quality. 1973. *Environmental quality—1973*. Washington, D.C.: U.S. Government Printing Office.
7. Allenby, B. R. 1999. *Industrial ecology: Policy framework and implementation*. Upper Saddle River, N.J.: Prentice-Hall.
8. Garner, G., and P. Sampat. 1999 (May). Making things last: Reinventing of material culture. *The Futurist*. pp. 24–28.
9. Relis, P., and H. Levenson. *Discarding solid waste as we know it: Managing materials in the 21st century*. Santa Barbara, Calif.: Community Environmental Council.
10. Young, J. E. 1991. Reducing waste-saving materials. In L. R. Brown, ed., *State of the world, 1991*, pp. 39–55. World Watch Institute. New York: W. W. Norton.
11. Steuteville, R. 1995. The state of garbage in America: Part I. *BioCycle* 36:54.
12. Gardner, G. 1998 (January–February). Fertile ground or toxic legacy? *World Watch*, pp. 28–34.
13. McGreery, P. 1995. Going for the goals: Will states hit the wall? *Waste Age* 26:68–76.
14. Brown, L. R. 1999 (March–April). Crossing the threshold. *World Watch*, pp. 12–22.
15. Rathje, W. L., and C. Murphy. 1992. Five major myths about garbage, and why they're wrong. *Smithsonian* 23:113–122.
16. Rathje, W. L. 1991. Once and future landfills. *National Geographic* 179(5):116–134.
17. Schneider, W. J. 1970. *Hydrologic implications of solid-waste disposal*. 135(22). U.S. Geological Survey Circular 601F. Washington, D.C.: U.S. Geological Survey.
18. Thomas, V. M., and T. G. Spiro. 1996. The U.S. dioxin inventory: Are there missing sources? *Environmental Science & Technology* 30:82A–85A.
19. Turk, L. J. 1970. Disposal of solid wastes—acceptable practice or geological nightmare? In *Environmental Geology*, pp. 1–42. Washington, D.C.: American Geological Institute Short Course, American Geological Institute.
20. Hughes, G. M. 1972. Hydrologic considerations in the siting and design of landfills. *Environmental Geology Notes*, no. 51. Urbana: Illinois State Geological Survey.
21. Bergstrom, R. E. 1968. Disposal of wastes: Scientific and administrative considerations. *Environmental Geology Notes*, no. 20. Urbana: Illinois State Geological Survey.
22. Cartwright, K., and Sherman, F. B. 1969. Evaluating sanitary landfill sites in Illinois. *Environmental Geology Notes*, no. 27. Urbana: Illinois State Geological Survey.
23. Rahn, P. H. 1996. *Engineering geology*, 2nd ed. Upper Saddle River, N.J.: Prentice-Hall. Chapter 7.
24. Bullard, R. D. 1990. *Dumping in Dixie: Race, class and environmental quality*. Boulder, Colo.: Westview Press.
25. Sadd, J. L., J. T. Boer, M. Foster, Jr., and L. D. Snyder. 1997. Addressing environmental justice: Demographics of hazardous waste in Los Angeles County. *Geology Today* 7(8):18–19.

26. Walker, W. H. 1974. Monitoring toxic chemical pollution from land disposal sites in humid regions. *Ground Water* 12:213–218.
27. Watts, R. J. 1998. *Hazardous wastes*. New York: John Wiley & Sons.
28. Wilkes, A. S. 1980. *Everybody's problem: Hazardous waste*. SW-826. Washington, D.C.: U.S. Environmental Protection Agency, Office of Water and Waste Management.
29. Elliot, J. 1980. Lessons from Love Canal. *Journal of the American Medical Association* 240:2033–2034, 2040.
30. Kufs, C., and C. Twedwell. 1980. Cleaning up hazardous landfills. *Geotimes* 25:18–19.
31. Albeson, P. H. 1983. Waste management. *Science* 220:1003.
32. New York State Department of Environmental Conservation. 1994. *Remedial chronology: The Love Canal hazardous waste site*. New York State.
33. Kirschner, E. 1994. Love Canal settlement: OxyChem to pay New York state $98 million. *Chemical & Engineering News* 72:4–5.
34. Westervelt, R. 1996. Love Canal: OxyChem settles federal claims. *Chemical Week* 158:9.
35. Whittell, G. 2000 (November 29). Poison in Paradise. (London) *Times 2*, p. 4.
36. Cooney, C. M. 1997. New EPA budget boosts particulate matter research, superfund cleanups. *Environmental Science and Technology*, 31(12):A555.
37. Bedient, P. B., H. S. Rifai, and C. J. Newell. 1994. *Ground water contamination*. Englewood Cliffs, N.J.: Prentice-Hall.
38. Huddleston, R. L. 1979. Solid-waste disposal: Land farming. *Chemical Engineering* 86:119–124.
39. McKenzie, G. D., and W. A. Pettyjohn. 1975. Subsurface waste management. In G. D. McKenzie and R. O. Utgard, eds., *Man and his physical environment: Readings in environmental geology*, 2nd ed., pp. 150–156. Minneapolis: Burgess Publishing.
40. National Research Council, Committee on Geological Sciences. 1972. *The earth and human affairs*. San Francisco: Canfield Press.
41. Cox, C. 1985. *The buried threat: Getting away from land disposal of hazardous waste*. No. 115-5. California Senate Office of Research.
42. Council on Environmental Quality. 1970. *Ocean dumping: A national policy: A report to the president*. Washington, D.C.: U.S. Government Printing Office.
43. Lenssen, N. 1989 (July–August). The ocean blues. *World Watch*, pp. 26–35.
44. U.S. Environmental Protection Agency. 2000. Forward pollution protection: The future look of environmental protection. Accessed August 12, 2000, at http://www.epa.gov/p2/p2case.htm#num4.

Chapter 28 Critical Thinking Issue References

Bruno, K. 1993 (February 23). E.P.A. plays a dioxin numbers game in Newark Bay dredging (letter). *New York Times*, p. A20.
Cernadas, A. 1993 (February 10). Why dredging in Newark Bay must proceed (letter). *New York Times*, p. A22.
Clark, S. L. 1993 (March 30). Testimony of the Environmental Defense Fund before the Subcommittee on Oceanography, Gulf of Mexico, and the Outer Continental Shelf of the Merchant Marine and Fisheries Commission, U.S. House of Representatives.
Muszynski, W. J., and J. A. Capo. 1993 (February 16). An awful problem gets dredged up (letter). *Wall Street Journal*, p. A15.
Schmidt, K. F. 1992. Dioxin's other face. *Science News* 141(2):24–27.
Strum, C. 1993 (January 27). U.S. suspends dredging permit for Newark Bay. *New York Times*, pp. B1, B4.
Strum, C. 1993 (January 30). From port to 17 Fathoms, views on dredging differ. *New York Times*, pp. 23, 27.
Strum, C. 1993 (February 13). Limited dredging allowed in Newark Bay. *New York Times*, p. 24.
U.S Army Corps of Engineers, New York District. 1989 (December). *Managing dredged material*. New York: U.S. Army Corps. of Engineers.
U.S. Environmental Protection Agency, Environmental Research Laboratory, Office of Research and Development. 1993 (March). *Interim report on data and methods for assessment of 2,3,7,8-tetra-*

chlorodibenzo-p-dioxin: Risks to aquatic life and associated wildlife. Duluth, Minn.: U.S. Environmental Protection Agency.

Chapter 29 Notes

1. Anonymous. 1998. *The Butchart Gardens*. Victoria, B.C.: The Butchart Gardens, Ltd.
2. McKelvey, V. E. 1973. Mineral resource estimates and public policy. In D. A. Brobst and W. P. Pratt, eds., *United States mineral resources*, pp. 9–19. U.S. Geological Survey Professional Paper 820.
3. U.S. Department of the Interior, Bureau of Mines. 1993. *Mineral commodity summaries, 1993*. I 28.149:993. Washington, D.C.: U.S. Department of the Interior.
4. Meyer, H. O. A. 1985. Genesis of diamond: A mantle saga. *American Mineralogist* 70:344–355.
5. Kesler, S. F. 1994. *Mineral resources, economics, and the environment*. New York: Macmillan.
6. Smith, G. I., C. L. Jones, W. C. Culbertson, G. E. Erickson, and J. R. Dyni. 1973. Evaporites and brines. In D. A. Brobst and W. P. Pratt, eds., *United States mineral resources*, pp. 197–216. U.S. Geological Survey Professional Paper 820.
7. Awramik, S. A. 1981. The pre-Phanerozoic biosphere—three billion years of crises and opportunities. In M. H. Nitecki, ed., *Biotic crises in ecological and evolutionary time*, pp. 83–102. Spring Systematics Symposium. New York: Academic Press.
8. Margulis, L., and J. E. Lovelock. 1974. Biological modulation of the Earth's atmosphere. *Icarus* 21:471–489.
9. Lowenstam, H. A. 1981. Minerals formed by organisms. *Science* 211:1126–1130.
10. Bateman, A. M. 1950. *Economic mineral deposits*, 2nd ed. New York: John Wiley & Sons.
11. Park, C. F., Jr., and R. A. MacDiarmid. 1970. *Ore deposits*, 2nd ed. San Francisco: W. H. Freeman.
12. Cornwall, H. R. 1973. Nickel. In D. A. Brobst and W. P. Pratt, eds., *United States mineral resources*, pp. 437–442. U.S. Geological Survey Professional Paper 820.
13. Van, N., J. Dorr, M. D. Crittenden, and R. G. Worl. 1973. Manganese. In D. A. Brobst and W. P. Pratt, eds., *United States mineral resources*, pp. 385–399. U.S. Geological Survey Professional Paper 820.
14. U.S. Secretary of the Interior. 1975. *Mining and mineral policy, 1975*. Washington, D.C.: U.S. Department of the Interior.
15. George, D. 1994. Framework for deep ocean mining fixed despite lack of technology. *Offshore* 54:52.
16. Brobst, D. A., W. P. Pratt, and V. E. McKelvey. 1973. *Summary of United States mineral resources*. U.S. Geological Survey Circular 682.
17. U.S. Geological Survey. 1984. *Yearbook, fiscal year 1983*. Washington, D.C.: U. S. Geological Survey.
18. Jeffers, T. H. 1991 (June). Using microorganisms to recover metals. *Minerals Today*. Washington, D.C.: U.S. Department of Interior, Bureau of Mines, pp. 14–18.
19. Haynes, B. W. 1990 (May). Environmental technology research. *Minerals Today*. Washington, D.C.: U.S. Bureau of Mines, pp. 13–17.
20. Sullivan, P. M., M. H. Stanczyk, and M. J. Spendbue. 1973. *Resource recovery from raw urban refuse*. U.S. Bureau of Mines Report of Investigations 7760.
21. Davis, F. F. 1972 (May). Urban ore. *California Geology*, pp. 99–112.
22. U.S. Geological Survey. 2000. *Minerals yearbook 1998—Recycling metals*. Accessed August 21, 2000, at http://minerals.usgs.gov.
23. Brown, L., N. Lenssen, and H. Kane. 1995. Steel recycling rising. In *Vital Signs 1995*. Worldwatch Institute.

Chapter 29 Critical Thinking Issue Reference

Debus, K. H. 1990 (August/September). Mining with microbes. *Technology Review* 93:50.

Chapter 30 Notes

1. Hill, J. B. 2000. *The legacy of Luna*. San Francisco: Harper.

2. Hawken, P., A. Lovins, and L. H. Lovins. 1999. *Natural capitalism*. Boston: Little, Brown.
3. Hubbard, B. M. 1998. *Conscious evolution*. Novato, Calif.: New World Library.
4. *Los Angeles Times*. 2001. Los Angeles Times Poll Alert, Study #458.
5. McHarg, I. L. 1971. *Design with nature*. Garden City, N.Y.: Doubleday.
6. Cohen, B. S. 1970. The constitution, the public trust doctrine and the environment. *Utah Law Review* 388.
7. McGinn, A. P. 2000 (April). POPs culture. *WorldWatch*, pp. 26–36.
8. Yannacone, V. J., Jr., B. S. Cohen, and S. G. Davison. 1972. *Environmental rights and remedies*. Rochester, N.Y.: Lawyers Co-operative Pub., pp. 39–46.
9. Bacow, L. S., and M. Wheeler. 1984. *Environmental dispute resolution*. New York: Plenum Press.
10. Council on Environmental Quality. 1979. *Environmental quality*. Washington, D.C.: Government Printing Office.
11. Brew, D. A. 1974. *Environmental impact analysis: The example of the proposed trans-Alaska pipeline*. U.S. Geological Survey Circular 695. Reston, Va.: U.S. Department of the Interior.
12. Parker, H. D. 1987. What is a geographic information system? In *GIS '87*, pp. 72–79. Second Annual International Conference, Exhibits and Workshops on Geographic Information Systems. American Society for Photogrammetry and Remote Sensing.
13. Star, J., and J. Estes. 1990. *Geographic information systems: An introduction*. Englewood Cliffs, N.J.: Prentice-Hall.
14. Callies, D. L. 1984. *Regulating paradise: Land use controls in Hawaii*. Honolulu: University of Hawaii Press.
15. Jain, R. K., L. V. Urban, G. S. Stacey, and H. E. Balbach. 1993. *Environmental assessment*. New York: McGraw-Hill.
16. Remy, M. H., T. A. Thomas, and J. G. Moose. 1991. *Guide to the California Environmental Quality Act*, 5th ed. Point Arena, Calif.: Solano Press Books.
17. National Park Service. 1974. *Cape Hatteras shoreline erosion policy statement*. Washington, D.C.: Department of the Interior.
18. Leatherman, S. P. 1983. Barrier dynamics and landward migration with Holocene sea-level rise. *Nature* 301:415–418.
19. National Park Service. 1984. *General management plan, development concept plan, and amended environmental assessment, Cape Hatteras National Seashore*.
20. McDonald, K. A. 1993. A geology professor's fervent battle with coastal developers and residents. *The Chronicle of Higher Education* 40:A8.
21. Tanji, K., A. Lauchli, and J. Meyer. 1986. Selenium in the San Joaquin Valley. *Environment* 28:6–11, 34–39.
22. U.S. Department of the Interior, Bureau of Reclamation. 1987. *Kesterson program*. Fact Sheet No. 4. Sacramento, Calif.
23. Rohse, M. 1987. *Land-use planning in Oregon: A no-nonsense handbook in plain English*. Corvallis: Oregon State University Press.
24. Curtin, D. J., Jr. 1991. *California land-use and planning law*, 11th ed. Point Arena, Calif.: Solano Press Books.
25. Abramovitz, J. N. 2001. Averting unnatural disasters. In World Watch Institute, *State of the world 2001*, pp. 123–142. New York: W. W. Norton.
26. White, G. F., and J. E. Hass. 1975. *Assessment of research on natural hazards*. Cambridge, Mass.: The MIT Press.
27. Advisory Committee on the International Decade for Natural Hazard Reduction. 1989. *Reducing disaster's toll*. Washington, D.C.: National Academy Press.
28. Steiner, F. 1983. Regional planning: Historic and contemporary examples. *Landscape Planning* 10:297–315.
29. Wilshire, H. G., and J. K. Nakata. 1976. Off-road vehicle effects on California's Mojave Desert. *California Geology* 29:123–132.
30. Wilshire, H. G., G. B. Bodman, D. Broberg, W. J. Kockelman, J. Major, H. E. Malde, C. T. Snyder, and R. C. Stebbs. 1977. *Impacts and management of off-road vehicles*. Report of the Committee on Environment and Public Policy. Boulder, Colo.: Geological Society of America.

31. Meadows, D. H., D. L. Meadows, J. Randers, and W. W. Behrens, III. 1972. *The limits to growth: A report for the Club of Rome's Project on the Predicament of Mankind.* New York: Universe Books (Potomac Associates).
32. Council on Environmental Quality and U.S. Department of State. 1980. *The global 2000 report to the president: Entering the twenty-first century.* Washington, D.C.: Government Printing Office.
33. Renner, M. 2002. Breaking the link between resources and repres-sion. In Worldwatch Institute State of the World 2002. New York, W. W. Norton

Chapter 30 Critical Thinking Issue References

Fautley, C. 2000 (October 15). Geology of the channel tunnel. *London Times.* Accessed November 9, 2000, at www.geologyshop.co.uk/chtung.htm.

Photo Credits

Chapter 1 Opener: Jeff Rotman/Stone/Getty Images. Page 2 (top): Frans Lanting/Minden Pictures, Inc. Page 2 (bottom): © Tom Bean. Page 6 (right): Corbis-Bettman. Page 7 (left): Peter Turnley/Black Star. Page 7 (right): Viviane Moos/Corbis Saba. Page 9: National Snow and Ice Data Center/Photo Researchers.

Chapter 2 Opener: Larry Goldstein/Stone/Getty Images. Page 15 (top): Gideon Mendel/Magnum Photos, Inc. Page 15 (bottom): D. Hudson/Corbis Sygma. Page 18 (top right): Jim Brandenburg/Minden Pictures, Inc. Page 18 (left): Alan Root/Okapia/Photo Researchers. Page 18 (bottom right): Norman Owen Tomalin/Bruce Coleman, Inc. Page 25 (left): U.S. Fish and Wildlife Service/J&K Hollingsworth. Page 25 (right): Joel Sartore/Grant Heilman Photography, Inc. Page 25 (bottom): ©Science/Visuals Unlimited. Page 26: ©Gary Braasch.

Chapter 3 Opener: Guido Alberto Rossi/The Image Bank/Getty Images. Page 33: M. Renaudeau/Hoa-Qui. Page 34: Chris Johns. Page 35 (top): Mark Segal/Stone/Getty Images. Page 35 (bottom) & 36: Courtesy Ed Keller. Page 39: Charles A. Mauzy/Stone/Getty Images. Page 40: Tom Till/DRK Photo. Page 44: ©AP/Wide World Photos.

Chapter 4 Opener: ©Corbis. Page 49 (top): TerraNova International/Photo Researchers. Page 49 (bottom): Courtesy Ed Keller. Page 60 (top): Kim Heacox/Peter Arnold, Inc. Page 60 (bottom): Manfred Gottschalk/Tom Stack & Associates. Page 69 (bottom): William Felger/Grant Heilman Photography.

Chapter 5 Opener: ©Still Pictures/Peter Arnold, Inc. Page 74: Chip Hires/Liaison Agency/Getty Images

Chapter 6 Opener: Stephen J. Krasemann/Photo Researchers. Page 95 (top left): Alvin E. Staffan/Photo Researchers. Page 95 (top center): Bill Ivy/Stone/Getty Images. Page 95 (top right): O. Spielman/CNRI/Phototake. Page 95 (bottom left): Jeanne Drake/Stone/Getty Images. Page 95 (bottom center): Runk/Schoenberger/Grant Heilman Photography. Page 95 (bottom right): Scott Nielsen/Bruce Coleman, Inc. Page 98: Farrell Grehan/Photo Researchers. Page 104 (left): Fridmar Damm/Leo de Wys, Inc. Page 104 (right): Jeff Gnass. Page 105 (top): Richard Hartmier/First Light, Toronto. Page 105 (bottom): Courtesy USDA Forest Service.

Chapter 7 Opener: Ferry/Liaison Agency/Getty Images. Page 110 (left): Robert Holmes/Corbis. Page 117 (top): Dr. Patricia Schultz/Peter Arnold, Inc. Page 117 (bottom): Kari Lounatmaa/Photo Researchers. Page 120 (left): Alvin E. Staffan/Photo Researchers. Page 120 (right): H. Reinhard/Bruce Coleman, Inc.

Chapter 8 Opener: ©AP/Wide World Photos. Page 131 (left & bottom right): Stephan C. White, Kansas Department of Agriculture. Page 131 (top right): Stephen G. Maka/DRK Photo. Page 133 (left): Hans & Judy Deste/Animals Animals. Page 133 (top right): Carl Purcell/Photo Researchers. Page 133 (bottom right): Stephen J. Krasemann/Stone/Getty Images. Page 138 (top left): Stephen J. Krasemann/DRK Photo. Page 138 (top center): Kim Heacox/DRK Photo. Page 138 (top right):

Tim Laman/The Wildlife Collection. Page 138 (bottom): Norman Tomalin/Bruce Coleman, Inc. Page 139 (left): Ferrero/Labat/Auscape International Pty. Ltd. Page 139 (top right): Toni Angermayer/Photo Researchers. Page 139 (bottom right): Martin Harvey/The Wildlife Collection. Page 144: Courtesy NASA. Page 145: John Shaw/Bruce Coleman, Inc. Page 146: Charles Glatzer/The Image Bank/Getty Images. Page 147 (top): Tom & Pat Leeson. Page 147 (bottom): Janis Burger/Bruce Coleman, Inc. Page 148: David Muench Photography. Page 149 (top): Frans Lanting/Minden Pictures, Inc. Page 149 (bottom): Steve Kaufman/Peter Arnold, Inc. Page 150 (top): David Muench Photography. Page 150 (bottom): Andre Jenny /Stock South. Page 151: William Johnson/Stock, Boston. Page 153: Photo by Dr. Alex Meisnesz.

Chapter 9 Opener: Runk/Schoenberger/Grant Heilman Photography, Inc. Page 157 (left): David Tomlinsin/Windrush Photos. Page 157 (right): Kim Heacox/Peter Arnold, Inc. Page 161: Courtesy Daniel Botkin.

Chapter 10 Opener: Albert Bierstadt, Sunset in the Yosemite Valley, 1868. The Haggin Museum. Page 170: ©Tom Bean. Page 172: Courtesy NASA Goddard Space Flight Center and the authors of Hall, F.G., D.B. Botkin, D.E. Strbel, K.D. Woods and S.J. Goetz, 1991; "Large Scale Patterns in Forest Succession As Determined by Remote Sensing," Ecology, 72: 628-640. Page 175 (left): Masha Nordbye/Bruce Coleman, Inc. Page 175 (right): ©Grant Heilman Photography. Page 176 (top): Breck Kent/Earth Scenes. Page 176 (center & bottom): Michael P. Gadomski/Earth Scenes. Page 177 (top): Terry Donnelly/Tom Stack & Associates. Page 177 (bottom): Courtesy Daniel Botkin. Page 184: David Tomlinsin/Windrush Photos.

Chapter 11 Opener: Herb Schmitz/Stone/Getty Images. Page 193 (top left): Kevin Morris/Stone/Getty Images. Page 193 (top right): JC Carton/Bruce Coleman, Inc. Page 193 (bottom left): Thomas Hovland/Grant Heilman Photography. Page 193 (bottom center): William E. Ferguson. Page 193 (bottom right): Hans Reinhard/Photo Researchers. Page 197: Doug Plummer/Photo Researchers. Page 199: J. Victolero, International Rice Research Institute. Page 204: Food and Agriculture Organization.

Chapter 12 Opener: Gary R. Zahm/DRK Photo. Page 212: Courtesy Steve Burr. Page 215 (top): From Trimble and Crosson, 2000; "U.S. Soil Erosion Rates - Myth and Reality," Science, 289: 248-250. Page 215 (bottom right): Jeri Gleiter/Peter Arnold, Inc. Page 215 (center right): Science Vu/Visuals Unlimited. Page 221 (top): Larry A. Hull. Page 221 (bottom): Matthew McVay/Stone/Getty Images. Page 223 (left): Simko/Visuals Unlimited. Page 223 (center): Arthur C. Smith III/Grant Heilman Photography, Inc. Page 223 (right): Bob Gurr/DRK Photo. Page 224 (top): Joy Sprurr/Bruce Coleman, Inc. Page 224 (bottom): Grant Heilman Photography. Page 225 (left): R. de la Harp/Biological Photo Service. Page 225 (right): Alex Von Koschembahr/Photo Researchers. Page 227 (top): Bill Bachman/Photo Researchers. Page 227 (center): Grant Heilman Photography. Page 227 (bottom): Corbis-Bettmann.

Chapter 13 Opener: C.C. Lockwood/Bruce Coleman, Inc. Page 233 (top): Larry Ulrich/DRK Photo. Page 235: David Muench Photography. Page 242: Inga Spence/Visuals Unlimited. Page 243: NASA/Science Source/Photo Researchers. Page 244 (top): Steve McCurry/Magnum Photos, Inc. Page 244 (bottom): Robert Frerck/Odyssey Productions. Page 246: Martin Miller/Visuals Unlimited. Page 249: ©Larry Ulrich.

Chapter 14 Opener: Stephen J. Krasemann/DRK Photo. Page 256: Wm. Grenfell/Visuals Unlimited. Page 267: John Cunningham/Visuals Unlimited. Page 271: ©PhotoDisc. Page 272 (top): Tom Bledsoe/DRK Photo. Page 272 (bottom): Catlin, George. Buffalo Chase, Mouth of the Yellowstone, 1832-1833. ©Smithsonian American Art Museum, Washington DC/Art Resource, NY. Page 274: Mike Price/Bruce Coleman, Inc. Page 276: Ken Lucas/Visuals Unlimited. Page 277 (top left): Richard Ellis/Photo Researchers. Page 277 (top right): Marc Epstein/DRK Photo. Page 277 (bottom left): ©AP/Wide World Photos. Page 277 (bottom right) Ross Frid/Visuals Unlimited. Page 278 (top): Richard Elliott/Stone/Getty Images. Page 278 (center): Tom & Pat Leeson/DRK Photo. Page 278 (bottom): Dianne Blell/Peter Arnold, Inc. Page 279 (left): Jeff Lepore/Photo Researchers. Page 279 (right): Gilbert Grant/Photo Researchers.

Chapter 15 Opener: Andy Levin/Photo Researchers. Page 284: ©AP/Wide World Photos. Page 286 (top left): T. Orban/Corbis Sygma. Page 286 (top right): David & Peter Turnley/Corbis. Page 286 (bottom): Courtesy Ed Keller. Page 287: Bill Brooks/Masterfile. Page 292: O. Franken/Corbis Sygma. Page 294: Martin Bond/Science Photo Library/Photo Researchers. Page 295: Michael Yamashita.

Chapter 16 Opener: Mark Gibson/Visuals Unlimited. Page 308: FPG International/Getty Images.

Chapter 17 Opener: Russell D. Curtis/Photo Researchers. Page 330 (left): George Hunter/Stone/Getty Images. Page 330 (right): Ken Graham/Bruce Coleman, Inc. Page 331: U.S. Fish and Wildlife Service/Liaison Agency/Getty Images. Page 332 (left): ©AP/Wide World Photos. Page 332 (right): U.S. Fish and Wildlife Service/Liaison Agency/Getty Images. Page 335: ©Grant Heilman Photography. Page 336: Courtesy Ed Keller. Page 337: William P. Hines/The Scranton-Times-Tribune Library.

Chapter 18 Opener: Courtesy Ed Keller. Page 347: ©Tom Bean. Page 349 (left): H. Gruyaeart/Magnum Photos, Inc. Page 349 (right): Courtesy Ed Keller. Page 350: T. J. Florian/Rainbow. Page 351: Courtesy Luz International. Page 355: C. Delis/Explorer. Page 356: Glen Allison/Stone/Getty Images. Page 360: Courtesy Pacific Gas & Electric Company.

Chapter 19 Opener: Hans Wolf/The Image Bank/Getty Images. Page 371 (left): Graham Finlayson/Stone/Getty Images. Page 371 (right): J. Schlegel/Leo de Wys, Inc. Page 372: Roger Ressmeyer/Corbis. Page 373: Courtesy the Princeton Plasma Physics Laboratory. Page 375: Corbis-Bettmann. Page 381: Igor Kostin/Corbis Sygma.

Chapter 20 Opener & 389: ©AP/Wide World Photos. Page 396: Jim Tuten/Black Star. Page 405 (top left): Stephen Krasemann/Stone/Getty Images. Page 405 (top right): Mike Price/Bruce Coleman, Inc. Page 405 (bottom left): Gregory G. Dimijian, M.D./Photo Researchers. Page 405 (bottom right): Sterling Dimmitt. Page 407: Courtesy Ed Keller. Page 408: Liu Liqun/Corbis. Page 410 (left): Sterling Dimmitt. Page 410 (right): Courtesy Jacksonville Corp of Engineers. Page 411 (top): Courtesy Ed Keller. Page 411 (bottom left): Stewart Halperin/Earth Scenes. Page 411 (bottom right): Cameron Davidson/Comstock, Inc. Page 412: ©AP/Wide World Photos. Page 415 (top): Rich Buzzelli/Tom Stack & Associates. Page 415 (bottom): Peter Cole/Bruce Coleman, Inc. Page 416: Courtesy Ed Keller.

Chapter 21 Opener: John Althouse/Liaison Agency/Getty Images. Page 421: ©AP/Wide World Photos. Page 424: Ben Osborne/Stone/Getty Images. Page 427 (top): Courtesy Ed Keller. Page 427 (bottom): McAllister/Liaison Agency/Getty Images. Page 428: William E. Ferguson. Page 430 (top): Courtesy Ed Keller. Page 430 (bottom): Michelle Barnes/Liaison Agency/Getty Images. Page 432: Charles Mason/Black Star. Page 433 (top): ©AP/Wide World Photos. Page 433 (bottom): Jim Strawser/Grant Heilman Photography. Page 434: John Cancalosi/DRK Photo. Page 435 (left): David Woodfall/DRK Photo. Page 435 (right): ©AP/Wide World Photos. Page 444 (top): Courtesy John Day, Louisiana State University. Page 444 (bottom): ©Integrated Water Systmen, Inc. Page 445: Courtesy Ed Keller.

Chapter 22 Opener: Johnny Johnson/DRK Photo. Page 459: Roger Ressmeyer/Corbis. Page 471: Shawn Henry/Corbis Saba. Page 478 (left): George D. Lepp/Corbis. Page 478 (right): Uwe Walz/Corbis.

Chapter 23 Opener: Hulton Archive/Getty Images. Page 486 (left): Courtesy NASA. Page 486 (right): Jason Reed/REUTERS/Liaison Agency/Getty Images. Page 488 (left): F. Hoffman/The Image Works. Page 488 (right): Courtesy Ed Keller. Page 489 (top): Jules Bucher/Photo Researchers. Page 489 (bottom): ©AFP/Corbis. Page 498: Bartholomew/Liaison Agency/Getty Images. Page 503 (left): Don & Pat Valenti/Stone/Getty Images. Page 503 (right): Don & Pat Valenti/Stone/Getty Images. Page 504: Bob Carey/Los Angeles Times. Page 507: Jim Mendenhall.

Chapter 24 Opener: ©AP/Wide World Photos. Page 522: Blanche/Liaison Agency/Getty Images. Page 523 (left): Andrew Syred/Photo Researchers. Page 523 (right): Oliver Meckes/Photo Researchers.

Chapter 25 Opener: Peter Cade/Stone/Getty Images. Page 545: Courtesy NASA.

Chapter 26 Opener: Ted Spiegel/PNI. Page 556 (top): Robert A. Lubeck/Animals Animals/Earth Scenes. Page 556 (bottom): Ken Wagner/Phototake. Page 566: C. Bradley Simmons/Bruce Coleman, Inc. Page 568: Photri.

Chapter 27 Opener: Paul Smith/Panos Pictures. Page 576 (top): Carlos Humerto T.D.C./Contact Press Images/PictureQuest. Page 582: Doug Courrance/FPG International/Getty Images. Page 584: Collection of ESL Information Services, Engineering Societies Library, New York City. Page 585: J. Messerschmidt/Leo de Wys, Inc. Page 587 (left): © & Courtesy Rhodeside & Harwell, Inc., Alexandria, VA. Page 589: Jeri Gleiter/Peter Arnold, Inc. Page 590 (top): Jim Richardson/Corbis. Page 590 (bottom): Robert Holmes/Corbis.

Chapter 28 Opener: Spencer Platt/Newsmakers/Getty Images. Page 599: Corbis Saba. Page 605: Alex Quesada/Matrix. Page 609: Courtesy John H. Kramer. Page 610: NY State Dept. of Environmental Conservation. Pages 611 & 612: Courtesy of Gloucester Fire Service & Sandhurst Area Action Group.

Chapter 29 Opener: Courtesy Ed Keller. Page 627: Helen Thompson/Animals Animals Earth Scenes. Page 631: Courtesy Rachel Haymon. Page 634: Royce Bair/TSS/ProFiles West, Inc. Page 635: Courtesy Ed Keller.

Chapter 30 Opener: ©AP/Wide World Photos. Page 643: Courtesy Florida House Learning Center, Sarasota County, Florida. Page 645 (left): Tom Van Sant/Photo Researchers. Page 645 (right): Konrad Wothe/Minden Pictures, Inc. Page 647: Alice Garick/Peter Arnold, Inc. Page 650 (top right): Courtesy NOAA. Page 650 (bottom right): A. Tannenbaum/Corbis Sygma. Page 650 (left): Courtesy NASA. Page 652: Bruce H. Morrison/Phototake. Page 653: ©Reuters NewMedia Inc./Corbis. Page 658: ©AP/Wide World Photos. Page 661: Courtesy Ed Keller.

Index

Barrier islands, 650-653, G-1
Bears. *See* Polar bears
Becquerel, 375, G-1
Bees, public service function, 565-566
Benthos, 152
Beta particles, 368, 369, G-2
Bioaccumulation, 289
Biochemical pxygen demand (BOD), 424-426, G-2
Biodiversity. *See* Biological diversity
Biogeochemical cycles:
    atmospheric questions, 53
    biological questions, 52-53
    in ecosystems, 60-62
    and environmental science, 52-53
    geological questions, 53
    glossary definition, G-2
    hydrological questions, 53
    illustrated, 54
    limiting factors, 53-55
    overview, 50, 55-56
Biogeography:
    defined, 124, 132
    effect on humans, 141-142
    glossary definition, G-2
    island-related, 139-141
    North American cross-sectional example, 135, 136
    purple loosestrife example, 131-132
    Wallace's realms, 132-134
Biohydrometallurgy, 637, G-2
Biological controls, 220-221, G-2
Biological diversity:
    and biological evolution, 111-114
    ethical basis for preserving, 127
    genetic, 115
    and genetic engineering, 114-115
    glossary definition, G-2
    habitat, 115
    how humans affect, 264
    influential factors, 124-128
    Mississippi Delta example, 110-111
    overview, 111, 115-117
    species, 115-117
Biological evolution:
    as a game, 115
    genetic drift process, 114
    glossary definition, G-2
    migration process, 112-114
    mutation process, 112
    natural selection process, 112
    overview, 111-112
    as story of life on earth, 114-115
Biological production. *See* Production, eco-logical
Biomagnification, 219, 289, G-2
Biomass. *See also* Biomass energy
    defined, 158
    in energy flow equations, 159
    glossary definition, G-2
    as renewable energy source, 344, 345, 346
Biomass energy:
    environmental impact, 359
    evaluating as energy source, 362
    future as power source, 359
    glossary definition, G-2
    list of examples, 358

overview, 357-358
    as renewable energy source, 344, 345, 346
    sources of fuel, 358
Biomass fuel, G-2. *See also* Biomass energy
Biomes:
    benthos, 152
    vs. biotic provinces, 135
    boreal forests, 145-146
    and convergent evolution, 135, 138
    defined, 135
    deserts, 149-150
    and divergent evolution, 138, 139
    fresh waters, 150-151
    glossary definition, G-2
    hydrothermal vents, 152, 167
    intertidal, 151-152
    list of types, 142
    maps of distribution, 142, 144, 145
    open ocean, 152
    overview, 142-143
    savannas, 149
    taiga, 145-146
    temperate forests, 146
    temperate grasslands, 148
    temperate rain forests, 146-147
    temperate shrub lands, 147-148
    temperate woodlands, 147
    tropical rain forests, 148-149
    tropical seasonal forests, 149
    tundra, 144-145
    upwellings, 152, 167
    wetlands, 150
Biopersistence, 615
Bioremediation, 436-437, G-2
Biosphere:
    and global warming, 476-479
    glossary definition, G-2
    overview, 42
Biota, 42, G-2
Biotechnology:
    biological diversity issues, 114-115
    genetically modified organisms, 207-208
    potential crop problems, 221-224
Biotic provinces. *See also* Wallace's realms
    vs. biomes, 135
    glossary definition, G-2
    overview, 134
Birth control, 90, 92
Birth rate:
    and demographic transition, 84, 85
    glossary definition, G-2
    vs. growth rate, 79, 81
    reducing, 90, 92
Bison, wildlife management issues, 272-273, 274
Black Death, 6
Black lung disease, G-2
Black smokers, 629, 631
Body burden, 288, G-2
Bogs, succession in, 175, 177, 178
Bone Valley, Florida, 68
Boreal forests, 145-146
Boundary Waters Canoe Area, Minnesota, 25-26, 172
Breeder reactors, 371-372, G-2
Brines, 628, G-2
British red squirrel, 119-121

British thermal unit, 312
Bubonic plague, 6
Buffalo, wildlife management issues, 272-273, 274
Buffers, 501, G-2
Buildings:
    and acid rain, 503-504
    design, and energy conservation, 315
    heating, ventilation, and air-conditioning systems, 525-526
    sick building syndrome, 521-522
Burner reactors, 367, G-2
Burr, Steve, 212
Bush fallow farming, 198
Butchart Gardens, Vancouver Island, Canada, 625
Butterfly (Julia Hill), 640, 641

C
Calcium carbonate, 501
Calcium cycle, in forest ecosystem, 61, 62
California:
    2001 energy crisis, 307
    growing rice in, 228-229
    land-use planning, 656
    Marble Mountain wilderness, 249
    San Joaquin Valley selenium toxicity, 653-655
California condor, 24-25
Canada:
    Butchart Gardens, 625
Canals, 407-408
Cape Hatteras National Seashore, North Canolina, 650-653
Capillary action, G-2
Carbon. *See also* Carbon cycle; Carbon dioxide (CO_2); Carbon monoxide (CO)
    defined, 63
    as one of "big six" building blocks of life, 54
    stored in vegetation, 20
Carbon cycle:
    carbon-silicate cycle, 65-67
    glossary definition, G-2
    illustrated, 63
    major pathways, 63
    overview, 63
    and photosynthesis, 63, 64
    and respiration, 63, 64
    unanswered questions, 65
Carbon dioxide (CO_2):
    glossary definition, G-2
    as greenhouse gas, 466-467
    Lake Nyos example, 285, 286
    monitoring atmospheric concentrations, 460
    and photosynthesis, 64
    as source of indoor air pollution, 524
Carbon monoxide (CO):
    ambient air quality standards, 513
    emission trends, 494
    environmental effects, 490, 493-494
    glossary definition, G-2
    major sources, 492
    overview, 493-494
    as primary pollutant, 491
    as source of indoor air pollution, 524

Food chain:
  in cities, 591, 592
  eating lower on, 166, 205
  and ecological communities, 97, 99-100
  and energy flow, 164-167
  glossary definition, G-7
  oceanic example, 100, 101, 102
  terrestrial example, 99-100
  Yellowstone hot springs example, 98-99
Food-chain concentration. *See*
      Biomagnification
Food supply:
  aquaculture as source of food, 196
  China example, 188-189
  crops as source of food, 192-194
  global perspective, 201-202
  limits to production, 203-205
  livestock as source of food, 194
  major crop species, 191-192
  mariculture as source of food, 196-197
  measures of availability, 203
  sources of food, 192-197
  ways to increase, 205-207
Food webs:
  glossary definition, G-7
  harp seal example, 100, 102
  oceanic examples, 100, 101, 102
  terrestrial example, 99-100
  trophic levels, 97
Force, 309, G-7
Forcing, 468, 470-473, G-7
Forests:
  and acid rain, 502
  alternatives to clear-cutting, 246
  boreal, 145-146
  calcium cycle, 61, 62
  clear-cutting, 245-246
  Coconino National Forest, 170
  economics of mahogany, 556-557
  and environmental unity, 39-40
  forestry as a profession, 242-243
  forestry issues, 236
  Fort Valley Experimental Forest, 170-171
  global perspective, 240-241
  global production and consumption of
      resources, 237-239
  harvesting, 7, 157, 241-242
  how trees grow, 239-240
  impact on atmosphere, 240-241
  managing, 242-243
  old-growth, 242
  public service functions, 236
  as resources, 235-237
  restoring, 170-171
  second-growth, 242
  secondary succession example, 176
  sulfur cycle, 61, 62
  sustainable, 241-242
  temperate, 146-147
  tropical, 148-149
  virgin, 242
  wetland, 233-234
Formaldehyde, as source of indoor air pollu-
      tion, 523, 524
Fort Valley Experimental Forest, Arizona,
      170, 171
Fortress city, 583, G-7

Fossil fuels:
  coal, 330, 333-338
  crude oil, 326-330
  glossary definition, G-7
  natural gas, 326, 328-330
  overview, 325-326
Fresh Kills landfill, New York, 599-600
Fresh waters, 151-152
Fuel cells, 353, G-7
Fugitive sources, 488, G-7
Fusion, nuclear, 366, G-7
Fusion reactors, 366, 372-373
Future value, 560-561

G

Gaia hypothesis, 8, 43, 45, G-7
Galapagos Islands, 139, 432-433
Game ranching, G-7
Gamma rays, 368-369, G-7
Garden cities, 585-586, G-7
Gases, hazardous, 498
Gasoline prices, 325
Gasoline tax, 340-341
Geese, threats to existence, 256-257
Genes, 112. *See also* Biotechnology
Genetic drift, 114, G-7
Genetic engineering:
  biological diversity issues, 114-115
  genetically modified organisms, 207-208
  potential crop problems, 221-224
Genetic risk, 268, G-7
Genetically modified organisms:
  glossary definition, G-7
  overview, 207-208
  potential crop problems, 221-224
Geochemical cycles, G-7. *See also*
      Biogeochemical cycles
Geographic Information System (GIS), 648,
      649, G-7
Geologic cycles:
  glossary definition, G-8
  how mineral deposits are formed, 626-630
  hydrologic cycle, 57-59
  overview, 56
  rock cycle, 59-60
  tectonic cycle, 56-57
Geologic record, role in studying global cli-
      matic change, 459
Geologic time scale, A-4
Geometric growth. *See* Exponential growth
Geopressurized systems, G-8. *See also*
      Geothermal energy
Geothermal energy:
  environmental impact, 360-361
  evaluating as energy source, 362
  future as power source, 361
  glossary definition, G-8
  as nonrenewable energy source, 344
  overview, 359
  system examples, 359-360
Gigawatt, 312
Glen Canyon Dam, 414, 415
Global circulation models (GCM), 461, G-8
Global extinction, 265, G-8
Global forecasting, 663-665, G-8
Global warming:
  adjustments to potential, 479-482

anthropogenic component, 463, 471-473
changes in biosphere, 476-479
changes in climate, 473-475
and electromagnetic radiation, 461-463
glossary definition, G-8
and Hudson Bay polar bears, 452
living with change, 479
mitigating, 479-481
negative and positive feedbacks, 469-470
overview, 461
potential effects, 473-479
and rise in sea level, 475-476
role of tree planting, 482
science of, 468-473
Grand Canyon, Arizona, 60, 414, 415
Granville Island, Vancouver, British
      Columbia, 658-659
Grasslands, 148
Gravel, 217, 392, 628, G-8
Gray squirrels, 119-121
Gray wolf, as endangered species, 280-281
Grays, defined, 375
Grazing. *See* Rangeland
Great Barrier Reef, Australia, 429, 430
Greece, ancient, 307-308
Green plans, 659, G-8
Green revolution, 205, G-8
Greenbelts, 586, G-8
Greenhouse effect, 463, 464-466, G-8
Greenhouse gases, 65, 463, 466-468, G-8.
      See also Global warming
Grizzly bear, wildlife management issues,
      271-272
Gross production (biology), 160, G-8
Groundwater:
  glossary definition, G-8
  interactions with surface water, 392-393
  Long Island, New York, 437-438
  overview, 392
  pollution, 436-438
  sources of pollution, 423
  and surface subsidence, 396
  sustainability, 402-403
  underground tank problem, 436-437
  use and problems, 395, 396
Growth efficiency, 165, G-8
Growth rate, 79, 82, 84, G-8
Guano, 68

H

Habitat:
  cities and wildlife, 591-592
  purple loosestrife example, 131-132
Habitats, 121, G-8
Half-life, 368, G-8
Hard path, 316-317, 318, G-8
Harp seal food web, 100, 102
Hawaiian goose, threats to existence, 256-257
Hazardous waste:
  alternatives to land disposal, 616-617
  comparison of reduction techniques, 614
  Comprehensive Environmental Response,
      Compensation, and Liability Act, 612
  demographics of nearby residents, 608
  glossary definition, G-8
  how land disposal works, 616
  incineration, 617